Power Electronics

Second Edition

CHHAYA BOOK CENTRE
Aurangpura, Aurangabad
☎ 0240- 2348245, Fax- 2351297

About the Authors

Dr M D Singh, former Principal at Laxmi Narayan College of Technology, Bhopal, worked as Assistant Director and Scientist at CEERI, Pilani, and has been Vice-President of the Indian Physics Association, BITS, Pilani. A former member of IEEE (USA) and a life member of IETE (India), Dr. Singh has contributed several research papers in many reputed national and international journals and has three patents to his credit. He obtained his B.E. from Jabalpur University and M.E. and Ph.D. from BITS, Pilani. He has widely travelled abroad and worked at several European institutions. Dr. Singh has been involved in research and teaching in power electronics, TV systems and satellite communication for over 40 years.

Dr K B Khanchandani is Professor, Department of Electronics and Telecommunication Engineering, SSGM College of Engineering, Shegaon. He has over 18 years of teaching and research experience in Power Electronics. He has published a number of research papers in journals, both national and international. He obtained his B.E. and M.E. from Amravati University and Ph.D. from National Institute of Technology, Bhopal. He has been involved in teaching and research in areas of power electronics, signal processing, analog and digital system design and VLSI design. He has worked on several research/development and laboratory modernization projects.

Power Electronics

Second Edition

M D Singh
Former Principal
Laxmi Narayan College of Technology
Bhopal

K B Khanchandani
Professor, Department of Electronics and Telecommunication Engineering,
SSGM College of Engineering
Shegaon, Maharashtra

Tata McGraw Hill Education Private Limited
NEW DELHI

McGraw-Hill Offices
New Delhi New York St Louis San Francisco Auckland Bogotá Caracas
Kuala Lumpur Lisbon London Madrid Mexico City Milan Montreal
San Juan Santiago Singapore Sydney Tokyo Toronto

Tata McGraw-Hill

Published by Tata McGraw Hill Education Private Limited,
7 West Patel Nagar, New Delhi 110 008

Copyright © 2007 by Tata McGraw Hill Education Private Limited

Fifth reprint 2009
RQXCRRQFRBDDC

No part of this publication may be reproduced or distributed in any form or by any means, electronic, mechanical, photocopying, recording, or otherwise or stored in a database or retrieval system without the prior written permission of the publishers. The program listings (if any) may be entered, stored and executed in a computer system, but they may not be reproduced for publication

This edition can be exported from India only by the publishers,
Tata McGraw Hill Education Private Limited

ISBN-13: 978-0-07-058389-4
ISBN-10: 0-07-058389-7

Head-Higher Education & School: *S. Raghothaman*
Executive Publisher—SEM & Tech Ed: *Vibha Mahajan*
Jr. Sponsoring Editor—SEM & Tech Ed: *Shalini Jha*

Deputy General Manager—Higher Education & Sales: *Michael J. Cruz*
Asst. Product Manager—SEM & Tech Ed: *Biju Ganesan*

Controller—Production: *Rajender P Ghansela*
Asst. General Manager—Production: *B L Dogra*
Senior Production Executive: *Anjali Razdan*

Information contained in this work has been obtained by Tata McGraw-Hill, from sources believed to be reliable. However, neither Tata McGraw-Hill nor its authors guarantee the accuracy or completeness of any information published herein, and neither Tata McGraw-Hill nor its authors shall be responsible for any errors, omissions, or damages arising out of use of this information. This work is published with the understanding that Tata McGraw-Hill and its authors are supplying information but are not attempting to render engineering or other professional services. If such services are required, the assistance of an appropriate professional should be sought

---

Typeset at Script Makers, 19, A1-B, DDA Market, Paschim Vihar, New Delhi 110 063, text and cover printed at AP Offset Pvt. Ltd., Delhi - 110 032

*In
Loving Memory of Our
Late Parents*

Contents

Preface to the Second Edition   xv
Preface to the First Edition   xix
Acknowledgements   xxi

1. POWER ELECTRONIC SYSTEMS: AN OVERVIEW   1

    1.1   Introduction   *1*
    1.2   History of Power Electronics Development   *2*
    1.3   Power Electronic Systems   *2*
    1.4   Power Semiconductor Devices   *4*
    1.5   Power Electronic Converters   *11*
    1.6   Power Electronic Applications   *12*
    1.7   Computer Simulation of Power Electronic Circuits   *14*

        Review Questions   *15*
        References   *16*

2. THYRISTOR: PRINCIPLES AND CHARACTERISTICS   17

    2.1   Introduction   *17*
    2.2   Principle of Operation of SCR   *18*
    2.3   Static Anode–Cathode Characteristics of SCR   *19*
    2.4   The Two-transistor Model of SCR (Two Transistor Analogy)   *23*
    2.5   Thyristor Construction   *25*
    2.6   Gate Characteristics of SCR   *27*
    2.7   Turn-on Methods of a Thyristor   *33*
    2.8   Dynamic Turn-on Switching Characteristics   *35*
    2.9   Turn-off Mechanism (Turn-off Characteristic)   *37*
    2.10   Turn-off Methods   *38*
    2.11   Thyristor Ratings   *55*
    2.12   Measurement of Thyristor Parameters   *63*
    2.13   Comparison between Transistors and Thyristors   *65*

        Review Questions   *66*
        Problems   *67*
        References   *69*

3. GATE TRIGGERING CIRCUITS 70

- 3.1 Introduction 70
- 3.2 Firing of Thyristors 71
- 3.3 Pulse Transformers 76
- 3.4 Optical Isolators (Optoisolators) 78
- 3.5 Gate Trigger Circuits 81
- 3.6 Unijunction Transistor 87
- 3.7 The Programmable Unijunction Transistor (PUT) 100
- 3.8 Phase Control using Pedestal-And-Ramp Triggering 106
- 3.9 Microprocessor Interfacing to Power Thyristor 108

  Review Questions 110
  Problems 111
  References 113

4. SERIES AND PARALLEL OPERATION OF THYRISTORS 114

- 4.1 Introduction 114
- 4.2 Series Operations of Thyristors 115
- 4.3 Need for Equalising Network 115
- 4.4 Equalising Network Design 118
- 4.5 Triggering of Series Connected Thyristors 121
- 4.6 Parallel Operation of Thyristors 124
- 4.7 Methods for Ensuring Proper Current Sharing 126
- 4.8 Triggering of Thyristors in Parallel 129
- 4.9 String Efficiency 130
- 4.10 Derating 131

  Review Questions 133
  Problems 133
  References 134

5. POWER SEMICONDUCTOR DEVICES 135

- 5.1 Introduction 135
- 5.2 Historical Perspective 137
- 5.3 Power Semiconductor Devices 139
- 5.4 Phase Controlled Thyristors 141
- 5.5 Inverter-Grade Thyristors 141
- 5.6 Asymmetrical Thyristor (ASCR) 142
- 5.7 Reverse Conducting Thyristor (RCT) 144
- 5.8 Bidirectional Diode Thyristor (Diac) 145
- 5.9 Bidirectional Triode Thyristor (TRIAC) 146
- 5.10 Silicon Unilateral Switch (SUS) 156
- 5.11 Silicon Bilateral Switch (SBS) 156
- 5.12 Silicon Controlled Switch (SCS) 156
- 5.13 Light-Activated Silicon-Controlled Rectifiers (LASCR) 158

- 5.14 Power MOSFETs *159*
- 5.15 Insulated Gate Bipolar Transistors (IGBTs) *187*
- 5.16 Gate Turn-off Thyristors (GTOs or Latching Transistors) *212*
- 5.17 Static Induction Devices *222*
- 5.18 MOS Controlled Thyristor (MCT) *224*
- 5.19 Integrated Gate-Commutated Thyristor (IGCT) *231*
- 5.20 MOS Turn-off Thyristor (MTO) *235*
- 5.21 Emitter Turn-Off Thyristor (ETO) *236*
- 5.22 Power Integrated Circuit (PICs) *244*
- 5.23 Comparison of Power Devices *247*
- 5.24 Silicon Carbide Devices *250*

   Review Questions 250
   Problems 254
   References 256

6. PHASE CONTROLLED CONVERTERS **258**

- 6.1 Introduction *258*
- 6.2 Control Techniques *259*
- 6.3 Single Phase Half-Wave Controlled Rectifier *263*
- 6.4 Single-Phase Full-Wave Controlled Rectifier (Two-quadrant Converters) *273*
- 6.5 Single-Phase Half Controlled Bridge-Rectifier *291*
- 6.6 Performance Factors of Line-commutated Converters *302*
- 6.7 The Performance Measures of Two-pulse Converters *303*
- 6.8 Three-Phase Controlled Converters *307*
- 6.9 Three-Pulse Converters (M_3 Connection) *308*
- 6.10 Six-Pulse Converters *323*
- 6.11 Three-Phase Fully Controlled Bridge Converter *329*
- 6.12 Three-Phase Half Controlled Bridge Converter (Three-Phase Semiconverters) *346*
- 6.13 The External Performance Measures of Six-Pulse Converters *359*
- 6.14 The Effect of Input Source Impedance *361*
- 6.15 Performance of Converter Circuits with Battery Load (Or Effect of Load Inductance) *378*
- 6.16 Selection of Converter Circuits *380*
- 6.17 Power Factor Improvement *380*
- 6.18 Microprocessor-Based Firing Scheme for Three-Phase Fully-Controlled Bridge Converter *395*

   Review Questions 402
   Problems 407
   References 411

7. DUAL CONVERTERS 412

7.1 Introduction *412*
7.2 Principle of Dual Converter (Ideal Dual Converter) *415*
7.3 Practical Dual Converter *416*
7.4 Dual Converter without Circulating Current Operation *417*
7.5 Dual Converter with Circulating Current Operation *421*
7.6 Dual-Mode Dual Converter *426*
7.7 Comparison between Non-Circulating Current Mode and Circulating Current Mode *428*
7.8 Microprocessor Based-Firing Scheme for a Dual Converter *429*

Review Questions 432
Problems 433
References 433

8. CHOPPERS 434

8.1 Introduction *434*
8.2 Basic Chopper Classification *436*
8.3 Basic Chopper Operation *437*
8.4 Control Strategies *444*
8.5 Chopper Configuration *447*
8.6 Thyristor Chopper Circuits *481*
8.7 Jones Chopper *496*
8.8 Morgan Chopper *502*
8.9 A.C. Choppers *504*
8.10 Source Filter *505*
8.11 Multiphase Chopper *508*
8.12 Flyback Converters [Switching Regulators] *510*

Review Questions 530
Problems 532
References 534

9. INVERTERS 535

9.1 Introduction *535*
9.2 Classification of Inverters *537*
9.3 Single-Phase Half-Bridge Voltage-Source Inverters *538*
9.4 Single-Phase Full-Bridge Inverters *545*
9.5 Performance Parameters of Inverters *551*
9.6 Voltage Control of Single-Phase Inverters *554*
9.7 Pulse-Width Modulated (PWM) Inverters *565*
9.8 Three-Phase Inverters *574*
9.9 Voltage Control of Three-Phase Inverters *593*
9.10 Thyristor-Based Inverters *593*

- 9.11 Series Inverters (Series Resonant Inverters) *594*
- 9.12 Self-Commutated Inverters *606*
- 9.13 Parallel Inverter *609*
- 9.14 The Single-Phase SCR Bridge Inverter *615*
- 9.15 Current Source Inverters *643*
- 9.16 Performance Comparisons of PWM, AVI and CSI *651*
- 9.17 Harmonic Reduction *653*
- 9.18 Harmonic Filters *657*

  Review Questions 664
  Problems 667
  References 669

10. CYCLOCONVERTERS 670

- 10.1 Introduction *670*
- 10.2 The Basic Principle of Operation *671*
- 10.3 Single-phase to Single-phase Cycloconverter *673*
- 10.4 Three-phase Half-wave Cycloconverters *680*
- 10.5 Cycloconverter Circuits for Three-Phase Output *686*
- 10.6 Output Voltage Equation *686*
- 10.7 Control Circuit *689*
- 10.8 Comparison between Cycloconverter and D.C. Link Converter *697*
- 10.9 Load-Commutated Cycloconverter *698*

  Review Questions 701
  Problems 702
  References 703

11. A.C. REGULATORS 704

- 11.1 Introduction *704*
- 11.2 Single-Phase A.C. Regulators *705*
- 11.3 Sequence Control of A.C. Regulators *723*
- 11.4 Three-phase A.C. Regulators *729*
- 11.5 A.C. Regulators to Feed Transformers *740*

  Review Questions 742
  Problems 744
  References 745

12. RESONANT CONVERTERS 746

- 12.1 Introduction *746*
- 12.2 Basic Resonance Circuit Concepts *747*
- 12.3 Classification of Resonant Converters *752*
- 12.4 Load Resonant Converters (Self-Commutating Converters) *753*
- 12.5 Parallel Resonant Inverters *767*

- 12.6 Class-E Resonant Inverters 768
- 12.7 Class-E Resonant Rectifier 770
- 12.8 Resonant-Switch Converters 772
- 12.9 ZVS Three-Level PWM-Converter 787
- 12.10 Resonant DC Link Inverters 793

  Review Questions 795
  Problems 796
  References 797

13. PROTECTION AND COOLING OF POWER SWITCHING DEVICES 798

- 13.1 Introduction 798
- 13.2 Overvoltage Conditions 799
- 13.3 Overvoltage Protection 803
- 13.4 Practical Overvoltage Protection in Naturally-Commutated Circuits 815
- 13.5 Overvoltage Protection in Forced-Commutated Circuits 816
- 13.6 Overcurrent Fault Conditions 816
- 13.7 Overcurrent Protection 821
- 13.8 Gate Protection 827
- 13.9 Heat Sinks 833
- 13.10 Thyristor Mounting Techniques 838
- 13.11 SCR Reliability 839

  Review Questions 843
  Problems 844
  References 845

14. CONTROL OF D.C. DRIVES 846

- 14.1 Introduction 846
- 14.2 Basic Machine Equations 847
- 14.3 Schemes for D.C. Motor Speed Control 849
- 14.4 Single-Phase Separately Excited Drives 851
- 14.5 Braking Operation of Rectifier Controlled Separately Excited Motor 870
- 14.6 Single-Phase Series D.C. Motor Drives 871
- 14.7 Three-Phase Separately Excited Drives 879
- 14.8 D.C. Chopper Drives 888
- 14.9 Phase-Locked Loop (PLL) Control of D.C. Drives 893
- 14.10 Microcomputer Control of D.C. Drives 895

  Review Questions 898
  Problems 899
  References 902

15. CONTROL OF A.C. DRIVES   903

    15.1   Introduction   *903*
    15.2   Basic Principle of Operation   *905*
    15.3   Speed Control of Induction Motors   *912*
    15.4   Stator Voltage Control   *913*
    15.5   Variable Frequency Control   *917*
    15.6   Rotor Resistance Control   *942*
    15.7   Slip Power Recovery Scheme   *949*
    15.8   Synchronous Motor Drives   *959*
    15.9   The Drive Selection   *978*

          Review Questions   *981*
          Problems   *983*
          References   *985*

16. POWER ELECTRONIC APPLICATIONS   987

    16.1   Introduction   *987*
    16.2   Uninterruptible Power Supply   *988*
    16.3   Switched Mode Power Supplies (SMPS)   *998*
    16.4   High Voltage D.C. Transmission   *1024*
    16.5   Static VAR Compensators   *1029*
    16.6   RF Heating   *1031*
    16.7   Switch-Mode Welding   *1041*
    16.8   Electronic Lamp Ballast   *1042*
    16.9   Battery Charger   *1042*
    16.10 Emergency Lighting System   *1044*
    16.11 Static Circuit Breaker   *1045*
    16.12 Time-Delay Circuit   *1047*
    16.13 Flasher Circuits   *1047*
    16.14 Integral Cycle Triggering (Or Burst Firing)   *1049*

          Review Questions   *1051*
          Problems   *1053*
          References   *1053*

APPENDIX A : Simulation Tools for Power Electronic Circuits   **1054**
APPENDIX B : High Voltage Gate Driver ICs (HVICs), Power Modules and Intelligent Power Modules   **1060**

INDEX   *1065*

Preface to the Second Edition

The field of electrical engineering is generally segmented into three major areas—electronics, power and control. Power electronics is a combination of these three areas. In broad terms, the function of power electronics is to process and control the electrical energy by supplying voltage and current in a form that is optimally suited to the load. The advent of power semiconductor devices, *thyristors,* in 1957, has been an exciting breakthrough in the art of electric power conversion and its control. Power electronics has undergone intense technological evolution during the last three decades. Starting with the conventional thyristor-type devices in early days, the recent availability of high-frequency, high-power, MOS-gated self-controlled devices are opening new frontiers in the field.

In modern power electronics equipment, there are essentially two types of semiconductor elements: the power semiconductors that can be considered as the muscles of the equipment, and the microelectronics control chips that provide power to the brain. Both the elements are digital in nature, except that one manipulates power upto gigawatts and the other handles only miliwatts. The close coordination of this end of the spectrum of electronics offers reduced size, cost advantages and high-level of performance.

The market demands on industry for productivity and quality are increasing. This has resulted in an increasing demand for automation in production processes and hence in the use of variable speed drives.

The use of electric cars, electric trams and electric subway trains can substantially reduce urban pollution problems. Power electronics permits generation of electric power from environmentally clean photovoltaic, fuel cells and wind energy sources. Widespread application of power electronics, with an eye for energy conservation and generation of power from environmentally clean sources, can help in solving problems like acid rains and greenhouse effects.

This textbook presents the basic tools for the analysis and design of power electronic circuits and provides methods and procedures suitable for a variety of power electronic applications. This text is suitable for undergraduate-level courses in power electronics and/or electrical machines and drives. It is also intended for practicing engineers who wish to gain knowledge of the recent developments in this rapidly expanding field.

In response to the comment and suggestions from students and teachers, several changes in the presentation of topics have been made and new topics have been included in this revised edition of the book.

The second edition offers several improvements over the first edition. Many sections have been rewritten to further simplify the presentation. Several chapters have been revised by adding new updated material. New problems have been formulated for many chapters.

The book consists of 16 chapters. Two chapters (Chapters 1 and 12) are newly added and the sequence of chapters has been modified for better flow of the content. The following is a brief description of the topics that are covered in each chapter with an emphasis on the revisions that have been made in the second edition.

Chapter 1 contains an overview of power electronics and its applications to make the reader aware of the meaning of power electronics and its importance.

Chapter 2 deals with the physical principle of thyristors, their structural details and switching characteristics. The section on turn-on switching characteristics has been completely revised.

Chapter 3 gives a brief description of gate triggering circuits. A new section on *microprocessor interfacing to power thyristors* has been added.

Chapter 4 contains series and parallel operations of thyristor.

Chapter 5 covers the various power semiconductor devices such as power MOSFETs, GTOs, IGBTs, SITs, SITHs, IGCTs, MCTs, MTOs and ETOs. Device structures and physical operations are described and typical terminal characteristics are shown. Many sections are completely revised. The section on power transistors along with solved examples has been moved over to the online learning centre (OLC) of the book. New sections and subsections on IGBTs and ETOs have been added.

Chapter 6 presents various phase-controlled rectifier circuits with their mathematical analysis and performance factors. A new section on power-factor improvement has been added. Many sections have been revised for better understanding. Sections on Firing Circuits for Line Commutated Converters and Triggering Circuits for Single-phase Fully Controlled Converters have been removed from the book. However, these are available on the OLC of the book.

Chapter 7 gives a detailed account of dual converters and their characteristics.

Chapter 8 is devoted to the discussion of the principles of choppers in D.C-to -D.C. conversion. A new section on *flyback converters* has been added. The section on chopper firing circuit has been removed and is available on the website.

Chapter 9 gives a comprehensive treatment of dc–ac inverters in which the various voltage-fed and current-fed inverter circuits are discussed and some typical forced-commutating circuits are investigated. A number of design examples are presented. Sections on three-phase bridge inverters with input circuit commutation and three-phase current source inverters are now available on the website.

Chapter 10 introduces the phase-controlled cycloconverters. The section on ring-connected cycloconverter circuits has been relegated to the website.

Chapter 11 presents a.c. regulators.

Chapter 12 deals with *resonant converters* where LC resonant circuits are utilized to improve the performance of converters. This chapter is newly added and covers the basic technology of resonant and soft-switching converters. Various forms of soft-switching techniques such as ZVS and ZCS are addressed.

Chapter 13 introduces the protection, cooling and mounting of power semi-conductor devices.

Chapter 14 discusses various schemes for D.C. motor speed control. The section on closed loop control of D.C. drives has been shifted to the website.

Chapter 15 introduces variable speed A.C. drives and briefly describes their benefits. It examines their classifications from different perspectives. The section on microprocessor controlled A.C. drive has been removed from the book and is available on the website.

Chapter 16 considers power electronics application circuits. Many sections in this chapter have been completely revised. Also, many new sections have been added.

Web Supplements

The Online Learning Centre of the book at (*http://www.mhhe.com/singh/pe2e*) has separate sections for students and instructors. Students' resource consists of PSPICE simulation examples and multiple-choice questions with answers. For instructors, the website offers the solutions manual wherein all the exercise problems in the book have been solved. It also has chapterwise PowerPoint slides which will help the teachers in preparing lectures and presentations.

M D Singh
K B Khanchandani

Preface to the First Edition

The field of electrical engineering is generally segmented into three major areas—electronics, power and control. Power electronics involves a combination of these three areas. In broad terms, the function of power electronics is to process and control the electrical energy by supplying voltage and current in a form that is optimally suited to the load. The advent of the power semiconductor device, *thyristor* in 1957, has been an exciting breakthrough in the art of electric power conversion and its control. Power electronics has undergone intense technological evolution during the last three decades. Starting with the conventional thyristor-type devices in the early days, the recent availability of high frequency, high power, MOS-gated self-controlled devices is opening new frontiers in power electronics.

In a modern power electronic equipment, there are essentially two types of semiconductor elements: the power semiconductors that can be considered as the muscle of the equipment, and the microelectronic control chips that provide the power to the brain. Both elements are digital in nature, except that one manipulates power up to gigawatts and the other handles only milliwatts. The close coordination of these end-of-the-spectrum electronics offers reduced size, cost advantages and high level of performance.

The market demands on industry for productivity and quality are increasing. This results in the increasing demand for automation in production processes and hence for the use of variable speed drives. Today, power electronics is an indispensable tool in any advanced country's industrial economy. Saving energy is an important aspect of power electronics applications.

The use of electric cars, electric trams and electric subway trains can substantially reduce urban pollution problems. Power electronics permits generation of electric power from environmentally clean photovoltaic, fuel cell and wind energy sources. Widespread application of power electronics, with an eye for energy conservation and generation of power from environmentally clean sources, can help in solving problems like acid rain and greenhouse effects.

This textbook is intended as an introduction to the basic theory and practice of modern power electronics, in particular it deals with the applications of power electronic techniques for d.c. and a.c. motor control. The text is suitable for degree-level and postgraduate courses in power electronics and/or electrical machines and drives. It is also intended for practicing engineers who wish to acquaint themselves with the recent developments in this rapidly expanding field.

The book consists of 14 chapters.

Chapter 1 deals with the physical principles of thyristors, their structural details and their switching characteristics. Chapter 2 gives a brief description of thyristor-firing circuits. Series and parallel operations of thyristors are discussed in Chapter 3.

Chapter 4 discusses all types of line-commutated phase-controlled converters with their mathematical analysis as well as performance factors and triggering circuits for these converters.

Chapter 5 is a comprehensive treatment of d.c.-a.c. inverters in which the various voltage-fed and current-fed inverter circuits are discussed, and some typical thyristor forced-commutating circuits are investigated.

Chapter 6 is devoted to the discussion of the principles of choppers in d.c. to d.c. conversion. Chapter 7 introduces the phase controlled cycloconverters. Chapter 8 gives a detailed account of dual converters and their characteristics. Chapter 9 deals with a.c. voltage controllers.

Chapter 10 discusses the various modern power semiconductor devices such as power transistors, power MOSFETs, GTOs, IGBTs, SITs and SITHs. Device structures and physical operations are described and typical terminal characteristics are shown.

Protection, cooling and mounting of SCRs is discussed in Chapter 11. Power electronics control of d.c. and a.c. motors is treated in Chapters 12 and 13 respectively. Chapter 14 considers power electronic application circuits.

Most importantly, it was the help and advice of Tata McGraw-Hill Publishing staff that made this whole project a reality. In particular, we wish to thank Ms Vibha Mahajan, Assistant Sponsoring Editor, Mini Narayanan, Copy Editor, and Anjali Razdan, Proofreader, of Tata McGraw-Hill. We are grateful to the authorities of SSGM College of Engineering, Shegaon, for providing all the facilities necessary for writing the book.

We would like to express our thanks for the many useful comments and suggestions provided by colleagues who reviewed this book during the course of its development, especially to Prof. J K Chatterjee of IIT Delhi and Dr Murugesh Mudaliar of BMS College of Engineering, Bangalore.

Finally, we would like to thank our family members for their love, patience and understanding during the preparation of the book.

M D Singh
K B Khanchandani

Acknowledgements

We are fortunate to have received many useful comments and suggestions from students, which have helped in improving the technical content and clarity of the book. We are grateful to all of them.

We are indebted to many readers in the academia and industry worldwide for their invaluable feedback and for taking the trouble to draw our attention to the improvements required and the errors in the first edition.

Our special thanks goes to the reviewers of the first edition who have helped tremendously in shaping the revised edition.

Prof Ms. Prerna Gaur (NSIT, Delhi) and Dr. Yaduvir Singh (TIET, Patiala) we very much hope that the above reviewers and others will keep the feedback coming.

Most importantly, it was the help and the advice of McGraw-Hill Education (India) staff that made this whole project a reality. In particular, we wish to thank Vibha Mahajan, Shalini Jha, Mini Narayanan and Anjali Razdan for their support, encouragement and hard work in bringing out this book. We are grateful to the authorities of SSGM College of Engineering, Shegaon, for providing all the facilities for writing this book.

Finally, the authors are grateful to their families for their love, tolerance, patience and support throughout this very time-consuming project. Readers of the book are welcome to send their comments and feedback.

M D SINGH
K B KHANCHANDANI

Chapter 1

Power Electronic Systems: An Overview

LEARNING OBJECTIVES:
- To become familiar with the power-electronic systems.
- To understand the overall systems view of power electronic converters.
- To introduce various power semiconductor devices.
- To consider the applications of power electronics.
- To introduce the simulation techniques for power electronic circuits.

1.1 INTRODUCTION

Generally, the electrical engineering field may be divided into three areas of specialization:

- Electronics • Power • Control

Electronics essentially deals with the study of semiconductor devices and circuits for the processing of information at lower power levels. In this rapidly developing society, electronics has emerged as the most important branch of engineering. The power area deals with both rotating and static equipment for the generation, transmission, distribution and utilisation of vast quantities of electrical power. The transmission and distribution system is a very vital link between generation and utilisation of electrical power and the relative strength of this system indicates the quality of electric power in a country. In India, after successfully establishing capabilities of generating electric power through various sources like hydel, thermal, nuclear, gas etc. a policy decision has been taken to give the highest priority to the transmission and distribution system in the 90's.

The control area deals with the stability and response characteristics of closed-loop systems using feedback on either a continuous or sampled-data basis.

Power electronics deals with the use of electronics for the control and conversion of large amounts of electrical power. The design of power electronics equipment involves interactions between the source and the load, and utilises small-signal electronic control circuits as well as power semiconductor devices. Therefore, power electronics draws as well as depends upon all other areas of electrical engineering.

Power electronics constitute a vast, complex and interdisciplinary subject that has gone through rapid technological evolution during the last four decades. As the technology is advancing and apparatus cost is decreasing along with the improvement of reliability, their applications are expanding in industrial, commercial, residential, military, aerospace and utility environments. Many innovations in power semiconductor devices, converter topologies, analytical and simulation techniques, electrical machine drives, and control and estimation techniques are contributing to this advancement. The frontier of the technology has been further advanced by the artificial intelligence (AI) techniques, such as fuzzy logic and artificial neural networks, thus bringing more challenge to power electronic engineers.

In the global industrial automation, energy generation, conservation of the 21st century, the widespread impact of power electronics is inevitable. In this chapter, we will overview the power devices, converters and applications of power electronics.

1.2 HISTORY OF POWER ELECTRONICS DEVELOPMENT

Until 1956, the application of semiconductors was confined to low power circuits and electronic engineering was also called as light current engineering. In September 1956, four engineers of the Bell Telephone Laboratory, USA, published a paper entitled "*PNPN* transistor switches" in the proceedings of the Institute of Radio Engineers. This paper triggered intensive research on *PNPN* devices. In 1957, Gordon Hall of General Electric Company, USA, developed the three terminal *PNPN* silicon based semiconductor device called as *silicon controlled rectifier* (SCR). Continuous modifications and improvement in its design as well as fabrication techniques have made it more and more economical and suitable for various control purposes. Later on, many other power devices having characteristics similar to that of an SCR were developed. Actually, the origin of power electronics can be traced back to the time when mercury arc devices were employed for the rectification of a.c. to d.c. or the inversion of d.c. to a.c. However, the rapidly increasing usage of power electronics nowadays has resulted from the development of solid state power devices.

1.3 POWER ELECTRONIC SYSTEMS

Block diagram of the generalised power electronics system is shown in Fig. 1.1 Power source may be an ac supply system or a dc supply system. In India, 1-phase and 3-phase 50 Hz ac supplies are readily available in most locations. Very low power drives (systems employed for motion control are called drives) are generally fed from 1-phase source. Rest of the drives are powered from 3-phase source. Low and medium power motors (tens of kilowatts) are generally fed from 400 V supply; for high ratings, motors may be rated at 3.3 kV, 6.6 kV, 11 kV and higher. In case of aircraft and space applications, 400 Hz ac supply is generally used to achieve high power to weight ratio for motors. In main line traction, a high voltage supply is preferred because of economy. In India, 25 kV, 50 Hz supply is employed.

Some loads are powered from a battery, e.g. fork lift trucks and milk vans. Depending on size, battery voltage may have typical values of 6 V, 12 V, 24 V, 48 V and 110 V dc. Solar powered drives which are used in space and water pumping applications are fed from a low voltage dc supply. Presently, though these drives are very expensive but have a great future for rural water pumping and low power transport applications.

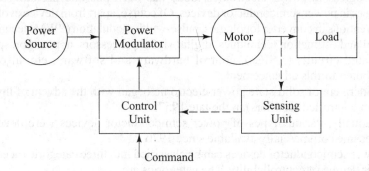

Fig. 1.1 *Block diagram of power electronic system*

Power modulator performs one or more of the following four functions:

(i) Converts electrical energy of the source as per the requirement of the load. For example, if the load is a dc motor, the modulator output must be adjustable direct voltage. In case the load is a 3-phase induction motor, the modulator may have adjustable voltage and frequency at its output terminals. When power modulator performs this function, it is known as converter.
(ii) Selects the mode of operation of the motor, i.e. motoring or braking.
(iii) Modulates flow of power from the source to the motor in such a manner that motor is imparted speed-torque characteristics required by the load.
(iv) During transient operations, such as starting, braking and speed reversal, it restricts source and motor currents within permissible values; excessive current drawn from source may overload it or may cause a voltage dip.

Motors commonly used in power electronic systems are:
(i) DC motors (shunt, series, compound and permanent magnet)
(ii) Induction motors (squirrel-cage, wound rotor and linear)
(iii) Synchronous motors (wound field and permanent magnet)
(iv) Brushless dc motors
(v) Stepper motors and
(vi) Switched reluctance motors

Power modulators are controlled by a control unit. Nature of the control unit for a particular system depends on the power modulator that is used. Control unit operates at much lower voltage and power levels. Sensing unit measures the load parameters, say speed in case of a rotating machine and compares it with the command. The difference of the two parameters processed by the control unit components now controls the turn-on of power semiconductor devices which are used in power modulators. As desired, the behaviour of the load circuit can be controlled over a wide range with the adjustment of the command.

1.4 POWER SEMICONDUCTOR DEVICES

The progress in power electronics today has been possible primarily due to advances in power semiconductor devices. Of course, apart from device evolution, the inventions in converter topologies, pulse-width modulation (PWM) techniques, control and estimation techniques, digital signal processors, application specific integrated circuits (ASICs), control hardware and software, etc. also have contributed to this advancement.

Modern era of solid-state power electronic began with the advent of thyristor (silicon controlled rectifiers) in the late 1957.

Gradually, various types of power semiconductor devices were developed and became commercially available since 1970.

Power semiconductor devices can be classified into three categories according to their degree of controllability. The categories are:

 (i) Uncontrolled turn-on and off devices (e.g. diode).
 (ii) Controlled turn-on and uncontrolled turn-off (e.g. SCR).
 (iii) Controlled turn-on and off characteristics [e.g. Bipolar junction transistors (BJTs), MOSFETs, Gate-turn-off thyristors (GTOs), Static induction thyristor (SITH), Insulated-gate bipolar transistors (IGBTs), static induction transistors (SITs), mos-controlled thyristors (MCTs)].

The on and off states of diodes are controlled by power circuit. Thyristors are turned-on by a control signal and are turned-off by the power circuit whereas the controllable switches are turned-on and off by controlled signals. The devices which behave as controllable switches are BJT, MOSFET, GTO, SITH, IGBT, SIT and MCT.

BJT, MOSFET, IGBT and MCT can withstand unipolar voltage whereas thyristors and GTOs can withstand bipolar voltages. BJT, MOSFET, IGBT and SIT requires continuous signal for keeping them in turn-on state but SCR, GTO, SITH and MCT requires pulse-gate signal for turning them ON and once these devices are ON, gate-pulse is removed.

Triac and RCT (reverse conducting thyristor) possess bidirectional current capability whereas all other remaining devices (diode, SCR, GTO, BJT, MOSFET, IGBT, SIT, SITH, MCT) are unidirectional current devices.

As the evolution of new and advanced devices continued, the voltage and current ratings and electrical characteristics of the existing devices began improving dramatically. In fact, the device evolution along with converter, control and system evolution was so spectacular in the last decade of 20th century, that we define it as the "decade of power electronics".

Thyristors are used for high power low frequency applications. Devices are available with 8000 V and 4000 A ratings. ABB recently introduced a monolithic ac switch that has the voltage ratings of 2.8 kV–6.5 kV and current ratings of 3000–6000 A. The advent of large GTOs push the thyristor voltage-fed inverter from the market. Currently, GTOs are available with 6000 V, 6000 A (Mitsubishi) ratings for large voltage-fed inverter applications.

Power MOSFET has grown in rating, but it's primary popularity is in high-frequency switching mode power supply and portable appliances. The BJT appeared and then fell into obsolescence due to the advent of IGBT at the higher end and power MOSFET at the lower end. The invention of IGBT is an important milestone in the history of power semiconductor devices. Commercial IGBTs are available with 3500 V, 1200 A, but upto 6.5 kV and 10 kV devices are under test in laboratory. Trench gate IGBT with reduced conduction drop is available upto 1200 V, 600A. IGBT intelligent power modules (IPM) from a number of vendors are available for 600 V, 50-300 A and 1200 V, 50-150 A to cover upto 150 hp ac drive applications.

Integrated Gate-Commutated Thyristor (IGCT) is basically a hard-driven GTO with built-in gate driver, and the device is available with 6000 V, 6000 A (10 kV devices are under test). Recently, ABB introduced reverse blocking IGCT (6000 V, 800 A) for use in current-fed inverter drives. Large band gap power semiconductor device with silicon carbide (SiC) that has high carrier mobility, high electrical and thermal conductivities and strong radiation hardness is showing high promise for next generation power devices. These devices can be fabricated for higher voltage, higher temperature, higher frequency and lower conduction drop. SiC diodes are commercially available, and other devices are expected in future. Table 1.1 lists some power devices and shows their respective V-I characteristics and symbolic representation. We will study all these devices in Chapter 5.

Table 1.1 Power devices, their characteristics and symbolic representation

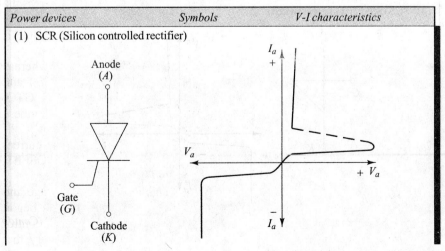

(Contd.)

Power devices	Symbols	V-I characteristics

(2) DIAC (Bidirectional diode thyristor)

(3) TRIAC (Bidirectional triode thyristor)

(4) SUS (Silicon unilateral switch)

(Contd.)

Power devices	Symbols	V-I characteristics

(5) SCS (Silicon controlled switch)

(6) SBS (Silicon bilateral switch)

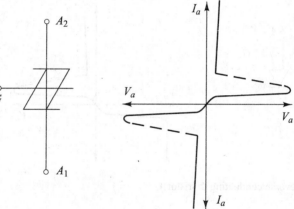

(7) LASCR (Light activated SCR)

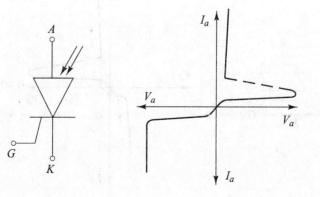

(Contd.)

Power devices	Symbols	V-I characteristics

(8) LASCS (Light activated SCS)

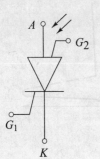 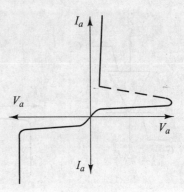

(9) PUT (Programmable unijunction transistor)

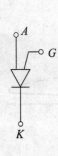 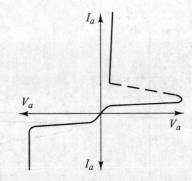

(10) RCT (Reverse conducting thyristor)

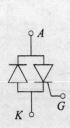 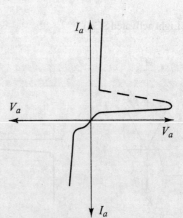

(Contd.)

Power devices	Symbols	V-I characteristics

(11) BJT (Bipolar Junction Transistor)

C
I_C
B
E

I_C
I_{Bn} $I_{Bn} > I_{B1}$
I_{B1}
O
V_{CE}

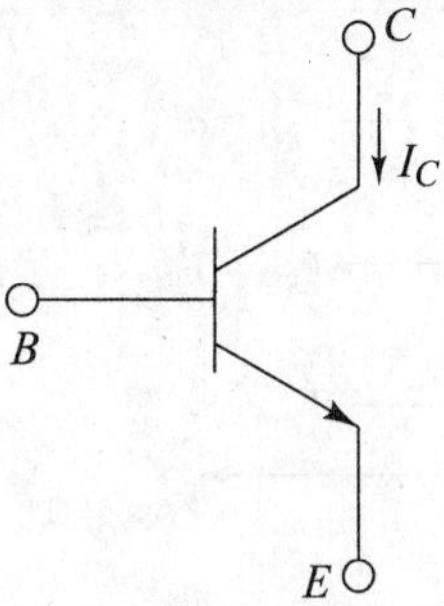

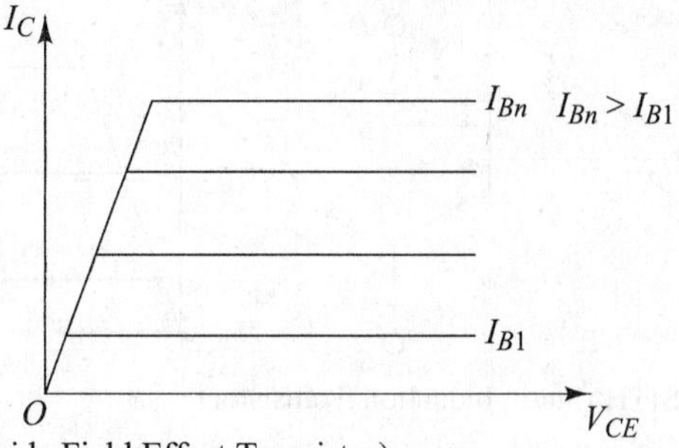

(12) N-channel MOSFET (Metal Oxide Field Effect Transistor)

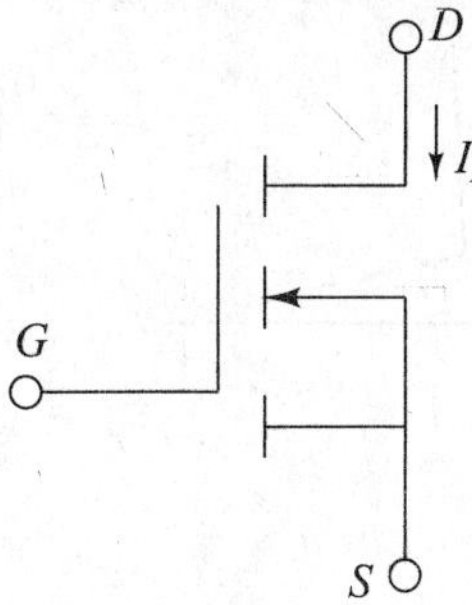

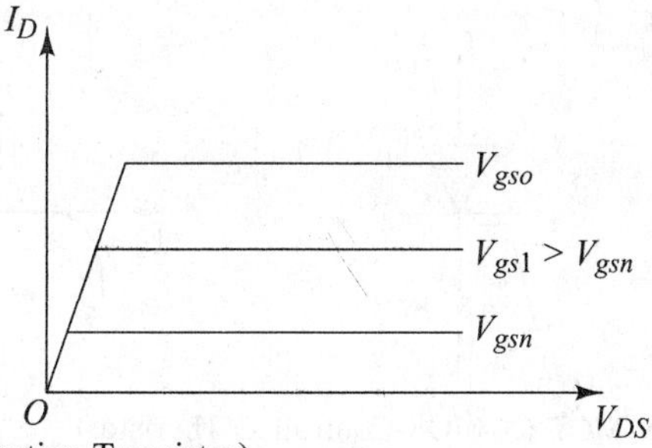

D
I_D
G
S

I_D
V_{gso}
$V_{gs1} > V_{gsn}$
V_{gsn}
O
V_{DS}

(13) IGBT (Insulated Gate Bipolar Junction Transistor)

C
I_C
G
I_E
E

I_C
V_{gsn}
V_{gs1} $V_{gsn} > V_{gs1}$
O
V_{CL}

(14) GTO (Gate-turn-off thyristors)

A
G
K

I_A
V_a
O
V_A
I_a

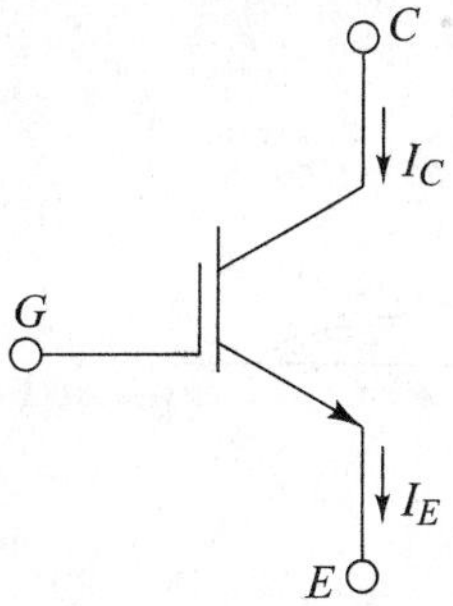

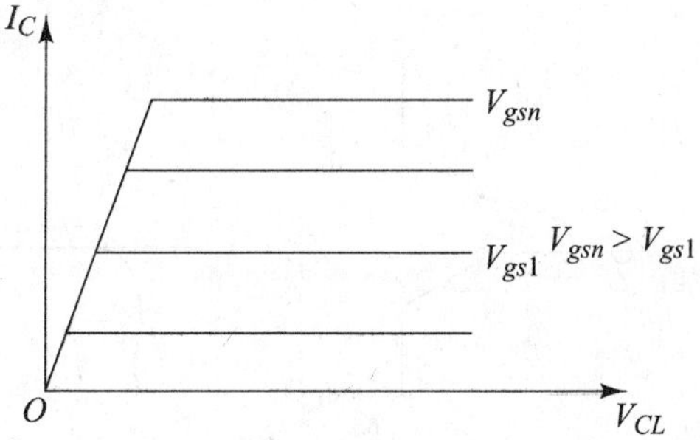

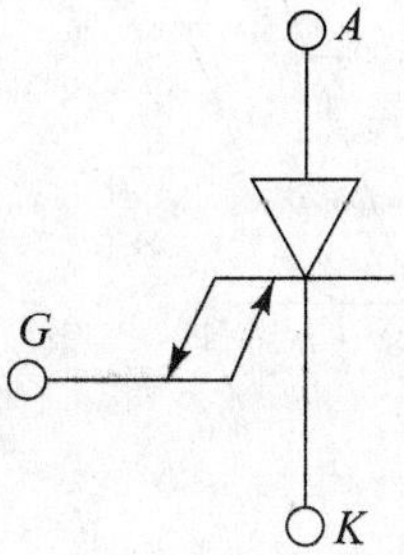

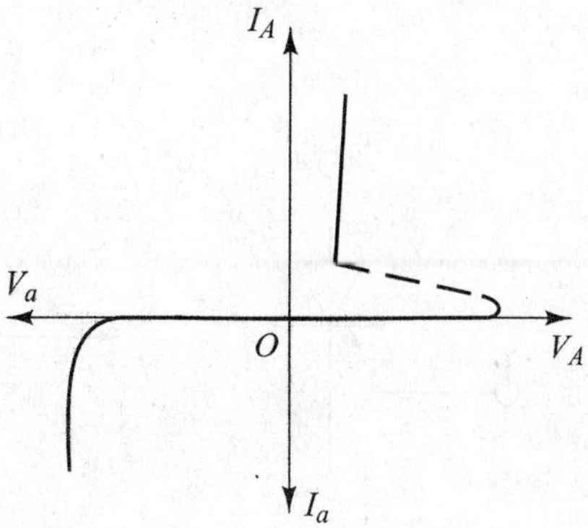

(Contd.)

| Power devices | Symbols | V-I characteristics |

(15) SIT (Static Induction Transistor)

(16) SITH (Static Induction Transistor)

(17) N-MCT (N-MOS-Controlled Thyristor)

(18) FCT (Field Controlled Thyristor)

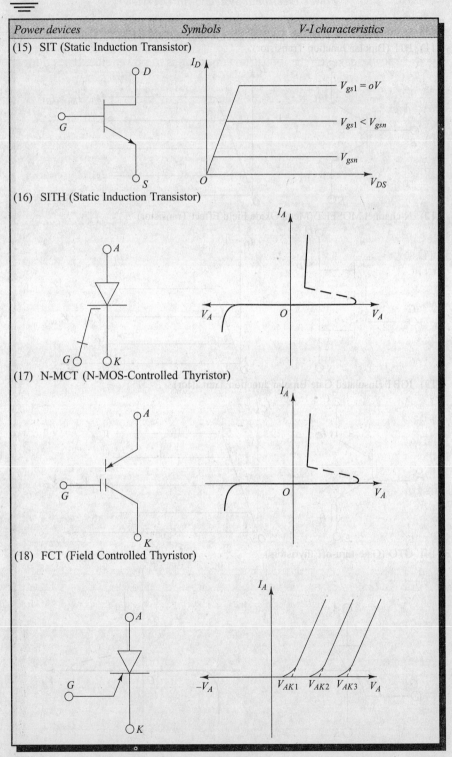

1.5 POWER ELECTRONIC CONVERTERS

The great strides taken in the industrial applications of power electronics during recent years have demonstrated that this versatile tool can be of great importance in increasing production, efficiency and control. Power Electronic Circuits are also called as power converters. A converter uses a matrix of power semiconductor switches to convert electrical power at high efficiency. The converter system is comprised of switches, reactive components L, C, and transformers. Switches include two terminal devices such as diodes and three terminal devices such as transistors or thyristors. These converters/controllers are generally classified into the following five broad categories:

1. Phase Controlled Rectifiers (AC to DC Converters) These controllers convert fixed ac voltage to a variable dc output voltage. These converters takes power from one or more ac voltage/current sources of single or multiple phases and delivers to a load. The output variable is a low-ripple dc voltage or dc current. These controller circuits use line voltage for their commutation. Hence they are also called as line commutated or naturally commutated ac to dc converters. These circuits include diode rectifiers and single/three phase controlled circuits. These controllers are discussed in detail in Chapter 6.

Applications: High voltage dc transmission systems
DC motor drives
Regulated dc power supplies
Static VAR compensator
Wind generator converters
Battery charger circuits

2. Choppers (DC to DC Converter) A chopper converts fixed dc input voltage to a variable dc output voltage. The dc output voltage may be different in amplitude than the input source voltage. Choppers are designed using semiconductor devices such as power transistors, IGBTs, GTOs, Power MOSFETs and thyristors. Output voltage can be varied steplessly by controlling the duty ratio of the device by low power signals from a control unit. Chopper has either a battery, a solar powered dc voltage source or a line frequency (50–60 Hz) derived dc voltage source. Choppers are discussed in Chapter 8.

Applications: DC drives
Subway cars
Battery driven vehicles
Electric traction
Switch mode power supplies

3. Inverters (DC to AC Converter) An inverter converts a fixed dc voltage to an ac voltage of variable frequency and of fixed or variable magnitude. A practical inverter has either a battery, a solar powered dc voltage source or a line frequency (50 Hz) derived dc voltage source (often unregulated). Inverters are widely used from very low-power portable electronic systems such as the flashlight discharge system in a photography camera to very high power industrial systems.

Inverters are designed using semiconductor devices such as power transistors, MOSFETs, IGBTs, GTOs and thyristors. Chapter 9 deals with the study of inverters in detail.

Applications: Uninterruptible power supply (UPS)
Aircraft and space power supplies
Induction and synchronous motor drives
High voltage dc transmission system
Induction heating supplies

4. Cycloconverters (AC to AC Converters) These circuits convert input power at one frequency to output power at a different frequency through one stage conversion. These are designed using thyristors and are controlled by triggering signals derived from a control unit.

The output frequency is lower than the source frequency. Output frequency in cycloconverter is a simple fraction such as $\frac{1}{3}, \frac{1}{5}$ and so on of the source frequency. These are mainly used for slow speed, very high power industrial drives. Cycloconverters are discussed in Chapter 10.

Applications: AC drives like rotary kilns multi-MW ac motor drives.

5. AC Voltage Controllers (AC Regulators) These converters convert fixed ac voltage directly to a variable ac voltage at the same frequency using line commutation. These converters employs a thyristorised voltage controller. Stepless control of the output voltage can be obtained by controlling firing angle of converter thyristors by low power signals from a control unit. This type of converters are briefly discussed in Chapter 11.

Applications: Lighting control
Speed control of large fans and pumps
Electronic tap changers

1.6 POWER ELECTRONIC APPLICATIONS

The importance of power electronics in industrial automation, energy systems, energy generation and conservation, and indirectly for environmental pollution control is tremendous. As the technology is maturing and cost is decreasing, power electronics is expanding in applications, such as switch mode power supplies (SMPS), UPS systems, electrochemical processes, heating and lighting, static VAR compensation, active filtering, high voltage dc system, photo-voltaic system, and variable frequency motor drives. The motor drives possibly constitute the most fascinating and complex applications of power electronics where the applications include computer peripherals, servos and robotics, pumps and fans, paper and textile mills, rolling mills, wind generation system, variable speed heat pump and air-conditioning, transportation system, ship propulsion etc.

The importance of power electronics is being increasingly visible now-a-days in the energy saving of electrical apparatus by more efficient use of electricity. The energy consumption in the world is increasing by leaps and bounds to improve the human living standard, particularly in industrialized countries. The major amount of this energy comes by burning fossil fuels, such as coal, natural gas and oil which create global warming effect besides urban pollution problem. The energy efficiency improvement of electrical apparatus with the help of power electronics not only reduce electricity consumption but the corresponding reduced power generation indirectly helps reduction of environmental pollution problem. It has been estimated that roughly 15% of electricity consumption can be saved by extensive application of power electronics. Table 1.2 list various applications of power-electronics. However, this list is not exhaustive.

Table 1.2 Applications of power-electronics in various sectors

Sectors	Applications
1. Home Appliances	Refrigerators, sewing machines, photography, airconditioning, food warming trays, washing machines, lighting, dryers, vaccum cleaners, electric blankets, grinders and mixers, cooking appliances
2. Games and entertainment	Games and toys, televisions, movie projectors
3. Commercial	Advertising, battery chargers, blenders, computers, electric fans, electronic ballasts, hand power tools, photocopiers, vending machines, light dimmers
4. Aerospace	Aircraft power systems, space vehicle power systems, satellite power systems
5. Automotive	Alarms and security systems, electric vehicles, audio and Rf amplifiers, regulators
6. Industrial	Blowers, boilers, chemical processing equipment, contactor and circuit breakers, conveyors, cranes and hoists, dryers, electric furnaces and ovens, electric, vehicles, electromagnets, electronic ignitions, elevators, flashers, gas-turbine starters, generator exciters, induction heating, linear induction motion control, machine tools, mining power equipments, motor drives and starters, nuclear reactor control, oil-well drilling equipment, paper mill machinery, power-supplies, printing press machinery, pumps and compressors, servo systems, steel mill instrumentation, temperature controls ultrasonic generators, uninterruptible power supplies (UPS), welding equipment
7. Medical	Fitness machines, laser power supplies, medical instrumentation
8. Security systems	Alarms and security systems, radar/sonar
9. Telecommunications	Uninterruptible power supplies (UPS), solar power supplies, VLF transmitters, wireless communication power supplies
10. Transportation	Magnetic levitation, trains and locomotives, motor drives, trolly buses, subways
11. Utility systems	VAR compensators, power factor correction, static circuit breakers, supplementary energy systems (solar, wind)

1.7 COMPUTER SIMULATION OF POWER ELECTRONIC CIRCUITS

Computer simulation plays a vital role in the analysis and design of power electronic circuits and systems. When a new converter circuit is to be developed, or a control strategy of a converter or drive system is formulated, it is often convenient to study the system performance by simulation before building the breadboard or prototype. The simulation not only validates the system's operation but also permits optimization of the system's performance by iteration of its parameters. Besides control and circuit parameters, the plant parameter variation effect can be studied. Thus, valuable time is saved in the design and development of a product, and the failure of components of poorly designed systems can be avoided.

The simulation techniques reduces the time to market the product. Simulation softwares are used to calculate the circuit waveforms, the steady state and dynamic performance of the systems, and the ratings of the various components used in the circuit. The tools used for simulation are generally classified either as circuit-oriented simulators or equation-solvers.

In circuit oriented simulator packages, the user needs to supply the circuit topology and the component values. The simulator internally generates the circuit equations that are completely transparent to the user. The user has the flexibility of selecting the details of the component models. With circuit oriented simulators, the initial set up time is small and it is easy to make changes in the circuit topology and control. Many built-in models for the components and the controllers are usually available. The focus is mainly on the circuit rather than the mathematical modelling. Circuit oriented simulators have a small control over the simulation process which results in maximum simulation time and may suffer from oscillation problems.

In equation solver packages, the user needs to supply the differential and algebraic equations which describes the circuits and the systems. These packages has the complete control over the simulation process which results in less simulation time.

Circuit-oriented simulators such as SPICE, EMTP, SABER, KREAN, etc. are available. The abbreviation SPICE stands for Simulation Program with Integrated Circuit Emphasis. It can handle nonlinearities and provides an automatic control on the time-step of integration. Many commercial versions of SPICE that operate on personal computers under several popular operating systems are available. Commercial versions can be divided into two types, i.e. mainframe versions and PC-based versions.

The mainframe versions are:
(i) PSpice (Microsim)
(ii) HSPICE (meta-software), which is designed for integrated circuit design with special device models.
(iii) RAD-SPICE (Metal Software), which simulates circuits subjected to ionizing radiation.
(iv) Precise (Electronic Engineering Software)

(v) Cadence-SPICE (Cadence Design)
(vi) Accusim (Mentor Graphics)
(vii) SPICE-Plus (Valid logic)

The PC-Versions are:

PSpice (Microsim)
AllSpice (Acotech)
IS-SPICE (Intusoft)
Z-SPICE (Z-Tech)
SPICE-plus (Analog Design Tools)
DSPICE (Daisy Systems)

In PSpice, many features are available which makes it a multilevel simulator. The controllers can be represented by their input-output behaviour-model. The input data can also be entered by drawing the circuit schematic. Free-evaluation (class-room) version of PSPICE is available. This evaluation version is very powerful for power electronic circuit simulations. PSPICE examples are provided on the on-line centre (OLC) of the book.

If we use an equation-solver, then we have to write differential and algebraic equations which describes various circuit states and the logical expressions within the controller. Then simultaneously solve these equations as a function of time. MATLAB is a very convenient equation solver. It is a software package for high-performance numerical computation and visualization. It provides an interactive environment with hundreds of built-in functions for technical computation, graphics and animation. The name MATLAB stands for MATrix LABoratory. SIMULINK is a powerful graphical user-interface to MATLAB which allows dynamic systems to be described in an easy block diagram form. A library of function blocks, such as sources, sinks, discrete, linear, nonlinear and connections can be used in the simulation. The power semiconductor device is normally simulated by the element switch, which is an element in nonlinear function block.

REVIEW QUESTIONS

1.1 Explain briefly the concept of power electronics.
1.2 State essential parts of power electronic systems.
1.3 With the help of neat block diagram, explain the functional elements of power electronic system.
1.4 What are the functions of power modulator?
1.5 Briefly explain the sources employed in power electronic systems.
1.6 State and explain the functions of various converters.
1.7 List all the motors employed in power-electronic system.
1.8 Discuss the basis of classifying power semiconductor devices into various categories.
1.9 List atleast two applications of power electronic converters.
1.10 Mention atleast two applications of power electronics in various energy activity sectors.

1.11 List the semiconductor devices which can withstand both unipolar as well as bipolar voltages.

1.12 What is the basic difference between cycloconverter and ac voltage controller?

1.13 List the devices which can withstand bidirectional current.

1.14 Write the basic difference between the equation solver packages and circuit oriented simulator packages?

REFERENCES

1. Bimal K. Bose, *Modern Power Electronics and AC Drives*, Pearson Education, Inc., 2002.
2. B.J. Baliga, "Trends in Power Semiconductor Devices, IEEE Trans. on Elec. Devices", Vol. 43, pp 1717–1731, 1996.
3. Ved Mohan, Tore M. Undeland and William P. Robbins, *Power Electronics: Converters, Applications and Design*, 2nd Ed., John Wiley & Sons, Inc., 1995.
4. B.K. Bose (Editor), *Power Electronics and Variable Frequency Drives: Technology and Applications*, IEEE Press, New York, 1997.
5. M.H. Rashid, *Power Electronics*; Pearson Education, 2002.

Chapter 2
Thyristor: Principles and Characteristics

LEARNING OBJECTIVES:
- To become familiar with the SCR.
- To define a thrysitor and its family.
- To study the V-I and dc gate-current control characteristics of an SCR.
- To explain the two-transistor analogy of silicon controlled rectifier.
- To explain the term commutation and also the different methods of commutation.
- To consider the turn-on and turn-off mechanisms of SCR.
- To establish the thyristor ratings.

2.1 INTRODUCTION

Thyristor is a general name given to a family of power semiconductor switching devices, all of which are characterised by a bistable switching action depending upon the PNPN regenerative feedback. The thyristor has four or more layers and three or more junctions. The SCR is the most widely used and important member of the thyristor family. This device has revolutionised the art of solid state power control. The SCR is almost universally referred to as the *thyristor*.

From the construction point of view, the thyristor (PNPN structure) can be best visualized as consisting of two transistors (a PNP and an NPN interconnected-to-form a regenerative feedback pair, as shown in Section 2.4). The name thyristor is derived by a combination of the capital letters from thyratron and transistor. Thus, a thyristor is a solid state device like a transistor and has characteristics similar to that of a thyratron tube.

In this chapter, we will study the operation, characteristics, ratings and commutation methods of thyristors.

2.2 PRINCIPLE OF OPERATION OF SCR

The structure and symbol of the thyristor (SCR) are shown in Fig. 2.1. It is a four layered *PNPN* switching device, having three junctions J_1, J_2 and J_3. It has three external terminals, namely, the anode (*A*), cathode (*K*) and gate (*G*). The anode and cathode are connected to the main power circuit. The gate terminal carries a low level gate current in the direction gate to cathode. Normally, the gate terminal is provided at the *P* layer near the cathode. This is known as cathode gate.

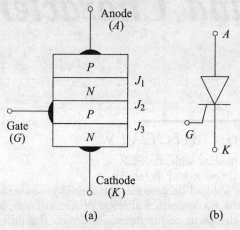

Fig. 2.1 *(a) Structure (b) symbol*

When the end *P* layer is made positive with respect to the end *N* layer, the two outer junctions, J_1 and J_3 are forward biased but the middle junction J_2 becomes reverse biased. Thus the junction J_2 because of the presence of depletion layer, does not allow any current to flow through the device. Only leakage current, negligibly small in magnitude, flows through the device due to the drift of the mobile charges. This current is insufficient to make the device conduct. The depletion layer, mostly of immobile charges do not constitute any flow of current. In other words, the SCR under the forward biased condition does not conduct. This is called as the *forward blocking state* or off-state of the device.

When the end *n* layer is made positive with respect to end *p* layer, the middle junction J_2 becomes forward biased, whereas the two outer junctions, J_1 and J_3 become reverse biased. The junctions J_1 and J_3 do not allow any current to flow through the device. Only a very small amount of leakage current may flow because of the drift of the charges. The leakage current is again insufficient to make the device conduct. This is known as the *reverse blocking state* or off-state of the device.

The width of the depletion layer at the junction J_2 decreases with the increase in anode to cathode voltage (since the width is inversely proportional to voltage). If the voltage between the anode and cathode is kept on increasing, a stage comes (corresponding to forward break-over voltage) when the depletion layer at J_2 vanishes. The reverse biased junction J_2 will breakdown due to the large voltage gradient across its depletion layer. This phenomenon is known as the *Avalanche breakdown*. Since the other junctions, J_1 and J_3 are already forward biased, there will be a free carrier movement across all the three junctions resulting in a large amount of current flowing through the device from anode to cathode. Due to the flow of this forward current, the device starts conducting and it is then said to be in the *conducting state* or on state.

2.3 STATIC ANODE–CATHODE CHARACTERISTICS OF SCR

An elementary circuit diagram for obtaining static *V–I* characteristics of a thyristor is shown in Fig. 2.2. Here, the anode and cathode are connected to the main source through a load. The gate and cathode are fed from another source E_g.

The static *V–I* characteristic of an SCR is shown in Fig. 2.3. Here, V_a is the anode–cathode voltage and I_a is the anode current. The thyristor *V-I* characteristics is divided into three regions of operation. These three regions of operation are described below.

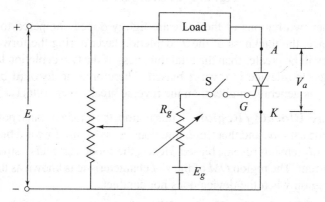

Fig. 2.2 *Elementary circuit*

1. Reverse Blocking Region When the cathode is made positive with respect to anode with the switch *s* open (Fig. 2.2), the thyristor becomes reverse biased. In Fig. 2.3, *OP* is the reverse blocking region. In this region, the thyristor exhibits a blocking characteristic similar to that of a diode. In this reverse biased condition, the outer junction J_1 and J_3 are reverse biased and the middle junction J_2 is forward biased. Therefore, only a small leakage current (in mA) flows. If the reverse voltage is increased, then at a critical breakdown level called reverse breakdown voltage V_{BR}, an avalanche will occur at J_1 and J_3 increasing the current sharply. If this current is not limited to a safe value, power dissipation will increase to a dangerous level that may destroy the device. Region *PQ* is the reverse-avalanche region. If the reverse voltage applied across the device is below this critical value, the device will behave as a high-impedance device (i.e., essentially open) in the reverse direction.

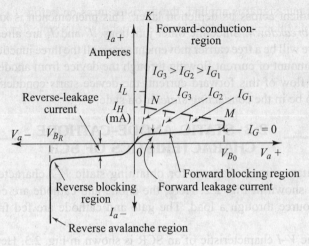

V_{B0} = Forward breakover voltage; V_{BR} = Reverse breakover voltage; I_G = Gate current; I_L = Latching current; and I_H = Holding current

Fig. 2.3 *V-I characteristics*

The inner two regions of the SCR are lightly doped compared to the outer layers. Hence, the thickness of the J_2 depletion layer during the forward biased conditions will be greater than the total thickness of the two depletion layers at J_1 and J_3 when the device is reverse biased. Therefore, the forward break–over voltage V_{BO} is generally higher than the reverse break–over voltage V_{BR}.

2. Forward Blocking Region In this region, the anode is made positive with respect to the cathode and therefore, junctions J_1 and J_3 are forward biased while the junction J_2 remains reverse biased. Hence, the anode current is a small forward leakage current. The region *OM* of the *V–I* characteristic is known as the forward blocking region when the device does not conduct.

3. Forward Conduction Region When the anode to cathode forward voltage is increased with the gate circuit kept open, avalanche breakdown occurs at the junction J_2 at a critical forward break-over voltage (V_{BO}), and the SCR switches into a low impedance condition (high conduction mode). In Fig. 2.3, the forward breakover voltage is corresponding to the point *M*, when the device latches on to the conducting state. The region *MN* of the characteristic shows that as soon as the device latches on to its *ON* state, the voltage across the device drops from say, several hundred Volts to 1–2 Volt, depending on the rating of the SCR, and suddenly a very large amount of current starts flowing through the device. The part *NK* of the characteristic is called as the forward conduction state. In this high conduction mode, the anode current is determined essentially by the external load impedance. Therefore when the thyristor conducts forward current, it can be regarded as a closed switch.

When a gate-signal is applied, the thyristor turns-on before V_{BO} is reached. The forward voltage at which the device switches to ON state depends upon the magnitude of gate current; higher the gate current, lower is the forward breakover voltage. Figure 2.3 shows that for gate current $I_G = 0$, the forward breakover voltage is V_{BO}. For I_{G1}, the forward breakover voltage is less than V_{BO} and for $I_{G2} > I_{G1}$, it is still further reduced. In practice, the magnitude of gate-current is more than the minimum gate current required to turn-on the SCR. The typical gate current magnitudes are of the order of 20 to 200 mA.

Once the SCR is conducting a forward current that is greater than the minimum value, called the *latching current*, the gate signal is no longer required to maintain the device in its ON state. Removal of the gate current does not affect the conduction of the anode current. The SCR will return to its original forward blocking state if the anode current falls below a low level, called the *holding current* (I_h). For most industrial applications, this holding current (typically 10 mA) can be regarded as being essentially zero. Note that latching current is associated with turn-on process and holding current with turn-off process. The holding current is usually lower than, but very close to the latching current. Hence, from the above discussion it becomes clear that the more convenient, reliable and efficient method of turning on the device employs the gate drive.

SOLVED EXAMPLES

Example 2.1 The latching current of a thyristor circuit in Fig. E2.1 is 50 mA. The duration of the firing pulse is 50 μs. Will the thyristor get fired?

Solution: As the SCR is triggered, the current will rise exponentially in the inductive circuit.

$$\therefore \quad i(t) = \frac{V}{R}\left(1 - e^{-t/\tau}\right)$$

where
$$\tau = \frac{L}{R} = \frac{0.5}{20} = 0.025 \text{s}$$

$$i(t) = \frac{100}{20}\left(1 - e^{-t/0.025}\right)$$

∴ At $t = 50$ μs,

$$i(50 \text{ μs}) = 5\left(1 - e^{\frac{-50 \times 10^{-6}}{0.025}}\right)$$

$$= 9.99 \text{ mA}$$

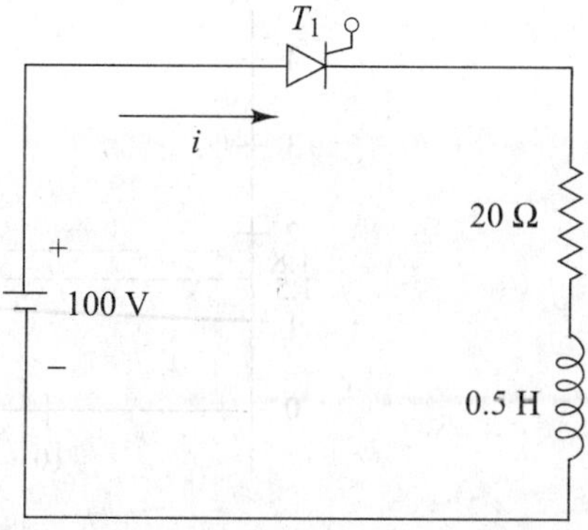

Fig. E2.1

Since the calculated circuit current value is less than the given latching current value of the SCR, it will not get fired.

Example 2.2 If the latching current in the circuit shown in Fig. E2.2 is 4 mA, obtain the minimum width of the gating pulse required to properly turn-on the SCR.

Solution: The circuit equation is,

$$V = L \frac{di}{dt}$$

where i is the latching current and t is the pulse-width.

$$\therefore \quad dt = \frac{L}{V} di$$

Integrating on both sides,

$$t = \frac{L}{V} i$$

$$t_{min} = \frac{0.1}{100} \times 4 \times 10^{-3} = 4 \, \mu s$$

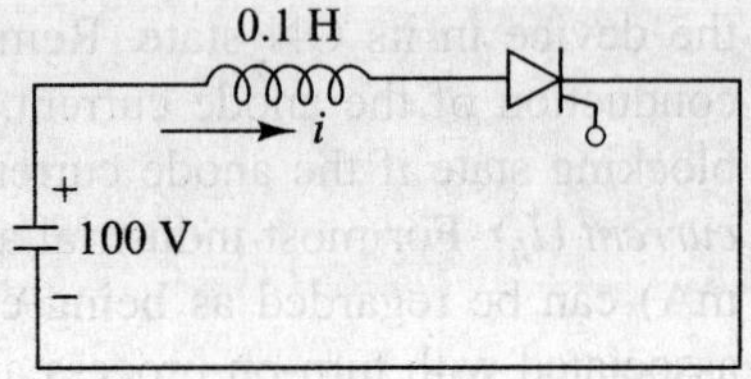

Fig. E2.2

The minimum width of the gating pulse required to properly turn-on the SCR is 4 μs.

Example 2.3 A typical V–I characteristic for a thyristor in ON state is shown in Fig. E2.3. Compute the average power-loss due to the rectangular current pulses of $I_{av} = 100A$, for conduction angles equal to
(a) 180° (b) 360°

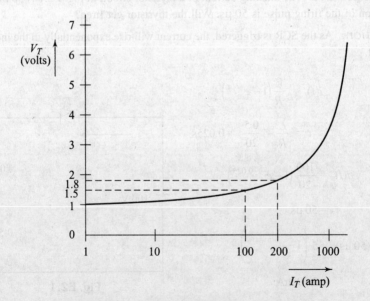

Fig. E2.3 *V–I characteristics in ON state*

Solution: Let I_m be the current during conduction and zero otherwise.

Therefore, $I_{av} = \dfrac{I_m \cdot \beta}{360°}$, where β = conduction angle.

∴ (a) For $\beta = 180°$, $I_m = \dfrac{I_{av} \times 360°}{\beta} = \dfrac{100 \times 360°}{180} = 200$ A

From the figure, the corresponding $V_T = 1.8$ V

∴ Average power loss $(P_{g_{av}}) = (1.8 \times 200) \times \dfrac{180°}{360°} = 180$ W

(b) For $\beta = 360°$, $I_m = I_{av} = 100$ A
From the figure, the corresponding $V_T = 1.5$ V

∴ Average power loss $= 1.5 \times 100 = 150$ W.

2.4 THE TWO-TRANSISTOR MODEL OF SCR (TWO TRANSISTOR ANALOGY)

The operation of an SCR can also be explained in a very simple way by considering it in terms of two transistors. This is known as the two transistor analogy of the SCR. The SCR can be considered as an *npn* and a *pnp* transistor, where the collector of one transistor is attached to the base of the other and *vice versa*, as shown in Fig. 2.4. This model is obtained by splitting the two middle layers of the SCR into two separate parts.

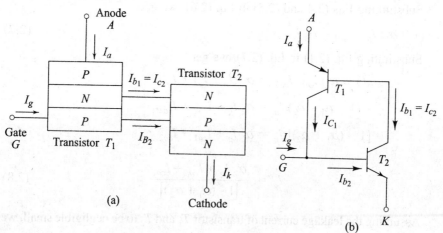

Fig. 2.4 *Two transistor analogy of SCR*

It is observed from the figure that the collector current of transistor T_1 becomes the base current of transistor T_2 and *vice versa*.

∴ $\quad I_{c1} = I_{b2}$ and $I_{b1} = I_{c2}$

Also, $\quad I_k = I_a + I_g \quad$ (2.1)

Now, we have the relation from transistor analysis,

$$I_{b1} = I_{e1} - I_{c1} \quad (2.2)$$

Also, $\quad I_{c1} = \alpha_1 I_{e1} + I_{co1} \quad$ (2.3)

where I_{co1} is the reverse leakage current of the reverse biased junction J_2 when the two outer layers are not present.

Substituting Eq. (2.3) in Eq. (2.2) we get

$$I_{b1} = I_{e1} - \alpha_1 I_{e1} - I_{co1}.$$

$$I_{b1} = (1 - \alpha_1) I_{e1} - I_{co1}$$

From Fig. 2.4, it is evident that the anode current of the device becomes the emitter current of transistor T_1 that is

$$I_a = I_{e1}$$

$\therefore \quad I_{b1} = (1 - \alpha_1) I_a - I_{co1} \quad$ (2.4)

Also, $\quad I_{c2} = \alpha_2 I_{e2} + I_{co2}$

From Fig. 2.4, it is also observed that the cathode current of the SCR becomes the emitter-current of transistor T_2.

$\therefore \quad I_k = I_{e2}$

$\therefore \quad I_{c2} = \alpha_2 I_k + I_{co2} \quad$ (2.5)

But $\quad I_{b1} = I_{c2} \quad$ (2.6)

Substituting Eqs (2.4) and (2.5) in Eq. (2.6), we get

$$(1 - \alpha_1) I_a - I_{co1} = \alpha_2 I_k + I_{co2} \quad (2.7)$$

Substituting Eq. (2.1) in Eq. (2.7), we get

$$(1 - \alpha_1) I_a - I_{co1} = \alpha_2 (I_a + I_g) + I_{co2}$$

$$(1 - \alpha_1 - \alpha_2) I_a = \alpha_2 I_g + I_{co2} + I_{co1}.$$

$$[1 - (\alpha_1 + \alpha_2)] I_a = \alpha_2 I_g + I_{co1} + I_{co2}$$

$\therefore \quad I_a = \dfrac{\alpha_2 I_g + I_{co_1} + I_{co_2}}{[1 - (\alpha_1 + \alpha_2)]} \quad$ (2.8)

Assuming the leakage current of transistor T_1 and T_2 to be negligible small, we have

$$I_a = \dfrac{\alpha_2 I_g}{1 - (\alpha_1 + \alpha_2)} \quad (2.9)$$

From Eq. (2.9), it can be analysed that if $(\alpha_1 + \alpha_2) = 1$, the value of anode current I_a becomes infinite, that is, the anode current suddenly attains a very high value, approaching infinity. In other words, we can say that the device suddenly latches into conduction (ON) state from the non-conduction (OFF) state. This characteristic of the device is known as its regenerative action. This can also be stated as the gate current I_g is of such a value that $(\alpha_1 + \alpha_2)$ approaches unity value, the device will trigger. This turn-on condition $\{(\alpha_1 + \alpha_2) \geq 1\}$ of the SCR can be satisfied in the following ways:

(a) If the temperature of the device is very high, the leakage current through it increase, which may then satisfy the required condition to turn it on.

(b) When the current through the device is extremely small, the alphas will be very small and the condition for breakover can be satisfied only by large values of hole multiplication factor M_p and electron multiplication factor M_n. Near the breakdown voltage of junction J_2, the multiplication factors are very high and the required condition for breakover can be obtained by increasing the voltage across the device to V_{BO}, which will close the breakdown voltage of junction J_2.

(c) The required condition for breakover can also be realised by increasing α_1 and α_2. In Fig. 2.4, if a current I_g is injected into the base P in the same direction as the current I_a across J_2, the current gain of the NPN transistor can now be increased independently of the anode to cathode voltage V_a and current I_a, because α_2 depends on $(I_a + I_g)$ and α_1 would still, depend on I_a. The total current gain will now depend on I_g and independent means of breakover is obtained. The presence of gate current modifies the static V-I characteristics as shown in Fig. 2.3.

2.5 THYRISTOR CONSTRUCTION

The successful and reliable operation of an SCR depends, to a large-extent on the design and fabrication of the device. The fabrication method chosen for a particular thyristor type therefore depends a great deal on the service expected from that type.

Basically, the *PNPN* device is a multilayered 'pellet' of alternate *P* and *N* type semiconductor material. This semiconductor is almost always silicon, the other element used being germanium. The pellets may be fabricated by anyone of the several methods, depending on the desired characteristics, complexity, and size of the finished device. The most popular pellet fabrication methods are:

(1) Planar diffused (all diffused)
(2) Alloy diffused.

The manufacture of both *alloy diffused* and *all diffused PNPN* pellets starts with the preparation of large area *PNP* wafers. These are formed by gaseously diffusing *P*-type impurities simultaneously into both faces of a thin wafer of *N*-type silicon. Where specific device characteristics are required, a second diffusion step is used to complete the final *PNPN* structure. To do this, each *PNP* wafer is selectively masked on a single side and subsequently diffused with *N*-type impurities through the windows in the mask. The finished *PNPN* wafers are then diced into individual pellets. Triacs and other more complex structures are fabricated using similar techniques. In the manufacture of some higher current SCR's, where only a limited number of pellets, sometimes only one, can be obtained from each wafer, the original *PNP* wafers are pelletised before adding the final *N*-region. Where this is the case, precision alloying techniques are used after pelletising to fuse a gold-antimony preform into each *PNP* pellet, thus forming the required *PNPN* structures.

2.5.1 Planar-Diffused (All Diffused)

The cross-sectional view of a typical all diffused SCR structure is shown in Fig. 2.5. As shown, the SCR consists of a four-layer pellet of *P* and *N*-type semiconductor materials. Silicon is used as the intrinsic semiconductor doped with proper impurities. The junctions are diffused type. A planar structure therefore describes a type of pellet where all the *PN* junctions come out to a single surface on the silicon pellet.

The principle advantage of planar construction is that junction information always takes place underneath a thin layer of silicon dioxide grown over the silicon wafer, before diffusion commences, which prevents contamination of silicon surfaces. As a result, planar, pellets are to a large degree protected from the outside environment. Disadvantages of planar construction are that more silicon is required per ampere of current carrying capability, and that more wafer processing steps are needed. Planar structures are best-suited, therefore, to low current devices where many pellets can be obtained from a single wafer, and to complex structures where photoresist techniques are required for geometry control.

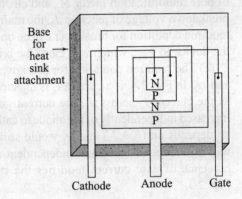

Fig. 2.5 *Cross-sectional view of planar-type SCR*

2.5.2 Alloy-Diffused [Mesa Type]

The cross-sectional view of a typical Mesa type construction is shown in Fig. 2.6. Here, the inner junction J_2 is obtained by diffusion, and then the outer two layers are alloyed to it. Since the *PNPN* pellet is required to handle large-currents, it is properly braced with tungsten or molybdenum plates to provide greater mechanical strength. One of these plates is hard soldered to a copper or an aluminum stud, that is threaded for attachment to a heat-sink. This provides an efficient thermal path for conducting the internal losses to the surrounding medium. The use of hard solder between the pellet and backup plates minimises thermal fatigue when the SCRs are subjected to temperature induced stresses. The gate or control electrode consists of a small aluminium wire that is connected to the silicon pellet. This is ohmic contact and not a rectifying junction. Control of the device is accomplished by applying a signal to the top PN junction; that is, between the gate and cathode leads.

When a larger cooling arrangement is required for high-power SCRs, the *press pack* or *hockey puck* construction is used, which provides for double-sided air or water cooling.

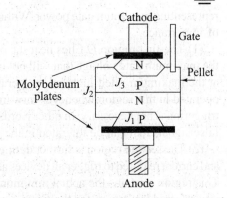

Fig. 2.6 *Cross-sectional view of alloy-diffused SCR (Mesa type)*

2.6 GATE CHARACTERISTICS OF SCR

In a thyristor, the gate is connected to the cathode through a PN junction and resembles a diode. Therefore, the V-I characteristic of a gate is similar to a diode but varies considerably in units. The circuit which supplies firing signals to the gate must be designed:

(1) to accommodate these variations,
(2) not to exceed the maximum voltage, and power capabilities of the gate,
(3) to prevent triggering from false signals or noise, and
(4) to assure desired triggering.

The design specification pertaining to gate characteristics are usually provided by the manufacturers. Figure 2.7 shows the gate characteristics of a typical SCR. Here, positive gate to cathode voltage V_g and positive gate to cathode current I_g represent d.c. values.

Applying gate drive increases the minority carrier density in the inner P layer and thereby facilitate the reverse breakdown of the junction J_2. There are maximum and minimum limits for gate voltage and gate current to prevent the permanent destruction of junction J_3 and to provide reliable triggering. Similarly, there is also a limit on the maximum instantaneous gate power dissipation ($P_{gmax} = V_g I_g$). The permissible maximum value of P_{gmax} depends on the type of gate drive. The gate signal can be d.c. or a.c. or a sequence of high frequency pulses. With pulse firing, a larger amount of instantaneous gate power-dissipation can be tolerated if the average-value of P_g is within the permissible limits. Hence, the gate can be driven harder (greater V_g and I_g) when pulse firing is used. This provides for reliable and faster turn-on of the device.

All possible safe operating points for the gate are bounded by the low and high current limits for the V-I characteristics, maximum gate voltage, and the hyperbola representing maximum gate power. Within these boundaries there are three regions of importance.

(1) The first region OA lies near the origin (shown hatched) and is defined by the maximum gate voltage that will not trigger any device. This value is obtained at the maximum rated junction temperature (usually 125°C). The gate must be operated in this region whenever forward bias is applied across the thyristor and triggering is not necessary. In other words, this region sets a limit on the maximum false signals that can be tolerated in the gate-firing circuit.

(2) The second region is further defined by the minimum value of gate-voltage and current required to trigger all devices at the minimum rated junction temperature. This region contains the actual minimum firing points of all devices. In a sense, it is a forbidden region for the firing circuit because a signal in this region may not always fire all devices or never fire any at all. In Fig. 2.7, OL and OV are the minimum gate-voltage and gate current limits respectively.

(3) The third region is the largest and shows the limits on the gate-signal for reliable firing. Ordinarily, a signal in the lower left part of this region is adequate for firing. For applications, where fast turn-on is required, a 'hard' firing signal in the upper right part of the region may be needed.

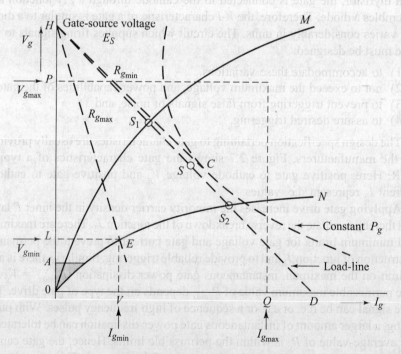

Fig. 2.7 *Gate characteristics*

In Fig. 2.7 curves ON and OM corresponds to the possible spread of the characteristic for SCRs of the same rating. For best results, the operating point S, which may change from S_1 to S_2, must be as close as possible to the permissible

P_g curve and must be contained within the maximum and minimum limits of gate voltage and gate current. This provides the necessary hard drive for the device. For selecting the operating point, usually a load line of the gate source voltage $E_s = OH$ is drawn as HD. The gradient of the load line HD (= OH/OD) will give the required gate source resistance R_g. The maximum value of this series resistance is given by the line HE, where E is the point of intersection of lines indicating the minimum gate voltage and gate current. The minimum value of gate source series resistance is obtained by drawing a line HC tangential to P_g curve.

A thyristor may be considered to be a charged controlled device. Thus, higher the magnitude of gate current pulse, lesser is the time needed to inject the required charge for turning on the thyristor. Therefore the SCR turn-on time can be reduced by using gate current of higher magnitude. It should be ensured that pulse width is sufficient to allow the anode current to exceed the latching current. In practice, the gate pulse width is usually taken as equal to or greater than SCR turn-on time, t_{on}. If T is the pulse width as shown in Fig. 2.8, then

$$T \geq t_{on}$$

With pulse firing, if the frequency of firing f is known, the peak instantaneous gate power dissipation P_{gmax} can be obtained as

$$P_{gmax} = V_g I_g = \frac{P_{gav}}{fT} \qquad (2.10)$$

where $\qquad f = \dfrac{1}{T_1}$ = frequency of firing or pulse repetition rate in Hz

and $\qquad T$ = pulse width in second

A duty cycle is defined as the ratio of pulse-on period to the periodic time of pulse. In the Fig. 2.8 pulse-on period is T and the periodic time is T_1. Therefore, duty-cycle is given by

$$\delta = \frac{T}{T_1} = fT \qquad (2.11)$$

From Eq. (2.10)

$$\frac{P_{gav}}{\delta} \leq P_{gmax} \qquad (2.12)$$

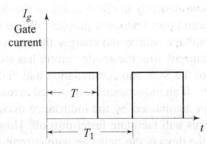

Fig. 2.8 Pulse gating

2.6.1 Gate Circuit Parameters

The gate cathode circuit with different circuit parameters is shown in Fig. 2.9. A series resistance R_g should be placed in series with the gate-source voltage E_g, to limit the magnitude of gate voltage and gate-current.

The shunt resistor R_{gc} is introduced to bypass the thermally generated leakage current across junction J_2 when the device is in the blocking state, in order to improve the thermal-stability of the device. This shunt resistance in turn will increase the required gate current and also the device holding and latching current levels.

Diode D_1 applies a negative voltage between the gate and the cathode when a reverse voltage is applied across the device. This negative gate-voltage reduces the reverse blocking current and improves the turn-off

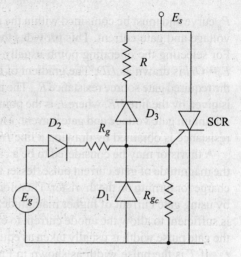

Fig. 2.9 Gate circuit

mechanism. This diode D_1 also serves to limit the reverse voltage applied between the cathode and the gate, if the gate, source voltage E_g is alternating. The negative gate current flows through the device while the SCR is ON because the diode D_1 will then be reverse biased. This will increase the dissipation of gate power. A series diode D_2 in the circuit will prevent the negative source current. Another diode D_3 is connected as shown in the figure to block the positive gate current coming from the supply when the device is forward biased.

A shunt capacitor C_s may be connected across gate to cathode to improve the dv/dt capability. However, pulse firing results in a larger portion of the gate drive being bypassed by the capacitor which will increase the delay time and consequently the di/dt rating of the device is also lowered. This shunt capacitor also poses one more problem. When the device is turned ON, the gate acts as a voltage source and charges the capacitor. This charge can provide enough gate current after the anode current has stopped thereby increasing the turn-off time of the SCR and commutation may fail.

If an inductance is connected across gate to cathode, the negative gate current is maintained by the inductance even after the anode-current has stopped, and this will facilitate faster turn-off. However, when pulse firing is used for gating the device, the negative gate current that continues to flow out of the gate can possibly turn-off the thyristor.

Thus, depending upon the specific requirements, gate circuit parameters are chosen. The use of a negative voltage bias between the gate and cathode is generally recommended. This will increase the forward breakover voltage and dv/dt withstanding capability. Similarly, the reverse leakage current will also be reduced by the negative gate bias. The only drawback is that a greater gate source voltage E_g is required to overcome this bias and turn-on the SCR.

SOLVED EXAMPLES

Example 2.4 An SCR has a V_g–I_g characteristics given as $V_g = 1.5 + 8\ I_g$. In a certain application, the gate voltage consists of rectangular pulses of 12 V and of duration 50 μs with duty cycle 0.2.

(a) Find the value of R_g series resistor in gate circuit to limit the peak power dissipation in the gate to 5 watts.

(b) Calculate average power dissipation in the gate.

Solution: During conduction,

$$V_{gs} = R_g I_g + V_g = R_g I_g + 1.5 + 8\ I_g$$

$$12 = (R_g + 8)\ I_g + 1.5 \qquad (i)$$

Peak power loss = $V_g I_g = 5$ ∴ $5 = (1.5 + 8I_g)\ I_g$

∴ $8 I_g^2 + 1.5 I_g - 5 = 0$ ∴ $I_g = \dfrac{-1.5 \pm \sqrt{(1.5)^2 - 160}}{16} = 0.7\ A$

Substitute I_g in Eq. (i), ∴ $12 = (R_g + 8)\ 0.7 + 1.5,\ R_g = 7\ \Omega$

Now, Average, power loss = peak power loss × duty cycle = 5 × 0.2 = 1W

Example 2.5 If the $V_g - I_g$ characteristics of an SCR is assumed to be a straight line passing through the origin with a gradient of 3×10^3, calculate the required gate source resistance. Given $E_{gs} = 10V$ and allowable $P_g = 0.012\Omega$.

Solution: The allowable $P_g = 0.012$ watt, ∴ $V_g I_g = 0.012$ (i)

Also, gradient = $\dfrac{V_g}{I_g} = 3 \times 10^3$, ∴ $V_g = 3 \times 10^3\ I_g$

Substituting V_g in Eq (i),

$(3 \times 10^3 \times I_g \times I_g) = 0.012, I_g = 2$ mA. ∴ $V_g = 3 \times 10^3 \times 2 \times 10^{-3} = 6$ V.

Example 2.6 A thyristor has a forward characteristic which may be approximated over its normal working range to the straight line shown in Fig. E2.6. Calculate the mean power-loss for—

(a) a continues on state current of 23 A.

(b) a half-sine wave of mean value 18 A.

(c) a level current of 39.6 A for one-half cycle.

(d) a level current of 48.5 A for one third cycle.

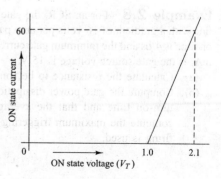

Fig. E2.6

Solution: (a) At a ON state current of 23A, the ON state voltage from Fig. E2.6 is

$$V_T = 1 + \frac{23 \times 1.1}{60} = 1.42 \text{ V}$$

∴ Power loss = $V_T I_T = 1.42 \times 23 = 32.7$ W

(b) The maximum value of the sine wave = 18π A

From the figure, at any current I, Voltage $V = 1 + \frac{1.1}{60} I$.

Over one cycle, the total base length is 2π from 0 to π, $I = 18\pi \sin x$, and from π to 2π, $I = 0$.

∴ Mean power = $\dfrac{1}{2\pi} \displaystyle\int_0^\pi V I \, dx = \dfrac{1}{2\pi} \displaystyle\int_0^\pi V\left(I + \dfrac{1.1}{60} 18\pi \sin x\right) 18\pi \sin x \, dx = 32.6$ W

(c) The mean power loss will be h the instantaneous power loss over the half cycle when the current is flowing.

∴ Mean power = $\left[39.6\left(1 + \dfrac{1.1}{60} 39.6\right)\right]/2 = 34.2$ W

(d) Now, the mean power loss for a level current of 48.5 A for one-third cycle is given by

Mean power = $\left[48.5\left(1 + \dfrac{1.1}{60} 48.5\right)\right]/3 = 30.5$ W.

Example 2.7 Compute the peak inverse voltage of thyristor connected in the three phase, six pulse bridge circuit having input voltage of 415 V. Voltage safely factor is 2.1.

Solution: We have the relation,

$$PIV = \sqrt{2} \times V_{in} \times V_f = \sqrt{2} \times 415 \times 2.1 = 1232.49 \text{ V}.$$

Example 2.8 For an SCR, the gate cathode characteristic is given by a straight line with a gradient of 16 volts per amp passing through the origin, the maximum turn-on time is 4 μs and the minimum gate current required to obtain this quick turn-on is 500 mA. If the gate source voltage is 15 V,
 (a) Calculate the resistance to be connected in series with the SCR gate.
 (b) Compute the gate power dissipation, given that the pulse width is equal to the turn-on time and that the average gate power dissipation is 0.3 W. Also, compute the maximum triggering frequency that will be possible when pulse firing is used.

Thyristor: Principles and Characteristics

Solution: (a) Given : $I_{g_{min}} = 500$ mA $= 0.5$ A, $\dfrac{V_g}{I_g} = 16$ V/A $\therefore V_g = 16 \times 0.5 = 8$ V

From Fig. E1.7,

$$R_S = \dfrac{E_{gs} - V_g}{I_g} = \dfrac{(15-8)}{0.5} = R_S = 14\ \Omega$$

(b)
Power dissipation,

$$P_g = V_g I_g$$
$$= 8 \times 0.5 = 4\ \text{W}$$

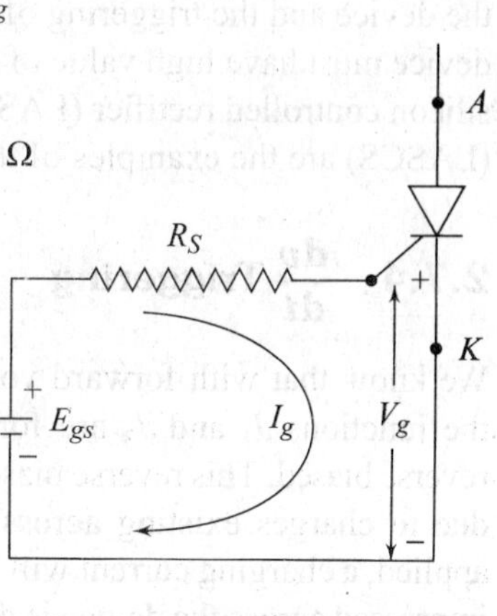

Now, $P_{g_{max}} = \dfrac{P_{g_{av}}}{f \cdot T_{on}}$ $\therefore$ $4 = \dfrac{0.3 \times 10^6}{f \times 4}$

$f = 18.75$ kHz $\therefore$ F ≈ 19 kHz.

Fig. E2.8

2.7 TURN-ON METHODS OF A THYRISTOR

A thyristor can be switched from a nonconducting state to a conducting state in several ways described as follows.

2.7.1 Forward Voltage Triggering

When anode-to-cathode forward voltage is increased with gate circuit open, the reverse biased junction J_2 will have an avalanche breakdown at a voltage called forward breakover voltage V_{BO}. At this voltage, a thyristor changes from OFF state (high voltage with low leakage current) to ON-state characterised by a low voltage across it with large forward current. The forward voltage-drop across the SCR during the ON state is of the order of 1 to 1.5 V and increases slightly with load current.

2.7.2 Thermal Triggering (Temperature Triggering)

Like any other semiconductor, the width of the depletion layer of a thyristor decreases on increasing the junction temperature. Thus, in a thyristor when the voltage applied between the anode and cathode is very near to its breakdown voltage, the device can be triggered by increasing its junction temperature. By increasing the temperature to a certain value (within the specified-limits), a situation comes when the reverse biased junction collapses making the device conduct. This method of triggering the device by heating is known as the thermal triggering process.

2.7.3 Radiation Triggering (Light Triggering)

In this method, as the name suggests, the energy is imparted by radiation. Thyristor is bombarded by energy particles such as neutrons or photons. With the help of this external energy, electron-hole pairs are generated in the device, thus increasing the number of charge carriers. This leads to instantaneous flow of current within the device and the triggering of the device. For radiation triggering to occur, the device must have high value of rate of change of voltage (dv/dt). Light activated silicon controlled rectifier (LASCR) and light activated silicon controlled switch (LASCS) are the examples of this type of triggering.

2.7.4 $\dfrac{dv}{dt}$ Triggering

We know that with forward voltage across the anode and cathode of a device, the junctions J_1 and J_3 are forward biased, whereas the junction J_2 becomes reverse biased. This reverse biased junction J_2 has the characteristics of a capacitor due to charges existing across the junction. If a forward voltage is suddenly applied, a charging current will flow tending to turn the device ON. If the voltage impressed across the device is denoted by V, the charge by Q and the capacitance by C_j, then

$$i_c = \frac{dQ}{dt} = \frac{d}{dt}(C_j V) = C_j \frac{dV}{dt} + V \frac{dc_j}{dt} \qquad (2.13)$$

The rate of change of junction capacitance may be negligible as the junction capacitance is almost constant. The contribution to charging current by the later term is negligible. Hence, Eq. (2.13) reduces to

$$i_c = C_j \frac{dV}{dt} \qquad (2.14)$$

Therefore, if the rate of change of voltage across the device is large, the device may turn-on even though the voltage appearing across the device is small.

2.7.5 Gate Triggering

This is the most commonly used method for triggering SCRs. In laboratories, almost all the SCR devices are triggered by this process. By applying a positive signal at the gate terminal of the device, it can be triggered much before the specified breakover voltage. The conduction period of the SCR can be controlled by varying the gate signal within the specified values of the maximum and minimum gate currents.

For gate triggering, a signal is applied between the gate and the cathode of the device. Three types of signals can be used for this purpose. They are either d.c. signals, pulse signals or ac signals.

1. D.C. Gate Triggering In this type of triggering, a d.c. voltage of proper magnitude and polarity is applied between the gate and the cathode of the device in such a way that the gate becomes positive with respect to the cathode. When the applied voltage is sufficient to produce the required gate current, the device starts conducting.

One drawback of this scheme is that both the power and control circuits are d.c. and there is no isolation between the two.

Another disadvantage of this process is that a continuous d.c. signal has to be applied, at the gate causing more gate power loss.

2. A.C. Gate Triggering a.c. source is most commonly used for the gate signal in all application of thyristor control adopted for a.c. applications. This scheme provides the proper isolation between the power and the control circuits. The firing angle control [discussed in Chapter 6] is obtained very conveniently by changing the phase angle of the control signal.

However, the gate drive is maintained for one half cycle after the device is turned ON, and a reverse voltage is applied between the gate and the cathode during the negative half cycle. The drawback of this scheme is that a separate transformer is required to step down the a.c. supply, which adds to the cost.

3. Pulse Gate Triggering This is the most popular method for triggering the device. In this method, the gate drive consists of a single pulse appearing periodically or a sequence of high frequency pulses. This is known as carrier frequency gating. A pulse transformer is used for isolation. The main advantage of this method is that there is no need of applying continuous signals and hence, the gate losses are very much reduced. Electrical isolation is also provided between the main device supply and its gating signals.

2.8 DYNAMIC TURN-ON SWITCHING CHARACTERISTICS

The static characteristics gives no indication as to the speed at which the SCR is capable of being switched from the forward blocking voltage to the conducting state and vice-versa. However, the transition from one state to the other does not take place instantaneously, it takes a finite period of time. This is illustrated in Fig. 2.10. As shown, the total turn-on time t_{on} of the SCR is subdivided into three distinct periods, called the *delay time, rise time* and *spread time*. These time periods are defined in terms of the waveforms of the anode voltage and current obtained in a circuit in which the anode-load consists of a pure-resistance.

(i) **Delay time (t_d):** This is the time between the instant at which the gate-current reaches 90% of its final value and the instant at which the anode current reaches 10% of its final value.

It can also be defined as the time during which anode voltage falls from V_a to $0.9\ V_a$, where V_a is the initial value of the anode voltage.

The gate current has non-uniform distribution of current density over the cathode surface due to the *p*-layer. Its value is much higher near the gate but decreases rapidly as the distance from the gate increases. It shows that during t_d, anode current flows in a narrow region near the gate where gate current density is the highest.

(ii) **Rise Time (t_r):** This is the time required for the anode current to rise from 10 to 90% of its final value. It can also be defined as the time required for the forward blocking off-state voltage to fall from 0.9 to 0.1 of its initial value-OP.

This time is inversely proportional to the magnitude of gate current and its build up rate. Thus, t_r can be minimized if high and steep current pulses are applied to the gate. For series *RL* circuit, the rate of rise of anode current is slow, therefore, t_r is more and for the *RC* series circuit, di/dt is high thus t_r is less. During rise-time, turn-on losses are the highest due to high anode voltage V_a and large anode current I_T occuring together in the thyristor.

(iii) **Spread-time (t_s):** The spread time is the time required for the forward blocking voltage to fall from 0.1 to its value to the on-state voltage drop (1 to 1.5 V). After the spread time, anode current attains steady-state values and the voltage drop across SCR is equal to the on-state voltage drop of the order of 1 to 1.5 V.

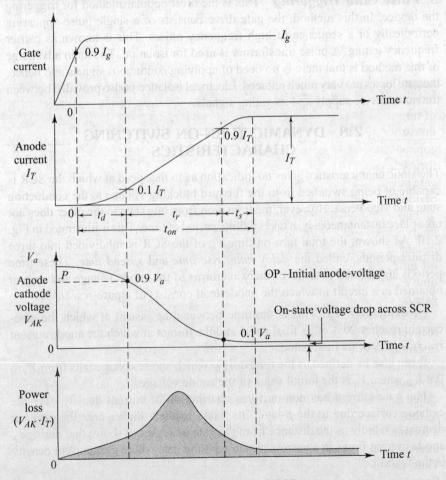

Fig. 2.10 *Waveforms during SCR turn-on*

(iv) **Turn-on Time (t_{on}):** This is the sum of the delay time, rise-time and spread time. This is typically of the order of 1 to 4 μs, depends upon the anode circuit parameters and the gate signal waveshapes.

The width of the firing pulse should, therefore, be more than 10 μs, preferably in the range of 20 to 100 μs. The amplitude of the gate-pulse should be 3 to 5 times the minimum gate current required to trigger the SCR.

From Fig. 2.10, it is noted that during rise-time, the SCR carries a large forward current and supports an appreciable forward voltage. This may result in high-instantaneous power dissipation creating local internal hot-spots which could destroy the device. It is, therefore, necessary to limit the rate of rise of current. Normally, a small inductor, called *di/dt* inductor is inserted in the anode circuit to limit the *di/dt* of the anode current.

The shadow area under the power-curve in Fig. 2.10 represents the switching loss of the device. This loss may be significant in high-frequency applications.

2.9 TURN-OFF MECHANISM (TURN-OFF CHARACTERISTIC)

Once the SCR starts conducting an appreciable forward current, the gate has no control on it and the device can be brought back to the blocking state only by reducing the forward current to a level below that of the holding current. Process of turn-off is also called as commutation. Various methods used for turning off thyristors will be discussed in Section 2.10. However, if a forward voltage is applied immediately after reducing the anode current to zero, it will not block the forward voltage and will start conducting again, although it is not triggered by a gate pulse. It is, therefore, necessary to keep the device reverse biased for a finite period before a forward anode voltage can be reapplied.

The turn-off time of the thyristor is defined as the minimum time interval between the instant at which the anode current becomes zero, and the instant at which the device is capable of blocking the forward voltage. The turn-off time is illustrated by the waveforms shown in Fig. 2.11. The total turn-off time t_{off} is divided into two time intervals the reverse, recovery time t_{rr} and the gate recovery time t_{gr}.

At the instant t_1, the anode forward current becomes zero. During the reverse recovery time, t_1 to t_3, the anode current flows in the reverse direction. At the instant t_2, a reverse anode voltage is developed and the reverse recovery current continues to decrease. At t_3, junction J_1 and J_3 are able to block a reverse voltage. However, the thyristor is not yet able to block a forward voltage because carriers, called *trapped charges*, are still present at the junction J_2. During the interval t_3 to t_4, these carriers recombine. At t_4, the recombination is complete and therefore, a forward voltage can be reapplied at this instant. The SCR turn-off time is the interval between t_4 and t_1. In an SCR, this time varies in the range 10 to 100 μs. Thus, the total turn-off time (t_q) required for the device is the sum of the duration for which the reverse recovery current flows after the application of reverse voltage, and the time required for the recombination of all excess carriers in the

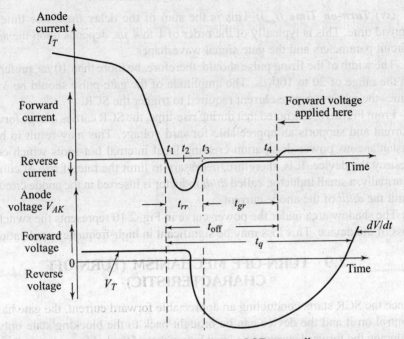

Fig. 2.11 *Waveforms during SCR turn-off*

inner two layers of the device. This may be noted that in case of highly inductive load circuit, the current cannot change abruptly at t_1. Also, the fast change in current at t_2 may give rise to high voltage surges in the inductance, which will then appears across the terminals of the thyristor.

In practical applications, the turn-off time required to the SCR by the circuit, called the circuit turn-off time t_q, must be greater than the device turn-off time t_{off} by a suitably safe margin, otherwise the device will turn-on at an undesired instant a process known as *commutation failure*. Thyristor having large turn-off time (50–100 μs) are called as slow switching or phase control type thyristors (or converter grade thyristors), and those having low turn-off time (10–50 μs) are called fast switching or inverter type thyristors. In high frequency applications, the required circuit turn-off time consumes an appreciable portion of the total cycle time and therefore, inverter grade thyristors must be used.

2.10 TURN-OFF METHODS

The term *commutation* basically means the transfer of current from one path to another. In thyristor circuits, this term is used to describe process of transferring current from one thyristor to another. As explained earlier, it is not possible for a thyristor to turn itself OFF; the circuit in which it is connected must reduce the thyristor current to zero to enable it to turn-off. 'Commutation' is the term to describe the methods of achieving this.

Commutation is one of the fundamental principles the use of thyristors for control purposes. A thyristor can only operate in two modes: it is either in the OFF state, i.e., open circuit, or in the ON state, i.e., short circuit. By itself it cannot control the level of current or voltage in a circuit. Control can only be achieved by variation in the time thyristors when switched ON and OFF, and commutation is central to this switching process. All thyristor circuits, therefore, involve the cyclic or sequential switching of thyristors. The two methods by which a thyristor can be commutated are as follows.

2.10.1 Natural Commutation

The simplest and most widely used method of commutation makes use of the alternating, reversing nature of a.c. voltages to effect the current transfer. We know that in a.c. circuits, the current always passes through zero every half cycle. As the current passes through natural zero, a reverse voltage will simultaneously appear across the device. This immediately turns-off the device. This process is called as *natural commutation* since no external circuit is required for this purpose. This method may use a.c. mains supply voltages or the a.c. voltages generated by local rotating machines or resonant circuits. The line commutated converters and inverters comes under this category.

2.10.2 Forced Commutation

Once thyristors are operating in the ON state, carrying forward current, they can only be turned OFF by reducing the current flowing through them to zero for sufficient time to allow the removal of charged carriers. In case of d.c. circuits, for switching off the thyristors, the forward current should be forced to be zero by means of some external circuits. The process is called *forced commutation* and the external circuits required for it are known as commutation circuits. The components (inductance and capacitance) which constitute the commutating circuits are called as commutating components. A reverse voltage is developed across the device by means of a commutating circuit that immediately brings the forward current in the device to zero, thus turning off the device. Producing reliable commutation is a difficult problem to be tackled while designing chopper and inverter circuits. The most important stage in the designing process is choosing a forced turn-off method and deciding its components.

The classification of the methods of forced commutation is based on the arrangement of the commutating components and the manner in which zero current is obtained in the SCR. There are six basic methods of commutation by which thyristors may be turned OFF.

1. Class A–self Commutation by Resonating the Load This is also known as resonant commutation. This type of commutation circuit using L-C components-in-series-with the load are shown in Fig. 2.12. In Fig. 2.12(a), load R_L is in parallel with the capacitor and in Fig. 2.12(b) load R_L is in series with the

L–C circuit. In this process of commutation, the forward current passing through the device is reduced to less than the level of holding current of the device. Hence, this method is also known as the current commutation method. The waveforms of the thyristor voltage, current and capacitor voltages are shown in Fig. 2.13.

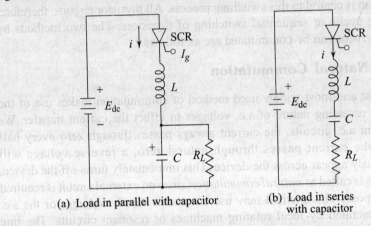

(a) Load in parallel with capacitor   (b) Load in series with capacitor

Fig. 2.12 *Class A commutation circuit*

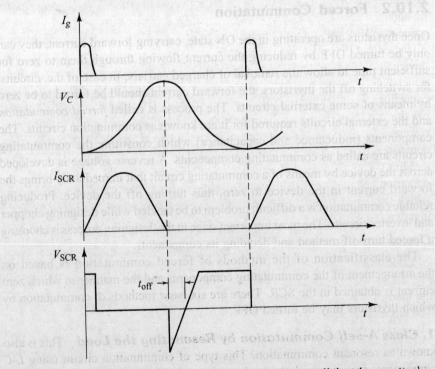

Fig. 2.13 *Voltages and currents in Class A (load is parallel with capacitor)*

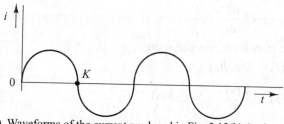

(a) Waveforms of the current produced in Fig. 2.12(b) (series capacitor)

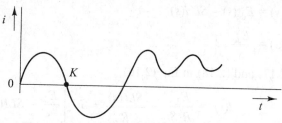

(b) Waveforms of the current produced in Fig. 2.12(a) (parallel capacitor)

Fig. 2.14

The load resistance R_L and the commutating components are so selected that their combination forms an underdamped resonant circuit. When such a circuit is excited by a d.c. source, a current of the nature shown in Fig. 2.14 will be obtained across the device. This current, as evident from its shape, has zero value at the point K where the device is automatically turned OFF. Beyond point K, the current is reversed in nature which assures definite commutation of the device. The thyristor when ON carries only the charging current of capacitor C which will soon decay to a valueless than the holding current of the device, when capacitor C is charged up to the supply voltage E_{dc}. This simultaneously switches off the thyristor. The time for switching off the device is determined by the resonant frequency which in turn depends on the values of the commutating components L and C, and the total load resistance.

This type of commutation circuits are most suitable for high frequency operation, i.e., above 1000 Hz, because of the need for an L–C resonant circuit which carries the full load current. This commutation circuit is used in series inverter.

Design Considerations

(a) Load in parallel with capacitor C Let us consider the resonant circuit of Fig. 2.12 (a). Let E_{dc} be the applied d.c. voltage, V be the load voltage, and i be the load current.
The circuit equation is

$$E_{dc} = L\frac{di}{dt} + V$$

and $$i = C\frac{dV}{dt} + \frac{V}{R}$$

By using Laplace transform, we can write

$$E_{dc}(s) - V(s) = S \cdot L\, I(s) \tag{2.15}$$

and $$I(s) = \frac{V(s)}{R} + S \cdot C \cdot V(s) \tag{2.16}$$

From Eq. (2.15), we can write

$$V(s) = E_{dc}(s) - SL\, I(s) \tag{2.17}$$

But $$E_{dc}(s) = \frac{E_{dc}}{S} \tag{2.18}$$

Substitute Eqs (2.17) and (2.18) in Eq. (2.16)

$$\therefore \quad I(s) = \frac{E_{dc}}{R \cdot S} - \frac{SLI(s)}{R} + SC\left[\frac{E_{dc}}{s} - SLI(s)\right]$$

$$I(s) + SL\frac{I(s)}{R} + S^2 CLI(s) = \frac{E_{dc}}{R \cdot S} + \frac{E_{dc}SC}{S}$$

$$\therefore \quad I(s)\left[1 + \frac{SL}{R} + S^2 CL\right] = \frac{E_{dc}}{S}\left[\frac{1}{R} + SC\right]$$

$$I(s)\left[\frac{R + LS + RCLS^2}{R}\right] = \frac{E_{dc}}{s}\left[\frac{1 + RCS}{R}\right]$$

$$I(s) = \frac{E_{dc}}{s}\left[\frac{1 + RCS}{R + LS + RCLS^2}\right]$$

$$I(s) = \frac{E_{dc}}{RLCS}\left[\frac{1 + RCS}{S^2 + \frac{1}{RC}S + \frac{1}{LC}}\right] \tag{2.19}$$

Taking inverse Laplace transform of Eq. (2.19), we get

$$i(t) = \frac{E_{dc}}{R}\left[1 + \frac{1}{\sqrt{1-\varepsilon^2}}\frac{W_n^2}{\varepsilon}e^{-t/RC}\sin(\omega t + \phi)\right]$$

where $$\varepsilon = \frac{1}{2R}\sqrt{\frac{L}{C}} \quad = \text{damping ratio}$$

$$W_n = \frac{1}{\sqrt{LC}} \quad = \text{undamped natural angular frequency.}$$

$$\omega = \omega_n \sqrt{1-\varepsilon^2}$$

or,
$$\omega = \frac{1}{\sqrt{LC}} \sqrt{1 - \frac{L}{4R^2C}} = \sqrt{\frac{1}{LC} - \frac{1}{4R^2C^2}}$$

$$\phi = \tan^{-1} \frac{2RC\omega}{-\varepsilon} - \tan^{-1} \frac{\sqrt{1-\varepsilon^2}}{-\varepsilon} = \tan^{-1} 2RC\omega$$

If $\quad i(t) = 0$ at $t = 0$,

$$\phi = -\sin^{-1} \frac{1}{A}$$

$$\therefore \quad i(t) = \frac{E_{dc}}{R}\left[1 + A e^{-t/2RC} \sin\left(\omega t - \sin^{-1}\frac{1}{A}\right)\right] \quad (2.20)$$

Now, load voltage from Eqs (2.15) and (2.16) can be written as

$$V(s) = \frac{E_{dc}}{LC\left(S^2 + \frac{1}{RC}S + \frac{1}{LC}\right)} \quad (2.21)$$

Taking inverse Laplace transform of Eq. (2.21), we get

$$V(t) = E_{dc}\frac{W_n}{\sqrt{1-\varepsilon^2}} e^{-t/2RC} \sin \omega t + E_{dc} \quad (2.22)$$

In this case, the triggering frequency of the thyristor must be less than W_n, so that the conduction cycle is completed.

(b) Load in series with capacitor C Let us consider the series resonant circuit of Fig. 2.12 (b). Let the thyristor be turned ON at $t = 0$ with the initial capacitor voltage zero.

The circuit equation is

$$E_{dc} = iR + L\frac{di}{dt} + \frac{1}{C}\int i\,dt \quad (2.23)$$

On differentiating and dividing by L, we get

$$\frac{d^2i}{dt^2} + \frac{R}{L}\frac{di}{dt} + \frac{i}{LC} = \frac{1}{L}\frac{d}{dt}E_{dc} \quad (2.24)$$

The corresponding homogeneous equation is of the second order and is as below.

$$\frac{d^2i}{dt^2} + \frac{R}{L}\cdot\frac{di}{dt} + \frac{1}{LC}i = 0 \quad (2.25)$$

The solution of this well known second order equation for under damped case is

$$i = e^{-\varepsilon t}[A_1 \cos \omega t + A_2 \sin \omega t] \quad (2.26)$$

where
$$\varepsilon = \frac{R}{2l} \quad (2.27)$$

and
$$\omega_o = \frac{1}{\sqrt{LC}} \quad (2.28)$$

$$\omega = \omega_o \sqrt{1-\varepsilon^2} = \sqrt{\frac{1}{LC} - \frac{R^2}{4L^2}} \quad (2.29)$$

When $i(0+) = i(0-) = 0$

$$A_1 = 0, \quad A_2 = \frac{E_{dc}}{L}$$

This gives
$$i(t) = e^{-\frac{R}{2L}t}\left[\frac{E_{dc}}{\omega L} \sin \omega t\right] \quad (2.30)$$

This equation shows that the thyristor-current i goes to zero at

$$\omega t = \pi$$

or
$$t = \frac{\pi}{\sqrt{\frac{1}{LC} - \frac{R^2}{4L^2}}} \quad (2.31)$$

Now,
$$\frac{di}{dt} = -e^{-\frac{\pi R}{2\omega L}}\left(\frac{E_{dc}}{l}\right)$$

Therefore, the capacitor voltage at the end of conduction, $V_c = E_{dc} - V_L$
where $V_L = L\,di/dt$

$$\therefore \quad V_c = E_{dc}\left[1 + e^{-\pi R/2\omega L}\right] \quad (2.32)$$

Now, if V_0 is the initial-state voltage of the capacitor then Eq. (2.30) becomes

$$i(t) = e^{-(R/2L)t}\left[\frac{E_{dc} - V_0}{\omega L} \cdot \sin \omega t\right] \quad (2.33)$$

and
$$V_c = E_{dc} + e^{-\pi R/2\omega t}(E_{dc} - V_0) \quad (2.34)$$

For $\omega > 0$, we now calculate the condition for underdamped.

$$\therefore \quad \frac{1}{LC} - \frac{R^2}{4L^2} > 0 \quad \text{i.e.,} \quad \frac{1}{LC} > \frac{R^2}{4L^2}$$

or
$$R < \sqrt{\frac{4L}{C}} \quad (2.35)$$

2. Class B—Self Commutation by an LC Circuit   In this method, the *LC* resonating circuit is across the SCR and not in series with the load. The commutating circuit is shown in Fig. 2.15 and the associated waveforms are shown in Fig. 2.16.

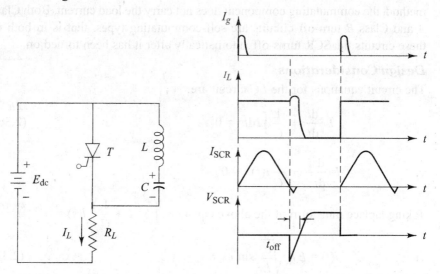

Fig. 2.15   *Class B commutation circuit*   **Fig. 2.16**   *Associated waveforms*

Initially, as soon as the supply voltage E_{dc} is applied, the capacitor *C* starts getting charged with its upper plate positive and the lower plate negative, and it charges up to the voltage E_{dc}.

When thyristor *T* is triggered, the circuit current flows in two directions: (1) The load current I_L flows through the path $E_{dc\,+} - T - RL - E_{dc\,-}$, and (2) Commutating current I_c.

The moment thyristor *T* is turned ON, capacitor *C* starts discharging through the path $C_+ - L - T - C_-$. When the capacitor *C* becomes completely discharged, it starts getting charged with reverse polarity. Due to the reverse voltage, a commutating current I_C starts flowing which opposes the load current I_L. When the commutating current I_C is greater than the load current I_L, thyristor *T* becomes turned OFF. When the thyristor *T* is turned OFF, capacitor *C* again starts getting charged to its original polarity through *L* and the load. Thus, when it is fully charged, the thyristor will be ON again.

Hence, from the above discussion it becomes clear that the thyristor after getting ON for sometime automatically gets OFF and after remaining in OFF state for sometime, it again gets turned ON. This process of switching ON and OFF is a continuous process. The desired frequency of ON and OFF states can be obtained by designing the commutating components as per the requirement. The main application of this process is in d.c. chopper circuits, where the thyristor is required to be in conduction state for a specified duration and then to remain in the OFF state also for a specified duration. Morgan chopper circuit using a saturable reactor in place of the ordinary inductor L is a modified arrangement for this process. The circuit has the advantage of longer oscillation period and therefore of more assurance of commutation. In this Class B commutation method, the commutating component does not carry the load current. Both Class A and Class B turn-off circuits are self-commutating types, that is in both of these circuits the SCR turns-off automatically after it has been turned on.

Design Considerations

The circuit equations for the LC circuit are:

$$L \frac{di}{dt} + \frac{1}{C} \int i\, dt = 0 \qquad (2.36)$$

$\therefore \qquad L \frac{d^2 i}{dt^2} + \frac{1}{C} i(t) = 0$

Taking laplace transform of the above equation, $\left(S^2 L + \frac{1}{C} \right) I(s) = 0$

$\therefore \qquad i(t) = E_{dc} \sqrt{\frac{C}{L}} \sin \omega_0 t \qquad (2.37)$

where $\qquad \omega_0 = \sqrt{\frac{1}{LC}} \qquad (2.38)$

Therefore, the peak commutation current is

$$I_{C_{(peak)}} = E_{dc} \sqrt{C/L} \qquad (2.39)$$

For this Class B commutation method, the peak discharge current of the capacitor is assumed to be twice the load-current I_L, and the time for which the SCR is reverse biased is approximately equal to one-quarter period of the resonant circuit.

Therefore, $\quad I_{C_{(peak)}} = 2\, I_L = E_{dc} \sqrt{C/L} \qquad (2.40)$

And $\qquad t_{off} = \frac{\pi}{2} \sqrt{LC} \qquad (2.41)$

3. Class C—Complementary Commutation (Switching a Charged Capacitor by a Load Carrying SCR) The class C commutation circuit is shown in Fig. 2.17. In this method, the main thyristor (SCR T_1) that is to be commutated is connected in series with the load. An additional thyristor (SCR T_2), called the complementary thyristor is connected in parallel with the main thyristor.

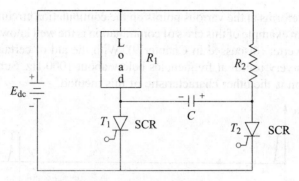

Fig. 2.17 *Class C–commutation circuit*

Circuit Operation

(a) Mode 0: [Initial-state of circuit] Initially, both the thyristors are OFF. Therefore, the states of the devices are –

$$T_1 \longrightarrow \text{OFF}, \quad T_2 \longrightarrow \text{OFF}, \quad \therefore \quad E_{c_1} = 0$$

(b) Mode 1: When a triggering pulse is applied to the gate of T_1, the thyristor T_1 is triggered. Therefore, two circuit current, namely, load current I_L and charging current I_C start flowing. Their paths are:

Load current I_L;

$$E_{dc}{+} - R_1 - T_1 - E_{dc-}$$

Charging current I_C;

$$E_{dc}{+} - R_2 - C_+ - C_- - T_1 - E_{dc-}$$

Capacitor C will get charged by the supply voltage E_{dc} with the polarity shown in Fig. 2.17. The states of circuit components becomes

$$T_1 \longrightarrow \text{ON}, \quad T_2 \longrightarrow \text{OFF}, \quad E_{c_1} = E_{dc}$$

(c) Mode 2: When a triggering pulse is applied to the gate of T_2, T_2 will be turned on. As soon as T_2 is ON, the negative polarity of the capacitor C is applied to the anode of T_1 and simultaneously, the positive polarity of capacitor C is applied to the cathode. This causes the reverse voltage across the main thyristor T_1 and immediately turns it off.

Charging of capacitor C now takes place through the load and its polarity becomes reverse. Therefore, charging path of capacitor C becomes

$$E_{dc+} - R_1 - C_+ - C_- - T_{2(a-k)} - E_{dc-}$$

Hence, at the end of Mode 2, the states of the devices are

$$T_1 \longrightarrow \text{OFF}, \quad T_2 \longrightarrow \text{ON}, \quad E_{c_1} = -E_{dc}$$

(d) Mode 3: Now, when thyristor T_1 is triggered, the discharging current of capacitor turns the complementary thyristor T_2 OFF. The state of the circuit at the end of this Mode 3 becomes

$$T_1 \longrightarrow \text{ON}, \quad T_2 \longrightarrow \text{OFF}, \quad E_{c_1} = E_{dc}$$

Therefore, this Mode 3 operation is equivalent to Mode 1 operation.

The waveforms at the various points on the commutation circuit are shown in Fig. 2.18. An example of this class of commutation is the well known McMurray–Bedford inverter (discussed in Chapter 9). With the aid of certain accessories, this class is very useful at frequencies below about 1000 Hz. Sure and reliable commutation is the other characteristic of this method.

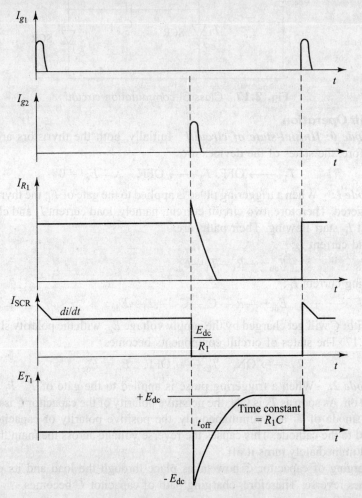

Fig. 2.18 *Circuit waveforms*

Design Considerations

As explained previously, when thyristor T_1 is conducting, capacitor C is charged to d.c. supply voltage E_{dc} through the resistor R_2. Now, when T_2 is triggered, a voltage twice the d.c. supply voltage E_{dc} is applied to the R_1C series circuit so that current through the circuit is,

$$i = \frac{2E_{dc}}{R_1} e^{-t/R_1C} \qquad (2.42)$$

Therefore, the voltage across the thyristor T_1 is

$$E_{T_1} = E_{dc} - iR_1 = E_{dc} - \frac{2E_{dc}}{R_1} e^{-t/R_1C} \cdot R_1 = E_{dc}\left(1 - 2e^{-t/R_1C}\right)$$

For making thyristor T_1 OFF, the capacitor voltage must be equal to the voltage E_{t_1}.

∴ $$E_c = E_{dc}\left(1 - 2e^{-t/R_1C}\right) \qquad (2.43)$$

Let $t = t_{off}$ when $E_c = 0$.
∴ Equation (2.43) becomes

$$0 = E_{dc}\left(1 - 2e^{-t_{off}/R_1C}\right) \quad \text{or} \quad 0 = 1 - 2e^{-t_{off}/R_1C} \qquad (2.44)$$

∴ $$t_{off} = 0.6931\, R_1 C \qquad (2.45)$$

or $$C = 1.44\, \frac{t_{off}}{R_1} \qquad (2.46)$$

So, from Eq. (2.45), R_1 and C must be such that the turn-off time of SCR T_1, t_{off}, is satisfied.

The maximum allowable $\frac{dV}{dt}$ rating for SCR T_1 may be obtained from the SCR T_1 data sheet.

The maximum $\frac{dV}{dt}$ across T_1 using the commutating components is given by

$$\frac{dV}{dt}_{(max)} > \frac{2E_{dc}}{R_1 C} \qquad (2.47)$$

4. Class D—Auxiliary Commutation (An Auxiliary SCR Switching a Charged Capacitor)
Figure 2.19 shows the typical Class D commutation circuit. In this commutation method, an auxiliary thyristor (T_2) is required to commutate the main thyristor (T_1), Assuming ideal thyristors and the lossless components, then the waveforms are as in Fig. 2.20. Here, inductor L is necessary to ensure the correct polarity on capacitor C.

Thyristor T_1 and load resistance R_L form the power circuit, whereas L, D and T_2 form the commutation circuit.

Circuit Operations

(a) Mode 0: [Initial Operation] When the battery E_{dc} is connected, no current flows as both thyristors are OFF. Hence, initially, the state of the circuit components becomes

$$T_1 \longrightarrow \text{OFF}, \quad T_2 \longrightarrow \text{OFF}, \quad E_C = 0$$

(b) Mode 1: Initially, SCR T_2 must be triggered first in order to charge the capacitor C with the polarity shown. This capacitor C has the charging path E_{dc+} —C_+ —C_- —T_2 —R_L—E_{dc-}. As soon as capacitor C is fully charged, SCR T_2 turns-off. This is due to the fact that, as the voltage across the capacitor increases, the current through the thyristor T_2 decreases since capacitor C and thyristor T_2 form the series circuit.

Hence the state of circuit components at the end of Mode 1 becomes,

$T_1 \longrightarrow$ OFF, $T_2 \longrightarrow$ OFF, $E_C = E_{dc}$

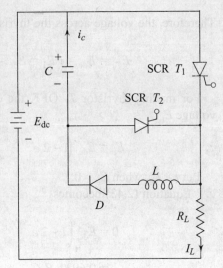

Fig. 2.19 Class D commutation circuit

(c) Mode 2: When thyristor T_1 is triggered, the current flows in two paths:
  (a) Load current I_L flows through E_{dc+} —T_1—R_L—E_{dc-}
  (b) Commutation current (Capacitor-discharges through) flows through C_+ —T_1—L—D—C_-.

After the capacitor C has completely discharged, its polarity will be reversed, i.e., its upper plate will acquire negative charge and the lower plate will acquire positive charge. Reverse discharge of capacitor C will not be possible due to the blocking diode D.

Therefore, at the end of Mode 2, the state of the circuit components becomes

$T_1 \longrightarrow$ ON, $T_2 \longrightarrow$ OFF, $E_C = -E_{dc}$

(d) Mode 3: When the thyristor T_2 is triggered, capacitor C starts discharging through the path C_+—$T_{2(A-K)}$ — $T_{1(k-A)}$—C_-. When this discharging current (commutating current I_C) becomes more than the load current I_L, thyristor T_1 gets OFF.

Therefore, at the end of Mode 3, the state of circuit component becomes

$T_1 \longrightarrow$ OFF, $T_2 \longrightarrow$ ON

Again, capacitor C will charge to the supply voltage with the polarity shown and hence SCR T_2 gets OFF. Therefore, thyristors T_1 and T_2 both get OFF, which is equivalent to Mode 0 operation.

This type of commutation circuit is very versatile as both time ratio and pulse width regulation is readily incorporated. The commutation energy may readily be transferred to the load and so high efficiency is possible. This method is used in Jone's chopper circuit.

Thyristor: Principles and Characteristics

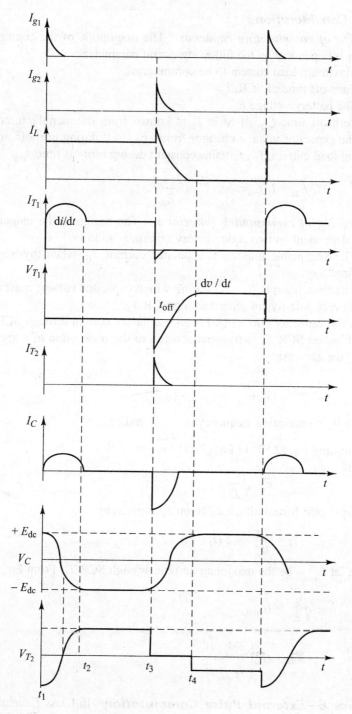

Fig. 2.20 *Associated waveforms*

Design Considerations

(a) Design of commutating capacitor The magnitude of the commutating capacitor is dependent on the following circuit parameters:
  (i) Maximum load current to be commutated
  (ii) Turn-off time of SCR, t_{off}
  (iii) The battery voltage E_{dc}

The turn-off time t_{off}, of SCR T_1 is known from the manufacturer's data sheet. The capacitor voltage changes from $-E_{dc}$ to 0 during turn-off time, t_{off}. Assuming load current, I_L, remains constant during turn-off time, t_{off},

$$C E_{dc} = I_L t_{off} \quad \therefore \quad C = \frac{I_L t_{off}}{E_{dc}} \tag{2.48}$$

(b) Designing of commutating inductor L The design of the inductor L is actually dependent on two contradictory criteria as follows:
  (i) The acceptable maximum capacitor current, I_C, when thyristor T_1 is fired.
  (ii) The time interval $(t_2 - t_1)$ during which capacitor voltage must reset to correct polarity for commutating SCR T_1.

Since the capacitor current (I_C) is an oscillatory current through SCR T_1, L, D, and C when SCR T_1 is triggered, therefore the peak value of current I_C is given by the expression,

$$I_{C_{(peak)}} = \frac{E_{dc}}{W_r L} \tag{2.49}$$

where W_r = oscillating frequency = $\dfrac{1}{\sqrt{LC}}$ rad/sec. $\tag{2.50}$

Substituting Eq. (2.50) in Eq. (2.49), we get

$$I_{C_{(peak)}} = E_{dc} \sqrt{\frac{C}{L}} \tag{2.51}$$

Also, periodic time during oscillation T_r, is given by

$$T_r = \frac{2\pi}{W_r} = 2(t_1 - t_2) \tag{2.52}$$

Now, let $I_{L_{(max)}}$ be the maximum current through SCR T_1. From Eq. (2.51),

$$E_{dc} \sqrt{\frac{C}{L}} \leq I_{L_{(max)}}$$

or

$$L \geq C \cdot \left(\frac{E_{dc}}{I_{L_{(max)}}}\right)^2 \tag{2.53}$$

5. Class E—External Pulse Commutation In Class E commutation method, the reverse voltage is applied to the current carrying thyristor from an external pulse source. A typical Class E commutation circuit is shown in

Fig. 2.21 and the associated waveforms are shown in Fig. 2.22. Here, the commutating pulse is applied through a pulse-transformer which is suitably designed to have tight coupling between the primary and secondary. It is also designed with a small air gap so as not to saturate when a pulse is applied to its primary. It is capable of carrying the load-current with a small voltage drop compared to the supply voltage. When the commutation of T_1 is desired, a pulse of duration equal to or slightly greater than the turn-off time specification of the thyristor is applied.

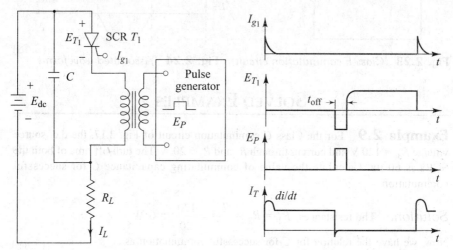

Fig. 2.21 Class E commutation circuit   **Fig. 2.22** Associated waveforms

When the SCR T_1 is triggered, current flows through the load R_L and the pulse transformer. When a pulse of voltage E_P from the pulse-generator is applied to the primary of the pulse transformer, the voltage induced in the secondary appears across thyristor T_1 as a reverse voltage $(-E_P)$ and turn it off. Since the induced pulse is of high frequency, the capacitor offers almost zero impedance. After T_1 is turned off, the load current decays to zero. Earlier to the commutation, the capacitor voltage remains at a small value of about 1 V.

This type of commutation method is capable of very high efficiency as minimum energy is required and both time ratio and pulse width regulation are easily incorporated. However, equipment designers have neglected this class for the designing of power circuits.

6. Class F—a.c. Line Commutation A typical line commutated circuit is shown in Fig. 2.23 and its associated waveforms are shown in Fig. 2.24. If the supply is an alternating voltages, load current will flow during the positive half cycle. During the negative half cycle, the SCR will turn-off due to the negative polarity across it. The duration of the half cycle must be longer than the turn-off time of the SCR. The maximum frequency at which this circuit can operate depends on the turn-off time of SCR.

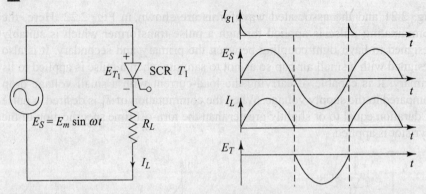

Fig. 2.23 Class F commutation circuit   **Fig. 2.24** Associated waveforms

SOLVED EXAMPLES

Example 2.9 For the Class C commutation circuit of Fig. 2.17, the d.c. source voltage E_{dc} = 120 V and current through R_1 and R_2 = 20 A. The turn-off time of both the SCRs is 60 μs. Calculate the value of commutating capacitance C for successful commutation.

Solution: The resistances $R_1 = R_2 = \dfrac{E_{dc}}{I} = \dfrac{120}{20} = 6$ W

Now, we have the relation for C for successful commutation as

$$C = 1.44 \cdot \dfrac{t_{off}}{R_1} = 1.44 \times \dfrac{60 \times 10^{-6}}{6} = 14.4 \,\mu F.$$

Example 2.10 For the Class D commutation circuit of Fig. 2.19, compute the value of the commutations capacitor C and commutating inductor L for the following data:

$E_{dc} = 50V, I_{L(max)} = 50$ A, t_{off} of SCR$_1$ = 30 μs

Chopping frequency f = 500 Hz and the load voltage variation required is 10 to 100%.

Solution: For reliable operation, let us assume 50% tolerance on turn-off time of SCR$_1$.

$$\therefore \quad t_{off} = \left(30 + \dfrac{50}{100} \times 30\right) = 45 \mu s$$

Now, we have the relation for the commutating capacitor, C as

$$C = \dfrac{I_L \, t_{off}}{E_{dc}} = \dfrac{50 \times 45 \times 10^{-6}}{50} = C = 45 \,\mu F.$$

The resetting time for capacitor voltage could be reduced by decreasing the value of L, but the peak capacitor current would increase as seen from Eq. (2.51). A large resetting time would limit the minimum voltage available at the load, which means the range of voltage available at the load is reduced.

Thyristor: Principles and Characteristics

Therefore, the minimum load voltage available is given by

$$V_{0(min)} = \frac{t_1 - t_2}{T} E_{dc}$$

where t is chopping time period, or

$$V_{0(min)} = \frac{\pi\sqrt{LC}}{T} E_{dc}$$

$$\therefore \quad L \leq \left(\frac{V_{0(min)}}{E_{dc}}\right)^2 \frac{T^2}{\pi^2 C} \quad \text{(i)}$$

Given $\quad V_{0(min)} = 10\%(50) = 5$ V

$$T = \frac{1}{f} = 2 \times 10^{-3} \text{ s}$$

$$\therefore \quad L \leq \frac{(2 \times 10^{-3})^2}{\pi^2 \times 45 \times 10^{-6}} \times \left(\frac{5}{50}\right)^2 \quad \text{or} \quad L \leq 90 \text{ }\mu\text{H}$$

Also, we have the relation

$$L \geq C \left(\frac{E_{dc}}{I_{L(max)}}\right)^2 \quad \text{or} \quad L \geq 45 \times 10^{-6} \left(\frac{50}{50}\right)^2$$

or $\quad L \geq 45$ μH

Hence, the range of commutating inductor is 45 μH < L < 90 μH
The choice of lower value of L would allow a larger voltage variation at the load.

2.11 THYRISTOR RATINGS

All semiconductor devices have definite limits to their capability and exceeding these even for short times will result in failure, loss of control, or irreversible deterioration. All thyristors, therefore, have to be used within their limits and this must include extreme conditions as may exist during circuit faults and it must take into account load, supply system, temperature and environmental variations. If extreme conditions are not precisely known and cannot be calculated, then appropriate safety margins have to be chosen to allow for the unknown factors. Correct safety margins can only be decided from practical operating experience. Therefore, the reliable operation of the device can be ensured only if its ratings are not exceeded under all operating conditions. The objective of this section is to discuss the various SCR ratings.

2.11.1 Voltage Ratings

It is essential that the voltage capability of a thyristor is not exceeded during operation even for a very short period of time. Therefore, the voltage rating of the device should be high enough to withstand anticipated voltage transients as well as the repetitive OFF state and reverse—blocking voltages. The various ratings related to the voltage are discussed in this section.

1. Working peak-off state forward voltage (V_{Dwm}) This is the maximum instantaneous value of the forward OFF state voltage that occurs across the thyristor excluding all repetitive and non-repetitive transient voltages.

2. Repetitive peak-off state forward voltage (V_{Drm}) It refers to the peak transient voltage that a thyristor in the OFF state can block repeatedly in the forward direction. This rating is specified at a maximum allowable junction temperature with gate circuit open or with a specified biasing resistance between the gate and cathode terminals.

3. Non-repetitive peak-off-state forward voltage (V_{DSM}) This is the maximum instantaneous value of any non-repetitive transient OFF state voltage that occurs across a thyristor.

4. Working peak reverse voltage (V_{RWM}) This is the maximum instantaneous value of the reverse voltage that occurs across the device excluding all repetitive and non-repetitive transient voltages.

5. Repetitive peak reverse voltage (V_{RRM}) This is the peak reverse transient voltage that may occur repeatedly in the reverse direction at the allowable maximum junction temperature. If this rating is exceeded, the SCR may be damaged due to excessive junction temperature.

6. Non-repetitive peak reverse voltage (V_{RSM}) This is the maximum transient reverse-voltage which can be safety blocked by the thyristor. The transient reverse voltage rating can be increased by inserting a diode of equal current rating in series with the thyristor.

7. On state voltage (V_T) This is the voltage drop between anode and cathode with specified forward ON state current and junction temperature. Its value is of the order of 1 to 1.5 V.

8. Gate trigger voltage (V_{GT}) This is the minimum gate voltage required to produce the gate trigger current.

9. Voltage safety factor (V_f) To avoid damage to a thyristor due to uncertain conditions, normal operating voltage is kept well below the V_{RSM} value of the device. The operating voltage and V_{RSM} (or peak inverse voltage V_{PIV}) value are related by the voltage safety factor V_f that is defined as

$$V_f = \frac{V_{RSM} \text{ (Peak Reverse Voltage)}}{\sqrt{2} \times \text{RMS Value of input voltage}} \qquad (2.54)$$

The normal value of this factor lies between 2 and 2.5.

10. Forward $\frac{dV}{dt}$ rating [rate of rise of OFF state voltage]

The $\frac{dV}{dt}$ rating of a thyristor indicates the maximum rate of rise of anode voltage that will not trigger the device without any gate signal. If the rate of rise of forward voltage is higher than the specified maximum value, it may cause switching from the OFF–state to the ON state. The mechanism for this phenomenon can be explained in terms of internal capacitance that the thyristor exhibits. When voltage across the device is increased, charges flow through it in a manner analogous to the charging current of a capacitor $\left(i = C\frac{dV}{dt}\right)$. The greater the rate of rise of applied voltage, the greater will be this flow of charges. As the rate of rise is increased, sufficient charges will eventually flow to act in the same manner as the charges injected when the gate is energised with a positive potential with respect to the cathode, and the thyristor will turn-on.

The $\frac{dV}{dt}$ rating also depends on the junction temperature. Higher the junction temperature, lower the $\frac{dV}{dt}$ of the device. Since the $\frac{dV}{dt}$ triggering causes random turn-on of a thyristor, it is never employed in practice. The circuit designer may often limit the maximum $\frac{dV}{dt}$ applied to an SCR by means of added suppressors or "snubber" networks (discussed in Chapter 13) placed across a device's terminals.

2.11.2 Current Ratings

The current carrying ability of the thyristor is determined by the temperature at its junction. Since the thyristor is made up of a semiconductor material, it's thermal capacity is, therefore, quite small. Hence, even for short over currents, the junction temperature may exceed the rated value and the device may be damaged. In this section, current ratings of SCR are discussed for both repetitive and non- repetitive type of current waveforms.

1. Average On-state Current (I_{TAV}) As the forward voltage drop across conducting SCR is very low, the power loss in an SCR depends mainly on forward average on state current I_{TAV}. The average current rating of a phase controlled thyristor (discussed in Chapter 6) results in different conduction angles. If the same average current is permitted for different conduction angles, the instantaneous current increases with the decrease in conduction angle.

This may result in an increased voltage drop across the device, and the average power dissipation may increase raising the junction temperature beyond the safe allowable limit. The permissible average current should, therefore, decrease with the decrease in conduction angle.

This current rating is of repetitive type and is specified at the maximum junction temperature. This rating varies with the case temperature. The manufacturers supply the data-sheet which shows the variation of the current rating with respect to the case-temperature.

2. RMS On-state Current (I_{RMS}) By the definition of rms values, the rms and average values are identical for a direct current. The rms current rating can be of importance when applying thyristors to high peak current, low duty cycle waveforms. Although, the average value of the waveform may be well within the ratings, it may be that the allowable rms rating is being exceeded. In order to prevent excessive heating in resistive elements of a thyristor, such as metallic joints, leads and interfaces, the rms value is given in the specification sheet of the SCR by all manufacturers. This current rating is also of repetitive type and is specified at the maximum junction temperature.

3. Surge Current Rating (I_{TSM}) (Non-repetitive, Peak-on State Current) When an SCR is working within its repetitive voltages and current ratings, its permissible junction temperature is never exceeded. However, a thyristor may be subjected to abnormal operating conditions due to faults or short circuits. In order to accommodate these unusual working conditions, the surge current rating is also specified by the manufacturers. The surge current rating indicates the maximum possible non-repetitive or surge current, the device can withstand and this may occur due to non-repetitive faults or short-circuits during the life-span of a thyristor.

Surge currents are assumed to be the sine waves with the frequency of 50 60 Hz depending upon the supply frequency. This rating is specified in terms of the number of surge cycles with corresponding surge current peak. The surge current rating is inversely proportional to the duration of the surge. It is usual to measure the surge duration in terms of the number of cycles of normal power frequency of 50 to 60 Hz.

The one cycle surge current rating is the peak value of an allowable non-recurrent half-sine wave of 10 ms (50 Hz) duration. For duration less than half cycle, that is 10 ms, a subcycle surge current rating is also specified. This rating for 50 Hz or 60 Hz supply is the peak value for a part of the half-sine wave. The subcycle surge current rating I_{sub} can be determined by equating the energies involved in one cycle surge and one subcycle surge as follows:

$$I_{SUB}^2 \, t = I^2 T \quad \text{or} \quad I_{SUB} = I \sqrt{\frac{T}{t}} \tag{2.55}$$

where T = time for one half-cycle of supply frequency (seconds)
  I = one cycle surge rating (ampere).
  I_{SUB} = subcycle surge current rating (ampere, A)
  t = duration of subcycle surge (seconds).
For 50 Hz supply $T = 10$ ms

$$\therefore \quad I_{SUB} = I \sqrt{\frac{1}{100t}} \tag{2.56}$$

4. I^2t Rating I^2t is the maximum allowable non-recurring value of the square of the instantaneous current integrated with respect to the time ($\int i^2 dt$). This rating is required to coordinate with the fast acting fuse for protecting the device during an overload or fault conditions. This rating is the measure of the thermal energy that the device can absorb for a short time before the fault is cleared by the fuse. It is usually specified for fault currents lasting for less than or equal to one half period at the supply frequency of 50 or 60 Hz.

5. $\dfrac{di}{dt}$ Rating The $\dfrac{di}{dt}$ rating of a thyristor indicates the maximum rate of rise in anode to cathode current. The maximum rate of change of current that the device can withstand during its ON state is called its critical rate of rise of current. The value of critical rate of rise in current for a particular device is always specified at its highest value of junction temperature that it can safely bear. If the rate of rise in anode current is very rapid compared to the spreading velocity of carriers across the junctions during the turn-on period, the local *hot spot* heating occurs due to high current density in the junction regions. This increases the junction temperature beyond the safe limit and the device may be damaged. Typical $\dfrac{di}{dt}$ values lies in the range 50 to 800 ampere per μ seconds.

The $\dfrac{di}{dt}$ value of the device mainly depends on the level of the gate current used to trigger the thyristor; within limits, the higher the gate current the higher the $\dfrac{di}{dt}$ capability. The need for high $\dfrac{di}{dt}$ switching levels has led to the developments of the amplifying gate thyristor (Chapter 10). An additional technique used to accomplish high $\dfrac{di}{dt}$ capability is to employ a hard-drive gate circuit.

6. Holding Current (I_H) The holding current may be defined as the minimum value of anode current below which the device stops conducting and returns to its OFF state. The value of this current is very small, usually in milliamperes. The holding current occurs when the current in a device in ON state is being decreased till it turns off. Therefore, this current is associated with the turn-off process.

7. Latching Current (I_L) The latching current of a device may be defined as the minimum ON state current required to keep the device in the ON state after the triggering pulse has been removed.

For an SCR to trigger, the anode current must be allowed to build up rapidly enough so that the latching current of the SCR is reached before the pulse is terminated. Therefore, for highly inductive anode circuits one must use a maintained type of trigger signal which assures gate drive until latching current has been attained. Generally, on the specification sheet the latching current is given two to three times the holding current.

8. Gate Current [$I_{g_{min}}$ and $I_{g_{max}}$] The current which is applied to the gate of the thyristor for control purposes, is known as its gate current. It may be of two types, the minimum gate-current ($I_{g_{min}}$) and maximum gate current ($I_{g_{max}}$). The minimum value of current required at the gate for triggering the device is called as the minimum gate current, $I_{g_{min}}$. The value of this current-depends on the rate of rise in current. ON the other hand, maximum gate current ($I_{g_{max}}$) is the maximum value of current that can be applied to the gate safety. Manufacturer's specification of $I_{g_{max}}$ must not be exceeded to avoid damage to the device. Within the safe-limits of $I_{g_{min}}$ and $I_{g_{max}}$, the conduction angle of thyristor can be controlled. More the gate current, earlier is the triggering of the device and *vice versa*.

2.11.3 Power Rating

The power-generated in the junction region of a thyristor in a normal operation consists of the following components of dissipation.

1. Forward-conduction.
2. Turn-on switching
3. Turn-off or commutation.
4. Forward and Reverse blocking.
5. Gate pulse triggering.

1. Forward Conduction Loss The average anode current multiplied by the forward voltage drop across the SCR is the average power dissipated in the thyristor. ON state conduction losses are the major source of junction heating for normal duty cycle and power-frequencies. Figure 2.25 illustrates the variation of the ON state average conduction loss in watts with the average current in amperes for various conduction angles for operation on from 50 Hz to 400 Hz. This type of information is generally supplied by the manufacturer. The curve marked d.c. is applicable for continuous direct-current. These curves are based on the current waveform which is the remainder of the half-sine wave which results when

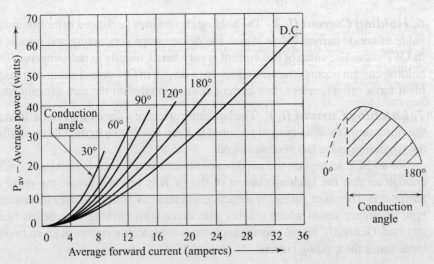

Fig. 2.25 *Average forward power dissipation for sinusoidal current waveform*

delayed angle triggering is used in a single phase resistive load circuit. These power curves are integrated product of the instantaneous anode current and ON state voltage drop. This integration can be performed graphically or analytically for conduction angles other than those listed, using the ON state voltage-current characteristic curves for the specific device. In the line commutated converters and a.c. regulator circuits, the forward conduction loss is the major source of junction heating.

2. Turn-on Losses Since the switching process takes a finite time, there is a relatively high voltage across the thyristor while a current flows. Therefore, this loss is rather higher than the turn-off loss. For example, by the time the current has reached 90 per cent of its final value, there may still be 10 per cent of the supply voltage across the device (see Fig. 2.10). Accordingly, appreciable power may be dissipated during this turn-on interval. Above 400 Hz switching, additional circuitry is used to reduce the switching losses or else some derating of the normal forward current is made to allow for the extra dissipation.

3. Turn-off Losses The turn-off power losses arises during the time of decay of reverse current, according to the product of the instantaneous values of reverse current and reverse voltage, may reach high peak values up to several kilowatts. It is possible during rapid turn-off for the reverse current to rise to a value comparable to the forward current. When the thyristor impedance starts to increase, dissipation occurs as the current falls and the reverse voltage builds up. To limit the rate of change of current at turn-off and hence the energy to be dissipated, circuit inductance is used. This also limits the rate of rise of forward current which is an advantage but the inductance can give rise to high reverse voltage transients during turn-off. In high frequency inverters, where the thyristors are switched ON and OFF several times in each supply cycle, the turn-on and turn-off losses may also have to be taken into consideration while selecting the device ratings since the switching loss may constitute a significant portion of the total loss.

4. Forward and Reverse Blocking Losses As mentioned in the previous section, a thyristor has different regions of operation. In the forward blocking region, anode is made positive with respect to cathode and the anode current is the small forward leakage current. Therefore, the forward blocking power loss is the integration of product of the forward blocking voltage and forward leakage current. Similarly, reverse power loss occurs in reverse blocking region. The forward power loss is generally small compared to the conduction-loss.

5. Gate Power Loss ($P_{g_{av}}$) The gate power loss is the mean power loss due to gate current between the gate and main terminals. Gate losses are negligible for pulse types of triggering signals. Losses may become more significant for gate signals with a high duty cycle or for SCRs in a small packages such as To–5, To–18 or Power Tab type packages.

2.11.4 Thermal Ratings

The following are the main thermal ratings of the device.

1. Junction Temperature (T_j) A thyristor's ability to block forward voltage applied to it can only be maintained within a specific junction temperature limit. If this junction temperature is exceeded, the thyristor will switch into the ON state even though no gate current may be flowing. This is usually the deciding factor in controlling the maximum current that can be carried by the thyristor for any significant period of time.

The operating junction temperature range of thyristors varies for the individual types. A low temperature limit may be required to limit the thermal stress in the silicon crystal to safe-values. This type of stress is due to the difference in the thermal coefficients of expansion of the materials used in fabricating the cell subassembly. The upper operating limit is imposed because of the temperature dependence of the breakover voltage, turn-off time and thermal stability considerations. The upper storage temperature limit in some cases may be higher than the operating limit. It is selected to achieve optimum reliability and stability of characteristics with time.

2. Transient Thermal Impedance (Z_{TH}) This is the resistance between the junction of a thyristor and its cooling surfaces. It is invariably expressed as degree centigrade temperature difference per watt of energy dissipated. The larger the thyristor, the smaller will be its thermal resistance value.

2.11.5 Turn-on and Turn-off Time Ratings

Selecting a thyristor for a particular application is very much dependent on the turn-on and turn-off time. Fast switching thyristors have very low values of turn-on and turn-off time. This is achieved by gold doping in silicon. Presence of gold in thyristors reduces minority carrier life-time but increases leakage currents. For faster devices leakage current is more and hence relatively lower values of voltages V_{RWM} and V_{OWM}.

SOLVED EXAMPLES

Example 2.11 A thyristor has an $\int i^2 \, dt$ rating of 15 amp^2 S and is being used to supply the circuit in Fig. E2.11 from a 120 V a.c. supply when a fault occurs, short circuiting the 10 Ω resistors to earth. What is the shortest fault clearance time to be achieved if damage to the thyristor is to be prevented?

Solution: The worst case fault occurs when the voltage is at maximum. Assume that the voltage is at the maximum value for the duration of the fault, then,

$$\int_0^{t_c} i^2\, dt = \int_0^{t_c} 120^2\, dt = 15$$

∴ Fault clearance time $= \dfrac{15}{(120)^2}$

$= 1.04$ ms.

Example 2.12 An SCR has half cycle surge current rating of 3000 A for 50 Hz supply. Calculate its one cycle surge current rating and $i^2 t$ rating.

Solution: Let I and I_{SUB} be the one cycle and sub-cycle surge current ratings of the SCR respectively. Then equating the energies involved in them, we get

$$I^2 t = I^2_{SUB}\, t$$

or $\quad i^2 \times \dfrac{1}{100} = (3000)^2 \times \dfrac{1}{200} \quad \therefore \quad i = 2121.32$ A

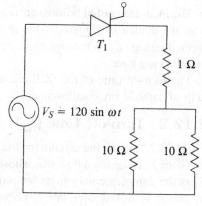

Fig. E2.11

Now, $i^2 t$ rating is given by the relation

$i^2 t =$ (rms value of one cycle surge current)$^2 \times$ time for one cycle $= i^2 \times \dfrac{1}{2f}$

$$= \left(\dfrac{3000}{5\sqrt{2}}\right)^2 \times \dfrac{1}{100} = 4500 \text{ A}^2 \text{ s}.$$

2.12 MEASUREMENT OF THYRISTOR PARAMETERS

2.12.1 Holding Current and Latching Current

The necessary circuit for the of holding current and latching current is shown in Fig. 2.26. For the measurement of latching current (I_L), the resistance R_S is varied slowly and the gate signal is applied repeatedly through switch S_1. Now, when the forward anode current is low, the SCR will go to the blocking state as soon as the gate signal is removed. If the anode current exceeds I_L, the device will continue to conduct even after the gate drive is removed. The reading of ammeter then indicates the value of latching current I_L.

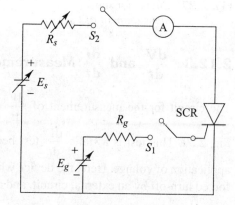

Fig. 2.26 *Circuit for measurement of I_L and I_H*

The measurement of holding current is obtained by reducing the anode current after the device is triggered. This is done by increasing R_S and decreasing the applied voltage E_S. When the anode current falls below I_H, the SCR will return to the blocking state.

The turn-on time of the SCR can also be measured from Fig. 2.26 with the help of cathode ray oscilloscope.

2.12.2 Turn-off Time (t_{off})

Figure 2.27 shows the circuit for the measurement of turn-off time, t_{off}. Here, thyristor T_1 is the main thyristor whose turn-off time is to be measured. Thyristor T_2 is the complementary thyristor required for turning thyristor T_1 OFF. When switch S_1 is closed, thyristor T_1 becomes ON and capacitor C will get charged through resistance R_S, with the polarity shown in the figure. Now, when switch S_2 is closed thyristor T_2 will get the gate pulse and becomes ON. Capacitor C will start discharging through the path. $C_+ —T_{2(A-K)} —R—T_{1(A-K)} —C_-$. Hence, thyristor T_1 becomes OFF. By measuring the voltage and current through the SCR T_1 with the help of CRO, the turn-off time of SCR T_1 can be calculated as shown in Fig. 2.28.

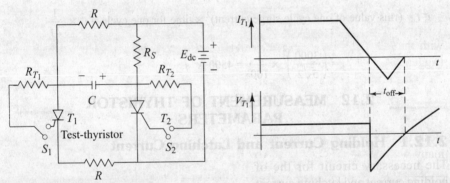

Fig. 2.27 Circuit for the measurement of t_{off}

Fig. 2.28 Related waveforms

2.12.3 $\dfrac{dV}{dt}$ and $\dfrac{di}{dt}$ Measurement

The circuit for the measurement of $\dfrac{dV}{dt}$ capability of the device is shown in Fig. 2.29. This test is a static $\dfrac{dV}{dt}$ test because the thyristor is OFF before the application of voltage. Here, the device which was initially ON, is subjected to a forced turn-off by an external circuit, and a forward voltage is then applied. Due to the presence of excess carriers in the inner layers, the reapplied $\dfrac{dV}{dt}$ rating is lower than the capability to withstand static $\dfrac{dV}{dt}$. In Fig. 2.29, the rate of application of the forward voltage is varied by changing the charging rate of capacitor C_1. The initial value of dv/dt, when switch S_1 is closed and S_2 is open, is given by

$$\frac{dV}{dt} = \frac{E_s R_s}{C_1} \qquad (2.57)$$

If this value is more than the $\dfrac{dV}{dt}$ rating of the SCR, it will conduct without any external gate signal. It is assumed that resistance R_1 is very small and the initial jump in voltage appearing across the SCR $\dfrac{E_s R_s}{(R_s + R_1)}$ is small enough to maintain the SCR in the OFF state.

For conducting the $\dfrac{di}{dt}$ test on the SCR, connect a small inductor in series with the anode circuit. The rate of change of anode current after the device is triggered, is controlled by varying the initial voltage on the capacitor.

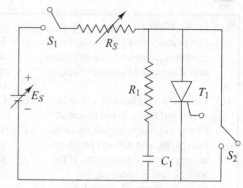

Fig. 2.29 Circuit for the measurement of $\dfrac{dV}{dt}$ and $\dfrac{di}{dt}$

2.13 COMPARISON BETWEEN TRANSISTORS AND THYRISTORS

Both transistors and thyristors are semiconductor devices, but they differ in many ways as under:

Transistors	Thyristors
(1) Transistor is a three-layer, two junction device.	(1) Thyristor is a four layer, three junction device.
(2) To keep a transistor in the conducting state, a continuous base current is required.	(2) Thyristors require a pulse to make it conducting and thereafter it remain conducting.
(3) When transistors (power transistor) conduct appreciable current, the forward voltage drop is of the order of 0.3 to 0.8 V.	(3) The forward voltage drop across the device is of the order of 1.2 to 2 V.
(4) The voltage and current ratings of transistors available at present are not as high as those of thyristors.	(4) Due to the difference in fabrication and operation, thyristors with very high voltage and current ratings are available.

(Contd.)

Transistors	Thyristors
(5) Power transistors have no surge current capacity and can withstand only a low rate of change of current.	(5) Thyristors have surge-current rating and therefore can withstand high rate of change of current compared to transistors.
(6) Commutation circuitry, which is costly and bulky, is not required.	(6) Commutation circuit is required.
(7) Power transistors switch on faster than SCRs, and turn-off problems are practically non existent. If the base current is removed, the transistor turns off. Therefore, power-transistors can be used in very high-frequency applications.	(7) Thyristors are used in comparatively low frequency applications.
(8) Circuits using power transistors will be smaller in size and less costly compared to circuits using thyristors.	(8) Comparatively larger in size and is costlier.
(9) There has been little operating experience in high power applications of transistors. Power transistors or Darlington pairs are more susceptible to failure.	(9) Thyristor circuits, on the other hand, have a proven record of many years of reliable operation.

REVIEW QUESTIONS

2.1 Describe the different modes of operation of a thyristor with the help of its static V–I characteristic.

2.2 Describe the holding current and latching current as applicable to an SCR with the help of its static V–I characteristic.

2.3 With the help of a neat diagram, explain the two transistor analogy of an SCR. Also discuss the triggering conditions of SCR.

2.4 Give the constructional details of an SCR. Sketch its schematic diagram and the circuit symbol.

2.5 Explain why
 (i) The inner two layers of an SCR are lightly doped and are wide.
 (ii) The inner n layer of an SCR is doped with gold.
 (iii) I_H is less than I_L.

2.6 Explain in detail the turn-off mechanism of an SCR.

2.7 Explain the various types of triggering methods of SCR briefly. Which is the universal method and why?

2.8 What are the different signals which can be used for turning on an SCR by gate control? Compare them.

2.9 Draw the gate characteristic of an SCR and explain it.

2.10 Draw the turn-off characteristic of an SCR and explain the mechanism of turn-off.

2.11 What are the different methods for turning off an SCR? Explain all methods in detail.

2.12 Define the following terms in connection with SCR:
  (i) Peak inverse voltage.
  (ii) Critical rate of rise of voltage
  (iii) Voltage safety factor
  (iv) Latching current.
  (v) Holding current

2.13 What do you mean by commutation of SCR? What are the different classes of forced commutation method? Explain the class C and class D methods.

2.14 Explain the following ratings of SCRs and their significance.
  (i) Peak working reverse voltage (V_{Rwm})
  (ii) Working peak OFF state forward voltage (V_{Dwm})
  (iii) Repetitive peak OFF state forward voltage (V_{DRM})
  (iv) Non-repetitive peak OFF state forward voltage (V_{RSM}).
  (v) Repetitive peak reverse voltage (V_{RRM})
  (vi) Non-repetitive peak reverse voltage (V_{RSM})
  (vii) On state voltage

2.15 Explain in detail the following current ratings of SCR in detail
  (i) Average ON state current
  (ii) Surge current rating
  (iii) RMS ON state current
  (iv) $I^2 t$ rating
  (v) $\dfrac{di}{dt}$ rating

2.16 Explain in detail the power rating of SCR.

2.17 What are $\dfrac{dV}{dt}$ and $\dfrac{di}{dt}$ ratings of SCRs? What happens if these ratings are exceeded? Explain.

2.18 Explain the following thermal ratings of SCRs.
  (i) Junction temperature.
  (ii) Transient thermal resistance.

2.19 Explain the methods of measurement of following SCR parameters
  (i) Holding and latching current
  (ii) Turn-off time
  (iii) $\dfrac{dV}{dt}$ and $\dfrac{di}{dt}$

2.20 Give the comparison between transistors and thyristors.

PROBLEMS

2.1 A typical thyristor circuit is shown in Fig. P2.1. If T_1 is switched ON at $t = 0$ determine the conduction time of thyristor T_1 and the capacitor voltage after T_1 is turned OFF. The inductor carries an initial current of $I_n = 250$A.
[*Ans*: $T_c = 38.45$ μs; $V_c = 297.97$ V]

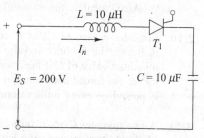

Fig. P2.1

2.2 The reverse biased junction capacitance of an SCR is 25 picofarads. The device can be turned ON if the charging current flowing through the junction capacitor is 5 mA. Calculate the $\dfrac{dV}{dt}$ capability of the device.

[*Ans*: 200 V/μs]

2.3 For an SCR, the gate-cathode characteristic has straight line slope of 130. For trigger source voltage of 15V, and allowable gate power dissipation of 0.5 watt, calculate the value of gate source resistance.

[Ans: $R_g \approx 112\ \Omega$]

2.4 The gate circuit of an SCR has a source voltage of 15 V and the load line has a slope of −120 V per ampere. The minimum gate current to turn-on the SCR is 25 mA. Calculate
  (i) Gate-source resistance R_s.
  (ii) The gate voltage and gate current for an average gate power dissipation of 0.4 watts.

[Ans: $R_S = 120\ \Omega$; $V_g = 4.63$ V; $I_g = 86.45$ mA]

2.5 The equivalent capacitance of the depletion layer of reversed biased junction of an SCR is 30 pt, fired with a $\dfrac{dV}{dt}$ of 150 V/μs. Calculate the capacitive current flowing through the junction.

[Ans: $I_C = 4.5$ mA]

2.6 If the latching current in the circuit shown in Fig. P2.6 is 100 mA, then compute the minimum width of the gating pulse required to properly turn-on the SCR.

2.7 A 3-phase, 6-pulse thyristor converter is connected to the 415 V main. If the peak inverse voltage of the thyristor is 1500 V compute the voltage safety factor.

[Ans: $V_f = 2.56$]

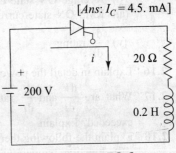

Fig. P2.6

2.8 A thyristor is triggered by a train of pulses of frequency 4 kHZ and of duty cycle 0.2. (a) Compute the gate pulse width (b) If the allowable average gate power is 1 W, find the maximum allowable gate power drive.

[Ans: (a) 50 μs = t_p; (b) $P_{g_{max}} = 5W$]

2.9 A thyristor V_g–I_g relation as $V_g = 1 + 10\ I_g$ and other gate characteristics as shown in Fig. 2.7. In an application, where the gating voltage is 20 V rectangular pulse of 180° duration, calculate the resistance R_g to be connected in series with the gate source voltage in order to limit the average gate power loss to 0.5 W. Also calculate the value of V_g and I_g during gating.

[Ans: $V_g = 3.7$ V; $I_g = 0.27$ A; $R_g = 60$]

2.10 Consider a case of Class C commutation circuit in Fig. 2.17. Let the load resistance be 5 ohms and the applied d.c. voltage be 120 V. Calculate the minimum value of C if the manufacturer's specified turn-off time of the thyristor for the forced commutation is 15 μs. What is the suitable value of R_2, if T_1 is pulsed on every milli second.

[Ans: $C = 3\mu Fi$; $R_2 = 55\ \Omega$].

2.11 A two thyristor Class C turn-off circuit is required to be designed for use as a blinker to turn-on and off a lamp of constant resistance of 10 ohms from a d.c. supply of 100 V. If the SCR used are of converter grade with turn-off time 50 μs. Find the value of the commutation capacitor so that communication failure may not occur. What economy is possible if inverter grade SCR of turn-off time 10 μs is available?

[Ans: $C \geq 7.15\ \mu F$].

REFERENCES

1. General Electric Co., *SCR Manual*, Sixth Edition, 1977.
2. M. Ramamoorthy, *An Introduction to Thyristor and their Applications*, East-West Press, 1977.
3. C.W. Lander, *Power Electronics*, McGraw-Hill Book Company, 1981.
4. B.M. Bird, of KG King, *Power Electronics*, John-Wiley & Sons, 1983.
5. B.R. Pelley, *Power-semiconductor Devices—A Status Review*, IEE Int Semiconductor Power Convertor Conf., 1982, 11–19.

Chapter 3

Gate Triggering Circuits

LEARNING OBJECTIVES:

- To define the basic requirements for the successful firing of thyristors.
- To examine the operation of optical isolators.
- To consider the operation of simple gate triggering circuits.
- To examine the operation of UJT relaxation oscillator.
- To consider the operation of PUT relaxation oscillator.
- To design the time delay circuits using PUT.
- To introduce microprocessor interfacing to power thyristor.

3.1 INTRODUCTION

The key factor in the widespread use of thyristors for controlling power is their ability to switch from non-conducting to conducting state in response to a small control signal. The proper triggering of a thyristor requires that the source of the trigger-signal should supply adequate gate current and voltage, without exceeding the thyristor gate ratings in accordance with the characteristics of the thyristor and the nature of its load and supply. Only a correctly designed firing circuit to supply the gate currents to thyristors will enable the full potential of both thyristors and equipment. The performance capabilities of a particular thyristor depend on the magnitude and wave-shape of the gate current; it will decide whether the device will fire over its full operating range, and whether the thyristor will be successful in accepting the circuit currents and voltage to which it is exposed.

Some systems operate successfully only when the gate current is flowing throughout the conduction period, while in others a single pulse is all that is required. The unbalance or harmonics in an equipment will be decided by the firing system used. The pattern of firing chosen for any thyristorised circuit will dictate the output waveform. The safe limits of operating range of the thyristor circuit are the result of the firing circuit design chosen. Since all thyristorised applications require some form of triggering, this chapter is devoted to the fundamentals of the gate triggering process, gate triggering devices and simple gate triggering circuits.

3.2 FIRING OF THYRISTORS

The basic requirements for the successful firing of a thyristor are that the current supplied to the gate should:
(i) be of adequate amplitude and sufficiently short rise time.
(ii) be of adequate duration.
(iii) occur at a time when the main circuit conditions are favourable to conduction.

3.2.1 Gate-Current Amplitude and Rise Time

The quoted firing current $I_{g(min)}$ in Chapter 2 expresses the minimum gate current required to fire all thyristors of a given type at a standard temperature. This is not normally a figure that can be applied directly in practice; a margin is required not only to cover possible errors and uncertainties but also to ensure that the turn-on time of the SCR, which becomes long and indeterminate as the gate-current approaches the critical firing level is acceptably short. Usually, the firing current is of about 1.4 $I_{g(min)}$ is commonly satisfactory so long as the main circuit conditions do not impose special requirements; a considerably higher current is often necessary, however, in order to reduce the turn-on time (usually to reduce the spread of turn-on times in a group of thyristors), to increase the di/dt rating of the thyristor or to reduce the turn-on switching loss. The gate voltage must exceed the quoted gate firing voltage, $V_{g(min)}$, by a corresponding margin.

The gate drive chosen must trigger the thyristor under the most adverse conditions, and, in particular, at the lowest thyristor junction temperature that is likely to occur. Both $V_{g(min)}$ and $I_{g(min)}$ exhibit a negative variation with temperature.

A specification of the gate current amplitude is incomplete unless the rise time of the pulse is also specified. To be useful, an adequate level of gate-current must be reached before the thyristor has turned ON—that is, within the turn-on delay time—and even given that this condition has met, the effectiveness of the pulse still increases somewhat as the rise time is further reduced. On the other hand, there is no particular advantage in a very short rise time if the amplitude is not sufficient to give a commensurately short turn-ON time. Hence, amplitude and rise time must be considered together in designing a firing circuit in accordance with any stipulated di/dt capability and switching performance.

The design of circuit for very short gate pulse rise times has to allow for the finite response time of the gate-cathode junction itself, which makes it necessary to apply a higher voltage than is apparent from the static gate characteristics. Commonly, an open-circuit source voltage of 15–20 Volt is specified in order to achieve the highest useful current amplitude with a commensurately short rise time.

3.2.2 Gate Pulse Duration

Under favourable conditions, a thyristor may be triggered successfully by a gate-pulse of a duration approximately equal to the turn-on time of the cell. However, a considerably longer pulse duration is desirable for one or more of the following reasons:

(a) A relatively long period may be required for the anode current, to rise to the latching current level.
(b) Oscillations, reflections or other disturbances may conspire to turn-off the thyristor shortly after it is first triggered.
(c) There may be uncertainty as to whether the anode circuit conditions are favourable to conduction when the firing pulse is initiated.

The relative importance of these factors, and possibly the ease with which they can be modified, determine the pulse duration required in a particular application. A certain way of ensuring that the pulse is of adequate duration is to extend it to the whole period for which the thyristor is intended to conduct; this is often done in cases where the moment, when the thyristor becomes forward-biased, is more or less unpredictable (as in many type of forced-commutation inverter with reactive load). Alternatively, the pulse is made to cover at least the period of uncertainty. Since, however, the generation of a very long pulse, particularly if high amplitude and a short-rise time are required, usually entails an unwelcome degree of complexity and cost, pulses of short duration are more often used where possible. An incidental advantage of a short-pulse is that if it coincides with the reverse anode voltage the resulting reverse loss reduces to insignificant.

In general, a pulse-duration of less than about 10 μs requires considerable care in the detailed design of the anode circuit, while a duration of 30–60 μs is usually sufficient to avoid problems so long as the anode circuit conditions are favourable to conduction.

Two alternatives to the use of a gate pulse long enough to cover the region of uncertain firing are:

(i) a control system in which the timing of the gate pulse is synchronized with the thyristor voltage or the current to zero, rather than with the supply voltage; and
(ii) an extended train of short pulses.

The first of these two possibilities is generally more appropriate for relatively simple applications where isolation is not required between the control system and thyristor main circuit. The second technique is of more general application, and has the additional advantage that it gives the thyristor repeated opportunities to turn-on. However, a disadvantage occur if the first-pulse of the train is not effective, since the actual instant of firing is then somewhat indeterminate.

3.2.3 Pulse Waveforms

Even if the anode circuit conditions require a gate pulse of high-amplitude, it is not necessary to maintain this for the whole duration of the pulse. If a thyristor starts to conduct at the beginning of the pulse, the conditions dictating the high amplitude disappear after the first few microseconds, while if it becomes forward biased within the duration of the pulse, at a dv/dt within the capability of the cell, neither a severe di/dt condition nor the necessity for a very short switching time can exist. On the other hand, a pulse of continuing high-amplitude has the disadvantages of making a bigger demand on the firing circuit, increasing the mean gate dissipation, and increasing the latching current. It is thus, a sound practice to use a pulse waveform with a leading edge of the required amplitude, and rise time followed by a tail of not more than the practical minimum amplitude required to fire the thyristor, namely, around 1.4 $I_{g(min)}$. The duration of the initial high amplitude pulse may depend on its amplitude since a high amplitude is associated with fast switching; if it is not a well defined square-pulse, an exponentially decaying pulse with a time constant of about 10 μs is usually adequate. Figure 3.1 illustrates a pulse waveform of this kind.

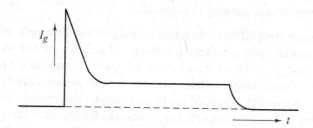

Fig. 3.1 *Extended gate firing pulse with high initial amplitude*

A negative overshoot at the end of the gate pulse is generally to be avoided since in some circumstances it may tend to turn-off the anode current that has been initiated in the thyristor.

3.2.4 Spurious Triggering

Spurious triggering, whether through stray pick-up on the gate in the firing control circuit, through excessive anode voltage or dv/dt in the main circuit, can damage a thyristor directly, apart from the possibility of indirect damage through a resulting malfunction of the circuit. This is so because the thyristor so triggered may not have the di/dt capability to withstand the anode circuit conditions. It follows that 'self-protecting' circuits depending on the automatic breakover of a thyristor in case of excessive voltage are to be avoided, unless the thyristor is specifically rated for such a duty.

3.2.5 Firing Angle Determination in Naturally Commutating Converters

The basic function of a firing angle control circuit attached to a naturally commutating converter is normally to determine the delay angle, in relation to the a.c. supply voltage, so that the desired voltage is applied to the load in accordance with a control input—i.e., in most cases to serve as a voltage-to-delay angle transducer. The systems principally employed are:

(a) Variable phase shift circuits
(b) Magnetic amplifiers
(c) Circuits in which the firing angle is determined by the intersection of a repetitive waveform, synchronous with the supply, and a reference or control level
(d) Digital timing circuits.

1. Variable Phase Shift A voltage or current is produced with a controllable phase relationship to the supply voltage and firing pulses, are generated at the zero-crossing points of the phase shifted waveforms. Most circuits of this type are suited less to electrical than to manual control, and do not readily provide the full 180° range that is commonly required.

2. Magnetic Amplifiers In suitable configurations, magnetic amplifiers can be made to produce output voltage waveforms closely resembling those of a half controlled bridge rectifier (discussed in Chapter 6), and their ideally linear relationship of mean output voltage to mean control voltage implies exactly the variation of 'firing angle' required to obtain a similar characteristic in a controlled rectifier. This makes them superficially attractive for thyristor firing circuits. In practice, the output from a conventional magnetic-amplifier is generally not suitable for firing a thyristor directly because of its poor rise time and variation of its initial amplitude with the firing angle. Even though this can be overcome by the addition of pulse forming networks, the disadvantages remain, that the characteristic tends to be seriously non-linear in the region of minimum firing angle, and that the virtually immediate response so easily provided by transistor circuits is not obtainable.

3. Waveform Intersections Systems Figure 3.2 illustrates the basic principle; the firing point is determined by the coincidence of the instantaneous level of a ramp or timing waveform and a triggering level at (i) the firing point is controlled by varying the slope of the ramp, while (ii) illustrates the more common and useful method of control in which the timing waveform is unvarying and control is effected by altering the triggering level. This principle is amenable to relatively easy predictable design employing normal analogue and digital techniques, and represents a great majority of systems in use at present.

The ramp, shown as linear in Fig. 3.2, may in some cases be exponential, or can, an advantage for some purposes, be a cosine function that yields a linear overall control characteristic with a fully or half controlled converter.

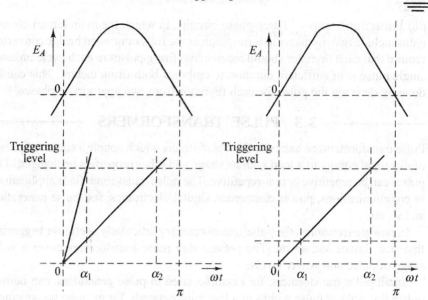

Fig. 3.2 *Firing angle control by waveform intersection* **(a)** *variable slope ramp* **(b)** *variable-triggering level*

The main problem in designing a circuit of this type is to synchronize the ramp function to the zero-crossings of the a.c. supply voltage in such a way that errors are not caused by transient disturbances or waveform distortion, without introducing unacceptable phase shifts.

4. Digital Systems Like the waveform intersection method, digital systems normally depend on measuring off the required time interval from the supply voltage zero, and considered simply as an alternative to the analogue approach to this particular requirement they do not appear to have any great advantage. They do, however, have considerable potential advantage as sections of wholly or mainly digital control systems, and lend themselves particularly well to microelectronic techniques.

3.2.6 Additional Design Features

Very often, the practical firing circuit design has to take into account additional aspects beyond the basic function of firing angle determination. These include:

(i) Firing angle limitation   A positive means of limiting to a value that will not lead to loss of commutation is commonly required in systems involving fully controlled converters (Chapter 6), and can be useful with a half controlled bridge as an alternative to a free-wheel diode.

(ii) Pulse distribution or gating   Depending on the design principles used, gating circuits are very essential to direct firing pulses to each thyristor in an appropriate region of the supply cycle and inhibit them in regions where they might lead to a circuit malfunction or other undesirable effects.

(iii) Pulse duplication Three-phase circuits, in which the main circuit current paths include two thyristors in series, such as the fully controlled bridge converter, require that each thyristor should receive two firing pulses in each cycle unless a single pulse is of sufficient duration to embrace both firing instant. This can be done by deriving the pulse for each thyristor from two appropriate phases.

3.3 PULSE TRANSFORMERS

Pulse transformers are basically the transformers which couple a source of pulses of electrical energy to a load with its shape and other properties unchanged. The pulses can be repetitive or non-repetitive. The pulse transformer finds applications in communications, power electronics, digital electronics, fast pulse generation and so on.

In power electronics, the pulse transformer is particularly useful for triggering thyristor, Triacs and so on. The present day pulse transformers cover a wide range of pulse and power levels.

Small pulse transformers, for example, used in pulse generators, can deliver only a few volts at pulse widths of a few microseconds. Large pulse transformers used in linear accelerators, radars and so on, can deliver power of the range 50 to 100 MW at 200 to 300 kV with pulse widths of the order of microseconds.

3.3.1 Pulse Transformer in Triggering Circuits

Pulse transformers are often used to couple a trigger pulse generator to a thyristor in order to obtain electrical isolation between the two circuits. The transformers commonly used for thyristor control are either 1:1 two-winding or 1:1:1 three-winding types. Figure 3.3 shows a complete output circuit to fire a thyristor correctly. The series resistor R either reduces the SCR holding current or balances gate currents in a three-winding transformer connected to two SCRs. The series diode D prevents reverse gate current in the case of ringing or reversal of the pulse transformer output voltage. The diode also reduces holding current of the SCR. In some cases where high noise levels are present, it may be necessary to load the secondary of the transformer with a resistor to prevent false triggering.

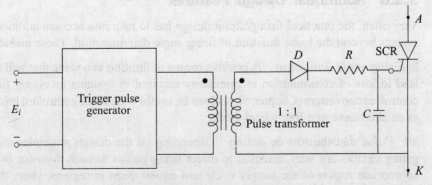

Fig. 3.3 *Complete output circuit*

Gate Triggering Circuits

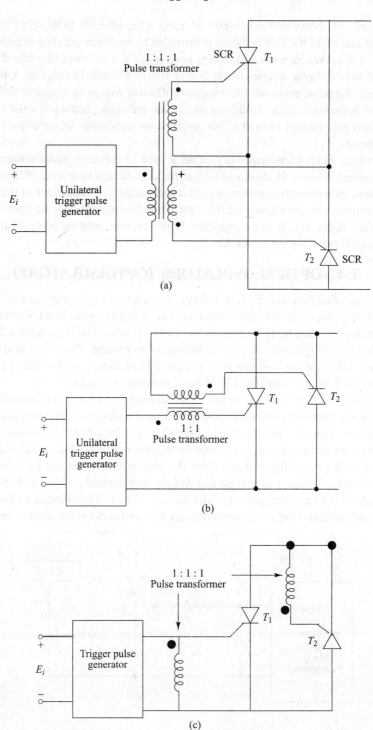

Fig. 3.4 *Pulse transformer connections for two SCRs*

Figure 3.4 shows different ways of using a transformer to drive an inverse parallel pair of SCRs. Full isolation is provided by the three-winding transformer in Fig. 3.4(a). Where such isolation is not required, a two-winding transformer may be used either in a series mode, Fig. 3.4(b) or a parallel mode Fig. 3.4(c). In any case, the pulse generator must supply sufficient energy to trigger both SCRs, and the pulse transformer (plus any additional balancing resistors) must supply sufficient gate current to both SCRs under worst conditions of unbalanced gate impedances.

A trigger pulse transformer is primarily used to enhance the efficiency. The simplest test is to use the designed trigger pulse generator to drive a 20 Ω resistor alone and then drive the same resistor through the pulse transformer. If the pulse waveforms across the resistor are the same under both conditions, the transformer is perfect. Some loss is to be expected, however, and must be compensated by increased drive from the generator.

3.4 OPTICAL ISOLATORS (OPTOISOLATORS)

A common situation where a low-voltage, low-current logic circuit controlling a high voltage, high current load is shown in Fig. 3.5. The logic circuit is usually an IC and is very often part of a process-control computer. The logic output E_0 will be either at 0 V (ground) or +5 V with respect to ground. When it is at 0 V, the SCR is "off" and the load receives no power from the ac source. When E_0 is +5 V, the SCR is "on", and the 230 V ac is switched to the load.

Generally it is not desirable to have a direct electrical connection between low-power control circuitry and the high-power load it controls, as done in this circuit. One reason is the possibility of noise being coupled from the high current circuit into the logic circuitry, especially over the common ground-line. Another reason is the possibility of high voltage from the load circuitry feeding back into the logic circuit as a result of component failure. For example, if the SCR became shorted out between the gate and cathode, 230 V would be coupled to the logic circuit output and could easily destroy many ICs, and perhaps the whole computer.

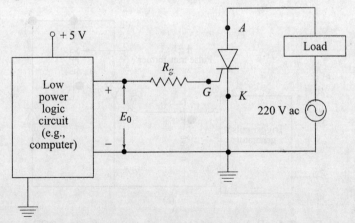

Fig. 3.5 *A low power logic circuit controlling a high-power load*

Thus, it is often necessary to electrically isolate the low-power control-circuitry and high power load circuitry. One means for doing this is to use an electro-mechanical relay (Fig. 3.6). The logic circuit drives the low current relay coil and the relay magnetically controls the switch contacts that connect power to the load.

While relays are widely used in industrial control applications, they do have certain shortcomings including—
(a) They are fairly expensive.
(b) Relays are bulkier than solid-state devices.
(c) They create magnetic fields and inductive "kick", which can be troublesome sources of electrical noise.
(d) Relays have shorter-life than semiconductor devices.
(e) The contacts create sparking upon opening; this is highly undesirable in many industrial environments.

These disadvantages are largely overcome by devices called *optical isolators* or *optoisolators*, which use light energy to couple the control signal to the load. Optoisolators consists of a light-source (usually an infra-red emitting diode, IRED), a light sensitive device (e.g., a phototransistor) and a switching device. In most cases the light sensor and switching device are one and the same.

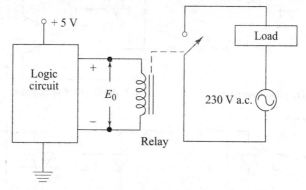

Fig. 3.6 *Electrical isolation provided by relay*

Figure 3.7 shows some of the available optoisolators. In each case, the devices inside the dotted lines are integrated into a single light-tight package with only input terminals x and y, and output terminals a and b accessible to the user. The input circuit is simply an IRED which emits IR radiation when it is sufficiently forward biased. This radiation is focused on a light sensitive device so that it switches "on" whenever sufficient current flows through the IRED. The optoisolators in Figs 3.7 (a), (b) and (c) are used to switch d.c. power to a load, while those in Figs 3.7 (d) and (e) can switch a.c. power.

The various opto-isolators have different output current capabilities. The IRED-phototransistor combination can switch output currents of only around 10 mA, while the IRED/photodarlington can typically switch 500–100 mA. The darlington transistor coupler in Fig. 3.7(b) can be used when increased output current capability is needed beyond that provided by the phototransistor output. The disadvantage is that the photodarlington has a switching speed less than that of the phototransistor.

An LASCR output coupler is shown in Fig. 3.7(c), while a phototriac output coupler is shown in (d). Both combinations can typically switch 500 mA currents. For heavier current loads, the device in (e) can be used. It uses a light-activated TRIAC to trigger a high current TRIAC that switches currents of over 1A.

Figure 3.8 illustrates how an optoisolator can be used to isolated low-power control circuitry from a high power load. The output voltage from the logic circuit provides the relatively low current needed to activate the IRED, which in turn controls the light activated TRIAC. When E_0 is 0 V, the IRED is nonconducting and the TRIAC is held "off", so the load receives no a.c. voltage. When E_0 is + 5 V, the IRED conducts and its radiant energy turns the TRIAC "on" switching the a.c. voltage across the load. The load receives a.c. power as long as E_0 remains at + 5 V. Note that there is no direct electrical connection between the logic circuit and load circuit.

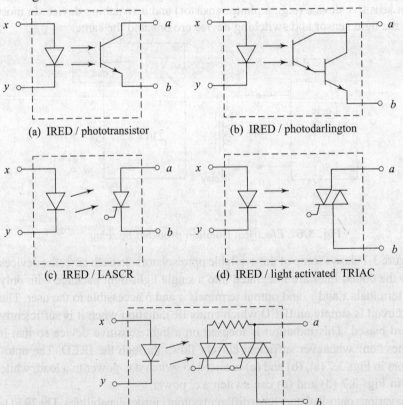

(a) IRED / phototransistor (b) IRED / photodarlington

(c) IRED / LASCR (d) IRED / light activated TRIAC

(e) IRED / Triac / Triac for high current a.c.

Fig. 3.7 *Common optoisolators*

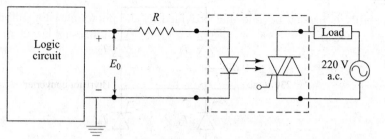

Fig. 3.8 *Electrical isolation provided by an optoisolator*

3.5 GATE TRIGGER CIRCUITS

The general block diagram of a gate-trigger circuit, for example, in a single phase converter is shown in Fig. 3.9. The thyristors are at line potential and the trigger circuit must be referenced within respect to a logic ground associated with

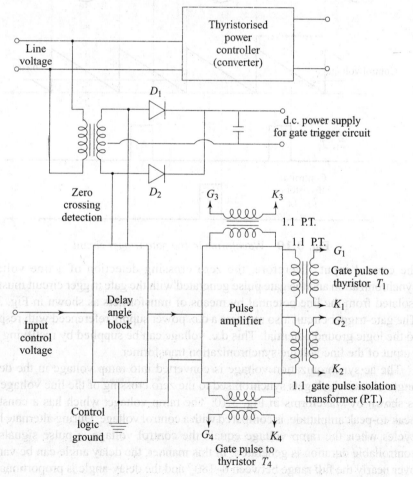

Fig. 3.9 *General block diagram of a thyristor gate trigger circuit*

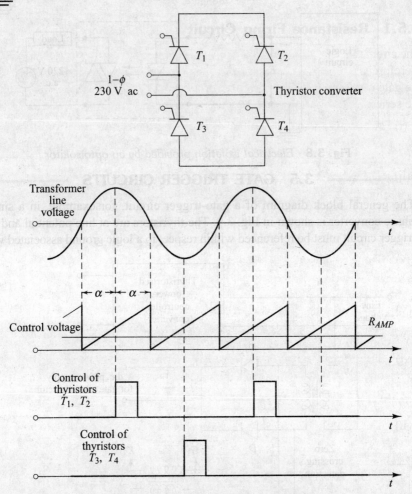

Fig. 3.10 *Waveforms in the gate trigger circuit*

the control input. Therefore, the zero crossing detection of a line voltage synchronization and the gate pulse generated with the gate trigger circuit must be isolated from the line potential by means of transformers as shown in Fig. 3.9. The gate-trigger circuit also requires a d.c. power supply referenced with respect to the logic ground potential. This d.c. voltage can be supplied by rectifying the output of the line voltage synchronization transformer.

The ac synchronization voltage is converted into ramp voltage in the delay angle block, which gets synchronised to the zero crossing of the line voltage, as is shown by waveforms in Fig. 3.10. The ramp voltage, which has a constant peak-to-peak amplitude, is compared with a control voltage. During alternate half cycles when the ramp voltage equals the control voltage, a pulse signals of controllable duration is generated. In this manner, the delay angle can be varied over nearly the full range between 0–180° and the delay-angle is proportional to the control voltage.

3.5.1 Resistance Firing Circuit

The circuit in Fig. 3.11 shows a simple method for varying the trigger angle and therefore, the power in the load. Instead of using a gate pulse to trigger the SCR, the gate current is supplied by an a.c. source of voltage e_s through R_{min}, R_v, and the series diode D. The circuit operates as follows:

(i) As e_s goes positive, the SCR becomes forward-biased from anode to cathode; however, it will not conduct ($e_L = 0$) until its gate current exceeds $I_{g(min)}$.

(ii) The positive e_s also forward biases the diode and the SCRs gate–cathode junction; this causes flow of a gate current i_g.

(iii) The gate current will increase as e_s increases towards its peak value. When i_g reaches a value equal to $I_{g(min)}$, the SCR turns "on" and e_L will approximately equal e_s (refer to point P on the waveform in Fig. 3.11).

(iv) The SCR remains "on" and $e_L \approx e_s$ until e_s decreases to the point where the load current is below the SCR holding-current. This usually occurs very close to the point until $e_s = 0$ and begins to go negative.

(v) The SCR now turns "off" and remains "off" while e_s goes negative since its anode–cathode is reverse biased, and since the SCR is now an open switch, the load voltage is zero during this period.

(vi) The purpose of the diode in the gate-circuit is to prevent the gate–cathode reverse bias from exceeding peak reverse gate voltage during the negative half-cycle of e_s. The diode is chosen to have peak reverse-voltage rating greater than the input voltage E_{max}.

(vii) The same sequence is repeated when e_s again goes positive.

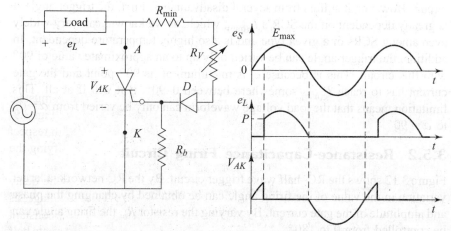

Fig. 3.11 *R-firing circuit and associated voltage waveforms*

The load-voltage waveform in Fig. 3.11 can be controlled by varying R_v which varies the resistance in the gate circuit. If R_v is increased, the gate current will reach its trigger value $I_{g(min)}$ at a greater value of e_s making the SCR to trigger at a latter point in the e_s positive half-cycle. Thus, the trigger angle α will increase. The opposite will occur if R_v is decreased. Of course, if R_v is made large enough the SCR gate current will never reach $I_{g(min)}$ and the SCR will remain "off". The minimum trigger angle is obtained with R_v equal to zero.

As shown in Fig. 3.11, the limiting resistor $R_{(min)}$ is placed between anode and gate so that the peak gate current of the thyristor I_{gm} is not exceeded. In the worst case, that is when the supply voltage has reached its peak, E_{max},

$$R_{min} \geq \frac{E_{max}}{I_{gm}} \tag{3.1}$$

The stabilising resistor R_b should have such a value that the maximum voltage drop across it does not exceed maximum possible gate voltage $V_{g(max)}$. From the voltage distribution,

$$R_b \leq \frac{(R_v + R_{min}) \cdot V_{g(max)}}{(E_{max} - V_{g(max)})} \tag{3.2}$$

The thyristor will trigger when the instantaneous anode voltage, e_s, is

$$e_s = I_{g(min)}(R_v + R_{min}) + V_d + V_{g(min)} \tag{3.3}$$

where $I_{g(min)}$ = minimum gate current to trigger the thyristor

V_d = voltage drop across the diode

$V_{g(min)}$ = gate-voltage to trigger, corresponding to $I_{g(min)}$.

The resistance trigger shown in Fig. 3.11 is the simplest and most economical circuit. However, it suffers from several disadvantages. First, the trigger angle α is greatly dependent on the SCR's $I_{g(min)}$, which, as we known, can vary widely even among SCRs of a given type and is also highly temperature dependent. In addition, the trigger angle can be varied only up to an approximate value of 90° with this circuit. This is because e_s is maximum at its 90° point and the gate current has to reach $I_{g(min)}$ somewhere between 0–90°, if it will if at all. This limitation means that the load voltage waveform can only be varied from $\alpha = 0°$ to $\alpha = 90°$.

3.5.2 Resistance-Capacitance Firing Circuit

Figure 3.12 shows the RC–half wave trigger circuit. By the RC network, a larger variation in the value of the firing angle can be obtained by changing the phase and amplitude of the gate current. By varying the resistor R_v, the firing angle can be controlled from 0 to 180°.

In the negative half-cycle, capacitor C charges through diode D_2 with lower plate positive to the peak supply voltage E_{max}. This capacitor voltage remains constant at $-E_{max}$ until supply voltage attains zero value. Now, as the SCR anode voltage passes through zero and becomes positive, capacitor C begins to charge through R_v from the initial voltage $-E_{max}$. When the capacitor charges to positive voltage equal to gate trigger voltage V_{gt} (= $V_{g(min)}$ + V_{D1}), SCR is triggered and after this, the capacitor holds to a small positive voltage, as shown in Fig. 3.12. During negative half-cycle, the diode D_1 prevents the breakdown of the gate to cathode junction. In the range of power-frequencies, the RC for zero output voltage is given by

$$R_v C \geq \frac{1.3T}{2} = \frac{4}{w} \qquad (3.4)$$

where $T = 1/f$ = period of ac line frequency in seconds.

As discussed above, the thyristor will turn ON when the capacitor voltage e_c equals ($V_{g(min)}$ + V_{D1}), provided the gate current $I_{g(min)}$ is available. Therefore, the maximum value of R_v is given by

$$e_s \geq I_{g(min)} R_v + e_c$$

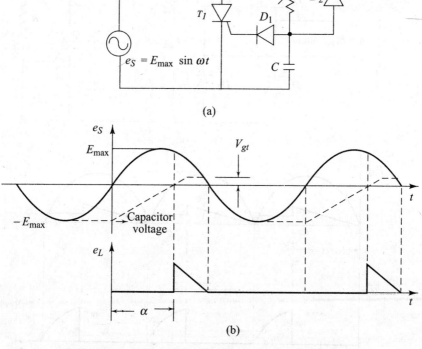

Fig. 3.12 *(a) RC firing circuit, (b) voltage-waveform*

$$= I_{g(min)} R_v + V_{g(min)} + V_{D1} \qquad (3.5)$$

or
$$R_v \leq \frac{e_s - V_{g(min)} - V_{D1}}{I_{g(min)}} \qquad (3.6)$$

where e_s is the instantaneous supply voltage at which the thyristor will turn ON. From Eqs 3.4 and 3.6, the suitable values of R_v and C can be obtained.

3.5.3 Resistor Capacitor-Full-Wave Trigger Circuit

Power can be delivered to the load in Fig. 3.12 only during the positive half-cycle of e_s because the SCR conducts only when it is forward biased. This limitation can be overcome in several ways, one of which is shown in Fig. 3.13. Here, the ac line voltage is converted to pulsating dc by the full-wave diode bridge. This allows the SCR to be triggered "on" for both half-cycle of the line voltage, which doubles the available power to the load.

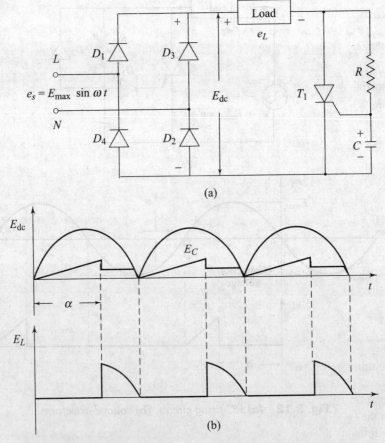

Fig. 3.13 (a) *RC full-wave trigger circuit* (b) *with waveforms*

Gate Triggering Circuits

In this circuit, the initial voltage from which capacitor C charges is almost zero. Capacitor C is set to this low positive voltage (upper plate positive) by the clamping action of SCR gate. When the capacitor charges to a voltage equal to V_{gt}, SCR triggers and rectified voltage E_{dc} appears across load as e_L. The value of R_{vC} is obtained from the following relation:

$$R_v C \geq 50 \frac{T}{2} = \frac{157}{w} \qquad (3.7)$$

As per Eq. 3.6, the value of R_v is given by

$$R_v \leq \frac{e_s - V_{gt}}{I_{g(\min)}} \qquad (3.8)$$

SOLVED EXAMPLES

Example 3.1 The circuit in Fig. 3.11 uses an SCR with $I_{g(\min)} = 0.1$ mA and $V_{g(\min)} = 0.5$ V. The diode is silicon and the peak amplitude of the input is 24 Volts. Determine the trigger angle α for $R_V = 100$ kΩ and $R_{\min} = 10$ kΩ.

Solution: The first step is to determine the instantaneous value of e_s at which triggering will occur. At the SCR trigger point, $V_{g(\min)} = 0.5$V and $I_{g(\min)} = 0.1$ mA. Using KVL around the gate circuit, we have

$$e_s = I_g (R_V + R_{\min}) + V_D + V_g$$

At the trigger point,

$e_{s(\text{trigger})} = 0.1$ mA $(110$ k$\Omega) + 0.7$V $+ 0.5$V $= 12.2$ V.

Since e_s is a sine-wave, it obeys the expression

$$e_s = E_{\max} \cdot \sin \omega t = E_{\max} \cdot \sin(2\pi f t)$$

where $2\pi f t$ is the phase angle at any instant of time. For our purposes, this angle is α. Thus, $E_{\max} = 24$ V,

$$e_s = 24 \sin \alpha. \quad \therefore \quad 12.2 = 24 \sin \alpha.$$

$$\sin \alpha = \frac{12.2}{24} \quad \therefore \quad \alpha = 30.6°.$$

3.6 UNIJUNCTION TRANSISTOR

The unijunction transistor, abbreviated UJT, is a three-terminal, single-junction device. The basic UJT and its variations are essentially latching switches whose operation is similar to the four-layer diode, the most significant difference being that the UJT's switching voltage can be easily varied by the circuit designer. Like the four-layer diode, the UJT is always operated as a switch and finds most frequent applications in oscillators, timing circuits and SCR/TRIAC trigger circuits.

3.6.1 Basic Operation

A typical UJT structure, pictured in Fig. 3.14, consists of a lightly doped, N-type silicon bar provided with ohmic contacts at each end. The two end connections are called base-1, designated B_1 and base-2, B_2. A small, heavily doped P-region is alloyed into one side of the bar closer to B_2. This P-region is the UJT emitter, E, and forms a P-N junction with the bar.

An interbase resistance, R_{BB}, exists between B_1 and B_2. It is typically between 4 kΩ and 10 kΩ, and can easily be measured with an ohmeter with the emitter open. R_{BB} is essentially the resistance of the N-type bar. This interbase resistance can be broken up into two resistances, the resistance from B_1 to emitter, called R_{B_1}, and resistance from B_2 to emitter, called R_{B_2}. Since the emitter is closer to B_2, the value of R_{B_1} is greater than R_{B_2} (typically 4.2 kΩ versus 2.8 kΩ).

The operation of the UJT can better be explained with the aid of an equivalent circuit. The UJT's circuit symbol and its equivalent circuit are shown in Fig. 3.15. The diode represents the P-N junction between the emitter and the base-bar (point x). The arrow through R_{B_1} indicates that it is variable since during normal operation it may typically range from 4 kΩ down to 10 Ω.

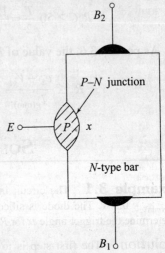

Fig. 3.14 Basic UJT structure

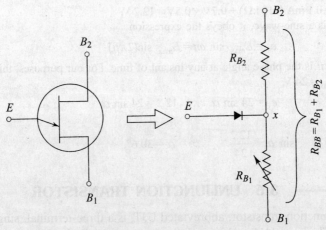

Fig. 3.15 UJT symbol and equivalent circuit

The essence of UJT operation can be stated as follows:
(a) When the emitter diode is reverse biased, only a very small emitter current flows. Under this condition, R_{B_1} is at its normal high-value (typically 4 kΩ). This is the UJT's "off" state.

(b) When the emitter diode becomes forward biased, R_{B1} drops to a very low value (reason to be explained later) so that the total resistance between E and B_1 becomes very low, allowing emitter current to flow readily. This is the "on" state.

Circuit-operation The UJT is normally operated with both B_2 and E biased positive relative to B_1 as shown in Fig. 3.16. B_1 is always the UJT reference terminal and all voltages are measured relative to B_1. The V_{BB} source is generally fixed and provides a constant voltage from B_2 to B_1. The V_{EE} source is generally a variable voltage and is considered the input to the circuit. Very often, V_{EE} is not a source but a voltage across a capacitor.

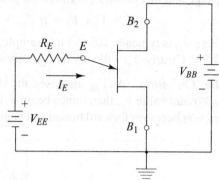

Fig. 3.16 Normal UJT biasing

We will analyze the UJT circuit operation with the aid of the UJT equivalent circuit, shown inside the dotted lines in Fig. 3.17(a). We will also utilize the UJT emitter-base-1 V_E-I_E curve shown in Fig. 3.17(b). The curve represents the variation of emitter current I_E, with emitter-base-1 voltage, V_E, at a constant B_2-B_1 voltage. The important points on the curve are labelled, and typical values are given in parentheses.

The "Off" state If we neglect the diode for a moment, we can see in Fig. 3.17(a) that R_{B_1} and R_{B_2} form a voltage divider that produces a voltage V_x from point x relative to ground.

$$\therefore \quad V_x = \frac{R_{B_1}}{R_{B_1} + R_{B_2}} \times V_{BB} = \underbrace{\frac{R_{B_1}}{R_{BB}}}_{\eta} \times V_{BB}$$

or simply,

$$V_x = \eta \, V_{BB} \tag{3.9}$$

where η (the greek letter "eta") is the internal UJT voltage divider ratio $\dfrac{R_{B_1}}{R_{BB}}$ and is called the *intrinsic stand off ratio*.

Values of η typically range from 0.5 to 0.8 but are relatively constant for a given UJT.

The voltage at point x is the voltage on the N-side of the P-N junction. The V_{EE} source is applied to the emitter which is the P-side. Thus, the emitter diode will be reverse-biased as long as V_{EE} is less than V_x. This is the "off" state and is shown on the V_E-I_E curve as being a very low current region. In the "off" state, then, we can say that the UJT has a very high resistance between E and B_1, and I_E is usually a negligible reverse leakage current. With no I_E, the drop across R_E is zero and the emitter voltage, V_E, equals the source-voltage.

The UJT "off" state, as shown on the V_E–I_E curve, actually extends to the point where the emitter voltage exceeds V_x by the diode threshold voltage, V_D, which is needed to produce forward current through the diode. The emitter voltage and this point, P, is called the *peak-point voltage*, V_p, and is given by

$$V_p = V_x + V_D = \eta V_{BB} + V_D \qquad (3.10)$$

where V_D is typically 0.5 V. For example, if $\eta = 0.65$ and $V_{BB} = 20$V, then $V_p = 13.5$ V. Clearly, V_P will vary as V_{BB} varies.

The "On" state As V_{EE} increases, the UJT stays "off" until V_E approaches the peak-point value V_p, then things begin to happen. As V_E approaches Vp, the P-N junction becomes forward biased and begins to conduct in the opposite direction.

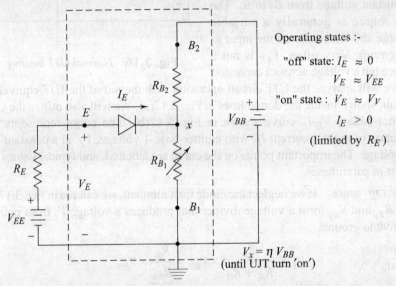

Fig. 3.17 *(a) Equivalent-circuit for UJT analysis*

Note on the V_E–I_E curve that I_E becomes positive near the peak point P. When V_E exactly equals V_p, the emitter current equals I_P, the *peak-point current*. At this point, holes from the heavily doped emitter are injected into the N-type bar, specially into the B_1 region. The bar, which is lightly doped, offers very little chance for these holes to recombine. As such, the lower half of the bar becomes replete with additional current carriers (holes) and its resistance R_{B_1} is drastically reduced. The decrease in R_{B_1} causes V_x to drop. This drop in turn causes the diode to become more forward biased, and I_E increases even further. The larger I_E injects more holes into B_1, further reducing R_{B_1}, and so on. When this *regenerative* or *snowballing* process ends, R_{B_1} has dropped to a very small value (2–25 Ω) and I_E can become very large, limited mainly by external resistance R_E.

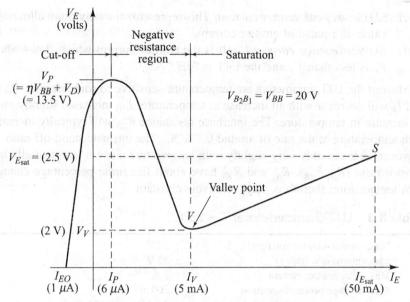

Fig. 3.17 (b) *Typical UJT V-I characteristic curve*

The UJT operation has switched to the low-voltage, high-current region of its $V_E - I_E$ curve. The slope of this "on" region is very steep, indicating a low resistance. In this region, the emitter voltage V_E, will be relatively small, typically 2 V, and remains fairly constant as I_E is increased up to its maximum rated value, $I_{E(\text{sat})}$. Thus, once the UJT is "on," increasing V_{EE} will serve to increase I_E while V_E remains around 2V.

Turning "Off" the UJT Once it is "on," the UJT's emitter current depends mainly on V_{EE} and R_E. As V_{EE} decreases, I_E will decrease along the "on" portion of the $V_E - I_E$ curve. When I_E decreases to point V, the valley point, the emitter current is equal to I_V, the *valley current*, which is essentially the holding current needed to keep the UJT "on". When I_E is decreased below I_V, the UJT turns "off" and its operation rapidly switches back to the "off" region of its $V_E - I_E$ curve, where $I_E \approx 0$ and $V_E - V_{EE}$. The valley current is the counterpart of the holding current in *PNPN* devices, and generally ranges between 1 and 10 mA.

3.6.2 UJT Parameters and Ratings

A set of parameter and ratings for a typical UJT (2N2646) are listed in Table 3.1. Some of the entries were defined earlier. Those which require explanation are:

(a) *Maximum reverse emitter voltage V_{B_2E}.* This is the maximum reverse bias which the emitter-base-2 junction can tolerate before breakdown occurs.

(b) *Maximum interbase voltage.* This limit is caused by the maximum power that the *N*-type base bar can safely dissipate.

(c) *Maximum peak emitter current.* This represents the maximum allowable value of a pulse of emitter current.

(d) *Emitter leakage current I_{E0}.* It is the emitter current which flows when V_E is less than V_P and the UJT is "off."

Most of the UJT parameters are temperature-sensitive to some degree. V_v, I_V and I_P will decrease with an increase in temperature. I_{E0} increases slightly with an increase in temperature. The interbase resistance R_{BB} will typically increase with temperature at the rate of around 0.8%/°C. The intrinsic stand-off ratio η, however, changes only very slightly with temperature since it is essentially the resistor ratio $(R_{B_1} R_{BB})$, R_{B_1} and R_{BB} have about the same percentage change with temperature; therefore, stays relatively constant.

Table 3.1 UJT characteristics at $T_j = 25°C$

Max. reverse-emitter voltage (V_{B_2E})	: 30 V
Max. interbase voltage (V_{BB})	: 35 V
Max. peak emitter current	: 2 A
Max. average power dissipation	: 300 mW
Interbase resistance (R_{BB})	: 4.7 kΩ to 9.1 kΩ
Intrinsic stand-off ratio (η)	: 0.56 to 0.75
Emitter-leakage current (I_{E0})	: 12 μA (max) at $V_{B_2E} = 30$ V
Valley current (I_v)	: 4 mA (min) at $V_{BB} = 20$ V
Valley voltage (V_v)	: 2 V (typical) at $V_{BB} = 20$ V
Peak-point current (I_P)	: 5 μA (max) at $V_{BB} = 25$ V

3.6.3 UJT Relaxation Oscillator

The UJT is often used as a trigger device for SCRs and TRIACs. Other applications include nonsinusoidal oscillators, sawtooth generators, phase-control, and timing circuits.

The most common UJT circuit in use today is the relaxation oscillator shown in Fig. 3.18. Also, this type of circuit is basic to other timing and trigger circuits. The operation is as follows:

Let us consider the situation in which the capacitor is at zero volts ($V_c = 0$) and the switch is suddenly closed at $t = 0$ applying E_{dc} to the circuit. Since $V_E = V_c = 0$, the UJT emitter diode is reverse-biased and the UJT is "off." The amount of reverse bias is V_x volts which can be obtained using the voltage divider rule:

$$V_x = \frac{(R_1 + R_{B_1})E_{dc}}{R_1 + R_{B1} + R_2 + R_{B2}} \tag{3.11}$$

In many cases, R_1 and R_2 are much smaller than R_{B_1} and R_{B_2}, and V_x becomes approximately equal to ηE_{dc} (Eq. 3.9).

Gate Triggering Circuits

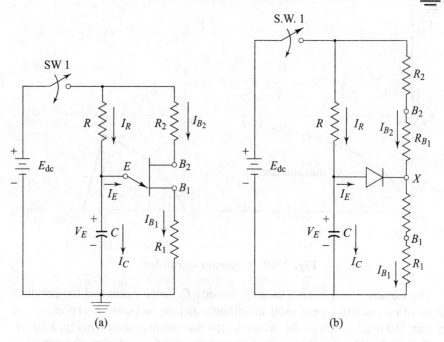

Fig. 3.18 *(a) UJT basic-relation oscillator (b) its equivalent circuit*

In this condition the only emitter current flowing will be small-reverse-leakage, I_{E0}. Also, R_{B_1} will be at its "off" value (typically 4 kΩ). Thus we can consider the emitter to be open ($I_E \approx 0$) and the capacitor will begin to charge toward the input voltage E_{dc} through resistor R. The capacitor voltage increases with a time constant of RC as illustrated in Fig. 3.19 (a). It will continue to increase until the voltage at the emitter reaches the peak-point value, V_{p_1} given by Eq. (3.10). At this time, the emitter diode becomes forward biased and the UJT turns 'on' with R_{B_1} dropping to a very low value (typically 10 ohms). Since the diode is now forward-biased, the capacitor will discharge through the low-resistance path containing the diode, R_{B_1} and R_1.

The capacitor discharge time constant is normally very short compared to its charging time constant (see Fig. 3.19(b)). An analytical expression for the discharge time constant is difficult to obtain since R_{B1} will continually change as the current I_E decreases. The discharging capacitor provides the emitter current needed to keep the UJT "on"; it will remain "on" until I_E drops below the valley current I_V, at which time the UJT will turn "off." This occurs at time T_2 when the capacitor voltage has dropped to the valley voltage V_v (typically 2–3 volts). At this time, R_{B1} returns to its "off" value, the diode is again reverse-biased and $I_E \approx 0$.

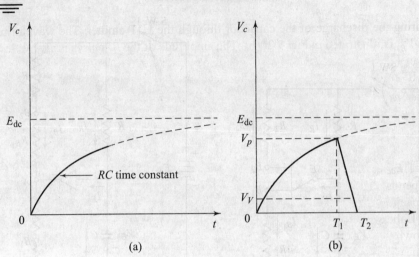

Fig. 3.19 *Capacitor waveform*

The capacitor will begin charging towards E_{dc} once again and the previous chain of events will repeat itself indefinitely as long as power is applied to the circuit. The result is a periodic sawtooth type waveform as shown in Fig. 3.20 (a).

To calculate the frequency of this waveform, we first calculate the period of one cycle. The length of one period, T_1, is essentially the time it takes for the capacitor to charge to V_p since the discharge time T_2 is usually relatively short. Thus $T \approx T_1$ and is given by

$$T = R.C. \log_e \left(\frac{E_{dc}}{E_{dc} - V_p} \right) \qquad (3.12)$$

In most cases, $V_p \approx \eta E_{dc} + V_0$ and the period can be written as

$$T \approx R.C. \log_e \left[\frac{E_{dc}}{E_{dc}(1 - \eta) - V_D} \right] \qquad (3.13)$$

The small diode drop V_D can often be ignored if $E_{dc} > 10$ V, resulting in the more approximate expression,

$$T \approx R.C. \log_e \left[\frac{1}{1 - \eta} \right] \qquad (3.14)$$

Examination of Eq. 3.14 brings out an important point, namely that T is relatively independent of supply voltage E_{dc}. This characteristic is important when designing a stable oscillator circuit. The oscillator frequency is given by $1/T$ and can be obtained by using either of the three previous equations for T.

Pulse outputs The UJT relaxation oscillator circuit can also supply pulse waveforms. If the output is taken from B_1 the result is a train of pulses occurring during the discharge of the capacitor through the UJT emitter. The waveforms of V_{B_1} is illustrated in Fig. 3.20(b). The amplitude of the B_1 pulses is always less than V_{pi} but is greater for larger values of C. The voltage at B_1 during the UJT "off" time will be very small and is determined by the voltage divider formed by R_1, R_{BB} and R_2 [see Fig. 3.18(b)] That is,

$$V_{B1} \text{ (off)} = \left(\frac{R_1}{R_1 + R_{BB} + R_2} \right) E_{dc} \qquad (3.15)$$

The rise time of the pulses at B_1 is very short (less than 1 μs), but the fall time depends on the values of C and R_1. A larger value of C or R_1 will cause a slower capacitor discharge and a longer fall-time.

If the output is taken at B_2, a waveform of negative going pulses is obtained as shown in Fig. 3.20(c). This results from the decrease in R_{B_1} when the UJT turns "on". This increases I_{B_2} which increases the drop across R_2 and thus reduces V_{B_2}. The amplitude of this pulses is usually about a couple of volts, but can be increased by increasing R_2.

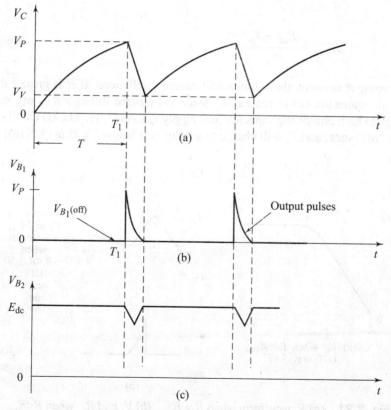

Fig. 3.20 *Waveform for UJT relation — oscillator*

The pulses at B_1 are usually the ones of most interest, they are of relatively high amplitude and are not affected by loading since they appear across a low-valued resistor R_1. These positive pulses are often used to trigger SCRs or other gated *PNPN* devices. The amplitude of these pulses is to some degree dependent on the value of C. For values of C of 1 µF or greater, the amplitude of the pulses is approximately equal to V_P (less than 2–3 V VJT drop). As C becomes smaller, the B_1 pulse decrease in amplitude. The reason for this is that the smaller value for C discharges a significant amount during the time that the UJT is making its transition from the "off" to "on" state. Thus, when the UJT finally reaches the "on" state, C has lost some of its voltage (V_p) and less voltage can appear across R_1 as the capacitor continues its discharge.

Varying the frequency The frequency of oscillations is normally controlled by varying the charging time constant RC. There are, however, limits on R. These limits are:

$$R_{min} = \frac{E_{dc} - V_v}{I_v} \tag{3.16}$$

$$R_{max} = \frac{E_{dc} - V_p}{I_p} \tag{3.17}$$

Keeping R between these limits will ensure oscillations. If R is greater than R_{max}, the capacitor never reaches V_p since the current through R is not large enough to both charge the capacitor and supply I_P to the UJT. The UJT will stay in the "off" state, and V_c will charge to a value just below V_P. (Fig. 3.21(a))

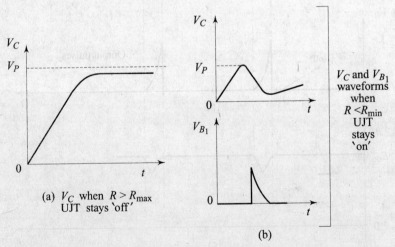

Fig. 3.21 *(a)* V_c *waveform when* $R > R_{max}$; *(b)* V_c *and* R_{B_1} *when* $R < R_{min}$

If R is smaller than R_{min}, the capacitor will reach V_P and discharge through the UJT, but the UJT will not turn "off" since the current through R is greater than the I_V needed to hold the UJT "on". The capacitor and V_{B_1} waveforms will consists of a single (Fig. 3.21(b)) representing one charge and discharge interval. This single pulse operation is sometimes used in time delay applications. The time delay is given by Eq. 3.12.

Examination of Eq. 3.16 indicates that to obtain a greater upper limit on frequency (a lower R_{min}) the value of I_V should be made larger. Similarly, to obtain a smaller lower limit on frequency (a higher R_{max}) the value of I_p should be made smaller. UJTs with I_v as high as 20 mA and I_p as low as 1 µA are presently available, resulting in a possible frequency range of 4000 : 1.

The frequency may also be varied by varying C. The lower limit on C is normally around 0.001 µF, while the upper limit depends on the size of R_1 (which limits on discharge current). In most applications of this circuit, the value of C is kept fixed and a variable resistor is used for R.

The temperature stability of the UJT relaxation oscillator frequency is normally very good. This is because η varies only slightly with temperature and the only variation in V_p is due to the small decrease in VD (2 mV/°C) with temperature. Its stability of frequency with variations in temperature and supply voltage coupled with its simplicity and low cost make the UJT oscillator a popular circuit for timing and pulsing applications.

3.6.4 The UJT as an SCR Trigger

The circuit examined in the previous section is often used as the gate trigger source in SCR applications. The basic circuit is shown in Fig. 3.22 when the B_1 pulse output is used to trigger the SCR a predetermined interval of time after the

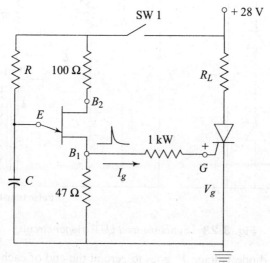

Fig. 3.22 *UJT oscillator as gate trigger source*

switch is closed. That is, the first B_1 pulse occurs T seconds after the 28 V is supplied to the UJT circuit. After the SCR has been triggered "on," subsequent pulses at its gate have no effect.

An important design consideration in this type of circuit concerns premature triggering of the SCR. The voltage at B_1 when the UJT is "off" (Eq. 3.15) must be smaller than the voltage needed to trigger the SCR, otherwise the SCR will be triggered immediately upon switch closure. Thus, we have the requirement

$$V_{B_1}(\text{off}) < (I_g \times 1 \text{ k}\Omega + V_g) \qquad (3.18)$$

3.6.5 Synchronized UJT-Triggering (Ramp Triggering)

Synchronized UJT triggering circuit is shown in Fig. 3.23. The diode bridge $D_1 - D_4$ rectifies a.c. to d.c. Resistor R_s lowers E_{dc} to a suitable value for the zener diode and UJT. The zener diode D_z is used to clip the rectified-voltage to a fixed voltage V_z. This voltage V_z is applied to the charging circuit RC. Capacitor C charges through R until it reaches the UJT trigger voltage V_p. The UJT then turns "on" and C discharges through the UJT emitter and primary of the pulse-transformer. The windings of the pulse transformer have pulse voltages at their secondary terminals. Pulses at the two secondary windings feed the same inphase pulse to two SCRs of a full wave circuit. SCR with positive anode voltage would turn ON. Rate of rise of capacitor voltage can be controlled by varying R. The firing angle can be controlled up to about 150°. This method of controlling the output power by varying charging resistor R is called as ramp control, open loop control or manual control.

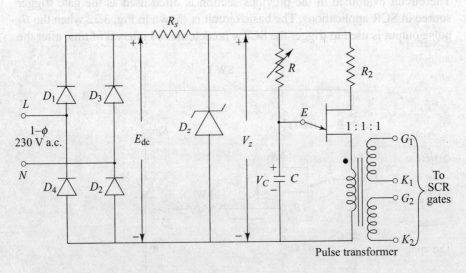

Fig. 3.23 *Synchronized UJT trigger-circuit*

As the zener diode voltage V_z goes to zero at the end of each half cycle, the synchronization of the trigger circuit with the supply voltage across SCRs is achieved. Thus the time t, equal to α/ω, when the pulse is applied to SCR for the first time, will remain constant for the same value of R. The various voltage waveforms are shown in Fig. 3.24.

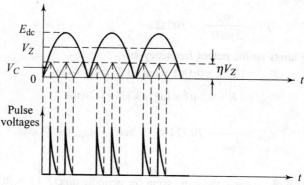

Fig. 3.24 *Generation of output pulses*

SOLVED EXAMPLES

Example 3.2 If $R_E = 1$ kΩ and $I_V = 5$ mA, determine the value of V_{EE} which will cause the UJT to turn "off."

Solution: At the valley point, $V_E = V_V = 2$ V and $I_E = I_V = 5$ mA.
Thus, using KVL, $V_{EE} = I_E R_E + V_E = 7$ V
is the value of V_{EE} below which the UJT will switch back to the "off" state.

Example 3.3 Design a UJT relaxation oscillator using UJT 2N2646, for triggering an SCR. The UJT has the following characteristics.

$$\eta = 0.7, I_P = 50 \,\mu A, V_v = 2V, I_V = 6 \text{ mA},$$
$$V_{BB} = 20 \text{ V}, R_{BB} = 7 \text{ K}\Omega, I_{E0} = 2 \text{ mA}.$$

Also, determine the limits for the output frequency of the oscillator.

Solution: Let us assume $C = 0.1$ μF.

$$\therefore R_{max} = \frac{20(1-0.7)}{50} \times 10^6 = 120 \text{ k}\Omega. \quad \text{and} \quad R_{min} = \frac{20-2}{6 \times 10^{-3}} = 3 \text{ k}\Omega.$$

The approximate value of R_2 is given by $R_2 = \dfrac{10^4}{\eta V_{BB}}$

$$\therefore \qquad R_2 = \frac{10^4}{0.7 \times 20} = 714.29 \,\Omega.$$

Now, $R_1 = \dfrac{V_{g(min)}}{I_{E0}}$

where $V_{g(min)}$ is the minimum gate voltage required to trigger SCR (≈ 0.2 V)

$\therefore \quad R_1 = \dfrac{0.2}{2 \times 10^{-3}} = 100 \, \Omega.$

Now, the limits on the output frequency are

$T_{max} = 120 \, K \times 0.1 \, \mu F \times \log_e (3.33) = 14.44$ ms (i)

$T_{min} = 3 \, K \times 0.1 \, \mu F \times \log_e (3.33) = 0.36$ ms. (ii)

$\therefore \quad f_{min} = \dfrac{1}{T_{max}} = 70.72$ Hz   and $f_{max} = 2.78$ kHz

Example 3.4 Derive the expression for periodic time T $\left[T = RC \log_e \left(\dfrac{1}{1-\eta} \right) \right]$

of the UJT relaxation oscillator.

Solution: Voltage across the capacitor is given by $V_p = V_{BB}(1 - e^{-t/RC})$

When $V_p = \eta V_{BB} + V_0$, the capacitor will discharge through R_1.

$\therefore \quad \eta V_{BB} + V_D = V_{BB}(1 - e^{-t/RC})$

Neglecting V_D, since it is small

$\eta V_{BB} = V_{BB}(1 - e^{-t/RC})$

$\therefore \quad e^{-t/RC} = 1 - \eta., \quad \therefore \quad e^{t/RC} = \dfrac{1}{1-\eta}$

or, $\dfrac{t}{RC} = \log_e \left(\dfrac{1}{1-\eta} \right) \quad \therefore \quad t = RC \log_e (1/1 - \eta),$

Here $t = T., \quad \therefore \quad T = RC \log_e \dfrac{1}{1-\eta}$

Therefore, frequency of oscillator f,

$f \approx \left(\dfrac{1}{RC \log_e (1/1-\eta)} \right)$

3.7 THE PROGRAMMABLE UNIJUNCTION TRANSISTOR (PUT)

The PUT is an improved version of a UJT. Actually, the PUT is a *PNPN* device, but its operation is so similar to the UJT that it is always considered with the UJT. As we shall see, the PUT behave like a UJT whose trigger voltage V_p can be set by the circuit designer via an external voltage divider.

Figure 3.25 shows the *PNPN* structure and the circuit symbol for the PUT. The anode (A) and cathode (K) are the same as for any *PNPN* device. The gate (G) is connected to the *N*-region next to the anode. Thus, the anode and gate constitute a *P-N* junction. It is this *P-N* junction which controls the "on" and "off" states of the PUT. The gate is usually positively biased relative to the cathode by a certain amount, V_g. When the anode voltage is less than V_g, the anode-gate junction is reverse-biased and the *PNPN* device is in the "off" state, acting as an open-switch between anode and cathode. When the anode voltage exceeds V_g by about 0.5 V, the anode gate junction conducts, causing the *PNPN* device to turn "on" in the same manner as does forward biasing the gate cathode junction of an SCR. In the "on" state, the PUT acts like any *PNPN* device between anode and cathode (low resistance and $V_{AK} \approx 1V$). The PUT is also referred to as a complementary SCR (CSCR).

The normal bias arrangement for the PUT is shown in Fig. 3.26. The voltage divider, R_1 and R_2 sets the voltage at the gate V_g. Note that R_1 and R_2 are external to the device and can therefore be chosen to produce any desired value of V_g. The anode cathode bias is provided by E_{dc}. As long as $E_{dc} < V_g$, the device is "off" with $I_A = 0$ and all of E_{dc} present across the anode cathode ($V_{AK} = E_{dc}$). The "off" state is summarized in part (a) of the figure.

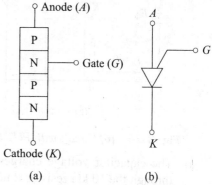

Fig. 3.25 *(a) PUT structure (b) circuit symbol*

If E_{dc} is increased to about 0.5V greater than the V_g bias value, the device turns "on". In other words, the peak-point voltage V_p for the PUT is given by

$$V_p = V_g = 0.5 \text{ V} \tag{3.19}$$

In the "on" state, the anode–cathode voltage, V_{AK}, drops to $\approx 1V$ and the anode current, I_A, is essentially equal to E_{dc}/R_1 being limited by R. In addition, V_g drops to a very low value (≈ 0.5 V) since R_2 is now shunted by the "on" *PNPN* structure. The PUT will remains in the "on"-state until the anode current is decreased below the valley-current, I_V. The "on" state is summarized in part (b) of the figure.

3.7.1 PUT Relaxation Oscillator

Figure 3.27 shows the relaxation oscillator, whose operation is very similar to the UJT oscillator. The various circuit waveforms are shown in part (b) of the figure. These waveforms reveal the following important points:

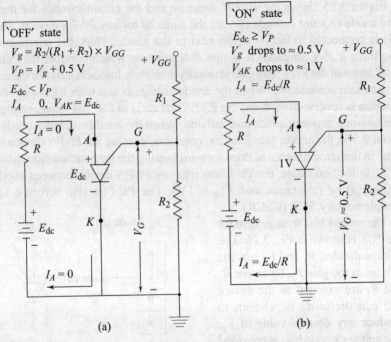

Fig. 3.26 *(a)* Circuit with PUT in "off" state *(b)* PUT in "on" state

(a) The capacitor voltage charges towards the 20 V supply voltage E_{dc} through the 10 kΩ resistor R until it reaches $V_p = 10.5$ V. At that point, the PUT turns "on" and the capacitor discharges rapidly through the PUT and the 100 Ω resistor R_s. The trigger voltage of 10.5 V is set by the voltage divider consisting of the two 20 kΩ resistors (R_1 and R_2) which bias V_g at + 10 V.

(b) The voltage at G remains at 10 V, while the capacitor charges and the PUT is "off." When the PUT turns "on," V_g drops to approximately zero. After the capacitor discharges, the PUT turns "off" (assuming current through 10 kΩ is less than I_V) and V_g returns to 10 V. This results in a negative going pulse at G.

(c) A positive pulse is produced across the 100 Ω resistor R_s as the capacitor discharges. The amplitude of this pulse is slightly lower than the capacitor peak voltage due to the anode cathode "on" voltage of ≈ 1V. The period of oscillator waveforms can be calculated from

$$T = R.C. \log_e \left(\frac{E_{dc}}{E_{dc} - V_p} \right) \qquad (3.20)$$

The frequency is simply $1/T$. Note that the above expression and the expression for the conventional UJT oscillator are exactly the same. The R_{min} and R_{max} limits for the charging resistor are determined just as they were for the UJT.

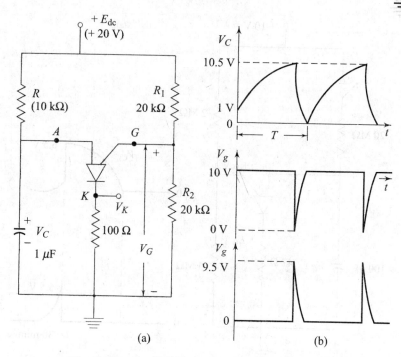

Fig. 3.27 *(a)* PUT oscillator *(b)* circuit waveforms

3.7.2 Advantages of the PUT

The PUT operation, though similar to the conventional UJT, has several important advantages over its predecessor. First, the switching voltage is easily varied by changing V_g through the voltage divider ratio. Second, the PUT can operate at low voltages (down to 3 V) making it compatible with integrated circuits.

Probably, the most important advantage of the PUT is its low peak point current, I_p. Recall from our earlier discussion of the UJT oscillator that the maximum charging resistor which could be used dependent on I_p. The PUT can be made to have a very low I_P (0.1 µA) by using large resistors for the gate bias voltage divider with a lower I_p, it is possible to use a much larger charging resistor. This is distinct advantage in long time-delay applications since a larger R would reduce the required value of C. The following example illustrates.

Figure 3.28 is a PUT time delay circuit similar to its UJT counterpart. Note that the voltage divider resistors are very large so that the PUT's I_p will be very low. Since $R_g = R_1 \| R_2 \approx 1 \text{ M}\Omega$, we can expect an I_p of around 0.1 µA. The value of V_g is easily seen to be

$$V_g = \frac{3 \text{ M}\Omega}{5 \text{ M}\Omega} \times 10 \text{ V} = 6 \text{ V}.$$

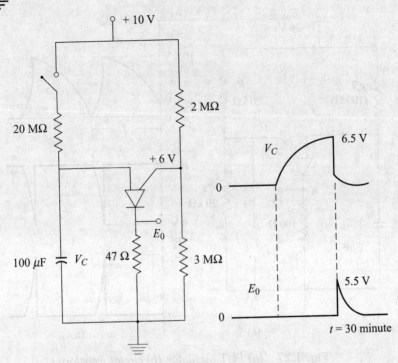

Fig. 3.28 *PUT-30 minute time-delay-circuit*

Thus V_p = 6.5 V. When the switch is closed, the capacitor will charge toward 10 V. When it reaches 6.5 V, it will discharge through the PUT, producing a pulse across the 47 Ω resistor. The time delay between closing the switch and the occurrence of the output pulse is to be designed for approximately 30 minutes.

To achieve such a long time delay requires a very large RC time constant. Equation 3.10 can be used to determine the required RC

$$T = RC \log_e \left(\frac{E_{dc}}{E_{dc} - V_p} \right)$$

$$1800 \text{ s} = RC \log_e \left(\frac{10}{10-6} \right) = RC \log_e (2.86) = RC (1.05)$$

So that $\qquad RC = \dfrac{1800 \text{ s}}{1.05} = 1713 \text{ s.}$

It is usually desirable to keep the value of C as low as possible. Large value of C usually requires electrolytic capacitors which suffer from high leakage and are physically quite large. Thus, it is necessary to make R very large. The largest R that can be used can be calculated using Eq. 3.17.

Gate Triggering Circuits

$$R_{max} = \frac{E_{dc} - V_p}{I_p} = \frac{(10-6)\text{ V}}{1.1\text{ μA}} = 40\text{ μΩ}$$

To be conservative, R should be kept well below R_{max} to ensure proper operation. The value of 20 mΩ is chosen. The value of C can now be found by satisfying

$$RC = 1713\text{ s} \quad \therefore \quad C = \frac{1713}{R} = \frac{1713}{20 \times 10^6} = 85.7 \times 10^{-6}\text{F} = 85.7\text{ μF}$$

Thus, a 100 μF capacitor can be used, and R can be varied to obtain the required delay.

If the same circuit were designed using a conventional UJT, the values of R and C might be 1 MΩ and 2000 μF. The higher I_p of the UJT results in a lower R_{max}.

Solved Example

Example 3.5 Design a free-running relaxation oscillator using a PUT to operate at a frequency of 5 to 50 Hz. DC supply is 12 V and I_p is 80 mA. It is used for triggering an SCR which will require 8 μs charge rate pulse. The voltage drop across PUT is 1 V.

Solution: Let us assume $R_s = 39\text{ Ω}$.

Since
$$T = R_s C = 8\text{ μs} \quad \therefore \quad C = \frac{8 \times 10^{-6}}{39} = 0.21\text{ μF}$$

The peak triggering current of 80 mA determines V_P, namely,

$$V_P = I_P R_s + 1\text{V} = (80\text{ mA})(39\text{ Ω}) + 1 = 4.12$$

where 1 V is the approximate PUT on-state voltage.

Now, $\quad V_P = \eta E_{dc} + V_D$

Neglecting V_D,

$$V_P = \eta E_{dc} \quad \therefore \quad \eta = \frac{V_p}{E_{dc}} = \frac{4.12}{12} = 0.34.$$

Now, the timing resistor R can be found from Eq. (3.20),

$$\therefore \quad R_{max} = \frac{T_{max}}{C \log_e\left(\frac{E_{dc}}{E_{dc} - V_p}\right)} = \frac{1}{C \log_e\left(\frac{E_{dc}}{E_{dc} - V_p}\right) f_{min}}$$

$$= \frac{1}{0.21 \times 10^{-6} \times \log_e\left(\frac{12}{12 - 4.12}\right) \times 5} = 2.26\text{ MΩ}$$

and $\quad R_{min} = \dfrac{1}{0.21 \times 10^{-6} \times 50 \times \log_e\left(\dfrac{12}{12-4.12}\right)} = 0.23\text{ MΩ}$

The maximum anode current occurs at the maximum frequency when R is a minimum:

$$I_{V(max)} \approx \frac{E_{dc}}{R_{min}} \approx \frac{12 \times 10^{-6}}{0.23} \approx 52.17 \, \mu A$$

$I_{V(min)}$ of the 2N6027 is 70 µA for $I_g = 1$ mA which should allow adequate safety margins. Therefore, to find R_1 and R_2, the following equations must be solved.

For $\quad \eta = 0.34, \quad I_g = \dfrac{2\eta E_{dc}}{R_g}$

$$1 \times 10^{-3} = \frac{2 \times 12 \times 0.34}{R_g} \quad \therefore \quad R_g = 8.16 \, k\Omega$$

Now, $\quad \eta = \dfrac{V_p}{E_{dc}} = \dfrac{R_2}{R_1 + R_2}$

The solutions for R_1 and R_2 are:

$$R_1 = \frac{R_g}{\eta} = \frac{8.16K}{0.34} = 24 \, k\Omega \quad \text{and} \quad R_2 = \frac{R_g}{1-\eta} = \frac{8.16K}{1-0.34} = 12.36 \, k\Omega.$$

3.8 PHASE CONTROL USING PEDESTAL-AND-RAMP TRIGGERING

Figure 3.29 shows the circuit for ramp-and-pedestal triggering of two thyristors connected in antiparallel for controlling power in an ac load. Ramp and pedestal triggering is an improved version of synchronized-UJT-oscillator triggering. The various voltage-waveforms are shown in Fig. 3.30.

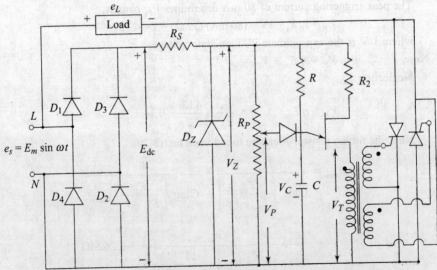

Fig. 3.29 *Ramp and pedestal trigger circuit for a.c. load*

Zener-diode voltage, V_z, is constant at its threshold-voltage. R_p acts as a potential divider. Wiper of R_p controls the value of pedestal voltage V_p. Diode D allows C to be quickly charged to V_p through the low-resistance of the upper portion of

R_p. The setting of wiper on R_p is such that, this value of V_p is always less than the UJT firing point voltage ηV_z. When wiper setting is such that V_P is small (See Fig. 3.3(a), voltage V_z charges C through R. When this ramp voltage V_C reaches ηV_z, UJT fires and voltage V_T, through the pulse transformer, is transmitted to the gate circuits of both thyristors T_1 and T_2. During first positive half-cycle, SCR T_1 is forward biased and is therefore, turned-on. After this, V_c reduces to V_P and then to zero at $\omega t = \pi$. As V_c is more than V_P, during the charging of capicitor C through charging resistor R, diode D is reverse-biased. Thus, V_P does not effect in anyway the discharge of C through UJT emitter and primary of pulse transformer. From period 0 to π, SCR T_1 is forward biased and is turned-on. From π to 2π, T_2 is turned-on. In this manner, load is subjected to alternating e_L, as shown in Fig. 3.30.

With the setting of wiper on R_P pedestal voltage V_P on C can be adjusted.

With low pedestal voltage across C, ramp charging of C to ηV_z takes longer time (Fig. 3.30(a)), and firing angle delay is, therefore, more and output voltage is low. With high pedestal on C, voltage-ramp charging of C through R reaches ηV_z faster, firing angle delay is smaller (Fig. 3.30(b)) and output voltage is high. This shows that output voltage is proportional to the pedestal voltage.

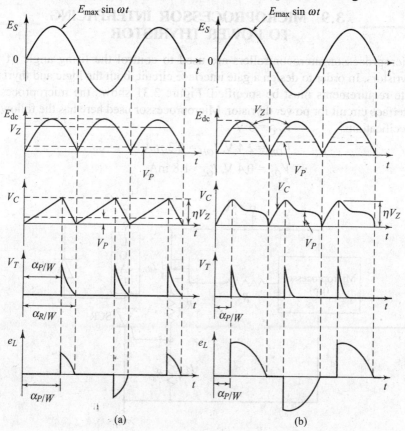

Fig. 3.30 *Waveform for ramp and pedestal control*

The time T required for capacitor to charge from pedestal voltage V_P to ηV_z can be obtained from the relation

$$\eta V_z = V_P + (V_z - V_P)(1 - e^{-T/RC})$$

Note that $(V_z - V_P)$ is the effective voltage that charges C from V_P to ηV_z. From above,

$$T = RC \ln \frac{V_z - V_P}{V_z(1-\eta)} \qquad (3.21)$$

and the firing angles are given by

$$\alpha_p = \omega R.C \ln \frac{V_z - V_p}{V_z(1-\eta)} \qquad (3.22)$$

and

$$\alpha_R = \omega \cdot RC \ln \frac{1}{1-\eta} \qquad (3.23)$$

where ω is the input frequency and η is the intrinsic stand off ratio of the UJT.

3.9 MICROPROCESSOR INTERFACING TO POWER THYRISTOR

Microprocessor/microcontrollers are used to control the firing angle of the thyristors. In order to design a gate interface circuit, both the logic and thyristor gate requirements must be specified. Figure 3.31 shows the microprocessor interface circuit for power thyristor. Microprocessor used here has the following specifications: $V_{CC} = 5$ V

$V_{OH} = 2.4$ V, $I_{OH} = 0.3$ mA

$V_{OL} = 0.4$ V, $I_{OL} = 1.8$ mA

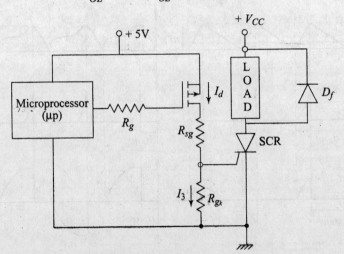

Fig. 3.31 *μp interfacing to SCR*

With these specifications, microprocessor cannot directly drive the power-thyristor. Therefore power interfacing circuitry is used. As shown, p-channel MOSFET is used as a interfacing device. MOSFET used has the following specifications: $V_{TH} = 3$ V, $I_d = 0.5$ A, $C_{gs} = 400$ pF

$$R_{ds(on)} = 10 \text{ ohms}$$

(i) Selection of Resistance Rg:
Resistor R_g limits the charging current of the capacitance C_{gs} and also decides the turn-on time of the MOSFET. From Fig. 3.31,

$$R_g = \frac{(V_{CC} - V_{OL})}{I_{OL}} \text{ ohms} = \frac{(5 - 0.4)}{1.8} = 2.7 \text{ k}\Omega$$

The MOSFET will not turn-on until C_{gs} has charged to 3V or, with a 5V supply, approximately one R-C time constant, that is

$$t_{\text{delay}} = R_g \cdot C_{gs} = 2.7 \times 10^3 \times 400 \times 10^{-12} = 1 \text{ μsec}$$

(ii) Section of resistance R_{gk} and R_{sg}:
The MOSFET must provide the thyristor gate current and the current through resistor R_{gk} when the gate is at 3V.

When $R_{gk} = \infty$, R_{sg} has the maximum value and is given by

$$R_{sg} \leq \frac{V_{CC} - V_{GT} - I_{GT} \cdot R_{ds(on)}}{I_{GT}} \Omega \leq \frac{5 - 3 - 75 \times 10^{-3} \times 10}{75 \times 10^{-3}} = 16.6 \text{ }\Omega$$

Choose $R_{sg} = 10$ ohms.

Since resistor R_{gk} provides a low cathode-to-cathode impedance in the off-state, hence improves SCR noise immunity.

When $V_{GT} = 3$V, $I_d = \dfrac{V_{CC} - V_{GT}}{R_{ds(on)} + R_{sg}} = 100$ mA

But $\qquad I_d = I_g + I_3$

∴ $\qquad 100 \text{ mA} = 75 \text{ mA} + I_3 \quad ∴ \quad I_3 = 25$ mA

∴ $\qquad R_{gk} = \dfrac{V_{GT}}{I_d - I_{GT}} = \dfrac{3}{25 \times 10^{-3}} = 120 \text{ }\Omega$

After turn-on, the gate voltage will be about 1 V. Therefore, the MOSFET current will be 200 mA.

∴ $I^2 R_{sg} = (200 \text{ mA})^2 \times 10 = 0.4$ W $\quad ∴ \quad$ Select $R_{sg} = 10 \text{ }\Omega/1$ W

REVIEW QUESTIONS

3.1 Explain the basic requirements for the successful firing of thyristor in detail.

3.2 With the help of neat diagram, explain the operation of resistance firing circuit. Also, draw and explain the associated waveforms. Also, for the same circuit, show that firing angle delay is proportional to the variable resistance.

3.3 Explain with the help of neat circuit diagram, the use of pulse-transformer in triggering circuits.

3.4 Explain in brief, the prime requirement of a trigger pulse transformer.

3.5 Under what circumstances should one consider using an optoisolator?

3.6 What advantages do opto-isolators have over electromechanical relays?

3.7 Why can't the circuit of Fig. 3.11 produce a trigger angle greater than 90°?

3.8 The circuit of Fig. 3.12 is adjusted to have a trigger angle of 90°?
   (a) Sketch e_L, V_{AK}, and V_c.
   (b) Sketch e_L if R_V is decreased.
   (c) Sketch e_L, if C_1 is increased
   (d) Sketch e_L if the amplitude of e_s is doubled.

3.9 Suppose the load voltage in Fig. 3.13(a) always has a trigger angle of 0° regardless of the setting of R. Which of the following could be a cause of this malfunction? Explain your choice (s):
   (a) C_1 is open.
   (b) SCR is shorted between anode and cathode.
   (c) One of the diodes in the bridge is shorted.
   (d) R is open.
   (e) The four layer diode is shorted.

3.10 After the UJT is turned "on," what happens to the value of R_{B1}? How can the value of V_P be varied for a given UJT? How is the UJT turned- "off"?

3.11 Consider the circuit of Fig. P.3.11. The oscillator is suppose to produce pulses at a rate of 10 kHz. When power is applied to the circuit, however, the circuit fails to oscillate. The technician testing the circuit notices that when the power is turned-off, the circuit temporarilly oscillates as the power-supply voltage drops to zero. Explain these observations and determine what is wrong with the circuit. How should the circuit be modified for proper operation?

3.12 Suppose that when the switch is closed in Fig. P. 3.12, the motor goes on instantly (no time delay). Which of the following reasons could be cause for the malfunction? (There may be more than one possible answer). Explain each choice:
   (a) The capacitor is shorted.
   (b) R_1 is too large.
   (c) The supply voltage is too large.
   (d) The SCR is too-sensitive.
   (e) The UJT is shorted from B_2 to B_1.

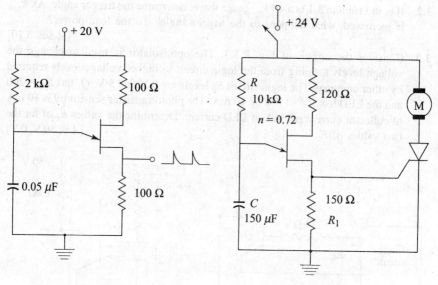

Fig. P.3.11 Fig. P. 3.12

3.13 Explain how the operation of PUT differ from that of the UJT. What are the advantages of PUTs over UJTs?

3.14 Draw and explain the general block-diagram of a thyristor trigger circuit.

3.15 Draw RC half-wave trigger-circuit for one SCR and discuss the function of the various components used. Describe with the help of waveforms how the output voltage is controlled by varying the resistance.

3.16 Explain with neat circuit diagram and waveforms the PUT relaxation oscillator.

3.17 Describe the RC full-wave trigger circuit for one SCR when the load is a.c. Also, draw the related voltage waveforms.

3.18 Explain the working of an oscillator employing an UJT. Derive expression for the frequency of triggering.

3.19 Draw and explain the equivalent circuit and V–I characteristic of the UJT in detail.

3.20 Draw and explain circuit diagram for the synchronized UJT triggering. Also, draw and explain the associated voltage waveforms.

3.21 Draw a circuit diagram for the ramp-and-pedestal trigger circuit used for the single-phase semiconverter. Describe its operation with appropriate waveforms.

3.22 Design the PUT 30 minute time-delay circuit.

3.23 Explain the PUT circuit operation in "on" state and "off" state.

3.24 With the help of neat diagram, explain the working of microprocessor interfacing to power thyristor.

PROBLEMS

3.1 The circuit of Fig. 3.11 utilizes an SCR with $I_{g(min)} = 0.1$ mA and $V_{g(min)} = 0.5$ V. If $R_2 = 10$ kΩ and the diode is silicon, determine the value of R_{min} needed to cause triggering when e_s reaches 3.2 V. [Ans. 10 kΩ]

3.2 If e_s in Problem 3.1 is a 40 V_{p-p} sine wave, determine the trigger angle. As R_{min} is increased, what happens to the trigger-angle? To the load-power?

[Ans. 5.10]

3.3 Consider the circuit of Fig. P.3.3. The optoisolator is used to change the voltage levels forming from the logic circuit to higher voltage levels required by other circuitry. The logic circuit E_o levels are 0 V and 5 V. Q_1 has $B_{dc} = 200$ and the LED has $V_F = 1.5$ V @ 10 mA. The phototransistor sensitivity is 10 mA of collector current per mA of LED current. Determine the values E_x of for the two values of E_0.

[Ans. 50 V, 0 V]

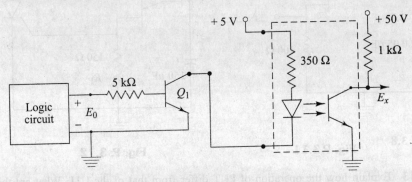

Fig. P.3.3

3.4 The optical coupler shown in Fig. P.3.4 is required to deliver at least 10 mA to the external load. If the current transfer ratio (it is the ratio of the output current to the input current through the LED. It is usually expressed as a percentage) is 60 per cent, how much current must be supplied to the input?

[Ans. 16.67 mA]

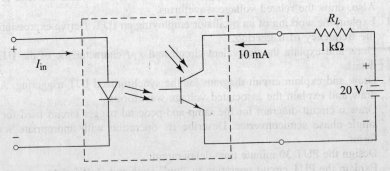

Fig. P.3.4

3.5 A UJT with the following parameters is used in the circuit of Fig. .3.16:

$$\eta = 0.66, V_D = 0.7\text{V}, I_V = 4 \text{ mA}$$
$$V_v = 1 \text{ V}, I_P = 10 \text{ } \mu\text{A}$$

(a) Assume that the UJT is initially "off". To what value must V_{EE} be raised to turn "on" the UJT? Use $V_{BB} = 20$ V.

(b) If $R_E = 1$ kΩ, to what value must V_{EE} be reduced before the UJT turns "off"?
[Ans. (a) 13.91 V (b) 5 V]

3.6 The UJT of Problem 3.5 is used in the oscillator circuit of Fig. 3.18. The circuit values are $R_1 = 100$ Ω, $R_2 = 50$ Ω , $R = 10$ kΩ, $C = 2$ μF and $E_{dc} = 24$ V.
  (a) Determine V_p
  (b) Determine whether the circuit will oscillate. If it oscillates, determine its frequency.
  (c) Accurately sketch and label the capacitor waveform.
[Ans. (a) 16.7 V. (b) ≈ 45 Hz]

3.7 For the oscillator of Example 3.6, determine the range of frequencies which can be obtained by varying R. How may this range of frequencies be changed without changing the UJT?

[Ans. ≈ 0.62 Hz to 68.6 Hz].

3.8 In Fig. P.3.12, how long after the switch is closed will the motor be energized?

[Ans. 1.9 Sec].

3.9 Determine the period and frequency of the oscillator of Fig. 3.38. Sketch the waveforms of V_c, V_g, and V_K, showing appr-oximate amplitude. Also, indicate the three waves to decrease the frequency of the oscillator of Fig. P.3.9.

[Ans. 5.2 kHz]

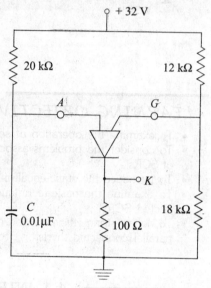

Fig. P.3.9

REFERENCES

1. General-Electric Company, *SCR manual*, 5th edition, N.Y., 1972.
2. C W Lander, *Power Electronics*, McGraw-Hill Book Company, 1981.
3. R Ramshaw, *Power Electronics*, ELBS, 1982.
4. F Csaki, *Power Electronics*, Akademiai Kiado, Budapest, 1980.
5. M. Ramamoorthy. *Thyristor and their Applications*, East-West-Press, 1977.
6. F E Gentry et.al., *Semiconductor Controlled Rectifiers*, Prentice Hall Inc, USA 1964.
7. S B Dewan and A Straughen, *Power Semiconductor Circuit*, John-Wiley, 1975.
8. C F Battersby, "Present techniques in gate firing", *IEE Conference Publication* 53, 146–153, 1969.

Chapter 4

Series and Parallel Operation of Thyristors

LEARNING OBJECTIVES:
- To examine the operation of series and parallel connected SCRs.
- To consider the problems associated with series and parallel connection of SCRs.
- To introduce the static-equalising and dynamic equalising networks.
- To examine the basic-techniques of triggering series and parallel connected SCRs.
- To define string-efficiency and derating factor in relation to a series and parallel connected thyristor circuits.

4.1 INTRODUCTION

The maximum power which can be controlled by a single solid-state power device (SCR) is determined by its rated blocking voltage and by its rated forward-current. To maximise either of these two ratings requires some compromise in the other rating. Also, laboratory devices can always be cited with capabilities considerably beyond those of devices which are commercially available at any particular time.

Thyristor ratings have considerably improved since its introduction in 1957. Presently, SCRs with voltage and current ratings of 10 kV and 3.5 kA are available. However, for many industrial applications and also in terminal of HVDC transmission-lines, a single SCR cannot meet the power-requirements. Therefore, for controlling the power, in high-voltage low-current circuits, low-voltage high-current circuits, SCRs have to be connected in series and parallel combinations. SCRs are connected in series for high-voltage operation, whereas they are connected in parallel for high-current operation. In this chapter, we will consider the problems associated with series and parallel connections of SCRs, and discuss the measures adopted to overcome these problems.

Series and Parallel Operation of Thyristors

4.2 SERIES OPERATIONS OF THYRISTORS

If the required voltage rating of the assembly or equipment incorporating thyristors is more than the voltage rating of a single thyristor, then several thyristors have to be united in a series connection. When we consider a series connection of thyristors, we mean thyristors of the same class to be connected in series. Like any other semiconductor device, characteristic properties of two thyristors of same make and rating are never same and this leads to the following two major problems during series connection of the devices.

(1) Unequal distribution of voltage across devices; and
(2) Difference in reverse-recovery characteristics.

4.3 NEED FOR EQUALISING NETWORK

Due to the difference in blocking currents, junction capacitances, delay times, forward-voltage drops as well as reverse-recovery for individual SCRs, external voltage equalisation networks and special considerations in gating circuit design are required.

4.3.1 Unequal Distribution of Voltage

The problem of unequal voltage sharing between series connected SCRs can be understood by referring to Fig. 4.1. Figure 4.1 shows the V–I characteristics of two identical SCRs. Here, V_{BO_1} and V_{BO_2} are the forward breakover voltages for thyristors T_1 and T_2, respectively. For SCR T_2, leakage-resistance $\left(=\dfrac{V_2}{I_0}\right)$ is high whereas for SCR T_1 it is low $\left(\dfrac{V_1}{I_0}\right)$. For the same leakage-current I_0 SCR T_2 supports rated voltage V_2 which is nearly on the point of breakdown, whereas SCR T_1 supports voltage V_1 which is less compared to V_2. Hence, thyristor having the highest leakage resistance (or low leakage current) will share a larger portion of the applied voltage.

This shows that when the thyristors of identical rating are connected in series, variations in their forward and reverse blocking characteristics causes unequal distribution of voltage in steady-state.

It is observed from Fig. 4.1 that the maximum voltage that the two-SCR string can block is only $(V_1 + V_2)$ and not the rated blocking voltage $2V_2$. Hence, in order to force equal sharing of voltages across the two SCRs under steady state condition, external resistors R have to be connected in parallel with each thyristor such that the parallel combination has the same resistance. The selection procedure for resistor R is discussed in more details in Section 4.4. The voltage sharing in series connected SCRs could also be achieved by connecting controlled avalanche rectifiers or metal oxide varistors across each thyristor in series string.

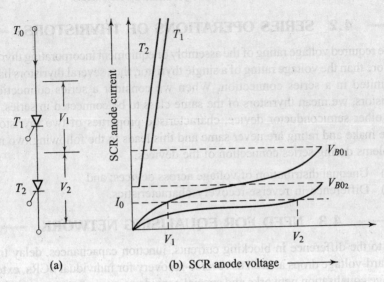

Fig. 4.1 *Distribution of voltage between two series connected SCRs*

4.3.2 Difference in Reverse Recovery Characteristics

The unequal distribution of voltages among thyristors in series also occurs during the transient conditions of turn-off, turn-on and high-frequency operation. A thyristor which has the largest turn-on time in the series string will have to momentarily support the full-string voltage. One method that can be used to minimise the unbalance caused by dissimilar turn-on delays is to apply high enough gate-drive with fast rise-time to minimise turn-on differences. If one thyristor in the string has short reverse-recovery time, it will alone support the string reverse-voltage. The problem of unequal voltage sharing among the series connected SCRs, due to difference in reverse recovery characteristics of the two unmatched SCRs, of the same type, can be understood by referring to Fig. 4.2.

As shown in Fig. 4.2, SCR T_1 is assumed to have less reverse-recovery time than that of SCR T_2. As a result, SCR T_1 recovers faster than SCR T_2 and it limits the reverse-current immediately. Therefore, SCR T_1 will share the maximum string voltage as SCR T_2 is not recovered. Hence, unequal voltage distribution occurs due to the difference in the reverse-recovery currents of the SCRs of the same time. Here, shaded area ΔQ represents the difference in reverse recovery charges of two thyristors. During turn-off, the capacitance of the reverse-biased junction determines the voltage distribution across SCRs in a series-connected string. As reverse-biased junctions are likely to have different capacitances, called self capacitances, the voltage distribution during turn-on and turn-off periods would be unequal. Voltage equalisation under these conditions can, however, be achieved by connecting shunt-capacitors as shown in Fig. 4.3. This capacitor removes inequalities in thyristor self-capacitances. Therefore, during turn-off periods, the resultant of self-capacitance and shunt capacitance of each thyristor in a string becomes equal.

Series and Parallel Operation of Thyristors 117

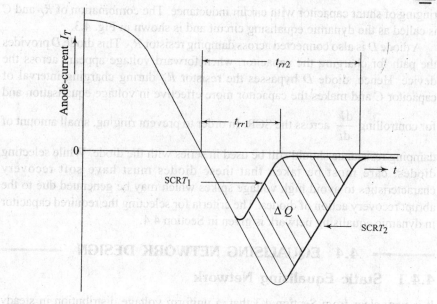

Fig. 4.2 *Reverse recovery currents of two SCRs (T_1, T_2) of the same type*

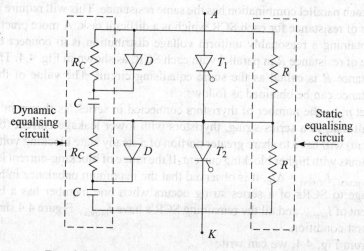

Fig. 4.3 *Equalising circuit for series operation*

During the period for which any of the SCR is in the blocking state, the corresponding capacitor will get charged to the voltage existing across that SCR. When this SCR is turned ON, the capacitor discharges heavy current through this SCR. To limit this discharge current, a damping resistor, R_C, is used in series with each capacitor as shown in Fig. 4.3. The value of these damping resistors much be kept to a reasonably low value in order not to reduce the effectiveness of the capacitors in equalising voltage during the reverse recovery interval. Resistor R_C also damps out the high-frequency oscillations that may arise due to the ringing of shunt capacitor with circuit inductance. The combination of R_C and C is called as the dynamic equalising circuit and is shown in Fig. 4.3.

A diode D is also connected across damping resistor R_C. This diode D provides the path for charging the capacitor, when forward voltage appears across the device. Hence, diode D bypasses the resistor R_C during charging interval of capacitor C and makes the capacitor more effective in voltage equalisation and for controlling $\dfrac{dV}{dt}$ across the SCR. In order to prevent ringing, small amount of damping resistance should still be used in series with the diode. While selecting diodes, care must be taken that these diodes must have soft recovery characteristics to avoid high voltage spikes which may be generated due to the abrupt recovery action of diodes. The criteria for selecting the required capacitor in dynamic equalising network is given in Section 4.4.

4.4 EQUALISING NETWORK DESIGN

4.4.1 Static Equalising Network

It is very clear from Section 4.3 that, a uniform voltage distribution in steady state can be achieved by connecting a suitable resistance across each SCR, such that each parallel combination has the same resistance. This will require different value of resistance for each SCR which is a difficult task. A more practical way of obtaining a reasonably uniform voltage distribution is to connect the same value of resistance R in parallel with each SCR, as shown in Fig. 4.4. This shunt resistance R is called as the static equalising circuit. The value of this shunt resistance can be obtained as follows:

Let n_s be the number of thyristors connected in series, as shown in Fig. 4.4. Recall that, in a series string, thyristors with lower leakage current (blocking-current) will have to share greater portion of a steady state blocking voltage than will units with higher blocking current. If the range of blocking-current is defined as $I_{b(max)} - I_{b(min)} = \Delta I_b$, it is observed that the maximum unbalance in blocking-voltage to SCRs of a series string occurs when one member has a blocking-current of $I_{b(min)}$ and all the remaining SCR's have $I_{b(max)}$. Figure 4.4 shows such a worst condition.

From Fig. 4.4, we can write

$$I_{b(min)} + I_1 = I_{b(max)} + I_2$$

or

$$I_{b(max)} - I_{b(min)} = I_1 - I_2 = \Delta I_b \qquad (4.1)$$

Let E_D be the maximum permissible blocking voltage, then

$$E_D = I_1 \cdot R \qquad (4.2)$$

Now, we can write string voltage E_s as

$$E_s = E_D + (n_s - 1) R I_2 \qquad (4.3)$$

From Eq. (4.1), we can write

$$I_2 = (I_1 - \Delta I_b)$$

Substitute above value of I_2 in Eq. (4.3)

$$\therefore \quad E_s = E_D + (N_s - 1) R \cdot (I_1 - \Delta I_b)$$

or

$$E_s = E_D + (n_s - 1) R I_1 - (n_s - 1) R \Delta I_b \quad (4.4)$$

Substitute Eq. (4.2) in Eq. (4.4).

or

$$E_s = E_D + (n_s - 1) E_D - (n_s - 1) R \Delta I_b.$$

$$E_s = n_s E_D - (n_s - 1) R \Delta I_b.$$

Now,

$$\therefore \quad R \leq \frac{n_s \cdot E_D - E_s}{(n_s - 1) \Delta I_b} \quad (4.5)$$

In general, only the maximum blocking currents for a particular thyristor type are provided by the manufacturer and the ΔI_b value is rarely provided. In such a case, it is usual to assume $\Delta I_b = I_{b(max)}$ with $I_{b(min)} = 0$. With this, the value of R calculated from Eq. (4.5) is lower than what is actually required.

Now, the resistor effective power-dissipation can be expressed as

$$P_R = \frac{(E_{RMS})^2}{R} \quad (4.6)$$

where E_{RMS} is the RMS voltage across R.

4.4.2 Dynamic Equalising Network

As mentioned in the previous section, shunt capacitors are required to limit the rate of rise of voltage on the SCR's and also during the reverse-recovery period, such capacitors provides a reverse recovery current path for slow SCR's around those SCR's which recover first. From Fig. 4.2, it is clear that the maximum unequal distribution of voltage under transient conditions occurs during turn-off due to the difference in the reverse-recovery characteristics. Therefore, the selection of capacitor value C is totally based on the reverse recovery characteristics. (Refer to Fig. 4.2).

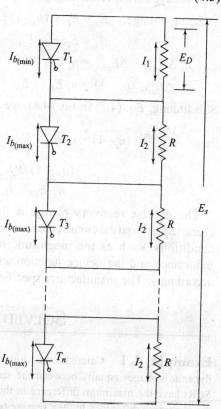

Fig. 4.4 Static equalising network

Due to the difference in the reverse-recovery characteristics, the shunt capacitors when charged to peak voltage are not charged equally.

Let n_s be the number of SCRs connected in series-string, and let us assume that SCR T_1 has the fastest recovery and remaining $(n_s - 1)$ thyristors have the slowest recovery.

Let $\Delta Q_{(max)}$ be the maximum permissible difference between reverse-recovery charge of SCRs of same time and $\Delta E_{(max)}$ be the maximum difference in voltage under this situation.

$$\therefore \quad \Delta E_{max} = \frac{\Delta Q_{(max)}}{C} \tag{4.7}$$

If E_D is the maximum permissible blocking voltage of fast SCR (SCR T_1), then voltage across remaining $(n_s - 1)$ SCRs will be $(E_D - \Delta E_{max})$.

$\therefore$ String-voltage, $\quad E_s = E_D + (n_s - 1)(E_D - \Delta E_{max})$.
$$= E_D + n_s E_D - n_s \Delta E_{max} - E_D + \Delta E_{max}$$

$$\therefore \quad n_s \Delta E_{max} - \Delta E_{max} = n_s E_D - E_s$$
$$\Delta E_{max}(n_s - 1) = n_s E_D - E_s \tag{4.8}$$

Substituting, Eq. (4.7) in Eq. (4.8), we get

$$\therefore \quad \frac{\Delta Q_{(max)}}{C}(n_s - 1) = n_s E_D - E_s$$

$$\therefore \quad C \geq \frac{(n_s - 1) \Delta Q_{max}}{n_s E_D - E_s} \tag{4.9}$$

The reverse recovery charge is the function of both the device design characteristics and the circuit commutation conditions. By varying the commutation conditions, such as the magnitude of forward conduction current, circuit inductance and the device junction temperature, the reverse-charge will vary accordingly. The manufacturer specifies the ΔQ for specific conditions.

SOLVED EXAMPLE

Example 4.1 Calculate the values of R and C that will divide the static and dynamic voltages equally between the series-connected SCRs in Example 4.1. These SCRs have the maximum difference in their off state leakage current, $\Delta I_b = 1$ mA and the maximum difference in their reverse-recovery charge, $\Delta Q = 30$ μC.

Solution: Using the relation, $R = \dfrac{n_s . E_D - E_s}{(n_s - 1) \Delta I_b}$

Given $\quad n_s = 18, \quad E_D = 500\,\text{V}, \quad E_s = 7500\,\text{V}, \quad \Delta I_b = 1\,\text{mA},$
$\quad \Delta Q = 30\,\mu\text{C}$

$$R = \frac{(18 \times 500 - 7500)}{[(81-1) \times 1 \times 10^{-3}]} = 88.24 \text{ k}\Omega$$

Now, $\quad C = \dfrac{(n_s - 1) \Delta Q_{max}}{(n_s E_D - E_s)} = \dfrac{(18-1) \times 30 \times 10^{-6}}{(18. \times 500 - 7500)} \quad C = 0.34 \text{ }\mu\text{F}.$

4.5 TRIGGERING OF SERIES CONNECTED THYRISTORS

Series operation of thyristors takes place satisfactorily only if all the thyristors are *fired* at the *same instant*. Even differences of few microseconds in the gate pulses to different thyristors can have a major influence on *the voltage* sharing in series operation.

Consequently, in most equipment using multiple-operation of thyristors, all the thyristors are *fired* from the same pulse-amplifier source. Usually, pulse transformers with separate secondary winding for each thyristor or separate pulse transformers per thyristor are used.

With series operation of thyristors, the following points need to be considered when selecting the firing system to be employed:

(1) All the thyristors must act as one. If one SCR of a series string is not *fired* when it should be, *the full circuit will be impressed across* it causing it to breakover and fail due to excessive di/dt.

(2) The thyristors will all be at different *voltage levels* with *respect to earth* and high voltage *insulation will be required* between all the gate circuits. This complicates pulse-transformer design, as more the insulation used, the slower the rate of rise of gate pulse, and this will affect the transient sharing of the total voltage.

This leads to the use of fiber-optic light guides as a means of obtaining the necessary insulation level. Unfortunately, the guides can only pass a small amount of energy to the gate and so a power-source local to the thyristor is required.

(3) All the thyristors must turn-off at the same instant or else the last to turn-off will be exposed to the full circuit voltage. Due to variation of holding currents, this can only be guaranteed if the gate-pulses continue for the whole of the conduction period.

The following are the primary methods in common use for triggering series-connected SCRs:
(1) Simultaneous triggering.
(2) Sequential triggering.
(3) Optical triggering.

4.5.1 Simultaneous Triggering

Figure 4.5 shows the circuit of simultaneous triggering of series connected SCRs. All the thyristors are triggered simultaneously and independently with the help of pulse-transforming. Hence, this method is also called as independent or individual *firing* method. Most of the trigger pulse transformers are provided with two secondary windings and these can be used for two series connected thyristors. For more than two thyristors, special triggering transformer has got to be made with sufficient number of secondaries.

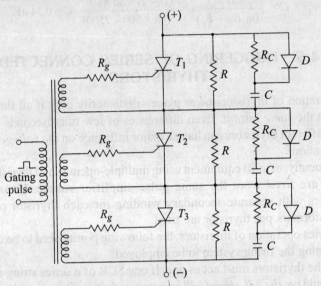

Fig. 4.5 *Simultaneous triggering*

The main triggering pulse is applied to the primary of the transformer. Each of the secondary winding is connected to the individual gates of respective SCRs in the string as shown in Fig. 4.5. Triggering requirements may differ quite widely between individual SCRs. To equalise the gate current in each SCR, a resistor R_g is connected in series with the secondary winding for swamping out any difference in a gate-to cathode impedance of individual units. Also, when using pulse transformer particular attention should be given to the insulation between windings. This insulation must be able to support at least the peak of the supply voltage.

4.5.2 Sequential Triggering

Figure 4.6 illustrates the sequential triggering technique. In this technique, one "master" SCR is triggered, and as its forward-blocking voltage begins to collapse, a gate signal is thereby applied to the "slave" SCR. Hence, this method is also referred to as slave-triggering method.

As shown in Fig. 4.6, the master SCR T_3 is turned-on by the external gate-pulse. The dynamic-equalising circuit is used for sequential turn-on. Initially, when supply voltage is applied to the SCR series string, all thyristors tranform in their forward-blocking state and all capacitors get charged with the polarity shown. Now, when gate pulse is applied to the master SCR T_3, turned-on. Capacitor C_3 starts discharging and the path of discharging becomes,

$$C_{3+} - R_{C_3} - R_{g2} - T_{2(G-K)} - T_{3(A-K)} - C_3$$

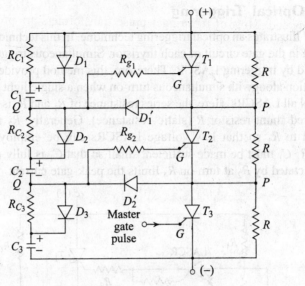

Fig. 4.6 *Sequential triggering*

Hence, thyristor T_2 turned ON. When T_2 turned-ON, its corresponding capacitor C_2 starts discharging and the path of discharging becomes,

$$C_{2+} - R_{C_2} - R_{g_1} - T_{1(G-K)} - T_{2(A-K)} - D_2^1 - C_{2-}.$$

Hence, thyristor T_1 turned-ON.

In this way, the discharge current of the shunt capacitor, through the SCR which is turned-on, will trigger the next SCR in the string. The process takes place very rapidly, and all the SCRs are turned-on in a very short time.

Here, the topmost SCR T_1 in the string will experience an increasing forward voltage due to the sequentially turning-on technique. This technique is generally used for generating impulse-voltages.

The minimum capacitance required to supply sufficient gate current to trigger under all conditions is given by:

$$C \geq \frac{10}{R_g + \dfrac{E_{GT(\max)}}{I_{GT(\max)}}} \mu F \qquad (4.10)$$

where $E_{GT(\max)}$ = maximum gate triggering voltage, $I_{GT(\max)}$ = maximum gate triggering current, and R_g = gate – source resistance.

Because of diodes D'_1 and D'_2, potential of points P and Q will not be the same. The resulting circulating currents may turn-on the SCRs. These currents must be minimised by selecting appropriate equalising circuit parameters so that the string is turned-on only when a gate signal is applied to the master SCR T_3.

4.5.3 Optical Triggering

Figure 4.7 illustrates an optical triggering technique. In this technique, LASCR is connected in the gate circuit of each thyristor. Simultaneous triggering of SCRs is achieved by triggering LASCR. Therefore, this method provides the required gate isolation alongwith simultaneous turn-on when a single light source is used to turn ON all LASCRs. Here, the series resistance of R_1 and R_2 is made equal to the required shunt resistor R (static–resistance). Generally, R_2 is made small compared to R_1 so that low voltage LASCRs can be employed. The time constant $R_1 C_1$ must be made sufficient small so that C_1 is fully charged to the voltage dictated by R_2 at turn-on R_4 limits the peak-gate current.

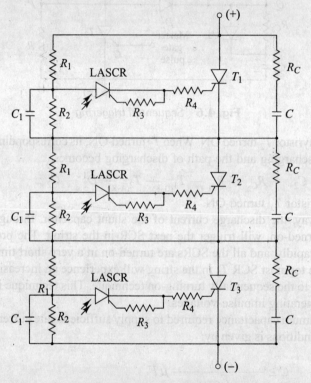

Fig. 4.7 *Optical triggering*

Therefore, in general for very high voltages, and particularly when trigger circuits is to be isolated from the power-circuit, triggering by a strong light source such as a Xenon flash tubes, phototransistor and fiber optic bundle is employed.

4.6 PARALLEL OPERATION OF THYRISTORS

The need to connect thyristors in parallel arises when the current or over current to be handled by the apparatus or equipment exceeds the rating of a single thyristor. Thyristors can be connected directly in parallel with each other if they have identical forward V–I characteristics. This is rarely the case unless very special selection of the thyristors is made to ensure good current sharing during normal load and under overload and faulty conditions.

The shape of the forward voltage of thyristors makes it difficult for them to achieve good sharing without assistance, as even quite a small voltage drop difference can result in a wide difference of load current, as shown in Fig. 4.8 shows the sharing of current between two parallel connected thyristors of same type.

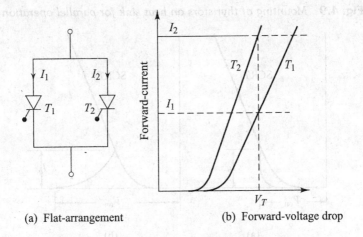

(a) Flat-arrangement (b) Forward-voltage drop

Fig. 4.8 *Parallel operation of two thyristors*

Figure 4.8(b) illustrates that for the same voltage drop V_T, SCR T_2 shares a rated current I_2, whereas SCR T_1 carries a current I_1 much less than the rated current I_2. Total rated current of the parallel unit is $(I_1 + I_2)$ instead of $2I_2$.

This unequal distribution of current in parallel connected thyristors leads to a thermal-runaway problem. For example, thyristor T_2 in Fig. 4.8 carries more current than the thyristor T_1. Due to large forward, current, its internal power dissipation will be more, thereby raising its junction temperature and reducing the dynamic resistance. This, in turn, will increase the current shared by SCR T_2 and the process becomes repetitive. The cumulative increase in current results in permanent damage to the SCR T_2, followed by the burning of SCR T_1. Therefore, all SCRs must operate at the same temperature when used in parallel connection. This can be done by a common heat-sink, as shown in Fig. 4.9.

The unequal distribution of current in a parallel unit is also caused by the inductive effect of current carrying conductors. When SCRs are arranged unsymmetrically, as shown in Fig. 4.9(a), the middle one will have more inductance due to more flux linkages. As a consequence, less current flows through the middle SCR as compared to outer two SCRs. This unequal current distribution can be avoided by mounting the SCRs symmetrically on the heat-sink as shown in Fig. 4.9(b).

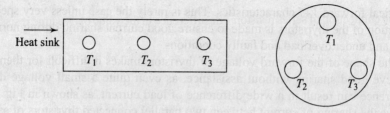

(a) Unsymmetrical arrangement   (b) Symmetrical arrangement

Fig. 4.9 *Mounting of thyristors on heat sink for parallel operation*

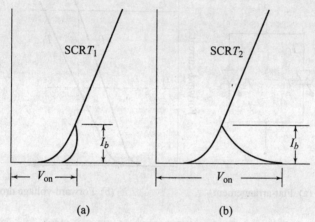

Fig. 4.10 *Static thyristor turn-on-behaviour*

The difference in the turn-on characteristics can also influence the operation of thyristors in parallel. Differences in finger-voltages (the minimum forward anode voltage at which an SCR can be successfully turned-on with a trigger signal of sufficient magnitude) may prohibit the SCR with highest turn-on voltages to trigger. This is explained in Fig. 4.10. If SCR T_1 is directly connected in parallel with SCR T_2 having identical characteristics, SCR T_2 will never turn-on in application requiring zero voltage triggering. When SCR T_1 is switched ON, the anode voltage of SCR T_2 would be that of the on-state voltage of SCR T_1 and consequently, it will never equal or exceed the minimum required anode voltage to trigger SCR T_2 even if the width of the trigger pulse is greater than the delay-time of the SCR T_2. Hence, it is very essential that in direct paralleling of SCRs, the forward characteristic of each and every SCR must be properly matched. Matched thyristors are generally available for direct parallel operation.

4.7 METHODS FOR ENSURING PROPER CURRENT SHARING

In d.c. circuits, a series resistor R may be added to each arm of the parallel arrangement to improve the sharing of current. To force steady state current sharing, the resistors are connected as shown in Fig. 4.11. In other words, due to the connection of series resistor. The difference in the value of dynamic resistance R_T of thyristors is compensated.

If the two thyristors are of identical rating, then the two external series resistors R_1 and R_2 are chosen such that the total voltage drops are equal, that is, $R_1 + R_{T_1} = R_2 + R_{T_2}$, where R_{T_1} and R_{T_2} are the corresponding dynamic resistances of the two SCRs, at the rated current.

If two SCRs of different forward current ratings I_{T_1} and I_{T_2} are to be operated in parallel, then the same resistance R can be used for both units to ensure proper current sharing by the SCRs. Let V_{T_1} and V_{T_2} be the respective voltage drops across the two SCRs for forward currents I_{T_1} and I_{T_2}. In parallel units, their anode-to-cathode voltage drops are same, therefore,

$$V_{T_1} + I_{T_1}(R + R_{T_1}) = V_{T_2} + I_{T_2}(R + R_{T_2}) \qquad (4.11)$$

Thus, from the knowledge of the spread of characteristics and the allowable power-loss in resistor R compatible with circuit conditions, the suitable value of R can be obtained from Eq. (4.11). Current equalisation by this method needs approximately 1.5 V voltage-drop across the series resistor at rated current. Obviously, such a addition of resistance affects overall efficiency and regulation.

Current sharing with reactors is a more efficient method than with resistors. In an a.c. circuit, the current distribution can be made more uniform by the magnetic coupling of the parallel paths as shown in Fig. 4.12. Figure 4.12 shows a 1:1 ratio reactor in bucking connection for two SCRs in parallel operation.

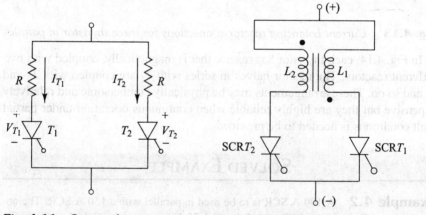

Fig. 4.11 *Current forcing using series resistor R*

Fig. 4.12 *Current sharing using reactor*

If currents I_{T_1} and I_{T_2} are equal, then the voltage drop in the reactor will be zero because of the mutual cancellation of flux linkages in the coils. If the current through SCR T_1 tends to increase above the current through SCR T_2 (i.e., $I_{T_1} > I_{T_2}$), a counter EMF will be induced proportional to the unbalanced current and tends to reduce the current through SCR T_1. At the same time, a boosting voltage is induced in series with SCR T_2, increasing the current flow through the SCR T_2.

The most important magnetic requirements of such a reactor are high saturation and low residual flux densities in order to provide as great a change in total flux each cycle as possible. Introducing a reactor also improves the transient sharing of current. Equalising reactors can be used in parallel more than two SCRs. Figure 4.13 shows the current equalising reactors connections for three thyristors in parallel, whereas, Fig. 4.14 shows the current equalising reactor connections for four thyristors in parallel.

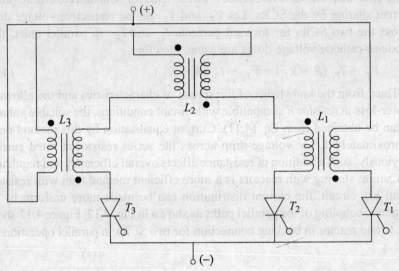

Fig. 4.13 *Current balancing reactor connections for three thyristor in parallel*

In Fig. 4.14, each thyristor has reactor that is magnetically, coupled with two different reactors. The reactor halves in series with T_1 are coupled with T_2 and T_4 and so on. These arrangements may be physically cumbersome and relatively expensive but they are highly reliable when continuous operation under partial fault conditions is needed to be provided.

SOLVED EXAMPLE

Example 4.2 A 100 A SCR is to be used in parallel with a 150 A SCR. The on-state voltage drops of the SCRs are 2.1 and 1.75 V, respectively. Calculate the series resistance that should be connected with each SCR if the two SCRs have to share the total current 250 A in proportion to their ratings.

Solution: Given

$$I_{T_1} = 100 \text{ A}, \quad V_{T_1} = 2.1 \text{ V}, \quad R = ?$$
$$I_{T_2} = 150 \text{ A}, \quad V_{T_2} = 1.75 \text{ V}$$

Using the relation,

$$V_{T_1} + I_{T_1}(R + R_{T_1}) = V_{T_2} + I_{T_2}(R + R_{T_2})$$

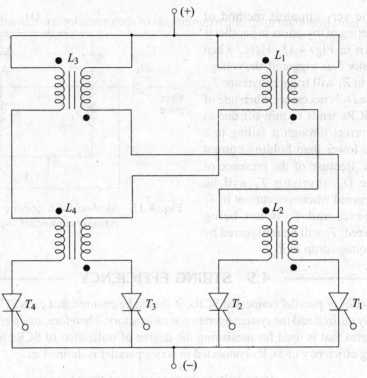

Fig. 4.14 *Current balancing reactor connections for four thyristors in parallel*

Let us assume that $(R + R_{T_1}) = R_1$
= resistance to be connected in series with each SCR.

$\therefore \quad V_{T_1} + I_{T_1} R_1 = V_{T_2} + I_{T_2} \cdot R_1$

$2.1 + 100 R_1 = 1.75 + 150 R_1 \quad \therefore R_1 = 0.007 \, \Omega$

4.8 TRIGGERING OF THYRISTORS IN PARALLEL

Parallel operation of thyristors can only takes place satisfactorily if all the thyristors are fired at the same time. Even differences of a few microseconds in the gate pulses to different thyristors can have a major influence on current balance in parallel operation. With parallel thyristors, all the thyristors must act as one and so the firing system employed must be highly reliable. The incorrect firing or misfiring of one thyristor will cause that thyristors to carry the full circuit current and this will be many times, its normal current level. Therefore the essential requirement for triggering SCRs connected in parallel is to use a common pulse generator for all. Also, it is necessary to drive the gates hard with sufficient peak and average power dissipation and with sufficient duration to ensure fast and complete turn-on.

The very simplest method of triggering of thyristors in parallel is shown in Fig. 4.15. Here, when thyristor T_1 is triggered, the voltage drop in R_1 will trigger thyristor T_2. Diode D_2 is necessary when one of the SCRs tends to turn-off due to the current through it falling to a value lower than holding current value. Because of the presence of diode D_2, thyristor T_1 will be retriggered whenever current in T_2 increases and T_1 is not being triggered. T_1 will be retriggered by the voltage drop in R_2.

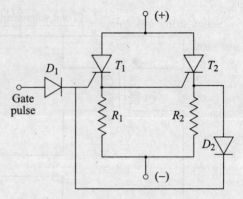

Fig. 4.15 *Method for triggering parallel-connected thyristors*

4.9 STRING EFFICIENCY

For series or parallel connected SCRs, it should be ensured that each SCR rating is fully utilized and the system operation is satisfactory. Therefore, string efficiency is a term that is used for measuring the degree of utilization of SCRs in string. String efficiency of SCRs connected in series/parallel is defined as

$$\text{String efficiency} = \frac{\text{(total voltage/current rating of the whole string)}}{\text{(individual voltage/current rating of SCR)} \times \text{(number of SCRs in series/parallel in the string)}} \quad (4.12)$$

In practice, this ratio is less than one. To obtaining the highest possible string efficiency, the SCRs connected in string must have identical characteristics. Since SCRs of the same ratings and specifications do not have identical characteristics, unequal current/voltage sharing is bound to occur for all SCRs in a string. Hence, the string efficiency can never be equal to one.

SOLVED EXAMPLE

Example 4.3 A thyristor string is formed by the series and parallel connection of thyristors. The voltage and current ratings of the string are 6 kV, and 4 kA respectively. Available thyristors have the voltage and current ratings of 1.2 kV and 1 kA, respectively. The string efficiency is 90% for both the series and parallel connections. Calculate the number of thyristors to be connected in series and parallel.

If the maximum blocking current is 15 mA and $\Delta Q_{max} = 25$ μC, calculate the values of R and C in Fig. 4.3.

Solution: Given: $E_s = 6$ kV, $I_m = 4$ kA, $E_D = 1.2$ kV, $I_T = 1$ kA, String efficiency = 90%.
Using the relation,

Series and Parallel Operation of Thyristors

Series-string efficiency = $\dfrac{\text{(Total voltage applied to whole string, } E_s)}{E_d \times n_s}$

$\therefore \quad n_s = \dfrac{E_s}{E_d \times \text{efficiency}} = \dfrac{6000}{1200 \times 0.9} = 5.56$

$n_s \approx 6 \approx$ number of SCRs in series

Similarly,

Number of SCRs in parallel, $n_p = \dfrac{I_m}{I_T \times \text{efficiency}} = \dfrac{4000}{1000 \times 0.9} = 4.44 \quad \therefore n_p = 5$

(ii) Given $I_{B\max} = 15$ mA, $\Delta Q_{\max} = 25$ μC

$\therefore \quad R \leq \dfrac{n_s E_D - E_s}{(n_s - 1) \Delta I_b}$

Here, assume $\Delta I_b = I_{B\max} = 15$ mA

$\therefore \quad R \leq \dfrac{6(1200) - 6000}{(6-1).(15 \times 10^{-3})} \quad \therefore R \leq 16\ \text{k}\Omega$

Now, $C \geq \dfrac{(n_s - 1).\Delta Q_{\max}}{(n_s . E_D - E_s)} = \dfrac{(6-1) \times (25 \times 10^{-6})}{(6 \times 1200) - 6000} \quad \therefore C \geq 0.104\ \mu\text{F}.$

4.10 DERATING

Several compensating networks used for voltage and current equalisation in series and parallel connected thyristors, respectively, have been discussed in previous sections. Although the differences in voltages/current will be reduced, these techniques do not entirely eliminate these differences. Therefore, in order to improve the reliability of the series/parallel string, an extra unit may be added so that the voltage/current applied to each device will be lower than its normal rating. Thus, there is an inherent derating of the units connected in series and parallel configuration.

The per cent derating will be given by the equations:

Per cent parallel derating = $\left(1 - \dfrac{I_m}{n_p I_T}\right) \times 100$ \hfill (4.13)

where I_m = total forward current, n_p = total number of SCRs in parallel; and I_T = forward-current rating of each SCR.

Similarly,

Per cent series derating = $\left(1 - \dfrac{E_s}{n_s E_D}\right) \times 100$ \hfill (4.14)

where E_s = total applied voltage to series-string; N_s = total number of SCRs in series; E_p = forward voltage rating of each device.

Derating factors are normally 8 to 20% for parallel connection of thyristors depending upon the number of devices in parallel. The other main reason for derating is poor cooling and heat-dissipation as number of devices have to appear in the same branch of the circuit.

SOLVED EXAMPLES

Example 4.4 Calculate the number of SCRs, each with rating of 500 V, 75 A required in each branch of a series and parallel combination for a circuit with the total voltage and current rating of 7.5 kV and 1000 A. Assume derating factor of 14%.

Solution: Given: E_s = 7.5 kV = total applied voltage of thyristor unit., I_m = 1000 A = total forward current. Derating factor = 14%, E_D = forward voltage rating of each device = 500 V,

I_T = forward current rating of each SCR = 75A

Now, using Eq. (4.14),

(i) % series derating = $\left(1 - \dfrac{E_s}{n_s E_D}\right) \times 100$

$14 = \left(1 - \dfrac{7500}{n_s \times 500}\right) \times 100 \quad \therefore n_s = 17.44$

$n_s \approx 18$ = number of SCRs connected in series-string.

(ii) % parallel derating = $\left(1 - \dfrac{I_m}{n_p I_T}\right) \times 100$

$14 = \left(1 - \dfrac{1000}{n_p \times 75}\right) \times 100 \quad \therefore n_p = 15.5 \approx 16.$

$\therefore$ Number of SCR connected in parallel = 16.

Example 4.5 Two SCRs are connected in parallel. One SCR has an approximate characteristic of $V = [0.9 + (2.4 \times 10^{-4} i)]$ V, and the other SCR $V = [1.0 + (2.3 \times 10^{-4} i)]$ V. Determine the current taken by each SCR if the total current is

(i) 500 A (ii) 1000 Amp (iii) 1500 Amp (iv) 2000 Amp

Determine the value of equal resistors which when placed in series with the SCRs will in (iv) bring the SCR current to within 10% of equal current sharing.

Solution: Due to parallel connection, voltage across each SCR will be identical, so if I_{T_1} and I_{T_2} be the respective SCR currents, then,

$0.9 + (2.4 \times 10^{-4}) I_{T_1} = 1.0 + (2.3 \times 10^{-4}) I_{T_2}$

and total current $I_T = I_{T_1} + I_{T_2}$.

Hence, for
(i) I_{T_1} = 457 A, I_{T_2} = 43 A (iii) I_{T_1} = 947 A, I_{T_2} = 553 A
(ii) I_{T_1} = 702 A, I_{T_2} = 298 A (iv) I_{T_1} = 1191 A, I_{T_2} = 809 A

In (iv) for 10% balance, the currents will be

I_{T_1} = 1100 A and I_{T_2} = 900A, then $0.9 + (2.4 \times 10^{-4} + R) 1100 = 1.0 + (2.3 \times 10^{-4} + R) 900$

$\therefore \qquad R = 0.215$ mΩ.

Review Questions

4.1 Describe the series and parallel operations of SCRs.

4.2 What is the necessity of connecting SCRs in series? What are the problems associated with series connection of SCRs? How are they eliminated?

4.3 Draw and explain the necessity of static and dynamic equalising circuit for series connected SCRs? Derive relations used for determining the values of shunt resistor R and capacitor C in this circuit.

4.4 Three SCRs of the same rating are connected in series. Will the voltage across each of them be same when—
 (i) all SCRs are in the forward-blocking state.
 (ii) all SCRs are conducting?

4.5 What are the different methods of triggering SCRs in series? Draw and explain sequential firing circuit for triggering of series connected SCRs.

4.6 Draw and explain the simultaneous triggering circuit of series connected SCRs?

4.7 Draw and explain the optical triggering circuit of series connected SCRs?

4.8 Why SCRs are required to be connected in parallel? What are the problems associated with parallel connection of SCRs? How they are eliminated?

4.9 Explain in brief why two SCRs of same rating when connected in parallel do not share equal currents. Suggest any method to equalise the currents and discuss its merits and demerits.

4.10 What do you understand by string efficiency? What is its significance.

4.11 Discuss thermal runaway process in parallel connected SCRs. How is it avoided?

4.12 What are the problems associated with firing of parallel connected SCRs? Draw and explain circuit for firing of parallel connected SCRs.

4.13 Series and parallel connected SCRs are derated, why? Define percentage derating of series and parallel connected SCRs.

4.14 Explain the terms:
 (i) String efficiency (ii) Derating factor.

4.15 What is the purpose of having parallel operation of SCRs? What care must be taken when paralleling the SCRs.

4.16 What will happen if one of the SCR has large delay time in parallel SCRs? Explain the convenient method of triggering parallel connected SCRs.

Problems

4.1 The voltage and current rating in a particular circuit are 3 kV and 750 A. SCRs with a rating of 800 V and 175 A are available. The recommended minimum derating factor is 15 per cent. Calculate the number of series and parallel units required. Also, obtain the required values of R and C to be used in the static and dynamic equalising circuits if the maximum forward leakage current for the SCRs is 10 mA and $\Delta Q = 20\ \mu C$.

[Ans $n_s = 5, n_p = 5, R = 25\ k\Omega, C = 0.08\ \mu F$]

4.2 Five thyristors, each of 500V and 500A are used in series and parallel circuit of 2 kV and 1.8 kA. Calculate voltage and current derating factor.
[*Ans* Voltage derating factor = 20%, Current derating factor = 28%]

4.3 The voltage rating in a particular circuit is 3.2 kV. SCRs with voltage rating of 60 V are available. Calculate the number of SCRs required to be connected in series? Also design the static equalising circuit for the above series connection if the maximum blocking current of the SCRs at the rated voltage is 6 mA.
[*Ans* $n_s = 54., R \approx 125 \, \Omega$]

4.4 A 'valve' in a HVDC circuit is a combination of SCRs in series and parallel. Calculate the number of SCRs, each with a rating 500 V, 50 A required in each branch of series parallel combination for a value with a total voltage and current rating of 7.5 kV and 1 kA. Assume derating factor 20%.
[*Ans* $n_s = 19; n_p = 25$]

4.5 Two thyristors, having a difference of 4 mA in latching current are connected in series in a circuit. Voltage across the devices are 500 V and 480 V. Calculate the required equalizing resistances.
[*Ans* $R = 5 \, k\Omega$]

4.6 A string of three thyristors connected in series as shown in Fig. 4.3 is designed to withstand an off-state voltage of 7.2 kV. If the compensating circuit components have values of $R_c = 30 \, \Omega$, $R = 24 \, k\Omega$ and $C = 0.008 \, \mu F$. Estimate the voltage across each thyristor in the off-state and the discharge current of each capacitor on turn-on. The leakage currents for the thyristors are $T_1 = 18$ mA, $T_2 = 24$ mA, and $T_3 = 16$ mA.
[*Ans* (i) $V_1 = 2.432$ kV, $V_2 = 2.29$ kV, $V_3 = 2.48$ kV
(ii) $I_{C_1} = 81.1$ A, $I_{C_2} = 76.27$ A, $I_{C_3} = 83.67$ A]

REFERENCES

1. *General-Electric Company, SCR Manual*, 5th edition Syracyse N.Y. 1972.
2. F Csaki, *et al, Power electronics*, Budapest: Akademiai Kiado, 1975.
3. M. Ramamoorthy, *Thyristors and their applications*, East-west Press, 1977.
4. S B Dewan, and A Straughen, *Power-semiconductor circuits*, John Wiley and Sons, 1975.
5. F E Gentry, *Semiconductor Controlled Rectifiers*, Prentice Hall 1964.
6. R S Ramshaw, *Power Electronics*, Chapman and Hall, London 1973.

Chapter 5

Power Semiconductor Devices

LEARNING OBJECTIVES:

- To introduce the major power devices like phase control and inverter grade SCRs, ASCRs, RCTs, DIACs, Triacs, SUS, SBS, SCS, LASCS, and LASCRs.
- To consider the switching performance, breakdown voltages, SOAs, thermal considerations, and applications of power MOSFET.
- To examine the operation, characteristics, latch-ups, switching performance, SOAs, gate-drive circuits, and snubbers of IGBT.
- To consider the operation of SITs.
- To examine the operation, switching performance, snubbers, characteristics, gate-drive circuits, over current protection, and applications of GTOs.
- To introduce the basic structure and operation of SITH and MCT.
- To introduce the PICs.
- To examine the operation of IGCT.
- To introduce the power device MTO.
- To examine the operation of ETO.

5.1 INTRODUCTION

Today, it is difficult to imagine a world of power conversion and control without silicon power devices. The power semiconductor device is the most important component in a power electronic system. In fact, the evolution of power-electronics has followed the evolution of devices.

Starting with the conventional thyristor-type devices in the early days, the recent availability of high frequency, high power, MOS-gated self-controlled devices are opening new frontiers in power electronics. It is interesting to note that in modern power electronics apparatus, there are essentially two types of semiconductor elements: the power semiconductors that can be defined as the muscle of the equipment, and microelectronic control chips, which have the power and intelligence of the brain. Both are digital in nature, except that one manipulates large power up to gigawatts, and the other handles power only of the order of milliwatts. Today's power electronics apparatus integrates both of these end-of-the-spectrum electronics, providing tremendous capability in size and cost reduction, as well as sophisticated system performance.

The progress in power-electronics today has been possible primarily due to advances in power-semiconductor devices. Of course, apart from device evolution, the inventions in converter topologies, PWM techniques, control and estimation techniques, digital signal processors, ASIC chips, control hardware and software, etc. have also contributed to this advancement. Modern era of solid-state power-electronics began with the advent of thyristor (silicon controlled rectifier) in the late 1950's. Gradually, other devices such as triac (1958), gate-turn-off thyristor (GTO-1958), bipolar power transistors (BPT-1975), Power-MOSFET (1975), insulated gate bipolar transistor (IGBT-1985), static induction transistor (SIT-1975) and integrated gate commutated thyristor (IGCT-1987) were introduced. As the evolution of new and advanced devices continued, the voltage and current ratings and electrical characteristics of the existing devices began improving dramatically. In fact, the device evolution along with converter, control and system evolution was so spectacular in the last decade of 29th century, we define it as the "decade of power-electronics.

Thyristors are used for high-power low frequency applications, such as HVDC systems, static phase-control type static VAR compensators, cycloconverters and load commutated inverter (LCI) synchronous machine drives. Currently, devices are available with 8 kV and 4 kA ratings ABB recently introduced a monolithic thyristor AC switch that has the voltage ratings of 2.8 kV–6.5 kV and current ratings of 3 kA–6 kA. The advent of large GTOs pushed the thyristor voltage-fed inverter from the market. GTOs are available with 6 kV, 6 kA (Mitsubishi) ratings for large voltage-fed inverter applications). Power MOSFET has grown in rating, but its primary popularity is in high-frequency switching mode power supply and portable appliances.

In 1998, Infineon Technology, Germany introduced Cool MOS (600 V) with only 20% conduction loss and reduced switching loss. However, there is difficulty of using its body diode. The BJT appeared and then fell into obsolescence due to the advent of IGBT at the higher end and power MOSFET at the lower-end. The invention of IGBT is an important milestone in the history of power-semiconductor devices. Commercial IGBTs are available with 3.5 kV, 1.2 kA, but upto 6.5 kV and 10 kV devices are under test in laboratory. Trench gate IGBT with reduced conduction drop is available up to 1.2 kV, 600 A. IGBT intelligent power modules (IPM) from number of vendors are available for 600 V, 50–300 A and 1200 V, 50–150 A to cover upto 150 hp ac drive applications. IGCT (also called GCT) is basically a hard-driven GTO with built-in gate driver, and the device is available with 6 kV, 6 kA (10 kV device, are under test). ABB introduced recently a reverse blocking IGCT (6 kV, 800 A) for use in current-fed inverter drives. Large-band gap power-semiconductor device with *silicon carbide* (SiC) that has high carrier mobility, high electrical and thermal conductivities and strong radiation hardness is showing high promise for next generation power devices. These devices can be fabricated for higher voltage, higher temperature, higher frequency and lower conduction drop. SiC diodes are commercially available, and other devices are expected in future.

Power semiconductor devices and their associated technology have come a long way from their beginnings with the invention of the bipolar transistor in the late 1940s. Presently, the spectrum of what are referred to as "power devices" span a very wide range of devices and technology from the massive 4 in, 3000-A thyristor to the high voltage integrated circuit and the power MOSFET, a device of VLSI complexity containing upto 1,50,000 separate transistors. In this chapter, the past, present and future of power devices will be reviewed. The main section of this chapter will review the technology and characteristics of bipolar power devices with separate subsections on thyristors, the gate turn-off thyristor (GTO), and the bipolar transistor. Within the thyristor subsection, the phase control thyristor, inverter thyristor, asymmetric thyristor (ASCR), reverse conducting thyristor (RCT), light-activated thyristors, and Triac are discussed. The other sections will deal with the new field of integrated power electronic devices. This will include a review of the evolution of power MOS technology together with a presentation of the characteristics of some of the power devices, such as the Insulated Gate Bipolar Transistor (IGBT), Static Induction Transistor (SIT), Static Induction Thyristors (SITH), and MOS-controlled thyristor (MCT).

5.2 HISTORICAL PERSPECTIVE

Nearly fifty years ago, the introduction of selenium rectifiers was heralded as a major advance over the thyratrons, ignitions, and even copper oxide devices of that time. However, the selenium era proved to be short lived. In the late 1940s, a number of important successive developments initiated the dawn of today's semiconductor era. Although point contact diodes were in use in the 1940s, most important was the discovery of the point contact transistor in 1947 and the subsequent invention of the junction transistor in 1949 at Bell Laboratories. Supported by countless process innovations, device concepts, and developments in physical understanding, the junction transistor formed the foundation for today's integrated circuits and power electronics business.

In 1952, Hall demonstrated the first significant power devices based upon semiconductor technology. The devices were fabricated using germanium mesa alloy junctions and resulted in a rectifier with impressive electrical characteristics for that time, i.e. a continuous forward current capability of 35 A and a reverse blocking voltage of 200 V. Hall also used the recently developed theories of carrier recombination, generation, and current flow to successfully model the electrical characteristics of power rectifiers and transistors.

In the mid 1950s, power device characteristics capitalized on the development of single-crystal silicon technology. The larger band gap of silicon rectifiers resulted in higher reverse voltage capability and higher operating temperatures. By the late 1950s, 500 V rectifiers were available in alloy junctions. The introduction of diffused junctions combined with mesa technology in the late 1950s proved to be the step necessary to realize reverse blocking capability of several kilovolts in later years. By the mid 1960s, theoretical avalanche breakdown voltages of upto 9000 V had been achieved by optimized contouring of mesa junctions. Increased current handling capability became a possibility by optimizing device packaging for minimum thermal and mechanical stress on the chip. Today, 77-mm diameter silicon rectifiers are available with continuous current ratings of 5000 A and a reverse voltage of 3000 V.

To put this development in perspective, while the first commercially available silicon transistors were announced by *TI* in 1954, it was almost a decade later that their practical application to high power conversion and control began. Emitter current crowding, reliability, and materials processing challenges precluded economic justification. The introduction of the planar process by Fairchild plus the application of photolithography techniques to wafer processing resulted in the birth of the power transistor business in the 1960s. A decade of effort in the industry related to second breakdown, power/speed performance and unique process such as epitaxy deposition, paid off. By the late 1970s, 200 A-500 V bipolar Darlington transistors with a gain of 50 were available together with 100 V–10 A transistors which could operate at frequencies upto 1 MHz. Since then, however, the application of MOS technology to power transistors has been a major focus of the industry due to the promise of high speed and high input impedance in many low voltage applications.

Beyond the important developments in material and process technology described above, there have been many significant device developments. One of the first of these was the publication of the *P-N-P-N* transistor switch concept in 1956 by Moll *et al*. Although, probably envisioned to be used for Bell's signal applications, engineers at General Electric quickly recognized its significance to power conversion and control, and within nine months announced the first commercial silicon controlled rectifier in 1957. This three-terminal power switch was fabricated using 5-mm square alloy-diffused mesa silicon chip and had a current rating of 25 A, and a blocking voltage capability of 300 V. The shorted emitter concept plus the planar process resulted in planar diffused SCRs in 1962. These processes resulted in high voltage blocking capability at junction temperatures of 125°C and made practical power control and conversion possible.

Since the early 1960s, thyristor producers have capitalized upon the process innovations of the signal industry while introducing new devices or structural improvements to existing devices. In 1961, a gate turn-off thyristor (GTO) was disclosed which combined the switching properties of a transistor with the low conduction losses of an SCR. In 1964, a bidirectional a.c. switch (TRIAC) was introduced by General Electric principally for 60-cycle consumer lighting and motor speed control. In 1965, light triggered thyristors were developed which later found significant application in optoelectronic couplers. In the late 1960s, a number of advances were made in the design of thyristor —cathode gate structures. Incorporation of interdigitated gates made possible high power 20 kHz inverters. Similarly, the inclusion of pilot gating techniques decreased the gating requirements as well as improved high frequency and pulse duty operation. The reverse conducting thyristor (RCT) and asymmetrical SCR were developed in 1970 to provide higher speed capability in those inverter applications where reverse blocking voltage was not required. In the mid 1970s, thyristor designers were intrigued with electric (or voltage) controlled thyristors which held promise for higher speed performance. The later, however, never came to fruition and the application of MOS concepts has proven to offer similar benefits but with greater ease in manufacture.

5.3 POWER SEMICONDUCTOR DEVICES

The power conversion world is a permanent quest for the ideal switch. Such a switch, in general terms, has the following requirements:

- high currents (turn-off, rms, average, peak, surge)
- high voltage (peak repetitive, surge, dc-continuous)
- fast switching (short on/off delays short rise/fall times, short turn-on/off times)
- low losses (conduction, switching)
- high frequency (fast switching, low switching losses)
- high reliability (low random failures, high power and temperature cycling, high blocking stability), low parts count)
- Compact constructions (low losses, low parts count).

Broadly, the power semiconductor devices can be classified into two main types according to the nature of their controllability:

1. Type I: Thyristors There are many different devices available today which can meet one or more of these requirements, but, as always, an improvement in one characteristic is usually gained at the expense of another. As a result, different devices have been optimized for different applications namely:

   (i) Phase-Control Thyristors
  (ii) Inverter-Grade Thyristors (fast-switching SCRs)
 (iii) Asymmetrical-Thyristors (ASCRs)
 (iv) Reverse-conducting Thyristors (RCTs)
  (v) Gate-Assisted Turn-off Thyristors (GATTs)
 (vi) Bidirectional Diode Thyristors (DIACs)

(vii) Bidirectional Triode Thyristors (Triacs)
(viii) Silicon Unilateral Switch (SUS)
(ix) Silicon Bilateral Switch (SBS)
(x) Silicon-Controlled Switch (SCS)
(xi) Light-Activated Silicon Controlled Rectifiers (LASCRs)

In general, the turn-on operation of the devices of this type is controllable using a trigger signal. However, the turn-off operation depends upon the condition of the power circuit. Hence, in this type, only turn-on switching is externally controllable.

2. Type II: Gate/Base Commutating Devices Both turn-on and turn-off operations of the device under this type are externally controllable by base or gate signals. High switching frequency devices which belongs to this type are:

(i) Power-BJT
(ii) Power-MOSFETs
(iii) Gate-Turn-off Thyristors (GTOs)
(iv) Static-Induction Thyristors (SITHs)
(v) Static-Induction Transistors (SITs)
(vi) Field-Controlled Thyristors (FCTs)
(vii) MOS-Controlled Thyristors (MCTs)
(viii) MOS-Turn-off Thyristors (MTOs)
(ix) Integrated Gate Commutated Thyristors (IGCTs)
(x) Emitter Turn-off thyristors (ETOs)

The power device design is pursued basically in two areas: that of the *transistor structure* and that of the *thyristor structure* whereby thyristors are generally preferred for their low conduction loses and transistors for their rugged turn-off capabilities. Numerous devices have been proposed, divided along these lines. Some strive to have the best of both worlds, exploiting the rugged on-state performanc of thyristors, while reverting to transistor-like behaviour prior to the critical turn-off phase.

Power BJT, MOSFET and IGBT has the transistor structure whereas GTO, FCT, MCT, IGCT, MTO and ETO has the thyristor structure.

As can be seen from the above paragraph, thyristor structures dominate in the number of proposed devices because of their inherent ability to conduct large current, with minimal losses. However, until recently, the only serious contenders for high power applications were the GTO (thyristor), with its cumbersome snubbers, and the IGBT (transistor), with its inherently high losses. Recent developments, however, have led to a device which successfully combines the best of thyristor and transistor characteristics, while fulfilling the additional requirements of manufacturability and high-reliability.

The salient design features, structures and applications of these different devices are discussed in the following sections.

5.4 PHASE CONTROLLED THYRISTORS

"Phase control" or "converter-grade" thyristors generally operate at line frequency. They are turned off by natural commutation and do not have special fast-switching characteristics. The turn-off time, t_q, is of the order of 50 to 100 µs. For a 77-mm diameter phase control thyristor, the conduction voltages vary from 1.5 V for 600 V to 2.5 V for 4000 V devices, and for a 5500 A, 1200 V thyristor it is typically 1.25 V. For a given size of a device with a given voltage design, the widths of the bases and the junction profiles are fixed by punch through and current gain considerations. Because of this, the forward drop becomes mainly dependent upon the diffusion length L (or lifetime, τ) in the wide n-base region and the ohmic contact losses at the end regions of the device and careful consideration of the diffusion profiles and diffusion technique must be taken to insure low forward drop.

Other key considerations of present day phase control thyristors are enhancement of the dynamic dv/dt capability and the improvement of the gate sensitivity. In case of the former, the rate of rise of the voltage reapplied at the end of the turn-off interval (dv/dt effect) in conjunction with the device capacitance produces a displacement current which can turn-on the device if sufficiently large. The use of emitter shorting structures which provide a shunting path for the displacement current has greatly improved dynamic dv/dt capabilities with typical dv/dt ratings of greater than 100 V/µs routinely achieved. The high temperature performance has been improved significantly by emitter shorting since the shorts will also by-pass thermally generated currents around the emitter.

In order to simplify the gate drive requirement and increase sensitivity, the use of amplifying gate, which was originally developed for fast switching "inverter" thyristors, is widely adopted in phase control SCR. The amplifying gate thyristor is shown in Fig. 5.1. As shown in the figure, an auxiliary thyristor T_{A_1} is turned ON by a gate signal and then the amplified output of T_{A_1} is applied as a gate signal to the main thyristor, T_1.

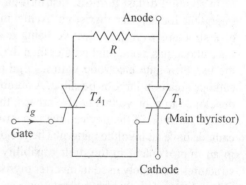

Fig. 5.1 *Amplifying gate thyristor*

The amplifying gate permits high dynamic characteristics with a typical dv/dt of 1000 V/µs and di/dt of 500 A/µs and simplifies the circuit design by reducing or minimizing dv/dt protection circuits and di/dt limiting inductor.

5.5 INVERTER-GRADE THYRISTORS

The most common feature of an inverter-grade thyristor which distinguishes it from a standard phase control type is that it has fast turn-off time, generally in the range of 5 to 50 μs, depending upon voltage rating. Therefore, these are used in high-speed switching applications with forced commutation (e.g., inverters in Chapter 9 and choppers in Chapter 8). The conduction voltages varies approximately as an inverse function of the turn-off time, t_q.

Inverter thyristors are generally used in circuits that operate from d.c. supplies, where current in the thyristor is turned off either through the use of auxiliary commutating circuitry, by circuit resonance, or by "load" commutation. Whatever be the circuit turn-off mechanism, fast turn-off is important because it minimizes size and weight of commutating and/or reactive circuit components.

For inverter thyristors, the turn-off time is dependent upon the stored charge at the time of current reversal, the amount of charge removed during the recovery phase, the rate for recombination during the recombination phase, and the amount of charge injected during reapplication of the forward voltage. The general practice for achieving short turn-off times reduction is to "kill" the lifetime by introducing recombination centres in the device structure *using gold or platinum doping* and electron irradiation. However, one of the problems in doing this is the adverse effect it has on the forward conduction drop, the turn-on time, and the plasma spreading properties of the device. As a result, a trade-off has to be made in selecting the appropriate lifetime. Incorporating cathode shorting will aid in this trade-off by reducing the initial stored charge and the amount of charge injected during the reapplication of the forward voltage. While it will not speed up the recovery and the recombination process, the advantage of the emitter shorting over lifetime reduction is that it has a much less adverse effect on forward drop.

In addition to fast turn-off, fast turn-on time and di/dt capability are usually important for inverter thyristors. Achieving this requires proper design of the gate structure with the objective being to design the device so that no part of the cathode has a potential greater than that of the gate electrode. This mandates the use of a gate electrode with a large perimeter ensuring that the adjacent cathode area is as large as possible. A number of interdigitated designs have been developed to achieve this goal. Perhaps, the most effective design in achieving good utilization of the device area is a structure in which both the gate and the cathode have an involute pattern. The use of involute gate structures has resulted in an improvement in the di/dt capability of inverter thyristors. Involute gate structures are widely used in inverter thyristors with 40 mm diameter and below. Other distributed gate structures, such as the "snow-flakes" pattern are being used for inverter thyristors with larger diameter.

These thyristors have high dv/dt of typically 1000 V/μs and di/dt of 1000 A/μs. The fast turn off and high di/dt are very important to reduce the size and weight of commutating and/or reactive circuit components. The conduction voltage of a 2200 A, 1800 V thyristor is typically 1.7 V.

5.6 ASYMMETRICAL THYRISTOR (ASCR)

As indicated above, the thyristor design involves a trade-off among various device parameters, such as forward and reverse voltage blocking capability, turn-on and turn-off times, and on-state voltage drop. The conventional thyristor may have a reverse blocking capability of thousands of volts, but this capability is not required for every application. In particular, the voltage-fed inverter circuit, which converts d.c. power to a.c., usually has rectifier diode connected in antiparallel across each thyristor to conduct reactive load currents and excess commutating current. In such circuits, the antiparallel diode clamps the thyristor reverse voltage to 1 or 2 V under steady circuit conditions.

If a high reverse voltage rating is unnecessary, the remaining thyristor characteristics can be optimized. One of the main characteristics of an asymmetrical thyristor (ASCR) is that they do not block significant reverse voltage. Therefore, an ASCR is specifically designed for applications where reverse blocking capability is unimportant. Typically, the reverse voltage rating is about 20 or 30 V and the forward voltage rating is of the range 400–2000 V. The switching times and on-state voltage drop of an ASCR are smaller than those of a conventional thyristor of the same rating. As already indicated, a fast turn off is important because it minimizes the size, weight, and cost of commutating circuit components, and permits operation at switching frequencies of 20 kHz, or more, with high-efficiency.

The conventional centre gate thyristor shown in Fig. 5.2 has a thick, lightly doped n-base region. Base thickness must be large enough to prevent the spreading of the depletion region during forward blocking into the anode p^+ region. When this spreading occurs due to the application of excessive forward voltage, a punch through conditions occurs and the thyristor immediately turns ON. Unfortunately, because the on-state voltage drop is also proportional to n-base thickness, a thyristor with a high peak repetitive forward voltage rating, VD_{RM}, will also have high on-state losses.

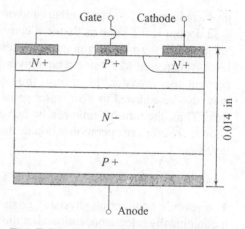

Fig. 5.2 *Conventional centre gate thyristor*

A narrower n-base is possible if a highly doped N^+ layer is introduced adjacent to the P^+ emitter, as shown in Fig. 5.3(a). The new N^+ layer acts as a buffer, preventing the depletion region from extending into the P^+ layer and allowing a higher average electric field in the lightly doped N-region. Consequently, a thinner N-base is obtained for the same forward blocking voltage capability. ASCR symbol is shown in Figure 5.3(b).

In the reverse blocking mode, the $N^+ P^+$ junction quickly avalanches at less than 50 V, so that the peak repetitive reverse voltage rating, V_{RRM}, is somewhat less than this value. However, the reduction in crystal thickness can be used to

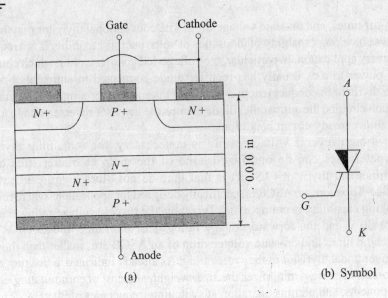

Fig. 5.3 *(a)* Asymmetrical thyristor (ASCR) *(b)* Symbol

give a reduction in on-state voltage and/or a faster turn-off time. If the amount of gold doping is the same as that in a conventional fast thyristor, turn-off time is unaltered but the on-state voltage drop is considerably reduced. If the gold doping level is increased to a value which gives the same on-state voltage as that in a conventional thyristor, the turn-off time is substantially reduced. Doping levels may also be adjusted to give lesser gains in both on-state voltage and turn-off time. Thus, the turn-off time can be halved and the on-state voltage reduced to allow more efficient operation at double the switching frequency of a conventional thyristor.

5.7 REVERSE CONDUCTING THYRISTOR (RCT)

The reverse conducting thyristor is simply an asymmetric thyristor with a monolithically integrated, antiparallel diode in a single silicon chip, as shown in Fig. 5.4 (a). The thyristor is turned OFF by passing a current pulse through the diode part of the chip. By combining the ASCR and the diode in one device, a more compact circuit layout is obtained and heat sinking is simplified. Stray loop inductance between the ASCR and the diode is also minimized, avoiding the generation of reverse voltage transients across the ASCR and making its turn-off behaviour more predictable. Isolation of thyristor and diode functions is important to ensure that charge carriers present in the diode during commutation do not diffuse into the thyristor part of the chip to cause retriggering when forward voltage is reapplied. Symbol of RCT is shown in Fig 5.4(b).

The forward blocking voltage varies from 400 to 2000 V and the current rating goes upto 500 A. The reverse blocking voltage is typically 30 to 40 V. A disadvantage of the RCT is that it is inflexible compared with two discrete devices because the current ratio between thyristor and diode parts of the chip is fixed for a given design. If the diode carries a commutation current pulse only, the greater part of the chip can be devoted to the thyristor to maximize its current capability. In the voltage source inverter, the load current is controlled by the thyristor and flows freely in the other direction through the diode. For such circuits, the RCT must have equal current ratings for thyristor and diode sections. Purpose-designed RCT devices are now being manufactured for high performance inverter and chopper circuits.

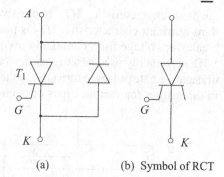

(a)   (b) Symbol of RCT

Fig. 5.4 *Reverse conducting thyristor equivalent circuit*

5.8 BIDIRECTIONAL DIODE THYRISTOR (DIAC)

A Diac is a two electrode, bidirectional avalanche diode which can be switched from the off-state to the on-state for either polarity of applied voltage. The schematic construction, voltage–current characteristics and circuit symbol of the Diac are shown in Fig. 5.5. Notice that the two leads are labelled as terminals T_1 and T_2 instead of the conventional anode–cathode designations. The term Diac is obtained from capital letters, *Diode that can work on a.c.*

Conduction occurs in the Diac when the breakover voltage is reached in either polarity across the two terminals. When T_1 is positive with respect to T_2, and if voltage V_{12} exceeds V_{BO_1}, then the structure *PNPN* conducts. The curve in Fig. 5.5(c) illustrates this characteristic. Similarly, when terminal T_2 is positive with respect to T_1 and if voltage V_{21} exceeds breakover voltage V_{BO_2}, the structure *PNPN* conducts. At voltages less than the breakover voltage, a very small amount of current called the leakage current flows through the device. Leakage current produced due to the drift of electrons and holes at the depletion region is not sufficient to cause conduction in the device. The device remains practically in nonconducting mode. This portion of the characteristics shown by region *OA* in Fig. 5.5(c) is called as the blocking state. At point '*A*', when the voltage level reaches the breakover voltage, the device starts conducting. During its conduction, the device exhibits negative resistance characteristics. The current flowing in the device starts increasing and the voltage across it starts decreasing. This portion of the characteristic shown by *AB* in Fig. 5.5(c) is known as the conduction state. The value of current corresponding to the point *A* is known as the breakover current. Similar explanation holds good for the negative half-cycle of triggering. The characteristic obtained in the third quadrant will be a replica of that obtained in the first quadrant. This is because the doping level is same at the two junctions of the device. Once the device starts conducting, the current flowing through it is very high which has to be limited by some external resistance. In the first-quadrant characteristic, MT_1 is positive with respect to MT_2, whereas in the third-quadrant characteristic, MT_2 is positive with respect to MT_1. The value of breakover voltage for a commonly used Diac type ST_2 is 30 volts.

Diac is mainly used as a trigger device for Triacs which require either positive or negative gate pulses to turn ON. In fact, matched Diac-Triac pairs are available in the market for various types of control circuits.

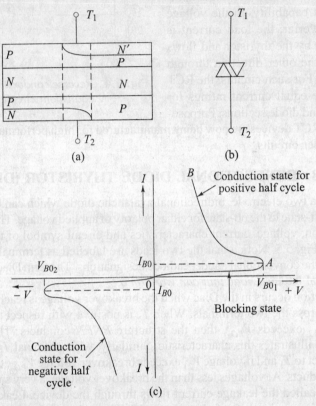

Fig. 5.5 *Schematic construction, circuit symbol and V–I characteristics of a diac*

5.9 BIDIRECTIONAL TRIODE THYRISTOR (TRIAC)

We have seen that the conventional thyristor, or SCR, has a reverse-blocking characteristic that prevents current flow in the cathode-to-anode direction. However, there are many applications, particularly in a.c.circuits, where bidirectional conduction is required. Two thyristors may be connected in inverse-parallel, but at moderate power levels the two antiparallel thyristors can be integrated into a single device structure, as shown in Fig. 5.6(a). This device, commonly known as *Triac* (triode a.c. switch) is represented by the circuit symbol shown in Fig. 5.6(b). Triac is the word derived by combining the capital letters from the

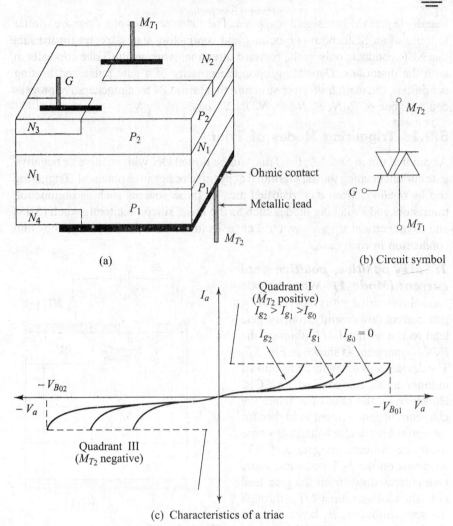

Fig. 5.6 *Triac structure and characteristic*

words *TRIode* and *a.c.* As the Triac can conduct in both the directions, the terms anode and cathode are not applicable to Triac. Its three terminals are usually designated as main terminals, MT_1, MT_2 and gate G, as in a thyristor. The terminal MT_1 is the reference point for measurement of voltages and currents at the gate terminal and at the terminal MT_2. The gate is near to terminal MT_1. The V–I characteristic of a Triac is shown in Fig. 5.6(c). This characteristics of the Triac are based on the terminal MT_1 as the reference point. The first quadrant is the region wherein MT_2 is positive with respect to MT_1 and *vice-versa* for the third-quadrant. The peak voltage applied across the device in either direction must be less than the breakover voltage in order to retain control by the gate. A gate current of specified amplitude of either polarity will trigger the Triac into conduction in either quadrant, assuming that the device is in a blocking condition initially before the gate signal is applied. The characteristics of a Triac are similar to those of an SCR, both in blocking and conducting states, except for the fact that SCR conducts only in the forward direction, whereas the Triac conducts in both the directions. Depending upon the polarity of a gate pulse and biasing conditions, the main four-layer structure that turns ON by a regenerative process could be one of $P_1 \, N_1 \, P_2 \, N_2$, $P_1 \, N_1 \, P_2 \, N_3$, or $P_2 \, N_1 \, P_1 \, N_4$.

5.9.1 Triggering Modes of Triac

As pointed out in Sec. 5.9, the Triac can be turned ON with positive or negative gate current keeping the MT_2 terminal at positive or negative potential. Triggering can be obtained from d.c., rectified a.c., or pulse sources such as unijunction transistors and switching diodes such as the Diac, silicon bilateral switch (SBS) and asymmetrical trigger switch. Let us examine how the Triac switches into conduction in each case.

1. MT_2 positive, positive gate current (Mode 1) When the gate current is positive with respect to MT_1, gate current flows mainly from the gate lead to the terminal MT_1 through the $P_2 - N_2$ junction, as shown in Fig. 5.7. The device turns on in the conventional manner as in the case of an SCR. However, in the case of a Triac, the gate current requirement is higher for turn-on at a particular voltage. Because of ohmic contacts of gate and MT_1 terminals on the P_2-layer, some more gate current flows from the gate lead G to the main terminal MT_1 through the semiconductor P_2 layer without passing through the P_2-N_2 junction. The main structure which ultimately turns ON through regenerative action is $P_1 \, N_1 \, P_2 \, N_2$. The P_2 layer is flooded with electrons when the gate current flows across the P_2–N_2 junction. These electrons diffuse to the edge of the junction J_2, and are collected by the N_1 layer. Therefore, the electrons build a space charge in the N_1 region and more holes from P_1 diffuse into N_1 to neutralize the negative space charge. These holes arrive at the junction J_2. They produce a positive space charge in the P_2 region which results in more electrons being injected from N_2 into P_2. This results in positive regeneration and ultimately the structure $P_1 \, N_1 \, P_2 \, N_2$ conducts the external current.

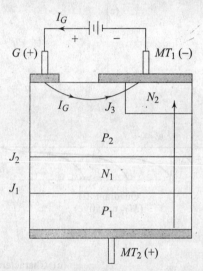

Fig. 5.7 MT_2 positive, positive gate current

2. MT_2 positive, negative gate current (Mode 2) A cross-sectional view of the structure is shown in Fig. 5.8.

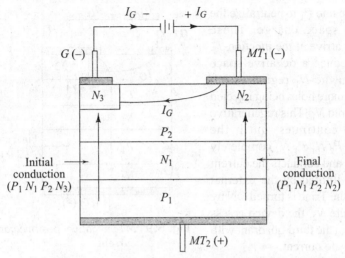

Fig. 5.8 MT_2 *positive, negative gate current*

When the terminal MT_2 is positive and gate terminal is negative with respect to terminal MT_1, gate current flows through P_2–N_3 junction and this gate current I_G forward biases the gate junction P_2–N_3 of the auxiliary $P_1\ N_1\ P_2\ N_3$ structure. As a result, Triac starts conducting through $P_1\ N_1\ P_2\ N_3$ layers initially. With the conduction of $P_1\ N_1\ P_2\ N_3$, the voltage drop across it falls but potential of layer between $P_2\ N_3$ rises towards the anode potential of MT_2. As the right hand portion of P_2 is clamped at the cathode potential of MT_1, a potential gradient exists across layer P_2, its left hand region being at higher potential than its right hand region. A current is thus established in layer P_2 from left to right which forward biased $P_2\ N_2$ junction and finally the main structure $P_1\ N_1\ P_2\ N_2$ begins to conduct. The device auxiliary structure, $P_1\ N_1\ P_2\ N_2$, may be considered as a pilot SCR, while the structure, $OP_1\ N_1\ P_2\ N_2$ may be regarded as the main SCR, both being built in one common structure. The anode current of the pilot SCR serves as the gate current for the main SCR. As compared with turn-on process discussed in above section, the device with MT_2 positive but gate current negative is less sensitive and therefore, more gate current is required.

3. MT_2 negative, positive gate current (Mode 3) When terminal MT_2 is negative and terminal MT_1 is positive, the device can be turned ON by applying a positive voltage between the gate and terminal MT_1. In this mode, the device operates in the third quadrant when it is triggered into conduction. The turn-on is initiated by remote gate control. The main structure that leads to turn-on is $P_2\ N_1\ P_1\ N_4$ with N_2 acting as a remote gate as shown in Fig. 5.9. The external gate current I_G forward biases $P_2\ N_2$ junction. Layer N_2 injects electrons into P_2 layer as shown by dotted arrows and are collected by the junction $P_2\ N_1$. The electrons from N_2 collected by $P_2\ N_1$ junction cause an increase of current through the junction $P_2\ N_1$. The holes injected from P_2, diffuse through N_1 and arrive in P_1. Hence, a positive space charge builds up in the P_1 region. More electrons from

N_4 diffuse into P_1 to neutralize the positive space charge. These electrons arrive at the junction J_2. They produce a negative space charge in the N_1 region which results in more holes being injected from P_2 into N_1. This regenerative process continues until the structure $P_2\ N_1\ P_1\ N_4$ completely turns ON and conducts the current which is limited by the external load. As the Triac is turned ON by remote gate N_2, the device is less sensitive in the third-quadrant with positive gate current.

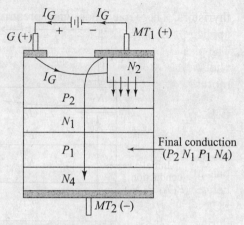

Fig. 5.9 MT_2 negative, positive gate current

4. MT_2 negative, negative gate current (Mode 4) A cross-sectional view of the structure is shown in Fig. 5.10. In this mode of operation, N_3 acts as a remote gate. The external gate current I_G forward biases $P_2\ N_3$ junction and electrons are injected as shown by the dotted arrows. These electrons from N_3 collected by $P_2\ N_1$ cause an increase of current across $P_1\ N_1$. The structure $P_2\ N_1\ P_1\ N_4$ turns ON by the regenerative action. The device turns ON due to the increased current in layer N_1. The device is more sensitive in this mode compared with the turn-on by the positive gate current as discussed above.

From the above four operating modes of Triac, it becomes clear that the sensitivity of the Triac is greatest in the first quadrant when turned ON with positive gate current and also in the third quadrant when turned ON with negative gate current. The sensitivity of the Triac is slightly lower in the first-quadrant when turned ON with negative gate current. Further, the Triac is much less sensitive in the third-quadrant with the positive gate current. Thus, the Triac is rarely operated in the first-quadrant with negative gate current and in the third-quadrant with positive gate current.

Because of the interaction between the two halves of the device, Triacs are limited in voltage, current, and frequency ratings as compared with conventional

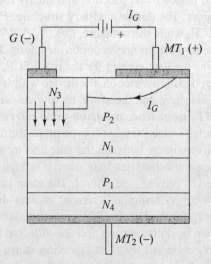

Fig. 5.10 MT_2 negative, negative gate current

thyristors. The Triac finds widespread use in consumer and light industrial appliances operating from 50 or 60 Hz a.c. supplies at moderate power levels. The plastic encapsulated Triac is a particularly cheap and compact device and is widely used for controlling the speed of single-phase a.c. series or universal motors, in such consumer appliances as food mixers and portable drills.

5.9.2 Commutation of Triacs

A Triac is equivalent to a pair of antiparallel connected SCRs. In an a.c. circuit the output voltage waveform with a Triac is the same as that obtainable with a pair of thyristors connected in antiparallel. However, one important difference between use of a pair of SCRs and use of a Triac in an a.c. circuit is that with SCRs, each SCR has an entire half cycle to turn-off, while a Triac must turn-off during the brief instant while the load current is passing through zero. For resistive loads, this is fairly simple to accomplish since the time available for a Triac to turn-off extends from the time the device current drops below holding current until the reapplied voltage exceeds the value of line voltage required to allow latching current. The task of commutating the Triac becomes more difficult with inductive loads.

The Triac circuit with an inductive load is shown in Fig. 5.11(a). The Triac voltage and current waveforms are also shown in Fig. 5.11(a). If we were to examine the waveforms at the current zero (i.e. at the turn-off point), a waveform such as Fig. 5.11(b) would be found.

From the current waveforms in Fig. 5.11 (b), it can be observed that the recovery current is active as a virtual gate current and trying to turn the device back ON. In addition there is a component to the reverse current which is due to the junction capacitance and the reapplied dv/dt. This component directly adds to the recovery current but does not appear until the Triac begins to block a voltage of opposite polarity. As the rate of removal of the current ($-di/dt$) decreases, the recovery current also decreases. This, then, implies that at lower values of di/dt, higher values of reapplied dv/dt are permissible for a given commutation capability. If by chance, the reapplied dv/dt is higher than the permissible value for a given (di/dt) rating, then additional protection circuits must be incorporated. The standard method is to use RC snubber such as R_s, C_s in Fig. 5.11(a). The values of R_s, and C_s depends upon the load current, line voltage and the reverse recovery charge of the Triac.

5.9.3 Advantages and Disadvantages of Triacs

As mentioned previously, a Triac is equivalent to a pair of antiparallel connected SCRs. In this section, we shall discuss the advantages and disadvantages of the Triac over an SCR.

(a) Advantages of Triac
1. Triacs can be triggered with positive or negative polarity voltages.

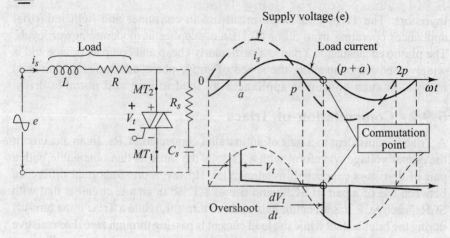

(a) Basic circuit and inductive load waveform

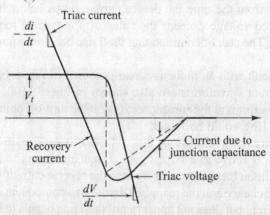

(b) Triac current and voltage at commutation

Fig. 5.11 *Commutation of Triac*

2. A Triac needs a single heat sink of slightly larger size, whereas antiparallel thyristor pair needs two heat sinks of slightly smaller sizes, but due to the clearance total space required is more for thyristors.
3. A Triac needs a single fuse for protection, which also simplifies construction.
4. In some d.c. applications, SCR is required to be connected with a parallel diode to protect against reverse voltage, whereas a Triac used may work without a diode, as safe breakdown in either direction is possible.

(b) Disadvantages of Triac
1. Triacs have low dv/dt rating compared to SCRs.
2. SCRs are available in larger rating compared to Triacs.
3. Since a Triac can be triggered in either direction, a trigger circuit with Triac needs careful consideration.
4. Reliability of Triacs is less than that of SCRs.

5.9.4 Phase-Control Using Triac

The versatility of the Triac and the simplicity of its use make it ideal for a wide variety of applications involving a.c. phase control. A phase-controlled circuit using Triac is shown in Fig. 5.12, where use is made of a Diac. Here, the Diac ensures that the Triac receives a clean, fast trigger pulse. During the positive half-cycle (when P is positive), the Triac requires a positive gate signal for turning it ON. This is provided by the capacitor C when its voltage is above the breakdown voltage of the Diac. The capacitor discharges through the Triac gate. When the Triac triggers, the voltage drop across PQ will be zero and the capacitor voltage will be reset to zero. A similar operation takes place in the negative half-cycle, and a negative gate pulse will be applied when the Diac breaks down in the reverse direction. The charging rate of capacitor C can be changed by varying the resistance R and hence the firing angle can be controlled.

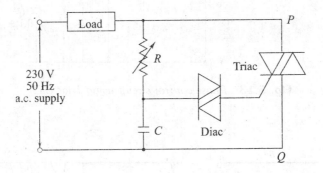

Fig. 5.12 *Basic Diac–Triac full-wave phase control*

In Fig. 5.12, the phase control is achieved by varying a resistor R. But, to have the control over the full range, from minimum to maximum power, the above simple RC timing circuit requires a very large change in the value of resistance R, presenting a low control gain. For manual control, this is most adequate. For systems which must perform a function, in response to some signal, the simple RC circuits are usually inadequate.

Figure 5.13 shows a conventional, manually controlled Triac circuit with a unijunction transistor. The control circuit is isolated from the mains. The diode bridge D_1–D_4 rectifies a.c. to dc. Resistor R_2 lowers rectified d.c. voltage to a suitable value for the zener diode and UJT. A zener diode clamps the control circuit voltage to a fixed level, as shown in Fig. 5.14. Since the peak-point voltage, e_p, of the unijunction transistor emitter is a fixed fraction of the interbase voltage, V_{BB}, as indicated by the dashed curve, the capacitor will charge on an exponential curve toward V_{BB} until its voltage reaches e_p. The UJT then turns ON and the capacitor discharges through the UJT emitter and primary of the pulse transformer. The resulting pulse via pulse-transformer and the current limiting resistor R_g triggers the Triac.

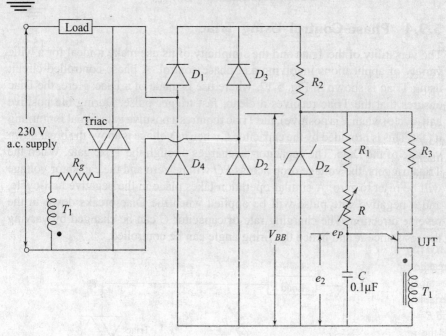

Fig. 5.13 *Phase-control circuit using Triac-UJT*

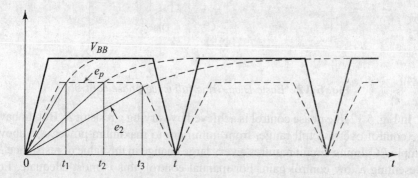

Fig. 5.14 *UJT voltage waveforms*

The instant in each half-cycle at which the Triac switches and the amount of power delivered to the load are determined by the setting of resistance R. The capacitor C should be chosen to suit the gate characteristic of the Triac in the range of 100 nF to 1 µf. In order to vary the firing point upto 180° in each half-cycle, the time constant $C(R + R_1)$ must exceed the time period of a half-cycle. For 50 Hz mains,

$$t = (R + R_1)\,C \geq 10 \text{ ms} \tag{5.1}$$

The resistance R_1 should limit the current to a value which prevents the UJT from remaining in conduction even after the capacitor has discharged.

$$\therefore \qquad R_1 > \frac{V_{BB\,max}}{I_{V\,min}} \qquad (5.2)$$

where I_V = valley current of the UJT.

Resistance R_3 provides a measure of temperature compensation for the point at which the UJT fires, and normally lies between 100 Ω and 1 Ω. Resistance R_g also helps making the circuit less sensitive to thermal variation and normally is below 100 Ω. This trigger method will fail with the inductive load which requires a maintained triggering (d.c. triggering), to built the current to a value exceeding the latching current of the Triac.

SOLVED EXAMPLE

Example 5.1 A Diac with a breakdown voltage of 40 V is used in a circuit such as that of Fig. 5.15, with R_1 variable from 1 kΩ to 25 kΩ, $C = 470$ nF₁ and $E = 240$ V RMS at 50 Hz. What will be the maximum and minimum firing delays with this arrangement?

Solution: Impedance of capacitor
$= 1/\omega C = 6773\ \Omega = |Z_c|$

The current through R_1 and C when the Diac is not conducting is

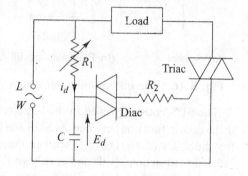

Fig. 5.15 *Triac firing circuit using a Diac*

$I_d = 240\sqrt{2}\ \sin(\omega t + \phi)/Z_d$.

where $\qquad Z_d = (R_1^2 + 1/\omega^2 C^2)^{1/2}$ and $\phi = \tan^{-1}(1/\omega RC)$

with $\qquad R_1 = 1000\ \Omega, Z_d = 6846\ \Omega$

The voltage across the capacitor is given by

$$V_c = I_d Z_c \qquad \therefore \qquad V_c = 335.8 \sin(\omega t - 8.4°)$$

When the Diac conducts, $V_c = 40$ V.

$\therefore$ Minimum delay $= \sin^{-1}\left(\dfrac{40}{335.8}\right) + 8.40 = 15.24°$

With $\qquad R_1 = 25$ kΩ, $Z_d = 25{,}901\ \Omega$.

The voltage across the capacitor is $I_d - Z_c$, as before

$$V_c = 88.76 \sin(\omega t - 74.84°)$$

When the Diac conducts, $V_c = 40$ V

Hence, Maximum delay $= \sin^{-1}(40/88.6°) + 74.84° = 101.6°$.

5.10 SILICON UNILATERAL SWITCH (SUS)

A silicon unilateral switch (SUS) is similar to a PUT, except for the fact that it has an internally built low-voltage avalanche diode between the gate and the cathode. The symbol for a SUS and its equivalent circuit are shown in Fig. 5.16. Its anode-to-cathode electrical characteristic is shown in Fig. 5.17 for no external connection to the gate terminal. Because of the presence of avalanche diode, SUS turns ON for a fixed anode gate voltage.

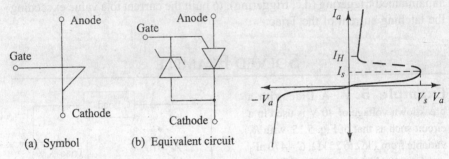

Fig. 5.16 *The silicon unilateral switch* **Fig. 5.17** *SUS characteristics curve*

The SUS is usually used in the basic relaxation oscillator circuit. The major difference in function between the SUS and UJT in relaxation oscillator circuitry is that the SUS switches at a fixed voltage, determined by its internal avalanche diode, rather than a fraction (η) of another voltage. Also, it should be noted that the switching current I_s is much higher in the SUS than in the UJT, and is also very close to I_H. These factors restrict the upper and lower limits of frequency or time delay which are practical with the SUS. For synchronization, lock-out, or forced switching, bias or pulse signals may be applied to the gate-terminal of the SUS.

5.11 SILICON BILATERAL SWITCH (SBS)

The silicon bilateral switch (SBS) is a device which essentially comprises two identical SUS structures, arranged in antiparallel, as shown in Fig. 5.18. As the name indicates, the device conducts in both directions when the applied voltage breaks the internal avalanche diode. The gate terminal is used only for external synchronization or for proper biasing. Since the device operates as a switch with both polarities of applied voltage, it is particularly useful for triggering the Triacs with alternate positive and negative gate pulses. The I–V characteristics of SBS is shown in Fig. 5.19.

5.12 SILICON CONTROLLED SWITCH (SCS)

The silicon controlled switch (SCS) is a commonly used low power *PNPN* device. It is essentially a miniature SCR with leads attached to all four of the semiconductor regions. The additional lead is connected to the *N* region below the anode *P* region. SCS has two gates, one anode gate like a PUT and another cathode gate like an SCR, as shown in Fig. 5.20.

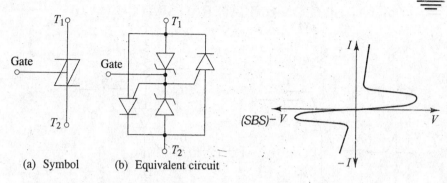

Fig. 5.18 *Silicon bilateral switch*   **Fig. 5.19** *SBS characteristic curve*

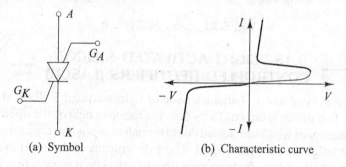

Fig. 5.20 *SCS characteristic curve*

The SCS can be triggered "on" or turned "off" by an appropriate signal at either gate. Refer to Fig. 5.21(a) in which a simple SCS circuit is shown. The cathode gate G_K is the normal gate used with SCRs. The anode gate G_A is the new gate. The SCS can be triggered "on" by either a positive pulse at G_K or a negative pulse at G_A. In the on-state, the SCS behaves like an SCR, namely, as a low-resistance with a voltage drop of typically 1V.

The SCS can be turned-off in any of the three ways:
(1) By reducing its anode current below I_H (same as SCR),
(2) By applying a negative pulse at G_K,
(3) By applying a positive pulse at G_A. Refer to Fig. 5.21(b).

Although the SCS can be turned-off at either gate, it suffers from relatively low turn-off gain similar to the gate-controlled switch. However, SCSs are low current devices designed for low current applications; therefore, this is not a serious drawback since low anode currents will require low gate currents for turn-off.

The SCS is used mainly in low power sensing circuits, timing and counting circuits and digital logic circuits. It is extremely versatile since it can be turned "on" and "off" by signals of either polarity.

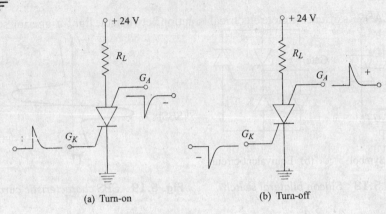

Fig. 5.21 *Basic SCS circuit*

5.13 LIGHT-ACTIVATED SILICON-CONTROLLED RECTIFIERS (LASCR)

The circuit symbol and V–I characteristic of light-activated SCRs is shown in Fig. 5.22. This device is turned ON by direct radiation of light on the silicon wafer. Electron-hole pairs which are created due to the radiation produce triggering current under the influence of electric field. The gate structure is designed to provide sufficient gate sensitivity for triggering from practical light sources (e.g. L_{ED} and to accomplish high di/dt and dv/dt capabilities). Once the LASCR is triggered to the "on" state, it behaves like a normal SCR. The LASCR will stay "on" even if the light disappears. It will turn "off" only if its anode current is decreased below I_H. The LASCR is most sensitive to light when its gate is open. This sensitivity can be varied by connecting a variable resistor between gate and cathode. In this way, the level of light at which the LASCR will trigger can be varied.

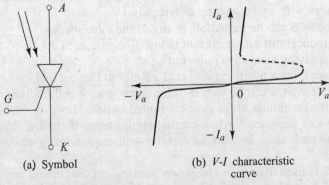

Fig. 5.22 *Light-activated SCRs*

The turn-on of a thyristor by optical means is an especially attractive approach for devices that are to be used in extremely high voltage circuits. LASCRs are used in high voltage and high current applications. [For example, high voltage d.c. (HVDC) transmission and static reactive power or volt-ampere reactive (VAR) compensation.]

LASCRs offer complete electrical isolation between the light-triggering source and the switching device of a power converter, which floats at a potential of as high as a few hundred kilovolts. The voltage rating of a LASCR could be as high as 4 kV at 1500 A with light triggering power of less than 100 mW. The typical di/dt is 250 A/ms and the dv/dt could be as high as 200 V/μs.

5.14 POWER MOSFETS

The power metal oxide semiconductor field-effect transistor (MOSFET) evolved from integrated circuit technology in the 1970s in response to the need to develop power transistors that can be controlled using much lower gate drive power levels compared to the existing power bipolar transistors. Power MOSFET is the most significant development in power semiconductor devices since the introduction of thyristors. It offers performances unavailable from bipolar transistors and thyristors, thus promises to not only offer better replacements for present-day devices but also lead to new circuit and systems concepts and bring power electronics into new areas of applications. Power MOSFETs have now gained a strong foothold in a multitude of applications, at power levels from a few watts to a few kilowatts in a few cases, to several hundreds of kilowatts. Typical uses include switching of linear power supplies, speed control of d.c. and a.c. motors, stepper motor controllers, relays, lighting controls, solenoid drivers, medical equipment, robotics, appliance controls, induction-heating, and instrumentation.

A power MOSFET is a voltage controlled device and requires only a small input current. In this device, the control signal is applied to a metal gate electrode that is separated from the semiconductor surface by an intervening insulator, typically silicon dioxide. The control signal required is essentially a bias voltage with no significant steady-state gate current flow in either the on-state or the off-state. Even during the switching of the devices between these states, the gate current is small at typical operating frequencies because it only serves to charge and discharge the input gate capacitance, the high input impedance is a primary feature of the power MOSFET that greatly simplifies the gate drive circuitry and reduces the cost of the power-electronics.

The power MOSFET is a unipolar device. Current conduction occurs through transport of majority carriers in the drift region without the presence of minority carrier injection required for bipolar transistor operation. In this device, during turn-off, no delays are observed as a result of storage or recombination of minority carriers. Their inherent switching speed is orders of magnitude faster than that for bipolar transistors. This feature is particularly attractive in circuits operating at high frequencies where switching power losses are dominant. The power MOSFET having operating frequencies are well above 100 kHz. The power MOSFETs switching timing is in order of 50–100 nanoseconds and can generate many kilowatts of power at frequencies up to 500 kHz.

5.14.1 Basic Structure

The MOSFET is the second category of field-effect transistor. It differs from the junction field-effect transistor (JFET) in that it has no *p-n* junction structure;

instead, the gate of the MOSFET is insulated from the channel by a silicon dioxide (SiO_2) layer.

There are two basic types of MOSFETs: depletion enhancement (DE) and enhancement only (E)

(1) Depletion enhancement MOSFET Figure 5.23 illustrates the basic structure of DE MOSFETs. The drain and source are diffused into the substate material and then connected by a narrow channel adjacent to the insulated gate. Both n-channel and p-channel devices are shown in figure.

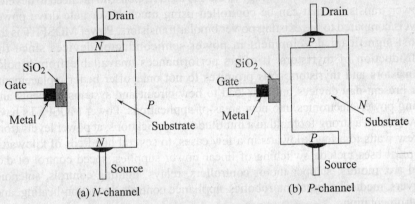

Fig. 5.23 *Basic structure of DE MOSFETs*

We will use the N-channel device to describe the basic operation. The P-channel operation is the same, except the voltage polarities are opposite to those of the N-channel.

The DE MOSFET can be operated in either of two modes: The depletion mode or the enhancement mode. Since the gate is insulated from the channel, either the positive or a negative gate voltage can be applied. The MOSFET operates in the depletion mode when a negative gate-to-source voltage is applied and in the enhancement mode when a positive gate-to-source voltage is applied.

(a) Depletion mode: Visualize the gate as one plate of a parallel plate capacitor and the channel as the other plate. The silicon dioxide insulating layer is the dielectric. With a negative gate voltage, the negative charges on the gate repel conduction electrons from the channel, leaving positive ions in their place. Thereby, the N-channel is *depleted* of some of its electrons, thus decreasing the channel conductivity. The greater the negative voltage on the gate, the greater the depletion of N-channel electrons. At a sufficiently negative gate-to-source voltage, V_{GS} (OFF), the channel is totally depleted and the drain current is zero. This depletion mode is illustrated in Fig. 5.24(a).

Like the N-channel JFET, the N-channel DE MOSFET conducts drain current for gate-to-source voltages between $V_{GS(off)}$ and zero. In addition, the DE MOSFET conducts for values of V_{GS} above zero.

(b) Enhancement mode: With a positive gate voltage, more conduction electrons are attracted into the channel, thus increasing (enhancing) the channel conductivity, as illustrated in Fig. 5.24(b).

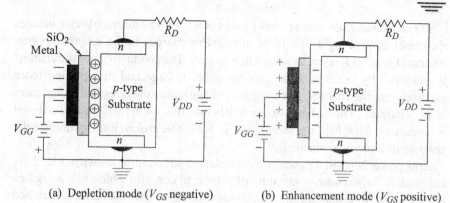

(a) Depletion mode (V_{GS} negative)   (b) Enhancement mode (V_{GS} positive)

Fig. 5.24 *Operation of n-channel DE MOSFET*

(c) Symbols: The schematic symbols for both the N-channel and the P-channel depletion enhancement MOSFETs are shown in Fig. 5.25. The substrate, indicated by the arrow, is normally (but not always) connected internally to the source. An inward substrate arrow is for N-channel, and an outward arrow is for P-channel.

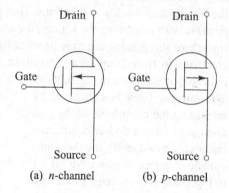

(a) *n*-channel   (b) *p*-channel

Fig. 5.25 *DE MOSFET schematic symbols*

(2) Enhancement MOSFET This type of MOSFET operates only in the enhancement mode and has no depletion mode. It defers in construction from the DEMOSFET in that it has *no physical channel*. Notice in Fig. 5.26(a) that the substrate extends completely to the SiO$_2$ layer.

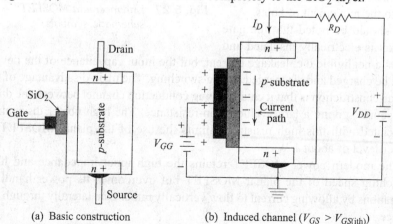

(a) Basic construction   (b) Induced channel ($V_{GS} > V_{GS(\text{ith})}$)

Fig. 5.26 *Enhancement MOSFET construction and operation*

For an n-channel device, a positive gate voltage above a threshold value induces a channel by creating a thin layer of negative charges in the substrate region adjacent to the SiO_2 layer, as shown in Fig. 5.26. The conductivity of the channel is enhanced by increasing the gate-to-source voltage and thus pulling more electrons into the channel. For any gate voltage below the threshold value, there is no channel. The schematic symbols for the n-channel and p-channel enhancement MOSFET are shown in Fig. 5.27. The broken lines symbolize the absence of a physical channel.

Low power MOSFETs usually have a planner structure, as shown in Fig. 5.26. Fabrication begins with a substrate of p-type silicon into which two n-regions are diffused. An insulating silicon dioxide layer, grown on the surface, is etched to allow the metallic source and drain connections to the n-regions. In the absence of gate bias, current cannot flow between the drain and source because of the two back-to-back p-n junctions. However, when the insulated gate is made positive with respect to the source, the electric field draws free electrons to the surface of the p-substrate, this process forms an n-type channel, which allows electrons to flow from source to drain and causes a lateral current flow, as indicated by the arrow in Fig. 5.26(b).

The current flow is enhanced by increasing the magnitude of the gate voltage to form a deeper conducting channel. Consequently, the n-channel enhancement mode planar MOSFET of Fig. 5.26(b) is a voltage-controlled current device in which conduction is entirely due to the movement of electrons. Conversely, it is possible to construct a p-channel MOSFET in which the conduction is entirely due to the movement of holes.

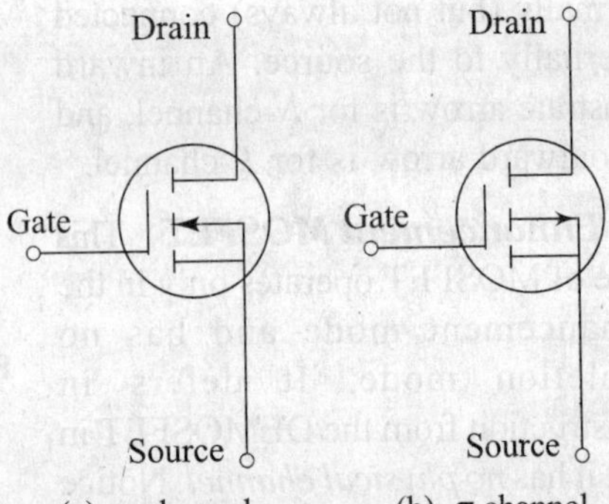

It should be noted that the gate contact is electrically insulated and

(a) v-channel   (b) π-channel

Fig. 5.27 *Enhancement MOSFET schematic symbols*

draws a negligible d.c. leakage current, but the input capacitance of the device must be charged and discharged during switching. The main disadvantage of the planar-construction is that it entails a long conduction channel between the drain and source, giving a large value of on-resistance. The high power dissipation associated with this high resistance limits the use of the planar MOSFET to power level of about 1 W.

The modern power MOSFET retains the high input impedance and high switching speed of the planar MOSFET but overcomes its power handling limitations by allowing current to flow vertically rather than laterally through the device.

5.14.2 Output Characteristics

The output characteristics, i.e. drain current I_D as a function of drain to source voltage V_{DS} with gate to source voltage V_{GS} as a variable parameter are as shown in Fig. 5.28. In this figure, the active, cut-off and ohmic regions are also shown. In power electronic applications, the MOSFET is used as a switch. Hence, device must be operated in the cut-off and ohmic region when turned-off and on respectively.

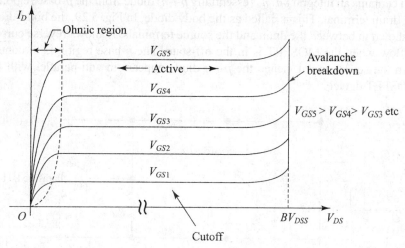

Fig. 5.28 *Output characteristics (I-V) of n-channel enhancement mode MOSFET*

The MOSFET is in the cut-off state when the gate-source voltage is less than the threshold voltage $V_{GS(th)}$. The device must withstand the applied voltage and to avoid the breakdown, the drain to source breakdown voltage should be greater than the applied voltage. When breakdown occurs, it is due to the avalanche breakdown of the drain body junction.

When a larger gate-to-source voltage is applied, the device is driven into the ohmic region where the drain to source voltage $V_{DS(on)}$ is small. In this operating region, the power dissipation can be kept reasonably low, by minimising the on-state voltage.

In the active region, the drain current I_D is almost independent of the drain-to-source voltage V_{DS}. As shown in Fig. 5.28, it is only dependent on the gate-to-source voltage V_{GS}. The power dissipation in the MOSFET is high in the active region.

The following points have been noted from the output characteristics:

(i) The MOSFETs are voltage controlled devices, i.e. the output current can be controlled by varying the gate-to-source voltage (V_{GS}).
(ii) The drain current will increase with increase in V_{GS}.
(iii) The gate-to-source voltage (V_{GS}) should be large enough to drive to MOSFET into ohmic region.

(iv) When the forward voltage applied to the MOSFET exceeds the breakdown voltage BV_{DSS}, the avalanche breakdown takes place. Operation above the breakdown voltage must be avoided.

(v) The second breakdown does not exist in MOSFETs.

5.14.3 Equivalent Circuit of Power MOSFET

The equivalent circuit of the power MOSFET is shown in Fig. 5.29. The MOSFET cell contains an integral $pn^- n^+$ (essentially p-i-n) diode from the p-base region to the drain terminal. This is called as the body diode. In Fig. 5.29, the body-diode is shown in between the drain and the source terminals. It allows reverse current to flow when the MOSFET is in the off-state. The p-base region is in contact with the source metal, hence the p-i-n diode appears in anti-parallel with the MOSFET device.

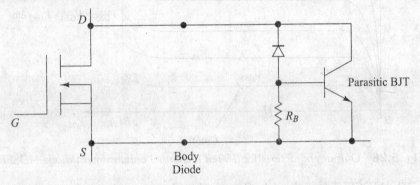

Fig. 5.29 *Equivalent circuit of power-MOSFET*

Often the integral diodes are required to conduct current that matches with the power-MOSFET rated current. The integral diode, however, exhibits very slow reverse recovery characteristics.

Besides the integral body diode, a parasitic BJT ($n^- pn^+$) also exists in the MOSFET structure. The p-region serves as the base of this parasitic BJT. The BJT must be kept cut-off at all times. This is done by shorting the base (p-body) to the emitter (n^+ source) by a common-metal covering.

The internal base of the transistor is connected to the source metal through a resistance R_B, the resistance of the lateral p-region. when the drain voltage is increased to near the avalanche breakdown voltage, current flows into the p-region of the BJT in addition to the normal MOSFET current through the n-channel. This avalanche current flows through the lateral p-region resistance R_B and produces a voltage drop that appears as the forward biasing voltage across the base to the emitter junction of the parasitic BJT. When this voltage increases beyond 0.7 V, the parasitic BJT is no longer capable of supporting the p-base/n^- collector breakdown voltage $BV_{DSS} = BV_{CBO}$.

A parasitic capacitance C_{gd} exists between the internal base of the BJT and the drain. The high rate of rise of the drain voltage dV_{DS}/dt causes a voltage transient between the base and the emitter of the parasitic BJT. The BJT turns-on and a latch-up condition develops. Thus, the parasitic BJT limits the maximum dV_{DS}/dt rating of the power MOSFET.

The reverse peak current through the body diode also flows through the resistance R_B during the turn-off transient in the MOSFET. The voltage drop across R_B forward biases the base to the emitter junction of the parasitic BJT, resulting in further increase in the reverse peak-current. In turn, the reverse-recovery time of MOSFET also increases.

5.14.4 Switching Performance

The power MOSFET are often used in switching applications because of their inherent high speed turn-on and turn-off capabilities. When used in this manner, they are maintained in either the on-state or forward blocking state for most of the time and must rapidly switch between these modes of operation. The high switching speed is accompanied by a very high rate of change of the drain voltage, which can cause an undesirable turn-on of the parasitic bipolar transistor in the power MOSFET. This can limit the switching speed and SOA of power MOSFET.

1. Transient analysis A typical switching circuit with a clamped inductive load (L_1) is shown in Fig. 5.30. As shown, the steady-state current flows through this inductive load. The inductance L_s is the stray inductance, not clamped by the diode D. First, consider the case of a step voltage V_{GA} applied at the gate-terminal G. The drain current flow is controlled by the gate voltage V_{GS} which is determined by the voltage across the gate-source capacitance C_{GS}. This voltage is governed by the charging of the capacitances C_{GS} and C_{GD} by resistor RG. No drain current will flow as long as the gate voltage V_{GS} remains below the threshold voltage V_T. The time taken for the gate voltage V_{GS} to reach the threshold voltage V_T represents a turn-on delay time. During this period, the input capacitance is simply ($C_{GS} + C_{GD}$) because the miller effect is operative only when the transistor is in its active region. The gate voltage which then rises exponentially is given by the equation:

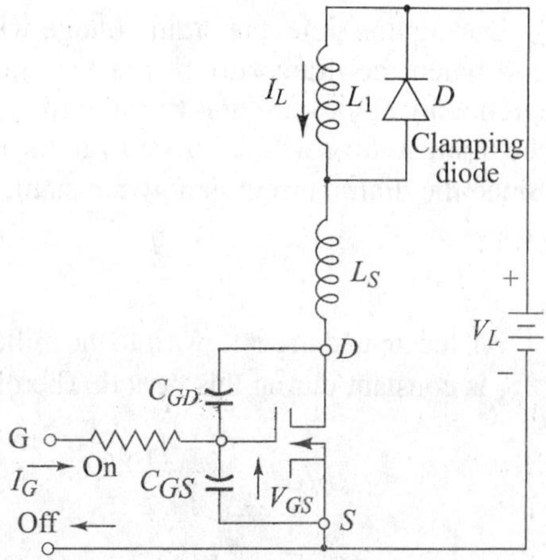

Fig. 5.30 Inductive switching circuit using power MOSFET for controlling load current

$$V_{GS} = V_{GA}\left\{1 - e^{-[t/R_G(C_{GS} + C_{GD})]}\right\} \quad (5.3)$$

From this expression, we can write the turn-on delay time as

$$t_{on} = R_G(C_{GS} + C_{GD})\ln\left[\frac{1}{1 - (V_T/V_{GA})}\right] \quad (5.4)$$

Beyond the turn-on delay time, the drain current flow begins to occur. Now, consider the linear transfer characteristics of power MOSFET which are used for transient analysis, shown in Fig. 5.31. As shown, the drain current will increase in proportion to the gate voltage. Since the device is now in its active region, the miller effect will determine the gate–drain capacitance that is being charged by the gate circuit. If the stray inductance is small, the Miller effect is small and the gate voltage will continue to rise exponentially as described by Eq. (5.5). The drain current will then have the form

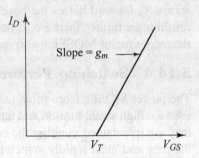

Fig. 5.31 Linear transfer-characteristics

$$I_D = g_m\left\{V_{GA}\left[1 - e^{-[t/R_G(C_{GS} + C_{GD})]}\right] - V_T\right\} \quad (5.5)$$

During this time, the drain voltage will be linearly constant. This period will end when the drain current reaches the load current I_L, that is, when all the circulating diode current is transferred to the power MOSFET. Beyond this point, the drain voltage will fall from V_L to the on-state voltage of the power MOSFET. Since the drain current is now constant, the gate voltage must be

$$V_{GS} = V_T + \frac{I_L}{g_m} \quad (5.6)$$

All the input current flows into the miller capacitance C_{GD}, since the gate voltage V_{GS} is constant during this period. Therefore, the input current is given by

$$I_G = \frac{V_G - V_{GS}}{R_G} \quad (5.7)$$

Thus,

$$\frac{dV_{GD}}{dt} = \frac{I_G}{C_{GD}} = \frac{V_G - (V_T + I_L/g_m)}{R_G \cdot C_{GD}} \quad (5.8)$$

Since the gate source voltage is constant during this period, the rate of change of the drain-source voltage is equal to the rate of change of the gate-drain voltage. Thus,

$$\frac{dV_D}{dt} = \frac{V_G - (V_T + I_L/g_m)}{R_G \cdot C_{GD}} \quad (5.9)$$

Integrating the above equation yields,

$$V_D = V_L - \left[\frac{V_G - (V_T + I_L/g_m)}{R_G \cdot C_{GD}}\right]t \quad (5.10)$$

The drain voltage will then decrease linearly with time. Figure 5.32 shows the composite gate and drain waveforms. As shown, the gate voltage may continue to rise beyond time t_3, but this will have no influence on the drain current or voltage because they have reached their steady-state values.

The above discussed analysis is based on the assumption that the ratio (L_s/R_G) must be small. During the interval t_2 and t_3, the high power dissipation due to the device turn-on transient occurs, and both high current and voltage are sustained by the device simultaneously. Also, during this period the device current–voltage locus must stay within its SOA.

For device turn-off, the corresponding analysis can be performed. In this case, an abrupt drop in applied gate voltage from V_G to zero is assumed to occur at the terminal G. The voltage V_{GS} then decreases exponentially with time as a result of discharging of the gate capacitance, until the gate voltage reaches the value required to obtain a saturated drain current equal to the load current I_L.

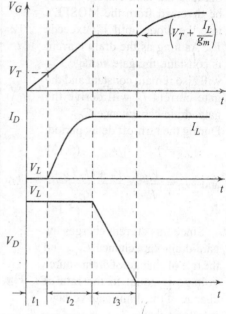

Fig. 5.32 Turn-on transients of power MOSFET for the case of small stray inductance (low LS/RG ratio)

During this period, the voltage V_{GS} is given by

$$V_{GS} = V_G \cdot e^{-(t/R_G \cdot C_G)}$$

Figure 5.33 shows the waveforms for turn-off transient of power MOSFET. The duration of this turn-off delay phase is given by the condition,

$$V_{GS}(t_4) = V_T + \frac{I_L}{g_m} = V_G \cdot e^{-(t_4/R_G \cdot C_G)} \quad (5.11)$$

From this equation, the turn-off delay time (t_4) is obtained as

$$t_4 = R_G C_G \ln\left[\frac{V_G}{V_T + (I_L/g_m)}\right] \quad (5.12)$$

Beyond the turn-off delay time, the drain current will remain at I_L, whereas the drain voltage will begin to rise toward V_L because the load current cannot be diversed from the MOSFET into the diode until V_D exceeds V_L. As long as the drain current is constant, the gate voltage V_{GS} will also remain constant and the gate current I_G will derive the gate–drain capacitance C_{GD}. During the turn-off delay period,

$$V_{GS} = V_T + I_L/g_m \quad (5.13)$$

and $\quad I_G = \dfrac{V_{GS}}{R_G} = \dfrac{V_T + (I_L/g_m)}{R_G} \quad (5.14)$

Since this current charges the gate-drain capacitance C_{GD} and the rate of change-of drain-source voltage is equal to the rate of change of the drain-gate voltage, it follows that

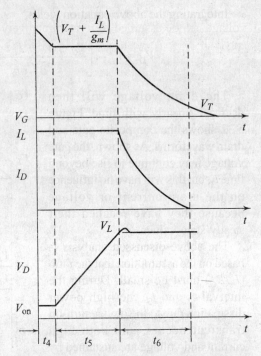

Fig. 5.33 *Turn-off transients of a power MOSFET for the case of small stray inductance (low L_s/R_G ratio)*

$$\frac{dV_{DS}}{dt} = \frac{dV_{DG}}{dt} = \frac{I_G}{C_{GD}} \quad (5.15)$$

Thus, $\quad V_{DS} = V_{on} + \dfrac{I_G}{C_{GD}} \cdot t \quad (5.16)$

Also, $\quad V_{DS} = V_{on} + \dfrac{1}{R_{GD} \cdot C_{GD}} \left(\dfrac{I_L}{g_m} + V_T \right) t \quad (5.17)$

During this period, from the on-state voltage drop V_{on} to the load voltage, the drain-voltage will rise linearly. The duration of this period can be obtained by equating V_{DS} to V_L:

$$t_5 = \frac{R_G \cdot C_{GD} \cdot (V_L - V_{on})}{(I_L/g_m) + V_T} \quad (5.18)$$

Freewheeling diode D will turn-on at this instant. To achieve this, it is essential that the MOSFET drain voltage exceeds the load supply voltage V_L. If the stray inductance L_S is small, the overshoot in the drain voltage will be small. However, if L_S is large and the rate of change of the drain current becomes large, the voltage on the MOSFET drain can exceed its breakdown voltage. The drain voltage can be assumed to remain relatively constant if the stray inductance is small. The gate voltage will then continue to decrease exponentially as given by the equation,

$$V_G = \left(\frac{I_L}{g_m} + V_T\right) e^{-(t/R_G \cdot C_G)} \tag{5.19}$$

The drain current will follow this change in the gate voltage, therefore,

$$I_D = (I_L + g_m V_T) e^{-(t/R_G \cdot C_G)} - g_m V_T \tag{5.20}$$

This period will extend until the gate voltage reaches the threshold voltage V_T and the drain current is reduced to zero. The time interval for this period is given by the expression

$$t_6 = R_G \cdot C_G \ln\left(\frac{I_L}{g_m V_T} + 1\right) \tag{5.21}$$

The gate voltage will continue to decay exponentially to zero after this interval. This will have no influence on the drain current of voltage, as they are at their steady-state voltages (off-state).

The power dissipation during the turn-off transients occurs primarily during the periods t_5 and t_6, where both the current and voltage are large. Once again, the current-voltage locus must remain within the SOA of the device. It is noted that during period T_6, the drain voltage overshoots. Its magnitude depends on the size of the stray inductance. If the stray inductance is large, the MOSFET can be forced into avalanche breakdown, causing destructive failure. The analysis presented here also indicates the need to keep the gate–drain capacitance small. This can be achieved by avoided the overlap of the gate electrode with the drift region or by using a thick oxide in the portion where the drift region comes to the surface.

2. (dv/dt) capability Power MOSFETs can be forced into current conduction, when a high dv/dt occurs at the drain terminal. In certain cases, this will lead to destructive failure of the devices. The various mechanisms that can lead to (dv/dt) induced turn-on are discussed in this section. Figure 5.34 shows the equivalent circuit used for analysis of (dv/dt) induced turn-on in power MOSFETs. As shown, ramp voltage is applied between drain and source. In addition to the device capacitances, the equivalent circuit shows the parasitic N-P-N bipolar transistor with a base emitter shunting resistance R_B. This shunting resistance is present in the device structure as a result of the overlapping of the source metal over both the N^+ emitter and the p-base regions.

Mode 1:   For the first mode of (dv/dt)-induced turn-on, consider the current I_1 flowing by means of the gate-to-drain capacitance C_{GD} into the gate circuit resistance R_G. If the drain voltage is much larger than the gate voltage, the voltage drop across the gate resistance will be approximately given by the equation:

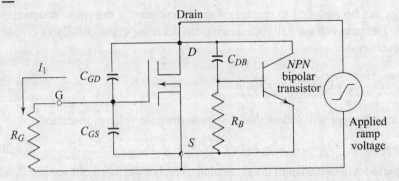

Fig. 5.34 *Equivalent circuit used for analysis of (dv/dt) induced turn-on in power MOSFETs*

$$V_{GS} = I_1 R_G \qquad (5.22)$$

$$= R_G \cdot C_{GD} \left(\frac{dv}{dt}\right) \qquad (5.23)$$

If the gate voltage V_{GS} exceeds the threshold voltage of the power MOSFET, the device will be forced into current conduction. The (dv/dt) capability set by this mode is given by

$$\left[\frac{dv}{dt}\right] = \frac{V_T}{R_G \cdot C_{GD}} \qquad (5.24)$$

The (dv/dt) capability can be increased by using a very low impedance gate drive circuit and raising the threshold voltage. Since the threshold voltage decreases with increasing temperature, this mode of turn-on can become aggravated with increasing power dissipation within the device. In general, it is non-destructive because the gate-voltage does not rise much above the threshold voltage and the device current is limited by the high device resistance.

Mode 2: This mode of (dv/dt)-induced turn-on occurs as a result of the existence of the parasitic bipolar transistor. At high rates of change of the drain voltage, a current I_2 flows by means of the capacitance C_{DB} into the base short-circuiting resistance R_B. If this current is sufficient for the base–emitter junction of the bipolar transistor to become forward-biased, it will turn-on the transistor. For low value of R_B, when the emitter junction is inactive, the breakdown voltage of the transistor, and hence the power MOSFET, approaches the collector–base breakdown voltage BV_{CBO} as shown in Fig. 5.35. But under a high (dv/dt) and a large values of R_B, the emitter–base junction becomes strongly forward-biased. The breakdown voltage then collapses to the level of the open-base transistor breakdown voltage, BV_{CEO}, which is about 60% of the collector–base breakdown voltage BV_{CBO}. If the applied drain voltage is greater than BV_{CEO}, the device will enter the avalanche breakdown or may be destroyed by second breakdown if the drain current is not externally limited.

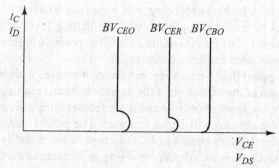

Fig. 5.35 *Breakdown characteristics of power MOSFET induced by turn-on of the parasitic n-p-n bipolar transistor*

The (dv/dt)-induced turn-on in this mode is dependent on the internal device structure. Consider the D_{mos} structure shown in Fig. 5.36 with one end of the N^+ emitter short circuited to the p-base region. The applied (dv/dt) creates a displacement current flow in capacitance C_{DB}. This current must flow laterally in the p-base region to the source metal. It produces a voltage drop along the resistance R_B that forward-biases the edge A of the N^+ emitter. When the edge A of the N^+ emitter becomes forward-biased by the lateral current flow, the bipolar transistor will turn-on, thus precipitating further current flow. An estimate of the (dv/dt) at which this will occur can be obtained by assuming that the bipolar transistor will turn ON at an emitter–base forward bias voltage of V_{BE}:

$$\frac{dv}{dt} = \frac{V_{BE}}{R_B \cdot C_{DB}} \qquad (5.25)$$

where R_B is a distributed base resistance.

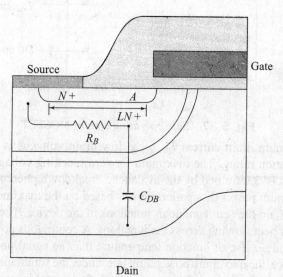

Fig. 5.36 *Cross section of power MOSFET with capacitance C_{DB} and base resistance R_B shown*

To obtain a high (dv/dt) capability, it is important to keep the resistance R_B small. This can be achieved by increasing the doping level of the p-base and keeping the length L_N^+ of the N^+ emitter as small as possible within the constraints of the lithography used for device fabrication.

It should be noted that p-base sheet resistance increases with increasing drain voltage as a result of the extension of the depletion layer, resulting in lowering of the dv/dt capability. In addition, increasing the temperature will lower the voltage V_{BE} at which the emitter will begin to inject. The p-base resistance will also increase with temperature because of a reduction in the mobility. These factors will cause a reduction in the (dv/dt) capability as temperature increases.

5.14.5 Safe Operating Area

Safe operating area (SOA) defines the limit of operation of the device on the I_D–V_{DS} characteristics, shown in Fig. 5.37. The following parameters determine the safe operating area:

$I_{D(max)}$: the maximum drain current

P_{max} : the maximum power dissipation and

V_{BDSS} : the forward blocking voltage.

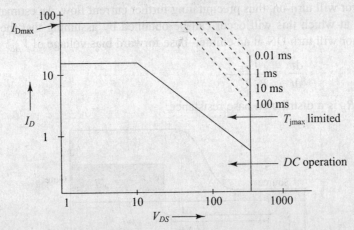

Fig. 5.37 *Safe operating area of MOSFET*

The maximum drain current $I_{D(max)}$ at low drain voltages is limited by the power dissipation rating. The maximum forward blocking voltage V_{BDSS} at low drain currents is determined by the avalanche breakdown phenomenon.

The maximum power dissipation limit is based on the maximum permissible temperature T_j on the semiconductor junctions in the device. The T_j depends on the amount of heat flowing across the junctions. A continuous power dissipation will cause a larger rise of junction temperature than an equal power for a short duration. Hence, the maximum power limit line, under the simultaneous application of high drain voltage and high drain current, is shifted upwards for shorter duration applications.

The SOAs in MOSFETs are large square regions. The large square region of SOA means a minimal snubber circuit.

5.14.6 Parallel Connection of MOSFETs

The power MOSFET is constructed like parallel connected MOSFET cells. Hence, it tends itself naturally for parallel connection for increasing the forward current-carrying capability. Also, MOSFET behaves like a resistance in the on-state and has a positive temperature coefficient. It means that in the parallel connection say two power MOSFETs if one of them draws more current than the other, its internal temperature will increase, this will increase its on-state resistance, to automatically reduce the current through it.

Hence, when a number of MOSFETs are connected in parallel for current sharing, dropping out of any device does not lead to more currents, the consequent thermal runaway and breakdown of the remaining devices. This phenomenon is the opposite of that encountered in power BJTs.

While paralleling the power MOSFETs it is necessary to take precautions to keep the parasitic oscillations as low as possible. For this, the resistance in series with gate-terminals must be chosen properly, also the gate driving circuit with a low output impedance must be used. Figure 5.38 shows the parallel connection of the power MOSFETs. In such direct paralleling of devices, inspite of all the precautions taken to select the devices with almost all identical parameters, the equal current sharing will not take place. Some of the devices will draw more current than the others. This is known as current unbalance. The factor which indicates the worst case current unbalance is known as the unbalance factor (α).

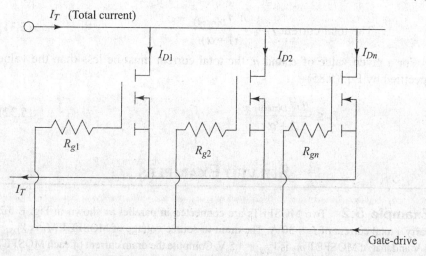

Fig. 5.38 *Parallel convection of power MOSFETs*

Unbalance Factor (α) Let n be the number of devices connected in parallel to share the total current I_T. Therefore, the average equal current shared by each device $I_{Daverage}$ is given by

$$I_{Dav} = I_T/n \tag{5.26}$$

To derive equation for α, the worst case condition is consider that is when any one device is taking maximum current say $I_{DC(max)}$, the factor of unbalance is then defined as

$$\alpha = \frac{I_{D(max)} - I_{D(av)}}{I_{D(av)}}, \text{ or } \alpha = \frac{I_{D(max)} - I_{D(av)}}{I_{D(av)}} \times 100\% \tag{5.27}$$

The value of α varies between 0 and 1 or between 0 and 100%. Ideally, the value of α must be 0. But, practically with increase in the value of α, the value of $I_{D(max)}$ increases. This will increase the possibility that the device carrying $I_{D(max)}$ may get damaged. In order to avoid the permanent damage to the device, the $I_{D(max)}$ must never exceed the maximum rated current of the device. That means,

$$I_{D(max)} \leq I_{D(rated)} \tag{5.28}$$

If α is known, then care can be taken to limit the total current to a lower value of increase the number of devices n

Substitute $I_{D(max)} = I_{D(rated)}$ in equation (5.81), we get

$$\alpha = \frac{I_{D(rated)} - I_{D(av)}}{I_{D(av)}} = \frac{I_{D(rated)}}{I_{D(av)}} - 1 \tag{5.29}$$

But $I_{D(av)} = \dfrac{I_T}{n} \quad \therefore \quad \alpha = \dfrac{I_{D(rated)}}{I_T/n} - 1 = \dfrac{n \cdot I_{D(rated)}}{I_T} - 1 \tag{5.30}$

$\therefore$ The total current $I_T = \dfrac{n \cdot I_{D(rated)}}{(1+\alpha)} \tag{5.31}$

For a given value of α and n, the total current must be less than the value specified by Eq. (5.85)

$$\therefore \quad I_T \leq \frac{n \cdot I_{D(rated)}}{(1+\alpha)} \tag{5.32}$$

SOLVED EXAMPLES

Example 5.2 Two MOSFETs are connected in parallel as shown in Fig. E 5.2, carry a total current of $I_T = 30$ A. The drain-to-source voltage of MOSFET M_1 is $V_{DS_1} = 4$ V and that of MOSFET m_2 is $V_{DS_2} = 4.5$ V. Compute the drain current of each MOSFET and differences in current sharing if the current sharing series resistances are:

(a) $R_{S_1} = 0.4\ \Omega$ and $R_{S_2} = 0.3\ \Omega$, and (b) $R_{S_1} = R_{S_2} = 0.7\ \Omega$.

Solution:

(a) Total current, $I_T = I_{D_1} + I_{D_2}$ and $V_{DS_1} + I_{D_1} R_{S_1} = V_{DS_2} + I_{D_2} R_{S_2}$

Also, $R_{S_2} I_{D_2} = R_{S_2}(I_T - I_{D_1})$ ∴ $V_{DS_1} + I_{D_1} R_{S_1} = V_{DS_2} + R_{S_2}(I_T - I_{D_1})$

or, $V_{DS_2} + I_{D_1}(R_{S_1} + R_{S_2}) = V_{DS_2} + R_{S_2} I_T$.

∴ $I_{D_1} = \dfrac{V_{DS_2} - V_{DS_1} + R_{S_2} \cdot I_T}{R_{S_1} + R_{S_2}} = \dfrac{4.5 - 4 + 0.3 \times 30}{(0.4 + 0.3)} = 13.57$ A or 45.23%

∴ $I_{D_2} = 30 - 13.57 = 16.43$ or 54.77%

Therefore, difference in current sharing,

$\Delta I = I_{D_2} - I_{D_1} = 16.43 - 13.57 = 2.86$ A

or $\Delta I = 54.77 - 45.23 = 9.54\%$

(b) $I_{D_1} = \dfrac{4.5 - 4 + 0.7 \times 30}{0.7 + 0.7} = 15.36$ A or 51.2%

∴ $I_{D_2} = 30 - 15.36 = 14.64$ A or 48.8%

∴ $\Delta I = 51.2 - 48.8 = 2.4\%$.

Fig. E 5.2

Example 5.3 Two MOSFETs are connected in parallel to share 120 A d.c. The device capacitances and estimated stray inductances are:

$C_{gs1} = 2200$ pf, $C_{gd1} = 390$ pf, $L_{g1} = 2$ nH

$C_{gs2} = 2700$ pf, $C_{gd2} = 330$ pf, $L_{g2} = 2.3$ nH

Determine the series resistance in the gate that would enable equal voltage sharing at high frequencies. Also, compute the safe highest frequency.

Solution:

The oscillating frequencies of the MOSFET is given by

$$f_1 = \dfrac{1}{2\pi\sqrt{L_{g1}(C_{gs1} + C_{gd1})}} = \dfrac{1}{2\pi\sqrt{2 \times 10^{-9}(2200 + 390) \times 10^{-12}}} = 6.99 \times 10^7 \text{ Hz}$$

and $$f_2 = \dfrac{1}{2\pi\sqrt{2.3 \times 10^{-9}(2700 + 330) \times 10^{-12}}} = 6.628 \times 10^7 \text{ Hz}$$

We would place a low-pass filter such that it could be at least a decade lower in frequency to effectively equalize the gate voltages at high frequencies.

Hence, the desired highest frequency of operation would be $f_n = 6$ MHz.

The series resistance in gate terminals of both MOSFETs is given by

$$R_g = \dfrac{1}{2\pi(C_{gs1} + C_{gd1})f_h} = \dfrac{1}{2\pi(2200 + 390) \times 10^{-12} \times 6 \times 10^6} = 10.24 \text{ ohm} \approx 11 \text{ ohm}$$

Example 5.4 In Fig. E 5.4 determine the power dissipated by each MOSFET.

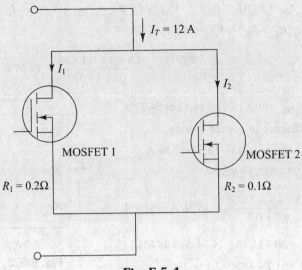

Fig. E 5.4

Solution:
Apply the current divider rule,

$$I_1 = \frac{R_2}{R_1 + R_2} \cdot I_T \quad \therefore \quad I_1 = \frac{0.1}{0.3} \cdot 12 = 4 \text{ A}$$

$$I_2 = 12 - 4 = 8 \text{ A}$$

$$\therefore \quad P_1 = I_1^2 \cdot R_1 = (4)^2 \times 0.2 = 3.2 \text{ W} \quad \text{and} \quad P_2 = I_2^2 \cdot R_2 = (8)^2 \times 0.1 = 6.4 \text{ W}$$

5.14.7 Gate Drive Design Considerations of the MOSFET

MOSFET is a voltage-controlled device; that is, a voltage of specified limits must be applied between gate and source in order to produce a current flow in the drain. Since the gate-terminal of the MOSFET is electrically isolated from the source by a silicon oxide layer, only a small leakage current flows from the applied voltage source into the gate. Thus, we can say that the MOSFET has as extremely high gain and high impedance. While designing gate-driving circuit, following factors are considered:

(i) In order to turn a MOSFET-on, a gate to source voltage pulse is needed to deliver sufficient current to charge the input capacitor in the desired time. The MOSFET input capacitance C_{iss} is the sum of the capacitors formed by the metal-oxide gate structure, from gate-to-drain (C_{gd}) and gate-to-source (C_{gs}). Thus, the driving voltage source impedance R_g must be very low in order to achieve high transistor speeds.

The driving impedance and required driving current can be obtained from the following equations:

$$R_g = \frac{t_r \ (or \ t_f)}{2.2 \ C_{iss}} \tag{5.33}$$

and
$$I_g = C_{iss}\frac{dv}{dt} \qquad (5.34)$$

where, R_g = generator impedance, Ω, C_{iss} = MOSFET input capacitance, pF,

dv/dt = generator voltage rate of change, V/ns

$t_r(t_f)$ = MOSFET rise time (fall time), ns

(ii) Excessive voltage applied to the gate of a power MOSFET may cause breakdown of oxide layer isolating gates and therefore permanent damage. The driving circuit must be designed so as not to cause any excessive voltage and any voltage transient.

(iii) The gate driving circuit must be able to sense and control the fault current, otherwise let-through fault currents can be much higher before self-limiting occurs (due to positive temperature coefficient) and can be destructive to the MOSFET.

(iv) When MOSFET is operated at very high frequencies, certain design precautions must be taken in order to minimize the problems, especially oscillations. Figure 5.39 shows a typical MOSFET driving a resistive load, working in the common-source mode.

There are basically two simple design rules which will present MOSFETs from oscillating when used in high frequencies. First, minimize all lead lengths going to the MOSFET terminals, especially the gate lead. If short leads are not possible, then designer may use a ferrite bead or a small resistor R_1 in series with the MOSFET, as shown in Fig. 5.39. Either, one of those elements when placed closed to the transistor gate will suppress parasitic oscillations.

Second, because of the extremely high input impedance of the MOSFET, the driving source impedance must be low in order to avoid positive feedback which may lead to oscillations. Also, we must note at this stage that while the d.c. input impedance of the MOSFET is very high, its dynamic or a.c. input impedance varies with frequency. Therefore, the rise and fall times of the MOSFET depend on the driving generator impedance.

Equation 5.33 is valid if $R_L \gg R_g$. This information, along with the fact that there are no storage or delay times associated with the MOSFET, allows the rise and fall times to be set by the designer. The resistor R_2 in the circuit of Fig. 5.39 is used to assist transistor turn-off.

(v) An optocoupler can be used for isolating the control-circuits from the power circuits.

(vi) A current sensing circuit can be used to protect the MOSFET against over drain current.

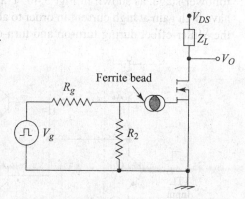

Fig. 5.39 MOSFET driving a resistive-load

Solved Example

Example 5.5 MOSFET is used for driving a resistive load with following parameters:

$$C_{iss} = 470 \text{ pf}, R_g = 200 \text{ }\Omega \quad \text{and} \quad R_L = 2.2 \text{ k}\Omega$$

Calculate the rise time of the driving waveform.

Solution:

We have the equation, $R_g = \dfrac{t_r}{2.2 \, C_{iss}}$

$\therefore \qquad t_r = (2.2)(200)(470 \times 10^{-12}) = 206.8$ ns

5.14.8 Gate Drive Circuits

A MOSFET does not require gate drive power; however, it require, the transfer of charge to and from the gate terminal for turning ON and OFF respectively.

5.14.8.1 Driving the MOSFET from TTL

Although it is possible to drive the MOSFET directly from the output of some transistor-transistor logic (TTL) families, direct driving is not recommended, since the transistor stays in the linear region for a long time before reaching saturation. Thus, the performance of the MOSFET may never reach its optimum point with such a gate drive.

A buffer circuit must be provided, in order to improve the switching performance (decreases rise and fall times) which will present very fast current sourcing and sinking to the gate-capacitances. Such a simple circuit is a complementary emitter follower-stage, as shown in Fig. 5.40. Transistors Q_1 and Q_2 must be chosen to have high gain at high current in order to able to deliver the current demanded by the Miller-effect during turn-on and turn-off.

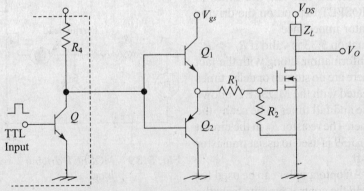

Fig. 5.40 *Emitter-follower driving circuit*

The current flowing in each buffer transistor at turn-on (Q_1) and turn-off (Q_2) is given by

$$I_{charge} = \frac{C_{gs} \cdot V_{Ds}}{t_r} \tag{5.35}$$

and
$$C_{gs} = C_{iss} - C_{rss} \tag{5.36}$$

where C_{gs} = gate-to-source capacitance, pF, C_{iss} = input capacitance, pF,
C_{rss} = reverse-transfer capacitance, pF, V_{gs} = gate-to source voltage, V
t_r = input pulse rise time, ns.

Assume that the gate-to-drain capacitance discharges at the same time.

$\therefore \quad t_r' = t_f$

Therefore, the discharge current is given by

$$I_{dis} = \frac{C_{rss} V_{Ds}}{t_r} \tag{5.37}$$

where V_{DS}-drain to source voltage, V.
The power-dissipated in each buffer-transistors is given by

$$P = V_{CE} I_C \, t_r f \tag{5.38}$$

where V_{CE} = buffer transistor saturation voltage, V
I_C = buffer transistor collector current, A
f = transistor switching frequency, kHz

A high current integrated buffer (DS0026) may be used to interface TTL levels to a MOSFET instead of discrete transistors, thus considerably improving the switching times, Fig. 5.41.

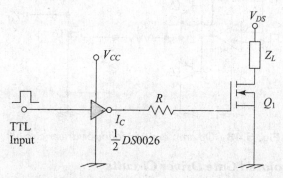

Fig. 5.41 *High current integrated buffer*

5.14.8.2 Driving the MOSFET from CMOS

Because of the MOSFET high input impedance, it may be directly driven by a CMOS gate, as shown in Fig. 5.42 (a). In order to achieve faster switching times, more than one CMOS gate may be paralleled, as shown in Fig. 5.42(b), to increase current availability to the MOSFET input capacitances.

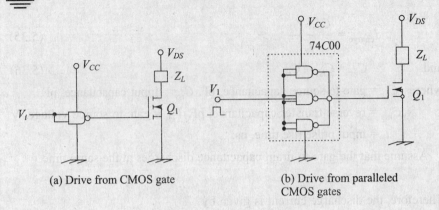

(a) Drive from CMOS gate (b) Drive from paralleled CMOS gates

Fig. 5.42 *CMOS drive circuits*

5.14.8.3 Driving the MOSFETs from Linear Circuits

It is conceivable that the MOSFET may be driven directly from the output of an operational amplifier, such as power op-amp capable of delivering high output current. However, the limiting factor is the slow slew rate of a power-op-amp, which limits the operating bandwidth to less than 25 kHz. In order to improve both bandwidth and slew rate to make the op-amp usable in driving the MOSFET, an emitter follower buffer may be used. Figure 5.43 shows the typical driving circuit using an op-amp.

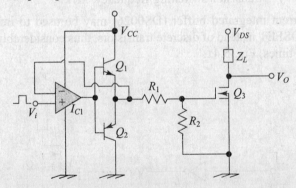

Fig. 5.43 *Op-amp emitter follower driver circuit*

5.14.8.4 Isolated Gate Driver Circuits

Figure 5.44 shows a pulse transformers based driver circuit. The pulse transformer provides the isolation needed to drive different MOSFETs at different voltage levels or to control an n-channel MOSFET driving a grounded or source-follower load. The size of the transformer reduces significantly when the operating frequency becomes very high (100 kHz or more).

Power Semiconductor Devices

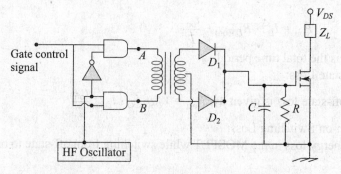

Fig. 5.44 *Pulse-transformer based drive circuit*

Since current (hence power) requirement for the gate-driver-circuit is very small at a steady state condition (few microwatts), a separate driver circuit and additional power supply are not required. However, during turn-on or turn-off transitions, a large current is required for fast switching. Another limiting factor is the switching capability of the diode. Thus Schottky diode can be used whose turn-off time is very small (0.23 μsec). Here high-frequency carrier signal reaches the logic gates when the input drive control signal is high. The output voltage of logic gates becomes alternately high and low. Therefore, current flows in the primary of the transformer from A to B and vice-versa. The voltage generated in the secondary winding of transformer is rectified and filtered. This d.c. voltage is used for driving a MOSFET.

Figure 5.45 shows an opto-coupler based gate driver circuit. Since the output current of the photo-transistor is not sufficient to drive a power-MOSFET, an additional transistor is used for amplification.

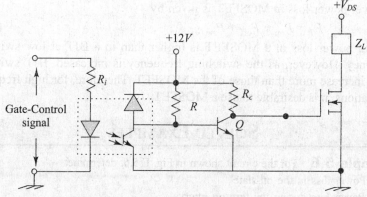

Fig. 5.45 *Opto-coupler based drive circuit*

5.14.9 MOSFET Losses

Following are the sources of power-losses in the switching MOSFET:
(i) On-state Loss (conduction loss):
   A MOSFET has relatively high-on-state losses given by:

$$P_{on} = I_D^2 \cdot R_{DS(on)} \cdot \frac{t_{on}}{T} \qquad (5.39)$$

where T is the total time period

(ii) Off-state Loss:

The off-state loss is given by, $P_{off} = V_{DS(max)} \cdot I_{DSS} \dfrac{t_{off}}{T}$ \qquad (5.40)

(iii) Turn-on Switching Loss:

The energy loss in the MOSFET while switching from off-state to on-state is given by

$$P_{sw_{on}} = \frac{V_{DS(max)} \cdot I_D \cdot t_r}{6} \qquad (5.41)$$

$$= [P_{sw_{on}} = (0.637) \times (1/2\, V_{DS(max)}) \cdot \left(\frac{1}{2} I_D\right) \cdot t_r\,]$$

(iv) Turn-off Switching Loss:

The energy loss in the MOSFET when it switches from the on-state to the off-state is given by

$$P_{sw_{off}} = \frac{V_{DS(max)} \cdot I_D \cdot t_f}{6} \qquad (5.42)$$

(v) Switching Power Loss:

The switching power loss is given by

$$P_{sw} = (P_s W_{on} + P_s W_{off}) \cdot f \qquad (5.43)$$

where f is the switching frequency

(vi) Total Power-Loss in MOSFET:

Tower power-loss in MOSFET is given by

$$P_T = P_{on} + P_{off} + P_{sw} \qquad (5.44)$$

Total power loss in a MOSFET is higher than in a BJT at low switching frequency. However, as the switching frequency is increased, BJT switching losses increase more than those of the MOSFET. Therefore, for high frequency applications, it is desirable to use a MOSFET.

SOLVED EXAMPLES

Example 5.6 For the circuit shown in Fig. E 5.6, determine:
(a) Power-loss in the on-state
(b) Power-loss during the turn-on interval
MOSFET parameters are: $t_r = 2\,\mu s$, $R_{DS(on)} = 0.2\,\Omega$, duty cycle $D = 0.7$ and $f = 30$ kHz.

Solution

Drain current given by, $I_D = \dfrac{V_{DS}}{R_L + R_{DS(on)}} = \dfrac{100}{12 + 0.2} = 8.2$ A

(a) Switching period $T = \dfrac{1}{f} = 1/30 = 33.33\ \mu sec$.

On-time is given by t_{on} = D.T. = $0.7 \times 33.33 \times 10^{-6} = 23.33\ \mu s$

Energy loss during on-time, $W_{on} = I_D^2\, R_{DS(on)} \cdot t_{on}$

$= (8.2)^2 \times (0.2)\,(23.33 \times 10^{-6}) = 313.74\ \mu J$

Now, power loss during on-time $P_{on} = W_{on} \cdot f$
$= 313.74 \times 10^{-6} \times 30 \times 10^3 = 9.41$ W

(b) Energy loss during turn-on, $W_{on} =$

$\dfrac{100 \times 8.2}{6} \times 2 \times 10^{-6} = 273.33\ \mu J$

Power loss during turn-on, $P_{on} = W_{on} \cdot f =$
$273.33 \times 10^{-6} \times 30 \times 10^3 = 8.2$ W

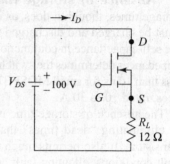

Fig. E 5.6

Example 5.7 Calculate the total power loss for the MOSFET having following para-meters:

$V_{DS} = 120$ V, $I_D = 4$ A, $t_r = 80$ ns, $t_f = 120$ ns, $I_{DSS} = 2$ mA, $R_{DS(on)} = 0.2\ \Omega$, duty-cycle $D = 50\%$, $F_{switching} = 45$ kHz.

Solution

(i) Total period $T = 1/f = 22.22\ \mu s$

(ii) Period $T_{on} = t_{off}$ (for 50% duty-cycle) = $11.11\ \mu sec.$

(iii) On-state loss, $P_{on} = \dfrac{(4^2) \times 0.2 \times 11.11 \times 10^{-6}}{22.22 \times 10^{-6}} = 1.6$ W

(iv) Off-state loss, $P_{off} = \dfrac{120 \times 2 \times 10^{-3} \times 11.11 \times 10^{-6}}{22.22 \times 10^{-6}} = 0.120$ W

(v) Turn-on switching power loss, $P_{W_{on}} = \dfrac{120 \times 4 \times 80 \times 10^{-9}}{6} \times 45 \times 10^3 = 0.288$ W

(vi) Turn-off switching power-loss, $P_{W_{off}} = \dfrac{120 \times 4 \times 120 \times 10^{-9}}{6} \times 45 \times 10^3 = 0.432$ W

(vii) Total power loss, $P_T = P_{on} + P_{off} + P_{W_{on}} + P_{W_{off}} = 1.6 + 0.120 + 0.288 + 0.432 = 2.44$ W

5.14.10 Comparison between Power MOSFETs and Bipolar Transistors

Power MOSFETs have a number of major performance advantages over bipolar transistors. These are discussed in the following sections:

1. *MOSFETs are voltage controlled* To switch a MOSFET ON, it is necessary simply to apply a voltage, typically 10 V for "full enhancement" between gate and source. The gate is isolated by silicon oxide from the body of the device, and the d.c. gain is virtually infinite. Drive power is negligible, and drive circuitry is generally considerably simpler than for a bipolar transistor. The drive currents for the bipolar must be supplied continuously during the whole conduction period, whereas gate current for the MOSFET flows only for the short periods needed to charge and discharge the self-capacitance, during the turn-on and turn-off conditions.

2. Absence of storage time The MOSFET has no inherent delay and storage times, though it does, as mentioned above, have self-capacitance which must be charged and discharged when switching. The time constant formed by the self-capacitance in conjunction with gate circuit and drain and source circuit impedances determines the switching times. Practical switching times range from less than 10 ns for small MOSFETs rated 1 A or less, to 50–100 ns for the largest ones, rated 10 to 30 A.

The absence of storage time increases circuit utilization factors by reducing or eliminating "dead times" that must be built into circuits that use bipolar transistors. It also permits much faster response when reacting to overload and fault conditions, allowing fault current to be arrested much more effectively as shown in Fig 5.46.

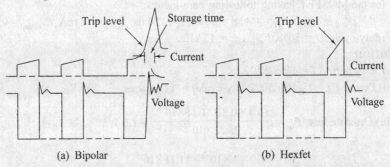

Fig. 5.46 *Response to overload in switching power supply (a) bipolar transistor (b) HEXFET*

3. High peak current capability The gain of a bipolar transistor decreases with increasing current, but the transconductance of a MOSFET increases with increasing current. High peak current in a bipolar transistor tends to "pull it out of saturation" and destroy it through overheating. The on-resistance of a MOSFET does increase with increasing current but the effect is much more being than with a bipolar, and the MOSFET has a much higher peak current carrying capability. For example, a 400 V HEXFET with a continuous drain current rating of 5.5 A has a usable peak-current of 20 A. A comparable bipolar transistor would have a usable peak current of perhaps 7.5 A. The underlying limitation on current handling capability of a HEXFET is junction heating. It is able to handle peak current well in excess of its continuous current rating, just so long as the rated maximum junction temperature is not exceeded.

4. Wide safe operating area The SOA of a power MOSFET is much better than that of a bipolar. The MOSFET, being a majority carrier device, has a positive temperature coefficient of resistance, and is immune from the hot-pot formation and second breakdown phenomenon that plague the bipolar transistor.

MOSFET are, therefore, generally much more rugged than bipolar, and snubber clamp circuitry can be smaller and less dissipative, as shown in Fig. 5.47.

MOSFET data sheets usually show SOA curves that cover current and voltage values upto the rated I_{DM} and V_{DS} values, respectively. Typical SOA data is based (for HEXFET) simply on junction temperature rise, and is for a case temperature of 25°C, and an internal dissipation that raises the junction to the rated maximum value of 150°C. SOA curves for each pulse duration follow a line of constant power, and are actually nothing more than a graphical statement of the absence of second breakdown.

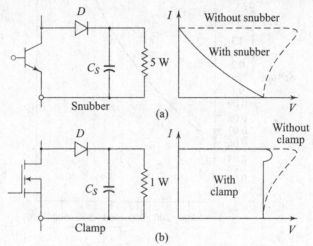

Fig. 5.47 *Comparison of snubber/clamp circuitry for* **(a)** *bipolar and* **(b)** *MOSFET*

5. At high frequency MOSFETs are more efficient The on-resistance of a MOSFET is one of its key characteristics because it determines the devices conduction losses in switching applications. For a given chip size, on-resistance increases with voltage rating. The 500 V rated power MOSFETs have on-resistance in the order of 0.4 Ω at 25°C. The corresponding conduction voltage drop at rated usable continuous current of 10 A is 4 V at 25°C, and about 8 V at a junction temperature of 150°C. Figure 5.48 shows a typical relationship between on-resistance and voltage rating for a HEXFET chip having dimensions of 6.5 mm × 6.5 mm.

Comparisons between power MOSFETs and bipolars often centre around the fact that the conduction voltage of a MOSFET is higher than for a bipolar, and becomes progressively more so as voltage rating increases. Conduction losses of a MOSFET when operating near rated current are therefore generally greater than for a bipolar. Switching losses of a MOSFET, on the other hand, are almost negligible, while the switching losses of a bipolar are often greater than its conduction losses, particularly at a high frequency. Therefore, the bipolar is usually more energy efficient at low frequency, while the MOSFET is more energy efficient at high frequency. The frequency crossover point depends upon specific circumstances, but is generally somewhere between 10 and 40 kHz.

It would not be correct to conclude from the above discussion that power MOSFETs will find use only in higher frequency applications, and will not be the preferred choice at lower frequency. Although MOSFETs certainly "shine" at high frequency, their higher conduction losses actually are often inconsequential when viewed from the standpoint of the overall system design. Considerations, such as circuit simplification, ruggedness, cost effectiveness and reliability of the overall system often favour the MOSFET, even in low frequency applications.

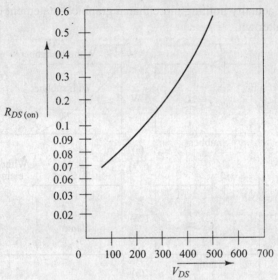

Fig. 5.48 *Relationship between on-resistance and voltage rating*

Today, MOSFET prices are close to those of bipolars, and overall system costs are frequently lower using MOSFETs because of the circuit simplification that result from their use. Price is much less a consideration than it is used to be and will become even less so in the future. Designers, no longer shackled by overriding cost considerations, are turning in droves to MOSFETs for the technical advantages they offer. **These advantages** are summarised as follows:

(a) Quality and reliability of MOSFETs are an order of magnitude better than for bipolar transistors, providing much better system reliability.
(b) Device circuitry is simpler and cheaper, and often can be standardized for a whole range of products.
(c) Overload and peak current handling capability are high. MOSFETs are generally much more rugged and "forgiving" than bipolars.
(d) MOSFETs are easy to parallel for higher current.
(e) MOSFETs are able to operate in hazardous radiation environments, and are suitable for space and military use.
(f) Its fast switching speeds permit much higher switching frequencies, much better efficiency at higher frequency, and often much smaller overall circuit size and weight.
(g) Absence of second breakdown reduces snubber circuitry in switching applications, and gives more power handling capability in linear applications.

(h) MOSFETs have more linear characteristics and have better temperature stability/giving better performance in linear applications, and reducing complexity of feedback circuitry.

(i) MOSFETs leakage current is relatively low, typically in the order of nanoamperes. This is important in typical applications such as automatic test equipment and relay switching.

(j) Drain source conduction threshold voltage is absent, eliminating electrical noise in sensitive a.c. switching applications.

(k) MOSFETs are easy to design. Their operation is "clean" and predictable, and is easily analyzed.

The above discussed major performance advantages of power MOSFET over bipolar transistors can be summarised in tabular form as:

Power BJT	Power MOSFET
1. BJT is a minority as well as majority carrier device.	1. MOSFET is a majority carrier device.
2. BJT is a current controlled device.	2. MOSFET is a voltage-controlled device.
3. BJT has negative temperature coefficient.	3. MOSFET has positive temperature coefficient.
4. BJT cannot operate at very high frequency.	4. MOSFET can operate at higher frequencies.
5. BJT has different shapes for the FBSOA and RBSOA.	5. FBSOA and RBSOA are identical. SOA is much better than that of BJT.
6. Second breakdown can take-place.	6. No possibility of second breakdown.
7. Peak-current capability is less than that of MOSFET.	7. Peak current capability of MOSFET is higher than that of BJT.
8. BJTs are less sensitive to voltage spikes than MOSFETs.	8. MOSFETs are more sensitive to voltage spikes than BJT.
9. The on-state voltage is lower than that of power-MOSFET. Therefore, the on-state loss is lower.	9. The on-state voltage is higher than that of power BJT. The on-state loss is higher than BJT.
10. Conduction losses of a BJT are less than that of MOSFET.	10. Conduction losses of a MOSFET are greater than BJT.
11. Switching losses are more.	11. Switching losses are less.
12. More energy efficient at low frequency	12. More energy efficient at high frequency.

5.15 INSULATED GATE BIPOLAR TRANSISTORS (IGBTs)

Another development in power MOS technology is the insulated gate bipolar transistor (IGBT). These devices combines the best qualities of power MOSFET and bipolar transistor. It has an input characteristics of a MOSFET and an output characteristic of a bipolar transistor. That means it has high input impedance and low on-state conduction loss. But it has no second breakdown problem like the bipolar transistors.

An IGBT is a voltage controlled device similar to power MOSFET, however, the turn-off time of an IGBT is significantly greater than that of a power MOSFET. It shares many of the appealing features of power MOSFET, such as ease of gate-drive, peak current capability, and ruggedness.

Other names for this device include conductivity modulated field effect transistor (COMFET), insulated gate transistor (IGT), GEMFET, bipolar mode MOSFET or bipolar MOS transistor.

5.15.1 Basic Structure of IGBT

Figure 5.49 (a) shows the vertical cross-section of n-channel IGBT. Like other devices, IGBT also uses the vertically oriented structure in order to maximize the area available for the current flow. This will reduce the resistance offered to the current flow and hence the on-state power loss taking place in the device. The IGBT also uses highly interdigitated gate-source structure in order to reduce the possibility of source/emitter current crowding. The doping levels used in different layers of IGBT are similar to those used in the comparable layers of the vertical MOSFET structure except for the body region. The main difference in the structure of IGBT as compared to that of a MOSFET is the existance of p^+ layer that forms the drain of the IGBT. It is also possible to make P-channel IGBT by changing the doping type in each of the layers of the device.

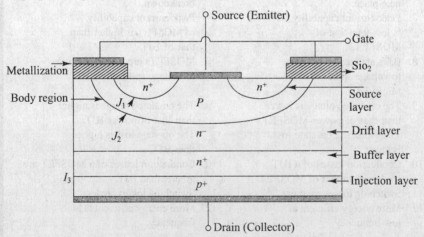

Fig. 5.49 (a) *Vertical cross section of IGBT*

The n^+ buffer layer between the p^+ drain contact and the n^- drift layer is not essential for the operation of the IGBT. IGBTs with buffer layer are termed as **punch-through (PT)** IGBTs whereas the IGBTs without buffer layer are termed as **non-punch-through (NPT)** IGBTs.

It is worth pointing out that the IGBT structure (Fig. 5.49(a)) contains a P-N-P-N thyristor structure between the collector and emitter terminals.

Figures 5.49(b) and 5.49(c) shows the circuit symbols of an IGBT. As shown, it has three terminals, namely, gate, collector (drain), and emitter (source).

In a p-channel IGBT, the directions of the arrowheads would be reversed.

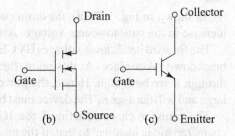

Fig. 5.49 (b) *Cross-sectional view and circuit symbols of an IGBT*

The symbol shown in Fig. 5.49(b) is essentially the same as that used for the n-channel power MOSFET, but with the addition of an arrowhead in the drain lead pointing into the body of the device indicating an injecting contact. The symbol shown in Fig. 5.49(c) is used if IGBT is considered to be basically a BJT with MOSFET gate input. This device has a collector and emitter rather than a drain and source. This symbol indicates that the IGBT has output characteristics similar to power BJT and input characteristics similar to the power MOSFET. We have used this symbol throughout the text book.

5.15.2 IGBT Voltage-Current Characteristics

The V-I characteristics of the n-channel IGBT is shown in Fig. 5.50 (a). In the forward direction they are similar to those of the logic level bipolar transistor, the only difference here is that the controlling parameter is the gate-to-source (emitter) voltage (V_{GS}), and the parameter being controlled is the drain (collector) current. When there is no voltage applied to the gate, the IGBT is in the off-state. In this state, the current [i_D or i_C] is zero and the voltage across the IGBT is equal to the source voltage. If a voltage $V_{GS} > V_{GS(th)}$ is applied to the gate, the device turns ON and allows current i_D to flow. This current is limited by the source-voltage and the load resistance. In the on-state, the voltage across the IGBT drops to zero.

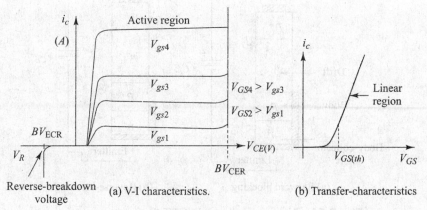

(a) V-I characteristics.   (b) Transfer-characteristics

Fig. 5.50 *IGBT-characteristics*

As shown in Fig. 5.50 (a), the drain current (collector current) increases with increase in the gate-to source voltage. Also, V_{GS} is positive.

The forward breakdown voltage (BVCER) is the voltage at which the avalanche breakdown takes place. At this point, the voltage across the device and current through it are both high. Hence, the power dissipated in the device will be very large and will damage it. The device must be therefore operated below this voltage.

The transfer characteristic of the IGBT is shown in Fig. 5.50 (b). This characteristic is identical to that of the power MOSFET. The characteristic curve is nonlinear for low values of collector current whereas it is fairly linear over most of the range where the collector current is higher. Therefore, in order to operate the IGBT in the linear region of this characteristics, it is necessary to apply V_{GS} greater than $V_{GS(th)}$.

5.15.3 Symmetric and Asymmetric IGBTs

IGBTs are classified on the basis of n^+ buffer layer. They are:
  (i) Symmetric IGBTs (Non punch-through IGBTs)
  (ii) Asymmetric IGBTs (punch-through IGBTs)

5.15.3.1 Symmetric IGBTs (Non Punch-Through IGBTs)

Figure 5.51 shows the structure of the non punch-through (NPT) IGBT. As shown, n^+ buffer layer is absent in the structure. In the forward blocking mode Fig. 5.51 (a), the collector (drain) is positive with respect to emitter (source). Therefore, the junction J_1 and J_3 are forward biased and hence can't decide the blocking capability of the device. The junction J_2 is reverse-biased and hence decides the blocking capability of the device. The junction J_2 is formed between moderately doped P (body) layer and lightly doped n^- (drift) layer. The forward blocking capacity is decided by the thickness of the n^- layer.

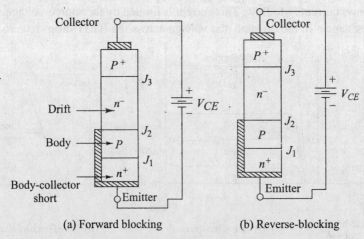

Fig. 5.51 *Non-punch-through IGBT (Buffer layer absent)*

When a reverse-voltage is applied between the collector and the emitter, Fig. 5.51 (b), junction J_2 is forward-biased and junction J_1 and J_3 are reverse-biased and the reverse-blocking capacity is decided by them. Since the source layer n^+ is heavily doped and p-layer (body-layer) is thin and moderately doped, the junction J_1 does not have a capacity to blocks voltage. However, the junction J_3 will block all the reverse-voltage due to the presence of n^- drift layer on its one side. Since the drift-layer is designed to block large voltages, the reverse blocking voltage of the non-punch through structure will be comparable to the forward blocking voltage. This way, the non-punch through IGBT has a symmetrical blocking capacity and they can be used in rectifier circuits.

5.15.3.2 Asymmetric IGBTs [Punch-through IGBTs]

Figure 5.52 shows the structure of the punch-through IGBT. As shown, the n^+ buffer layer is used between the p^+ collector injection layer and n^- drift layer.

If the forward voltage is applied, Fig. 5.52 (a), junctions J_1 and J_3 are forward biased and junction J_2 is reverse-biased. The forward blocking capacity is therefore decided by junction J_2. As this junction is formed between body layer (p) and drift layer (n^-), the thickness of the drift layer will decide the forward blocking capacity. This is identical to non-punch-through structure.

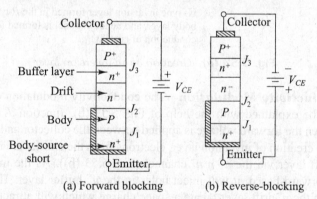

(a) Forward blocking (b) Reverse-blocking

Fig. 5.52 *Punch-through IGBTs*

When a reverse voltage is applied between collect and emitter, Fig. 5.52 (b), junction J_2 is forward biased and junctions J_1 and J_3 are reverse-biased and will have to block the reverse-voltage. But J_1 cannot block large voltages as explained in Section 5.15.3.1 and now due to the presence of n^+ buffer layer J_3 also cannot block a large reverse-voltage. Therefore, the device will breakdown at low reverse-voltages. Thus, the punch-through structure will have an asymmetric blocking capacity. They have faster turn-off times and are used for inverter applications.

Punch-through IGBT structures are more popular and are widely used.

5.15.4 Operating Principle

The operating principle of an IGBT is similar to that of MOSFET. The operation can be divided into two parts as follows:

(a) Creation of an Inversion Layer The operation of an IGBT is based on the principle of creation of an inversion layer. When the gate-to-source voltage (V_{GS}) greater than $V_{GS\,(threshold)}$ is applied, n-type inversion layer is created, beneath the SiO_2 layer, as shown in Fig. 5.53 (a). Due to the formation of n-type of layer in the p-type body layer, a channel ($n^+\,n\,n^-$) is formed which helps to establish the current as shown in Fig. 5.53 (a).

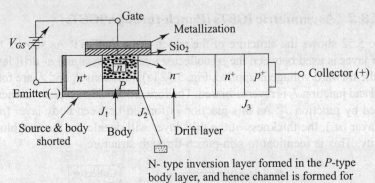

Fig. 5.53 (a) *Creation of an inversion layer*

(b) Conductivity Modulation The conductivity modulation of the drift-layer can be explained with the help of Fig. 5.53 (b). Junction J_3 is forward-biased when the forward voltage is applied between the collector and the emitter. Due to the creation of inversion layer, electrons from the emitter are injected into the n^- drift layer via the $n^+\,n\,n^-$ channel (Fig. 5.53 (b)). As the junction J_3 is already forward-based, it will inject holes in the n^+ buffer layer. The electrons injected in the n^- drift layer create a space charge which will attract holes from the n^+ buffer layer which were injected by the p^+ layer. This way, double injection takes-place into the n^- drift region from both sides as shown. This increases the

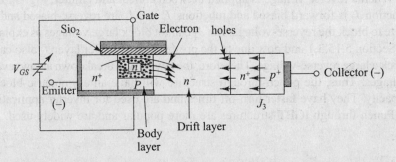

Fig. 5.53 (b) *Conductivity modulation of the drift layer in IGBT*

conductivity of the drift region and reduces the resistance to its minimum. In this way, the conductivity modulation reduces the on-state voltage across the IGBT.

The main difference between the MOSFET and the IGBT is that, there is no conductivity modulation of drift layer in MOSFET. Therefore, the on-state resistance $R_{DS(on)}$ and hence the on-state power-loss is very high in MOSFET. However, in IGBTs the conductivity modulation of the drift layer reduces the on-state resistance and hence the on-state power-loss. Hence, the on-state losses in IGBT are less than that in MOSFET.

5.15.5 Equivalent Circuit

An equivalent circuit for modelling the operation of the IGBT is shown in Fig. 5.54. This equivalent circuit consists of a coupled P-N-P and N-P-N transistor pair representing the **four layer parasitic thyristor structure** with a MOSFET shunting the upper-N-P-N transistor. This structure may give rise to latching up of the device to either a high-current mode or a zero current mode. The emitter of transistor Q_1 is termed the collector of the IGBT. The collector of Q_1, which is also the emitter of the transistor Q_2, is termed the emitter of the IGBT. The main collector current is carried by the PNP transistor Q_1. The input side of the IGBT is like MOSFET therefore, the MOSFET has been included in the equivalent circuit.

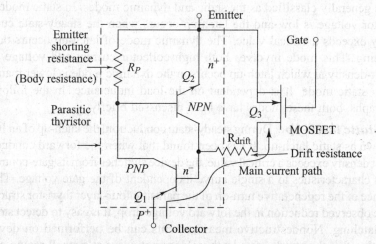

Fig. 5.54 *Equivalent circuit of an IGBT*

The two resistances that appear in the equivalent circuit of the IGBT are the drift region resistance and the body spreading resistance R_P. The drift region resistance is considerably large as the drift layer is very lightly doped and therefore has low conductivity. The resistance R_P is due to the later component of the current flowing through the IGBT structure.

If the body spreading resistance (R_P) is significant then it will turn-on the n-p-n transistor which will further turn-on the parasitic thyristor structure and get latched into the on-state. In this way, the latch-up takes place. Once the parasitic thyristor gets latched, it cannot be turned-off as the gate does not have any control over it. The only way to turn-off the IGBT is by way of forced-commutation, exactly similar to the conventional SCR. If the latch-up is not terminated quickly, the IGBT will be destroyed due to the excessive power-dissipation.

In order to prevent the turn-on of the parasitic-thyristor, R_P should be kept minimum. For this, various modified structures of the IGBT have been suggested. One of them is to use a heavy doping of the body region to decrease the resistance. The second one is to use a hole current bypass structure.

5.15.6 Latch-Ups in IGBTs

As discussed in previous section, that, the IGBT structure contains a parasitic *P-N-P-N* thyristor structure between the collector and the emitter terminals. The presence of this parasitic four-layer thyristor structure in the IGBT creates the possibility of the device latching up by regenerative action. This mode of operation is highly undesirable because it leads to loss of control of the collector current by the applied gate voltage. Once the device has latched up, it can be turned off only by either externally turning off the collector voltage or reversing its polarity. Latch-up usually produces catastrophic failure of the devices as a result of excessive heat dissipation in d.c. circuits. The latch-up modes of IGBT can be generally classified as the static and dynamic modes. In static mode, the collector voltage is low and the latch-up occurs when the steady-state current density exceeds a critical value. The dynamic mode of latch-up occurs during switching. This mode involves both high collector current and voltage. The current-density at which latch-up occurs in the dynamic mode is lower than that for the static mode. It is dependent on the load inductance. In the following paragraphs, both these modes have been discussed briefly.

(i) Static latch-up During steady-state conduction, the latch-up of an IGBT is known as static latch-up. It has been found that when the forward conduction current density exceeds a critical value, the device switches from its gate-controlled output characteristics to a single curve independent of the gate voltage. This is evidence of the regenerative turn-on of the parasitic four-layer thyristor structure. By the observed reduction in the forward voltage drop, it is easy to detect steady-state latching. Nondestructive measurements can be performed on devices, indicating that the steady-state latch-up does not occur at a small region of the device but it is distributed over most of the active area.

It has been found that the steady-state latch-up current is given by the relation

$$I_{L,ss} = \frac{0.7}{\alpha_{PNP} \cdot R_p} \tag{5.45}$$

where $I_{L,ss}$ = steady-state latching current,
 0.7 = forward voltage drop across the N^+-P junction (J_1),
 α_{PNP} = current gain of the *P-N-P* transistor.
 R_p = resistance of the *P*-base under the N^+ emitter.

From Eq. (5.45), it becomes clears that the latching current can be increased by reducing the current gain α_{PNP} and the resistance of the P-base region. It has also been found experimentally that the latching current decrease nearly inversely with increasing emitter length L_E.

If the P-N-P-transistor current gain is high, the hole current (I_h) passing through the P-base is a large fraction of the collector current. The latching current will then be low. When the lifetime in the N-base is reduced to increase switching speed, the α_{PNP} also decreases. This raises the current I_h at which α_{NPN} will be sufficiently high to create-latch-up. Thus, lifetime reduction has two effects on the static latch-up:
(1) it reduces the magnitude of the hole-current flowing into the P-base region (i.e. the ratio (I_h/I_c) decreases), and
(2) it reduces the current gain of the P-N-P transistor. These two effects combine to raise the static latching current density.

(ii) Dynamic latch-up Under switching conditions, IGBTs have been observed to latch-up at a lower current density. It has been found experimentally that, if the IGBT is subjected to rapid turn-on but does not undergo gate-controlled turn-off, its latching current density is identical to the static latching current density. Figure 5.55 shows the gate and collector current-voltage waveforms for this case. It should be noted that, to maintain this mode of operation, the anode voltage must be independently turned-off while the IGBT is in its steady-state forward conduction mode prior to turning OFF the gate voltage. This type of operation is encountered in an a.c. circuits where the device is turned-on during the positive half-cycle but does not have to undergo gate controlled turn-off. In

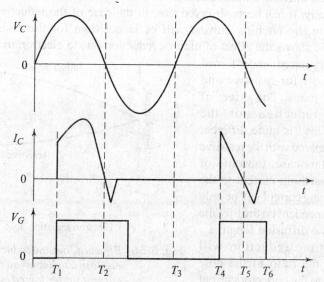

Fig. 5.55 *Switching waveforms for an IGBT operating in an a.c. circuit without undergoing forced gate turn-off*

such type of application, the IGBT can be used to control the rate of rise of the collector current (load current) at times T_1 and T_4 to minimize the generation of electrical noise in the power circuits. The gate voltage is maintained on beyond times T_2 and T_5, at which point the collector voltage reverses; this allows the IGBT to undergo reverse-recovery. It is not subjected to gate-controlled turn-off conditions. The time (T_1 and T_4) at which the IGBT is turned ON can be adjusted to control the power delivered to the load.

The dynamic latching current density is found to be less than the static latching current density if the IGBT is turned-off when the collector voltage is positive by switching off the gate drive voltage.

It has been found that the condition for latch-up during turn-off with resistive-load is given by the equation

$$\alpha_{NPN,\ DR} = (1 - \alpha_{PNP,\ DR}) \tag{5.46}$$

since $\alpha_{PNP,\ DR}$ (DR refers to dynamic turn-off under a resistive load) is larger than $\alpha_{PNP,ss}$, the hole current I_h at which latch-up occurs and hence, the steady-state collector current I_C prior to turn-off will be less than for static latch-up. It has been also observed that the degradation in latching current due to turn-off under a resistive load will become worse as the d.c. supply voltage increases. The degradation in latching current will also be more pronounced for devices with smaller P-N-P transistor current gain α_{pnp}. Thus, although the use of lifetime reduction techniques to increase the switching speed may at first sight be expected to raise latching current density because of a reduction in the current gains of the coupled transistors, in practice lifetime reduction can even cause a decrease in the dynamic latching current density.

Practically, it has been observed that, in the case of the inductive load, the decrease in the latching current will be larger than for the resistive load. Figure 5.56 shows the effect of lifetime reduction due to electron irradiation on the latching current under forced gate turn-off for resistive and inductive loads. The effect of lifetime reduction for the inductive load is quite different when compared with the resistive load. In this case, the ratio of dynamic latching current to the static latching current will be only weakly dependent on the lifetime through the diffusion length L_a. The lifetime reduction will decrease the ratio of the hole current I_h to the collector current I_C and raise the static latching current density. This will result in an increase in the absolute dynamic latching current as shown in Fig. 5.56.

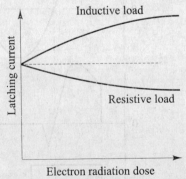

Fig. 5.56 *Effect of lifetime reduction due to electron irradiation on the latching current under forced gate turn-off for resistive and inductive loads*

Also, it has been observed that the dynamic latching current density is a function of the gate series resistance R_G. Practically, it has been found that the dynamic latching current will increase with increasing series gate resistance. The rate of rise of the collector voltage decreases when the gate resistance increases. Since IGBTs are typically used in circuit that take advantage of their gate turn-off capability, it is necessary to design the structure such that the dynamic latching current density exceeds the maximum operating current density. This is a much more stringent design requirement than that needed to meet a static latch-up criterion.

An ideal design for the IGBT is one whereby the current will saturate at a level three to five times higher than the steady-state operating current level (to allow for surge conditions) for the specified gate drive voltage. At the same time, the dynamic latching current level must exceed the collector saturation current. Such a design would prevent IGBTs latch-up under all circuit operating conditions and prevent destructive failure of devices.

5.15.7 Switching Characteristics

The IGBT is designed such that the turn-on and turn-off times of the device can be controlled by the gate-to-emitter source impedance. Its equivalent input capacitance is lower than a power MOSFET with a comparable current and voltage rating. The device is turned ON by applying a positive voltage between the gate and emitter terminals. When V_{GE} is greater than $V_{GE(th)}$, collector current flows. In switching applications where $V_{GE} \gg V_{GE(th)}$, the device saturates.

During turn-on, the IGBT is similar to a power MOSFET and does not exhibit a significant storage time during turn-off. However, during turn-off, it exhibits a fall time that consists of two distinct time intervals designated as T_{f_1} and T_{f_2}. The two time intervals are very distinct for the slow device and hardly noticeable for a fast device. The turn-off delay is caused by the discharge time constant of the effective gate-to-emitter capacitance and R_{GE}.

1. Controlling current fall time By the use of external circuitry, the current fall-time of the IGBT can be controlled (Fig. 5.57). This dependence is shown in Fig. 5.58. T_{f_2} is not controllable and is an inherent characteristics of the type of IGBT that is selected (that is, slow, medium or fast switching types). Since T_{f_1} and T_{f_2} contribute significantly to switching losses, the control feature of T_{f_1} by a resistor (R_{GE}) is a definite advantage. For example, in the case of an inductive load, the fall time can be controlled to the extent that snubber less operation is possible, since $E_L = -LdI/dt$.

Figures (5.58) and (5.59) are idealized representations of the two phases of the device turn-off. That is, a flow device can be used for d.c. and low frequency applications with minimal gate turn-off current or a fast device can be used with a linear turn-off characteristics [$R_{GE} = N$ per unit (p.u.)], and $N > 1$, Fig. 5.59. For higher frequency operation, a fast device with $R_{GE} = 1$ p.u. will minimize switching losses due to T_{f_1} and T_{f_2}.

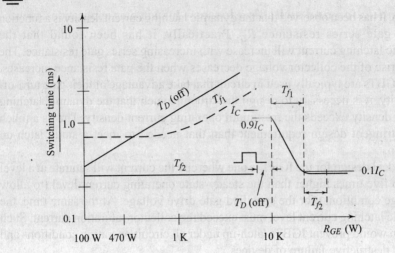

Fig. 5.57 *Typical turn-off time V_s R_{GE}*

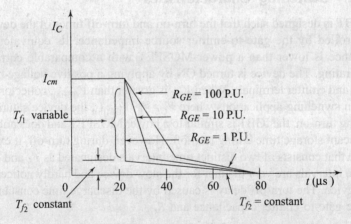

Fig. 5.58 *Full time control for a slow device*

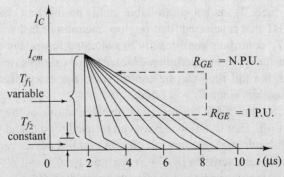

Fig. 5.59 *Full time control for a fast device*

2. Conduction and switching losses

In switching applications, power losses in the IGBTs will consists of:

(i) Drive losses
(ii) Conduction losses
(iii) Off-state losses, and
(iv) Switching losses.

In d.c. or low frequency applications, where the total switching times are much less than the period, switching losses are generally negligible, Also, if the ambient temperature extremes are limited, the off-state losses are generally insignificant.

Figure 5.60(a) shows the idealized switching waveforms for a resistive load, whereas Fig. 5.60(b) shows the waveforms for an inductive load. Switching losses can be determined as a function of time, current and voltage by using the waveforms of Fig. 5.60 as a reference. For purely inductive loads, turn-on losses are very small because the transistor turns ON at essentially zero current each cycle. Therefore, for a purely inductive load, switching losses will be determined by the turn-off losses.

Note the deviation from normal rise and fall-time definitions (i.e. instead of 10% to 90%, 90% to 10%, 0 to 100% and 100 to 0% is used to determine power dissipation).

From Fig. 5.60, following points have been noted:

Rise-time $(T_r) = 0 - t_1$,   Conduction time $(T_c) = t_2 - t_1$
Turn-off delay $(T_d) = t_3 - t_2$,   Fall-time one $(T_{f_1}) = t_4 - t_3$
Fall-time two $(T_{f_2}) = t_5 - t_4$,   Period $(T) = 1/f$
Maximum-collector current $= I_{cm}$
Collector-to-emitter saturation volts $= V_{CE(SAT)}$
Gate-to-emitter volts $= V_{GE}$
Collector-to-emitter power supply voltage $= V_{CC}$.

When gate-voltage is applied to an IGBT operating in a saturated switching modes, the collector-to-emitter voltage decreases and goes into hard saturation if sufficient gate voltage is available. The sequence of events for a resistive load is shown in Fig. 5.60(a). $V_{CE(SAT)}$ and $I_{C(MAX)}$ defines the conduction region during time interval $t_3 - t_1$. The switching losses are contained in intervals 0 to t_1 and $t_3 - t_4$ as shown in Fig. 5.60(a). The switching power losses are directly proportional to frequency and are independent of duty cycle or pulse width. Conduction losses are proportional to duty cycle $D (= T_1/T)$, while off-state losses are proportional to 1-duty cycle. It has been observed that T_{f_2} adds another component to the switching losses and will be a limiting parameter for high speed switching applications.

The energy loss in the IGBT during the turn-on process is given by

$$W_{ON} = \frac{V_{CE(max)} \cdot I_{C(max)} \cdot t_{on}}{6} \quad (5.47)$$

The average power dissipated due to the turn-on process is given by

$$P_{ON} = W_{ON} \cdot f_s \quad (5.48)$$

where f_s is the IGBT switching frequency.

200   *Power Electronics*

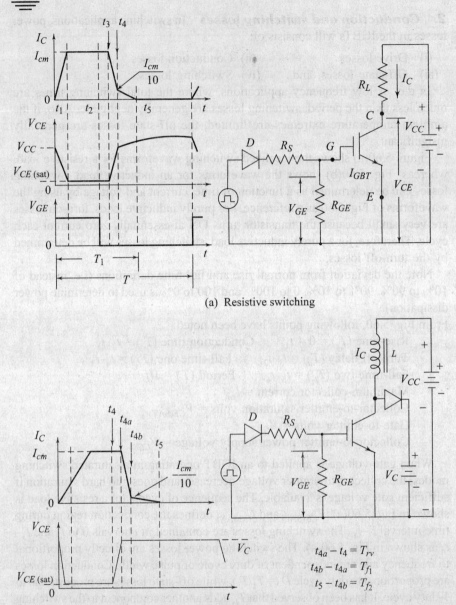

Fig. 5.60 *Idealized switching waveforms*

The energy loss during turn-off (t_{off}) is the same as that for the turn-on is given by

$$W_{\text{OFF}} = \frac{V_{CE(\max)} \cdot I_{C(\max)} \cdot t_{\text{off}}}{6} \qquad (5.49)$$

SOLVED EXAMPLE

Example 5.8 The IGBT used in the circuit of Fig. E 5.8 has the following data:
$t_{ON} = 3$ μs, $t_{off} = 1.2$ μs, $D = 0.7$,
$V_{CE(sat)} = 2$ V and $f_s = 1$ kHz
Determine: (i) average load current
(ii) conduction power loss
(iii) switching power loss during turn-on and turn-off.

Solution

(i) From Fig. E 5.8, $I_{C(max)} = \dfrac{V_{CC} - V_{CE(sat)}}{R_L}$

$= \dfrac{200 - 2}{10} = 19.8$ A

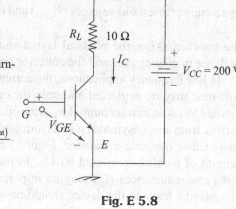

Fig. E 5.8

∴ $I_{C(avg)}$ = Duty cycle $(D) \times I_{Cmax} = 0.7 \times 19.8 = 13.86$ A

(ii) Conduction power-loss = $V_{CE(sat)} \cdot I_{C(avg)} = 2 \times 13.86 = 27.72$ W

(iii) Switching power loss:

(a) During turn-on = $W_{ON} \cdot f_s = \dfrac{V_{CE(max)} I_{C(max)} t_{on}}{6} \cdot f = \dfrac{200 \times 19.8 \times 3 \times 10^{-6}}{6} \times 1 \times 10^3$

$= 1.98$ W

(b) During turn-off = $W_{off} \cdot f = \dfrac{V_{CE(max)} I_{C(max)} t_{off}}{6} \cdot f_s = 0.792$ W

5.15.8 Parallel Operation of IGBT

The parallel operation of IGBT also suffers from the static and dynamic current sharing problems. Similar to other devices like BJT and MOSFET. It is well known from the previous discussion that in static case, there should be equal load sharing by the individual device but in dynamic case, the turn-on time, turn-off time and the magnitude of the collector current must be balanced. Therefore, the on-time, the off-time and the on-state voltage as a function of collector current should be determined over the operating temperature in order to determine the performance of the IGBT during parallel-operation.

With collector current, the $V_{CE(sat)}$ varies and it may be as high as 2.5 V. At collector current upto about 70% of its rated value, the device has a negative temperature coefficient and about this value it shows positive temperature coefficient. For example, as IGBT of 20 A rating has a negative temperature coefficient of V_{CE} upto a current of 14 A (70% 20) and shows a positive temperature coefficient upto 20 A. ΔV_{CE} varies from $- 0.5$ mV/°C at 1.0 A to about 0.75 mV/°C at current of about 18 A. The temperature coefficient becomes almost zero at about 14 A. Hence, an IGBT behaves like a BJT and a power-MOSFET.

In order to achieve good current sharing in parallel-connected IGBTs, their gate-emitter threshold voltages (V_{GEth}) and transconductance, $\left(g = \dfrac{I_C}{V_{GE}}\right)$ should be matched. Also, the physical layout should be such as to obtain geometric balance in gate-emitter and the collector-emitter areas.

In low frequency applications, the current unbalance during turn-on and turn-off time may be neglected and only the current unbalance during the on-state should be taken into account. But, as the operating frequency increases, problems arises from the unsymmetry of layout and the lead resistance and inductance may cause dynamic unbalance. Figure 5.61 shows high-frequency equivalent circuit of parallel-connected IGBTs. In parallel operation, the lead resistances (R_s) and inductances (L_s) play an important role for dynamic unbalance. All connecting leads to the device should be symmetrical and as short as possible. To prevent oscillations, R_{s1}, R_{s2}, R_{GE1} and R_{GE2} are recommended.

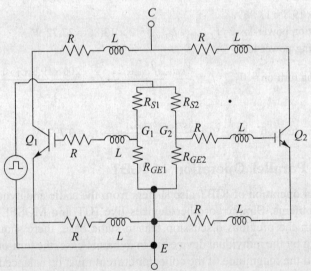

Fig. 5.61 *High-frequency equivalent circuit*

There may be a large-unbalance of current due to unequal turn-on and turn-off times of IGBTs Q_1 and Q_2. This can be minimized by matching the trans conductance and designing a drive circuitry to provide equal turn-on and turn-off times. The turn-on times can be made equal by adjusting either R_{s1} or R_{s2} and turn-off times by R_{GE1} or R_{GE2}. This can be further improved by bucking inductors in addition to matching transconductance.

5.15.9 Safe Operating Areas

Both during turn-on and turn-off, the IGBTs have robust safe operating areas. Figure 5.62(a) shows the forward-bias safe operating area (FBSOA), whereas Fig. 5.62(b) shows the reverse-bias safe operating area (RBSOA). As shown, FBSOA is square for short switching times, identical to the FBSOA of the power MOSFET. For longer switching times the IGBTs are thermally limited, as shown in FBSOA, and this is also identical to the FBSOA behaviour of the power MOSFET. The RBSOA is somewhat different than the FBSOA, as shown in Fig. 5.62(b). The upper right-hand corner of the RBSOA is progressively cut-out and the RBSOA becomes smaller as the rate of change of reapplied collector-to-emitter voltage dV_{CE}/dt becomes larger. The reason for this restriction on the RBSOA as a function of reapplied dV_{CE}/dt is to avoid latch-up. Too large a value of dV_{CE}/dt during turn-off will cause latch-up of the IGBT exactly as it can in thyristors. This value, fortunately, is quite large comparing favourably with other power devices. By proper choice of V_{GE} and gate drive resistance, the device user can easily control the reapplied dV_{CE}/dt.

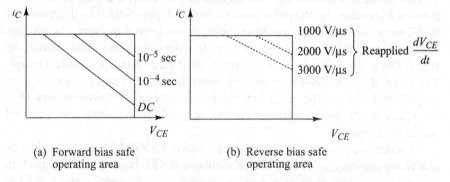

(a) Forward bias safe operating area

(b) Reverse bias safe operating area

Fig. 5.62 *Safe operating area of an IGBT*

The latch-up is avoided by proper setting the maximum collector current, I_{CM}. It is usually determined on the basis of dynamic latch-up conditions. There is also a maximum permissible gate–emitter voltage, $V_{GE(max)}$. As long as this voltage is not exceeded, an external circuit fault that tries to force the collector current to be larger than I_{CM} will cause the IGBT to leave the on-state and enter the active region where the current becomes a constant independent of the collector–emitter voltage. Under these conditions, the IGBT must be turned-off as quickly as possible because of the excessive power dissipation. This behaviour is desirable because latch-up will not occur and the gate control over the collector current is maintained. It was observed that, when V_{GE} is 10 to 15 V, collector currents of 4 to 10 times the nominal rated current are obtained. It is indicated by the recent measurements that the device can withstand such currents for 5 to 10 μs, depending on the value of V_{CE} and can be turned-off by V_{GE}.

The breakdown voltage of the *PNP* transistor sets the maximum permissible collector–emitter voltage. The beta of the *PNP* transistor is quite low, so its breakdown voltage is essentially BV_{CBO}, the breakdown voltage of the drift-body junction. In commercially available IGBTs, maximum permissible junction temperature is 150°C. A very desirable feature of the IGBT is the fact that the on-state voltage $V_{CE(on)}$ varies very little between room temperatures and the maximum junction temperature. Due to the combination of the positive temperature coefficient of the MOSFET voltage drop position of $V_{CE(on)}$ and the negative temperature coefficient of the voltage drop across the drift region, IGBT has the flat temperature characteristics.

5.15.10 IGBT Snubbers

The square SOA of the IGBT for switch mode operation minimizes the need for snubber circuits in most situations. The ability to control the turn-on and turn-off times by controlling the gate current through appropriate sizing of the series gate resistance further minimizes the need for turn-on and turn-off snubbers. The peak current handling capability of the IGBT, which is greater than that of most power MOSFETs, is another factor that makes the use of snubbers unnecessary in most applications. Therefore, in the case of the IGBT, the need for snubbers must be determined for the particular application. Figure 5.63 is an illustration of IGBT turn-off characteristics with and without a polarized snubber. The snubber is used for load-line shaping and functions to reduce switching losses within the IGBT. The snubber also reduces turn-off dv/dt of the collector-to-emitter voltage. The controllable collector current capability is also increased since it varies proportional to V_{CE}. That is, at low values of collector-to-emitter voltage ($V_{CE} \ll V_{CER}$), controllable collector current is much greater than the specified maximum value.

It has been said earlier that the switching losses in the IGBT increase significantly due to the presence of T_{f_2}. By use of snubbers, device heating is minimized. In Fig. 5.63, operation at point P with no snubber, T_{rv_1}, T_{f_1} and T_{f_2}, the IGBT has switching losses equal to:

$$P_{av(sw)} = V_{CC} \cdot I_{cm} \cdot f \frac{T_{rv_1} + 1.1 \cdot T_{f_1} + 0.1 T_{f_2}}{2} \qquad (5.50)$$

When the polarized snubber is used (cases Q and R with T_{rv_2} and T_{rv_3}, respectively), these losses are reduced. One can write equations for each case to get an exact expression or use graphical analysis. It is readily observed that snubbering definitely reduces the peak power and average power that the IGBT must dissipate.

5.15.11 Gate Drive Design Consideration of IGBTs

The gate-drive considerations for an IGBT and the MOSFET are almost identical due to the fact that the input characteristics of the IGBTs are similar to that of MOSFET. While designing the gate-drive circuits, following aspects are to be considered:

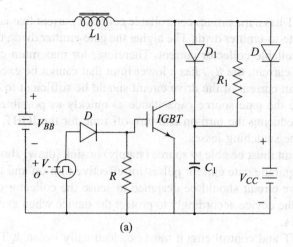

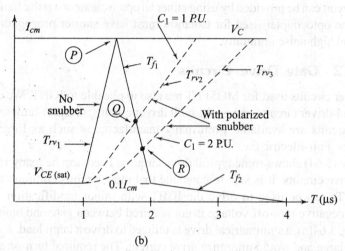

Fig. 5.63 *(a) and (b) IGBT turn-off with/without polarized snubber*

(i) The IGBT is a voltage controlled device like MOSFET. That is, it has a gate-to-emitter threshold voltage and a capacitive input impedance. In order to turn the device 'on', the input capacitance must be charged upto a value greater than $V_{GE(th)}$ before collector current can begin to flow. The collector to emitter saturation voltage decreases with an increase in magnitude of V_{GE}.

That is, for lowest value of 'on'-state voltage/the gate drive must be able to supply V_{GE} much greater than $V_{GE(th)}$.

(ii) In order to turn-'off' the IGBT, the gate drive circuit must provide a resistance between gate and emitter (R_{GE}). This resistance provides the path for the gate capacitance to discharge and turns-off the device. R_{GE} determines discharge time constant and therefore dv/dt during turn-off (discharge).

The IGBT has a maximum controllable collector current that is dependent on the gate-to-emitter dv/dt. The higher the gate-emitter dv/dt, the lower is the controllable collector current. Therefore, for maximum controllable collector current, the R_{GE} has a lower limit that cannot be exceeded.

(iii) The output current of the drive circuit should be sufficient to charge and discharge the gate source capacitance as quickly as possible. This will help in reducing the turn-on and turn-off time for the IGBT, and hence reduce the switching losses.

(iv) Gate circuit must be able to source (supply) or sink (draw) short duration, high magnitude gate current pulses for effective turn-on and turn-off.

(v) Gate drive circuit should be designed to sense the collector current and turn-off the device accordingly to protect the device when excessive current flows.

(vi) The IGBT and control circuit must be electrically isolated. The isolated circuit can be provided by using either an optoisolator or a pulse transformer. The optocoupler used for isolation must have shorter propagation delay and high noise immunity.

5.15.12 Gate Drive Circuits

The driver circuits used for MOSFET are also applicable to IGBT. We can have fabricated-driver circuits or custom-built driver circuits. Custom-built integrated driver circuits are available from many manufacturers such as International Rectifiers, Fuji-Electric etc.

Figure (5.64) shows some typical drive circuits. There can be many variations in the drive circuits. It is safe to conclude that most circuits common to power MOSFETs can be used to drive the IGBT with minor modification. In most cases, a negative turn-off voltage is not required between gate and emitter.

In Fig. 5.64(a), a symmetrical drive is utilized to drive a lamp load. R_{GE} would be quite large and would minimize drive current. The required turn-on and turn-off times would be slow, tens of microseconds.

In Fig. 5.64(b), a high speed asymmetrical drive is utilized. Isolation between primary and secondary is realized by the transformer T_1. A diode is used in conjunction with R_{S_1} to provide a fast turn-on time without affecting the turn-off time. Therefore, R_{S_1} and R_{GE} would determine turn-on and turn-off times, respectively.

In Fig. 5.64(c), an asymmetrical drive is utilized for a relay or solenoid driver. R_S determines turn-on time, while R_{GE} determines turn-off time. D_1 is normally required for this type of application in order to prevent excessive collector-to-emitter voltage due to effects of $L\dfrac{di}{dt}$ during turn-off of Q_1. Since T_{f_1} can be controlled, it is possible to eliminate D_1 in some applications or, in many cases, the need for critical layouts is minimized when D_1 must be used.

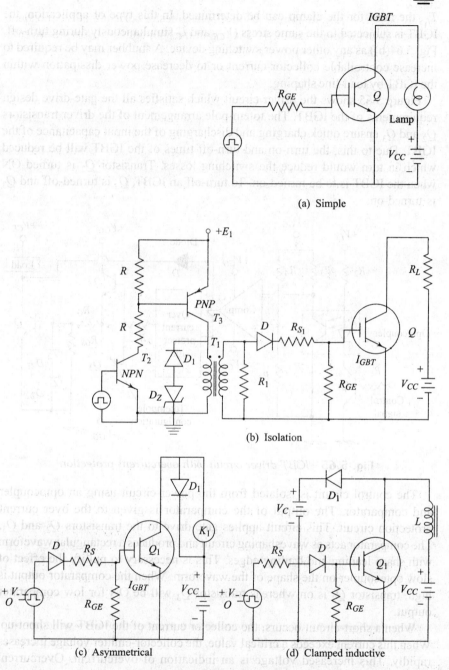

Fig. 5.64 *IGBT drive circuits*

In Fig. 5.64(d), a high speed drive would be used. The drive is asymmetrical since R_S determines turn-on time and R_{GE} determines turn-off time. The collector-to-emitter voltage is clamped by D_1 and V_c. Depending on the value of L, I_c and

T_{f_1}, the need for the clamp can be determined. In this type of application, the IGBT is subjected to the same stress (V_{CE} and I_C simultaneously during turn-off, Fig. 5.64(b)) as any other power switching device. A snubber may be required to increase controllable collector current or to decrease power dissipation within the IGBT by load-line shaping.

Figure 5.65 shows the driver circuit which satisfies all the gate-drive design requirements of the IGBT. The totem-pole arrangement of the driver transistors Q_2 and Q_3 ensure quick charging and discharging of the input capacitance of the IGBT. Due to this, the turn-on and turn-off times of the IGBT will be reduced which in turn would reduce the switching losses. Transistor Q_2 is turned ON when the IGBT is to be turned-on. To turn-off an IGBT, Q_2 is turned-off and Q_3 is turned-on.

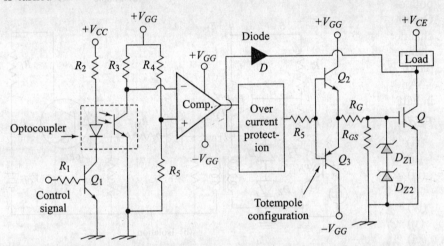

Fig. 5.65 *IGBT driver circuit with overcurrent protection*

The control circuit is isolated from the power-circuit using an optocoupler and comparator. The output of the comparator is given to the over current protection circuit. This circuit applies gate-drive to the transistors Q_2 and Q_3. The comparator acts as waveshaping circuit and produces a rectangular waveform with sharp leading and trailing edges. This is necessary to nullify the effect of slow optocoupler on the shape of the waveform. When the comparator output is high, transistor Q_2 is on whereas transistor Q_3 will be ON for low comparator output.

When a short-circuit occurs, the collector current of the IGBT will shoot-up. When this current exceeds a critical value, the collector-emitter voltage increases rapidly. This increased voltage is an indication of overcurrent. Overcurrent protection feature has been incorporated in this circuit.

Under the normal operating conditions ($I_C < I_{Cmax}$) diode D will conduct and the comparator output makes the transistor Q_2-ON and hence IGBT will be ON.

But as soon as the overcurrent condition arises, the collector-emitter voltage increases rapidly which will reverse-bias the diode D. As soon as diode D is turned-off, the comparator output is not allowed to pass through, to the totempole driver circuit and hence IGBT quickly turned-off. This way, IGBT is protected against overcurrent condition.

Custom-Built Driver IC Custom built IGBT (or MOSFET) driver ICs are available in various modules, for driving single IGBT, for driving IGBTs in a half-bridge, single-phase full bridge and 3-phase bridge circuits. The IC IR 2110 from international rectifier is a high voltage and high speed dual driver for IGBTs (or MOSFETs). The device inputs are compatible with standard CMOS outputs or LSTTL outputs using pull-up resistors. It provides low impedance output for both of its output channels, this is essential for driving the IGBTs or MOSFETs. Out of the two output channels one is a floating channel and the other is a fixed channel. The floating channel can be configured to drive an IGBT or an n-channel MOSFET whose source voltage can be upto 500 volts from the common pin of IR 2110.

An IR 2100 is used to drive high voltage switching IGBT or power MOSFET in half-bridge, dual forward or other topologies.

Features: Following are the important features of such drivers:
  (i) 500 V rated floating supply offset voltage.
  (ii) 2 A peak output current capability per channel.
  (iii) 25 ns switching time with IGBT or MOSFET (with input capacitance 1000 pF).
  (iv) 100 ns propagation delay time.
  (v) CMOS compatible Schmitt trigger inputs.
  (vi) 10-20 V output drive operating voltage range.
  (vii) 15 mW total quiescent power dissipation with 15 V supply.
  (viii) Under voltage lockout (shutdown).
  (ix) High dv/dt and negative transient immunity.

Functional Block Diagram: Figure 5.66 shows the functional block-diagram for *IC IR* 2110. It is a 14-pin IC available in DIP package. As shown, it is capable of driving two IGBTs-simultaneously. There are two channels, the upper-one is the floating channel and lower one is fixed one. Schmitt triggers are used at the inputs with hysteresis. The driver supply voltage V_{CC} (pin 3) is sensed by an under voltage sensing circuit, and if it is found lower than a critical value, then the outputs will be inhibited to avoid operation of devices in active region. The two driver are the totem-pole drivers required to drive the IGBTs.

Typical Connection: Figure 5.67 shows the typical connections for *IR* 2110 to two IGBTs. The power supply for the upper channel is generated by the boot-straping circuit consisting of elements D_1, R_3 and C_1. The diode D_1 is a fast recovery diode which must have a voltage with standing capability higher than the peak voltage (V_{BUS}). The value of R_3 (= 10 Ω) and C_1 (= 0.047 μf) depends o the operating frequency (above 10 kHz), duty cycle, gate-charge requirements of the device and the start up transients of the bootstrapped floating supply.

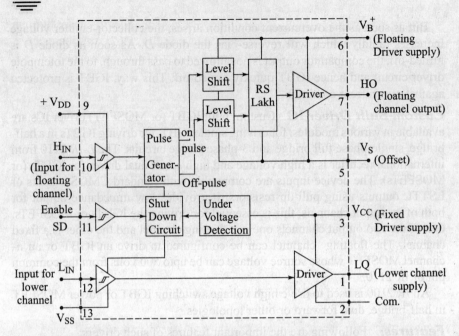

Fig. 5.66 *Functional block-diagram (Courtesy: International Rectifier)*

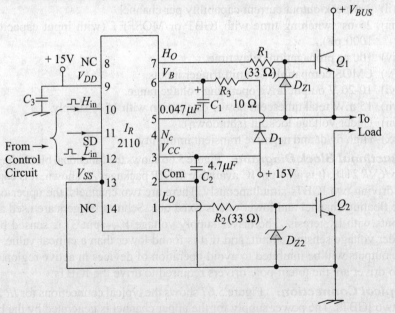

Fig. 5.67 *Typical connections*

Initially IGBT Q_2 is turned-on and Q_1 is off. Capacitor C_1 will charge through D_1, R_3, load, Q_2 to 15 V with the polarity shown in Fig. 5.67. As soon as Q_1 is turned-on, the potential at its source increases to V_{BUS}. This will boost the potential at the negative terminal of capacitor C_1 to V_{BUS}. The initial voltage on the capacitor is 15 V. Therefore, the voltage at pin 6 with respect to power ground becomes (V_{BUS} + 15 V), this voltage then gets connected to the gate of Q_1 through the upper driver. Thus, the gate emitter voltage of Q_1 is maintained at 15 V.

However, the capacitor C_1 discharges through the upper device. The discharge time of C_1 should be longer than the conduction period of Q_1. Therefore the values of R_3 and C_1 should be properly selected depending on the frequency of operation. If the capacitor C_1 discharges quickly even before the completion of the conduction interval of Q_1, then V_{GS} of Q_1 will drop down and Q_1 will turn-off.

Capacitors C_2 and C_3 are the supply bypass capacitors which are required to supply the transient current needed to switch the capacitive load. The capacitors C_1, C_2 and C_3 must be connected close to the actual device. A 0.047 µf ceramic disc capacitor in parallel with a 4.7 µf tantalum capacitor is recommended for the V_{CC} bypass. A 0.01 µf ceramic disc capacitor is usually sufficient to bypass the logic supply (V_{DD}). SD (pin 11) can be used to disable the outputs by connecting it to ground.

V_{DD} (pin 9) and V_{CC} (pin 3) can be connected together and operated from a single power supply.

Typical waveforms of IC 2110 are shown in Fig. 5.68 and the recommended dc operating conditions are given in Table 5.1.

Table 5.1 Recommended dc operating conditions

Symbol	Parameter	Min.	Max.	Unit
V_{BS}	Floating supply voltage	10	20	Volts
V_{HO}	High side channel o/p voltage	0	V_B	Volts
V_{CC}	Fixed supply voltage	10	20	Volts
V_{LO}	Low side channel output voltage	0	V_{CC}	Volt
V_{DD}	Logic supply voltage	3	V_{CC}	Volt
V_{IN}	Logic input voltage (H_{in}, L_{in}, S_D)	0	V_{DD}	Volt
V_{SS}	Logic supply offset voltage	−1.0	1.0	Volt

Applications: The high voltage bridge driver IC IR2110 can be used in the following applications:
  (i) AC and DC motor drives.
  (ii) High frequency switch mode power supplies.
  (iii) Inverters
  (iv) Choppers
  (v) Battery charger
  (vi) Induction heating and welding.

These driver ICs are designed for general and specific applications
(i) Current type: IR 2110, IR 2111, IR 2112
(ii) Special type:
- For square-wave and high voltage: IR 2111
- For resonant and phase shifted PWM: IR 2112, IR 2113
- For buck-boost converters: IR 2125
- For 3-phase, six step, PWM: IR 30, IR 2132
- For oscillating application converters: IR 2155.

5.15.13 Comparison of IGBT and MOSFET

MOSFETs	IGBTs
1. In the power MOSFET, the decrease in the electron mobility with increasing temperature results in a rapid increase in the on-state resistance of the channel and hence the on-state drop.	1. In IGBTs, this increase in voltage drop is very small.
2. The on-state voltage drop increases by a factor of 3 between room temperature and 200°C.	2. Here with the identical conditions, the increment in the on-state voltage drop is very small.
3. All highest temperature, maximum current rating goes down to 1/3 value.	3. At high ambient temperature; IGBT is extraordinarily well suited.
4. Current sharing in multiple paralleled MOSFETs is comparatively poor than IGBTs.	4. Current sharing in multiple paralleled IGBTs is far better than power MOSFET.
5. The turn-on transients are identical to IGBTs.	5. Turn-on transients are identical to MOSFETs.
6. Power MOSFET is suited for applications that require low blocking voltages and high operating frequencies.	6. IGBT is the preferred device for applications that require high blocking voltages and lower operating frequencies.

5.16 GATE TURN-OFF THYRISTORS (GTOs OR LATCHING TRANSISTORS)

Previous chapters describes the thyristor and their use in power electronic applications. Also, we have seen that thyristors can block high voltage (several thousand volts) in the off-state and conduct large currents (several thousand amperes) in the on-state with only a small on-state voltage drop (a few volts). The most useful of all is their capability of being switched ON when desired by means of a control signal applied at the gate of the thyristors. However, as we know that once an SCR is turned ON by the gate signal, the gate loses control and it can be brought back to the blocking state only by reducing the forward current to a level below that of the holding current. This is the serious deficiency in thyristors that prevent their use in switch mode applications. This section describes the structure and operation of thyristors that have a gate turn-off capability, the so-called gate turn-off thyristors or GTOs.

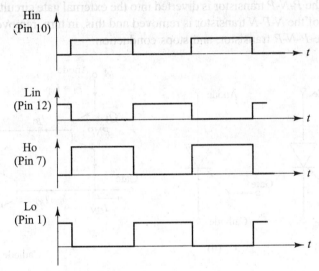

Fig. 5.68 *Typical waveforms for IC 2110*

5.16.1 Basic Structure

The gate turn-off thyristor (GTO) incorporates many of the advantages of the conventional thyristor and the high-voltage switching transistor. It is a *PNPN* device that can be triggered into conduction by a small positive gate-current pulse, but also has the capability of being turned-off by a negative gate-current pulse. However, the turn-off current gain is low (typically 4 or 5). For example, a 4000 V, 3000 A device may need −750 A gate current to turn it OFF. This facility allows the construction of inverter circuits without the bulky and expensive forced commutating components associated with conventional thyristor circuitry. The GTO also has a faster switching speed than the regular thyristor, and it can withstand higher voltage and current than the power transistor or MOSFET.

The GTO is a three-terminal device with anode, cathode and gate terminals. The various circuit symbols are shown in Fig. 5.69. The two-way arrow convention (Fig. 5.69(i)) on the gate lead distinguishes the GTO from the conventional thyristor. Figure 5.70 shows the two-transistor analogy of the GTO. Like the conventional thyristor, the GTO switches regeneratively into the on-state when a positive gating signal is applied to the base of the *N-P-N* transistor. In a regular thyristor, the current gains of the *N-P-N* and *P-N-P* transistors are large in order to maximize gate sensitivity at turn-on and to minimize on-state voltage drop. But this pronounced regenerative. latching effect means that the thyristor cannot be turned-off at the gate. Internal regeneration is reduced in the GTO by a reduction in the current gain of the P-N-P transistor, and turn-off is achieved by drawing sufficient current from the gate. The turn-off action may be explained as follows. When a negative bias is applied at the gate, excess carriers are drawn from the base region of the N-P-N transistor, and the collector current of the P-N-P transistor is diverted into the external gate circuit. Thus, the base drive of the N-P-N transistor is removed and this, in turn, removes the base drive of the P-N-P transistor, and stops conduction.

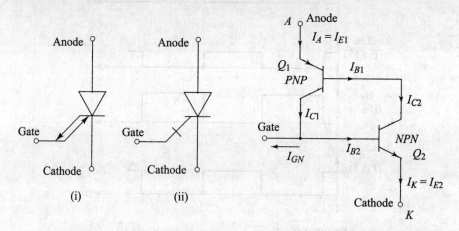

Fig. 5.69 Circuit symbols **Fig. 5.70** Two-transistor analogy of GTO

The reduction in gain of the P-N-P transistor can be achieved by the diffusion of gold or other heavy metal to reduce carrier lifetime, or by the introduction of anode to N-base short-circuiting spots, as in Fig. 5.71, or by a combination of these two techniques. Device characteristics are influenced by the particular technique used. Thus, the gold-doped GTO retains its reverse-blocking capability but has a high on-state voltage drop. The shorted anode emitter construction has a lower on-state voltage, but the ability to block reverse voltage is sacrificed. Large GTOs also have an interdigitated gate-cathode structure in which the cathode emitter consists of many parallel connected N-type fingers diffused into the P-type gate region, as in Fig. 5.71. This configuration ensures a simultaneous turn-on or turn-off of the whole active area of the chip.

GTOs are available with symmetric or asymmetric voltage blocking capabilities. A symmetric blocking device cannot have anode shorting and, therefore, is somewhat slower. The use of asymmetrical GTOs requires the connection of a diode in series with each GTO to gain the reverse blocking capability, whereas symmetrical GTOs have the ability to block a reverse voltage. In symmetrical GTOs, N-base is doped with a heavy metal to reduce the turn-off time. The asymmetrical GTOs offer more stable temperature characteristics and lower on-state voltage compared to symmetrical GTOs.

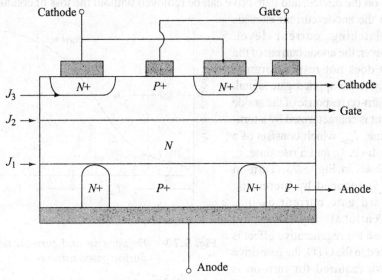

Fig. 5.71 *Basic GTO structure showing anode to N-base short-circuiting spots*

5.16.2 Switching Performance

The simplified gate drive circuit of Fig. 5.72 shows separate d.c. supplies for turn-on and turn-off. The GTO is gated into conduction by means of transistor T_1 in the turn-on circuit. The switching device in the turn-off circuit should have a high peak current capability. An auxiliary thyristor or MOSFET is appropriate for this duty. Figure 5.72 shows an auxiliary SCR, TH_1, which is gated to initiate the turn-off process. Turn-off performance may be enhanced by the presence of some series inductance, L, as shown. The voltage supply for the turn-off circuit is in the region of 10 to 20 V and the gate current at turn-off, applied for few microseconds, is typically about one-fifth of the anode current prior to turn-off. Consequently, the energy required to turn-off the GTO is much less than that needed to turn-off a conventional thyristor.

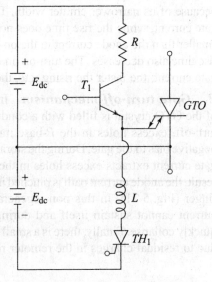

Fig. 5.72 *Basic gate-drive circuit for a GTO*

1. Gate turn-on mechanism The gate turn-on mechanism of the GTO is similar to that of a conventional thyristor. A steep-fronted pulse of gate current turns-on the device, and gate drive can be removed without the loss of conduction when the anode current exceeds the latching current level. However, the anode current of the GTO does not respond immediately to the applied gate signal. The turn-on response of the anode current is characterized by a turn-on time, T_{on}, which consists of a delay time, t_d, and a rise time, t_r, as shown in Fig. 5.73. Turn-on time is reduced by increasing forward gate current as in a conventional thyristor, but because the regenerative effect is reduced in the GTO, the gate drive current required for turn-on is larger. To ensure conduction of all cathode fingers and a reduction in on-state voltage, some manufactures recommend a continuous gating current during the entire conduction period.

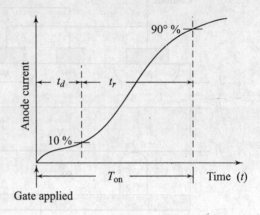

Fig. 5.73 Delay, rise and turn-on times during gated turn-on

For high frequency applications, a short turn-on time, especially a short rise time is required to reduce the switching power loss. The power loss dissipated during the delay-time is negligible because of the low anode current.

The GTO is expected to have a faster turn-on time than the conventional thyristor because of its narrower emitter width. The delay time decreases with increasing gate current, while the rise time does not vary so far as the gate current is much smaller than the anode current in the on-state. However, for larger gate drive, the rise time also decreases. The turn-on time also depends upon the rising rate of the gate current; the faster the rising rate, the shorter the turn-on time.

2. Gate turn-off mechanism In the conducting state, the central region of the GTO crystal is filled with a conducting electron-hole plasma. To achieve turn-off, excess holes in the P-base must be removed by the application of a negative bias to the gate. During the storage phase of the turn-off process, negative gate current extracts excess holes in the P-base through the gate terminal. As a result, the anode current path is pinched into a narrow filament under each cathode finger (Fig. 5.71). In this non-regenerative three-layer section of the crystal, current cannot sustain itself and during the full period, the current filaments quickly collapse. Finally, there is a small but slowly decaying tail of anode current due to residual charges in the remoter regions of the crystal.

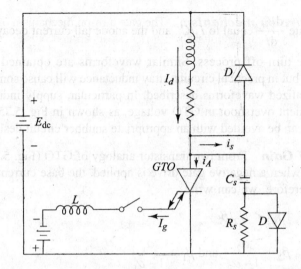

Fig. 5.74 Basic GTO switching circuit with a clamped inductive load

Figure 5.74 shows the basic GTO switching circuit with a clamped inductive load. Figure 5.75 shows the associated voltage and current waveforms at turn-off for the simplified circuit of Fig. 5.74. As shown, the d.c. supply feeds an inductive load through a series connected GTO. A freewheeling diode is connected across the load to allow circulation of load current during the off-period of the GTO. The snubber capacitor reduces the rate of rise of forward voltage at turn-off, thereby improving the current interrupt capability of the GTO and also limiting the turn-off losses in the device.

Assume the GTO is conducting a steady load current, I_d, when the gate

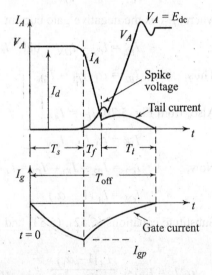

Fig. 5.75 Voltage and current waveforms during turn-off of a GTO

turn-off is initiated at time zero by the application of a reverse bias between gate and cathode. A reverse gate current, I_g, builds up at a rate determined by the inductance of the gate circuit, but the anode current remains constant throughout the storage time T_s. Anode current, I_A, then decreases rapidly to the residual, or tail, current during the fall time, T_f, and the load current is diverted into the snubber capacitance, C_s. As a result, forward voltage, V_A, builds up across the device at a rate $\dfrac{dV_A}{dt}$ equal to I_d/C_s, and the anode tail current decays to zero to complete the turn-off process. Similar waveforms are obtained in inverter applications, but in practical circuits, stray inductance will cause some departure from the idealized waveforms described. In particular, supply inductance will cause a transient overshoot in GTO voltage, as shown in Fig. 5.75. Excessive overvoltage can be avoided with an appropriate snubber circuit design.

3. Turn-off Gain From two transistor analogy of GTO (Fig. 5.70), we can observe that when a negative gate drive is applied, the base current I_{B2} will be reduced. Therefore, we can write,

$$I_{B2} < I_{C2}/\beta_2 \qquad (5.51)$$

where $\beta_2 = \dfrac{\alpha_2}{1-\alpha_2}$ and $\beta_1 = \dfrac{\alpha_1}{1-\alpha_1}$ $\qquad (5.52)$

But $I_{C1} = I_{B2} + I_{GN}$

where, I_{GN} is the negative gate current.

$\therefore \qquad I_{B2} = I_{C1} - I_{GN},$ but $I_{C1} = \alpha_1 I_{E1}$

Thus, $I_{B2} = \alpha_1 \cdot I_{E1} - I_{GN}$

Also, from Fig. 5.70, $I_{E1} = I_A$

$$I_{B2} = \alpha_1 I_A - I_{GN} \qquad (5.53)$$

Now, $I_{C2} = I_{B2} = I_{E1} - I_{C1} = I_A - \alpha_1 I_A$

$$I_{C2} = I_A(1-\alpha_1) \qquad (5.54)$$

Substitute equations (5.52), (5.53) and (5.54) in equation (5.51), we get

$$\alpha_1 I_A - I_{GN} < \dfrac{I_A(1-\alpha_1)}{\alpha_2/(1-\alpha_2)} \quad \therefore \quad \alpha_1 I_A - \dfrac{I_A(1-\alpha_1)}{\alpha_2/(1-\alpha_2)} < I_{GN}$$

$$\therefore \dfrac{\alpha_1\alpha_2 I_A - I_A(1-\alpha_1)(1-\alpha_2)}{\alpha_2} < I_{GN}$$

$$\therefore \dfrac{\alpha_1\alpha_2 I_A - I_A(1-\alpha_2-\alpha_1+\alpha_1\alpha_2)}{\alpha_2} < I_{GN}, \quad \therefore \dfrac{I_A(\alpha_1+\alpha_2-1)}{\alpha_2} < I_{GN}$$

$$\therefore \quad I_{GN} > \frac{I_A}{\alpha_2/(\alpha_1+\alpha_2-1)}, \quad I_{GN} > \frac{I_A}{\beta_{off}} \qquad (5.55)$$

where, $\quad \beta_{off} = \dfrac{\alpha_2}{(\alpha_1+\alpha_2-1)} \qquad (5.56)$

The parameter β_{off} is known as the turn-off gain. From Eq. (5.55) it becomes clear that to turn-off the GTO, the negative gate current must be greater than the anode current divided by the turn-off gain.

Magnitude of I_{GN} for Reliable Turn-OFF of GTO

(i) From Eq. (5.56) it is clear that the magnitude of I_{GN} for reliable turn-off depends on the value of turn-off gain β_{off}. Higher the β_{off}, lower will be the required value of I_{GN}.

(ii) α_1 and α_2 are the transportation factors of two transistors Q_1 and Q_2 and their values are always between 0 and 1. Therefore, the turn-off gain of a practical GTO is of the order 3 to 5. For example, if GTO carries an anode current of 900 A, then the required negative gate current of the GTO is of the order of 180 A to 300 A, which is very high and the cost of the turn-off circuit will be very high.

(iii) In GTO, there is always a maximum value of the controllable anode current. If the anode current exceeds this value, then no value of gate, current can turn-off the GTO. If an attempt is made to turn-off a GTO carrying an anode current which is higher than the maximum controllable value then it will be damaged. This is because if anode current is higher than the maximum value, the value of negative gate-current required to turn it OFF also increases. Due to this the gate cathode junction becomes reverse-biased and will eventually breakdown.

5.16.3 GTO Characteristics

In order to realize the gate-turn-off capability of the GTO, basic design trade-offs are necessary and therefore, some device characteristics are inferior to those of a conventional thyristor of comparable ratings. Because of low internal regeneration, there is an increase in latching and holding current levels, and there is also an increase in the on-state voltage drop and the associated power loss. In a GTO with a shorted anode emitter, the reverse voltage rating is appreciably less than the forward-blocking voltage but many inverter circuits do not require the capability withstand reverse voltage. However, the GTO retains many of the advantages of the thyristor and has a faster switching speed. Its surge current capability is comparable to that of a conventional thyristor, so that device protection is possible with a fast semiconductor fuse. Because of the interdigitated gate-cathode structure of the GTO, the di/dt limitation at turn-on is less stringent than that in a conventional thyristor. In general, the GTO has the high blocking voltage and large current capability that are characteristic of thyristor devices. Consequently, the GTO can be used in equipment operating directly from three-phase a.c.supplies at 440 V and above. A wide range of GTO devices is now available, and a single device can be obtained with a present power capability of 4500 V and 3500 A.

The GTO inverter has a number of advantages over the conventional thyristor inverter. In particular, the GTO circuit has about 60 per cent of the size and weight of the thyristor unit and has a higher efficiency because the increase in gate-drive power and on-state powerloss is more than compensated by the elimination of forced commutation losses. Several manufacturers have adopted GTO devices as switching elements in a range of packaged adjustable-frequency inverter drives, and the use of GTOs in inverters is growing rapidly.

The I–V characteristics of a GTO in the forward direction is identical to that of a conventional thyristor. However, in the reverse direction, the GTO has virtually no blocking capability because of the anode short structure. The latching current for large power GTOs is several amperes (here, 2A) as compared to 100–500 mA for conventional thyristors of the same rating.

5.16.4 GTO Snubber Circuits

The GTO turn-off characteristics will depend on the snubber circuit. The function of the snubber circuit is three fold:
(1) With inductive load, when the GTO is switched OFF, the current which was flowing through the GTO is diverted into the snubber circuit which would otherwise have caused a loss (power dissipation) in it.
(2) The snubber capacitor suppresses the rate of rise of forward voltage (dv/dt) immediately following a turn-off, thereby preventing the possibility of maltriggering of GTO.
(3) It helps in increasing the peak gate controllable current capability.

The requirements of a snubber circuit are as follows:
(i) The snubber circuit must be wired as short as possible to reduce wiring inductance.
(ii) Fast recovery diode is necessary to decrease the dv/dt and GTO and turn-off power dissipation.
(iii) The smallest internal inductance of the capacitor will be also desirable.

The power dissipation (W_s) of the snubber circuit is approximately expressed as follows:

$$W_s = \frac{1}{2} C_s \cdot V_c^2 \cdot f \tag{5.57}$$

where C_s is snubber capacitor, V_c the capacitor voltage, f the operating frequency.

The snubber capacitor of GTO should be larger, say 2 µF, than that of a conventional thyristor, say 0.5 µF. Therefore, GTO snubber circuit power dissipation will be larger by about four times than the thyristor one. But a commutation circuit power dissipation of conventional thyristor equipment will be larger by about 50 times than off-gating circuit one of GTO. For instance, at operating frequency of 1000 Hz, the calculated results are as follows:

	GTO	SCR
Snubber + switching	680 W	190 W
Commutating (turn-off)	40 W	2140 W
On-state	600 W	580 W
Total	1320 W	2910 W

Therefore, the total efficiency of a GTO inverter will increase more than the conventional thyristor.

5.16.5 Overcurrent Protection of GTOs

In common with other semiconductor devices, GTOs are most susceptible to damage by excessive voltage and current than most other non-solid-state electrical components. Many of the protection requirements of conventional thyristors are also applicable to GTO devices; however, the gate turn-off capability offers not only a different method for overcurrent protection but also opens a new avenue for possible failure.

Many electric power converters or other apparatus employing power semiconductors require the equipment to be so designed as to deliver a certain amount of overcurrent to the load for a certain time and to protect the load against higher currents for longer durations. Such an operation must be regarded as normal for the switching devices and must be within their repetitive rating. However, if such an overcurrent control system is not provided or fails to operate properly, or if there is an internal failure in the converter equipment, fault conditions will prevail and the devices must be protected as much as possible. For conventional thyristors, the surge current should be limited to less than the rated capability of the device by suitable surge limiting impedances, eventual natural or forced commutation of the surge, fuses, or fast circuit breakers.

Because GTO devices also have a significant surge capability, they may also be protected in the same way; however, any attempt to extinguish a high surge current by gate turn-off control is almost certain to result in destruction of the device. The limit of safety is the noncurrent peak turn-off capability, $I_{ATO(max)}$. The usual failure mechanism of the GTO under on overcurrent condition is reverse bias second breakdown, just as in a power transistor. For most practical circuits, the reapplied dv/dt will increase in direct proportion to the anode current because the snubber capacitor is usually a fixed value. Thus, the peak power dissipated by the GTO will increase very rapidly with the anode current.

From the above discussion, the following design techniques are recommended:

(i) Some means should be provided to monitor the anode current in a GTO device and to suppress the generation of any gate pulses if a level exceeding I_{ATO} (non-respective) is detected. The conventional surge limit will then be in effect.

(ii) If a current level between the repetitive and non-repetitive turn-off rating is sensed, turn-off action should be initiated at once, on a one-shot basis. The current sensing device is already available in many power conversion circuits, to perform other functions. In any event, the current sensor should have a suitably fast response, so corrective action can be taken to gate the GTO OFF at less than I_{ATO} (non-repetitive) in the event of a fault or to suppress gating if the anode current is greater than I_{ATO} (non-respective).

5.17 STATIC INDUCTION DEVICES

For high frequency applications generally static devices are used. The members of static-induction device family are: Static induction transistors (SITs), Static induction thyristors, (SITH), Static-induction diodes (SID), Static induction transistor logic (SITL), and static induction MOS transistors. However, the first two devices are commercially available devices.

5.17.1 Static Induction Transistors (SITs)

Static induction transistor is a higher power high frequency device and is essentially the solid-state version of the triode vacuum tube. SIT was commercially introduced by Tokin Corp. in 1987. Figure 5.76(a) shows the cross-sectional view of the basic static induction transistor and Fig. 5.76(b) shows its symbol. It is a short N-channel vertical device where the gate electrodes are buried within the drain and source N-type epi-layers. An SIT is identical to a JFET except for vertical and buried gate construction, which gives a lower channel resistance causing a lower drop. An SIT has a short channel length, low gate series resistance, low gate source capacitance, and small thermal resistance. It has a low noise, low distortion, and high audio-frequency power capability. The turn-on and turn-off times are very small, typically 0.25 µs.

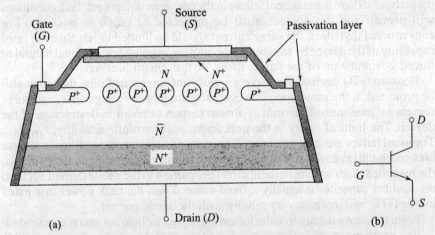

Fig. 5.76 *(a) Basic structure of SIT, and (b) circuit symbol*

An SIT is a normally on-device, and a negative gate voltage holds it off. The on-state drop is high, typically 90 V for a 180 A device and 18 V for an 18 A device. The normally on-characteristics and the high on-state drop limit its applications for general power conversions. In fact, a faster-than-MOSFET switching speed is possible because of lower equivalent gate source capacitance and resistance. The current rating of SITs can be upto 300 A, 1200 V. SITs are most suitable for high power, high frequency applications (e.g. audio, VHF/UHF, and microwave amplifiers). Japanese universities and industries have used SITs in AM/FM transmitters, induction heating, high-voltage low-current power supplies, ultrasonic generators and linear power amplifiers.

Since it is a majority carrier device, the SOAs are limited by junction temperature. Device paralleling is easy because of positive temperature coefficient characteristics of channel resistance.

5.17.2 Static Induction Thyristors (SITHs)

A SITH, or SI-thyristor, is a self controlled GTO-like on–off device. SITH was commercially introduced in 1988 by Toyo Electric Company of Japan. It is turned ON by applying a short positive gate voltage pulse like normal thyristors and is turned-off by application of negative voltage pulse to its gate. Figure 5.77 shows the basic structure of SITH and the device symbol. It is essentially a $P^+N^+N^+$ diode with a buried P^+ grid like gate structure. The structure is analogous to SIT except that a P^+ layer has been added to the anode side similar to SIT, SITH is a normally-on device with the N-region saturated with minority carrier. If the gate is reverse-biased with respect to the cathode, a depletion layer will block the anode current flow. The device does not have reverse blocking capability due to the emitter shorting that is needed for high speed operation. The turn-off behaviour of SITH is similar to that of GTO, i.e. the negative gate current is large and the anode circuit shows a tail current. The general comparison with GTO can be summarized as follows:

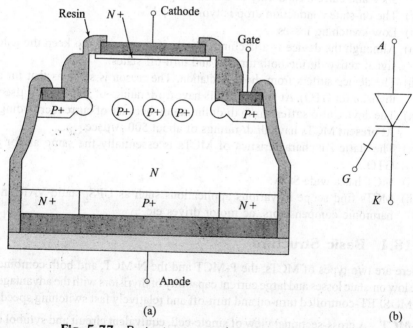

Fig. 5.77 *Basic structure and circuit symbol of SITH*

(1) Unlike GTO, SITH is a normally-on device.
(2) The conduction drop is higher.
(3) The turn-off current gain is lower, typically 1 to 3 instead of 4 to 5 for a GTO.
(4) Both devices show a long tail current.

(5) The switching frequency is higher.
(6) The dv/dt and di/dt ratings are higher.
(7) The SOA is improved.

5.18 MOS CONTROLLED THYRISTOR (MCT)

The MOS controlled thyristor is a thyristor with an insulated gate terminal. The insulated gate simplifies the drive circuits. The device combines the advantages of MOSFET and the thyristor. MCT has low on-state losses and high current capabilities of thyristors and the simpler drive characteristics and faster switching speed of MOSFETs. The main features of the device are:

(i) There is a separate MOSFET for turn-on and turn-off. Hence, faster turn-on and turn-off times.
(ii) Typical values of switching times are of the order of 1 μs.
(iii) MCTs are designed for asymmetrical blocking and little reverse-blocking capability, typically only about 25 V.
(iv) MCTs can easily be connected in series-parallel combination for a higher power requirement. Device with maximum voltage capability of 2 kV–3 kV and current capability of 200 A are presently available.
(v) The on-state conduction drop is typically 1.1 V.
(vi) Low switching losses.
(vii) Although the device is a latching switch, it is necessary to keep the gate signal active during both turn-on and turn-off states.
(viii) The device suffers from dv/dt limitation. The reason is same as that for a thyristor (or GTO). At present MCTs have dv/dt ratings of 500–1000 V/μsec.
(ix) The device also suffers from di/dt limitation because of current crowding. At present MCTs have di/dt ratings of about 500 A/μsec.
(x) The static i-v characteristics of MCTs is essentially the same as for a GTO.
(xi) MCT has a wide SOA.
(xii) MCTs find scope in various applications such as UPS, static VAR and harmonic compensators, ac motor drives etc.

5.18.1 Basic Structure

There are two types of MCTs, the P-MCT and the N-MCT, and both combines the low on-state losses and large current capability of thyristors with the advantages of MOSFET-controlled turn-on and turn-off and relatively fast switching speeds.

N-MCT A cross-sectional view of single-cell, equivalent circuit and symbol of N-MCT is shown in Fig. 5.78. A complete N-MCT is constructed as the parallel-connection and many small MCT cells are fabricated integrally on the same silicon wafer. The thyristor portion of the device has the same *pnpn* structure as a conventional thyristor.

As shown, the MCT structure has four inherent transistors. The two transistors pnp-(Q_1) and npn-(Q_2) constitute the thyristor and the two MOSFETs, n-channel

MOSFET (Q_3) and p-channel MOSFET (Q_4) are connected to the base of Q_2 and Q_1 respectively. The on-FET is a *n*-channel MOSFET and the off-FET is a *p*-channel MOSFET. These MOSFETs are located around the cathode and hence shares the same side of the silicon wafer as the cathode. The lightly doped region which must contain the depletion layer of the blocking junction is placed in the *n*-type region neared to the anode.

N-MCT Turn-On For an *N*-MCT, the gate voltage is applied with respect to the cathode. There are two MOSFETs, one for turn-on and the other for turn-off. If a positive voltage is applied between gate and cathode, an N-channel MCT turns-on and P-channel turns-off. If a negative voltage is applied between gate and cathode, the P-channel MOSFET turns-on and N-channel turns-off. Hence, *N*-MCT can be turned-on by applying positive voltage and turned-off by applying negative voltage.

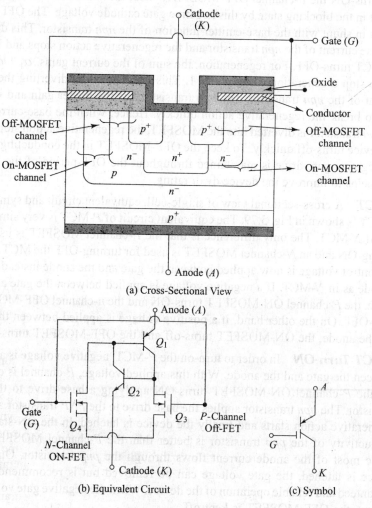

Fig. 5.78 *N*-MCT

When a positive voltage is applied between gate and cathode, the anode-current flows through the N-channel of the ON-MOSFET. Hence the ON-FET is activated which will simultaneously ensure that the OFF-FET is driven into its blocking state. The anode current flows through the N-channel of the ON MOSFET. This applies base current to the *pnp* transistor, the collector current of which turns-on the *npn* transistor. The regenerative action starts and the device latches. The gate-drive can be removed once the device latches. However, the gate drive is recommended for reliable operation.

Once the device latches due to the better conduction properties of the *npn* transistor, than the parallel ON-MOSFET carries the major portion of the current.

N-MCT Turn-Off In order to turn-OFF the N-MCT, a negative voltage is applied between the gate and the cathode. This negative voltage creates P-channel and turns-ON the P-channel OFF-MOSFET. Also, the n-channel ON-MOSFET is kept in the blocking state by this negative gate cathode voltage. The OFF-FET comes in shunt with the base-emitter junction of the *npn* transistor. This diverts the base current of the *npn* transistor and the regenerative action stops and hence the MCT turns-OFF. For regeneration, the sum of the current gains, $\alpha_1 + \alpha_2 \approx 1$ and to stop regeneration, $\alpha_1 + \alpha_2 \ll 1$. This is achieved by diverting the base current of the *npn* transistor. The *npn* transistor has the higher gain and it will help to break the regenerative action quickly. Hence, when the base-current of the *npn* transistor is diverted into the MOSFET, the regenerative action stops and the device turns-off quickly. To keep the OFF-MOSFET in the conducting state, the negative gate-drive is maintained throughout the OFF-period of the MCT. This helps to improve the device dv/dt rating.

P-MCT A cross-sectional view of single-cell, equivalent circuit and symbol of P-MCT is shown in Fig. 5.79. The equivalent circuit of P-MCT is very similar to that of N-MCT. The only difference is that the P-channel MOSFET is used for turning-ON and an N-channel MOSFET is used for turning-OFF the MCT. Also, the control voltage is now applied between the gate and the anode instead of the cathode as in N-MCT. If a negative voltage is applied between the gate and the anode, the P-channel ON-MOSFET turns-ON and the n-channel OFF-MOSFET turns-OFF. On the other hand, if a positive voltage is applied between the gate and the anode, the ON-MOSFET turns-off and the OFF-MOSFET turns-ON.

P-MCT Turn-ON In order to turn-on the P-MCT, negative voltage is applied between the gate and the anode. With this applied voltage, P-channel is created and the P-channel ON-MOSFET turns-ON applying a base-drive to the *npn* transistor. The *npn* transistor applies the base drive to the *PNP* transistor and the regenerative action starts and finally the device is latched. In the ON-state, the conductivity of the *pnp* transistor is better than the P-channel MOSFET and hence most of the anode current flows through the *pnp* transistor. Once the device is latched, the gate voltage can be removed but is recommended for guaranteed and reliable operation of the device. When the negative gate voltage is present, the OFF-MOSFET is kept-off.

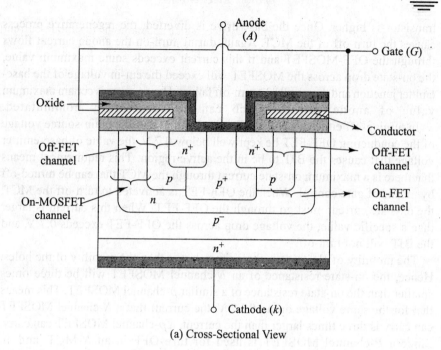

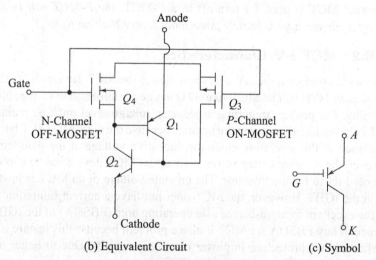

Fig. 5.79 P-MCT

P-MCT Turn-Off The P-MCT is turned-off by applying a positive voltage between the gate and the anode. This turns-off the ON-MOSFET and an N-channel is created and the N-channel off-MOSFET turns-ON. This MOSFET is in shunt with the base-emitter junction of the p-n-p transistor. This diverts the base drive of the transistor through the OFF MOSFET and breaks the regenerative process. For this purpose, pnp transistor was chosen since the gain of this transistor is higher. Once the base drive is diverted, the regenerative process causes the turn-off of the MCT. Again, during turn-on the anode current flows through the OFF-MOSFET and if this current exceeds some maximum value, the on-state drop across the MOSFET will exceed the cut-in voltage of the base-emitter junction and there will be a turn-off failure. Hence, there is certain maximum value of anode current which can be successfully commutated.

In turning-off either type of MCT, it is essential that the drain-source voltage of the conducting OFF-FET be kept well below 0.7 V, the value of base-emitter voltage that causes the BJT to be in the active region. This requirement means that there is a maximum on-state current through the MCT that can be turned-off by means of gate control. When the OFF-FET is activated to turn-off the MCT, the on-state current must go through the OFF-FET. When this current is larger than a specific value, the voltage drop across the OFF-FET exceeds 0.7 V, and the BJT will not turn-off.

The mobility of electrons is three times larger than the mobility of the holes. Hence, the on-state resistance of an N-channel MOSFET will be three times smaller than the on-state resistance of a similar p-channel MOSFET. This means that for the same voltage drop of 0.7 V, the current that a N-channel MOSFET can carry is three times larger than the current, a p-channel MOSFET can carry. Since a P-channel MOSFET is used for turn-OFF in an N-MCT, and an N-channel MCT is used for turn-off in a P-MCT, *the P-MCT will be able to turn-off a current approximately three-times larger than an N-MCT.*

5.18.2 MCT I-V Characteristics

In the static condition, the I-V characteristics of the MCT are similar to those of thyristors and GTOs. The MCT, like GTO has negligible reverse voltage blocking capability. For proper triggering, a pedestal voltage level must be maintained. MCT behaves like thyristor in the on-state because the OFF-MOSFET has a high impedance in this operating condition (negative voltage at the gate-terminal). However, the on-state voltage drop of an actual MCT device has been observed to exceed those for the thyristor. The on-state voltage of an MCT is lower than that in the IGBT. However, the MCT does not have a current saturation feature and the excellent forward-biased safe operating area (FBSOA) of the IGBT. The absence of any FBSOA for MCT is also a problem because this feature is useful for short-circuit protection in power electronic circuits. Due to better turn-off characteristics, MCT is preferable to GTO.

The dv/dt limitation of MCT is identical to those in thyristors and GTO. However, it is essential to ensure that the OFF-MOSFET is in the conducting state prior to turning off. It is accomplished by using a pedestate of positive voltage at the gate of MCT to maintain it in the conducting state. If the gate signal is allowed to go to zero during the blocking state, then even a moderate value of dv/dt can turn-on the MCT. The di/dt limitations of MCT are the result of current crowding due to the decrement in the on-state voltage with increasing temperature.

5.18.3 Switching Characteristics

MCT is capable of switching from the ON-state to the OFF-state and vice-versa with typical switching times of 1 μsec. A step-down converter using N-MCT is shown in Fig. 5.80(a) and switching waveforms are shown in Fig. 5.80(b). An important advantage of the MOS gate structure for turning ON the thyristor is that the current used to trigger the device is not provided by the gate circuit. This current is derived from the anode circuit. Consequently, a high dv/dt capability can be achieved without influencing the gate-drive requirements.

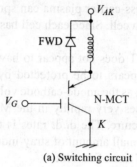

(a) Switching circuit

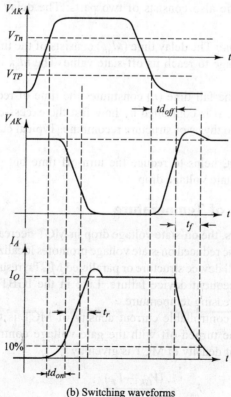

(b) Switching waveforms

Fig. 5.80 *N-MCT switching behaviour*

The turn-on time consists of two parts: The turn-on delay time (td_{on}) and the current rise time (t_r).

(i) *Turn-on delay time (upto 10% I_{max}):* The delay time consists of the time for the injection of excess carriers before the regenerative action begins. Typical values of the turn-on delay time are about 0.5 µs.

(ii) *Rise-time:* The rise time is the time of spreading the plasma of the excess carrier under the gate electrode. The MCT current rises very rapidly due to the regenerative action. Typical values of t_r are 0.5 µs and are governed by how fast the excess-carrier plasma can spread across the entire cross-sectional area of each cell. Since each cell has a relatively small area, this time is quite small.

During turn-on, the MCT does not appear to have the typical hard-switched waveforms but instead appears to be protected by a turn-on snubber circuit because the current is rising as the anode-cathode voltage is falling. This is due to the fact that the current rises very rapidly in the MCT due to the regenerative action of the thyristor structure. The di/dt rates is the range of 1000 A/µs are easily achievable. Hence, small amount of stray-inductance can act as a turn-on snubber to limit the rate of rise of current.

The turn-off time also consists of two parts: The delay time td_{off} and the current fall time t_f.

(i) *The delay time:* The delay time (td_{off}) consists of the time for the anode to cathode voltage to reach its off-state value. The td_{off} is typically 1 µs or less.

(ii) *Fall time:* The fall time (t_f) constitutes the time of recombination in the base-regions. It is controlled by how fast the excess-carriers in the base regions of the thyristor structure recombine. Typical current fall-times are 0.5–1 µs.

The anode short, helps to reduce the turn-off time but only at the cost of increasing the on-state voltage drop.

5.18.4 Effect of Temperature

Similar to thyristors, the on-state voltage drop in MCT decreases with the rise in the temperature. The reduced on-state voltage promotes localization of the current flow in the multicell device structure or paralleled MCTs, resulting in the thermal runaway and consequent device failure. Like in the IGBT, the turn-off time increases with increasing temperature.

The maximum controllable current density of MCT is the highest current density that can be turned-off with the gate voltage control. The maximum controllable current density of MCT is given by

$$J_{mcc} \propto \frac{(V_{GK} - V_T)}{\alpha_1} \tag{5.58}$$

where V_{GK} is the gate to cathode voltage, V_T is the threshold gate voltage for turn-on and α_1 is the current gain of the *pnp* transistor. It has been found that the J_{mcc} decreases very rapidly with the increase in the anode voltage if no snubbers are used for the MCT. Also, with an increase in temperature, J_{mcc} decreases rapidly. This factor is the most important limitation on the high temperature operation of MCTs.

5.18.5 Safe Operating Area

Figure 5.81 shows the SOA of the MCT. The SOA is limited by the maximum controllable anode current, maximum power dissipation and device temperature. As the anode-cathode voltage increases, the maximum current starts decreasing due to power dissipation and internal device avalanche mechanism. Since the 50 A is larger enough, the snubberless operation is possible.

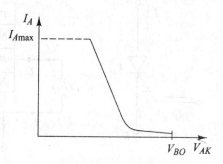

Fig. 5.81 *FBSOA for MCT*

5.19 INTEGRATED GATE-COMMUTATED THYRISTOR (IGCT)

IGCT, developed by Mitsubishi and ABB, is basically a GTO with hard turn-off, which in combination with other advances in the packaging, achieves a fast turn-off and low turn-off switching losses. Development of GTO technology has led to a wide variety of applications in the range of 1 to 20 MVA, mainly adjustable speed drives and railway interties. Although these applications have proven the reliability and cost effectiveness of the GTO thyristor, there are still cumbersome disadvantages of this technology. The inhomogeneous switching mode of the GTO gives rise to the need for dv/dt limiting snubber capacitors, and large dispersions in turn-off times complicate the series connection of GTOs.

A breakthrough in GTO technology was achieved by the introduction of the "Hard-Driven-GTO", leading to the possibility of snubberless switching. This novel switching mode is characterised by an extremely fast commutation of the cathode current to the gate (Gate-Commutated Thyristor) whereby the cathode emitter is turned-off before the voltage at the main-blocking junction rises. Thus, the IGCT combines in a unique-manner the advantages of a thyristor, low-on-state voltage drop and high blocking voltage ratings, with the rugged switching behaviour of the transistor.

5.19.1 Principle of Operation

IGCT is a power semiconductor switch for power conversion which sets performance standards with regards to power, reliability, speed, efficiency, cost, weight and volume. IGCT symbol is shown in Fig. 5.82(a).

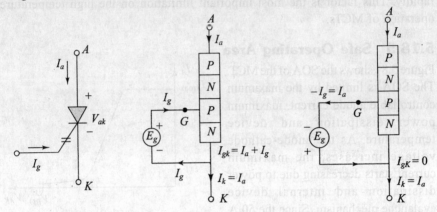

Fig. 5.82 *(a) Symbol (b) Conducting GCT (c) Blocking GCT*

In the conducting state, IGCT is a regenerative thyristor switch like an SCR or GTO as shown in Fig. 5.82 and as such, is characterised by high current capability and low on-state voltage. In the blocking state, the gate cathode junction is reverse-biased and is effectively 'out of operation' and the resultant device is that of Fig. 5.83.

Figures 5.82(b) and 5.82(c) represent the conducting and blocking states of GTOs with one major difference, namely that the GCT can transit from 5.82(b) to 5.82(c) instantaneously whereas a GTO must do so via an intermediate and indeterminate state which is neither one thing nor the other (known as 'GTO phase'). It is because of this GTO phase, that GTO requires large-snubbers to minimise the rate of reapplication of voltage (dv/dt). Thyristor devices, unlike simple transistors, are sensitive to dv/dt.

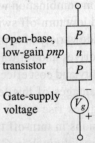

Fig. 5.83 *Equivalent circuit of blocking GCT.*

In IGCT-technology, elimination of the 'GTO zone' is achieved by quickly diverting (commutating) the entire anode current away from the cathode and out of the gate before charge is removed from the anode n-base, i.e. before the anode p-n-p transistor 'realises' that the cathode n-p-n transistor has stopped injecting. At the instant that the cathode ceases injection the device is described by the circuits of Figs. 5.82 (c) and Fig. 5.83. The device becomes a transistor prior to it having to sustain any blocking voltage at all. Because turn-off occurs after the device has become a transistor, no external dv/dt limitation is required and the

IGCT may operate snubberlessly as does a MOSFET or an IGBT. This is illustrated in Fig. 5.84—which shows IGCT turn-off. In this Fig. 5.84, we see the classical turn-off waveform of a transistor. In this case, it is the *p-n-p* transistor of the GCTs regenerative transistor pair which blocks automatically after extinguishing the *npn* transist with unity gain. Because of this, it is necessary for the gate drive circuit to handle the full-anode current and to do so quickly "before the anode transistor knows what's happened". Thus the key to IGCT design lies in very low inductance gate circuits, which, depending on the device rating, may require coaxial devices and multi-layer circuit boards.

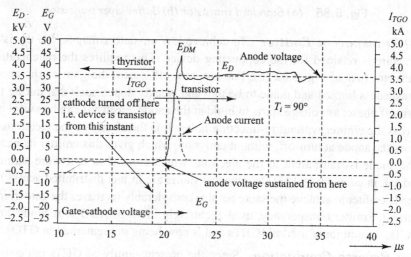

Fig. 5.84 *Snubberless Turn-off of a 3 kA/4.5 kV IGCT*

GCTs exploits several technologies which enable significant loss reductions and ultimately component reductions to appear the requirements of ideal switch and these are:

(i) Buffer Layer The electrical field in the blocking *pn*-junction has a triangular distribution as shown in Fig. 5.85 (a). The total blocking voltage of the device is the integral of this field over distance (thickness of the silicon *n*-base); hence the higher the required blocking voltage the higher the silicon thickness and the greater the conduction and switching losses. Figure 5.85 (b) shows that, were the field to have a constant rather than a diminishing profile, the same blocking voltage could be achieved with a thinner silicon resulting in a more efficient device. This type of field-distribution is exactly that which is achieved by the introduction of the *buffer-layer*, an additional intrinsic *n*-layer which redistributes the otherwise triangular field. This arrangement is also known as the *pin-structure*. For example, the use of buffer layer in a 4.5 kV GCT, reduces the required water thickness by about 40% which allows a corresponding reduction of the conduction and switching losses.

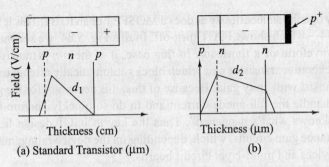

Fig. 5.85 *(a)* Standard transistor *(b)* Buffer-layer transistor

(ii) Transparent Emitter To achieve low on-state voltages, the thyristor structure is retained for the conducting device. This requires the regeneration transistor pair of Fig. 5.82 (b). For low turn-off losses, the gain of the anode transistor is limited and made to be thin and weakly doped such that during the turn-off phase, as voltage starts to build at the anode, electrons can be swept out across the emitter without re-injecting holes. It is the ability of the electrons to cross the anode at turn-off, without emission, which gives this emitter its name. The anode is transparent to the electrons which cross it as if it were shorted. Indeed, in conventional GTOs, physical shorts are in fact distributed across the anode emitter to achieve the same but this considerably increases the gate trigger current. Emitter transparency, used exclusively in all of ABB semiconductor's GCTs, is commonly used in IGBTs and is now being implemented in GTOs.

(iii) Reverse Conduction Since the present family of GCTs is destined primarily for voltage source inverters, they are asymmetric and intended for use with fast-antiparallel diodes. In the past (1970's), fast thyristors were successfully fabricated with monolithically integrated fast diodes. This has also been achieved with GTOs, but prior to the application of buffer layer and transparent emitter technology, the GTO substrate silicon material was nearly twice the thickness of that of an appropriate antiparallel free-wheeling diode (FWD). Thus, integration was achieved at the cost of device performance. However, the present GCTs fully exploit the benefits of the above described technologies and thus allow optimised FWDs to monolithically integrate on the same-wafer. They are thus all reverse conducting.

(iv) Controlled Diode Recovery As illustrated by Fig. 5.84, IGCTs are capable of snubberless turn-off, characteristics of most transistors. This implies that the integrated FWD must also turn-off without a snubber and this at high commutation rates (di_r/dt). To allow this, the diode sections of the reverse conducting GCTs are given a profiled (as opposed to uniform) recombination centre distribution through proton irradiation. Diode reverse recovery characteristics are then controlled by this "lifetime profile" and ensure that there are no discontinuities in the reverse current as it fails to zero (no snap-off).

IGCT allows reliable, cost effectively, efficient and compact power-electronics to be designed over the power range of fractions to hundreds of megawatts and is the gateway to new applications in such areas such as medium voltage drives and transmission and distribution.

5.20 MOS TURN-OFF THYRISTOR (MTO)

Silicon Power Corporation (SPCO) has developed the MTO, which is a combination of a GTO and MOSFETs, which together overcome the limitations of the GTO regarding its gate-drive power, snubber circuits, and dv/dt limitations. Unlike IGBT, the MOS structure is not implanted on the entire device surface, but instead the MOSFETs are located on the silicon, all around the GTO to eliminate need for high-current GTO turn-off pulses. The GTO structure is essentially retained for its advantages of high voltage, high current and lower forward conduction losses than IGBTs. With the help of these MOSFETs, and tight packaging to minimize the stray inductance in the gate-cathode loop, the MTO becomes significantly more efficient than conventional GTO, requiring drastically smaller gate-drive while reducing the charge-storage time on turn-off, providing improved performance and reduction of system costs. As before, the GTO is still provided with double-sided cooling, and lend to itself to thin packaging technology for even more efficient removal of heat from the GTO.

5.20.1 Principal of Operation

Figure 5.86 shows the structure, symbol and the equivalent circuit of the MTO. The structure shown (Fig. 5.86 (a)) is that of a thyristor with four layers, and has two gate structures, one for turn-on and the other for the turn-off. For both gates, the metal is directly bonded on the p-layer.

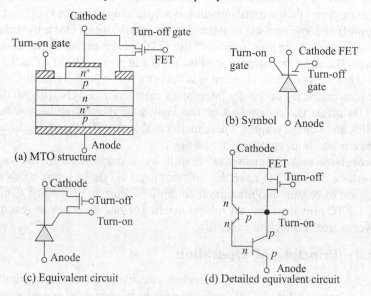

Fig. 5.86 *MOS-turn-off Thyristor*

As in GTO, turn-on is achieved by turn-on current pulse of about one-tenth of the main current for 5–10 μsec followed by a small back-porch current. Turn-on pulse turns-on the upper *npn* transistor, which is turn turns-on the lower *pnp* transistor leading to a latched-turn-on.

Turn-off is carried out by application of just a voltage pulse of about 15 V to the MOSFET gates, thereby turning the MOSFETs ON, which shorts the emitter and the base of the *n-p-n* transistor, shunting-off the latching process. In contrast, the contentional GTO turn-off is carried out by sweeping enough current out of the emitter base of the upper *n-p-n* transistor with a large negative pulse to stop the regenerative latching action. What is equally important with the new approach is that the turn-off can be much faster, (1–2 μsec as against 20–30 μsec) and the losses corresponding to the storage time are almost eliminated. This also means high dv/dt and much smaller snubber capacitors and elimination of the snubber resistor.

Small turn-off time also means that MTOS can be connected in series without matching of devices because virtually cell devices turn-off simultaneously resulting in cell the devices taking their share of current. Since MOSFETs are essentially turned-on in parallel to the GTOs gate-cathode, the rapid turn-off requires MOSFETs with a very low forward voltage drop. MOSFETs are small inexpensive, and commercially produced in large quantities. Fast turn-off of MTOs and other advanced GTOs have essentially overcome the disadvantages of GTOs compared to the IGBTs with regard to the over-current protection.

5.21 EMITTER TURN-OFF THYRISTOR (ETO)

ETO was developed at Virginia Power Electronics Center in collaboration with Silicon Power Corporation (SPCO), is another variation on the GTO, and incorporates low voltage transistors in series with a high voltage GTO to achieve fast turn-off and low turn-off switching losses. Basically ETO is a hybrid power semiconductor device that turns-off the GTO under the unity turn-off gain condition. Based on the integration of the GTO and power-MOSFET technology, the ETO is a high power device that is suitable for use in high-frequency and high power converters. By optimally integrating commercial GTOs with MOSFETs, the ETOs offers the advantages of fast switching speed, snubberless turn-off capability and voltage control. These merits enable the ETO the ability to achieve high power and high switching frequency.

Theoretical and experimental results have demonstrated the superior performance of the ETO over that of traditional GTOs in terms of switching speed, ease of control and maximum controllable current capability. Compared to IGCT, ETO provides clear advantages in terms of gate-drive power-requirement and over current protection capability.

5.21.1 Principle of Operation

Figure 5.87 shows the principle equivalent circuit and the circuit symbol of the ETO. As shown, MOSFET Q_E (emitter switch) is connected in series with the

GTO and a second MOSFET Q_G (gate-switch) is connected across this series MOSFET and the GTO gate. The two MOSFETs are operating as a complementary pair to help the GTOs turn-off. During a normal turn-off transient, Q_E is turn-off and Q_G is turned-on. The turn-off of the emitter switch Q_E cuts off the GTOs cathode current path and all of the cathode currents are quickly transfered to the gate-path. In this way, the latch-up mechanism of the GTO is broken and the ETO is turned-off under a unity-gain turn-off condition (also known as a hard driven turn-off condition).

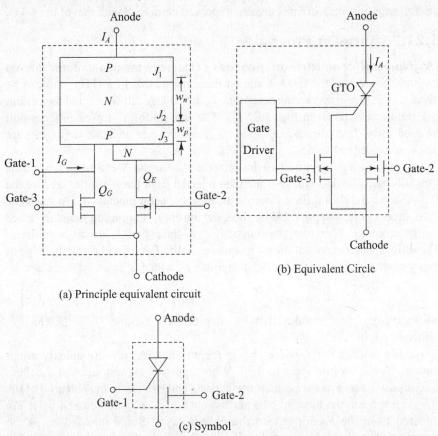

Fig. 5.87 *Emitter-turn-off thyristor*

The benefit of the unity-gain turn-off is two fold. First, the storage time, which is the time needed for the gate current to remove all minority charges in the p-base region, is now significantly decreased to about 1 μs. Second, unlike the traditional GTO, the turn-off of an ETO is changed to an open-base p-n-p process that further ensures the current uniform distribution among GTO cells over the turn-off transient. Therefore, ETO has a wide reverse biased safe operation area (RBSOA) and snubberless turn-off capability.

During the turn-on transient, Q_E is turned-on and Q_G is turned-off. A high-current pulse is injected into the GTO gate to reduce the turn-on delay time and improve the turn-on di/dt rating. The built in *PNP* and *NPN* transistors inside the GTO latch up quickly, and the anode voltage of the ETO collapses to a low voltage. So the turn-on process of the ETO is similar to that of the GTO. Depending on the amplitude and rise rate of the gate-current as well as the structure of the GTO, an ETO can be uniformly or non-uniformly turned-on. Since the gate driver of the ETO is tightly integrated with the ETO, the ability to provide the desired gate-turn-on current is greatly improved compared with that of the GTO.

5.21.2 Turn-On Process

(A) Non-uniform turn-on process Generally the non-uniform turn-on transient process of the ETO is almost the same as that of a GTO, and can be divided into three-intervals the delay time t_d, the voltage fall-time t_f and the current rise time t_r, as shown in Fig. 5.87 (a). The non-uniform turn-on process will proceed if the gate current amplitude (I_{Gm}) and the rate of increase (dI_G/dt) are below some critical values.

By applying a pulse current to the gate, the *P*-base and *N*-emitter junction J_3 is forward-biased, excess electrons are then injected from the *N*-emitter layer to the *P*-base layer and then diffuse towards the base-collector junction J_2. At the same time, excess holes are provided by the gate-current I_G in order to maintain space charge neutrality. Therefore, the conductivity of the *P*-base is heavily modulated. The diffusing electrons that do not recombine in the *P*-base will reach the edge of the J_2 space-charge region after the delay time t_P, which is given by

$$t_P = W_{P(\text{undep.})}^2 / 2D_n \tag{5.59}$$

where $W_{P(\text{undep})}$ is the width of the un-depleted *P*-base and D_n is the electron diffusion coefficient.

Having reached the J_2 space charge region, the electrons are quickly swept through by the built-in electric field. When approaching the junction J_1, these electrons will serve as excess majority carriers and will lower the potential of the *N*-base layer. So, the forward bias across junction J_1 is increased and holes are injected from the *P*-emitter into the *N*-base layer. Some injected holes are recombined in the *N*-base, while offers will diffuse through the un-depleted *N*-base towards junction J_2. These holes will reach the edge of the J_2 space charge layer after a transient time interval t_n, which is given by

$$t_n = W_{n(\text{undep.})}^2 / 2D_p \tag{5.60}$$

where $W_{n(\text{undep.})}$ is the width of the un-depleted *N*-base and D_P is the hole-diffusion coefficient. The holes are then instantaneously swept across junction J_2 and into the *P*-base region. These injected holes serve as excess majority carriers and will cause the increased injection of electrons from the *N*-emitter into the *P*-base. As a result, positive feedback is formed.

The delay time t_d is defined as the transient time between the moment the gate current increases to 10% of the final value and the time when the anode voltage decreases to 90% of the initial value, as shown in Fig. 5.88. It is the time from the application of the gate-current to the moment the J_2 space charge region starts to be discharged. In the case of the GTO, the rate at which the gate-current is applied in generally low, and the delay time t_d is generally larger than the transient time for the carriers diffusing through the base-region. So,

$$t_d \geq t_n + t_p = W_{p(\text{undep.})}^2/2D_n + W_{n(\text{undep.})}^2/2D_p \tag{5.61}$$

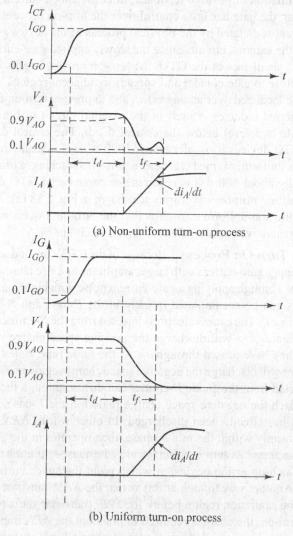

Fig. 5.88 *ETOS turn-on process*

In the next phase, during the voltage fall-time, the J_2 space charge region is quickly flooded with excess carriers. The positive space-charge region in the N-base is discharged by the electrons injected from the N-emitter over junction J_3, and the negative space charge region in the P-base is discharged by the holes injected from the P-emitter over J_1. So the thickness of the J_2 space charge region is reduced, resulting in both the fall of the anode voltage and the increase of the anode current. The voltage fall-time t_f is defined as the time it takes for the anode voltage to be reduced from 90% to 10% of its initial value, as shown in Fig. 5.88.

With the formation of positive feedback, after the anode current exceeds the latching current, the gate has little control over the turn-on process. The rise of the anode current is dictated by the physical process of the device as well as the constraints of the external circuit. Since the slowly applied gate-current initially only turns-on a small area of the GTO, the turn-on process begins with a small cathode area close to gate contact and spreads to adjacent regions.

To avoid the localized overheating effect and to prevent turn-on failure in the device, an external inductor is used in the external circuit to limit the anode current rise rate to a level below the critical di_A/dt. The critical di_A/dt for the commercial GTO devices is usually below 1000 $A/\mu s$. To ensure that the entire junction area is uniformly turned ON before turn-off switching, a minimum turn-on time (usually about 100 µs) is required for commercial GTO devices. The typical non-uniform turn-on waveforms are shown in Fig. 5.88 (a). Another way of describing the non-uniform turn-on is that the turn-on process relies heavily on the self regenerative process of the thyristor latch-up.

(B) Uniform Turn-On Process Because of the ETO's low inductance (about 10 nH) gate-loop, a gate-current with larger amplitude and rise rate can be applied during turn-on. Through applying a gate-current pulse with higher amplitude, the excess carriers can be accumulated quickly in the P-base and N-base regions during delay-time t_p. The excess electrons injected from the N-emitter will diffuse towards junction J_2 and will discharge the positive space charge region in the N-base once they have passed through J_2. At the same time, holes provided by the gate-current will discharge the negative space charge region in P-base. Before the holes injected from the anode P-emitter can diffuse through the un-depleted N-base and reach the negative space charge region in the P-base, the J_2 space charge region has already been discharged. In other words, NPN transistor is turned-on uniformly within the initial phase, then operates in the active region, quickly moving across its active region towards the quasi-saturation region before the thyristor latch-up action begins, at which point the anode current increases significantly. Another way to look at this is that the NPN transistor is turned-on quickly into the saturation region before the PNP transistor starts to turn-on. In this type of turn-on, the charges required to maintain the NPN transistor within the saturation region are provided by the gate-drive unit. Little charges are injected from the anode side. In this case, the turn-on delay-time t_d is larger than t_p and is less than $t_p + t_n$.

To achieve this type of turn-on, during the delay time t_d, the gate current should deliver enough charge to discharge the J_2 space charge region. So the gate current required to uniformly turn-on the ETO is given as

$$I_G t_d > Q_{GT} \tag{5.62}$$

where Q_{GT} is the charge needed to turn-on the GTO device and discharge the space charge region. Therefore,

$$Q_{GT} \geq A \cdot w \cdot n* \cdot q/r_n \tag{5.63}$$

where r_n is the J_3 emitter injection efficiency,

A is the device area, w is the total base and n-base and P-base width and

$n*$ is the modulated carrier plasma level, which is much higher than the N-base background doping N_D.

So,
$$I_G > A \cdot w \cdot n* \cdot q/r_n/t_d \tag{5.64}$$

It can be seen from Eq. (5.64) that a larger ETO device requires a higher gate-current in order to achieve uniform turn-on. The gate current rise-rate (dI_G/dt) should satisfy the following:

$$\frac{dI_G}{dt} \gg \frac{I_G}{t_d} \tag{5.65}$$

During the voltage fall phase, the angle voltage collapses quickly, while the anode current stays at a low level, as shown in Fig. 5.88 (b). The anode voltage continues to decrease until reaching a saturation voltage level inspite of the increase in anode current. The turn-on loss is therefore reduced since there is not much overlap between the anode voltage and current. Under this condition, the GTO is uniformly turned-on without current crowding problems. Therefore, the critical anode current rise rate can be increased. The minimum on-time can be reduced since the turn-on process of the GTO in this case is more similar to that of a transistor than it is to a thyristor. Under this condition, ETO's turn-on performance is greatly improved.

It should be mentioned that at a higher anode current rise rate, an anode voltage bump may occur, indicating either that there is still an unmodulated N-base-region (or that the $n*$ level) provided by the gate current is still too low. To eliminate this bump, a gate current with an even higher amplitude is required to increase $n*$ in Eq. (5.64). In other words, one indication of uniform turn-on is that the voltage fall phase has no bump during the high current rise-phase.

5.21.3 ETO Gate Drive Circuit

The ETOs gate-drive circuit and control signals are shown in Fig. 5.89. Following the turn-on command, switch Q_P is turned-on and the 18 V voltage source begins to charge the inductor L_P through Q_P, L_P and Q_G. After a charging time t_P, Q_P is turned-off and the current in L_P freewheels through Q_G and D_P. Then, Q_G is turned-off and Q_E is turned-on. The current in Q_G will be transferred instantaneously to the GTOs gate, since the inductor L_P serves as a current source at this moment. The current decrease rate in Q_G, which is also the gate current rise rate, is dictated only by the breakdown voltage of Q_G (about 55 V) and the gate loop inductance (about 10 nH). Thus, the gate current rise rate can be as high as 5 kA/μs, which is high enough to uniformly turn-on the ETO. Once the current in L_P is totally commutated to GTOs gate, it will freewheel through emitter switch Q_E and diode D_P and then slowly drop down to zero, as shown in Fig. 5.89 (b). By varying the charging time t_P, gate-current pulses with different amplitudes can be obtained and the ETOs corresponding turn-on performance can be characterized.

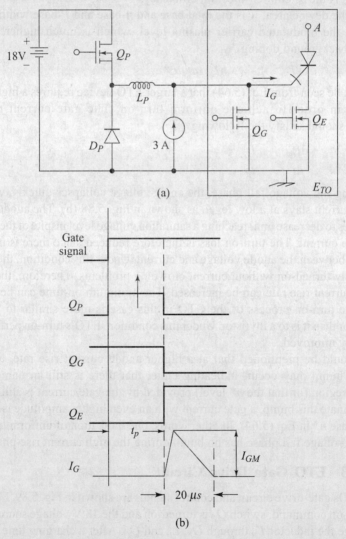

Fig. 5.89 *(a)* ETOs gate drive circuit and *(b)* control signals

5.21.4 Snubberless Turn-Off Process

A typical snubberless ETO turn-off characteristic is shown in Fig. 5.90, which defines several important time instances during a snubberless turn-off. The turn-off operation begins from the time zero point t_0. Thereafter, the emitter switch Q_E is turned-off and the gate-switch Q_G is turned-on. In the period t_0–t_1, the voltage on the emitter switch rises and stays at a clamped value. Because the internal structure of the GTO gate-cathode is a *PN* junction, the potential of the GTO gate will increase correspondingly, and the current from the upper *PNP* transistor in the GTO will be commutated to the gate switch Q_G until finally the cathode current becomes zero.

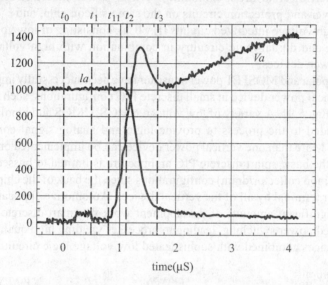

Fig. 5.90 *ETO turn-off characteristic*

At t_1, the ETO achieves the so-called unity turn-off gain. In the time period t_1–t_2, the anode current remains constant, flowing through the *PNP* transistor to the GTO gate. The minority carriers in the *p*-base region are pulled out by this current until t_{11} when the main junction J_2 is recovered. From t_{11} to t_2, the minority carriers in the N⁻–region are swept out while the voltage increases to the d.c. link voltage value. The GTOs emitter (gate-cathode) *PN* junction is turned-off from t_1 to t_2.

In the time period t_2–t_3, because there is no more base current injected into the base of the *PNP* transistor, the anode current falls to its tail value rapidly and voltage spike appears due to the di/dt applied to stray inductances. Finally in period after t_3, the anode voltage remains high while the anode current decreases slowly, determined by the minority carriers recombination in the undepleted N⁻- region.

5.22 POWER INTEGRATED CIRCUIT (PICs)

Power integrated circuit technology has developed rapidly during recent years, making it possible to combine substantial power handling and sophisticated control functions on the same silicon chip. A PIC is loosely defined as "smart power." The motivation for power circuit integration is reduction of cost and improvement of reliability. The main problems in a PIC synthesis are isolation between high voltage and low voltage devices and cooling.

The different PIC processes and architectures that have been developed around the world can generally be classified into two major broad categories:

(1) Smart discrete PICs, combining vertical power device structures with driver and protection circuits on the same silicon chip, and
(2) High voltage integrated circuits (HVICs), consisting of low voltage analog and digital logic circuitry in combination with high voltage lateral power devices.

Both bipolar and MOSFET power transistor have been successfully implemented as the vertical power device in small discrete PIC configurations, such as the one shown in Fig. 5.91. A variety of low voltage NMOS, CMOS and bipolar devices can be added to the process to provide logic and analog signal conditioning functions. More than one vertical power device can be implemented on the same chip, but the basic smart discrete PIC architecture is limited to bussed terminal (e.g., common collector/drain) configurations since the back of the chip is shared as a power terminal by all of the vertical devices. Automotive applications have spurred significant amounts of development work on smart discrete PIC's for multiplexed power switching, leading to processes yielding high quality vertical power devices combined with sophisticated low voltage logic circuitry.

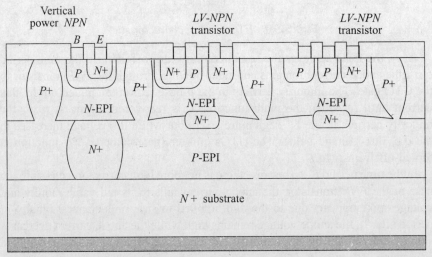

Fig. 5.91 *Smart discrete PIC incorporating high-voltage vertical devices*

HVICs incorporate lateral high voltage device in place of the vertical structures so that the back side of the IC is no longer used as an active electrical terminal for the PIC. Three basic isolation techniques, self, junction, and dielectric isolation are widely used to permit HVIC subcircuits to operate at widely separated voltage potentials. Figure 5.92 illustrates a junction-isolated structure, which uses reverse-biased P-N-junction to form "island" in this n-epitaxial layer for the isolated HVIC subcircuits, with the P-substrate held at the most negative potential in the circuit.

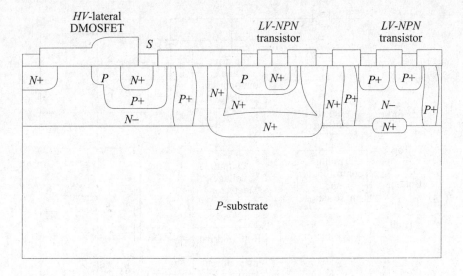

Fig. 5.92 *High voltage integrated circuit (HVIC) including high voltage lateral devices*

Charge control techniques have been developed for distributing the electric field in the HVIC epitaxial layer so that lateral power transistors can be flexibly designed with breakdown voltages of 500 V or more set by the device physical dimensions. A wide variety of bipolar, CMOS, and I^2L (integrated injected logic) devices can be accommodated in the same HVIC process to form analog and digital logic subcircuits operating at different potentials but communicating via high-voltage level shifting lateral devices.

Recently, a large family of PICs that include power MOSFET smart switch, half-bridge inverter driver, H-bridge inverter, two-phase step motor driver, one-quadrant chopper for d.c. motor drive, three-phase brushless d.c. motor driver, three-phase diode rectifier–PWM inverter are being available. Evidently, the majority of PICs are being targetted for motion control applications. Recently, application specific PICs (ASPIC) are also being available in the market.

Figure 5.93 shows a half-bridge power MOSFET driver manufactured by Harris (Gs 6000/1). It is more appropriate to define it as an HVIC. If the ground of the HVIC driver chip is referred to the negative power bus terminal, as in indicated in Fig. 5.93, the gate drive command for the upper switch S_1 must be level shifted in order to accommodate the S_1s floating source terminal reference potential. Supplying gate charge to the upper S_1 switch is further complicated by the fact that the on-state gate voltage for S_1 must be higher than the positive power bus potential $+E_{dc}$ by 10 to 15 V.

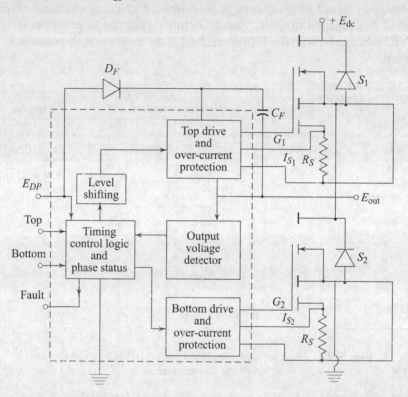

Fig. 5.93 *Block diagram of HVIC driving two power MOSFETs with integrated current sensors in half brige configuration*

The most common technique for supplying the modest gate charge requirements of S_1 is by means of a simple diode–capacitor charge pump. As illustrated in Fig. 5.94, capacitor C_f is charged by the low voltage supply E_{DD} when S_2 is on and then delivers the charge to the S_1 gate at a voltage above E_{dc} when S_1 is ON. Switch S_2 must be turned ON often enough to refresh the charge extracted from C_f, placing an upper limit on the duty cycle of the upper switch. Alternatively, more complicated charge pump configurations using independent free-running oscillators can be adopted to circumvent these duty-cycle limitations.

HVIC half bridge drivers using 500 V junction isolation PIC technology are commercially available with various combinations of features. The block diagram of half-bridge integrated power MOSFET driver in Fig. 5.93 illustrates key features of Harris chip, which is representative of the most sophisticated of the available HVIC half bridge driver offerings.

In addition to the level shifting functions, the HVIC logic includes half-bridge lockout protection to ensure that S_1 and S_2 are never gated ON at the same time. Other protection features include overcurrent for both S_1 and S_2 (using integrated current sensors, if desired), logic supply E_{DD} under voltage detection, and output state monitor to detect when E_{out} is not following the input logic command. The detected faults are reported to the control controller, which is then given as a logic output. HVIC half bridge drivers are available with sufficiently high voltage and current drive capabilities to be used in integral horsepower motor drives rated at 10 HP or higher.

PICs have the following advantages:

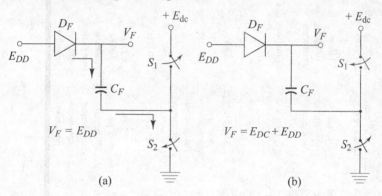

Fig. 5.94 Operating principles of charge pump for upper switch gate drive supply: **(a)** capacitor charges when S_2 is closed, **(b)** capacitor supplies charge to S_1 gate when S_2 is open

(1) Lower assembly and inventory costs from parts count reduction.
(2) Improved reliability by eliminating package interconnections.
(3) Major volume and weight reductions via circuit integration.
(4) Faster dynamic/protection response times from on-chip logic.
(5) Improved features protection possible using integrated sensors.

PICs have the following disadvantages:

(1) PIC chip cost may be higher than sum of replaced components.
(2) Lack of long reliability track record for new PIC technology.
(3) Limited PIC power handling capability in existing packages.
(4) Performance compromises due to integrated power/logic process.
(5) High engineering cost of developing a full-custom PIC design.

5.23 COMPARISON OF POWER DEVICES

Power semiconductors can be classified as unipolar or bipolar according to their primary feature, the charge carriers involved in the transport of current. In unipolar devices (SIT, MOSFET) only one type of charge carrier is involved in current transport; high blocking voltages and a high current capacity cannot therefore be attained at one and the same time. However, they can be controlled with very

Table 5.2 Power devices comparison

S.No.	Parameters	Thyristor	BJT	Power MOSFET	GTO	IGBT	SIT	SITH	MCT
1.	Voltage and current ratings	10 kV/4 kA.	2 kV/1 kA	500 V/200 A	5 kV/3 kA	1200 V/400 A	1200 V/300 A	1200 V/300 A	3 kV/300 A
2.	Linear/Trigger	Triggered Current	Linear Current	Linear Voltage	Triggered Current	Linear Voltage	Linear Voltage	Linear Current	Triggered Voltage
3.	Gating	Symmetric/ Asymmetric	Asymmetric	Asymmetric	Symmetric/ Asymmetric	Asymmetric	Asymmetric	Asymmetric	Asymmetric
4.	Voltage blocking								
5.	SOA	Does not apply	Second Break (down (SB) and T_j limited	T_j limited	T_j limited SB at turn-off	T_j limited	T_j limited	T_j limited	T_j limited
6.	Conduction drop	< 2 Volts	< 2 Volts	4–6 Volts	> 4 Volts	3.3 Volts	18 V	4 V	1–2 V
7.	Temp. coeff. of resistivity	Negative	Negative	Positive	Negative	Approximately flat (positive at high current)	Positive	Negative	Negative
8.	Turn-off gain	—		—	3 to 5			≤3	—
9.	Switching frequency	500 Hz	10 kHz	upto 100 kHz	2 kHz	10 kHz	70 kHz	10 kHz	10 kHz
10.	Turn-on time	2 μs	1 μs	100 ns	2–4 μs	<1 μs	0.25 μs	2 μs	1 μs
11.	Turn-off time	200 μs	2–5 μs	150–200 ns	10 μs	2 μs	300 ns	9–10 μs	2 μs
12.	Snubber	Unpolarized	Polarized	Polarized (snubberless operation is possible)	Polarized or unpolarized	Polarized (snubberless operation possible)	Polarized (snubberless operation possible)	Polarized	Polarized (snubberless operation possible)

(Contd.)

Table 5.2 (Contd.)

S.No.	Parameters	Thyristor	BJT	Power MOSFET	GTO	IGBT	SIT	SITH	MCT
13.	Reapplied $\frac{dV}{dt}$ (V/µs)	30	5 device	Limit for device loss	Limit for Miller effect and SOA	Limited by device loss	Limit for high	Very	2000	5000
14.	Turn-on $\frac{di}{dt}$ (A/µs)	200	21A/µs	loss 300	100	Very high	Very high	Very high	900	1000
15.	Leakage current (mA)	3	1	30.	2.5	0.2	1	0.1	25	1.0
16.	Protection	Gate inhibit or fast fuse	Gate inhibit or fast fuse	Gate inhibit or fast fuse.	Base control.	Gate control.	Gate control.	Gate control.	Fast fuse or gate inhibit.	Fast fuse or gate inhibit.
17.	Applications	d.c./a.c. motor drives, large-power supplies, lighting and heating, static var compensators, electronic circuit breakers	Lamp dimming, heating control, zero voltage switched a.c. relay, a.c. motor starting and control	Motor drives, UPS systems, static var compensators, induction heating	Motor drives, UPs systems, static var and harmonic compensators, Switched mode power supplies	Switching mode power Supply, Brushless d.c. motor drive, electronic d.c. relay	A.C. motor drives, Ups systems, static var and harmonic compensators, switching mode power supplies	Induction heating, ultrasonic generators, Am/Fm generators. Linear power amplifiers	Induction heating static var compensation	a.c. motor drives, Ups systems, static var and harmonic compensators.
18.	Comments	Large surge current capability	Large surge current capability	High uncontrollable surge, current. Large snubber to limit device loss switching low frequency at high power	Current gain varies with °C and collector current.	Built-in body diode can carry full current, but slow recovery. Becomes higher with higher voltage and current ratings.	Possibility of the device latching up by regenerative action. Combines the feature like low forward voltage with low switching loss.	Off leakage current is more difficult to define	This device is extremely process sensitive, and small perturbation in the manufacturing process would produce major change in the characteristics	−196 to +250 T_j range possible. Characteristics comparable to IGBT excepted lower drop

small amounts of energy (field controlled) and are, thanks to their short switching times, suitable for very high switching frequencies. MOSFETs and SITs are restricted to the low power range.

In the case of bipolar devices (BJT, IGBT, GTO, SITH, FCT) charge carriers are injected at a forward biased *P-N* junction from the heavily doped zone (emitter) to the lightly doped zone. The conductivity of the lightly doped zone can be raised by several powers of ten (conductivity modulation), leading to feasible devices with both high blocking and current carrying capabilities. BJTs, and IGBTs, are suitable for applications in the medium and lower power ranges, and GTOs and SITHs for the upper power range. Due to their switching behaviour, which is caused by their physical characteristics, bipolar devices are restricted to low to medium switching frequencies. This also applies to power BJTs, especially in a Darlington circuit.

5.24 SILICON CARBIDE DEVICES

Recent analysis has shown that very high performance FETs and Schottky rectifiers can be obtained by replacing silicon with gallium arsenide, silicon carbide or semiconducting diamond. Among these silicon carbide is the most promising material because its technology is more mature than for diamond and the performance of the silicon carbide devices is expected to be an order of magnitude better than that of gallium arsenide devices. In these devices, the high breakdown electric field strength of silicon carbide leads to a 200 fold reduction in the resistance of the drift region. As a consequence of this low drift region resistance, the on-stack voltage drop for even the high voltage FET is much smaller than for any unipolar or bipolar silicon device. These switches can be expected to switch-off in less than 10 nanoseconds and have superb FBSOA. The analysis also indicates that having voltage Schottky barrier rectifiers with on-state voltage drops close to 1 volt may be feasible with no reverse recovery transients.

If the silicon power switch and rectifier are replaced with the silicon carbide devices, it becomes possible to reduce the turn-off time 10 nanoseconds because both the switch and the rectifier behave as nearly ideal devices. It is also observed that the power losses have been drastically reduced at all switching frequencies. Also, power loss obtained with silicon carbide devices is smaller than that for silicon devices. In addition, it is anticipated that these devices can be operated at higher junction temperatures due to the large band gap of SiC, making the heat sink requirement less stringent.

REVIEW QUESTIONS

5.1 Consider each of the following descriptions. For each one indicate which *PNPN* device or devices the statement describes. A given statements can pertain to more than one *PNPN* device.
  (i) A bilateral two-terminal device.
  (ii) Turned "on" only by exceeding anode–cathode switching voltage.

(iii) A bilateral device which is triggered by a gate signal.
(iv) A unilateral device that can be turned "off" by a negative gate signal.
(v) Can be turned "on" or "off" at either gate.
(vi) A low-current, four-terminal device.
(vii) A unilateral two-terminal device with V_s usually ≤ 10 V.
(viii) Can be turned "off" by reducing current below I_H.
(ix) Can be turned "on" by a negative pulse.
(x) Once it is turned "on" it remains in the low resistance even after the signal that produced turn "on" has been removed.
(xi) Triggered "on" by applying light energy.

5.2 List the different members of the thyristor family. Draw their characteristics and explain in brief.

5.3 Explain briefly the difference between the phase control thyristors and inverter grade thyristors.

5.4 Draw and explain the construction of an asymmetric thyristor. Also, explain why the doping levels are adjusted in ASCRs.

5.5 Explain the difference between ASCRs and RCTs.

5.6 Explain the disadvantages of the RCT in detail.

5.7 Discuss in brief the concepts of gate-assisted turn-off thyristors.

5.8 Draw the V–I characteristics of a Diac and explain its working principles. Explain the term breakover voltage of the Diac.

5.9 Explain with the help of a circuit diagram, how Diac is used as a triggering agent for a Triac.

5.10 Discuss the advantages of ASCR and RCT over conventional thyristors.

5.11 Explain with the help of layer diagram the construction of a Triac.

5.12 Draw the V–I characteristics of a Triac and explain its working principle.

5.13 Explain the difference between a Triac and an SCR.

5.14 Explain the various triggering modes of a Triac. Compare their sensitivity.

5.15 Draw and explain the a.c. phase controller circuit using Triac and Diac. Which modes are used here? Discuss the best triggering mode recommended.

5.16 List the advantages and disadvantages of Triac over SCR.

5.17 If an SCR and a Triac of same rating are available, which will you prefer as a control device? Give reasons.

5.18 Explain why the Diac–Triac matched pair should be used in a control circuit.

5.19 Draw a neat circuit diagram of a simple light dimmer circuit using Triac and draw the waveforms of voltage across the bulb and current passing through it for $\alpha = 0°$ and $\alpha = 90°$.

5.20 Explain why silicon unilateral switch (SUS) turns ON by a fixed anode to cathode voltage. Draw relaxation oscillator circuit using SUS and explain its working.

5.21 With the help of equivalent circuit, explain the operation of silicon-bilateral switch.

5.22 Draw and explain the V–I characteristics of SCS. Also, differentiate between an SCS and SCR.

5.23 Differentiate between LASCRs and SCRs.

5.24 Draw and explain the operation of cross-sectional structure of power MOSFET.

5.25 Draw and explain the output characteristics of n-channel enhancement mode MOSFET.

5.26 Draw electrical equivalent circuit of a power MOSFET and explain why are they preferred in the inverter application.

5.27 Draw and explain the switching behaviour of power-MOSFET.

5.28 What is the significance of safe-operating area of a power MOSFET? Name other operating limitations of a power MOSFET.

5.29 Discuss briefly the parallel operation of MOSFETs.

5.30 Briefly discuss the gate-drive design considerations of the MOSFET.

5.31 Briefly discuss the following gate-drive circuits for power MOSFETs:
  (i) Driving the MOSFET from TTL
  (ii) Emitter-follower driving circuit
  (iii) CMOS gate-drive circuits
  (iv) Op-amp emitter-follower driver circuit
  (v) Isolated-gate drive circuit

5.32 Briefly explain the sources of power-losses in the switching MOSFET.

5.33 Compare power BJTs with power MOSFETs.

5.34 Justify the following statements:
  (i) Power MOSFET is operated at high enough gate-source voltage to minimize the conduction losses.
  (ii) Parallel operation of MOSFETs can be done more easily as compared to thyristors.
  (iii) Antiparallel diode is connected across MOSFET.
  (iv) Operating frequency of power MOSFET is higher than that of a power BJT.

5.35 With the help of neat structural diagram and suitable waveforms, explain the operation of insulated-gate BJT. (IGBT).

5.36 Explain the following points with respect to IGBT:
  (i) Forward blocking capability.
  (ii) Reverse blocking capability, and
  (iii) Forward conduction mode.

5.37 Describe briefly the latch-ups in an IGBT.

5.38 Discuss the static latch-up and dynamic latch-ups in an IGBT.

5.39 Discuss the switching characteristics of the IGBT with the help of neat circuit diagrams and waveforms.

5.40 Explain the safe operating areas of an IGBT.

5.41 Briefly describe the gate-drive circuits for IGBT.

5.42 Justify the following statements:
  (i) IGBT uses a vertically oriented structure
  (ii) IGBT combines the advantages of MOSFET and power BJT
  (iii) IGBT is preferred as a power switch over both the power BJT and MOSFET
  (iv) Punch-through IGBT structures are more popular and are widely used.

5.43 Compare SCR, Power BJT, MOSFET and IGBT on the basis of following parameters:
  (i) operating frequency   (ii) trigger circuit
  (iii) drop   (iv) snubbers
  (v) V-I rating   (vi) Applications

5.44 Briefly explain the V-I characteristics of an IGBT.

5.45 Explain the basic difference between non-punch-through IGBTs and punch-through IGBTs.
5.46 Explain the operating principle of an IGBT on the basis of:
  (i) creation of an inversion-layer
  (ii) conductivity modulation of the drift layer
5.47 Draw and explain the equivalent circuit for modelling the operation of the IGBT.
5.48 Explain the parallel operation of IGBT. Also, highlight the problems faced while parallel-operation.
5.49 Briefly explain the gate-drive design considerations of IGBTs.
5.50 Draw and explain the IGBT driver circuit with overcurrent protection.
5.51 Draw and explain the functional block-diagram of IC IR 2110. Also, draw the typical waveforms.
5.52 How does a GTO differ from a conventional thyristor. Give its circuit symbol.
5.53 With the help of a neat structural diagram, explain the operation of GTO.
5.54 Explain briefly the switching behaviour of a GTO.
5.55 Describe the turn-off process in a GTO with the help of appropriate voltage and current waveforms.
5.56 Briefly discuss the overcurrent protection of GTOs.
5.57 Derive the expression for turn-off gain of GTO. Also, discuss on the magnitude of this parameter for reliable turn-off of GTO.
5.58 With the help of basic structural diagram explain the operation of static induction transistor. Also, list the applications of SIT.
5.59 Give the merits and demerits of a GTO as compared to a conventional SCR.
5.60 With the help of a neat structural diagram, explain the operation of static induction thyristors.
5.61 With the help of structure diagram and equivalent circuit, explain the behaviour of a MOS-controlled thyristor.
5.62 Justify: MCT is a GTO with MOS-controlled gate.
5.63 Explain the operating principle of N-MCT by neatly sketching the cross-sectional view and equivalent circuit diagram.
5.64 Explain the operating principle of P-MCT by neatly sketching the cross-sectional view and equivalent circuit diagram.
5.65 Briefly discuss the turn-on and turn-off behaviour of N-MCT
5.66 Briefly explain the turn-on and turn-off behaviour of P-MCT.
5.67 Comment on the I-V characteristics of MCT.
5.68 Draw and explain the switching characteristics of MCT.
5.69 Explain the following timing parameters related to MCT:
  (i) turn-on delay time and current rise time
  (ii) turn-off delay time and fall time
5.70 Justify: P-MCT can turn-off a current approximately three-times larger than an N-MCT.
5.71 Explain the effect of temperature on the operation of MCT.
5.72 Explain briefly the safe-operating area of MCT.
5.73 Briefly describe the operation of IGCT.

5.74 Explain the following terms with reference to IGCT:
  (i) Buffer layer
  (ii) Transparent emitter
  (iii) Reverse conduction
  (iv) Controlled diode recovery
5.75 Explain with the help of neat diagrams the basic principle of operation of MTO.
5.76 Describe the basic principle of operation of ETO.
5.77 Explain briefly the nonuniform and uniform turn-on process of ETO with the help of neat waveforms.
5.78 Draw and explain the operation of ETO gate-drive circuit.
5.79 With the help of neat waveforms, explain the snubberless turn-off process of ETO.
5.80 Compare MTO and IGCT briefly.
5.81 Compare MTO and ETO on the basis of:
  (i) speed
  (ii) gain
  (iii) efficiency
  (iv) control method
  (v) maximum ratings

PROBLEMS

5.1 The circuit shown in Fig. P 5.1 is used to control the power dissipated in the 10Ω load resistor. Assume that the bilateral switching diode has a breakdown voltage of ± 2.8 V and that the holding voltage of the diode and Triac are negligible.

The applied sinusoidal voltage is 300 V RMS, at 60 Hz.
  (i) Compute the conduction angle.
  (ii) Draw the waveform of the voltage applied to the load.
  (iii) Compute the total power dissipated by the load resistor.

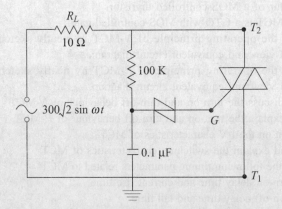

Fig. P 5.1

[Ans (i) Conduction angle = 103.5°; (iii) Total power dissipated = 5.81 kW]

5.2 Figure P 5.2 shows an SCR used in a over-heater detector circuit. The thermistor used in the circuit is a temperature sensitive resistor; its resistance increases with temperature. The SUS shown in the circuit has $V_s = 8$ V.

(a) Explain what happens as the temperature of the thermistor increases.
(b) At what value of R_T will the SCR trigger "on" and active the indicator light? (Assume $I_s = 0$).
(c) How can we vary the temperature at which triggering occurs?

[*Ans* (b) 1 kΩ]

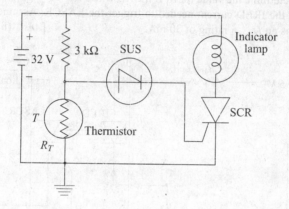

Fig. P 5.2

5.3 Figure P 5.3 shows a LASCR being used in a flame-detector circuit. Since LASCRs are relatively low current devices, the LASCR cannot drive the alarm directly. Here, the LASCR is used to trigger the high current Triac, which switches power to the alarm.
  (a) Describe the circuit operation.
  (b) Determine an appropriate value for R_1 if the LASCR has $I_H = 10$ mA and the Triac has an $I_{G(max)}$ rating of 50 mA. Assume $I_{GT} = 5$ mA, $V_{GT} \approx 0$ for the Triac.
  (c) Why wouldn't a photoconductive cell or phototransistor be used instead of the LASCR?

[*Ans* (b) 5.75 kΩ]

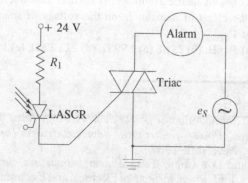

Fig. P 5.3

5.4 Figure P 5.4 shows how a computer might control a high-power load without direct electrical connection. The computer output is either at 0 V or +5 V relative to ground.
   (a) Describe the circuit operation for the two possible value of V_{cmp}.
   (b) Determine the value for R_1 if the IRED has $C_F = 1$ V @ $I_F = 10$ mA, which is the IRED current needed to trigger the LASCR. Assume the transistor has an $I_{C(max)}$ rating of 30 mA. [Ans (b) 133 – 400 Ω]

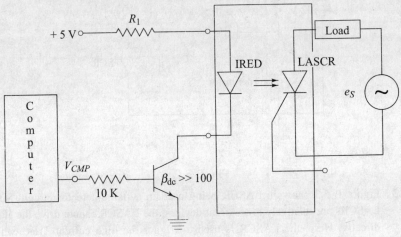

Fig. P 5.4

5.5 A MOSFET is operating as a chopper switch at a frequency of 60 kHz, having the circuit arrangement similar to Fig. 5.48. The d.c. input voltage of the chopper is $E_{dc} = 40$ V and the load curent is 50 A. The switching times are $t_a = 70$ ns and $t_f = 20$ ns. Compute the values of:
(a) L_s; (b) C_s, (c) R_s for critically damped condition; (d) R_S, if the discharge time is limited to one-third of the switching period; (e) R_S, if peak discharge current is limited to 6% of load current; and (f) power loss due to R_C snubber, P_s, neglecting the effect of inductor L_s on the voltage of snubber capacitor, C_s. Assume that $V_{CE(sat)} = 0$.
   [Ans (a) 56 nH; (b) 25 nF (c) 2.99 Ω; (d) 222.22 Ω; (e) 13.33 Ω; (f) 1.2 W.]

REFERENCES

1. S K Gandhi, *Semiconductor power devices*, John Wiley, New York, 1987.
2. B W Williams, *Power electronics, devices, drivers, and applications*, John Wiley, New York 1987.
3. B J Baliga and D Y Chen, (Eds.), *Power transistors, device, design and applications*, IEEE Press, Institute of Electrical and Electronic Engineers, New York, 1984.

4. S Adler Michad King W Owyang, BJ Baliga and RA Kokosa, "*The evolution of power device technology*" IEEE Transactions on Electron Devices, 1984.
5. SCR Manual, 6th Ed., General Electric Company, New York, 1979.
6. B K Bose, "Recent advances in power electronics," IEEE Transactions on Power Electronics, 1992.
7. B R Pelley, "Power MOSFETs—A status review," International Power Electronics Conference Record, 1983.
8. V A K Temple, "MOS controlled thyristors a new class of power devices," IEEE Transactions on Electron Devices, 1986.
9. N Y Seki, Tsuruta, and K Ichiawa, "Gating circuit developed for high power thyristors", IEEE Power Electronics Specialists Conference, 1981.
10. B J Baliga, M Chang, P Shafer, and MW Smith, "The insulated gate transistor – A new power switching device," IEEE industry Applications Society Meeting, 1983.
11. M Thomas Johns, "Designing intelligent muscle into industrial motion control" IEEE Trans Industrial Electronics, OCT., 1990.
12. Mohan/Undeland/Robins, "Power-Electronics," Second Edition, Wiley, 1995.
13. Zhenxue, X., Y. Bai, B. Zhang, and A. Huang, "The uniform turn-on of ETO", IEEE 2002.

Chapter 6
Phase Controlled Converters

LEARNING OBJECTIVES:
- To examine the operation of naturally commutated converters.
- To understand the operation of converters in the rectifying and inverting modes.
- To understand the operation of single-phase and three-phase half controlled and fully controlled converters.
- To develop the general equations describing converter behaviour.
- To consider the function and operation of the freewheeling diode.
- To examine the effect of source inductance on commutation and to define overlap.
- To understand the various firing schemes for converters.
- To Understand the various techniques of improving power factor in phase controlled converters.

6.1 INTRODUCTION

Rectification is a process of converting an alternating current or voltage into a direct current or voltage. This conversion can be achieved by a variety of circuits based on and using switching devices. The widely used switching devices are diodes, thyristors, power-transistors, power MOS, etc. The rectifier circuits can be classified broadly into three classes: uncontrolled, fully-controlled and half-controlled. An uncontrolled rectifier uses only diodes and the d.c. output voltage is fixed in amplitude by the amplitude of the a.c. supply. The fully-controlled rectifier uses thyristors as the rectifying elements and the d.c. output voltage is a function of the amplitude of the a.c. supply voltage and the point-on-wave at which the thyristors are triggered (called firing-angle α). The half-controlled rectifier contains a mixture of diodes and thyristors, allowing a more limited control over the d.c. output voltage-level than the fully-controlled rectifier. The half-controlled rectifier is cheaper than a fully-controlled rectifier of the same ratings but has operational limitations.

Uncontrolled and half-controlled rectifiers will permit power to flow only from the a.c. system to the d.c. load and are, therefore, referred to as unidirectional converters. However, with a fully-controlled rectifier it is possible, by control of the point-on-wave at which switching takes place, to allow power to be transferred from the d.c. side of the rectifier back into the a.c. system. When this occurs, operation is said to be in the inverting mode. The fully controlled converter may therefore be referred to as a bidirectional converter.

Hence it is possible for the phase-controlled converters to provide either a one- quadrant, two-quadrant or four-quadrant operation at its d.c. terminals. Figure 6.1(a) shows the diagrammatic representation of one-quadrant converter which contains controlled and uncontrolled rectifiers in different circuit positions. With one-quadrant converters only one polarity of voltage and current at d.c. terminals is possible. Figure 6.1(b) shows the two-quadrant converter diagrammatic representation, which contains controlled rectifiers in all circuit positions. With two quadrant converters, the power can be made to flow either from the a.c. to the d.c. side of the converter or *vice versa*. Figure 6.1(c) shows the diagrammatic representation of four-quadrant converters which contains two 2- quadrant converters, connected back-to-back with one another, thus providing the facility for bidirectional current-flow through the load.

Rectifiers are often described by their pulse-number. This is the number of discrete switching operations involving load-transfer (commutation) between individual diodes, thyristors, etc., during one cycle of the a.c.-supply waveform. The pulse-number is, therefore, directly related to the repetition period of the d.c. voltage waveform and it is sometimes expressed in terms of the ripple-frequency of this waveform. For example, a six-pulse converter operating from a 50 Hz-supply produces a 300 Hz (6×50) ripple in the output d.c. voltage-waveform.

The converter type depends on the power to be handled and how much voltage ripple will be tolerated. For low-powers, below 20 kW, single-phase circuits are adequate but they themselves can take different forms. For high powers, above 20 kW, three-phase circuits are used.

This chapter presents the analysis of different configurations of phase controlled rectifiers. The output voltage in all the configurations is controlled by delaying the firing-angle of the thyristor.

6.2 CONTROL TECHNIQUES

Figure 6.2 shows the technique of controlled conversion from ac to dc for a half-wave circuit which uses a unidirectional switch. When this switch S is turned-on, it conducts current in the direction of the arrow. The output voltage waveform depends on the switch control waveform and the pulse-triggered switch, such as SCR, GTO and MCTs or level-triggered switch such as BJT, MOSFET and IGBTs. Current pulses are required for triggering SCRs and GTOs whereas voltage pulses are required for MCTs, MOSFETs and IGBTs. Usually the source has an inductance, such as the ac line inductance or leakage inductance of transformer. For analysis purpose we neglect this inductance.

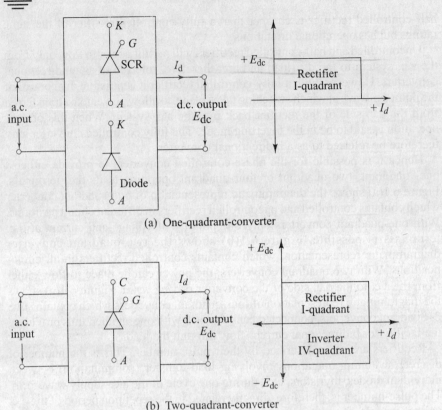

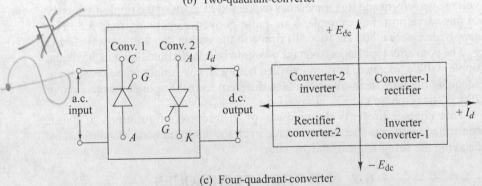

(c) Four-quadrant-converter

Fig. 6.1 *Types of phase-controlled converters*

6.2.1 Phase Angle-Control [Firing Angle Control]

In a.c. circuits, the SCR can be turned "on" by the gate at any angle, with respect to the applied voltage. This firing angle is measured with respect to a given reference, at which the firing pulses are applied to the thyristor gates. The reference point is the point at which the application of the gate pulses results in the maximum mean positive d.c.-terminal voltage of which the converter is capable. In other words, a firing-angle of 0° corresponds to the conditions when each thyristor in the circuit is fired at the instant its anode voltage-first becomes positive in each cycle, under this condition, therefore, the converter operates in exactly the same manner as if it was an uncontrolled rectifier circuit. The 'α' is the symbol for the firing-angle. Hence, the most efficient method to control the turning "on" of a thyristor is achieved by varying the firing-angle of thyristor. Such a method of control is called as phase-angle control. This phase-angle control is a highly efficient means of controlling the verage-power to loads such as lamps, heaters, motors, d.c. suppliers, etc.

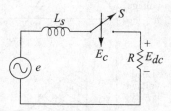

Fig. 6.2 Half-wave controlled converter using unidirectional switch

Those converters which work on the principle of natural commutation are called "line-commutated converters." There are many forms of phase-control possible with the thyristor. These various forms are discussed in the subsequent sections.

Figure 6.3 shows various waveforms obtained using phase-angle control for half-wave-controlled converter circuit. In a pulse-triggered switch, the falling edge of the control pulse (V_C) lies at an angle $\alpha + \delta$, as shown in Fig. 6.3 (c), where δ is a short angle. Once the switch is turned-on, it remains ON even if the triggering pulse has subsided. It can be turned-off only when the current through it is reduced below the holding current (in case of SCR).

In a level triggered switch, the falling edge lies at an angle π, as shown in Fig. 6.3 (d).

6.2.2 Extinction Angle Control

Figure 6.4 shows the output voltage waveform and the control pulses for the level-triggered and pulse-triggered switches. The rising edge of the control pulse coincides with the beginning of the input voltage waveform. The falling edge lies at a controllable angle ($\pi - \beta$). Angle β is called as the extinction angle. In a pulse triggered switch, the control pulse consists of two short pulses: one for turning-on and the other for forced-turn-off.

6.2.3 Pulse Width Modulation (PWM) Control

The output voltage waveform and control pulses using PWM control are shown in Fig. 6.5. As shown, the control pulse is symmetrically positioned with respect to the positive and negative peaks of the input voltage waveform. Pulse width δ is the control parameter. Like extinction angle control, the control pulse consists of two short pulses in case of pulse-triggered switch.

262   *Power Electronics*

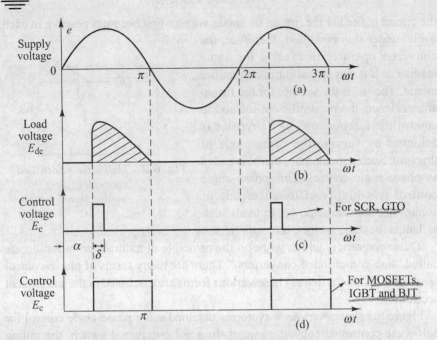

Fig. 6.3 *Phase-angle control*

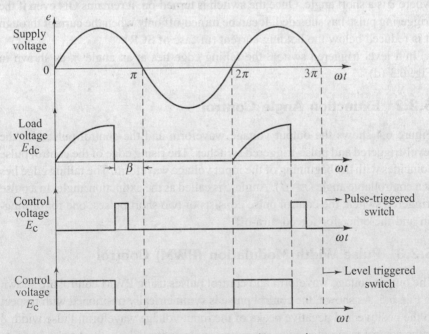

Fig. 6.4 *Extinction angle control*

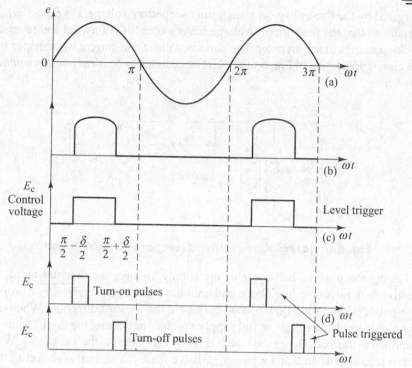

Fig. 6.5 *PWM control*

In multiple pulse width modulation (MPWM) control, control signals consists of p-pulses per half-period of the input voltage waveform. Both the pulse-width δ and the number of pulses p can be used as the control parameters. The two widely used methods are:

(a) Uniform multiple PWM [p-pulses of uniform pulse-width] and
(b) Sine-PWM [p-pulses with sinusoidal variation of pulse-width]

This type of control is suited for level triggered switches.

6.3 SINGLE PHASE HALF-WAVE CONTROLLED RECTIFIER

In its simplest form, phase control can be described by considering the half-wave thyristor circuit. In a half-wave single-phase controlled rectifier only one SCR is employed in the circuit. It is included in between the a.c. source and the load. The performance of the controlled rectifier very much depends upon the type and parameters of the output (load) circuit.

6.3.1 With Resistive Load

Figure 6.6(a) shows the circuit-diagram of a single-phase half-wave converter with resistive load. Triggering circuit is not shown in the figure. The circuit is energized by the line voltage or transformer secondary voltage, $e = E_m \sin \omega t$. It is assumed that the peak supply voltage never exceeds the forward and reverse-blocking ratings of the thyristor. The various voltage and current waveshapes for this circuit are shown in Fig. 6.6(b). *SCR is assumed to be ideal one throughout the chapter.*

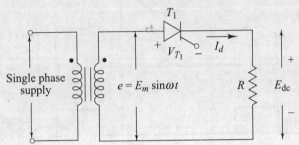

Fig. 6.6 *(a) Halfwave-controlled rectifier with resistive load*

During the positive half-cycle of the supply voltage, the thyristor anode is positive with respect to its cathode and until the thyristor is triggered by a proper gate-pulse, it blocks the flow of load current in the forward direction. When the thyristor is fired at an angle α, full supply voltage (neglecting the thyristor drop) is applied to the load. Hence the load is directly connected to the a.c. supply. With a zero reactance source and a purely resistive load, the current waveform after the thyristor is triggered will be identical to the applied voltage wave, and of a magnitude dependent on the amplitude of the voltage and the value of load resistance R. As shown in Fig. 6.6(b), the load current will flow until it is commutated by reversal of supply voltage at $\omega t = \pi$. The angle $(\pi - \alpha = \beta)$ during which the thyristor conducts is called the conduction angle. By varying the firing angle α, the output voltage can be controlled. During the period of conduction, voltage drop across the device is of the order of one volt.

During the negative half-cycle of the supply voltage, the thyristor blocks the flow of load current and no voltage is applied to the load R.

The voltage and current relations are derived as follows:

(a) Average Load Voltage The average value of the load-voltage can be derived as

$$E_{dc} = \frac{1}{2\pi} \int_{\alpha}^{\pi} E_m \cdot \sin \omega t \, d(\omega t)$$

where E_m is the peak value of the a.c. input voltage

$$= \frac{1}{2\pi} E_m [-\cos \omega t]_{\alpha}^{\pi} \quad E_{dc} = \frac{E_m}{2\pi}[1 + \cos \alpha]. \tag{6.1}$$

The maximum output voltage is obtained when $\alpha = 0$.

$$\therefore \quad E_{dcmax} = \frac{E_m}{\pi} \tag{6.2}$$

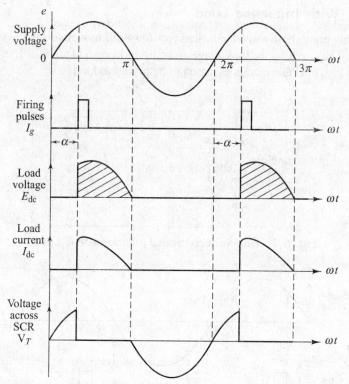

Fig. 6.6 (b) *Waveforms for a half-wave circuit*

(b) **Average load current** With resistive load, the average load current is directly proportional to the average load voltage divided by the load resistance:

$$\therefore \quad I_d = \frac{E_m}{2\pi R}[1+\cos\alpha] \tag{6.3}$$

(c) **RMS load voltage** The RMS load voltage for a given firing angle α is given by

$$E_{rms} = \left[\frac{1}{2\pi}\int_\alpha^\pi (E_m \sin\omega t)^2 \, d(\omega t)\right]^{1/2} = \left[\frac{E_m^2}{2\pi}\int_\alpha^\pi \sin^2\omega t \, d(\omega t)\right]^{1/2}$$

$$= E_m\left[\frac{1}{2\pi}\int_\alpha^\pi \left(\frac{1-\cos 2\omega t}{2}\right) d(\omega t)\right]^{1/2} = E_m\left[\frac{1}{4\pi}\left(\omega t - \frac{\sin 2\omega t}{2}\right)_\alpha^\pi\right]^{1/2}$$

$$E_{rms} = E_m\left[\frac{\pi - \alpha}{4\pi} + \frac{\sin 2\alpha}{8\pi}\right]^{1/2} \tag{6.4}$$

For firing angle $\alpha = 0$, $E_{rms} = \dfrac{E_m}{2}$ \hfill (6.5)

6.3.2 With Inductive Load

The single phase half-wave controlled rectifier with inductive-load is shown in Fig. 6.7. The waveshapes for voltage and current in case of an inductive load are given in Fig. 6.8. The load is assumed to be highly inductive.

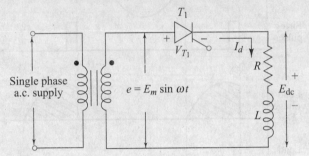

Fig. 6.7 *Half-wave controlled rectifier with R-L load*

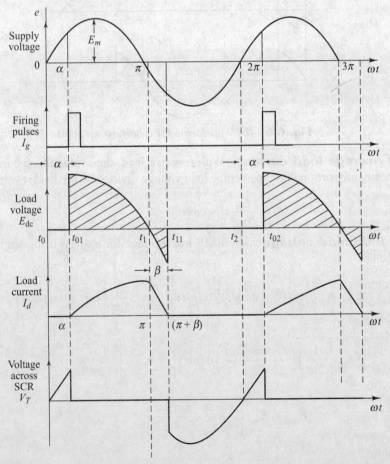

Fig. 6.8 *Waveforms for a half-wave controlled rectifier with RL load*

The operation of the circuit on inductive loads changes slightly. Now at instant t_{01}, when the thyristor is triggered, the load-current will increase in a finite-time through the inductive load. The supply voltage from this instant appears across the load. Due to inductive load, the increase in current is gradual. Energy is stored in inductor during time t_{01} to t_1. At t_1, the supply voltage reverses, but the thyristor is kept conducting. This is due to the fact that current through the inductance cannot be reduced to zero.

During negative-voltage half-cycle, current continues to flow till the energy stored in the inductance is dissipated in the load-resistor and a part of the energy is fed-back to the source. Hence, due to energy stored in inductor, current continuous to flow upto instant t_{11}. At instant, t_{11}, the load-current is zero and due to negative supply voltage, thyristor turns-off.

At instant t_{02}, when again pulse is applied, the above cycle repeats. Hence the effect of the inductive load is increased in the conduction period of the SCR.

The half-wave circuit is not normally used since it produces a large output voltage ripple and is incapable of providing continuous load-current.

The average value of the load-voltage can be derived as:

$$E_{dc} = \frac{1}{2\pi} \int_{\alpha}^{\pi+\alpha} E_m \cdot \sin \omega t \, d(\omega t)$$

Here, it has been assumed that in negative half-cycles, the SCR conducts for a period of α.

$$\therefore \quad E_{dc} = \frac{E_m}{2\pi} \left[-\cos \omega t\right]_{\alpha}^{\pi+\alpha} \quad \text{Or,} \quad E_{dc} = \frac{E_m}{\pi} \cos \alpha \tag{6.6}$$

From Eqs (6.1) and (6.6), it is clear that the average load-voltage is reduced in case of inductive load. This is due to the conduction of SCR in negative cycle.

6.3.3 Effect of Freewheeling Diode

Many circuits, particularly those which are half or uncontrolled, include a diode across the load as shown in Fig. 6.9. This diode is variously described as a commutating diode, flywheel diode or by-pass diode. This diode is commonly described as a commutating diode as its function is to commutate or transfer load current away from the rectifier whenever the load-voltage goes into a reverse-state.

This diode serves two main functions:
(i) It prevents reversal of load voltage except for small diode voltage-drop.
(ii) It transfers the load current away from the main rectifier, thereby allowing all of its thyristors to regain their blocking states.

Figure 6.10 shows a half-wave controlled rectifier with a freewheeling diode D_f connected across R–L load. The load-voltage and current waveforms are also shown in Fig. 6.11.

With diode D_f, thyristor will not be able to conduct beyond 180°.

268 Power Electronics

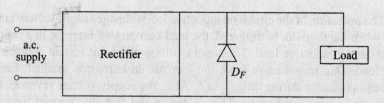

Fig. 6.9 *Position of commutating diode D_F*

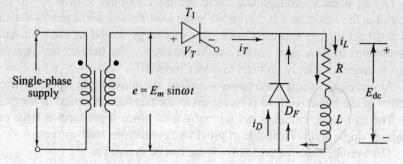

Fig. 6.10 *Half-wave rectifier with a freewheeling diode*

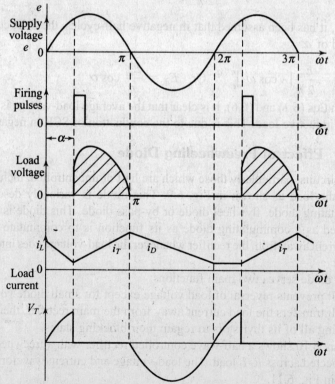

Fig. 6.11 *Waveforms for half-wave controlled-rectifier with inductive load and freewheeling diode*

Phase Controlled Converters

We know that during the positive half-cycle, voltage is induced in the inductance. Now, this induced voltage in inductance will change its polarity as the di/dt changes its sign and diode D_f will start conducting as soon as the induced voltage is of sufficient magnitude, thereby enabling the inductance to discharge its stored energy into the resistance.

Hence, after 180°, the load current will freewheel through the diode and a reverse-voltage will appear across the thyristor. The power flow from the input takes place only when the thyristor is conducting. If there is no freewheeling diode, during the negative portion of the supply voltage, thyristor returns the energy stored in the load inductance to the supply line. With diode D_f, the freewheeling action takes place and no power will be returned to the source. Hence, the ratio of the reactive power flow from the input to the total power consumed in the load is less for the phase-control circuit with a freewheeling diode. In other words, the freewheeling diode improves the input power-factor.

Mathematically:

$$\frac{EI \sin \phi}{EI} = \text{less} \quad \therefore \sin \phi = \text{less} \quad (6.7)$$

Since, $\phi = $ less $\therefore$ Power-factor $\cos \phi = $ more (6.8)

Hence it is clear that the freewheeling diode helps in improvement of power-factor of the system.

SOLVED EXAMPLES

Example 6.1 If the half-wave controlled rectifier has a purely resistive load of R and the delay angle is $\alpha = \pi/3$. Determine:
(a) Rectification efficiency (b) Form factor (c) Ripple factor
(d) Transformer utilization factor (e) Peak inverse voltage for SCR T_1

Solution:

(a) Rectification efficiency $\eta = \dfrac{p_{dc}}{p_{ac}}$ where, p_{dc} = dc load power = E_{dc}^2/R and

p_{ac} = rms load power = $\dfrac{E_{rms}^2}{k}$. We have the relation, from Eq. (6.1),

$$E_{dc} = \frac{E_m}{2\pi}(1 + \cos \alpha), \text{ Since, } \alpha = \pi/3 \quad \therefore E_{dc} = 0.239 E_m$$

Also, from equation (6.4), we have

$$E_{rms} = E_m \left[\frac{\pi - \alpha}{4\pi} + \frac{\sin 2\alpha}{8\pi} \right]^{1/2}$$

For firing angle $\alpha = \pi/3$, $E_{rms} = 0.485 E_m$

$\therefore$ Rectification efficiency $\eta = \dfrac{(0.239 E_m)^2}{(0.485 E_m)^2} \quad \therefore \eta = 24.28\%$

(b) Form factor $(ff) = \dfrac{E_{rms}}{E_{dc}} = \dfrac{0.485 E_m}{0.239 E_m} = 2.033 = 203.3\%$

(c) Ripple factor $(Rf) = (FF^2 - 1)^{1/2} = (2.033^2 - 1)^{1/2} = 1.77 = 177\%$

(d) Transformer utilisation factor $(TUF) = \dfrac{P_{dc}}{E_s \cdot I_s}$

where E_s and I_s are the rms secondary voltage and current respectively.

Now, $\quad E_s = E_m/\sqrt{2} = 0.707 E_m$

Also, $\quad I_s = $ rms load current $= \dfrac{E_{rms}}{R} = \dfrac{0.485 E_m}{R}$

$\therefore \quad TUF = \dfrac{E_{dc}^2/R}{E_s I_s} = \dfrac{(0.239 E_m)^2/R}{0.707 E_m \times 0.485 E_m/R} = 0.166 \text{ or } 16.6\%$

(e) Peak inverse voltage $= E_m$

Example 6.2 An SCR is used to control the power of 1 kW, 230 V, 50 Hz heater. Determine the heater power for firing angles of 45° and 90°.

Solution:

Please refer to Fig. 6.6 (a) Here R is the heater resistance.
rms current is the heat producing component of load current.

$\therefore$ From equation (6.4), we have $E_{rms} = E_m \left[\dfrac{\pi - \alpha}{4\pi} + \dfrac{\sin 2\alpha}{8\pi} \right]^{1/2}$

(i) At $\alpha = \pi/4$, $E_{rms} = 155$ V $\therefore$ Heat power $W = E_{rms}^2/R$

Now, $\quad R = \dfrac{230^2}{1 \text{ kW}} = 52.90 \, \Omega$

$\therefore \quad W = (115)^2/52.90 = 454.15$ watts.

(ii) At $\quad \alpha = \pi/2$, $E_{rms} = 115$ Volts.

$\therefore \quad W = (115)^2/52.90 = 250$ watts.

Example 6.3 A single-phase half-wave controlled converter is operated from a 120 V, 50 Hz supply. Load resistance $R = 10 \, \Omega$. If the average output voltage is 25% of the maximum possible average output voltage, determine:
(a) firing angle, (b) rms and average output currents (c) average and rms SCR currents.

Solution:

(a) We have, E_{dc} (average output voltage) $= \dfrac{E_m}{2\pi}(1 + \cos \alpha)$

The maximum output voltage is obtained when $\alpha = 0$ $\therefore E_{dc_{max}} = \dfrac{E_m}{\pi}$

Given $\quad E_{dc} = 25\% \left(\dfrac{E_m}{\pi}\right) = 0.25 \dfrac{E_m}{\pi}$

$$\therefore \qquad 0.25 \frac{E_m}{\pi} = \frac{E_m}{2\pi} (1 + \cos \alpha)$$

$\therefore$ Firing angle $\alpha = \dfrac{\pi}{3}$ or $60°$ and $E_{dc} = 0.238 E_m$

(b) Average output current $I_{dc} = \dfrac{E_{dc}}{R}$

$$= \frac{0.238 E_m}{R} = \frac{0.238 \times 120 \times \sqrt{2}}{10} = 4.04 \text{ Amp}$$

RMS output voltage $E_{rms} = E_m \left[\dfrac{\pi - \alpha}{4\pi} + \dfrac{\sin 2\alpha}{8\pi} \right]^{1/2}$

$$= 0.4484 E_m = 0.4484 \times 120 \times \sqrt{2} = 76.096 \text{ Volts}$$

$\therefore$ RMS load current $I_{rms} = \dfrac{E_{rms}}{R} = 7.61 \text{ A}$

(c) Since the thyristor current waveform is same as the load current waveform, therefore, average and rms SCR current will be same as that of average and rms load current values.

Example 6.4 A d.c. battery is charged through a resistor R as shown in Fig. 6.12. Derive an expression for the average value of charging current in terms of E_m, E_b, R, etc. on the assumption that SCR is fired continuously.

(i) For an a.c. source voltage of 230 V, 50 Hz, find the value of average-charging current for $R = 5 \, \Omega$ and $E_b = 150$ V.
(ii) Find the power supplied to battery and that dissipated in the resistor.
(iii) Calculate the supply power factor.

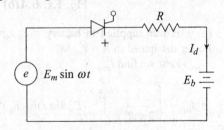

Fig. Ex. 6.4(a)

Solution: By neglecting SCR drop, we can have the following equation from Fig. Ex. 6.4(a),

$$E_m \sin \omega t = E_b + I_d R \quad \therefore \quad I_d = \frac{E_m \sin \omega t - E_b}{R}$$

The related waveforms of Fig. Ex. 6.4(a) are shown in Fig. Ex. 6.4(b).
From Fig. Ex. 6.4(b), it is clear that SCR is turned ON when $E_m \sin \alpha_1 = E_b$ and is turned OFF when $E_m \sin \alpha_2 = E_b$, where $\alpha_2 = \pi - \alpha_1$.
The battery charging requires only the average current, I_d, given by

$$I_d = \frac{1}{2\pi R} \left[\int_{\alpha_1}^{(\pi - \alpha_1)} (E_m \sin \omega t - E) \, d\omega t \right] = \frac{1}{2\pi R} [2 E_m \cos \alpha_1 - E(\pi - 2\alpha_1)]$$

(i) We have, $E_m \sin \alpha_1 = E_b$.

$$\therefore \quad \alpha_1 = \frac{\sin^{-1} E_b}{E_m} = \sin^{-1} \frac{150}{230\sqrt{2}} = 27.46°$$

$$\therefore \quad I_d = \frac{1}{2\pi \times 5} \left[\begin{array}{l} (2 \times 230 \times \sqrt{2} \cos 27.46) \\ -150(\pi - \frac{2 \times 27.46 \times \pi}{180}) \end{array} \right] = 7.96 \text{ A}.$$

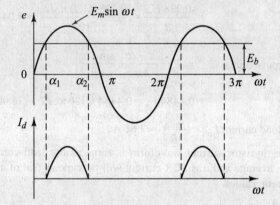

Fig. Ex. 6.4(b) *Waveforms*

(b) Power supplied to battery = $E_b.I_d = 7.96 \times 150 = 1.94$ kW
Power dissipated in $R = I_{rms}^2 R$
First we find I_{rms}

$$\therefore I_{rms} = \left[\frac{1}{2\pi R^2} \int_{\alpha_1}^{(\pi - \alpha_1)} (E_m \sin \omega t - E_b)^2 \, d\omega t \right]^{1/2}$$

$$= \left[\frac{1}{2\pi R^2} \int_{\alpha_1}^{\pi - \alpha_1} E_m^2 . \sin^2 \omega t - 2E_b^2 - 2 E_m E_b \sin \omega t \, d\omega t \right]^{1/2}$$

$$= \left[\frac{1}{2\pi R^2} \left\{ (E_b^2 + E_m^2)(\pi - 2\alpha_1) + E_m^2 . \sin 2\alpha_1 - 4E_m.E_b \cos \alpha_1 \right\} \right]^{1/2}$$

Therefore, by substituting the values, we get

$$I_{rms} = \left[\frac{1}{2\pi.25} \left\{ \begin{array}{l} (150^2 + 230^2)\left(\pi - 2 \times 27.46.\frac{\pi}{180}\right) + (230)^2. \\ (\sin 2 \times 27.46) - (4.\sqrt{2}.230.150.\cos 27.466) \end{array} \right\} \right]^{1/2} = 14.81 \text{ A}.$$

$\therefore$ Power dissipated in resistor = $(14.81)^2. 5 = 1096.68$ W

(c) Supply power factor = $\dfrac{I_{rms}^2 \cdot R + I_d \cdot E_b}{E_{rms} \cdot I_{rms}}$

$= \dfrac{1096.68 + 1194}{230 \times 14.81} = 0.672$ lagging.

6.4 SINGLE-PHASE FULL-WAVE CONTROLLED RECTIFIER (TWO-QUADRANT CONVERTERS)

There are two basic configurations of full wave controlled rectifiers. Their classification is based on the type of SCR configuration employed. They are—

(1) Mid-point converters.   (2) Bridge converters.

6.4.1 Mid-point Converters (M-2 Connection)

In a single phase full-wave controlled-rectifier circuit with mid-point configuration two SCRs (M-2) and a single-phase-transformer with centre-tapped secondary windings are employed. These converters are also referred to as two pulse converters as two triggering pulses or two sets of triggering pulses are to be generated during every cycle of the supply to trigger the various SCRs. Single-phase full-wave circuit with transformer mid-point configuration are generally used for rectifiers of low ratings.

1. With Resistive Load Figure 6.12 illustrates a 2-pulse mid-point converter circuit with resistive-load. This type of full-wave rectifier circuit uses two SCRs connected to the centre-tapped secondary of a transformer, as shown in Fig. 6.12. The input signal is coupled through the transformer to the centre-tapped secondary.

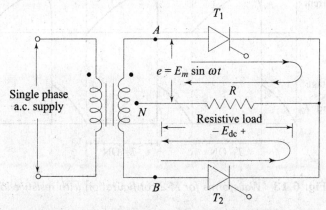

Fig. 6.12 *Full-wave mid-point circuit with resistive load*

During the positive half-cycle of the a.c. supply, i.e. when terminal A of the transformer is positive with respect to terminal B, or the secondary-winding terminal A is positive with respect to N, SCR$_1$ (T_1) is forward-biased and SCR$_2$

(T_2) is reverse-biased. Since no triggering pulses are given to the gates of the SCRs, initially they are in off-state. When SCR$_1$ is triggered at a firing-angle α, current would flow from terminal A through SCR$_1$, the resistive load R and back to the centre-tap of the transformer (i.e. terminal N). This current path is also shown in Fig. 6.12. This current continuous to flow up to angle π when the line voltage reverses its polarity and SCR$_1$ is turned-off. Depending upon the value of α and the load circuit parameters, the conduction angle of SCR$_1$ may be any value between 0 and π.

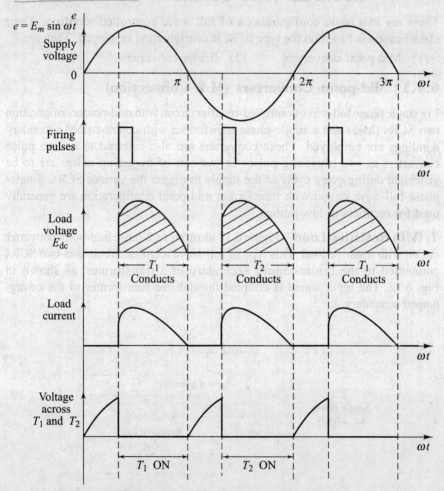

Fig. 6.13 *Waveforms for M-2 configuration with resistive-load*

During the negative half-cycle of the a.c. supply, the terminal B of the transformer is positive with respect to N. SCR$_2$ is forward-biased. When SCR$_2$ is triggered at an angle $(\pi + \alpha)$, current would flow from terminal B, through SCR$_2$, the resistive load and back to centre-tap of the transformer. This current continues till angle 2π, then SCR_2 is turned off. Here it is assumed that both thyristors are triggered with the same firing angle, hence they share the load current equally.

Each half of the input-wave is applied across the load. Thus, across the load, there are two pulses of current in the same direction. Hence the ripple frequency across the load is twice that of the input supply frequency. The voltage and current waveforms of this configuration is shown in Fig. 6.13. It is clear from Fig. 6.13 that with purely resistive load, the load current is always discontinuous.

The voltage and current relations are derived as follows:

(a) Average d.c. Output Voltage The output d.c. voltage, E_{dc}, across the resistive load is given by

$$E_{dc} = \frac{1}{\pi}\int_{\alpha}^{\pi} E_m \cdot \sin \omega t \, d(\omega t) = \frac{E_m}{\pi}[-\cos \omega t]_{\alpha}^{\pi} = \frac{E_m}{\pi}[1+\cos \alpha] \qquad (6.9)$$

(b) Average-load Current The average-load current is given by

$$I_{dc} = \frac{E_m}{\pi \cdot R}[1+\cos \alpha] \qquad (6.10)$$

(c) RMS Load-voltage The RMS load-voltage for a given firing angle α is given by

$$E_{rms} = \left[\frac{1}{\pi}\int_{\alpha}^{\pi} E_m^2 \sin^2 \omega t \, d\omega t\right]^{\frac{1}{2}} = E_m \cdot \left[\frac{1}{\pi}\int_{\alpha}^{\pi} \sin^2 \omega t \, d\omega t\right]^{\frac{1}{2}}$$

$$= E_m \cdot \left[\frac{1}{\pi}\int_{\alpha}^{\pi}\left(\frac{1-\cos 2\omega t}{2}\right) d\omega t\right]^{\frac{1}{2}} = E_m \cdot \left[\frac{1}{2\pi}\left(\omega t - \frac{\sin 2\omega t}{2}\right)_{\alpha}^{\pi}\right]^{\frac{1}{2}}$$

$$= E_m \cdot \left[\frac{1}{2\pi}\left(\pi - \alpha + \frac{\sin 2\alpha - \sin 2\pi}{2}\right)\right]^{\frac{1}{2}} = E_m \cdot \left[\frac{1}{2\pi}\left(\pi - \alpha + \frac{\sin 2\alpha}{2} - 0\right)\right]^{\frac{1}{2}}$$

$$E_{rms} = E_m \cdot \left[\frac{\pi - \alpha}{2\pi} + \frac{\sin 2\alpha}{4\pi}\right]^{\frac{1}{2}} \qquad (6.11)$$

2. With inductive Load The circuit diagram of the single-phase full wave, or bi-phase half-wave controlled rectifier with R_L load is shown in Fig. 6.14. The various voltage and current waveforms are shown in Fig. 6.15.

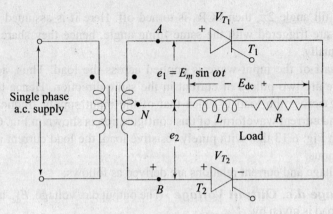

Fig. 6.14 *Bi-phase half-wave circuit*

With reference to Fig. 6.14, thyristor T_1 can be fired into the on-state at any time after e_1 goes positive. Once thyristor T_1 is turned-on, current builds up in the inductive load, maintaining thyristor T_1 in the on-state up to the period when e_1 goes negative. However, once e_1 goes negative, e_2 becomes positive, and the firing of thyristor T_2 immediately turns on thyristor T_2 which takes up the load current, placing a reverse voltage on thyristor T_1, its current being commutated (transferred) to thyristor T_2. The thyristor voltage, V_T, waveform in Fig. 6.15 shows that it can be fired into conduction at anytime when V_T is positive. The peak reverse (and forward) voltage that appears across the thyristor is $2\,E_m$, that is, the maximum value of the complete transformer secondary voltage.

The load-current may be continuous or discontinuous, depending on the inductance value. The load current is continuous if inductance value is greater than its critical value. It is discontinuous if inductance value is less than its critical value. The analysis given here assumes that the inductance is sufficiently large, so that each thyristor conducts for a period of 180° (conduction of current is continuous). Also, both thyristors are triggered with the same delay angle, hence they share the load current equally. As shown in Fig. 6.15, due to large inductance in the circuit and continuous current conduction, the thyristors continue to conduct even when their anode voltages are negative with respect to the cathode. The load current is shown to be constant d.c.

Now, the output d.c. voltage E_{dc} can be obtained as

$$E_{dc} = \frac{1}{\pi} \int_{\alpha}^{\pi+\alpha} E_m \cdot \sin \omega t \, d(\omega t) = \frac{E_m}{\pi} \left[\cos \alpha - \cos(\pi + \alpha) \right]$$

$$E_{dc} = \frac{2 E_m}{\pi} \cos \alpha. \tag{6.12}$$

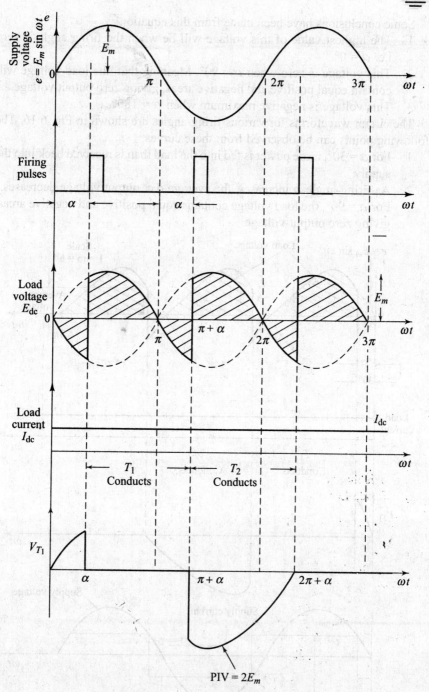

Fig. 6.15 *Waveforms for M-2 connection with R-L load*

Some conclusions have been made from this equation—
1. The highest value of this voltage will be when the firing angle is zero i.e. $\alpha = 0°$.
2. This voltage is zero when $\alpha = 90°$. Meaning that, the load voltage will contain equal positive and negative areas, giving zero output voltage.
3. This voltage is negative maximum when $\alpha = 180°$.

The circuit waveforms for various firing-angles are shown in Fig. 6.16. The following points can be observed from these curves:
1. For $\alpha = 30°$, more power is fed into the load than is received back into the supply.
2. As firing-angle α increases, the average d.c. output voltage decreases. For $\alpha = 90°$, the load voltage contains equal positive and negative areas, giving zero output-voltage.

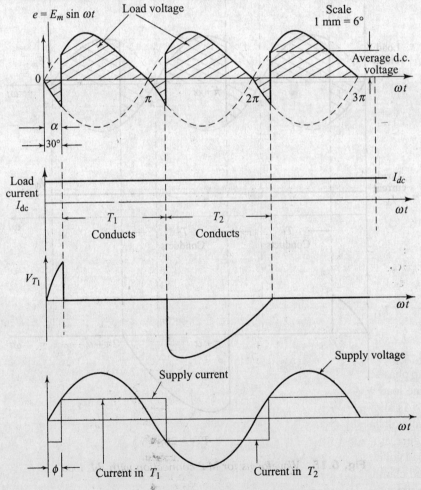

Fig. 6.16 *(a) Waveforms for $\alpha = 30°$*

3. For firing-angle $\alpha > 90°$, the d.c. voltage goes negative. At $\alpha = 180°$, the negative d.c. voltage is maximum.
4. The period for which a thyristor is reverse-biased reduces as the firing angle increases to 180°. To turn OFF a thyristor, it must be reverse-biased for greater than its turn-off time.

Therefore, the maximum firing-angle must always be less than 180°.

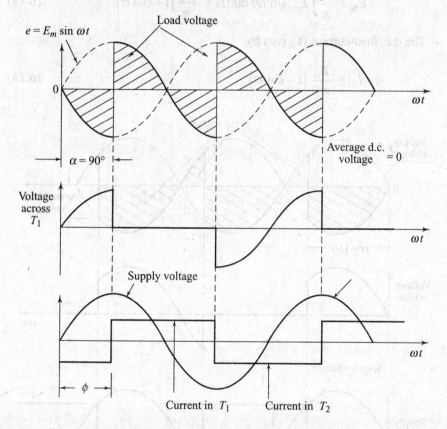

Fig. 6.16 *(b) Waveforms for $\alpha = 90°$*

3. Effect of Freewheeling Diode Figure 6.17 shows the full-wave centre tap phase-controlled thyristor-circuit with inductive load and freewheeling diode. The load voltage and current-waveforms are also shown in Fig. 6.18.

As shown in Fig. 6.18, the thyristors are triggered at angle α. The variable d.c. voltage at the load is obtained by varying this firing angle α. From the same figure, it is also clear that as the supply voltage goes through zero at 180°, the load voltage cannot be negative since the freewheeling diode, D_f, starts conducting and clamps the load voltage to zero volts. A constant load current is maintained by freewheeling current through the diode. The conduction period of thyristors and diode is also shown in Fig. 6.18. The stored-energy in the inductive load circulates current through the feedback-diode in the direction shown in Fig. 6.17. The rate of decay of this current depends upon the time-constant of the load.

The average d.c. output voltage can be calculated as

$$E_{dc} = \frac{1}{\pi} \int_{\alpha}^{\pi} E_m \cdot \sin \omega t \, d(\omega t) = \frac{E_m}{\pi} [1 + \cos \alpha] \qquad (6.13)$$

The d.c. load-current is given by

$$I_{dc} = \frac{E_m}{\pi R} [1 + \cos \alpha] \qquad (6.14)$$

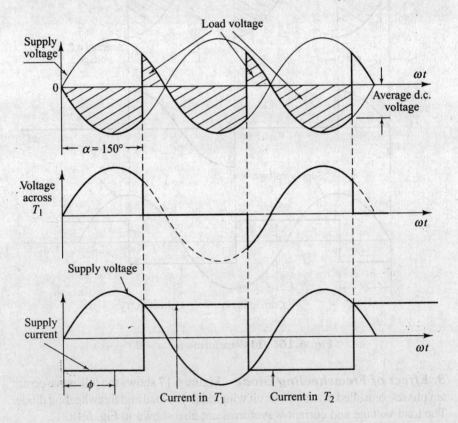

Fig. 6.16 *(c) Waveforms for $\alpha = 150°$*

It is also observed from Fig. 6.18, that the freewheeling diode, D_F, carries the load-current during the firing-angle α when the thyristors are not conducting. Hence, the current through the diode D_F is given by

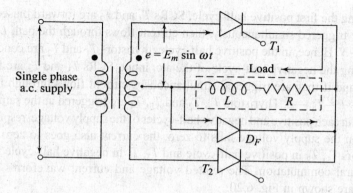

Fig. 6.17 *M-2 configuration with freewheeling diode D_f*

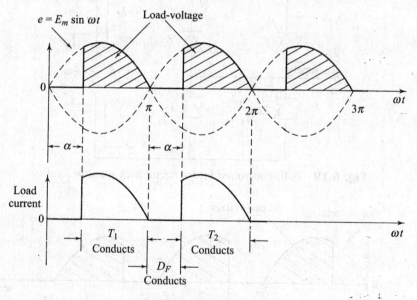

Fig. 6.18 *Waveforms*

$$ID_f = I_{dc}\frac{\alpha}{\pi} = \frac{E_m}{\pi R}(1+\cos\alpha)\frac{\alpha}{\pi}$$

$$ID_f = \frac{E_m}{\pi^2 R}(\alpha + \alpha'\cos\alpha) \qquad (6.15)$$

6.4.2 Bridge-Configurations (B-2 Connection)

An alternative-circuit arrangement of a two-quadrant converter, operating from a single-phase supply, is a fully controlled bridge-circuit as shown in Fig. 6.19. The operation of this circuit is in principle similar to that of the 2-pulse midpoint circuit of Fig. 6.12. In the bridge circuit, diagonally opposite pair of thyristors are made to conduct, and are commutated, simultaneously.

During the first positive half-cycle, SCRs T_1 and T_2 are forward biased and if they are triggered simultaneously, then current flows through the path L—T_1—R—T_2—N. Hence, in the positive half cycle, thyristors T_1 and T_2 are conducting.

During the negative half-cycle of the a.c. input, SCRs T_3 and T_4 are forward biased and if they are triggered simultaneously, current flows through the path N—T_3—R—T_4—L. Thyristors T_1, T_2 and T_3, T_4 are triggered at the same firing angle α in each positive and negative half-cycles of the supply voltage, respectively.

When the supply voltage falls to zero, the current also goes to zero. Hence thyristors T_1, T_2 in positive half-cycle and T_3, T_4 in negative half-cycle turn-off by natural commutation. The related voltage and current waveforms for this circuit are shown in Fig. 6.20.

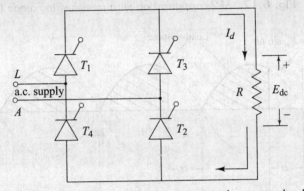

Fig. 6.19 *Fully-controlled bridge-circuit with resistive-load*

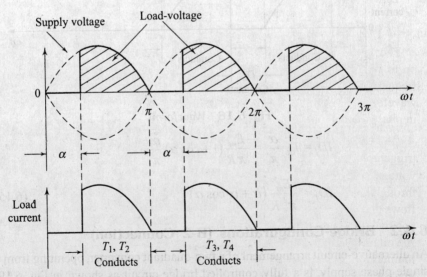

Fig. 6.20 *Waveforms for fully-controlled bridge with resistive-load*

When the supply voltage falls to zero, the current also goes to zero. Hence thyristors T_1, T_2 in positive half-cycle and T_3, T_4 in negative half-cycle turn-off by natural commutation. The related voltage and current waveforms for this circuit are shown in Fig. 6.20. The relations for E_{dc}, I_{dc} and E_{rms} for this bridge configuration is similar to the Eqs (6.9), (6.10) and (6.11), respectively.

1. Fully Controlled Bridge Circuit with Inductive Load [R–L Load]

The single phase fully controlled bridge circuit with R–L load is shown in Fig. 6.21. Conduction does not takes place until the thyristors are fired and, in order for current to flow, thyristors T_1 and T_2 must be fired together, as must thyristor T_3 and T_4 in the next half-cycle. To ensuring simultaneous firing, both thyristors T_1 and T_2 are fired from the same firing circuit. Inductance L is used in the circuit to reduce the ripple. A large value of L will result in a continuous steady current in the load. A small value of L will produce a discontinuous load-current for large-firing angles. The waveforms with two different firing-angles are shown in Fig. 6.22.

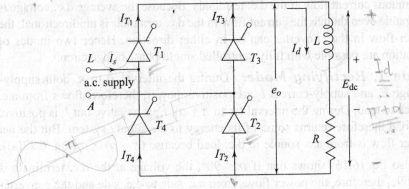

Fig. 6.21 *Fully controlled single-phase bridge with R–L load*

The voltage waveform at the d.c. terminals comprises a steady d.c. component on to which is superimposed an a.c. ripple component, having a fundamental frequency equal to twice that of a.c. supply. The input line-current has a square-waveform of amplitude I_d, and the fundamental component of this waveform is in phase with the input-voltage.

As shown in Fig. 6.22(a), at firing angle $\alpha = 60°$, thyristors T_1 and T_2 are triggered. Current flows through the path $L—T_1—A—L—R—B—T_2—N$. Supply voltage from this instant appears across output terminal and forces the current through load. This load-current, I_d, is assumed to be constant. This current also flows through the supply and the direction is from line to neutral, which is taken positive, as shown in Fig. 6.22(a) along with the applied voltage. Now, at instant π, voltage reverses. However, because of very large inductance L, the current is maintained in the same direction at constant magnitude I_d which keeps the thyristors T_1 and T_2 in conducting state and hence, the negative supply voltage appears across output terminals.

At an angle $\pi + \alpha$, thyristors T_3 and T_4 are fired. With this, the negative line voltage reverse biases thyristors T_1 through T_3, and T_2 through T_4 of commutating thyristors T_1 and T_2. The current flows through the path N—T_3—A—L—R—B—T_4—L. This continues in every half cycle and we get the output voltage as shown in the figure. As shown, the line current is positive when T_1, T_2 are conducting and negative when T_3, T_4 are conducting.

The average output d.c. voltage can be obtained as

$$E_{dc} = \frac{1}{\pi} \int_{\alpha}^{\pi+\alpha} E_m \sin \omega t \, d(\omega t) = \frac{E_m}{\pi} \left[-\cos \omega t\right]_{\alpha}^{\pi+\alpha}$$

$$= \frac{E_m}{\pi} \left[\cos \alpha - \cos(\pi + \alpha)\right] \quad E_{dc} = \frac{2E_m}{\pi} \cos \alpha \qquad (6.16)$$

By controlling the phase-angle of firing pulses, applied to the gates of the thyristors in the range 0°–180°, the average-value of the d.c. voltage can be varied continuously from positive maximum to negative maximum, assuming continuous current flow at the d.c. terminals. Because the average d.c. voltage is reversible even though the current flow in the d.c. terminals is unidirectional, the power-flow in the converter can be in either direction. Hence two modes of operation are possible with fully controlled single-phase bridge circuit.

Mode 1, Rectifying Mode: During the interval α to π, both supply-voltage E_s and supply-current I_s are positive; power, therefore, flows from a.c. source to load. During the interval (π to $\pi + \alpha$), E_s is negative but I_s is positive, the load therefore returns some of its energy to the supply system. But the net power flow is from a.c. source to d.c. load because $(\pi - \alpha) > \alpha$ (Fig. 6.22(a)).

Also Eq. (6.16) shows that if $\alpha < 90°$, the voltage at the d.c. terminals is positive, therefore, the power flows from a.c. side to d.c. side and the converter operates as a rectifier.

Mode 2, Inverting Mode: In Fig. 6.22(b), the firing pulses are retarded by an angle of 135°. The d.c. terminal voltage waveforms now contains a mean negative component, and the fundamental component of the a.c. line-current waveforms lags the voltage by an angle of 135°. Since the mean d.c. terminal voltage is negative [$\alpha > 90°$], the d.c. power, and hence also the mean a.c. power, must also be negative. In other words, power is now being delivered from the d.c. side of the converter to the a.c. side, and the converter is operating as a "line-commutated inverter." This is confirmed by the fact that the a.c. line current is displaced from the a.c. voltage by an angle greater than 90°, which indicates that a mean component of power is being delivered to the a.c. side of the circuit.

In order to achieve this situation in practice, it is necessary for a source of d.c. voltage E whose voltage equals to the average d.c. voltage (negative) of the converter must be connected in the output, as shown in Fig. 6.23. It is this source of external voltage which drives the direct-current into the converter against the "counter" voltage produced at its d.c. terminals. If this d.c. voltage-source is not connected, the conduction will cease somewhere before the angle (180° + α).

Phase Controlled Converters

Such an operation is used in the regenerative breaking mode of d.c. drives and in high-voltage direct-current (HVdc) transmission.

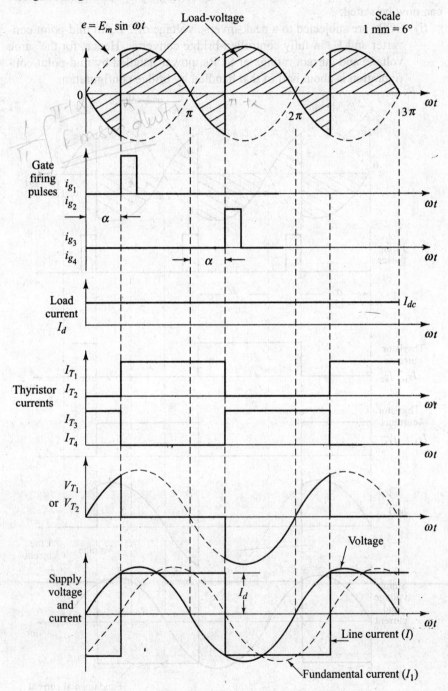

Fig. 6.22 (a) Waveforms for $\alpha = 60°$

As both the types of phase-controlled converters have been studied, the advantages of single-phase bridge converter over single-phase mid-point converter can now be stated:

(i) SCRs are subjected to a peak-inverse voltage of $2\,E_m$ in mid-point converter and E_m in fully controlled-bridge converter. Hence, for the same voltage and current ratings of SCRs, power handled by mid-point configuration is about half of that handled by bridge configuration.

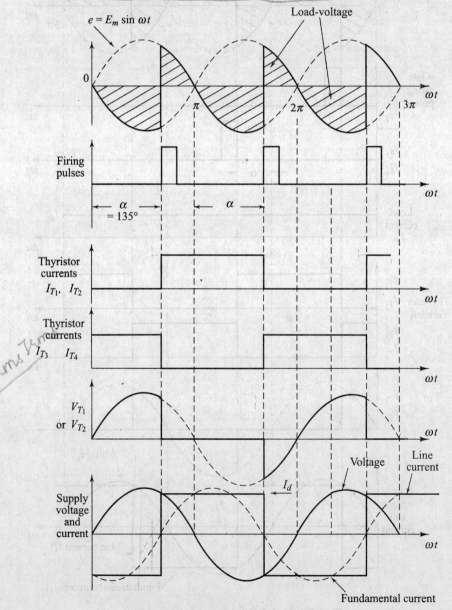

Fig. 6.22 *(b) Waveforms for $\alpha = 135°$*

(ii) In mid-point configuration, each secondary should be able to supply the load-power. As such, the transformer rating in mid-point converter is double the load-rating. This, however, is not the case in the single-phase bridge converter.

Hence, bridge configuration is preferred over mid-point configuration. However, the choice between these two types depends primarily on cost of the various components, available source voltage and the load-voltage requirements. Mid-point converter is used in case the terminals on d.c. side have to be grounded.

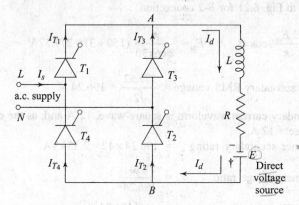

Fig. 6.23 *Fully-controlled bridge-converter operating as a "line-commutated inverter"*

SOLVED EXAMPLES

Example 6.5 A highly inductive d.c. load requires 12 A at 150 V from a 230 V single-phase a.c. supply. Give design details for this requirement using
 (a) *M*-2 connection.   (b) *B*-2 connection, for $\alpha = 30°$.
Assume each SCR to have a voltage drop of 1.5 V. Also, compare the two designs.

Solution: (a) Refer to Fig. 6.14.
For *M*-2 connection with inductive load, we have the relation:

$$E_{dc} = \frac{2 E_m}{\pi} \cos \alpha$$

Now with 1.5 V SCR drop, we can write, $E_{dc} = \frac{2 E_m}{\pi} \cos \alpha - 1.5$

∴ $150 = \frac{2 E_m}{\pi} \cos 30 - 1.5$, $E_m = \frac{\pi}{2 \cos 30}(150 + 1.5) = 274.8$ V.

Hence, each section of the transformer secondary requires an RMS voltage of

$$E_{rms} = \frac{274.8}{\sqrt{2}} = 194.31 \text{ V and carrier RMS current}$$

$$I_{rms} = \frac{12}{\sqrt{2}} = 8.49 \text{ A}$$

∴ Transformer secondary rating $= 2 \times 194.31 \times 8.49 = 3.3$ kVA

Transformer voltage ratio $= \dfrac{230}{194.31}$

Transformer primary current is a sequence wave of RMS value $= 12 \left(\dfrac{194.31}{230}\right)$
$= 10.14$ A

PIV for each SCR $= 2E_m \approx 550$ V, with $I_{rms} = 12/\sqrt{2} = 8.5$ A.

(b) Refer to Fig. 6.21 for B-2 connection.

Now, $E_{dc} = \dfrac{2E_m}{\pi} \cos \alpha$ ∴ $E_m = \dfrac{\pi}{2\cos 30}(150 + 3) = 277.52$ V

Transformer secondary RMS voltage $= \dfrac{277.52}{\sqrt{2}} = 196.24$ V.

The secondary current waveform is square-wave, 12 A and, as the current is level, its RMS value = 12 A.

Transformer secondary rating $= 196.24 \times 12 = 2.35$ kVA.

Transformer voltage ratio $= \dfrac{230}{196.24}$

Transformer RMS primary current $= 12 \left(\dfrac{196.24}{230}\right) = 10.24$ A

For each SCR, PIV $= E_m = 277.52$ V, $I_{rms} = \dfrac{12}{\sqrt{2}} = 8.5$ A

(c) The SCR loss in (a) $= \dfrac{1.5}{150} = 1\%$ of the load power

The SCR loss in (b) $= \dfrac{3}{150} = 2\%$ of the load power

Comparing the two circuits, the B–2 connection is superior on transformer size and SCR voltage rating requirements.

Example 6.6 A single-phase fully-controlled bridge circuit shown in Fig. 6.21 is used for obtaining a regulated d.c. output voltage. The RMS value of the a.c. input voltage is 230 V, and the firing angle is maintained at $\dfrac{\pi}{3}$, so that the load-current is 4 A.

(i) Calculate the d.c. output voltage and the active and reactive power input.
(ii) Assuming that the load resistance remains the same, calculate the quantities in (i) if a freewheeling diode is used at the output. The firing angle is maintained at $\dfrac{\pi}{3}$.
(iii) If SCR$_3$ is damaged and gets open-circuited, calculate the average d.c. output voltage for the average direct-current output. For this case, a freewheeling diode is connected. The firing angle is $\dfrac{\pi}{3}$.

Phase Controlled Converters

Solution: From Eq. (6.16), we have the d.c. output voltage,

$$E_{dc} = \frac{2 E_m}{\pi} \cos \alpha = \frac{2 \times \sqrt{2} \times 230}{\pi} \cos\left(\frac{\pi}{3}\right) = 103.54 \text{ V} \tag{i}$$

From Eq. (6.37), we have Active power input $P_i = \frac{2 E_m}{\pi} I_d \cos \alpha$

$$= \frac{2 \times \sqrt{2} \times 230}{\pi} \times 4 \cos\left(\frac{\pi}{3}\right) = 414.15 \text{ W}$$

From Eq. (6.38), we have, Reactive power input $Q_i = \frac{2 E_m}{\pi} I_d \sin \alpha$

$$= \frac{2 \times \sqrt{2} \times 230 \times 4}{\pi} \sin\left(\frac{\pi}{3}\right) = 717.32 \text{ Vars.} \tag{ii}$$

(b) The load resistance $R = \frac{E_{dc}}{I_d} = \frac{103.54}{4} = 25.89 \, \Omega$

When the freewheeling diode DF is connected across the load, then average load-voltage is given by Eq. (6.13)

$$\therefore \quad E_{dc} = \frac{E_m}{\pi} (1 + \cos \alpha)$$

$$= \frac{\sqrt{2} \times 230}{\pi} (1 + \cos \pi/3) = 155.30 \text{ V} \tag{iii}$$

It can be observed from Eq. (i) and (iii) that, for the same firing angle, the average d.c. output voltage E_{dc} will be higher when the freewheeling diode is used.

With freewheeling diode D_F, the phase angle Q of the fundamental component of input-current is given by

$$\theta = \alpha/2 \tag{iv}$$

Therefore, the fundamental value of the alternating line current I_1 is

$$I_1 = \frac{2\sqrt{2} I_d}{\pi} \cos(\alpha/2) \tag{v}$$

where $I_d = \frac{E_{dc}}{R} = \frac{155.30}{25.89} = 6 \text{ A} \quad \therefore I_1 = \frac{2\sqrt{2} \times 6}{\pi} \cos(\pi/6) = 4.68 \text{ A}$

$\therefore$ Active power input $(P_i) = E_{rms} I_1 \cos(\alpha/2) = 230 \times 4.68 \times 0.866$

$$= 932.16 \text{ watts.} \tag{vi}$$

Reactive power input (Q_i)

$$= E_{rms} I_1 \sin(\alpha/2) = E_{rms} \frac{2\sqrt{2} I_d}{\pi} \cos(\alpha/2) \cdot \sin(\alpha/2)$$

$$= \frac{E_{rms} \sqrt{2}}{\pi} I_d \cdot \sin \alpha. \tag{vii}$$

$$\therefore \qquad Q_i = \frac{230 \times \sqrt{2} \times 6 \times \sin(\pi/3)}{\pi} = 538 \text{ Vars}$$

From Eqs (ii) and (vii), it becomes clear that the freewheeling diode DF reduces the reactive-power-input for any given firing-angle with the same-direct by 50%.

(c) If SCR_3 is damaged and gets open-circuited, then the fully-controlled bridge circuit will reduce to a controlled circuit, which behaves as a half-wave controlled circuit. Assuming continuous load current I_d, we write

$$E_{dc} = \frac{1}{2\pi}\int_\alpha^\pi E_m \sin \omega t\, d(\omega t) = \frac{E_m}{2\pi}(1+\cos \alpha)$$

$\therefore$ For $\alpha = \pi/3$, $E_{dc} = \dfrac{230\sqrt{2}}{2\pi}(1+\cos \pi/3) = 77.65$ V.

$\therefore$ Load current $I_d = \dfrac{E_{dc}}{R} = \dfrac{77.65}{25.89} \approx 3$ A

Example 6.7 A single phase fully controlled thyristor bridge converter supplies a load consisting of R, L and V_c. The inductance L in the circuit is so large that the output current may be considered to be virtually constant. Assume the SCR to be ideal with following data:

RMS supply voltage = 220 V, load resistance = 0.5 Ω, output current i_{dc} = 10 A. Determine:
(a) Firing angle α if $E_C = 135$ V (b) α if $E_C = -145$
(c) Which source (ac or dc) is supplying power in (a) and (b).
(d) Draw the load voltage waveform for (a) and (b).

Solution:

(a) Average load voltage, $E_{dc} = \dfrac{2E_m}{\pi}\cos\alpha$ $E_{dc} = \dfrac{2\times\sqrt{2}\times 220}{\pi}\cos\alpha = 198\cos\alpha$

From Fig. Ex. 6.7 (a), $E_{dc} = I_{dc}\cdot R + E_c = (10\times 0.05) + 135 = 140$ V
$\therefore E_{dc} = 140 = 198\cos\alpha$ $\therefore \alpha = 45°$

(b) From Fig. Ex. 6.7 (b), $E_{dc} = I_{dc}\cdot R - V_C = (10\times 0.5) - 145 = -140$ V

(a)   (b)

Fig. Ex. 6.7

$\therefore \qquad E_{dc} = -140 = 198 \cos \alpha \quad \therefore \quad \alpha = 135° \text{ or } \dfrac{3\pi}{4}.$

(c) In Fig. Ex. 6.7 (a), the converter works as rectifier since power flows from ac source to the load. In Fig. Ex. 6.7 (b), converter works as inverter since power flows from dc to ac circuit.

(d) Load voltage waveforms are shown in Fig. Ex. 6.7 (c) and (d) respectively.

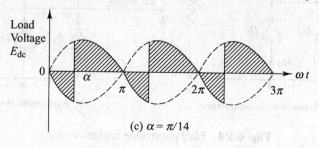

(c) $\alpha = \pi/14$

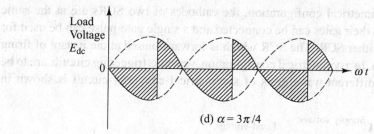

(d) $\alpha = 3\pi/4$

Fig. Ex. 6.7 *Load voltage waveforms*

6.5 SINGLE-PHASE HALF CONTROLLED BRIDGE-RECTIFIER

The phase-controlled converter circuits discussed in previous sections are capable of operating with both positive and negative mean voltages at the d.c. terminals. Many applications actually require operation only with a positive voltage, that is only in the rectifying-mode. In such cases, it is generally advantageous to connect uncontrolled diodes into certain parts of the circuit.

When one pair of SCRs is replaced by diodes in single-phase fully controlled bridge-circuit, the resultant circuit obtained is called as a half-controlled bridge-circuit. With this type of circuit, it is possible to provide a continuous control of the mean d.c. terminal voltage, from maximum to virtually zero, but reversal of the mean voltage is not-possible. In other words, only a one-quadrant operation can be obtained.

6.5.1 Half Controlled Bridge Rectifier with Resistive Load

The two different versions of half-controlled bridge-circuit with resistive load are shown in Fig. 6.24. Half-controlled converters are also known as "semi-converter."

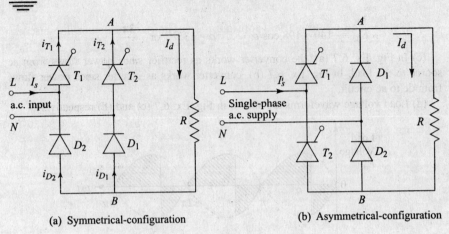

(a) Symmetrical-configuration (b) Asymmetrical-configuration

Fig. 6.24 *Half controlled-bridge circuits*

In a symmetrical configuration, the cathodes of two SCRs are at the same potential so their gates can be connected and a single gate-pulse can be used for triggering either SCR. The SCR which is forward-biased at the instant of firing will turn-on. In asymmetrical configuration, separate-triggering circuits are to be used. The different-waveforms of symmetrical converter-circuit is shown in Fig. 6.25.

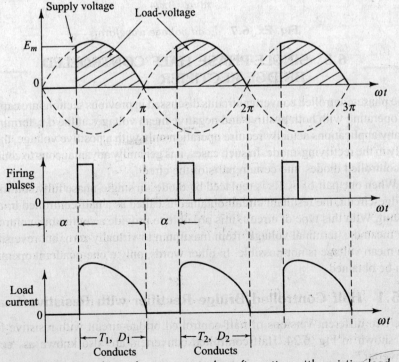

Fig. 6.25 *Waveforms for symmetrical configuration with resistive load*

Phase Controlled Converters

Now, consider the symmetrical configuration of half-controlled bridge-circuit. During the positive half-cycle of the a.c. supply, thyristor T_1 and diode D_1 are forward-biased and are in the forward-blocking mode. When the SCR T_1 is triggered, at a firing-angle α, the current flow through the path L—T_1—R—D_1—N. As shown in Fig. 6.25, the load-current will flow until it is commutated by reversal of supply voltage at $\omega t = \pi$.

During the negative half-cycle of the a.c. supply, thyristor T_2 and diode D_2 are forward-biased. When SCR T_2 is triggered at an angle $(\pi + \alpha)$, the current would flow through the path N—T_2—A—R—B—D_2—L. This current is continuous till angle 2π, when SCR T_2 is turned-off.

The voltage and current-relations are derived as follows:

(i) Average d.c. load-voltage:

$$E_{dc} = \frac{1}{\pi} \int_{\alpha}^{\pi} E_m \cdot \sin \omega t \, d(\omega t)$$

$$= \frac{1}{\pi} E_m [-\cos \omega t]_{\alpha}^{\pi} = \frac{E_m}{\pi}[1 + \cos \alpha] \quad (6.17)$$

(ii) Average load-current:

$$I_d = \frac{E_m}{\pi R}[1 + \cos \alpha]$$

(iii) RMS load-voltage: The RMS load voltage for a given firing angle α is given by

$$E_{rms} = \left[\frac{E_m^2}{\pi} \int_{\alpha}^{\pi} \sin^2 \omega t \, d(\omega t)\right]^{1/2} = E_m \left[\frac{1}{\pi} \int_{\alpha}^{\pi} \left(\frac{1 - \cos 2\omega t}{2}\right) \cdot d(\omega t)\right]^{1/2}$$

$$= E_m \left[\frac{1}{2\pi}\left(\omega t - \frac{\sin 2\omega t}{2}\right)_{\alpha}^{\pi}\right]^{1/2} = E_m \left[\frac{\pi - \alpha}{2\pi} + \frac{\sin 2\alpha}{4\pi}\right]^{1/2} \quad (6.18)$$

6.5.2 Half Controlled Bridge Rectifier with R-L Load

Figure 6.26 shows two alternative arrangements of 2-pulse half-controlled bridge converters with inductive load. The various voltage and current waveforms for both symmetric and asymmetric configurations are shown in Fig. 6.27. Consider the symmetrical circuit configuration. As shown in Fig. 6.27(a), thyristor T_1 is turned-on at a firing angle α in each positive half-cycle. From this instant α, supply voltage appears across output terminals AB, through thyristor T_1 and diode D_1. Current flows through the path L—T_1—A—L—R—B—D_1—N. Here, the filter inductance L is assumed to be sufficiently large as to produce continuous load current. This current I_d is taken to be constant. Hence during positive half-cycle, thyristor T_1 and diode D_1 conducts.

Now, when the supply voltage reverses at $\omega t = \pi$, the diode D_2 is forward-biased since diode D_1 is already conducting. The diode D_2 then turns ON, and the load current passes through D_2 and T_1. The supply voltage reverse-biases D_1 and turns it off. Thus, the load-current freewheels through the path R—B—D_2—T_1—A—L during the interval from π to $(\pi + \alpha)$ in each supply-cycle.

During the negative half-cycle, at the instant $(\pi + \alpha)$, a triggering-pulse is applied to the forward-biased thyristor T_2. Thyristor T_2 is turned ON. As thyristor T_2 is turned ON, the supply voltage reverse-biases T_1 and then turns it OFF by the line commutation. Therefore, the load-current flows through T_2 and D_2, the above-cycle repeats and the waveforms obtained are as shown in Fig. 6.27(a), which are similar to that of fully-controlled converter with a freewheeling diode. Here, we have seen that the conduction period of thyristors and diodes are equal, therefore this circuit is called as the symmetrical configuration.

Now consider the circuit of Fig. 6.26(b). During the positive half-cycle of the a.c.-supply, thyristor T_1 and diode D_1 are forward-biased. As shown in Fig. 6.27(b), thyristor T_1 is turned ON at firing angle α. Current flows through the path L—T_1—A—L—R—B—D_1—N. Hence, T_1 and D_1 conduct from α to π. Similarly, T_2 and D_2 conduct from $(\pi + \alpha)$ to 2π in each negative half-cycle of the a.c. supply. The freewheeling action is provided by diodes D_1 and D_2 from 0 to α and from π to $(\pi + \alpha)$ in each supply cycle. In this converter configuration, the conduction periods of thyristors and diodes are unequal. Hence this circuit configuration is known as the asymmetrical configuration.

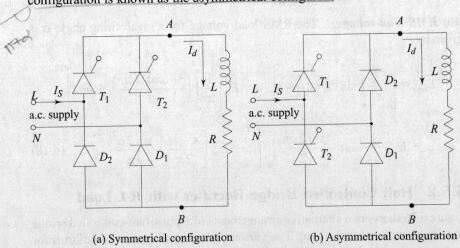

(a) Symmetrical configuration   (b) Asymmetrical configuration

Fig. 6.26

In the above discussion, we have seen that the thyristor conducts for a longer interval in the symmetrical circuit configuration. Therefore, the thyristors used in this circuit must have a higher average current-rating compared to those in the asymmetrical-circuit configuration.

The average d.c. output voltage, neglecting diode and SCR drop, is given by

$$E_{dc} = \frac{1}{\pi} \int_{\alpha}^{\pi} E_m \sin \omega t \, d(\omega t) = \frac{E_m}{\pi} [1 + \cos \alpha]$$

In comparing the operation of the half-controlled 2-pulse circuit with that of the fully-controlled circuit, the following points are evident:

(i) Since half the thyristors are replaced by diodes, a half-controlled converter costs less than a fully-controlled converter.
(ii) The periods of the negative-voltage "swing" at the d.c. terminals obtained with fully-controlled bridge-circuit are replaced by "freewheeling" periods of zero voltage in the half-controlled circuit. The elimination of the negative swings of voltage at the d.c. terminals is advantageous because it results in a reduction of the ripple voltage, with correspondingly reduced filtering requirements.

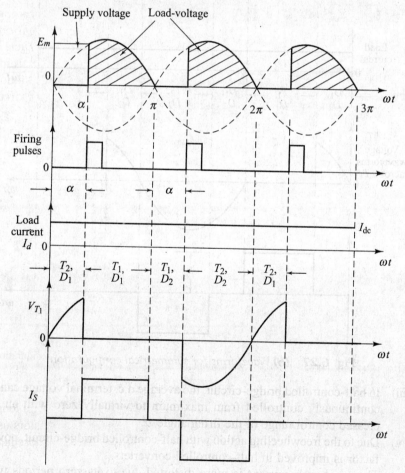

Fig. 6.27 (a) Waveforms for symmetrical configuration

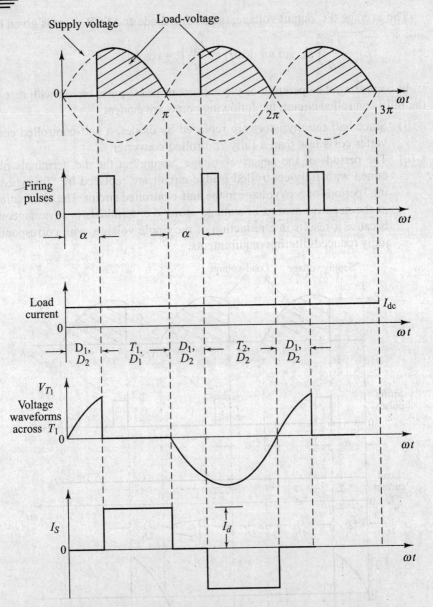

Fig. 6.27 *(b) Waveforms for symmetrical configuration*

- (iii) In half-controlled bridge-circuit, the average d.c. terminal voltage can be continuously controlled from maximum to virtually zero with an increased control range of the firing angle α.
- (iv) Due to the freewheeling action with half-controlled bridge-circuit, power factor is improved in half-controlled-converters.
- (v) The a.c. supply-current is more distorted due to its zero periods with half-controlled circuit, compared to fully-controlled bridge-circuit.

Half-controlled converters are widely used in main line a.c. traction where large d.c. motors are supplied from a single-phase a.c. supply.

6.5.3 Operation with Practical R-L Loads

In practical circuits, the load will always be a combination of some finite resistance R with inductance L. The actual values of R and L decides the amount of energy stored by the load and the time taken to release that energy in the process of freewheeling. Hence, the operation of the semiconverter (with R-L load), is divided into two modes, depending on the nature of the load current.

(a) Continuous current mode
(b) Discontinuous current mode

Figures 6.28 (a) and 6.28 (b) show the load voltage and current waveforms for these modes. The operation of the semiconverter remains same.

Continuous Conduction Mode As shown in the continuous conduction mode, the load current increases and decreases gradually and has a finite ripple but due to the higher value of load inductance, the load current does not go to zero. The conduction angle for the devices in the symmetrical configuration will remain π radians. During the freewheeling intervals, the load current decreases from I_{max} to I_{min} as shown in Fig. 6.28 (a). This mode is preferred in the motor control applications as it yields better results.

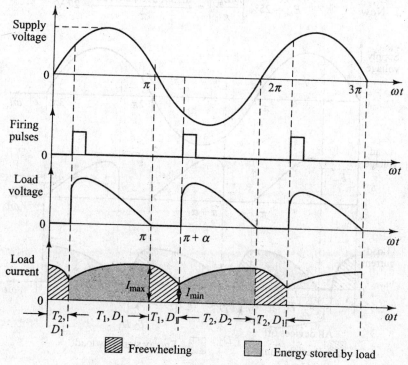

Fig. 6.28 *(a) Continuous conduction mode*

Discontinuous Condition Mode As shown in Fig. 6.28(b), in discontinuous mode, the load current decreases to zero in the freewheeling interval even before the next SCR is turned-on. The load current ripple content increases and the conduction angle for the devices decreases. This mode is not suitable for applications like dc motor speed control.

SOLVED EXAMPLES

Example 6.8 A single-phase semiconverter is operated from 120 V, 50 Hz ac supply. The load resistance is 10 Ω. If the average output voltage is 25% of the maximum possible average output voltage, determine

(a) firing angle
(b) rms and average output current
(c) rms and average thyristor current

Solution:

(a) For single-phase semiconverter

We have $E_{dc} = \dfrac{E_m}{\pi}[1 + \cos \alpha]$

For $\alpha = 0$, average output voltage will be maximum, $\therefore E_{dc_{max}} = \dfrac{E_m}{\pi}$

$\therefore$ Now $E_{dc} = 25\% \left[\dfrac{2E_m}{\pi}\right] = \dfrac{0.25 \times 2 \times \sqrt{2} \times 120}{\pi} = 27$ V

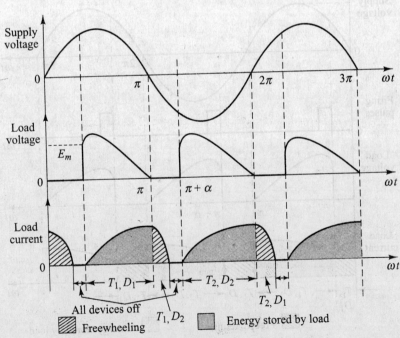

Fig. 6.28 *(b) Discontinuous conduction mode*

$$\therefore \quad 27 = \frac{120 \times \sqrt{2}}{\pi}(1+\cos\alpha), \quad \therefore \quad \alpha = 120°$$

(b) We have, rms load voltage $= E_{rms}$

$$= E_m \left[\frac{\pi - \alpha}{2\pi} + \frac{\sin 2\alpha}{4\pi} \right]^{1/2} = 53 \text{ V}$$

$$\therefore \text{ rms load current} = I_{rms} = \frac{E_{rms}}{R} = 5.3 \text{ Amp.}$$

Average load current $= I_{dc} = E_{dc}/R = 2.7$ Amp.

(c) Average thyristor current

$$I_{TH(av)} = \frac{1}{2\pi} \int_\alpha^\pi I_m \sin\omega t \, d\omega t = \frac{I_m}{2\pi}(1+\cos\alpha)$$

$$I_m = \frac{E_m}{R} = 16.97 \text{ Amp} \quad \therefore \quad I_{m(av)} = \frac{16.97}{2\pi}\left(1+\cos\frac{2\pi}{3}\right) = 1.35 \text{ Amp.}$$

RMS thyristor current $= I_{TH(rms)} = \left\{ \frac{1}{2\pi} \int_\alpha^\pi I_m^2 \sin^2\omega t \, d\omega t \right\}^{1/2}$

$$= \left\{ \frac{I_m^2}{2\pi} \int_\alpha^\pi \frac{1-\cos 2\omega t}{2} d\omega t \right\}^{1/2}$$

$$= \left\{ \frac{I_m^2}{2\pi} \int_\alpha^\pi \frac{1-\cos 2\omega t}{2} d\omega t \right\}^{1/2} = \frac{16.97}{2}\left[\frac{1}{\pi}\left(\frac{\pi}{3} - 0.433\right)\right]^{1/2} = 3.75 \text{ Amp}$$

Example 6.9 For the circuit shown in Fig. (Ex. 6.9),
  (a) draw the current waveforms for each device assuming load to be highly inductive
  (b) obtain the expression for the average load voltage.
  (c) for an ac source of 230 V/50 Hz, and firing angle of $\pi/4$, find the average output current and power delivered to battery with $R = 10\,\Omega$, $L = 10$ mH and $E_C = 120$ V.

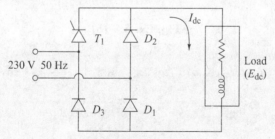

Fig. Ex. 6.9 (a)

Solution:
  (a) The load voltage and current waveforms through all devices is shown in Fig. [Ex. 6.9(b)]
  (b) From the load voltage waveform average load voltage is given by

$$E_{dc} = \frac{1}{2\pi}\left[\int_\alpha^\pi E_m \sin\omega t \, d(\omega t) + \int_\pi^{2\pi} E_m \sin\omega t \, d\omega t \right]$$

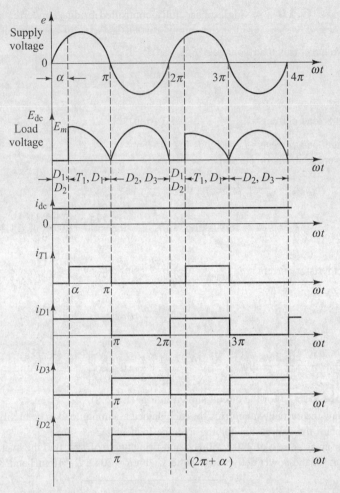

Fig. Ex. 6.9 (b)

$$= \frac{-E_m}{2\pi} \left[(\cos \pi - \cos \alpha) + (\cos 2\pi - \cos \pi) \right] = \frac{E_m}{2\pi} \left[(1 + \cos \alpha) + 2 \right]$$

$$= \frac{E_m}{2\pi} [3 + \cos \alpha]$$

(c) $$E_{dc} = \frac{\sqrt{2} \times 230}{2\pi} [3 + \cos 45°] = 194.67$$

$$I_{dc} = \frac{(194.67 - 120)}{10} = 7.47 \text{ A}$$

Now, power delivered to battery = $E_C \cdot I_{dc}$ = 120 × 7.47 = 896.40 W

Phase Controlled Converters

Example 6.10 A single-phase fully-controlled bridge converter supplies an inductive load. Assuming that the output-current is virtually constant, and is equal to I_d, determine the following performance measures, if the supply voltage is 230 V and if the firing angle is maintained at $(\pi/6)$ radians.
(i) Average output voltage. (ii) Supply RMS current.
(iii) Supply fundamental RMS current. (iv) Fundamental power factor.
(v) Supply power factor. (vi) Supply harmonic factor.
(vii) Voltage ripple factor.

Solution:

(i) Average output voltage, $E_{dc} = \dfrac{2E_m}{\pi} \cos \alpha = \dfrac{2\sqrt{2} \times 230}{\pi} \cos(\pi/6) = 179.33$ V

(ii) Supply RMS current = I_d

(iii) Supply fundamental RMS current = $\dfrac{2\sqrt{2}}{\pi} I_d$

(iv) Fundamental power factor or displacement factor = $\cos \alpha = 0.86$

(v) Supply power factor = $\dfrac{2\sqrt{2}}{\pi} \cos \alpha = 0.78$

(vi) Supply harmonic factor = $\sqrt{\dfrac{\pi^2}{8} - 1} = 0.483$

(vii) Voltage ripple factor $KV = \left(\dfrac{\pi^2}{8\cos^2 \alpha} - 1 \right)^{1/2} = \left(\dfrac{\pi^2}{8\cos^2 \pi/6} - 1 \right)^{1/2} = 0.803$.

Example 6.11 A single-phase semiconverter is operated from 120 V, 50 Hz ac supply. The load current with an average value I_{dc} is continuous and ripple free firing angle $\alpha = \pi/6$. Determine:
(a) Displacement factor (b) Harmonic factor of input current
(c) Input power factor

Solution:

Supply rms current $= I_{rms} = I_{dc} \cdot \left(1 - \dfrac{\alpha}{\pi}\right)^{1/2} = I_{dc} \left(1 - \dfrac{\pi/6}{\pi}\right)^{1/2} = 0.91 \, I_{dc}$

Now, the rms value of the supply fundamental component of input current

$= I_{rms1} = \dfrac{2\sqrt{2} \, I_{dc}}{\pi} \cos \alpha/2 = \dfrac{2\sqrt{2}}{\pi} I_{dc} \cos 15 = 0.869 \, I_{dc}$

(a) Displacement factor $= \cos \phi_1 = \cos(-\alpha/2) = \cos(-15°)$ DF = 0.9659
(b) Harmonic factor (Hf) of input current

$= \left[\left(\dfrac{I_{rms}}{I_{rms1}} \right)^2 - 1 \right]^{1/2} = \left[\left(\dfrac{0.91}{0.869} \right)^2 - 1 \right]^{1/2} = 30.80\%$

(c) Input power factor $= \dfrac{I_{rms1}}{I_{rms}} \cos \alpha/2 = 0.922$ (lagging)

6.6 PERFORMANCE FACTORS OF LINE-COMMUTATED CONVERTERS

The above discussions have revealed that even though the a.c. input supply is purely sinusoidal, the input current, output voltage and output current are complex waveforms. The current at the a.c. input terminal of the converters consists of a fundamental component with superimposed harmonic components. Therefore, depending upon the load-current waveform, a phase-difference between the input voltage and current may be introduced by the controlled-rectifier, which leads to a power-factor different from that of the load. Therefore in this section, it is first necessary to define the important external performance parameters of the phase-controlled converters. These are as follows:

1. Input Displacement Angle (ϕ_1) The input-displacement angle, denoted by ϕ, is defined as the angular displacement between the fundamental component of the a.c. line current and the associated line to neutral voltage. In all phase-controlled converter circuits, the fundamental component is either in phase or lags behind the voltage by an angle which depends upon the firing angle.

2. Input Displacement Factor (cos ϕ_1) The input displacement factor is defined as the cosine of the input displacement angle.

3. Input Power Factor (PF) The input power factor is defined as the ratio of the total mean input power to the total RMS input volt-amperes (apparent power). Since only the fundamental component contributes to the mean input power, the power factor may be defined as

$$PF = \frac{E_1 I_1 \cos \phi_1}{E_{rms} I_{rms}} = \frac{I_1}{I_{rms}} \cos \phi_1 \qquad (6.19)$$

where, $E_1 = E_{rms}$ = phase-voltage, I_1 = fundamental component of the supply current.

ϕ_1 = input displacement angle, I_{rms} = supply rms current

4. D.C. Voltage Ratio (r) The d.c. voltage ratio, denoted r, is defined as the ratio of the mean d.c. terminal voltage at a given firing-angle α, to the maximum possible d.c. terminal voltage, that is, the voltage obtained when the firing angle is zero degree.

5. Input Current Distortion Factor The distortion factor of the current in a given input-line is defined as ratio of the RMS amplitude of the fundamental component, to the total RMS amplitude.

6. Input Harmonic Factor (I_H) The input harmonic factor is defined as the ratio of the total harmonic content to the fundamental component.

$$\therefore \quad I_H = \frac{(I_{\text{rms}}^2 - I_1^2)^{1/2}}{I_1} \qquad (6.20)$$

7. Voltage Ripple Factor (K_V) The voltage-ripple factor is defined as the ratio of the net harmonic-content of the output voltage to the average output voltage

$$K_V = \frac{\sqrt{E_{\text{dc}_{\text{rms}}}^2 - E_{\text{dc}}^2}}{E_{\text{dc}}} \qquad (6.21)$$

where E_{dc} = average value of output voltage,
$E_{\text{dc}_{\text{rms}}}$ = RMS value of output voltage.

8. Current Ripple Factor (K_I) The current ripple factor is defined as the ratio of the net harmonic content of the output voltage to the average output current.

$$K_I = \frac{\sqrt{I_{d\,\text{rms}}^2 - I_d^2}}{I_d} \qquad (6.22)$$

where I_d = average value of output-current
$I_{d\,\text{rms}}$ = RMS value of the output-current.

6.7 THE PERFORMANCE MEASURES OF TWO-PULSE CONVERTERS

Here, we consider the 2-pulse-converter of Fig. 6.21, that is, fully-controlled single-phase bridge with R–L load. The various output and input waveforms of this type of converter are shown in Fig. 6.22.

The current in an a.c. supply connected to a converter can be represented generally by Fourier Series. That is,

$$I = \sum_{n=1}^{\infty} (a_n \cos n\,\omega t) + (b_n \sin \omega t) \qquad (6.23)$$

$$= \sum_{n=1}^{\infty} C_n \cdot \sin(n\,\omega t + \phi_n) \qquad (6.24)$$

where

$$C_n = \sqrt{a_n^2 + b_n^2} \qquad (6.25)$$

$$\phi_n = \tan^{-1}\left(\frac{a_n}{b_n}\right) \qquad (6.26)$$

The fundamental component can be obtained by first determining the coefficients a_n, b_n, and c_n.

Now, Fourier coefficients of the supply-current (I) waveform, assuming constant load current I_d can be obtained from Fig. 6.22 as

$$a_1 = \frac{2}{\pi} \int_{\alpha}^{(\pi+\alpha)} I_d \cos \omega t \, d(\omega t) = -\frac{4}{\pi} I_d \sin \alpha \qquad (6.27)$$

$$b_1 = \frac{2}{\pi} \int_{\alpha}^{(\pi+\alpha)} I_d \sin \omega t \, d(\omega t) = \frac{4}{\pi} I_d \cos \alpha \qquad (6.28)$$

$$\therefore \quad C_1 = \sqrt{a_1^2 + b_1^2} = \sqrt{\left(\frac{-4}{\pi} I_d \sin \alpha\right)^2 + \left(\frac{4}{\pi} I_d \cos \alpha\right)^2}$$

$$C_1 = \frac{4 I_d}{\pi} \qquad (6.29)$$

The fundamental component of the supply current and its phase angle with reference to the corresponding phase voltage are:

$$I_1 = \frac{C_1}{\sqrt{2}} \quad I_1 = \frac{2\sqrt{2} \, I_d}{\pi} \qquad (6.30)$$

$$\phi_1 = \tan^{-1} \frac{a_1}{b_1} = \tan^{-1} \left[\frac{\frac{-4}{\pi} \cdot I_d \sin \alpha}{\frac{4}{\pi} I_d \cos \alpha} \right] = \tan^{-1}(-\tan \alpha) \quad \phi_1 = -\alpha \qquad (6.31)$$

This negative sign for ϕ_1 shows that the current lags behind the voltage. The expressions for the external performances of a 2-pulse converter can be derived as follows:

(1) Input Displacement Factor (DSF)

$$\text{DSF} = \cos \alpha \qquad (6.32)$$

(2) Input Power Factor (PF)

$$PF = \left(\frac{I_1}{I_{rms}}\right) \cos \phi_1$$

$$= \left(\frac{\frac{2\sqrt{2} \, I_d}{\pi}}{I_d}\right) \cos \alpha \quad PF = \frac{2\sqrt{2}}{\pi} \cos \alpha \qquad (6.33)$$

(3) D.C. Voltage Ratio

$$r = \frac{\dfrac{1}{\pi}\displaystyle\int_{\alpha}^{(\pi+\alpha)} E_m \sin\omega t\, d(\omega t)}{\dfrac{1}{\pi}\displaystyle\int_{0}^{\pi} E_m \sin\omega t\, d(\omega t)}, \quad r = \cos\alpha \qquad (6.34)$$

(4) Input Current Distortion Factor

$$= \frac{I_1}{I_{rms}} = \frac{2\sqrt{2}\, I_d}{\pi \cdot I_d} = \frac{2\sqrt{2}}{\pi} \qquad (6.35)$$

(5) Input Harmonic Factor (I_H)

We know that

$$I_H = \frac{\left(I_{rms}^2 - I_1^2\right)^{1/2}}{I_1}$$

Substituting the values of I_{rms} and I_1,

$$I_H = \frac{\left[I_{d^2} - \left(\dfrac{2\sqrt{2}\, I_d}{\pi}\right)^2\right]^{1/2}}{2\sqrt{2}\, I_d/\pi} = \frac{\left[I_{d^2} - \dfrac{8 I_{d^2}}{\pi^2}\right]^{1/2}}{2\sqrt{2}\, I_d/\pi}$$

$$= \frac{\left[\dfrac{\pi^2 I_{d^2} - 8 I_{d^2}}{\pi^2}\right]^{1/2}}{2\sqrt{2}\, Id/\pi} = \left[\dfrac{\pi^2 - 8}{8}\right]^{1/2} \quad I_H = \left[\dfrac{\pi^2}{8} - 1\right]^{1/2} \qquad (6.36)$$

(6) Voltage Ripple Factor (k_V)

We have the equation for k_V as

$$k_V = \frac{\sqrt{E_{dc\,rms}^2 - E_{dc}^2}}{E_{dc}} = \frac{\left[\dfrac{E_m^2}{2} - \left(\dfrac{2E_m}{\pi}\cos\alpha\right)^2\right]^{1/2}}{\dfrac{2E_m}{\pi}\cos\alpha}$$

$$= \frac{E_m\left[\dfrac{1}{2} - \dfrac{4}{\pi}\cos^2\alpha\right]^{1/2}}{2E_m \cos\alpha/\pi} = \frac{\pi\left[\dfrac{1}{2} - \dfrac{4}{\pi^2}\cos^2\alpha\right]^{1/2}}{2\cos\alpha}$$

$$= \frac{\pi \left[\dfrac{\pi^2 - 8\cos^2\alpha}{2\pi^2}\right]^{1/2}}{2\cos\alpha} = \left[\dfrac{\pi^2 - 8\cos^2\alpha}{2\cos^2\alpha}\right]^{1/2}$$

$$k_V = \left(\dfrac{\pi^2}{8\cos^2\alpha} - 1\right)^{1/2} \qquad (6.37)$$

(7) Active Power Input (P_i) Only the fundamental component in the series of Eq. (6.23), contributes to the mean a.c. input power and the mean power as the harmonic components of current is obviously zero.

Therefore, the mean a.c. input power in a given line is given by:

P_i = RMS line voltage × RMS fundamental component of current × displacement factor.

$= E_{rms} \times I_1 \times \cos\alpha$.

$$= E_{rms} \times \dfrac{2\sqrt{2}\, I_d}{\pi} \cos\alpha = \dfrac{2 E_m}{\pi} I_d \cos\alpha \qquad (6.38)$$

(8) Reactive Power Input (Q_i) Reactive power input is given by

$$Q_i = E_{rms} \cdot I_1 \sin\alpha = \dfrac{2 E_m}{\pi} I_d \sin\alpha \qquad (6.39)$$

6.7.1 Comparison between Semiconverter and Full Converter

The performance of the single-phase semiconverter and full converter with R-L load is compared based on the various performance parameters, as shown in Table 6.1.

Table 6.1

S.No.	Parameter	Full-converter	Semiconverter
1.	Average-load voltage	$E_{dc} = \dfrac{2 E_m}{\pi}\cos\alpha$	$E_{dc} = \dfrac{E_m}{\pi}(1+\cos\alpha)$
2.	RMS Load Voltage	$E_{rms} = \dfrac{E_m}{\sqrt{2}}$	$E_{rms} = \dfrac{E_m}{\sqrt{2}}\left[\dfrac{1}{\pi}\left(\pi - \alpha + \dfrac{\sin 2\alpha}{2}\right)\right]^{1/2}$
3.	Form factor (FF)	$F.F. = \dfrac{\pi}{2\sqrt{2}\cos\alpha}$	$F.F. = E_{rms}/E_{dc}$
4.	Ripple Factor (RF)	$RF = \left[\dfrac{\pi^2}{8\cos^2\alpha} - 1\right]^{1/2}$	$RF = [FF^2 - 1]^{1/2}$
5.	Rectification Efficiency (η)	$\eta = \dfrac{8\cos^2\alpha}{\pi^2}$	$\eta = 1/FF^2$

(Contd.)

Table 6.1 (Contd.)

S.No.	Parameter	Full-converter	Semiconverter
6.	Operation	Two quadrant converter (Rectification and inversion)	Single quadrant converter (Rectification only)
7.	Fundamental Power Factor (FPF)	$FPF = \cos \alpha$	$FPF = \cos(\alpha/2)$
8.	Input Power Factor (PF)	$PF = \dfrac{2\sqrt{2}}{\pi} \cos \alpha$	$PF = \dfrac{\sqrt{2}\,(1+\cos\alpha)}{[\pi(\pi-\alpha)]^{1/2}}$
9.	Harmonic Factor (HF)	$HF = \left[\dfrac{\pi^2}{8} - 1\right]^{1/2}$	$HF = \left[\dfrac{\pi(\pi-\alpha)}{4(1+\cos\alpha)} - 1\right]^{1/2}$
10.	Free wheeling Mode	Absent	Present
11.	Active Power (P_i)	$P_i = \dfrac{2E_m I_{dc}}{\pi} \cos \alpha$	$P_i = \dfrac{2E_m I_{dc}}{\pi} \cos^2\left(\dfrac{\alpha}{2}\right)$
12.	Reactive Power (Q_i)	$Q_i = \dfrac{2E_m}{\pi} I_{dc} \sin \alpha$	$Q_i = \dfrac{E_m I_d}{\pi} \sin \alpha$
13.	Supply current	Square-wave	Quasi-square wave
14.	Harmonic present in supply current	Only odd harmonics	Only odd harmonics
15.	Power Flow	Bidirectional	Unidirectional

+ Note: Reactive power input to the semiconverter is half of the reactive power input to the full converter.

6.8 THREE-PHASE CONTROLLED CONVERTERS

The converter operating from a single-phase supply produces a relatively high proportion of a.c. ripple-voltage at its d.c. terminals. This ripple is generally undesirable because of its heat producing effect. Therefore, a large outlay of smoothing reactor is necessary to smoothen the output voltage as well as to reduce the possibility of discontinuous operation. The need for smoothing can be minimised by increasing the number of pulses. A three phase a.c. supply with a suitable transformer connection permits an increase in the pulse number. When the number of pulses of the converter is increased, the number of segments that fabricate the output voltage also increases and consequently the ripple content decreases. Higher the pulse number, smoother is the output voltage.

Three-phase rectifier circuits are used for large power applications. Generation of the three-phase a.c. power is now universal and in some countries, only generation frequencies may be different. Now-a-days, 11 kV, 33 kV, 66 kV three-phase a.c. supply is available to the industries. These voltages are suitably stepped down using transformers. These transformers are generally delta-connected on primary side and star-connected on the secondary side.

Three-phase controlled converter circuits can be studied under following categories:
(1) Three-pulse converters
(2) Six-pulse converters
(3) Twelve-pulse converters.

6.9 THREE-PULSE CONVERTERS (M_3 CONNECTION)

Three pulse converters are also known as the three-phase half-wave controlled rectifier. The simplest type of phase-controlled converter operating from a three-phase supply is the three-pulse midpoint converter.

6.9.1 Three-Phase Half-Wave Controlled Rectifier with Resistive Load

Figure 6.29 shows the power-diagram of a three-phase half-wave controlled rectifier with resistive load. This configuration is called as the mid-point configuration because all the phase emfs can have a common terminal which may be considered as the neutral point or the mid-point. As shown in figure, the primary is connected in a delta fashion and secondary in star. The load is connected to the neutral point. For the analysis of the circuit, the leakage inductance and on state SCR drops are assumed to be zero. The vector diagram of the three-phase voltage is shown in Fig. 6.30.

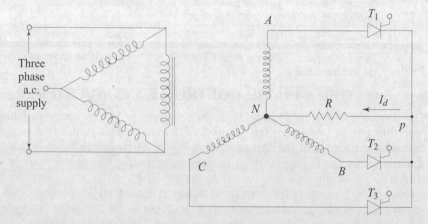

Fig. 6.29 *Three-phase half-wave controlled rectifier with resistive load*

The circuit functions in a manner such that only one SCR is conducting at any given instant, the one which is connected to the phase having the highest instantaneous positive value. Here, no SCR can be triggered below a phase angle of 30° because it remains reverse-biased by the other conducting phase. The firing angle α for a particular thyristor connected in a particular phase is therefore measured from 30° with respect to the corresponding phase voltage.

The various waveforms for the three-phase half-wave controlled rectifier with resistive load are shown in Fig. 6.30. As shown in the vector diagram, phase A and phase C are equally positive with respect to the neutral. The SCR T_1 connected to phase A cannot be triggered below an angle of 30° since it is already reverse-biased by the already conducting SCR T_3. Therefore, the minimum firing

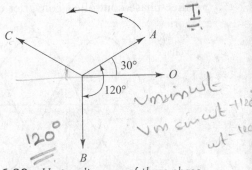

Fig. 6.30 Vector diagram of three-phase voltage

angle is $\pi/6$. The firing angle α is measured from the cross-over points of the voltage waveform as these are the points at which the equivalent thyristors would start to conduct. This gives a conduction period for each thyristor equal to one-third of the supply cycle-period (120° or $2\pi/3$).

As shown in Fig. 6.31 (a), SCR T_1 will start conducting from $\omega t = 30°$ to $\omega t = 150°$, as this SCR T_1 is the most positive as compared to the other two SCRs T_2, T_3 during this interval. SCR T_2 will conduct from $\omega t = 150°$ to 270° and SCR T_3 from $\omega t = 270°$ to 390°. When an SCR is conducting, the common cathode terminal p rises to the highest positive voltage of that phase and the other two blocking SCRs are reverse-biased. The voltage across the load follows the positive supply voltage envelop and has the waveforms as shown in Fig. 6.31(b).

With resistive load, there are two modes of conduction. They are:
(1) Continuous conduction mode for $a \leq 30°$.
(2) Discontinuous conduction mode for $a > 30°$.

When the firing angle a is taken in between 0° to 30°, from the cross-over point, the load current is continuous. This is due to the fact, that the maximum value of the conduction angle of an SCR is 120°. Therefore, for firing angle $a \leq 30°$, we are having the continuous conduction mode of the operation. This continuous conduction mode is shown in Fig. 6.31(c).

If the firing angle is kept more than 30°, then the conduction angle will be less than 120°. Therefore, the output voltage and current pulses become discontinuous, that is, during some part of the cycle the voltage and current remains at zero. This discontinuous conduction mode of operation is shown in Fig. 6.31(d).

Now, we derive the various voltage and current relationships for both continuous and discontinuous conduction modes.

(1) Continuous conduction Mode (for $0 \leq a \leq 30°$)

(i) Average Load Voltage:

$$E_{dc} = \frac{1}{2\pi/3} \int_{\alpha+30°}^{\alpha+150°} E_m \cdot \sin \omega t \, d(\omega t) = \frac{3 E_m}{2\pi} \left[-\cos \omega t\right]_{\alpha+30°}^{\alpha+150°}$$

$$= \frac{3\sqrt{3}}{2\pi} E_m \cdot \cos \alpha \qquad (6.40)$$

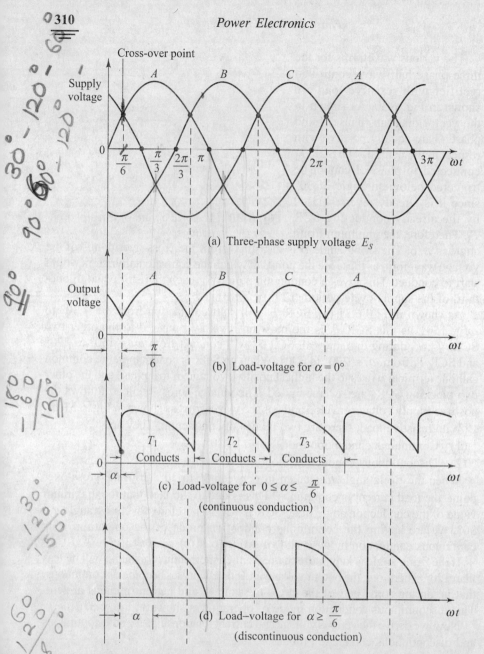

Fig. 6.31 *Waveforms for 3-pulse converter with resistive-load R*

(ii) Average-load Current:

$$I_d = \frac{3\sqrt{3}}{2\pi R} E_m . \cos \alpha \qquad (6.41)$$

(iii) RMS Load Voltage: The RMS load voltfage for a given firing angle α is given by:

$$E_{rms} = \left[\frac{3}{2\pi}\int_{\alpha+30°}^{\alpha+150°} E_m^2 \sin^2 \omega t \, d(\omega t)\right]^{\frac{1}{2}} = E_m\left[\frac{3}{2\pi}\int_{\alpha+30°}^{\alpha+150°}\left(\frac{1-\cos 2\omega t}{2}\right)d(\omega t)\right]^{\frac{1}{2}}$$

$$= E_m\left[\frac{3}{4\pi}\int_{\alpha+30°}^{\alpha+150°} d(\omega t) - \frac{3}{4\pi}\int_{\alpha+30°}^{\alpha+150°}\cos 2\omega t \, d(\omega t)\right]^{\frac{1}{2}}$$

$$= E_m\left[\frac{3}{4\pi}(\omega t)\Big|_{\alpha+30°}^{\alpha+150°} - \frac{3}{4\pi}\left(\frac{\sin 2\omega t}{2}\right)\Big|_{\alpha+30°}^{\alpha+150°}\right]^{\frac{1}{2}}$$

$$= E_m\left[\frac{3}{4\pi}[\alpha+150° - \alpha - 30°] - \frac{3}{8\pi}\begin{bmatrix}\sin 2(\alpha+150°)\\-\sin 2(\alpha+30°)\end{bmatrix}\right]^{\frac{1}{2}}$$

$$= E_m\left[\frac{1}{2} - \frac{3}{8\pi}2(-\cos 2\alpha)\frac{\sqrt{3}}{2}\right]^{\frac{1}{2}} = E_m\left[\frac{1}{2} + \frac{3\sqrt{3}}{8\pi}\cos 2\alpha\right]^{\frac{1}{2}} \quad (6.42)$$

(2) Discontinuous conduction Mode (for $30° \leq \alpha \leq 150°$)

(i) Average load voltage:

$$E_{dc} = \frac{1}{2\pi/3}\int_{\alpha+30°}^{180°} E_m \cdot \sin \omega t \, d\omega t = \frac{3E_m}{2\pi}[-\cos \omega t]_{\alpha+30°}^{180°}$$

$$= \frac{3E_m}{2\pi}[1 + \cos(\alpha+30°)] \quad (6.43)$$

(ii) Average load current:

$$I_d = \frac{3E_m}{2\pi R}[1 + \cos(\alpha+30°)] \quad (6.44)$$

(iii) RMS load voltage:

$$E_{rms} = \left[\frac{1}{2\pi/3}\int_{\alpha+30°}^{180°} E_m^2 \cdot \sin^2 \omega t \, d(\omega t)\right]^{\frac{1}{2}}$$

$$= \frac{\sqrt{3}\,E_m}{2\sqrt{2}}\left[\frac{5\pi - 3\alpha}{3\pi} + \frac{\sin(2\alpha + \pi/3)}{\pi}\right]^{\frac{1}{2}} \quad (6.45)$$

6.9.2 Three-Phase Half-Wave Controlled Rectifier with Inductive Load (R–L)

Figure 6.32 shows the power diagram of three-phase half-wave controlled rectifier with R–L load. The primary of the transformer is connected in delta fashion and secondary in star. For the simplification of the analysis, it is assumed that the current waveform for each transformer secondary winding is rectangular.

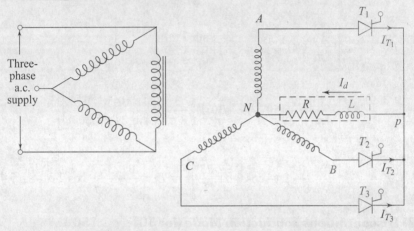

Fig. 6.32 *Three-pulse converter with R-L load*

Typical waveforms associated with the three-pulse converter, assuming a perfectly smooth direct-current, are shown in Fig. 6.33. In Fig. 6.33(a), the firing angle is 0°. Each thyristor conducts for a period of 120° of the input a.c. supply, and the fundamental frequency of the ripple voltage at the d.c. terminals is 3-times the input frequency, that is, there are three-pulses at the output. The current waveform in each thyristor consists of a unidirectional rectangular block, having a duration of 120°.

Figure 6.33(b) shows the waveforms for $\alpha = 45°$. Therefore, each thyristor blocks forward voltage for a 45° period, prior to the instant at which it is fired. The average voltage at the d.c. terminals has been reduced and the fundamental component of the a.c. input current waveforms now lags the voltage by 45°.

Figure 6.33(c) shows the waveforms for $\alpha = 90°$. Each thyristor now blocks forward and reverse voltage for equal periods of time. The average voltage at the d.c. terminals is zero, and the fundamental component of the a.c. input current waveform lags the voltage by 90°. Therefore, as the firing angle is increased from zero, the average d.c. output voltage decreases and is zero when $\alpha = 90°$. Because of the large smoothing reactor, the current remains continuous, even when the instantaneous unsmoothed voltage is negative. The conducting period of each thyristor in the steady-state condition with controlled operation is 120°. Hence, by varying the firing angle α from 0 to 90°, rectifier operation is obtained with Fig. 6.33.

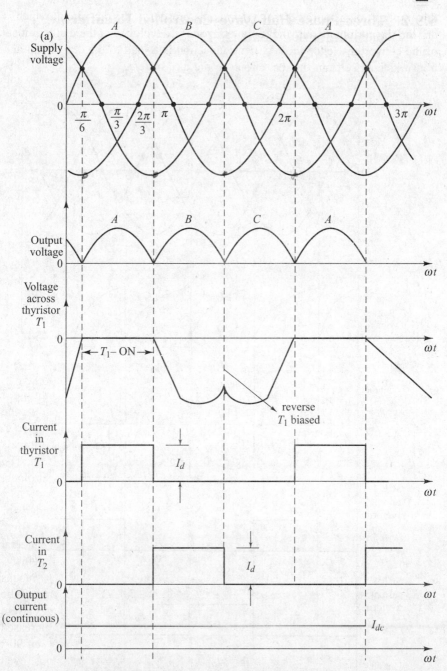

Fig. 6.33 *(a) Voltage and current waveforms for $\alpha = 0°$*

Figure 6.33(d) shows the waveform for $\alpha = 135°$. The blocking voltage waveform developed across each thyristor is now predominantly in the forward direction. Therefore, the mean d.c. voltage at the d.c. terminals is negative and this implies that a source of voltage must be present in the d.c. circuit which is driving the current against the "counter" voltage developed at the d.c. terminals of the converter. Hence the rectifier connection is a load to the d.c. circuit and transmits energy from the d.c. circuit to the a.c. system.

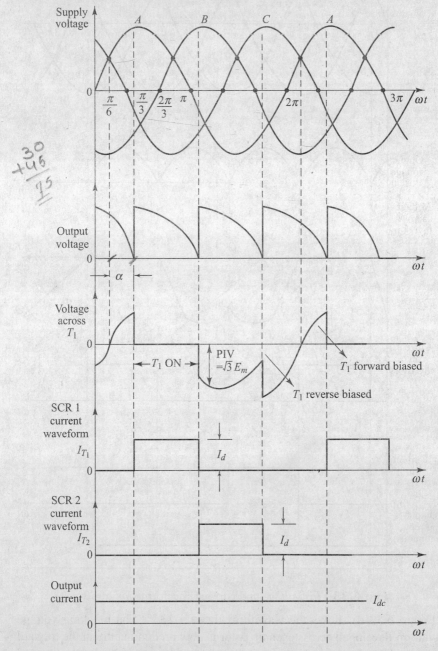

Fig. 6.33 *(b) Voltage and current waveforms for $\alpha = 45°$*

Figure 6.33(e) shows the waveforms for $\alpha = 180°$. The thyristor blocking voltage is now almost entirely in the forward direction, and this voltage swings 'negative' only for a short "turn off" period immediately following the conduction interval. The mean voltage at the d.c. terminals has its maximum negative value, which is almost equal and opposite to the maximum positive value obtained at $\alpha = 0°$. Therefore, the converter is now operating at "full-inversion" and the phase-displacement between the fundamental component of the a.c. input current waveform and the voltage is almost 180°.

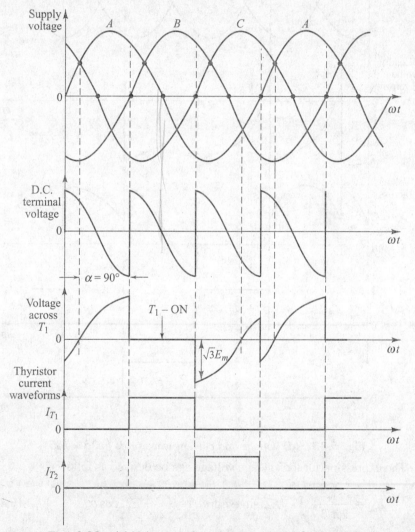

Fig. 6.33 *(c) Voltage and current waveforms for $\alpha = 90°$*

During the inverter operation, to maintain the current continuously, a voltage-source of reverse polarity should be connected on the d.c. side, otherwise the discontinuous conduction operation results. Figure 6.34 shows the graph of average-output voltage versus firing angle α.

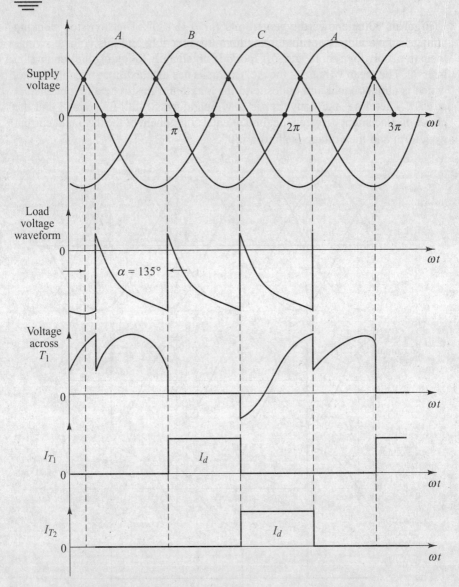

Fig. 6.33 *(d) Voltage and current waveforms for $\alpha = 135°$*

The expression for the average voltage can be derived as follows:

$$E_{dc} = \frac{1}{2\pi/3} \int_{\alpha+30}^{\alpha+150} E_m \cdot \sin \omega t \, d(\omega t) = \frac{3\sqrt{3}}{2\pi} E_m \cdot \cos \alpha \qquad (6.46)$$

Following points are noted from this expression:
(i) When $\alpha = 0°$, the maximum value of output-voltage is obtained.

$$\therefore \qquad E_{dc \, max} = \frac{3\sqrt{3}}{2\pi} E_m.$$

(ii) This voltage is zero when $\alpha = 90°$.
(iii) This voltage is negative maximum when $\alpha = 180°$.

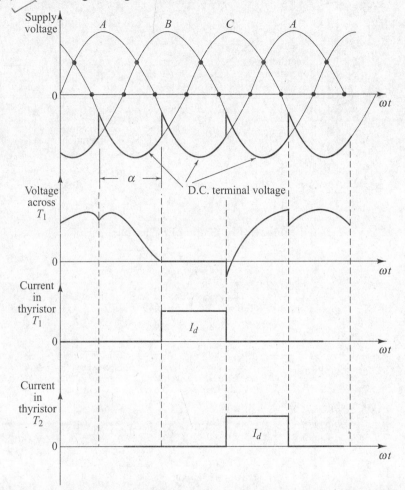

Fig. 6.33 *(e) Voltage and current waveforms for $\alpha = 180°$*

The product of voltage and current is the instantaneous value of power. For the phase angle between 0 and 90°, the average value of power is positive, indicating a flow of energy from the a.c. system to the d.c. circuit. With $\alpha = 90°$, the positive and negative value of power are in balance so that effective power becomes zero. For $\alpha > 90°$, the average power is negative, so that energy is transmitted from the d.c. circuit to the a.c. system.

In the above analysis, we have assumed that the current waveform in each transformer secondary, which is the same as the associated thyristor current waveform consists of a unidirectional or "half-wave" rectangular "block" having a duration of 120°. But this waveform contains a direct component having an amplitude of $I_d/3$. Therefore, if a simple star connected transformer secondary is used to

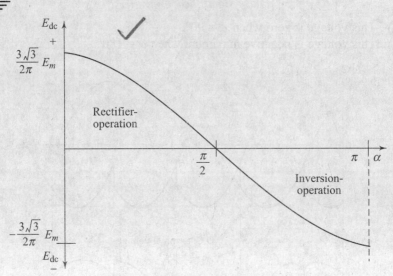

Fig. 6.34 *Graph of E_{dc} vs. α*

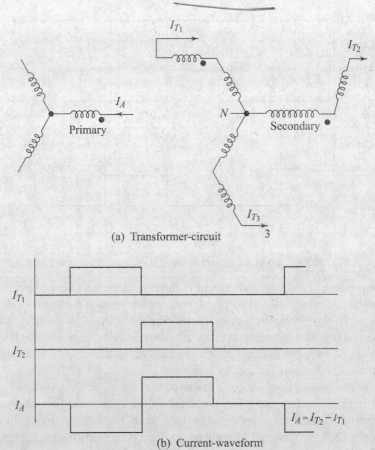

(a) Transformer-circuit

(b) Current-waveform

Fig. 6.35 *Zig-zag connection*

carry this current, then a d.c. magnetisation of the transformer core results. To avoid this problem, the interconnected star called as the zig-zag connection, shown in Fig. 6.35(a), is used as the secondary of the supply-transformer. The current which is reflected into the primary is now a.c. as shown in Fig. 6.35(b), being as much positive as negative, hence avoiding any d.c. component in the core mmf.

6.9.3 Effect of Freewheeling Diode

The circuit of three-pulse mid-point converter with freewheeling diode is shown in Fig. 6.36. The related waveforms, which assumes a perfectly smooth load-current, are shown in Fig. 6.37. For firing angles less than 30°, the d.c. terminal voltage of the converter is always positive, and the freewheeling diode does not come into operation. As the firing angle is retarded beyond this point, so the load current starts to freewheel through the diode for certain periods, thus cutting off the input line current, and preventing the d.c. terminal load voltage from swinging into the negative direction.

Hence, the effect of the freewheeling diode is to cause a reduction of ripple voltage of the d.c. terminals, and, at the same time, to divert the load current away from the input lines. The total range of firing angle control required for this circuit is 150°.

SOLVED EXAMPLES

Example 6.12 A three-phase half-wave controlled rectifier has a supply of 200 V/phase. Determine the average load voltage for firing angle of 0°, 30°, 60°, assuming a thyristor volt drop of 1.5 V and continuous load current.

Solution: From Eq. (6.45), we have

$$E_{dc} = \frac{3\sqrt{3}}{2\pi} E_m \cdot \cos \alpha - \text{drop across SCR} = \frac{3\sqrt{3}}{2\pi} 200 \sqrt{2} \cos \alpha - 1.5$$

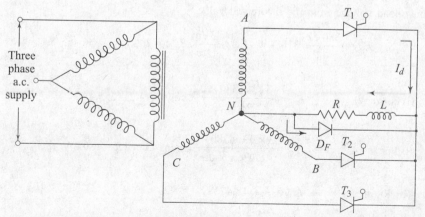

Fig. 6.36   *Three-pulse mid-point converter with a freewheeling diode*

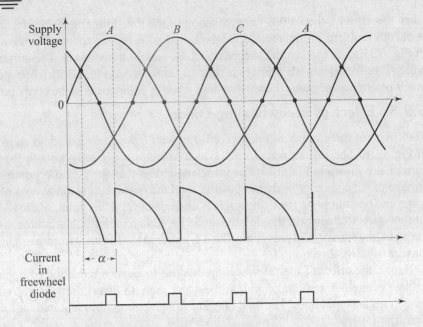

Fig. 6.37 *Waveforms with a freewheeling diode for $\alpha = 60°$*

∴ (i) For $\alpha = 0°$, $E_{dc} = 232.41$ V (ii) For $\alpha = 30°$, $E_{dc} = 201.07$ V
(iii) For $\alpha = 60°$, $E_{dc} = 115.45$ V

Example 6.13. A three-phase, half-wave controlled converter is connected to a 380 V (line) supply. The load-current is constant at 32 A, and is independent of firing angle. Find the average load voltage at firing angle of 0° and 45°, given that the thyristors have a forward voltage-drop of 1.2 V. What value of current and peak reverse voltage rating will the thyristor require and what will be the average power dissipation in each thyristor?

Solution: (a) The effect of the thyristor voltage drop (V_t) will be to reduce the mean-load voltage from the theoretical value.

∴ $$E_m = \frac{380\sqrt{2}}{\sqrt{3}} \quad 310.3 \text{ V.} \quad E_{dc} = \frac{3\sqrt{3}}{2\pi} E_m \cos \alpha - V_t$$

∴ (i) For $\alpha = 0$, $$E_{dc} = \frac{3\sqrt{3} \times 380 \times \sqrt{2}}{2\pi \times \sqrt{3}} \cos 0° - 1.2 = 255.4 \text{ V.}$$

(ii) For $\alpha = 45°$, $$E_{dc} = \frac{3 \times \sqrt{3} \times 380 \times \sqrt{2}}{2 \times \pi \times \sqrt{3}} \cos 45° - 1.2 = 180.2 \text{ V.}$$

(b) Ratings $$I_{rms} = \frac{32}{\sqrt{3}} \quad 18.47 \text{ A.}$$

Phase Controlled Converters

For three-phase half-controlled converters, the reverse-voltage can be seen to be the difference between the two-phase voltages, i.e. the line voltage of the three-phase supply.

The peak inverse voltage (PIV) is therefore the peak value of the a.c. line voltage.

$$PIV = \sqrt{2}\, V_{line} = \sqrt{2} \times \sqrt{3} \times E_{phase} = \sqrt{2} \times 380 = 537.4\ V.$$

Now, the average power dissipated in the thyristor is obtained by averaging the instantaneous power dissipation over one cycle.

$$\therefore \text{Average power} = \frac{1}{2\pi} \int_{\alpha}^{\alpha + 2\pi/3} V_{TH.}\, I_{TH.}\, d\omega t = \frac{V_{TH.}\, I_{TH.}}{3} = \frac{1.2 \times 32}{3} = 12.8\ W.$$

Example 6.14 For a 3-ϕ half-wave converter driving a large inductive load, derive an equation for
 (i) Form factor and ripple factor (ii) Thyristor and supply current ratings
Also, calculate thyristor current ratings for $R = 10\ \Omega$ and $\alpha = 30°$.

Solution:

(i) Form factor (FF) = $\dfrac{E_{dc_{rms}}}{E_{dc(avg)}}$, $E_{dc_{rms}} = \left[\dfrac{3}{2\pi} \int_{30+\alpha}^{150+\alpha} E_m^2 \sin^2(\omega t)\, d\omega t\right]^{1/2}$

$$= \left[\frac{3E_m^2}{4\pi} \int_{30+\alpha}^{150+\alpha} (1 - \cos 2\omega t)\, d\omega t\right]^{1/2} = E_m \left(\frac{1}{2} + 0.2 \cos 2\pi\right)^{1/2}$$

$$= E_m (0.5 + 0.2 \cos 2\alpha)^{1/2}$$

$\therefore$ We have $E_{dc} = \dfrac{3\sqrt{3}}{2\pi} \cdot E_m \cos \alpha$

$$\therefore\quad FF = \frac{E_m (0.5 + 0.2 \cos 2\alpha)^{1/2}}{3\sqrt{3}/2\pi\, E_m \cos \alpha} = \sqrt{\frac{0.73 + 0.2925 \cos 2\alpha}{\cos^2 \alpha}}$$

$$\therefore\quad RF = \sqrt{FF^2 - 1}$$

(ii) Let the average output current be I_{dc}. Here, the thyristor and supply current ratings are same.

Since the current waveform has peak value of I_{dc} and the duty cycle is 1/3, we get

$$I_{T(avg)} = I_{S(av)} = \frac{I_{dc}}{3} \quad \therefore I_{dc} = \frac{E_{dc(av)}}{R_L} = 24.3\ Amp$$

$$I_{T(rms)} = I_{S(rms)} = I_{dc}/\sqrt{3},\ I_{T(peak)} = I_{S(peak)} = I_{dc}$$

PIV rating of SCR = $\sqrt{3}\, E_m = \sqrt{2}\, E_{line}$.

$$\therefore\quad I_{T(avg)} = \frac{I_{dc}}{3} = 8.1\ A,\ I_{T(rms)} = 14\ Amp,$$

$$I_{Tpeak} = I_{dc} = 24.3\ A,\ PIV = \sqrt{2} \times 415 = 590\ V$$

Example 6.15 A three-phase half-wave converter is operated from a 3-ϕ-γ-connected 220 V, 50 Hz supply and load resistance is $R = 10$ ohm. If the average output voltage is 25% of the maximum possible average voltage, determine:
  (i) delay-angle
  (ii) rms and average output currents
  (iii) average and rms thyristor currents
  (iv) rectification efficiency
  (v) transformer utilization factor and
  (vi) input power factor

Solution:

Given = Line-voltage $E_L = 220$ V, $f = 50$ Hz, $R_L = 10$ ohm, Phase voltage $E_s = 220/\sqrt{3} = 127.02$ V $\cong 127$ V.

$$E_m = \sqrt{2} \cdot E_s = 179.63 \text{ V}$$

(i) $E_{dc} = \dfrac{3\sqrt{3} \, E_m}{2\pi} \cos \alpha$

For $\alpha = 0$, maximum possible output average voltage will be obtained.

$$\therefore \quad E_{dc_{max}} = \dfrac{3\sqrt{3}}{2\pi} \cdot E_m = \dfrac{3\sqrt{3} \times 179.63}{2\pi} = 148.55 \text{ V}$$

$\therefore$ Average output voltage $\cdot E_{dc} = 25\% \, E_{dc_{max}} = 0.25 \times 148.55 = 37.14$ V

Normalized average output voltage is given by

$$E_{dc_n} = \dfrac{E_{dc}}{E_{dc_{max}}} = \cos \alpha \quad \quad (1)$$

We know that the load current is continuous if $\alpha \leq \pi/6$. i.e. if

$$E_{dc_n} \geq \cos(\pi/6) = 86.6\%$$

Since given $E_{dc_n} = 25\%$, which is less than 86.6%, hence the load current is discontinuous.

$\therefore$ Now with resistive load $\alpha \geq \pi/6$, $E_{dc} = \dfrac{3 E_m}{2\pi} [1 + \cos(\alpha + 30°)]$

$$37.14 = \dfrac{3 \times 179.63}{2\pi} [1 + \cos(\alpha + 30°)] \quad \therefore \quad \alpha = 94.5°$$

(ii) Average output current, $I_{dc} = \dfrac{E_{dc}}{R} = \dfrac{37.14}{10} = 3.714$ Amp. Now, from equation (6.45), $E_{rms} = 62.12$ V and $I_{rms} = \dfrac{E_{rms}}{I_{dc}} = \dfrac{62.12}{10} = 6.21$ A

(iii) Thyristor average current, $I_{T_{avg}} = \dfrac{I_{dc}}{3} = \dfrac{3.714}{3} = 1.24$ A

and rms thyristor current $= I_{T_{rms}} = \dfrac{I_{rms}}{\sqrt{3}} = \dfrac{6.21}{\sqrt{3}} = 3.59$ A

(iv) Rectification efficiency, $\eta = \dfrac{P_{dc}}{P_{ac}} = \dfrac{37.14 \times 3.714}{62.1 \times 6.21} = 35.75\%$

(v) RMS input line current is same as thyristor rms current, and
Input volt-ampere rating = E.I. = $3\,E_s I_s = 3 \times 127 \times 3.59 = 1367$ W

$$\text{TUF} = \dfrac{P_{dc}}{E_s I_s} = \dfrac{37.14 \times 3.714}{1367} = 10.09\%$$

(vi) Output power, $P_o = I_{rms}^2 \cdot R = (6.21)^2 \times 10 = 385$ W

$$\text{Input power factor} = \dfrac{P_o}{\text{Input } VA} = \dfrac{385}{1367} = 0.28 \text{ (lagging)}$$

6.10 SIX-PULSE CONVERTERS

The three-pulse converters described in the above section have not become very popular because of the fact that they require special types of converter transformer to prevent d.c. magnetisation. Therefore, a three-phase converter with higher pulse number is developed to provide large-output and least ripple content. Six-pulse and twelve-pulse converters have been developed. Above 12-pulse connections, that is, 18-pulse and 24-pulse connections, are rarely used. Six-pulse connections are most widely used for industrial applications, whereas for transmission lines, 12-pulse connections are preferred.

The six-pulse connections have the following advantages compared to three-pulse-converters:
 (i) Commutation is made very easier.
 (ii) Distortion on the a.c. side is reduced due to the reduction in lower-order harmonics.
 (iii) Inductance required in series is considerably reduced.

The six-pulse converters can be realised in the following forms:
 (i) Simple six-pulse converter.
 (ii) Six-pulse mid-point converter with interphase-transformers.
 (iii) Six-pulse bridge-converter.

6.10.1 Simple Six-Pulse Mid-Point Converter with Inductive Load (or Six-Pulse Double-Star Circuit or M-6 Connection)

The power diagram of a six-phase half-wave controlled rectifier with inductive load is shown in Fig. 6.38. The converter transformer used converts three-phase supply to six-phase one and makes it available to the converter terminals. The secondary of the transformer has six-phases. The secondary voltages are displaced from each other by 60° and are supplied to the anodes of the six thyristors. The common cathode connection becomes a d.c. terminal. The load is connected between the star-point of the secondaries of the transformer and common cathode point. The vector diagram of the six-phase voltage is shown in Fig. 6.39. The related voltage and current waveforms are shown in Fig. 6.40.

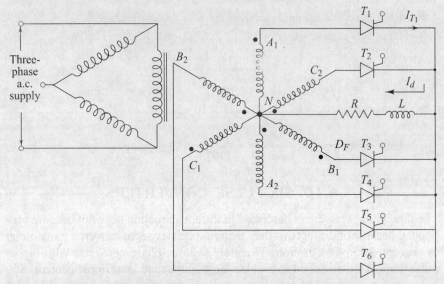

Fig. 6.38 *Six-phase half-wave controlled rectifier with R-L load (M-6 connection)*

Although a six-pulse voltage waveform is obtained at the d.c. terminals, this circuit has the practical disadvantage that each thyristor conducts for a period of only 60° during each cycle. This results in a relatively high ratio of RMS to average current in the thyristors, as well as in the windings of the transformer. Consequently, the "utilization factor" of the circuit is relatively poor. For this reason, this six-pulse circuit is not often used in practice.

The average value of the d.c. voltage is given by

$$E_{dc} = \frac{1}{2\pi/6} \int_{\alpha+\pi/3}^{\alpha+\frac{2\pi}{3}} E_m \cdot \sin\omega t \, d(\omega t) = \frac{6E_m}{2\pi}[-\cos\omega t]_{\alpha+\pi/3}^{\alpha+2\pi/3}$$

$$= \frac{3E_m}{\pi}\cos\alpha \qquad (6.47)$$

By comparing Eqs (6.46) and (6.47), it is clear that the average value of the d.c. voltage available at the output terminals of six-pulse converters is greater than that of three-pulse converter.

The variation of the firing angle controls the average voltage. For angles $0 < \alpha < 90°$, the load voltage varies from positive maximum to zero and the

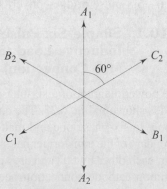

Fig. 6.39 *Vector diagram*

converter operates in the rectification mode. For $90° < \alpha < 180°$, the voltage varies from zero to negative maximum and the converter operates in the inversion mode. If a negative d.c. source is available at the load, then the power is transferred from the d.c. side to the a.c. side. The mean d.c. voltage is superimposed by harmonics.

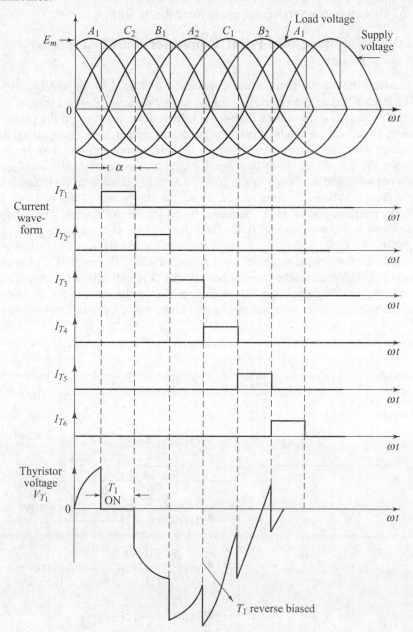

Fig. 6.40 *Simple six-pulse mid-point converter waveforms*

The minimum order of the harmonics present is six-times the supply frequency. Because of this, the smoothing inductance required in this case is smaller than the one required for the three-pulse converter.

Now, if the load is purely resistive, then two modes of operation are possible:
(i) Continuous conduction mode for $0 \leq \alpha \leq \pi/3$.
(ii) Discontinuous conduction mode for $\pi/3 \leq \alpha \leq 2\pi/3$.

6.10.2 Six-Pulse Mid-Point Converter with Interphase Reactor

In practice, the simple transformer connection of Fig. 6.38, in which current flows in each leg of the primary winding for only one third of a cycle, is not normally used, as it introduces high-level of harmonic current into the primary system. To reduce the levels of primary harmonic current, a six-pulse mid-point converter with interphase reactor is used. By interposing an interphase-reactor between the d.c. terminals of two 3-pulse circuits, having mutually displaced a.c. input voltages, as shown in Fig. 6.41, a better utilization of thyristors and transformer winding can be achieved. Here, the star-points of the two sets of secondary winding have been connected by an interphase reactor. The related waveforms are shown in Fig. 6.42. Each three-pulse commutating group of thyristors (T_1, T_3, T_5 and T_2, T_4, T_6) now operates independently of the other and therefore, each thyristor conducts for a period of 120°. The mean d.c. terminal voltages of each group is the same as one another. Theoretically, since there is no mean voltage difference across the interphase-reactor, the value of the mean d.c. terminal voltage of the combination is equal to the mean d.c. terminal voltage of

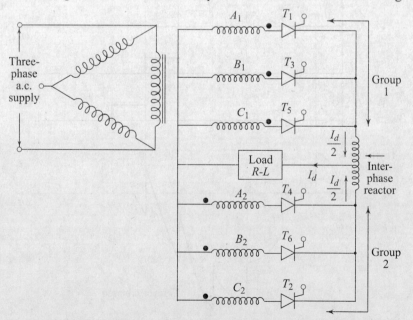

Fig. 6.41 *Six-pulse mid-point converter with interphase reactor*

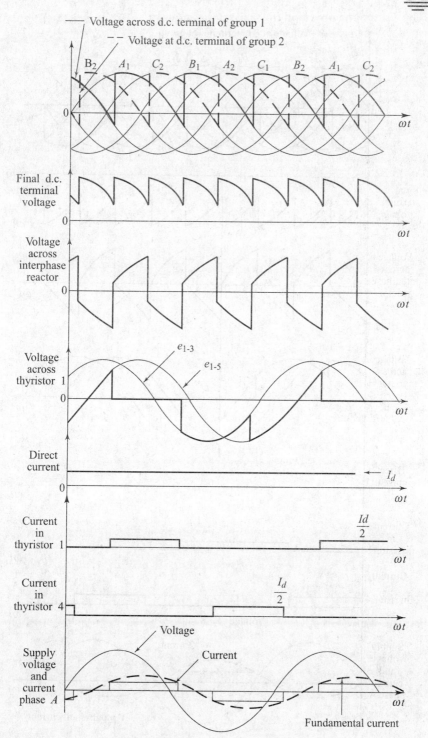

Fig. 6.42 *(a)* $\alpha = 45°$, waveforms

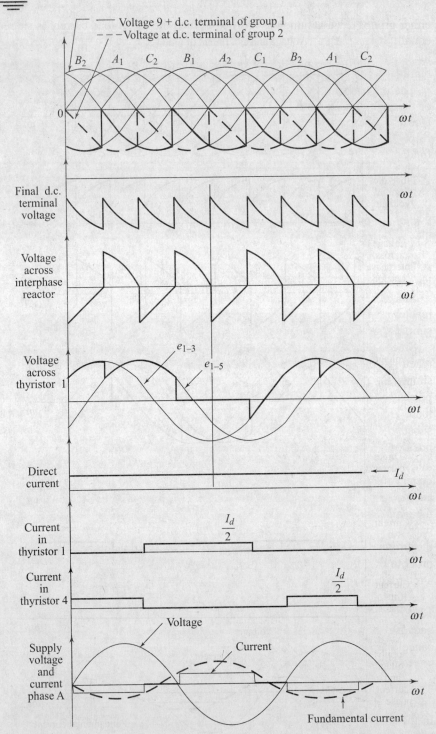

Fig. 6.42 *(b)* $\alpha = 150°$, *waveforms*

either group. If we assume the ideal case, then the direct load current is shared equally between the two groups and therefore, there is no d.c. magnetisation of the core of the interphase reactor. Practically, some current unbalance may exist but this can be kept relatively small.

The instantaneous voltage at the mid-point of the interphase reactor consists of the arithmetic mean of the instantaneous voltages of the individual groups. Due to the phase-displacement between the individual group voltages, the resulting ripple-voltage waveform at the mid-point of the interphase reactor has a fundamental frequency of 6 times the input-frequency.

The mode of operation described above is achieved only if the exciting current of the reactor is free to flow in both directions. The independent operation of the two groups is maintained over the major portion of the load range because of the peak-exciting current of the interphase reactor which might be typically about 1 or 2% of the full load-direct current.

At no load, the operation of the circuit becomes similar to that of the simple six-pulse converter circuit of Fig. 6.38, whereas at light load, the operation of the circuit reverts to a more complex "intermittent conduction mode."

Since, the firing-angle control range for the six-pulse converter with interphase-reactor is 30° advanced with respect to the firing-angle control range for the simple six-pulse converter.

Since the firing-angle control range for the 6-pulse converter with interphase reactor is 30° advanced with respect to the firing-angle control range for the simple 6-pulse-converter, this means, with a fixed phase-position of the firing pulses, that the firing angle effectively advances by 30° between light load and no load. Thus, depending upon the particular operating point in the control range, a substantial increase in the d.c. terminal voltage may be obtained, if the phase position of the firing pulses remains unaltered as the load is removed. Even if a 30° adjustment of firing angle is made, to compensate against this effect, the mean value of the d.c. terminal voltage still rises by about 15% between light load and no load. This is due to the fact that the voltage at the d.c. terminals follows the actual input voltage waves at no load, whereas it follows the mean voltage waves resulting from the two 60° displaced input waves at increased load, and the amplitudes of these latter waves are only 0.866 of the input waves. Hence, in order to maintain a constant d.c. terminal voltage between light load and no load, a retardation of the firing angle by slightly more than 30° is required.

Although the individual input line-current waveforms consists of unidirectional 120° rectangular "blocks," there is no d.c. magnetization of the transformer core because the d.c. mmf's of the 2-windings on each phase of the transformer cancel one another.

6.11 THREE PHASE FULLY CONTROLLED BRIDGE CONVERTER

In the previous section, it has been shown that a six-pulse converter circuit is obtained by connecting a d.c. terminal of two 3-pulse converters in parallel. If we connect the d.c. terminals of two 3-pulse converters in series, the converter so formed is called as 6-pulse bridge converter, shown in Fig. 6.43 (a). This converter is most widely used in industrial applications upto the 120 kW level, where two quadrant operation is required.

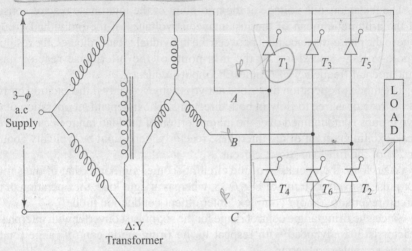

Fig. 6.43 *(a) 3-φ Full converter*

The load is fed via a three-phase half-wave connection to one of the three-supply lines, no neutral being required. Hence transformer connection is optional. However, for isolation of output from the supply source, or for higher output requirements, the transformer is to be connected. If transformer is used, then one winding is connected in delta because the delta connection gives the circulating path for third harmonic current. Therefore, third harmonic current does not appear in line which is an advantage. This circuit consists of two groups of SCRs, positive group and negative group. Here, SCRs T_1, T_3, T_5 forms a positive group, whereas SCRs T_4, T_6, T_2 forms a negative group. The positive group SCRs are turned-on when the supply voltages are positive and negative group SCRs are turned-on when the supply voltages are negative.

When connected to the a.c. supply, firing gate pulses will be delivered to the SCRs in the correct sequence. But, if only a single gate-pulse is used, no current will flow, as the other SCR in the current path will be in the off-state. Hence, in order to start the circuit functioning, two thyristors must be fired at the same time in order to commence current-flow, one of the upper-arm and one of the lower arm.

For describing the operation of the circuit, one should remember the following points:

(i) Each device should be triggered at a desired firing angle 'α'.
(ii) Each SCR can conduct for 120°.
(iii) SCRs must be triggered in the sequence T_1, T_2, T_3, T_4, T_5, and T_6.
(iv) The phase shift between the triggering of the two adjacent SCRs is 60°.

Phase Controlled Converters

(v) At any instant, two SCRs can conduct and there are such six pairs. The six pairs are $(T_6, T_1), (T_1, T_2), (T_2, T_3), (T_3, T_4), (T_4, T_5)$ and (T_5, T_6).

(vi) Each SCR conducts in two pairs and each pair conducts for 60°.

(vii) The incoming SCR commutates the outgoing SCR, i.e. SCR T_1 commutates SCR T_5, SCR T_2 commutates SCR T_6 and so on.

Table 6.2 gives the conducting pair, the incoming and outgoing SCRs.

Table 6.2 Firing sequence of SCRs

S.No.	ωt	Incoming SCR	Conducting pair	Outgoing SCR	Line voltage across the load
1.	$30° + \alpha$	T_1	(T_6, T_1)	T_5	E_{AB}
2.	$90° + \alpha$	T_2	(T_1, T_2)	T_6	E_{AC}
3.	$150° + \alpha$	T_3	(T_2, T_3)	T_1	E_{BC}
4.	$210° + \alpha$	T_4	(T_3, T_4)	T_2	E_{BA}
5.	$270° + \alpha$	T_5	(T_4, T_5)	T_3	E_{CA}
6.	$330° + \alpha$	T_6	(T_5, T_6)	T_4	E_{CB}

(viii) When the two SCRs are conducting, i.e. one from positive (upper) group and one from negative (lower) group, the corresponding line voltage is applied across the load. For example, when T_6 and T_1 are conducting, the line voltage V_{AB} is applied across the load, as shown in Table 6.2.

(ix) Phase A is connected to first half bridge (T_1/T_4), B-phase to second half-bridge (T_3/T_6) and C-phase to third half bridge (T_5/T_2).

(x) When the upper SCR of a half bridge conducts, the current of that phase is positive whereas when the lower SCR conducts, the current is negative.

(xi) From the vector diagram 6.43 (b), it is clear that, when the phase voltage makes an angle of 30° with neutral, the upper SCR in that half bridge is forward biased. Similarly, the lower SCR is forward biased when its phase makes an angle of 210° with neutral. Hence, $\alpha = 0°$ when $\omega t = 30°$.

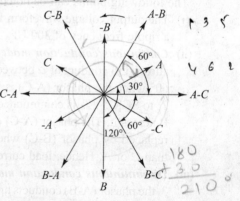

Fig. 6.43 (b) Vector diagram

The defining equations for phase and line voltages are:

$E_{AN} = E_m \sin(\omega t)$

$E_{BN} = E_m \sin(\omega t - 120°)$

$E_{CN} = E_m \sin(\omega t + 120°)$

$E_{AB} = \sqrt{3}\, E_m \sin(\omega t + 30°)$

$E_{AC} = \sqrt{3}\, E_m \sin(\omega t - 30°)$

$E_{BC} = \sqrt{3}\, E_m \sin(\omega t - 90°)$

$E_{BA} = \sqrt{3}\, E_m \sin(\omega t - 150°)$

$E_{CA} = \sqrt{3}\, E_m \sin(\omega t + 150°)$

$E_{CB} = \sqrt{3}\, E_m \sin(\omega t + 90°)$

6.11.1 With Resistive Load (R-load)

Three-phase fully controlled bridge rectifier with resistive load is shown in Fig. 6.44. For six-pulse operation, each SCR has to be fired twice in its conduction cycle, that is firing intervals should be 60°. As shown in Fig. 6.43 (b), there are six line to line phasors, each having a maximum conduction angle of 60°. The output voltage waveform for different values of 'α', i.e. $\alpha = 0°, 30°, 60°, 90°$ and 120° are shown in Fig. 6.45 (a).

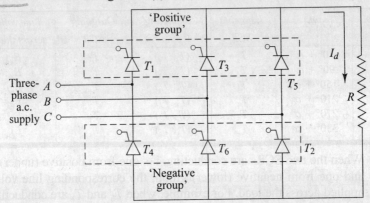

Fig. 6.44 *3-phase fully controlled bridge rectifier with resistive load*

The following points can be noted from Fig. 6.45 (a).
(i) The output voltage waveform for any value of α is a six pulse wave with a ripple frequency of 300 Hz.
(ii) **Continuous conduction mode ($0 \leq \alpha \leq \pi/3$)** When the phasor (A-B) is allowed to conduct at α between zero to $\pi/3$, it continuous to conduct by 60° when the phasor (A-C) is fired. The conduction is shifted from SCR T_6 to SCR T_2. T_6 is commutated off by the reverse-voltage of phase C and B across it. The phasor (A-C) conducts after another 60° after which it is replaced by phasor (B-C) when phase B voltage assumes greater value than C or A. Hence, load current is continuous for α between 0 to $\pi/3$.
(iii) **Discontinuous conduction mode: ($\pi/3 \leq \alpha \leq 2\pi/3$)** When $\pi/3 \leq \alpha \leq 2\pi/3$, the phasor (A-B) conducts upto an angle π after which both the thyristors T_1 and T_6 are commutated off because phase B becomes positive with respect to phase C and after 60°, when T_2 and T_1 are fired, phase (A-C) conducts also upto angle π, hence load current remains zero from angle π to the next firing pulse and becomes discontinuous, therefore, the fully-controlled bridge circuit produces a ripple frequency of six-times the supply frequency at all trigger angles.
(iv) For $\alpha = 120°$, the output voltage is zero and hence $\alpha_{max} = 120° (2\pi/3)$.

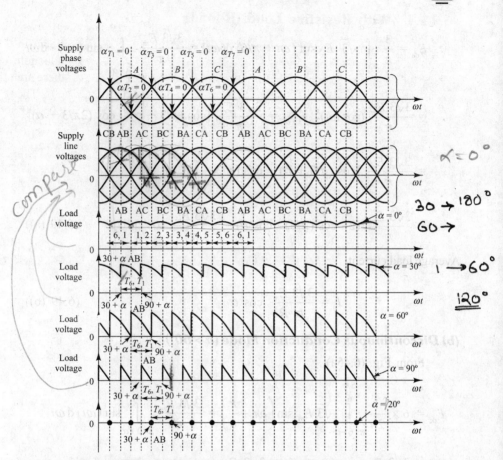

Fig. 6.45 *(a) Voltage waveforms for various firing angles*

The voltage and current relations for both the modes can be derived as follows:

(a) Continuous Conduction Mode: ($\alpha < 60°$) The general equation for the average load voltage is given by

$$E_{dc} = \frac{1}{2\pi} \int_0^{2\pi} E_{dc}(\omega t) \cdot d(\omega t)$$

Now for $\alpha < 60°$, and from Fig. 6.45 (a), we can write

$$E_{dc} = 6 \times \frac{1}{2\pi} \int_{\frac{\pi}{6}+\alpha}^{\frac{\pi}{2}+\alpha} E_{AB(\omega t)} \cdot d(\omega t)$$

where the line-to line-voltage E_{AB} is given by:

$$E_{AB} = \sqrt{3} \ E_m \sin(\omega t + \pi/6) \tag{6.48}$$

$$\therefore E_{dc} = \frac{3}{\pi} \int_{\frac{\pi}{6}+\alpha}^{\frac{\pi}{2}+\alpha} \sqrt{3}\, E_m \sin(\omega t + \pi/6)\, d(\omega t) = \frac{3\sqrt{3}\, E_m}{\pi} \int_{\frac{\pi}{3}+\alpha}^{\frac{2\pi}{3}+\alpha} \sin(\omega t) \cdot d\omega t$$

$$= \frac{3\sqrt{3}\, E_m}{\pi} (\cos \omega t)_{2\pi/3+\alpha}^{\pi/3+\alpha} = \frac{3\sqrt{3}\, E_m}{\pi} [\cos(\pi/3 + \alpha) - \cos(2\pi/3 + \alpha)]$$

$$= \frac{3\sqrt{3}\, E_m}{\pi} [\cos(\pi/3 + \alpha) + \cos(\pi/3 - \alpha)]$$

$$= \frac{3\sqrt{3}\, E_m}{\pi} (2 \cdot \cos(\pi/3) \cdot \cos \alpha) = \frac{3\sqrt{3}\, E_m}{\pi} \cos \alpha \qquad (6.49\ (a))$$

Average load current:

$$I_d = \frac{3\sqrt{3}\, E_m}{\pi \cdot R} \cos \alpha \qquad (6.49\ (b))$$

(b) Discontinuous Conduction Mode ($\alpha > 60°$)

From Fig. (6.45a),

$$E_{dc} = 6 \times \frac{1}{2\pi} \int_{\pi/6+\alpha}^{5\pi/6} \sqrt{3}\, E_m \sin\left(\omega t + \frac{\pi}{6}\right) = \frac{3\sqrt{3}\, E_m}{\pi} \int_{\pi/3+\alpha}^{\pi} \sin(\omega t)\, d\omega t$$

$$= \frac{3\sqrt{3}\, E_m}{\pi} (\cos \omega t)_{\pi}^{\pi/3+\alpha} = \frac{3\sqrt{3}\, E_m}{\pi} [1 + \cos(\alpha + \pi/3)] \qquad (6.50)$$

For α_{max}, $E_{dc} = 0$, $\therefore \frac{3\sqrt{3}\, E_m}{\pi} [1 + \cos(\alpha + \pi/3)] = 0$

Hence, $\alpha_{max} = 120°$

Average load current:

$$I_d = \frac{3\sqrt{3}\, E_m}{\pi R} [1 + \cos(\alpha + \pi/3)] \qquad (6.51)$$

Detailed Waveforms (Operation) with R-Load for $\alpha = 30°$ Figure 6.45(b) shows the detailed waveforms for output voltage, output current, supply current, SCR current and voltage with $\alpha = 30°$. From these waveforms, following points can be noted:
  (i) Output voltage and current waveforms are six pulse with ripple frequency of 300 Hz.

(ii) For $\alpha \leq 60°$, the supply current waveforms are of 120° in each half cycle and for $\alpha > 60°$ (not shown in figure) it is less than 120°, i.e. $(180° - \alpha)$ in each half-cycle.

(iii) For $\alpha \leq 60°$, the SCR waveform is 120° wide and for $\alpha > 60°$, it is $(180° - \alpha)°$ wide.

(iv) PIV rating of SCR is $\sqrt{3}\, E_m$.

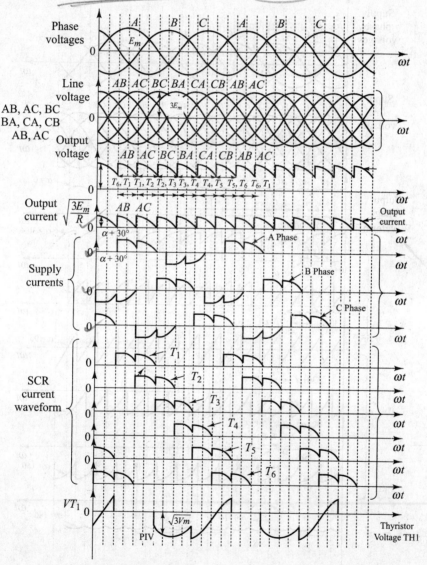

Fig. 6.45 *(b) Waveform with $\alpha = 30°$*

6.11.2 With Inductive Load (R-L)

The power diagram of the three-phase fully-controlled converter with R-L load is shown in Fig. 6.45. The output voltage waveforms with firing angles are shown in Fig. 6.46 (a). The load inductance is assumed to be very large so as to produce a constant load current.

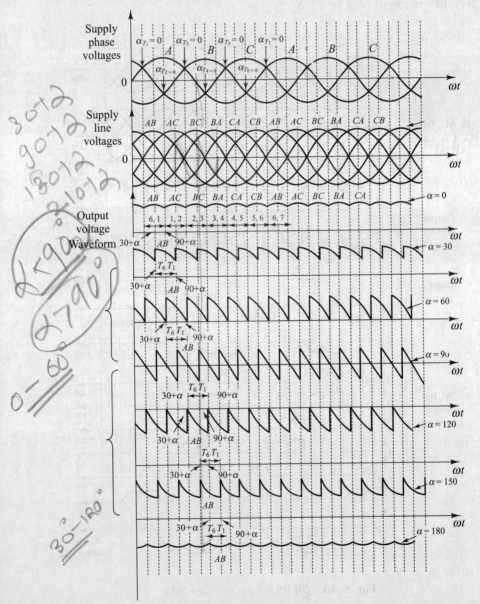

Fig. 6.46 *(a) Output voltage waveforms with R-L load*

Following points can be noted from the waveforms of Fig. 6.46.
(i) Waveforms are similar with R-load for $\alpha = 0°$, $30°$, and $60°$.
(ii) For $\alpha > 60°$, the waveforms are different. The voltage goes negative due to the inductive nature of the load. The previous thyristor pair continuous to conduct till the next SCR is triggered. For examples T_6 and T_1 continuous to conduct upto $(90 + \alpha)$ till thyristor T_2 is triggered and when T_2 is triggered it commutates the SCR T_6 and then T_1 and T_2 starts conducting.
(iii) For $\alpha = 90°$, the area under the positive and the negative cycle are equal and the average voltage is zero.
(iv) For $\alpha < 90°$, average output voltage is positive and for $\alpha > 90°$, the average output voltage is negative.
(v) The maximum value of α is $180°$.
(vi) The output is always a six pulse, i.e. ripple frequency is 300 Hz irrespective of the value of α.

Expression for Average Output Voltage and RMS Output Voltage:

(i) Average Output Voltage

From Fig. 6.46, we can write:

$$\text{Average output voltage, } E_{dc} = 6 \times \frac{1}{2\pi} \int_{30+\alpha}^{90+\alpha} E_{Ry(\omega t)} \, d\omega t$$

$$= \frac{3}{\pi} \int_{30+\alpha}^{90+\alpha} \sqrt{3} \, E_m \sin(\omega t + 30) \, d\omega t = \frac{3}{\pi} \int_{60+\alpha}^{120+\alpha} \sqrt{3} \, E_m \sin(\omega t) \, d\omega t$$

$$= \frac{3\sqrt{3} \, E_m}{\pi} [\cos(\omega t)]_{120+\alpha}^{60+\alpha} = \frac{3\sqrt{3}}{\pi} E_m [\cos(60+\alpha) - \cos(120+\alpha)]$$

$$E_{dc} = \frac{3\sqrt{3} \, E_m}{\pi} \cos\alpha \text{ for } 0 \leq \alpha \leq 180° \qquad (6.52)$$

where E_m is the peak-value of the live-to neutral voltage.

As the firing angle changes from 0 to $90°$, the voltage also changes from maximum to zero and the converter is said to be in *rectification mode*. For the angles in the range $90°$ to $180°$, the voltage varies from 0 to negative maximum and the converter is in the *inversion mode*. It can transfer power from d.c. side to a.c. if there is a negative d.c. source available at the d.c. terminals. The mean value of the d.c. voltage is superimposed by ripple content.

(ii) R.M.S. Output Voltage

$$E_{rms} = \left[\frac{1}{2\pi} \int_0^{2\pi} E_{dc}^2(\omega t) \, d\omega t \right]^{1/2} = \left[6 \times \frac{1}{2\pi} \int_{30+\alpha}^{90+\alpha} E_{AB}^2(\omega t) \, d\omega t \right]^{1/2}$$

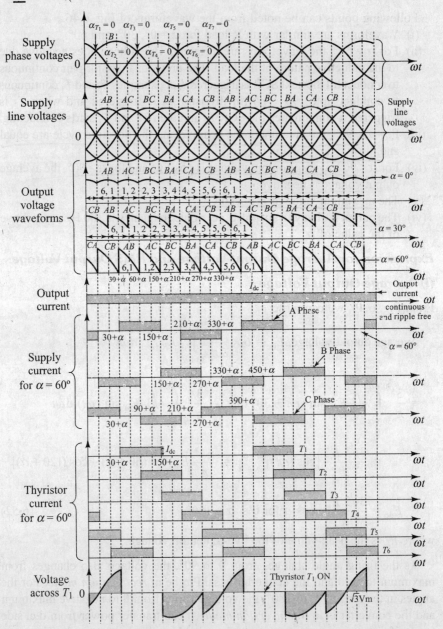

Fig. 6.46 *(b) Waveforms for rectifier mode of operation ($\alpha < 90°$)*

$$= \left[\frac{3}{\pi} \int_{30+\alpha}^{90+\alpha} \left(\sqrt{3}\, E_m \sin(\omega t + 30) \right)^2 d\omega t \right] = \left[\frac{9 E_m^2}{\pi} \int_{60+\alpha}^{120+\alpha} \sin^2(\omega t)\, d\omega t \right]^{1/2}$$

$$= \left[\frac{9E_m^2}{2\pi}\int_{60+\alpha}^{120+\alpha} 1-\cos(2\omega t)\,d\omega t\right]^{1/2} = \left[\frac{9E_m^2}{2\pi}\left\{60° + \frac{1}{2}(\sin(2\omega t))_{120+\alpha}^{60+\alpha}\right\}\right]^{1/2}$$

$$= \left[\frac{9E_m^2}{2\pi}\left\{60° + \frac{1}{2}(\sin(120+2\alpha) - \sin(240+2\alpha))\right\}\right]^{1/2}$$

$$= \left[\frac{9E_m^2}{2\pi}\left\{60° + \frac{1}{2}(\sqrt{3}\cos 2\alpha)\right\}\right]^{1/2}$$

$$= \frac{3E_m}{2}\left[\frac{2}{3} + \frac{\sqrt{3}}{\pi}\cos 2\alpha\right]^{1/2}, \quad \text{for } 0 \leq \alpha \leq 180° \qquad (6.53)$$

(A) Rectifier Operation ($\alpha < 90°$) The waveforms for rectifier mode of operation are shown in Fig. 6.46 (b). Following points can be noted from these waveforms:
  (i) The output voltage waveforms are shown for $\alpha = 0°, 30°$, and $60°$ whereas the waveforms for supply current, SCR current and SCR voltage are shown only for $\alpha = 60°$.
  (ii) Supply waveforms are quasi-square-wave for α upto $90°$.
  (iii) SCR current waveforms are rectangular waves of duration $120°$.
  (iv) PIV rating of thyristor is ($\sqrt{3}\ E_m$).
  (v) The three supply currents are phase shifted by $120°$ and the six thyristor currents are phase shifted by $60°$.
  (vi) The load current waveform is continuous and ripple free and hence the average, rms and peak values are all equal to I_{dc}.
  (vii) The circuit works in the first quadrant since E_{dc} and I_{dc} are both positive and hence $P_{dc(av)}$ is positive and power flows from source to load.

(B) Inverter Operation ($90° \leq \alpha \leq 180°$) The waveforms for inverter mode of operation are shown in Fig. 6.46 (c). Following points can be noted from these waveforms:
  (i) Voltage waveforms are shown for $\alpha = 90°$ and $\alpha = 120°$ whereas the supply current and thyristor current waveforms are shown for $\alpha = 120°$.
  (ii) Average output voltage is negative for inverting operation.
  (iii) Waveforms for supply currents are quasi-square-waves with a mutual phase-shift of $120°$.
  (iv) SCR current waveforms are rectangular waves with $120°$ wide and the mutual phase shift is $60°$.
  (v) The average power flow is from load to source since (E_{dc} = negative and I_{dc} = positive) average output power $P_{dc}(av)$ is negative. Hence, circuit works in the fourth quadrant.

Solved Examples

Example 6.16 (a) Derive a general expression for the average load voltage of P-pulse fully-controlled converter. (b) Also determine the RMS value of current that flows for $1/p$ of each cycle.

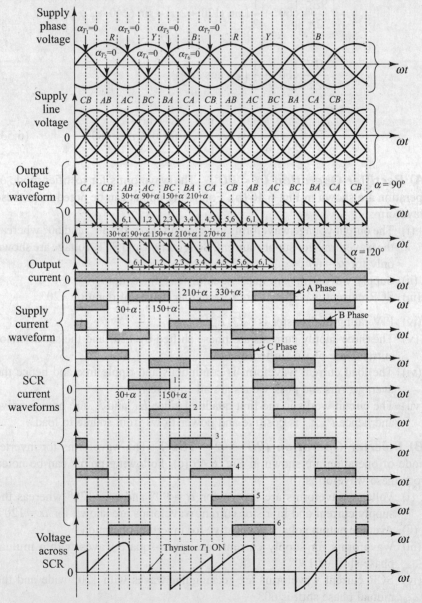

Fig. 6.46 *(c) Waveforms for inversion operation ($\alpha > 90*$)*

Phase Controlled Converters

Solution: (a) Fig. Ex. 6.16 shows the general waveform for the P-pulse fully controlled converter, where P is the pulse number of the output

From the waveform of Fig. Ex. 6.16, we can write,

$$E_{dc} = \frac{1}{2\pi/p} \int_{-\frac{\pi}{p}+\alpha}^{+\frac{\pi}{p}+\alpha} E_m \cos \omega t \, d(\omega t) = \frac{p}{2\pi} E_m \left[\sin\left(\frac{\pi}{p}+\alpha\right) - \sin\left(-\frac{\pi}{p}+\alpha\right) \right]$$

$$= \frac{p \cdot E_m}{2\pi} \left[\begin{array}{l} \sin\dfrac{\pi}{p} \cos \alpha + \cos\dfrac{\pi}{p} \sin \alpha - \\ \sin\left(-\dfrac{\pi}{p}\right) \cos \alpha + \cos\left(-\dfrac{\pi}{p}\right) \sin \alpha \end{array} \right]$$

$$= \frac{p \cdot E_m}{2\pi} \sin \frac{\pi}{p} \cos \alpha.$$

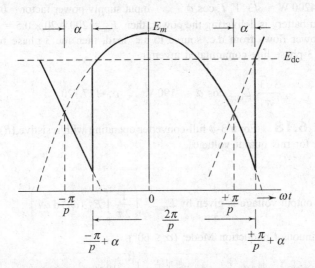

Fig. Ex. 6.16 *General waveform for P-pulse converter*

(b) Let us divide the cycle into p intervals, then the current in one interval is I and zero in the other interval. Therefore, the sum of the squares of each interval becomes I^2.

∴ RMS value of current = $\left(\dfrac{I^2}{p}\right)^{1/2} = \dfrac{1}{\sqrt{p}}$.

Example 6.17 (a) A three-phase fully-controlled converter charges a battery from a three-phase supply of 230 V, 50 Hz. The battery emf is 200 V and its interval resistance is 0.5 Ω. On account of inductance connected in series with the battery, changing current is constant at 20 A. Calculate (i) firing angle (ii) supply power factor

(b) In case it is desired that power flows from d.c. source to a.c. load in part (a), find the firing angle for the same current.

Solution: (a) (i) The battery terminal voltage is

$$E_{dc} = 200 + 20 \times 1.5 = 210 \text{ V}.$$

But $E_{dc} = \dfrac{3\sqrt{3}\,E_m}{\pi}\cos\alpha = 210 \text{ V}$ ∴ $\alpha = \cos^{-1}\dfrac{210 \times \pi}{3\sqrt{2}\times 230} = 47.45°$

(ii) For constant load-current $I_d = 20$ A, the supply current is of rectangular wave or square wave with a amplitude of 20 A. For this converter, we know that the supply current flow, for 120° over a positive half-cycle of 180°.

∴ RMS value of supply current over one half-cycle is,

$$I_s = \left[\frac{1}{\pi}(20)^2 \cdot \frac{2\pi}{3}\right]^{1/2} = 20\sqrt{2/3} = 16.33 \text{ A}.$$

RMS value of output current $I_{dc\,rms} = 20$ A.

Power-delivered to load $= E_{dc} \cdot I_{dc} + I^2_{dc\,rms} \cdot R = 200 \times 20 + (20)^2 \times 0.5 = 4200$ W

Now $4200 \text{ W} = \sqrt{3}\, V_s I_s \cos\phi$ ∴   Input supply power factor $= 0.646$ (lag).

(b) When battery is delivering the power, then $E_{dc} = 200 - 20 \times 0.5 = 190$ V.

When power flows from d.c. source to a.c. load, then the 3-phase full converter works as a 3-phase line commutated inverter.

∴ $\dfrac{3\sqrt{3}}{\pi} E_{mph} \cos\alpha = -190 \text{ V}$. ∴ $\alpha = 127.720$

Example 6.18 For a 3-ϕ full-converter operating with resistive (R) load, derive an equation for rms output voltage.

Solution:

The rms output voltage is given by $E_{rms} = \left(\dfrac{1}{2\pi}\displaystyle\int_0^{2\pi} E_{dc}^2(\omega t)\, d\omega t\right)^{1/2}$

(i) Continuous Conduction Mode: ($\alpha \leq 60°$)

$$\therefore E_{rms} = \left[\frac{6}{2\pi}\cdot \int_{30+\alpha}^{90+\alpha} E_{AB}^2(\omega t)\,d\omega t\right]^{1/2} = \left[\frac{6}{2\pi}\int_{30+\alpha}^{90+\alpha}\left[\sqrt{3}\,E_m \sin(\omega t + 30)\right]^2 d\omega t\right]^{1/2}$$

$$= \left[\frac{E_m^2}{\pi}\int_{60+\alpha}^{120+\alpha}\sin^2(\omega t)\,d\omega t\right]^{1/2} = \left[\frac{9 E_m^2}{2\pi}\left\{\int_{60+\alpha}^{120+\alpha}(1-\cos(2\omega t)\,d\omega t)\right\}\right]^{1/2}$$

$$= \left[\frac{9 E_m^2}{2\pi}\left\{60° + \frac{1}{2}(\sin(2\omega t)\big|_{120+\alpha}^{60+\alpha}\right\}\right]^{1/2}$$

$$= \left[\frac{9 E_m^2}{2\pi}\left\{\frac{\pi}{3} + \frac{1}{2}(\sin(120+2\alpha) - \sin(240+2\alpha))\right\}\right]^{1/2}$$

$$E_{\text{rms}} = \left[\frac{9E_m^2}{2\pi}\left\{\frac{\pi}{3}+\frac{1}{2}\sqrt{3}\cos 2\alpha\right\}\right]^{1/2} \quad E_{\text{rms}} = \frac{3}{2}E_m\left[\frac{\frac{2\pi}{3}+\sqrt{3}\cos 2\alpha}{\pi}\right]^{1/2} \quad (1)$$

(ii) Discontinuous Conduction Mode ($\alpha > 60°$)

$$E_{\text{rms}} = \left[6 \times \frac{1}{2\pi}\int_{\alpha+30}^{150} E_{AB}^2(\omega t)\,d\omega t\right]^{1/2} = \left[\frac{6}{2\pi}\int_{\alpha+30}^{150}\left(\sqrt{3}\,E_m \sin(\omega t + 30)\right)^2 d\omega t\right]^{1/2}$$

$$= \left[\frac{9E_m^2}{2\pi}\left(\int_{\alpha+60}^{180}\sin^2(\omega t)\,d\omega t\right)\right]^{1/2} = \left[\frac{9E_m^2}{2\pi}\left(\int_{\alpha+60}^{180}(1-\cos(2\omega t))\,d\omega t\right)\right]^{1/2}$$

$$= \left[\frac{9E_m^2}{2\pi}\left(120° - \alpha + \frac{1}{2}(\sin(2\omega t))\Big|_{180°}^{\alpha+60°}\right)\right]^{1/2}$$

$$= \left[\frac{9E_m^2}{2\pi}\left(\frac{2\pi}{3} - \alpha + \frac{1}{2}(\sin(120° + 2\alpha) - \sin(360))\right)\right]^{1/2}$$

$$= \left[\frac{9E_m^2}{2\pi}\left(\frac{2\pi}{3} - \alpha + \frac{1}{2}\sin(120 + 2\alpha)\right)\right]^{1/2} \quad E_{\text{rms}} = \frac{3}{2}E_m\left[\frac{4\pi/3 - \alpha + \sin(120 + 2\alpha)}{\pi}\right]^{1/2}$$

(2)

Example 6.19 For the 3-ϕ converter operating from 3ϕ, 415 V/50 Hz supply, find out the SCR rating if the load resistance is 100 Ω in series with a large smoothing inductor.

Solution:

The SCR current ratings are $I_{T(av)}$, $I_{T(rms)}$ and $I_{T(peak)}$

$$I_{T(av)} = \frac{1}{2\pi}\int_0^{2\pi} I_{T(\omega t)}\,d\omega t = \frac{1}{2\pi}\int_{30+\alpha}^{150+\alpha} I_{dc}\cdot d\omega t = \frac{I_{dc}}{2\pi}(150 + \alpha - 30\cdot\alpha) = \frac{I_{dc}}{3}$$

$$I_{T(rms)} = \left[\frac{1}{2\pi}\int_0^{2\pi} i_T^2(\omega t)\cdot d\omega t\right]^{1/2} = \left[\frac{1}{2\pi}\int_{30+\alpha}^{150+\alpha} I_{dc}^2\,d\omega t\right]^{1/2}$$

$$= \left[\frac{I_{dc}^2}{2\pi}\cdot 2\pi/3\right]^{1/2} = \frac{I_{dc}}{\sqrt{3}},\ I_{T(peak)} = I_{dc}\ \text{Now},\ I_{dc} = \frac{E_{dc}}{R}$$

From equation (6.52),

For $\quad \alpha = 0,\ E_{dc} = \dfrac{3\sqrt{3}\times\sqrt{2/3}\times 415}{\pi}\cos 0 = 560.5$ V

$\therefore \quad I_{dc} = \dfrac{560.5}{100} = 5.605$ A

∴ Thyristor current ratings are

$$I_{T(avg)} = \frac{I_{dc}}{3} = \frac{5.605}{3} = 1.868 \text{ A} \quad I_{T(rms)} = \frac{I_{dc}}{\sqrt{3}} = 3.235 \text{ A} \quad I_{T(peak)} = I_{dc} = 5.605 \text{ A}$$

Now, Peak Inverse Voltage, PIV $\geq \sqrt{3} \; E_m, \geq \sqrt{3} \times \sqrt{\frac{2}{3}} \times 415 \geq 590$ V

Example 6.20 For 3-ϕ full converter operating from 3-ϕ, 415/50 Hz supply. Determine, and plot the following:
(a) Fundamental component of the supply current
(b) 5th, 7th, 11th, and 13th harmonics for $\alpha = 0°, 30°, 60°, 90°, 120°, 150°$ and $180°$. Assume large inductive load with $R_L = 10 \; \Omega$.

Solution:
(i) The nth harmonic component of the supply current is given by

$$I_{s(n)} = \frac{C_n}{\sqrt{2}} = \frac{2\sqrt{2} \; I_{dc}}{\pi} \frac{\cos(30\,n)}{n} \tag{1}$$

∴ Fundamental component of the supply current is given by (substitute $n = 1$)

$$\therefore \; I_{s(fund)} = 0.9 \; I_{dc} \cos 30 = 0.78 \; I_{dc} = \frac{\sqrt{6}}{\pi} I_{dc} \tag{2}$$

Now $I_{dc} = \dfrac{E_{dc}}{R_L} = \dfrac{6\sqrt{3} \; E_m \cos \alpha}{2\pi R_L}$, Eq. (2) implies, $I_{s(fund)} = \dfrac{0.78 \times 6\sqrt{3} \; E_m}{2\pi} \cdot \dfrac{|\cos \alpha|}{R_L}$

But, $E_m = \sqrt{\dfrac{2}{3}} \times 41.5 = 338.85$ V ∴ $I_{s(fund)} = \dfrac{0.78 \times 6\sqrt{3} \times 338.95}{2\pi} \dfrac{|\cos \alpha|}{100}$

$$= 4.372 \; |\cos \alpha| \tag{3}$$

and $I_{s(n)} = \dfrac{5.372 \; |\cos \alpha|}{n}$, where $n = 5, 7, 11,$ and 13. $\tag{4}$

(ii) Table Ex. 6.20 shows fundamental and harmonics for different values of α.

S.No.	α (degrees)	$I_{s(fund)} =$ $4.372 \,\|\cos \alpha\|$	$I_{s(5)} =$ $\dfrac{4.372\|\cos \alpha\|}{5}$ (Amp)	$I_{s(7)} =$ $\dfrac{4.372\|\cos \alpha\|}{7}$ (Amp)	$I_{s(11)} =$ $\dfrac{4.372\|\cos \alpha\|}{11}$ (Amp)	$I_{s(13)} =$ $\dfrac{4.372\|\cos \alpha\|}{13}$ (Amp)
1.	0°	4.372	0.875	0.625	0.4	0.336
2.	30°	3.786	0.757	0.54	0.345	0.29
3.	60°	2.186	0.4372	0.3122	0.198	0.168
4.	90°	0	0	0	0	0
5.	120°	2.186	0.4372	0.3122	0.198	0.168
6.	150°	3.786	0.757	0.54	0.345	0.29
7.	180°	4.372	0.875	0.625	0.4	0.339

Figure Ex. 6.20 shows the plot of RMS fundamental and harmonic current with α.

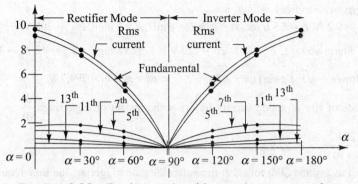

Fig. Ex. 6.20 *Fundamental and harmonics current with* α

Example 6.21 For the three phase full-converter operating from 3ϕ /415 V/ 50 Hz supply with resistive load, determine the average output voltage for $\alpha = 0°$, 30°, 60°, 90° and 120°. Plot E_{dc} Vs α.

Solution:

Case I: for $\alpha < 60°$, $E_{dc} = \dfrac{6\sqrt{3}\, E_m}{2\pi} \cos \alpha = \dfrac{6\sqrt{3} \times \sqrt{2/3} \times 415}{2\pi} \cos \alpha = 560.5 \cos \alpha$

∴ For $\alpha = 0$, $E_{dc} = 560.5$ V, For $\alpha = 30°$, $E_{dc} = 485.4$ Volts

For $\alpha = 60°$, $E_{dc} = 280.25$ Volts

Case II For $\alpha > 60°$, $E_{dc} = \dfrac{6\sqrt{3}\, E_m}{2\pi} [1 + \cos(60 + \alpha)] = 560.5\,[1 + \cos(60 + \alpha)]$

For $\alpha = 90°$, $E_{dc} = 75.1$ V, For $\alpha = 120°$, $E_{dc} = 0$ V

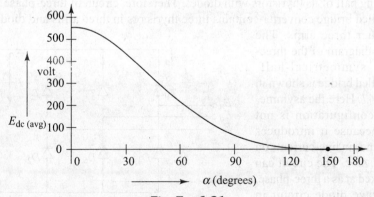

Fig. Ex. 6.21

Example 6.22 The holding current of SCRs in 3-ϕ full-wave converter is 200 mA and the delay time is 2.5 μ sec. Converter is supplied from a star-connected 280V/50 supply and has a load of $R = 2$ ohm and $L = 8$ mH. It is operated with a delay angle of $\alpha = 60°$ (= $\pi/3$). (i) Determine the minimum width of gate-pulse width, t_p. (ii) If $L = 0$, determine t_p.

Solution:

(i) $I_H = 0.2$ A, $t_d = 2.5\ \mu$ sec, $\alpha = \pi/3$, $L = 8$ mH, $R = 2$ ohm, $E_{line} = 208$ V, $F = 50$ Hz

∴ Phase voltage, $E_p = 208/\sqrt{3} = 120$ V & Peak voltage, $= E_m = \sqrt{2}\ E_p = 169.7$ V

Now, $e = \sqrt{3}\ E_m \sin(\omega t + \pi/6)$, ∴ at $\omega t = \pi/3$, $e_1 = 294.2$ V.

Rate of rise of anode current (di/dt) at the instant of triggering, $\dfrac{di}{dt} = \dfrac{P_{-1}}{L} = $

$\dfrac{294.2}{8 \times 10^{-3}} = 36.77$ kA/sec.

Let us assume di/dt for short time after the gate triggering, the time required for the anode current to rise to the level of holding current, t_1 becomes:

$$I_H = t_1 \cdot (di/dt) = t_1 \times 36770 = 0.2 \quad \therefore t_1 = \dfrac{0.2}{36770} = 5.439\ \mu\text{ sec.}$$

∴ Minimum width of gate pulse is

$$t_{p1} = t_1 + t_d = 5.439 + 2.5 = 7.933\ \mu \text{ sec.}$$

(ii) Rate of rise of anode current, di/dt at the instant of triggering is

$$di/dt = E_1/L = \dfrac{294.2}{0} = \infty \text{ A/sec}, \quad \therefore t_1 = I_H/di/dt = \dfrac{0.2}{\infty} = 0\ \mu \text{ sec}$$

∴ Minimum width of gate-pulse is $t_{p2} = t_1 + t_d = 0 + 2.5 = 2.5\ \mu$ sec.

6.12 THREE-PHASE HALF CONTROLLED BRIDGE CONVERTER (THREE-PHASE SEMICONVERTERS)

Freewheeling mode of operation of bridge connected rectifiers can be realized by replacing half of its thyristors with diodes. Therefore, circuit of three-phase half-controlled bridge converter contains three thyristors in three arms and diodes in the other three arms. The circuit diagram of the three-phase symmetrical-half-controlled bridge is shown in Fig. 6.47. Here the asymmetrical configuration is not used because it introduces imbalance in line-currents on the a.c. side. The circuit can be looked at as a three-phase, half-wave diode circuit in series with a three-phase, half-wave, phase-controlled thyristor circuit.

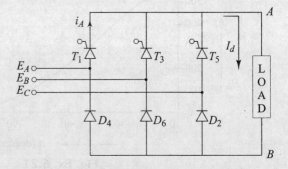

Fig. 6.47 Half-controlled three-phase bridge (3-ϕ semiconverters)

Three-phase semiconverters are used in industrial applications upto the 120 kW level, where one quadrant operation is required. The power-factor of this converter decreases as the delay angle increases, but is better than that of three-phase half-wave converters. This converter has inherent free-wheeling action which improves its power-factor. The free-wheeling action takes place between the two devices (SCR and diode) in the same arm, i.e. between $T_1 D_4$, $T_3 D_6$ and $T_5 D_2$. With this free-wheeling action, the voltage drop across the load ($V_{TH} + V_D$) becomes approximately 2 V. Hence, this leads to following drawbacks:

(i) This voltage drop reduces the average output voltage.
(ii) During the free-wheeling period, the conduction losses increases which decreases the efficiency of the converter. This also makes the load current less continuous for the same operating conditions.
(iii) SCR's continues to conduct in the negative cycle of the line voltage and hence the conduction period of each SCR is 180° (or π). This increases the average and rms current ratings of the SCRs.

All these drawbacks can be reduced by placing an external freewheeling diode across the load, as shown in Fig. 6.48. This reduces the on-state voltage drop across the load and conduction losses and makes the load current more continuous. The free-wheeling diode also guarantees the commutation of each SCR at the end of the corresponding half-cycle and hence the average and rms current rating of the devices (SCR's and diodes) decreases. This helps to decrease the cost and cooling requirements of the converter.

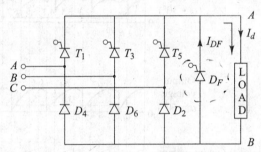

Fig. 6.48 *Three-phase semiconverter with free-wheeling diode D_f*

6.12.1 Operation with Resistive (R) Load

The related voltage and current waveforms for the 3-ϕ semiconverter (Fig. 6.47) are shown in Fig. 6.49 (a). In Fig. 6.49 (a), the output voltage waveforms (for $\alpha = 0°$, 30°, 60°), supply current waveforms. (for $\alpha = 0°$ and 90°), and devices current waveforms (for $\alpha = 90°$) are shown. The following points can be observed from Fig. 6.49 (a):

(i) For $\alpha = 0$, the output voltage waveform is a six pulse output.
(ii) For $\alpha \geq 30°$, the output is only a 3-pulse and hence this converter is known as 3-pulse converter.
(iii) The output voltage waveforms goes to zero after every pulse for $\alpha = 60°$ and for $\alpha > 60°$, it remains zero for a finite time and is thus discontinuous. The pulse width for $\alpha \leq 60°$ is 120° and for $\alpha \geq 60°$ is $(180 - \alpha)$.
(iv) The three-phase current waveforms are of 120° width in each half-cycle for any $\alpha \leq 60°$ and for any $\alpha \geq 60°$, the pulse-width in each half-cycle starts decreasing and is given by $(180° - \alpha)$.
(v) The SCR and diode current waveforms are also 120° wide in each cycle for $\alpha \leq 60°$ and of duration $(180° - \alpha)$ for $\alpha > 60°$.

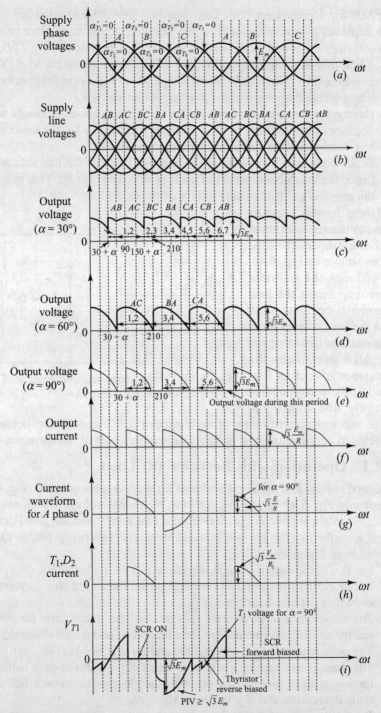

Fig. 6.49 *(a) Voltage and current waveforms for 3-φ semiconverter for $\alpha = 30°, 60°$ and $90°$*

Phase Controlled Converters

Operation: Operation of the circuit can be explained with the help of waveforms in Fig. 6.49 (a) and the phasor diagram of Fig. 6.43 (b). The following points will be helpful to understand the operation.

(i) Diodes starts conducting as soon as they are forward-biased.
(ii) Line voltages conducts in the sequence E_{AB}, E_{AC}, E_{BC}, E_{BA}, E_{CA} and E_{CB}.
(iii) A line voltage which has the highest value compared to others will conduct, i.e. when it makes an angle of 60° with neutral.
(iv) Table 6.3 shows the devices which conducts during each line voltages.

When a particular pair is conducting the output instantaneous voltage during that period is its instantaneous line voltage. For example, during the period when D_6 and T_1 are conducting, output voltage is E_{AB} and similarly when T_5 and D_6 are conducting, the output voltage is E_{CB}.

Table 6.3

S.No.	Conducting Line Voltages	Conducting Devices
(i)	E_{AB}	(D_6, T_1)
(ii)	E_{AC}	(T_1, D_2)
(iii)	E_{BC}	(D_2, T_3)
(iv)	E_{BA}	(T_3, D_4)
(v)	E_{CA}	(D_4, T_5)
(vi)	E_{CB}	(T_5, D_6)

(v) When a line voltage is 9 + 60° with neutral, its phase-voltage is at 30° with neutral. There are six devices. Each device is forward-biased when its phase voltage is at 30° with neutral. Table 6.4 gives $\alpha = 0$ for each device.

Table 6.4

S.N	Devices	Phase voltage at 30° with neutral	Angle from $\omega t = 0$ $\omega t = 0$ is with reference to phase R
(i)	T_1	E_A	$\omega t = 30°$
(ii)	D_2	E_C	90°
(iii)	T_3	E_B	150°
(iv)	D_4	$-E_A$	210°
(v)	T_5	$-E_C$	270°
(vi)	D_6	$-E_B$	330°

(vi) Since SCR conducts only when it is triggered and a diode conducts as soon it is forward biased the phasor, E_{AB}, E_{BC} and E_{CA} starts conducting only when the respective SCRs T_1, T_3, and T_5 are triggered. While the phasors E_{AC}, E_{BA} and E_{CB} starts conducting as soon as their respective diodes D_2, D_4 and D_6 are forward biased.

It should also be noted that only one line phasor can conduct at a time. When next phasor comes into conduction, it commutates the previous conducting phasor. If a new diode conducts, it commutates the previously conducting diode while if a new SCR starts conducting it commutates the previously conducting SCR.

For a = 60°, each line phasor and hence each pair conducts for 60°. Table 6.5 gives the instant at which each phasor starts conducting, the incoming device, the outgoing device, the pair conduction, and the conducting period of each pair. The reference point is $\omega t = 0$, in phase with 'A'-phase. It should be noted that the outgoing device is commutated by the incoming device.

Table 6.5 (Incoming, outgoing and conducting devices)

S.No.	Triggering instant	Incoming device	Outgoing device	Pair conducting and phasor	Conducting period of each pair	Conducting period of outgoing device
1.	$\omega t = 30 + \alpha$	T_1	T_5	D_6 and T_1 (E_{AB})	$90 - (30 + \alpha)$ $= 60 - \alpha$	$360 + 30 + \alpha -$ $(270 + \alpha) = 120°$
2.	90°	D_2	D_6	T_1 and D_2 (E_{AC})	$(150 + \alpha - 90)$ $= (60 + \alpha)$	$360 + 90 - 330$ $= 120$
3.	$150 + \alpha$	T_3	T_1	D_2 and T_3 (E_{BC})	$210 - (150 + \alpha)$ $= (60 - \alpha)$	$150 + \alpha$ $- (30 + \alpha)$ $= 120°$
4.	210	D_4	D_2	T_3 and D_4 (E_{BA})	$270 + \alpha - 210$ $= (60 + \alpha)$	$210 - 90 = 120°$
5.	$270 + \alpha$	T_5	T_3	D_4 and T_5 (E_{CA})	$330 - (270 + \alpha)$ $= (60 - \alpha)$	$270 + \alpha$ $- (150 + \alpha)$ $= 120°$
6.	330	D_6	D_4	T_5 and D_6 (E_{CB})	$360 + (30 + \alpha)$ $-330 = (60 + \alpha)$	$330 - 210$ $= 120°$

(vii) From Table 6.5, we can observe the following:
   (a) Phasors E_{AC}, E_{BA} and E_{CB} always starts conducting at $\omega t = 90°$, 210° and 330° respectively at any α while phasors E_{AB}, E_{BC} and E_{CA} start conducting at $\omega t = 30 + \alpha$, $150 + \alpha$, and $270 + \alpha$ respectively. Each device conducts for 120°.
   (b) Phasors E_{AB}, E_{BC} and E_{CA} conducts for $(60 - \alpha)$ while phasors E_{AC}, E_{BA} and E_{CB} conducts for $(60 + \alpha)$. The total conduction period of $(E_{AB} + E_{AC})$, $(E_{BC} + E_{BA})$ and $(E_{CA} + E_{CB})$ is 120°.
(viii) For $\alpha = 0$, all phasors conducts for 60° while for any $\alpha \leq 60°$, Phasor E_{AB}, E_{CB} and E_{CA} conducts for period less than 60° while phasors E_{AC}, E_{BA} and E_{CB} conducts for period greater than 60°. The total conduction period of $(E_{AB} + E_{AC})$, $(E_{BC} + E_{BA})$ and $(E_{CA} + E_{CB})$ is still 120°.
(ix) For $\alpha = 60°$, phasors E_{AB}, E_{BC} and E_{CA} conducts for 0° and phasors E_{AC}, E_{BA} and E_{CB} conducts for 120°, i.e. $(180° - \alpha)$.

(x) For $\alpha > 60°$, phasor E_{AB}, E_{BC}, and E_{CA} are unable to conduct and phasors E_{AC}, E_{BA}, and E_{CB} conducts for less than 120°, i.e.
$$180 - \alpha \ (\alpha > 60°)$$
(xi) For $\alpha = 180°$, no phasor can conduct and the output voltage is zero.

Expression for Output Voltage For 3-ϕ semiconverter, the nature of output voltage waveform is different for $\alpha \leq 60°$ and $\alpha \geq 60°$. We derive the expression for output voltage for these two cases.

Case-I $\alpha \leq 60°$.

The average output voltage is given by

$$E_{dc} = 3 \times \frac{1}{2\pi} \left[\int_{30+\alpha}^{90} E_{AB\,(\omega t)} \, d\omega t + \int_{90}^{150+\alpha} E_{AC\,(\omega t)} \, d\omega t \right]$$

Substituting the value of $E_{AB}(\omega t)$ and $E_{AC}(\omega t)$ from Eq. (6.48), we get

$$E_{dc} = \frac{3}{2\pi} \left[\int_{30+\alpha}^{90} \sqrt{3}\, E_m \sin(\omega t + 30) \, d\omega t + \int_{90}^{150+\alpha} \sqrt{3}\, E_m \sin(\omega t - 30) \, d\omega t \right]$$

$$E_{dc} = \frac{3\sqrt{3}\, E_m}{2\pi} \left[\left(\cos(\omega t + 30)\right)_{90}^{30+\alpha} + \left(\cos(\omega t + 30)\right)_{150+\alpha}^{90} \right]$$

$$= \frac{3\sqrt{3}\, E_m}{2\pi} [\cos(60 + \alpha) - \cos(120) + \cos(60) - \cos(120 + \alpha)]$$

$$= \frac{3\sqrt{3}\, E_m}{2\pi} [1 + \cos(60 + \alpha) - \cos(120 + \alpha)] = \frac{3\sqrt{3}\, E_m}{2\pi} [1 + \cos \alpha]$$

(6.54 (a))

Case II $\alpha \geq 60°$, $E_{dc} = 3 \times \frac{1}{2\pi} \left[\int_{30+\alpha}^{210} E_{Ac}(\omega t) \, d\omega t \right]$

Substitute the value of E_{AC}, we get,

$$E_{dc} = \frac{3}{2\pi} \int_{30+\alpha}^{210} \sqrt{3}\, E_m \sin(\omega t - 30) \, d\omega t = \frac{3\sqrt{3}\, E_m}{2\pi} [\cos(\omega t - 30)]_{210}^{30+\alpha}$$

$$= \frac{3\sqrt{3}\, E_m}{2\pi} [\cos(\alpha) - \cos(180)] = \frac{3\sqrt{3}\, E_m}{2\pi} (1 + \cos \alpha) \quad (6.54 \text{ (b)})$$

Hence, from both cases we can observe that the average output voltage for any α, $(0 \leq \alpha \leq 180)$ is same and is given by Eq. (6.54).

RMS Output Voltage

$$E_{rms} = \left(\frac{1}{2\pi} \int_0^{2\pi} E_{dc}^2(\omega t)\, d\omega t \right)^{1/2}$$

Case I $\alpha \le 60°$,
RMS output voltage is given by

$$E_{rms} = \left\{ \frac{3}{2\pi} \left[\int_{30+\alpha}^{90} E_{AB}^2(\omega t)\, d\omega t + \int_{90}^{150+\alpha} E_{Ac}^2(\omega t)\, d\omega t \right] \right\}^{1/2}$$

$$= \left\{ \frac{3}{\pi} \left[\int_{30+\alpha}^{90} \left(\left(\sqrt{3}\, E_m \sin \omega t + 30 \right) \right)^2 d\omega t \right. \right.$$

$$\left. \left. + \int_{90}^{150+\alpha} \left(\sqrt{3}\, E_m \sin(\omega t - 30) \right)^2 d\omega t \right] \right\}^{1/2}$$

$$= \left\{ \frac{9 E_m^2}{2\pi} \left[\int_{30+\alpha}^{90} \sin^2(\omega t + 30)\, d\omega t + \int_{90}^{150+\alpha} \sin^2(\omega t - 30)\, d\omega t \right] \right\}^{1/2}$$

$$= \left\{ \frac{9 E_m^2}{2\pi} \left[\int_{60+\alpha}^{120} \sin^2(\omega t)\, d\omega t + \int_{60}^{120+\alpha} \sin^2 \omega t\, d\omega t \right] \right\}^{1/2}$$

$$= \left\{ \frac{9 E_m^2}{4\pi} \left[\int_{60+\alpha}^{120} (1 - \cos 2\omega t)\, d\omega t + \int_{60}^{120+\alpha} (1 - \cos 2\omega t)\, d\omega t \right] \right\}^{1/2}$$

$$= \left\{ \frac{9 E_m^2}{4\pi} \left[(60 - \alpha) + \frac{1}{2} (\sin 2\omega t)\big|_{120}^{60+\alpha} + (60 + \alpha) + \frac{1}{2} (\sin 2\omega t)\big|_{120+\alpha}^{60} \right] \right\}^{1/2}$$

$$= \left\{ \frac{9 E_m^2}{4\pi} \left[120° + \frac{1}{2} (\sin(120 + 2\alpha) - \sin(240) \right. \right.$$

$$\left. \left. + \sin(120) - \sin(240 + \alpha) \right] \right\}^{1/2}$$

$$= \left\{ \frac{9 E_m^2}{4\pi} \left[120° + \frac{1}{2} \left(\sqrt{3} + \sin(120 + 2\alpha) - \sin(240 + 2\alpha) \right) \right] \right\}^{1/2}$$

Since $\sin(120 + 2\alpha) - \sin(240 + 2\alpha) = \sqrt{3}\cos 2\alpha$ and $120° = 2\pi/3$, we get

$$E_{rms} = \left\{\frac{9E_m^2}{4\pi}\left[\frac{2\pi}{3} + \frac{\sqrt{3}}{2}(1+\cos 2\alpha)\right]\right\}^{1/2} = \frac{3}{2}E_m\left[\frac{2}{3} + \frac{\sqrt{3}}{2\pi}(1+\cos 2\alpha)\right]^{1/2}$$

(6.55)

Case II $\alpha > 60°$, RMS output voltage is given by:

$$E_{rms} = \left[\frac{3}{2\pi}\int_{30+\alpha}^{210} E_{AC}^2(\omega t)\,d\omega t\right]^{1/2} = \left[\frac{3}{2\pi}\int_{30+\alpha}^{210}\left(\sqrt{3}\,E_m\sin(\omega t - 30)\right)^2 d\omega t\right]^{1/2}$$

$$= \left[\frac{9E_m^2}{2\pi}\int_{30+\alpha}^{210}\sin^2(\omega t - 30)\,d\omega t\right]^{1/2} = \left[\frac{9E_m^2}{2\pi}\int_{\alpha}^{180}\sin^2\omega t\,d\omega t\right]^{1/2}$$

$$= \left[\frac{9E_m^2}{4\pi}\int_{\alpha}^{180}(1-\cos 2\omega t)\,d\omega t\right]^{1/2}$$

$$= \frac{3}{2}E_m\left\{\frac{1}{\pi}\left[(180-\alpha) + \frac{1}{2}(\sin(2\omega t))\right]_{180}^{\alpha}\right\}^{1/2}$$

$$= \frac{3}{2}E_m\left\{\frac{1}{\pi}\left[(180-\alpha) + \frac{1}{2}(\sin 2\alpha - \sin(360°))\right]\right\}^{1/2}, \pi = 180°$$

$$\therefore \quad E_{rms} = \frac{3E_m}{2}\left[\frac{\pi - \alpha + 1/2\sin 2\alpha}{\pi}\right]^{1/2} \quad (6.56)$$

Devices Ratings: SCRs and diode conducts only for 120° for $\alpha \leq 60°$ and for $(180° - \alpha)$ for $\alpha \geq 60°$.

(a) Case I $\alpha \leq 60°$
  (i) SCR and diode rms current ratings:
  $$I_{T(rms)} = I_{D(rms)} = \frac{I_{dc(rms)}}{\sqrt{3}} = \frac{\text{rms value of output current}}{\sqrt{3}}$$
  (ii) Average current ratings:
  $$I_{D(avg)} = I_{T(avg)} = \frac{I_{dc(avg)}}{3} = \frac{\text{Average value of output current}}{3}$$

(b) Case II $\alpha > 60°$, $I_{T(rms)} = I_{D(rms)} = \frac{I_{dc(rms)}}{\sqrt{3}}$

$$I_{T(avg)} = I_{d(avg)} = \frac{I_{dc}}{3}, \quad I_{T(peak)} = I_{D(peak)} = \frac{E_m}{R_L} \quad (6.57)$$

$$\text{PIV} > \sqrt{2}\; E_m, \text{ and } I_{dc(rms)} = \frac{E_{rms}}{R_L} = \frac{\text{RMS value of ouput voltage}}{R}$$

6.12.2 Operation with Inductive Load

Waveforms for semiconverter with large inductive load are shown in Fig. 6.49 (b). The following points can be observed from the Fig. 6.49 (b):

(i) The voltage waveform is same as that with resistive (R) load. Hence, the average and rms values of the output voltage waveform are same and are given by Eqs. (6.54), (6.55) and (6.56).

(ii) **Continuous conduction mode:** The output current waveform is continuous for $\alpha < 60°$, and is ripple free as shown for $\alpha = 30°$. Thus, the form-factor (FF) of current waveform is unity and the ripple factor is zero. The peak value of current I_{dc}.

(iii) **Discontinuous conduction mode:** For firing angles $\alpha > 60°$, the discontinuous mode occurs, as shown for $\alpha = 90°$. It can be observed from the waveforms that the output voltage becomes zero during a part of the output voltage period because of the freewheeling action.

(iv) Supply current waveform is a quasi-square wave for any $\alpha \leq 60°$. For any $\alpha > 60°$, the pulse width per half-cycle decreases to $(180 - \alpha)$, i.e. less than 120°. The peak value of the current waveform is I_{dc}.

(v) SCR and diode current waveform for any α, $0 \leq \alpha \leq 180°$ is a rectangular wave of duration 120°, i.e. of duty-cycle 1/3. The peak value of the waveform is I_{dc}.

(vi) PIV rating of SCRs and diode is $\sqrt{3}\; E_m$.

(vii) The maximum value of α is 180°

(viii) For $\alpha > 60°$, freewheeling action takes place when the output voltage is zero.

(ix) Operation of the circuit can be understand with the help of following points:
   (a) The source current is equal to load current I_{dc} when either the SCR or the diode is conducting otherwise the source current is zero.
   (b) In general, if SCR conducts means phase current is positive and if diode conducts, the phase current is negative. Phase current is zero when neither device is conducting.
   (c) Devices conducting current is equal to load current I_{dc}.

(x) **Half-waving effect:** For large firing angle delays, commutation failure may take place due to the limited time available in the symmetrical 3-ϕ semiconverter circuit configuration, if the current is assumed to be continuous. This may result "in-half-waving effect". That is, the converter operates as an uncontrolled half wave converter. This effect is shown in Fig. 6.49 (c), with the thyristor. T_1 in conduction and the firing pulses to

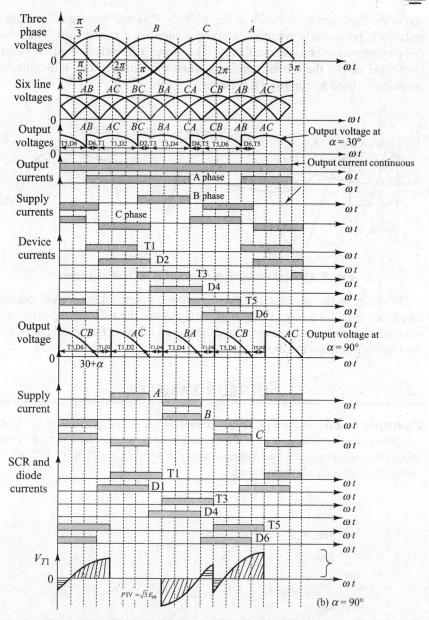

Fig. 6.49 (b) Waveforms for $\alpha = 30°$ and $\alpha = 90°$

SCRs T_3 and T_5 are inhibited. For the condition shown in Fig. 6.49 (c), the output voltage becomes uncontrolled being E_{AB} or E_{AC} for 120° each.

Prior to the half-waving effect, the waveforms are shown for $\alpha = 120°$ in Fig. 6.49 (c). The ripple in the output current becomes quite appreciable. The thyristor current is the same as the load current, since the thyristor conducts all the times. This effect can be eliminated by connecting a freewheeling diode across the load circuit, as shown in Fig. 6.49(c). This half-waving effect leads to instability, particularly when the converter supplies a motor load. The motor drive system may become oscillatory. Therefore, a freewheel diode is usually connected across the output terminals when the symmetrical half-controlled converter is used to control a d.c. motor.

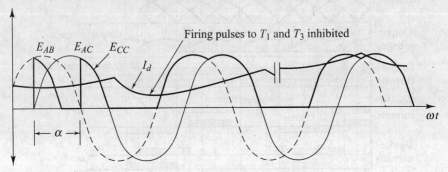

Fig. 6.49 *(c) Waveforms illustrating half-waving effect*

However, in the asymmetrical configuration, commutation failure does not take place even for a large firing angle delay. Also, the half-waving effect does not arise with this configuration, since there is an inherent freewheel path for the load current through the diodes.

SOLVED EXAMPLES

Example 6.23 A three-phase half-controlled bridge converter shown in Fig. 6.7, is supplied at a line-voltage of 15 V. Plot a curve relating average load voltage to firing delay angle, and sketch the load-voltage waveform at firing angles of 0°, 30°, 60°, 90° and 120°, neglecting all voltage drop.

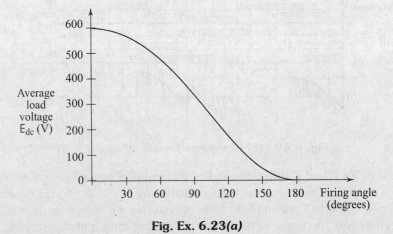

Fig. Ex. 6.23(a)

Solution: Average load voltage is given by

$$E_{dc} = \frac{3}{2\pi} 415\sqrt{2}\ (1 + \cos \alpha).$$

Figure 6.23(a) shows the plot for different value of a and E_{dc}. Figure 6.23(b) shows the load voltage waveforms.

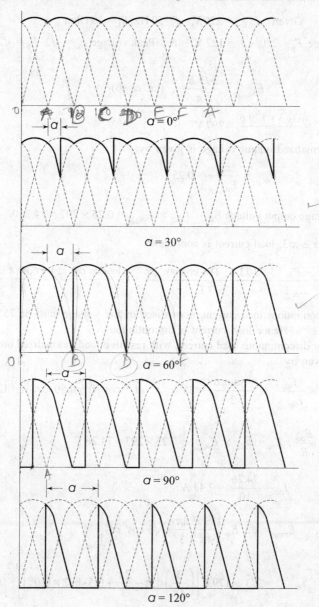

Fig. Ex. 6.23(b)

Example 6.24 The three-phase semiconverter is operated from a three phase Y-connected 220 V, 50 Hz supply and load resistance is $R_L = 10$ ohms. If the average output voltage is 25% of the maximum possible average output voltage, determine.
(a) delay angle (b) rms and average output currents
(c) average and rms thyristor currents (d) rectification efficiency and
(e) transformer utilization factor

Solution: Given $E_{Line} = 220$ V, $f = 50$ Hz, $R_L = 10$ Ohm
Phase voltage $E_p = 220/\sqrt{3} = 127$ V, & Peak voltage $E_m = \sqrt{2}\ E_p = \sqrt{2} \times 127 = 179.6$ V

$\therefore$ $$E_{dc_{max}} = \frac{3\sqrt{3}\ E_m}{\pi} \quad \text{(i.e. } \alpha = 0\text{)}$$

$\therefore$ $E_{dc_{max}} = \dfrac{3\sqrt{3} \times 179.6}{\pi} = 297$ V

Now, normalized output voltage is given by

$\therefore$ $$E_{dc_n} = \frac{E_{dc}}{E_{dc_{max}}} = 0.25$$

$\therefore$ Average output voltage $E_{dc} = E_{dc_n} \times E_{dc_{max}} = 0.25 \times 297 = 74.26$ V

(a) For $\alpha \geq \pi/3$, load current is continuous

$$E_{dc_n} = \frac{E_{dc}}{E_{dc_{max}}} = \frac{3\sqrt{3}\ E_m (1+\cos\alpha)/2\pi}{3\sqrt{3}\ E_m/\pi} = \frac{(1+\cos\alpha)}{2} = \frac{1+\cos(\pi/3)}{2} = 75\%$$

For continuous load current, normalised output voltage must be 75%. But here it is 25%. Hence load current is discontinuous.

For discontinuous load current, with resistive-load, normalized output voltage is given by

$$E_{dc_n} = \frac{E_{dc}}{E_{dc_{max}}} = \frac{1+\cos\alpha}{2} = 25\% \quad \therefore (1+\cos\alpha) = 0.50 \quad \therefore \alpha = 120°$$

(b) $I_{dc} = \dfrac{E_{dc}}{R}$, $E_{dc} = E_{dc_n} \times E_{dc_{max}} = 0.25 \times 297 = 74.26$ V

$\therefore$ $I_{dc} = \dfrac{74.26}{10} = 7.43$ A

Now, $E_{rms} = \sqrt{3}\ E_m \left[\dfrac{3}{4\pi}\left(\dfrac{2\pi}{3} + \sqrt{3}\cos^2\alpha \right) \right]^{1/2}$

$= \sqrt{3} \times 179.6 \left[\left(\dfrac{3}{4\pi}\right) \cdot \left(\pi - \dfrac{2\pi}{3} + 0.5\sin 2\times 120°\right) \right]^{1/2} = 119.1$ V

$\therefore$ $I_{rms} = \dfrac{E_{rms}}{R_L} = \dfrac{119.1}{10} = 11.91$ A

(c) Average thyristor current $I_{T(avg)} = \dfrac{I_{dc}}{3} = \dfrac{7.426}{3} = 2.46$ A

RMS thyristor current $= I_{Trms} = \dfrac{I_{rms}}{\sqrt{3}} = \dfrac{11.91}{\sqrt{3}} = 6.88$ A

(d) Rectification efficiency $(\eta) = \dfrac{E_{dc} \cdot I_{dc}}{E_{rms} \cdot I_{rms}} = \dfrac{74.26 \times 7.426}{119.1 \times 11.91} \times 100 = 38.8\%$

(e) Transformer utilization factor (TUF)

$= \dfrac{P_d}{E_s \cdot I_s} = \dfrac{E_{dc} \cdot I_{dc}}{E_s \cdot I_s}$ where, $I_s = I_{rms}\sqrt{2}/3 = 9.73$ A

$E_s \cdot I_s = 3 \times 127 \times 9.73 = 3705$ VA $\therefore$ TUF $= \dfrac{74.26 \times 7.439}{3705} = 0.149$

6.13 THE EXTERNAL PERFORMANCE MEASURES OF SIX-PULSE CONVERTERS

The current-waveforms for the three-phase fully controlled bridge converter are shown in Fig. 6.50. This current-waveform is an even function. Therefore, the expansion into Fourier-series contains only cosine terms.

$\therefore$
$$I = \sum_{n=1}^{\infty} a_n \cos n\omega t \qquad (6.58)$$

where,
$$a_n = \dfrac{I_d}{\pi}\left[\int \cos \omega t\, d\omega t - \int \cos \omega t\, d\omega t + \int \cos \omega t\, d\omega t\right]$$

$$= \dfrac{2I_d}{\pi}\int \cos n\omega t\, d\omega t = \dfrac{4I_d}{n\pi}\sin\dfrac{n\pi}{3} \qquad (6.59)$$

The series is, therefore,

$$I = \dfrac{4I_d}{\pi}\dfrac{\sqrt{3}}{2}\left[\cos \omega t + \dfrac{1}{5}\cos 5\omega t\right]$$

$\therefore$ RMS value of fundamental $= \dfrac{2I_d\sqrt{3}}{\sqrt{2}\pi}$ \qquad (6.60)

RMS value of current $= \sqrt{\dfrac{2}{3}}I_d$ \qquad (6.61)

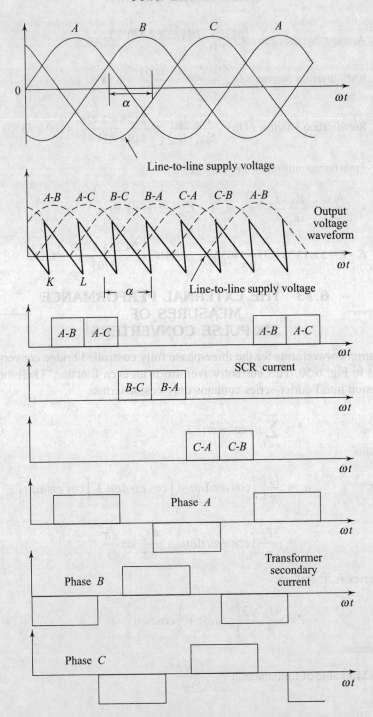

Fig. 6.50 *Voltage and current waveforms*

Solved Example

Example 6.25 For the 3-ϕ full converter, operating from 3-ϕ/415 V/50 Hz supply and deriving a large inductive load with $R_L = 100$ ohm, determine and plot the following for $\alpha = 0°, 30°, 60°, 90°, 120°, 150°$ and $180°$.

(a) Displacement power factor (DPF) (b) Distortion factor (DF)
(c) Harmonic factor   (d) Supply power factor

Solution: The nth harmonic component of supply current is given by

$$I_{s(n)} = \frac{c_n}{\sqrt{2}} = \frac{2\sqrt{2} \, I_{dc}}{\pi} \frac{\cos(30\,n)}{n} = 0.9 \, I_{dc} \cdot \frac{\cos 30n}{n}$$

$\therefore$ Fundamental component of supply current is given by (substitute $n = 1$),

$$I_{s(fund)} = 0.9 \, I_{dc} \cos 30 = 0.78 \, I_{dc}$$

(a) Displacement power factor $= \cos(\phi_1) = \cos \alpha$

(b) $DF = \dfrac{I_{s(fund)}}{I_{s(rms)}}$, where $I_{s(rms)} = \sqrt{\dfrac{2}{3}} \, I_{dc} = 0.816 \, I_{dc}$ $\therefore DF = \dfrac{0.78 \, I_{dc}}{0.816 \, I_{dc}} = 95.6\%$

(c) $HF = \sqrt{\dfrac{1}{DF^2} - 1} = \sqrt{\left(\dfrac{1}{0.956}\right)^2 - 1} = 30.8\%$

(d) Supply PF $= \dfrac{3}{\pi} \cos \alpha = 0.956 \cos \alpha$

Table Ex. 6.25 shows all above parameter value for different values of α and Fig. Ex. 6.25 shows related graphs.

S.N	α (degrees)	DPF $= \cos \alpha$	DF $= 95.6\%$	HF $= 30.8\%$	PF $= 0.956 \cos \alpha$
1	0	1	95.6%	30.8%	0.956
2	30	0.866	95.6%	30.8%	0.828
3	60	0.5	95.6%	30.8%	0.478
4	90	0	95.6%	30.8%	0
5	120	-0.5	95.6%	30.8%	-0.478
6	150	-0.866	95.6%	30.8%	-0.828
7	180	-1	95.6%	30.8%	-0.956

6.14 THE EFFECT OF INPUT SOURCE IMPEDANCE

The circuit analyses and formulae expressed in the previous sections are based on the simplifying assumptions that, even with inductive loads, the transfer or commutation of the current from one thyristor to the next took place

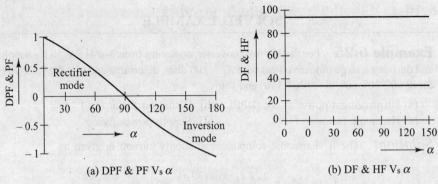

Fig. Ex. 6.25

instantaneously. This assumption greatly simplifies calculations and at low frequencies and commercial voltages, generally does not substantially effect the results of these calculations.

Actually, in practice, inductance and resistance must be present in the supply source, and time is required for a current change to take place. The net result is that the current commutation is delayed, as it takes a finite time for the current to decay to zero in the outgoing thyristor, while the current will rise at the same rate in the incoming thyristor. Thus, in practice, the commutation process may occupy a quite significant period of time, during which both the "incoming" and "outgoing" thyristors are simultaneous in conduction. This period, during which both the outgoing and incoming thyristors are conducting, is known as the overlap period and the angle for which both devices share conduction is known as the overlap angle (μ) or commutation angle.

During this "commutation overlap" period, the waveforms of the voltage at the output terminals of the converter, as well as the current and voltage at the input terminal are different from those obtained with zero-source inductance. This has the modifying effect upon the external performance characteristics of the converter. At the output terminals, the effect of the input source inductance is to cause a loss of mean voltage, as well as modification to the harmonic distortion terms, while at the input terminals, a slight reduction of displacement factor, with respect to the a.c. source voltage, as well as the modification to the distortion-terms in the current waveform, takes place.

The inductive reactance of the a.c. supply is normally much greater than its resistance and, as it is the inductance which delays to current change, it is reasonable to neglect the supply resistance. However, if the source impedance is resistive, then there will be a drop across this resistance and the average output voltage of a converter gets reduced by an amount equivalent to the average drop. Since the source resistance is usually small, the commutation angle during which the load current is transferred from the outgoing to the incoming thyristors is generally neglected. The effect of source inductance is investigated in this section for both single-phase and three-phase fully-controlled converters.

6.14.1 Single-Phase Fully-Controlled Converters

The commutation overlap is more predominant in full converters than semiconverters. The single-phase fully-controlled bridge converter is shown in Fig. 6.51(a). L_s is the source inductance. The load current is assumed constant. The a.c. supply may be represented by its Thevenin's equivalent circuit, each phase being a voltage source in series with its inductance. The major contributor to the supply impedance is the transformer leakage reactance. The equivalent circuit of fully controlled single phase converter is shown in Fig. 6.51(b). When the terminal L of the source voltage, e_s, is positive, then the current flows through the path $L - L_s - T_1 - R - L - T_2 - N$. This is shown as e_1, L_s, T_1, T_2, and load in Fig. 6.51(b). Similarly, when terminal N of source voltage, e_s, is positive, load current flows through the path $N - T_3 - \text{load} - T_4 - L_s - L$. This is shown as e_2, L_s, T_3, T_4 and load in Fig. 6.51(b). The related circuit voltage and current waveforms are shown in Fig. 6.51(c).

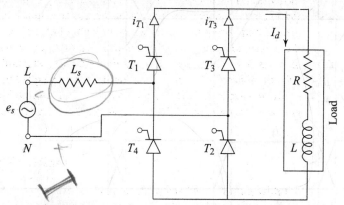

Fig. 6.51 *(a) Single-phase fully-controlled converter with source inductance L_s*

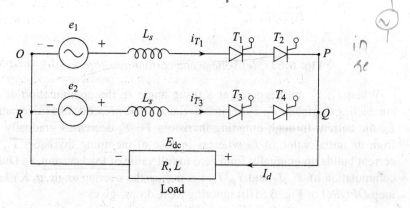

Fig. 6.51 *(b) Equivalent circuit*

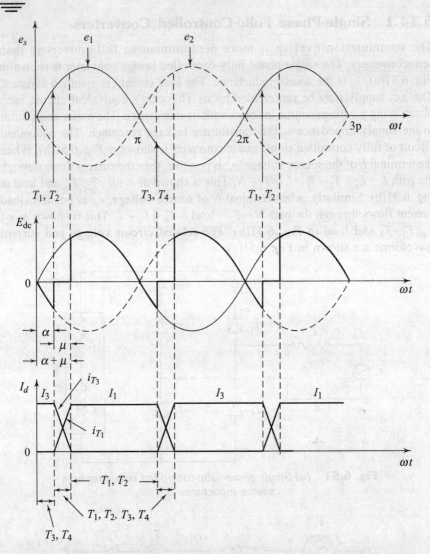

Fig. 6.51 *(c) Voltage and current-waveforms with L_s*

When T_1, T_2 are triggered at a firing angle α, the commutation of already conducting thyristors T_3, T_4 begins. Because of the presence of source inductance L_s, the current through outgoing thyristors T_3, T_4 decreases gradually to zero from its initial value of I_d, whereas in case of incoming thyristors T_1, T_2, the current builds up gradually from zero to full value of load current, I_d. During the commutation of T_1, T_2 and T_3, T_4, i.e. during the overlap angle μ, KVL for the loop *OPQRO* of Fig. 6.51(b) ignoring SCR drops, gives

$$e_1 - L_s \frac{d_{iT_1}}{dt} = e_2 - L_s \frac{d_{i_3}}{dt}$$

or
$$e_1 - e_2 = L_s \left(\frac{d_{iT_1}}{dt} - \frac{d_{i_3}}{dt} \right) \tag{6.62}$$

But from Fig. 6.51(c), we have the relations
$e_1 = E_m \sin \omega t$ and $e_2 = -E_m \sin \omega t$.
Substituting in Eq. (6.62), we get

$$L_s \left(\frac{d_{iT_1}}{dt} - \frac{d_{i_3}}{dt} \right) = 2E_m \times \sin \omega t \tag{6.63}$$

Since the load current is assumed constant, we can write

$$i_{T_1} + i_{T_3} = I_d \tag{6.64}$$

∴ Differentiating Eq. (6.64), with respect to t.

$$\frac{d_{iT_1}}{dt} = -\frac{d_{iT_3}}{dt} \tag{6.65}$$

Substitute Eq. (6.65) in Eq. (6.63), we get

$$L_s \left(2 \frac{d_{iT_1}}{dt} \right) = 2 E_m \cdot \sin \omega t$$

∴ $$\frac{d_{iT_1}}{dt} = \frac{E_m}{L_s} \sin \omega t \tag{6.66}$$

If the overlap angle is μ, then the current through thyristor pair T_1, T_2 builds up from zero to I_d during this interval.
Therefore, at $\omega t = \alpha, i_{T_1} = 0$ and at $\omega t = (\alpha + \mu), i_{T_1} = I_d$
∴ Therefore, from Eq. (6.66), we can write

$$\int_0^{I_d} d_{iT_1} = \frac{E_m}{L_s} \int_{\alpha/\omega}^{(\alpha+\mu)/\omega} \sin \omega t \, d(\omega t)$$

∴ $$I_d = \frac{E_m}{\omega L_s} [\cos \alpha - \cos(\alpha + \mu)] \tag{6.67}$$

It can be observed from Fig. 6.51(c) that the output voltage is zero during the interval μ. There are two commutations in each cycle. Thus, the average output-voltage is given by

$$E_{dc} = \frac{E_m}{\pi} \int_{\alpha+\mu}^{\pi+\alpha} \sin \omega t \, d(\omega t) = \frac{E_m}{\pi} [-\cos \omega t]_{\alpha+\mu}^{\pi+\alpha}$$

$$= \frac{E_m}{\pi} [\cos(\alpha + \mu) - \cos(\alpha + \pi)]$$

$$\therefore \quad E_{dc} = \frac{E_m}{\pi}[\cos\alpha + \cos(\alpha + \mu)] \quad (6.68)$$

Now, from Eq. (6.66), we have

$$\cos(\alpha + \mu) = \cos\alpha - \frac{\omega L_s}{E_m} I_d$$

Substituting this value of $\cos(\alpha + \mu)$ in Eq. (6.68), we get

$$E_{dc} = \frac{E_m}{\pi}\left[\cos\alpha + \cos\alpha - \frac{\omega L_s}{E_m} I_d\right] \quad E_{dc} = \frac{2E_m}{\pi}\cos\alpha - \frac{\omega L_s}{E_m} I_d \quad (6.69)$$

Also, from Eq. (6.67), we have $\cos\alpha = \frac{\omega L_s}{E_m} I_d + \cos(\alpha + \mu)$

Substituting this value of $\cos\alpha$ in Eq. (6.68), we get

$$E_{dc} = \frac{2E_m}{\pi}\cos(\alpha + \mu) + \frac{\omega L_s}{\pi} I_d \quad (6.70)$$

With the help of Eq. (6.69), a d.c. equivalent circuit for a two-pulse single-phase fully-controlled converter can be drawn, as shown in Fig. 6.51(d).

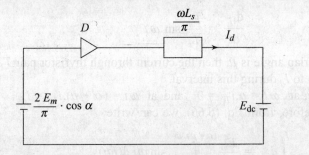

Fig. 6.51 *(d) D.C. equivalent circuit*

Diode D in Fig. 6.51(d) indicates that load current is unidirectional. This equivalent circuit shows that the effect of source inductance is to present an equivalent resistance of magnitude $\frac{\omega L_s}{\pi}$ in series with interval voltage of rectifier $\frac{2E_m}{\pi}\cos\alpha$. With the load current I_d, the voltage drop is $\frac{\omega L_s I_d}{\pi}$. Hence it becomes clear that with source inductance, the output voltage of a converter is reduced by $\frac{\omega L_s I_d}{\pi}$ value. The variation of output-voltage with output-current is shown in Fig. 6.51(e). From this Fig. 6.51(e), it is also clear that as load current I_d (or source inductance) increases, the commutation interval or the overlap angle increases and as a consequence, the average output voltage decreases.

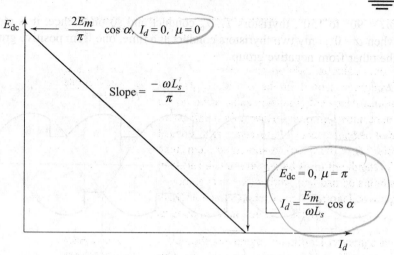

Fig. 6.51 *(e) Variation of output voltage E_{dc} with I_d*

Thus, in single-phase fully-controlled converter, as long as commutation angle μ is less than π, the output voltage is given by Eq. (6.68) and when μ is equal to π, the load will be permanently short-circuited by SCRs and the output voltage will be zero because during the overlap period, all SCRs will be conducting.

6.14.2 Three-Phase Fully-Controlled Bridge Converter

Three-phase fully-controlled bridge converter with a source inductance L_s in each line is shown in Fig. 6.52. In this case also, the load current I_d is assumed to be constant. Figure 6.53 shows the various voltage and current waveforms for a six-pulse bridge converter.

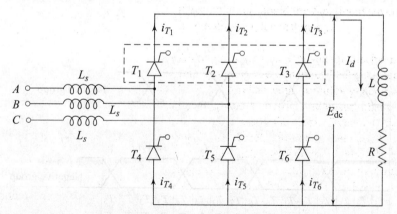

Fig. 6.52 *Three-phase fully-controlled bridge with source inductance L_s*

The conduction of various SCRs with firing angle $\alpha = 0$ is shown in Fig. 6.53(b). As shown in Fig. 6.53(b), thyristors T_3, T_5 conduct up to $\omega t = 30°$. From $\omega t = 30°$ to $90°$, thyristors T_1, T_5 conduct. Similarly, from instant

$\omega t = 90°$ to $150°$, thyristors T_1, T_6 conduct and so on. Hence, it is clear that when $\alpha = 0°$, only two thyristors conduct at a time, one from positive group and the other from negative group.

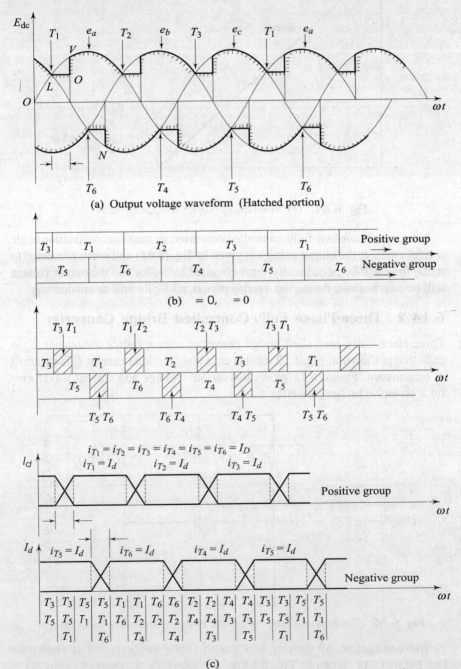

Fig. 6.53 *(a), (b) and (c) Voltage and current waveform*

Phase Controlled Converters

The effect of overlap is shown in Fig. 6.53(c) from instant $\omega t = 0°$ to $30°$, thyristors T_3, T_5 conduct. At instant $\omega t = 30°$, T_3 is outgoing thyristor and thyristor T_1 is incoming and both T_3, T_1 belong to the same group (positive group). As thyristor T_1 is triggered, current through T_3 starts decaying, while through T_1 current begins to build up. Therefore, at $\omega t = 30° + \mu$, current i_{T_3} is zero, whereas current $i_{T_1} = I_d$. Therefore, from the period $\omega t = 30°$ to $30° + \mu$, three thyristors T_3, T_5, T_1 conduct. After $\omega t = 30° + \mu$, thyristors T_5, T_1 conduct. At instant $\omega t = 90°$, as thyristor T_6 is triggered, current i_{T_5} begins to decrease and current i_{T_6} starts to build up. Therefore, from $\omega t = 90°$ to $90° + \mu$, three thyristors T_5, T_1, T_6 conduct. At $\omega t = 90° + \mu$, current i_{T_5} becomes zero and current $i_{T_6} = I_d$. After $\omega t = 90° + \mu$, only two thyristors T_1, T_6 conduct. This sequence of operation repeats with other SCRs of the fully-controlled bridge converter.

From the above sequence of operation, it can be observed that when positive group of SCRS are undergoing commutation, two SCRs from the positive group and one SCR from the negative group conduct. After the commutation of positive group is completed, only two SCRs conduct, one SCR from positive-group and the one from negative-group. In a similar way, when negative group of SCRs are undergoing commutation, three SCRs conduct, two SCRs from the negative group and one SCR from the positive group. After the commutation of negative group is completed, only two SCRs conduct, one from positive group and one from negative group. The sequence of conduction of thyristors is shown in Fig. 6.53(c), which is as follows:

T_3–T_5, T_3–T_5–T_1, T_5–T_1, T_5–T_1–T_6, T_1–T_6,
T_1–T_2–T_6, T_6–T_2, T_6–T_2–T_4, T_2–T_4, T_2–T_4–T_3,
T_4–T_3, T_4–T_3–T_5, T_3–T_5 and so on.

Hence it becomes clear that three thyristors and two thyristors conduct alternatively. From Fig. 6.53(c), it is also observed that for a six-pulse converter, there are six shaded regions, which indicate six commutations in each cycle of the supply voltage.

Now, during the transfer of current from T_3 to T_1 (i.e. commutation of T_3 and T_1), the output voltage is obtained by taking average of corresponding phase voltage e_c and e_a of the positive group. This means that voltage from $\omega t = 30°$ to $30° + \mu$ follow the curve $\frac{e_a + e_c}{2}$ from the positive group, as shown by the region LO in Fig. 6.53(c). Similarly, during commutation of T_5, T_6, the voltage waveform from the negative group is $\frac{e_b + e_c}{2}$, as indicated by region ED.

In Fig. 6.53(c), firing delay angle α has been taken as zero ($0°$), just to have more concentration on the effect of source inductance L_s. However, similar procedure is applicable for any firing-angle (α), provided that overlap angle μ must be less than $60°$. In Fig. 6.53(a), the output voltage is shown by the hatched portion.

As mentioned earlier, the effect of source inductance, L_s, is to reduce the average d.c. output voltage. This reduced d.c. output voltage is shown by the region LOV in Fig. 6.53(a). This reduced voltage region is almost proportional to the triangular area. Therefore, the average value of this reduced output voltage due to overlap is equal to the triangular area LOV divided by the periodicity of this triangular area, which is equal to $\dfrac{\pi}{3}$. Hence,

Reduction in output voltage due to overlap

$$= \frac{1}{\pi/3}\int_0^\mu e_L \cdot d(\omega t) = \frac{3}{\pi}\int_0^\mu L_s \cdot \frac{di}{dt} d(\omega t) = \frac{3L_s}{\pi}\int_0^{\mu/\omega} \omega \cdot \frac{di}{dt} dt$$

$$= \frac{3\omega L_s}{\pi}\int_0^{I_d} di = \frac{3\omega L_s}{\pi} I_d \qquad (6.71)$$

Thus, output voltage with overlap = interval voltage of 3-phase fully-controlled converter (Eq. 6.52) – fall in output voltage due to overlap

$$= \frac{3\sqrt{3}}{\pi} E_{mph} \cdot \cos\alpha - \frac{3\omega L_s}{\pi} I_d \qquad (6.72)$$

In general, for P-pulse converter, reduction in output voltage due to overlap

$$= \frac{p}{2\pi}\int_0^\mu L_s\left(\frac{di}{dt}\right) d(\omega t) = \frac{p\omega L_s}{2\pi}\int_0^{\mu/\omega}\left(\frac{di}{dt}\right) dt = \frac{p\omega L_s}{2\pi}\int_0^{I_d} d_i = \frac{p\omega L_s \cdot I_d}{2\pi}$$

$$(6.73)$$

Output voltage for a 3-phase fully-controlled converter, similar to that given for single-phase fully-controlled converter in Eq. (6.70) is given by

$$E_{dc} = \frac{3\sqrt{3}\, E_{mph}}{\pi} \cos(\alpha + \mu) + \frac{30\,\omega L_s}{\pi} I_d \qquad (6.74)$$

SOLVED EXAMPLES

Example 6.26 A single-phase full-wave mid-point converter with freewheeling diode, as shown in Fig. 6.72, is supplied from a 120 V, 50 Hz supply with a source inductance of 0.33 mH. Assuming the load-current is continuous at 4 A, find the overlap angles for
(a) transfer of current from a conducting thyristor to the commutating diode, and
(b) from the commutating diode to a thyristor when the firing angle is 15°.

Solution: (a) Suppose, commutation from thyristor to diode begins at time $t = 0$, the instant when the load voltage starts to reverse (neglecting device voltage-drops). Hence, at the onset of commutation and referring to Fig. Ex. 6.26(a), we can write

$$e = -E_m \sin \omega t = -L\frac{di}{dt}, \quad \therefore \quad di = E_m \sin\omega t\, \frac{dt}{L}$$

Integrating, $i = \dfrac{E_m}{L}\int_0^t \sin\omega t\, d(\omega t) = \dfrac{E_m}{\omega L}(1 - \cos\omega t)$

Fig. Ex. 6.26 *Commutation from thyristor to diode*

Commutation is complete when current $i = I_d$, at which point $\omega t = \mu_1$.

Hence, $\qquad I_d = \dfrac{E_m}{\omega L}(1 - \cos\mu)$

∴ Substituting the values, we get

$$4 = \dfrac{120}{2\times\pi\times 50\times 0.333\times 10^{-3}}(1 - \cos\mu_1) \quad \therefore\ \mu_1 = 6.786°.$$

(b) Now, commutation from the diode to the thyristor begins at the instant of firing the thyristor substituting $t = 0$, at the instant of firing then, referring to Fig. Ex. 6.27, we can write

$$e = E_m \sin(\omega t + \alpha) = L\dfrac{di}{dt}$$

After integrating, as before, this gives

$$i = \dfrac{E_m[\cos\alpha - \cos(\omega t - \alpha)]}{\omega L}$$

Commutation is complete when $i = I_d$ and $\omega t = \mu_2$, when

$$I_d = \dfrac{E_m[\cos\alpha - \cos(\mu_2 - \alpha)]}{\omega L}$$

After substituting the values, we get
$\mu_2 = 0.536°.$

Example 6.27 A three-phase, half-wave converter is operating in the inverting mode connected to a 415 V (line) supply. If the angle of firing advance is 18° and the overlap 3.8°, find the mean voltage at the load.

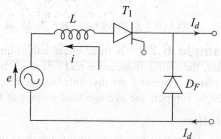

Fig. Ex. 6.27 *Commutation from diode to thyristor*

Solution: Refer to Eq. 6.90, $E_{dc} = \dfrac{3\sqrt{3}}{4\pi} E_m \cdot [\cos\beta + \cos(\beta - \mu)]$

Given: $\beta = 180°$ and $\mu = 3.8°$ $E_{dc} = \dfrac{3\sqrt{3}}{4\pi} \times \dfrac{415\sqrt{2}}{\sqrt{3}} \times \cos(18° + \cos 14.2°) = 269.1$ V.

Example 6.28 A three-phase fully-controlled bridge converter is connected to three-phase a.c. supply of 400 V, 50 Hz and operates with a firing angle $\alpha = \pi/4$. The load current is maintained constant at 10 A and the load voltage is 360 V. Compute:

(i) Source inductance, L_s. (ii) Load resistance, R.
(iii) Overlap angle, μ.

Solution: (i) Refer to Eq. (6.77),

$$E_{dc} = \dfrac{3\sqrt{3}}{\pi} E_{mph} \cdot \cos\alpha - \dfrac{3\omega L_s}{\pi} I_d$$

Given: $E_{dc} = 360$ V, $E_{mph} = 400 \cdot \left(\dfrac{\sqrt{2}}{\sqrt{3}}\right)$ $\alpha = \pi/4$, $I_d = 10$ A

$\therefore \qquad 360 = \dfrac{3\sqrt{3}}{\pi} \times 400 \left(\dfrac{\sqrt{2}}{\sqrt{3}}\right) \cos\left(\dfrac{\pi}{4}\right) - \dfrac{3 \times 2\pi \times 50 \times L_s \times 10}{\pi}$

$360 = 381.17 - 3000 L_s$, $L_s = 7.3$ mH.

Source inductance per phase will be 7.3 mH

(ii) Load resistance, $R = \dfrac{E_{dc}}{I_d} = \dfrac{360}{10} = 36\ \Omega$

(iii) Overlap angle can be determined from Eq. (6.79).

$\therefore \qquad E_{dc} = \dfrac{3\sqrt{3}}{\pi} E_{mph} \cos(\alpha + \mu) + \dfrac{3\omega L_s}{\pi} I_d$

$\therefore \qquad 360 = \dfrac{3\sqrt{3}}{\pi} 400 \left(\dfrac{\sqrt{2}}{\sqrt{3}}\right) \cos(\alpha + \mu) + \dfrac{3 \times 2\pi \times 50 \times 7 \times 10^{-3} \times 10}{\pi}$

$\therefore \qquad \cos(\alpha + \mu) = 0.63$ $\alpha + \mu = 50.95$ $\mu = 50.95 - 45°$ $\mu \approx 6°$.

Example 6.29 A three-phase half-controlled rectifier is supplied at 150 V/ph, 50 Hz, the source inductance and resistance being 1.2 mH and 0.07 Ω, respectively, per phase. Assuming the thyristor voltage drop of 1.5 V and continuous load current of 30 A, compute the average load voltage at firing angles of 0°, 30° and 60°.

Solution:
(i) There will be a thyristor voltage drop of 1.5 V all the times.
(ii) Also, resistance voltage drop = $I_d \cdot R = 30 \times 0.07 = 2.1$ V

(iii) The source reactance also leads to a voltage drop given by the equation,

$$= \frac{3\omega L_s}{2\pi} I_d = \frac{(3 \times 2\pi \times 50 \times 1.2 \times 10^{-3} \times 30)}{2\pi} = 5.4 \text{ V}$$

(iv) The average d.c. output voltage neglecting the voltage drops is given by,

$$E_{dc} = \frac{3\sqrt{3} E_m}{2\pi} \cos \alpha = \frac{3\sqrt{3} \times 150 \times \sqrt{2}}{2\pi} \cos \alpha = 175.4 \cos \alpha.$$

(v) Now, calculate the E_{dc} by considering all voltage drops

∴ At $\alpha = 0°$, $E_{dc} =$ $175.4 \cos 0° = 1.5 - 2.1 - 5.4 = 166.4$ V
   At $\alpha = 30°$, $E_{dc} =$ $175.4 \cos 30° - 1.5 - 2.1 - 5.4 = 142.9$ V
   At $\alpha = 60°$, $E_{dc} =$ $175.4 \cos 60° - 1.5 - 2.1 - 5.4 = 78.7$ V.

Example 6.30 A three-phase fully-controlled bridge converter is connected to 415 V supply, having a reactance of 0.3 Ω phase and resistance of 0.05 Ω/phase. The converter is working in the inversion-mode at a firing advance angle of 35°. Compute the average generator voltage. Assume $I_d = 60$ A and thyristor drop = 1.5 V.

Solution: For three-phase fully-controlled bridge converter, the voltage drop due to overlap is given by $= \dfrac{3\omega L_s}{\pi} I_d = \dfrac{3 \times 0.3 \times 60}{\pi} = 17.19$ V.

Voltage drop due to SCRs $= 2 \times 1.5 = 3$ V
Voltage drop due to supply resistance $= 2 \times 0.05 \times 60 = 6$ V.
The no-load average voltage ($\mu = 0$), at advance firing angle β is given by

$$E_{dc(no\ load)} = \frac{3\sqrt{3}}{\pi} E_{mph} \cdot \cos \alpha = \frac{3\sqrt{3}}{\pi} \left(415 \frac{\sqrt{2}}{\sqrt{3}}\right) \cos(\mu - 35°) = -459.1 \text{ V}$$

(Negative sign indicates inversion)
Since the generator has to supply all the voltage drops, therefore
Average generator voltage $= 459.1 + 17.19 + 3 + 6 = 485.29$ V.

Example 6.31 Draw the thyristor voltage waveform for a six-pulse converter operating in the inverter mode for a firing advance angle of 30° and overlap 10°, for (a) six-phase half-wave connection, (b) bridge connection.

Solution:
(a) The waveforms for six-phase half-wave connection is shown in Fig. 6.31(a). These waveforms are drawn with reference to Fig. 6.38.
(b) The waveforms for bridge-connection are shown in Fig. Ex. 6.31(b). These waveforms are drawn with reference to Fig. 6.41. In these waveforms, thyristor T_1 voltage is represented by hatched region. During commutations, the line-to-neutral voltage at the bridge terminals is midway between the line and the other commutating line.

Example 6.32 Three-phase fully-controlled bridge converter to a supply voltage of 230 V per phase and frequency of 50 Hz. The source inductance is 3 mH.

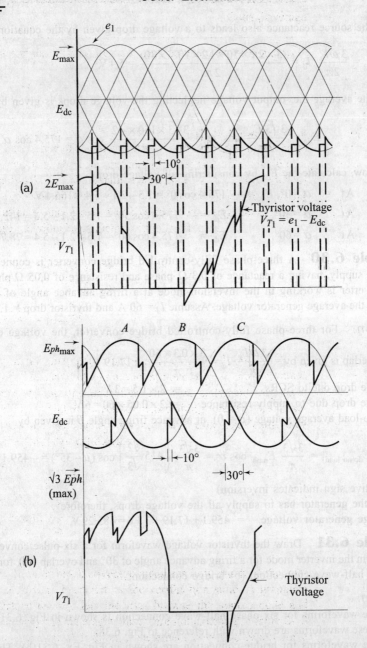

Fig. Ex. 6.31 *(a) Waveforms for six-phase half-wave connection*
(b) Waveforms for bridge connection

The load current on d.c. side is constant at 15 A. If the load consists of a d.c. source voltage of 400 V having an internal resistance of 1 Ω, compute the following:

(i) Firing angle. (ii) Overlap angle.

Solution:

(i) Converter output voltage

$$E_{dc} = E_{\text{load source}} + I_d \cdot R = 400 + 15 \times 1 = 415 \text{ V}.$$

Now, using the relation

$$\therefore \quad E_{dc} = \frac{3\sqrt{3}}{\pi} E_{mph} \cos\alpha - \frac{3\omega L_s}{\pi} I_d$$

$$415 = \frac{3\sqrt{3}}{\pi} \cdot 230 \left(\sqrt{2}\right) \cos\alpha - \frac{3 \times 2\pi \times 50 \times 3 \times 10^{-3} \times 15}{\pi}$$

Firing angle $\alpha = 37.20°$.

(ii) Using Eq. (6.79), we get

$$415 = \frac{3\sqrt{6} \times 230}{\pi} \cos(\alpha + \mu) + \frac{3(2\pi\, 50)\, 3 \times 10^{-3} \times 15}{\pi} \quad \therefore \mu = 6.53°$$

$\therefore$ Overlap angle $\mu = 4.53°$.

Example 6.33 Design the 3-ϕ bridge converter for following specifications:

(i) 2.3 kV/50 Hz, 3-ϕ power supply

(ii) Delta-star transformer with 10:1 ratio

(iii) DC load current = 90 A

(iv) DC load voltage, varying from +500 V to −500 V

(v) Commutating inductance per phase = 50 μH

Solution:

Step 1 Calculation of firing angle α

Reduction in output voltage due to overlap is given by

$$E_{dc(red)} = \frac{3\omega \cdot L_s}{\pi} \cdot I_d = \frac{3 \cdot 2 \cdot \pi \cdot f \cdot L_s \cdot I_d}{\pi} = 6 \cdot F \cdot L_s \cdot I_d = 6 \times 50 \times 50 \times 10^{-6} \times 90$$

$$= 1.35 \text{ V}$$

Assume SCR drop $V_T = 1.5$ V

$\therefore$ Average output voltage $E_{dc} = E_{dc1} + E_{dc(red)} + 2 \cdot V_T = 500 \text{ V} + 1.35 + 2 \cdot (1.5)$

$$E_{dc} = 504.35 \text{ V}$$

Transformer turn ratio = 10 : 1

$\therefore$ Line rms voltage $E_L = \left(\dfrac{2300}{10}\right)\sqrt{3} = 398.4$ V Now $E_{dc} = 1.35\, E_L \cos\alpha = 504.35$ V

$$\therefore \quad \cos\alpha = \frac{504.35}{1.35 \times 398.4} = 0.94 \quad \therefore \quad \alpha = 19.9°$$

Step 2 Selection of SCRs

SCRs withstand a peak value E_m in reverse direction. Allow a 50% voltage margin for line voltage fluctuation and snubber overshoot. Therefore, voltage rating is

$$E_{DRM} = E_{RRM} = 398.2 \sqrt{2} \times 1.5 = 845 \text{ V}$$

Thyristor average and rms currents are

$$I_{avg} = \frac{90}{3} = 30 \text{ A}, \quad I_{rms} = \frac{90}{\sqrt{3}} = 51.96 \text{ A}$$

∴ Select SCRs having ratings 900 V and 63 A (rms)
In the inverting mode, $E_{dc} = -500 + 1.35 + 3 = 495.65$ V
which gives cos $\alpha = -0.921$

$$\therefore \quad \cos(\alpha + \mu) = \cos \alpha - \frac{\omega L_s \cdot I_d}{E_m} = \cos \alpha - \frac{\omega L_s \cdot I_d \sqrt{3}}{\sqrt{2} \, E_L}$$

$$= -0.921 - \frac{2 \times \pi \times 50 \times 10^{-6} \times 90 \times \sqrt{3}}{\sqrt{2} \times 398.4} = -0.926 \quad \therefore \quad \alpha + \mu = 157.8°$$

$$\therefore \quad \gamma = 180 - (\alpha + \mu) = 22.21° \quad \text{or,} \quad t_{off} = \frac{22.21}{2 \times \pi \times 50 \times 57.3} = 1.028 \text{ ms}$$

which is adequate. Assuming with a 33% duty cycle for $I_{av} = 30$ A. $T_{c_{max}} = 104°$C and $P_{av} = 43$ W.

Step 3 Calculation of heat—sink parameters

Thermal resistance, $\theta_{CA} = \theta_{CS} + \theta_{SA} = \dfrac{(T_{c_{max}} - T_A)}{P_{av}}$

Let us assume $T_A = 25°$C and $\theta_{CS} = 0.675°$C/W

$$\therefore \quad \theta_{SA} = \frac{104 - 25}{43} - 0.075 = 1.76°\text{C/W}$$

Step 4 Transformer selection

RMS secondary current = $\sqrt{\dfrac{2}{3}} \, I_d = \sqrt{2/3} \cdot 90 = 73.48$ A

RMS primary current = $\sqrt{\dfrac{2}{3}} \dfrac{I_d}{n} = \sqrt{2/3} \dfrac{90}{10} = 7.35$ A.

Transformer VA rating = $3 \, E_{line} \, T_{ph} = \dfrac{3 \times 398.4}{\sqrt{3}} \times 73.48 = 50.7$ kVA

Line rms fundamental curernt is

$$I_{Lf} = \frac{3}{\pi} \cdot \frac{2 I_d}{n} \cdot \frac{1}{\sqrt{2}} = \frac{3 \times \sqrt{2} \times 90}{\pi \times 10} = 12.16 \text{ A}$$

and the rms current is

$$I_L = \frac{\sqrt{2}\, I_d}{n} = \frac{\sqrt{2} \times 90}{10} = 12.73 \text{ A}.$$

Step 5 Calcualtion of converter parameters

In rectification mode, Displacement factor = $\cos \alpha = 0.94$, DF = $\dfrac{12.16}{12.73} = 0.96$ and PF = $0.94 \times 0.96 = 0.90$.

Fundamental input power $P_i = \sqrt{3} \times 2300 \times 12.16 \times 0.94 = 45.53$ kW

output power $= E_{dc}\, I_{dc} = 500 \times 90 = 45$ kW.

∴ Efficiency, $\eta = \dfrac{45}{45.53} = 98.8\%$ where the losses due to conduction drop only has been considered.

Example 6.34 A line commutated ac to dc converter is shown in Fig. Ex. 6.34. It operates from a three phase, 50 Hz, 580 V (line-to-line) supply. It supplies a load current, I_0 of 3464 A. Assume I_0 to be ripple free and neglect source impedance.

(a) Calcualte the delay angle α of the converter if its average output voltage is 648 V.
(b) Calculate the power delivered to the load R in kW.
(c) Calculate the fundamental reactive power drawn by converter from the supply in kVAR.

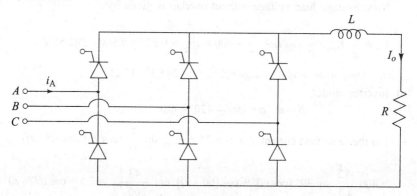

Fig. Ex. 6.34

Solution:

(a) $E_{dc} = \dfrac{3\sqrt{3}}{\pi} E_m \cos \alpha = \dfrac{3\sqrt{2}}{\pi} E_{line} \cos \alpha$

∴ $648 = \dfrac{3 \times \sqrt{2} \times 580}{\pi} \cos \alpha$, $\alpha = 34.18°$.

(b) Power delivered, $P_{dc} = 64.8 \times 3464 = 2244.67$ kW.

(c) AC terminal voltage,

$$P_{AC} = 1.05\, P_{dc} = 1.05 \times 2245 = 2357.25 \text{ kVA}$$

$$\therefore \quad Q_{(KVAR)} = \sqrt{(2357.25)^2 - (2245)^2} = 718.75$$

Example 6.35 A 3-phase fully controlled thyristor bridge converter is operated from an ac supply of 400 V_{rms} line to line. When the converter is operated in the rectifier mode at a control angle $\alpha = 30°$, the overlap angle (r) due to the line reactance is 15°. Calculate the reduction in dc output voltage due to the overlap. If the converter operates in the inverter mode with $\alpha = 120°$ and without any change in the dc load current, what will be the overlap angle (r)?

Solution:

(a) Rectifier mode: $\alpha = 30°$, $r = 15°$, Reduction in output voltage = ?

For a P-pulse fully controlled rectifier with an overlap angle r, the average voltage:

$$E_{dc} = \frac{P}{2\pi} E_{max} \sin\frac{\pi}{p} [\cos\alpha + \cos(\alpha+r)], \; P = \text{number of pulses}$$

where $P = 6$, $E_{max} = \sqrt{\frac{2}{3}} \cdot 400$, $\therefore E_{dc} = \frac{6}{2\pi} 400 \sqrt{\frac{2}{3}} \cdot \sin\left(\frac{180}{6}\right) [\cos 30 + \cos(30+45)]$

$$= 245.3 \text{ V}$$

Now, average load voltage without overlap is given by

$$E_{dc} = \frac{P}{\pi} E_{max} \sin\frac{\pi}{P} \cos\alpha = \frac{6}{\pi} \times 400 \times \sqrt{\frac{2}{3}} \times 0.523 \times 0.866 = 282.51 \text{ V}$$

$\therefore$ Drop in dc output voltage = $282.51 - 245.3 = 37.21$

(b) Inverter mode:

$$\beta = \pi - \alpha = 180° - 120° = 60°$$

For the same load current, $E \times I_0 = 25 = E_{max} \sin\frac{\pi}{P} [\cos\beta + \cos(\beta - r)]$

$$= 400 \sqrt{\frac{2}{3}} \sin 30° [\cos 60° + \cos(60-r)] \text{ or } \frac{43.3}{200\sqrt{2}} = 2.5 + \cos(60-r)$$

or $0.5 + \cos(60 - r) = 0.153$, $\boxed{r = -50°}$

6.15 PERFORMANCE OF CONVERTER CIRCUITS WITH BATTERY LOAD (OR EFFECT OF LOAD INDUCTANCE)

In the previous-section, the operation of the converter-circuits have been discussed by assuming the load-currenzt constant. This assumption will be valid only if the value of the load inductance L is made very high. For converters used as regulated d.c. power supplies, an output filter is required to reduce the ripple in the direct-current and voltage of the load. Therefore, inductance L is made reasonably large to act as filter choke. For such applications, this assumption is valid. But, there are other applications of controlled-converters in which the output voltage is not filtered and only the rectified voltage is used (e.g. battery charges and speed controllers for d.c. motors). Under this condition, the load current will not be constant.

We have seen the voltage and current waveforms for single-phase fully-controlled bridge with purely resistive load, that is, L is zero (Fig. 6.20). It is observed from this Fig. 6.20, that the load current is discontinuous since the SCRs turn off by natural commutation at the end of every half-cycle.

The circuit diagram of a fullwave controlled bridge-rectifier with resistive and battery load (active load) is shown in Fig. 6.54. The voltage and current waveforms of this type of power circuit are shown in Fig. 6.55.

As shown in Fig. 6.55, the load current waveform is not only decided by the firing angle α, but also by the battery load E. Here, for the simplification purpose, source impedance is neglected.

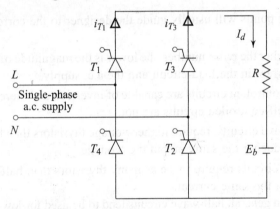

Fig. 6.54 *Single-phase fully-controlled bridge with resistive and battery load*

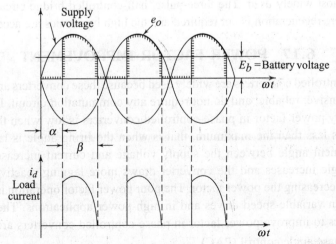

Fig. 6.55 *Voltage and current waveforms with resistive and battery load*

The average d.c. output voltage will be,

$$e_{dc} = \frac{1}{\pi} \int_{\alpha}^{\alpha+\beta} E_m \sin \omega t \, d(\omega t) = \frac{E_m}{\pi}[-\cos \omega t]_{\alpha}^{\alpha+\beta}$$

$$= \frac{E_m}{\pi}[\cos \alpha - \cos(\alpha + \beta)] \text{ where } \beta \text{ is known as the conduction angle.}$$

(6.75)

From Fig. 6.55, we can write

$$\alpha + \beta = \pi - \sin^{-1}\left(\frac{E_b}{E_m}\right) \quad (6.76)$$

6.16 SELECTION OF CONVERTER CIRCUITS

The following points will usually guide the designer to the correct choice of a circuit:

(1) The higher the pulse number, the lower is the magnitude of ripple voltage and current in the load-circuit and the a.c. supply.
(2) Fully-controlled circuits are capable of inversion and therefore regeneration, half-controlled circuits are not.
(3) Half-wave circuits, require higher-voltage thyristors than bridge-circuits for producing the same output d.c. voltage.
(4) Bridge circuits require twice as many thyristors than half-wave circuits to carry the same current.

Therefore, in general, half-wave circuits tend to be used for low-voltage, high-current use and where the minimum of thyristors are required. Bridge converters are the most widely used. The three-pulse, half-controlled bridge circuit is only used where regeneration is not required and the high harmonics are acceptable.

6.17 POWER FACTOR IMPROVEMENT

Phase-controlled converters are widely used because these converters are simple, less expensive, reliable, and do not require any commutation circuit. However, the supply power factor in phase-controlled converters is low when the output voltage is less than the maximum, that is when the firing angle is large. The displacement angle between the supply voltage and current increases as the firing angle increases and the converter draws more lagging reactive power, thereby decreasing the power factor. The poor power factor operation is a major concern in variable-speed drives and in high-power applications. The various techniques to improve power factor in phase-controlled converters are:

1. Phase angle control (PAC)

2. Semiconverter operation of full converters
3. Asymmetrical firing.

In forced-commutated converters, each thyristor is provided with its own commutation circuitry and therefore, can be commutated at any desired instant. The various control schemes implementing forced commutation and which improves the power factor and other performance characteristics are:
1. Extinction angle control (EAC)
2. Symmetrical angle control (SAC)
3. Pulse-width modulation (PWM) control.

6.17.1 Phase Angle Control (PAC)

Various performance perameters, such as supply power factor, displacement factor and harmonic factor of phase-controlled converters, were discussed in previous sections. In both semiconverter and full converter systems, power factor and displacement factor decrease as the output voltage decreases (or as firing angle increases). The semiconverter provides some improvement in power factor, displacement factor and lower order harmonics.

6.17.2 Semiconverter Operation of Full Converters

A full converter system is commonly used where regeneration is required. However, a semiconverter is used where regeneration is not required. It is possible with some complexity in the control logic circuit to operate a full converter as a semi-converter in both rectifying and inverting modes in order to exploit the better performance characteristics of semiconverters. This is illustrated in Fig. 6.56(a).

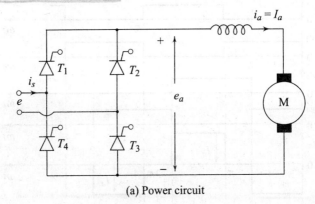

(a) Power circuit

Fig. 6.56(a)

In the rectification mode (Fig. 6.56(b)) SCRs T_3 and T_4 are triggering during the whole of the positive half-cycle and negative half-cycle, respectively, of the supply voltage. Thyristors T_3 and T_4, therefore, behave as diodes during the respective half-cycles. The firing angle α for SCRs T_1 and T_2 is varied to change the output voltage.

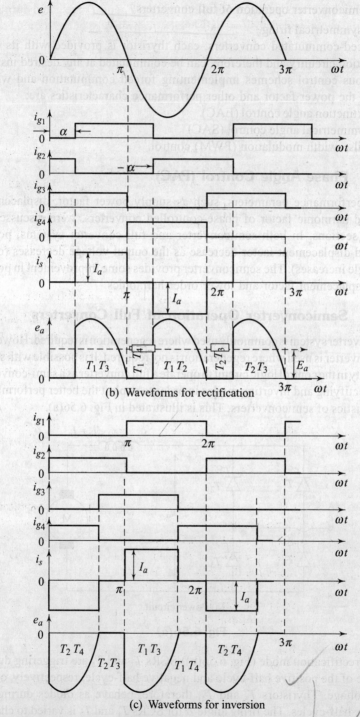

Fig. 6.56 *(b)* and *(c)* Semiconverter operation of a full converter

In the inversion mode (Fig. 6.56(c)), thyristors T_1 and T_2 are gated during the whole of the negative and positive half-cycles, respectively, of the supply voltage so that they behave as diodes during their respective half-cycles. The firing angle for SCRs T_3 and T_4 is varied to change the output voltage.

When two SCRs of the same leg (such as T_1 and T_4 or T_2 and T_3) conduct, the motor current freewheels through them producing zero output voltage. The nature of the motor terminal voltage e_a and the supply current i_s is the same as that in a semiconverter system. Therefore, performance is improved at the cost of complexity in the control logic circuit.

Another interpretation of the foregoing control scheme is as follows: In the rectification mode, the firing of the negative group of thyristors (T_3 and T_4) is kept at zero. Therefore, when the firing angle of the positive group of thyristors T_1 and T_2 is zero, the output voltage is maximum; when it is 180°, the output voltage is zero. In the inverting mode, the triggering of the positive group of thyristor is kept at 180° and when the firing angle for the negative group is also 180°, the output voltage has its maximum negative value. In practice, the firing angle must be limited to less than 180° in all cases to allow a margin for commutation. As the firing angle for T_3 and T_4 is reduced from 180°, the output voltage becomes less negative. This interpretation helps in understanding the application of this control scheme in three-phase full converters, where it is called sequential control. The firing angle for the positive group of SCRs is kept constant at zero and that for the negative group of thyristors is varied for the rectification mode. For the inversion mode, the firing angle for the positive group is varied, and that for the negative group is kept at 180°. This control scheme, although it improves power factor in both single-phase and three-phase converters, is not recommended for three-phase converters because of three major disadvantages:

(a) Even harmonic currents are present in the supply line current.
(b) Third harmonic ripple is present in the output.
(c) There is a danger of commutation failure.

6.17.3 Asymmetrical Firing

The improvement in power factor can be achieved by a technique known as asymmetrical firing, shown in Fig. 6.57(c). In this scheme thyristors are triggered at different angles, whereas in the symmetrical firing scheme (Fig. 6.57(b) both thyristors are triggered at the same firing angle. For example, if 0.5 p.u. average output voltage is required, SCRs are triggered at $\alpha = 90°$ in the symmetrical firing scheme. To obtain the same average output voltage in the asymmetrical firing scheme, if SCR T_1 is triggered at $\alpha_1 = 60°$, then SCR T_2 is triggered at $\alpha_2 = 120°$. Since one SCR is triggered at a lower value of the firing angle, the power factor is slightly improved in asymmetrical firing. However, asymmetrical firing produces several disadvantages. It generates d.c. and even harmonic currents in the supply line current.

The d.c. component can, however, be avoided if the firing angles are alternated in successive cycles. If sufficient inductance is not present in the motor armature

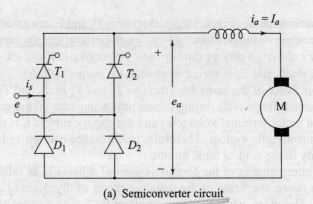

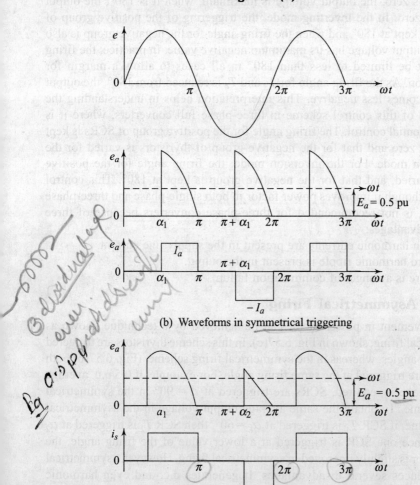

Fig. 6.57 *(b)* and *(c)* Symmetrical and asymmetrical triggering

circuit, asymmetrical firing makes the motor current very peaky and discontinuous. The disadvantages outweigh the advantages of minor improvement in power factor and therefore, asymmetrical firing is only of theoretical interest.

6.17.4 Extinction Angle Control (EAC)

Figure 6.58 shows the single-phase semiconverter, where thyristors T_1 and T_2 are replaced by switches S_1 and S_2. The switching actions of S_1 and S_2 can be either performed by SCRs or by GTOs. A dotted square of Fig. 6.58 represents an SCR(GTO) and its commutation circuitry. For the sake of simplicity, commutation circuits are not shown. Figure 6.59 shows the waveforms for the extinction angle control scheme.

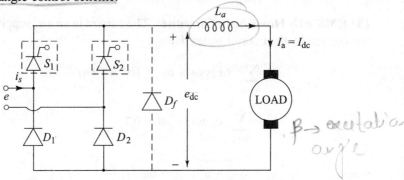

Fig. 6.58 *Power circuit employing forced commutation*

In this scheme, as shown in Fig. 6.59, switch S_1 is turned ON at $\omega t = 0$ and turned OFF by forced commutation at $\omega t = \beta$. Switch S_2 is turned ON at $\omega t = \pi$ and is turned OFF at $\omega t = (\pi + \beta)$. The output voltage is controlled by varying the extinction angle, β. The fundamental component i_1 of the supply current i_s,

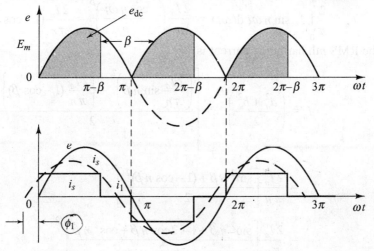

Fig. 6.59 *Extinction angle control scheme waveforms*

shown in Fig. 6.59 leads the supply voltage. The displacement factor is leading, and this feature may be desirable to compensate for line voltage drop.

(1) Average Output Voltage The average output voltage is

$$E_{dc} = \frac{1}{\pi} \int_0^\beta E_m \sin \omega t \, d(\omega t) = \frac{E_m}{\pi} \left[-\cos \omega t\right]_0^\beta = \frac{E_m}{\pi} (1 - \cos \beta) \quad (6.77)$$

(2) RMS Supply Current The RMS supply current I is

$$I = \left[\frac{1}{\pi} \int_0^\beta I_a^2 \, d(\omega t)\right]^{1/2} = I_a \sqrt{\beta/\pi} \quad (6.78)$$

(3) RMS nth Harmonic Current The current in an a.c. supply may be represented by the Fourier series that:

$$I = \sum_{n=1}^\infty (a_n \cos n\omega t + b_n \sin \omega t)$$

$$= \sum_{n=1}^\infty c_n \sin(n\omega t + \phi_n)$$

where $\quad c_n = \sqrt{a_n^2 + b_n^2}, \quad \phi_n = \tan^{-1}\left(\frac{a_n}{b_n}\right)$

The Fourier coefficients are:

$$a_n = \frac{2}{\pi} \int_0^\beta I_a \cos n\omega t \, d(\omega t) = \frac{2 I_a}{\pi}\left(\frac{\sin n\omega t}{n}\right)_0^\beta = \frac{2 I_a}{\pi n}(\sin n\beta)$$

$$b_n = \frac{2}{\pi} \int_0^\beta I_a \sin n\omega t \, d(\omega t) = \frac{2 I_a}{\pi}\left(-\frac{\cos n\omega t}{n}\right)_0^\beta = \frac{2 I_a}{\pi n}(1 - \cos n\beta)$$

The RMS nth harmonic current is

$$I_n = \left(\frac{a_n^2 + b_n^2}{2}\right)^{1/2} = \left[\frac{\left(\frac{2 I_a}{\pi n} \sin n\beta\right)^2 + \left\{\frac{2 I_a}{\pi n}(1 - \cos \beta)\right\}^2}{2}\right]^{1/2}$$

$$= \frac{2 I_a}{\pi n}\left[\frac{\sin^2 n\beta + (1 - \cos n\beta)^2}{2}\right]^{1/2}$$

$$= \frac{2 I_a}{\pi n}\left[\frac{\sin^2 n\beta + 1 - 2\cos n\beta + \cos^2 n\beta}{2}\right]^{1/2}$$

$$= \frac{2I_a}{\pi n}\left[\frac{1+1-2\cos n\beta}{2}\right]^{1/2} = \frac{2I_a}{\pi n}[1-\cos n\beta]^{1/2}$$

$$= \frac{2I_a}{\pi n}\left[1-\left(1-2\sin^2\frac{n\beta}{2}\right)\right]^{1/2}$$

$$= \frac{2I_a}{\pi n}\left[1-1+2\sin^2\frac{n\beta}{2}\right]^{1/2} = \frac{2\sqrt{2}\,I_a}{\pi n}\left[\sin^2\frac{n\beta}{2}\right]^{1/2}$$

$$= \frac{2\sqrt{2}\,I_a}{\pi n}\sin\frac{n\beta}{2} \quad \text{where } n \text{ is odd} \quad (6.79)$$

(4) Displacement Angle The displacement angle of the nth harmonic is

$$\phi_n = \tan^{-1}\frac{a_n}{b_n} = \tan^{-1}\left[\frac{2I_a \sin n\beta/\pi n}{2I_a(1-\cos n\beta)/\pi n}\right]$$

$$= \tan^{-1}\left[\frac{\sin n\beta}{1-\cos n\beta}\right] = \tan^{-1}\left[\frac{2\sin n\beta/2 \cos n\beta/2}{1-(1-2\sin^2 n\beta/2)}\right]$$

$$= \tan^{-1}\left[\frac{\cos n\beta/2}{\sin n\beta/2}\right] = \tan^{-1}\left[\frac{\sin(n\pi/2 - n\beta/2)}{\cos(n\pi/2 - n\beta/2)}\right]$$

$$= \tan^{-1}[\tan(n\pi/2 - n\beta/2)]$$
$$= n\pi/2 - n\beta/2$$

$$= n\left(\frac{\pi}{2}-\frac{\beta}{2}\right) \quad \text{where } n \text{ is odd} \quad (6.80)$$

(5) Supply Power Factor The supply power factor is

$$P_F = \left(\frac{I_1}{I}\right)\cos\phi_1 = \frac{2\sqrt{2}\,I_a \sin\beta/2 \big/ \pi}{I_a\sqrt{\beta/\pi}}\cos\left(\frac{\pi}{2}-\frac{\beta}{2}\right)$$

$$= \frac{2\sqrt{2}\,\sin\beta/2}{\sqrt{\pi}\sqrt{\beta}}\sin\beta/2 = \frac{2\sqrt{2}\,\sin^2\beta/2}{\sqrt{\pi\beta}}$$

$$= \frac{2\sqrt{2}}{\sqrt{\pi\beta}}(1-\cos^2\beta/2) = \frac{2\sqrt{2}}{\sqrt{\pi\beta}}\left[1-\left(\frac{1+\cos\beta}{2}\right)\right]$$

$$= \frac{2\sqrt{2}}{\sqrt{\pi\beta}}\left(\frac{2-1-\cos\beta}{2}\right) = \frac{\sqrt{2}(1-\cos\beta)}{\sqrt{\pi\beta}} \quad (6.81)$$

(6) Displacement Factor The displacement factor is

$$D_F = \cos \phi_1 = \cos\left(\frac{\pi}{2} - \frac{\beta}{2}\right) = \sin \beta/2, \text{ leading} \quad (6.82)$$

(7) Harmonic-Factor The harmonic factor is

$$H_F = \frac{(I^2 - I_1^2)^{1/2}}{I_1} = \frac{\left(I_a^2 \beta/\pi - 8 I_a^2 \sin^2 \beta/2 \big/ \pi^2\right)^{1/2}}{2\sqrt{2}\, I_a \sin \beta/2 \big/ \pi}$$

$$= \left[\frac{(\pi I_a^2 \beta - 8 I_a^2 \sin^2 \beta/2)\big/\pi^2}{8 I_a^2 \sin^2 \beta/2 \big/ \pi^2}\right]^{1/2} = \left[\frac{\pi \beta - 8 \sin^2 \beta/2}{8 \sin^2 \beta/2}\right]^{1/2}$$

$$= \left[\frac{\pi \beta}{8 \sin^2 \beta/2} - 1\right]^{1/2} \quad \text{But } \sin^2 \beta/2 = \left(\frac{1 - \cos \beta}{2}\right)$$

$$= \left[\frac{\pi \beta}{4(1 - \cos \beta)} - 1\right]^{1/2} \quad (6.83)$$

In a semiconverter system, performance of the converter under extinction angle control is similar to that of phase-angle control, with the exception that the displacement factor is leading in extinction angle control, whereas it is lagging in phase-angle control.

6.17.5 Symmetrical Angle Control (SAC)

Figure 6.60 shows the voltage and current waveforms for symmetrical angle control scheme. As shown in Fig. 6.60, switch S_1 is triggered at angle α and turned off at $(\pi - \alpha)$.

Expressions for the various performance parameters are as follows:

(1) Average Output Voltage The average output voltage is given by

$$E_{dc} = \frac{1}{\pi} \int_{\alpha}^{\pi - \alpha} E_m \sin \omega t \, d(\omega t) = \frac{E_m}{\pi} \left[-\cos \omega t\right]_{\alpha}^{\pi - \alpha}$$

$$= \frac{E_m}{\pi} [\cos \alpha - \cos(\pi - \alpha)] = \frac{2 E_m}{\pi} \cos \alpha \quad (6.84)$$

(2) RMS Supply Current The RMS supply current is given by

$$I = \left[\frac{1}{\pi} \int_{\alpha}^{\pi - \alpha} I_a^2 \, d(\omega t)\right]^{1/2} = \left[\frac{I_a^2}{\pi}(\pi - \alpha - \alpha)\right]^{1/2} = I_a \left(1 - \frac{2\alpha}{\pi}\right)^{1/2} \quad (6.85)$$

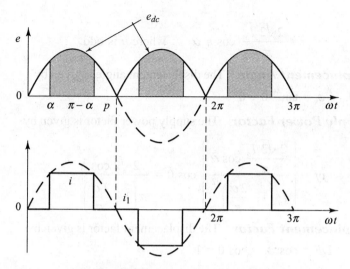

Fig. 6.60 Symmetrical angle control

(3) RMS nth Harmonic Current

The Fourier coefficients are:

$$a_n = \frac{2}{\pi} \int_{\alpha}^{\pi-\alpha} I_a \cos n\omega t \, d(\omega t) = \frac{2 I_a}{\pi} \left(\frac{\sin n\omega t}{n} \right)_{\alpha}^{\pi-\alpha}$$

$$= \frac{2 I_a}{\pi n} [\sin n(\pi - \alpha) - \sin n\alpha]$$

$$= \frac{2 I_a}{\pi n} [\sin n\alpha - \sin n\alpha] = 0$$

$$b_n = \frac{2}{\pi} \int_{\alpha}^{\pi-\alpha} I_a \sin n\omega t \, d(\omega t) = \frac{2 I_a}{\pi} \left(-\frac{\cos n\omega t}{n} \right)_{\alpha}^{\pi-\alpha}$$

$$= \frac{2 I_a}{\pi n} [\cos n\alpha - \cos n(\pi - \alpha)]$$

$$= \frac{2 I_a}{\pi n} [\cos n\alpha + \cos n\alpha] = \frac{4 I_a}{\pi n} \cos n\alpha$$

The RMS nth harmonic current is

$$I_n = \left(\frac{a_n^2 + b_n^2}{2} \right)^{1/2} = \left(\frac{\dfrac{16 I_a^2}{\pi^2 n^2} \cos^2 n\alpha}{2} \right)^{1/2}$$

$$= \frac{2\sqrt{2}\, I_a}{\pi n} \cos n\alpha \quad \text{where } n \text{ is odd.} \quad (6.86)$$

(4) Displacement Angle The displacement angle is given by

$$\phi_n = \tan^{-1} 0 = 0° \quad (6.87)$$

(5) Supply Power Factor The supply power factor is given by

$$pf = \left(\frac{\dfrac{2\sqrt{2}\, I_a}{\pi} \cos \alpha}{I_a \left(1 - \dfrac{2\alpha}{\pi}\right)^{1/2}} \right) \cos 0 = \frac{2\sqrt{2} \cos \alpha}{\pi \left(1 - \dfrac{2\alpha}{\pi}\right)^{1/2}} \quad (6.88)$$

(6) Displacement Factor The displacement factor is given by

$$DF = \cos \phi_1 = \cos 0 = 1 \quad (6.89)$$

(7) Harmonic Factor The harmonic factor is given by

$$HF = \frac{\left[I_a^2 \left(1 - \dfrac{2\alpha}{\pi}\right) - 8 I_a^2 \cos^2 \alpha / \pi^2 \right]^{1/2}}{2\sqrt{2}\, I_a \cos \alpha / \pi} \equiv \frac{\left[\left(1 - \dfrac{2\alpha}{\pi}\right) - 8\cos^2 \alpha / \pi^2 \right]^{1/2}}{8 \cos^2 \alpha / \pi^2}$$

$$\equiv \left[\frac{\left(1 - \dfrac{2\alpha}{\pi}\right)}{8 \cos^2 \alpha / \pi^2} - \frac{8 \cos^2 \alpha / \pi^2}{8 \cos^2 \alpha / \pi^2} \right]^{1/2} \equiv \left[\frac{\pi(\pi - 2\alpha)}{8 \cos^2 \alpha} - 1 \right]^{1/2} \quad (6.90)$$

In this control scheme, the supply current pulse is placed symmetrically with respect to the supply voltage peak and therefore, the fundamental current i is in phase with the supply voltage. This makes the displacement factor unity and improves the power factor.

6.17.6 Pulse-Width Modulation (PWM) Control

In phase angle control, extinction angle control, and symmetrical angle control schemes, the supply current consists of one pulse per half-cycle and the lowest order harmonic is the third. It is very difficult to filter out the lowest order harmonic current. The lowest order harmonics can be eliminated and/or reduced if the supply current has more than one pulse per half-cycle. The pulse-width modulation scheme is shown in Fig. 6.61. In this scheme, as the switch is turned ON and OFF several times during each half-cycle, the width of the pulses is varied to change the output voltage.

Lowest order harmonics can be eliminated or reduced by selecting the type of modulation for the pulse widths and the number of pulses per half-cycle. Higher order harmonics may increase, but these are of concern because they can be eliminated easily by filters.

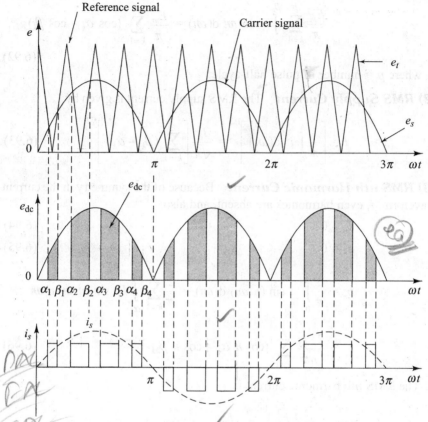

Fig. 6.61 *Sinusoidal pulse-width modulation control*

Figure 6.61 shows the sinusoidal pulse-width control technique. In this scheme, as shown, firing signals for the thyristors are obtained by comparing a triangular voltage e_t with a rectified sinusoidal voltage e_s, which is in phase with the supply voltage e. The output voltage e_a is varied by changing the amplitude of e_s or the modulation index m. The modulation index is defined as

$$E_m = \frac{\text{amplitude of } e_s}{\text{amplitude of } e_t} \qquad (6.91)$$

To determine the expressions for the output voltage and the various performance parameters, the instants of turn-on (i.e. α'_s) and turn-off (i.e. β'_s) are obtained using Newton's method to solve for the intersecting points between the signals e_s and e_t. Once the values of α'_s and β'_s are known, the expressions for the performance parameters can be easily obtained.

(1) Average Output Voltage The average output voltage is given by

$$E_{dc} = \frac{1}{\pi} \int_0^\pi E_m \sin \omega t \, d(\omega t)$$

$$= \frac{E_m}{\pi} \sum_{k=1}^{P} \int_{\alpha_k}^{\beta_k} \sin \omega t \, d(\omega t) = \frac{E_m}{\pi} \sum_{k=1}^{P} (\cos \alpha_k - \cos \beta_k)$$

(6.92)

where p = number of pulse half-cycle.

(2) RMS Supply Current The RMS supply current is given by

$$I = \left[\frac{1}{\pi} \int_0^\pi i^2 \, d(\omega t) \right]^{1/2} = \frac{I_a}{\sqrt{\pi}} \left[\sum_{k=1}^{P} (\beta_k - \alpha_k) \right]^{1/2}$$

(6.93)

(3) RMS nth Harmonic Current Because of the symmetry in the current waveform, i, even harmonics are absent, and also:

$$I_o = 0$$ (6.94)

$$a_n = 0$$ (6.95)

$$b_n = \frac{2}{\pi} \int_0^\alpha i \sin(n \omega t) \, d(\omega t) = \frac{2}{\pi} \sum_{k=1}^{P} \int_{\alpha_k}^{\beta_k} I_a \cdot \sin(n \omega t) \, d\omega t$$

$$= \frac{2 \cdot I_a}{n \pi} \sum_{k=1}^{P} [\cos n \alpha_k - \cos n \beta_k]$$

(6.96)

The RMS nth harmonic current is

$$\therefore \quad I_n = \left(\frac{a_n^2 + b_n^2}{2} \right)^{1/2} = \frac{b_n}{\sqrt{2}}$$

$$= \frac{\sqrt{2} \cdot I_a}{\pi} \sum_{k=1}^{P} [\cos n \alpha_k - \cos n \beta_k]$$

(6.97)

$$\therefore \quad I_{s(\text{fund})} = \frac{b_1}{\sqrt{2}} = \frac{\sqrt{2} \, I_a}{\pi} \sum_{k=1}^{P} [\cos \alpha_k - \cos \beta_k]$$

(4) Displacement Angle

$$\phi_n = 0°$$ (6.98)

(5) Supply Power Factor The supply power factor is

$$\text{PF} = \frac{I_1}{I} \cos \phi_1 = \frac{I_1}{I}$$

(6.99)

(6) Displacement Factor The displacement factor is given by

$$\text{DF} = \cos \phi_1 = \cos 0 = 1$$ (6.100)

(7) Harmonic Factor
The harmonic factor is given by

$$\text{HF} = \left[\frac{I^2 - I_1^2}{I_1^2}\right]^{1/2} \quad (6.101)$$

In a sinusoidal PWM control, the displacement factor is unity and the power factor is improved. The lowest order harmonic is the fifth for four pulses per half-cycle, and the seventh for six pulses per half-cycle. Therefore, lower-order harmonics that are difficult to filter out are eliminated or reduced by selecting the number of pulses per half-cycle. Note that sinusoidal modulation is maintained as long as the modulation index m is limited to less than unity.

Various techniques for improving the supply power factor of phase-controlled drives have been discussed in the preceding sections. Information about the total harmonic control is important only if input filters are not used. Current with high harmonic content may distort the supply voltage. In some control schemes, the harmonic factor is high in the low-speed region. This is due to high contents of higher order harmonics, which are easily filtered out if input filters are used. The important current harmonics that the filter designer needs to consider are those of the lowest order. In this respect, the PWM control scheme has an advantage because by a proper choice of the number of pulses per half-cycle, the lowest order harmonics can be eliminated.

An input filter can eliminate most of the harmonic currents from the line, thereby making the line current essentially sinusoidal. A higher number of pulses per half-cycle increases the ripple frequency of the motor current. The armature circuit inductance may be sufficient to smooth out the motor current, and additional inductance may not be necessary in the armature circuit. The number of pulses per half-cycle should not be too large. If it is, the switching loss of the thyristors increases and special, costly thyristors having low turn-off time are required. Six pulses per half-cycle appear to be a good choice, in which case harmonic current below the seventh are eliminated. The PWM control scheme is receiving serious consideration in single-phase traction systems.

Power factor and displacement factor improve in forced commutation schemes. E_{AC} and P_{AC} schemes provides similar characteristics. However, E_{AC} makes power factor leading, which may be desirable in supply systems that feed other large inductive loads.

SOLVED EXAMPLE

Example 6.36 A single-phase semiconverter is operated from 230 V/50 Hz supply and is driving a dc motor load. Determine the following if the armature current is continuous and ripple free and equal to 10 A.
  (i) RMS Supply Current (I_{rms})
  (ii) Fundamental Supply Current $I_{s(\text{fund})}$
  (iv) Distortion Factor (DF)
  (v) Harmonic Factor (HF)

(iii) Displacement Factor (DPF) (vi) Supply Power Factor (PF)
to get an average output voltage of 175 V, if the converter is operated in
(a) Extinction Angle Control (EAC), also find β.
(b) Symmetrical Angle Control (SAC), also find α.
(c) PWM control with 3-pulses per half cycle and
$\alpha_1 = 33°$, $\beta_1 = 62°$, $\alpha_2 = 78°$, also obtain E_{dc}.

Solution:
(a) *EAC control:*

Average O/P voltage $= E_{dc} = \dfrac{E_m}{\pi}(1 - \cos\beta)$, $175 = \dfrac{\sqrt{2} \times 230}{\pi}(1 - \cos\beta)$

$\therefore \qquad \beta = 133.65°$

(i) RMS Supply Current $I_{srms} = I_a\sqrt{\beta/\pi} = 10\sqrt{\dfrac{133.65}{180}} = 8.62$ Amp

(ii) $I_{S(fund)} = \dfrac{2\sqrt{2}\, I_a}{\pi}\sin\dfrac{\beta}{2} = \dfrac{2\sqrt{2}\,(10)}{\pi}\sin\left(\dfrac{133.65}{2}\right) = 8.28$ Amp

(iii) Displacement factor, DPF $= \sin\beta/2 = \sin\dfrac{133.65}{2} = 0.92$ (lagging)

(iv) DF $= \dfrac{I_{s(fund)}}{I_{s(rms)}} = \dfrac{8.28}{8.62} = 0.96 = 96\%$

(v) HF $= \sqrt{\dfrac{1}{DF^2} - 1} = 29.4\%$

(vi) PF $= $ DF $\times$ DPF $= 0.96 \times 0.92 = 0.8832$ (lagging)

(b) *SAC control:*

$E_{dc} = \dfrac{2E_m}{\pi}\cos(\alpha)$, $\therefore \; 175 = \dfrac{2\sqrt{2}\,(230)}{\pi}\cos(\alpha) \; \therefore \; \alpha = 32.31°$

(i) $I_{s(rms)} = I_a\left(1 - \dfrac{2\alpha}{\pi}\right)^{1/2} = 10\left(1 - \dfrac{2(32.31)}{\pi}\right)^{1/2} = 8$ Amp

(ii) $I_{s(fund)} = \dfrac{2\sqrt{2}\,I_a}{\pi}\cos\alpha = \dfrac{2\sqrt{2}\times 10}{\pi}\cos(32.31) = 7.6$ Amp

(iii) DPF $= $ Unity (1)

(iv) DF $= \dfrac{I_{s(fund)}}{I_{s(rms)}} = \dfrac{7.6}{8} = 0.95 = 95\%$

(v) HF $= \sqrt{\dfrac{1}{DF^2} - 1} = \sqrt{\dfrac{1}{(0.95)^2} - 1} = 0.328 = 32.8\%$

(vi) PF $= $ DF $\times$ DPF $= 0.95$

(c) *PWM control:*

$\alpha_1 = 33°$, $\beta_1 = 62°$, $\alpha_2 = 78°$, $\beta_3 = 180 - \alpha_1$, $\alpha_1 = 180 - \beta_1$,
$\beta_2 = 180 - \alpha_2$, $\beta_3 = 147$, $\alpha_3 = 118°$, $\beta_2 = 102°$

$$E_{dc} = \frac{E_m}{\pi}\sum_{k=1}^{p}(\cos\alpha_k - \cos\beta_k) = \frac{E_m}{\pi}\begin{bmatrix}\cos(\alpha_1) - \cos(\beta_1) + \cos(\alpha_2) \\ -\cos(\beta_2) + \cos(\alpha_3) - \cos(\beta_3)\end{bmatrix}$$

$$= \frac{\sqrt{2}\,(230)}{\pi}\begin{bmatrix}\cos(33) - \cos(62) + \cos(78) \\ -\cos(102) + \cos(118) - \cos(147)\end{bmatrix} = 119.5\text{ V}$$

(i) $I_{s(\text{rms})} = \frac{I_a}{\sqrt{\pi}}\left(\sum_{k=1}^{p}\beta_k - \alpha_k\right)^{1/2} = \frac{10}{\sqrt{180}}\left(\beta_1 - \alpha_1 + \beta_2 - \alpha_2 + \beta_3 - \alpha_3\right)^{1/2}$

$$= \frac{10}{\sqrt{180}}(62 - 33 + 102 - 78 + 147 - 118)^{1/2} = 6.75\text{ Amp}$$

(ii) $I_{s(\text{fund})} = \frac{\sqrt{2}\cdot I_a}{\pi}\sum_{k=1}^{p}(\cos\alpha_k - \cos\beta_k)$

$$= \frac{\sqrt{2}}{\pi}\times 10\begin{bmatrix}\cos 33 - \cos(62) + \cos(78) \\ -\cos(102) + \cos(118) - \cos(147)\end{bmatrix} = 5.2\text{ Amp}$$

(iii) DPF $= \cos\phi_1 = 1$ (unity)

(iv) DF $= \dfrac{I_{s(\text{fund})}}{I_{s(\text{rms})}} = \dfrac{5.2}{6.75} = 77\%$

(v) HF $= \sqrt{\dfrac{1}{DF^2} - 1} = 82.76\%$

(vi) PF $=$ DF $\times$ DPF $= 0.77$

6.18 MICROPROCESSOR-BASED FIRING SCHEME FOR THREE-PHASE FULLY-CONTROLLED BRIDGE CONVERTER

The six-pulse bridge converter is one of the most widely used types of solid-state converters. In recent years, the microprocessor technique has been applied to general d.c. motor control systems for high performance. The microprocessor control of thyristor converter has the advantage of less hardware, system flexibility, low-maintenance cost, improved performance and control capability.

Phase control of thyristor converter has the following three basic operations:
1. Line synchronization

2. Delay control
3. Pulse distribution to the thyristors sequentially.

These functions are implemented using a microprocessor to the converter bridge, shown in Fig. 6.62. The main purpose of the firing circuit is to convert the angle control input into a corresponding phase-angle between the triggering pulses and the a.c. supply voltage.

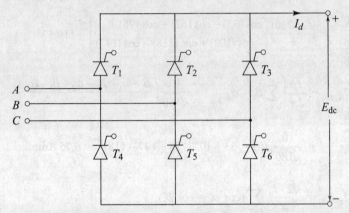

Fig. 6.62 *Three-phase bridge-converter*

6.18.1 Line Synchronization

Since the firing angle is referred to the a.c. supply voltage, the control must be synchronized to the a.c. supply. Here synchronization can be accomplished for every one-sixth of the supply period (60°) and a fast response is achieved. The interrupt handling capability of the microprocessor provides an effective means to synchronize with the line-voltages. It detects the zero crossings and generates corresponding interrupt signals in order to call necessary service routines. By this approach, the interface between the microprocessor and the supply line is reduced to a simple zero crossing detector. In addition, these detectors also provide information regarding the status of the phases of the supply voltage. This information called line status is necessary for the distribution of triggering pulses.

The a.c. supply is given as the input signals for the zero crossing detectors through a delta–star connected step-down transformer as shown in Fig. 6.63(a). The output of the transformer will be lagging 30° with reference to the a.c. supply voltage so the output of the zero-crossing detector will be 30° lagging with supply voltage as shown in Fig. 6.63(b). The output of the ZCD will be in high level for the positive cycle of the input and in zero level for the negative cycle of the input.

The zero crossing output of the A and B phases are given as input to an exclusive-OR gate and its output will be a square-wave with ON-period 120° and OFF-period 60° as shown in Fig. 6.63(c). This wave and the zero crossing output of the phase C is given as input to another exclusive-OR gate and its output, a square-wave of width 60°, as shown in Fig. 6.63(d). Using an edge detecting circuit, the edges, both rising and falling, are detected. Figures 6.63(e) and 6.63(f) show the output of the monostable, multivibrators triggered at the rising and falling edges, respectively. Figure 6.63(g) shows the synchronization signal IRQ_1, generated by OR-ing the monostable output using an OR-gate. Thus, a synchronization signal IRQ_1, is obtained at every 60° of the supply voltages.

6.18.2 Delay Control

The relation between the firing angle and the control input may be linear or inverse cosine, depending on the application. The most commonly used method

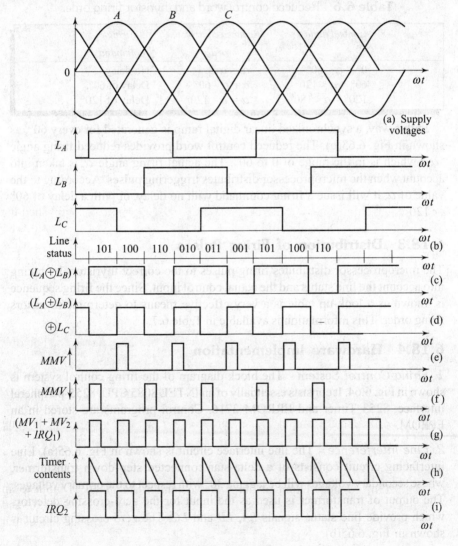

Fig. 6.63 *Waveforms*

to obtain controllable delay is the ramp technique since it can be readily complemented using analog or digital circuits.

In this approach, a ramp synchronized to the supply voltage, is generated and compared to the control input. Firing angle is determined by the instant when the ramp attains the value of the control input.

In this scheme, a controllable delay is provided by a programmable timer, operation of which is under the control of the microprocessor. Every one-sixth of the supply period, on the arrival of the interrupt signal $IRQ1$, the microprocessor initializes the timer with a reduced control word. This reduced control word is calculated from the angle-control input according to Table 6.6.

Table 6.6 Reduced control word and thyristor firing order

Control input	Reduced control word	Firing command
$0° < \alpha < 60°$	$\alpha_t = \alpha$	No-delay.
$60° < \alpha < 120°$	$\alpha_t = \alpha - 60°$	Delay of $60°$.
$120° < \alpha < 180°$	$\alpha_t = \alpha - 120°$	Delay of $120°$.

In this way, a synchronized linear digital ramp is generated for every $60°$, as shown in Fig. 6.63(b). The reduced control word provides reduced firing angle 'α_R' which is in the range of 0 to $60°$. The actual firing angle α_t is taken into account when the microprocessor distributes triggering pulses. According to the value of α, it will issue a firing command with no delay, or with a delay of $60°$ or $120°$.

6.18.3 Distribution of Firing Pulses

The microprocessor distributes firing pulses to the correct thyristors by taking into account the line status and the actual control input. Since the firing sequence is known as a look-up table is a very effective means to determine thyristors firing order. This information is available in Table 6.7.

6.18.4 Hardware Implementation

1. Firing Control System The block diagram of the firing control system is shown in Fig. 6.64. It consists essentially of an INTEL 8085 CPU, 8255-peripheral interface, 8253 Timer and ERROM 2716. Control programs are stored in an EPROM.

2. Line Interference The line interface circuit is shown in Fig. 6.65(a). Line interfacing circuit consists of a delta-star connected step-down transformer, whose secondary voltages will be lagging $30°$ with respect to the primary voltages. The output of transformer is used as the input for the zero-crossing detectors which provide line status signals *LA*, *LB* and *LC*. The zero-crossing circuit is shown in Fig. 6.65(b).

Table 6.7 Firing command table

Control input	Line status			Firing command					
	A	B	C	T_1	T_6	T_2	T_4	T_3	T_5
(1) $0 < \alpha < 60°$									
	0	0	0	0	0	0	0	1	1
	0	1	0	0	0	1	1	0	0
	0	1	1	0	0	0	1	1	0
	1	0	0	1	1	0	0	0	0
	1	0	1	1	0	0	0	0	1
	1	1	0	0	1	1	0	0	0
(2) $60° < \alpha < 120°$	0	0	1	0	0	0	1	1	0
	0	1	0	0	1	1	0	0	0
	0	1	1	0	0	1	1	0	0
	0	0	0	1	0	0	0	0	1
	1	0	1	0	0	0	0	1	1
	1	1	0	1	1	0	0	0	0
(3) $120° < \alpha < 180°$									
	0	0	1	0	0	1	1	0	0
	0	1	0	1	1	0	0	0	0
	0	1	1	0	1	1	0	0	0
	1	0	0	0	0	0	0	1	1
	1	0	1	0	0	0	1	1	0
	1	1	0	1	0	0	0	0	1

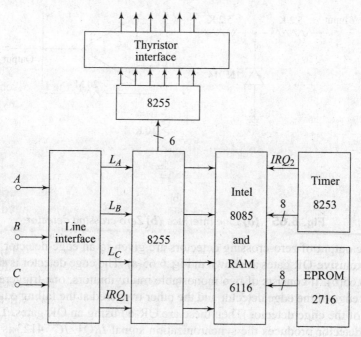

Fig. 6.64 *Firing control system*

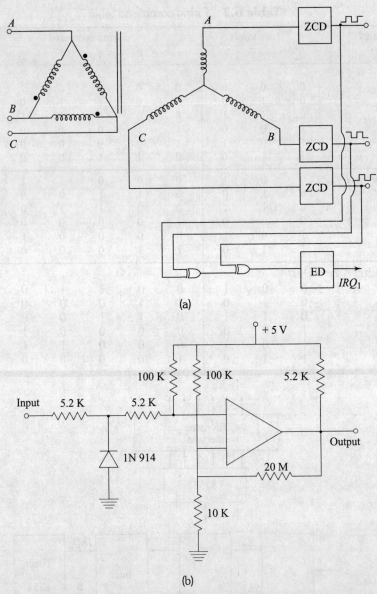

Fig. 6.65 *(a) Line interface (b) Zero crossing detector*

The output of zero-crossing detectors are given to an edge-detector through two exclusive-OR gates as shown in Fig. 6.65(a). The edge detector is shown in Fig. 6.65(b). It consists of two monostable multivibrators, one triggered at the rising edge of the edge detector and the other triggered at the falling edge of the input of the edge detector. Then these are OR-ed using an OR gates. Thus, the edge detector produces the synchronization signal *IRQ*1. *IC* 74123 is used as the monostable-multivibrator and its configuration and pin details are given in Fig. 6.65(d).

3. Driver Stage Figure 6.66 shows the interface between one thyristor and the microprocessor. Similarly all other thyristors are connected with the processor. The power level of the output signal from the microprocessor is not sufficient to trigger the thyristor. Hence, the power-level is raised using a Darlington-pair, as shown in Fig. 6.66. To isolate the control circuitry from the power circuitry, amplified gate-signal is given to the thyristor gate through a pulse-transformer. The gate-pulse width is reduced to 100 ms in the software itself to avoid the saturation of transformer. The diode connected in the primary side of the pulse-transformer acts as the freewheeling diode. The resistor connected across the secondary terminals protect the thyristor from getting triggered due to spurious pulses.

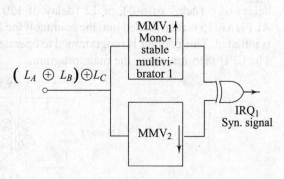

Fig. 6.65 *(c)* Edge detector

R_D	Input		Output	
$\bar{A}$	$\bar{A}$	B	Q	$\bar{Q}$
L	X	X	L	H
X	H	X	L	H
X	X	L	L	H
H	L	↑	⊓	⊔
H	↓	H	⊓	⊔
↑	L	H	⊓	⊔

Pin configuration:
- $\bar{A}_1$ — 1 — V_{CC}
- B_1 — 2 — Rextr.1/Cext.1
- R_{D1} — 3 — Cext.1
- $\bar{Q}_1$ — 4 — Q_1
- Q_2 — 5 — $\bar{Q}_2$
- Cextr2 — 6 — R_{D2}
- Rextr2 — 7 — B_2
- GND — 8 — $\bar{A}_2$

Fig. 6.65 *(d)* IC 74123

6.18.5 Software Implementation

The firing control software consists of two subroutines which are executed every one-sixth of the supply period. Timing diagram and control subroutine flow charts are shown in Figs 6.67(a) and (b), respectively. Subroutine SUB_1 is called and executed on reception of a synchronization pulse IRQ_1. First, the CPU reads the line status "LSTAT" and the control-input "ALPHA." It then calculates the reduced control word ALPHAR. According to Table 6.6, firing delay 'FDEL' is determined. Depending on the value of ALPHA, FDEL is 0 (no delay) or 6 (delay of 60°), or 12 (delay of 120°). The reduced control word ALPHAR 15 is then loaded into the counter if the 8253 timer and down-counting is initiated. This counter is programmed to operate in the single shot timer mode. The CPU then returns to the main program.

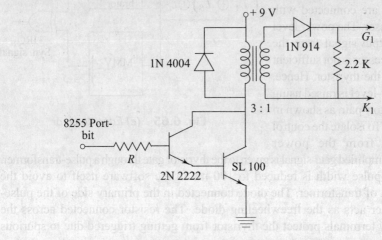

Fig. 6.66 *Driver interface*

When the counter has reached the final state (time out), an interrupt signal IRQ_2 is issued to the CPU. Subroutine SUB_2 is then called and executed. First, the CPU determines the firing-command from firing command look-up Table 6.7, taking into account the line status LSTAT and firing delay FDEL (from SUB_1). Then it sends the firing command to the thyristor gates through the 8255 peripheral interface and returns to the main program.

In the proposed firing control scheme, the control input is read and processed at every zero crossing of supply voltages and conse-quently changes in the control input is taken into account at the next zero crossing. The new firing angle is effectively obtained within one-sixth of the supply period. Hence the res-ponse time is always less than one-sixth of the supply period, i.e. 3.33 ms for 50 Hz supply.

REVIEW QUESTIONS

6.1 Explain with the help of neat power-diagram and associated waveforms, the operation of a single-phase half-wave controlled converters with
   (i) Resistive load (ii) Inductive load.
6.2 Derive an expressions for the
   (i) average load voltage (iii) RMS load voltage,
   (ii) average load current
   for single-phase half-controlled converter with resistive load and inductive load.

Phase Controlled Converters

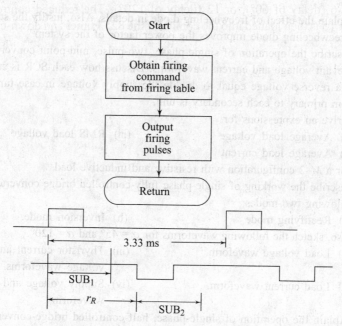

Fig. 6.67 *(a) Firing control software timing diagram*

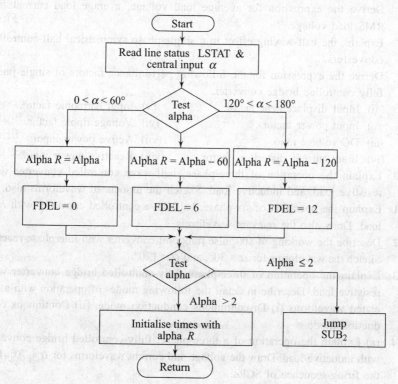

Fig. 6.67 *(b) Firing control subroutine flow chart*

6.3 Explain the effect of freewheeling diode in details. Also, justify the statement "Freewheeling diode improves the power factor of the system."

6.4 Describe the operation of single-phase, two-pulse, mid-point converter with relevant voltage and current waveforms. Discuss how each SCR is subjected to a reverse voltage equal to double the supply voltage in case turns ratio from primary to each secondary is unit.

6.5 Derive an expressions for
   (i) Average load voltage
   (ii) Average load current
   (iii) RMS load voltage
   for a M–2 configuration with resistive and inductive load.

6.6 Describe the working of single-phase fully-controlled bridge converter in the following two-modes:
   (a) Rectifying mode
   (b) Inversion mode.
   Also, sketch the following waveforms for $\alpha = 45°$ and $\alpha = 120°$.
   (i) Load voltage waveform.
   (ii) Load current waveform.
   (iii) Thyristor current and voltage waveforms.
   (iv) Supply voltage and current waveforms.

6.7 Explain the operation of single-phase, half-controlled bridge-converter with resistive load and inductive load with the associated waveforms.
Derive the expression for average load voltage, average load current and RMS load voltage.

6.8 Explain, the half-waving effect in a single-phase symmetrical half-controlled converters.

6.9 Derive the expression for the following performance factors of single-phase fully controlled bridge converter.
   (i) Input displacement factor.
   (ii) Input power factor.
   (iii) DC voltage ratio.
   (iv) Input-current distortion factor.
   (v) Input harmonic factor.
   (vi) Voltage ripple factor.
   (vii) Active power input.
   (viii) Reactive power input.

6.10 Explain the operation of three-phase, half-wave controlled converter with resistive load, and inductive load. Sketch the associated waveforms also.

6.11 Explain the operation of six-phase, half-wave controlled converter with R–L load. Draw also the relevant waveforms.

6.12 Describe the working of six-pulse midpoint converter with interphase reactor. Sketch the waveforms for $\alpha = 30°$, and $\alpha = 120°$.

6.13 Explain the operation of three-phase fully-controlled bridge converter with resistive load. Describe in detail the following modes of operation with associated waveforms (i) Discontinuous conduction mode. (ii) Continuous conduction mode.

6.14 (a) Explain the operation of a three-phase, fully-controlled bridge converter with inductive load. Draw the voltage and current waveforms for $\alpha = 70°$. List the firing-sequence of SCRs.
(b) Derive the expression for average load voltage.

6.15 Explain the operation of a three-phase, half-controlled bridge converter with associated waveforms.

6.16 Derive an expressions for all external performance measures of six-pulse-converters.

6.17 Explain the effect of source inductance on the performance of a three-phase, fully-controlled bridge converter.

6.18 Derive an expression for output voltage of a three-phase, fully-controlled bridge converter by conducting the following factors: (i) overlap-angle, (ii) source-inductance.

6.19 For a 3-phase thyristor-controlled half-wave-rectifier, with resistive load, show that the average output voltage are given by

$$E_{dc} = \frac{3\sqrt{3}}{2\pi} E_m \cos \alpha, \text{ for } 0 < \alpha < \frac{\pi}{6}$$

and $\quad E_{dc} = \frac{3}{2\pi} E_m [1 + \cos(\alpha + \pi/6)] \text{ for } \frac{\pi}{6} < \alpha < \frac{5\pi}{6}$

where E_m is the maximum value of phase-voltage and α is the firing angle.

6.20 A three-phase, fully-controlled converter is connected to a resistive load. Show that the average output voltage is given by

$$E_{dc} = \frac{3\sqrt{3}}{2\pi} E_m \cos \alpha, \quad \text{for } 0 < \alpha < \frac{\pi}{3}$$

and $\quad E_{dc} = \frac{3\sqrt{3}}{2\pi} E_m [1 + \cos(\alpha + \pi/3)] \quad \text{for } \frac{\pi}{3} < \alpha < \frac{2\pi}{3}$

6.21 Discuss the effect of source-inductance on the performance of a single-phase fully-controlled converter, indicating clearly the conduction of various thyristors during one cycle.

Derive an expression for its output voltage in terms of
(a) maximum voltage E_m, firing angle α and overlap angle μ.
(b) E_m, α, L_s and I_d.

Show that the effect of source inductance is to present an equivalent resistance of $\frac{\omega L_s}{\pi}$ Ω in series with the internal rectifier voltage.

6.22 Explain the effect of battery load on the performance of single-phase fully-controlled bridge converter.

6.23 Develop a microprocessor based firing scheme for three-phase fully controlled bridge circuits.

6.24 For a 3-ϕ semiconverter operating from 3-ϕ ideal supply and delivering power to a purely resistive load, derive an equation for the average output voltage if the triggering angle is α.

6.25 For the semiconverter with triggering angle α and 3ϕ 415 V/50 Hz supply and with inductive load, determine:

(i) rms output voltage (iv) transformer utilization factor
(ii) rms source current (v) SCR and diode ratings
(iii) rectification efficiency

6.26 For the 3-ϕ semiconverter, operating from 3-ϕ/415 V/50 Hz supply with inductive loads. Derive:
(i) Distortion Factor (DF) (iii) Supply Power Factor (SPF)
(ii) Displacement Factor (DPF)

6.27 For the 3-ϕ full-converter operating from 3-ϕ, 415 V/50 Hz supply with inductive load, derive the followings:
(i) average output voltage (v) Distortion Factor (DF)
(ii) rms output voltage (vi) Harmonic factor (HF)
(iii) FF and PF (vii) Transformer utilisation factor (TUF)
(iv) Displacement Power Factor (DPF)

6.28 A commutating diode is placed across the load of three-phase half-wave controlled rectifier. Plot a graph of average load voltage against delay angle α by taking supply voltage of 100 V/phase and neglecting device voltage drop.

6.29 Show that for a generated ac to dc converter, operating from a polyphase supply with a highly inductive load, the average output voltage E_{dc} α cos α, where α is the firing angle.

6.30 For a 3-ϕ full-converter operating from a 3ϕ/415 V/50 Hz supply, derive an equation for the average output voltage in terms of the overlap angle 'μ'. Also, find out the reduction in output voltage.

6.31 A 3-ϕ full converter is driving a large inductive load and load current is equal to I_{dc}. The triggering angle of the SCRs is α. Derive an equation for the following:
(a) Total rms input power (c) Total reactive power input
(b) Total active power input Assume $L_c = 0$.

6.32 Show that the performance of a three-phase full-converter as influenced by source inductance is given by the relation:

$$\cos(\alpha + \mu) = \cos \alpha - \frac{\omega L_s I_d}{E_m}.$$

The symbols used have their usual meanings.

6.33 List the various techniques of improving power factor in phase-controlled converters.

6.34 With the help of circuit diagram and waveforms, explain the semiconverter operation of full converters.

6.35 With the help of waveform, explain how improvement in power factor can be achieved by a asymmetrical triggering.

6.36 With the help of circuit diagram and their associated waveforms, explain the extinction angle control scheme for improvement of power factor.

6.37 Derive the expressions for average output voltage, RMS supply current, RMS nth harmonic current, displacement angle, supply power factor, displacement factor, and harmonic factor in case of extinction angle control scheme.

6.38 Explain with associated waveform, how power factor can be improved with symmetrical angle control scheme.

6.39 Derive the expressions for average output voltage, RMS supply current, RMS nth harmonic current, displacement angle, supply power factor, displacement factor, and harmonic factor in case of symmetrical angle control scheme.

6.40 Explain in detail the sinusoidal pulse-width modulation control scheme for power factor improvement.

6.41 Derive the expressions for average output voltage, RMS supply current, RMS nth harmonic current, displacement angle, supply power factor, displacement factor, harmonic factor, in sinusoidal pulse-width modulation control scheme for improvement of power factor.

PROBLEMS

6.1 A single-phase half-wave rectifier is used to supply power to a load of impedance 10 Ω from 230 V, 50 Hz a.c. supply at the firing angle of 30°. Calculate
(a) Average load voltage. (c) Load current.
(b) Effective value.

[Ans (a) 96.35 V, (b) 159.8 V, (c) 9.636 A]

6.2 A voltage source $e = 100 \sin 377\,t$, supplies a resistive load of 100 Ω through a thyristor, which performs half-wave controlled rectification. Calculate, the average power in the load, if the firing angle is fired at 45° with respect to the supply voltage waveform.

[Ans. 22.7 W]

6.3 A highly inductive load, such that load current can be assumed constant, is to be supplied from a 230 V, 50 Hz, single-phase supply by a fully-controlled and a half-controlled bridges. Compare the average load. Voltage provided by each bridge at firing angles of 30° and 90°. Neglect device voltage drops.

[Ans. (i) Fully controlled bridge–
E_{dc}, 30° = 179.33 V, E_{dc}, 90° = 0
(ii) Half-controlled bridge–
E_{dc} 30° = 193.199 V, E_{dc} 90° = 103.54 V]

6.4 A single-phase 230 V, 1 kW heater is connected across a single-phase, 230 V, 50 Hz supply through an SCR. For firing angle of 45° and 90°, calculate the power absorbed by the heater-element.

[Ans. (i) $\alpha = 45°$, power = 454.57 W
(ii) $\alpha = 90°$, power = 250 W]

6.5 Calculate the average output voltage of a three-phase half-controlled bridge operating with a triggering angle of $\pi/2$ and connected to three-phase a.c. supply of 400 V and 50 Hz. The load current i_d is assumed to be continuous.

[Ans. $E_{dc} = 270$ V]

6.6 A single-phase fully-controlled bridge is connected to an a.c. supply of 230 V and 50 Hz is used for the speed control of d.c. motor with separate field excitation. The full-load average armature current is 10 A and the converter operates at a firing angle $\alpha = \pi/4$. Neglecting the inductance and resistance of both armature and source, calculate the minimum value of series inductance, L_d, required in the armature circuit to provide for continuous current conduction.

[Ans. $L_d = 46.5$ mH]

6.7 A fully-controlled three-phase bridge converter is working in the inversion mode with a firing advance angle of 25°. If the a.c. supply is 220 V with a reactance of 0.1 Ω/phase, determine the maximum current that can be commutated, allowing a recovery angle of 5°. Neglect device voltage drops.

[Ans. $I_d = 140$ A]

6.8 A six-pulse thyristor converter, connected to 230 V, 50 Hz, 3-ϕ a.c. supply. If the reactance on a.c. side is 0.1502 and the commutation angle is 15°, calculate the d.c. load current.

[Ans. $I_d = 90.47$ A]

6.9 Calculate d.c. output voltage from a six-pulse double-star circuit connected to an a.c. supply of 415 V, three-phase, 50 Hz at a firing angle of 30°.

[Ans. $E_{dc} = 161.38$V]

6.10 A three-phase, half-wave converter is supplying a load with a continuous constant current of 40 A over a firing angle from 0 to 75°. What will be the power dissipated by the load at these limiting values of firing angle? The supply voltage is 415 V (line).

[Ans. for $\alpha = 0°$, Power = 11.2 kW
$\alpha = 75°$, Power = 2.9 kW]

6.11 A single-phase, half-controlled bridge is constructed as shown in Fig. P. 6.11. Sketch the load voltage waveform for an R–L load at a firing angle of 30°.

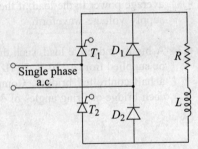

Fig. P. 6.11

6.12 A three-phase, fully-controlled bridge converter is supplying a d.c. load of 400 V, 60 A from a three-phase, 50 Hz, 660 V (line) supply. If the thyristors have a voltage drop of 1.2 V when conducting then, neglecting overlap, compute

(i) the firing angle of thyristor.

(ii) RMS current in thyristors.

(iii) the mean power loss in thyristors.

(iv) If the a.c. supply has an inductance per phase of 3.6 mH, what will be the new value of firing angle required to meet the load requirements?

[Ans. (i) $\alpha = 63.10°$, (ii) $I_{rms} = 34.64$ A, (iii) Power = 24 W, (iv) $\alpha = 58.23°$]

6.13 (a) A three-phase, fully-controlled bridge converter is connected to a three-phase, 50 Hz, 415 V (line) supply and is operating in the inverting mode at a firing advance angle of 30°. If the a.c. supply has a resistance and inductance per-phase of 0.09 Ω and 1 mH, respectively. Find

(i) d.c. source voltage.

(ii) overlap angle.

(iii) recovery angle, when the d.c. current is constant at 52 A.

Thyristors have a forward voltage drop when conducting of 1.8 V.

(b) In case (a), what will be the maximum d.c. current that can be accommodated at a firing advance angle of 22.5°, allowing for a recovery angle of 5° ?

[Ans. (a) (i) 506.6 V; (ii) $\mu = 8.17°$; (iii) $\gamma = 21.43°$, (b) $I_d = 67.53$A]

6.14 A three-phase fully-controlled converter is employed to change a battery with an emf of 95 V and an interval resistance of 0.1 Ω. The supply RMS voltage is 110 V line-to-line and sufficient inductance is induced in the output-circuit to maintain the current virtually constant at 20 A. Determine

    (i) the firing angle α,   (ii) power factor of supply.

[Ans. (i) $\alpha = 49.22°$, (ii) p.f. = 0.62]

6.15 A three-phase fully-controlled bridge converter operates in both rectification and inversion mode. The leakage inductance of each phase of input transformer winding is 0.002 H. The three-phase input voltage is balanced and sinusoidal, and has an RMS magnitude of 230 V per phase and frequency 50 Hz. The load current of d.c. side is 15 A.
  (a) Calculate the drop in the d.c. output voltage caused by the internal reactance drop.
  (b) When the bridge is working in rectification mode and the d.c. output voltage is 200 V, then compute
    (i) firing angle α
    (ii) overlap angle for the phase currents.
  (c) When the bridge is to be functioning as an inverter with a load current 15 A and d.c. output voltage and 200 V then calculate the recovery angle or margin angle.

[Ans. (a) 9 V, (b) (i) = 67°, (ii) = 2.48°, (c) 67°]

6.16 (a) A single-phase semiconverter delivers power to *RLE* load with $R = 5$ Ω, $L = 10$ mH and $E = 80$ V. The a.c. supply voltage is 230 V, 50 Hz. For the continuous conduction, find the average value of output-current for a firing angle of 50°. (b) If one of the two main SCRs is damaged and open-circuited, find the new value of average output current on the assumption of continuous conduction. Draw the output voltage waveforms and indicate the conducting periods of devices.

[Ans. (a) 18.013 A; (b) 1 A]

6.17 A three-phase fully-controlled thyristor bridge converter supplies a d.c. voltage source of 400 V having an interval resistance of 1.8 Ω. Assume highly inductive load with a content load current of 20 A. The supply RMS load voltage per phase is 230 V and source inductance in each phase is 0.005 H. Compute the following by ignoring the source resistance:
  (i) Firing angle α for an output voltage of 436 V
  (ii) Overlap angle.

[Ans. (i) 30°; (ii) 11°]

6.18 A single phase fully controlled bridge rectifier is given 230 V, 50 Hz supply. The firing angle is 45° and the load is highly inductive. Determine:
  (i) Average output voltage   (iii) Power factor
  (ii) Voltage ripple factor   (iv) Form factor

Ans. [(i) 146.42 V, (ii) 1.21, (iii) 0.637 (lagging), (iv) 1.57]

6.19 A single phase fully controlled bridge is operated with a resistive load $R = 10\ \Omega$, the input voltage to the bridge is 230 V. The firing angle is 60°. Determine
 (i) Average load voltage
 (ii) Average and rms load current
 (iii) Form factor and ripple factor
 (iv) Average output power
 (v) Rectifier efficiency, and
 (vi) SCR ratings

Ans. [(i) 155.3 V, (ii) 15.53 A, 20.63 A, (iii) 1.328, 0.874 (iv) 2.41 kW,
(v) 56.66%, (vi) $I_{TH(av)} = 7.77$A, PIV = 325.26 V]

6.20 A single phase full converter supplies an inductive load. Supply voltage is 230 V/50 Hz and the firing angle is 60°. Assuming that the output current is continuous and ripple free and equal to 5 Amp. Determine:
 (i) Average output voltage
 (ii) Supply rms current
 (iii) Fundamental power factor
 (iv) Distortion factor
 (v) Input power factor
 (vi) Harmonic factor
 (vii) Active and reactive power
 (viii) Average and rms values of thyristor current

Ans. [(i) 103.53, (ii) 15 Amp, (iii) 0.5, (iv) 0.9, (v) 0.45 lagging, (vi) 0.4834,
(vii) 1.553 kW, 2.69 kVar, (viii) $I_{TH(av)} = 7.5, I_{Th(rms)} = 10$]

6.21 A 3-ϕ full converter is operated from a Δ:Y connected transformer whose secondary ratings are 3-ϕ, 415 V, 50 Hz. Derive an equation for the transformer utilisation factor

Ans. TUF = 0.956 $|\cos \alpha|$

6.22 For a 3-ϕ semiconverter operating from 3-ϕ, 415 V/50 Hz supply,
 (a) derive the following:
  (i) RMS value of supply current (I_{rms})
  (ii) Distortion factor (DF)
  (iii) Displacement factor (DPF)
  (iv) Supply Power Factor (PF)
  (v) Transformer Utilisation Factor (TUF)
 (b) Find out the values of DF, DPF and PF

[Ans. (a) (i) $I_{rms} = 22.88(1 + \cos \alpha)$, for $\alpha < 60°$,

$$= 8.025\,(1 + \cos \alpha) \left(\frac{\pi - \alpha}{\pi}\right)^{1/2}, \alpha > 60°$$

(ii) DF = 0.96 cos ($\alpha/2$) for $\alpha \leq 60°$

$$= \frac{[0.78 \cos (\alpha/2)\, I_{dc}]}{\left(\frac{\pi - \alpha}{\pi}\right)^{1/2}} \text{ for } \alpha > 60°$$

(iii) DPF = cos ($\alpha/2$)
(iv) PF = DF × DPF = 0.96 cos² ($\alpha/2$), for $\alpha \leq 60°$

$$= \frac{0.78 \cos^2 \alpha}{\left(\pi - \frac{\alpha}{\pi}\right)^{1/2}} \text{ for } \alpha > 60°$$

(v) TUF = 0.48 (1 + cos α) for $\alpha \leq 60°$

$$= \frac{0.39\,(1 + \cos \alpha)}{\left(\frac{\pi - \alpha}{\pi}\right)^{1/2}} \text{ for } \alpha > 60°$$

6.23 A three phase full-wave converter is operated from a 3-ϕ Y-connected 208 V/ 50 Hz supply and the load resistance is $R = 10\ \Omega$. If it is required to obtain an average output voltage of 50% of the maximum possible output voltage, determine:

(a) firing angle α
(b) rms and average output currents
(c) average and rms thyristor currents
(d) rectification efficiency
(e) transformer utilisation factor, and
(f) input power-factor (PF)

Ans. (a) $\alpha = 60°$ (b) $I_{dc} = 14.05$ Amp, $I_{rms} = 15.93$ A
(c) $I_{Tavg} = 4.68$ A, $I_{Trms} = 9.2$ Amp.
(d) $\eta = 77.8\%$ (e) TUF = 42.1%
(f) 0.542 (lagging).

6.24 The load current of a three-phase full converter is continuous with a negligible ripple content

(a) Express the input current in fourier series, and determine the harmonic factor HF of input current, displacement factor DF, and the input power factor PF.

(b) If the delay angle $\alpha = \pi/3$, determine normalized output voltage, HF, DF, and PFZ

Ans. (a) $i_{s(t)} = I_{dc} + \sum_{n=1,2,\ldots}^{\alpha} (a_n \cos n\omega t + b_n \sin n\omega t)$

HF = 31.08%, DF = cos(−α), PF = 0.9549 DF (b) $E_n = 0.5$ P4
HF = 31.08%, DF = 0.5, PF = 0.478 (lagging)

6.25 A three-phase full-converter feeds power to a resistive load of 10 Ω. For a firing angle of 30°, the load takes 5 kW. Find the magnitude of per phase input supply voltage.

Ans. $E_{ph} = 108.59$ V

REFERENCES

1. B R Pelley, *Thyristor Phase Controlled-Converters and Cycloconverters*, Wiley-Interscience, New York, 1971.
2. P C Sen, *Thyristorised DC Drives*, Wiley Interscience, New York.
3. C W Lander, *Power Electronics*, McGraw-Hill Book Company (UK).
4. S B Dewan, *Power-Semiconductor-Circuits*, John Wiley and Sons, New York, 1975.
5. F Csaki, *Power Electronics*. Akademi ai Kidao, Budapest, 1975.
6. M Ramamoorthy, *Thyristors and their Applications*, East-West Press Pvt Ltd, 1977.

Chapter 7
Dual Converters

LEARNING OBJECTIVES:
- To describe the need and function of a dual converter.
- To understand what is meant by an ideal dual converter and a non-ideal dual converter.
- To examine the operation of a dual converter bridge.
- To classify the dual converters in terms of their operating modes.
- To examine the operation of dual converters with circulating current.
- To examine the operation of dual converters with non-circulating current.
- To introduce the dual mode, dual converter.
- To develop some basic control schemes for dual converters.

7.1 INTRODUCTION

The fully controlled converter, as described in Chapter 6, can produce a reversible direct output voltage with output current in one direction, and in terms of a conventional voltage/current diagram (Fig. 7.1) is said to be capable of operation in two quadrants, the first and fourth. Such a range of operation is useful for certain purposes, examples being the control of a d.c. torque motor, i.e. a motor used to provide unidirectional torque with reversible rotation (Fig. 7.2), and a d.c. transmission link between two a.c. systems in which power can be transmitted in either direction according to the polarity of the voltage with current flows always in one direction (Fig. 7.3). Equally, a converter may be used under steady-stage conditions in the first-quadrant only but transiently in the second-quadrant in order to extract energy from the load quickly and thereby improve the response of the system to changing command signals.

Dual Converters

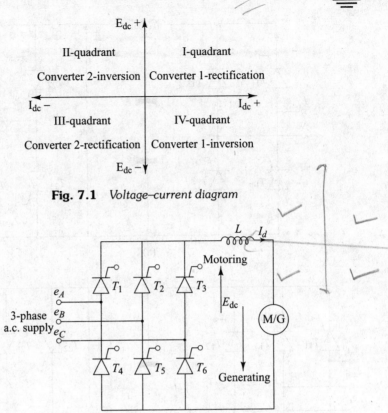

Fig. 7.1 Voltage–current diagram

Fig. 7.2 Fully controlled three-bridge converter controlling a torque motor in two-quadrant

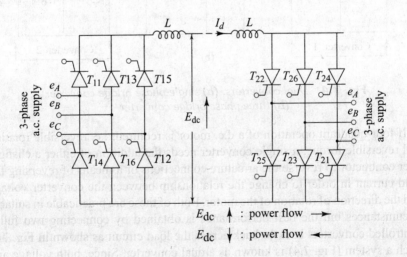

Fig. 7.3 Two fully controlled two-quadrant converter in a d.c. transmission link for reversible power flow

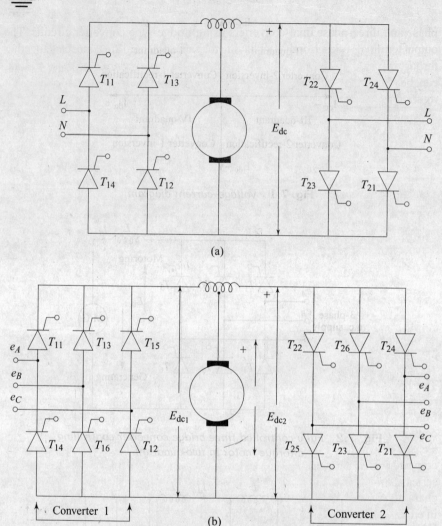

Fig. 7.4 *Dual-converters.* **(a)** *single-phase bridge converter,* **(b)** *three-phase bridge converter*

If four-quadrant operation of a d.c. motor is required, i.e. reversible rotation and reversible torque, a single converter needs the addition of either a change-over contractor to reverse the armature connections or a means of reversing the field current in order to change the relationship between the converter voltage and the direction of rotation of the motor. Both of these are practicable in suitable circumstances but the best performance is obtained by connecting two fully-controlled converters back-to-back across the load circuit as shown in Fig. 7.4. Such a system (Fig. 7.4) is known as a dual converter. Since, both voltage and current of either polarity are obtained with a dual converter, therefore the system will provide the four-quadrant operation. Figures 7.4(a) and (b) show the single-phase and three-phase dual-converters using bridge-type converter circuits. The output terminals of each converter having the same potential are connected together through a reactor. The four possible quadrants of converter operation thus resulting can be translated into four (steady-state) combinations of motor-torque and rotation as shown in Fig. 7.5.

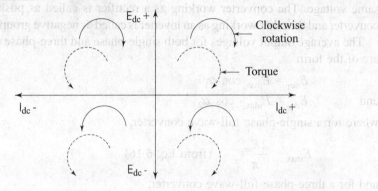

Fig. 7.5 *Torque-rotation diagram*

7.2 PRINCIPLE OF DUAL CONVERTER (IDEAL DUAL CONVERTER)

The basic principle of operation of the dual converter can be explained with reference to the simplified equivalent diagram of the d.c. circuit shown in Fig. 7.6. In this simplified representation, assumption is made that the dual converters are ideal and they produce pure d.c. output voltage, that is, there is no a.c. ripple at the d.c. output terminals. As shown in Fig. 7.6, each two-quadrant converter is assumed to be a controllable direct voltage source, connected in series with a diode. Diodes D_1 and D_2 represent the unidirectional current flow characteristics of the converters. The current in load circuit can, however, flow in either direction.

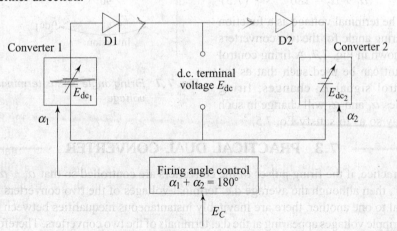

Fig. 7.6 *Ideal dual converter equivalent circuit*

The firing angles of the individual converters of the dual converter are regulated by a control voltage E_c so that their d.c. voltages are equal in magnitude but opposite in polarity. Therefore, they can drive the current in opposite directions through the load. Thus, when one converter operates as a rectifier having a given d.c. terminal voltage, the other converter operates as an inverter with exactly the same voltage. The converter working as a rectifier is called as positive group converter and the other working as an inverter is called as negative group converter.

The average output voltages for both single-phase and three-phase converters are of the form

$$E_{dc_1} = E_{max} \cos \alpha_1 \quad (7.1)$$

and $\quad E_{dc_2} = E_{max} \cdot \cos \alpha_2 \quad (7.2)$

where for a single-phase full-wave converter,

$$E_{max} = \frac{2E_m}{\pi} \quad \text{(from Eq. 6.16)}$$

and for a three-phase full-wave converter,

$$E_{max} = \frac{3\sqrt{3} E_{mph}}{\pi}$$

For an ideal converter,

$$E_{dc} = E_{dc_1} = -E_{dc_2} \quad (7.3)$$

Substitution of E_{dc_1} and E_{dc_2} from Eqs (7.1) and (7.2) in Eq. (7.3) gives,

$$E_{max} \cos \alpha_1 = -E_{max} \cos \alpha_2 \quad (7.4)$$

or $\quad \cos \alpha_1 = -\cos \alpha_2$
$\quad\quad\quad\quad = \cos(180 - \alpha_2)$
or $\quad \alpha_1 = 180 - \alpha_2$
or $\quad \alpha_1 + \alpha_2 = 180° \quad (7.5)$

The terminal voltage as a function of firing angle for the two converters is shown in Fig. 7.7. A firing control circuit can be used such that as the control signal E_c changes, firing angles α_1 and α_2 will change in such a way so as to satisfy Eq. 7.5.

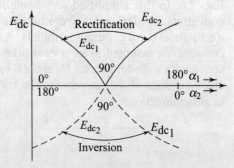

Fig. 7.7 Firing angle versus terminal voltage

7.3 PRACTICAL DUAL CONVERTER

In practice, if the firing pulses of the converters are controlled so that $\alpha_1 + \alpha_2 = 180°$, then although the average d.c. terminal voltages of the two converters are equal to one another, there are inevitably instantaneous inequalities between the a.c. ripple voltages appearing at the d.c. terminals of the two converters. Therefore, with this simple control method if the two converters are solidly connected, the instantaneous voltage difference at the terminals of the two converters will produce enormous circulating currents between the two converters that will not flow through the load. Therefore, in practice, it is necessary to use some means for controlling the flow of circulating current between the two converters. Generally, two following modes are used for the above purpose.

(1) Dual converter without circulating current mode.
(2) Dual converter with circulating current mode.

Both of these operating modes are discussed in the following sections.

7.4 DUAL CONVERTER WITHOUT CIRCULATING CURRENT OPERATION

In a dual converter without circulating current operating mode, the flow of circulating current is completely inhibited through appropriate automatic control of the firing pulses, so that only that converter which carries the load current is in conduction and the other converter is temporarily blocked. Since only one converter operates at a time and the other is in blocking state, no reactor is required between the converters.

At a particular instant, suppose converter 1 is operating as a rectifier and is supplying the load-current while pulses to second converter are blocked. For the inversion operation, converter 1 is first blocked by removing its firing pulses and load current is reduced to zero. Converter 2 is made to conduct by applying the firing pulses to it. The current in converter 2 would now build up through the load in the reverse direction. So long as converter 2 is in operation, converter 1 is in the blocking state since the firing pulses are withdrawn from it. The pulses to converter 2 are applied after a delay time (current-free safety interval) of 10 to 20 ms. This delay time ensures reliable commutation of thyristors in converter 1. If the converter 2 is triggered before the converter 1 has been completely turned-off, a large circulating current would flow between the two converters.

Irregular "jumps" in the level of the d.c. terminal voltage at the point of current reversal must be avoided in order to achieve a smooth change-over of current from one converter to the other. Thus, the firing pulse control should, ideally, be such that the mean d.c. terminal voltage of the converter 2, at the instant of current reversal, is the same as that of the converter 1.

From the above discussion, it becomes clear that such a mode of operation requires sophisticated control system which automatically "blocks" and "deblocks" the individual converters in accordance with the direction of load current suitably with a safety intervals. The load current, under steady-state conditions, may be continuous or discontinuous. Therefore, the control circuit for this mode of operation of a dual converters is so designed as to give satisfactory operation under both continuous and discontinuous load current.

Several control schemes for a circulating current-free dual converter can be evolved, and a few basic schemes are discussed here. In this scheme, since only one converter conducts at a time, it is possible to have only one firing unit and to switch the firing pulses to the appropriate converter.

7.4.1 Converter Selection by Control Signal Polarity

Figure 7.8 shows the basic block diagram of this scheme. Here, the control signal E_c represents the difference between a set speed and the actual speed of the d.c. motor in a closed-loop feedback scheme or it may represent the speed and direction of rotation in an open loop control scheme. The firing pulses for both converters are generated by the control signal such that the sum of their firing angles (α_1, α_2), is 180°. These pulses, however, are 'ANDed' with a signal that represents the polarity of the control signal. Hence, only one converter receives firing pulses and the other does not. AND gates are used for inhibiting α_1 or α_2. This is done by developing M and N inputs. M input is high and N input is low when control voltage E_c is positive. The output of AND gate-1 becomes high since its both inputs are at high level. Therefore, converter 1 receives the firing pulses and the firing pulses to converter-2 are blocked. The converter-1 operates as a rectifier and the d.c. motor rotates in the positive direction.

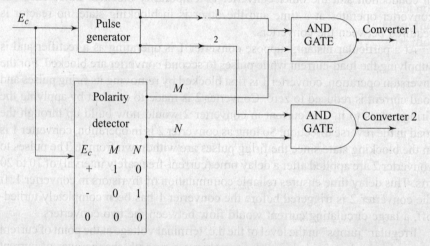

Fig. 7.8 *Basic block diagram*

If the control voltage polarity is negative, M input becomes low and N input becomes high. AND gate-2 becomes enabled and converter 2 receives the firing pulses. Therefore, converter-1 becomes in the blocking state and converter-2 becomes ON. When the control voltage E_c is zero, both M and N inputs becomes zero and the firing pulses are blocked from both converters.

In the steady-state, this converter selection scheme works satisfactorily. However, when control voltage E_c changes from positive to negative, circulating current may flow. As discussed above, when control voltage E_c is switched from positive to negative, converter 2 starts conducting. It may happen that conduction in converter 1 has not totally stopped; this results in a circulating current between the converters.

Dual Converters

Figure 7.9 shows the transfer characteristics with dead-band. The transfer characteristics is plotted between the output voltage and the control voltage. The problem of circulating current between the converters can be overcome by introducing dead-band in the transfer characteristics or by a slow change of control voltage E_c.

7.4.2 Converter Selection by Load-Current-Polarity

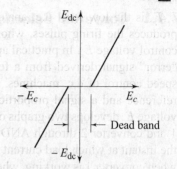

Fig. 7.9 Transfer characteristics

A control scheme for a circulating current-free mode of dual converter in a simplified diagrammatic form is shown in Fig. 7.10. This scheme consists of a firing pulse generator, comparators and a current transducer which acts in conjunction to block the firing pulses to the idle converter until the current of the conducting converter has reached zero. In accordance with the direction of current there exists an automatic blocking and deblocking of the individual converters.

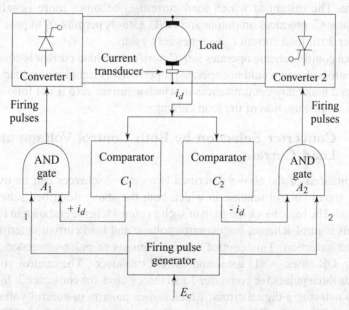

Fig. 7.10 Control scheme for a dual converter with no circulating current

The function of the firing pulse generator is to supply firing pulses to the thyristors within the associated converters, provided that the associated AND gate "A_1" or "A_2" is closed. Since these gates are controlled by the comparator outputs, it is inherent that one of these must always be closed, and the other open, and hence only one of the two converters has firing pulses applied at any- one time.

E_c is the low level d.c. analog control voltage. The firing pulse generator produces the firing pulses, whose phase is controlled in accordance with the control voltage E_c. In practical applications, the control voltage E_c might be the "error" signal derived from a feedback control loop. For example, in case of speed control of d.c. machines, E_c might be the difference between the speed reference and a signal proportional to the actual machine speed. The control voltage E_c develops two graphs of pulses α_1 and α_2, each one going to converter 1 and converter 2 through AND gates A_1 and A_2. The current transducer sense the instant at which load current i_d becomes zero. The load-current i_d is positive when converter 1 is working, whereas it is negative when converter 2 is working.

Suppose that a situation exists in which the positive converter supplies current to the load at a given positive, voltage-level. Under this condition, as long as load current i_d is positive, comparator C_1 produces high output. Therefore, AND gate $-A_1$ becomes enabled and applies the firing pulses to the converter 1. The moment at which the positive converter current reaches zero, comparator C_1 goes low and therefore, AND gate A_1 inhibits the firing pulse α_1. This process remains continuous, till load-current i_d becomes positive again.

If the polarity of the control signal E_c is changed, the current i_d becomes negative. The instant at which load current i_d becomes more negative, the comparator C_2 produces an output and AND gate-A_2 permits α_2 to pass through converter 2 till load current i_d becomes zero again.

Such a control scheme operates satisfactorily if the load current is continuous. This control scheme would not operate properly with discontinuous load current, nor indeed under any circumstances in which a current zero is not followed by a reversal in the direction of the load current.

7.4.3 Converter Selection by Both Control Voltage and Load Current

The limitations of the above described two control schemes can be overcome using a signal control scheme in which both the above described schemes are combined. The basic block diagram of such a control scheme is shown in Fig. 7.11.

In this control scheme, both control voltage and load current determines the converter selection. This control scheme consists of pulse-generator, polarity detector, OR gates, AND gates and current transducer. The control voltage E_c develops three pulses for converter 1 and three pulses for converter 2. In voltage polarity detector, a digital circuit which senses polarity of control voltage E_c is employed. The truth-table of this circuit is also shown in Fig. 7.11. The current polarity detector senses the load current i_d and develops three currents as shown in Fig. 7.11. These currents are ORed in OR gates O_1 and O_2. OR gate O_1 develops high only when load-current i_d is positive. OR gate O_2 develops high only when current i_d is negative. The three signals, namely, α_1 (α_2), E_c and i_d are ANDed in AND gates A_1 and A_2. Therefore, we get α_1 high only when E_c is high and i_d is high. Thus, converter 1 receives the firing pulses if E_c is positive and load current i_d is zero (discontinuous load current) or positive.

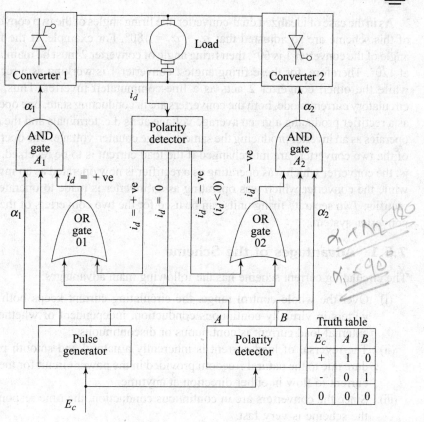

Fig. 7.11 *Basic block diagram*

On the other hand, converter 2 receives the firing pulses if E_c is negative and load current is zero or negative. If the control voltage E_c is zero, both AND gates, A_1 and A_2, will block the firing pulses. The transition from one converter to another is accompanied by a delay which is very minimum. Therefore, this control scheme is most useful for the control of dual converters.

7.5 DUAL CONVERTER WITH CIRCULATING CURRENT OPERATION

It has been observed from the above discussion that certain difficulties may arise due to discontinuous load current with the dual-converter operating in the circulating current free mode. To overcome these difficulties, very sophisticated control schemes may be necessary. An alternative operating technique, with which the difficulties associated with the circulating current free dual converter in discontinuous conduction do not arise, is to use a circulating current operating scheme. In this scheme, a current limiting reactor is inserted between the d.c. terminals of the two converters as shown in Fig. 7.4. The firing angles of the individual converters are regulated in such a way that a controlled amount of current is allowed to circulate between the converters.

As in the case of idealized dual-converter, the firing angles of the two converters of this scheme are so adjusted that $\alpha_1 + \alpha_2 = 180°$. For example, if the firing angle of the converter 1 is 60°, then firing angle of converter 2 must be maintained at 120°. Therefore, for these firing angles, converter 1 is working as a rectifier while the other, converter 2 acts as a line-commutated inverter. Thus, with circulatory current mode, both the converters are in conducting state, one operates as a rectifier producing a given average voltage at its d.c. terminals and the other operates as an inverter producing the same average counter-voltage. The operation of the two converters are interchanged if the load current is to be reversed. That is, the converter which was operating as a rectifier is now operated as an inverter while the converter which was operating as an inverter is made to operate as a rectifier. Two separate firing units can be used for the two converters of the dual converter system.

7.5.1 Advantages of the Scheme

The circulating current scheme has the following main advantages:
 (i) Over the whole control range, the circulating current keeps both converters in virtually continuous conduction, independent of whether the external load current is continuous or discontinuous.
 (ii) The reversal of load-current is inherently a natural and smooth procedure due to the natural freedom provided in the power circuit for the load current to flow in either direction at anytime.
 (iii) Since the converters are in continuous conduction, the time response of the scheme is very fast.
 (iv) The current sensing is not required and the normal delay period of 10 to 20 ms as in the case of a circulating current free operation is eliminated.
 (v) Linear transfer characteristics are obtained.

7.5.2 Disadvantages of the Scheme

The circulating current scheme has the following main disadvantages:
 (i) Since the current limiting reactor is required in this scheme, the size and cost of this reactor may be quite significant at high power levels.
 (ii) Since the converters have to handle load as well as circulating currents, the thyristors with high current ratings are required for these converters.
 (iii) The efficiency and power factor are low because of circulating current which increases losses.

In spite of these drawbacks a dual converter with circulating current mode is preferred if load current is to be reversed quite frequently and a fast response is desired in the four-quadrant operation of the dual converter.

7.5.3 Operation with Waveforms

The following assumptions are made for describing the operation of dual converter with circulating current.

(i) The reactor is lossless.
(ii) The firing angles of two converters are controlled so that there sum is 180° (i.e. $\alpha_1 + \alpha_2 = 180°$).

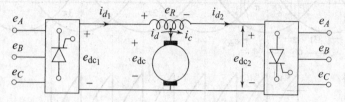

Fig. 7.12 *Dual converter basic block diagram*

The basic power block diagram of a dual converter with circulating current mode is shown in Fig. 7.12. The various voltage and current polarities are also shown in Fig. 7.12. Figure 7.13 shows the associated voltage and current waveforms. In Fig. 7.13(a), the supply line voltages, e_A, e_B and e_C, are shown. Here, converter-1 operates as a rectifier while converter 2 operates as inverter. As shown in Fig. 7.13(a), converter 1 is triggered at point P with firing angle $\alpha_1 = 60°$. Converter 2 is triggered at point Q with firing angle $\alpha_2 = 120°$. The output voltage waveforms of converter 1 is shown in Fig. 7.13(b), whereas for converter 2, it is shown in Fig. 7.13(c). The instantaneous d.c. output voltage shown in Fig. 7.13(d) is the average of the instantaneous converter voltage. It is to be noted that this output voltage waveform has the shape different from that of either of the converter voltages. The average value of output voltage is, however, the same as the average of each converter voltage. Figure 7.13(e) shows the instantaneous voltage across the reactor (e_R). This instantaneous voltage is the difference between the instantaneous converter voltages. In Fig. 7.13(f), the circulating current waveform is shown this circulating current shown is obtained from the time integral of the voltage across the reactor. Both at no-load and load conditions, both converters are kept in a continuous conduction state, because of the circulating current.

Now, let us consider the phase voltages as follows:

$$e_A = E_m \sin \omega t \tag{7.6}$$

$$e_B = E_m \sin \left(\omega t - \frac{2\pi}{3} \right) \tag{7.7}$$

$$e_C = E_m \sin \left(\omega t + \frac{2\pi}{3} \right) \tag{7.8}$$

and $\quad e_A + e_B + e_C = 0 \tag{7.9}$

From Figs 7.13 (a), (b), (c) and (e), we have the following relations.

During the interval $(\pi/6 + \alpha_1) < \omega t < \left(\frac{\pi}{6} + \alpha_1 + \frac{\pi}{3} \right)$,

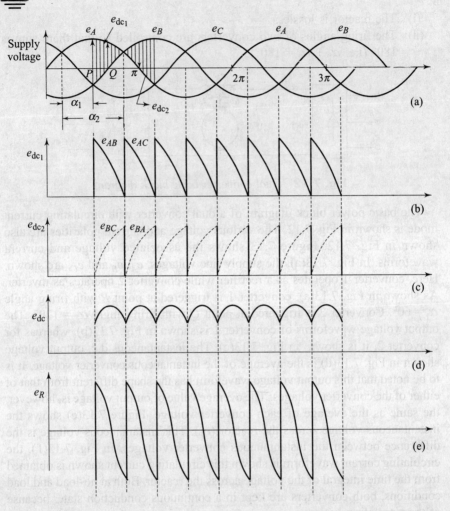

Fig. 7.13 *(a) Voltage and current waveforms for a circulatory current type dual converter*

$$e_{dc_1} = e_A - e_B = e_{AB} \tag{7.10}$$

$$e_{dc_2} = -(e_C - e_B) = e_B - e_C = e_{BC} \tag{7.11}$$

$$e_R = e_{dc_1} - e_{dc_2} = e_A + e_C - 2e_B \tag{7.12}$$

Eq. (7.12) can also be written as

$$e_R = e_A + e_C + e_B - e_B - 2e_B$$

$$= e_A + e_B + e_C - 3e_B \tag{7.13}$$

Substituting Eq. (7.9) in Eq. (7.13),

$$e_R = -3e_B \tag{7.14}$$

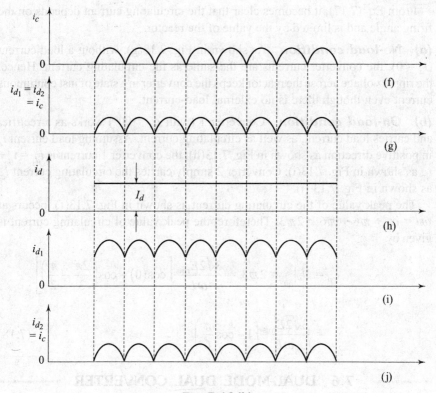

Fig. 7.13(b)

Since, the circulating current i_c is the time integral of the voltage across the reactor,

$$\therefore \quad i_c = \frac{1}{\omega L} \int_{\alpha_1 + \pi/6}^{\omega t} e_R \cdot d(\omega t) \tag{7.15}$$

Substituting Eqs (7.7) and (7.14) in Eq. (7.15),

$$\therefore \quad i_c = \frac{1}{\omega L} \int_{\alpha_1 + \pi/6}^{\omega t} -3E_m \cdot \sin\left(\omega t - \frac{2\pi}{3}\right) d(\omega t)$$

$$i_c = \frac{3E_m}{\omega L}\left[\cos\left(\omega t - \frac{2\pi}{3}\right) - \cos\left(\alpha_1 - \frac{\pi}{2}\right)\right] \tag{7.16}$$

or

$$i_c = \frac{3\sqrt{2} E_{rms}}{\omega L}\left[\cos\left(\omega t - \frac{2\pi}{3}\right) - \cos\left(\alpha_1 - \frac{\pi}{2}\right)\right] \tag{7.17}$$

From Eq. (7.17), it becomes clear that the circulating current depends on the firing angle and is limited by the value of the reactor.

(a) No-load condition As shown in Fig. 7.13(g), without a load current ($i_d = 0$), the converter currents are the same as the circulating current. Hence, the ripple voltage across the reactor keeps the converter in a state of just continuous current even though there is no external load-current.

(b) On-load condition Converter 1 (with $\alpha_1 < 90°$) works as a rectifier and carries load current, as well as circulating current. Assuming load current i_d in positive direction as shown in Fig. 7.13(h), the converter 1 current is $i_{d_1} = i_d + i_c$, as shown in Fig. 7.13(i). Converter 2 simply carries the circulating current i_c, as shown in Fig. 7.13 (i).

The peak value of the circulating current as shown in Fig. 7.13(f), occurs at $\omega t = \alpha_1 + \pi/6 + \pi/6 = 2\pi/3$. Therefore, the peak value of circulating current is given by

$$i_{cp} = i_c\big|_{\omega t\, =\, 2\pi/3} = \frac{3\sqrt{2}E_{rms}}{\omega L}\left[\cos(0) - \cos\left(\frac{2\pi}{3} - \frac{\pi}{2}\right)\right]$$

$$= \frac{3\sqrt{2}E_{rms}}{\omega L}\left(1 - \cos\frac{\pi}{6}\right) \qquad (7.18)$$

7.6 DUAL-MODE DUAL CONVERTER

The foregoing study reveals that both the dual converter operating modes have certain merits and demerits. It is possible that the converters will be loaded to an extent much higher than the maximum load in the circulating current mode. This make the presence of the circulating current undesirable. If the load current tends to be discontinuous, however, it is necessary to operate in the circulating current mode. These contradictory requirements load to the concept of dual mode operation in which the dual converter will operate in both the circulating current and noncirculating current modes.

Figure 7.14 shows the basic block diagram for a dual-mode dual converter. This basic block diagram consists of comparators, current sensor, AND gates, pulse-generator, error amplifier and integrator. In this control scheme, the dual converter is operated mostly in the noncirculating current mode. However, if the load current falls below a level called "threshold" level, the dual converter operates in the circulating current mode. It is simplified that below the threshold level, there is the possibility of discontinuous load current.

The load current is first sensed and is then fed to the two comparators, C_1 and C_2, as shown in Fig. 7.14. If the load current i_d is more than the threshold voltage, $-e_{TH}$, then the comparator C_1 produces an output e_{d_1}. If the load current i_d is less than the threshold voltage, $+e_{TH}$, then the comparator C_2 produces an output e_{d_2}.

The comparator outputs e_{d_1} and e_{d_2} are used as the inputs for the AND gates A_1 and A_2, respectively. The second input for both AND gates is the pulses generated by the pulse-generator. E_C is the control voltage. The output e_{d_1} and e_{d_2} allow the firings pulses to be applied to the converters. Both converters receive the firing pulses when the load current is below the threshold value and hence, circulating current flows through the dual converter. Otherwise, only one converter, receives the firing pulses, and the dual converter operates in the noncirculating current mode. Figure 7.15 illustrates the associated waveforms. The circulating current in this control scheme flows only when the load current is small. Thus, the size of the reactor is small. When only one converter conducts, the core of the reactor can saturate at high load currents and the reactor does not have to perform the circulating current limiting function.

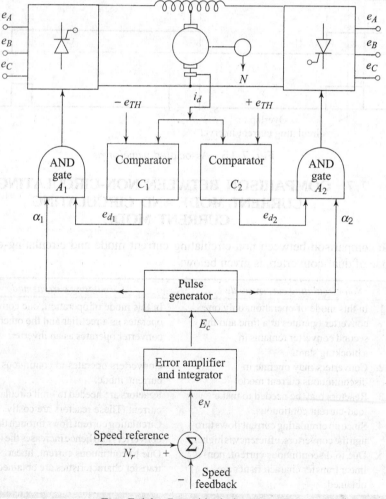

Fig. 7.14 *Basic block diagram*

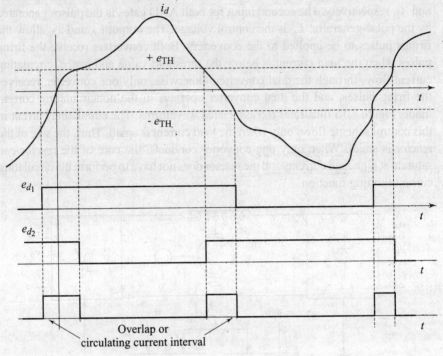

Fig. 7.15 *Associated waveforms*

7.7 COMPARISON BETWEEN NON-CIRCULATING CURRENT MODE AND CIRCULATING CURRENT MODE

The comparison between non-circulating current mode and circulating-current mode of dual-converters is given below:

Non-circulating current mode	Circulating current mode
1. In this mode of operation, only one converter operates at a time and the second converter remains in a blocking state.	In this mode of operation, one converter operates as a rectifier and the other converter operates as an inverter.
2. Converters may operate in discontinuous current mode.	Converters operates in continuous current mode.
3. Reactors may be needed to make load-current continuous.	Reactors are needed to limit circulating current. These reactors are costly.
4. Since no circulating current flows through the converters, efficiency is higher.	Circulating current flows through the converters and hence increases the losses.
5. Due to discontinuous current, non-linear transfer characteristics are obtained.	Due to continuous current, linear transfer characteristics are obtained.

(Contd.)

Dual Converters

Non-circulating current mode	Circulating current mode
6. Due to discontinuous current, response is sluggish.	Due to continuous-current in the converters, response is fast.
7. Due to spurious firing, faults between converters results in dead short-circuit conditions.	Due to spurious firing, fault currents between converters are restricted by the reactor.
8. In this mode of operation, the crossover technique is complex.	In this mode of operation, the crossover technique is simple.
9. Loss of control for 10 to 20 ms is observed in this mode of operation.	Since converters do not have to pass through blocking unlocking and safety intervals of 10 to 20 ms, hence control is never lost in this mode of operation.
10. The control scheme needs command module to sense the change in polarity.	As both the converters are operating at the same time, the control scheme does not require command module.
11. The complete scheme is cheaper compared to circulating current mode.	The complete scheme is expensive.
12. In this mode of operation, the converter loading is the same as the output load.	In this mode of operation the converter loading is higher than the output load.

7.8 MICROPROCESSOR BASED-FIRING SCHEME FOR A DUAL CONVERTER

Figure 7.16 illustrates a scheme of firing control of a dual converter using a microprocessor. This scheme consists of the following:

(a) Thyristor power converter In this case, a dual converter used for four-quadrant operation of the d.c. motor. There are two converter groups, one positive group and the other negative group. When two thyristors of a particular group conduct simultaneously, then the load current flows.

(b) Power signal operation From the three-phase voltages, digitised signals are obtained. An isolation is provided by means of the filament single-phase transformers. The output of the transformers is amplified. The amplified signals are saturated to obtain the digitized signals. These signals are used to obtain the base interrupt.

(c) Frequency doubler stage In this stage, a base interrupt signal is obtained and this signal frequency is double to that of the supply. This is obtained by combining the two sets of pulse signals obtained at the rising and falling edges of the above digitized signals. The interrupt signals is available at each 60° interval so that six firing pulses are available for firing the converter. At the falling edge of the interrupt, signal starts a new firing cycle.

(d) Firing angle control A microprocessor is used for the firing angle control of the dual converter. To perform the functions of firing range selection, firing pulse generation, cross over protection, etc. the microprocessor can be

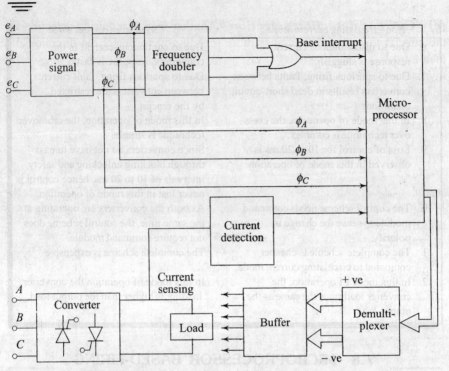

Fig. 7.16 Block diagram for firing scheme of a dual converter using microprocessor

programmed using a suitable software. The thyristor of the converter in conduction should receive the firing pulses for either modes of continuous and discontinuous operation. Two thyristors should receive the pulses at anytime. The firing signal contains several pulses. The firing pulses so obtained turn the thyristor ON reliably. A proper software facilitates the implementation of the firing pulses with minimum hardware. A second requirement is firing angle control. This can be achieved by storing the angle command in two registers of the processor. To indicate the firing pulses for the positive group of thyristors or negative group of thyristors, information should also be available. This information can be made available again from the digitized power signals and also power source voltages. A logic can be developed to define the firing angle range and also the firing pulses to a given converter. Using proper look-up tables, the firing angle selection can be suitably implemented. The following steps are involved:

(i) Read in the digitized signals.
(ii) The range indicater and a look-up table provide the firing-code.
(iii) After checking the protections, the gates of the thyristors are triggered.
(iv) At the end of firing process, the microprocessor will take care of interrupted program. The microprocessor can be programmed to do the other jobs of servo-system.

Dual Converters

(e) Current detection selection Another feature requiring detection is the current detection. It becomes more meaningful to detect the direction of load-current rather than the detection of thyristor currents. Positive group of thyristors conduct if the current is in the positive direction. However, they may be conducting if the current direction is not positive. This is due to the turn-off characteristics of the SCRs. Therefore, it is necessary to device special methods to make sure that no SCR is conducting when the current direction signals are zero. This is necessary to avoid the firing of the thyristors in the other converter.

One of the protection features to be implemented using a microprocessor is failure of any phase. This can be implemented using digitized power signals, cross-over conditions, etc. The multi-level protection is possible and this increases the reliability.

Figure 7.17 illustrates the block-diagram for microprocessor based speed control of a d.c. motor using a dual converter.

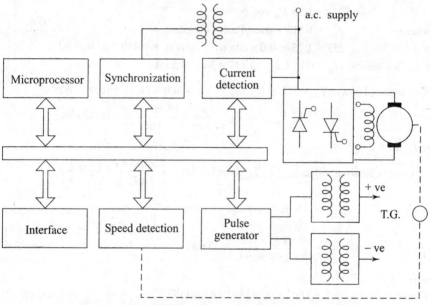

Fig. 7.17 *Basic block diagram*

SOLVED EXAMPLES

Example 7.1 Compute the peak value of the circulating current for the 3-ϕ circulatory current type dual converter consisting of two three-phase fully controlled bridges for the given data.

Per phase supply RMS voltage = 230 V, ω = 315 rad/s, L = 12 mH

$$\alpha_1 = 60°, \alpha_2 = 120°.$$

Solution: The peak value of the circulating current from Eq. (7.18) is given by

$$i_{cp} = \frac{3\sqrt{2}E_{rms}}{\omega L}(1 - \cos \pi/6), = \frac{3\sqrt{2} \times 230}{315 \times 12 \times 10^{-3}}(1 - \cos \pi/6) = 34.58 \text{ A}.$$

Example 7.2 Design a dual converter to achieve a four-quadrant operation of the separately excited d.c. motor. Motor and converter specifications are given by
  (i) Motor specifications
      $E_a = 220$ V, $I_a = 30$ A, $N = 1500$ rpm.
  (ii) Converter specifications
      Supplied from 3-ϕ, 400 V, 50 Hz supply
      Assume drop in the circuit is 15%.

Solution: Consider that dual converter consist of six-pulse converters to achieve a four-quadrant operation.
  (i) Step 1 Rectifier operation:
Total drop in the system = $220 \times 0.15 = 33$ V.
$\therefore$ Total d.c. voltage, $E_{dc\alpha} = E_{dc}$ + drop, = $220 + 33 = 253$ V.
For six-pulse bridge converter, we have the relation

$$E_{dc\ \alpha} = 1.35\ E_{ac} \cos \alpha_1.$$

where $\quad E_{ac}$ = RMS value of a.c. voltage.
$\therefore \quad 253 = 1.35 \times 400 \times \cos \alpha_1 \therefore \cos \alpha_1 = 0.469 \therefore \alpha_1 = 62°$.

A.C. line current $I_{ac} = 0.817. I_{dc} = 0.817 \times 30 = 24.51$ A.

A.C. terminal power, $P_{ac} = \sqrt{3} \times E_{ac} \times I_{ac} = \sqrt{3} \times 400 \times 24.51 = 16.98$ kW.

$$P_{ac} = 1.05\ P_{dc}, \quad \therefore P_{dc} = \frac{P_{ac}}{1.05} = \frac{16.98 \times 10^3}{1.05} = 16.17 \text{ kW}.$$

(ii) Step 2

Current limiting inductance L_C is given by, $L_C = \dfrac{2 \times 1.35 \times E_{ac}}{6\omega I_{ripple}} \left[\dfrac{1}{7} + \dfrac{1}{5}\right]$

where $\quad I_{ripple} = \dfrac{I_d}{5}$ for six-pulse converter = $\dfrac{30}{5} = 6$ A.

$\therefore \quad L_C = \dfrac{2 \times 1.35 \times 400}{6 \times 2\pi \times 50 \times 6} = 33$ mH

(iii) Step 3
Firing angle $\quad \alpha_2° = 180° - \alpha_1 = 180° - 62 = 118°$

(iv) Selection of SCR
  (a) Voltage rating, PIV = $2\sqrt{2}\ E_{ac} = 2\sqrt{2} \times 400 = 1131.37$ PIV = 1200 V.
  (b) Current rating

$$I_T = 2\sqrt{2} \times I_{ac} = 2\sqrt{2} \times 24.51 = 69.32 \text{ A} \cong 70 \text{ A}.$$

REVIEW QUESTIONS

7.1 Explain with a neat circuit diagram the basic principle of a dual converter.
7.2 Compare the ideal dual converter mode with non-ideal dual converter mode.
7.3 Describe in detail the operation of dual converter without circulating current.
7.4 With the help of basic block diagram, explain how converter selection is achieved by control signal polarity.

Dual Converters 433

7.5 With the help of suitable control scheme, explain how converter selection in achieved by load current polarity.

7.6 Explain by giving basic block diagram, the converter selection by both control voltage and load current approach.

7.7 Discuss the advantages and disadvantage of the converter selection method of both control voltage and load current.

7.8 Explain in detail the operation of dual converter with circulating current. List the advantages and disadvantages of the same scheme.

7.9 Draw the basic block diagram of a dual converter operating in circulating current mode and describe the operation with associated waveforms.

7.10 Derive the expression for peak value of the circulating current.

7.11 By giving the basic block diagram, explain the operation of a dual mode dual converter in detail.

7.12 Give the comparison between non-circulating current mode and circulating current mode.

7.13 Draw and explain in detail the microprocessor based firing scheme for a dual converter.

PROBLEMS

7.1 Calculate the peak value of the circulating current for the 3-ϕ circulatory current type dual converter consisting of two three-phase fully controlled bridges for the given data.
Per phase supply RMS voltage = 230 V, f = 15 Hz, L = 15 mH, α_1 = 60°, α_2 = 120° [Ans ic_p = 27.74 A.]

7.2 Two three-phase full-converters are connected in antiparallel to form a three phase dual converter of the circulating current type. The input to the dual-converter is 3-ϕ, 400 V, 50 Hz. If the peak value of the circulating current is to be limited to 20 A, find the value of inductance needed for the reactor for firing angle of 60°. [Ans L_C = 20.89 mH.]

7.3 Design a dual converter to achieve at four-quadrant operation for I_d = 10 A at 200 V. The converter is supplied from 400 V, 3-ϕ and 50 Hz supply.
I_{ripple} = 2 A.
[Ans (i) α_1 = 68°, α_2 = 112° (ii) $L_c = I_{ac}$ = 8.17, P_{ac} = 5.66 kW, P_{dc} = 5.39 kW.
(iii) L_c = 98 mH (iv) PIV = 1200 V, I_T = 22 Amp]

REFERENCES

1. P C Sen, *Thyristor D.C. drives*, John Wiley, 1981.
2. C W Lander, *Power electronics*, McGraw-Hill.
3. B R Pelley, *Thyristor phase controlled converters and cycloconverters*, John Wiley, 1971.
4. G De, *Principles of thyristorised converters*, Oxford and IBH Publishing Co., Calcutta, 1982.

Chapter 8
Choppers

LEARNING OBJECTIVES:
- To describe the need and function of a chopper.
- To consider the operation of a d.c. chopper.
- To explain the different chopper control techniques.
- To examine the operation of step-up and step-down choppers.
- To classify the d.c. choppers in terms of their operating envelopes.
- To consider the operation of buck, boost, buck-boost and cuk-switching regulators.
- To explain the working principle and circuit analysis of Type A chopper.
- To explain the working principles of Type B, Type C, Type D and Type E chopper circuits.
- To consider the operation of various chopper commutation circuits.
- To examine the operation of Jones and Morgan chopper circuits.
- To explain the working principles of an a.c. chopper and a multiphase chopper.
- To consider the operation of buck, boost, buck-boost and cuk-switching regulators.

8.1 INTRODUCTION

To produce quality goods in any industry, the processes necessarily require the use of variable speed drives. Variable speed d.c. and a.c. drives are being increasingly used in all industries. These drives and processes take power from d.c. voltage sources. In many cases, conversion of the d.c. source voltage to different levels is required. For example, subway cars, trolley buses, or battery operated vehicles require power from a fixed voltage d.c. source. However, their speed control requires conversion of fixed voltage d.c. source to a variable-voltage d.c. source for the armature of the d.c. motor.

Generally, following techniques are available for obtaining the variable d.c. voltage from a fixed d.c. voltage:

(1) Line Commutated Converters (Conversion of AC supply to variable DC supply using controlled rectifiers; covered in Chapter 6).

(2) AC Link Chopper (inverter-rectifier) In this method the d.c. is first converted to a.c, by an inverter (d.c. to a.c. converter). The obtained a.c. is then stepped up or down by a transformer and then rectified back to d.c. by a rectifier. As the conversion is in two stages, d.c. to a.c. and a.c. to d.c., this technique is therefore, costly, bulky and less efficient. However, the transformer provides isolation between load and source. Figure 8.1 illustrates the conversion processes.

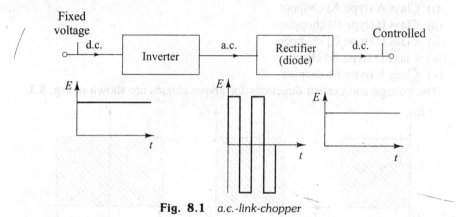

Fig. 8.1 *a.c.-link-chopper*

(3) DC Chopper (d.c. to d.c. power converters) A d.c. chopper is a static device (switch) used to obtain variable d.c. voltage from a source of constant d.c. voltage, Fig. 8.2. Therefore, chopper may be thought of as d.c. equivalent of an a.c. transformer since they behave in an identical manner. Besides, the saving in power, the d.c. chopper offers greater efficiency, faster response, lower maintenance, small size, smooth control, and, for many applications, lower cost, than motor-generator sets or gas tubes approaches.

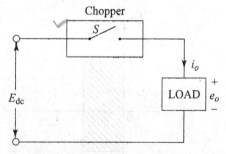

Fig. 8.2 *Basic chopper configuration*

Solid-state choppers due to various advantages are widely used in trolley cars, battery-operated vehicles, traction-motor control, control of a large number of d.c. motors from a common d.c. bus with a considerable improvement of power factor, control of induction motors, marine hoists, forklift trucks and mine haulers. The objective of this chapter is to discuss the basic principles of chopper operation and more common types of chopper configuration circuits.

8.2 BASIC CHOPPER CLASSIFICATION

DC choppers can be classified as:

(A) According to the Input/Output Voltage Levels

  (i) **Step-down chopper:** The output voltage is less than the input voltage.

  (ii) **Step-up chopper:** The output voltage is greater than the input voltage.

(B) According to the Directions of Output Voltage and Current

  (i) Class A (type A) chopper
  (ii) Class B (type B) chopper
  (iii) Class C (type C) chopper
  (iv) Class D (type D) chopper
  (v) Class E (type E) chopper

The voltage and current directions for above classes are shown in Fig. 8.3.

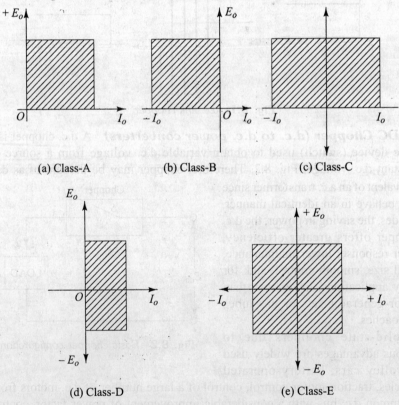

(a) Class-A   (b) Class-B   (c) Class-C

(d) Class-D   (e) Class-E

Fig. 8.3 *Chopper configurations*

(C) According to Circuit Operation

  (i) *First-quadrant chopper:* The output voltage and both must be positive. (Type A).

(ii) *Two-quadrant chopper:* The output voltage is positive and current can be positive or negative (class-C) or the output current is positive and the voltage can be positive or negative (class-D).

(iii) *Four-quadrant chopper:* The output voltage and current both can be positive or negative (class-E).

(D) According to Commutation Method

(i) Voltage-commutated choppers   VCLI
(ii) Current-commutated choppers
(iii) Load-commutated choppers   VLCI
(iv) Impulse-commutated choppers

8.3 BASIC CHOPPER OPERATION

8.3.1 Principle of Step-Down Chopper (Buck-Converter)

In general, d.c. chopper consists of power semiconductor devices (SCR, BJT, power MOSFET, IGBT, GTO, MCT, etc., which works as a switch), input d.c. power supply, elements (R, L, C, etc.) and output load. (Fig. 8.4). The average output voltage across the load is controlled by varying on-period and off-period (or duty cycle) of the switch.

A commutation circuitry is required for SCR based chopper circuit. Therefore, in general, gate-commutation devices based choppers have replaced the SCR-based choppers. However, for high voltage and high-current applications, SCR based choppers are used. The variations in on- and off periods of the switch provides an output voltage with an adjustable average value. The power-diode (D_P) operates in freewheeling mode to provide a path to load-current when switch (S) is OFF. The smoothing inductor filters out the ripples in the load current. Switch S is kept conducting for period T_{on} and is blocked for period T_{off}. The chopped load voltage waveform is shown in Fig. 8.5.

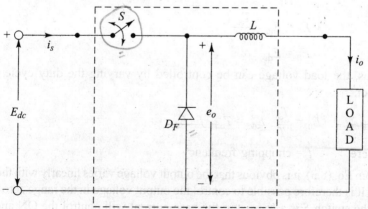

Fig. 8.4 *Basic chopper circuit*

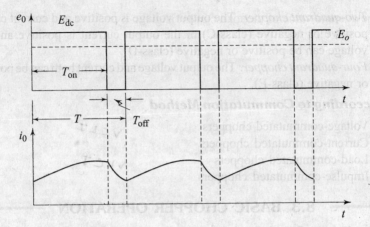

Fig. 8.5 *Output voltage and current waveforms*

During the period T_{on}, when the chopper is on, the supply terminals are connected to the load, terminals. During the interval T_{off}, when the chopper is off, load current flows through the freewheeling diode D_F. As a result, load terminals are short circuited by D_F and load voltage is therefore, zero during T_{off}. In this manner, a chopped d.c. voltage is produced at the load terminals.

From Fig. 8.5, the average load-voltage E_0 is given by

$$E_0 = E_{dc} \frac{T_{on}}{T_{on+off}} \qquad (8.1)$$

where T_{on} = on-time of the chopper, T_{off} = off-time of the chopper

$T = T_{on} + T_{off}$ = chopping period

If $\alpha = \dfrac{T_{on}}{T}$ be the duty cycle, then above equation becomes,

$$E_0 = E_{dc} \cdot \frac{T_{on}}{T} \qquad (8.2)$$

$$E_0 = E_{dc} \cdot \alpha \qquad (8.3)$$

Thus, the load voltage can be controlled by varying the duty cycle of the chopper.

Also, $\qquad E_0 = \dfrac{T_{on}}{T} E_{dc} = T_{on} \cdot f \cdot E_{dc} \qquad (8.4)$

where f = chopping frequency

From Eq. (8.3), it is obvious that the output voltage varies linearly with the duty-cycle. It is therefore possible to control the output voltage in the range zero to E_{dc}.

If the switch S is a transistor, the base-current will control the ON and OFF period of the transistor switch. If the switch is GTO thyristor, a positive gate pulse will turn-it ON and a negative gate pulse will turn it OFF. If the switch is an SCR, a commutation circuit is required to turn it OFF.

The average value of the load current is given by

$$I_0 = \frac{E_o}{R} = \frac{\alpha \cdot E_{dc}}{R} \qquad (8.5)$$

The effective (RMS) value of the output voltage is given by

$$E_{o\,(RMS)} = \sqrt{\frac{E_{dc}^{\,2} \cdot T_{on}}{T}} = E_{dc} \cdot \sqrt{\frac{T_{on}}{T}}$$

$$= E_{dc}\sqrt{\alpha} \qquad (8.6)$$

SOLVED EXAMPLES

Example 8.1 A d.c. chopper circuit connected to a 100 V d.c. source supplies an inductive load having 40 mH in series with a resistance of 5 Ω. A freewheeling diode is placed across the load. The load current varies between the limits of 10 A and 12 A. Determine the time ratio of the chopper.

Solution: The average value of the load current $= \dfrac{I_1 + I_2}{2} = \dfrac{10 + 12}{2} = 11$ A.

The maximum value of the load current $= \dfrac{100}{5} = 20$ A

Now, the average value of the voltage, $E_{0av} = 100 \times \dfrac{11}{20} = 55$ V

Also, $\quad E_{dc} \cdot \dfrac{T_{on}}{T_{on} + T_{off}} = E_{0av} \quad$ or $\quad \dfrac{T_{on}}{T_{on} + T_{off}} = \dfrac{E_{0av}}{E_{dc}}$

$$\dfrac{T_{on}}{T_{on} + T_{off}} = \dfrac{55}{100} = 0.55 \quad \therefore T_{on} = 0.55\,(T_{on} + T_{off})$$

$$\therefore \qquad \dfrac{T_{on}}{T_{off}} = \dfrac{0.55}{0.45} = 1.222.$$

Example 8.2 For the chopper circuit shown in Fig. Ex. 8.2, express the following variables as functions of E_{dc}, R, and duty cycle α.
  (i) Average output voltage and current.
  (ii) Output current at the instant of commutation.
  (iii) Average and RMS freewheeling diode currents.
  (iv) RMS value of the output voltage.
  (v) RMS and average load currents.

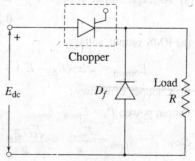

Fig. Ex. 8.2

Solution: With resistive load, load current waveforms are similar to load voltage waveforms.

∴ (i) Average output voltage $E_0 = E_{dc} \dfrac{T_{on}}{T} = E_{dc} \cdot \alpha$.

Average output current, $I_{0av} = \dfrac{E_0}{R} = \dfrac{E_{dc}}{R}\alpha$.

(ii) Output current at the instant of commutation = $\dfrac{E_{dc}}{R}$.

(iii) Freewheeling diode does not come into picture for a resistive load. Hence, average and RMS values of freewheeling diode currents are zero.

(iv) RMS value of output voltage

$$= \left[\dfrac{T_{on}}{T}E_{dc}^2\right]^{1/2} = \sqrt{\alpha}\cdot E_{dc}$$

(v) Now, average thyristor current

$$= \dfrac{E_{dc}}{R}\cdot\dfrac{T_{on}}{T} = \alpha\dfrac{E_{dc}}{R}$$

RMS thyristor current $= \left(\alpha\cdot\left(\dfrac{E_{dc}}{R}\right)^2\right)^{1/2} = \sqrt{\alpha}\cdot\dfrac{E_{dc}}{R}$

Example 8.3 A step-down dc chopper has a resistive load of $R = 15$ ohm and input voltage $E_{dc} = 200$ V. When the chopper remains ON, its voltage drop is 2.5 V. The chopper frequency is 1 kHz. If the duty cycle is 50%, determine:
(a) Average output voltage
(b) RMS output voltage
(c) Chopper efficiency
(d) Effective input resistance of chopper

Solution:

Given: Input voltage $E_{dc} = 200$ V, duty cycle $\alpha = 0.5$

$R = 15\ \Omega$, $F = 1$ kHz, Chopper drop $E_d = 2.5$ V

(a) Average output voltage $E_0 = \alpha\cdot(E_{dc} - E_d)$
$= 0.5(200 - 2.5) = 98.75$ V

(b) RMS output voltage

$E_{0(rms)} = \sqrt{\alpha}\ (E_{dc} - E_d) = \sqrt{0.5}\ (200 - 2.5) = 139.653$ V

(c) Chopper efficiency

Output power, $P_0 = E_{0rms}\cdot I_{0rms}$

Now, $I_{0rms} = \dfrac{E_{0rms}}{R} = \dfrac{\sqrt{\alpha}\cdot E_{dc}}{R}$

∴ $P_0 = \sqrt{\alpha}\cdot E_{dc}\cdot\dfrac{\sqrt{\alpha}\cdot E_{dc}}{R} = \dfrac{\alpha E_{dc}^2}{R}$

If E_d is the chopper drop, then

$$P_0 = \frac{\alpha(E_{dc} - E_d)^2}{R} = \frac{0.5(200 - 2.5)^2}{15} = 1300.21 \; \omega$$

Now, the input power to the chopper is given by

$$P_i = \frac{1}{T}\int_0^T E_{dc} \, i_s \, dt = \frac{1}{T}\int_0^{T_{on}} E_{dc} \frac{(E_{dc} - E_d)}{R} dt = \frac{1}{T}\int_0^{\alpha \cdot T} \frac{E_{dc}(E_{dc} - E_d)}{R} dt$$

$$= \frac{E_{dc}(E_{dc} - E_d)}{T \cdot R}(t)_0^{\alpha_T} = \frac{\alpha E_{dc}(E_{dc} - E_d)}{R} = \frac{0.5(200)(200 - 2.5)}{15} = 1316.67 \; W$$

$\therefore$ Chopper efficiency, $\eta = \dfrac{P_o}{P_i} = \dfrac{1300.21}{1316.67} = 0.9874 = 98.74\%$

8.3.2 Principle of Step-up Choppers

The chopper configuration of Fig. 8.4 is capable of giving a maximum voltage that is slightly smaller than the input d.c. voltage (i.e. $E_0 < E_{dc}$). Therefore, the chopper configuration of Fig. 8.4 is called as step-down choppers. However, the chopper can also be used to produce higher voltages at the load than the input voltage (i.e., $E_0 \geq E_{dc}$). This is called as step-up chopper and is illustrated in Fig. 8.6.

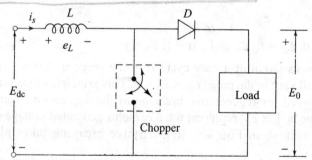

Fig. 8.6 Step-up chopper or boost choppers

When the chopper is ON, the inductor L is connected to the supply E_{dc}, and inductor stores energy during on-period, T_{on}.

When the chopper is OFF, the inductor current is forced to flow through the diode and load for a period T_{off}. As the current tends to decrease, polarity of the emf induced in inductor L is reversed to that of shown in Fig. 8.6, and as a result voltage across the load E_0 becomes

$$E_0 = E_{dc} + L\frac{di_s}{dt}$$

that is, the inductor voltage adds to the source voltage to force the inductor current into the load. In this manner, the energy stored in the inductor is released to the load. Here, higher value of inductance L is preferred for getting lesser ripple in the output.

During the time T_{on}, when the chopper is ON, the energy input to the inductor from the source is given by

$$W_i = E_{dc} I_s T_{on} \tag{8.7}$$

Equation 8.7 is based on the assumption that the source current is free from ripples.

Now, during the time T_{off}, when chopper is OFF, energy released by the inductor to the load is given by

$$W_0 = (E_0 - E_{dc}) I_s T_{off} \tag{8.8}$$

Considering the system to be lossless, and, in the steady-state, these two energies will be equal.

$$\therefore \quad E_{dc} \cdot I_s T_{on} = (E_0 - E_{dc}) I_s T_{off}$$

or $$E_0 = E_{dc} \frac{T_{on} + T_{off}}{T_{off}} \tag{8.9}$$

or $$E_0 = E_{dc} \frac{T}{T - T_{on}} \tag{8.10}$$

or $$E_0 = E_{dc} \frac{1}{T/T - T_{on}/T}, \text{ But, } \frac{T_{on}}{T} = \alpha$$

$$\therefore \quad E_0 = \frac{E_{dc}}{1 - \alpha} \tag{8.11}$$

For $\alpha = 0$, $E_0 = E_{dc}$; and $\alpha = 1$, $E_0 = \infty$.

Hence, for variation of a duty cycle α in the range $0 < \alpha < 1$, the output voltage E_0 will vary in the range $E_{dc} < E_0 < \infty$. This principle of step-up chopper can be employed for regenerative breaking of the d.c. motors even at lower operating speeds. Let E_{dc} represent the d.c. motor generated voltage and E_0 the d.c. source voltage in Fig. 8.6. Regenerative breaking takes place when $\left(E_{dc} + L\dfrac{di_s}{dt}\right)$ exceeds E_0. Even at decreasing motor speeds, duty cycle α can be so adjusted that $\left(E_{dc} + L\dfrac{di_s}{dt}\right)$ is more than the fixed supply voltage E_0.

SOLVED EXAMPLE

Example 8.4 A step-up chopper is used to deliver load voltage of 500 V from a 220V d.c. source. If the blocking period of the thyristor is 80 μs, compute the required pulse width.

Solution: From Eq. (8.10) we have, $E_0 = E_{dc} \dfrac{T_{on} + T_{off}}{T_{off}}$

$$\therefore \quad 500 = 220 \frac{T_{on} + 80 \times 10^{-6}}{80 \times 10^{-6}}, \quad \therefore T_{on} = 101.6 \times 10^{-6} = 101.6 \text{ μs}.$$

8.3.3 Principle of Step-Up/Down Choppers

A chopper can also be used both in step-up and step-down modes by continuously varying its duty cycle. The principle of operation is illustrated in Fig. 8.7. As shown, the output, voltage polarity is opposite to that of input voltage E_{dc}.

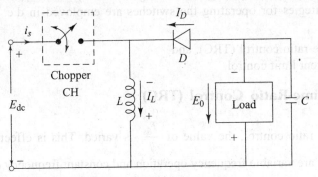

Fig. 8.7 *Step-up/down chopper*

When the chopper is ON, the supply current flows through the path E_{dc+} – CH – L – E_{dc-}. Hence, inductor L stores the energy during the T_{on} period.

When the chopper CH is OFF, the inductor current tends to decrease and as a result, the polarity of the emf induced in L is reversed as shown in Fig. 8.7. Thus, the inductance energy discharges in the load through the path,

$$L_+ - \text{Load} - D - L_-.$$

During T_{on}, the energy stored in the inductance is given by

$$W_i = E_{dc} I_s T_{on} \tag{8.12}$$

During T_{off}, the energy fed to the load is

$$W_o = E_0 I_s T_{off} \tag{8.13}$$

For a lossless system, in steady-state: Input energy, W_i = output energy, W_o

$$\therefore \quad E_{dc} \cdot I_s \cdot I_{on} = E_0 I_s T_{off}, \text{ or } E_0 = E_{dc} \cdot \frac{T_{on}}{T_{off}} \tag{8.14}$$

or $\qquad E_0 = E_{dc} \cdot \dfrac{T_{on}}{T - T_{on}} = E_{dc} \dfrac{1}{T/T_{on} - T_{on}/T_{on}}$

Substituting $\dfrac{T_{on}}{T} = \alpha$, we get, $E_0 = E_{dc} \cdot \dfrac{1}{1/\alpha - 1}$

or $\qquad E_0 = E_{dc} \dfrac{\alpha}{1-\alpha} \tag{8.15}$

For $0 < \alpha < 0.5$, the step-down chopper operation is achieved and for $0.5 < \alpha < 1$, step-up chopper operation is obtained.

8.4 CONTROL STRATEGIES

It is seen from Eq. (8.3), that, average value of output voltage, E_0 can be controlled by periodic opening and closing of the switches. The two types of control strategies for operating the switches are employed in d.c. choppers. They are:

(1) Time-ratio control (TRC), and
(2) Current limit control.

8.4.1 Time-Ratio Control (TRC)

In the time-ratio control, the value of $\dfrac{T_{on}}{T}$ is varied. This is effected in two ways. They are variable frequency operation and constant frequency operation.

1. Constant Frequency System In this type of control strategy, the on-time T_{on}, is varied but the chopping frequency f ($f = 1/T$, and hence the chopping period T) is kept constant. This control strategy is also called as the *pulse-width modulation control*.

Figure 8.8 illustrates the principle of pulse-width modulation. As shown, chopping period T is constant. In Fig. 8.8(a), $T_{on} = \frac{1}{4} T$, so that duty cycle $\alpha = 25\%$. In Fig. 8.8 (b), $T_{on} = \frac{3}{4} T$, so that duty cycle $\alpha = 75\%$. Hence, the output voltage E_0 can be varied by varying the on-time T_{on}.

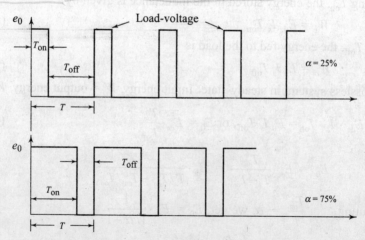

Fig. 8.8 *Pulse-width modulation control (constant frequency f)*

2. Variable Frequency System In this type of control strategy, the chopping frequency f is varied and either–

(a) ON-time, T_{on}, is kept constant or (b) OFF-time, T_{off}, is kept constant. This type of control strategy is also called as *frequency modulation control*.

Figure 8.9 illustrates the principle of frequency modulation. As shown in Fig. 8.9(a), chopping period T is varied but on-time T_{on} is kept constant. The output voltage waveforms are shown for two different duty cycles. In Fig. 8.9(b), chopping period T is varied but T_{off} is kept constant.

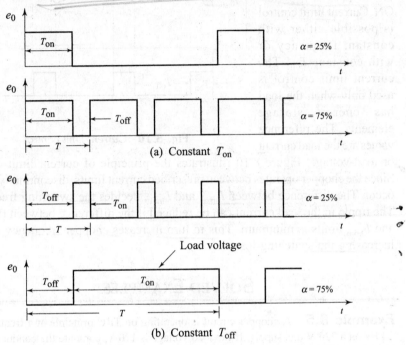

Fig. 8.9 *Output voltage waveforms for variable frequency system*

Frequency modulation control strategy has the following major disadvantages compared to pulse-width modulation control.

(i) The chopping frequency has to be varied over a wide range for the control of output voltage in frequency modulation. Filter design for such wide frequency variation is, therefore, quite difficult.
(ii) For the control of duty cycle, frequency variation would be wide. As such, there is a possibility of interference with signalling and telephone lines in frequency modulation technique.
(iii) The large OFF-time in frequency modulation technique may make the load current discontinuous, which is undesirable.

Thus, the constant frequency system (PWM) is the preferred scheme for chopper drives.

8.4.2 Current Limit Control

In current limit control strategy, the chopper is switched ON and OFF so that the current in the load is maintained between two limits. When the current exceeds upper limit, the chopper is switched OFF. During OFF period, the load current freewheels and decreases exponential. When it reaches the lower limit, the chopper is switched ON. Current limit control is possible either with constant frequency or with constant T_{on}. The current limit control is used only when the load has energy storage elements. The reference values are the load current

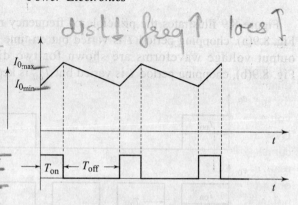

Fig. 8.10 Current limit control

or load-voltage. Figure 8.10 illustrates the principle of current limit control. Since the chopper operates between prescribed current limits, discontinuity cannot occur. The difference between I_{0max} and I_{0min}, decides the switching frequency. The ripple in the load current can be reduced if the difference between the I_{0max} and I_{0min} limits is minimum. This in turn increases chopper frequency thereby increasing the switching losses.

SOLVED EXAMPLES

Example 8.5 A chopper circuit is operating on TRC principle at a frequency of 2 kHz on a 220 V d.c. supply. If the load voltage is 170 V, compute the conduction and blocking period of thyristor in each cycle.

Solution: From Eq. (8.2), $E_0 = E_{dc} \cdot T_{on} \cdot f$
Given : $f = 2$ kHz, $E_{dc} = 220$ V, $E_o = 170$.

Conduction period, $T_{on} = \dfrac{E_0}{E_{dc} \cdot f} = \dfrac{170}{220 \times 2 \times 10^3} = T_{on} = 0.386$ ms.

But, chopping period, $T = \dfrac{1}{f} = \dfrac{1}{2 \times 10^3} = 0.5$ ms

∴ Blocking period of SCR, $T_{off} = T - T_{on} = 0.5 - 0.386 = 0.114$ m sec.

Example 8.6 In a 110 V dc chopper drive using the CLC scheme, the maximum possible value of the accelerating current is 300 A, the lower-limit of the current pulsation is 140 A. The ON- and OFF periods are 15 ms and 12 ms, respectively. Calculate the limit of current pulsation, chopping frequency, duty cycle and the output voltage.

Solution: Given: $T_{on} = 15$ ms, $T_{off} = 12$ ms, $I_{0max} = 300\ A$, $I_{0min} = 140\ A$
Now, maximum limit of current pulsation = 300 − 140 = 160 A.

Chopping frequency = $\dfrac{1}{T} = \dfrac{1}{15 + 12} = 37$ Hz & ratio, $\alpha = \dfrac{T_{on}}{T} = \dfrac{15}{27} = 0.56$

Output voltage, $E_0 = \alpha E_{dc} = 0.56 \times 110 = 61.60$ V

8.5 CHOPPER CONFIGURATION

Choppers may be classified according to the number of quadrants of the $E_0 - I_0$ diagram in which they are capable of operating. A classification that is convenient for the discussion that follows is shown in Fig. 8.11. The polarity of the output voltage and the direction of energy flow cannot be changed in Figs 8.4 and 8.6. By various combination of connections it is possible to realize any combination of output voltage and current polarity. With reference to the combination shown in Fig. 8.11, if the load is a separately excited motor of constant field, then the positive voltage and positive current in the first quadrant give rise to a "forward drive." Changing the polarity of both the armature voltage and the armature current results in a "reverse" drive (quadrant III). In II and IV quadrants, the direction of energy flow is reversed and the motor operates as a *generator braking* rather than driving.

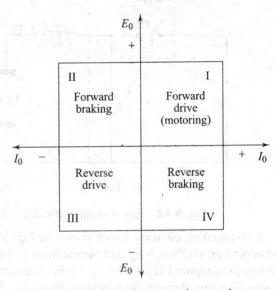

Fig. 8.11 *Polarities of output voltage and current*

In regenerative breaking, most of the breaking energy is returned to the supply. The condition for regeneration is that the rotational emf must be more than the applied voltage so that the current is reversed and the mode of operation changes from motoring to generating. It was observed that about 35% of the energy put into an automotive vehicle during typical urban traction is theoretically recoverable by regenerative breaking. However, the exact value of the recoverable energy is a function of the type of driving, the efficiency of the drive train, gear ratios in the drive/train etc. Therefore, the choppers which gives this regenerative breaking facility are widely used compared to systems without regenerative breaking.

D.C. chopper circuits are combined in accordance with the quadrants, in which a d.c. motor assumed as a load is required to operate. In the first and third quadrants, for instance, a resistance may also serve as a load, but a generating mode can be maintained over any significant span of time only, if the load is capable of delivering sustained power. This section describes the classification of various chopper configurations.

8.5.1 First Quadrant or Class A Chopper [Step-down Chopper with R–L Load]

Figure 8.12 illustrates the basic power circuit of first quadrant chopper. The term 'first quadrant' signifies that circuit parameters E_0 and I_0 occur only in the first quadrant of $E_0 - I_0$ diagram.

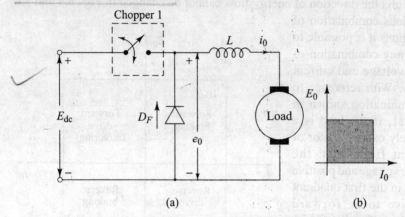

Fig. 8.12 *Type A chopper circuit and $E_0 - I_0$ characteristic*

Commutating circuitry is not shown in Fig. 8.12 for simplicity. When the chopper CH_1 is ON, $e_0 = E_{dc}$ and current flows in the direction shown in Fig. 8.12. When the chopper CH_1 is OFF, $e_0 = 0$ but the current i_0 flows in the load in the same direction through freewheeling diode D_f. Therefore, both average load voltage E_0 and current I_0 are positive and thus power flows from source to load. This operation is shown by the hatched area in Fig. 8.12(b). Therefore, this configuration is used for motoring operation of d.c. motor load. Class A chopper circuit is also called as step-down chopper as average output voltage E_0 is always less than the d.c. input voltage E_{dc}. Due to the motoring operation, this chopper is also called as motoring chopper.

1. Steady-state Time-domain Analysis Class A chopper circuit of Fig. 8.12 can also be drawn in terms of three separate circuit elements, as shown in Fig. 8.13. Here, load is R–L E_b type load. E_b is the load voltage which may be a d.c. motor or a battery.

The operation of this system may be understood from the consideration of the waveforms of the circuit variables shown in Fig. 8.14. This Fig. 8.14 shows the two modes of circuit operation.

(i) When chopper CH_1 is ON, the supply voltage E_{dc} appears at the terminals of the armature circuit and, the current i_0 would increase until it reached the steady-state value expressed by

$$i_0 = \frac{E_{dc} - E_b}{R} \qquad (8.16)$$

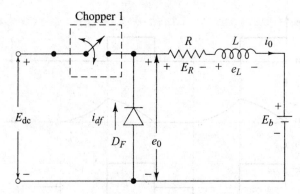

Fig. 8.13 *First quadrant chopper with R – L load*

The average current I_0 in the circuit can be controlled by commutating chopper CH_1 before the current has reached the value given by Eq. (8.16), and allowing it to decay through diode DF either to zero, as shown in Fig. 8.14(a), or to some lower value than it had attained while CH_1 was conducting, as shown in Fig. 8.14(b). If this process of turning chopper CH_1 ON and OFF is repeated at regular intervals, then average value of i_0 is controlled.

As shown in Fig. 8.14(a), the turn-on time of chopper T_{on} is shorter in relation to chopping period T, which results in a discontinuous current. Therefore, the current waveform consists of a series of pulses and these pulses become identical when steady-state conditions have been reached.

If turn-on time T_{on} is longer in relation to T, the load current will not decay to zero during the interval $T_{on} < t < T$, but will merely decrease until CH_1 is again turned-on. Therefore, in the steady state, the current will flow continuously as shown in Fig. 8.14(b).

Mode 1: $0 \leq t \leq T_{on}$ When the chopper is ON, current flows through the path $E_{dc+} - R - L - E_b - E_{dc-}$. For this mode of operation, the differential equation governing its performance is given by

$$E_{dc} = R \cdot i_0 + L\frac{di_0}{dt} + E_b, \text{ for } 0 \leq t \leq T_{on} \tag{8.17}$$

Mode 2: $T_{on} \leq t \leq T$ When chopper is OFF, the load current continuously flowing through the freewheeling diode D_f. For this mode of operation, the differential equation governing its performance is given by

$$0 = R \cdot i_0 + L\frac{di_0}{dt} + E_b, \text{ for } T_{on} \leq t \leq T \tag{8.18}$$

From Fig. 8.14(b), it is observed that the initial value of current for Eq. (8.17) is I_{0min} and I_{0max} for Eq. (8.18).

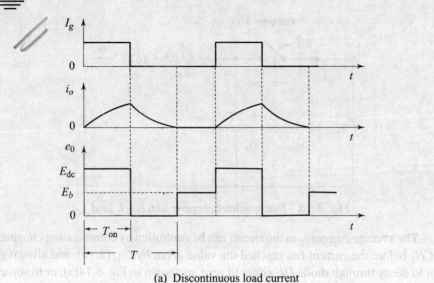

(a) Discontinuous load current

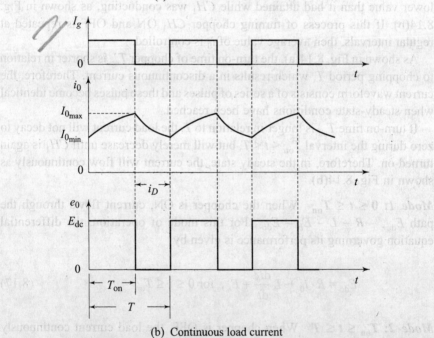

(b) Continuous load current

Fig. 8.14 *First quadrant chopper, two modes operation*

Now, by taking Laplace-transforms of Eqs (8.17) and (8.18), we can write

$$\frac{(E_{dc} - E_b)}{s} = R\,I_{0(s)} + L[sI_{0(s)} - I_{0min}] \tag{8.19}$$

and

$$\frac{-E_b}{s} = RI_{0(s)} + L[sI_{0(s)} - I_{0max}] \tag{8.20}$$

Equation 8.19 can also be written as

$$I_0(s) = \frac{(E_{dc} - E_b)}{s(R + Ls)} + \frac{LI_{0min}}{(R + Ls)}$$

or

$$I_0(s) = \frac{(E_{dc} - E_b)}{sL(s + R/L)} + \frac{L \cdot I_{0min}}{L(s + R/L)}$$

or

$$I_0(s) = \frac{(E_{dc} - E_b)}{sL(s + R/L)} + \frac{I_{0min}}{(s + R/L)} \tag{8.21}$$

Now, by taking inverse-Laplace of Eq. 8.21, we get

$$i_0(t) = \frac{(E_{dc} - E_b)}{R}\left(1 - e^{-R/L \cdot t}\right) + I_{0min} \cdot e^{-R/L \cdot t}, \ 0 \le t \le T_{on} \tag{8.22}$$

Let us define time-constant, $\tau = L/R$, so Eq. (8.22) becomes

$$i_0(t) = \frac{(E_{dc} - E_b)}{R}\left(1 - e^{-t/\tau}\right) + I_{0min} \cdot e^{-t/\tau}, 0 \le t \le T_{on} \tag{8.23}$$

When chopper is commutated at $t = T_{on}$, $i_0(t) = I_{0max}$.
Thus, Eq. (8.23) becomes

$$I_{0max} = \frac{(E_{dc} - E_b)}{R}\left(1 - e^{-T_{on}/\tau}\right) + I_{0min} \ e^{-T_{on}/\tau} \tag{8.24}$$

From Eq. (8.20), we can write, $I_{0(s)}[R + Ls] = \frac{-E_b}{s} + L \cdot I_{0max}$

or $I_{0(s)} = \frac{-E_b}{s(R + Ls)} + \frac{L \cdot I_{0max}}{(R + Ls)}$ = or $I_{0(s)} = \frac{-E_b}{s \cdot L \cdot (s + R/L)} + \frac{L \cdot I_{0max}}{(s + R/L)}$

or

$$I_{0(s)} = \frac{-E_b}{sL(s + R/L)} + \frac{I_{0max}}{(s + R/L)} \tag{8.25}$$

Now, by taking inverse-Laplace transform of Eq. (8.25) we get

$$i_{0(t)} = \frac{-E_b}{R}\left(1 - e^{-t/\tau}\right) + I_{0max} e^{-t/\tau}, T_{on} \le t \le T \tag{8.26}$$

For interval $T_{on} \le t \le T$, let us define $t' = t - T_{on}$, so that when
$t = T_{on}, t' = 0$, and for $t = T, t' = T - T_{on} = T_{off}$.
Substituting t' in Eq. (8.26), we get

$$i_0(t') = \frac{-E_b}{R}\left(1 - e^{-t'/\tau}\right) + I_{0max} e^{-t'/\tau}, T_{on} \le t \le T \tag{8.27}$$

Now, at $t' = T - T_{on} = T_{off}, i_0(t') = I_{0min}$.
Thus, Eq. (8.27) becomes

∴
$$I_{0min} = \frac{-E_b}{R}\left(1 - e^{-(T - T_{on})/\tau}\right) + I_{0max} e^{-(T - T_{on})/\tau} \tag{8.28}$$

Equations (8.24) can be solved for I_{0max} and I_{0min}, as follows:
From Eq. (8.24), we can write

$$I_{0max} = \frac{E_{dc}}{R}\left(1 - e^{-T_{on}/\tau}\right) - \frac{E_b}{R}\left(1 - e^{-T_{on}/\tau}\right) + I_{0min} e^{-T_{on}/\tau} \tag{8.29}$$

Substituting Eq. (8.28) for I_{0min} in Eq. (8.29), gives

$$I_{0max} = \frac{E_{dc}}{R}\left(1 - e^{-T_{on}/\tau}\right) - \frac{E_b}{R}\left(1 - e^{-T_{on}/\tau}\right)$$

$$- \frac{E_b}{R} e^{-T_{on}/\tau}\left(1 - e^{-(T-T_{on})/\tau}\right) + I_{0max} e^{-(T-T_{on})/\tau} \cdot e^{-T_{on}/\tau}$$

$$= \frac{E_{dc}}{R}\left(1 - e^{-T_{on}/\tau}\right) - \frac{E_b}{R} + \frac{E_b}{R} e^{-T_{on}/\tau} - \frac{E_b}{R} e^{-T_{on}/\tau}$$

$$+ \frac{E_b}{R} e^{-(T-T_{on})/\tau} e^{-T_{on}/\tau} + I_{0max} e^{-T/\tau} e^{+T_{on}/\tau} e^{-T_{on}/\tau}$$

$$= \frac{E_{dc}}{R}\left(1 - e^{-T_{on}/\tau}\right) - \frac{E_b}{R}\left(1 - e^{-T/\tau}\right) + I_{0max} e^{-T/\tau}$$

or,

$$I_{0max} - I_{0max} e^{-T/\tau} = \frac{E_{dc}}{R}\left(1 - e^{-T_{on}/\tau}\right) - \frac{E_b}{R}\left(1 - e^{-T/\tau}\right)$$

$$I_{0max}(1 - e^{-T/\tau}) = \frac{E_{dc}}{R}\left(1 - e^{-T_{on}/\tau}\right) - \frac{E_b}{R}\left(1 - e^{-T/\tau}\right)$$

or, $$I_{0max} = \frac{E_{dc}}{R}\left[\frac{1 - e^{-T_{on}/\tau}}{1 - e^{-T/\tau}}\right] - \frac{E_b}{R} \tag{8.30}$$

Now, substitute value of I_{0max} from Eq. (8.30) into Eq. (8.28),

$\therefore$ $$I_{0min} = \frac{-E_b}{R} + \frac{E_b}{R} e^{-(T-T_{on})/\tau} + \frac{E_{dc}}{R}\left[\frac{1 - e^{-T_{on}/\tau}}{1 - e^{-T/\tau}}\right]$$

$$\times e^{-(T-T_{on})/\tau} - \frac{E_b}{R} e^{-(T-T_{on})/\tau}$$

$$= \frac{E_{dc}}{R}\left[\frac{1 - e^{-T_{on}/\tau}}{1 - e^{-T/\tau}}\right]\frac{e^{T_{on}/\tau}}{e^{T/\tau}} - \frac{E_b}{R}$$

or, $$I_{0min} = \frac{E_{dc}}{R}\left[\frac{e^{+T_{on}/\tau} - 1}{e^{+T/\tau} - 1}\right] - \frac{E_b}{R} \tag{8.31}$$

When chopper CH_1 is continuously turned-on, then, $T_{on} = T$ and both I_{0max} and I_{0min} have the value given by Eq. (8.16), i.e.

$$I_{0max} = I_{0min} = \frac{E_{dc} - E_b}{R} \qquad (8.32)$$

Thus, for given value of E_{dc}, E_b, R, α and τ, the maximum I_{0max} and minimum I_{0min} value of load current can be obtained from Eqs (8.30) and (8.31) respectively.

2. Steady-state Ripple From Fig. 8.14(b), it is observed that load-current i_o varies between the value I_{0max} and I_{0min}. Therefore, the ripple current $(I_{0max} - I_{0min})$ can be calculated from Eqs (8.30) and (8.31) as follows:

$$\therefore \ (I_{0max} - I_{0min}) = \frac{E_{dc}}{R}\left[\frac{(1-e^{-T_{on}/\tau})}{(1-e^{-T/\tau})} - \frac{e^{T_{on}/\tau} - 1}{e^{T/\tau} - 1}\right]$$

$$= \frac{E_{dc}}{R}\left[\frac{(1-e^{-T_{on}/\tau})}{(1-e^{-T/\tau})} - \frac{(1-e^{-T_{on}/\tau})e^{T_{on}/\tau}}{(1-e^{-T/\tau})e^{T/\tau}}\right]$$

$$= \frac{E_{dc}}{R}\left[\frac{(1-e^{-T_{on}/\tau}) - (1-e^{-T_{on}/\tau})e^{(T_{on}-T)/\tau}}{(1-e^{-T/\tau})}\right]$$

$$= \frac{E_{dc}}{R}\left[\frac{(1-e^{-T_{on}/\tau})(1-e^{-(T-T_{on})/\tau})}{(1-e^{-T/\tau})}\right] \qquad (8.33)$$

It is seen from Eq. (8.33) that, ripple-current is independent of back emf E_b. Now, we know that $\alpha = T_{on}/T$ $\therefore T_{on} = \alpha \cdot T$ (8.34)
Also $\quad T - T_{on} = (1 - \alpha)T$ (8.35)

Substitute Eqs (8.34) and (8.35) into Eq. (8.33) gives

$$(I_{0max} - I_{0min}) = \frac{E_{dc}}{R}\left[\frac{(1-e^{-\alpha T/\tau})(1-e^{-(1-\alpha)T/\tau})}{(1-e^{-T/\tau})}\right]$$

$$\therefore \quad \text{Per unit (PU) ripple current} = \frac{(I_{0max} - I_{0min})}{E_{dc}/R}$$

$$= \frac{(1-e^{-\alpha T/\tau})(1-e^{-(1-\alpha)T/\tau})}{(1-e^{-T/\tau})} \qquad (8.36)$$

For duty cycle $\alpha = 50\%$ (i.e. $\alpha = 0.5$) and $T/\tau = 5$, per unit ripple current become 0.848. Similarly, for $\alpha = 0.5$ and $T/\tau = 25$, per unit ripple current = 1. In this way, the variation of per unit ripple current as a function of duty cycle α and ratio T/τ can be plotted, as shown in Fig. 8.15. The PU ripple current has the maximum value when $\alpha = 0.5$. As the inductance L is increased, time-constant τ also increases and ratio T/τ reduces, therefore PU ripple current decreases, as shown in Fig. 8.15.

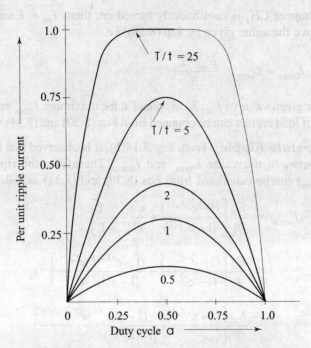

Fig. 8.15 PU ripple current as a function of a and T/τ

3. Fourier Analysis of Output Voltage

Figure 8.14(b) illustrates the load voltage waveform e_0, for continuous load current. As shown, this voltage waveform is periodic in nature and is independent of load-circuit parameters. This voltage waveform may be described by the Fourier-series as

$$e_0 = E_0 + \sum_{n=1}^{\infty} C_n \tag{8.37}$$

where C_n = value of nth harmonic voltage

$$= \frac{2E_{dc}}{n\pi} \sin n\pi\alpha \sin(n\omega_0 t + \theta_n) \tag{8.38}$$

E_0 = average voltage of output voltage = αE_{dc}

$$\alpha = \text{duty cycle} = \frac{T_{on}}{T}, \text{ and } \theta_n = \tan^{-1}\frac{\sin 2\pi n\alpha}{1 - \cos 2\pi n\alpha}$$

The average output voltage can be controlled by varying the duty-cycle α. From Eq. (8.38), it becomes clear that the amplitude of the harmonic voltage $\left(=\frac{2E_{dc}}{n\pi}\sin n\pi\alpha\right)$ depends on the order of harmonic n and also on the duty cycle α. The maximum value of the nth harmonic occurs when $\sin n\pi\alpha = 1$ and its value is

$$\frac{2E_{dc}}{n\pi} = \frac{0.6366\,E_{dc}}{n} \text{ V, and its RMS value is}$$

$$\frac{2E_{dc}}{\sqrt{2}\,n\pi} = \frac{0.45\,E_{dc}}{n} \text{ V} \tag{8.39}$$

Now, the harmonic current in the load is given by $i_n = \dfrac{C_n}{Z_n}$

where Z_n is the load impedance at harmonic frequency nf Hz and is given by

$$Z_n = \sqrt{R_L^2 + (n\omega_0 L)^2}$$

For negligible load resistance, R_L, $i_n = \dfrac{C_n}{n\omega_0 L}$ or $i_n \propto \dfrac{1}{n^2}$ \hfill (8.40)

Thus, the harmonic current decreases, as the order of the harmonic (n) increases. Another technique for measure of the harmonic content of a waveform, without calculating its harmonic components, is the a.c. ripple voltage E_r. It is defined as,

$$E_r = \sqrt{E_{rms}^2 - E_0^2} \tag{8.41}$$

In the above equation, E_{rms} and E_0 are the RMS and average value of the output-voltage respectively. Now, the RMS value of output voltage is given by

$$E_{rms} = \left[\alpha \cdot E_{dc}^2\right]^{1/2} = \sqrt{\alpha} \cdot E_{dc}$$

Therefore, Eq. (8.41) becomes

$$\therefore \quad E_r = \sqrt{\alpha \cdot E_{dc}^2 - \alpha^2 E_{dc}^2} = E_{dc}\sqrt{\alpha - \alpha^2} \tag{8.42}$$

Ripple factor (RF) is defined as the ratio of a.c. ripple voltage to average voltage, and is given by

$$RF = \frac{E_r}{E_0}, \text{ or, } RF = \frac{E_{dc}\sqrt{\alpha - \alpha^2}}{\alpha \cdot E_{dc}} = \frac{E_{dc}\sqrt{\alpha}\sqrt{1-\alpha}}{\alpha \cdot E_{dc}}$$

$$= \frac{\sqrt{1-\alpha}}{\sqrt{\alpha}} = \sqrt{\frac{1-\alpha}{\alpha}} \tag{8.43}$$

4. Continuous Conduction Limit With pulse-width modulation control in class A chopper circuit, the "on" time (T_{on}) can be reduced by increasing "off" time T_{off}. At a particular low value of T_{on}, the off-time is large and current i_0 may go to zero. As discussed previously, that current i_0 in class A chopper cannot reverse, hence it stays at zero. Therefore, the limit of continuous conduction is reached when current I_{0min} given by Eq. (8.31) goes to zero.

The duty-cycle (α') and the limit of continuous conduction is given by Eq. (8.31) to zero. Therefore,

$$I_{0\min} = \frac{E_{dc}}{R}\left[\frac{e^{T_{on}/\tau}-1}{e^{T/\tau}-1}\right] - \frac{E_b}{R} = 0$$

or

$$\frac{e^{T_{on}/\tau}-1}{e^{T/\tau}-1} = \frac{E_b}{E_{dc}}$$

Let us define, $\quad g = \dfrac{E_b}{E_{dc}} \quad \therefore \dfrac{e^{T_{on}/\tau}-1}{e^{T/\tau}-1} = g$

or $\quad (e^{T_{on}/\tau}-1) = g(e^{T/\tau}-1)$ or $e^{T_{on}/\tau} = 1 + g(e^{T/\tau}-1)$

or $\quad \alpha' = \dfrac{T_{on}}{T} = \left(\dfrac{\tau}{T}\right)\ln\left[1 + g(e^{T/\tau}-1)\right]$ \hfill (8.44)

The load current is continuous if the actual-duty cycle (α) is more than the duty-cycle (α') obtained from Eq. (8.44), and, it becomes discontinuous if α is less than α'. The limit of continuous conduction for one-quadrant chopper can also be explained with the help of graph shown in Fig. 8.16. This graph shows the variation of duty cycle α' with respect to normalised back emf (g). The duty cycle α' can be calculated for various values of α' from 0 to 1 and by considering $\tau/T = 2$. For example, for $g = 0.4$,

$\therefore \quad \alpha' = 2\ln[1 + 0.4(e^{0.5} - 1)] = 0.4614$

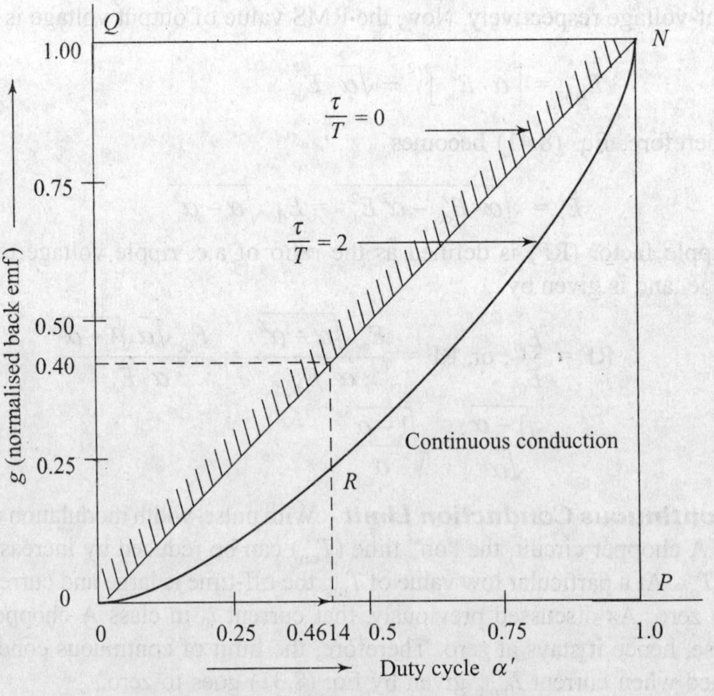

Fig. 8.16 *Continuous conduction limit*

As shown, for the value of $\tau/T = 2$, the region ORNPO represents the continuous conduction region and the other region ORNQO represents the discontinuous conduction region. Point marked R in Fig. 8.16 gives the limit of continuous conduction with duty-cycle $\alpha'(= 0.4614)$, for $\tau/T = 2$, and $g = 0.4$. For these values of $\dfrac{\tau}{T}$ and g, if actual duty cycle (α) is more than $\alpha'(= 0.4614)$, then the point will lie in the continuous conduction region, and if α is less than α', then the point will lie in the discontinuous conduction region. The straight line ON corresponds to $\tau/T = 0$ value.

5. Expression for Average Load Current Figure 8.17 shows the various voltage and current waveforms for one-quadrant chopper with R-L E_b load. Now, the average-load current (I_{0Av}) over a complete cycle can be obtained by adding the average value of the input current (or thyristor current I_{TAV}) and the average value of the freewheeling diode current (I_{DF}). Therefore, first we calculate the values of I_{TAV} and I_{DF}.

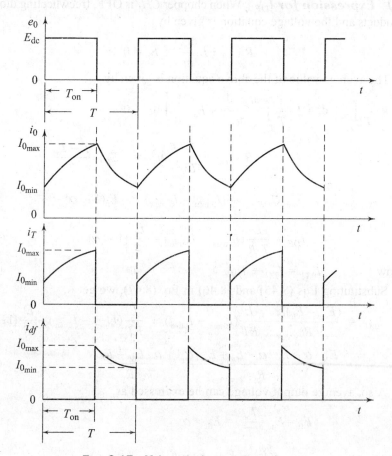

Fig. 8.17 *Voltage and current waveforms*

(a) Expression for I_{TAV} When chopper CH_1 is ON, the voltage equation for the chopper circuit of Fig. 8.13 by taking source or thyristor current i_t can be written as, $E_{dc} = Ri_T + L\dfrac{di_T}{dt} + E_b$ or, $(E_{dc} - E_b)dt = Ri_T\,dt + L\dfrac{di_T}{dt}\cdot dt$

Now, the average value of above equation is,

$$\dfrac{(E_{dc} - E_b)}{T}\int_0^{T_{on}} dt = R\cdot\dfrac{1}{T}\int_0^{T_{on}} i_t\cdot dt + \dfrac{1}{T}\int_0^{T_{on}} L\cdot di_T$$

∴ $$(E_{dc} - E_b)\dfrac{T_{on}}{T} = R\cdot I_{TAV} + \dfrac{1}{T}\int_{I_{0min}}^{I_{0max}} L\,di_T$$

or, $$(E_{dc} - E_b)\alpha = R\cdot I_{TAV} + \dfrac{L}{T}(I_{0\max} - I_{0\min})$$

or $$\dfrac{(E_{dc} - E_b)\alpha}{R} - \dfrac{L}{RT}(I_{0\max} - I_{0\min}) = I_{TAV} \qquad (8.45)$$

(b) Expression for I_{DF} When chopper CH_1 is OFF, freewheeling diode D_f conducts and the voltage equation is given by

$$R\cdot i_{df} + L\dfrac{di_{df}}{dt} + E_b = 0$$

The average value of the above equation is given by

$$R\cdot\dfrac{1}{T}\int_{T_{on}}^{T} i_{df}\cdot dt + L\cdot\dfrac{1}{T}\int_{T_{on}}^{T}\dfrac{di_{df}\cdot dt}{dt} + E_b\cdot\dfrac{1}{T}\int_{T_{on}}^{T} dt = 0$$

or $$R\cdot I_{DF} + \dfrac{L}{T}\int_{I_{0max}}^{I_{0min}} di_{df} = -E_b\cdot\left(\dfrac{T - T_{on}}{T}\right)$$

or $$RI_{DF} + \dfrac{L}{T}(I_{0\min} - I_{0\max}) = -E_b(1 - \alpha)$$

∴ $$I_{DF} = \dfrac{L}{T\cdot R}(I_{0\max} - I_{0\min}) - \dfrac{E_b}{R}(1 - \alpha) \qquad (8.46)$$

Now $$I_{0AV} = I_{TAV} + I_{DF} \qquad (8.47)$$

Substituting Eqs (8.45) and (8.46) in Eq. (8.47), we get

$$I_{0AV} = \dfrac{(E_{dc} - E_b)\alpha}{R} - \dfrac{L}{RT}(I_{0\max} - I_{0\min}) + \dfrac{L}{RT}(I_{0\max} - I_{0\min}) - \dfrac{E_b}{R}(1-\alpha)$$

$$= \dfrac{E_{dc}\cdot\alpha - E_b\cdot\alpha - E_b + E_b\cdot\alpha}{R} = \dfrac{\alpha\cdot E_{dc} - E_b}{R} \qquad (8.48)$$

Also, average output voltage can be expressed as,

$$E_0 = E_{dc}\cdot\dfrac{T_{on}}{T} = E_{dc}\cdot\alpha$$

Therefore, $$I_{0AV} = \dfrac{\alpha E_{dc} - E_b}{R} = \dfrac{E_0 - E_b}{R} \qquad (8.49)$$

Thus, the average output current can only be the average output voltage minus the back emf divided by the d.c. impedance of the load, i.e. its resistance.

SOLVED EXAMPLES

Example 8.7 For the ideal type A chopper circuit, following conditions are given: E_{dc} = 220 V, chopping frequency, f = 500 Hz; duty cycle α = 0.3 and R = 1 Ω; L = 3 mH; and E_b = 23 V. Compute the following quantities.
  (i) Check whether the load current is continuous or not.
  (ii) Average output current.
  (iii) Maximum and minimum values of steady-state output current.
  (iv) RMS values of first, second and third harmonics of load current.
  (v) Average value of source current.
  (vi) The input power, power absorbed by the back emf E_b and power loss in the resistor.
  (vii) RMS value of output current using the result of (ii) and (iv).
  (viii) The RMS value of load current using the results of (iv). Compare the result with that obtained in part (vii) above.

Solution:
  (i) We know from the chopper theory, that the load current is continuous only when actual value of duty cycle α is greater than α'. Therefore, first calculate α'. From Eq. (8.44), we write, $\alpha' = \left(\dfrac{\tau}{T}\right) \ln\left[1 + g\left(e^{T/\tau} - 1\right)\right]$

  where $\tau = L/R = \dfrac{3 \times 10^{-3}}{1} = 3 \times 10^{-3}$ s.

  $T = \dfrac{1}{f} = \dfrac{1}{500} = 2000$ μs · $g = \dfrac{E_b}{E_{dc}} = \dfrac{23}{220} = 0.105$.

  $\therefore \alpha' = \left(\dfrac{3 \times 10^{-3}}{2000 \times 10^{-6}}\right) \ln\left[1 + 0.105\left(e^{0.67} - 1\right)\right] = 1.5 \ln[1.100]$ $\alpha' = 0.143$

  Since $\alpha > \alpha'$, load current is continuous.
  (ii) Average output current,

  $$I_{0av} = \dfrac{\alpha \cdot E_{dc} - E_b}{R} = \dfrac{0.3 \times 220 - 23}{1} = 43 \text{ A}.$$

  (iii) From Eq. (8.30), maximum value of output current is given by

  $$I_{0max} = \dfrac{E_{dc}}{R}\left[\dfrac{1 - e^{-T_{on}/\tau}}{1 - e^{-T/\tau}}\right] - \dfrac{E_b}{R}$$

  Now, $\alpha = \dfrac{T_{on}}{T}$ $0.3 = \dfrac{T_{on}}{2000 \times 10^{-6}}$ $\therefore T_{on} = 600$ μs.

  Also, $\dfrac{T_{on}}{\tau} = \dfrac{600 \times 10^{-6}}{3 \times 10^{-3}} = 200 \times 10^{-3}$.

$$\therefore \quad I_{0max} = \frac{220}{1}\left[\frac{1-e^{-200\times 10^{-3}}}{1-e^{-0.67}}\right] - \frac{23}{1} \quad \therefore I_{0max} = 58.64 \text{ A}.$$

From Eq. (8.31), $I_{0min} = \frac{E_{dc}}{R}\left[\frac{e^{T_{on}/\tau}-1}{e^{T/\tau}-1}\right] - \frac{E_b}{R}$

$$= \frac{220}{1}\left[\frac{e^{0.2}-1}{e^{0.67}-1}\right] - \frac{23}{1} = 28.05 \text{ A}.$$

(iv) The RMS value of first harmonic voltage is given by

$$E_1 = \frac{2E_{dc}}{\sqrt{2}\pi}\sin\pi = \frac{2\times 220}{\sqrt{2}\pi}\sin(\pi\times 0.3) = 80.121 \text{ V}.$$

Now, $\quad Z_1 = \sqrt{R^2+(WL)^2} = \sqrt{(1)^2+(2\pi\times 500\times 3\times 10^{-3})^2} = 9.48\ \Omega.$

$$\therefore \quad I_1 = \frac{E_1}{Z_1} = \frac{80.121}{9.48} = 8.452 \text{ A}.$$

Similarly, $\quad I_2 = \frac{2\times 220}{2\sqrt{2}\pi}\sin 2\pi\alpha \cdot \frac{1}{\sqrt{1^2+(2\pi\times 500\times 2\times 3\times 10^{-3})^2}} = 2.494 \text{ A}$

$$I_3 = \frac{2\times 220}{3\sqrt{2}\pi}\sin(162°) \cdot \frac{1}{\sqrt{1^2+(2\pi\times 3\times 500\times 3\times 10^{-3})^2}} = 0.624 \text{ A}.$$

(v) From Eq. (8.45), the average value of source current is given by

$$I_{TAV} = \frac{(E_{dc}-E_b)\alpha}{R} - \frac{L}{RT}(I_{0max} - I_{0min})$$

$$= \frac{(220-23)0.3}{1} - \frac{3\times 10^{-3}}{1\times 2000\times 10^{-6}}(58.64-28.05)$$

$$= 59.1 - 1.5(30.59) = 13.215$$

(vi) Input power = $E_{dc}\times$ average source current = $220\times 13.215 = 2907.3$ W

Power absorbed by load emf = $E_b\times$ average load current = $23\times 43 = 989$ W.

Power loss in resistor R = Input power – power absorbed by load emf

$$= 2907.3 - 989 = 1918.3 \text{ W}.$$

(vii) $I_{rms} = \sqrt{I_{0av}^2+I_1^2+I_2^2+I_3^2} = \sqrt{(43)^2+(8.452)^2+(2.494)^2+(0.624)^2}$

$$= 1927.05 \text{ A} = 43.89 \text{ A}.$$

(viii) Power loss in resistor $I^2 R = 1918.3$ W

$$\therefore \quad I_{rms} = \sqrt{\frac{1918.3}{1}} = 43.798 \text{ A}$$

The value of I_{rms} in both parts is nearly the same.

Example 8.8 An ideal chopper operating at a chopping period of 2 ms supplies a load of 4 Ω having an inductance of 8 mH from a 80 V battery. Assuming the load is shunted by a perfect commutating diode, and battery to be lossless,

(a) compute the load current waveform $f_{on}\dfrac{T_{on}}{T_{off}}$ values of
  (ii) 1/1 (ii) 4/1 (iii) 1/4.
(b) Also, calculate the mean value of load voltage and current at each setting.

Solution:

(a) During the on-period, the battery is switched to a series R_L load, having an initial current I_{0min}.

During the off-period, the load current decays in the R–L load through the diode, having an initial value of I_{0max}. Fig. Ex. 8.8 shows the waveforms for each setting.

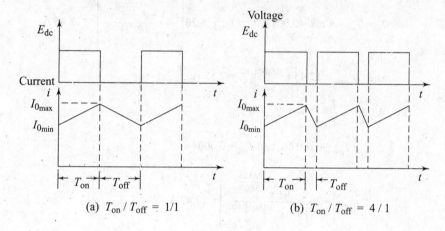

(a) $T_{on}/T_{off} = 1/1$ (b) $T_{on}/T_{off} = 4/1$

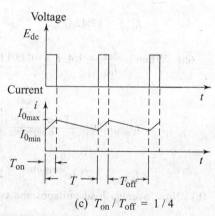

(c) $T_{on}/T_{off} = 1/4$

Fig. Ex. 8.8 *Three different waveforms*

When chopper is ON, $i = I_{0min}\cdot\left(\dfrac{E_{dc}}{R} - I_{0min}\right)\left(1 - e^{-T_{on}/\tau}\right) = I_{0max}$ (a)

When chopper is OFF, $i = I_{0max}\, e^{-T_{off}/\tau} = I_{0min}$ (b)

From Eqs (a) and (b), we can write $I_{0\max} = \dfrac{E_{dc}}{R}\left(\dfrac{1-e^{-T_{on}/\tau}}{1-e^{-T/\tau}}\right)$ (c)

and $\quad I_{0\min} = I_{0\max}\, e^{-T_{off}/\tau}$ (d)

Now, $\quad \dfrac{E_{dc}}{R} = \dfrac{80}{4} = 20$ A. $\tau = L/R = \dfrac{0.008}{4} = 0.002$ s. $T = 2$ ms. $= 0.002$ s

(i) When $\dfrac{T_{on}}{T_{off}} = 1/1$,

$T_{on} = T_{off} = 1$ ms $= 0.001$ s. $\quad \therefore I_{0\max} = 20\left(\dfrac{1-e^{-\tfrac{0.001}{0.002}}}{1-e^{-0.002/0.002}}\right) = 12.45$ A.

and $I_{0\min} = 12.45\, e^{-0.001/0.002} = 7.55$ A.

(ii) When $\dfrac{T_{on}}{T_{off}} = 4/1, T_{on} = 4\, T_{off}$

$\therefore \quad T_{on} = 0.0016$ s, and $T_{off} = 0.0004$ s

$\therefore \quad I_{0\max} = 20\left(\dfrac{1-e^{-\tfrac{0.0016}{0.002}}}{1-e^{-0.002/0.002}}\right) = 17.42$ A.

$I_{0\min} = 17.42\,(e^{-0.004/0.002}) = 2.36$ A

(iii) When $\dfrac{T_{on}}{T_{off}} = 1/4, T_{off} = 0.0016$ s, $T_{on} = 0.0004$ s.

$\therefore \quad I_{0\max} = 20\left(\dfrac{1-e^{-\tfrac{0.004}{0.002}}}{1-e^{-0.002/0.002}}\right) = 5.73$ A.

$I_{0\min} = 5.73\,(e^{-0016/0.002}) = 2.57$ A.

(b) Now, average load voltages and currents are given by

(i) $E_0 = E_{dc}\left(\dfrac{T_{on}}{T}\right) = 80\left(\dfrac{1}{2}\right) = 40$ V, $I_{0av} = \dfrac{E_0}{R} = \dfrac{40}{4} = 10$ A

(ii) $E_0 = 80(4/5) = 64$ V, $I_{0av} = \dfrac{64}{4} = 16$ A

(iii) $E_0 = 80\,(1/5) = 16$ V, $I_{0av} = \dfrac{16}{4} = 4$ A.

Choppers

Example 8.9 An R–L E_b type load is operating in a chopper circuit from a 400 volts d.c. source. For the load, $L = 0.05$ H and $R = 0$. For a duty cycle of 0.3, find the chopping frequency to limit the amplitude of load current excursion to 8 A.

Solution: The related circuit diagram is shown in Fig. Ex. 8.9.

The average output voltage is given by, $E_0 = \alpha \cdot E_{dc}$
As the average value of voltage drop across inductance L is zero,

$$E_b = E_0 = \alpha \cdot E_{dc} = 0.3 \times 400 = 120 \text{ V}$$

During the on-period of the chopper T_{on}, the difference in source voltage E_{dc} and load back emf E_b, i.e. $(E_{dc} - E_b)$, appears across L, as shown in Fig. Ex. 8.9.

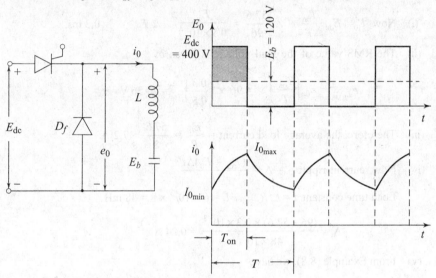

Fig. Ex. 8.9 *Chopper circuit and waveforms*

∴ During T_{on}, volt–time area applied to inductance = $(400 - 120) T_{on} = 280 T_{on}$ V–s (a)

As shown in Fig. Ex. 8.9, the current through inductance L rises from $I_{0\,min}$ to $I_{0\,max}$. From this, volt–time area across L during current change is given by

$$\int_0^{T_{on}} E_L dt = \int_0^{T_{on}} L \cdot \frac{di_0}{dt} dt = \int_{T_{0max}}^{T_{0max}} L \cdot di = L(I_{0\,max} - I_{0\,min}) = L \Delta I_0 \quad \text{(b)}$$

During T_{on}, the volt–time areas given by Eqs (a) and (b) must be equal.

∴ $\qquad\qquad\qquad 280 T_{on} = L \Delta I_0$

∴ $\qquad T_{on} = \dfrac{0.05 \times 8}{280} = 1.43$ ms

∴ Chopping frequency, $f = 1/T = \dfrac{\alpha}{T_{on}} = \dfrac{0 \cdot 3}{1.43 \times 10^{-3}} = 209.79$ Hz.

Example 8.10 A simple d.c. chopper is operating at a frequency of 2 kHz from a 96 V d.c. source to supply a load resistance of 8 Ω. The load time constant is 6 ms. If the average load voltage is 57.6 V, find the T_{on} period of the chopper, the average load current, the magnitude of the ripple current and its RMS value.

Solution: Chopping period, $T = 1/f = \dfrac{1}{2000} = 0.5$ ms.

Given load time constant = 6 ms

∴ Load time constant = 12 T, therefore treat as a linear current variation.

(i) Now, $E_0 = E_{dc} \cdot \dfrac{T_{on}}{T}$ ∴ $\dfrac{57.6}{96} = \dfrac{T_{on}}{0.5 \times 10^{-3}}$ ∴ $T_{on} = 0.3$ ms.

(ii) The RMS value of the load voltage is given by

$$F_{L_{RMS}} = E_{dc}\left(\dfrac{T_{on}}{T}\right)^{1/2} = 96 \times \left(\dfrac{0.3}{0.5}\right)^{1/2} = 74.36 \text{ V}$$

(iii) Therefore, the average load current = $\dfrac{E_0}{R} = \dfrac{57.6}{8} = 7.2$ A

(iv) Now, current ripple $= \Delta_i = \dfrac{(E_{dc} - E_0)\Delta t}{L}$

Load time constant $\tau = L/R$, ∴ $L = 6 \times 10^{-3} \times 8 = 48$ mH.

∴ $\Delta_i = \dfrac{(96 - 57.6) \times 0.3 \times 10^{-3}}{48 \times 10^{-3}} = 0.24$ A

(v) From Example (8.8), we have

$$I_{0\,max} = \dfrac{E_{dc}}{R}\left(\dfrac{1 - e^{-T_{on}/\tau}}{1 - e^{-T/\tau}}\right) = \dfrac{96}{8}\left(\dfrac{1 - e^{-\frac{0.3\,ms}{6\,ms}}}{1 - e^{-0.5/6\,ms}}\right) = 7.32 \text{ A}.$$

Similarly,

$$I_{0\,min} = I_{0\,max}\, e^{-T_{off}/\tau} = 7.32\, e^{-0.2\,ms/6\,ms} = 7.08 \text{ A}$$

The RMS value of the ripple current is given by

$$I_{r,\,RMS} = \dfrac{(I_{0\,max} - I_{0\,min})}{2\sqrt{3}} = \dfrac{(7.32 - 7.08)}{2\sqrt{3}} = 0.0693 \text{ A}.$$

Example 8.11 A d.c. motor with armature resistance $R_a = 0.4$ Ω and armature inductance $t_a = 8$ mH, is having a back emf of 80V while carrying a current of 10 A. The motor is connected to a d.c. source of 180 V by the main SCR of the chopper. If the SCR turns off after 1 ms, compute the current in the motor
  (i) at the instant the thyristor turns off, and
  (ii) 8 ms after SCR turns off.

Solution: The differential equation with the given chopper conditions is given by

$$E_{dc} = R_a \cdot i_a + L_a \frac{di_a}{dt} + E_b, \text{ Now, } \tau = L/R = 8 \text{ mH}/0.4 = 20 \text{ ms.}$$

The solution of the above equation is given by

$$i(t) = \frac{E_{dc} - E_b}{R_a}\left(1 - e^{-t/\tau}\right) + I_0 \cdot e^{-t/\tau} = \frac{180 - 80}{0.4}\left(1 - e^{-t/20 \times 10^{-3}}\right) + 10(e^{-t/20 \times 10^{-3}})$$

$$= 250\left(1 - e^{-t/20 \times 10^{-3}}\right) + 10\, e^{-t/20 \times 10^{-3}}$$

(i) At $t = 1 \times 10^{-3}$ s, $I(t) = 250\left(1 - e^{-1 \times 10^{-3}/20 \times 10^{-3}}\right) + 10\, e^{-1/10^{-3}/20 \times 10^{-3}}$

$$I(t) = 12.193 + 9.512\, I(t) = 21.705 \text{ A.}$$

(ii) Current is freewheeled through the load for the period of 9 ms.

$$\therefore \quad i_f = I(t) \cdot e^{-t/\tau} = i_f = 21.705\, e^{-9 \times 10^{-3}/20 \times 10^{-3}} = 13.84 \text{ A}$$

Example 8.12 A DC chopper operates on 230 V dc and frequency of 400 Hz, feeds an R-L load. Determine the ON time of the chopper for output of 150 V.

Solution:

Given: $E_{dc} = 230$ V, $f = 400$ Hz, $E_0 = 150$ V

We have, $E_0 = \alpha \cdot E_{dc}$ $\therefore 150 = \alpha \cdot 230$, $\therefore \alpha = 0.65$

Time period of output voltage wave is given by

$$T = 1/F = 1/400 = 2.5 \times 10^{-3} \text{ sec}$$

Now, $\alpha = \frac{T_{on}}{T}$, $\therefore t_{on} = \alpha \cdot T = 0.65\,(2.5 \times 10^{-3})$

$\therefore$ On-time of chopper, $t_{on} = 1.6305$ m sec

Example 8.13 A single-quadrant type A chopper is operated with the following specifications:

(i) ideal battery of 220 V (ii) on-time $t_{on} = 1$ msec (iii) off-time $t_{off} = 1.5$ msec

Determine: (a) Average and RMS output voltages (b) Ripple and form factor

Solution:

Time period $T = t_{on} + t_{off} = (1 + 1.5) = 2.5$ msec, Duty cycle $\alpha = \frac{T_{on}}{T} = \frac{1}{2.5} = 0.4$

(a) Average output voltage, $E_0 = \alpha \cdot E_{dc} = (0.4)(220) = 88$ V.

RMS output voltage $E_{0rms} = \sqrt{\alpha} \cdot E_{dc} = \sqrt{0.4}\,(220) = 139.14$ V

(b) Form-factor (FF) = $\dfrac{\text{RMS Value}}{\text{Average Value}} = \dfrac{\sqrt{\alpha} \cdot E_{dc}}{\alpha \cdot E_{dc}} = \dfrac{1}{\sqrt{\alpha}} = \dfrac{1}{\sqrt{0.4}} = 1.58$

Ripple factor (RF) = $\sqrt{(FF)^2 - 1} = \sqrt{\dfrac{1}{\alpha} - 1} = \sqrt{\dfrac{1-\alpha}{\alpha}} = \sqrt{\dfrac{1-0.4}{0.4}} = 1.23$

8.5.2 Second-Quadrant or Class B Chopper [Step-up Chopper with R–L Load]

Figure 8.18(a) shows the basic power circuit of the second-quadrant chopper. The term 'second-quadrant' signifies that circuit parameters E_0 and I_0 occur only in the second-quadrant of E_0–I_0 diagrams. Figure 8.18(b) shows the chopper with R–L load.

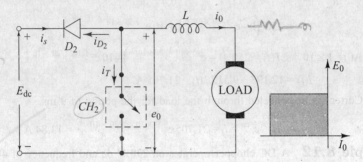

Fig. 8.18(a) Type-B chopper circuit and E_0–I_0 characteristic

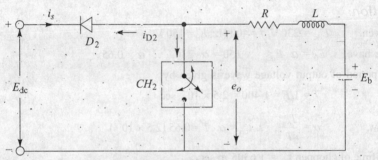

Fig. 8.18(b) Class B with R–L load

If the chopper is turned ON and OFF during regular intervals of period T, emf E_b stores energy in inductance L whenever chopper is conducting and part of that stored energy is delivered to source E_{dc} by current through diode D_2 when CH_2 is commutated. Therefore, when the chopper is ON, $e_0 = 0$ and when chopper is OFF and diode D_2 conducting, $e_0 = E_{dc}$. When the chopper is OFF, load current i_0 has the direction opposite to that shown in Fig. 8.18. Hence, average load voltage E_0 is positive and average load current I_0 is negative. It means that, power flow takes place from load to source. Since active load is capable of providing continuous power output, the reversal power-flow is possible in this type of chopper configuration. Because of power flow from load of lower-voltage e_0 to the source of higher-voltage E_{dc}, this configuration is also referred to as step-up chopper. This configuration is used for regenerative breaking of d.c. motors. A typical application is the chopper drive of a subway-train.

The related circuit waveforms are shown in Fig. 8.19. Here, the interval during which diode D_2 conducts is designated as T_{off}. Hence, the cycle of operation starts at $t = 0$ in Fig. 8.19, at the instant when chopper CH_2 is commutated.

Choppers 467

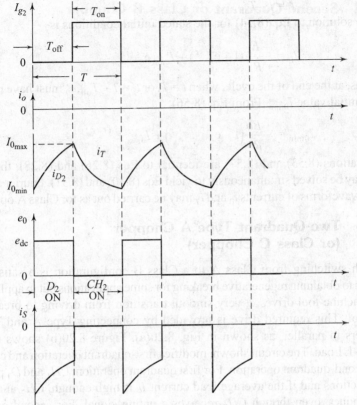

Fig. 8.19 *Second-quadrant chopper waveform*

As shown in Fig. 8.19, let i_0 has value $I_{0\min}$ at $t = 0$. For interval $0 < t < T_{\text{off}}$, diode D_2 conducts and $e_0 = E_{\text{dc}}$. During this interval, the voltage equation becomes

$$L\frac{di_0}{dt} + Ri_0 = E_{\text{dc}} - E_b \qquad (8.50)$$

or

$$\frac{di_0}{dt} + \frac{R}{L}i_0 = \frac{E_{\text{dc}} - E_b}{L} \qquad (8.51)$$

The solution of the above Eq. (8.51) for the stated initial conditions is

$$I_{0(t)} = \left(\frac{E_{\text{dc}} - E_b}{R}\right)\left(1 - e^{-t/\tau}\right) + I_{0\min} \cdot e^{-t/\tau} \qquad (8.52)$$

At $t = T_{\text{off}}$, I_0 has reached a magnitude $I_{0\max}$, where $I_{0\min} < I_{0\max} < 0$. Thus, from Eq. (8.52), $I_{0\max} = \left(\dfrac{E_{\text{dc}} - E_b}{R}\right)\left(1 - e^{-T_{\text{off}}/\tau}\right) + I_{0\min} \cdot e^{-T_{\text{off}}/\tau} \qquad (8.53)$

At $t = T_{\text{off}}$, chopper CH_2 is turned-on and at T_{off}, e_0 becomes zero and $i_0 = I_{0\max}$. During the interval $T_{\text{off}} < t < T$, voltage equation becomes

$$\frac{di_0}{dt'} + \frac{R \cdot i_0}{L} = \frac{-E_b}{L} \qquad (8.54)$$

where $t' = t - T_{\text{off}}$ (8.55)

The solution to Eq. (8.54) for the stated initial conditions is

$$I_0 = \frac{-E_b}{R}(1 - e^{-t'/\tau}) + I_{0\max} e^{-t'/\tau} \qquad (8.56)$$

Thus, at the end of the cycle, when $t = T$, or $t' = T - T_{\text{off}}$, i_0 must have returned to its initial value $I_{0\min}$. From Eq. (8.56),

$$I_{0\min} = \frac{-Eb}{R}\left(1 - e^{-(T-T_{\text{off}})/\tau}\right) + I_{0\max} e^{-(T-T_{\text{off}})/\tau} \qquad (8.57)$$

Equations (8.53) and (8.57) are identical to Eqs (8.26) and (8.28), therefore, they may be solved simultaneously to yield Eqs (8.30) and (8.31). Fourier analysis of the waveforms of currents i_0 and i_T may be carried out as for Class A operation.

8.5.3 Two-Quadrant Type A Chopper (or Class C Chopper)

Though switching from Class A to a Class B configuration is a satisfactory method to obtaining regenerative breaking for some applications but in applications like machine-tool drives, a very smooth transition from driving to breaking is essential. This required drive is provided by connecting Type A and Type B choppers in parallel, as shown in Fig. 8.20(a). Figure 8.20(b) shows class C with R–L Load. The circuit shown, modifies first-quadrant operation and converts it to second-quadrant operation. For first quadrant operation, CH_1 and D_1 perform the functions and if the average load current I_0 is high enough, CH_2 and D_2 do not conduct, even though CH_2 receives a gating signal. For second quadrant operation, CH_2 and D_2 perform the functions and if the average load current I_0 has a sufficiently large negative value, CH_1 and D_1 do not conduct, even though CH_1 receives a gating signal. Figure 8.21 shows the gate current, load voltage and supply current waveforms.

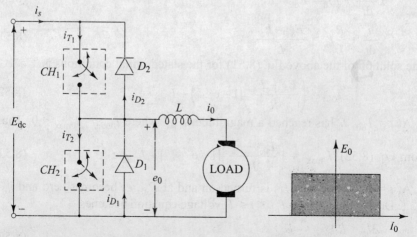

Fig. 8.20(a) *Type-C chopper circuit and $E_0 - I_0$ characteristics*

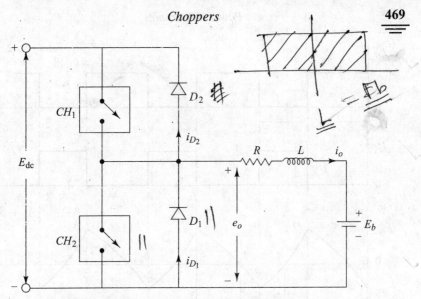

Fig. 8.20(b)   *Class-C chopper with R-L load*

Initially, when both the choppers are OFF, both the diodes D_1 and D_2 become also OFF, and therefore, load is isolated from the supply. As shown in Fig. 8.21, at point P, chopper CH_1 is triggered and it starts to conduct. The load current i_0 is positive and the load receives power from the supply. Therefore, the output voltage $e_0 = E_{dc}$ when chopper CH_1 or diode D_2 conducts. At point Q, chopper CH_1 is turned OFF and inductance L forces the load current to flow through diode D_1 till the value of $L\dfrac{di_s}{dt}$ becomes equal to the back emf (E_b) of the load and the load-current i_0 becomes zero. Therefore, diode D_1 conducts from point Q to point R, as shown in Fig. 8.21. At this point R, if the gate signal to chopper CH_2 is available the back emf (E_b) of the motor forces current in the opposite direction through L and CH_2. This continues until CH_2 is turned-OFF and CH_1 is turned-ON. Now, when CH_2 is turned-OFF, the energy of the inductance forces current through diode D_2 to the supply. The input current becomes negative. During this period, CH_1 cannot conduct due to reverse bias but comes into conduction when the input current reduces to zero, provided the gate signal is available to CH_1 and both the load and input current becomes positive.

Hence, it becomes clear that load voltage $e_0 = 0$ if chopper CH_2 or diode D_1 conducts, and $e_0 = E_{dc}$ if chopper CH_1 or diode D_2 conducts. Therefore, average load voltage E_0 is positive. However, load current i_0 have both positive and negative directions. It is positive if CH_1 is ON or D_1 conducts and negative if CH_2 is ON or D_2 conducts. Since average load voltage E_0 is positive and average load current I_0 is reversible, power flow is reversible. It is also clear that both thyristors may not be turned-ON simultaneously because that would short-circuit source E_{dc}. They are turned-ON alternately, as shown by the gating signal waveforms in Fig. 8.21. This chopper configuration is, therefore, used for both motoring and regenerative breaking of d.c. motor.

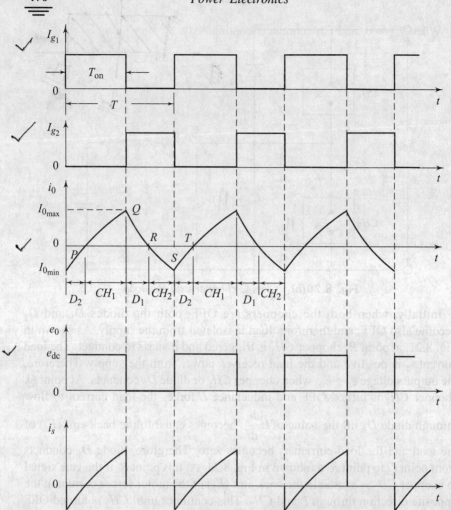

Fig. 8.21 Associated waveforms

The Class A chopper analysis may be applied directly to this Class C chopper circuit, the only new feature being that $I_{0\,min}$ and $I_{0\,max}$ in Eqs (8.30) and (8.31) may be positive or negative. The quadrant in which the chopper is operating may be determined by Eq. (8.39), which states

$$E_0 = \frac{T_{on}}{T} E_{dc} \qquad (8.58)$$

If $E_0 > E_b$, then $I_0 > 0$ and resultant energy is delivered to the load circuit (armature circuit). On the other side, if $E_0 < E_b$, then $I_0 < 0$ and the net energy-flow is to the source E_{dc}.

When $T_{on} = T$, and CH_1 conducts continuously,

$$I_0 = I_{0min} = I_{0max} = \frac{E_{dc} - E_b}{R} \qquad (8.59)$$

When $T_{on} = 0$, and CH_2 conducts continuously,

$$I_0 = I_{0min} = I_{0max} = \frac{-E_b}{R} \qquad (8.60)$$

but this is the state in which all regenerated energy is dissipated in the load-circuit resistance (armature circuit resistance).

8.5.4 Two-Quadrant Type B Chopper (or Class D Chopper)

Figure 8.22(a) shows the basic power circuit of two-quadrant Type B chopper or Type D chopper. Two-quadrant chopper permits either to change voltage polarity while maintaining current polarity, or *vice versa*. Fig. 8.22(b) shows the Class-D chopper with R–L load.

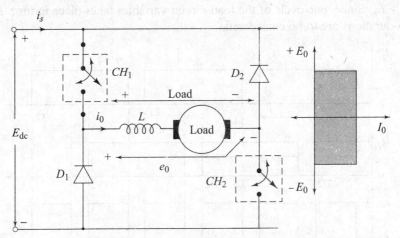

Fig. 8.22(a) Type D chopper and E_0–I_0 characteristics

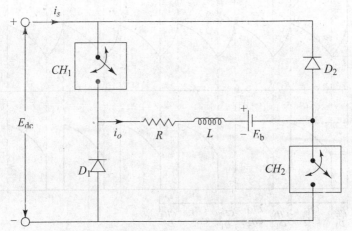

Fig. 8.22(b) Class-D chopper with R–L load

The circuit operation of a Type D chopper can be considered into two different modes of operation; one mode for which $T_{on} > T/2$ and the two gating signals overlap, the other mode for which $T_{on} < T/2$ and only one chopper is turned-ON at any instant or none of the chopper is turned-ON. Here, the load current is assumed to be continuous for circuit analysis.

(a) Mode 1 Operation: $T_{on} > T/2$ When both the choppers are turned-ON, the current flows through the path, $E_{dc} - CH_1 - \text{load} - CH_2 - E_{dc}$. Therefore, both the diodes D_1 and D_2 are turned-OFF. The supply voltage E_{dc} is applied to the load circuit and load current i_0 increases.

When only one chopper is turned-ON, that chopper and one of the diodes short-circuit the load branch and provide a path in which some of the energy stored in inductance L may be dissipated in maintaining a decreasing load current i_0.

Circuit analysis: The associated waveforms for this mode of operation is shown in Fig. 8.23. For steady-state operation, it is necessary in this mode that $E_{dc} > E_b$. Since, one cycle of the load-circuit variables takes place in time $T/2$, two-durations are to be considered.

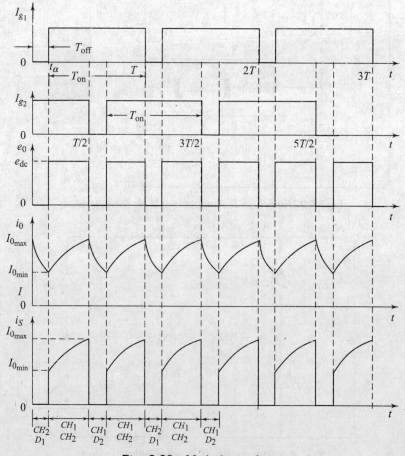

Fig. 8.23 Mode 1 waveforms

(i) Interval $0 < t < t_\alpha$ Let us assume that load current is continuous. Therefore, at $t = 0$, let $i_0 = I_{0\max}$.

For this duration, as shown in Fig. 8.23, chopper CH_2 is only turned-on and current flows through CH_2 and D_1.

Now, voltage equation becomes (R–L–E_b type load)

$$R\,i_0 + L\frac{di_0}{dt} + E_b = 0 \tag{8.61}$$

The solution of Eq. (8.61) with initial conditions becomes

$$i_0 = -\frac{E_b}{R}\left(1 - e^{-t/\tau}\right) + I_{0\max} e^{-t/\tau},\ 0 < t < t\alpha \tag{8.62}$$

where $\tau = L/R$ seconds

As shown in Fig. 8.23, at $t = t_\alpha$, $i_0 = I_{0\min}$.
Substituting these conditions in Eq. (8.61), we get

$$I_{0\min} = -\frac{E_b}{R}\left(1 - e^{-t_\alpha/\tau}\right) + I_{0\max} e^{-t_\alpha/\tau} \tag{8.63}$$

(ii) Interval $t_\alpha < t < T/2$ Let us consider,

$$t' = t - t_\alpha \tag{8.64}$$

Now, both the choppers are turned-on. Voltage-equation becomes

$$R \cdot I_0 + L\frac{di_0}{dt'} + E_b = E_{dc} \tag{8.65}$$

The solution of Eq. (8.65) with stated initial conditions becomes

$$i_0 = \frac{(E_{dc} - E_b)}{R}\left(1 - e^{-t'/\tau}\right) + I_{0\min} e^{-t'/\tau},\ t_\alpha < t < t/2 \tag{8.66}$$

At $t = T/2$ or $T' = T/2 - T_\alpha$, $i_0 = I_{0\max}$. Substituting these conditions in Eq. (8.66), we get,

$$I_{0\max} = \left(\frac{E_{dc} - E_b}{R}\right)\left(1 - e^{-(T/2 - t_\alpha)/\tau}\right) + I_{0\min} e^{-(T/2 - t_\alpha)/\tau} \tag{8.67}$$

After solving Eqs (8.63) and (8.67), as in Section 8.5.1, we get

$$I_{0\max} = \frac{E_{dc}}{R} \cdot \frac{\left(1 - e^{-(T/2 - t_\alpha)/\tau}\right)}{(1 - e^{-T/2\tau})} - \frac{E_b}{R} \tag{8.68}$$

And,

$$I_{0\min} = \frac{E_{dc}}{R} \cdot \frac{\left(e^{+(T/2 - t_\alpha)/\tau} - 1\right)}{(e^{+(T/2\tau - 1)})} - \frac{E_b}{R} \tag{8.69}$$

(b) Mode 2 Operation: $T_{on} < T/2$ The related waveforms for this mode of operation is shown in Fig. 8.24. In this mode, it is necessary that $E_b < 0$ and $-E_b > E_{dc}$, for steady state operation. When both the choppers CH_1 and CH_2 are turned-off, the load voltage has the reverse-polarity to that shown in Fig. 8.22. Hence, load current starts to flow through the path.

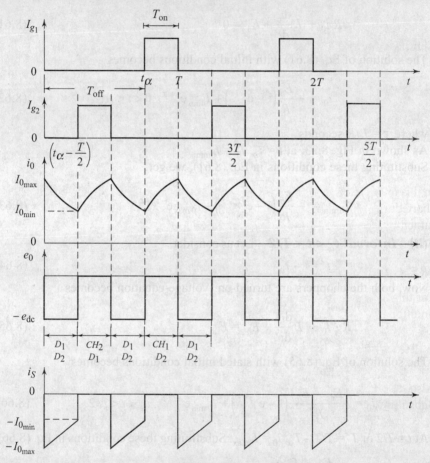

Fig. 8.24 *Mode 2 waveforms*

Load$_+$ – D_2 – E_{dc+} – E_{dc} – D_1 – Load. The diodes D_1 and D_2 are called as energy recovery diodes because they provide the path for flow of load current. Again, two durations are considered in this mode for circuit analysis.

Circuit Analysis

(i) Interval $0 < t < (t_\alpha - T/2)$: Both the choppers CH_1 and CH_2 are turned-off. Again, assume continuous current operation. Therefore,

at $\quad t = 0, i_0 = I_{0max}$.

Now, write the voltage equation,

$$R\,i_0 + L\frac{di_0}{dt} + E_b = -E_{dc} \qquad (8.70)$$

With the stated initial conditions, the solution of Eq. (8.70) becomes

$$i_0 = \frac{-(E_{dc}+E_b)}{R}(1-e^{-t/\tau}) + I_{0\max}e^{-t/\tau}, \quad 0 < t < t_\alpha - T/2 \qquad (8.71)$$

At $t = t_\alpha - T/2$, $i_0 = I_{0\min}$.
Substituting the above conditions, in Eq. (8.61), we get,

$$I_{0\min} = \frac{-(E_{dc}+E_b)}{R}(1-e^{-(t_\alpha-T/2)/\tau}) + I_{0\max}e^{-(t_\alpha-T/2)/\tau} \qquad (8.72)$$

(ii) Internal $(t_\alpha - T/2) < t < T/2$ Let us define,

$$t' = t - \left(t_\alpha - \frac{T}{2}\right) \qquad (8.73)$$

In this interval, either chopper is turned on. Therefore, load circuit terminals are shorted and the load current builds up because $E_b < 0$. When chopper CH_2 is turned on, the load current flows through the path CH_2 and D_1. Now, write the mesh equation for this interval,

$$R\,i_0 + L\frac{di_0}{dt'} + E_b = 0 \qquad (8.74)$$

For this interval, at $t' = 0$, $i_0 = I_{0\min}$.
Solution to Eq. (8.74) with stated conditions is,

$$i_0 = \frac{-E_b}{R}(1-e^{-t'/\tau}) + I_{0\min}e^{-t'/\tau}, \quad (t_\alpha - T/2) < t < T/2 \qquad (8.75)$$

Also, at $t = T/2$, or $t' = T - t_\alpha$, $i_0 = I_{0\max}$.
Substituting the above conditions in Eq. (8.75), we get

$$I_{0\max} = \frac{-E_b}{R}\left(1-e^{-(T-t_\alpha)/\tau}\right) + I_{0\min}e^{-(T-t_\alpha)/\tau} \qquad (8.76)$$

Equations (8.72) and (8.76) can be solved as in Section 8.5.1 to yield

$$I_{0\max} = \frac{-E_{dc}}{R}\frac{\left(e^{(t_\alpha - T/2)/\tau}-1\right)}{\left(e^{T/2\tau}-1\right)} - \frac{E_b}{R} \qquad (8.77)$$

and,

$$I_{0\min} = \frac{-E_{dc}}{R}\frac{\left(1-e^{-(t_\alpha - T/2)/\tau}\right)}{\left(1-e^{-T/2\tau}\right)} - \frac{E_b}{R} \qquad (8.78)$$

Hence, from the above two modes of operation, it becomes clear that in two quadrant Type B chopper, output voltage $e_0 = +E_{dc}$ if both the choppers are on and $e_0 = -E_{dc}$ if both the choppers are off and both diodes D_1 and D_2 are conducting. Also, it becomes clear that average output voltage E_0 is positive when $t_{on} > t_{off}$ and becomes negative when $t_{on} < t_{off}$. In this configuration, since average load current I_0 is positive and average load voltage E_0 is reversible, power-flow is also reversible. This Type D chopper configuration may be used for both motoring and regenerative breaking of the d.c. motor.

1. Fourier analysis of output voltage The load voltage waveform (e_0) of Figs 8.23 and 8.24 may be described by the Fourier series as

$$e_0 = E_0 + \sum_{n=1}^{\infty} C_n \sin(n\omega_0 t + \theta_n) \tag{8.79}$$

where n is a positive integer and ω_0 is defined by

$$\omega_0 = \frac{4\pi}{T} \text{ rad/s} \tag{8.80}$$

Equation (8.80) is based on the fact that one cycle of e_0 takes place in $T/2$ seconds. Also, the average value of e_0 is given by

$$E_0 = \left(1 - \frac{2t_\alpha}{T}\right) E_{dc} \tag{8.81}$$

or

$$E_0 = \left(\frac{2T_{on}}{T} - 1\right) E_{dc} \tag{8.82}$$

$$C_n = \frac{4E_{dc}}{n\pi} \sin \frac{n\omega_0 t_\alpha}{2} = \frac{4E_{dc}}{n\pi} \sin \frac{2\pi n t_\alpha}{T} \tag{8.83}$$

and

$$\theta_n = \frac{\pi}{2} - \frac{n\omega_0 t_\alpha}{2} = \frac{\pi}{2} - \frac{2n\pi t_\alpha}{T} \text{ rad} \tag{8.84}$$

The amplitude of the fundamental frequency in the waveform of e_0 is given by

$$C_1 = \frac{4E_{dc}}{\pi} \sin \frac{\omega_0 t_\alpha}{2} = \frac{4E_{dc}}{\pi} \sin \frac{2\pi t_\alpha}{T} \tag{8.85}$$

Amplitude C_1 has its maximum value when $\omega_0 t_\alpha = \pi$ or 3π

That is, $\dfrac{t_\alpha}{T} = 0.25$ or 0.75

The second value in above two cases corresponds to operation in the fourth quadrant, as shown in Fig. 8.24.

The average value of load-current I_0 is given by

$$I_0 = \frac{E_0 - E_b}{R} = \frac{1}{R}\left[\left(1 - \frac{2t_\alpha}{R}\right) E_{dc} - E_b\right] \tag{8.86}$$

Assuming $\omega_0 L \gg R$, the RMS value of the fundamental component of current is

$$I_{R_1} = \frac{2\sqrt{2} E_{dc}}{\pi \omega_0 L} \frac{\sin 2\pi t_\alpha}{T} = \frac{E_{dc}}{\sqrt{2} \cdot \pi^2 L} \sin \frac{2\pi t_\alpha}{T} \tag{8.87}$$

The RMS load current is given by

$$I_{RMS} = \sqrt{(I_0)^2 + (I_{R_1})^2} \qquad (8.88)$$

On the assumption that load current is constant at magnitude I_0, the source current waveform is similar to that of e_0, and, the pulses of source current is having magnitude I_0. Therefore, with the similar analysis of e_0, the average value of source current is given by

$$I_s = \left(1 - \frac{2t_\alpha}{T}\right) I_0 \qquad (8.89)$$

The output power P_0 is given by, $P_0 = E_0 \cdot I_0$

$$= \left(1 - \frac{2t_\alpha}{T}\right) E_{dc} \cdot I_0 = E_{dc} \cdot I_s = p_{in} \qquad (8.90)$$

SOLVED EXAMPLES

Example 8.14 A Class C chopper is operated from a 220 V battery. The load is a dc motor with $R = 0.1\ \Omega$, $L = 10$ mH and $E_b = 100$ V. Determine the following:
 (i) Duty cycle for the motoring mode
 (ii) Critical duty-cycle for the regenerative mode
 (iii) Duty cycle to achieve regenerative braking at the rated current of 10 Amp
 (iv) Power returned to the source during braking

Solution:
Given $E_{dc} = 220$ V, $R = 0.1\ \Omega$, $L = 10$ mH, $E_b = 100$
(i) The average load current is given by

$$i_0 = \frac{E_o - E_b}{R} = \frac{\alpha \cdot E_{dc} - E_b}{R}$$

$$\therefore\ \alpha = \frac{i_0 \cdot R + E_b}{E_{dc}} = \frac{(10 \times 0.1) + 100}{220}, \quad \therefore\ \alpha = 0.459$$

(ii) Critical duty cycle for regenerative braking $= \dfrac{E_b}{E_{dc}} = 0.4545$

(iii) Duty cycle to achieve regenerative braking at the rated current of $i_0 = 10$ Amp.

$$\therefore\ \text{Rated load current, } i_0 = \frac{\alpha \cdot E_{dc} - E_b}{R}$$

For regeneration this current should be negative.

$$\therefore\ -10 = \frac{(\alpha \cdot 220) - 100}{0.1}, \quad \therefore\ D = 0.45$$

(iv) Power returned to source during braking
$= E_b\, i_0 - i_0^2\, R = 100 \times 10 - (10^2 \times 0.1) = 990$ Watts

Example 8.15 A two-quadrant chopper operating in the first and fourth quadrant is operated from a 220 V battery. The load is dc motor with $R = 0.1\ \Omega$, $L = 10$ mH and $E_b = 100$ V, determine:
  (i) Duty cycle α_m for motoring mode
  (ii) Critical duty cycle for regenerative braking
  (iii) Duty-cycle to achieve regenerative braking at the rated current of 10 Amp
  (iv) Power returned to the source during braking
  (v) The switching frequency of the devices if the output frequency is 5 kHz.

Solution:
  Given: $E_{dc} = 220$ V, $R = 0.1\ \Omega$, $L = 10$ mH, $E_b = 100$ V, $i_0 = 10$ Amp
  Class-D chopper operates in first and fourth quadrant
  (i) Duty-cycle for motoring mode (α_m):
  From equation (8.82), $E_0 = (2 \cdot \alpha_m - 1)\, E_{dc}$

  Average-load current $I_0 = \dfrac{E_0 - E_b}{R}$

  $10 = \dfrac{(2\alpha_m - 1)\cdot(220) - (110)}{0.1}$ $\quad \therefore \alpha_m = 0.73$

  (ii) Critical duty-cycle for regenerative braking is given by
  $$\alpha_c = (1 - \alpha_m) = (1 - 0.73) = 0.27$$
  (iii) Duty-cycle for regeneration of rated armature current (α_R):

  Rated average armature current $= \dfrac{E_b - E_0}{R}$

  But $\quad E_0 = -(2\alpha_R - 1)$ during braking

  $\therefore\quad 10 = \dfrac{100 + (2\alpha_R - 1)\, 220}{0.1}$ $\quad \therefore \alpha_R = 0.275$

  (iv) Power-returned to the source during braking interval:
  $P = E_b \cdot i_0 - i_0^2 R_a = (100 \times 10) - (100 \times 0.1) = 990$ W
  (v) Switching frequency of device:
  In Class-D chopper, the switching frequency of power switch is half the output frequency.
  $\therefore \qquad f_s = 2.5$ kHz.

8.5.5 Four-Quadrant Chopper (or Class E Chopper)

Figure 8.25(a) shows the basic power circuit of Type E chopper. From Fig. 8.25, it is observed that the four-quadrant chopper system can be considered as the parallel combination of two Type C choppers. In this chopper configuration, with motor load, the sense of rotation can be reversed without reversing the polarity of excitation. In Fig. 8.25, CH_1, CH_4 D_2 and D_3 constitute one Type C chopper and CH_2, CH_3, D_1 and D_4 form another Type C chopper circuit. Figure 8.25(b) shows Class-E with R–L load.

Choppers

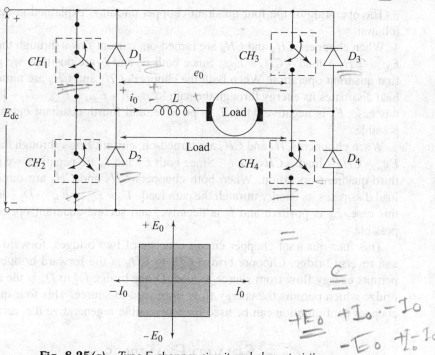

Fig. 8.25(a) *Type E chopper circuit and characteristic*

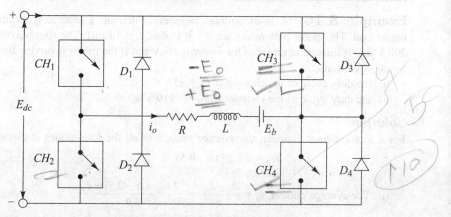

Fig. 8.25(b) *Class E chopper with R–L load*

If chopper CH_4 is turned on continuously, the antiparallel connected pair of devices CH_4 and D_4 constitute a short-circuit. Chopper CH_3 may not be turned on at the same time as CH_4 because that would short circuit source E_{dc}.

With CH_4 continuously on, and CH_3 always off, operation of choppers CH_1 and CH_2 will make E_0 positive and I_0 reversible, and operation in the first and second quadrants is possible. On the other hand, with CH_2 continuously on and CH_1 always off, operation of CH_3 and CH_4 will make E_0 negative and I_0 reversible, and operation in the third and fourth quadrants is possible.

The operation of the four-quadrant chopper circuit is explained in detail as follows:

When choppers CH_1 and CH_4 are turned-on, current flows through the path, $E_{dc+} - CH_1$ – load – $CH_4 - E_{dc-}$. Since both E_0 and I_0 are positive, we get the first quadrant operation. When both the choppers CH_1 and CH_4 are turned-off, load dissipates its energy through the path load–$D_3 - E_{dc+} - E_{dc-} - D_2$ – load. In this case, E_0 is negative while I_0 is positive, and fourth-quadrant operation is possible.

When choppers CH_2 and CH_3 are turned-on, current flows through the path, $E_{dc+} - CH_3$ – load – $CH_2 - E_{dc-}$. Since both E_0 and I_0 are negative, we get the third-quadrant operation. When both choppers CH_2 and CH_3 are turned-off, load dissipates its energy through the path load– $D_1 - E_{dc+} - E_{dc-} - D_2$ – load. In this case, E_0 is positive and I_0 is negative, and second-quadrant operation is possible.

This four-quadrant chopper circuit consists of two bridges, forward bridge and reverse bridge. Chopper bridge CH_1 to CH_4 is the forward bridge which permits energy flow from source to load. Diode bridge D_1 to D_4 is the reverse bridge which permits the energy flow from load-to-source. This four-quadrant chopper configuration can be used for a reversible regenerative d.c. drive.

SOLVED EXAMPLES

Example 8.16 A four-quadrant chopper is driving a separately excited dc motor load. The motor parameters are $R = 0.1$ ohm, $L = 10$ mH. The supply voltage is 200 V d.c. If the rated current of the motor is 10 A and if the motor is driving the rated torque. Determine:

(i) the duty cycle of the chopper if $E_b = 150$ V.
(ii) the duty cycle of the chopper if $E_b = -110$ V.

Solution:
For a four-quadrant chopper, the average voltage in all the four-modes is given by
$$E_0 = 2 E_{dc} \cdot (\alpha - 0.5)$$

(i) The average current, $i_0 = \dfrac{E_0 - E_b}{R} = \dfrac{2 E_{dc} \cdot (\alpha - 0.5) - E_b}{R}$

$$10 = \frac{2 \times 200 (\alpha - 0.5) - 150}{0.1} \quad \therefore \alpha = 0.876$$

Since, $\alpha > 0.5$, this mode is forward-motoring

(ii) Now, $10 = \dfrac{2 \times 200 (\alpha - 0.5) - 110}{0.1}$, $\therefore \alpha = 0.228$

As $\alpha < 0.5$, this mode is reverse motoring mode.

8.6 THYRISTOR CHOPPER CIRCUITS

In the previous sections, a chopper has been considered as an ideal on-off switch. The power switch used can be any power semiconductor devices like, SCR, power BJT, Power MOSFET, IGBT, MCT, etc. the power devices used as a switch must satisfy the required voltage and current ratings. The use of these devices however decides the performance of chopper in the following manner:

(i) As the operating frequency increases, the size of the inductance required to make the current continuous decreases. Hence, the size and cost of the chopper decreases. The efficiency and reliability increases. The operating frequency of the power devices is minimum for SCR and maximum for power MOSFET.

(ii) The power-devices have different power-capabilities, operating frequency and different gating requirements. The power-level of SCR is highest and power-level for MOSFET is the lowest. Gating requirements of transistor are complex while for MOSFET, IGBT and MCT are simplest. Hence, different power-devices are suitable for different requirements, e.g. for high-power-levels, SCR is the choice whereas for low power-levels, MOSFET can be used so that the gating requirements are simplest and operating frequency can be made higher to reduce the size of the chopper.

SCR based chopper circuit consists of a main power SCR switch together with the commutation circuitry to turn it OFF. There are several ways in which an SCR can be turned-off. The commutation techniques used for dc choppers are:

1. Forced Commutation In this type of commutation, current through the thyristor is forced to become zero to turn it OFF. This can be accomplished in two ways:

(a) Voltage commutation: In this scheme, a charged capacitor momentarily reverse-biases the conducting SCR and turns it OFF.

(b) Current commutation: In this scheme, a current pulse is forced in the reverse direction through the conducting SCR. As the net current becomes zero, the thyristor is turned OFF.

2. Load Commutation In this type of commutation, the load current flowing through the SCR either becomes zero (as in natural or line commutation employed in converters) or is transferred to another device from the conducting thyristors.

8.6.1 Voltage or Impulse Commutated Chopper

Figure 8.26 shows the basic power circuit diagram of voltage commutated chopper. This commutation circuit comprises an auxiliary SCR T_2, a diode D, inductor L and capacitor C; the complete chopper circuitry has been outlined with a dotted box. The main power switch is SCR T_1.

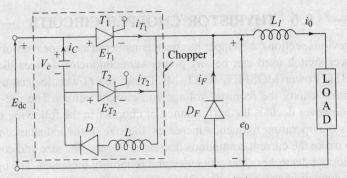

Fig. 8.26 *Voltage commutated chopper*

To start the circuit, capacitor C is initially charged with the polarity shown, by triggering the auxiliary SCR T_2. Capacitor C gets charged through the path $E_{dc+} - C_+ - C_- - T_2 - L_1$ load $- E_{dc-}$. As the charging current decays to zero, thyristor T_2 will be turned OFF. Figure 8.27 shows the associated voltage and current waveforms. For convenience, the chopper operation is divided into certain modes and is explained as under:

(i) Mode I Operation $(0 < t < t_2)$: The main SCR T_1 is triggered at t_0. Source current flows in two paths, load current i_o constitutes one path and commutation current i_c, the other path. With the triggering of SCR T_1, load gets connected to the supply and the load voltage, $e_0 = E_{dc}$.

Load current i_0 flows through the path, $E_{dc+} - T_1$ - load $- E_{dc-}$, whereas the commutating current i_c flows through the path, $C_+ - T_1 - L - D - C_-$. The capacitor current first rises from zero to a maximum value when voltage across capacitor C is zero at $t = t_1/2$. As i_c decreases to zero, capacitor is charged to a reverse voltage $(-E_{dc})$ at $t = t_1$ as shown in Fig. 8.27. This reverse voltage on capacitor is held constant by diode D.

At $t = 0$, voltage across auxiliary thyristor T_2 is $(-E_{dc})$, whereas it is zero at $t_1/2$ and E_{dc} at t_1. This voltage variation is shown as cosine wave in Fig. 8.27. Therefore, at $t = t_1$, $i_{T_1} = I_0$, $V_c = -E_{dc}$, $E_{T_2} = E_{dc}$, $e_0 = E_{dc}$, as shown in Fig. 8.27. These conditions continue upto the period t_2.

(ii) Mode II Operation $(t_2 < t < t_3)$: At a desired instant t_2, the auxiliary SCR T_2 is to be triggered for turning-off the main SCR T_1. With the turning-on of T_2, reverse capacitor voltage $(-E_{dc})$ appears across T_1, which reverse-biases it, and turns it OFF. Since the capacitor voltage commutates the main SCR T_1, the given circuit of Fig. 8.26 is called as voltage commutated chopper. Current i_{T_1} becomes zero at t_2.

After the SCR T_1 is turned-off, the capacitor C and SCR T_2 provide the path for load current i_0 through $E_{dc} - C - T_2$ - load. The load voltage is the sum of source voltage and the voltage across the capacitor. Hence, at instant t_2, load voltage is $e_0 = E_{dc} + E_{dc} = 2E_{dc}$, and it decreases linearly as the voltage across the capacitor decreases. During this mode, $V_c = E_{T_1}$, since the capacitor is directly connected across T_1 through T_2. As the capacitor discharges through the load, V_c and E_{T_1} change from $(-E_{dc})$ to zero at $(t_2 + t_q)$. Load-voltage e_0 changes from $2E_{dc}$ at t_2 to E_{dc} at $(t_2 + t_q)$. After $(t_2 + t_q)$, V_c and E_{T_1} start rising from zero towards E_{dc}, whereas e_0 starts falling towards zero. Hence, in this mode, V_c and E_{T_1} change linearly from $(-E_{dc})$ at t_2 to E_{dc} at t_3, since load current i_0 is assumed constant. Similarly, e_0 changes linearity from $2E_{dc}$ at t_2 to zero at t_3.

(iii) Mode III Operation ($t_3 < t < t_4$) : For this mode, $t_3 < t < T$. At t_3, $V_c = E_{T1} = E_{dc}$, $e_0 = 0$ and capacitor current decays to zero, therefore, SCR T_2 turns-off naturally. At t_3, since capacitor is slightly overcharged, freewheeling diode gets forward-biased. The load current after t_3 freewheels through this diode D_F. Note that during freewheeling period from t_3 to T, E_{T_2} is slightly negative as capacitor C is somewhat overcharged.

Thus, during this mode, $i_c = 0$, $i_{T_1} = 0$, $i_f = I_{0m}$, $E_{T_1} = E_{dc}$, $e_0 = 0$, $i_{T_2} = 0$.

Now, at $t = T$, the main thyristor T_1 is triggered again and the cycle as described from $t = 0$ to $t = T$ repeats.

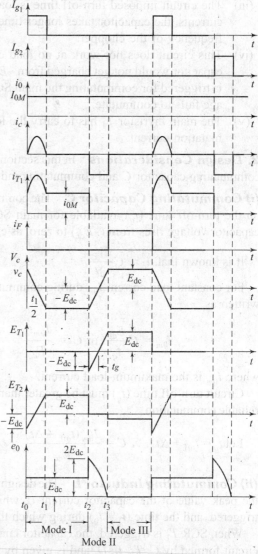

Fig. 8.27 *Related voltage and current waveforms*

1. Disadvantages Since voltage commutated chopper circuit is a simple chopper circuit, it is therefore widely used. However, it suffers from the following disadvantages:
  (i) A starting circuit is required, and the starting circuit (such as logic circuit) should be such that it triggers auxiliary SCR T_2 first.
  (ii) Load voltage jumps to twice the supply voltage when the commutation is initiated.

(iii) The circuit imposed turn-off time is load dependent. At very low load currents, the capacitor takes longer time to discharge, thus limiting the frequency of the chopper.

(iv) This circuit does not work at no load conditions, because, at no load, capacitor would not get charged from $-E_{dc}$ to E_{dc} when auxiliary SCR T_2 is triggered for commutating the main SCR T_1, that is, the capacitor voltage fails to commutate.

(v) The main thyristor T_1 has to carry the load current as well as the commutation current.

2. Design Considerations In this section, we see the selection method of commutating capacitor C and commutating inductor L.

(i) Commutating Capacitor C The commutating capacitor value is based on the turn-off time (t_q) available for main SCR T_1. During this time t_q, the capacitor voltage rises from $(-E_{dc})$ to zero, as explained in Mode II operation.

It is known that, $i_c = C \dfrac{dV_c}{dt}$

For constant load current I_{0m} during commutation, the above relation can be written as

$$I_{0m} = C \frac{E_{dc}}{t_q} \text{ or } C = \frac{I_{0m} t_q}{E_{dc}} \tag{8.91}$$

where I_{0m} is the maximum-load current.

Circuit turn-off time (t_q) must be greater than the SCR turn-off time (t_{off}), for reliable commutation.

Let $t_q = t_{off} + \Delta t$ $\therefore C = \dfrac{I_{0m}(t_{off} + \Delta t)}{E_{dc}}$ \hfill (8.92)

(ii) Commutating Inductor L The design criteria for inductor L is based on the peak value of the capacitor current i_c, which flows through T_1 when it is triggered, and the time $(t_1 - t_0)$ during which the capacitor current pulse lasts.

When SCR T_1 is triggered, the capacitor current i_c flows through the ringing circuit formed by $C, T_1, L, D,$ and is given by

$$i_c = \frac{E_{dc}}{\omega_r \cdot L} \sin \omega_r \cdot t \tag{8.93}$$

where $\omega_r = \dfrac{1}{\sqrt{LC}} = \dfrac{2\pi}{T_r}$ is the ringing frequency in radians per second and $T_r = 2(t_1 - t_0)$.

$$i_c = E_{dc} \cdot \sqrt{\frac{C}{L}} \sin \omega_r t \tag{8.94}$$

Also, the peak capacitor current (I_{cp}) is

$$I_{cp} = \frac{E_{dc}}{\omega_r \cdot L} = E_{dc} \cdot \sqrt{\frac{C}{L}} \qquad (8.95)$$

When T_1 is turned-on, it carries both, load current and the peak capacitor current. It is usual to take I_{cp} less than, or equal to, load current I_{0m}, so that the peak current through T_1 is not unnecessarily large, i.e.

$$I_{cp} \leq I_{0m} \quad \text{or} \quad E_{dc}\sqrt{\frac{C}{L}} \leq I_{0m} \qquad (8.96)$$

or
$$L \geq C \left(\frac{E_{dc}}{I_{0m}}\right)^2 \qquad (8.97)$$

By selecting a small value of the commutating inductance L, the resetting time $(t_1 - t_0)$ can be reduced. It can be seen from Eq. (8.95), that the small value of inductance L increases the peak value of the capacitor current. Also, if the duration $(t_1 - t_0)$ is large, the range of load voltage is reduced.

Therefore, the minimum load voltage is given by

$$E_{0(\min)} = \frac{(t_1 - t_0)}{T} \cdot E_{dc} \qquad (8.98)$$

where T is the chopping period of the chopper

or
$$E_{0(\min)} = \frac{\pi\sqrt{LC}}{T} E_{dc} \qquad (8.99)$$

The minimum value of load, voltage should be equal to or less than 10% of the supply voltage E_{dc} in order to vary the load voltage over a wide range.

Therefore, $\dfrac{(t_1 - t_0)}{T} E_{dc} \leq 0.1\, E_{dc}$ or $\dfrac{(t_1 - t_0)}{T} \leq 0.1$

or $\dfrac{\pi\sqrt{LC}}{T} \leq 0.1$ or $L \leq \dfrac{0.01 T^2}{\pi^2 C}$ $\qquad (8.100)$

If duty-cycle has to be reduced, inductance L value has to be decreased but the peak capacitor current increases. Hence, a compromise has to be made between the Eqs (8.97) and (8.99).

SOLVED EXAMPLES

Example 8.17 A voltage commutated chopper circuit of Fig. 8.26 controls the battery-powered electric cars. The battery voltage is 80 V, starting current is 80 A and thyristor turn-off time is 20 μs, chopping period 2000 μs. Compute the values of the commutating capacitor C and the commutating inductor L.

Solution: For reliable operation, the circuit turn-off time must be greater than the SCR turn-off time (t_{off}).

Let $\quad t_q = t_{\text{off}} + \Delta t$

Also, let $\quad \Delta t = t_{\text{off}} = 20 \, \mu s \quad \therefore t_q = 40 \, \mu s$.

From Eq. (8.92), $\quad C = \dfrac{I_{0m}(t_{\text{off}} + \Delta t)}{E_{dc}} = \dfrac{80(40 \times 10^{-6})}{80} = 40 \, \mu F$.

From Eq. (8.97), $L \geq C \left(\dfrac{E_{dc}}{I_{0m}}\right)^2 \geq 40 \times 10^{-6} \left(\dfrac{80}{80}\right)^2 \geq 40 \, \mu H$

Also given, chopping period = 2000 µs from Eq. (8.100).

$\therefore \quad L \leq \dfrac{0.01 T^2}{\pi^2 C} \quad L \leq \dfrac{0.01(2000 \times 10^{-6})^2}{\pi^2 \times 40 \times 10^{-6}} \leq 101.32 \, \mu H$.

The value of the commutating inductor is in the range $40 < L < 101.32 \, \mu H$. A value of L close to 40 µH will be a good choice because it will allow the minimum load voltage be about 5% of the supply voltage, hence giving a wider range of variation in the load voltage.

8.6.2 Current or Resonant–Pulse Commutated Chopper

Figure 8.28 shows the basic power-circuit for current commutated chopper. Here, T_1 is the main thyristor. The commutation circuit consists of auxiliary thyristor T_2, capacitor C, inductor L, diodes D_1 and D_2. D_F is the freewheeling diode and R is the charging resistor. The main SCR T_1 is commutated by a current pulse generated in the commutation circuitry. The important feature of this type of chopper is that the reverse voltage across the device is applied through a diode connected in antiparallel to the SCR. Since this is limited to about one volt, the turn-off time of SCR increases in comparison with the voltage commutations.

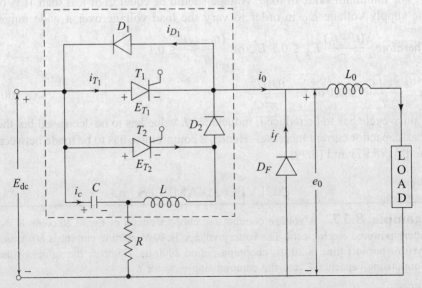

Fig. 8.28 *Current commutated chopper circuit*

As in voltage commutated chopper, here also the energy for current commutation comes from the energy stored in a capacitor. To start the circuit, the capacitor is charged to a voltage E_{dc} through the path $E_{dc+} - C - R - E_{dc-}$. The main thyristor T_1 is triggered at $t = t_0$, so that load voltage $e_0 = E_{dc}$ and load current $i_0 = I_{0m}$, up to $t = t_i$, as shown in Fig. 8.29.

For convenience, the commutation process is divided into certain modes and is explained as follows.

(i) Mode I Operation At time $t = t_1$, auxiliary SCR T_2 is triggered to commutate the main thyristor T_1. When thyristor T_2 is turned-on, an oscillatory current $\left(i_c = \dfrac{E_{dc}}{\omega_r L} \sin \omega_r t \right)$ is set up in the circuit consisting of C, T_2, and L. At t_2, the capacitor current i_c reverses, therefore, SCR T_2 gets turned-off due to natural commutation, and at t_2, $V_c = -E_{dc}$. In this mode, main thyristor T_1 remains unaffected and hence load voltage and load current remains E_{dc} and I_{0m} respectively.

(ii) Mode II Operation Since T_2 is turned-off at t_2, oscillatory current i_c flows through C, L, D_2 and T_1. As shown in Fig. 8.29, after t_2, current i_c would flow through thyristor T_1 and not through D_1, because D_1 is reverse biased by a small voltage drop across conducting thyristor T_1. Hence, after t_2, i_c would flow through T_1 and not through D_1. As current i_c flows in the opposite direction in T_1, it decreases the current i_{T_1}.

At t_3, $i_c = i_{T_1}$ and so the net current through T_1 is zero and it turns-off. As the oscillatory current through T_1 turns it OFF, it is called as current commutated chopper. During this mode, load voltage remains E_{dc} through T_1.

(iii) Mode III Operation Since T_1 is turned-off at t_3, i_c becomes more than i_0. After t_3, i_c supplies load current i_0 and diode D_1 begins to conduct the current $(i_c - i_0)$ and the drop in D_1 due to this current keeps the thyristor T_1 reverse-biased for the time t_q $(= t_4 - t_3)$.

(iv) Mode IV Operation As shown in Fig. 8.29, at t_4, $i_c = i_0$ and $iD_1 = 0$; therefore, diode D_1 is reverse-biased. After t_4, a constant current equal to i_0 flows through $E_{dc} - c - L - D_L$ – load and therefore, capacitor C is charged linearly to source-voltage E_{dc} at t_5. Therefore, during the period (t_4-t_5), $i_c = i_0$.

As D_1 is turned-off at, t_4, $v_c = E_{t_1}$. Now, the load voltage $e_0 = E_{dc} - V_c$ at t_4 and at t_5, $V_c = E_{dc}$. Hence, at t_5 load voltage become $e_0 = E_{dc} - E_{dc} = 0$. During the interval (t_4-t_5), V_c increases linearity and therefore, load voltage e_0 decreases to zero linearity, during this period.

(v) Mode V Operation As shown in Fig. 8.29, at t_6, capacitor C is actually overcharged to a voltage somewhat greater than source voltage E_{dc}. Therefore at t_5, the freewheeling diode D_F becomes forward biased and starts to conduct the load current i_0. As i_c is not zero at t_5, the capacitor C is still connected to load through source $E_{dc} - C - L$ and D_2, and as a result C is overcharged by the transfer of energy from L to C. Therefore, at t_6, $i_c = 0$ and V_c becomes more than E_{dc}.

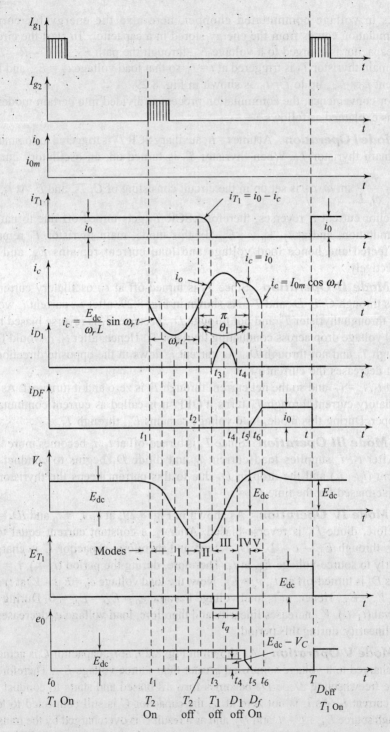

Fig. 8.29 *Voltage and current waveforms for current commutated chopper*

During interval t_5 to t_6, $i_0 = i_c + i_F$ and therefore, as i_c decays, i_F builds up. At t_6, $i_F = i_0$ and $i_c = 0$. Commutation process is completed at t_6 and the commutation interval is $(t_6 - t_1)$.

From t_6 onwards, load current freewheels through D_F and decays. As i_c is zero and D_2 is open circuited, capacitor voltage V_c decays through R for the freewheeling period of the chopper.

At $t = T$, the main thyristor T_1 is triggered again and the cycle repeats.

This chopper was developed by the Hitachi-Electric Company, Japan and is widely used in traction cars.

1. Advantages The following are the main advantages of current commutated chopper:
 (i) The capacitor always remains charged with the correct polarity.
 (ii) Commutation is reliable as long as load current is less than the peak commutation current I_{cp}.
 (iii) The auxiliary thyristor T_2 is naturally commutated as its current passes through zero value.

2. Design Considerations The values of the commutating components L and C are selected to satisfy the following conditions, for reliable commutation of main SCR T_1.
 (a) The peak commutating current I_{cp} must be greater than the maximum possible load current I_{0m}. This condition is very essential for reliable commutation of main SCR T_1.

From Eq. (8.95), the oscillating current is given by

$$i_c = E_{dc}\sqrt{\frac{C}{L}} \sin \omega_r t = I_{cp} \cdot \sin \omega_r t$$

From condition (a), we make the relation

$$I_{cp}\left(=E_{dc}\sqrt{\frac{C}{L}}\right) > I_{0m}. \text{ or, } E_{dc}\sqrt{\frac{C}{L}} = x \cdot I_{0m} \qquad (8.101)$$

where I_{0m} = maximum load current and x is greater than 1. Therefore,

$$x = \frac{I_{cp}}{I_{0m}} \qquad (8.102)$$

 (b) The circuit turn-off time (t_q) must be greater than the thyristor T_1 turn-off time (T_{off}), therefore,

$$t_q = t_{off} + \Delta t.$$

From Fig. 8.29 with the current waveform, we can write,

$$t_q = t_4 - t_3 \quad \text{or} \quad \omega_r t_q = \pi - 2\theta_1 \qquad (8.103)$$

Also, $I_{cp} \sin \theta_1 = I_{0m}$ or, $\theta_1 = \sin^{-1}\left(\dfrac{I_{0m}}{I_{cp}}\right) = \sin^{-1}\left(\dfrac{1}{x}\right)$ \hfill (8.104)

Circuit turn-off time for main SCR T_1 from Eq. (8.103),

$$t_q = \dfrac{1}{\omega_r}(\pi - 2\theta_1) \tag{8.105}$$

Substitute the value of θ_1 from Eq. (8.104),

$$t_q = \dfrac{1}{\omega_r}\left[\pi - 2\sin^{-1}\left(\dfrac{I_{0m}}{I_{cp}}\right)\right] \tag{8.106}$$

From Eq. (8.106), it becomes clear that as load current I_{0m} increases, circuit turn-off time for main SCR T_1 decreases. Thus, a certain value of ratio $\left(\dfrac{I_{0m}}{I_{cp}}\right)$ must be maintained for ensuring necessary turn-off time t_q.

Substituting $\omega_r = \dfrac{1}{\sqrt{LC}}$ in Eq. (8.106), we get

$$t_q = \sqrt{LC}\left[\pi - 2\sin^{-1}\left(\dfrac{1}{x}\right)\right] \tag{8.107}$$

or $\qquad \sqrt{C} = \dfrac{t_q}{\sqrt{L}\left[\pi - 2\sin^{-1}(1/x)\right]}$ \hfill (8.108)

Substituting this value of $\sqrt{C}$ in Eq. (8.101), we get

$$\dfrac{E_{dc}}{L}\dfrac{t_q}{\left[\pi - 2\sin^{-1}(1/x)\right]} = x \cdot I_{0m} \text{ or } L = \dfrac{E_{dc} t_q}{x \cdot I_{0m}\left[\pi - 2\sin^{-1}(1/x)\right]} \tag{8.109}$$

From Eq. (8.107), we have

$$\dfrac{1}{\sqrt{L}} = \dfrac{\sqrt{C}}{t_q}\left[\pi - 2\sin^{-1}(1/x)\right]$$

Substituting the above value of $\sqrt{L}$ in Eq. (8.101) gives

$$\dfrac{E_{dc}\sqrt{C}\sqrt{C}\left[\pi - 2\sin^{-1}(1/x)\right]}{t_q} = x \cdot I_{0m} \text{ or } C = \dfrac{x \cdot I_{0m} \cdot t_q}{t_q\left[\pi - 2\sin^{-1}(1/x)\right]} \tag{8.110}$$

3. Peak Capacitor Voltage From capacitor voltage waveform of Fig. 8.29, it is clear that the maximum capacitor voltage is reached at t_8. Therefore, voltage at $t_6 = V_{cp}$ = voltage at t_5 + voltage rise due to the energy transferred from L to C during t_5 to t_8.

At instant t_5, energy stored in L is $\frac{1}{2}LI_{0m}^2$ and at t_6, this energy is transferred to C. Thus, the voltage rise of capacitor C due to this transfer to energy is,

$$\frac{1}{2}CV_c^2 = \frac{1}{2}LI_{0m}^2 \qquad (8.111)$$

or $\qquad V_c = I_{0m}\sqrt{\dfrac{L}{C}} \qquad (8.112)$

Now, substituting Eq. (8.112) in Eq. (8.103),

$\therefore \qquad V_{cp} = E_{dc} + I_{0m}\sqrt{\dfrac{L}{C}} \qquad (8.113)$

The peak voltage and current value of capacitor are also the peak ratings of both the main and auxiliary thyristors.

The charging resistor R is selected such that the chopping period T is much greater than $3\,RC$.

SOLVED EXAMPLES

Example 8.18 For a current commutated chopper, peak commutating current is thrice the maximum possible load current. The source voltage is 220 V d.c. and main SCR turn-off time is 20 μs. For a maximum load current of 180 A, compute
(a) the value of commutating components L and C
(b) maximum capacitor voltage, and
(c) the peak commutating current.

Solution: Given: $x = 3$, $t_{off} = 20$ μs, $\therefore t_q = t_{off} + \Delta t = 20 + 20 = 40$ μs

(a) From Eq. (8.109), $L = \dfrac{E_{dc} \cdot t_q}{x \cdot I_{0m}\left[\pi - 2\sin^{-1}(1/x)\right]}$

$= \dfrac{220 \times 40 \times 10^{-6}}{3 \times 180\,[\pi - 2\sin^{-1}(1/3)]} = 6.62$ μH

From Eq. (8.110), $C = \dfrac{x \cdot I_{0m} \cdot t_q}{E_{dc}\left[\pi - 2\sin^{-1}(1/x)\right]}$

$= \dfrac{3 \times 180 \times 40 \times 10^{-6}}{220 \times [\pi - 2\sin^{-1}(1/s)]} \quad \therefore \quad C = 39.88$ μF

(b) From Eq. (8.113) the peak capacitor voltage is given by

$V_{cp} = E_{dc} + I_{0m}\sqrt{\dfrac{L}{C}} = 220 + 180\sqrt{\dfrac{6.62 \times 10^{-6}}{39.88 \times 10^{-6}}} = 293.34$ V.

(c) Peak commutating current
$i_{cp} = x \cdot I_0 = 3 \times 220 = 660$ A

8.6.3 Load Commutated Chopper

Figure 8.30 shows the power-circuit of a load commutated chopper. This chopper circuit consists of four thyristors T_1, T_2, T_3 and T_4, and one commutating capacitor C. Here, thyristors T_1, T_2 form one pair and thyristors T_3, T_4 form another pair which conduct the load current alternatively. When one thyristor pair T_1, T_2 functions as main thyristors, at the same time other thyristor pair T_3, T_4 functions as auxiliary thyristors, and *vice-versa*. Again, operation of the chopper circuit is divided into different operating modes. These modes are described with the associated waveforms as shown in Fig. 8.31.

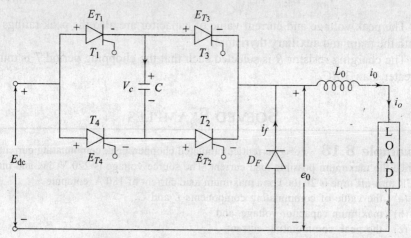

Fig. 8.30 *Load commutated chopper circuit*

The waveforms shown in Fig. 8.31 starts at the instant $t = t_0$. It is assumed that prior to this instant t_0, capacitor C was charged to the reverse voltage ($-E_{dc}$) due to the conduction of thyristors T_3 and T_4. Therefore, before the instant t_0, the capacitor upper plate becomes negative and lower plate positive.

(i) Mode I Operation As shown in Fig. 8.31, at $t = t_0$ both thyristors T_1 and T_2 are triggered. Therefore, load current flows through the path $E_{dc} - T_1$, C, T_2 and the load. Load voltage e_0 now becomes, $e_0 = E_{dc} - V_c$, i.e., 2 E_{dc}. The capacitor C is charged linearly by a constant load current i_0 from $(-E_{dc})$ at $t = 0$ to E_{dc} at t_1.

When the capacitor is charged fully positive at $t = t_1$, the current through the conducting thyristors T_1, T_2 becomes zero and these go into the blocking mode. The load voltage e_0 falls linearly. The freewheeling diode D_f becomes forward biased, and the load current is transferred from T_1 and T_2 to D_f.

(ii) Mode-II Operation As shown in Fig. 8.31, from t_1 to T, the freewheeling diode D_F conducts the load current. For the period t_1 to T, $V_c = E_{dc}$, $i_c = 0$, $i_f = i_0$ and the load voltage $e_0 = 0$. Now, at $t = T$, the second pair of thyristors T_3, T_4 is triggered. This places the fully charged capacitor across thyristors T_1 T_2, reverse biasing them and turning them off. The cycle now repeats.

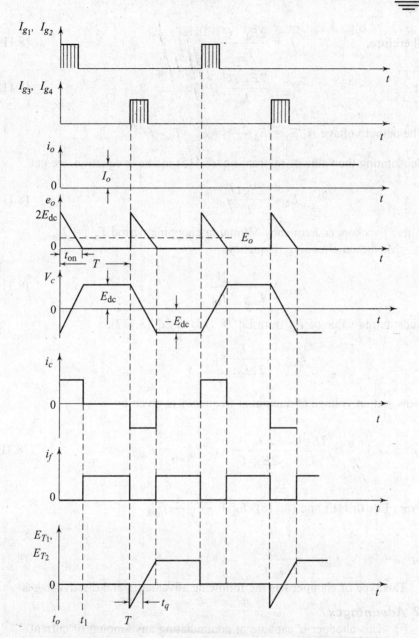

Fig. 8.31 *Voltage and current waveforms of load-commutated chopper*

1. Design Considerations The average value of the chopper output voltage is controlled by changing the firing frequency of the choppers. Thus, it is a frequency modulated chopper.

As shown in Fig. 8.31, for a constant load current I_0, capacitor voltage changes from $(-E_{dc})$ to E_{dc} in time T_{on}, i.e. total change in voltage is $2E_{dc}$ in time t_{on}.

Therefore,
$$I_0 = \frac{2E_{dc} \cdot C}{T_{on}} \tag{8.114}$$

or
$$T_{on} = \frac{2E_{dc} \cdot C}{I_0} \tag{8.115}$$

The output voltage is, $E_0 = E_{dc} \dfrac{T_{on}}{T} = E_{dc} \cdot T_{on} \cdot f$

Substituting the value of t_{on} from Eq. (8.115) in above equation, we get

$$E_0 = \frac{2E_{dc}^2 cf}{I_0} \tag{8.116}$$

where f = chopper frequency Minimum chopping period $T_{min} = T_{on}$.
$\therefore$ Maximum chopping frequency,

$$f_{max} = \frac{1}{T_{min}} = \frac{1}{T_{on}} \tag{8.117}$$

Substituting value of T_{on} from Eq. (8.115) in Eq. (8.117),

$$f_{max} = \frac{I_0}{2 \cdot E_{dc} \cdot c}$$

Now, output voltage at maximum frequency is given by

$$E_0|_{f_{max}} = \frac{2E_{dc}^2 C}{I_0} \cdot \frac{I_0}{2E_{dc} \cdot C} \quad \therefore E_0|_{f_{max}} = E_{dc} \tag{8.118}$$

From Eqs (8.116) and (8.118), $E_{dc} = \dfrac{2E_{dc}^2 \cdot C}{I_0} f_{max}$

or
$$f_{max} = \frac{I_0}{2E_{dc} \cdot C} \tag{8.119}$$

This type of chopper has the following advantages and disadvantages.

2. Advantages
   (i) This chopper is capable of commutating any amount of current.
   (ii) No commutating inductor is required in this chopper circuit which is normally costly, bulky and noisy.
   (iii) This circuit can operate at high frequencies of the order of kHz, and therefore filtering requirements to smooth out load current are minimal.

3. Disadvantages
This chopper circuit has some minor disadvantages, as follows:

(i) The peak load voltage is twice the supply voltage. However, this peak can be reduced by filtering.
(ii) Because of higher switching losses at high frequencies and losses in the two conducting thyristors in series with the load, efficiency may become low for high power applications.
(iii) Since freewheeling diode D_f is subjected to twice the supply voltage ($2 E_{dc}$) in a short time, a fast recovery type diode must be used.
(iv) The commutating capacitor has to carry the load current at a frequency half the chopping period.
(v) One thyristor pair should be turned-on only when the other pair is commutated. This can be realised by sensing the capacitor current that is alternating.

SOLVED EXAMPLE

Example 8.19 A load commutated chopper, fed from a 230 V d.c. source has a constant load current of 50 A. For a duty cycle of 0.4 and a chopping frequency of 2 kHz, calculate
(a) the value of commutating capacitance. (b) average output voltage.
(c) circuit turn-off time for one SCR pair. (d) total commutation interval.

Solution:

(a) From Eq. (8.115), we have $C = \dfrac{t_{on} \cdot I_0}{2 E_{dc}}$

Given: $\alpha = 0.4 = \dfrac{T_{on}}{T}$, $T = \dfrac{1}{f} = \dfrac{1}{2 \times 10^3} = 0.5$ ms

$\therefore 0.4 = \dfrac{T_{on}}{0.5 \text{ ms}}$, $\therefore T_{on} = 0.2$ ms

$C = \dfrac{0.2 \times 10^{-3} \times 50}{2 \times 230}$, $C = 21.74$ μF

(b) From Eq. (8.118), $E_0 = \dfrac{2 E_{dc}^2 \cdot cf}{I_0}$

$E_0 = \dfrac{2 \times 230 \times 230 \times 21.74 \times 10^{-6} \times 2 \times 10^3}{50} = 92$ V

(c) For load commutated chopper circuit, the circuit turn-off time for each SCR is

$t_q = \dfrac{1}{2} T_{on} = \dfrac{1}{2} \cdot \dfrac{C \cdot 2 E_{dc}}{I_0} = \dfrac{C \cdot E_{dc}}{I_0} = \dfrac{21.74 \times 10^{-6} \times 230}{50} = 100$ μs

(d) Total commutation interval $= \dfrac{2 C E_{dc}}{I_0} = \dfrac{2 \times 21.74 \times 10^{-6} \times 230}{50} = 200$ μs

8.7 JONES CHOPPER

Figure 8.32 shows the basic power circuit of Jones chopper. This chopper circuit is another example of Class D commutation. In this circuit, SCR T_1 is the main thyristor, whereas SCR T_2, capacitor C, D_2, and autotransformer (T) forms the commutating circuit for the main thyristor T_1. Therefore, the special feature of this circuit is the tapped autotransformer T through a portion of which the load current flows. Here, L_1 and L_2 are closely coupled so that the capacitor always gets sufficient energy to turn-off the main SCR T_1.

If the main thyristor T_1 is ON for a long period, then the motor will reach the maximum steady-state speed determined by the battery voltage, the motor and the mechanical load characteristics. If thyristor T_1 is OFF, the motor will not rotate. Now, if thyristor T_1 is alternatively ON and OFF in a cyclic manner, the motor will rotate at some speed between maximum and zero. Figure 8.33 shows the voltage and current waveforms of the chopper circuit.

Let us assume that initially capacitor C is charged to a voltage E_{dc} with the polarity shown in Fig. 8.32. As shown in Fig. 8.33, SCR T_1 is triggered at time $t = t_1$, current flows through the path $C_A - T_1 - L_2 - D_1 - C_B$ and capacitor C charges to opposite polarity, i.e. plate B is positive and plate A is negative. However, diode D_1 prevents further oscillation of the resonating $L_2 C$ circuit. Hence, capacitor C retains its charge until SCR T_2 is triggered. In Fig. 8.33, the capacitor voltage waveforms are drawn at bottom plate B of capacitor.

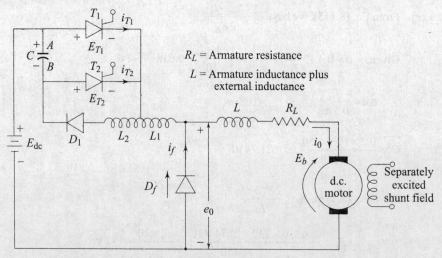

Fig. 8.32 *Freewheeling diode d.c. motor drive using Jones chopper*

Now, at time $t = t_3$, SCR T_2 is triggered. Current flow through the path $C_B - T_2 - T_1 - C_A$. Therefore, discharge of capacitor C reverse-biases SCR T_1 and turns it OFF. The capacitor again charges up with plate A positive and SCR T_2 turns-off because the current through it falls below the holding current value when capacitor C is recharged.

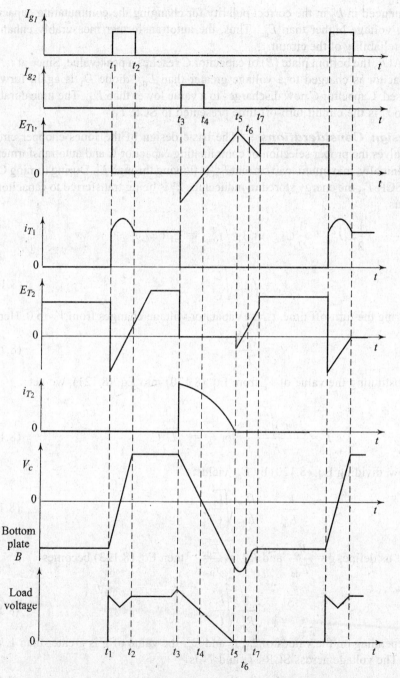

Fig. 8.33 *Jones chopper voltages and current waveforms*

The cycle repeats when SCR T_1 is again triggered. The use of autotransformer insures that whenever current is delivered from dc source to the load, a voltage is induced in L_2 in the correct polarity for changing the commutating capacitor to a voltage higher than E_{dc}. Thus, the autotransformer measurably enhances the reliability of the circuit.

At t_5, the bottom plate (B) of capacitor C reaches a peak value. Since at t_5, the capacitor is charged to a voltage greater than E_{dc}, diode D_1 is again forward biased. Capacitor C now discharges to a value lower than E_{dc}. The time duration t_3 to t_4 is the circuit turn-off time presented to SCR T_1.

Design Considerations : The basic design of the Jones chopper circuit involves the proper selection of commutating capacitor C and autotransformer T.

Initially, maximum load current I_{0m} is flowing through L_1. During turning OFF of SCR T_1, the energy stored in inductance L_1 is being transferred to capacitor C. Thus,

$$\frac{1}{2} L_1 I_{0m}^2 = \frac{1}{2} C V_c^2 \text{ or } V_c^2 / I_{0m}^2 = L_1/C$$

or
$$V_c = I_{0m} \cdot \sqrt{\frac{L_1}{C}} \qquad (8.120)$$

During the turn-off time, t_q, the capacitor voltage changes from V_C to 0. Hence,

$$t_q = \frac{V_c \cdot C}{I_{0m}} \qquad (8.121)$$

Substituting the value of V_c from Eq. (8.120) into Eq. (8.121), we get

$$t_q = \frac{I_{0m} \cdot \sqrt{\frac{L_1}{C}} \cdot C}{I_{0m}} \text{ or } t_q = \sqrt{L_1 C} \qquad (8.122)$$

Now, dividing Eq. (8.120) by E_{dc} yields:

$$\frac{V_c}{E_{dc}} = \frac{I_{0m}}{E_{dc}} \sqrt{\frac{L_1}{C}} \qquad (8.123)$$

Let us define, $g = \dfrac{V_c}{E_{dc}}$ and $R_m = \dfrac{E_{dc}}{I_{0m}}$ then Eq. (8.123) becomes

$$g = \frac{1}{R_m} \sqrt{\frac{L_1}{C}} \qquad (8.124)$$

Depending on the values of L_1, C and R_m, the value of g is greater than 1.

The voltage across SCRs T_1 and T_2 is

$$V_c = g \cdot E_{dc} \qquad (8.125)$$

Hence, a large value of g would require an increase in the voltage rating of the thyristor in the circuit.

Choppers

In this type of chopper circuit, only dissipative elements are winding resistance and the forward conducting resistance of the SCRs and diodes. Therefore, this circuit is basically very efficient. The inductance L maintains the load current through diode D_f when SCR T_1 is not conducting. Hence, the motor torque which is proportional to the load current is smooth rather than pulsating. If the inductance in the load circuit is small, the change in the load current is substantial, which would result in substantial torque ripple in the motor load.

SOLVED EXAMPLES

Example 8.20 The Jones chopper of Fig. 8.32 controls the electric car. Compute the value of commutating capacitor C and transformer inductance L_1 and L_2 for the following data:

$$E_{dc} = 60V, t_q = 20 \, \mu s, I_{0m} = 140 \, A, g = 4$$

Solution: From Eq. (8.124), $g = \dfrac{1}{R_m} \sqrt{\dfrac{L_1}{C}}$

But, $R_m = \dfrac{E_{dc}}{I_{0m}} = \dfrac{60}{140} = 0.43 \, \Omega$ ∴ $\sqrt{\dfrac{L_1}{C}} = 4 \times 0.43 = 172$ \hfill (a)

Also, from Eq. (8.122), $t_q = \sqrt{L_1 C}$ ∴ $20 \times 10^{-6} = \sqrt{L_1 C}$ \hfill (b)

Dividing Eq. (b) by Eq. (a), we get, $11.63 \times 10^{-6} = C$ ∴ $C = 11.63 \, \mu F$

and $L_1 = 34.4 \, \mu H$

Normally, L_2 is designed to be equal to L_1.

Example 8.21 A d.c. series motor is used in an electric rapid transit system, as shown in Fig. 8.32. The Jones chopper is used as a speed controller for this purpose.
(a) For the following data, compute the inductance, L, required to limit the current swing in the armature under the worst condition to 6 amperes.
Given data :
(i) Motor performance parameters:
n = rated speed in rpm = 1750, HP = horse power rating = 150 HP
R_a = armature circuit resistance in ohm = 0.0099 Ω., E_{ff} = efficiency of motor = 90%
E_a = armature voltage = 1200 V, L = inductance in series with armature
L = (armature inductance + series field inductance + external inductance)
(ii) Chopping period $T = 2500 \, \mu s$.
(b) Also calculate the steady-state speed and the current swing in the armature for $\alpha = 0.10$ and $L = 125$ mH.
Assume rated current in the armature for all values of α, and $E_{dc} = 1200$ V, $\alpha = 0.1$

Solution: Average output voltage, $E_0 = \dfrac{T_{on}}{T} \cdot E_{dc} = \alpha E_{dc}$

The voltage across the inductor L is given by $(E_{dc} - \alpha \cdot E_{dc})$

$$\therefore \quad L \cdot \frac{di_a}{dt} = (E_{dc} - \alpha E_{dc}) \quad \text{or} \quad \frac{di_a}{dt} = \frac{(E_{dc} - \alpha E_{dc})}{L}$$

Here, $dt = T_{on}$ $\therefore di_a = \frac{(E_{dc} - \alpha E_{dc})}{L} T_{on}$ \hfill (a)

We can write the above equation as

$$di_a = \frac{(E_{dc} - \alpha E_{dc})}{L} \frac{T_{on}}{T} \cdot T = \frac{(E_{dc} - \alpha E_{dc})}{L} \alpha \cdot T \tag{b}$$

Now, differentiating Eq. (b) w.r.t α,

$$\therefore \quad \frac{di_a}{d\alpha} = \frac{E_{dc}}{L} T - \frac{2\alpha E_{dc}}{L} T$$

For worst case condition, $\frac{di_a}{d\alpha} = 0$

$$\therefore \quad 0 = \frac{E_{dc}}{L} T(1 - 2\alpha)$$

$$\therefore \quad (1 - 2\alpha) = 0$$

$\alpha = 0.5$ is the worst condition. Using $\alpha = 0.5$, in Eq. (b),

$$di_a = 6 = \frac{(1200 - 0.5 \times 1200)}{L} \times 0.5 \times 2500 \times 10^{-6}$$

$$\therefore \quad L = 125 \text{ mH}$$

(b) Power input to the motor $= P_{in.} = \frac{\text{power output}}{\text{efficiency}} = \frac{150 \times 746}{0.9} = 124.33 \text{ kW}$

Therefore, rated current in the armature $I_a = \frac{P_{in}}{E_{dc}} = \frac{124.33 \times 10^3}{1200} = 103.61$ A

$\therefore$ Armature voltage under rated torque condition is

$$E_{a(rated)} = E_{dc} - I_a \cdot R_a = 1200 - 103.1 \times 0.0099 = 1198.97 \text{ V}$$

For the above armature voltage, the speed is 1750.
For $\alpha = 0.1$, voltage at the armature is given by

$$\alpha \times 1200 - I_a \cdot R_a = 0.1 \times 1200 - 103.61 \times 0.0099 = 118.97 \text{ V}.$$

$\therefore$ Motor speed, $N = \frac{118.97}{1198.97} \times 1750 = 173.65$ rpm.

Current swing $= \Delta i_a = \frac{(E_{dc} - \alpha E_{dc})}{L} \times \alpha T$

$$= \left(\frac{1200 - 0.1 \times 1200}{125 \times 10^{-3}} \right) \times 0.1 \times 2500 \times 10^{-6} = 2.16 \text{ A}.$$

Example 8.22 Design the Jones chopper circuit for optimum frequency considerations to meet the following specifications:

Source voltage, E_{dc} = 200 V, Load current, I_0 = 50 A and t_q = 200 µs

Solution:

(i) *Selection of commutating capacitor C.* For obtaining optimum frequency so as to have lesser commutation losses and smaller commutation components, the commutation capacitor C can also be given by the following equation,

$$C \geq \frac{\pi}{2} \frac{t_q}{E_{dc}} I_0 \geq \frac{\pi}{2} \frac{200 \times 10^{-6}}{200} \times 50 \geq 78.54 \text{ µF}.$$

Capacitor voltage rating = safety factor $\times E_{dc}$ = 1.5 × 200 = 300 V.

(ii) *Selection of inductance L_1.* From Eq. (8.122), we have the relation

$$t_q = \sqrt{L_1 C}$$

$$\therefore L_1 = \frac{(t_q)^2}{C} = \frac{200 \times 200 \times 10^{-12}}{78.54 \times 10^{-6}} = 0.51$$

$$\therefore L_1 = L_2 = 0.51 \text{ mH}.$$

(iii) *Selection of SCR T_1 (main SCR)*

$$V_{BO} = \text{safety factor} \times E_{dc} = 1.5 \times 200 = 300 \text{ V}$$
$$I_T = \text{safety current} \times I_0 = 1.5 \times 50 = 75 \text{ A}.$$

(iv) *Selection of auxiliary SCR T_2*

Turn-off time, $t_q \leq \frac{\pi}{2}\sqrt{L_1 C} \leq \frac{\pi}{2}\sqrt{0.51 \times 10^{-3} \times 78.54 \times 10^{-6}} \leq 315$

or $t_q \cong 250$ µs and V_{Bo} = 300 V and I_T = 75 A.

(v) *Selection of diode D_1*
PIV rating of diode = V_{BO} of SCR = 300 V and $I_D = I_T$ = 75 A.

(vi) *SCR dynamic characteristics*

(a) *Main* SCR T_1, $\dfrac{dV}{dt} = \dfrac{I_{0\max}}{C}$ volts per µs

$$= \frac{50}{78.54 \times 10^{-6}} = 0.64 \text{ volts per µs}.$$

Initial $\dfrac{di}{dt} = \dfrac{E_{RRm}}{L_1}$ amp-per Hz

$$= \frac{\text{safety factor} \times E_{dc}}{L_1} = \frac{1.5 \times 200}{0.51 \times 10^{-3}} = 0.59 \text{ A/µs}.$$

(b) *Main* SCR T_2, $\dfrac{dV}{dt} = \dfrac{E_{peak}}{\sqrt{L_1 C}}$ volts per µs

$$= \frac{1.5 \times 200}{\sqrt{0.51 \times 10^{-3} \times 78.54 \times 10^{-6}}} = 1.49 \text{ V/µs}.$$

Initial $\dfrac{di}{dt} = \dfrac{E_{peak}}{L_s}$ A/μs

Let us assume, L_1 = stray inductance during the discharging loop = 2 μH.

∴ $\dfrac{di}{dt} = \dfrac{1.5 \times 200}{2\mu H} = 150$ A /μs.

8.8 MORGAN CHOPPER

Figure 8.34 shows the power-circuit of Morgan chopper. In this circuit, T_1 is the main thyristor, whereas capacitor C, saturable reactor SR and diode D_1 forms the commutating circuit. The exciting current of the saturable reactor is assumed to be negligible small. When the saturable reactor is saturated, it has very low inductance. The voltage and current waveforms of the Morgan-chopper is shown in Fig. 8.35.

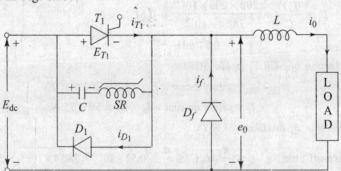

Fig. 8.34 Morgan chopper

When the main SCR T_1 is OFF, capacitor C will charge to the supply voltage E_{dc} with the polarity as shown in Fig. 8.34 and the saturable reactor is placed in the positive saturation condition. The capacitor charging path is $E_{dc+} - C - S_R -$ Load $- E_{dc-}$.

As shown in Fig. 8.35, thyristor T_1 is triggered at time $t = t_1$. When thyristor T_1 is turned-on, the capacitor voltage appears across the saturable reactor and the core flux is driven from the positive saturation towards negative saturation. The capacitor voltage remains essentially constant with the same polarity, till the negative saturation point is reached. This is due to the negligible exciting current of the SR.

When the core flux reaches the negative saturation, the capacitor discharges through the SCR T_1 and the post-saturation inductance of SR. This forms a resonant circuit with a discharging time of $\pi\sqrt{L_s C}$ seconds, where L_s is the post-saturation inductance of the reactor. Thus, the discharging time of the capacitor is comparatively small and the reversal of the polarity of the capacitor takes place very quickly. After this, the capacitor voltage which is now $-E_{dc}$ is impressed on the saturable reactor in the reverse direction and the core is driven from negative saturation towards positive saturation.

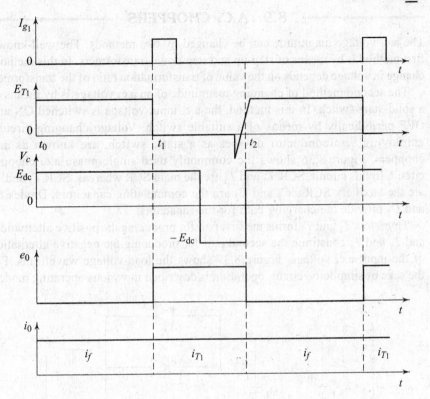

Fig. 8.35 *Morgan chopper voltage and current waveform*

After a fixed interval of time, the core flux reaches the positive saturation after which the capacitor discharges very quickly through SCR T_1 in the reverse direction and the post-saturation inductance as before. The discharge current first passes through SCR T_1, turning it OFF and then through diode D_1.

When SCR T_1 is turned-off the load current flows through the freewheeling diode D_f. Since the volt–time integral to saturate the core is constant, the ON period of SCR T_1 is fixed. The ON period is a function of L_s, C and the average output voltage can be altered only by varying the operating frequency. Output voltage is lowered by lowering the frequency and increases by increasing the frequency. The ON period, however, can be controlled by varying the volt time product of the saturable reactor by means of d.c. controlled current through it. Also, the total ON time of SCR T_1 is determined by the time required for the saturable reactor to move from positive saturation to negative saturation and back to positive saturation again. Hence, the use of saturable reactor in place of linear reactor is advantageous in two ways: At the time of turn-off and charging of the capacitor, the inductance (saturated) is low and for on-time, it is high (unsaturated). The circuit cost is low due to the use of only one thyristor.

8.9 A.C. CHOPPERS

The a.c. voltage magnitude can be changed by two methods. The well-known first method is by means of step-up and step-down transformers. In this method, change in voltage depends on the value of transformation ratio of the transformer.

The second method of changing magnitude of an a.c. voltage is by means of a solid-state switch. In this method, the a.c. input voltage is switched ON and OFF periodically by means of a suitable switch. Voltage changing circuits employing semiconductor devices as a static switch, are known as a.c. choppers. Figure 8.36 shows the commonly used single-phase a.c. chopper circuit. In this circuit, SCR T_1 and T_2 are the main SCR whereas, SCR T_3 and T_4 are the auxiliary SCRs. C_1 and C_2 are the commutating capacitors. Diodes D_1 and D_2 provide the charging path for the capacitors.

Thyristors T_1 and T_3 forms the first pair for producing the positive alternation, and T_2 and T_4 constitute the second pair for producing the negative alternation of the input a.c. voltage. Figure 8.37 shows the load-voltage waveforms. For the sake of simplicity, circuit operation is described in various operating modes.

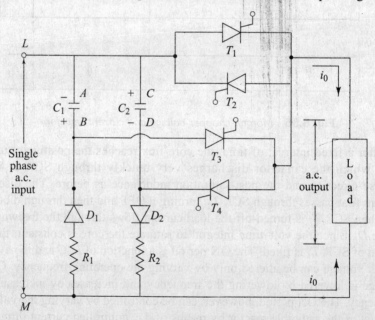

Fig. 8.36 *Single phase a.c. chopper*

(i) Mode 0 operation Initially, during the positive half-cycle of the supply voltage, capacitor C_2 charges through the path $L - C_2 - D_2 - R_2 - M$, with the polarity shown in Fig. 8.36. Similarly, during the negative half-cycle of the supply voltage, capacitor C_1 charges through the path $M - R_1 - D_1 - C_1 - L$, with the polarity shown in Fig. 8.36. The voltage across these capacitors is used for commutation of main SCRs T_1 and T_2.

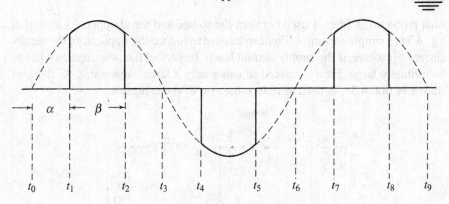

Fig. 8.37 *Load voltage waveform*

(ii) Mode I Operation As shown in Fig. 8.37, during the first positive half-cycle of the supply voltage, thyristor T_1 is triggered at instance t_1 with a firing angle α. The current flows through the path L – SCR T_1 – Load – M. When the instantaneous voltage reaches the instant T_2, auxiliary thyristor T_3 is triggered. As soon as thyristor T_3 is triggered, capacitor C_1 will start discharging through the path $C_B - T_3 - T_1 - C_A$. When the discharging current of capacitor C_1 becomes more than the forward-current of the SCR T_1, SCR T_1 becomes turned-off. The auxiliary SCR T_3 will be automatically turned-off at instant t_3 because of the zero current at this instant. Hence, SCRs T_1 and T_3 forms the first pair for producing the positive alternation of the input a.c. voltage.

(iii) Mode II Operation For the formation of the negative alternation, second pair of thyristors T_2 and T_4 are used. The main SCR T_2 is triggered at the instant t_4, as shown in Fig. 8.37, during the first negative half-cycle of the input voltage. The current flows through the path M – Load – T_2 – L. When the instantaneous voltage reaches the instant t_5, SCR T_4 is triggered. As soon as thyristor T_4 is triggered, capacitor C_2 will start discharging through the path $C_c - T_2 - T_{4(A-K)} - C_D$. When this discharging current is more than the load current, SCR T_2 becomes turned-off. At instant t_6, SCR T_4 is automatically turned-off as the current passing through it becomes zero.

Again at instant t_7, SCR T_1 is triggered to produce the next positive alternation. This is a continuous process and repeated again and again to generate an a.c. voltage across the load. The load power can be changed simply by varying the pulse-width (or conduction angle) β. The main advantage of this type of a.c. chopper is that, whatever the pulse width β, the fundamental input power factor is always unity. The circuit is generally used for obtaining a regulated a.c. output voltage.

8.10 SOURCE FILTER

A disadvantage of the d.c. to d.c. chopper is that the d.c. supply current is pulsating because of the chopping operation. Therefore, the supply current has harmonics that will produce undesirable effects, such as source voltage fluctuation, signal interference, supply distortion, additional heating and so on. In order to overcome such problems, a filter is used between the source and the chopper. As shown in Fig. 8.38, a simple capacitor filter can be used to reduce the ripple from the supply current. However, if the supply current has to be perfect d.c., the capacitor has to be infinitely large. Hence, instead of using only a large capacitor, L–C filter, as shown in Fig. 8.39, is used to reduce the size of the capacitor.

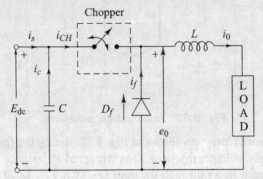

Fig. 8.38 *Basic chopper circuit with capacitor input filter*

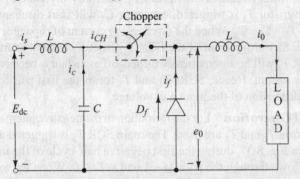

Fig. 8.39 *Basic chopper circuit with input L–C filter*

Design Considerations Figure 8.40 shows the equivalent circuit for the nth harmonic of Fig. 8.39. The filter capacitor C makes it possible for the chopper to draw pulsed currents and it reduces the over voltage present so as to bring them to an acceptable level for the choppers input. The inductor L reduces the ripple current in the supply to an amplitude low enough such that it may not interfere with other users. This is particularly important in electric traction where the presence of a.c. currents flowing in the line and rails can upset the signalling or remote control circuits.

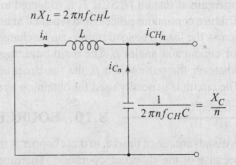

Fig. 8.40 *Equivalent circuit for nth harmonic*

Referring to the equivalent circuit of Fig. 8.40, the nth harmonic current in the supply is given by

$$I_n = \frac{X_C/n}{(nX_L)-(X_C/n)} I_{CH_n} \qquad (8.126)$$

where
$$X_L = 2\pi f_{CH} \cdot L \qquad (8.127)$$

$$X_C = \frac{1}{2\pi f_{CH} C} \qquad (8.128)$$

where, f = order of harmonic, f_{CH} = chopper frequency
I_n = RMS nth harmonic of the supply current,
I_{CH_n} = RMS nth harmonic of the chopper current

Equation (8.126) can also be written as

$$I_n = \frac{X_C}{n^2 X_L - X_C} I_{CH_n} \qquad (8.129)$$

Substituting the value of X_C and X_L from Eqs (8.127) and (8.128) into Eq. (8.129),

$$I_n = \frac{1}{2\pi f_{CH} \cdot C} \left[\frac{1}{n^2 2\pi f_{CH} \cdot L - \frac{1}{2\pi f_{CH} \cdot C}} \right] I_{CH_n}$$

$$= \frac{1}{4\pi^2 n^2 f_{CH}^2 LC - 1} I_{CH_n} \qquad (8.130)$$

$$= \frac{1}{\left(\dfrac{n \cdot f_{CH}}{f_r}\right)^2 - 1} I_{CH_n} \qquad (8.131)$$

where f_r = resonant frequency of filter circuit.

The chopper frequency f_{CH} and the resonant frequency f_r should be different, otherwise, due to resonance, large voltage oscillations will build up in the supply voltage. To avoid this resonance phenomenon, normally f_{CH} is two or three times more than f_r.

Thus, the supply harmonic current is approximately given by

$$I_n = \left(\frac{f_r}{n f_{CH}}\right)^2 \cdot I_{CH_n} \qquad (8.132)$$

From the above expression it becomes clear that the supply harmonic can be reduced by the following considerations.

(i) By increasing the chopper frequency for the same filter elements.
(ii) By decreasing f_r (i.e. increasing the values of L and or C) for the same chopper frequency.
(iii) By decreasing ripple amplitude of the chopper current, which will reduce I_{CH_n}.

8.11 MULTIPHASE CHOPPER

A multiphase chopper consists of two or more choppers operating at the same frequency but with a proper phase shift. This type of operation enables the load and power supply to be subjected to an effective frequency which is a multiple of the chopping frequency. As a result, the supply harmonic current is reduced and the ripple amplitude decreases.

The two chopper configuration of Fig. 8.41, is called as two-phase chopper. Similarly, three choppers connected in parallel will constitute a three phase chopper.

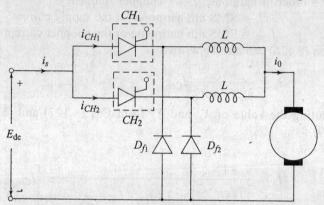

Fig. 8.41 *Two phase chopper*

The main operating modes of the multiphase chopper are
(i) In-phase operating mode and
(ii) Phase shifted operating mode.

In-phase operating mode, all the parallel connected choppers are ON and OFF at the same instant whereas in the phase-shifted operating mode the choppers are ON and OFF at different instants of time. Inductance L in series with each chopper is sufficiently large so that each chopper operates independent of each other.

Figure 8.42(a) shows the input current waveforms for in-phase operation for a duty-cycle of 30%. The load current I_0 be assumed to be ripple free. As shown in the figure, both choppers CH_1 and CH_2 are

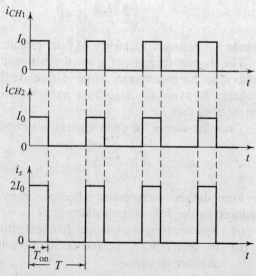

Fig. 8.42(a) *In-phase operation with α = 30%*

ON and OFF at the same instant. Also, the input current i_s, which is the sum of currents i_{CH_1} and i_{CH_2}, is seen to be doubled as shown. The in-phase operation corresponds to the case of single chopper operation.

Figure 8.42(b) shows the input current waveforms for phase-shifted operation for a duty-cycle of 30%. As shown in Fig. 8.42(b), the conducting periods of the two chopper units never overlap. As shown, the two choppers are operated at the same frequency and duty-ratio but with a phase difference of 180°. In this mode, chopper CH_1 is ON for $0.3T$ from $t = 0$, whereas chopper CH_2 is made ON such that the input current obtained from $i_{CH_1} + i_{CH_2}$ is periodic in nature.

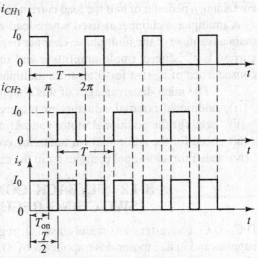

Fig. 8.42(b) *Phase-shifted operation with $\alpha = 30\%$*

From Figs 8.42(a) and (b), it becomes clear that for phase-shifted operation, the frequency of input current is doubled and its ripple current amplitude is halved as compared to the in-phase operation of the chopper. Figure 8.42(a) shows that in case of in-phase operation, frequency of harmonics in the input current is equal to switching frequency $(1/T)$ of each chopper, whereas Fig. 8.42(b) shows that in case of phase-shifted mode, frequency of harmonics in the input current is twice the switching frequency of each chopper. As the frequency of harmonics in the input current is twice the switching frequency, the size of filter is reduced in the phase-shifted chopper. Hence, the phase-shifted operation of multiphase choppers is usually preferred.

In phase shifted mode, for 50% duty-cycle, the supply current is continuous ripple free. Also, the supply current will be half of the load current. For 60% duty cycle, the supply current will be continuous having half of the load current as amplitude and

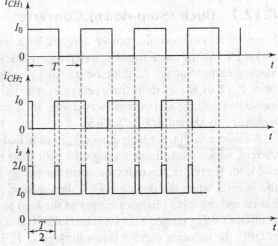

Fig. 8.42(c) *Phase-shift operation with $\alpha = 60\%$*

containing a pedestal of half the load current as shown in Fig. 8.42(c).

A multiphase chopper is used where load-current requirement is large. The main advantage of the multiphase-chopper over a single chopper is that its input current has reduced ripple amplitude and increased ripple frequency. As a consequence of it, size for filter for a multiphase chopper is reduced.

The main disadvantages of this chopper circuit are–
(i) additional external inductors are required
(ii) need for the additional motor connections
(iii) need for the additional commutating components
(iv) need for additional complexity in the control logic.

8.12 FLYBACK CONVERTERS [SWITCHING REGULATORS]

D.C.–D.C. converters are widely used in regulated switch-mode d.c. power supplies and in d.c. motor drive applications. Often, the input to these converters is an unregulated d.c. voltage, which is obtained by rectifying the line voltage and therefore, it will fluctuate due to changes in the line voltage magnitude. Switch-mode, d.c.-to-d.c. converters are used to convert the unregulated d.c. input into a controlled d.c. output at a desired voltage level.

The following d.c.–d.c. converters are discussed in this section:
1. Buck (step-down) converter.
2. Boost (step-up) converter.
3. Buck-Boost (step-down/up) converter.
4. Cuk converter.

Of these four converters, only the step-down and step-up are the basic converter topologies. Both the buck-boost and the cuk converters are combinations of the two basic topologies. The name "flyback converter" is descriptive of the inductive energy flyback action typically encountered in this type of converter operation.

8.12.1 Buck (Step-down) Converter

Figure 8.43(a) shows the power diagram of a buck converter using a power MOSFET. As the name implies, a step-down (buck) converter produces a lower average output voltage E_0 than the d.c. input voltage E_{dc}. By varying the duty-ratio T_{on}/T of the switch, the average output voltage can be controlled. The associated voltage and current waveforms for a continuous current flow in the inductor L is shown in Fig. 8.43(b).

As shown in Fig. 8.43(b), device T_1 is switched ON at time $t = 0$. The supply current, which rises, flows through the path filter inductor L, filter capacitor C, and load. Therefore, the inductor stores the energy during the T_{on} period. During the interval when the device is ON, the diode in Fig. 8.43(a) becomes reverse biased and the input provides energy to the load as well as to the inductor. Now, at instant $t = T_{on}$, device T_1 is switched OFF. During the interval when the device is OFF, the inductor current flows through L, C, load, and freewheeling diode D_F and hence diode D_F conducts. The output voltage fluctuations are very much

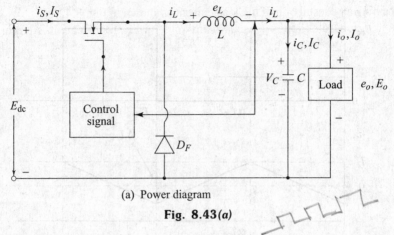

(a) Power diagram

Fig. 8.43(a)

diminished, using a low-pass filter, consisting of an inductor and a capacitor. The corner frequency f_c of this low-pass filter is selected to be much lower than the switching frequency, thus essentially dominating the switching frequency ripple in the output voltage. Depending on the switching frequency, filter inductance, and capacitance the inductor current could be discontinuous.

In general, the voltage across the inductor L is given by

$$e_L = L \frac{di}{dt}$$

In time T_{on}, assuming that the inductor current rises linearly from I_1 to I_2, we can write,

$$\therefore \quad E_{dc} - E_0 = L \left(\frac{I_2 - I_1}{T_{on}} \right) \quad (8.133)$$

Let us define, the change in current as
$\Delta I = I_2 - I_1$ (i.e. peak-to-peak ripple current of L)
$\therefore$ Equation (8.133) becomes

$$E_{dc} - E_0 = \frac{\Delta I \cdot L}{T_{on}} \quad (8.134)$$

or,

$$T_{on} = \frac{\Delta I \cdot L}{E_{dc} - E_0} \quad (8.135)$$

As shown in Fig. 8.43(b), during time T_{off}, inductor current falls linearly from I_2 to I_1,

$$-E_0 = -L \frac{\Delta I}{T_{off}} \quad (8.136)$$

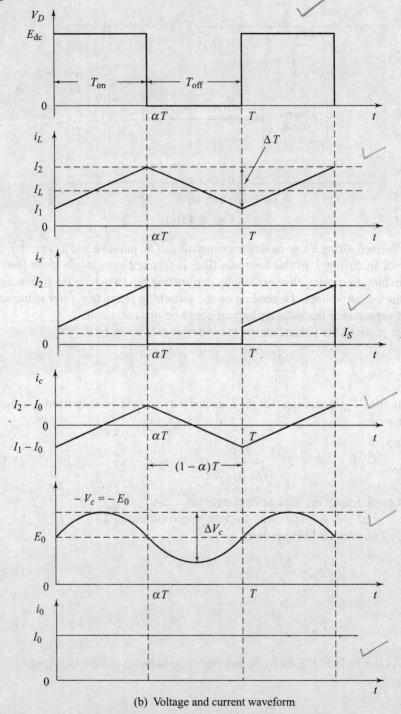

(b) Voltage and current waveform

Fig. 8.43(b) *Buck-chopper with continuous i_L*

or $\quad T_{\text{off}} = \dfrac{\Delta I \cdot L}{E_0}$ \hfill (8.137)

Equating the values of ΔI in Eqs (8.134) and (8.137) gives

$$\Delta I = \dfrac{(E_{dc} - E_0)}{L} T_{on} = \dfrac{T_{off} \cdot E_0}{L}$$

$\therefore E_{dc} T_{on} - E_0 T_{on} = T_{off} E_0$
$\quad E_0 T_{off} + E_0 T_{on} = E_{dc} T_{on}$
$\quad E_0 (T_{off} + T_{on}) = E_{dc} T_{on}$
$\therefore \quad E_0 T = E_{dc} T_{on}$

$\therefore \quad E_0 = \dfrac{E_{dc} \cdot T_{on}}{T}$

Let us define, duty ratio $= \dfrac{T_{on}}{T} = \alpha$.

$\therefore \quad E_0 = \alpha \cdot E_{dc}$ \hfill (8.138)

By assuming a lossless circuit, we can write,

$$E_{dc} I_s = E_0 \cdot I_0 = \alpha E_{dc} \cdot I_0$$

Therefore, the average input current is given by

$$I_s = \alpha \cdot I_0 \hfill (8.139)$$

Now, the switching period T can be calculated as,

$$T = T_{on} + T_{off} \hfill (8.140)$$

By substituting the values of T_{on} and T_{off} from Eqs (8.135) and (8.137) in Eq. (8.140), we get

$$T = \dfrac{\Delta I L}{E_{dc} - E_0} + \dfrac{\Delta I L}{E_0} \text{ or } T = \dfrac{1}{f} = \dfrac{\Delta I L E_{dc}}{E_0 (E_{dc} - E_0)} \hfill (8.141)$$

From Eq. (8.141), we can write the equation for peak-to-peak ripple current as

$$\Delta I = \dfrac{E_0 (E_{dc} - E_0)}{L \cdot E_{dc}} \cdot T \hfill (8.142)$$

or $\quad \Delta I = \dfrac{E_0 (E_{dc} - E_0)}{f \cdot L \cdot E_{dc}}$ \hfill (8.143)

or $\quad \Delta I = \dfrac{E_0 (E_{dc}/E_{dc} - E_0/E_{dc})}{f \cdot L \cdot E_{dc}/E_{dc}}$

From Eqs. (8.138), we have

$\quad \alpha = \dfrac{E_0}{E_{dc}}, \quad \therefore \Delta I = \dfrac{\alpha \cdot E_{dc}(1 - \alpha)}{f \cdot L}$ \hfill (8.144)

We can write the inductor current i_L by applying Kirchhoff's current law as,

$$i_L = i_C + i_O$$

By assuming that the load ripple current Δi_O is very small and therefore, neglected, we can write the above equation as

$$\Delta i_L = \Delta i_C$$

The average capacitor current, which flows into for $T_{on}/2 + T_{off}/2 = T/2$, is

$$I_C = \frac{\Delta I}{4}$$

Now, the capacitor voltage is expressed as

$$V_c = \frac{1}{C}\int i_C \, dt + V_c(t=0)$$

and the peak-to-peak ripple voltage of the capacitor is

$$\Delta V_c = V_c - V_c(t=0) = \frac{1}{C}\int_0^{t/2} \frac{\Delta I}{4} dt$$

$$= \frac{\Delta I \cdot T}{8C} = \frac{\Delta I}{8 f_c} \tag{8.145}$$

Substituting the value of ΔI from Eq. (8.143) or (8.144) in Eq. (8.145) yields

$$\Delta V_c = \frac{E_0 \cdot (E_{dc} - E_0)}{8 L C f^2 E_{dc}} \tag{8.146}$$

or

$$\Delta V_c = \frac{E_{dc}\alpha(1-\alpha)}{8 L C f^2} \tag{8.147}$$

Since the buck chopper circuit requires only one transistor, it is a simple one and has high efficiency, greater than 90%. The inductor L limits the di/dt of the load-current. This type of chopper circuit provides one polarity of output voltage and unidirectional output current. In case of possible short circuit across the diode path, it requires a protection circuitry. Its main application is in regulated d.c. power supplies and d.c. motor speed control.

SOLVED EXAMPLES

Example 8.23 The buck-converter in Fig. 8.43 has an input voltage of $E_{dc} = 14$ V. The required average output voltage is $E_0 = 6$ V and the peak-to-peak output ripple voltage is 15 mV. The switching frequency is 30 kHz. If the peak-to-peak ripple curent of inductor is limited to 0.6 A. Determine: (a) the duty cycle α, (b) the filter inductance L, and (c) the filter capacitor C.

Solution:

Given: $E_{dc} = 14$ V, $E_0 = 6$ V, $\Delta VC = 15$ mV, $\Delta I = 0.6$ A, $f = 30$ kHz

(a) From Eq. (8.138), duty-cycle α is given by

$$\alpha = \frac{E_0}{E_{dc}} = \frac{6}{14} = 0.4285 = 42.85\%$$

(b) Now, from Eq. (8.143), we can write,

$$L = \frac{E_0(E_{dc} - E_0)}{f \cdot E_{dc} \cdot \Delta I} = \frac{6(14 - 6)}{30 \times 10^3 \times 14 \times 0.6} = 190.48 \, \mu H$$

(c) From Eq. (8.145), we have

$$C = \frac{\Delta I}{8 \times f \times \Delta V_C} = \frac{0.6}{8 \times 30 \times 10^3 \times 15 \times 10^{-3}} = 166.67 \, \mu F.$$

Example 8.24 A buck converter operating at 50 kHz is fed from a 12 V battery and supplies 5 V to load. Neglecting switch and device-losses, determine:
(a) The maximum on-period of MOSFET switch given that battery voltage varies from 13.5 V in fully charged state to 10 V at the end of discharge.
(b) Battery drain current under nominal condition with 10 Amp. load.
(c) The value of choke required to maintain continuous current operation for a ripple current of 500 mA and worst case battery voltage conditions.

Solution:

Given: $E_o = 5$ V, $E_{dc_{max}} = 13.5$, $E_{dc_{min}} = 10$ V, $I_o = 10$ A, $\Delta I = 500$ mA

(a) From Eq. (8.138), $\alpha = \dfrac{E_o}{E_{dc}}$ $\therefore \dfrac{T_{on}}{T} = \dfrac{E_o}{E_{dc}}$

$$\therefore \quad t_{on(max)} = \frac{E_o}{E_{dc_{min}} \cdot f} = \frac{5}{10 \times 50 \times 10^3} = 10 \, \mu$$

$\therefore$ Maximum period of conduction for switch $= 10 \, \mu$ sec.

(b) If the switch and device losses are neglected, then output power will be equal to input power, i.e.

$$E_{dc} \cdot I_s = E_0 \cdot I_0 \quad \therefore \quad 12 \times I_s = 5 \times 10 \quad \therefore \quad I_s = 4.16 \text{ A}$$

(c) From Eq. 8.143, $L = \dfrac{5(10 - 5)}{50 \times 10^3 \times 10 \times 500 \times 10^{-3}} = 100 \, \mu H$

8.12.2 Boost (Step-up) Converter

Figure 8.44(a) shows the circuit diagram of a boost chopper using a power MOSFET. As the name implies, the output voltage is always greater than the input voltage. When the power device is ON, the inductor L is connected to the supply E_{dc}, and inductor stores energy during on-period, T_{on}. Hence, diode D_F is reverse biased and isolates the output stage. When the power device is OFF, the output stage receives energy from the inductor as well as from the input. The current which was flowing through the transistor would now flow through L, D_F, C and load. The associated voltage and current waveforms are shown in Fig. 8.44(b).

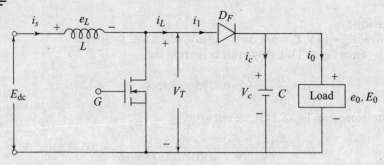

Fig. 8.44(a) Circuit diagram

During T_{on}, by assuming that the inductor current rises linearly from I_1 to I_2, we can write,

$$E_{dc} = L\frac{(I_2 - I_1)}{T_{on}} = L\frac{\Delta I}{T_{on}} \quad (8.148)$$

or
$$T_{on} = \frac{\Delta I \cdot L}{E_{dc}} \quad (8.149)$$

During time T_{off}, the inductor current falls linearly from I_2 to I_1, therefore, we can write,

$$E_{dc} - E_0 = -L \cdot \frac{\Delta I}{T_{off}} \quad (8.150)$$

or
$$T_{off} = \frac{\Delta I \cdot L}{E_0 - E_{dc}} \quad (8.151)$$

From Eqs (8.148) and (8.150), the peak-to-peak ripple current of inductor L can be written as,

$$\Delta I = \frac{E_{dc} \cdot T_{on}}{L} = \frac{(E_0 - E_{dc})T_{off}}{L} \quad (8.152(a))$$

Substituting $T_{on} = \alpha T$ and $T_{off} = (1 - \alpha)T$, yields the average output voltage,

$$E_0 = E_{dc} \cdot \frac{T}{T_{off}} = \frac{E_{dc}}{1 - \alpha} \quad (8.152(b))$$

Assuming a lossless circuit, $P_i = P_o$.

$$\therefore \quad E_{dc} \cdot I_s = E_0 I_0 = \frac{E_{dc} I_0}{(1-\alpha)}$$

Therefore, the average input current becomes

$$I_s = \frac{I_0}{1-\alpha} \quad (8.153)$$

Now, the switching period T can be obtained as

$$T = \frac{1}{f} = T_{on} + T_{off} = \frac{\Delta I \cdot L}{E_{dc}} + \frac{\Delta I \cdot L}{E_0 - E_{dc}} = \frac{\Delta I \cdot L \cdot E_0}{E_{dc}(E_0 - E_{dc})} \quad (8.154)$$

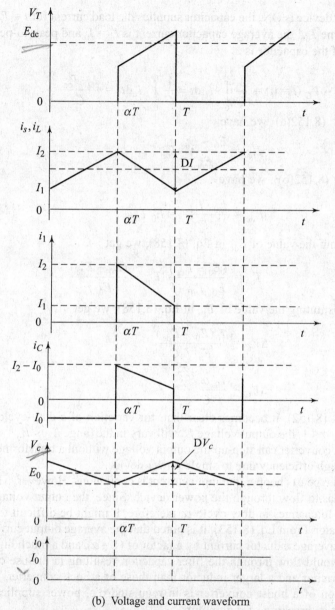

(b) Voltage and current waveform

Fig. 8.44(b) *Boost chopper with continuous i_L*

From above equation, the peak-to-peak ripple current becomes

$$\Delta I = \frac{E_{dc}(E_0 - E_{dc})}{f \cdot L \cdot E_0} \qquad (8.155)$$

or

$$\Delta I = \frac{E_{dc} \cdot \alpha}{f \cdot L} \qquad (8.156)$$

When the device is ON, the capacitor supplies the load current for $t = T_{on}$ period. During time T_{on}, the average capacitor current is $I_C = I_0$ and peak-to-peak ripple voltage of the capacitor is

$$\Delta V_C = V_C - V_C(t=0) = \frac{1}{C}\int_0^{T_{on}} I_C dt = \frac{1}{C}\int_0^{T_{on}} I_0 dt = \frac{I_0 \cdot T_{on}}{C} \qquad (8.157)$$

From Eq. (8.152(a)) we have

$$T_{on} = \frac{E_0 - E_{dc}}{E_{dc}} T_{off} \qquad (8.158)$$

From Eq. (8.152(b)), we have

$$E_{dc} = \frac{E_0 \cdot T_{off}}{T} = E_0 \, T_{off} \cdot f$$

Substituting the value of E_{dc} in Eq. (8.158), we get

$$T_{on} = \frac{(E_0 - E_{dc})}{E_0 \cdot T_{off} \cdot f} \cdot T_{off} = \frac{(E_0 - E_{dc})}{E_0 \cdot f}$$

Now, substituting the value of T_{on} in Eq. (8.158), we get

$$\Delta V_C = \frac{I_0 (E_0 - E_{dc})}{E_0 \cdot f \cdot C} \qquad (8.159)$$

or

$$\Delta V_C = \frac{I_0 \cdot \alpha}{f \cdot C} \qquad (8.160)$$

From Eq. (8.152), it becomes clear that, for variation of a duty-cycle α in the range $0 < \alpha < 1$, the output voltage E_0 will vary in the range $E_{dc} < E_0 < \infty$. Hence, the boost converter can step-up the output voltage without a transformer. Again, it has a high efficiency due to single power device.

In this type of chopper, the input current is continuous. However, a high peak current has to flow through the power device. Since, the output voltage is very sensitive to changes in duty-cycle α, therefore, it might be difficult to stabilize the regulator. From Eq. (8.153), it is noted that the average output current is less than the average inductor current by a factor of $(1 - \alpha)$, and a much higher RMS current would flow through the filter capacitor, resulting in the use of a larger filter capacitor and a larger inductor than those of a buck converter. The main application of a boost converter is in regulated d.c. power supplies and the regenerative breaking of d.c. motors.

SOLVED EXAMPLES

Example 8.25 Consider the boost converter of Fig. 8.44. The input voltage to this converter is 6 V. The average output voltage $E_0 = 18$ V and the average load current $I_0 = 0.4$ A. The switching frequency is 20 kHz of $L = 250$ μH and $C = 420$ μF. Determine: (a) the duty-cycle α, (b) the ripple current of inductor, ΔI, (c) the peak current of inductor, I_2, and (d) the ripple voltage of filter capacitor, ΔV_C.

Solution:

Given: $E_{dc} = 6$ V, $E_0 = 18$ V, $I_0 = 0.4$ A, $f = 20$ kHz, $L = 250$ μH, and $C = 420$ μF.

(a) From Eq. (8.152),

$$18 = \frac{6}{1-\alpha}, \quad \text{or} \quad 1 - \alpha = \frac{6}{18} \quad \text{or} \quad \alpha = 0.6667 = 66.67\%$$

(b) From Eq. (8.155),

$$\Delta I = \frac{6 \times (18-6)}{20 \times 10^3 \times 250 \times 10^{-6} \times 18} = 0.80 \text{ A}$$

(c) From Eq. (8.153),

$$I_s = \frac{0.4}{1-0.667} = 1.20 \text{ A}$$

∴ Peak inductor current,

$$I_2 = I_s + \frac{\Delta I}{2} = 1.20 + \frac{0.80}{2} = 1.6 \text{ A}$$

(d) From Eq. (8.160),

$$\Delta V_C = \frac{0.4 \times 0.667}{20 \times 10^3 \times 420 \times 10^{-6}} = 31.76 \text{ mV}.$$

8.12.3 Buck-boost Converter (Choppers)

Figure 8.45(a) shows the circuit diagram of a buck-boost converter. As shown, a buck-boost converter is nothing but cascade connection of the two basic converters: the step-down converter and the step-up converter. The main application of such a converter is in regulated d.c. power supplies, where a negative polarity output may be desired with respect to the common terminal of the input voltage, and the output voltage can be either higher or lower than the input voltage. For a continuous load current, the waveforms for the steady-state voltage and currents are shown in Fig. 8.45(b).

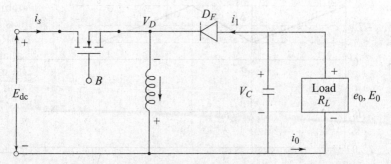

Fig 8.45(a) Circuit diagram

When the power MOSFET is switched ON, the supply current flows through the path $E_{dc+} - T_1 - L - E_{dc-}$. Hence, inductor L stores the energy during the T_{on} period.

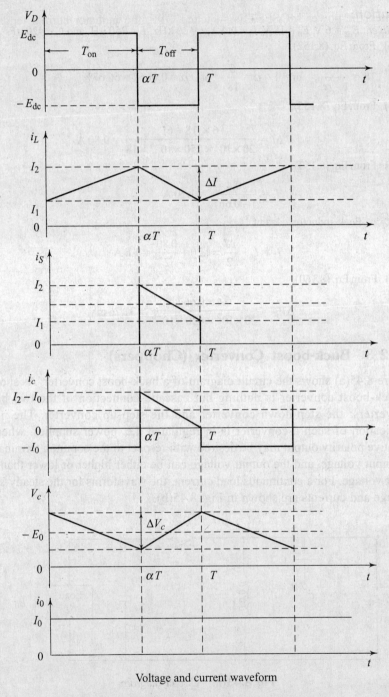

Voltage and current waveform

Fig 8.45(b) Buck-boost converter with continuous i_L

Choppers

When the power MOSFET is switched OFF, the inductor current tends to decrease and as a result, the polarity of the emf induced in L is reversed as shown in Fig. 8.45(a). Thus, the inductance energy discharges in the load through the path $L_+ -$ Load $- D - L_-$.

During time T_{on}, by assuming that the inductor current rises linearly from I_1 to I_2, we can write,
$$E_{dc} = L \cdot \frac{I_2 - I_1}{T_{on}} = L \cdot \frac{\Delta I}{T_{on}} \tag{8.161}$$

or,
$$T_{on} = \frac{\Delta I L}{E_{dc}} \tag{8.162}$$

Now, during time T_{off}, the inductor current falls linearly from I_2 to I_1, therefore, we can write

$$E_0 = -L \cdot \frac{\Delta I}{T_{off}} \tag{8.163}$$

or
$$T_{off} = \frac{-\Delta I \cdot L}{E_0} \tag{8.164}$$

where the peak-to-peak ripple current of inductor L is given by
$$\Delta I = I_2 - I_1$$

From Eqs (8.161) and (8.163), we can write,

$$\Delta I = \frac{E_{dc} \cdot T_{on}}{L} = \frac{-E_0 \cdot T_{off}}{L}$$

or
$$E_0 = -E_{dc} \cdot \frac{T_{on}}{T_{off}} = -E_{dc} \cdot \frac{T_{on}/T}{T_{off}/T} = \frac{-E_{dc} \cdot \alpha}{(T - T_{on})/T}$$

or
$$E_0 = -E_{dc} \cdot \frac{\alpha}{(1-\alpha)} \tag{8.165}$$

For a lossless system, in a steady-state,

$$E_{dc} \cdot I_s = -E_0 \cdot I_0 \text{ or } E_{dc} I_s = E_{dc} \cdot \frac{\alpha}{1-\alpha} I_0$$

Therefore, the average input current is given by

$$I_s = \frac{\alpha}{1-\alpha} I_0 \tag{8.166}$$

Now, the switching period, T, can be calculated as

$$T = \frac{1}{f} = T_{on} + T_{off}$$

Substituting the values of T_{on} and T_{off} from Eqs (8.162) and (8.164), we get

$$T = \frac{\Delta IL}{E_{dc}} - \frac{\Delta I \cdot L}{E_0} = \frac{\Delta IL(E_0 - E_{dc})}{E_{dc} \cdot E_0} \tag{8.167}$$

From Eq. (8.167), the peak-to-peak ripple current can be written as

$$\Delta I = \frac{E_{dc} \cdot E_0}{f \cdot L \cdot (E_0 - E_{dc})} \tag{8.168}$$

or

$$\Delta I = \frac{E_{dc} \cdot \alpha}{f \cdot L} \tag{8.169}$$

During the period T_{on}, when the device is ON, the filter-capacitor supplies the load current. Therefore, the average discharging current of the capacitor is $I_C = I_0$.

Also, the peak-to-peak ripple voltage of the capacitor is given by

$$\Delta V_C = \frac{1}{C} \int_0^{T_{on}} I_C dt = \frac{1}{C} \int_0^{T_{on}} I_0 dt = \frac{I_0 \cdot T_{on}}{C} \tag{8.170}$$

From Eq. (8.165), we can write, $E_0(1 - \alpha) = -E_{dc} \cdot \alpha$, $E_0 = \alpha(E_0 - E_{dc})$

or

$$E_0 = \frac{T_{on}}{T}(E_0 - E_{dc}), \text{ or } T_{on} = \frac{E_0}{(E_0 - E_{dc})f}$$

Substituting the above value of T_{on} in Eq. (8.170), we get

$$\Delta V_C = \frac{I_0 \cdot E_0}{(E_0 - E_{dc})f \cdot C} \tag{8.171}$$

or

$$\Delta V_C = \frac{I_0 \cdot \alpha}{f \cdot C} \tag{8.172}$$

From Eq. (8.165), it becomes clear that the buck-boost converter provides output voltage polarity reversal without a transformer. This type of chopper has a high efficiency. The inductor L limits the di/dt of the fault current when the device is under the fault condition. In this type of converter circuit, the output short circuit protection would be easy to implement.

Example 8.26 Consider the buck-boost converter of Fig. 8.45. The input voltage to this converter is $E_{dc} = 14$ V. The duty cycle $\alpha = 0.6$ and the switching frequency is 25 kHz. The inductance $L = 180$ μH and filter capacitance $C = 220$ μF. The average load current $I_0 = 1.5$ A. Compute:
 (a) the average output voltage, E_0;
 (b) the peak-to-peak output voltage ripple, ΔV_c;
 (c) the peak-to-peak current of inductor, ΔI; and
 (d) the peak current of the device I_p.

Solution:

(a) From Eq. (8.165), $E_0 = -\dfrac{14 \times 0.6}{1 - 0.6} = -21$ V.

(b) From Eq. (8.172), peak-to-peak output ripple voltage is

$$\Delta V_c = \dfrac{1.5 \times 0.6}{25 \times 10^3 \times 220 \times 10^{-6}} = 0.16 \text{ V}.$$

(c) From Eq. (8.169), the peak-to-peak inductor ripple is

$$\Delta I = \dfrac{14 \times 0.6}{25 \times 10^3 \times 180 \times 10^{-6}} = 1.87 \text{ A}$$

(d) From Eq. (8.166),

$$I_s = \dfrac{0.6 \times 1.5}{1 - 0.6} = 2.25 \text{ A}$$

Now, since the average input current I_s is the average of duration $\alpha \cdot T$, the peak-to-peak current of the MOSFET is given by

$$I_p = \dfrac{I_s}{\alpha} + \dfrac{\Delta I}{2} = \dfrac{2.25}{0.6} + \dfrac{1.87}{2} = 4.69 \text{ A}$$

Example 8.27 A buck-boost converter is operated from a 24 V battery and supplies an average load current of 2 Amp. Its switching frequency is 50 kHz. Neglecting diode and switch drop, determine

(a) Range of duty-cycle variation required to maintain the output voltage at 15 V, given that the battery voltage ranges from 26 V in the fully charged state to 21 V in the discharged state.
(b) The peak to peak choke ripple current for the nominal supply voltage, given that the choke value is 500 μH.
(c) Average supply current drawn from the battery under nominal condition.

Solution:

(a) From Eq. (8.165),

(i) when $E_{dc} = 26$ V, $\dfrac{E_o}{E_{dc}} = \dfrac{\alpha}{1-\alpha}$

$\therefore \quad \dfrac{15}{26} = \dfrac{\alpha}{1-\alpha} \quad \therefore \alpha = 0.366$

(ii) When $E_{dc} = 21$ V

$\therefore \quad \dfrac{15}{21} = \dfrac{\alpha}{1-\alpha}, \quad \therefore \alpha = 0.417$

$\therefore$ Duty cycle varies from 0.366 to 0.417

(b) Nominal supply voltage = 24 V

we have, $\dfrac{E_o}{E_{dc}} = \dfrac{\alpha}{1-\alpha}, \dfrac{15}{24} = \dfrac{\alpha}{1-\alpha}, \quad \therefore \alpha = 0.385$

From (8.169),

$$\Delta I = \dfrac{24 \times 0.385}{50 \times 10^3 \times 500 \times 10^{-6}} = 0.3696 \text{ A} \quad = \Delta I = 369.6 \text{ mA}$$

(c) Assuming switch and the device losses to be zero,
∴ Power supplied by the battery = Load power
$$24 \times I_s = 15 \times 2, \quad \therefore I_s = 1.25 \text{ Amp.}$$

8.12.4 Cuk Converter

Figure 8.46(a) shows a Cuk converter, after the name of the inventor. The key difference between this converter and the previously discussed converters, from the operation point of view, is that a capacitor, rather than an inductor, is used for

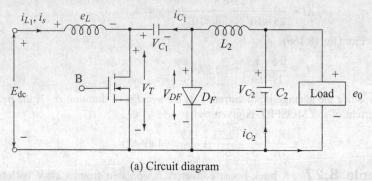

(a) Circuit diagram

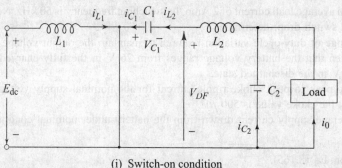

(i) Switch-on condition

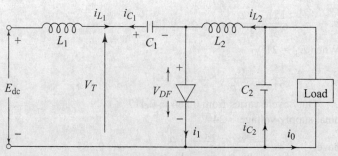

(ii) Switch-off condition

(b) Equivalent circuits

Fig. 8.46(a) and (b)

energy storage and transfer to accomplish power transformation. From this point of view, the Cuk converter is a capacitive energy flyback converter. In fact, the Cuk converter and the buck-boost converter are electrically duals of each other. The point connected MOSFET-inductor-diode combination is replaced by the dual, serially connected diode-capacitor-MOSFET combination. Similar to the buck-boost converter, the Cuk converter provides an output voltage which is less than or greater than the input voltage but the output voltage polarity is opposite to that of the input voltage.

When the input voltage is applied to the circuit and the MOSFET T_1 is switched-off, the inductor currents i_{L_1} and i_{L_2} flow through the diode D_F. The equivalent circuit is shown in Fig. 8.46(b). Therefore, capacitor C_1 is charged through L_1, D_F and the input supply E_{dc}. Current i_{L_1} decreases, because V_{C_1} is larger that E_{dc}.

Energy stored in inductor L_2 feeds the output. Therefore, current i_{L_2} also decreases.

Now, when the power MOSFET is ON, capacitor voltage V_{C_1} reverse biases the diode D_F and turns it OFF. The inductor currents i_{L_1} and i_{L_2} flow through the device as shown in Fig. 8.46(b). Since $V_{C_1} > E_{dc}$, C_1 discharges through the device, transferring energy to the output and L_2, therefore i_{L_1} increases. The input feeds energy to L_1 causing i_{L_1} to increase. The waveforms for steady-state voltages and currents are shown in Fig. 8.46(c) for a continuous load current.

During time T_{on}, by assuming that the current of inductor L_1 rises linearly from $I_{L_{11}}$ to $I_{L_{12}}$, we can write

$$E_{dc} = L_1 \cdot \frac{I_{L_{12}} - I_{L_{11}}}{T_{on}} = L_1 \cdot \frac{\Delta I_1}{E_{dc}} \qquad (8.173)$$

or
$$T_{on} = \frac{\Delta I_1 \cdot L_1}{E_{dc}} \qquad (8.174)$$

During time T_{off}, the current of inductor L_1 falls linearly from $I_{L_{12}}$ to $I_{L_{11}}$ due to the charged capacitor C_1.

$$\therefore \quad E_{dc} - V_{C_1} = -L_1 \cdot \frac{\Delta I_1}{T_{off}} \qquad (8.175)$$

or
$$T_{off} = \frac{-\Delta I_1 \cdot L_1}{E_{dc} - V_{C_1}} \qquad (8.176)$$

where V_C is the average value of capacitor C_1, and

$$\Delta I_1 = I_{L_{12}} - I_{L_{11}}$$

From Eqs (8.173) and (8.175), we can write

$$\Delta I_1 = \frac{E_{dc} \cdot T_{on}}{L_1} = \frac{-(E_{dc} - V_{C_1})}{L_1} T_{off}$$

or $\quad E_{dc} T_{on} = -E_{dc} T_{off} + V_{C_1} T_{off}$

$E_{dc}(T_{on} + T_{off}) = V_{C_1} T_{off}$

or
$$V_{C_1} = \frac{E_{dc} \cdot T}{T_{off}} = \frac{E_{dc}}{1 - \alpha} \qquad (8.177)$$

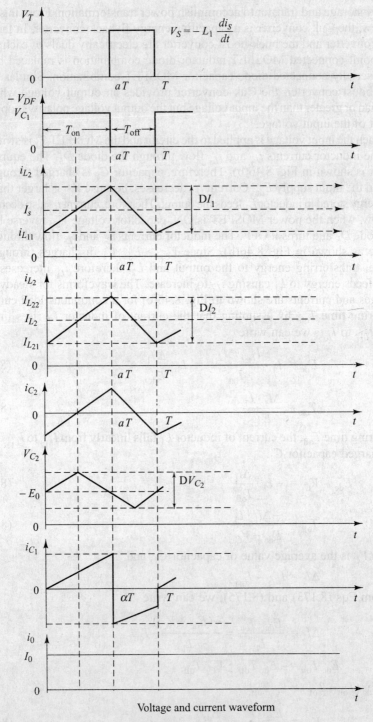

Voltage and current waveform

Fig. 8.46(c) Cuk converter

Also, during time T_{on}, by assuming that the current of filter inductor L_2 rises linearly from $I_{L_{21}}$ to $I_{L_{22}}$, we get

$$V_{C_1} + E_0 = L_2 \cdot \frac{I_{L_{22}} - I_{L_{21}}}{T_{on}} = L_2 \cdot \frac{\Delta I_2}{T_{on}} \qquad (8.178)$$

or
$$T_{on} = \frac{\Delta I_2 \, L_2}{V_{C_1} + E_0} \qquad (8.179)$$

Also, during time T_{off}, current of inductor L_2 falls linearly from $I_{L_{22}}$ to $I_{L_{21}}$, therefore, $E_0 = -L_2 \dfrac{\Delta I_2}{T_{off}}$ \hfill (8.180)

or
$$T_{off} = -\frac{\Delta I_2 L_2}{E_0} \qquad (8.181)$$

where $\Delta I_2 = I_{L_{22}} - I_{L_{21}}$.

Now, from Eqs (8.178) and (8.180), we can write

$$\Delta I_2 = \frac{(V_{C_1} + E_0)T_{on}}{L_2} - \frac{E_0 \cdot T_{off}}{L_2}$$

$$V_{C_1} T_{on} = -E_0 \, T_{off} - E_0 \, T_{on}$$

$$V_{C_1} = \frac{-E_0(T_{off} + T_{on})}{T_{on}} = \frac{-E_0}{T_{on}/T} \quad V_{C_1} = \frac{-E_0}{\alpha} \qquad (8.182)$$

We can calculate the average output voltage by equating Eqs (8.176) to Eq. (8.182) as,

$$V_{C_1} = \frac{E_{dc}}{1-\alpha} = \frac{-E_0}{\alpha} \quad \text{or} \quad E_0 = \frac{-\alpha \cdot E_{dc}}{1-\alpha} \qquad (8.183)$$

For a lossless system, in a steady-state,

$$E_{dc} \, I_s = -E_0 \cdot I_0 = E_{dc} \, I_s = \frac{\alpha \cdot E_{dc}}{1-\alpha} I_0$$

∴ Average input current,

$$I_s = \frac{\alpha \cdot I_0}{1-\alpha} \qquad (8.184)$$

Now, the switching period T can be calculated as

$$T = \frac{1}{f} = T_{on} + T_{off}$$

Substituting the value of T_{on} and T_{off} from Eqs (8.174) and (8.176), we get

$$T = \frac{\Delta I_1 \cdot L_1}{E_{dc}} - \frac{-\Delta I_1 \cdot L_1}{E_{dc} - V_{C_1}} = \frac{-\Delta I_1 \cdot L_1 V_{C_1}}{E_{dc}(E_{dc} - V_{C_1})} \qquad (8.185)$$

From Eq. (8.185), the peak-to-peak ripple current of inductor L_1 becomes

$$\Delta I_1 = \frac{-E_{dc} \cdot (E_{dc} - V_{C_1})}{f \cdot L_1 V_{C_1}} \tag{8.186}$$

or
$$\Delta I_1 = \frac{E_{dc} \cdot \alpha}{f \cdot L_1} \tag{8.187}$$

Also, the switching period T can be obtained from Eqs (8.179) and (8.181) as,

$$T = \frac{1}{f} = T_{on} + T_{off} = \frac{\Delta I_2 \cdot L_2}{V_{C_1} + E_0} - \frac{\Delta I_2 \cdot L_2}{E_0}$$

$$= \frac{-\Delta I_2 \cdot L_2 \cdot V_{C_1}}{E_0 (V_{C_1} + E_0)} \tag{8.188}$$

From Eq. (8.188), the peak-to-peak ripple current of inductor L_2 becomes

$$\Delta I_2 = \frac{-E_0 (V_{C_1} + E_0)}{f \cdot L_2 \cdot V_{C_1}} \tag{8.189}$$

or
$$\Delta I_2 = \frac{-E_0 (1-\alpha)}{f \cdot L_2} = \frac{\alpha \cdot E_{dc}}{f \cdot L_2} \tag{8.190}$$

As discussed previously, when the power-device is OFF, the energy transfer capacitor C_1 is charged by the input current for time $t = T_{off}$. Therefore, the average charging current for C_1 becomes

$$I_{C_1} = I_s$$

Also, the peak-to-peak ripple voltage of the capacitor C_1 is given by

$$\Delta V_{C_1} = \frac{1}{C_1} \int_0^{T_{off}} I_{C_1} dt = \frac{1}{C_1} \int_0^{T_{off}} I_s dt = \frac{I_s \cdot T_{off}}{C_1} \tag{8.191}$$

From Eq. (8.183), we can write, $E_0(1 - \alpha) = -E_{dc}\,\alpha$

$E_0 = \alpha(E_0 - E_{dc})$, or $E_0 = T_{on}/T(E_0 - E_{dc})$

or
$$\frac{T_{on}}{T} = \frac{E_0}{(E_0 - E_{dc})}$$

$$\therefore \quad \frac{T - T_{off}}{T} = \frac{E_0}{E_0 - E_{dc}}, \quad 1 - \frac{T_{off}}{T} = \frac{E_0}{E_0 - E_{dc}}$$

$$\frac{-T_{off}}{T} = \frac{E_0}{E_0 - E_{dc}} - 1, \quad \frac{-T_{off}}{T} = \frac{E_{dc}}{E_0 - E_{dc}}$$

or
$$T_{off} = \frac{E_{dc}}{(E_{dc} - E_0) f}$$

Substituting the value of T_{off} in Eq. (8.191), we get

$$\Delta V_{C_1} = \frac{I_s \cdot E_{dc}}{(E_{dc} - E_0) f \cdot C_1} \tag{8.192}$$

or $$\Delta V_{C_1} = \frac{I_s \cdot (1-\alpha)}{f \cdot C_1} \tag{8.193}$$

If we assume that the load current ripple Δi_0 is negligible, then we have

$$\Delta i_{L_2} = \Delta i_{C_2}$$

Therefore, the average charging current of capacitor C_2, which flows for time $T/2$, becomes

$$I_{C_2} = \frac{\Delta I_2}{4}$$

and peak-to-peak ripple voltage of capacitor C_2 is given by

$$\Delta V_{C_2} = \frac{1}{C_2} \int_0^{T/2} I_{C_2} \cdot dt = \frac{1}{C_2} \int_0^{T/2} \frac{\Delta I_2}{4} \cdot dt = \frac{\Delta I_2}{8 f C_2} \tag{8.194}$$

or $$\Delta V_{C_2} = \frac{-E_0(1-\alpha)}{8 C_2 L_2 f^2} = \frac{\alpha \cdot E_{dc}}{8 C_2 L_2 f^2} \tag{8.195}$$

A unique feature of the Cuk converter is that both the input and the output current are non-pulsating. In this circuit it is also possible to simultaneously eliminate the ripple in i_{L_1} and i_{L_2} completely, leading to lower external filtering requirements. The significant disadvantages is the requirement of a capacitor C_1 with a large ripple current carrying capability.

SOLVED EXAMPLE

Example 8.28 Consider the cuk converter of Fig. 8.46. The input voltage to this converter is 15 V. The duty cycle is $\alpha = 0.4$ and the switching frequency is 25 kHz. The filter inductance is $L_2 = 380$ μH and filter capacitance, $C_2 = 220$ μF. The energy transfer capacitance $C_1 = 400$ μF and inductance $L_1 = 250$ μH. The average load current is $I_0 = 1.25$ A. Determine:
(a) the average output voltage E_0,
(b) the average input current I_s,
(c) the peak-to-peak ripple current of inductor L_1, ΔI_1,
(d) the peak-to-peak ripple voltage of capacitor C_1, ΔV_{C_1}
(e) the peak-to-peak ripple current of inductor L_2, ΔI_2,
(f) the peak-to-peak ripple voltage of capacitor C_2, ΔV_{C_2}, and
(g) the peak current of the device I_P.

Solution: Given: $E_{dc} = 15$ V, $\alpha = 0.4$, $I_0 = 1.25$ A, $f = 25$ kHz, $L_1 = 250$ μH, $L_2 = 380$ μH, $C_1 = 400$ μF, $C_2 = 220$ μF.

(a) From Eq. (8.183), $E_0 = -\dfrac{0.4 \times 15}{(1-0.4)} = -10$ V

(b) From Eq. (8.184), $I_s = \dfrac{0.4 \times 1.25}{(1-0.4)} = 0.83$ A

(c) From Eq. (8.187), $\Delta I_1 = \dfrac{15 \times 0.4}{25 \times 10^3 \times 250 \times 10^{-6}} = 0.96$ A

(d) From Eq. (8.193), $\Delta V_{C_1} = \dfrac{0.83(1-0.4)}{25 \times 10^3 \times 400 \times 10^{-6}} = 49.8$ mV

(e) From Eq. (8.190), $\Delta I_2 = \dfrac{0.4 \times 15}{25 \times 10^3 \times 380 \times 10^{-6}} = 0.63$ A

(f) From Eq. (8.194), $\Delta V_{C_2} = \dfrac{0.63}{8 \times 25 \times 10^3 \times 220 \times 10^{-6}} = 14.31$ mV

(g) The average voltage across the diode can be obtained from

$$V_{DF} = -\alpha V_{C_1} = -E_0 \cdot \alpha \dfrac{1}{-\alpha} = E_0$$

For a lossless circuit,

$$I_{L_2} \cdot V_{DF} = E_0 \cdot I_0$$

and the average value of the current in inductor L_2 is

$$I_{L_2} = \dfrac{I_0 \cdot E_0}{V_{DF}} = I_0 = 1.25 \text{ A}$$

Therefore, the peak current of the device is,

$$I_p = I_s + \dfrac{\Delta I_1}{2} + I_{L_2} + \dfrac{\Delta I_2}{2} = 0.83 + \dfrac{0.96}{2} + 1.25 + \dfrac{0.63}{2} = 2.88 \text{ A}$$

REVIEW QUESTIONS

8.1 Draw the schematics of step-down and step-up choppers and derive an expression for output voltage in terms of duty-cycle for a step-up and step-down chopper.

8.2 With the circuit diagram and output voltage waveforms, explain the working of Jones chopper.

8.3 With the circuit diagram and output voltage waveform, explain the principle of operation of a chopper.

8.4 Explain the time ratio control and current limit control, and control strategies used for chopper.

8.5 With the help of circuit diagram, explain the working of step-up/step-down chopper.

Choptres

8.6 Draw the circuit of a two-quadrant chopper and explain its working.

8.7 With the help of voltage and current waveforms, explain the working of first-quadrant chopper. Give the complete time domain analysis of class A chopper.

8.8 Derive the expressions for $I_{0\,max}$ and $I_{0\,min}$ for class A chopper. Also, derive the expression for per unit ripple current.

8.9 Explain the continuous conduction mode and non-continuous conduction mode of class A chopper.

8.10 Derive the expression for average load current for class A chopper.

8.11 With the help of circuit diagram and associated waveforms, explain the principle of working of class C chopper.

8.12 With the help of a circuit diagram, explain the working of class D chopper.

8.13 Give the detailed circuit analysis of class D chopper.

8.14 Derive the expression for output power for class D chopper.

8.15 Draw the schematic of class E chopper and explain the working of the same.

8.16 Give the classification of chopper commutation.

8.17 Describe the voltage commutated chopper with associated voltage and current-waveforms as a function of time.

8.18 Derive the expressions for commutating components L and C for a voltage-commutated chopper. Also, discuss the assumptions made for designing the components.

8.19 With the help of basic power circuit diagram, explain the working of a current commutated chopper. Also, draw the associated waveforms.

8.20 Mention the advantages of Jones chopper circuit over other chopper circuits. Give the application of this chopper.

8.21 Explain in brief how average voltage across the load is made more than d.c. supply voltage using chopper. Derive the expression for the average voltage.

8.22 Draw a schematic diagram of a single-phase a.c. chopper and discuss in brief with output voltage and current waveforms.

8.23 Draw the single SCR chopper circuit for the control of a d.c. series motor. Explain its working with voltage and current waveforms.

8.24 Derive the expression for commuting components L, C and turn-off time t_q for the current commutated chopper circuit.

8.25 Discuss the working of load commutated chopper with associated voltage and current waveforms. Show voltage variation across each pair of SCRs as a function of time.
Derive an expression from which the value of commutating capacitor of this chopper can be calculated.

8.26 Enumerate the merits and demerits of load commutated chopper.

8.27 Describe a Morgan chopper with associated voltage and current waveforms. Enumerate the demerits of Morgan chopper compared to Jones chopper.

8.28 Discuss the design considerations of the source filter.

8.29 What is a multiphase chopper? Bring out clearly, with appropriate waveforms, the difference between the in-phase operation and phase-shifted operation of a multiphase chopper. Also, explain why phase shifted operation is always preferred.
List the merits and demerits of a multiphase choppers.

8.30 Draw and explain the working of any chopper firing circuit.

8.31 With the help of neat circuit diagram and associated waveforms discuss the operation of a Buck converter.

8.32 Derive the expressions for peak-to-peak ripple current of inductor and peak-to-peak ripple voltage of capacitor in terms of circuit components, supply voltage, frequency and duty-ratio, for a Buck converter.

8.33 List the advantages and disadvantages of the Buck chopper.

8.34 Discuss the operation of Boost converter with the help of neat circuit diagram and waveforms.

8.35 For a Boost converter, derive the expressions for peak-to-peak ripple current and ripple voltage in terms of circuit components, frequency, supply voltage and duty ratio.

8.36 List the advantages and disadvantages of the Boost chopper.

8.37 With the help of a neat circuit diagram and associated waveforms, discuss the operation of Buck–Boost converter. List the advantages and disadvantages of this type of converter.

8.38 Derive the expression for peak-to-peak ripple current and ripple voltage in case of Buck–Boost converter.

8.39 Briefly discuss the operation of Cuk converter with the help of a circuit diagram and voltage and current waveforms.

PROBLEMS

8.1 An 80 V battery supplies an R–L load through a chopper circuit. The load inductance is 40 mH and resistance is 6 Ω. The load has a freewheeling diode D_f across it. It is required to vary the current between the limits 10 A and 12 A. Calculate the time ratio of the chopper?

$$\left[Ans \quad \frac{T_{on}}{T_{off}} = 4.12 \right]$$

8.2 For the ideal type A chopper circuit, with a R-LE_b type load, following operating conditions are given
E_{dc} = 220 V, chopping period = 1×10^{-3} s,
T_{on} = 400 µs, R = 1.5 Ω, L = 6 mH and E_b = 44 V.
Compute the following quantities
   (i) Check whether the load current is continuous.
   (ii) Average output current.
   (iii) Maximum and minimum values of steady-state output current.
   (iv) RMS values of first, second and third harmonics of load current.
   (v) Average value of source current.
   (vi) The input power, the power absorbed by the back emf E_b and power loss in the resistor.
   (vii) RMS value of output current using the results of (ii) and (iv).
   (viii) The RMS value of load current using the resuls of (iv), compute the result with that obtained in part (vii) above
   [Ans (i)α' = 0.221, current is continuous. (ii) 29.33 A (iii)33.77 A, 24.98 A
   (iv)2.496 A, 0.772 A, –0.1716 A. (v) 11.781 A (vi) 2591.82 W,
   1290.52 W, 1301.3 W. (vii) 29.45 A (viii) 29.45 A.]

Choppers

8.3 A chopper circuit is operating on TRC principle at a frequency of 1 kHz on a 220 V d.c. supply. If the load voltage is 180 V, calculate the conducting and blocking period of thyristor in each cycle.

[*Ans* $T_{on} = 0.82$ μs, $T_{off} = 0.18$ μs.]

8.4 A step-up chopper with a pulse width of 150 μs is operating on 220 V d.c. supply. Compute the load voltage if the blocking period of the device is 40 μs.

[*Ans* $E_0 = 14045$ V]

8.5 A d.c. on–off chopper operating at 1 kHz and duty cycle of 10% is supplied from a 200 V source. If the load inductance is 10 mH and resistance 10 Ω. Compute the maximum and minimum circuit in the load.

[*Ans.* $I_{0max} = 2.22$ A, $I_{0min} = 0.22$ A]

8.6 An ideal d.c. chopper operating at a frequency of 600 Hz supplies a load resistance of 5Ω, inductance 1 g mH from a 110 V d.c. source. If the source has zero impedance and the load is shunted by an ideal diode as shown in Fig. 8.6, calculate average load voltage and current at mark/space $\left(\dfrac{T_{on}}{T_{off}}\right)$ ratios of

(i) 1/1 (ii) 5/1 and (iii) 1/3

[*Ans* (i) $E_0 = 55$ V, $I_0 = 11$ A (ii) $E_0 = 91.67$ V, $I_0 = 18.33$ A (iii) $E_0 = 27.5$ V, $I_0 = 5.5$ A.]

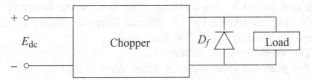

Fig. P.8.6

8.7 A current-commutated chopper controls a battery powered electric car. The battery voltage is 100 V, starting current is 100 A, thyristor turn-off time is 20 μs, chopper frequency is 400 Hz. Compute the values of commutating capacitor and commutating inductor. Assume $\dfrac{I_{cp}}{I_{0m}} = 3$.

[*Ans* $C = 48.74$ μF; $L = 4.4158$ μH.]

8.8 For a current-commutated chopper, peak commutating current is twice the maximum possible load current, the source voltage is 230 V d.c. and main SCR turn-off time is 30 μ sec. For a maximum load current of 200 A, compute
(a) the values of commutating inductor and capacitor
(b) maximum capacitor voltage
(c) the peak commutating current.

[*Ans* (a) $L = 13.178$ μH (b) 345 V (c) 400 A $C = 39.86$ μF.]

8.9 A chopper fed from a 220 V d.c. source is working at a chopping period of 20 ms and is connected to an R–L, Load of $R = 5$ Ω and $L = 40$ mH.
(a) compute the value of α at which the minimum load current will be
 (i) 5 A (ii) 10 A (iii) 20 A (iv) 30 A
(b) For the values of duty-cycle obtained in part (a) above, calculate the values of maximum currents and ripple factors.

[Ans. (a) 0.328, 0.506, 0.722, 0.862 (b) 26.823 A, 34.405 A, 40.051 A, 42.379 A. ripple factors 1.4313, 0.988, 0.6205, 0.40.]

8.10 A d.c. battery is charged from a constant d.c. source of 220 V through a chopper. The d.c. battery is to be charged from its internal emf of 90 V to 122 V. The battery has internal resistance of 1 Ω. For a constant charging current of 10 A, calculate the range of duty cycle.

[Ans. 0.45 to 0.6.]

8.11 The Buck-converter in Fig. 8.43 has an input voltage of E_{dc} = 16 V. The required average output voltage is E_0 = 8 V, and the peak-to-peak output ripple voltage is 10 mV. The switching frequency is 25 kHz. If the peak-to-peak ripple current of inductor is limited to 0.7 A. Determine: (a) duty-cycle α, (b) filter inductance, L, and (c) the filter capacitor C.

[Ans (a) 50%; (b) 228.57 µH; (c) 350 µF.]

8.12 Consider the boost converter of Fig. 8.44. The input voltage to this converter is 8 V. The average output voltage E_0 = 16 V and the average load current I_0 = 0.5 A. The switching frequency is 30 kHz. If L = 160 µH and C = 380 µF. Compute (a) the duty-cycle α; (b) the ripple current of inductor, ΔI; (c) the peak current of inductor, I_2, and (d) the ripple voltage of filter capacitor, ΔV_c.

[Ans. (a) 50%; (b) 0.83 A; (c) 1.415 A; (d) 21.93 mv.]

8.13 Consider the buck–bost converter of Fig. 8.45. The input voltage to this converter is E_{dc} = 10 V. The duty cycle α = 0.3 and the switching frequency is 25 kHz. The inductance I = 150 µH and filter-capacitance C = 220 µF. The average load current I_0 = 1.2 A. Determine:

(a) The average output voltage, E_0,
(b) The peak-to®-peak output voltage ripple, ΔV_C,
(c) The peak-to-peak current of inductor, ΔI, and
(d) the peak current of the transistor I_p. [Ans (a) – 4.29 V; (b) 65.45 mV; (c) 0.8 A; (d) 2.1 A.]

REFERENCES

1. S. B. Dewan, *Power Semiconductor Circuits*, John Wiley, New York, 1975.
2. R. S. Ramshaw, *Power Electronics*, Chapman and Hall, London 1973.
3. P. Sen, *Thyristor d.c. drives*, John Wiley, 1981.
4. F. Csaki et al., *Power Electronics*, Akademiaikiado Budapest, 1975.
5. B. D. Bedford, and R. G. Hoft, *Principles of Inverter Circuits*, John Wiley, 1964
6. N W Mapham and J C Hey 'Jones chopper circuit' *IEEE Inst. Conf. Rec.*, p. 124, (1964).
7. W. McMurray, "Thyristor commutation in d.c. choppers a comparative study" *IEEE Trans on IA*, vol. IA-14, No. 6, Nov/Dec. 1978, p. 547.
8. M.H. Rashid, *"Power-Electronics"* Pearson Education 2002.

Chapter 9

Inverters

LEARNING OBJECTIVES:

- To describe the need and function of an inverter.
- To consider the operation of single-phase half-bridge and full-bridge transistorised inverters.
- To introduce the performance parameters of inverters.
- To examine the operation of unipolar and bipolar pwm inverters.
- To consider the operation and design of series inverter with different circuit arrangements.
- To examine the operation of a three-phase series inverter.
- To consider the operation of a high frequency series inverter.
- To consider the operation and design considerations of self-commutated inverters.
- To consider the operation and design details of parallel inverter with different circuit arrangements.
- To examine the operation and detailed design aspects of various voltage source bridge inverter circuits.
- To examine the operation of three-phase bridge inverters with different conduction modes.
- To consider the means of controlling a variable frequency inverter output-voltage.
- To introduce various schemes of pulse width modulated inverters.
- To examine the basic techniques of harmonic filtering and to introduce filter types.
- To examine the operation of current source inverters as a means of producing a variable frequency supply.

9.1 INTRODUCTION

The d.c. to a.c. power converters are known as inverters. In other words, an inverter is a circuit which converts a d.c. power into an a.c. power at desired output voltage and frequency. The a.c. output voltage could be fixed at a fixed or variable frequency. This conversion can be achieved either by controlled turn-on and turn-off devices (e.g. BJTs, MOSFETs, IGBTs, MCTs, SITs, GTOs, SITHs) or by forced commutated thyristors, depending on applications. For low and medium power outputs, the above-mentioned power devices are suitable but for high power outputs, thyristors should be used. The output voltage waveforms of an ideal inverter should be sinusoidal. The voltage waveforms of practical inverters are, however, nonsinusoidal and contain certain harmonics. Square wave or quasi-square wave voltages may be acceptable for low and medium power applications, and for high power applications low-distorted, sinusoidal waveforms are required. The output frequency of an inverter is determined by the rate at which the semiconductor devices are switched *on* and *off* by the inverter control circuitry and consequently, an adjustable frequency a.c. output is readily provided. The harmonic contents of output voltage can be minimized or reduced significantly by switching techniques of available high speed power semiconductor devices. The filtering of harmonics is not feasible when the output frequency varies over a wide range, and the generation of a.c. waveforms with low harmonic content is important. When the a.c. output voltage of an inverter is given to a transformer or a.c. motor, this output voltage must be varied in conjunction with frequency to maintain the proper magnetic conditions. Therefore, the output voltage control is an essential feature of an adjustable frequency system, and various techniques for achieving voltage control are discussed in this chapter.

The d.c. power input to the inverter may be battery, fuel cell, solar cells or other d.c. source. But in most industrial applications, it is fed by a rectifier. This configuration of a.c. to d.c. converter and d.c. to a.c. inverter is called a d.c. link converter because it is a two-stage static frequency converter in which a.c. power at network frequency is rectified and then filtered in the d.c. link before being inverted to a.c. at an adjustable frequency. Rectification is achieved by standard diode or thyristor converter circuits, and inversion is achieved by the circuit techniques described in this chapter.

Inverters are mainly classified as voltage source inverters and current source inverters. A voltage fed inverter (VFI), or voltage source inverter (VSI), is one in which the d.c. source has small or negligible impedance. In other words, a voltage source inverter has stiff d.c. voltage source at its input terminals. Because of a low internal impedance, the terminal voltage of a voltage source inverter remains substantially constant with variations in load. It is, therefore, equally suitable to single motor and multi-motor drives. Any short-circuit across its terminals causes current to rise very fast, due to the low time constant of its internal impedance. The fault current cannot be regulated by current control and must be cleared by fast-acting fuse links. On the other hand, the current fed, or current source, inverter (CSI) is supplied with a controlled current from a d.c. source of high impedance. Typically, a phase controlled thyristor rectifier feeds the inverter with a regulated current through a large series inductor. Thus, load current rather than load voltage is controlled, and the inverter output voltage is dependent upon the load impedance. Because of a large internal impedance, the terminal voltage of a current source inverter changes substantially with a change in load. Therefore, if used in a multi-motor drive, a change in load on any motor affects other motors. Hence, current source inverters are not suitable for multi-motor drives. Since the inverter current is independent of load impedance, it has inherent protection against short-circuits across its terminals. Some of the important industrial applications of inverters are:

- Variable speed a.c. motor drives.
- Induction heating.
- Aircraft power supplies.
- Uninterruptible power supplies (UPS).
- High voltage d.c. transmission lines.
- Battery-vehicle drives.
- Regulated voltage and frequency power supplies, etc.

9.2 CLASSIFICATION OF INVERTERS

Inverters can be classified on the basis of a number of factors:

(a) Classification According to the Nature of Input Source Based on the nature of input power source, inverters are classified as
  (i) Voltage source inverters (VSI)
  (ii) Current source inverters (CSI)

In case of VSI, the input to the inverter is provided by a ripple free dc voltage source whereas in CSI, the voltage source is first converted into a current source and then used to supply the power to the inverter.

(b) Classification According to the Waveshape of the Output Voltage The inverters can be classified according to the nature of output voltage waveform as:
  (i) Square-wave inverter
  (ii) Quasi-square wave inverter
  (iii) Pulse-width modulated (PWM) inverters

A square-wave inverter produces a square-wave ac voltage of a constant magnitude. The output voltage of this type of inverter can only be varied by controlling the input dc voltage. Square-wave ac-output voltage of an inverter is adequate for low and medium power applications. However, the sine-wave output voltage is the ideal waveform for many high-power applications. Two methods can be used to make the output closer to a sinusoid. One is to use a filter circuit on the output side of the inverter. This filter must be capable of handling the large power output of the inverter, so it must be large and will therefore add to the cost and weight of the inverter. Moreover, the efficiency will be reduced due to the additional power-losses in the filter.

The second method, pulse-width modulation (PWM) uses a switching scheme within the inverter to modify the shape of the output voltage waveform.

9.2.1 Thyristor Inverter Classification

The thyristor inverters can be classified in the following categories:
1. According to the method of commutation.
2. According to the connections.

(a) Classification According to the Method of Commutation According to the method of commutation, the SCR inverters can mainly be categorised in two types, viz.
1. Line commutated inverters
2. Forced commutated inverters.

1. Line Commutated Inverters In case of a.c. circuits, a.c. line voltage is available across the device. When the current in the SCR goes through a natural zero, the device is turned-off. This process is known as natural commutation process and the inverters based on this principle are known as line commutated inverters.

2. Forced Commutated Inverters In case of d.c. circuits, since the supply voltage does not go through the zero point, some external source is required to commutate the device. This process is known as the forced commutation process and the inverters based on this principle are called as forced commutated inverters. As the device is to be commutated forcefully, these types of inverters require complicated commutation circuitries. These inverters are further classified as: (i) Auxiliary commutated inverters and (ii) Complementary commutated inverters.

(b) Classifications According to Connections According to the connections of the thyristors and commutating components, the inverters can be classified mainly in three groups. These are:
1. Series inverters.
2. Parallel inverters.
3. Bridge inverters.: Bridge inverters are further classifed as: (i) Half-bridge and (ii) Full-Bridge.

9.3 SINGLE-PHASE HALF-BRIDGE VOLTAGE-SOURCE INVERTERS

Figure 9.1 shows the basic configuration of single-phase half-bridge inverter. Switches S_1 and S_2 are the gate-commutated devices such as power BJTs, MOSFETs, GTO, IGBT, MCT, etc. When closed, these switches conducts and current flows in the direction of arrow. The operation of this inverter for different types of load is explained in the following sections:

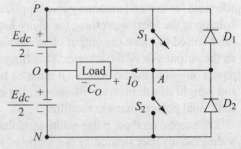

Fig. 9.1 *Half-bridge inverter*

9.3.1 Operation with Resistive Load

The operation of the circuit can be divided into two periods:
 (i) Period-I, where switch S_1 is conducting from $0 \leq 1 \leq T/2$ and
 (ii) Period-II, where switch S_2 is conducting from $T/2 \leq t \leq T$,
where $T = 1/f$ and f is the frequency of the output voltage waveform. Figure 9.2 shows the waveforms for the output voltage and switch currents for a resistive-load.

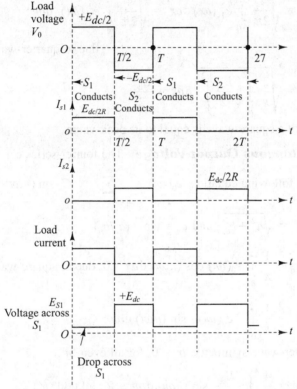

Fig. 9.2 *Voltage and current waveforms*

Switch S_1 is closed for half-time period ($T/2$) of the desired ac output. It connects point p of the dc source to point A and the output voltage e_0 becomes equal to $+E_{dc}/2$.

At $t = T/2$, gating signal is removed from S_1 and it turns-off. For the next half-time period ($T/2 < t < T$), the gating signal is given to S_2. It connects point N of the dc source to point A and the output voltage reverses. Thus, by closing S_1 and S_2 alternately, for half-time periods, a square-wave ac voltage is obtained at the output. With resistive load, waveshape of load current is identical to that of output voltage. Simply by controlling the time periods of the gate-drive signals, the frequency can be varied. Here diodes D_1 and D_2 do not play any role. The voltage across the switch when it is OFF is E_{dc}. Gating circuit should be designed such that switches S_1 and S_2 should not turn-on at the same time.

Circuit Analysis

(i) RMS Output Voltage
The average value of the output voltage is given by

$$E_{0(av)} = \frac{1}{2\pi} \int_0^{2\pi} e_0(\omega t) \, d\omega t$$

Now, rms value of the output voltage is given by

$$E_{0(rms)} = \sqrt{\frac{1}{2\pi} \int_0^{\pi/2} e_0^2(\omega t) \, d\omega t} = \sqrt{\frac{4}{2\pi} \int_0^{\pi/2} e_0^2(\omega t) \, d\omega t}$$

due to quarter-wave symmetry

$$= \sqrt{\frac{2}{\pi} \int_0^{\pi/2} \left(\frac{E_{dc}}{2}\right)^2 d\omega t} = \frac{E_{dc}}{2} \quad (9.1)$$

RMS value of a square-wave is equal to its peak-value.

(ii) Instantaneous Output-Voltage
The fourier-series can be found out by using the following equation:
$$e_0(\omega t) = \sum_{n=1,2,3,\ldots}^{\infty} C_n \sin(n\omega t + \phi_n)$$

where, $C_n = \sqrt{a_n^2 + b_n^2}$, and $\phi_n = \tan^{-1}(a_n/b_n)$

and $\quad a_n = \dfrac{1}{\pi} \displaystyle\int_0^{2\pi} e_0(\omega t) \cos(n\omega t) \, d\omega t = 0$, due to square-wave symmetry

and $\quad b_n = \dfrac{1}{\pi} \displaystyle\int_0^{2\pi} e_0(\omega t) \cdot \sin(n\omega t) \, d\omega t$

Due to quarter-wave symmetry, $b_n = 0$, for all even 'n'.

$\therefore \quad b_n = \dfrac{4}{\pi} \displaystyle\int_0^{\pi/2} \dfrac{E_{dc}}{2} \sin(n\omega t) \, d\omega t,$ for all odd 'n'.

$\therefore \quad b_n = \dfrac{2 E_{dc}}{n\pi},$ for odd value of n.

$\therefore \quad C_n = \sqrt{a_n^2 + b_n^2} = \dfrac{2 E_{dc}}{n\pi}$ and $\phi_n = \tan^{-1}\left(\dfrac{a_n}{b_n}\right) = 0$

Therefore, the instantaneous output voltage of a half-bridge inverter can be expressed in fourier-series form as

$$e_0(\omega t) = \sum_{n=1,3,5,\ldots}^{\infty} \frac{2 E_{dc}}{\pi \, n} \sin(n\omega t) \quad (9.2)$$

$$= 0, \quad \text{for } n = 2, 4, \ldots \text{ (even values of } n\text{)}$$

The nth harmonic-component is given by

$$e_0(n) = \frac{c_n}{\sqrt{2}} = \frac{2 E_{dc}}{n\pi \cdot \sqrt{2}} = \frac{\sqrt{2}}{n} \frac{E_{dc}}{n} \quad \text{for } n = 1, 3, 5, \ldots \quad (9.3)$$

RMS value of fundamental components is obtained by substituting $n = 1$ in Eq. (9.3)

$$\therefore \quad E_{1_{(rms)}} = \frac{\sqrt{2}}{\pi} E_{dc} = 0.45 E_{dc} \quad (9.4)$$

(iii) Switch (Device) Voltage and Current Ratings From Fig. 9.2,

$$V_{CEO(transistor)} \geq 2 \frac{E_{dc}}{2} \geq E_{dc} \quad (9.5)$$

The current waveform for switch is a square-wave with a peak value of $E_{dc}/2R$.

$$\therefore \quad I_{T_{avg}} = \frac{1}{T} \int_0^{T/2} \frac{E_{dc}}{2R} \, dt = \frac{E_{dc}}{4R} \quad (9.6)$$

$$\therefore \quad I_{T_{rms}} = \sqrt{\frac{1}{T} \cdot \int_0^{T/2} \left(\frac{E_{dc}}{2R}\right)^2 dt} = \frac{E_{dc}}{2\sqrt{2} R} \quad (9.7)$$

and

$$I_{T_{peak}} = \frac{E_{dc}}{2R} \quad (9.8)$$

9.3.2 Operation with RL Load

With an inductive-load, the output voltage waveform is similar to that with a resistive-load, however the load-current cannot change immediately with the output voltage. The operation of half-bridge inverter with RL load is divided into four distinct modes. Voltage and current waveforms are shown in Fig. 9.3. D_1 and D_2 are known as the feedback diodes.

Mode I ($t_1 < t < t_2$): S_1 is turned-on at instant t_1, the load voltage is equal to $+E_{dc}/2$ and the positive load current increases gradually. At instant t_2, the load-current reaches the peak value. Switch S_1 is turned-off at this instant. Due to same-polarity of load voltage and load current, the energy is stored by the load [Fig. 9.4(a)].

Mode II ($t_2 < t < t_3$): Due to inductive load, the load current direction will be maintained even-after S_1 is turned-off. The self-induced voltage in the load will be negative. The load current flows through lower half of the supply and D_2 as shown in Fig. 9.4(b). In this mode, the stored energy in load is feedback to the lower half of the source and the load voltage is clamped to $-E_{dc}/2$.

Mode III ($t_3 < t < t_4$): At instant t_3, the load-current goes to zero, indicating that a_1, the stored energy, has been returned back to the lower half of supply. At

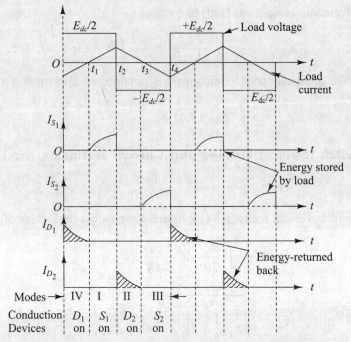

Fig. 9.3 *Voltage and current waveforms with RL load*

instant t_3, S_2 is turned-on. This will produce a negative load voltage $e_0 = -E_{dc}/2$ and a negative load current. Load current reaches a negative peak at the end of this interval (Fig. 9.4(c)].

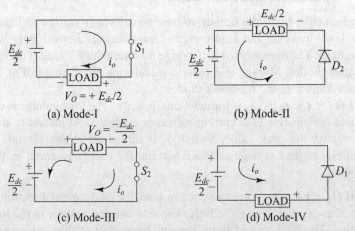

Fig. 9.4 *Operating modes*

Mode IV ($t_0 < t < t_1$): Switch S_2 is turned-off at instant t_4. The self induced voltage in the inductive load will maintain the load current. The load voltage changes its polarity to become positive $E_{dc}/2$, load current remains negative and the stored energy in the load is returned back to the upper half of the dc source (Fig. 9.4(d)).

Inverters

At t_5, the load current goes to 0 and S_1 can be turned-on again. This cycle of operation repeats.

Circuit-Equations

(i) Instantaneous Current i_0:- With RL load, the equation for instantaneous current i_0 can be obtained from Eq. (9.2), as

$$i_0(t) = \sum_{n=1,3,5,\ldots}^{\infty} \frac{2 E_{dc}}{n\pi \sqrt{R^2 + (n\omega L)^2}} \sin(n\omega t - \theta_n) \quad (9.9)$$

Here, $Z_n = \sqrt{R^2 + (n\omega L)^2}$ is the impedance offered by the load to the nth harmonic component, $\dfrac{2 E_{dc}}{n\pi}$ is the peak amplitude of nth harmonic voltage, and

$$\theta_n = \tan^{-1}\left(\frac{n\omega L}{R}\right) \quad (9.10)$$

(ii) Fundamental Output Power: The output power at fundamental frequency ($n = 1$) is given by

$$P_{1_{rms}} = E_{1_{rms}} \cdot I_{1_{rms}} \cdot \cos\theta_1 = I_{1_{rms}}^2 \cdot R \quad (9.11)$$

where $E_{1_{rms}}$ = RMS value of fundamental output voltage.

$I_{1_{rms}}$ = RMS value of fundamental output current

$\theta_1 = \tan^{-1}(\omega L / R)$

But

$$I_{1_{rms}} = \frac{2 E_{dc}}{\sqrt{2} \cdot \pi \cdot \sqrt{R^2 + (\omega L)^2}} \quad (9.12)$$

$$P_{1_{rms}} = I_{1_{rms}}^2 \cdot R = \left[\frac{2 E_{dc}}{\pi \cdot \sqrt{2} \sqrt{R^2 + (\omega L)^2}}\right]^2 \cdot R \quad (9.13)$$

$$= \left[\frac{4 E_{dc}^2 \cdot R}{2\pi^2 (R^2 + \omega^2 L^2)}\right] = \left[\frac{2 E_{dc}^2 \cdot R}{\pi^2 (R^2 + \omega^2 L^2)}\right] \quad (9.14)$$

Importance of fundamental power is that in many applications such as electric motor drives, the output power due to fundamental current is generally the useful power and the power due to harmonic current is dissipated as heat and increases the load dissipation.

9.3.3 Cross Conduction or Shoot through Fault

In the half-bridge inverter circuit, each switch conducts for a period of $T/2$ secs. At any particular instant, one switch is turned-on and the other is turned-off. However, the outgoing switch does not turn-off instantaneously due to its finite turn-off delay. Due to this, both switches (incoming and outgoing) conduct simultaneously for a short-time. This is known as cross-conduction or shoot-through-fault.

When both switches conducts simultaneously, the input dc supply is short-circuited and with this switches get damaged. Cross conduction can be avoided by allowing the outgoing switch to turn-off completely first and then applying the gate-drive to the incoming device. A dead-band or delay is introduced between the trailing-edge of the base-drive of outgoing device and the leading-edge of the base-drive of the incoming device. Therefore, during the dead-band interval, no device receives base-drive. Hence, the dead-band should be longer than the turn-off time of the power-devices used in the inverter circuit.

SOLVED EXAMPLES

Example 9.1 The single-phase half-bridge inverter has a resistive load of 10 Ω and the center-tap dc input voltage is 96 V. Compute:
(i) RMS value of the output voltage.
(ii) Fundamental component of the output voltage waveform.
(iii) First five harmonics of the output-voltage waveform.
(iv) Fundamental power consumed by the load.
(v) RMS power consumed by the load.
(vi) Verify that the rms value determined by harmonic summation method is nearly equal to the value determined by integration method.

Solution:

(i) From Eq. (9.1), $E_{0(rms)} = \dfrac{E_{dc}}{2} = 96$ volts

(ii) From Eq. (9.4), $E_{1(fund)} = \dfrac{\sqrt{2}}{\pi} E_{dc} = 0.9 \times 96 = 86.40$ volts

(iii) From Eq. (9.3), first five harmonics are given by

$$E_{0(3)} = \frac{E_1}{3} = \frac{86.4}{3} = 28.8 \text{ V}, \quad E_{0(5)} = \frac{E_1}{5} = \frac{86.4}{5} = 17.28 \text{ V}$$

$$E_{0(7)} = \frac{E_1}{7} = \frac{86.4}{7} = 12.34 \text{ V}, \quad E_{0(9)} = \frac{E_1}{9} = \frac{86.4}{9} = 9.6 \text{ V}$$

$$E_{0(11)} = \frac{E_1}{11} = \frac{86.4}{11} = 7.85 \text{ V}$$

(iv) Fundamental power, $P_{0(fund)} = \dfrac{E_{1(fund)}^2}{R} = \dfrac{(86.0)^2}{10} = 746.5$ W

(v) RMS power, $P_{0(rms)} = \dfrac{E_{0(rms)}^2}{R} = \dfrac{(96)^2}{10} = 921.6$ W

(vi) RMS value by harmonic summation method is

$$E_{0(\text{rms})} = \sqrt{E_1^2 + E_3^2 + E_5^2 + E_7^2 + E_9^2 + E_{11}^2} = 94.34 \text{ V}$$

Thus, the two values are equal. The value obtained by harmonic summation method is always less than that found by direct integration method.

9.4 SINGLE-PHASE FULL-BRIDGE INVERTERS

Figure 9.5 shows the power-diagram of the single-phase bridge inverter. The inverter uses two pairs of controlled switches (S_1S_2 and S_3S_4) and two pairs of diodes (D_1D_2 and D_3D_4). The devices of one pair operate simultaneously. In order to develop a positive voltage (+ E_0) across the load, switches S_1 and S_2 are turned-on simultaneously whereas to have a negative voltage ($-E_0$) across the load, we need to turn-on the switches S_3 and S_4. Diodes D_1, D_2, D_3 and D_4 are known as the feedback diodes.

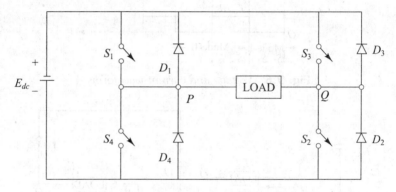

Fig. 9.5 *Single-phase full-bridge inverter*

9.4.1 Operation with Resistive Load

Voltage and current waveforms with resistive-load are shown in Fig. 9.6. The bridge-inverter operates in two-modes in one-cycle of the output.

(i) Mode-I (0 < t < T/2): In this mode, switches S_1 and S_2 conducts simultaneously. The load voltage is $+E_{dc}$ and load current flows from P to Q. The equivalent circuit for mode-I is shown in Fig. 9.7(a). At $t = T/2$, S_1 and S_2 are turned-off and S_3 and S_4 are turned-on.

(ii) Mode-II (T/2 < t < T): At $t = T/2$, switches S_3 and S_4 are turned-on and S_1 and S_2 are turned-off. The load voltage is $-E_{dc}$ and load current flows from Q to P. The equivalent circuit for mode-II is shown in Fig. 9.7 (b). At $t = T$, S_3 and S_4 are turned-off and S_1 and S_2 are turned-on again.

As the load is resistive, it does not store any energy. Therefore, feedback diodes are not effective here.

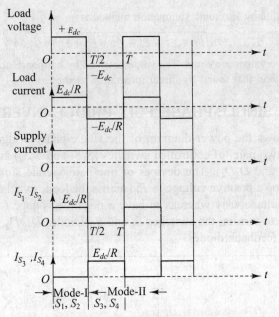

Fig. 9.6 *Voltage and current waveforms*

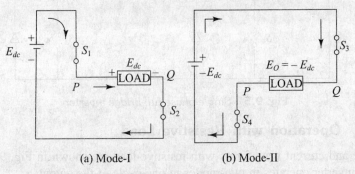

(a) Mode-I (b) Mode-II

Fig. 9.7 *Equivalent circuit*

Circuit-Analysis The analysis of the full-bridge inverter with resistive-load can be carried-out on similar lines of half-bridge inverter with resistive-load. Hence all equations of half-bridge are valid with $E_{dc}/2$ replaced by E_{dc}.

(i) RMS output voltage, $E_{0(rms)} = E_{dc}$ \hfill (9.15)

(ii) Fourier series, $e_{0(\omega t)} = \sum\limits_{n=1,3,5,\ldots}^{\infty} \dfrac{4 E_{dc}}{n\pi} \sin(n\omega t)$ \hfill (9.16)

(iii) Fundamental output voltage, $E_{0(fund)} = \dfrac{2\sqrt{2}}{\pi} \cdot E_{dc}$ \hfill (9.17)

(iv) n^{th} harmonic voltage $E_0(n) = \dfrac{E_{0(fund)}}{n}$ \hfill (9.18)

(v) Transistor (switch) ratings:

$$V_{CE_0} \geq E_{dc}, \quad I_{T(av)} = \frac{E_{dc}}{2R}$$

$$I_{T(rms)} = \frac{E_{dc}}{\sqrt{2}\,R}, \quad I_{T(peak)} = \frac{E_{dc}}{R} \tag{9.19}$$

SOLVED EXAMPLES

Example 9.2 A single-phase full-bridge inverter is operated from a 48V battery and is supplying power to a pure resistive load of 10 Ω. Determine:
 (i) the fundamental output voltage and the first five harmonics.
 (ii) RMS value by direct integration method and harmonic summation method.
 (iii) Output rms power and output fundamental power.
 (iv) Transistor switch ratings.

Solution: Given: $E_{dc} = 48$ V, $R = 10$ ohm

(i) From (9.17), $E_{0(fund)} = \dfrac{2\sqrt{2}}{\pi} E_{dc} = \dfrac{2\sqrt{2}}{\pi}(48) = 43.22$ V

Now n^{th} harmonic voltage, $E_{0(n)} = \dfrac{E_{0\,(fund)}}{n}$

$\therefore \quad E_{0(3)} = \dfrac{43.22}{3} = 14.40$ V, $\quad E_{0(5)} = \dfrac{43.22}{5} = 8.64$ V

$E_{0(7)} = \dfrac{43.22}{7} = 6.17$ V, $E_{0(9)} = \dfrac{43.22}{9} = 4.80$ V, $E_{0(11)} = \dfrac{43.22}{11} = 3.92$ V

(ii) $E_{0(rms)} = E_{dc} = 48$ V $\quad$ also, $\quad E_{0(rms)} = \sqrt{E_1^2 + E_3^2 + E_5^2 + E_7^2 + E_9^2 + E_{11}^2}$

$= \sqrt{(43.22)^2 + (14.40)^2 + (8.64)^2 + (6.17)^2 + (4.8)^2 + (3.92)^2} = 47.18$ V

Hence, the two values are nearly equal.

(iii) Output rms power, $P_{0rms} = \dfrac{E_{0(rms)}^2}{R} = \dfrac{48^2}{10} = 230.4$ W

Output fundamental power, $P_{0(fund.)} = \dfrac{E_{0(fund)}^2}{R} = 186.624$ W

(iv) Switch (Transistor) Ratings:

$$V_{CE_0} \geq E_{dc} \geq 48 \text{ V}, \quad I_{T(peak)} \geq \frac{E_{dc}}{R} \geq 4.8 \text{ A}$$

$$I_{T(rms)} \geq \frac{E_{dc}}{\sqrt{2}\cdot R} \geq 3.394 \text{ A}, \quad I_{T(avg)} \geq \frac{E_{dc}}{\sqrt{2}\,R} \geq 2.4 \text{ A}$$

9.4.2 Operation with RL Load

Voltage and current waveforms for single-phase bridge inverter with RL load are shown in Fig. 9.8. The operation of the circuit is explained in four-modes.

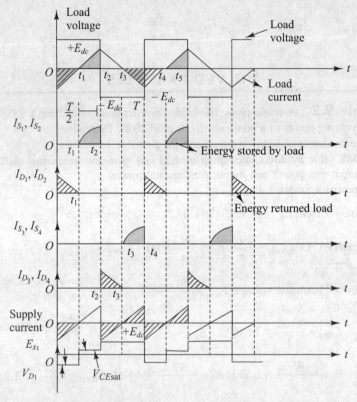

Fig. 9.8 *Voltage and current waveforms*

(i) Mode-I ($t_1 < t < t_2$): At instant t_1, the switch S_1 and S_2 are turned-on. Switches are assumed to be ideal switches. Point P gets connected to positive point of d.c. Source E_{dc} through S_1 and point Q gets connected to negative point of input supply. The output voltage, $e_0 = +E_{dc}$, Fig. 9.9(a). The load current starts increasing exponentially due to the inductive nature of the load. The instantaneous current through S_1 and S_2 is equal to the instantaneous load current. During this interval, energy is stored in inductive load.

(ii) Mode-II ($t_2 < t < t_3$): Both the switches Q_1 and Q_2 are turned-off at instant t_2. Due to the inductive nature of the load, the load current does not reduce to zero instantaneously. There is a self-induced voltage across the load which maintains the flow of current in the same-direction. The polarity of this voltage is exactly opposite to that in mode-1, The output voltage becomes $-E_{dc}$, but the load current continues to flow in the same direction, through D_3 and D_4 as shown in Fig. 9.9(b). Thus, in this mode, the stored energy in the load inductance is returned back to the source. Load current decreases exponentially and goes to 0 at instant t_3 when all the energy stored in the load is returned back to supply. D_3 and D_4 are turned-off at t_3.

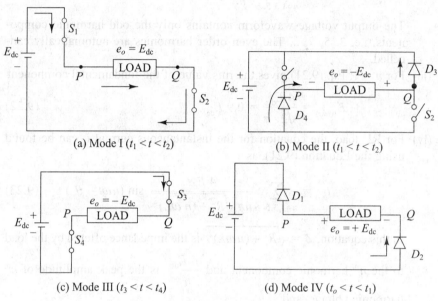

Fig. 9.9 Equivalent circuits

(iii) Mode III ($t_3 < t < t_4$): Switches S_3 and S_4 are turned-on simultaneously at instant t_3. Load voltage remains negative ($-E_{dc}$) but the direction of load current will reverse. The current increases exponentially in the other direction and the load again stores the energy.

(iv) Mode IV ($t_0 < t < t_1$): Switches S_3 and S_4 are turned-off at instant t_0 (or t_4). The load inductance tries to maintain the load current in the same direction by inducing the positive-load voltage. This will forward-bias the diodes D_1 and D_2. The load energy is returned back to the input dc supply. The load voltage becomes $e_0 = +E_{dc}$ but the load current remains negative and decreases exponentially towards 0. At t_1 (or t_5), the load current goes to zero and switches S_1 and S_2 can be turned-on again. The conduction period with a very highly inductive load, will be $T/4$ or 90° for all the switches as well as the diodes. The conduction period of switches will increase towards $T/2$ or 180° with increase in the load power-factor.

Circuit Analysis

(i) RMS output voltage can be obtained from

$$E_{0rms} = \left[\frac{2}{T/2} \int_0^{T/2} E^2 \, dt \right]^{1/2} \quad \therefore \quad E_{0rms} = E_{dc} \qquad (9.20)$$

(ii) The instantaneous output voltage can be expressed in fourier series as

$$e_{0(\omega t)} = \sum_{n=1,3,5,\ldots}^{\infty} \frac{4 E_{dc}}{n\pi} \sin n\omega t \qquad (9.21)$$

The output voltage waveform contains only the odd harmonic components, i.e. 3, 5, 7, ... The even order harmonics are automatically cancelled.

(iii) For $n = 1$, Eq. (9.21) gives the rms value of the fundamental component

$$E_{1(rms)} = \frac{4 E_{dc}}{\sqrt{2} \cdot \pi} = 0.9 E_{dc} \qquad (9.22)$$

(iv) For RL load, the equation for the instantaneous current i_0 can be found using the Equation (9.21), as

$$i_{0(t)} = \sum_{n=1,3,5,\ldots}^{\infty} \frac{4 E_{dc}}{n\pi \sqrt{R^2 + (n\omega L)^2}} \sin(n\omega t - \theta_n) \qquad (9.23)$$

In this equation, $Z_n = \sqrt{R^2 + (n\omega L)^2}$ is the impedance offered by the load to the nth harmonic component and $\frac{4 E_{dc}}{n\pi}$ is the peak amplitude of nth harmonic voltage, and

$$\theta_n = \tan^{-1}(n\omega L/R) \qquad (9.24)$$

SOLVED EXAMPLES

Example 9.3 The full-bridge inverter of Fig. 9.5 has a source voltage $E_{dc} = 220$ V. The inverter supplies an RLC load with $R = 10$ ohm, $L = 10$ mH and $C = 52$ μF. The inverter frequency is 400 Hz. Determine:
 (a) the rms load current at fundamental frequency
 (b) the rms value of load current
 (c) the power output and   (d) the average supply current

Solution:
Given: $E_{dc} = 220$ V, $R = 10$ Ω, $L = 10$ mH, $C = 52$ μF, $f = 400$ Hz
The inductive reactance of the fundamental voltage
$$X_L = 2\pi f L = 2 \times \pi \times 400 \times 10 \times 10^{-3} = 25.13 \ \Omega$$
The capacitive reactance for the fundamental voltage,
$$X_C = \frac{1}{2\pi f.c} = \frac{1}{2\pi \times 400 \times 52 \times 10^{-6}} = 7.7 \ \Omega$$
Impedance offered to the nth harmonic-component
$$Z_n = \sqrt{R^2 + \left(nX_L - \frac{X_c}{n}\right)^2}$$

Impedance offered to different harmonic components are obtained by putting different values of n,

$$\therefore\ Z_1 = \sqrt{(10)^2 + (25.13 - 7.7)^2} = 20.09,\ Z_3 = \sqrt{100 + \left(3 \times 25.13 - \frac{7.7}{3}\right)^2} = 73.51$$

Similarly,

$$Z_5 = \sqrt{100 + \left(5 \times 25.13 - \frac{7.7}{5}\right)^2} = 124.51,\ Z_7 = \sqrt{100 + \left(5 \times 25.13 - \frac{7.7}{7}\right)^2} = 175.09$$

$$Z_9 = \sqrt{100 + \left(9 \times 25.13 - \frac{7.7}{9}\right)^2} = 225.54$$

Now, the rms value of the nth harmonic component of the output voltage is given by

$$E_n = \frac{0.9\ E_{dc}}{n} = \frac{0.9 \times 220}{n} = \frac{198}{n}$$

Also, rms value of the nth harmonic component of the current is given by

$$I_n = \frac{E_n}{Z_n} = \frac{198}{nz_n} \quad (9.25)$$

(a) RMS value of the fundamental component of load current,

$$I_1 = \frac{198}{1 \times 20.05} = 9.86\ A$$

(b) Substituting the values of n and Z_n in Eq. (9.25), rms values of different harmonic components may be obtained as, $I_3 = 0.897\ A$

$$I_5 = 0.318\ A,\ I_7 = 0.161\ A,\ I_9 = 0.097\ A$$

$\therefore$ RMS value of the load current,

$$I = \sqrt{I_1^2 + I_3^2 + I_5^2 + I_7^2 + I_9^2}$$

$$= \sqrt{(9.86)^2 + (0.897)^2 + (0.318)^2 + (0.161)^2 + (0.097)^2}$$

$$= \sqrt{97.21 + 0.804 + 0.101 + 0.025 + 0.009} = 9.90\ A$$

(c) Output power, $P_0 = I^2 \cdot R = (9.90)^2 \times 10 = 980.10\ W$

(d) Average supply current, $I_{av} = \dfrac{P_0}{E_{dc}} = \dfrac{980.10}{220} = 4.46\ A$

9.5 PERFORMANCE PARAMETERS OF INVERTERS

Ideally, an inverter should give a sinusoidal voltage at its output. However, outputs of practical inverters are non-sinusoidal and may be resolved into fundamental and harmonic components. Performance of an inverter is usually evaluated in terms of the following performance parameters:

(a) Harmonic Factor of nth Harmonic (HF_n) The harmonic factor is a measure of the individual harmonic contribution in the output voltage of an inverter.

It is defined as the ratio of the rms voltage of a particular harmonic component to the rms value of fundamental component.

$$\therefore \quad \mathrm{HF}_n = \frac{E_{n_{rms}}}{E_{1_{rms}}} \tag{9.26}$$

(b) Total Harmonic Distortion (THD) A total harmonic distortion is a measure of closeness in a shape between the output voltage waveform and its fundamental component. It is defined as the ratio of the rms value of its total harmonic component of the output voltage and the rms value of the fundamental component. Mathematically,

$$\mathrm{THD} = \sqrt{\sum_{n=2,3,\ldots}^{\infty} E_{n_{rms}}^2} \Big/ E_{1_{rms}} \tag{9.27}$$

$$= \sqrt{\frac{E_{0_{rms}}^2 - E_1^2}{E_1}} \tag{9.28}$$

(c) Distortion Factor (DF) A distortion factor indicates the amount of harmonics that remain in the output voltage waveform, after the waveform has been subjected to second-order attenuation (i.e. divided by n^2). It is defined as

$$\mathrm{DF} = \sqrt{\frac{\sum_{n=2,3,\ldots}^{\infty} \left(\frac{E_{n_{rms}}}{n^2}\right)^2}{E_{1_{rms}}}} \tag{9.29}$$

(d) Lowest-Order Harmonics (LOH) The lowest frequency harmonic, with a magnitude greater than or equal to three-per cent of the magnitude of the fundamental component of the output voltage, is known as *lowest-order harmonic*. Higher the frequency of the LOH, lower will be the distortion in the current waveform.

SOLVED EXAMPLES

Example 9.4 A single-phase half-bridge inverter has a resistive load of $R = 3\ \Omega$ and the dc input voltage $E_{dc} = 24$ Volts. Determine:
  (a) IGBT ratings (b) Total harmonic distortion THD
  (c) The distortion factor DF
  (d) The harmonic factor and the distortion factor of the lowest order harmonic

Solution:

(a) IGBT ratings: (i) Average I_{avg} current $= \dfrac{1}{T} \displaystyle\int_0^{T/2} \dfrac{E_{dc}}{2R}\, dt = \dfrac{E_{dc}}{2RT}(T/2) = \dfrac{E_{dc}}{4R}$

$\therefore$ Average IGBT current $= \dfrac{24}{3 \times 4} = 2$ Amp

(ii) IGBT peak current = $I_{peak} = \dfrac{E_{dc}/2}{R} = 4$ Amp

(iii) Peak reverse blocking voltage V_{BR} of each IGBT,

$$V_{BR} = 2 \times \dfrac{E_{dc}}{2} = 24 \text{ Volts}$$

(b) Total harmonic distortion (THD):

$$\text{THD} = \dfrac{1}{E_{1rms}}\left[\sum_{n=2,3}^{\infty} E_{n_{rms}}^2\right]^{1/2}, \quad \therefore E_{1rms} = \dfrac{2E_{dc}}{\sqrt{2}\,\pi} = 10.8 \text{ V}$$

$$\text{RMS harmonic voltage} = \left[\sum_{n=3,5,7}^{\infty} E_{n_{rms}}^2\right]^{1/2}$$

$$= \sqrt{E_0^2 - E_{1rms}^2} = [12^2 - (10.8)^2]^{1/2} = 5.23 \text{ V}$$

$\therefore \quad \text{THD} = \dfrac{5.23}{10.8} = 0.484 = 48.4\%$

(c) Distortion Factor DF: $= \dfrac{1}{E_{1rms}}\left[\sum_{n=3,5,7}^{\infty}\left(\dfrac{E_{n\,rms}}{n^2}\right)^2\right]^{1/2}$

To determine $\dfrac{E_{n\,rms}}{n^2}$, we have to find $E_{n\,rms}$,

$\therefore \quad E_0 = \sum\limits_{n=1,3,5}^{\infty} \dfrac{2E_{dc}}{n\pi} \sin n\omega t = 0$, for $n = 2, 4, 6$.

$\therefore \quad E_0 = \dfrac{2E_{dc}}{\pi}\sin\omega t + \dfrac{2E_{dc}}{3\pi}\sin 3\omega t + \dfrac{2E_{dc}}{5\pi}\sin 5\omega t + \dfrac{2E_{dc}}{7\pi}\sin 7\omega t + ...$

$$E_{3rms} = \dfrac{2E_{dc}}{3\pi\sqrt{2}} = 3.6 \text{ V}, E_{5rms} = \dfrac{2E_{dc}}{5\pi\sqrt{2}} = 2.16 \text{ V}$$

$$E_{7rms} = 1.54 \text{ V}, E_{9rms} = 1.2 \text{ V},$$

$$E_{11rms} = 0.982 \text{ V}, E_{13rms} = 0.83 \text{ V}$$

$$\left[\sum_{n=3,5,7}^{\infty}\left(\dfrac{E_{n\,rms}^2}{n^2}\right)^2\right]^{1/2} = \left[\left(\dfrac{E_3}{3^2}\right)^2 + \left(\dfrac{E_5}{5^2}\right)^2 + \left(\dfrac{E_7}{7^2}\right)^2 + ...\right]^{1/2}$$

$$= [0.16 + 0.0348 + 2.3 \times 10^{-3} + ...]^{1/2} = 0.44 \text{ V}$$

$\therefore \quad \text{DF} = \dfrac{0.44}{10.8} = 0.041 = 4.1\%$

(d) The lowest harmonic is third harmonic.

$\therefore$ HF for the third harmonic = $\text{HF}_3 = \dfrac{E_{3rms}}{E_{1rms}} = \dfrac{3.6}{10.8} = 33.33\%$

DF of the third harmonic $DF_3 = \dfrac{\left(E_{3\text{rms}}/3^2\right)}{E_{1\text{rms}}} = \dfrac{3.6/9}{10.8} = 0.037 = 3.7\%$.

Example 9.5 A single-phase transistorized bridge inverter has a resistive load of $R = 3\ \Omega$ and the dc input voltage of $E_{dc} = 48$ Volts. Determine:
(a) Transistor ratings (b) Total harmonic distortion
(c) Distortion factor DF
(d) Harmonic factor and distortion factor at the lowest order harmonic

Solution:
Given: $E_{dc} = 48$ V, $R = 3\ \Omega$

(a) Transistor Ratings: (i) Peak transistor current $I_{p1} = \dfrac{E_{dc}}{R} = \dfrac{48}{3} = 16$ Amp

(ii) Average-transistor current $I_{av} = \dfrac{I_P}{2} = 8$ Amp

(iii) Peak reverse blocking voltage, $V_{BR} = 48$ V $= E_{dc}$

(b) The rms harmonic voltage is

$$E_n = \left[E_{0_{\text{rms}}}^2 - E_{1_{\text{rms}}}^2\right]^{1/2}$$

where $E_{0_{\text{rms}}} = 48$ V, $E_{1_{\text{rms}}} = 0.9 \times 48 = 43.2$ V

$\therefore\ E_n = \sqrt{(48)^2 - (43.2)^2} = 20.92$ V and $\therefore$ THD $= \dfrac{20.92}{43.2} = 48.43\%$

(c) $DF = \dfrac{1}{E_{1_{\text{rms}}}} \cdot \left[\displaystyle\sum_{n=3,5,7}^{\infty}\left(\dfrac{E_{n_{\text{rms}}}}{n^2}\right)^2\right]^{1/2}$, Here, $\left[\displaystyle\sum_{n=3,5,7}^{\infty}\left(\dfrac{E_{n_{\text{rms}}}}{n^2}\right)^2\right]^{1/2} = 0.0342$ V

$DF = \dfrac{0.0342}{0.9} = 3.8\%$

(d) The lowest order harmonic is the third harmonic. The rms value of the third harmonic is $E_{3_{\text{rms}}} = E_{1_{\text{rms}}}/3$

$\therefore\ HF_3 = \dfrac{E_{3_{\text{rms}}}}{E_{1_{\text{rms}}}} = 33.33\%$ and $DF_3 = \dfrac{\left(E_{3_{\text{rms}}}/3^2\right)}{E_{1_{\text{rms}}}}$

$= 1/27 = 3.704\%$

9.6 VOLTAGE CONTROL OF SINGLE-PHASE INVERTERS

In many industrial applications, it is often required to vary the output voltage of the inverter due to the following reasons:
  (i) to compensate for the variations in the input voltage.
  (ii) to compensate for the regulation of inverters.
  (iii) to supply some special loads which need variation of voltage with frequency, such as an induction motor.

The various methods for the control of output voltage of inverters are as under:
(a) External control of a.c. output voltage.
(b) External control of d.c. input voltage.
(c) Internal control of inverter.

The first two methods require the use of peripheral components, whereas the third method requires no peripheral components. These methods are now briefly discussed.

9.6.1 External Control of A.C. Output Voltage

In this type of control, as shown in Fig. 9.10, an a.c. voltage controller is inserted between the output terminals of inverter and the load terminals. Through the firing angle control of a.c. voltage controller, the voltage input to the a.c. load is regulated. This method gives rise to higher harmonic content in the output voltage, particularly when the output voltage from the a.c. voltage controller is at low level. Therefore, this method is rarely employed except for low power applications.

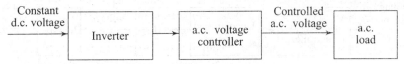

Fig. 9.10 *External control method*

9.6.2 External Control of D.C. Input Voltage

When the available voltage source is a.c., then d.c. voltage input to the inverter is controlled through a fully-controlled rectifier as shown in Fig. 9.11(a), through an uncontrolled rectifier and a chopper as shown in Fig. 9.11(b), or through an a.c. voltage controller and an uncontrolled rectifier as shown in Fig. 9.11(c). In case the available voltage source is d.c., then d.c. voltage input to the inverter is controlled by means of a chopper as shown in Fig. 9.11(d). The main advantages of voltage control schemes of Fig. 9.11 are as follows:
  (i) As the inverter output voltage is controlled through the adjustment of d.c. input voltage to the inverter, output voltage waveform and its harmonic contents are not affected appreciably.
  (ii) If the d.c. input to the inverter is varied to compensate for source voltage fluctuations, the inverter can be designed for a very limited voltage range. Such an inverter is most efficient, both in terms of power loss and component utilisation.

This method of voltage control, however, suffers from the following disadvantages:
  (i) For reducing the ripple content of d.c. voltage input to the inverter, a filter circuit is required in all type of schemes of Fig. 9.11. Filter circuit increases the cost, weight and size, and at the same time reduces efficiency and makes the transient response sluggish.

(ii) In control schemes of Fig. 9.11, the number of power converters used for the control of inverter output voltage varies from two to three. More power handling converter stages result in more losses and reduced efficiency of the entire scheme.

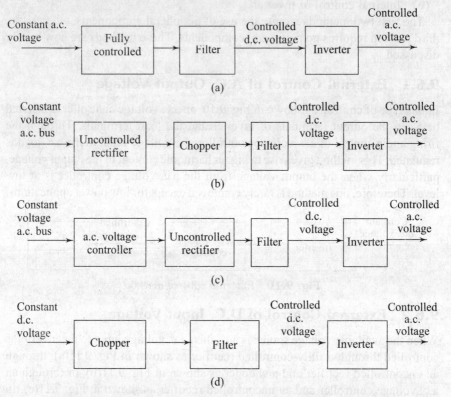

Fig. 9.11 *Voltage control by controlling d.c. input voltage*

(iii) The commutating capacitor voltage decreases as the d.c. input voltage is reduced. This has the effect of reducing the circuit turn-off time for the SCR for a constant load current. Therefore, for a large variation of output voltage for a constant load current, control of d.c. input voltage is not desirable. This limitation can, however, be overcome by a separated fixed d.c. source for charging the commutating capacitor, but this makes the scheme costly and complicated.

9.6.3 Internal Control of Inverter

Inverter output voltage can also be adjusted by exercising a control within the inverter itself. The two possible ways of doing this are:
(a) Series inverter control, and
(b) Pulse-width modulation control.

1. Series Inverter Control This method of voltage control involves the use of two or more inverters in series. Figure 9.12(a) illustrates how the output voltage of two inverters can be summed up with the help of transformers to obtain an adjustable output voltage. In this figure, the inverter output is fed to two transformers whose secondaries are connected in series. Phasor sum of the two voltages E_{L_1}, E_{L_2} gives the resultant voltage E_L as shown in Fig. 9.12(b). The voltage E_L is given by

$$E_L = [E_{L_1}^2 + E_{L_2}^2 + 2 E_{L_1} E_{L_2} \cos \theta]^{1/2}$$

It is essential that the frequency of output voltages E_{L_1}, E_{L_2} from the two inverters is the same. When θ is zero, $E_L = E_{L_1} + E_{L_2}$ and for $\theta = \pi$, $E_L = 0$ in case $E_{L_1} = E_{L_2}$. The angle θ can be varied by the firing angle control of two inverters. The series connection of inverters, called multiple inverter control, does not augment the harmonic content even at low output voltage levels.

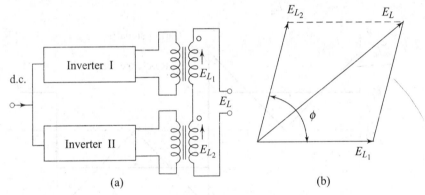

Fig. 9.12 *Internal control of inverters by their series connection*

2. Pulse-width Modulation Control The most efficient method of controlling the output voltage is to incorporate pulse width modulation control (PWM control) within the inverters. In this method, a fixed d.c. input voltage is supplied to the inverter and a controlled a.c. output voltage is obtained by adjusting the on-and-off periods of the inverter devices. The PWM control has the following advantages:
  (i) The output voltage control can be obtained without any additional components.
  (ii) With this type of control, lower order harmonics can be eliminated or minimised along with its output voltage control. The filtering requirements are minimised as higher order harmonics can be filtered easily.

The main drawback of this method is that the SCRs used in this method must have very low turn-on and turn-off times (inverter-grade SCRs), therefore, they are very expensive.

The commonly used PWM control techniques are:
(a) Single-pulse width modulation (SPWM)
(b) Multiple-pulse width modulation (MPWM)
(c) Sinusoidal pulse width modulation (sin PWM)

9.6.3.1 Single-pulse Width Modulation

In single-pulse width modulation control, there is only one pulse per half-cycle and the width of the pulse is varied to control the inverter output voltage. The generation of gating signals and output voltage of single-phase full bridge inverters is shown in Fig. 9.13. As shown in Fig. 9.13, the gating signals are generated by comparing a rectangular reference signal of amplitude, E_R, with a triangular carrier wave of amplitude E_c. The fundamental frequency of output voltage is determined by the frequency of the reference signal. The pulse-width, P, can be varied from $0°$ to $180°$ by varying E_R from 0 to E_c. The ratio of E_R to E_c is the control variable and is defined as the *amplitude modulation index*. The amplitude modulation index, or simply modulation index is

$$M = \frac{E_R}{E_c} \tag{9.30}$$

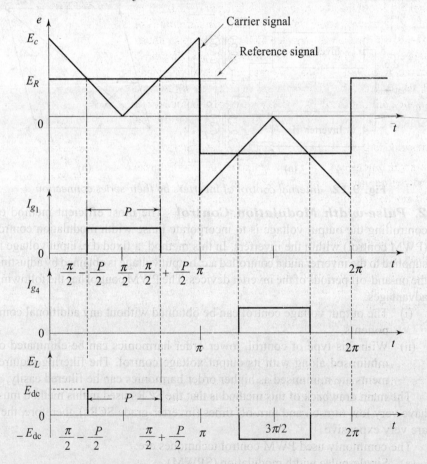

Fig. 9.13 *Single pulse-width modulation*

Inverters

The following Fourier-series describes the waveform of E_L as

$$E_L = \sum_{n=1,3,5,\ldots}^{\infty} A_n \sin n\omega t + \sum_{n=1,3,5,\ldots}^{\infty} B_n \sin n\omega t \qquad (9.31)$$

where
$$A_n = \frac{2}{\pi}\int_0^{\pi} E_{dc} \sin n\omega t \, d(\omega t) = \frac{2}{\pi}\int_{(\pi/2-p)}^{(\pi/2+p)} \sin n\omega t \, d(\omega t).$$

$$= \frac{4 E_{dc}}{5\pi} \sin \frac{np}{2}$$

and
$$B_n = \frac{2 E_{dc}}{\pi}\int_{(\pi/2-p)}^{(\pi/2+p)} \cos n\omega t \, d(\omega t) = 0 \qquad (9.32)$$

Thus,
$$E_L = \sum_{n=1,3,5,\ldots}^{\infty} \frac{4 E_{dc}}{n\pi} \sin \frac{np}{2} \sin n\omega t \qquad (9.33)$$

When pulse-width P is equal to its maximum value of π radians, then the fundamental component of output voltage E_L, from Eq. (9.33), has the peak value of

$$E_{L1m} = \frac{4 E_{dc}}{\pi} \qquad (9.34)$$

The RMS output voltage can be found from

$$E_{Lrms} = \left[\frac{2}{\pi}\int_{(\pi-p)/2}^{(\pi+p)/2} E_{dc}^2 \, d(\omega t)\right]^{1/2} = E_{d.c} \cdot \sqrt{\frac{P}{\pi}} \qquad (9.35)$$

The peak value of the nth harmonic component from Eq. (9.33) is given by

$$E_{Lnm} = \frac{4 E_{dc}}{n\pi} \sin \frac{np}{2} \qquad (9.36)$$

From Eqs (9.34) and (9.36), $\dfrac{E_{Lnm}}{E_{L1m}} = \dfrac{\sin \frac{np}{2}}{n} \qquad (9.37)$

The ratio as given by Eq. (9.37) is plotted in Fig. 9.14 for $n = 1$, $n = 3$, $n = 5$, 7 for different pulse widths. From these curves it may be observed that when the fundamental component is reduced to nearly 0.33, the amplitude of the third harmonic is also 0.33. When fundamental component is reduced to about 0.143, all the three harmonics (3, 5, 7) become almost equal to the fundamental. This shows that in this type of voltage control scheme, as great deal of harmonic content is introduced in the output voltage, particularly at low output voltage levels.

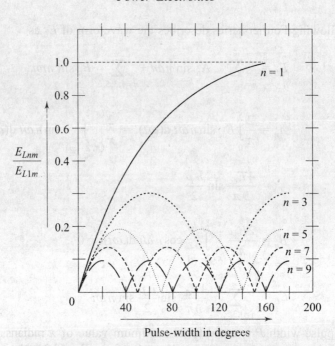

Fig. 9.14 *Harmonic content in SPWM*

9.6.3.2 Multiple Pulse-width Modulation

In this method of pulse-width modulation, the harmonic content can be reduced using several pulses in each half-cycle of output voltage. By comparing a reference signal with a triangular carrier wave, the gating signals are generated for turning-on and turning-off of a thyristor, as shown in Fig. 9.15(a). The carrier frequency, f_c, determines the number of pulses per half-cycle, m, whereas the frequency of reference signal sets the output frequency, f_o. The modulation index controls the output voltage. This type of modulation is also known as symmetrical pulse width modulation (SPWM). The number of pulses N_p per half-cycle is found from the expression

$$N_p = \frac{f_c}{2f_0} = \frac{m_f}{2} \qquad (9.38)$$

where $m_f = \dfrac{f_c}{f_0}$ is the frequency modulation ratio. The variation of modulation index (M) from 0 to 1 varies the pulse width from 0 to π/N_p and the output voltage from 0 to $E_{d.c.}$. For SPWM, the output voltage for single-phase bridge inverters is shown in Fig. 9.15(b).

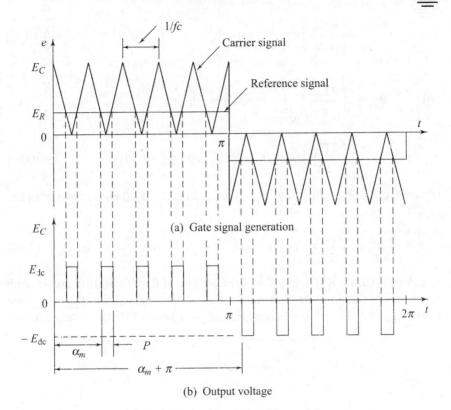

Fig. 9.15 *Multiple-pulse width modulation*

If P is the width of each pulse, the RMS output voltage can be obtained from the following expression:

$$E_{L(rms)} = \left[\frac{2N_p}{2\pi} \int_{(\pi/N_p - P)/2}^{(\pi/N_p + P)/2} E_{dc}^2 \, d(\omega L) \right]^{1/2} = E_{dc} \sqrt{\frac{N_p \cdot P}{\pi}} \qquad (9.39)$$

The general expression for various harmonics in the output voltage is obtained by deriving an expression for a general pair of pulses, such that the positive pulse of duration P starts at $\omega t = \alpha$ and the negative one of the same width starts at $\omega t = \pi + \alpha$. This is shown in Fig. 9.15(b). The effects of all pulses can be combined together to obtain the effective output voltage. Thus for this pair of pulses,

$$A_{nm} = \frac{2E_{dc}}{\pi} \int_{\alpha_{m-p/2}}^{\alpha_{m+p/2}} \sin n\,\omega t \, d(\omega t)$$

$$= \frac{2E_{dc}}{2\pi} [\cos n\,(\alpha m + p/2) - \cos n\,(\alpha_m - p/2)]$$

$$= \left(\frac{4E_{dc}}{n\pi}\right) \sin n\frac{p}{2} \cdot \sin n\alpha_m \qquad (9.40(a))$$

and $\quad B_{nm} = \left(\dfrac{2E_{dc}}{\pi}\right) \displaystyle\int_{\alpha_{m-p/2}}^{\alpha_{m+p/2}} \cos n\,\omega t\, d(\omega t)$

$$= \left(\frac{2E_{dc}}{2\pi}\right)[\sin(\alpha_m + p/2) - \sin(\alpha_m - p/2)] \qquad (9.40(b))$$

If there are K pulses situated at $\alpha_1, \alpha_2, \alpha_3, \ldots, \alpha_m, \ldots, \alpha_k$, then we have from Eq. (9.40),

$$A_n = \frac{4E_{dc}}{n\pi}\sin n\frac{p}{2}\sum_{m=1}^{k}\sin n\alpha_m \quad \text{and} \quad B_n = \frac{4E_{dc}}{n\pi}\cos n\frac{p}{2}\sum_{m=1}^{k}\sin n\alpha_m \qquad (9.41)$$

In Fig. 9.16, for $K = 3$ and $K = 10$, amplitude of first, third, fifth and seventh harmonics as a ratio of the maximum value of fundamental (i.e. E_{Lnm}/E_{L1m}) are plotted against the pulse width expressed as a ratio of distance between two

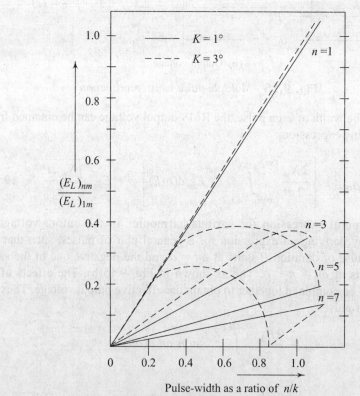

Fig. 9.16 *Harmonic content of MPWM*

adjacent pulses, i.e. ($P/\pi/k$). As the number of pulses per half-cycle (i.e. k) is increased, the considerable reduction in lower order harmonics is achieved, as shown in Fig. 9.16. For example, with $K = 10$, substantial reduction in third, fifth and seventh harmonics is achieved in the complete range of the output voltage. With larger values of N_p, the amplitudes of lower order harmonics would be lower, but the amplitudes of some higher order harmonics would increase. However, such higher order harmonics produce negligible ripple or can easily be filtered out.

With this method, since voltage control is achieved with a simultaneous reduction of lower order harmonics, this scheme is comparatively advantageous over single-pulse modulation. However, due to larger number of pulses per half-cycle, frequent turning-on and turning-off of thyristors is required which increases the switching losses. Also, for this scheme inverter-grade thyristors are required which are costly.

9.6.3.3 Sinusoidal Pulse Width Modulation

In this method of modulation, several pulses per half-cycle are used as in the case of multiple pulse width modulation. Instead of maintaining the width of all pulses the same as in the case of multiple-pulse modulation, the width of each pulse is varied proportional to the amplitude of a sine-wave evaluated at the centre of the same pulse. By comparing a sinusoidal reference signal with a triangular carrier wave of frequency, f_c, the gating signals are generated, as shown in Fig. 9.17(a). The frequency of reference signal, f_r, determine the inverter output frequency, f_o, and its peak amplitude, E_r, controls the modulation index, M, and then in turn the RMS output voltage, E_L. The number of pulses per half-cycle depends on the carrier frequency. Within the constraint that two thyristors of the same arm (T_1, T_4) cannot conduct at the same time, the instantaneous output voltage is shown in Fig. 9.17(a). The same gating signals can be generated using unidirectional triangular carrier-wave as shown in Fig. 9.17(b).

By varying the modulation index M, the RMS output voltage can be varied. It can be observed that the area of each pulse corresponds approximately to the area under the sine-wave between the adjacent midpoints of OFF periods on the gating signals. If P_m is the width of the mth pulse, Eq. (9.39) can be extended to find the rms output voltage

$$E_L = E_{dc} \left(\sum_{m=1}^{N_p} \frac{P_m}{\pi} \right)^{1/2} \qquad (9.42)$$

Harmonic analysis of the output modulated voltage wave reveals that SPWM has the following important features:

(i) For modulation index less than one, the largest harmonic amplitudes in the output voltage are associated with harmonics of order $f_c/f_r \pm 1$ or $2N_p \pm 1$, where N_p is the number of pulses per half-cycle. Thus, by increasing the number of pulses per half-cycle, the order of dominant

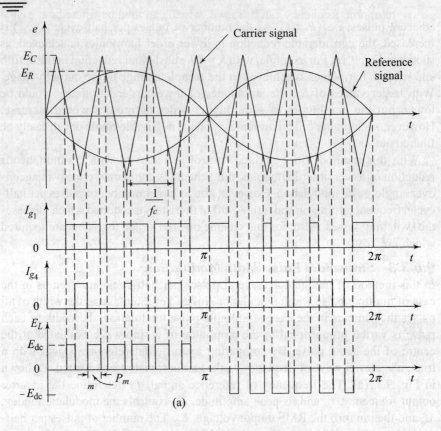
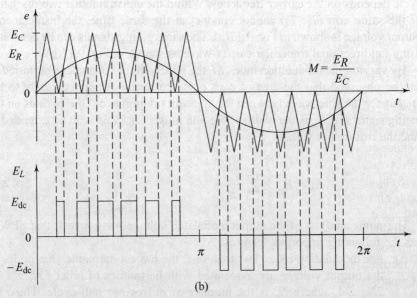

Fig. 9.17 *Sinusoidal pulse-width modulation*

harmonic frequency can be raised, which can then be filtered out easily. For $N_p = 5$, harmonics of the order of 9 and 11 become significant in the output voltage. It may be noted that the highest order of significant harmonic of modulated voltage-wave is centred around the carrier frequency, f_c.

(ii) For modulation index greater than one, lower order harmonics appear since for modulation index greater than one, pulse width is no longer a sinusoidal function of the angular position of the pulse.

9.7 PULSE-WIDTH MODULATED (PWM) INVERTERS

Square-wave inverters suffers from two major drawbacks:

(i) The output voltage of the inverter cannot be controlled for a fixed-source voltage. To achieve voltage control, the inverter must be fed either from a controlled ac-dc or dc-dc converter.

(ii) The output voltage contains appreciable harmonics (low-frequency range). Also, THD is very high.

Due to these drawbacks, square-wave inverter is rarely used in practice. To achieve voltage control within the inverter and to reduce the harmonic contents in the output voltage, PWM inverters are used. In PWM inverters, width of the output pulses are modulated to achieve the voltage control. PWM technique allows:

(i) Variation of output voltage within the inverter by varying the gain of the inverter. This allows the input d.c. voltage to be of fixed amplitude.

(ii) Variation of output frequency either by varying the number of pulses per half cycle of the output or by varying the period for each half-cycle with fixed number of pulses in each half-cycle.

(iii) Simultaneous variation of output voltage and frequency is also possible. So that V/F ratio can be kept constant. This feature is required in induction motor-drives.

(iv) Control of harmonics at the output of the inverter.

9.7.1 Pulse-Width Modulated Half-Bridge Inverters

With a half-bridge inverter, an output voltage of zero is not possible—the output voltage can be only positive or negative. Therefore, the output voltage is allowed to reverse instead of being zero. Figure 9.18 shows a PWM-waveform for a half-bridge inverter. We can control the output voltage by controlling the width 2δ.

Sinusoidal PWM (Fig. 9.19) is commonly used with half-bridge inverters. A rectified sinusoidal reference signal is compared with a triangular carrier-wave. For the time during which the reference

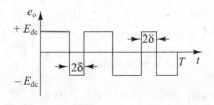

Fig. 9.18 PWM in half-bridge inverter

signal is higher than the carrier-wave, the switches are operated to produce positive-going pulses; otherwise, negative-going pulses are produced:

When $\quad\quad\quad E_R > E_C$, S_1 is ON and $e_0 = + E_{dc}$
When $\quad\quad\quad E_R < E_C$, S_2 is ON and $e_0 = - E_{dc}$

Switch conduction is also shown in Fig. 9.19.

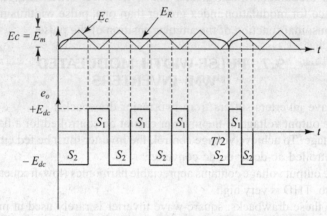

Fig. 9.19 *Sinusoidal PWM (S_1, S_2 are switches in half-bridge inverter)*

Let us define a parameter M_f, called the *carrier frequency ratio* as

$$M_f = \frac{\text{Frequency of the carrier signal}}{\text{Frequency of the modulating signal}} = \frac{f_c}{f_m} \geq 1 \quad\quad (9.43)$$

Circuit Analysis:-

(i) Fundamental Output Voltage The fundamental output voltage can be very easily found by assuming that the carrier ratio is quite high. Fundamental output voltage is proportional to the instantaneous modulation index and to the peak-value of the output voltage ($E_{dc}/2$).

$$\therefore \quad E_{0_{(\text{fund.})}} = \text{Instantaneous modulation index} \times \text{Peak value of output voltage}$$

$$= \frac{E_m \sin w_m t}{E_c} \cdot \frac{E_{dc}}{2} = \frac{E_m}{E_c} \cdot \frac{E_{dc}}{2} \sin w_m t$$

but $\quad\quad\quad M = \dfrac{E_m}{E_c}$

$$\therefore \quad E_{0_{(\text{fund.})}} = M \cdot \frac{E_{dc}}{2} \cdot \sin w_m t, \, M \leq 1 \quad\quad (9.44)$$

But, $\quad\quad\quad E_{0_{(\text{fund.})}} = M \cdot \dfrac{E_{dc}}{2\sqrt{2}} \quad\quad\quad\quad\quad\quad\quad\quad (9.45)$

$$\therefore \quad E_{0_{(\text{fund.})}} = \sqrt{2} \, E_{0_{(\text{fund.})}} \cdot \sin(w_m t) \quad\quad (9.46)$$

Maximum value of fundamental output voltage occurs at
$M = 1$, $\therefore$ Equation (9.45) $\Rightarrow$

$$\therefore \quad E_{0_{(fund.)}} = \frac{E_{dc}}{2\sqrt{2}} = 0.707 \cdot \frac{E_{dc}}{2} \qquad (9.47)$$

(ii) Inverter Gain

$$\text{Gain} = \frac{\text{Fundamental output voltage}}{\text{D.C. Input voltage}} = \frac{0.707 \cdot M \, E_{dc}/2}{E_{dc}/2}$$

$$E_G = 0.707 \cdot M \qquad (9.48)$$
$$E_{G(max)} = 0.707 \text{ when } M = 1 \qquad (9.49)$$

(iii) RMS Output Voltage
RMS output voltage is given by

$$E_{0(rms)} = \frac{E_{dc}}{2} \qquad (9.50)$$

The general equation for the rms value is given by

$$E_{0(rms)} = \sqrt{E_{0(fund)}^2 + E_{0(2)}^2 + E_{0(3)}^2 \ldots} = \sqrt{E_{0(fund)}^2 + \sum_{n=2}^{\infty} E_0^2(n)}$$

$$\therefore \quad \sum_{n=2}^{\infty} E_{0(n)}^2 = \sqrt{E_{0(rms)}^2 - E_{0(fund)}^2} \qquad (9.51)$$

Substitute the $E_{0(rms)}$ value from (9.50) and $E_{0(fund)}$ value from (9.45) in Eq. (9.51), we get

$$E_n = \sqrt{\left(\frac{E_{dc}}{2}\right)^2 - \left(\frac{M}{\sqrt{2}}\right)^2 \cdot \left(\frac{E_{dc}}{2}\right)^2} = \frac{E_{dc}}{2}\sqrt{\frac{2-M^2}{2}} \qquad (9.52)$$

where $\quad E_n = \sqrt{\sum_{n=2}^{\infty} E_{0(n)}^2} \qquad (9.53)$

Thus the amplitude of the harmonics are independent of M_F.

(iv) Distortion and Harmonic Factor

$$\text{Distortion factor} \quad DF = \frac{E_{0(fund)}}{E_{0(rms)}} = \frac{0.707 \cdot M \cdot E_{dc}/2}{E_{dc}/2}$$

$$= 0.707 \, M \qquad (9.54)$$
$$DF_{max} = 0.707 \text{ when } M = 1 \qquad (9.55)$$

$$\text{Harmonic Factor} \quad HF = \sqrt{\frac{1}{DF^2} - 1} = \sqrt{\frac{2}{M^2} - 1} \qquad (9.56)$$

$$HF_{max} = 100\% \text{ when } M = 1 \qquad (9.57)$$

(v) Harmonics at the Output The carrier ratio as defined earlier is given by
$$M_F = f_c/f_m = 2 \cdot p$$
where p is the number of pulses per half-cycle. The harmonic frequencies present at the output can be expressed as
$$f_n = k_1 \cdot f_c + k_2 \cdot f_m$$
where f_n = frequency of the nth harmonic

$$\therefore \quad \frac{f_n}{f_m} = k_1 \cdot \frac{f_c}{f_n} + k_2$$

The order of the harmonic 'n' can be written as
$$n = \frac{f_n}{f_m} = k_1 \cdot m_f + k_2 \qquad (9.58)$$

The carrier ratio is usually chosen as odd number. The waveform then will have a quarter-wave symmetry and only odd harmonics are present. This is one of the requirements of a PWM signal. Now only odd harmonics are present hence if k_1 is odd then k_2 is even and vice versa. Therefore, the harmonic present at the output are

$$\left.\begin{array}{l} n = M_f,\ M_{f\pm 2},\ M_{f\pm 4},\ M_{f\pm 6},\ \dots \\ n = 2M_{f\pm 1},\ 2M_{f\pm 3},\ 2M_{f\pm 5},\ \dots \\ n = 3M_f,\ 3M_{f\pm 2},\ 3M_{f\pm 4},\ 3M_{f\pm 6}\ \dots \end{array}\right\} \qquad (9.59)$$

Frequency spectrum of the PWM signal is shown in Fig. 9.20. The sidebands can be clearly seen. The centre frequencies are $M_f, 2M_f, 3M_f, \dots$ The amplitude

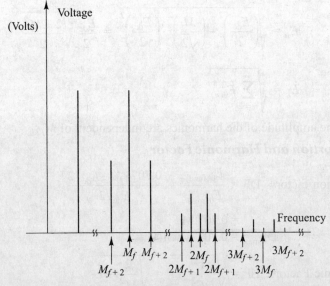

Fig. 9.20 *Frequency-spectrum for bipolar sinusoidal PWM-output in half-bridge-inverter*

Inverters

of the harmonics corresponding to the centre frequency M_f, $3M_f$, $5M_f$, are noticeable, since they are odd harmonics. However, since $2M_f$, $4M_f$ ... etc. are even harmonics, their amplitudes are negligible. For any switching frequency or carrier-ratio M_f, the amplitude of the harmonics is given by the same Eq. (9.53) and the harmonic factor is independent of M_f. However, the order of the most significant harmonics M_f, $M_{f\pm2}$ is raised and the filtering becomes easier. The cost and size of the filter decreases. However, increasing the carrier ratio M_f, increases the switching frequency of the transistor, which increases the switching loss and the cost of the transistor. This decreases the inverter efficiency.

SOLVED EXAMPLES

Example 9.6 A single-phase half-bridge bipolar PWM inverter is operated from a centre tap 240 volts DC supply. The fundamental output frequency is adjusted to 50 Hz. The carrier frequency used is 1.2 kHz. Modulation index is adjusted to 0.8. Determine:
(a) Carrier ratio (M_f) and the number of pulses per cycle
(b) Fundamental output voltage
(c) Distortion and the harmonic factor of the output voltage waveform.
(d) The order of the first five significant harmonics.

Solution:

Given: $\dfrac{E_{dc}}{2} = 240$ V, $f_m = 50$ Hz, $F_c = 1.2$ kHz, $M = 0.8$

(a) $2p = M_f = \dfrac{f_c}{f_m} = \dfrac{1.2 \times 10^3}{50} = 24$

(b) From Eq. (9.45), $E_{0(\text{fund})} = \dfrac{1}{\sqrt{2}} \cdot M \cdot \dfrac{E_{dc}}{2} = 0.707 \, M \, \dfrac{E_{dc}}{2}$

$\qquad = 0.707 \times 0.8 \times 240 = 135.74$ V

(c) From Eq. (9.50), $E_{0_{rms}} = \dfrac{E_{dc}}{2} = 240$ V

Distortion factor DF $= \dfrac{E_{0(\text{fund})}}{E_{0(\text{rms})}} = \dfrac{135.74}{240} = 0.5655 = 56.55\%$

Harmonic factor, HF $= \sqrt{\dfrac{1}{DF^2} - 1} = \sqrt{\dfrac{1}{(0.5655)^2} - 1} = 1.458 = 145.8\%$

$\qquad = 1.458 = 145.8\%$

(d) Harmonics are given by Eq. (9.59)

$\qquad n = M_f, M_{f\pm2}, M_{f\pm4}, M_{f\pm6}, \ldots$
$\qquad n = 2M_{f\pm1}, 2M_{f\pm3}, 2M_{f\pm5}, \ldots$
$\qquad n = 3M_f, 3M_{f\pm2}, 3M_{f\pm4}, 3M_{f\pm6}, \ldots$

$$n = 4M_{f\pm 1}, 4M_{f\pm 3}, 4M_{f\pm 5}, \ldots$$
$$n = 5M_f, 5M_{f\pm 2}, 5M_{f\pm 4}, 5M_{f\pm 6}, \ldots$$

Thus, the significant harmonics are:

$$n = 24, 20, 22, \quad n = 47, 49, 45, 51, \quad n = 72, 70, 74, \quad n = 95, 97$$
$$n = 120, 118, 122 \text{ etc.}$$

Example 9.7 A half-bridge inverter is operated in a bipolar PWM scheme and is fed from a centre-tap dc supply of $E_{dc}/2$ each. The fundamental output at modulation index of 0.8 must be 220 V. Compute the dc and the switch voltage ratings.

Solution:

Given: $\quad E_{0(\text{fund.})} = 220 \text{ V}$

From Eq. (9.39), the fundamental output voltage = $0.707 \, M \, \dfrac{E_{dc}}{2}$

$$\therefore \quad \frac{E_{dc}}{2} = \frac{E_{0(\text{fund})}}{0.707 \cdot M} = \frac{220 \, V}{0.707 \times 0.8} = 388.97 \text{ Volts}$$

Switch voltage rating, $V_{CEO} \geq 2 \times \dfrac{E_{dc}}{2} \geq 778 \text{ V}$

9.7.2 PWM Full-Bridge Inverters

In a PWM inverter, the output voltage waveform has a constant amplitude whose polarity reverses periodically to provide the output fundamental frequency. The source voltage is switched at regular intervals to produce a variable output voltage. The output voltage of the inverter is controlled by varying the pulse-width of each cycle of the output voltage. A full-bridge inverter is shown in Fig. 9.21. As shown, the bridge inverter can be considered to be a combination of two half-bridge circuits. The first half-bridge consists of two switches S_1 and S_4 whereas the second-one consists of the switches S_2 and S_3. The load voltage E_{PQ} is the difference between the output voltage E_{PO} and E_{QO} of the individual half-bridge circuits.

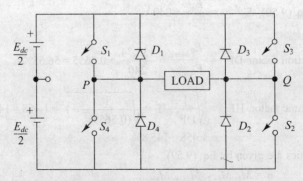

Fig. 9.21 *Full-bridge inverter*

9.7.2.1 PWM with Bipolar Voltage Switching

The PWM bipolar switching scheme used with half-bridge inverter can be used more efficiently with the full-bridge inverter.

As shown in Fig. 9.22, a triangular wave of peak-amplitude 'E_c' is compared with a sine-wave of peak amplitude 'E_m' to generate the base-drives for the two devices in the half-bridge circuit (S_1 and S_4). Base-drives for S_2 and S_3 are exactly 180° out of phase to those of S_4 and S_1 respectively. Thus, S_1 and S_2 conduct simultaneously to make the instantaneous load voltage $+E_{dc}$ and S_3, S_4 then conduct simultaneously to make the instantaneous load voltage $-E_{dc}$. The load-voltage waveform is a bipolar PWM waveform with a peak voltage of $\pm E_{dc}$ volts as shown in Fig. 9.22.

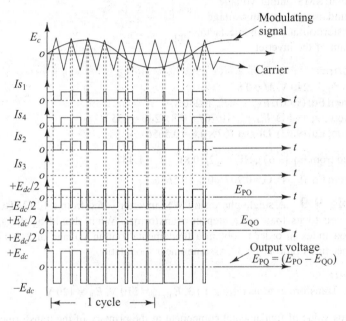

Fig. 9.22 *PWM with Bipolar-voltage switching*

Circuit Analysis If the load voltage waveform for half-bridge and full-bridge PWM inverters are compared, then it will be observed that they are identical except for the fact that peak-voltage for full-bridge inverter is $+E_{dc}$ volts instead of ($E_{dc}/2$). Therefore, the analysis will proceed on the same lines as that for the half-bridge inverter to yield the following results.

(i) RMS value of output $E_{0(rms)} = E_{dc}$ (9.60)
(ii) RMS value of the fundamental $E_{0(fund)} = 0.707\, M \cdot E_{dc}$ (9.61)
(iii) Distortion Factor, DF = $0.707 \cdot M$ (9.62)
(iv) Harmonic-factor, Hf = $\sqrt{2/M^2 - 1}$ (9.63)
(v) Gain of inverter, $G = 0.707 \cdot M$ (9.64)

(vi) Dominant harmonics: Dominant harmonics for M_f = odd are again similar to half-bridge inverter and are

$$\left. \begin{array}{l} n = M_f, M_{f\pm 2}, M_{f\pm 4}, M_{f\pm 6}, \ldots \\ = 2M_{f\pm 1}, 2M_{f\pm 3}, 2M_{f\pm 5}, \ldots \\ = 3M_f, 3M_{f\pm 2}, 3M_{f\pm 4}, \ldots \end{array} \right\} \qquad (9.65)$$

SOLVED EXAMPLES

Example 9.8 A full-bridge bipolar PWM inverter is fed from a 240 V battery and is driving an RL load. Compute the following if modulation index $M = 0.8$:
  (i) Total RMS output voltage
  (ii) Fundamental output voltage
  (iii) Distortion and harmonic factor
  (iv) Gain of the inverter

Solution:
  Given: $E_{dc} = 240$ V, $M = 0.8$
  (i) From Eq. (9.60), $E_{0(rms)} = E_{dc} = 240$ V
  (ii) From Eq. (9.61), $E_{0(fund)} = 0.707 \times 0.8 \times 240 = 135$ V
  (iii) From Eq. (9.62), DF $= 0.707 \times 0.8 = 0.5656$

  and from Eq. (9.63), HF $= \sqrt{2/(0.8)^2 - 1} = 1.457$
  (iv) From Eq. (9.64), G $= 0.707 \times 0.8 = 0.5656$

Example 9.9 A single-phase sin PWM inverter is fed from a 120 V DC bus and is connected to its load via a step-up transformer. What should be the amplitude modulation index in order to obtain a fundamental output voltage of 210 V_{rms} if the transformer turns ratio is 1 : 3? Assume linear modulation.

Solution:
  Given: Transformer turns ratio is 1 : 3, $E_{01rms} = 210$ V, $E_{dc} = 120$ V

  The rms value of fundamental component at the primary of the transformer $= \dfrac{210}{3}$
  $= 70$.
  Now from Eq. (9.61), $E_{0(fund)} = 0.707 M \cdot E_{dc}$, $70 = 0.707 \times M \times 120$ $\therefore$ $M = 0.83$

9.7.2.2 PWM with Unipolar PWM Switching

Waveforms for unipolar PWM bridge inverter are shown in Fig. 9.22. In unipolar PWM switching, the polarity of the PWM output voltage remains positive or negative for half-cycle period of output as shown in Fig. 9.22. For example, in positive half-cycle the output voltage polarity is either ($+E_{dc}$) or zero. Similarly, in the negative half-cycle, the output polarity is either ($-E_{dc}$) or zero. This is called as three-level PWM, since output takes three voltage levels, $+E_{dc}, 0, -E_{dc}$, Fig. 9.22.

Unipolar PWM waveform is generated as follows:
(i) In full-bridge unipolar PWM-inverter, the two half-bridges are given two separate control signals.
(ii) Bipolar triangular wave of peak-amplitude is compared with two sinusoidal modulating signals which are 180° out of phase, as shown in Fig. 9.23.

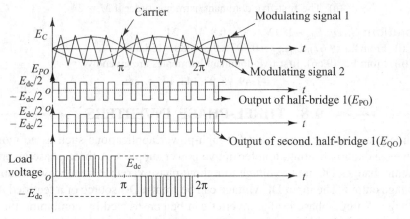

Fig. 9.23 *Waveforms for unipolar PWM bridge inverter*

(iii) The base-driving waveforms for the switches S_1 and S_4, which forms the first-half-bridge inverter are generated by comparing the triangular waveform with the first sinusoidal modulating signal. Due to this, a bipolar PWM waveform is generated at the output of the first half-bridge (V_{P0}).
(iv) The base-driving signals for switches S_2 and S_3 which forms the second half-bridge inverter are generated by comparing the triangular waveform with second modulating signal (out of phase modulating signal). Due to this a bipolar PWM signal is generated at the output of second half-bridge inverter (E_{QO}).
(v) The output voltage of the bridge inverter = $E_{PO} - E_{QO} = E_{PQ}$

The output voltage is generated by using the following logic:
If switches S_1 and S_2 are ON, $E_{PQ} = + E_{dc}$
If switch S_3 and S_4 are ON, $E_{PQ} = - E_{dc}$
If S_1 and S_3 are ON, $E_{PQ} = 0$
If S_4 and S_2 are ON, $E_{PQ} = 0$

Harmonic Spectrum and Fundamental Output In unipolar PWM, due to the 180° phase-shift between two reference signals, the carrier frequency at the output is effectively doubled. Hence, the most significant harmonic is twice that of bipolar PWM. The most significant harmonics selecting M_f as even are given by

$$n = 2M_{f \pm 1}, 4M_{f \pm 1}, 6M_{f \pm 1} \qquad (9.66)$$

and the fundamental output (RMS) value is given by

$$E_{0(\text{fund})} = 0.707 \cdot M \cdot E_{dc}, \text{ for } 0 \leq M \leq 1 \qquad (9.67)$$

SOLVED EXAMPLES

Example 9.10 A full-bridge inverter is operated from a dc supply of 280 V and in unipolar PWM mode.
Determine: (i) Fundamental output voltage at $M = 0.8$
(ii) The first-five dominant harmonic order if $M_f = 24$

Solution: Given: $E_{dc} = 280$ V, $M = 0.8$, $M_f = 24$
(i) From Eq. (9.67), $E_{0(fund)} = 0.707 \times 0.8 \times 280 = 158.36$ V
(ii) From Eq. (9.66), first five dominant harmonics are given by
$$n = 47, 49, 95, 97, 143, 145, \ldots \text{etc.}$$

9.8 THREE-PHASE INVERTERS

Three-phase inverters are used for high-power applications such as ac motor drives, induction heating, uninterruptive power supplies. A three-phase inverter circuit changes DC input voltage to a three-phase variable frequency, variable-voltage output. The input DC voltage can be from a DC source or a rectified AC voltage. A three-phase bridge inverter can be constructed by combining three-single-phase half-bridge inverters. Figure 9.24 shows the basis circuit of three-

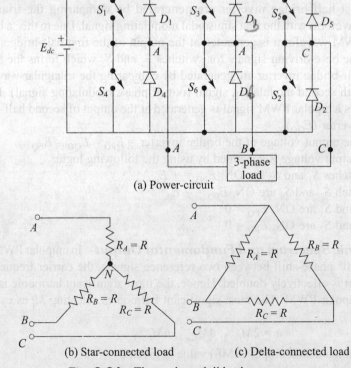

(a) Power-circuit

(b) Star-connected load

(c) Delta-connected load

Fig. 9.24 *Three-phase full-bridge inverter*

phase bridge inverter. As shown, it consists of six power-switches with six associated freewheeling diodes. The switches are opened and closed periodically in the proper sequence to produce the desired output waveform. The rate of switching determines the output frequency of the inverter.

Basically, there are two possible schemes of gating the devices. In one scheme, each device (switch) conducts for 180° and in the other scheme, each device conducts for 120°. But in both these schemes, gating signals are applied and removed at 60° intervals of the output voltage waveform. These modes of device conduction are described in the following subsection:

9.8.1 180°–Conduction Mode with Resistive Load

In this control scheme, each switch conducts for a period of 180° or half-cycle electrical. Switches are triggered in sequence of their numbers with an interval of 60°. At a time, three switches (one from each leg) conduct. Thus, two switches of the same leg are prevented from conducting simultaneously. One complete cycle is divided into six modes, each of 60° intervals. The operation of the circuit can be understood from the waveforms shown in Fig. 9.25 and the operation Table 9.1.

Switch pair in each leg, i.e. S_1, S_4, S_3 S_6, and S_5, S_2 are turned-on with a time interval of 180°. It means that switch S_1 conducts for 180° and switch S_4 for the next 180° of a cycle. Switches, in the upper group, i.e. S_1, S_3, S_5 conduct at an interval of 120°. It means that if S_1 is fired at 0°, then—S_3 must be triggered at 120° and S_5 at 240°. Same is true for lower group of switches. On the basis of this gating scheme, Table 9.1 is prepared.

Table 9.1 Operation Table

S.No.	Interval	Device conducting	Incoming device	Outgoing device
1	I	5, 6, 1	1	4
2	II	6, 1, 2	2	5
3	III	1, 2, 3	3	6
4	IV	2, 3, 4	4	1
5	V	3, 4, 5	5	2
6	VI	4, 5, 6	6	3

The following points can be noted from the wave forms (Fig. 9.25) and the operating Table 9.1,
  (i) Each switch conducts for a period of 180°.
 (ii) Switches are triggered in the sequence 1, 2, 3, 4, 5, and 6.
(iii) Phase shift between triggering the two adjacent switches is 60°.
(iv) From table, it is observed that in every step of 60° duration, only three switches are conducting—two from upper group and one from the lower group and vice-versa.

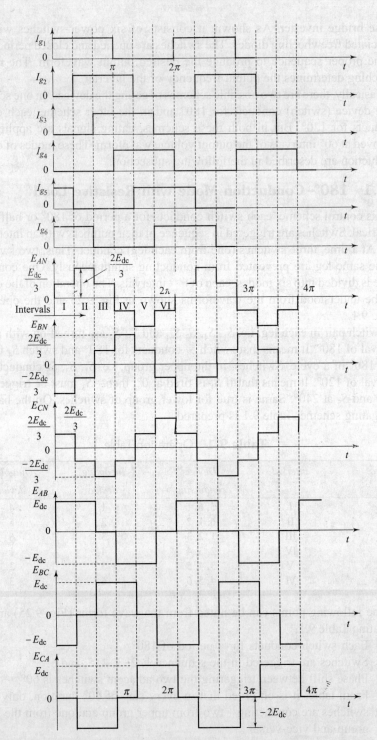

Fig. 9.25 *Voltage waveforms for 180° conduction*

(v) The output voltage waveforms (E_{AB}, E_{BC}, E_{CA}) are quasi-square-wave with a peak-value of E_{dc}. The three-line voltages are mutually phase-shifted by 120°.

(vi) The three-phase-voltages E_{AN}, E_{BN}, and E_{CN} are six-step saves, with step heights $\dfrac{E_{dc}}{3}$ and $\dfrac{2}{3} E_{dc}$.

(vii) Line voltage E_{AB} is leading the phase-voltage E_{AN} by 30°.

In Fig. 9.25, phase voltages E_{AN}, E_{BN} and E_{CN} have also been drawn for star-connected resistive load. For a star-connected load, the line-to-neutral voltages must be determined to find the line or phase currents. There are three modes of operation in a half-cycle and the equivalent circuits are shown in Fig. 9.26 for a star-connected load.

From Figs (9.25) and (9.26), it is observed that

(i) During interval I for $0 \le \omega t < \pi/3$,

$$Req = R_B + (R_A \parallel R_C)$$

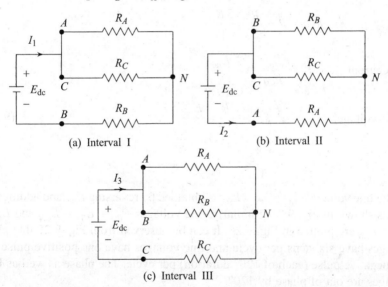

Fig. 9.26 *Equivalent circuits for star-connected resistive load*

$$= R + \frac{R}{2} = \frac{3R}{2} \quad (\because R_A = R_B = R_C)$$

Current, $\quad I_1 = \dfrac{E_{dc}}{Req} = \dfrac{2 E_{dc}}{3R}$

Now, $\quad E_{AN} = E_{CN} = \dfrac{I_1 \cdot R}{2} = \dfrac{E_{dc}}{3}$ \hfill (9.68)

Also, $\quad E_{BN} = -I_1 \cdot R = \dfrac{-2 E_{dc}}{3}$

(ii) During interval II, for $\pi/3 \leq \omega t < 2\pi/3$,

$$Req = R + R/2 = \frac{3R}{2}$$

Current, $\quad I_2 = \dfrac{E_{dc}}{Req} = \dfrac{2E_{dc}}{3R}$

Now, $\quad E_{AN} = I_2 R = \dfrac{2E_{dc}}{3}$ \hfill (9.69)

Also, $\quad E_{BN} = E_{CN} = \dfrac{-I_2 R}{2} = \dfrac{-E_{dc}}{3}$

(iii) During interval III, for $\dfrac{2\pi}{3} \leq \omega t < \pi$,

$$Req = R + \frac{R}{2} = \frac{3R}{2}$$

Current, $\quad I_3 = \dfrac{E_{dc}}{Req} = \dfrac{2E_{dc}}{3R}$

Phase-voltage, $\quad E_{AN} = E_{BN} = \dfrac{I_3 \cdot R}{2} = \dfrac{E_{dc}}{3}$ \hfill (9.70)

and $\quad E_{CN} = -I_3 R = \dfrac{-2E_{dc}}{3}$

The line voltage $E_{AB} = E_{AN} - E_{BN}$ is obtained by reversing E_{BN} and adding it to E_{AN} as shown in Fig. 9.25. Similarly, line voltages $E_{BC} = E_{BN} - E_{CN}$ and $E_{CA} = E_{CN} - E_{AN}$ are plotted in Fig. 9.25. It can be observed from Fig. 9.25 that phase voltages have six steps per cycle and line voltages have one positive pulse and one negative pulse (each of 120° duration) per cycle. The phase as well as line-voltages are out of phase by 120°.

The instantaneous line-to-line voltage, E_{AB}, in Fig. 9.25, can be expressed in a Fourier-series, recognizing that E_{AB} is shifted by $\pi/6$ and even harmonics are zero,

$$E_{AB} = \sum_{n=1,3,5,\ldots}^{\infty} \frac{4E_{dc}}{n\pi} \cos\frac{n\pi}{6} \sin n(\omega t + \pi/6) \quad (9.71)$$

E_{BC} and E_{CA} can be found from Eq. (9.71) by phase shifting E_{AB} by 120° and 240°, respectively,

$$E_{BC} = \sum_{n=1,3,5,\ldots}^{\infty} \frac{4E_{dc}}{n\pi} \cos\frac{n\pi}{6} \sin n(\omega t - \pi/2) \quad (9.72)$$

$$E_{CA} = \sum_{n=1,3,5...}^{\infty} \frac{4E_{dc}}{n\pi} \cos\frac{n\pi}{6} \sin n\left(\omega t - \frac{7\pi}{6}\right) \quad (9.73)$$

For $n = 3, 9, 15 \cdots, \cos\frac{n\pi}{6} = 0$.

Hence, it is noted from Eqs (9.71), (9.72) and (9.73) that the triple harmonics ($n = 3, 9, 15,...$) would be zero in the line-to-line voltages.

The line-to-line RMS voltage can be found from

$$E_L = \left[\frac{2}{\pi}\int_0^{2\pi/3} E_{dc}^2 \, d(\omega t)\right]^{1/2} = \sqrt{\frac{2}{3}} E_{dc} = 0.8165 E_{dc} \quad (9.74)$$

From Eq. (9.69), the RMS nth component of the line voltage is

$$E_{Ln} = \frac{4E_{dc}}{\sqrt{2}\, n\pi} \cos\frac{n\pi}{6} \quad (9.75)$$

which, for $n = 1$, gives the fundamental line voltage.

$$E_{L_1} = \frac{4E_{dc}\cos 30}{\sqrt{2}\,\pi} = 0.7797\, E_{dc} \quad (9.76)$$

The RMS value of line-to-neutral voltages can be found from the line voltage,

$$E_p = \frac{E_L}{\sqrt{3}} = \frac{\sqrt{2}\,E_{dc}}{3} = 0.4714\, E_{dc} \quad (9.77)$$

SOLVED EXAMPLES

Example 9.11 A three-phase bridge inverter is operated in 180° conduction mode. Draw the output line voltage waveforms and obtain:
(a) Fourier series for the line voltage
(b) RMS value of the nth harmonic-line voltage
(c) RMS value of the fundamental component of the line voltage
(d) First four harmonics present at the output in the line voltage waveform
(e) The rms value of the line voltage
(f) Distortion and harmonic factor

Solution:
(a) The line voltage waveform of a 3-ϕ bridge inverter operating in 180° mode is a quasi-square wave. The waveform is shown in Fig. Ex. 9.11.

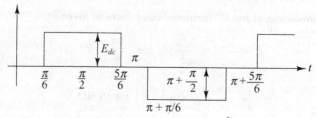

Fig. Ex. 9.11 *Output waveform*

Since the waveform has a quarter-wave symmetry,

∴ $a_n = 0$, for all n, and $b_n = 0$, for all even 'n'.

Also, $b_n = \dfrac{4}{\pi} \displaystyle\int_0^{\pi/2} e(\omega t) \cdot \sin(n\omega t)\, d\omega t$ for all odd 'n'.

From Fig. Ex. (9.11), we can write, $b_n = \dfrac{4}{\pi} \displaystyle\int_{\pi/6}^{\pi/2} E_{dc} \sin(n\omega t)\, d\omega t$

$= \dfrac{4 E_{dc}}{n\pi} \left[\cos \cdot (n\omega t)_{\pi/2}^{\pi/6} \right] \quad b_n = \dfrac{4 E_{dc}}{n\pi} \left[\cos\left(\dfrac{n\pi}{6}\right) - \cos\left(\dfrac{n\pi}{2}\right) \right]$

Since $\cos\left(\dfrac{n\pi}{2}\right) = 0$ for all odd 'n', we get, $b_n = \dfrac{4 E_{dc}}{n\pi} \cos(n\pi/6)$

Now,

$$c_n = \sqrt{a_n^2 + b_n^2} = \dfrac{4 E_{dc}}{n\pi} \cos\left(\dfrac{n\pi}{6}\right) \tag{9.78}$$

and $\phi_n = \tan^{-1}\left(\dfrac{a_n}{b_n}\right) = 0 \tag{9.79}$

Average value of the waveform is given by

$$E_{avg} = \dfrac{1}{2\pi} \int_0^{2\pi} e(\omega t)\, d\omega t = \dfrac{1}{2\pi} \left[\int_{\pi/6}^{5\pi/6} \cdot E_{dc}\, d\omega t + \int_{\pi+\pi/6}^{\pi+5\pi/6} (-E_{dc}) \cdot d\omega t \right]$$

$$= \dfrac{E_{dc}}{2\pi} \left[\dfrac{5\pi}{6} - \dfrac{\pi}{6} + \pi + \dfrac{\pi}{6} - \pi - \dfrac{5\pi}{6} \right] = 0$$

Now, the fourier-series for any periodic waveform is given by

$$e(\omega t) = E_{avg} + \sum_{1,3,5,\ldots}^{\infty} c_n \sin(n\omega t + \phi_n)$$

$$= 0 + \sum_{1,3,5,\ldots}^{\infty} \dfrac{4 E_{dc}}{n\pi} \cos\left(\dfrac{n\pi}{6}\right) \sin(n\omega t + 0)$$

$$= \sum_{1,3,5,\ldots}^{\infty} \dfrac{4 E_{dc}}{n\pi} \cos\left(\dfrac{n\pi}{6}\right) \sin(n\omega t) \tag{9.80}$$

(b) The rms value of the nth harmonic component is given by

$$E_n = \dfrac{c_n}{\sqrt{2}} = \dfrac{2\sqrt{2}}{n\pi} E_{dc} \cos\left(\dfrac{n\pi}{6}\right)$$

$$= \dfrac{0.9}{n} E_{dc} \cdot \cos\left(\dfrac{n\pi}{6}\right) \text{ for all odd '}n\text{'.} \tag{9.81}$$

Inverters

(c) The rms value of the fundamental can be obtained by substituting $n = 1$, in Eq. (9.81),

$$\therefore \quad E_{1(\text{line})} = 0.9\, E_{dc} \cdot \cos\left(\frac{n\pi}{6}\right) = 0.78\, E_{dc} = 78\%\, E_{dc} \tag{9.82}$$

(d) We have

$$E_{n(\text{line})} = \frac{0.9\, E_{dc}}{9} \cos\left(\frac{n\pi}{6}\right) \text{ for all odd '}n\text{'}.$$

The order of the harmonics and their amplitudes for the first four-non zero harmonics is given by Table 9.2.

Table 9.2

S.No.	Harmonic order	Amplitude (RMS value)
1.	3	0
2.	5	$-\dfrac{0.78}{5} E_{dc}$
3.	7	$-\dfrac{0.78}{7} E_{dc}$
4.	9	0
5.	11	$+\dfrac{0.78}{11} E_{dc}$
6.	13	$+\dfrac{0.78}{11} E_{dc}$
7.	15	0

From Table (9.2), we can observe that the tripplen harmonics (3, 5, 11, ...) are zero. Hence, the first four harmonics present are given by,

$$h = 1, 5, 7, 11, 13, \ldots$$
$$\text{or} \quad h = 6k \pm 1,\, k = 1, 2, 3, \ldots \tag{9.83}$$

(e) The fourier series is given by

$$E_{\text{line}} = 0.78\, E_{dc} \left[\frac{\sin(1\,\omega t)}{1} - \frac{1}{5}\sin(5\omega t) - \frac{1}{7}\sin(7\,\omega t) + \frac{1}{11}\sin(11\,\omega t) + \frac{1}{13}\sin(13\,\omega t) + \ldots \right] \tag{9.84}$$

The rms value is given by

$$E_{\text{rms(line)}} = \sqrt{\frac{1}{2\pi} \int_0^{2\pi} e^2\, (d\omega t)}$$

Since the waveform has quarter-wave symmetry, we can write,

$$E_{\text{rms}} = \sqrt{\frac{4}{\pi} \int_0^{\pi/2} e^2\,(\omega t)\, d\omega t}$$

From Fig. (9.26), $E_{\text{rms(line)}} = \sqrt{\dfrac{2}{\pi} \int_{\pi/6}^{\pi/2} E_{dc}^2 \, d\omega t} = \sqrt{\dfrac{2}{\pi} E_{dc}^2 \left(\dfrac{\pi}{2} - \dfrac{\pi}{6}\right)}$

$$= \sqrt{\dfrac{2}{3}} \, E_{dc} = 81.6\% \, E_{dc} \qquad (9.85)$$

(f) Distortion factor, $DF = \dfrac{E_{1(\text{line})}}{E_{\text{rms(line)}}} = \dfrac{0.78 \, E_{dc}}{0.816 \, E_{dc}} = 0.9558 = 95.58\%$

and Harmonic factor, $HF = \sqrt{\dfrac{1}{DF^2} - 1} = \sqrt{\dfrac{1}{(0.9558)^2} - 1} = 30.76\%$ \qquad (9.86)

Example 9.12 A three-phase bridge inverter is fed from a 500 V dc source. The inverter is operated in 180° conduction mode and it is supplying a purely resistive, star-connected load. Determine the following:

(i) RMS value of the output line and phase voltages.
(ii) RMS value of the fundamental component of the line and phase voltages.

Solution:

Given: $E_{dc} = 500$ V

(i) From Eq. (9.85),

$$E_{\text{line(rms)}} = 0.816 \, E_{dc} = 0.816 \times 500 = 408 \text{ V}$$

and $E_{\text{phase(rms)}} = \dfrac{E_{\text{line(rms)}}}{\sqrt{3}} = \dfrac{408}{\sqrt{3}} = 235.56$ V

(ii) From Eq. (9.82), $E_{\text{line(fund)}} = 0.78 \, E_{dc} = 0.78 \times 500 = 390$ V

and $E_{\text{phase(fund)}} = \dfrac{E_{\text{line(fund.)}}}{\sqrt{3}} = \dfrac{390}{\sqrt{3}} = 225.16$ V

Example 9.13 A three-phase bridge inverter is fed from a dc source of E_{dc} volts. The inverter is operated in 180° conduction mode and it is supplying a purely resistive star connected load with $R\Omega$/phase. Determine the ratings of the transistor.

Solution:

The transistor current waveform is shown in Fig. Ex. 9.13.

Transistor peak current, $I_{\text{peak}} = \dfrac{2}{3} \dfrac{E_{dc}}{R}$

$$I_{\text{rms}} = \sqrt{\dfrac{1}{2\pi} \left[\int_0^{\pi/3} \left(\dfrac{E_{dc}}{3R}\right)^2 d\omega t + \int_{\pi/3}^{2\pi/3} \left(\dfrac{2 E_{dc}}{3R}\right)^2 d\omega t + \int_{2\pi/3}^{\pi} \left(\dfrac{E_{dc}}{3R}\right)^2 d\omega t\right]}$$

$$= \sqrt{\dfrac{1}{2\pi} \left(\dfrac{E_{dc}}{3R}\right)^2 \cdot \left[\dfrac{\pi}{3} + 4 \times \dfrac{\pi}{3} + \dfrac{\pi}{3}\right]} = \dfrac{E_{dc}}{3R}$$

Transistor average current, $I_{avg} = \dfrac{1}{2\pi}\left[\displaystyle\int_0^{\pi/3} \dfrac{E_{dc}}{3R} \cdot d\omega t + \int_{\pi/3}^{2\pi/3} \dfrac{2E_{dc}}{3R} d\omega t + \int_{2\pi/3}^{\pi} \dfrac{E_{dc}}{3R} d\omega t\right]$

$= \dfrac{1}{2\pi} \cdot \dfrac{E_{dc}}{3R}\left[\dfrac{\pi}{3} + \dfrac{2\cdot\pi}{3} + \dfrac{\pi}{3}\right] = \dfrac{1}{3} \cdot \dfrac{2 E_{dc}}{3R}$

The maximum voltage, the device will handle = E_{dc}

∴ $V_{CEO} \geq E_{dc} \geq 1.5\, E_{dc}$

∴ Device Ratings are:

$I_T \geq 1.5\, I_{peak} = 1.5 \cdot \dfrac{2}{3} \cdot \dfrac{E_{dc}}{R} \geq \dfrac{E_{dc}}{R}$

and $V_{CEO} \geq 1.5\, E_{dc}$
where 1.5 is the safety factor

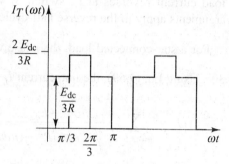

Fig. Ex. 9.13 *Transistor current waveform*

9.8.2 180° Conduction Mode with RL-Load

From the above discussion, it becomes clear that with resistive loads, the diodes across the switch have no functions. If the load is inductive, then the current in each arm of the load will be delayed to its voltage as shown in Fig. 9.27.

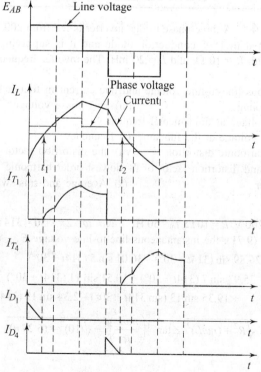

Fig. 9.27 *Waveforms for 180° firing with an induction (RL) load*

When switch S_1 is triggered, S_4 is turned-off but, because the load current cannot reverse, the only path for this current is through diode D_1 (see Fig. 9.24). Hence, the load phase is connected to the positive end of the d.c. source but, until the load current reverses at t_1, switch S_1 will not take up conduction. Similar arguments apply in the reverse half-cycle at instant t_2.

For a star-connected load, the phase voltage is $E_{AN} = \dfrac{E_{AB}}{\sqrt{3}}$ with a delay of 30°. Using Eq. (9.69), the line current I_L for an RL load is given by

$$I_L = \sum_{n=1,3,5\ldots}^{\infty} \left[\frac{4E_{dc}}{\sqrt{3}\cdot n\pi\sqrt{R^2+(n\omega L)^2}} \cos\frac{n\pi}{6}\right] \sin(n\omega t - \theta_n)$$

where $\theta_n = \tan^{-1}\left(\dfrac{n\omega L}{R}\right).$ \hfill (9.87)

For a delta-connected load, the phase currents can be obtained directly from the line-to-line voltages. Once the phase currents are known, the line currents can be determined.

SOLVED EXAMPLES

Example 9.14 A three-phase bridge inverter is fed from 200 V dc source. The inverter is operated in 180° conduction mode and it is supplying inductive, star-connected load with $R = 10\,\Omega$, and $L = 20$ mH. The inverter frequency is $f_0 = 50$ Hz. Determine:
  (i) instantaneous line-to-line voltage and line current in fourier series
  (ii) rms line voltage   (iii) rms phase voltage
  (iv) rms line voltage at fundamental frequency
  (v) rms phase voltage at the fundamental frequency
  (vi) the total harmonic distortion   (vii) the distortion factor
 (viii) harmonic and distortion factor of the lowest order harmonic
  (ix) load power   (x) Average and rms switch current

Solution:
  Given: $E_{dc} = 200$ V, $R = 10\,\Omega$, $f_0 = 50$ Hz $\quad\therefore\quad \omega = 2\pi\times 50 = 314$ rad/s.
  (i) Using Eq. (9.71), the instantaneous line-to-line voltage E_{AB}, can be written as
$$E_{AB} = 226.89\sin(314t+30°) - 36.01\sin 5(314t+30°)$$
$$- 35.93\sin 7(314t+30°) + 10.51\sin 11(314t+30°)$$
$$+ 19.35\sin 13(\sin 314t+30°) - 2.34\sin 17(314t+30°) + \ldots$$

Now, $Z_L = \sqrt{R^2+(n\omega L)^2}\,\angle\tan^{-1}\left(\dfrac{n\omega L}{R}\right) = \sqrt{(10)^2+(6.28\,n)^2}\,\angle\tan^{-1}\left(\dfrac{6.28}{10}\right)$

Instantaneous line (or phase) current from Eq. (9.87) is given by

$I_L = 10.62 \sin(314t - 32.12) - 0.56 \sin(5 \times 314t - 72.33)$
$\quad - 0.402 \sin(7 \times 314t - 77.18) + 0.076 \sin(11 \times 314t - 81.76)$
$\quad + 0.118 \sin(13 \times 314t - 83) - 0.01 \sin(17 \times 314t - 84.69) - \ldots$

(ii) From Eq. (9.74), $E_L = 0.8165 \times 200 = 163.30$ V
(iii) From Eq. (9.77), $E_P = 0.4714 \times 200 = 94.28$ V
(iv) From Eq. (9.76), $E_{L1} = 0.7797 \times 200 = 155.94$ V
(v) $E_{P1} = E_{L1}/\sqrt{3} = 90.03$ V
(vi) From Eq. (9.76), $E_{L1} = 0.7797 E_{dc}$

$$\left(\sum_{n=5,7,11,\ldots}^{\infty} E_{L_n}^2\right)^{1/2} = \left(E_L^2 - E_{L1}^2\right)^{1/2} = 48.47 \text{ V}$$

Now, $\quad$ THD $= \dfrac{48.47}{155.94} = 31.08\%$

(vii) RMS harmonic line voltage is

$$EL_n = \left[\sum_{n=5,7,11,\ldots}^{\infty} \left(\frac{EL_n}{n^2}\right)^2\right]^{1/2} = 0.00666 \, E_{dc} = 1.332$$

Now, $\quad$ DF $= \dfrac{1.332}{155.94} = 0.854\%$

(viii) The lowest-order harmonic is fifth, $E_{L5} \equiv \dfrac{E_{L1}}{5}$

$\therefore$ Harmonic factor, $\text{HF}_4 = \text{HF}_5 = E_{L5}/E_{L1} = 1/5 = 20\%$

$\therefore \quad \text{DF}_5 = \left(\dfrac{E_{L5}/5^2}{E_{L1}}\right) = \dfrac{1.247}{155.94} = 0.8\%$

(ix) The line current is the same as the phase current for star-connected loads and the rms line current,

$$I_L = \frac{(10.62^2 + 0.56^2 + 0.402^2 + 0.076^2 + 0.118^2 + 0.01^2)^{1/2}}{\sqrt{2}} = 7.53 \text{ A}$$

Now, load power $P_0 = 3 I_L^2 R = 3 \times (7.53)^2 \times 10 = 1.70$ kW

(x) The average supply current, $I_s = \dfrac{P_0}{200} = 8.50$ A

$\therefore$ The average switch (device) current $I_{T_{avg}} = \dfrac{8.5}{3} = 2.83$ A

Since the line-current is shared by two switches, the rms value of a switch current is

$$I_{T_{rms}} = \frac{I_L}{\sqrt{2}} = \frac{7.53}{\sqrt{2}} = 5.32 \text{ A}$$

9.8.3 120° Conduction Mode with Resistive Load

In this type of conduction mode, each switch conducts for 120°. At any instant of time, only two switches remain on. Like in the above case, here also gate pulse indicates the conduction period of each switch. In this case also, six commutations per cycle are required. The gating signals and various voltage waveforms of three-phase bridge inverters with 120° conduction for each switch is shown in Fig. 9.28. In this figure, one period of inverter operation has been divided into six intervals. The firing sequence of six switches is prepared, as shown in Table 9.3. Like 180° mode, 120° mode inverter also requires six steps, each of 60° duration, for completing one cycle of the output a.c. voltage.

Table 9.3 Operation Table

S.No.	Interval	Conducting devices	Incoming device	Outgoing device
1	I	S_6, S_1	S_1	S_5
2	II	S_1, S_2	S_2	S_6
3	III	S_2, S_3	S_3	S_1
4	IV	S_3, S_4	S_4	S_2
5	V	S_4, S_5	S_5	S_3
6	VI	S_5, S_6	S_6	S_4

Following points can be noted from the waveforms (Fig. 9.28) and the operating Table 9.3:

(i) The base drives of two switches in the same-half-bridge have an inherent dead band of 60°. Hence, there is no possibility of cross conduction or shoot-through fault.

(ii) Conduction period for each switch is 120°.

(iii) The phase-shift between the triggering of every two adjacent switches is 60°.

(iv) Three line voltages, E_{AB}, E_{BC} and E_{CA} are six-step waves, with step heights $E_{dc}/2$ and E_{dc}. The three-line voltages are mutually phase shifted by 120°.

(v) The three-phase voltages E_{AN}, E_{BN} and E_{CN} are quasi-square-waves with peak values of $E_{dc}/2$. They are also mutually phase-shifted by 120°.

(vi) The line-voltage E_{AB} is leading the phase-voltage E_{AN} by 30°.

From Fig. 9.28 and Table 9.3, it is observed that two switches conduct at a time, one from the upper group and the other from the lower group. There are three modes of operation in one half-cycle and the equivalent circuits for a star-connected load are shown in Fig. 9.29.

(i) **During interval I, for $0 \leq \omega t < \pi/3$**, switches S_1 and S_6 conduct.

$$\therefore \quad E_{AN} = \frac{E_{dc}}{2}, \quad E_{BN} = \frac{-E_{dc}}{2}, \quad E_{CN} = 0. \quad (9.88)$$

(ii) **During interval II,** $\left(\text{for } \frac{\pi}{3} \leq \omega t \leq \frac{2\pi}{3}\right)$, switches S_1 and S_2 conduct.

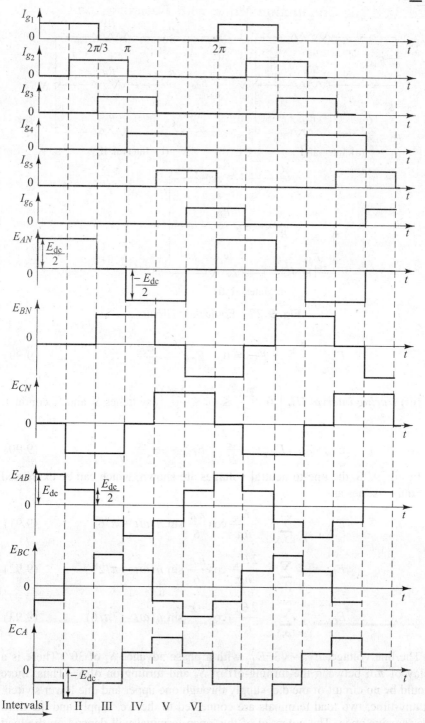

Fig. 9.28 *Gating signals and voltage waveform for 120° conduction*

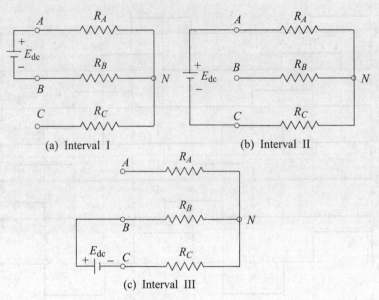

Fig. 9.29 *Equivalent circuits*

$$\therefore \quad E_{AN} = \frac{E_{dc}}{2}, \quad E_{BN} = 0, \quad E_{CN} = \frac{-E_{dc}}{2}. \tag{9.89}$$

(iii) During interval III, $\left(\text{for } \dfrac{2\pi}{3} \leq \omega t \leq \dfrac{3\pi}{3} \right)$, switches S_2 and S_3 conduct.

$$\therefore \quad E_{AN} = 0, \quad E_{BN} = \frac{E_{dc}}{2}, \quad E_{CN} = \frac{-E_{dc}}{2}. \tag{9.90}$$

In Fig. 9.28, the line-to-neutral voltages are shown, which can be expressed in Fourier-series as

$$E_{AN} = \sum_{n=1,3,5,\ldots}^{\infty} \frac{2E_{dc}}{n\pi} \cos\frac{n\pi}{6} \sin n(\omega t + \pi/6) \tag{9.91}$$

$$E_{BN} = \sum_{n=1,3,5,\ldots}^{\infty} \frac{2E_{dc}}{n\pi} \cos\frac{n\pi}{6} \sin n(\omega t - \pi/2) \tag{9.92}$$

$$E_{CN} = \sum_{n=1,3,5,\ldots}^{\infty} \frac{2E_{dc}}{n\pi} \cos\frac{n\pi}{6} \sin n(\omega t - 7\pi/6) \tag{9.93}$$

The line voltage, $E_{AB} = \sqrt{3}\, E_{AN}$ with a phase advance S_4 of 30°. There is a delay of $\pi/6$ between the turning-off of S_1 and turning-on of S_4. Thus, there should be no circuit of the d.c. supply through one upper and one lower switch. At any-time, two load terminals are connected to the d.c. supply and the third one remains open. The potential of this open terminal will depend on the load characteristics and would be unpredictable.

Solved Examples

Example 9.15 A three-phase bridge inverter is operated in 120° conduction mode. Determine
(a) fourier series for the line voltage
(b) the nth harmonic component
(c) the fundamental component
(d) the rms value
(e) distortion and harmonic factor
(f) first four harmonics

Solution:
Line-voltage waveform is a six-step waveform, shown in Fig. 9.28.

(a) Since the waveform has quarter-wave symmetry, we get
$$a_n = 0, \text{ for all } n. \text{ and } b_n = 0, \text{ for all even } n.$$

and
$$b_n = \frac{4}{\pi} \int_0^{\pi/2} e(\omega t) \sin(n\omega t) \, d\omega t \text{ for all odd } n.$$

$\therefore$
$$c_n = b_n = \frac{4}{\pi} \left[\int_0^{\pi/3} \frac{E_{dc}}{2} \cdot \sin(n\omega t) \, d\omega t + \int_{\pi/3}^{\pi/2} E_{dc} \cdot \sin(n\omega t) \, d\omega t \right]$$

$$= \frac{4 E_{dc}}{2n\pi} \left[(\cos n\omega t)_{\pi/3}^{0} + 2 (\cos (n\omega t))_{\pi/2}^{\pi/3} \right]$$

$$= \frac{E_{dc}}{n\pi} \left[1 - \cos(n\pi/3) + 2 \cos(n\pi/3) - 2 \cos(n\pi/2) \right]$$

but $\cos(n\pi/2) = 0$, for all odd n,

$$c_n = \frac{2 E_{dc}}{n\pi} [1 + \cos(n\pi/3)] \qquad (9.94)$$

and
$$\phi_n = 0 \qquad (9.95)$$

$\therefore$
$$E_{line}(\omega t) = \sum_{n=1,3,5,\ldots}^{\infty} \frac{2 E_{dc}}{n\pi} [1 + \cos(n\pi/2) \sin(n\omega t)] \qquad (9.96)$$

(b) The nth harmonic component,
$$E_{n(line)} = \frac{c_n}{\sqrt{2}} = \frac{\sqrt{2} \, E_{dc}}{n\pi} [1 + \cos(n\pi/2)] \qquad (9.97)$$

(c) Fundamental component,
$$E_{1(line)} = \frac{\sqrt{2} \, E_{dc}}{\pi} (1.5) = 67.52\% \, E_{dc} \qquad (9.98)$$

(d) Since the waveform has quarter-wave-symmetry,

$\therefore$
$$E_{rms(line)} = \sqrt{\frac{4}{2\pi} \left[\int_0^{\pi/3} \left(\frac{E_{dc}}{2} \right)^2 d\omega t + \int_{\pi/3}^{\pi/2} E_{dc}^2 \, d\omega t \right]}$$

$$= \sqrt{\frac{4}{\pi} \left(\frac{E_{dc}}{2} \right)^2 \cdot \left(\frac{\pi}{3} + 4 \left(\frac{\pi}{2} - \frac{\pi}{3} \right) \right)} = \frac{E_{dc}}{\sqrt{2}} = 70.7\% \, E_{dc} \qquad (9.99)$$

(e) $\text{DF} = \dfrac{E_{1(\text{line})}}{E_{\text{rms(line)}}} = 95.5\%$ \hfill (9.100)

and $\quad \text{HF} = \sqrt{\dfrac{1}{\text{DF}^2} - 1} = 30.8\%$ \hfill (9.101)

(f) The first four non-zero harmonics are 5, 7, 11, 13.

$\therefore \quad E_5 = \dfrac{\sqrt{2}\, E_{\text{dc}}}{5\pi}\left[1 + \cos\left(\dfrac{5\pi}{3}\right)\right] = 0.135\, E_{\text{dc}}$

$\quad E_7 = \dfrac{\sqrt{2}\, E_{\text{dc}}}{7\pi}\left[1 + \cos\left(\dfrac{7\pi}{3}\right)\right] = 0.096\, E_{\text{dc}}$

$\quad E_9 = \dfrac{\sqrt{2}\, E_{\text{dc}}}{9\pi}\left[1 + \cos\left(\dfrac{9\pi}{3}\right)\right] = 0$

$\quad E_{11} = \dfrac{\sqrt{2}\, E_{\text{dc}}}{11\pi}\left[1 + \cos\left(\dfrac{11\pi}{3}\right)\right] = 0.06\, E_{\text{dc}}$

$\therefore$ Fourier series is

$$e(\omega t) = E_{\text{dc}}\left[\sin \omega t + \dfrac{1}{5}\sin 5\omega t + \dfrac{1}{7}\sin 7\omega t + \dfrac{1}{11}\sin 11\omega t + \dfrac{1}{13}\sin 13\omega t + \ldots\right]$$

Example 9.16 A three phase transistorized bridge inverter is operating in 120° conduction mode. Draw the current waveform for transistor and obtain its ratings also.

Solution:
Current waveform for the device is shown in Fig. Ex. 9.16.

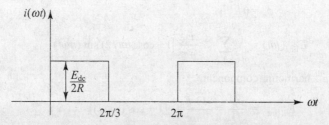

Fig. Ex. 9.16

(a) Transistor current ratings,

(i) $I_{\text{rms}} = \sqrt{\dfrac{1}{2\pi}\displaystyle\int_0^{2\pi} i^2(\omega t)\, d\omega t} = \sqrt{\dfrac{1}{2\pi}\displaystyle\int_0^{2\pi/3}\left(\dfrac{E_{\text{dc}}}{2R}\right)^2 d\omega t}$

$\qquad = \sqrt{\dfrac{1}{2\pi}\times\left(\dfrac{E_{\text{dc}}}{2R}\right)^2 \cdot \dfrac{2\pi}{3}} = \dfrac{E_{\text{dc}}}{2R}\dfrac{1}{\sqrt{3}}$ \hfill (9.102)

(ii) $I_{\text{avg(rms)}} = \dfrac{1}{2\pi}\displaystyle\int_0^{2\pi/3}\left(\dfrac{E_{\text{dc}}}{2R}\right) d\omega t = \dfrac{E_{\text{dc}}}{2R}\cdot\dfrac{1}{3}$ \hfill (9.103)

(iii) $I_{peak} = \dfrac{E_{dc}}{2R}$ \hfill (9.104)

(b) The maximum voltage across the device is E_{dc}.

∴ $V_{CEO} \geq E_{dc} \geq 1.5\, E_{dc}$ [where 1.5 is safety-factor] \hfill (9.105)

Example 9.17 For a three-phase bridge inverter operating in 120° conduction mode, determine:
  (i) dc voltage for a fundamental line voltage of 415 V
  (ii) rms line and phase voltage   (iii) Device voltage rating.

Solution:
  (i) From Eq. (9.98),

$$E_{1(line)} = \dfrac{\sqrt{2}}{\pi} E_{dc}(1.5)$$

∴ $E_{dc} = \dfrac{415\,\pi}{\sqrt{2}(1.5)} = 614.60\text{ V} \cong 615\text{ V}$

(ii) $E_{rms(line)} = 70.7\%\, E_{dc} = 434.5\text{ V}$

Now, $E_{rms(phase)} = \sqrt{\dfrac{4}{2\pi}\displaystyle\int_{\pi/6}^{\pi/2}\left(\dfrac{E_{dc}}{2}\right)^2 d\omega t} = \sqrt{\dfrac{2}{\pi}\left(\dfrac{E_{dc}}{2}\right)^2\cdot\left(\dfrac{\pi}{2}-\dfrac{\pi}{6}\right)} = \sqrt{\dfrac{2}{3}}\dfrac{E_{dc}}{2}$

$= 40.8\%\, E_{dc} = 40.8\%\,(615) = 250.92\text{ V}$

(iii) Device voltage rating, $V_{CEO} \geq 1.5\, E_{dc} \geq 1.5 \times 615 = 922.5\text{ V}$

9.8.4 Comparison of Two Conduction Modes

In 180° mode conduction, when gate signal I_{g_1} is removed to turn-off switch S_1 at $\omega t = 180°$, gating signal I_{g_4} is simultaneously applied to turn-on switch S_4 in the same leg. In practice, a commutation interval must exist between the removal of I_{g_1} and application of I_{g_4} for proper and reliable operation of the inverter circuit. Since enough time may not be provided for the commutation of switch and two switches in series may simultaneously conduct, resulting into short circuit of the source by these switches.

This problem is overcome considerably in 120° mode inverter. In this inverter, there is a 60° interval between the turning-off of S_1 and turning-on of S_4. During this 60° interval, switch S_1 can be commutated safely. Thus, enough time is made available for the outgoing switch to commutate before the switch in series is turned-on. Therefore, commutation is more reliable and the possibility of two series switches conducting simultaneously is much less.

The second important difference is in terms of utility of devices. The comparison of the two patterns is done using a figure of merit termed as utility factor (UF) and is defined as

$$UF = \dfrac{P_0}{P_T} \quad\quad (9.106)$$

where P_0 is the rated output power of the inverter and P_T is the measure of total power handling capability of the devices employed in the inverter, and is defined as

$$P_T = N \cdot V_{DRM} \cdot I_{rms} \tag{9.107}$$

where N = number of thyristors, V_{DRM} = repetitive peak forward-voltage
I_{rms} = rated RMS forward current

For the purpose of comparison of the two control schemes, it is usual to calculate the utility factor for ideal resistive load.

(i) Inverter with 180°-Conduction For star-connected load, if R is the resistance per phase, then

$$P_0 = 3 \left(\frac{1}{\pi} \int_0^\pi \frac{E_{AN}^2}{R} \, d(\omega t) \right) \tag{9.108}$$

Substituting from Eq. (9.69) gives

$$P_0 = \frac{2 E_{dc}^2}{3 R} \tag{9.109}$$

Device rating will be chosen such that the RMS current flowing through it will be equal to its RMS current rating I_{rms}. During a cycle, current of any phase is shared between two device in series; for example, for phase A, device S_1 carries load current during positive half-cycle and device S_4 during the negative half-cycle. Therefore, under the rated conditions, the RMS value of load phase current I_p is given by, $\quad I_p = \sqrt{2} \, I_{rms}$ \hfill (9.110)

Load power P_0 can also be written as

$$P_0 = 3 I_p^2 R = 6 I_{rms}^2 R \tag{9.111}$$

From Eqs (9.109) and (9.111), $I_{rms} = \dfrac{E_{dc}}{3R}$ \hfill (9.112)

For this load, the repetitive peak forward voltage rating of the device should be equal to E_{dc}. Thus,

$$P_T = 6 \cdot E_{d.c.} \cdot \frac{E_{dc}}{3R} = \frac{2 E_{dc}^2}{R} \tag{9.113}$$

Now, from Eqs (9.106), (9.109) and (9.113),

$$UF = \frac{P_0}{P_T} = \frac{1}{3} = 33.33\% \tag{9.114}$$

Equation (9.114) indicates that in 180° conduction, the rated power output of the inverter is only 33.33 per cent of the combined maximum power rating of the six switches.

(ii) Inverter with 120° Conduction For star-connected load, with a per phase resistance of R, output power P_0 will be given by Eq. (9.108). Substituting Eq. (9.88) in Eq. (9.108), we get

$$P_0 = \frac{E_{dc}^2}{2R} \tag{9.115}$$

Here also, the relation between I_{rms} and I_p is given by Eq. (9.110), and P_0 in terms of I_{rms} is given by Eq. (9.111). Now, from Eqs (9.111) and (9.115),

$$I_{rms} = \frac{1}{2\sqrt{3}} \cdot \frac{E_{dc}}{R} \tag{9.116}$$

From Eq. (9.107)

$$P_T = 6 \cdot E_{d.c.} \cdot \frac{1}{2\sqrt{3}} \cdot \frac{E_{dc}}{R} \quad P_T = \sqrt{3} \frac{E_{dc}^2}{R} \tag{9.117}$$

Also, from Eq. (9.106),

$$UF = \frac{1}{2\sqrt{3}} = 28.9\% \tag{9.118}$$

In this case, the rated inverter output is 28.9 per cent of the combined maximum power rating of six switches. From Eqs (9.114) and (9.118), it becomes clear that the utility factor is more in case of 180°-conduction scheme.

9.9 VOLTAGE CONTROL OF THREE-PHASE INVERTERS

A three-phase inverter may be considered as three single-phase inverters and the output of each single-phase inverter is shifted by 120°. The voltage control techniques discussed in above section are applicable to three-phase inverters. For an example, the generation of gating signals with sinusoidal pulse-width modulation are shown in Fig. 9.30. There are three-sinusoidal reference waves, each shifted by 120°. A carrier wave is compared with the reference signal corresponding to a phase to generate the gating signals for that phase. The output voltage, as shown in Fig. 9.30, is generated by eliminating the condition that two switching devices in the same arm cannot conduct at the same time.

9.10 THYRISTOR-BASED INVERTERS

Gate-commutated power devices, like BJT, MOSFET, IGBTs, etc., are used for low- and medium-power applications. For high power applications, it is necessary to connect them in series and/or parallel combinations, which increases the circuit complexity. Therefore, for high power applications, fast-switching thyristors (Inverter-grade) which are available in high voltage and current ratings are more suitable.

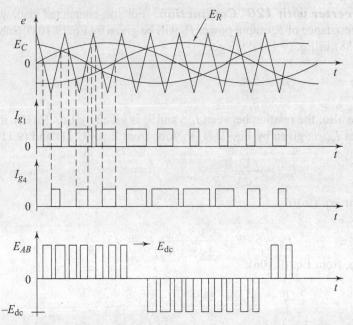

Fig. 9.30 *Sinusoidal pulse-width modulation for three-phase inverter*

9.11 SERIES INVERTERS (SERIES RESONANT INVERTERS)

In some inverters, the commutating elements may come in series with the load or in parallel with the load during operation. In this type of inverters, as indicated by the name, the commutating elements, viz. L and C are connected in series with the load. This constitutes a series R-L-C resonant circuit. If the load is purely resistive, it only has resistance in the circuit. In case of load being inductive or capacitive in nature, its inductance or capacitance part is added to the commutating elements (being in series). This type of thyristorised inverter produces an approximately sinusoidal waveform at a high output frequency, ranging from 200 Hz to 100 kHz, and is commonly used in relatively fixed output applications such as ultrasonic generators, induction heating, sonar transmitter, fluorescent lighting, etc. Due to the high-switching frequency, the size of commutating components is small.

9.11.1 Basic Series Inverter

Figure 9.31(a) shows the circuit diagram of basic series inverter. Two thyristors T_1 and T_2 are used to produce the two halves (positive and negative respectively) in the output. As shown, the commutating elements L and C are connected in series with the load R to form the series R-L-C circuit. The values of L and C are chosen such that, they form an underdamped circuit. This is necessary to produce the required oscillations. This condition is fulfilled by selecting L and C such that

$$R^2 < \frac{4L}{C} \tag{9.119}$$

The operation of a basic series inverter circuit can be divided into following three operating modes.

Mode 1: This mode begins when a d.c. voltage E_{dc} is applied to the circuit and thyristor T_1 is triggered by giving external pulse to its gate. As soon as SCR T_1 is triggered, it starts conducting and resulting in some current to flow through the R-L-C series circuit. Capacitor C gets charged up to voltage, say, E_c, with positive polarity on its left plate and negative polarity on its right plate. The load current is of alternating nature. This is due to the underdamped circuit formed by the commutating elements. It starts building up in the positive half, goes gradually to its peak-value, then starts returning and again becomes zero, as shown in Fig. 9.31(b). When the current reaches its peak-value, the voltage across the capacitor is approximately the supply voltage E_{dc}. After this, the current starts decreasing but the capacitor voltage still increases and finally the current becomes zero but the capacitor retains the highest voltage, i.e. $(E_{dc} + E_c)$, where E_c is the initial voltage across the capacitor at the instant SCR T_1 was turned-on. At P, SCR T_1 is automatically turned-off because the current flowing through it becomes zero.

Mode 2: During this mode, the load current remains at zero for a sufficient time (T_{off}). Therefore, both the thyristors T_1 and T_2 are OFF. During this period PQ, capacitance voltage will be held constant.

Mode 3: Since the positive polarity of the capacitor C appears on the anode of SCR T_2, it is in conducting mode and hence triggers immediately. At Q, SCR T_2 is triggered. When SCR T_2 starts conducting, capacitor C gets discharged through it. Thus, the current through the load flows in the opposite direction forming the negative alternation. This current builds up to the negative maximum and then decreases to zero at point R. SCR T_2 will then be turned-off. Now, the capacitor voltage reverses to some value depending upon the values of R, L and C.

Again, after some time delay (T_{off}), SCR T_1 is triggered and in the same fashion other cycles are produced. This is a chain of process giving rise to alternating output almost sinusoidal in nature. One important point to be noted here is that the supply from the d.c. source is intermittent in nature. Positive alternation of the a.c. output is drawn from the d.c. input source, whereas for the negative alternation the current is drawn from the capacitor.

It is necessary to maintain a time delay between the point when one SCR is turned-off and other SCR is triggered. If this is not done, both the SCRs will start conducting simultaneously resulting in a short circuit of the d.c. input source. This time delay (T_{off}) must be more than the turn-off time of the SCRs. The output frequency is given by

$$F = \left[\frac{1}{T/2 + T_{off}}\right] \text{ Hz} \tag{9.120}$$

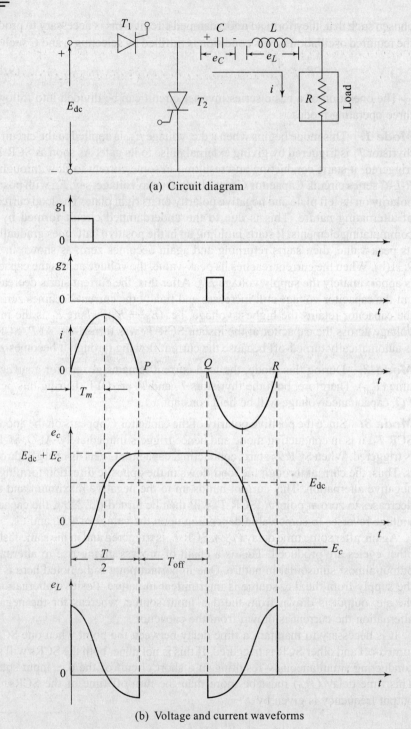

(a) Circuit diagram

(b) Voltage and current waveforms

Fig. 9.31 *Series inverter*

where T is the time period for oscillations and is given by

$$\frac{T}{2} = \frac{\pi}{\sqrt{1/LC - R^2/4L^2}} \tag{9.121}$$

and T_{off} is the time-delay between turn-off of one SCR and turn-on of the other SCR. Thus, by changing the value of T_{off}, frequency can be changed without changing the commutating elements. The basic series inverter has the following drawbacks:

1. The load voltage waveform has more distortion due to the time delay. This distortion is specially high for frequencies less than the resonance frequency.
2. The maximum inverter frequency is limited to a value that is slightly less than the circuit ringing frequency. If the inverter frequency exceeds the circuit ringing frequency, the d.c. source will be short-circuited.
3. The commutating element must have high rating because these components carry the load current continuously and the capacitor supplies the load current in every alternate half-cycle.
4. Load current is drawn from the d.c. source only during one half-cycle and this increases the peak current rating of the d.c. source. Since the current drawn from the d.c. source is not continuous in nature, more ripples are present in it.
5. The peak amplitude and duration of the load current in each half-cycle depends on load parameters, resulting in poor output regulation for the inverter.

Some of the above limitations can be overcome by little modifications in the circuit. Such a circuit has been described in Section 9.11.4.

9.11.2 Circuit Analysis

Again, for the sake of simplicity, the circuit analysis is carried out in the same operating mode:

Mode 1: This mode 1 operation can be described by the following differential equation:

$$E_{dc} = \frac{1}{C}\int i\, dt - E_c + L\frac{di}{dt} + iR. \tag{9.122}$$

The initial conditions at $t = 0$ are
(i) Initial current $i(0) = 0$.
(ii) Initial voltage on capacitor $E_c(t=0) = -E_c$

Since the circuit is underdamped, the solution of Eq. (9.122) yields

$$i(t) = A \cdot e^{-\frac{Rt}{2L}} \sin \omega_r t \tag{9.123}$$

where ω_r is the resonant frequency and is given by

$$\omega_r = \sqrt{\frac{1}{LC} - \frac{R^2}{4L^2}} \tag{9.124}$$

The constant A, in Eq. (9.123) can be determined from the initial condition:

From Eq. (9.122), $\left.\dfrac{di}{dt}\right|_{t=0} = \dfrac{E_{dc} + E_c}{L}$ \hfill (9.125)

From Eq. (9.123), $\left.\dfrac{di}{dt}\right|_{t=0} = A\,\omega_r$ \hfill (9.126)

From Eqs (9.125) and (9.126), we get

$$A = \frac{E_{dc} + E_c}{\omega_r \cdot L} \tag{9.127}$$

Substituting Eq. (9.127) into Eq. (9.123) we get,

$$i(t) = \frac{E_{dc} + E_c}{\omega_r \cdot L}\, e^{-\frac{Rt}{2L}} \sin \omega_r t \tag{9.128}$$

The time T_m when the current $i(t)$ in Eq. (9.128) becomes maximum can be found from the condition

$$\frac{di}{dt} = 0$$

$\therefore \quad \omega_r \dfrac{-RT_m}{e^{2L}} \cos \omega_r \cdot T_m - \dfrac{R}{2L} e^{-\frac{RT_m}{2L}} \sin \omega_r T_m = 0$

or $\quad \omega_r \cdot \dfrac{-RT_m}{e^{2L}} \cos \omega_r \cdot T_m = \dfrac{R}{2L} e^{-\frac{RT_m}{2L}} \sin \omega_r T_m$

or $\quad \dfrac{\omega_r \cdot 2L}{R} = \dfrac{\sin \omega_r T_m}{\cos \omega_r T_m} \quad$ or $\quad \dfrac{\omega_r \cdot 2L}{R} = \tan \omega_r T_m$

or
$$T_m = \frac{1}{\omega_r} \tan^{-1} \left(\frac{\omega_r 2L}{R} \right) \tag{9.129}$$

The current $i(t)$ becomes zero at $t = T/2$. Therefore, SCR T_1 will be turned-off. Capacitor C will be charged to voltage V_c.

Mode 2: During this mode, thyristors T_1 and T_2 are OFF. Therefore,

$$i(t) = 0 \quad \text{and} \quad V_c = E_c + E_{dc}.$$

Inverters

Mode 3: This mode begins when thyristor T_2 is turned-on and a reverse resonant current flows through the load. This mode of circuit operation can be described by

$$V_c = \frac{1}{C}\int i\, dt + L\frac{di}{dt} + iR. \tag{9.130}$$

with the initial condition $i(t = 0) = 0$.

Under steady state conditions, the positive and negative load current waveforms must be identical. For this purpose, Eqs (9.122) and (9.130) have to be same. Thus, the necessary condition for the steady state is

$$V_c = E_{dc} + E_c \tag{9.131}$$

The value of E_c in the steady state can be obtained using Eq. (9.131). The capacitor voltage, at the end of a half-cycle is given by

$$V_c = \frac{1}{C}\int_0^{T/2} i\, dt - E_c \tag{9.132}$$

Substituting Eq. (9.131) in Eq. (9.132), we get

$$E_{dc} + E_c = \frac{1}{C}\int_0^{T/2} i\, dt - E_c$$

$$E_{dc} = \frac{1}{C}\int_0^{T/2} i\, dt - 2E_c$$

$$\therefore \quad E_{dc} = \frac{1}{C}\left[\int_0^{T/2} \frac{E_{dc} + E_c}{\omega_r L} e^{-\frac{Rt}{2L}} \sin \omega_r t\, dt\right] - 2E_c$$

After solving the above equation, we get

$$E_{dc} = \frac{E_{dc} + E_c}{C(R^2 + 4W_r^2 L^2)} \times 4L\left\{i + e^{-\frac{R\pi}{2\omega_r L}}\right\} - 2E_c \tag{9.133}$$

From Eq. (9.133), E_c is obtained as

$$E_c = \frac{E_{dc}\left\{1 - \left[\dfrac{4L}{c(R^2 + 4W_r^2 L^2)}\right]\left[1 + e^{-\frac{R\pi}{2\omega_r L}}\right]\right\}}{\left(\dfrac{4L}{C(R^2 + 4W_r^2 L^2)}\right)\left(1 + e^{-\frac{R\pi}{2\omega_r L}}\right) - 2} \tag{9.134}$$

Substituting the value of W_r from Eq. (9.124) in above equation, we get

$$E_c = E_{dc} \left[\frac{e^{-\frac{R\pi}{2\omega_r L}}}{1 - e^{-R\pi/2\omega_r L}} \right] \tag{9.135}$$

9.11.3 Design Aspects

1. Design of Inductance L A suitable value of L is chosen on the basis of the attenuation factor $e^{-Rt/2L}$. When $W_r \cdot t = \pi/2$, the peak-value of A will be reduced by the factor $e^{-R\pi/4\omega_r L}$. The optimum value of this attenuation factor (AF) is 0.5. Therefore, we can write,

$$\text{AF} = \frac{-R\pi}{e^{4\omega_r \cdot L}} \tag{9.136(a)}$$

or,

$$L = \frac{-R}{8 f_r \, l_{\ln} \, (\text{A.F.})} \tag{9.136(b)}$$

2. Design of Capacitance C

(i) Capacitance C is obtained from the expression for W_r. From Eq. (9.124), we have

$$\omega_r = \sqrt{\frac{1}{LC} - \frac{R^2}{4L^2}} \quad \therefore \quad \omega_r^2 = \frac{1}{LC} - \frac{R^2}{4L^2}$$

$$\therefore \quad C = \frac{1}{L} \left[\frac{1}{\omega_r^2 + R^2/4L^2} \right] \tag{9.137}$$

If the load is variable, the maximum value of R must be used in Eq. (9.137) so that the circuit remains underdamped for all load conditions.

(ii) For proper operation of the circuit, the voltage rating of capacitor must be greater than $(E_c + E_{dc})$ value.

The value of E_c in the steady-state is obtained from Eq. (9.135). This voltage E_c must be of sufficient magnitude to turn-off the SCRs within time T_{off}. Therefore, the minimum value of R is used in this equation.

3. Selection of Thyristor

(i) The forward blocking voltage rating of the SCR must be greater than the value $(E_c + E_{dc})$.

(ii) The current rating of the SCR must be greater than the peak-load current value. The peak-load current value is given by the expression

$$I_{\text{peak}} = \frac{E_c + E_{dc}}{W_r \, L} \, e^{-\frac{R\pi}{4\omega_r L}} \tag{9.138}$$

Here also, the minimum value of resistance is used.

(iii) The ringing frequency ω_r must be so chosen as to approximate the value of the desired output frequency ω_0, such that the off-time $\left(T_{\text{off}} = \pi\left(\dfrac{1}{\omega_0} - \dfrac{1}{\omega_r}\right)\right)$ of the SCRs is greater than their turn-off time [i.e. $T_q \ll T_{\text{off}}$].

9.11.4 Modified Series Inverter

Some of the limitations of basic series inverter circuit of Fig. 9.31 can be overcome by making some modifications in the basic form. One such modification is to replace the normal inductance by a mutually-coupled inductance as shown in Fig. 9.32. Inductors L_1 and L_2 have the same inductance and are closely coupled. It can be seen that even if SCR T_2 is triggered a little before SCR T_1 is turned-off, there will not be any possibility of short circuit at the d.c. input source. This is explained as follows:

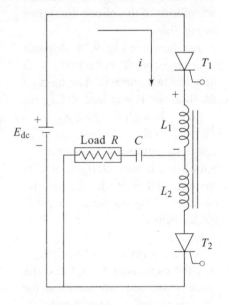

Let us suppose that SCR T_2 is triggered shortly before SCR T_1 is turned-off. At the instant when SCR T_2 is triggered, the voltage across the capacitor will be slightly less than $(E_c + E_{dc})$ and the load voltage and current will be closed to zero. Hence, a voltage equal to the voltage across the capacitor minus the load voltage will appear across L_2. Since L_1 is closely coupled to L_2, the same voltage will appear across L_1. The

Fig. 9.32 *Series inverter with coupled inductors*

voltage across L_1 will tend to increase the cathode potential of SCR T_1 more than its anode potential and therefore, SCR T_1 will be reverse-biased and turn-off. Thus, even if SCR T_2 is turned-on before SCR T_1 is switched-off, it will not result into short circuiting of the d.c. source. A similar operation will take place if SCR T_1 is triggered before SCR T_2 is turned-off.

Hence, we can increase the limit of output frequency to more than the resonance frequency of the R-L-C resonant circuit.

Further, the drawback of high pulsed current from the d.c. supply can be overcome in a half bridge configuration as shown in Fig. 9.33, where $L_1 = L_2$ and $C_1 = C_2$. The power is drawn from the d.c. source during both half-cycles of output voltage. One-half of the load current is supplied by capacitor C_1 or C_2 and the other-half by the d.c. source.

Initially, capacitor C_2 is assumed to be charged to voltage E_c with upper plate negative and lower plate positive. As the capacitors C_1 and C_2 together are connected across the battery, the total voltage across C_1 and C_2 should be equal to E_{dc}. Therefore, C_1 will be charged to $(E_{dc} + E_c)$ value with upper plate positive. The various voltage and current waveforms of this circuit are shown in Fig. 9.34.

As shown in Fig. 9.34, thyristor T_1 is triggered at instant '0'. With this, the two currents flow through the thyristor T_1 and load R. Current i_1 flows through the path $E_{dc}+ - T_1 - L_1 - R - C_{2+} - C_{2-} - E_{dc-}$, thus charging C_2. Capacitor C_1, which is already charged at this instant will provide the second current i_2. This discharge current i_2 has the path $C_{1+} - T_1 - L_1 - R - C_1 -$.

Since the driving voltage $(E_{dc} + E_c)$, the capacitors C_1, C_2 and the initial conditions are identical for both these paths, the two currents i_1 and i_2 will always be equal. Hence, 50% of the load current is drawn from the input source and 50% from the discharge of the capacitor.

At the end of the half-cycle (instant P), when the load current becomes zero, SCR T_1 will be turned-off and the voltage across the capacitors reversed. In the steady state, capacitor C_2 will be charged to voltage $(E_{dc} + E_c)$ in the opposite direction and capacitor C_1 to E_c. When SCR T_2 is triggered at instant Q, identical operations will take place in the following negative half-cycle.

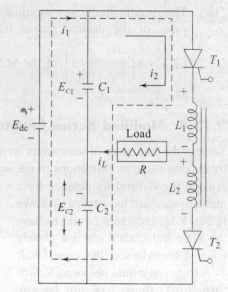

Fig. 9.33 *Half-bridge improved series inverter*

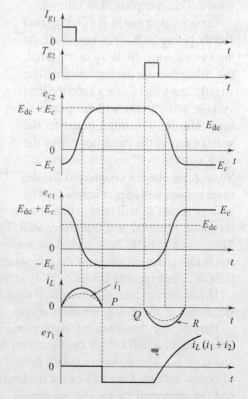

Fig. 9.34 *Voltage and current waveforms*

Again, 50% of the load current is obtained from the d.c. input source and the rest 50% from the discharge of capacitor C_2. Hence, the input d.c. supply no more remains intermittent in nature and the ripples are reduced to the minimum.

The design criteria for the commutating components (L_1, L_2, C_1, C_2) are the same as those discussed in Section 9.11.3. The peak forward off-state voltage for the SCRs is ($E_{dc} + E_c$) and E_c is the peak reverse voltage. In the steady state, the voltage E_c is given by Eq. (9.135).

9.11.5 Three-Phase Series Inverter

A three-phase series inverter can be developed using three separate single-phase inverters and connecting them in a manner as shown in Fig. 9.35. Each phase can function as an independent series inverter if the capacitors across the d.c. supply are large enough so as to establish a constant neutral voltage. The elements, L, C and R connected in series with each thyristor and are so selected that the current response is underdamped as in the basic series inverter.

The thyristors are triggered in the sequence T_1, T_6, T_2, T_4, T_3 and T_5. The firing frequency being six times the output frequency. Therefore, the interval between successive firing will be $T/6$, where T is the period of the output. If coupled inductors similar to those in the modified inverter were used in each leg of the inverter, then T_4 can be turned-on before the current in T_1 has gone to zero, as explained in Section 9.11.4. Similarly, the thyristors in the other leg can likewise be turned-on. However, if the inductors in each leg are not coupled, one thyristor in a leg can be turned-on only when the current through the other thyristor in the same leg has decreased to zero. The design considerations are the same as those for a single-phase series inverters.

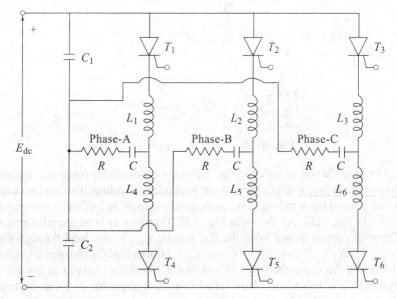

Fig. 9.35 *Three-phase series inverter* ($L_1 = L_2 = L_3 = L_4 = L_5 = L_6 = L$)

9.11.6 High Frequency Series Inverter

As mentioned earlier, the output frequency of the series inverter is limited to its resonant frequency. However, using a coupled-inductor, it is possible to increase the range of inverter frequency slightly above the circuit resonant frequency. But, ultimately the device turn-off time specification limits the maximum possible frequency at which a series inverter can be designed and operated. The high frequency output, up to 30 kHz, can be obtained by using the inverter grade thyristors. One method of designing an inverter which can operate at a high-frequency with thyristors having normal turn-off time specifications (i.e. converter-grade SCRs) is to employ the *time sharing principle*. In time sharing inverter, a number of series inverters are used in parallel as shown in Fig. 9.36, and are operated one at a time in a specified sequence. In an *n*-stage series inverter, each inverter will be active and will supply the load current for $(1/n)^{th}$ of the whole period.

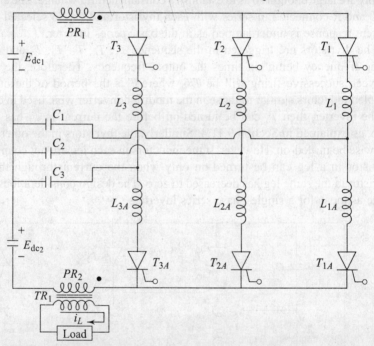

Fig. 9.36 *Time sharing inverter circuit*

The time sharing inverter of Fig. 9.36 consists of a high-frequency transformer having two primary windings and one secondary winding. The load is connected to the secondary winding. The associated voltage and current waveforms are shown in Fig. 9.37. As shown in Fig. 9.37, thyristor T_1 is triggered at instant '0'. Charging current flows from the d.c. source E_{dc_1} to the load through the path $E_{dc_1+} - PR_1 - T_1 - L_1 - C_1 - E_{dc_1-}$. Capacitor C_1 charges to voltage E_c which will be more than E_{dc_1}. Capacitor C_1 is fully charged at instant P and thyristor T_1 is turned-off. Now, after instant P, thyristor T_3A can be triggered which will give the charging path for capacitor C_3. Capacitor C_3 therefore, charges through the path $E_{dc_2+} - C_3 - L_3A - T_3A - PR_2 - E_{dc_2-}$. This current will flow into the dot in second-primary PR_2 and so load current in secondary will be in the direction opposite to earlier one. After SCR $T_3 A$ is turned-off, SCR T_2 is turned-on. Thus, the SCRs are triggered in the sequence T_1, T_3A, T_2, T_1A, T_3 and $T_2 A$, producing alternate positive and negative half cycles of current in the load.

Voltage E_c across every capacitor, in the steady state, at the end of the conduction period of the corresponding SCR, will be the same and therefore, both the positive and negative half-cycle load current waveforms will be identical, as shown in Fig. 9.37. The main advantage of the circuit is that after each SCR is turned-off a reverse voltage appears across the thyristor until its complementary SCR is triggered since the voltage E_c across the capacitor will be more than the supply voltage. For example, SCR T_1 will have a reverse voltage after it is turned-off until SCR T_1A is triggered.

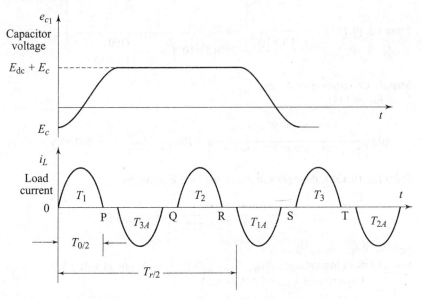

Fig. 9.37 *Voltage and current waveforms*

The reverse-bias on thyristor T_1 will exist during the time SCRs T_3A and T_2 conduct, assuming that the output frequency is equal to the ringing frequency. In other words, if each thyristor has a ringing frequency fr $(1/T_r)$ and the output frequency is f_0 $(1/T_0)$ [for optimum design, $T_r = n\, T_0$ for an n-stage inverter, where T_0 is the period of each resonant circuit], then the SCRs will be reverse-biased for approximately a duration of $(T_r - T_0)/2$. This specifies the turn-off time requirement of each SCR. By selecting a high value of T_r, T_0 can be made as small as possible for a given turn-off time of the SCR and hence, high frequency output can be obtained.

SOLVED EXAMPLE

Example 9.18 Design a series inverter circuit for operation in the frequency range 1 to 5 kHz. The load resistance may vary from 25 Ω to 100 Ω. The peak load current is limited to 3 A and the supply voltage is 100 V.

Solution:
Step 1: *Design of inductance L*
Let us assume attenuation factor = 0.5

From Eq. 9.136(b), and $fr = 5$ kHz $\therefore L = \dfrac{-100}{8 \times 5 \times 10^3 \times \ln(0.5)} = 3.6$ mH.

Step 2: *Design of capacitance C*

From Eq. (9.137), $C = \dfrac{1}{3.6 \times 10^{-3}} \left[\dfrac{1}{\left(2\pi \times 5 \times 10^3\right)^2 + \dfrac{(100)^2}{4 \times (3.6 \times 10^{-3})^2}} \right]$ $C = 0.24$ μF.

Step 3: *Calculation of E_c and I_{peak}*
From Eq. (9.135),

$$E_c = 100 \left[\dfrac{e^{-25\pi/2 \times 2\pi \times 5 \times 10^3 \times 3.6 \times 10^{-3}}}{1 - e^{-25\pi/2 \times 2\pi \times 5 \times 10^3 \times 3.6 \times 10^{-3}}} \right] = 100 \left[\dfrac{0.7066}{1 - 0.7066} \right] = 240.83 \text{ V}$$

From Eq. (9.138), the peak-current in the load is given by

$$I_{peak} = \dfrac{100 + 240.83}{2\pi \times 5 \times 10^3 \times 3.6 \times 10^{-3}} \times 0.5 = 1.5 \text{ A.}$$

Step 4: *Selection of thyristor*
Forward blocking voltage rating, V_{Bo}, $V_{Bo} \geq E_c + E_{dc} \geq 340.83 \approx 400$ V.
Current $I_T \geq I_{peak} \geq 1.5$ A $I_T \approx 2$ A

Now $T_{off} = \pi \left(\dfrac{1}{\omega_0} - \dfrac{1}{\omega_r} \right) = \pi \left(\dfrac{1}{2\pi \times 1 \times 10^3} - \dfrac{1}{2\pi \times 5 \times 10^3} \right) \approx 0.4$ ms.

$\therefore \ t_q \leq T_{off}, t_q \approx 0.3$ ms.

9.12 SELF-COMMUTATED INVERTERS

Inverters discussed in Section 9.11 are also called as self-commutated inverters since they do not use any separate commutation circuits. The self commutated inverter, which uses Class-A type of commutation circuit, is shown in Fig. 9.38. In this circuit, capacitor C and inductor L_1 form an underdamped resonant circuit with the load. Here, thyristors T_1, T_2 and T_3, T_4 are triggered in pairs. When thyristors T_1 and T_2 are triggered, the supply current flows through the path E_{dc+} — T_1 — L_1 — C — L_1 — T_2 — E_{dc-}. The capacitor C will get charged with the polarity shown. When capacitor C is completely charged, current through thyristors T_1 and T_2 reduces and both thyristors become turned-off. The voltage across the capacitor will be more than the supply voltage, at the instant of zero-current. Therefore, the capacitor C will begin discharging into the supply E_{dc} through feedback diodes D_{1A} and D_{2A}, and the discharging path becomes $C_+ - L_1 - D_{1A} - E_{dc} - D_{2A} - L_1 - C_-$. These diodes apply a reverse voltage across thyristors T_1 and T_2. Now, thyristors T_3 and T_4 will be triggered at the end of the half-period of the output. Diodes D_{1A} and D_{2A} become reverse biased and the current will shift to the SCRs T_3 and T_4. Capacitor C will then begin charging in the opposite direction. As in series-inverters of Section 9.11, here also it is necessary that the output frequency f_o must be lower than the resonant frequency $fr \left[= \dfrac{1}{2\pi\sqrt{2L_1 C}} \right]$ of the commutating circuit. In other words, the effective load power-factor must be leading so that the SCR current may become zero before the load voltage polarity reverses.

Since the capacitor is connected in parallel with the load, the load current waveform is continuous over a wide range of output frequency. This is an advantage over the series-inverter of Section 9.11. The design steps for commutating components L_1 and C are as follows:

1. The resonant frequency fr is given by the formula

$$fr = \frac{1}{2\pi\sqrt{2L_1 C}} \qquad (9.139)$$

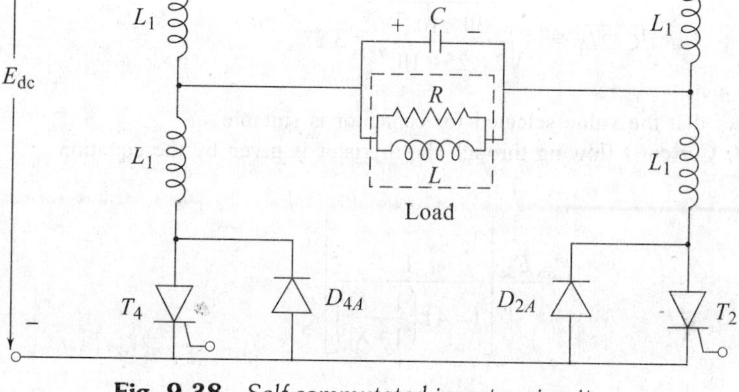

Fig. 9.38 *Self-commutated inverter circuit*

2. The ratio fr/fo is chosen properly. This ratio generally used is 1.39. A lower value of this ratio gives a shorter turn-off time and higher values produce waveform distortion.
3. The ratio L/L_1 is chosen very properly. The ratio generally used is 200. A lower value produces greater component voltage and high values cause waveform distortion.
4. Capacitor C is chosen such that $R\sqrt{C/2L_1}$ lies between 3 and 5.
5. Capacitor C can also be chosen from the following equation

$$C = \frac{5\pi}{6\omega_r \cdot R \ln 2} \tag{9.140}$$

where
$$\omega_r = \sqrt{\frac{1}{LC} - \frac{1}{4R^2C^2}} \tag{9.141}$$

SOLVED EXAMPLES

Example 9.19 Design a self-commutated inverter circuit to operate at a frequency of 3 kHz with an optimum distortion. The load specifications are as follows: $R = 5\,\Omega$, $L = 5$ mH, $E_{dc} = 100$ V. Also compute the output power.

Solution:
Step 1: Resonant frequency, $f_r = 1.35 f_o = 1.35 \times 3$ kHz $= 4.05$ kHz.

Step 2: Since $\quad \dfrac{L}{L_1} = 200 \Rightarrow L_1 = \dfrac{L}{200} = \dfrac{5\times 10^{-3}}{200} = 25\,\mu H$

Step 3: Selection of capacitor C

From Eq. (9.139), $f_r = 4.05 \times 10^3 = \dfrac{1}{2\pi\sqrt{2 \times 25 \times 10^{-6} \times C}} \quad \therefore C \approx 30\,\mu F.$

Now, $\quad R\sqrt{C/2L_1} = 5\sqrt{\dfrac{30 \times 10^{-6}}{2 \times 25 \times 10^{-6}}} = 3.87$

employs that the value selected for capacitor is suitable.

Step 4: Current I flowing through the thyristor is given by the equation

$$I = \frac{E_{dc}}{Z_0}\left[\frac{1}{1 - 4\tau\left(\dfrac{1-K}{1+K}\right)}\right]$$

where constant $\tau = RC = 150\,\mu s.$

Attenuation factor, $K = e^{-T/2\tau}$

where $T = \dfrac{1}{f_0} = \dfrac{1}{3 \times 10^3} = 0.33$ msec $\quad \therefore K = 0.333$

Output impedance, $Z_0 = \dfrac{R \times j\omega L}{R + j\omega L} = \dfrac{5 \times j \times 2\pi \times 3 \times 10^3 \times 5 \times 10^{-3}}{5 + j \times 2\pi \times 3 \times 10^3 \times 5 \times 10^{-3}} \approx 5 \angle 3°$

$\therefore \quad I = \dfrac{100}{5} \left[\dfrac{1}{1 - 4 \times 150 \times 10^{-6} \left(\dfrac{1 - 0.333}{1 + 0.333} \right)} \right] = 20$ A

Step 5: Maximum output voltage ($E_{o(\text{max})}$) is given by the relation,

$$E_{o(\text{max})} = I \cdot Z_o \left(1 - \dfrac{2K}{1+K} \right) = 20 \times 5 \left(1 - \dfrac{2 \times 0.333}{1 + 0.333} \right) = 50.03 \text{ V}$$

$$E_{o(\text{rms})} = 50.03/\sqrt{2} = 35.38 \text{ V}$$

Step 6: Selection of thyristor

$V_{Bo} \geq 2 (E_{dc} + E_{o(\text{max})}) \geq 2 (100 + 50.03) \geq 300.06$ V

$I_T \geq 2.I \geq 2 \times 20 \geq 40$ A

$t_q \leq \quad t \cdot \ln \left[\dfrac{2}{1+K} \right] \leq 150 \times 10^{-6} \ln \left[\dfrac{2}{1 + 0.333} \right] \leq 60.85 \text{ μs}.$

Capacitor voltage, $V_c = 2 E_{o(\text{max})} = 2 \times 50.03 = 100.06$ V

Now, output power, $P_o = E_o I_o \cos \phi = 50.03 \times \dfrac{35.38}{5} \cos (3°) = 353.53$ W.

9.13 PARALLEL INVERTER

A parallel inverter is used to produce a square-wave from a d.c. supply. Basically, by alternately switching the two thyristors, the d.c. source is connected in alternative sense to the two halves of the transformer primary, thereby inducing a square-wave voltage across the load in the transformer secondary.

Figure 9.39 shows the parallel inverter cirucit. Two diodes are connected as shown in Fig. 9.39 to feedback the stored load energy during those periods when the load current reverses relative to the voltage. Figure 9.40 shows the load voltage and current waveforms of parallel inverter.

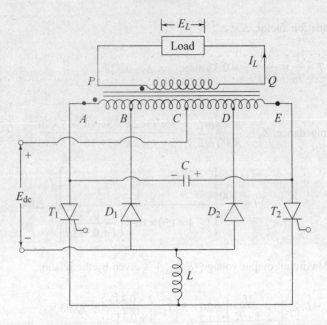

Fig. 9.39 *Parallel inverter with feedback diodes*

Thyristors T_1 and T_2 are the main load carrying thyristors. Inductor L and capacitor C are the commutating components. Diodes D_1 and D_2 are the feedback diodes, which permit the load reactive power to be fed back to the d.c. supply. The circuit operation can be divided into different operating modes.

Mode 1: During this mode, thyristor T_1 is triggered at instant R, as shown in Fig. 9.40. Battery voltage now forces the current to the primary section CA through the path $E_{dc+} - C - A - T_1 - L - E_{dc-}$. Thus, neglecting the small voltage drop across L, the supply voltage E_{dc} will appear across the left-half of the transformer primary winding CA. Terminal C is positive with respect to terminal A. The flux produced due to this current induces the voltage in all sections of transformer winding as number of turns in section CA are assumed to be equal to number of turn in secondary. The load voltage (secondary voltage) is nearly equal to E_{dc} and is in such a direction so as to force a current into the dot at terminal P. This is because, the current in primary leaves the dot at terminal A.

Now, due to autotransformer action, voltage E_{dc} is induced in CE section of primary winding. Therefore, terminal E will be at a potential of $2\,E_{dc}$ with respect to A. Thus, capacitor C will get charged to twice the supply voltage (i.e. $2\,E_{dc}$) with the polarity as shown in Fig. 9.39. Since the load is inductive, load current increases gradually from instant R, as shown in Fig. 9.40.

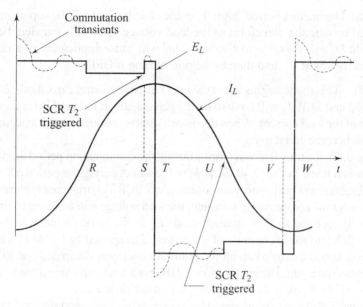

Fig. 9.40 *Load voltage and current waveforms*

Mode 2: This mode begins when thyristor T_2 is switched ON at instant S, as shown in Fig. 9.40. When T_2 is turned-on, capacitor C will immediately apply a reverse voltage of 2 E_{dc} across SCR T_1 and turns it OFF. When SCR T_1 is turned-off, the capacitor will discharge through SCR T_2, inductor L, diode D_1, and a portion of a transformer primary winding BA. Thus, the energy stored in the capacitor will be fed back to the load through the transformer coupling of windings BA and PQ. During this period, the potential of point B will be fixed by the d.c. input supply and the load voltage will still be positive but more than E_{dc}.

The load current, which earlier was flowing through SCR T_1, will now flow through CB and diode D_1 to the negative input terminal. This can happen only if diode D_1 is forward biased and the capacitor discharge current is more than the load current. As the potential of point B increases sufficiently to reverse bias diode D_1, the capacitor will no longer discharge through SCR T_2 and point B will not get connected to the supply negative terminal. The current through inductor L will now flow through diode D_2, DE and SCR T_2, and the trapped energy in inductor L will be fed back to the load.

The load current T_1, which earlier was flowing through CB, will now flow from D to C through diode D_2, and the load reactive energy will be returned to the d.c. supply. Since point D is now connected to the negative supply terminal, the load voltage polarity will be reversed and more than E_{dc}. Also, the capacitor C will be charged in the opposite direction to slightly more than twice the supply voltage. Thyristor T_2 will stop conducting after all the energy in the commutating inductor L has been completely dissipated in the load. Immediately following the commutation of SCR T_1, energy is transferred from the capacitor and inductor to the load. During this period, high frequency oscillations will be superimposed on normal rectangular waveform of the load voltage. After this transient period, only diode D_2 will continue to conduct. This will cause application of a reverse voltage across SCR T_2, and thereby help in turning it OFF.

Mode 3: This mode begins when the load current becomes zero, diode D_2 will be blocked and SCR T_2 will have to be triggered again at instant U to reverse the direction of the load current. When thyristor T_2 starts conducting, the load voltage will again become equal to E_{dc}.

In Fig. 9.40, the transient waveforms during the commutation period are shown by the dashed lines. In Fig. 9.40, SCR T_1 will conduct during the period RS, when both load voltage and load current are positive. At S, SCR T_2 is triggered to commutate SCR T_1. After the commutation transient, the load voltage will be reversed and the current will continue to flow through diode D_2 in the same direction as before. SCR T_2 will be turned-off because of the reverse bias applied by D_2. At U, the load current will become zero when all the inductive energy is dissipated and SCR T_2 will be triggered again. During the period UW, both load voltage and load current are reversed. At W, SCR T_1 is triggered to turn-off SCR T_2.

Since the SCRs have to be triggered twice in each half cycle, and since interval SU is load dependent, it is necessary that the SCRs be gated by a train of pulses for a minimum duration of a quarter cycle. It is noted that the load voltage will rise above E_{dc} during period SU when the feedback diodes conduct. In Fig. 9.39, if these feedback diodes are connected to points A and E, then the load voltage waveforms will be rectangular. But, such a connection will require the energy trapped in the commutating components to be dissipated as heat in the SCRs and diodes, thereby necessitating the de-rating of the components. Thus, the efficiency of the circuit can be increased by connecting the diodes D_1 and D_2 to the tap points.

1. Design Aspects

Design of commutating components: The design of the commutating components of the parallel inverter circuit of Fig. 9.39 is based on the following assumptions:
1. The load-current waveform should be rectangular with an amplitude I_L.
2. The duration for which the capacitor applies reverse voltage across the turned-off SCR should be slightly more than the turn-off time t_q of the SCR.
3. The peak discharge current from the capacitor through SCR T_2, inductor L and diode D_1 should be twice the normal load current.
4. The load is assumed to be purely resistive so that load current along with the load voltage also reverse polarity.

In Fig. 9.39, when SCR T_2 is turned on, the capacitor discharge current will be given by the equation

$$i(t) = A \sin \omega t + B \cos \omega t \tag{9.142}$$

where $\omega = 1/\sqrt{LC}$, and L and C are the commutating components A and B are the constants which are to be determined from the initial conditions, namely,

Inverters

$$i(0) = I_L' \text{ and } \left.\frac{di}{dt}\right|_{t=0} = \frac{2E_{dc}}{L}$$

By substituting first condition in Eq. (9.142), we get, $B = I_L'$ (9.143)

Now, differentiating Eq. (9.142) w.r.t. t, we get

$$\frac{di(t)}{dt} = A\omega \cos \omega t - BW \sin \omega t$$

and $\left.\dfrac{di}{dt}\right|_{t=0} = AW$ (9.144)

Substituting second condition in Eq. (9.144), we get

$$AW = \frac{2E_{dc}}{L} \quad \therefore A = \frac{2E_{dc}}{LW} = \frac{2E_{dc}}{L}\sqrt{LC}$$

$$A = 2E_{dc}\sqrt{\frac{C}{L}} \tag{9.145}$$

Substituting value of A and B constants from Eqs (9.143) and (9.145) into Eq. (9.142), we get, $i(t) = 2E_{dc}\sqrt{\dfrac{C}{L}} \sin \omega t + I_L' \cos \omega t$ (9.146)

Now, the peak discharge current $\left[2E_{dc}\sqrt{\dfrac{C}{L}}\right]$ from the capacitor should be twice the reflected load current, i.e.

$$2E_{dc}\sqrt{C/L} = 2I_L', \text{ or, } E_{dc}\sqrt{C/L} = I_L' \tag{9.147}$$

If the transformer turns ratio is unity, then the reflected load current becomes equal to normal load current and Eq. (9.147) becomes

$$E_{dc}\sqrt{C/L} = I_L \tag{9.148}$$

Diode D_1 conducts till current $i(t)$ reaches the peak value, and SCR T_1 will be reverse-biased as long as diode D_1 conducts. This duration must be at least equal to the turn-off time t_q of SCR T_1. With these considerations, t_q is given by

$$t_q \approx \frac{\pi}{3}\sqrt{LC} \tag{9.149}$$

The duration (t_{off}) for which the SCRs will be reverse biased is given by

$$t_{off} = \frac{1}{\tau} \ln 2 \tag{9.150}$$

where constant $\tau = \dfrac{1}{4R_L C}$ (9.151)

The duration (t_{off}) must be greater than the turn-off time (t_q) of the SCR, for proper commutation.

From Eqs (9.147) and (9.149), the values of commutating components can be obtained. The voltage rating of capacitor should be greater than twice the supply voltage E_{dc} and the current rating of inductor should be more than twice the reflected load current, i.e.

Voltage rating of capacitor $(V_c) \gg 2\,E_{dc}$
and current rating of inductor $\gg 2\,I_L'$.

Selection of SCR For proper operation, the forward blocking voltage rating of the SCR should be greater than twice the supply voltage, i.e. $V_{BO} \geq 2\,E_{dc}$ and, the peak-current rating of SCR should be greater than twice the reflected load current, i.e. $I_T \geq 2\,I_L'$

SOLVED EXAMPLE

Example 9.20 Design a parallel inverter to feed a load at 200 V, 50 Hz and peak-load current is 2 A, E_{dc} = 40 V. Specify the rating of SCRs, transformer and commutating components.

Solution:
Step 1: Calculation of commutating components from Eq. (9.148),

$$40\sqrt{C/L} = \frac{200}{40}2, \sqrt{C/L} = 0.25 \tag{a}$$

Now, from Eq. (9.149), and by assuming t_q = 40 µs, we can write

$$\sqrt{CL} = \frac{3 \times 40 \times 10^{-6}}{\pi} \sqrt{CL} = 38.2 \times 10^{-6} \tag{b}$$

Solving Eqs (a) and (b), we get L = 152.8 µH. C = 9.55 µF.
Now, voltage rating of capacitor, $V_c \geq 2\,E_{dc} \geq 2 \times 40 \geq 80$ V.

Step 2: *Design of output transformer*
The output transformer has a centre-tapped primary winding with a voltage rating of 2 E_{dc} = 80 V, for each side and secondary winding with a voltage rating of about 500 V. The turns ratio of the secondary winding to one-half of the primary winding is 6.

The design of output-transformer is based on the fundamental frequency terms. A square wave of amplitude 200 V has a fundamental frequency amplitude given by 415 V. Assume a core flux density,
B_{max} = 1.0 Wb/m^2 and a core-cross section = 25 cm^2.
∴ Number of turns on the secondary side,

$$N_2 = \frac{415}{\sqrt{2} \times 4.44 \times 50 \times 25 \times 10^{-4}} = 530$$

∴ $\dfrac{N_2}{1/2 N_1} = 6, \quad \therefore N_1 = \dfrac{2 N_2}{6} = \dfrac{2 \times 530}{6}$ N_1 = 176.67.

Step 3: *Selection of thyristor*

$$V_{Bo} \geq 2\,E_{dc} \geq 2 \times 40\text{ V} \geq 80\text{ V}$$

and $\quad I_T \geq I_L' \geq 2 \times \dfrac{200}{4} \times 2 \geq 20$ A.

9.14 THE SINGLE-PHASE SCR BRIDGE INVERTER

There are various configurations in inverters and among them the bridge inverters are popular because they can be easily extended for multi-phase operation, pulse-width modulation, etc. Also, the output transformer is not essential in bridge inverters. In this section, certain bridge inverter configurations and their principle of operation will now be discussed.

9.14.1 Single-Phase Half-Bridge Inverter

Figure 9.41(a) gives the circuit configuration of a single-phase half-bridge inverter. For this basic circuit configuration, the triggering circuit and the commutation circuit are not shown for simplicity. The gating signals for the thyristors and the resulting output voltage waveform are shown in Fig. 9.41(b).

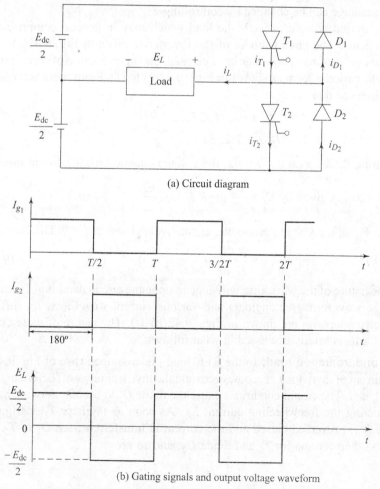

Fig. 9.41 *Single phase half-bridge inverter*

As shown in Fig. 9.41(b), for the interval $0 < t \le T/2$, thyristor T_1 is conducting and load is subjected to a voltage $E_{dc}/2$ due to the upper voltage source $E_{dc}/2$. At instant $t = T/2$, thyristor T_1 is commutated and T_2 is turned-on. During the interval $T/2 < t \le T$, thyristor T_2 conducts and the load is subjected to a voltage $\left(-\dfrac{E_{dc}}{2}\right)$ due to the lower voltage source $\dfrac{E_{dc}}{2}$. Each thyristor is gated at frequency $f = 1/T$ of the a.c. supply desired. The gating signal of the two thyristors have a phase angle of 180°. From Fig. 9.41(b), the output is easily seen to be rectangular a.c. waveform of frequency,

$$\omega = \frac{2\pi}{T} \text{ rad/s}$$

where T is the triggering period of the thyristors. Frequency of the inverter output voltage can be changed by controlling T.

The output waveform feeds the load which may in general comprise RLC components. The circuit model of the inverter is given in Fig. 9.42(a). After several cycles of source voltage have elapsed, the time variation of current settles down to periodic form as shown in Figs. 9.42(c) to (f). From these waveforms, it is observed that

$$i_L = -I_L;\ t = 0,\ T,\ 2T,\ \ldots$$

and
$$i_L = I_L;\ t = T/2,\ 3T/2,\ \ldots$$

During the interval $0 < t < t/2$, the voltage equation for the circuit model of Fig. 9.42(a) is given by
$$E_L = \frac{E_{dc}}{2} = R\,i_L + L\frac{di_L}{dt} + \frac{1}{C}\int_0^t i_L\,dt + V_{co} \qquad (9.152)$$

where V_{co} is the voltage across the capacitive element at $t = 0$. Differentiating Eq. (9.152), we get,
$$\frac{d^2 i_L}{dt^2} + \frac{R}{L}\frac{di_L}{dt} + \frac{1}{LC} i_L = 0 \qquad (9.153)$$

The nature of the waveform will depend upon the circuit damping. The output voltage waveform (rectangular) and various current waveforms for different load characteristics are drawn in Figs. 9.42(b)–(f). The nature of these current waveforms is briefly discussed in what follows:

RLC underdamped load: In the RLC load underdamped case of Fig. 9.42(c), the current of thyristor T_1 becomes zero and the thyristor turns-off before thyristor T_2 is gated. The circuit conditions cause the diode D_1 to become forward biased, conducting the freewheeling current i_{D1}. As soon as thyristor T_2 is triggered, diode D_1 is reversed-biased and the current is transferred from D_1 to T_2. This process then repeats for T_2 and diode D_2, and so on.

Inverters

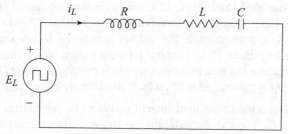

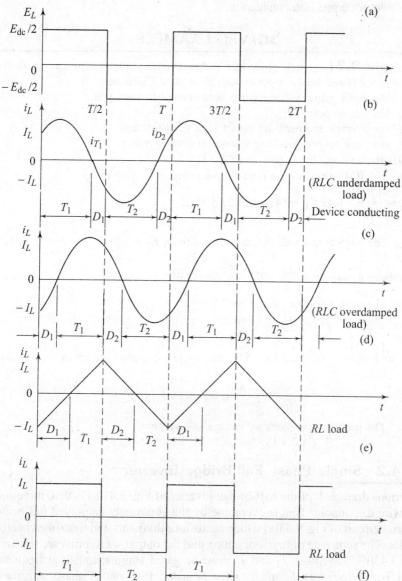

Fig. 9.42 *Circuit model and waveforms*

RL and RLC overdamped load: In the RLC load overdamped case of Fig. 9.42(d) and the *RL* load of Fig. 9.42(e), T_1 is forced to switch OFF (at $T/2$) while it is still carrying current. As a consequence, the voltage across the *L*-component of the load reverses, causing diode D_2 to become forward-biased which then conducts the freewheeling current. As the circuit current tends to reverse, diode D_2 switches OFF and the current is conducted by T_2 which has already been gated at $t = T/2$.

R load: For the ideal *R* load case, load current waveform i_L is identical with load-voltage waveform E_L, there is theoretically no freewheeling current and diodes D_1–D_2 do not come into conduction.

Solved Example

Example 9.21 Single-phase half-bridge-inverter of Fig. 9.41(a) has a resistive-load of $R = 3\ \Omega$ and the d.c. input voltage $E_{dc} = 50$ V. Calculate
  (i) the RMS output voltage at the fundamental frequency E_1.
 (ii) the output power, P_o.
(iii) the average and peak current of each thyristor, and
 (iv) the peak reverse-blocking voltage of each thyristor.

Solution: Given: $E_{d.c.} = 50$ V and $R = 3\ \Omega$.
  (i) The RMS value of the fundamental component is given by

$$E_1 = \frac{2\,E_{dc}}{\sqrt{2}\,\pi} = 0.45\,E_{dc} = 0.45 \times 50 = 22.5\text{ V}.$$

 (ii) The RMS output voltage can be found from, $E_L = \dfrac{E_{dc}}{2} = \dfrac{50}{2} = 25$ V

$\therefore\ $ Output power, $P_o = \dfrac{E_L^2}{R} = \dfrac{(25)^2}{3} = 208.33$ W.

(iii) The peak thyristor current

$$I_{peak} = \frac{E_{dc}/2}{R} = \frac{50/2}{3} = 8.33\text{ A}.$$

Since each thyristor conducts for a 50% duty cycle, the average current of each thyristor is

$$I_{av} = \frac{50}{100} \times I_{peak} = 0.5 \times 8.33 = 4.165\text{ A}.$$

 (iv) The peak reverse blocking voltage is given by
$$V_{BR} = 2 \times E_{dc}/2 = 2 \times 50/2 = 50\text{ V}$$

9.14.2 Single Phase Full-Bridge Inverter

A serious drawback of the half-bridge inverter of Fig. 9.41(a) is that, it requires a 3-wire d.c. supply. This is overcome by the commonly employed full-bridge inverter circuit of Fig. 9.43(a) which needs four thyristors and four-freewheeling diodes. The sequence of thyristor gating and the output waveforms are shown in Fig. 9.43(b). Thyristors T_1 and T_2 must be gated simultaneously at frequency $f = 1/T$ and thyristors T_3 and T_4 must be gated 180° out of phase with these. Frequency of output voltage can be controlled by varying the periodic time *T*.

When thyristors T_1 and T_2 conduct, the load voltage will be positive, i.e. E_{dc} and when thyristors T_3 and T_4 conduct, the output voltage will be negative, i.e. $(-E_{dc})$, as shown in Fig. 9.43(b). Diodes D_1 to D_4 serve to feed the load reactive power back to the d.c. supply. As shown in Fig. 9.43(b), the load voltage waveform is fairly rectangular and is not affected by the nature of the load. This is an advantage of bridge inverter over the series inverter. The circuit model of single-phase full-bridge inverter is same as shown in Fig. 9.42(a). The load voltage and current waveforms for single-phase full-bridge inverter will be same as that of Figs. 9.42(b)–(f), but the components conducting period will be different. In place of SCR T_1, here two thyristors, T_1 and T_2 conduct. Similarly, in place of SCR T_2, thyristors, T_3 and T_4 conduct and in place of D_1, diodes D_1, D_2 conduct, whereas instead of D_2, here D_3, D_4 conduct.

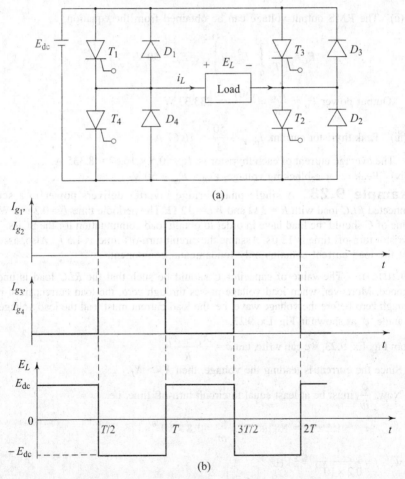

Fig. 9.43 *Single-phase full-bridge inverter*

Some of the commonly used commutation circuits for bridge inverters will be described in the following sections.

SOLVED EXAMPLES

Example 9.22 Single-phase full-bridge inverter of Fig. 9.43(a) has a resistive load of $R = 3\ \Omega$ and the d.c. input voltage $E_{dc} = 50$ V. Compute
(i) the RMS output-voltage at the fundamental frequency E_1
(ii) the output power P_0
(iii) the average and peak currents of each thyristor
(iv) the peak reverse-blocking voltage of each thyristor.

Solution: Given: $E_{dc} = 50$ V, $R = 3\ \Omega$

(i) The RMS value of fundamental component is given by

$$E_1 = \frac{4 E_{dc}}{\sqrt{2}\,\pi} = 0.90\,E_{d.c.} = 0.90 \times 50 = 45\text{ V}.$$

(ii) The RMS output voltage can be obtained from the equation

$$E_L = \left(\frac{2}{T}\int_0^{T/2} E_{dc}^2\, dt\right)^{1/2} = E_{dc} = 50\text{ V}.$$

∴ Output power $P_o = \dfrac{E_{dc}^2}{R} = \dfrac{(50)^2}{3} = 833.33$ W.

(iii) Peak thyristor current, $I_{peak} = \dfrac{50}{3} = 16.67$ A

∴ The average current of each thyristor is, $I_{av} = 0.5 \times 16.67 = 8.335$ A.

(iv) Peak reverse-blocking voltage, $V_{BR} = E_{dc} = 50$ V.

Example 9.23 A single phase-bridge inverter delivers power to a series connected RLC load with $R = 3\ \Omega$ and $WL = 12\ \Omega$. The periodic time $T = 0.2$ ms. What value of C should the load have in order to obtain load commutation for the SCRs. The thyristor turn-off time is 12 μs. Assume the circuit turn-off time as $1.5\,t_q$. Also, assume that the load current contains only fundamental component.

Solution: The value of capacitor C should be such that the RLC load is underdamped. Moreover, when load voltage passes through zero, the load current must pass through zero before the voltage wave, i.e. the load current must lead the load voltage by an angle 'ϕ' as shown in Fig. Ex. 9.23.

From Fig. Ex. 9.23, we can write, $\tan\phi = \dfrac{X_C - X_L}{R}$

Since the current is leading the voltage, then $X_C > X_L$.

Now, $\dfrac{\phi}{\omega}$ must be at least equal to circuit turn-off time, i.e.

$$1.5 \times 12 = 1.8\ \mu s.\quad \therefore\ \frac{\phi}{\omega} = 1.8 \times 10^{-6}\text{ s}.$$

Now, $f = \dfrac{1}{0.2 \times 10^{-3}} = 5$ kHz.

∴ $\phi = \omega \times 1.8 \times 10^{-6} = 2 \times \pi \times 5 \times 10^3 \times 1.8 \times 10^{-6} = 0.0565$ rad $\phi = 3.23°$.

∴ $\tan 3.23 = \dfrac{X_C - 12}{3}\ \ X_C = 12.17 = \dfrac{1}{2\pi \times 5 \times 10^3 \times C}\ \ \therefore\ C = 2.62\ \mu F.$

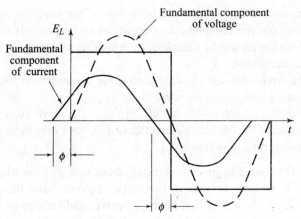

Fig. Ex. 9.23 *Phasor diagram*

9.14.3 The McMurray Inverter (Auxiliary Commutated Inverter)

The McMurray inverter is an impulse commutated inverter which relies on an LC circuit and an auxiliary thyristor for commutation in the load circuit. The impulse is derived from the resonating LC circuit and is applied to turn off a thyristor carrying the load current. A single-phase full-bridge thyristor inverter using auxiliary commutation is shown in Fig. 9.44.

The commutation scheme of Fig. 9.44 was first reported by W. McMurray and therefore, it is called "McMurray scheme." It is also described as "auxiliary commutation scheme" because it requires an additional thyristor for the turning-off of each main thyristor. The circuit consists of the main thyristors T_1, T_2, T_3 and T_4, the freewheeling diodes D_1, D_2, D_3 and D_4, the auxiliary thyristors T_{A1}, T_{A2}, T_{A3} and T_{A4}, and the commutating components L and C. When the thyristor pair T_1 and T_2 conducts, load is connected across the d.c. source causing a positive voltage across the load. When thyristor pair T_3 and T_4 conducts, a negative voltage is produced across the load. Thus, alternatively making the pair of thyristors to conduct, an alternating voltage is produced across the load.

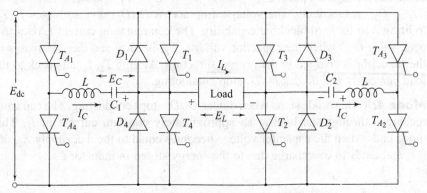

Fig. 9.44 *Single-phase McMurray inverter*

In any inverter, understanding the operation of the commutation process is very important and the commutating circuit has to be properly designed. The operation of the circuit will be described for a lagging pF load with the following simplifying assumptions:

(1) Load current remains constant during the commutation interval.
(2) Thyristors and diodes are ideal switches.
(3) Inductor L and capacitors are ideal in that they have no resistance.

The operation of the McMurray inverter of Fig. 9.44 may be subdivided into various operating modes as follows:

Mode 1: This mode begins when the thyristor pair T_1/T_2 is triggered. When thyristors T_1/T_2 become turned-on, the supply current flows through the path $E_{dc+} - T_1 - Z_L - T_2 - E_{dc-}$ and hence, positive load voltage E_L is obtained.

The commutating capacitors C_1/C_2 are already charged to a voltage E_c with the polarities shown in Fig. 9.44 because of the commutation of the previously conducting thyristors T_4/T_3.

Mode 2: This mode begins when thyristors T_{A_1}/T_{A_2} are triggered to turn-off the main thyristors T_1/T_2 which were conducting. Figure 9.45 depicts the operations after T_{A_1}/T_{A_2} have been turned-on to initiate commutation at time t_o. When thyristors T_{A_1}/T_{A_2} have been turned-on, capacitors C_1/C_2 start discharging. Capacitor C_1 forms the discharging loop $C_{1+} - T_1 - T_{A_1} - L - C_{1-}$, and capacitor C_2 forms the discharging loop $C_{2+} - L - T_{A_2} - T_2 - C_{2-}$, and therefore, current I_c rises taking part of the load current from T_1/T_2. Voltage drop across $T_1(T_2)$ reverse-biases $D_1(D_2)$, current I_C can, therefore, flow only through $T_1(T_2)$ and not through $D_1(D_2)$. As load current I_L is constant, an increase in I_c causes a corresponding decrease in I_{T_1} (I_{T_2}) so that $I_{T_1} = I_L - I_c$ (KCL at node P) and $I_{T_2} = I_L - I_c$ (KCL at node Q).

At t_1, the capacitor current I_c rises to I_L and, therefore, currents I_{T_1} and I_{T_2} become zero. As a result, main thyristors T_1 and T_2 become turned-off at time t_1.

Mode 3: After t_1 as resonant current I_c exceeds I_L, the excessive current $I_C - I_L = I_{D_1}(I_{D_2})$ circulates through feedback diodes $D_1(D_2)$. The resonating oscillation continues through the paths $C_{1+} - D_1 - T_{A_1} - L - C_{1-}$ and $C_{2+} - L - T_{A_2} - D_2 - C_{2-}$, respectively. The voltage drop across $D_1(D_2)$ reverse biases $T_1(T_2)$ to bring it to forward blocking capability. The commutating current I_c rises to a peak value (I_{cp}) when the capacitor voltage (E_c) is zero, and then decreases as the capacitor is charged in the reverse direction. At time T_2, I_c falls back to the load current I_L and diodes D_1/D_2 stop conducting.

Mode 4: This mode starts when diodes D_1/D_2 stops conducting. The capacitor recharges through the load at an approximately constant current of I_L. This mode ends when the capacitor voltage becomes equal to the d.c. supply E_{dc} at $t = t_3$ and tends to overcharge due to the energy stored in inductor L.

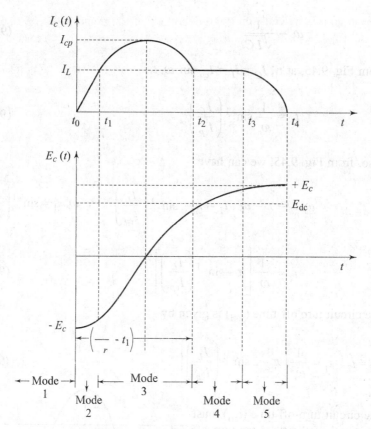

Fig. 9.45 *Voltage and current waveforms*

Mode 5: This mode begins when the capacitor voltage tends to be greater than E_{dc}, and diodes $D_4(D_3)$ become forward-biased. The energy stored in inductor L is transferred to the capacitor, causing it to be overcharged with respect to supply voltage E_{dc}. This mode ends when the capacitor current falls again to zero and the capacitor voltage is reversed to that of original polarity and now the circuit is ready for the next cycle of operation.

During the next half cycle, thyristor pair T_3/T_4 is triggered and a negative half cycle of voltage is produced across the load.

Design Considerations

Design of Commutating Components (L and C) From mode 2 operation, it becomes clear that when thyristors T_{A_1}/T_{A_2} has been turned-on, capacitors C_1/C_2 start discharging. This discharging current is given by the expression

$$I_c = E_c \sqrt{\frac{C}{L}} \sin \omega_r \cdot t = I_{cp} \sin \omega_r \cdot t \tag{9.154}$$

where $\qquad I_{cp} = E_c \sqrt{\dfrac{C}{L}}$ \hfill (9.155)

and
$$\omega_r = \frac{1}{\sqrt{LC}} \qquad (9.156)$$

From Fig. 9.45, at t_1, $I_C = I_L = I_{cp} \sin \omega_r t_1$.

$$\therefore \quad t_1 = \frac{1}{\omega_r} \sin^{-1}\left(\frac{I_L}{I_{cp}}\right) \qquad (9.157)$$

Also, from Fig. 9.45, we can have

$$\omega_r t_2 = \pi - \omega_r \cdot t_1 = \pi - \sin^{-1}\left(\frac{I_L}{I_{cp}}\right) \quad \left[\because \omega_r \cdot t_1 = \sin^{-1}\left(\frac{I_L}{I_{cp}}\right)\right]$$

$$\therefore \quad t_2 = \frac{1}{\omega_r}\left[\pi - \sin^{-1}\left(\frac{I_L}{I_{cp}}\right)\right] \qquad (9.158)$$

The circuit turn off-time (t_{off}) is given by

$$t_{off} = t_2 - t_1 = \frac{1}{\omega_r}\left[\pi - 2\sin^{-1}\left(\frac{I_L}{I_{cp}}\right)\right] \qquad (9.159)$$

The circuit turn-off time (t_{off}) must be greater than the thyristor turn-off time t_q. In practice, this condition, however can be realised by various combinations of values of C and L as shown in Fig. 9.46. The current commutation pulse I_C that gives the required turn-off time with the minimum amount of capacitor energy $\frac{1}{2}CV^2$, is the optimum pulse. This can be achieved from Eqs (9.157) and (9.159) as under.

From Eq. (9.159),

$$\frac{\omega_r \cdot t_{off}}{2} = \frac{\pi}{2} - \sin^{-1}\left(\frac{I_L}{I_{cp}}\right)$$

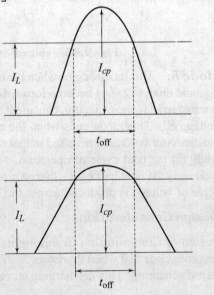

Fig. 9.46 *Pertaining to the choice of L and C for the inverter circuit of Fig. 9.47*

or $\dfrac{I_L}{I_{cp}} = \sin\left(\dfrac{\pi}{2} - \dfrac{\omega_r \cdot t_{off}}{2}\right) = \cos\left(\dfrac{\omega_r \cdot t_{off}}{2}\right)$

Now, let $x = \dfrac{I_{cp}}{I_L}$, so that

$$\cos\left(\dfrac{\omega_r \cdot t_{off}}{2}\right) = 1/x \qquad (9.160)$$

or $\dfrac{t_{off}}{\sqrt{LC}} = 2\cos^{-1}(1/x) = m(x)$ \qquad (9.161)

The energy W that the commutating circuit must provide in order to commutate the main thyristor is, $W = \dfrac{1}{2} C E_c^2 = \dfrac{1}{2} L I_L^2$

Substituting the value of $E_c = I_{cp}\sqrt{\dfrac{L}{C}}$ from Eq. (9.155) in the above relation,

we get, $W = \dfrac{1}{2} C E_c I_{cp} \sqrt{\dfrac{L}{C}} = \dfrac{1}{2}\sqrt{LC}\, E_c I_{cp}$

From Eq. (9.161), $\sqrt{LC} = \dfrac{t_{off}}{2\cos^{-1}(1/x)}$. Substituting this value of $\sqrt{LC}$ in

the above relation, we get, $W = \dfrac{1}{2} \dfrac{t_{off}\, E_c \cdot I_{cp}}{2\cos^{-1}(1/x)} = \dfrac{t_{off}\, E_c \cdot x \cdot I_L}{4\cos^{-1}(1/x)}$

Here, E_c, I_L, t_{off} has the dimensions of the energy. In normalised form, the above relation is

$$\dfrac{W}{E_C \cdot I_L\, t_{off}} = \dfrac{x}{4\cos^{-1}(1/x)} = n(x) \qquad (9.162)$$

On plotting $n(x)$ against x from Eq. (9.162), it is found that normalized commutation energy $n(x)$ has a minimum value of 0.446 when $x = 1.5$.

From Eq. (9.161), $m(x) = 2\cos^{-1}(1/1.5) = 1.68$. The design is carried out on the basis of worst operating conditions which consists of minimum supply voltage $E_{dc(min)}$ and the maximum load current $I_{L(max)}$.

From Eq. (9.155), $E_{dc(min)}\sqrt{\dfrac{C}{L}} = I_{cp} = x\, I_{L(max)} = 1.5\, I_{L(max)}$

or, $\sqrt{\dfrac{C}{L}} = 1.5\, \dfrac{I_{L(max)}}{E_{dc(min)}}$ \qquad (9.163)

From Eq. (9.161), $\sqrt{LC} = \dfrac{t_{off}}{m(x)} = \dfrac{t_{off}}{1.68}$ \qquad (9.164)

Multiplication of Eqs (9.163) and (9.164) gives

$$C = \frac{1.5\, t_{\text{off}} \cdot I_{L(\max)}}{1.68\, E_{\text{dc(min)}}} = 0.893\, \frac{t_{\text{off}} \cdot I_{L(\max)}}{E_{\text{dc(min)}}} \qquad (9.165)$$

Substitute value of C in Eq. (9.164), we get

$$L = \frac{0.397\, E_{\text{dc(min)}}\, t_{\text{off}}}{I_{L(\max)}} \qquad (9.166)$$

However, as explained above, the McMurray circuit requires several commutations to achieve an increased capacitor voltage, following an increment in load current, and this slow build-up may result in a commutation failure for a sudden increase in load. Consequently, it has been suggested that the capacitor voltage should be stabilized at a high value, independent of load, by advanced firing of main thyristor T_4. If T_4 is gated while excess discharge current I_C, is still flowing in diode D_1, a step voltage E_{dc}, is applied to the LC circuit which permits a controlled build-up of the steady-state capacitor voltage to a value that is adequate for commutation of the maximum load current.

The inherent build-up of capacitor voltage in the McMurray circuit may result in a steady-state voltage that is appreciably higher than the d.c. supply voltage. In high voltage inverters, the resulting voltage stresses on the inverter components are unacceptably large, and a modified version of the McMurray circuit is used which sets the capacitor voltage equal to the d.c. supply voltage, E_{dc}, prior to each commutation.

SOLVED EXAMPLE

Example 9.24 For the McMurray single-phase inverter circuit of Fig. 9.45, obtain the value of commutating components L and C for the following parameter values:

(i) Maximum load current = 80 A (ii) t_q of SCRs = 40 μs (iii) $E_{\text{dc(min)}}$ = 200 V

Solution: The circuit turn-off time, t_{off}, must be greater than turn-off time of the SCRs for the reliable commutation. ∴ t_{off} = 60 μs.

From Eq. (9.165), $C = \dfrac{0.893 \times 60 \times 10^{-6} \times 80}{1.68 \times 20} = 127.57\, \mu F.$

From Eq. (9.164), $L = \dfrac{0.397 \times 20 \times 60 \times 10^{-6}}{80}$ $L = 5.96\, \mu H.$

9.14.4 The Modified McMurray Inverter

The modified McMurray full-bridge inverter circuit is shown in Fig. 9.47. A resistor R, and auxiliary diodes $D_{A_1}, D_{A_2}, D_{A_3}, D_{A_4}$ are added to the basic McMurray circuit. These components permit the removal of the capacitor overcharge at the end of the commutation process.

After commutation of thyristor T_1, the excess energy in capacitor C_1, less the energy dissipated in R, is returned to the d.c. supply (E_{dc}) by a current flowing through L, R, D_{A_1}, the d.c. source, and D_4. Similarly, after commutation of T_2, the excess energy in capacitor C_2, less the energy dissipated in R, is returned to the d.c. supply by a current flowing through L, D_3, E_{dc}, D_{A2} and R. The voltage drop across R and D_{A_1} commutates T_{A_1}. Similarly, voltage drop across R and D_{A_2} commutates T_{A_2}. Resistor R limits the undershoot of capacitor voltage during this discharge interval. If the resistor is large enough to overdamp the series circuit, the capacitor is discharged to the d.c. supply voltage, E_{dc}, for all values of load current. For critical damping of resonant circuit, consisting of R, L, C in series in Fig. 9.47,

$$\sqrt{\frac{1}{LC} - \left(\frac{R}{2L}\right)^2} = 0, \quad \text{or} \quad R = 2\sqrt{\frac{L}{C}} \qquad (9.167)$$

Here, $C_1 = C_2 = C$.

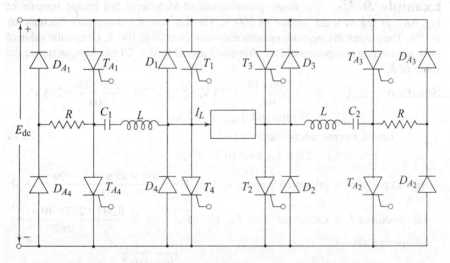

Fig. 9.47 *Modified McMurray full-bridge inverter*

Improved operation of the modified McMurray circuit can also be obtained by means of advanced firing of the main thyristor T_4, which must be gated just before D_1 stops conducting. Practical circuit design must take into consideration the influence of the thyristor snubber circuits (protection circuits) and di/dt limiting inductors on the commutating circuit.

Figure 9.48 shows the power diagram of a modified McMurray half-bridge inverter circuit. Commutation circuit operation is similar to that of the full-bridge inverter circuit described above.

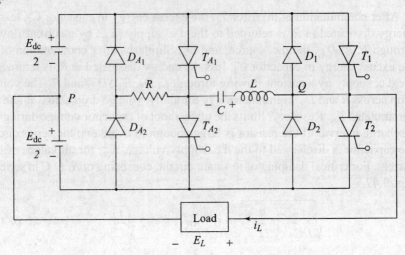

Fig. 9.48 *Modified McMurray half-bridge inverter*

Example 9.25 The single-phase modified McMurray full-bridge inverter of Fig. 9.47, is fed by a d.c. source of 300 V. The d.c. source voltage may fluctuate by ± 15%. The current during commutation may vary from 20 to 100 A. Obtain the value of the commutating components if the thyristor turn-off time is 20 μs. Also, compute the value of R.

Solution: $E_{dc(min)} = 300 - \dfrac{300 \times 15}{100} = 255$ V, $E_{dc(max)} = 300 + \dfrac{300 \times 15}{100} = 345$ V

$I_{L(min)} = 20$ Amp and $I_{L(max)} = 100$ A

Let us assume safety factor = 2

∴ $T_{off} = 2\, t_q = 2 \times 20 \times 10^{-6} = 40$ μs

(i) Capacitor C is calculated from Eq. (9.165), $C = \dfrac{0.893 \times 40 \times 10^{-6} \times 100}{255} = 14$ μF.

(ii) Inductor L is calculated from Eq. (9.166), ∴ $L = \dfrac{0.397 \times 255 \times 40 \times 10^{-6}}{100}$

$= 40.5$ μH.

(iii) R is calculated from Eq. (9.167), ∴ $R = 2\sqrt{\dfrac{40.5 \times 10^{-6}}{14 \times 10^{-6}}} = 3.4\ \Omega$.

Example 9.26 A three-phase bridge inverter is supplied from a 600 V source. For a star connected resistive load of 15 Ω/phase, find the RMS load current, the load-power and the thyristor ratings for, (i) 120° conduction and (ii) 180° conduction.

Solution: Load is star connected. [Fig. Ex. 9.26(a)]

(i) *120° conduction*

Load current amplitude = $\dfrac{600}{2 \times 15} = 20$ A.

RMS load current = $\left\{ \dfrac{1}{2\pi} \left[\int\limits_{0}^{2\pi/3} 20^2\, d\theta + \int\limits_{\pi}^{5\pi/6} 20^2\, d\theta \right] \right\}^{1/2}$

$= [(20^2 + 20^2)/3]^{1/2} = 16.33$ A.

∴ Load power $= (16.33)^2 \times 15 \times 3 = 12$ kW.

Thyristor RMS current $= (20^2/3)^{1/2} = 11.5$ A.

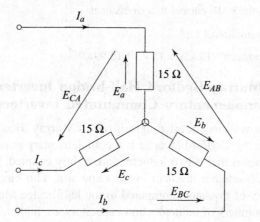

Fig. Ex. 9.26(a) *Star connected load*

(ii) *180° conduction*

At any instant, load on inverter [Fig. Ex. 9.26(b)] $= 15 + \dfrac{15}{2} = 22.5$ Ω.

$I_1 = \dfrac{E_{dc}}{22.5} = 26.67$ A, $I_2 = \dfrac{I_1}{2} = 13.33$ A

Phases are connected in parallel for two-third of a cycle, therefore:

RMS load current $= \left\{ \dfrac{1}{2\pi} \left[\int_0^{2\pi/3} (13.33)^2 \, d\theta + \int_{2\pi/3}^{2\pi/3} (26.76)^2 \, d\theta + \int_{4\pi/3}^{2\pi} (13.13)^2 \, d\theta \right] \right\}^{1/2}$

Fig. Ex. 9.26(b) *Effective-load for 180° conduction*

$$= \left(\frac{2 \times (13.13)^2 + (26.67)^2}{3} \right)^{1/2} = 18.85 \text{ A}.$$

Thyristors carry a current of 26.67 A for one-sixth of a cycle and 13.13 for a half-cycle. Therefore, the RMS current in a thyristor is

RMS load current 2 = 13.13 A.

Load power = $(18.85)^2 \times 15 \times 3 = 15.99$ kW.

9.14.5 McMurray–Bedford Half-bridge Inverter (Complementary Commutated Inverters)

Figure 9.49 shows the circuit diagram of McMurray–Bedford half-bridge inverter. McMurray–Bedford inverter is a complementary impulse commutated inverter. This means that if two inductors are tightly coupled, triggering of one thyristor turns-off another thyristor in the same arm. This circuit configuration uses less number of thyristors compared to the half-bridge McMurray inverter. However, the numbers of inductors and capacitors are higher.

The McMurray–Bedford inverter circuit of Fig. 9.49 consists of main thyristors T_1, T_2 and feedback diodes D_1, D_2. Commutation circuitry consists of two capacitors C_1, C_2 and magnetically coupled inductors L_1 and L_2, having the same inductance L and the same number of turns. Inductors L_1 and L_2 are wound on a core with an air gap so as to avoid saturation. Values of these inductors are of the order of microheneries.

The simplifying assumption for this inverter circuit are the same as for the inverter discussed in the previous section. The circuit operation can be divided into five operating modes and the equivalent circuits for modes are shown in Fig. 9.50. The voltages and currents waveforms for the inverter circuit of Fig. 9.49 is shown in Fig. 9.51.

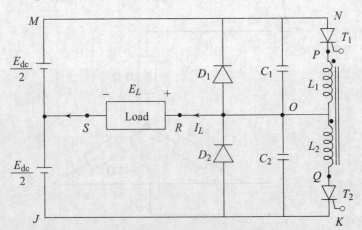

Fig. 9.49 *McMurray–Bedford half-bridge inverter*
($C_1 = C_2 = C$ and $L_1 = L_2 = L$)

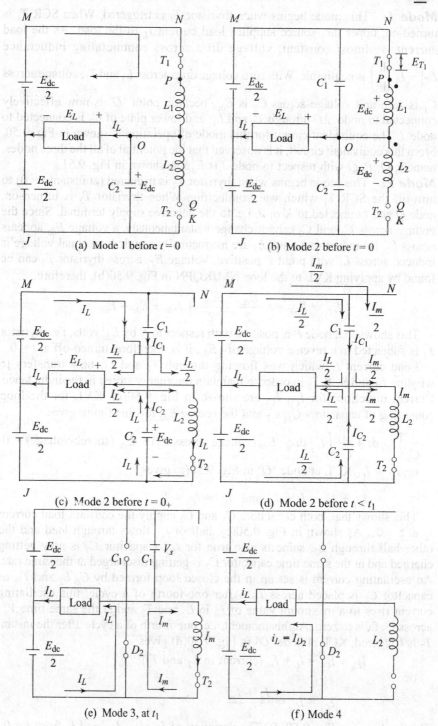

Fig. 9.50 *Equivalent circuits*

Mode 1: This mode begins when thyristor T_1 is triggered. When SCR T_1 is turned-on, upper d.c. source supplies load current I_L to the load. As the load current is almost constant, voltage drop across commutating inductance $L_1 \left(= L_1 \dfrac{di}{dt} \right)$ is negligible. With zero voltage drop across L_1 and T_1, voltage across C_1 is zero and voltage across C_2 is E_{dc}, because point 'O' is now effectively connected to node 'M' through T_1 and L_1, and lower plate of C_2 is connected to node J. The equivalent circuit for this mode of operation is shown in Fig. 9.50. From this equivalent circuit, it is observed that the potential of all the three nodes, namely, O, P, Q with respect to node J is E_{dc}, as shown in Fig. 9.51.

Mode 2: This mode begins when thyristor T_2 is triggered (at instant $t = 0$) to turn-off the SCR T_1 which was conducting. When thyristor T_2 is turned-on, node Q gets connected to K or J, i.e. to the negative supply terminal. Since the voltage across C_1 and C_2 cannot change instantaneously, a voltage E_{dc} appears across L_2. As inductors L_1 and L_2 are magnetically coupled, an equal voltage is induced across L_1 with point P positive. Voltage E_{T_1} across thyristor T_1 can be found by applying KVL to the loop NMJKQPN in Fig. 9.50(b), therefore,

$$E_{T_1} = E_{NP} = -\frac{E_{dc}}{2} - \frac{E_{dc}}{2} + E_{dc} + E_{dc} = E_{dc}$$

This shows that node P is positive with respect to N by E_{dc} volts, i.e. thyristor T_1 is subjected to a reverse voltage of $-E_{dc}$, it is, therefore, turned-off at $t = 0_+$.

Load current I_L which was flowing through T_1 and L_1, now transfers to winding L_2 and SCR T_2 in order to maintain the energy stored in the inductance. Current directions for I_{C_1}, I_{C_2} are shown in Fig. 9.50(c). KVL for the loop consisting of capacitors C_1, C_2 and the source E_{dc} for this figure gives

$$-\frac{1}{C_1} \int I_{C_1} dt + \frac{1}{C_2} \int I_{C_2} dt - E_{dc} \text{ (voltage across } C_2\text{)} + E_{dc} \text{ (source-voltage)} = 0$$

or, $I_{C_1} = I_{C_2}$ KCL at node 'O' in Fig. 9.50(c) gives

$$I_{C_1} + I_{C_2} = I_L + I_L. \text{ and } I_{C_1} = I_{C_2} = I_L.$$

This shows that, both capacitors C_1 and C_2 supply the constant load current I_L at $t = 0_+$. As shown in Fig. 9.50(c), half of I_{C_1} flows through load and the other half through L_2, same is also true for I_{C_2}. Capacitor C_1 is now getting charged and at the same time capacitor C_2 is getting discharged at the same rate. An oscillating current is set up in the closed loop formed by C_2 L_2 and T_2, as capacitor C_2 is placed across L_2. After one-fourth of a cycle, this oscillating current rises to a maximum value of I_m in L_2 and T_2 and at the same time V_{C_2} across C_2 falls to zero. At this moment, i.e. one-fourth of a cycle after the instant T_2 is triggered, KCL at node 'O' in Fig. 9.50(d) gives

$$I_{C_1} + I_{C_2} = I_L + I_m \text{ (current in } L_2 \text{ and } T_2\text{)}$$

or

$$I_{C_1} = I_{C_2} = \frac{I_L + I_m}{2}$$

This is shown in Fig. 9.50(d). The variation of I_{C_1}, I_{C_2}, I_{T_2}, and I_L from $t = 0$ to t_1 is shown in Fig. 9.51.

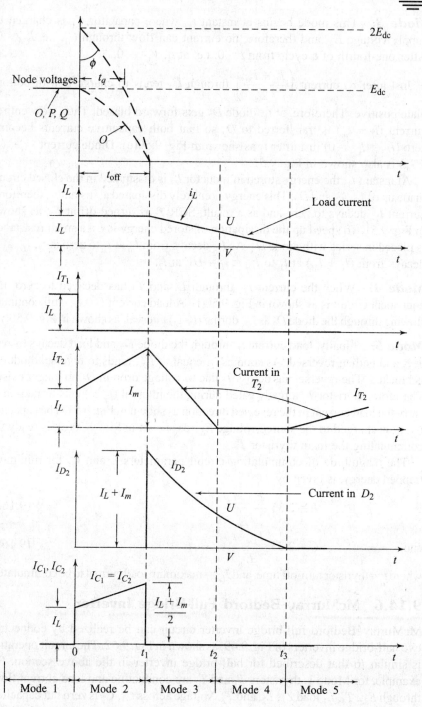

Fig. 9.51 *Voltage and current waveforms for McMurray–Bedford inverter circuit*

Mode 3: This mode begins at instant t_1, where capacitor C_1 is charged to supply voltage E_{dc} and therefore, no current can flow through C_1, i.e. $I_{C_1} = 0$. After one-fourth of a cycle from $t = 0$, i.e. at t_1, $V_{C_2} = 0$.

Just after t_1, current $\left(\dfrac{I_L + I_m}{2}\right)$ through C_2 tends to charge it with bottom plate positive. Therefore, at t_1, diode D_2 gets forward biased. Thus, now entire-current $(I_L + I_m)$ is transferred to D_2 so that both capacitive currents become zero $(I_{C_1} = I_{C_2} = 0)$ just after t_1 as shown in Fig. 9.50(e). Diode current $I_{D_2} = (I_0 + I_m)$ is also shown in Fig. 9.51.

At instant t_1, the energy stored in inductor L_2 is dissipated in the closed circuit made up of L_2, T_2 and D_2. This energy is entirely dissipated at instant t_2, therefore, current I_{T_2} decays to zero and as a result, SCR T_2 is turned-off at T_2; as shown in Fig. 9.51. To speed up the dissipation of stored energy is L_2, a small resistance is placed in series with diode D_2. As I_{T_2} decays from I_m at t_1 to zero at t_2, I_{D_2} also decays from $(I_L + I_M)$ at t_1 to $I_{D_2} = i_L = UV$ at t_2.

Mode 4: When the current I_{T_2} through L_2 and T_2 has decayed to zero, the equivalent circuit is as shown in Fig. 9.50(f). A load current $I_L = I_{D_2}$ still continues flowing through the diode D_2 as I_{D_2} during $(t_3 - t_2)$ interval, as shown in Fig. 9.50(f).

Mode 5: Finally, load current i_L through the diode D_2 and load decays to zero at t_3 and is then reversed. As soon as i_L, equal to I_{D_2}, tends to reverse, diode D_2 is blocked. The reverse bias across T_2 due to voltage drop in D_2 no longer exists. Therefore, thyristor T_2 already gated during the interval $(t_3 - t_2)$ gets turned-on to carry the load current in the reversed direction as shown in Fig. 9.51. The capacitor C_1, now charged to the source voltage E_{dc}, as shown in Fig. 9.50(e), is ready for commutating the main thyristor T_2.

The magnitude of commutating circuit parameters L and C, for minimum trapped energy, is given by

$$L = 2.35 \frac{E_{dc} \cdot t_q}{I_{Lm}} \qquad (9.168)$$

and

$$C = 2.35 \frac{I_{Lm} \cdot t_q}{E_{dc}} \qquad (9.169)$$

where t_q = thyristor turn-off time, and I_{Lm} = maximum load current to be commutated.

9.14.6 McMurray–Bedford Full-Bridge Inverter

McMurray–Bedford full-bridge inverter circuit can be realized by connecting two half-bridge inverters of Fig. 9.49, as shown in Fig. 9.52. The circuit operation is similar to that described for half-bridge inverter in the above section. For example, for Mode 1, thyristors T_1 and T_2 are conducting and load current flows through E_{dc}, T_1, L_1 load Z_L, L_2, and T_2. Voltage across C_1, C_2 is zero but capacitors C_3, C_4 are charged to voltage E_{dc}. For initiating commutation of T_1, T_2, thyristors T_3, T_4 are triggered. This reverse-biases T_1, T_2 by voltage $(-E_{dc})$ and makes them turned-off.

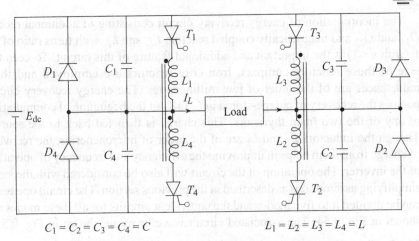

Fig. 9.52 *McMurray–Bedford full-bridge inverter*

9.14.7 Modified McMurray–Bedford Half-Bridge Inverter

In the inverter configuration discussed above, the energy trapped in commutating inductor is wasted and this increases the temperature of the devices, as well as decreases the efficiency of the inverter. A part of the trapped energy can be recovered and fed back to the d.c. source by having the modification as shown in Fig. 9.53. This configuration uses a feedback-winding in the commutation circuitry. The number of SCRs required here is less than that compared to the half-bridge McMurray inverter. However, the numbers of inductors and capacitors are more. Thyristors T_1 and T_2 are the main thyristors and diodes D_1, D_2 are the feedback diodes. Commutation circuit consists of magnetically coupled inductors L_1 and L_2 having the same number of turns and the same inductance L, and capacitors C_1 and C_2.

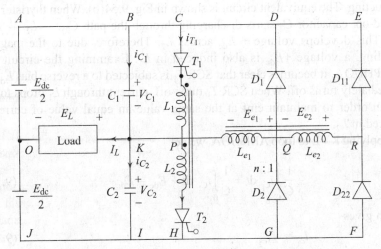

Fig. 9.53 *Modified McMurray–Bedford half-bridge inverter*

The incorporation of energy recovery circuit consisting of additional diodes D_{11} and D_{22} and magnetically coupled reactors L_{e_1} and L_{e_2} with turns ratio of $n : 1$ (with $n < 1$) is the important and additional feature of this circuit. To keep the size of these reactors compact, iron core reactors are employed and their inductances are of the order of few milliheneries. The energy recovery circuit pursues the recovery of charge-stored in L_1 or L_2 at the beginning of commutation of any of the two main thyristors. This charge is then fed back to the source. Though the inductors L_1 and L_2 are of the order of microheneries, the recovery of charge from them helps in improving the efficiency and frequency of operation of the inverter. The operation of the circuit will also be considered with the same simplifying assumptions as described in the previous section. The circuit operation can be divided into five modes and the equivalent circuits for all these modes are shown in Fig. 9.54. The associated circuit waveforms are shown in Fig. 9.55.

Mode 1 : This mode begins when thyristor T_1 is conducting and upper d.c. source supplies load current I_L to the load. As the load current I_L is almost constant, voltage drop across L_1 is negligible. With zero voltage drop across L_1 and T_1, voltage across C_1 is zero and voltage across C_2 is E_{dc}. The equivalent circuit for this mode of operation is shown in Fig. 9.54(a).

Energy recovery transformer will be feeding the energy back to the source only when secondary voltage E_{e_2} is either greater than E_{dc} or less than E_{dc}. Since the ratio of turns of L_{e_1} and L_{e_2} is $n : 1$, this statement implies that for energy recovery E_{e_1} should either be greater than $n\,E_{dc}$ or less than $-n\,E_{dc}$. In the situation under consideration, D_1 is reverse biased and E_{e_2} is zero. Thus none of the diodes is conducting and the currents through L_{e_1} and L_{e_2} are also zero. The current through L_2 is also zero as T_2 is not conducting. As before, load current will be assumed constant during the commutation interval.

Mode 2 : This mode begins when T_2 is fired to turn-off T_1 which was conducting. The equivalent circuit is shown in Fig. 9.54(b). When thyristor T_2 is turned-on, capacitor C_2 starts discharging through the path $C_{2+} - L_2 - T_2 - C_{2-}$. This develops voltage $+ E_{dc}$ across L_2. Therefore, due to the magnetic coupling, a voltage $+E_{dc}$ is also induced in L_1. Examining the circuit loop ABCPHIJOA, it becomes clear that SCR T_1 is subjected to a reverse bias E_{dc} and immediately turns off. When SCR T_1 turns-off, current through L_1 drops to zero and in order to maintain emf at the same value an equal value of current is induced in L_2.

Applying KVL to loop ABKIJOA, we get

$$\frac{1}{C}\int_0^t i_{C_1}\,dt + \frac{1}{C}\int_0^t i_{C_2}\,dt + E_{dc} = E_{dc} \qquad (9.170)$$

which gives

$$i_{C_1} = -i_{C_2} \qquad (9.171)$$

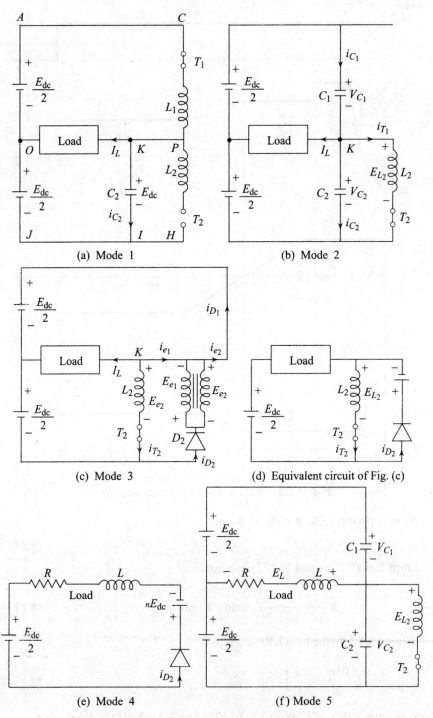

Fig. 9.54 *Equivalent circuits*

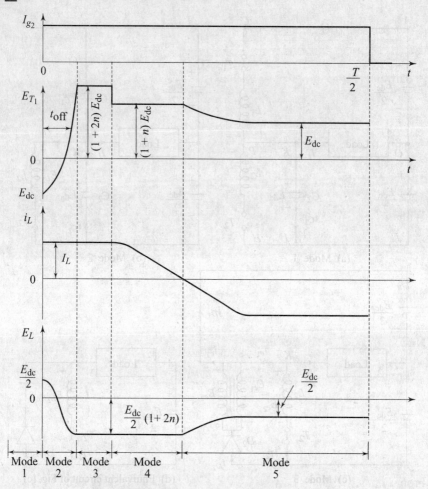

Fig. 9.55 *Voltage and current waveforms*

Now, applying KCL at node K, we get

$$I_L + i_{T_2} + i_{c_2} = i_{c_1} \tag{9.172}$$

From Eqs (9.171) and (9.172), we get

$$i_{c_1} = \frac{I_L + i_{T_2}}{2} \quad \text{and} \quad i_{c_2} = -\left(\frac{I_L + i_{T_2}}{2}\right) \tag{9.173}$$

Writing KVL to the loop KPHI, we get

$$L\frac{di_{T_2}}{dt} = \frac{1}{C}\int_0^t i_{c_2}\,dt + E_{dc} \tag{9.174}$$

Substitute value of i_{c_2} from Eq. (9.173) into Eq. (9.174), we get

Inverters 639

$$L \frac{di_{T_2}}{dt} = \frac{1}{C}\int_0^t -\left(\frac{I_L + i_{T_2}}{2}\right) dt + E_{dc} \qquad (9.175)$$

Differentiating Eq. (9.175), we get

$$\frac{d^2 i_{T_2}}{dt^2} + \frac{di_{T_2}}{2LC} + \frac{I_L}{2LC} = 0 \qquad (9.176)$$

This is second order differential equation with the following initial conditions,

$$i_{T_2(0)} = I_L \text{ and } \frac{di_{T_2}}{dt}(0) = \frac{E_{dc}}{L} \qquad (9.177)$$

With the above initial conditions, the solution of the Eq. (9.176), is

$$i_{T_2} = 2 I_L \cos \omega_r \cdot t + \frac{E_{dc}}{\omega_r \cdot L} \sin \omega_r \cdot t - I_L \qquad (9.178)$$

where

$$\omega_r = \frac{1}{\sqrt{2LC}} \qquad (9.179)$$

Now, $\quad E_{L_2} = L \dfrac{di_{T_2}}{dt} = -2\, \omega_r\, L\, I_L \sin \omega_r\, t + E_{dc} \cos \omega_r\, t \qquad (9.180)$

From the loop consisting of L_1, T_1, L_2 and the source, and noting

$$E_{L_1} = E_{L_2},\ E_{T_1} = E_{dc} - 2\, E_{L_2}$$

Substituting from Eq. (9.180),

$$E_{T_1} = E_{dc} + 4\, \omega_r\, L\, I_L \sin \omega_r\, t - 2\, E_{dc} \cos \omega_r\, t \qquad (9.181)$$

SCR T_1 will be subjected to a reverse-bias as long as E_{T_1} is less than zero. From Eq. (9.180), the circuit turn-off time t_{off} of SCR T_1 is given by

$$\cos \omega_r\, t_{\text{off}} - \frac{2\omega_r . L I_L}{E_{dc}} \sin \omega_r . t_{\text{off}} = \frac{1}{2} \qquad (9.182)$$

This mode of operation comes to an end at instant t_1 when $E_{e_1} = n \cdot E_{dc}$.

Mode 3 : This mode begins at instant t_1 where diodes D_{11} and D_2 start conducting and energy recovery starts. From Eq. (9.180), the value of t_1 is obtained as follows:

$$-2\, \omega_r\, L\, I_L \sin \omega_r . t_1 + E_{dc} \cos \omega_r\, t_1 = n\, E_{dc} \qquad (9.183)$$

The equivalent circuit for this mode of operation is shown in Fig. 9.54(c). Voltages E_{e_1} and E_{e_2} are clamped at nE_{dc} and E_{dc}, respectively, since diodes D_2 and D_{11} are conducting. Therefore, E_{L_2} and V_{c_2} are also damped at $-n\, E_{dc}$ and the load voltage is clamped at $-\left(\dfrac{E_{dc}}{2}\right)(2n + 1)$. In the equivalent circuit, the transformer formed by L_{e_1} and L_{e_2} has a connection in which secondary is connected across a voltage E_{dc}. When the voltage is transferred to the primary side, the equivalent circuit part consisting of the transformer and the secondary voltage E_{dc} can be replaced by a source of voltage $-n\,E_{dc}$ in series with a diode and the equivalent circuit shown in Fig. 9.54(d) is obtained. The diode is incorporated to emphasize that the circuit permits the flow of current only in one direction.

Now, $$\frac{di_{T_2}}{dt'} = -n\,E_{dc} \qquad (9.184)$$

where t' is the time measured from the beginning of this mode, i.e.

$$t' = t - t_1$$

From Eq. (9.178), by substituting $t = t_1$, the initial value $i_{T_2(0)} = I_{T_2}$ can be obtained. Using this initial condition, the solution of Eq. (9.184) yields

$$i_{T_2} = I_{T_2} - \frac{n\,E_{dc}}{L} t' \qquad (9.185)$$

From Eq. (9.185), it is clear that i_{T_2} decreases linearly. From node k,

$$i_{e_1} = -(i_{T_2} + I_L) \qquad (9.186)$$

$$= \frac{n\,E_{dc}\,t'}{L} - I_{T_2} - I_L \qquad (9.187)$$

Now $i_{D_{11}} = i_{e_2} = -n\,i_{e_1}$

$$i_{D_2} = i_{e_2} - i_{e_1} = -(n+1)\,i_{e_1} \qquad (9.188)$$

When i_{T_2} becomes zero, this mode comes to an end at $t' = t_1'$.
From Eq. (9.185),

$$t_1' = \frac{L}{n\,E_{dc}} I_{T_2}$$

Let $\quad t_1 + t_1' = t_2 \qquad (9.189)$

This is also the end of the commutation interval.

Mode 4: Since diodes D_2 and D_{11} are still conducting in mode 3, voltages E_{e_1}, E_{e_2}, V_{c_1} and V_{c_2} are all still clamped to their values in Mode 3, however, i_{T_2} and E_{L_2} have fallen to zero values. Fig. 9.54(e) shows the equivalent circuit for this mode of operation.

From the equivalent circuit, we can write

$$\frac{di_{D_2}}{dt''} + \frac{R}{L} i_{D_2} = -\frac{E_{dc}}{2L}(1 + 2n) \qquad (9.190)$$

where t'' is the time measured from the beginning of Mode 4 and is given by

$$t'' = t - t_2 \qquad (9.191)$$

Initial value of i_{D_2} is obtained from Eqs (9.187) and (9.189) as

$$i_{D_2} = (n+1)\,I_L \qquad (9.192)$$

This mode is over at $t'' = t_1'$ when $i_{D_2} = 0$.

Mode 5: Let $t_2 + t'' = t_3$ (9.193)

Equivalent circuit for Mode 5 is shown in Fig. 9.54 (f). All circuit currents are momentarily zero. Thyristor T_2 is forward biased and will conduct again if gate pulse is present. This explains the reason for using a gate pulse of duration of 180°. At the beginning of this mode, capacitors C_2 and C_1 are charged to voltages $-nE_{dc}$ and $(n+1)E_{dc}$, respectively. Further, transients in the commutation circuit will occur until C_1 and C_2 are charged to voltages of E_{dc} and 0, respectively. The load current will start flowing in the negative direction and build up in accordance with the time constant of the load. If the steady state has already been reached in the circuit operation, the load current will reach the value $-I_L$ when commutation of T_2 is initiated by turning-on T_1.

Design of commutation circuit The circuit turn-off time of thyristors T_1 and T_2 is given by Eq. (9.182), which is rewritten below.

$$\cos \omega_r t_{off} - \frac{2\omega_r \cdot L I_L}{E_{dc}} \cdot \sin \omega_r t_{off} = 1/2$$

or
$$t_{off} = \frac{1}{\omega_r}\left[\cos^{-1}\frac{1}{2\gamma} - \theta\right]$$ (9.194)

where
$$\gamma = \left(1 + \frac{4\omega_r^2 L^2 I_L^2}{E_{dc}^2}\right)^{1/2}$$ (9.195)

and
$$\theta = \tan^{-1}\frac{2\omega_r \cdot L I_L}{E_{dc}}$$ (9.196)

As E_{dc} decreases or I_L increases, the circuit turn-off time decreases. The design is, therefore, carried out for the worst situation, i.e. for the maximum value of I_L and the minimum value of E_{dc}.

9.14.8 Modified McMurray–Bedford Full-Bridge Inverter

Modified McMurray–Bedford full-bridge inverter circuit is shown in Fig. 9.56. The number of thyristors, diodes, and other components in a full-bridge inverter is double of those in half-bridge inverter of Fig. 9.53. Here, thyristors T_1, T_2 conduct in one half-cycle and thyristors T_3, T_4 conduct in another half-cycle. When thyristors T_1, T_2 are triggered, load current I_L flows through the path E_{dc+} — T_1 — L_1 — P Z_L — Q — L_2 — T_2 — E_{dc-}. Capacitor C_1 discharges through the path C_{1+} — T_1 — L_1 — C_{1-} and C_2 discharges through the path C_{2+} — L_2 — T_2 — C_{2-}. The voltage across the capacitors C_1 and C_2 becomes zero, whereas capacitors C_3 and C_4 will be charged to the voltage E_{dc}. Figure 9.57 shows the load voltage waveform and the various conduction periods, transients during commutation are not shown. SCRs T_1 and T_2 will conduct during period JK. At K, thyristors T_3 and T_4 will be triggered. When T_3, T_4 will be turned-on, capacitor C_4 starts discharging through the path C_{4+} — L_4 — T_4 — C_{4-}. This develops voltage E_{L_4} across L_4. Therefore, due to the magnetic coupling voltage

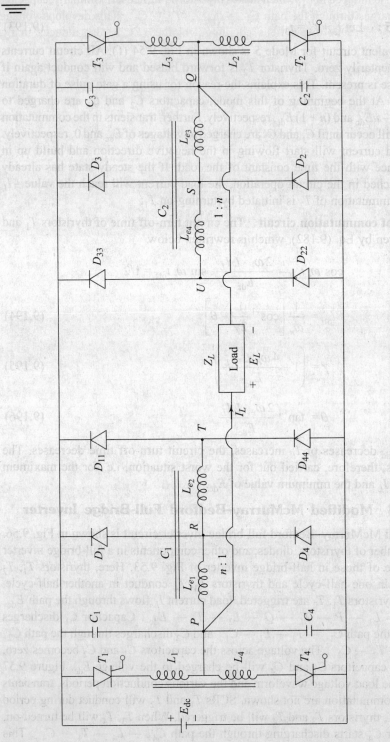

Fig. 9.56 Modified McMurray–Bedford full-bridge inverter circuit

$+E_{dc}$ is also induced in L_1, which makes thyristor T_1 turned-off. Similarly, capacitor C_3 discharges through the path $C_{3+} - L_3 - T_3 - C_{3-}$. This develops voltage E_{L_3} across L_3. Due to magnetic coupling, voltage $+E_{dc}$ is also induced in L_2, which makes thyristor T_2 turned-off. Now, energy recovery starts when E_{Le_1} (= E_{Le_2}) is greater than E_{dc} and the energy recovery path becomes $P - L_{e1} - R - D_1 - E_{dc} - D_2 - S - L_{e_3} - Q$. When E_{Le_1} (= E_{ie_2}) is less than E_{dc}, diodes D_1 and D_2 stops conducting and diodes D_{11}, D_{22} starts conducting and the recovery path become $P - L_{e_1} - R - L_{e_2} - T - D_{11} - E_{dc} - D_{22} - L_{e_4} - S - L_{e_3} - Q$. At the end, the load current becomes zero and then it reverses.

In Fig. 9.56, the freewheeling diodes D_1 and D_4 are connected to the tap points R and S on coils C_1 and C_2. Therefore, diode D_4 will start conducting only when the potential of point P becomes $-n E_{dc}$. Similarly, diode D_3 will not conduct until the potential of Q becomes $(E_{dc} + n E_{dc})$. When these diodes begin conducting, the potential of points R and S, as also of P and Q, will be fixed, so the capacitor discharge current will become zero and the load current will be supplied through these diodes. The load voltage during the period the freewheeling diodes conduct will be $(2n E_{dc} + E_{dc})$. This duration is designated L_M in Fig. 9.57, where the load current is positive and the load voltage is negative. Hence, power will flow from the load to the supply. After this recovery period, i.e. when all the load inductive energy has been fed back, the current will go to zero. Since thyristors T_3 and T_4 are in the off-state, they will have to be retriggered to start the negative cycle. These SCRs will conduct during the period MN. At N, SCRs T_1 and T_2 will be fired and similar operations as just described will take place. The design of commutating components L and C may be obtained from Eqs (9.194), (9.195) and (9.196).

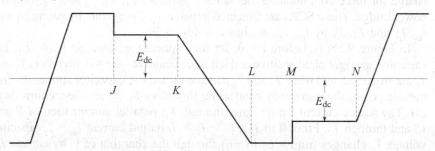

Fig. 9.57 *Load voltage waveform*

9.15 CURRENT SOURCE INVERTERS

The inverters that have so long been discussed are described as voltage source inverters because the output voltage of the inverter is maintained constant independent of the load. However, the load current is forced to fluctuate from positive to negative, and *vice versa* and to cope with inductive loads, thyristors with freewheeling diodes are required. Normally, for maintaining inverter output voltage constant, a large capacitor is connected at the d.c. input side of the inverter.

In a current-source inverter (CSI), the current from the d.c. source is maintained at an effectively constant level, irrespective of load or inverter conditions. This is achieved by inserting a large inductance in series with the d.c. supply to enable changes of inverter voltage to be accommodated at low values of di/dt. The d.c. input to current-source inverter is obtained from a fixed voltage a.c. source through a controlled rectifier-bridge, or through a diode bridge and a chopper. In order that current input to CSI is almost ripple free, L-filter is used before CSI. As it is a constant-current system, the current sourced inverter is used typically to supply high-power factor loads whose impedance either remains constant or decreases at harmonic frequencies in order to prevent problems either on switching or with harmonic overvoltages.

The current source inverters find their use in the following applications:
(i) Speed control of a.c. motors
(ii) Induction heating
(iii) Lagging VAR compensation.
(iv) Synchronous motor starting, etc.

Current source inverters may be either load commutated or force commutated. Load commutation is possible when load Pf is leading. For lagging Pf loads, forced commutation is essential. Use of commutating capacitor is an important feature of force-commutated current-source inverters.

9.15.1 Single-Phase Capacitor-Commutated Current-Source Inverters with Resistive Load

Figure 9.58(a) shows a single-phase, current-source bridge inverter circuit with resistance load R. Capacitor C in parallel with the load is used for storing the charge for force commutating the SCRs. Thyristors T_1, T_2, T_3 and T_4 form the power bridge. These SCRs are triggered in pairs; T_1, T_2 together by gating signals I_{g_1}, I_{g_2} and T_3, T_4 by I_{g_3}, I_{g_4}, as shown in Fig. 9.59.

In Figure 9.58(a), before $t = 0$, let the capacitor voltage be $V_c = -E_1$, i.e. capacitor has right plate positive and left plate negative. At $t = 0$, thyristors T_1 and T_2 are triggered, and when T_1 and T_2 become turned-on, capacitor applies reverse voltage across the previously conducting thyristors T_3, T_4 and hence turn them off. The source current I now flows through T_1, parallel combination of R and C, and through T_2. From 0 to $T/2$, $I_{T_1} = I_{T_2} = I$, output current $I_{ac} = I$, capacitor voltage V_c changes from $-E_1$ to $+E_1$ through the charging of C by current I_C. Note that, here load voltage $E_L = V_C$. Thus, the waveform of $I_L = E_L/R = V_c/R$ has the same nature as that of V_C, as shown in Fig. 9.59. When T_3, T_4 are gated at $t = T/2$, $V_c = E_1$ reverse-biases T_1, T_2; these are therefore turned-off immediately. The source current now flows through T_3, parallel combination of R, C and T_4. From instant $\dfrac{T}{2}$ to T, $I_{T_3} = I_{T_4} = I$, but $I_{ac} = -I$. The variation of a.c. current I_{ac} is shown in Fig. 9.58(b). The current I_{ac} is a square wave of amplitude I.

Circuit analysis From $0 < t < T/2$, equivalent circuit for the CSI of Fig. 9.58(a) is as shown in Fig. 9.58(c). The capacitor is initially charged to a voltage $-E_1$. Traversing the closed path PQRSP, we get

Inverters

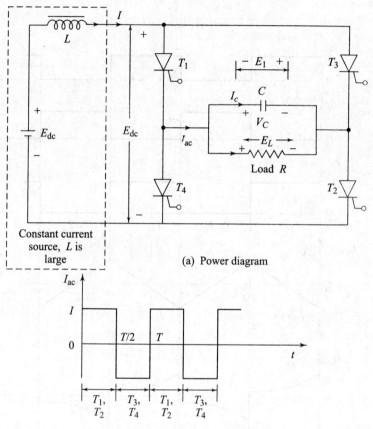

(a) Power diagram

(b) a.c. output current waveform

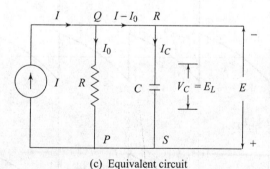

(c) Equivalent circuit

Fig. 9.58 *Single-phase current-source inverters*

$$RI_L - \frac{1}{C}\int (I - I_L)\, dt + E_1 = 0,\ 0 \le t \le \frac{T}{2}$$

or,

$$V_C = E_L = RI_L = \frac{1}{C}\int (I - I_L)\, dt - E_1 \qquad (9.197)$$

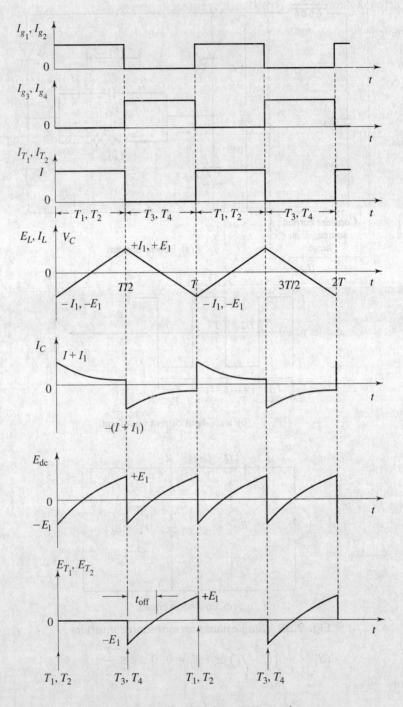

Fig. 9.59 Voltage and current waveforms

Inverters

Differentiation of Eq. (9.197) with respect to t gives,

$$R \cdot \frac{dI_L}{dt} + \frac{I_L}{C} = \frac{I}{C} \tag{9.198}$$

or

$$\left(RP + \frac{1}{C}\right) I_L = \frac{I}{C} \tag{9.199}$$

Complementary function (CF) of the solution is obtained from force-free equation,

$$\left(RP + \frac{1}{C}\right) I_{cp} = 0, \text{ or } P = -\frac{1}{RC}, \qquad \therefore I_{cp} = A \cdot e^{-t/RC}$$

For particular integral (PI), substitute $P = 0$ in Eq. (9.199)

$$\therefore \qquad \frac{I_L}{C} = \frac{I}{C} \qquad \text{or} \qquad I_L = I$$

Hence, complete solution for load current I_L, from Eq. (9.199), is

$$I_L = \text{P} \cdot \text{I} + \text{C.F.} = I + A \cdot e^{-t/RC} \tag{9.200}$$

The load current at $t = 0$, under steady state operation from Fig. 9.58, is

$$I_L = -I_1.$$

Therefore, from Eq. (9.200),

$$-I_1 = I + A \qquad \text{or} \qquad A = -(I + I_1)$$

Substitute this value of A in Eq. (9.200),

$$\therefore \qquad I_L = I - (I + I_1) e^{-t/RC}$$

or

$$I_L = I(1 - e^{-t/RC}) - I_1 \cdot e^{-t/RC} \; / \; 0 < t < \frac{T}{2} \tag{9.201}$$

As only steady solution is desired, current I_L at $t = T/2$ becomes I_1 (Fig. 9.58). Therefore, Eq. (9.201) becomes

$$I_1 = I(1 - e^{-T/2RC}) - I_1 \cdot e^{-T/2RC} = \text{or } I_1 = I \left[\frac{1 - e^{-T/2RC}}{1 + e^{-T/2RC}}\right] \tag{9.202}$$

$$= I \text{ if } T \gg 2RC$$

Substitution of I_1 from Eq. (9.202) in Eq. (9.201), gives

$$I_L = I\left(1 - e^{-t/RC}\right) - I\left[\frac{1 - e^{-T/2RC}}{1 + e^{-T/2RC}}\right]$$

or

$$I_L = I\left[1 - \frac{2 e^{-t/RC}}{1 + e^{(-T/2RC)}}\right] \tag{9.203}$$

The turn-off time t_{off} provided by the circuit to each SCR is obtained from the condition that when $t = t_{off}$, $V_C = I_L R = 0$. Therefore, from Eq. (9.203),

$$V_C = I_L R = RI\left[1 - \frac{2e^{-t_{off}/RC}}{1 + e^{-T/2RC}}\right] = 0 \text{ or } e^{(-t_{off}/RC)} = \frac{1}{2}\left[1 + e^{-T/2RC}\right]$$

or,
$$t_{off} = RC \ln\left[\frac{2}{1 + e^{-T/2RC}}\right] \quad (9.204)$$

The average value of the input voltage E_{dc} can be calculated from the equation

$$E_{dc} = \frac{1}{T/2}\int_0^{T/2} I_L R \, dt = \frac{2}{T} IR \int_0^{T/2}\left[1 - \frac{2e^{-t/RC}}{1 + e^{-T/2RC}}\right] dt$$

or,
$$E_{dc} = IR\left[1 - \frac{4RC}{T}\left(\frac{1 - e^{-T/2RC}}{1 + e^{-T/2RC}}\right)\right] \quad (9.205)$$

When input-power ($E_{dc} \cdot I$) is positive, power is delivered to load R.

Design Aspects:
(i) From Eq. (9.204), it is observed that as the inverter frequency ($1/T$) is increased, the turn-off time provided by the circuit reduces. But, the circuit commutating time t_{off} should be more than the SCR turn-off time (t_q) for reliable operation. It means that there is an upper limit to the inverter frequency beyond which inverter SCRs will fail to commutate.

(ii) When the inverter frequency ($1/T$) is low, the plot of I_L, or E_L, versus time, from Eq. (9.203), becomes flatter as shown by dotted curve in Fig. 9.60. As this curve shape is nearer to a square-wave, it can be inferred that for low inverter frequencies, inverter has square-wave output for load current I_L or load voltage E_L.

When inverter frequency is high, or when T is small, waveform of E_L or I_1 is shown by full line curve in Fig. 9.60. As this full line curve is closer to a sine-wave, it can be said that for high inverter frequency, CSI has sinusoidal waveshape for output load current or load voltage.

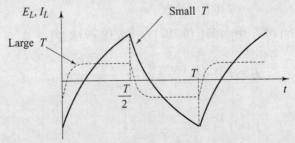

Fig. 9.60 Waveforms for single-phase CSI with R-load

(iii) *Square-wave output*: It has been found that for obtaining square-wave of the load current,

$$\frac{T}{2RC} > 5.0$$

If t_q is the turn-off time for the SCRs used as CSIs, then from Eq. (9.204), assuming $t_q = t_{off}$, $t_q = RC \ln \frac{2}{1+e^{-5}} \approx RC \ln_2 = 0.69\, RC$

or, $$C = \frac{t_q}{0.69\, R} \qquad (9.206)$$

For $\frac{T}{2RC} = 5$ or for $T = 10\, RC$,

maximum inverter frequency is given by,

$$f_{max} = \frac{1}{T} = \frac{1}{10\, RC}$$

Substituting the value of C from Eq. (9.206),

$$f_{max} = \frac{1}{10} \cdot \frac{0.69\, R}{t_q} \quad \therefore f_{max} = \frac{0.069\, R}{t_q} \qquad (9.207)$$

(iv) *Sinusoidal wave output*: For obtaining sinusoidal waveshape for load current, the capacitive reactance X_c at three times the minimum frequency f_{min}, should be less than $R/2$, i.e.

$$X_c \text{ at } 3 f_{(min)} \leq R/2$$

or $\dfrac{1}{3.2\pi \cdot f_{min}\, C} \leq R/2$ or $C \geq \dfrac{0.106}{R\, f_{min}} \qquad (9.208)$

The inverter should be therefore operated at frequencies higher than f_{min} in order to obtain the sinusoidal waveshape.

SOLVED EXAMPLES

Example 9.27 A capacitor commutated 1-ϕ bridge inverter is operated at 50 Hz with load resistance of 5 ohm. Thyristor turn-off time is 62 μsec. Determine:
(i) commutating capacitor C for successful commutation of SCR.
(ii) load current I_L (iii) $F_{critical}$ for reliable commutation. (iv) $R_{critical}$

Solution:
Given: $f = 50$ Hz, $t_q = 62\, \mu$sec, $R = 5$ ohm.

(i) From Eq. (9.204), $t_{off} = R.C. \ln \left[\dfrac{2}{1+e^{-T/2RC}} \right]$

Let the circuit turn-off time (t_{off}) be five times greater than the SCR turn-off time (t_q) for successful commutation.

$\therefore \quad t_{off} = 5 \cdot t_q = 5 \times 62 = 310 \, \mu sec$ and $T = 1/f = 20$ msec.

Now, if $T/2RC > 5$, then
$$t_{off} = 0.69 \, RC$$
$\therefore \quad 390 \times 10^{-6} = 0.69 \times 5 \times C, \quad \therefore \quad C = 89.85 \, \mu f \, C = 100 \, \mu f$ (std. value)

(ii) From Eq. (9.201),

Load current, $I_1 = I \left[\dfrac{1 - e^{-T/2RC}}{1 + e^{-T/12RC}} \right]$, Now, $\dfrac{T}{2RC} = \dfrac{20 \times 10^{-3}}{2 \times 5 \times 100 \times 10^{-6}} = 20$

$\therefore \quad I_1 = I \left[\dfrac{1 - e^{-20}}{1 + e^{-20}} \right] \cong I$

(iii) For reliable operation, $t_{off} \geq 1.5 \, t_q \geq 1.5 \times 62 = 93 \, \mu sec$

Now, from (9.204), $93 \, \mu s = 500 \, \mu s \cdot \ln \left(\dfrac{2}{1 + e^{-T/1000 \times 10^{-6}}} \right)$

$T_{critical} = 1.514$ msec, $\quad \therefore \quad f_{critical} = 660.5$ Hz

Hence, if $f > f_{critical}$, SCR will fail to commutate.

(iv) Since $T_{off} = 0.69 \, RC$
For successful commutation, $t_{off} = 1.5 \, t_q = 1.5 \times 62 = 93 \, \mu sec$
$\therefore \quad 93 \, ms = 0.69 \times R \times 100 \, \mu f \quad \therefore \quad R_{critical} = 1.34 \, \Omega$

9.15.2 Comparison between Voltage Source Inverter (VSI) and Current Source Inverter (CSI)

(i) In VSI_1 input voltage is maintained constant and input current may or may not be constant. In CSI, input current is maintained constant and input voltage may change with load.
(ii) In VSI, the misfiring of switching devices may cause short circuit across the source and create serious problems, whereas in CSI, since input current is maintained constant, misfiring of switching devices or short circuit across source would not be a serious problem.
(iii) In VSI, the amplitude of the output voltage does not depend on the load. However, the amplitude and waveform of output current depends upon the load. In CSI, the amplitude of output current is independent of load. However, the amplitude and waveform of output voltage depends upon load.
(iv) In VSI, the peak current of power devices depends upon circuit conditions and may not be limited under worst conditions, whereas in CSI, since input current is maintained constant, peak current of power devices is limited.
(v) The commutation circuits for SCR in CSI are comparatively simpler than that in VSI.
(vi) CSI has the ability to handle reactive or regenerative load without freewheeling diodes, hence no freewheeling diodes are needed. However, in VSI, free-wheeling diodes are essential to handle reactive power of load.

(vii) In CSI, to obtain constant-current source, a larger reactor is required in series with voltage-source, as compared with that in VSI.

(viii) In VSI, to control input voltage, in extra converter stage may be used. Similarly, in CSI, to control input current, an extra converter stage may be used.

9.16 PERFORMANCE COMPARISONS OF PWM, AVI AND CSI

Each of these types of inverters are usually designed to allow operation within the motor constraints. The differences are in the technique used to generate volts/hertz adjustable frequency, and any inherent design limitations on minimum and maximum frequency.

It is not the intent of this book to take sides or to pass judgement on the relative merits or pros and cons of the three types of inverters. They all do a good job when properly applied and properly designed. Some manufacturers offer, two design and in some cases, all three, depending on horsepower ranges and application requirements.

For the moment, however, the most frequently stated advantages and disadvantages of these three most commonly used types of inverters are listed as follows:

(1) AVI [Adjustable Voltage Inverters] :

Advantages:
 (i) *Basic simplicity*: It has simple logic and can be operated open loop (no feedback of amperes or volts is required for steady-state operation).
 (ii) A single controller can be used with more than one motor.
 (iii) Reliability is good, somewhat better than for PWM types.
 (iv) Voltage stresses on motor insulation are relatively low.
 (v) Can be designed for up to 500 Hz operation.

Disadvantages:
 (i) Speed range is limited because of motor cogging at 6 Hz and below.
 (ii) D.C. link stability can be problem at low speeds because of motor interaction with d.c. link filter elements.
 (iii) Requires an additional set of power devices in the input stage if regeneration back to the a.c. line is desired.
 (iv) To obtain extended ride through capability on incoming power loss, a d.c. chopper must be added to the d.c. link.
 (v) Input power factor is poor below base speed.

(2) PWM [Pulse-Width Modulated] :

Advantages:
 (i) A wider speed range (below rated frequency) is possible.
 (ii) Can be used with more than one motor.
 (iii) Input power factor is good at all frequencies.
 (iv) Diode input stage allows ride through on input power interruption.

Disadvantages:
  (i) Logic circuitry is relatively complex
  (ii) Operation above 120 to 150 Hz is difficult.

(3) CSI [Current Source Inverters] :

Advantages:
  (i) Capable of regeneration back to the a.c. line because d.c. link polarity can be reversed.
  (ii) Large d.c. link filter inductor and regulated power supply acts as a current limiter, making it easier to apply protective fuses.
  (iii) Ability to ride through power-link interruptions.
  (iv) The converter-grade thyristors are sufficient.
  (v) Current source inverters does not generate radio-frequency interference.

Disadvantages:
  (i) Cogging can occur at speeds below 6 Hz.
  (ii) D.C. link filter inductor is large, costly, and contributes to losses and enclosure size.
  (iii) Can cause high voltage spikes on motor terminals.
  (iv) Usually not possible to use with more than one motor.
  (v) Motor power factor appears on the incoming line to the controller.
  (vi) Voltage clamping devices lower overall efficiency.
  (vii) May require special tuning to motor parameters.

Figure 9.61 shows a generalized frequency controller with a d.c. link. The input uses six semiconductors to provide the d.c., and the output uses six semiconductors to provide the adjustable frequency. Table 9.4 is a summary of power devices as used in these three most common types of frequency controllers.

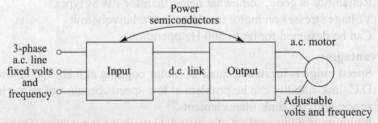

Fig. 9.61 *General adjustable frequency controller with d.c. link*

Table 9.4 Summary of power semiconductors as used with the three basic types of adjustable frequency controllers

Type of controller	Input devices	d.c. link voltage	Output devices
PWM	Diodes	Constant	Thyristors or
AVI	IGBTs/Thyristors/MCTs	Adjustable	IGBTs/MOSFETs/MCTs or
CSI	IGBTs/Thyristor/GTOs	Adjustable	GTOs

9.17 HARMONIC REDUCTION

The power-electronics equipments, such as rectifier, inverters and choppers have switching devices and their operation produces current and voltage harmonics into the system from which they are working. These harmonics affect the operation of other equipments connected to the same system through conduction or by radio interference. The harmonics present in the inverter system lead to the following disadvantages:

(i) Harmonic currents will lead to excessive heating in the induction motors connected with the thyristor system. This will reduce the load carrying capacity of the motor.
(ii) If the control and regulating circuits are not properly shielded, harmonics from power side can affect their operation and malfunctioning can result.
(iii) On critical loads, torque pulsation produced by the harmonic current can be harmful.
(iv) Harmonic currents cause losses in the a.c. system and can even sometime produce resonance in the system. Under such resonant conditions, the metering and instrumentation may also be affected because of the distortion, etc.

These effects can be minimized by reducing the harmonic content.

There are various methods available to reduce the harmonic content. The reduction of harmonic contents or the improvement in waveshape can be done by following methods:

(i) Single-pulse width modulation
(ii) Transformer connections
(iii) Multiple commutation in each half-cycle
(iv) Stepped wave-inverters

Normally, a single-phase bridge inverter produces a square wave. This square-waveform contains $33\frac{1}{3}$ per cent third harmonic, 20 per cent fifth harmonic and $14\frac{1}{2}$ per cent seventh harmonic. In some applications, harmonics at the output should be less than 5 per cent. It is customary to reduce the lower order harmonics by some technique and use filter for higher order harmonics. This procedure not only decreases the cost of the filter but also improves the transient response very much.

9.17.1 Harmonics Reduction by Single-pulse Width Modulation

In the case of single-pulse width modulation, the width of the pulse is adjusted to reduce the harmonic. In general, the RMS value of the amplitude of harmonic voltage of a single pulse modulated wave is given by (From Eq. (9.36)).

$$E_{Ln} = \frac{4E_{dc}}{\sqrt{2}\,n\pi} \sin \frac{np}{2} = \frac{2\sqrt{2}\,E_{dc}}{n\pi} \sin \frac{np}{2}$$

where P is the width of the pulse and E_{dc} is the supply d.c. voltage. For example, if the third harmonic is to be eliminated, $E_{L_3} = 0$,

$$\therefore \quad E_{L_3} = \frac{4E_{dc}}{3\sqrt{2}\,\pi} \sin \frac{3p}{2} = 0 \quad \therefore P = 120°$$

Similarly, to eliminate fifth harmonic, $P = 72°$.

By this method, therefore, only one harmonic can be eliminated at a time.

9.17.2 Harmonic Reduction by Transformer Connections

To have a net output voltage with reduced harmonic content, output voltage from two or more inverters can be combined by means of transformers. The essential condition of this scheme is that the output voltage waveforms from the inverters must be similar but phase shifted from each other. Figure 9.62(a) shows a scheme for connecting two inverters and two transformers for harmonic elimination. Their output voltages, E_{L_1} from inverter 1 and E_{L_2} from inverter 2, are shown in Fig. 9.62(b). As shown in this figure, waveform E_{L_2} is taken to have a phase-shift of $\pi/3$ radians with respect to E_{L_1} waveform. By adding the vertical ordinates of E_{L_1} and E_{L_2}, resultant output voltage E_L is obtained. In this scheme, it is assumed that the transformers have a turns ratio of 1 : 1.

The absence of third harmonic in the output waveform E_L can be explained by writing the Fourier-series for E_{L_1} and E_{L_2}.

$$\therefore \quad E_{L_1} = A_1 \sin \omega t + A_3 \sin 3\omega t + A_5 \sin 5\omega t + \ldots$$

Since the output of second inverter, E_{L_2}, is delayed by $\pi/3$.

$$E_{L_2} = A_1 \sin(\omega t - \pi/3) + A_3 \sin(\omega t - \pi/3) + A_5 \sin(\omega t - \pi/3) + \ldots$$

The resultant voltage E_L is obtained by vector addition.

$$E_L = E_{L_1} + E_{L_2} = \sqrt{3}\,[A_1 \sin(\omega t - \pi/6) + A_5 \sin 5(\omega t + \pi/6) + \ldots] \quad (9.209)$$

From the above expression of E_L, it is observed that with the phase shifting of $\pi/3$ and combining voltages by transformer connection, it is possible to eliminate third harmonics. Along with third harmonics, the multiples of third harmonics, such as 9, 12, are also eliminated. It should be noted that the resultant fundamental component is not twice the individual voltage, but it is $\sqrt{3}/2$ (= 0.866) of that for individual output voltage and the effective output has been reduced by $(1 - 0.866) = 13.4\%$. The main disadvantage of this method of harmonic reduction is the need for more number of inverters and transformers of similar ratings.

Inverters

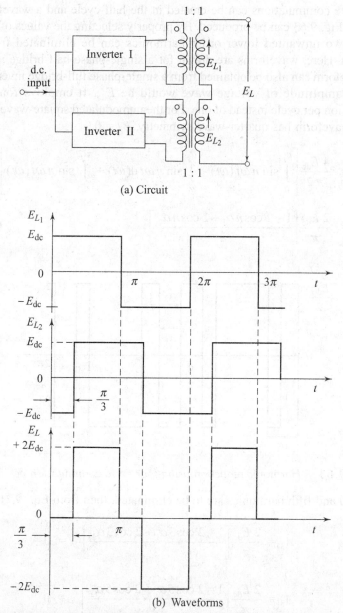

Fig. 9.62 *Harmonic reduction by transformer connection*

9.17.3 Harmonic Reduction by Multiple Commutation in Each Half-cycle

This method is explained with respect to a single-phase-inverter. Normally, there is one commutation per half-cycle at the end of each half-cycle and this produces a square-wave output. Instead of having commutation at the end of half-cycle, some more commutations can be created in the half-cycle and a waveform as shown in Fig. 9.63 can be produced. By properly selecting the values of α_1 and α_2, any two unwanted lower order harmonics can be eliminated from the waveform. Here, waveforms are drawn for a single-phase half-bridge inverter. This waveform can also be obtained from a single-phase full-bridge inverter, but then the amplitude of voltage wave would be E_{dc}. It employs four extra commutation per cycle instead of one. For the unmodulated square-wave, as this voltage waveform has quarter-wave symmetry, $B_n = 0$,

$$\therefore \quad A_n = \frac{4 E_{dc}}{\pi 2} \left[\int_0^{\alpha_1} \sin n\omega t\, (\omega t) - \int_{\alpha_1}^{\alpha_2} \sin n\omega t\, d(\omega t) + \int_{\alpha_2}^{\pi/2} \sin n\omega t\, (\omega t) \right]$$

$$= \frac{2 E_{dc}}{\pi} \left[\frac{1 - 2\cos n\alpha_1 + 2\cos n\alpha_2}{n} \right] \quad (9.210)$$

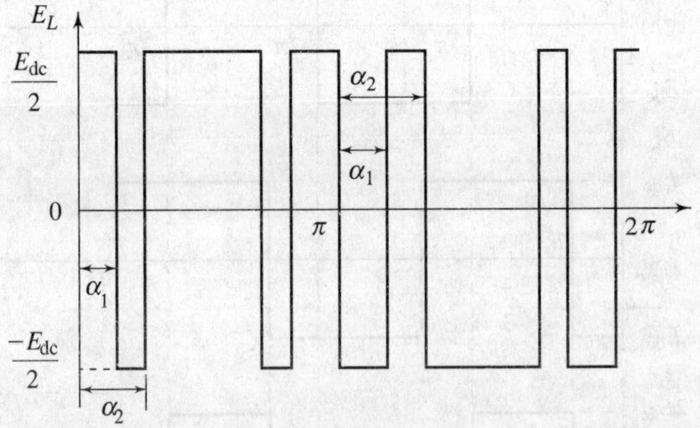

Fig. 9.63 *Harmonic reduction using four extra commutation per cycle*

If third and fifth harmonics are to be eliminated, then from Eq. (9.210),

$$A_3 = \frac{2 E_{dc}}{\pi} \left[\frac{1 - 2\cos 3\alpha_1 + 2\cos 3\alpha_2}{3} \right] = 0$$

and

$$A_5 = \frac{2 E_{dc}}{\pi} \left[\frac{1 - 2\cos 5\alpha_1 + 2\cos 5\alpha_2}{5} \right] = 0$$

Or $\quad\quad 1 - 2\cos 3\alpha_1 + 2\cos 3\alpha_2 = 0$
and $\quad\quad 1 - 2\cos 5\alpha_1 + 2\cos 5\alpha_2 = 0.$

The above two simultaneous equations can be solved numerically to calculate α_1 and α_2 under the condition that

$\quad\quad 0 < \alpha_1 < 90°\quad$ and $\quad \alpha_1 < \alpha_2 < 90°,$
which gives $\quad \alpha_1 = 23.62°\quad$ and $\quad \alpha_2 = 33.6°.$

Similarly, any two harmonics can be eliminated by calculating the corresponding values of α_1 and α_2. This method produces a fundamental voltage of 83.9% or 0.839 times the amplitude of fundamental component of unmodulated voltage wave. Thus, with this method of harmonic reduction, inverter is derated by (100 – 83.9) 16.1%. Another disadvantage of this method is that, there are additional four commutations per cycle and this leads to more switching losses which decreases the efficiency of operation.

9.17.4 Harmonic Reduction Using Stepped Wave Inverters

This method may be called as wave-stepping, in which pulses of different widths and heights are added to produce a resultant stepped wave with reduced harmonic content. Figure 9.64(a) shows two stepped-wave inverters fed from a common d.c. supply. These inverters are connected to a common load through transformers having turns ratio of 1 : 3 and 1 : 1, respectively.

The inverter I is so operated that its output voltage is E_{L_1}, as shown in Fig. 9.64(b). The output-voltage level is either zero or positive during the first half-cycle. During second half-cycle (not shown in the figure), the output voltage would be either zero or negative. This type of modulation in which the output voltage has only two levels during any half-cycle, is called as *two-level modulation*.

The inverter II is so operated that its output voltage is E_{L_2}, as shown in Fig. 9.64(c). It is observed front E_{L2} waveform that the level of output voltage is positive, zero and negative during the first half-cycle. Therefore, inverter II is operated with *three-level modulation*. The resultant output voltage waveform from a series combination of inverter I and inverter II is obtained by superimposing the waveforms of Figs 9.64(b) and (c), as shown in Fig. 9.64(d). The load-voltage waveform of Fig. 9.64(d) shows that the amplitude of output voltage is $4E_{dc}$ and waveform has four steps. Fourier analysis of Fig. 9.64(d) would give harmonics whose amplitudes would depend upon the values of P_1, P_2, P_3, P_4 and amplitude of E_L. By a proper choice of these parameters, third, fifth and seventh harmonics can be reduced considerably and the fundamental component optimised. It is noted that the three-level modulation of inverter II helps achieving the required wave-stepping of the resultant output voltage waveform.

9.18 HARMONIC FILTERS

The inverters described in above sections have output waveforms of either the quasi-square waveform or pulse-width modulated waveform. In order to attenuate the harmonic content of these waveforms, it is necessary to pass them through a filter. A wide variety of filters are available to improve the output waveform. The following are the normally used filters:
  (i)   LC filters
  (ii) Resonant-arm filter
  (iii) OTT filter

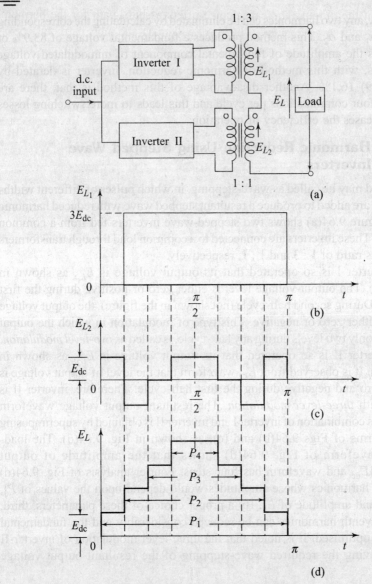

Fig. 9.64 Stepped wave inverters

9.18.1 LC Filters

The simplest form of a low-pass LC filter is shown in Fig. 9.65. The inductance offers a high impedance to harmonic voltage; higher the harmonic number, higher will be the impedance and lower will be the magnitude of the

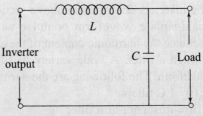

Fig. 9.65 Low pass filter (single-state)

harmonic at the output. The capacitance offers a shunt path for the harmonic current. The higher the frequency, the lower will be X_C and more harmonic current will be by-passed. The L_C filter of Fig. 9.65 has a no-load output/input ratio as

$$\frac{1/\omega C}{\omega L - 1/\omega C} = \frac{1}{\omega^2 LC - 1} \qquad (9.211)$$

If the inverter output is designed for elimination of the low-order harmonics, then this simple low-pass filter is adequate provided $\omega^2 LC$ is not excessively high at the fundamental frequency.

Since, the impedance of capacitor C will decrease as the frequency increases and the impedance of inductance L will increase as frequency increases, therefore, this filter will attenuate the higher frequencies more easily.

Figure 9.66 shows the cascaded LC filter. Cascaded filters are used for higher attenuations. For the same KVA rating of L and C, cascaded filters give a better attenuation but the extra components involved increases the cost. For most of the practical inverters, not more than two stages are used.

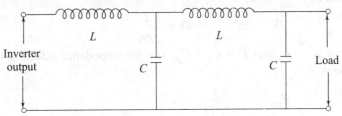

Fig. 9.66 *Cascaded LC filter*

9.18.2 Resonant-Arm Filter

The output voltage of the inverter, will have a certain harmonic content. However, an approximately sinusoidal output may be required, e.g. in the case of so-called uninterruptible current sources designed to take-over during network failures. A sinusoidal output voltage can be realised by means of a combined series-parallel resonant filter circuit tuned to the fundamental of the output voltage. Of course, the sinusoidal output-current driven by the sinusoidal output-voltage will give rise to an a.c. component on the d.c. side of the inverter and, therefore, it will augment the ripple in the input current.

The resonant-arm filter of Fig. 9.67 is more appropriate for situations where low order harmonics are present. Both the series-arm, L_s C_s, and the parallel-arm, L_P C_P, are turned to the inverter output frequency. The series-arm presents zero impedance to the fundamental frequency, but finite increasing impedance to higher frequencies. The parallel-arm presents infinite impedance at the fundamental frequency, but a reducing impedance to higher frequencies.

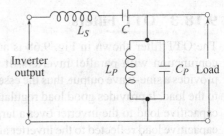

Fig. 9.67 *Resonant-arm filter*

The effectiveness of a filter is characterized by the distortion factor,

$$K = \left(\frac{\sum_{n=2}^{\infty} E_n^2}{\sum_{n=1}^{\infty} E_n^2} \right)^{1/2} \tag{9.212}$$

Its usual allowable value for standby inverters is 5 per cent. In order to estimate the parameters of the resonant circuit one may calculate the average rated power of the inductor and capacitor of the inverter.

At no-load, and assuming for simplicity the impedance of the elements in the series and parallel circuits to be equal, the filter factor of the filter with respect to the nth harmonic can be expressed (neglecting losses in the filter) approximately in terms of the inductance and capacitance involved. Let X denote the reactance of the capacitor and reactor with respect to the angular frequency ω of the fundamental:

$$jX = J\omega L, \quad -jX = \frac{1}{j\omega C} \tag{9.213}$$

where $L = L_s = L_p$ and $C = C_s = C_p$. The net impedance referring to the nth harmonic is

$$Z_{pn} = -j \frac{nx}{n^2 - 1}, \tag{9.214}$$

in the parallel resonant circuit, and

$$Z_{sn} = j \times (n - 1/n) \tag{9.215}$$

in the case of the series-resonant circuit. The ratio of the output voltage of the filter to its input voltage is

$$\frac{E_L}{E_i} = \frac{Z_{pn}}{Z_{sn} + Z_{pn}} = \frac{1}{(n - 1/n)^2 - 1} \tag{9.216}$$

We have assumed the inductors to be of equal impedance. The series inductor carries the output current of the inverter: the parallel inverter carries its output voltage. The same holds for the capacitors in the resonant circuits. Accordingly, at a distortion factor of about 5%, the average rated power of each circuit element will equal the inverter output.

9.18.3 OTT Filters

The OTT filter shown in Fig. 9.68 is an extremely useful circuit when used in conjunction with parallel inverters. It performs three important functions. It provides a sine-wave output, thus the essential elimination of the harmonic content to the load. It provides good load regulation while at the same time maintaining a capacitive load to the inverter over a large load range of load power factor. This capacitive load reflected to the inverter aids SCR commutation, as well as inverter output regulation. The OTT filter has the following advantages:

(1) The input impedance remains capacitive in spit
   in the load power factor and impedance magn
(2) As long as the normalized load impedance
   input impedance is always capacitive.
(3) The OTT-filter has the further advantage of
   ance of 4.5 for open, i.e. infinite load in
   which decrease input impedance with increasing
(4) The input impedance of the filter reflects the output impeda
   output becomes short-circuited, i.e. the load and the input zeros are
   and the same, thus short-circuit input current is theoretically infinite; this
   being ideal for tripping protective devices under faulted load conditions.

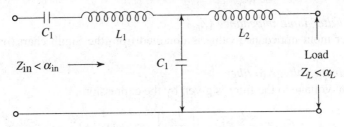

Fig. 9.68 OTT-filter

The following is the design procedure for OTT-filter to produce sinusoidal voltage.

Step 1: *Load resistance*

Load resistance, $R_L = \dfrac{E_L^2 \times Pf^2}{P_o}$ Ω (9.217)

where E_L is the output voltage (volts RMS), P_o the output power (watts), and Pf the rated load power factor.

Step 2: *Load reactance*
The load reactance, X_L, is given by, $X_L = \dfrac{R_L}{Pf}\sqrt{1 - Pf^2}$ (9.218)

Step 3: *Load impedance*
Load impedance is given by the expression

$$|Z_L| = \sqrt{R_L^2 + X_L^2} \;(\Omega) \quad (9.219)$$

and $\quad \angle Z_L = \cos^{-1} Pf$ (degrees) (9.220)

Step 4: *Filter-design impedance*
Filter-design impedance is given by the expression

$$Z_P \leq \dfrac{|Z_L|}{2} \;(\Omega) \quad (9.221)$$

Step 5: *Design radian frequency*

$$W_D = 2\pi f \text{ (rad/s)} \quad (9.222)$$

where f is the output frequency (Hz).

ter-element value
e of capacitances and inductances used in the filter are as follows:

$$C_1 = \frac{1}{6Z_D W_D} \quad \text{(farad)} \tag{9.223}$$

$$C_2 = \frac{1}{3Z_D W_D} \quad \text{(farad)} \tag{9.224}$$

$$L_1 = \frac{9Z_D}{2W_D} \quad \text{(henry)} \tag{9.225}$$

$$L_2 = \frac{Z_D}{W_D} \quad \text{(henry)} \tag{9.226}$$

Step 7: *Filter-input impedance* (Z_{IN})
The filter-input impedance value is obtained from the Smith chart (not shown here).

Step 8: *Input voltage to filter*
The input voltage to the filter is given by the expression

$$F_{SQ} = \frac{\sqrt{2}}{4} \pi |Z_{IN}| \left(\frac{P_o}{R_{IN}}\right)^{1/2} \quad \text{(volts)} \tag{9.227}$$

SOLVED EXAMPLES

Example 9.28 Design an OTT-filter required for a parallel inverter to meet the following specifications:
Output voltage, $E_L = 230$ V, frequency, $f = 50$ Hz load current, $I_L = 1.5$ A
$E_{dc} = 40$ V

Solution: Here, $f = 50$ Hz
∴ Design radian frequency, $W_D = 2\pi f = 2\pi \times 50 = 314$ rad/s

The load impedance $Z_L = \dfrac{E_L}{I_L} = \dfrac{230}{1.5} = 153.33\ \Omega$.

Filter design impedance, $Z_p = \dfrac{|Z'_L|}{2} = \dfrac{153.33}{2} = 76.67\ \Omega$.

Calculation of filter-components:

$$C_1 = \frac{1}{6Z_D W_D} = \frac{1}{6 \times 76.67 \times 314} = 6.92\ \mu F.$$

$$C_2 = \frac{1}{3Z_D W_D} = \frac{1}{3 \times 76.67 \times 314} = 13.85\ \mu F.$$

$$L_1 = \frac{9Z_D}{2W_D} = \frac{9 \times 76.67}{2 \times 314} = 1.098\ H.$$

$$L_2 = \frac{Z_D}{W_D} = \frac{76.67}{314} = 0.244\ H.$$

Example 9.29 Design a 400 Hz parallel inverter with sine-wave output to meet the following specifications:

$$\text{Output power } (P_o) = 360 \text{ W}$$
$$\text{Output voltage } (E_L) = 120 \text{ V (RMS)}$$
$$\text{Output frequency } (f) = 400 \text{ Hz}$$
$$\text{Rated load power-factor } (PF) = 0.7 \text{ lagging}$$
$$\text{Available d.c. supply } (E_{dc}) = 28 \text{ V.}$$

Solution:
(A) *Part I: OTT-filter design*

(i) Load resistance from Eq. (9.217) is, $R_L = \dfrac{(120)^2 \times (0.7)^2}{360} = 20 \, \Omega$

(ii) Load reactance X_L from Eq. (9.218) is, $X_L = \dfrac{20}{0.7}\sqrt{1-(0.7)^2} = 20 \, \Omega$

(iii) Load impedance (Z_L) from Eq. (9.219) is given by

$$|Z_L| = \sqrt{(20)^2 + (20)^2} = 28.3 \, \Omega$$

and $\quad \angle Z_L = \cos^{-1}(0.7) = \dfrac{\pi}{4} = 45°.$

(iv) Filter design impedance from Eq. (9.221) is, $Z_D \leq \dfrac{28.3}{2} \approx 15 \, \Omega$

(v) Design radian frequency from Eq. (9.222) is $W_D = (2)(3.14)(400) = 2500$ rad/s.

(vi) Filter element values from Eqs (9.223) to (9.226) is,

$$C_1 = \dfrac{1}{6 \times 15 \times 2500} = 4.5 \times \mu F = 4.5 \times 10^{-6} \, F$$

$$C_2 = \dfrac{1}{3 \times 15 \times 2500} = 9 \, \mu F = 9 \times 10^{-6} \, F$$

$$L_1 = \dfrac{9 \times 15}{2 \times 2500} = 27 \times 10^{-3} \, H$$

and $\quad L_2 = \dfrac{15}{2500} = 6 \text{ mH} = 6 \times 10^{-3} \, H$

(vii) Let us assume filter input impedance Z_{in} from Smith chart (not shown here) as
$Z_{in} = 80 - j23., \quad \therefore P_{in} = 80 \, \Omega$

and $\quad X_{in} = 23 \, \Omega, \quad \therefore |Z_{in}| = 83 \, \Omega.$

(viii) From Eq. (9.227), the input-voltage to filter is given by

$$E_{SQ} = \dfrac{\sqrt{2}}{4} \times \pi \times 83 \times \left(\dfrac{360}{80}\right)^{1/2} = 195 \, V.$$

(B) *Part II: Inverter design*

(i) Transformer turns ratio, $n = \dfrac{E_{SQ}}{E_{dc}} = 7$

(ii) Assuming 85% efficiency, the input power becomes

$$P_i = \frac{P_o}{\text{efficiency}} = \frac{360 \times 100}{85} = 424 \text{ W}$$

(iii) Average thyristor current is given by

$$I_{T(av)} = \frac{P_o \times |Z_{in}|}{2 \times E_{dc} \times P_{in}} = \frac{360 \times 83}{2 \times 28 \times 80} = 6.8 \text{ A}$$

$$V_{Bo} \approx 2.5 \, E_{dc} = 2.5 \times 28 = 70 \text{ V}.$$

(iv) For the parallel inverter, the commutating components can also be obtained from the following relations,

$$L = \frac{6 \times E_{dc} \cdot T_{off}}{\pi \, I_{P(SCR)}} \quad \text{(i)}$$

and

$$C = \frac{3 \, T_{off} \cdot I_{peak}}{8 \pi \, E_{dc}} \quad \text{(ii)}$$

For this design problem, SCR C 141 is selected, which has the following ratings, t_q = 10 μs.

$$\left(\frac{dv}{dt}\right)_{max} = 200 \text{ V/μs}$$

Therefore, choose T_{off} = 12 μs and $I_{peak(SCR)}$ = 14 A.

$$\therefore \quad L = \frac{6 \times 28 \times 12}{14 \times 3.14} = 45 \text{ μH}$$

and

$$C = \frac{3 \times 12 \times 10^{-6} \times 14}{8 \times 3.14 \times 28} = 0.75 \text{ μF}.$$

Also, turn-on

$$\left.\frac{di}{dt}\right|_{t=0} = \frac{2 \, E_{dc}}{L} = \frac{2 \times 28}{45} = 1.25 \text{ A/μs},$$

REVIEW QUESTIONS

9.1 Explain the principle of operation of an inverter
9.2 How are inverters classified?
9.3 With the help of neat circuit diagram and associated waveforms, explain the operation of single phase half bridge MOSFET based voltage-source inverter with
   (i) Resistive load and
   (ii) Inductive load
9.4 Derive the following expressions for 1-ϕ half-bridge transistorised VSI
   (i) RMS output voltage   (ii) instantaneous output voltage
   (iii) n^{th} harmonic component   (iv) switch voltage and current ratings

9.5 Explain the cross conduction or shoot through fault in inverters. How will you overcome it?

9.6 With the help of neat circuit diagram and associated waveforms, explain the operation of single-phase full-bridge MOSFET based voltage source inverter with
    (i) Resistive load   (ii) Inductive load

9.7 Explain the function of feedback diodes used in antiparallel with transistors in inverters.

9.8 Explain operation of a square-wave transistorised inverter. Assume RL load. Draw the following waveforms:
    (i) Input voltage   (ii) Input current
    (iii) Output voltage and current
    (iv) Transistor and feedback diode current waveform
    (v) Voltage across one transistor

9.9 Explain the following performance parameters of inverters
    (i) Harmonic factor of nth harmonic   (ii) Total harmonic distortion
    (iii) Distortion factor   (iv) Lowest order harmonic

9.10 Explain why a PWM inverter is superior to a square-wave inverter.

9.11 Explain why a PWM technique is becoming increasingly popular and simpler with advances in
    (i) Microprocessor/microcontroller
    (ii) Power-semiconductor devices

9.12 List different voltage control and PWM techniques used in $1-\phi$ inverter.

9.13 Briefly explain bipolar and unipolar PWM full-bridge inverters.

9.14 Compare transistorised and thyristorised inverters.

9.15 How is the output voltage and frequency of a PWM inverter varied?

9.16 What are the advantages and drawbacks of PWM inverter over square-wave or quasi-square inverter?

9.17 Why voltage control is needed in inverter circuits? State the various methods of voltage control in inverter circuits and explain each of them briefly.

9.18 For a single-phase-transistorised PWM half-bridge inverters, derive the expressions for the following:
    (i) Fundamental output voltage   (ii) Inverter gain
    (iii) RMS output voltages   (iv) Distortion and harmonic factor
    (v) Harmonics at the output

9.19 Derive the following for $1-\phi$ transistorised bipolar PWM inverter:
    (i) RMS value of output voltage
    (ii) Fundamental RMS output voltage
    (iii) Distortion factor
    (iv) Harmonic factor
    (v) Gain of inverter
    (vi) Dominant harmonics

9.20 With the help of neat circuit diagram and waveforms, explain briefly the operation of transistorised three-phase bridge inverter with resistive load in
    (i) 180° conduction mode   (ii) 120° conduction mode

9.21 Compare 180° and 120° conduction mode of $3-\phi$ transistorised bridge inverter.

9.22 Compare between voltage source and current-source inverters.

9.23 Draw and explain the simple SCR series inverter circuit employing Class A type commutation. Draw and discuss the important waveforms. State the limitations of this series inverter.

9.24 Draw a modified series inverter circuit. Explain qualitatively how you can have output frequency higher than series resonance frequency.

9.25 Give the detailed design aspects of series inverter.

9.26 Draw and explain the operation of three-phase series-inverter circuit.

9.27 Draw and explain the operation of the time-sharing inverter circuit. Also draw the related voltage and current waveforms.

9.28 Explain the operation of self-commutated inverter circuit with the help of a neat circuit diagram.

9.29 Draw a neat diagram of parallel inverter employing feedback-diodes. Explain the working of inverter with the help of voltage and current waveforms. What care should be taken to avoid commutation failure?

9.30 Give the design aspects of parallel inverter employing feedback diodes.

9.31 Explain the operation of single-phase bridge inverter with the help of load, voltage and load current waveforms.

9.32 A single-phase half-bridge inverter may be connected to a load consisting of (i) R (ii) R_L, and RLC overdamped (iii) RLC underdamped. For all these loads, draw the load voltage and load current waveforms under-steady-state operating conditions. Also, indicate the conduction of the various elements of the inverter-circuit.

9.33 Explain the operation of single-phase bridge inverter with the help of voltage waveforms.

9.34 Draw a circuit of auxiliary-commutated (McMurray inverter) single-phase bridge inverter and explain its operation by drawing voltage and current-waveforms.

9.35 Give the design considerations of single-phase auxiliary commutated inverter.

9.36 Draw and explain the operation of single-phase modified McMurray full-bridge inverter.

9.37 Describe modified McMurray half-bridge inverter with appropriate voltage and current waveforms. For this circuit, find an expression that gives the circuit turn-off time for the main thyristor in terms of load, current, peak capacitor current, etc. Discuss how commutating circuit components can be designed on the basis of minimum commutation energy. Also draw the related voltage and current waveforms.

9.38 Describe modified McMurray–Bedford half-bridge inverter circuit with related voltage and current waveforms.

9.39 Draw and explain the operation of single-phase McMurray–Bedford full-bridge inverter.

9.40 Draw and explain the operation of modified McMurray–Bedford inverter half-bridge circuit. Draw the related voltage and current waveform. Also, give the design of commutation circuit.

9.41 Draw and explain the operation of modified McMurray–Bedford full-bridge inverter circuit. Also draw the related voltage waveforms.

9.42 With an appropriate power diagram, discuss the principle of working of a three-phase bridge inverter. Draw phase and line voltage waveforms on the assumption that each thyristor conducts for 120° and the resistive load is star connected. Also, prepare a table which shows the sequence of firing of various SCRs.

9.43 Repeat the above question, in case each thyristor conducts for 180°.

9.44 What is the need for controlling the output at the output terminals of an inverter? Discuss briefly and compare the various methods employed for the control of output voltage of inverters.

9.45 What is pulse-width modulation? List the various PWM techniques. How do these differ from each other?

9.46 Draw the waveforms and discuss the performance (in harmonic reduction) of following methods of PWM control used in inverters.
   (i) SM—single-pulse modulation.
   (ii) SMM—symmetrical multiple-pulse modulation.
   (iii) sin m—sinusoidal pulse modulation

9.47 For single-pulse modulation, show that output voltage can be expressed as

$$E_L = \sum_{n=1,3,5,\ldots}^{\infty} \frac{4 E_{dc}}{n\pi} \sin \frac{np}{2} \sin n\omega t$$

where P is the pulse-width.

Sketch the variation of $\dfrac{E_{Lnm}}{E_{L1m}}$ as a function of pulse width P. Here E_{Lnm} is the peak value of the nth harmonic component and E_{L1m} is that of the fundamental component.

9.48 Explain sinusoidal pulse modulation as used in PWM inverters. Write the important features of the same.

9.49 Explain in brief the voltage control of three phase inverters.

9.50 State the need for reduction of harmonics in inverters. Outline the various methods for reduction of harmonics or the improvement in waveshape.

9.51 Explain the single-phase PWM, transformer connection, multiple-commutation and filter-methods used for reduction of harmonics in inverter output.

9.52 Explain how the harmonic reduction using stepped wave inverters is done.

9.53 Discuss briefly the resonant arm filter.

9.54 Draw and explain the Ott-filter circuit. Give the detailed design procedure for Ott-filters to produce sinusoidal voltage.

9.55 Draw and explain the operation of single-phase capacitor commutated current source-inverters with resistive-load. Draw also the related voltage and current-waveform.

9.56 Give the circuit analysis of CSI with resistive load.

9.57 Give the performance-comparison of PWM, AVI and CSI.

PROBLEMS

9.1 Compute the output frequency of a series inverter with the following specifications: $L = 10$ mH; $C = 0.1$ μF; $R = 150$ Ω; $T_{off} = 0.2$ ms.
Also, find the attenuation factor (AF).

[Ans. $F_0 = 4.89$ kHz, AF = 0.68.]

9.2 Design a series inverter to meet the following specifications:
Maximum output frequency = 4 kHz
Load resistance varies from 400 to 100 Ω
Supply voltage = 120 V.
[Ans. $Fr = 5$ kHz, $L = 14.43$ mH, $C = 58.7$ μF, $E_c = 289.71$ V. $I_{peak} \approx 0.76$ Amp. $I_T = 1$ A, $V_{BO} \approx 600$ V. $T_q \approx 20$ μs.]

9.3 Design a self-commutated inverter to operate at a frequency of 2 kHz with a optimum distortion. The load specifications are as follows:
$R = 2$ Ω, $L = 3$ μH, $E_{dc} = 80$ V.
Also, compute the output-power
[Ans. $f_r = 2.7$ kHz, $L_1 = 15$ μH, $C = 115.8$ μF $I = 40$ A, $E_{omax} = 39.40$ V, $V_{BO} \geq 238.8$, $I_T = 80$ A, $V_c = 78.8$ V, $t_q \leq 93$ μs, $P_o = 548.5$ W]

9.4 Obtain the value of the commutating components L and C of a parallel inverter for an output voltage of 230 V, 50 Hz and peak load current 1 A. The d.c. input voltage is 30 V. Assume turn-off time of SCR is 40 μs.
[Ans. $L = 149$ μH, $C = 9.75$ μF.]

9.5 Design a parallel inverter, assuming 100% efficiency transformer to supply 50 W at a 240 V into a resistive load, using a square wave of frequency 1 kHz. A 12 V d.c. supply is available. Also, $I_L = 1$A.
[Hint: For square wave, $t_q = t_{on} = 500$ μs]
[Ans. $L = 0.28$ mH, $C = 790$ μF. $V_{BO} = 30$ V, $I_L = 4$ A.]

9.6 The single-phase full-bridge inverter of Fig. 9.16(a) has the resistive-load of 2.4 Ω and the d.c. input voltage of 48 V. Determine
(i) the RMS output voltage at the fundamental frequency, E_1
(ii) the output power, P_o.
(iii) the average and peak currents of each thyristor, and
(iv) the peak reverse-blocking voltage of each thyristor.
[Ans (i) 43.2 V (ii) 960 ω (iii) 20 A, 10 A. (iv) 48 V.]

9.7 A single-phase bridge inverter delivers power to a series connected RLC load with $R = 2$ Ω and $\omega L = 10$ Ω. The period time T is 0.1 ms. What value of capacitor C should the load have in order to obtain load commutation for the SCRs. The thyristor turn-off time is 10 μs. Assume circuit turn-off time $T_{off} = 1.5\ t_q$. Also, assume that the load current contains only fundamental component.
[Ans. $C = 1.248$ μF.]

9.8 The single-phase modified McMurray half-bridge inverter of Fig. 9.51, is fed by a d.c. source of 230 V. The d.c. source voltage may fluctuate by ±10%. The current during commutation may vary from 30 to 120 A. Determine the values of the commutating components if the thyristor turn-off time is 10 μs. Also, compute the value of R.
[Ans. $C = 10.35$ μF, $L = 13.7$ μH $R = 2.3$ Ω]

9.9 A single-phase bridge inverter of Fig. 9.46 is used to supply a load of 10 Ω resistance, 24 mH inductance from a 360 V d.c. source. If the inverter is operating at 60 Hz, determine the steady-state power delivered to the load for
(a) square wave operation
(b) quasi-square wave operation with an on-period of 0.6 of a cycle.
[Ans (a) 5.8 kW (b) 3.35 kW.]

9.10 A three-phase bridge-inverter is fed from a d.c. source of 200 V. If the load is star-connected of 10 Ω/phase pure resistance, determine the RMS load current, the required RMS current rating of the thyristors and the load power for (i) 120° firing, and (ii) 180° firing.

[Ans (i) 120° firing. $I_{L(rms)}$ = 8.16 A, $I_{T(rms)}$ = 9.8 A.] Load power = 2000 W.
(ii) 180° firing $I_{L(rms)}$ = 9.43 A, $I_{T(rms)}$ = 6.67 A. Load power = 2667 W.]

REFERENCES

1. B D Bedford, and RG Hoft, *Principle of Inverter Circuits*, John Wiley and Sons, Inc. New York, 1964.
2. C W Lander, *Power Electronic*, McGraw-Hill Book Company (UK) Ltd.
3. S B Dewan, *Power-semiconductor Circuits*, John Wiley and Sons, New York, 1975.
4. F F Mazda, *Thyristor Control*, John Wiley and Sons, 1973.
5. J M D Murphy, *Thyristor Control of A.C. Motors*, Pergamon Press, 1973.
6. *SCR Manual*, Sixth edition, GE Company, 1979.
7. S K Datta, *Power Electronics and Controls*, PHI, 1989.
8. B M Bind, and K G King, *Power Electronics*, John Wiley and Sons, 1983.
9. M Ramamoorthy, *Thyristors and Their Applications*, East-West Press, 1977.
10. H S Patel, 'Generalized technique of harmonic-elimination and voltage control in thyristor converter; *IEEE Transactions on Industry Applications*, Vol. IA9, No. 3, 1973, pp. 310–317.
11. S C Bansal, and Rao, 'Evaluation of PWM-Inverter Schemes', *Proc. IEE*, 125, 4, 328–334, 1978.
12. J R Bows, and A Midoun, 'New PWM Switching Strategy for Microprocessor Controlled Inverter Drives, *Proc. IEE*, 133B, 4, 237–254, 1986.
13. S B Dewan, and K Gallant, 'Analysis of a Single-phase Current Source Inverter', *Conf. Record Annual Meeting EEF-IAS*, Sept. 1974, p. 1025.
14. M.H. Rashid, *Power-Electronics*, Pearson Education, 2002.

Chapter 10

Cycloconverters

LEARNING OBJECTIVES:
- To consider the basic principle of operation of a cycloconverter.
- To examine the operation of various configurations of a single phase to single phase cycloconverter.
- To consider the operation of a three phase to single phase cycloconverter.
- To introduce the various control schemes for a cycloconverter.
- To develop the general equation describing cycloconverter behaviour.
- To introduce the line-commutated cycloconverter.

10.1 INTRODUCTION

A cycloconverter is a type of power controller in which an alternating voltage at supply frequency is converted directly to an alternating voltage at load frequency without any intermediate d.c. stage. In a line-commutated cycloconverter, the supply frequency is greater than the load frequency. The operating principles were developed in the 1930s when the grid controlled mercury arc rectifier became available. The techniques were applied in Germany, where the three phase 50 Hz supply was converted to a single phase a.c. supply at $16\frac{2}{3}$ Hz for railway traction. In the United States, a 400 HP scheme in which a synchronous motor was supplied from a cycloconverter comprising 18 thyratrons was in operation for several years as a power station auxiliary drive. However, because these early schemes were not sufficiently attractive technically or economically, they were discontinued.

Subsequent invention of the thyristor and the development of reliable transistorized and microprocessor based control circuitry have led to a revival of interest in the cycloconversion principle. Sophisticated control circuits permit the conversion of a fixed input frequency to an adjustable output frequency at an adjustable voltage, and such schemes are attractive for a.c. motor drives. The ruggedness and low weight of the solid-state cycloconverter also make it attractive for aircraft electrical systems, which require the production of a constant output frequency from a variable-speed alternator.

A cycloconverter is controlled through the timings of its firing pulses, so that it produces an alternating output voltage. By controlling the frequency and depth of phase modulation of the firing angles of the converters, it is possible to control the frequency and amplitude of the output voltage. Thus, a cycloconverter has the facility for continuous and independent control over both its output frequency and voltage. This frequency is normally less than 1/3 of the input frequency. The quality of the output voltage wave and its harmonic distortion also impose the restriction on this frequency. The distortion is very low at low output frequencies. The cycloconverters are normally used to provide either a variable frequency from a fixed input frequency or a fixed frequency from a variable input frequency.

A cycloconverter can handle load of any power factor and allows power flow in both the directions. The output voltage waveshape inevitably contains harmonic distortion components in addition to the required sinusoidal component. These distortion terms are produced as a necessary outcome of the basic mechanism of the cycloconverter, whereby the output voltage is fabricated from segments of the input voltage waves. These distortions can be minimised by adequate filters at the output. The distortion of the output voltage increases if the ratio of the output and input frequency increases.

In this chapter, the basic principles of operation of the phase-controlled cycloconverter are explained. The various cycloconverter circuits and controls are also discussed.

10.2 THE BASIC PRINCIPLE OF OPERATION

The basic principle of operation of a cycloconverter can be explained with reference to an equivalent circuit shown in Fig. 10.1. Each two-quadrant converter is now represented as an alternating voltage source, which corresponds to the fundamental or wanted voltage component generated at its output terminals. The diodes connected in series with each voltage source shows the unidirectional conduction of each two-quadrant converter. If the ripple in the output voltage of each converter is neglected, then it becomes ideal and represents the desired output voltage.

The basic control principle of an ideal cycloconverter is to continuously modulate the firing angles of the individual converters, so that each produces the same sinusoidal a.c. voltage at its output terminals. Thus, the voltages of the two generators in Fig. 10.1 have the same amplitude, frequency and phase, and the voltage at the output terminals of the cycloconverter is equal to the voltage of either of these generators. It is possible for the mean power to flow either "to" or "from" the output terminals, and the cycloconverter is inherently capable of operation with loads of any phase angle, within a complete spectrum of 360°.

Because of the unidirectional current carrying property of the individual converters, it is inherent that the positive half-cycle of load current must always be carried by the positive converter, and the negative half-cycle by the negative converter, regardless of the phase of the current with respect to the voltage. This means, each two-quadrant converter operates both in its rectifying and in its inverting region during the period of its associated half-cycle of current.

Figure 10.2 shows the output voltage and current waveforms illustrating the operation of the idealized cycloconverter circuit with loads of various displacement angles. In Fig. 10.2(a), the displacement angle of the load is 0°. In this case, each converter carries the load current only when it operates in its rectifying region, and it remains idle throughout the whole period in which its terminal voltage is in the inverting region of operation. In Fig. 10.2(b), the displacement angle of the load is 60° lagging. During the first 120° period of each half-cycle of load current, the associated converter operates in its rectifying region, and delivers power to the load. During the latter 60° period of each half-cycle of load current, on the other hand, the associated converter operates in its inverting region, and under this condition the load is regenerating power back into the cycloconverter output terminals, and hence into the a.c. system at the input side. In Fig. 10.2(c), the displacement angle of the load is 60° leading. In this case, during the first 60° period of the load current half-cycle, the associated converter operates in its inverting region; and during the latter 120° period, in its rectifying region. In Fig. 10.2(d), the displacement angle of the load is 180°. In this case, the load is fully regenerative and it continuously delivers power into the output terminals of the cycloconverter over the whole period of each output cycle. Thus, during each half-cycle of current, the associated converter operates permanently in its inverting region.

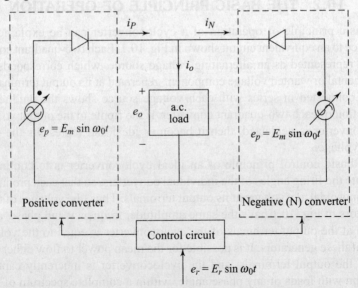

Fig. 10.1 *Equivalent circuit of cycloconverter*

10.3 SINGLE-PHASE TO SINGLE-PHASE CYCLOCONVERTER

It is a single-phase cycloconverter whose input and output are single phase a.c. The input a.c. voltage of supply frequency 50 Hz is converted into lower frequency a.c. output. There are mainly two configurations for this type of cycloconverter, viz. centre-tapped transformer configuration and bridge configuration.

10.3.1 Centre-tapped Transformer Configuration

Figure 10.3 shows the power circuit of a single-phase to single-phase cycloconverter employing a centre-tapped transformer. There are four thyristors, namely, P_1, N_1, P_2, and N_2. Out of the four SCRs, SCRs P_1 and P_2 are responsible

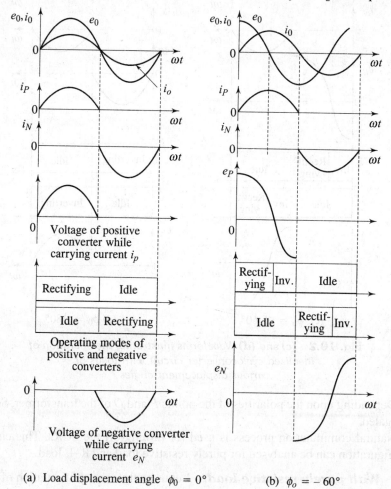

(a) Load displacement angle $\phi_0 = 0°$ (b) $\phi_o = -60°$

Fig. 10.2 *(a) and (b)*

for generating the positive halves forming the positive group. The other two SCRs, N_1 and N_2, are responsible for producing the negative halves forming the negative group. This configuration is meant for generating 1/3 of the input frequency, i.e. this circuit generates a frequency of $16\frac{2}{3}$ Hz at its output.

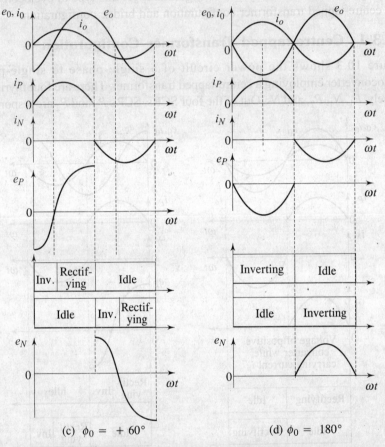

Fig. 10.2 *(c)* and *(d)* *Waveforms illustrating the operation of idealized cycloconverter circuit with loads of various displacement angles*

Depending upon the polarities of the points P and Q of the transformer, SCRs are gated.

Natural commutation process is used for turning-off the SCRs. This circuit configuration can be analysed for purely resistive load and R–L load.

(1) With purely resistive load: Let us analyse the configuration of Fig. 10.3 for a purely resistive load. During the first positive half-cycle, when point P is positive and point Q is negative, SCR P_1 being in conducting mode is gated. The current flows through positive point P, SCR P_1, load and the negative point

O. In the negative half-cycle, when point Q is positive and point P is negative, SCR P_1 is automatically turned-off and SCR P_2 is triggered simultaneously. Path for the current flow in this condition will be from positive point Q, SCR P_2, load and the negative point O. Direction of flow of current through the load remains the same as in the positive half-cycle. Next moment, again point P becomes positive and point Q becomes negative, thus, SCR P_2 is automatically line commutated. SCR P_1 is gated simultaneously. The current path again becomes as in the previous case when SCR P_1 was conducting. Thus, it is seen that the direction of flow of current through the load remains same in all the three half-cycle, or, in other words, the three positive half-cycles are being obtained across the load to produce one combined positive half-cycle as output.

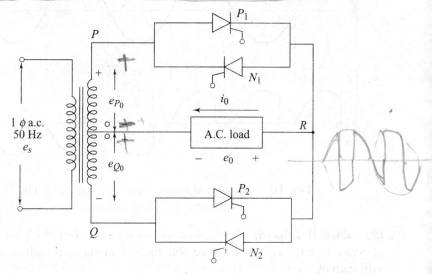

Fig. 10.3 *Single phase to single phase cycloconverter circuit*

Similarly, in the next negative half-cycle of the a.c. input, when point Q is again positive and point P is negative, SCR P_1 is automatically switched OFF. Now, instead of SCR P_2, SCR N_1 (which is also in conducting mode) is gated. The path for the current flow will be from point Q, load, SCR N_1 and back to negative point P. Thus, the direction of flow of current through the load is reversed. In the next positive half-cycle, point P is positive and point Q is negative. SCR N_1 is automatically turned-off. SCR N_2 which is in the conducting mode is simultaneously turned-on. The path for the current flow becomes from positive point P, load, SCR N_2 to the negative point Q. Thus, the direction of flow of current through the load remains the same. For the next negative half-cycle of the a.c. input when point Q is positive and point P negative, SCR N_2 is automatically switched OFF and SCR N_1 is gated. The current flow through the load again remains in the same direction. We can thus analyse it as producing one negative half-cycle at the output by combining three negative halves of the input. In other words, it can be said that, three cycles of the input a.c. have been combined to produce one cycle at the output, i.e. three positive half cycles at the output by the SCRs P_1 and P_2 whereas, three negative half-cycle of the input a.c. are combined to produce one negative half cycle at the output by SCRs N_1 and N_2. This clearly indicates that the input frequency 50 Hz is reduced to 1/3rd (16.2/3 Hz) at the output across the load. The input and output waves are shown in Fig. 10.4. The output voltage magnitude can be changed by varying the firing angle of the SCRs.

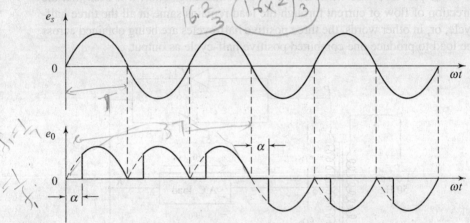

Fig. 10.4 *Input and output waveforms of a $16\frac{2}{3}$ Hz cycloconverter (pure resistance load)*

(2) With R–L load: Let us now analyse the case of an R–L load. This type of cycloconverter will be described both for discontinuous as well as continuous load current.

(a) Discontinuous load current: When point P is positive with respect to point O in Fig. 10.3, forward-biased SCR P_1 is triggered at $\omega t = \alpha$. With this, load current starts building up in the positive direction from point R to point O. Load current i_0 becomes zero at $\omega t = \beta > \pi$ but less than $(\pi + \alpha)$, as shown in Fig. 10.5(c). Thyristor P_1 is thus naturally commutated at $\omega t = \beta$ which is already reverse biased after π. After half a cycle, point Q is positive with respect to O. Now, forward-biased thyristor P_2 triggered at $\omega t = (\pi + \alpha)$. Load current is again positive from point R to O and builds up from zero as shown in Fig. 10.5(c). At $\omega t = (\pi + \beta)$, current decays to zero and SCR P_2 is naturally commutated. At $(2\pi + \alpha)$, SCR P_1 is again turned-on. Load current in Fig. 10.5(c) is seen to be discontinuous. After four positive half-cycles of load voltage and load current, thyristor N_2 (after P_2, N_2 should be fired) is gated at $(4\pi + \alpha)$ when point O is positive with respect to point Q.

As SCR N_2 is forward-biased, it starts conducting but load current direction is reversed, i.e. it is now from point O to R. After SCR N_2 is triggered, load current builds up in the negative direction as shown in Fig. 10.5(c). In the next

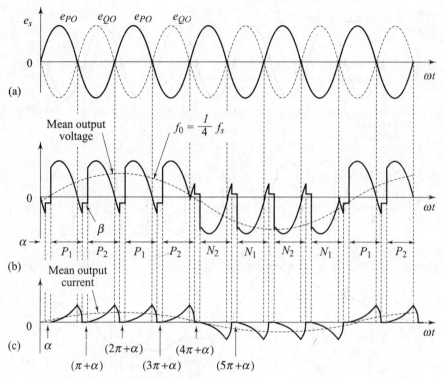

Fig. 10.5 *Voltage and current waveforms for a single-phase to single-phase cycloconverter with discontinuous load current*

half-cycle, point O is positive with respect to point P but before SCR N_1 is fired, current i_0 decays to zero and SCR N_2 is naturally commutated. Now, when SCR N_1 is gated at $(5\pi + \alpha)$, current i_0 again builds up but it decays to zero before SCR N_2 in sequence is again gated. In this manner, four negative half-cycles of load voltage and load current, equal to the number of four positive half-cycles, are generated. Now SCR P_1 is again triggered to fabricate further four positive half-cycles of load voltage and so on. For discontinuous load current, natural commutation is achieved, i.e. SCR P_1 goes to blocking state before SCR P_2 is gated and so on.

In Fig. 10.5, mean output voltage and current waves are also shown. It can be observed from this figure that the frequency of output voltage and current is $f_0 = 1/4 f_s$.

(b) Continuous load current: When point P is positive with respect to point O in Fig. 10.3, SCR P_1 is triggered at $\omega t = \alpha$, positive output voltage appears across load and load current starts building up, as shown in Fig. 10.6(c). At $\omega t = \pi$, supply and load voltages are zero. After $\omega t = \pi$, SCR P_1 is reverse-biased. As load current is continuous, SCR P_1 is not turned-off at $\omega t = \pi$. When SCR P_2 is triggered in sequence at $(\pi + \alpha)$, a reverse voltage appears across SCR P_1; it is therefore turned-off by natural commutation. When SCR P_1 is commutated,

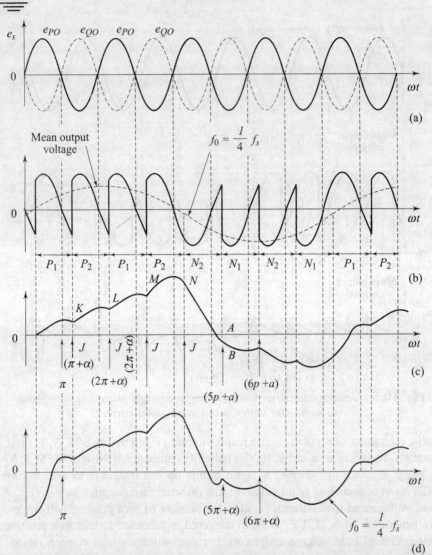

Fig. 10.6 *Voltage and current waveforms for single phase to single phase cycloconverter with continuous load current*

load current has built up to a value equal to KJ, as shown in Fig. 10.6. With the turning-on of SCR P_2 at $(\pi + \alpha)$, output voltage is again positive as it was with SCR P_1 ON. As a consequence, load current builds up further than KJ as shown in Fig. 10.6. At $(2\pi + \alpha)$, when SCR P_1 is again turned-on, SCR P_2 is naturally commutated and load current through SCR P_1 builds up beyond KL as shown. At the end of four positive half-cycles of output voltage, load current is KN. When SCR N_2 is now triggered after SCR P_2, load is subjected to a negative voltage cycle and load current i_0 decreases from positive KN to negative AB (say) as shown in Fig. 10.6(1). Now, SCR N_2 is commutated and SCR N_1 is gated at $(5\pi + \alpha)$. Load current i_0 becomes more negative than AB at $(6\pi + \alpha)$, this is because with SCR N_1 ON, load voltage is negative. For four negative half-cycles of output voltage, current i_0 is shown in Fig. 10.6(c). Load current waveform is redrawn in Fig. 10.6(d) under steady-state conditions. It is seen from load current waveform that current i_0 is symmetrical about ωt axis in Fig. 10.6(d). The positive group of voltage and current wave consists of four pulses and same is true for negative group of wave. One positive group of pulses along with one negative group of identical pulses constitute one cycle for the load voltage and load current. The supply voltage has, however, gone through four cycles. The output frequency is, therefore, $f_0 = 1/4\, f_s$ in Fig. 10.6.

10.3.2 Bridge Configuration

The input transformer used for the midpoint connection shown in Fig. 10.3 will not be required if the bridge configuration shown in Fig. 10.7 is used. Here, two single phase fully-controlled bridges are connected in opposite directions. Bridge 1 supplies load current in the positive half of the output cycle and bridge 2 supplies load current in the negative half of the output cycle. The two bridges should not conduct together as this will produce a short-circuit at the input.

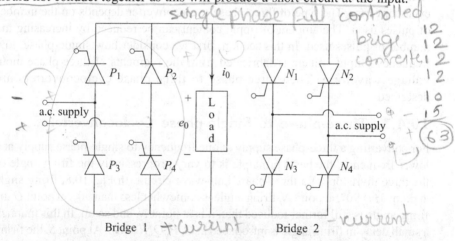

Fig. 10.7 Bridge configuration single phase cycloconverter

Instead of one thyristor in the centre-tap transformer configuration, two thyristors come in series with each voltage source in the bridge configuration. For resistive loads, the SCRs undergo natural commutation and produce discontinuous current operation as shown in Fig. 10.4. For inductive loads, the load current may be continuous or discontinuous, depending upon the firing angle and load power factor. The load voltage and current waveforms would be similar to Fig. 10.5 for discontinuous load current and as in Fig. 10.6 for continuous load current.

When the load current is positive, the firing pulses to the SCRs of bridge 2 will be inhibited and bridge 1 will be gated. Similarly, when the load current is negative, bridge 2 will be gated and the firing pulses will not be applied to the SCRs in bridge 1. This is the circulating current free mode of operation.

Thus, the firing angle control scheme must be such that one converter can conduct at a time, and the change-over of firing pulses from one converter to the other should be periodic according to the output frequency. However the firing angles of SCRs of both the converters should be the same to produce a symmetrical output. When a cycloconverter operates in the noncirculating current mode, the control scheme becomes complicated if the load current is discontinuous. The control scheme becomes somewhat simplified if some amount of circulating current is allowed to flow between them. In this case, a circulating current limiting reactor is connected between the positive and negative converters. This circulating current by itself keeps both the converters in virtually continuous conduction over the whole control range. This type of operation is called as the circulating-current mode of operation.

10.4 THREE-PHASE HALF-WAVE CYCLOCONVERTERS

A three-phase cycloconverter works on the same principles as single-phase cycloconverter. The type of three-phase cycloconverter depends on the number of pulses used. The amount of ripple content can be reduced by increasing the number of pulses used. In this section, first we consider how single-phase, low frequency output voltage is fabricated from the segments of three-phase input voltage waveform. Then three-phase to three-phase cycloconverters are described.

10.4.1 Three-phase to Single-phase Cycloconverters

For converting a three-phase supply at one frequency to single-phase supply at a lower frequency, the basic principle is to vary progressively the firing angle of the three thyristors of a three-phase half-wave circuit. In Fig. 10.8, firing angle at point M is 90°, at point N, firing angle is somewhat less than 90°, at point O the firing angle is still further reduced than it is at point N, and so on. In this manner, a small delay in firing angle is introduced at O, P, Q, R and S. At point S, the firing angle is zero and the mean output voltage, given by $E_0 = E_{do} \cos \alpha$ (E_{do} is the average output voltage with zero firing delay) is maximum at point S. At point M, the mean output voltage is zero as $\alpha = 90°$. After point S, a small delay in firing angle is further introduced progressively at point T, U, V, W, X and Y. At point Y, the firing angle is again 90° and the value of mean output voltage is zero.

The gating circuitry is suitably designed to introduce progressive firing angle delay as discussed. In Fig. 10.8, the single-phase output voltage, fabricated from three-phase input voltage, is shown by thick curve. Mean output voltage wave is obtained by joining points pertaining to average voltage values. For example, at point M, $\alpha = 90°$, $E_0 = 0$; at point S, $\alpha = 0°$, therefore E_0 has maximum mean output voltage, and so on. It can be observed from Fig. 10.8 that fabricated output voltage given by thick curve can be resolved into fundamental frequency output voltage plus several other harmonic components. The load inductance can, however, filter out the high frequency unwanted harmonics. Figure 10.8 reveals that for one half-cycle of fundamental frequency output voltage (marked mean output voltage in this figure), there are eight half-cycles of supply frequency voltage. This shows that the output frequency $f_0 = \frac{1}{8} f_s$, where f_s is the supply frequency.

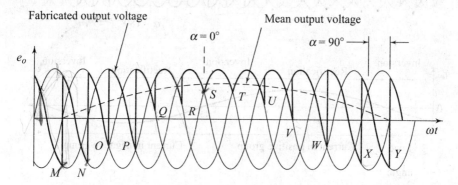

Fig. 10.8 *Fabricated and mean output voltage waveform for a single phase cycloconverter*

It is obvious from Fig. 10.8 that for obtaining positive half-cycle of low-frequency output voltage, firing angle is varied from 90° to 0° and then to 90°. For obtaining one cycle (consisting of one positive half-cycle and one negative half-cycle) of low frequency output voltage, the firing angle should be varied from 90° to 0° to 90° for positive half-cycle and from 90° to 180° and back to 90° for negative half-cycle. This is illustrated in Fig. 10.9.

Thus, it has been observed from above that a complete cycle of low frequency output voltage can be fabricated from the segments of three-phase input voltage waveform by the use of phase-controlled converters. The cycloconverter can be made to deliver any p_f load. In Fig. 10.9, the device is shown to deliver a lagging p_f load. In a thyristor converter circuit, current can only flow in one direction. For allowing the flow of current in both the directions during one complete cycle of load current, two three-phase half-wave converters must be connected in antiparallel as shown in Fig. 10.10. The converter circuit that permits the flow of current during positive half-cycle of low-frequency output current is called as positive-group converter. The other group permitting the flow of current during the negative half-cycle of output current is called as negative group converter. For a three phase to single phase cycloconverter, schematic diagram is shown in Fig. 10.10(a) and the basic circuit configuration in Fig. 10.10(b). This figure uses two-three-phase half-wave converters in antiparallel, the positive group for the conduction of positive load current and negative group for the flow of negative load current.

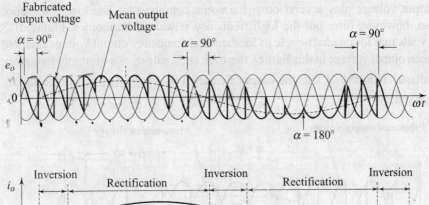

Fig. 10.9 *Voltage and current waveforms for a three-phase half-wave cycloconverter*

Examination of Fig. 10.9 reveals that when output current is positive (above the reference line ωt), positive converter conducts. Under this condition, positive converter acts as a rectifier when output voltage is positive, and as an inverter when output voltage is negative. When output current is negative, the negative converter conducts; under this condition, negative converter acts as a rectifier when output voltage is negative and as an inverter when output voltage is positive.

It can thus be inferred, in general, that one of the two component converters in Fig. 10.10 would operate as rectifier if the output voltage and current have the same polarity and as an inverter if these are of opposite polarity.

Though the output voltages of the two converters in the same phase have the same average value, their output voltage waveforms as a function of time are, however, different and as a result, there will be a net potential difference across the two converters of Fig. 10.10(a). This net voltage would cause a circulating current in the two converters. This circulating current can be avoided by removing the gating signals from idle converter or can be limited to a low value by inserting an intergroup reactor (IGR) between the positive and negative group converters, as shown in Fig. 10.10(b). In order that the average value of the two converters are equal in magnitude and opposite in sign, the sum of their firing angles must be 180°. In other words, if α_p and α_n are the firing angles for positive and negative group converters, respectively, then these firing angles should be so controlled as to satisfy the condition $\alpha_p + \alpha_n = 180°$.

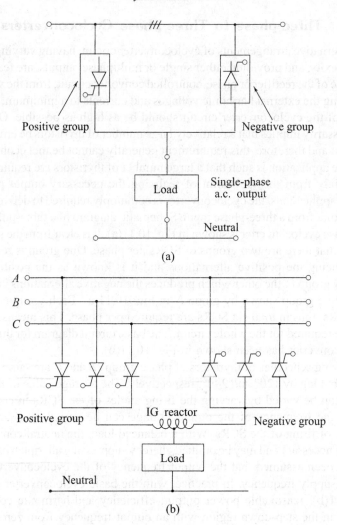

Fig. 10.10 *Three-phase to single-phase cycloconverter: **(a)** schematic diagram **(b)** basic circuit configuration with I_G reactor*

The continuous current of each group in the circulating current mode imposes a higher loading on each group compared to the non-circulating current mode of operation. In practice, the circulating current mode would only be used when the load current is low, so that continuous load current with a better waveform can be maintained. At the higher levels of load current, the groups would be blocked to prevent circulating current. Control circuits would be used to sense the level of the load current, allowing firing pulses to each group at low current levels, but blocking firing to one or other group at the higher current levels. The reactor would be designed to saturate at the higher current levels when the cycloconverter is operating in the non-circulating current mode, thus permitting a smaller core.

10.4.2 Three-phase to Three-phase Cycloconverters

Many alternative arrangements of cycloconverter circuit, having varying degrees of complexity, and providing either single or multiphase outputs, are feasible. As in the case of the rectifier or phase-controlled converter circuit, from the viewpoint of reducing the external harmonic voltages and currents to a minimum, the pulse number of the cycloconverter circuit should be as high as possible. Of-course, this necessarily implies that a relatively large number of thyristors be employed in the circuit and therefore, this requirement generally cannot be met economically, unless the application is such that a large number of thyristors are required in any case, purely from the viewpoint of realizing the necessary output power. In practical applications, the cycloconverter is commonly required to deliver a three-phase output from a three-phase input. Schematic diagram of a three-pulse phase, three-phase cycloconverter is shown in Fig. 10.11(a). It is clear from the schematic diagram that there are two groups of SCRs per phase. One group is responsible for producing the positive alternations and it is known as the positive group shown by group P, the other which produces the negative alternations is called as the negative group shown by group N in Fig. 10.11(a). Each group consists of three SCRs, thus in total six SCRs are required per phase. This means, in all 18 SCRs are required for the whole circuit. The basic circuit diagram for this scheme of cycloconverter has been shown in Fig. 10.11(b).

The firing schedules of thyristors of phase groups B and C are same as that of group A but lag by 120° and 240°, respectively. The average value of the output voltage can be varied by varying the firing angles of the SCRs in conduction, whereas the frequency of the output voltage can be varied by changing the sequence of firing of the SCRs. With a balanced load, the neutral connection is no longer necessary and may be omitted, thereby suppressing all triplen harmonics.

It has been assumed that the output frequency of the cycloconverter is less than the supply frequency. In practice, with the basic cycloconverter circuit of Fig. 10.11(b), reasonable power output, efficiency and harmonic content are obtained in the step-down region with an output frequency from zero to about one-third of the input frequency. As the cycloconverter output frequency approaches the supply frequency, harmonic distortion in the output voltage increases because the output voltage waveform is composed of fewer segments of the supply voltage. As a result, losses in the cycloconverter and in the a.c. motor become excessive, and there is a drop in overall efficiency. By using more complex converter circuits with higher pulse numbers, the output voltage waveform is improved and the maximum useful ratio of output to input frequency is increased to about one-half. The a.c. motor normally presents a high impedance at the ripple frequency and hence, the output current is nearly sinusoidal and no additional filtering is necessary. Systems designed for aircraft power supplies that operate with a fixed output frequency (400 Hz) are provided with an output filter.

Cycloconverters   685

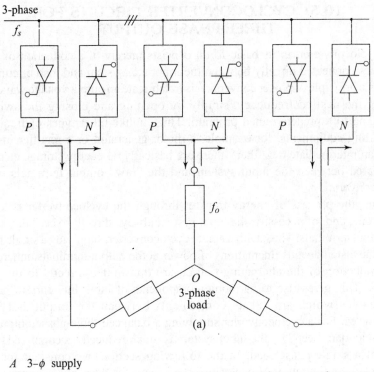

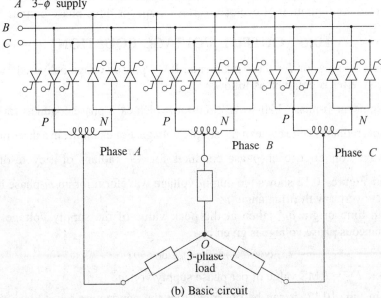

Fig. 10.11 *Three-phase to three-phase cycloconverter **(a)** schematic diagram **(b)** basic circuit*

10.5 CYCLOCONVERTER CIRCUITS FOR THREE-PHASE OUTPUT

The cycloconverter, in its basic form, consists merely of a collection of static switches connected directly between the input a.c. system and load circuit, and the basic principle of power conversion is to fabricate an output voltage waveform having the desired frequency, simply by opening and closing the switches according to a predetermined program. Thus, unlike other types of frequency converting equipments, for example, a motor-generator, or a rectifier inverter (with an intermediate d.c. filter), there are basically no energy storage elements connected between the input system and the "raw" output terminals of the cycloconverter.

Thus, the process of energy transfer through the cycloconverter is a very direct one, and, of necessity, the input system always directly "sees" the load at the output terminals. Thus, with a single cycloconverter, supplying a single phase load, the instantaneous fluctuations of power at the output terminals, inherent in the production of the alternating output, are transmitted directly to the input system. This gives rise to harmonic components of input line current having frequencies which are "beat" components between the output and input frequencies. For a cycloconverter supplying a balanced three-phase output, the harmonic load "seen" by the input system is much reduced, as compared to the case of a single-phase load. In the following section, we consider the ring connected configuration of cycloconverter circuit for three-phase output.

10.6 OUTPUT VOLTAGE EQUATION

A general expression for the fundamental RMS phase voltage delivered by the cycloconverter is obtained as follows:

Assume an m-phase, half-wave circuit in which each phase conducts for $\frac{2\pi}{m}$ electrical radians in one cycle of the supply voltage. For example, in a three phase half-wave converter, each phase conducts for $\frac{2\pi}{3}$ radians of a cycle of 2π radians. Figure 10.12 shows the output voltage waveform for an m-phase half-wave converter with firing angle α.

With time origin pp' taken at the peak value of the supply voltage, the instantaneous phase voltage is given by

$$e = E_m \cos \omega t = \sqrt{2}\, E_{ph} \cdot \cos \omega t$$

where E_{ph} = RMS value of per phase supply voltage.

From Fig. 10.12, it can be observed that the conduction takes place from $(-\pi/m)$ to (π/m) if the firing angle is zero. For firing angle α, the phase conducts from $[(-\pi/m) + \alpha]$ to $[(+\pi/m + \alpha]$, and the average output voltage, E_{dc}, equals the average height of the shaded area in Fig. 10.12:

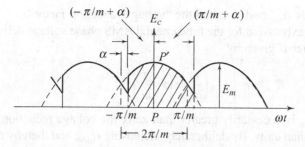

Fig. 10.12 *Output voltage waveform for m-phase half-wave converter with firing angle α*

$$\therefore\ E_{dc} = \frac{m}{2\pi} \int_{(-\pi/m+\alpha)}^{(\pi/m+\alpha)} \sqrt{2} E_{ph} \cos \omega t\, dt(\omega t) = \sqrt{2} E_{ph} \left(\frac{m}{\pi}\right) \sin\left(\frac{\pi}{m}\right) \cos \alpha \tag{10.1}$$

When the delay angle is zero, E_{dc} has the maximum value E_{do}:

$$E_{do} = \sqrt{2} E_{ph} \left(\frac{m}{\pi}\right) \sin\left(\frac{\pi}{m}\right) \tag{10.2}$$

If the cycloconverter delay angle is slowly varied, the output phase voltage at any point of the low frequency cycle may be calculated as the average output voltage for the appropriate delay angle. This calculation ignores the rapid fluctuations superimposed on the average low frequency waveform. Assuming continuous current conduction, the average output voltage is given by

$$E_{dc} = E_{do} \cos \alpha$$

If E_{or} is the fundamental RMS value of per phase output voltage of cycloconverter, then the peak output voltage for zero firing angle is

$$\sqrt{2} E_{or} = E_{do} = \sqrt{2} E_{ph} \left(\frac{m}{\pi}\right) \sin\left(\frac{\pi}{m}\right)$$

or

$$E_{or} = E_{ph} \left(\frac{m}{\pi}\right) \sin\left(\frac{\pi}{m}\right) \tag{10.3}$$

However, the firing angle of the positive group α_p cannot be reduced to zero, for this value corresponds to a firing angle of 180° in the negative group ($\alpha_p = \pi - \alpha_n$). In practice, inverter firing cannot be delayed by 180 degrees because sufficient margin must be allowed for commutation overlap and thyristor turn-off time. Consequently, the delay angle of the positive group cannot be reduced below a certain finite value, α_{min}. Therefore, the maximum output voltage per phase is,

$$E_{dc\,max} = E_{do} \cos \alpha_{min} = r \cdot E_{do} \tag{10.4}$$

where $r = \cos \alpha_{min}$ and is called the "voltage reduction factor".

Thus, the expression for the fundamental RMS phase voltage delivered by the cycloconverter is given by

$$E_{or} = r \cdot \left[E_{ph} \left(\frac{m}{\pi} \right) \sin \left(\frac{\pi}{m} \right) \right] \tag{10.5}$$

Since α_{min} is necessarily greater than zero, the voltage reduction factor, r, is always less than unity. By deliberately increasing α_{min}, and thereby reducing the range of variation of α about the 90 degree value, the output voltage, E_{or}, can be reduced, and a static method of voltage control is obtained. In practice, the output voltage is less than the theoretical value given by Eq. (10.5), due to the influence of commutation overlap and the circulating currents between positive and negative groups.

Equation (10.5) gives the RMS value of the per phase output voltage for a three phase to three phase or three phase to single phase cycloconverter employing m-phase half-wave circuits as shown in Figs 10.10 or 10.11.

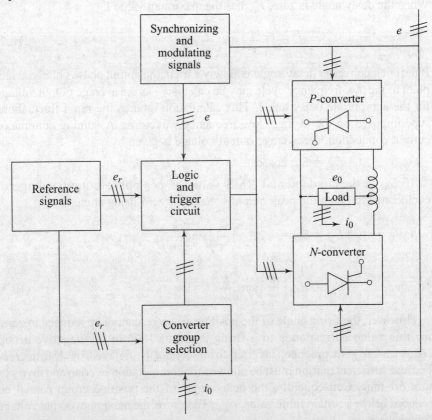

Fig. 10.13 *Control circuit block-diagram for a cycloconverter with non-circulating current mode*

10.7 CONTROL CIRCUIT

The function of the control circuit is to deliver correctly timed, properly shaped, firing pulses to the gates of the thyristors in the power converter so as to generate a voltage of the desired wave shape at the output terminals of a cycloconverter. The functional block diagram of the control circuit of a cycloconverter with the non-circulating current operation is shown in Fig. 10.13. The same control circuit is applicable to a cycloconverter with the circulating current operation but the block designated as converter group selection will not be present in this case. The control circuit can be arranged in four functional blocks:

1. Synchronizing circuit.
2. Reference voltage sources.
3. Logic and triggering circuit.
4. Converter group selection circuit.

10.7.1 Synchronizing Circuit

The main function of the synchronizing circuit is to derive low voltage signals to the control circuit which operates at low voltages. These low voltage signals must be synchronized to the voltages supplied to the main power circuit. Step-down transformers may be used for this purpose with filter circuit to avoid waveform distortion if any. While deriving the modulating voltages at the supply frequency, the phase shifting network may also be required. To determine the instants at which the firing pulses are to be produced to the thyristors in the two converter groups, the modulating voltages are compared with the reference voltages.

10.7.2 Reference Voltage Signals

The reference signal is designed to control the output voltage in the sense that the output voltages tends to follow the reference signal. It means that if the amplitude and frequency of the reference signal is varied, then the amplitude and frequency of the output voltage varies automatically. In the case of three phase to three-phase cycloconverter, the reference signal does additional function of shifting e_{OA}, e_{OB}, e_{OC} by phase shift of 120°. Three phase variable-frequency, variable-voltage sine wave reference voltage sources can be designed in various ways.

The reference voltage generator block diagram is shown in Fig. 10.14. In one of the design approach, a fixed high frequency oscillator and resistor–capacitor phase shifting networks are used to generate a three phase sine wave. In this case, UJT is used as a relaxation oscillator at a variable frequency to trigger athree-stage ring counter. The output frequency of the UJT relaxation oscillator is controlled by a d.c. voltage. The ring counter develops three, three phase, square-wave output at a frequency f_d and at a phase shift of 120°. Fixed frequency oscillator at f_c gives three outputs, ϕ_A, ϕ_B and ϕ_C. Three mixers (transistorised chopper), one for each phase, are used to combined the fixed and variable frequencies. The output of the mixer stage is a square wave with

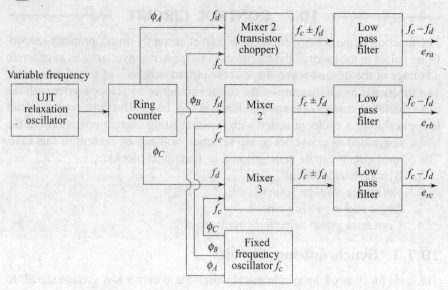

Fig. 10.14 *Reference voltage generator block-diagram*

half-wave symmetry. It generates a fundamental and a series of odd harmonics. If all higher order harmonics are neglected, it can be shown that the sum $(f_c + f_d)$ and the difference $(f_c - f_d)$ of fixed frequency and chopper fundamental frequency (f_d) would result at the output of the chopper. By using low-pass filters, the frequency component $(f_c - f_d)$ is selected and the component $(f_c + f_d)$ is eliminated. Finally, the reference signals can be shown to be

$$e_{ra} = r \cdot E_m \sin 2\pi (f_c - f_d) t \tag{10.6}$$

$$e_{rb} = r \cdot E_m \sin (2\pi (f_c - f_d) t - 120°) \tag{10.7}$$

$$e_{rc} = r \cdot E_m \sin (2\pi (f_c - f_d) t + 120°) \tag{10.8}$$

where E_m is the peak of the cosine modulating wave.

From Eqs (10.6) to (10.8), it becomes clear that the amplitude and frequency of the reference waves are controlled by varying r and f_d, respectively, while the phase sequence of the three phase output is controlled by setting f_d greater or less than f_c. The amplitude is controlled by varying r from 0 to 1.

10.7.3 Logic and Triggering Circuit

The block diagram of the main logic and triggering circuit for the phase A output of a cycloconverter using the reference voltage e_{ra} is shown in Fig. 10.15. Two similar blocks, one with the reference voltage e_{rb} and the second with the reference voltage e_{rc} are used for the other two phases of the output in the case of a three phase to three phase cycloconverter circuit.

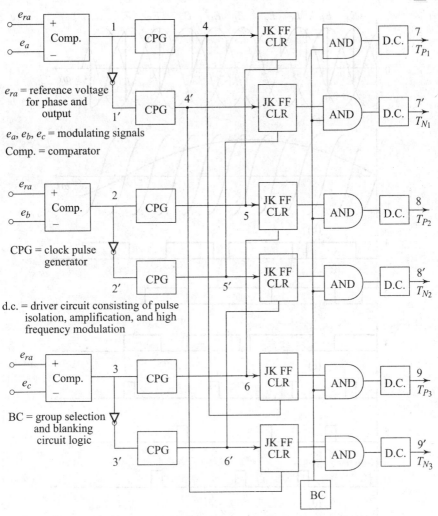

Fig. 10.15 *Block diagram of logic and trigger circuit*

The reference voltage e_{ra} is compared with the modulating voltages e_a, e_b and e_c corresponding respectively to supply voltages e_A, e_B and e_C. Comparators (comp) produces short pulses. These pulses drive clock-pulse generator (c_p), which is a positive edge triggered monostable multivibrator. Therefore, we get six pulses (4, 4′, 5, 5′, 6, 6′). Now, these short pulses act as clock-pulses to the flip-flops. The clearing of the pulses is done when not required. This clearing is done in the sequence $6 \to 5 \to 4$. This limits the pulse presence only when that thyristor is supposed to conduct, otherwise pulse is blocked. The pulse-output of the flip-flop is ANDed with the input coming from the blanking circuit. The output of the AND gate is given to the driver circuit which amplifies and isolates the pulses and drives the respective thyristors. With the circulating current operation, no blanking circuit is needed. Hence, this input may be set at the logic '1' permanently.

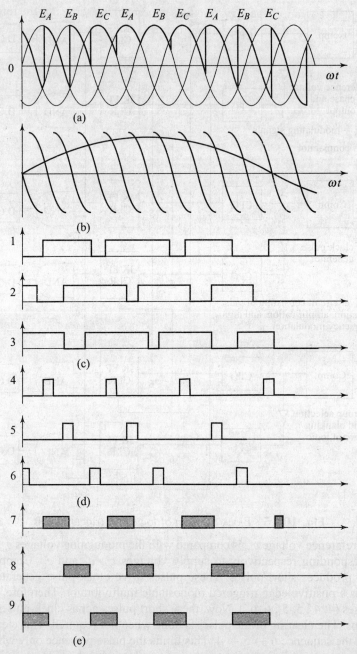

Fig. 10.16 *Various waveforms of logic circuit for one phase only. (a) supply voltage of 3-pulse cycloconverter showing output voltage of one half-cycle. (b) modulating signals for positive converter and reference voltage (100%). (c) outputs of comparators used in the positive group. (d) clock pulse for positive group. (e) gate pulses to SCRs*

Cycloconverters

The block diagram of the logic circuit is general in the sense that the reference signals and modulating signals could be of any waveshape. In most applications, sinusoidal reference signals and consinusoidal modulating signals are commonly employed. The basic principle of the consinusoidal modualtion is the same as that of cosine wave crossing-pulse timing control which is explained in Chapter 9. Figure 10.16 illustrates typical waveforms at differnt points of the logic and triggering circuit of Fig. 10.15 assuming sinusoidal reference voltages and consinusoidal modulating signals. The waveforms are drawn for one half-cycle of the output voltage for the positive converter with unity power factor load. The same procedure can be followed for developing the output voltage for the other half cycle of the positive converter. For the negative converter, different modulating signals have to be taken into account. This is taken care of by using an inverter at the output of each comparator (Fig. 10.15).

10.7.4 Converter Group Selection

The block diagram of the converter bank selection is shown in Fig. 10.17. The load current is allowed to flow through the P-converter or the N-converter through suitable logic. D is the delay during which period, the firing pulses to both the converters are inhibited. The delays are not introduced in some control schemes where a small circulating current is permitted during the cross-over instants of the fundamental load current only. The scheme is still recognized as the non-circulating current operation since during a major portion of the output cycle, it operates in the non-circulating current mode.

The circuit is an essential part of the control scheme of a cycloconverter with the non-circulating current operation. The function of this circuit is to see that only one converter operates at a time depending upon the polarity of the current. The positive converter is operated when the load current is positive and the negative converter is operated when the load current is negative. The converter group selection is not straight forward primarily due to non-ideal nature of the output current waveform. Since the actual load voltage waveform itself is far from sinusoidal, the load current is also non-sinusoidal. Depending upon load circuit parameters, converter pulse number, the load current may become zero before the fundamental half-period. If the group selection and blanking circuit were to operate at each current zero instant, it may cause erratic switching of converters. The result of this is to further distort the output voltage. One possible solution to this problem is to see that the blanking circuit operates at the zero crossings of the fundamental current. The fundamental component of the load current is extracted and the converter bank selection is made to occur at the zero crossings of this fundamental components of the current. There are, however, some operational difficulties, in the design of filter components specifically when the cycloconverter is required to operate over a range of the output frequency and with variable load. Filters which operates satisfactorily over the desired range of frequency will have to be used. Thus, the envelope distortion of the output current and the output voltage are reduced and the possible steady state discontinuous conduction within the fundamental period does not cause any erratic switching of converters.

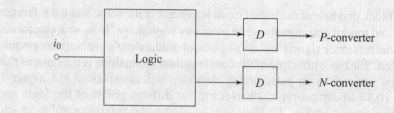

Fig. 10.17 *Converter group selection in non-circulating current scheme*

Because of filters, certain amount of phase shift may be introduced between the zero-crossings of the fundamental output current and the actual load current. In order to eliminate the waveform distortion, certain amount of circulating current may be allowed to flow during these short overlap periods in some control schemes. The presence of the circulating current reactor is a must in such a design. However, if no circulating current is permitted to flow, some distortion in the output voltage is to be tolerated. This distortion arises due to the delays introduced at the zero-crossings of the load current to ensure turn-off of thyristors in the outgoing group before the thyristors in the incoming group are turned-on.

A fundamental different approach to the converter group selection is to employ a closed loop control of the output voltage, arranged in such a way that it automatically selects the converter banks. The firing pulses are permanently applied to both the converters, but a bias voltage is introduced between their voltage transfer characteristics, so that there is no possibility for appreciable circulating current to flow. The action of the closed-loop voltage control is such that it automatically counteracts the bias voltage of whichever converter, for the time being, fabricates the output voltage wave. In this way, the current carrying converter is automatically 'pulled' into operation, while, in the meantime, the idle converter is pushed to one side. At the current cross-over point, the action of the voltage feedback loop is inherently such that the error signal seeks the incoming converter, and thus, the bank selection function is furnished automatically. A control scheme of this type is shown in simplified functional form in Fig. 10.18.

The function of each of the positive and negative firing pulse generators is to produce firing pulses for the thyristors of the associated converters. The phase of these firing pulses is controlled in accordance with the analog input voltages e_{cp} and e_{CN}, so that (with continuous conduction) the mean output voltage of the associated converter is directly proportional to the respective input voltage.

Each of the firing pulse generators have equal and opposite fixed bias voltages E applied to their inputs. The polarities of these voltages are such as to tend to retard the converter firing angles, and thus to "bias-off" the converters, with respect to one another, thereby preventing the flow of circulating current. A second, common control voltage is applied to the inputs of both firing pulse generators; this is the output signal of the error voltage amplifier.

The operation of the scheme can be explained with reference to the waveforms shown in Fig. 10.19. For simplicity, the presence of ripple voltage at the output of the cycloconverter is neglected, and continuous conduction (except at the current crossover) into a load of unity displacement factor, is assumed.

Consider the operation during the time period t_1–t_2, during which the positive converter is in conduction. The action of the voltage feedback loop during this time is automatically such as to force the output voltage of the positive converter to follow the reference input voltage. This necessarily implies that the error voltage amplifier produces at its output, in addition to a sinusoidal modulating signal, a steady bias voltage of E. The presence of this component of voltage is necessary in order to counterbalance the steady bias voltage applied to the positive firing pulse generator, thereby reducing the net bias voltage at the input of this firing pulse generator to zero. In the meantime, the total bias voltage appearing at the input of the negative firing pulse generator is $2E$, and therefore this converter is "biased off."

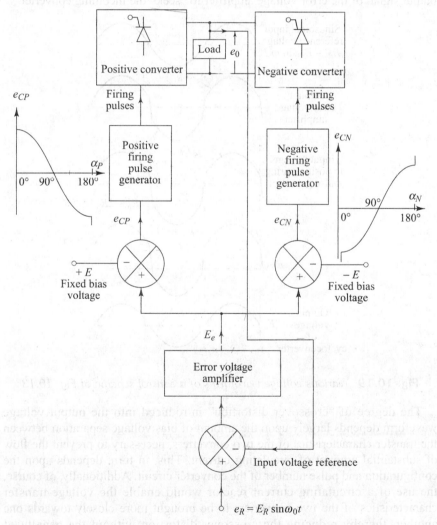

Fig. 10.18 *Control scheme for a cycloconverter using voltage-sensing principle of converter group selection*

The current in the positive converter falls to zero at time t_2. The output voltage now remains momentarily at zero (since, with no current, it cannot reverse) and hence, the polarity of the actual voltage error becomes negative. The output signal of the error voltage amplifier now moves in a negative going direction, thus seeking the negative converter, and pulling this into conduction, while, at the same time, the positive converter is automatically pushed to one side.

This type of control scheme is attractive because of its simplicity, and because it provides a naturally robust and reliable mode of system operation. However, it should be appreciated that the principle relies for its operation upon the deliberate production of an error between the input reference voltage and the mean output voltage, in the region of current crossover. It is this error which forces the output signal of the error voltage amplifier to "seek" the incoming converter.

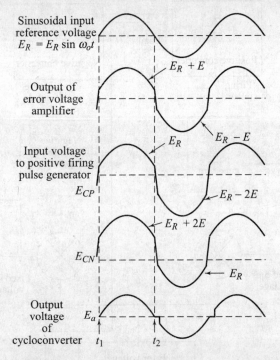

Fig. 10.19 *Various voltage waveforms of a control scheme of Fig. 10.13*

The degree of "crossover distortion" introduced into the output voltage waveform depends largely upon the amount of bias voltage separation between the transfer characteristics of the two converters, necessary to prevent the flow of substantial amount of circulating current. This, in turn, depends upon the configuration and pulse number of the converter circuit. Additionally, of course, the use of a circulating current reactor would enable the voltage-transfer characteristics of the two converters to be brought more closely towards one another, thereby reducing the crossover distortion without the penalty of introducing high amplitude circulating ripple currents.

10.8 COMPARISON BETWEEN CYCLO-CONVERTER AND D.C. LINK CONVERTER

A comparison between cycloconverter and d.c. link converter is given in a tabular form as follows:

Cycloconverter	D.C. Link Coverter
1. In a cycloconverter, a.c. power at one frequency is converted directly to a lower frequency in a single conversion stage.	d.c. link converter has two power controllers and the full output power is converted in two stages.
2. Cycloconverter functions by means of phase commutation and no auxiliary forced commutation circuits are necessary. This results in a more compact power circuit and also eliminates circuit losses associated with forced commutation.	d.c. link converter, on the other hand, requires forced commutation for the inverter even though the rectifier operates on phase control principle.
3. Cycloconverter is inherently capable of power transfer in either direction between source and load. It can supply power to loads at any power factor and is also capable of regeneration at full power over the complete speed range, down to stand-still. This feature makes a cycloconverter preferable for large reversing drives requiring rapid acceleration and deceleration. This type of application occurs principally in the metal rolling industry.	This feature is slightly difficult and is involved to incorporate in a d.c. link converter.
4. Commutation failure causes a short circuit of the a.c. supply. But if an individual thyristor fuse blows off, a complete shutdown is not necessary and the cycloconverter continues to function with somewhat distorted waveforms. A balanced load is presented to the a.c. supply even with unbalanced output conditions.	The d.c. link converter cannot provide this feature.
5. Cycloconverter delivers a high quality sinusoidal waveform at low output frequencies since it is fabricated from a large number of segments of the supply waveform. This is often preferable for very low speed applications.	The d.c. link converter, on the other hand, generates a stepped waveform which may cause a nonuniform rotation of an a.c. motor at very low frequencies (< 10 Hz). The distorted waveform also causes system instability at low frequencies.
6. For reasonable power output and efficiency, the output frequency is limited to 1/3 input frequency.	The frequency can be varied from zero to rated value. The upper frequency limit is divided by the device turn-off time.

(Contd.)

Cycloconverter	D.C. Link Coverter
7. Requires larger number of thyristors and its control circuitry is more complex. This is not justified for small installations but it is economical for units 20 kVA and more.	The d.c. link converter requires only 12 thyristors and control circuits are less involved.
8. Has very low power factor, particularly at reduced output voltages.	The d.c. link converter has high input power factor if diode rectifier is used. With phase controlled p_f depends upon phase angle.
9. Extremely attractive for large power low speed reversing drives.	Extremely suitable for high frequencies.

10.9 LOAD-COMMUTATED CYCLOCONVERTER

The load-commutated cycloconverter differs from the previously described line-commutated cycloconverter in that the thyristors can be commutated by the reversal of the load voltage. In this case, the load must possess a generated or back, emf that is independent of the source voltage. The most common example of such a load is a wound-field or permanent magnet synchronous machine. In such cases, the load frequency can be equal to or greater than the source frequency and still allow natural thyristor commutation. The thyristor gating is based on two control signals: first, with respect to the source voltage, to control the load voltage; and second, with respect to the synchronous machine-generated emf, to ensure that current will flow in the correct phase of the load machine at the correct time. For optimum torque per ampere in a synchronous machine, commutation should take place with 180° degree of firing delay with respect to the machine voltage. This, however, leaves no margin for commutation overlap and the thyristor turns-off. As in the other types of thyristor-based synchronous motor drive, the commutation must be advanced from the optimum position. It is the function of the shaft position sensor and the control electronics to insure that the commutation is sufficient in advance of the line-to-line voltage crossing. This condition results in a phase current that leads the phase voltage.

In aircraft applications, load-commutated cycloconverters are used to start and accelerate a gas turbine through its high frequency synchronous a.c. generator operating as a motor. After the turbine has started and reached its operating speed, the generator cycloconverter delivers 400 Hz power to the aircraft.

SOLVED EXAMPLES

Example 10.1 A three pulse cycloconverter feeds a single phase load of 190 V, 45 A at a power factor of 0.7 lagging. Determine:
(a) the required supply voltage (b) thyristor rating, and
(c) power factor of the supply current.

Cycloconverters

Neglect device and supply impedance volt-drops.

Solution: Using Eq. 10.3, we can write

(a) $$190 = \left(\frac{3}{\pi}\right) \times E_{ph} \cdot \sin(\pi/3) = 229.74 \text{ V/ph}$$

∴ $E_{ph} \approx 230$ V/ph (RMS supply voltage per phase)

(b) To determine the thyristor ratings, the worst case will be when the output frequency is so low that a converter group is acting as a rectifier feeding a d.c. load for a long period, with a current equal to the maximum value of the cycloconverter load.

∴ Maximum value of the cycloconverter load = $45\sqrt{2}$ A.

∴ Thyristor RMS current $I_{rms} = \dfrac{45\sqrt{2}}{\sqrt{3}} = 36.74$ A

$$\text{PIV} = \sqrt{3}\ E_{ph(max)} = \sqrt{3} \times 230\sqrt{2} = 563.38 \text{ V}.$$

(c) For each cycle of load current, each input line will conduct short blocks of current for one-third of the time.

The RMS value of a complete sine wave = 45 A, hence the RMS value for one-third period-conduction equals

$$I_{rms} = \sqrt{(45^2/3)} = 25.98 \text{ A}$$

Input power per phase = 1/3 × load power = 1/3 × 190 × 45 × 0.7 = 1995 W.

$$\text{Power factor} = \frac{\text{power}}{E_{rms} \cdot I_{rms}} = \frac{1995}{230 \times 25.98} = 0.33$$

Example 10.2 A six-pulse, blocked group cycloconverter is fed from a three phase, 600 V (line), 50 Hz supply. The supply has an inductance of 1.146 mH/phase. If the cycloconverter is supplying a variable resistive load with a current of 28 A, estimate the peak and RMS value of load voltage for firing angles of 0°, 30°, and 60°.

Solution: From Eq. 9.183 (Chapter 9), we can write the peak voltage which is the mean voltage of the equivalent rectifier, as

$$E_{peak} = \frac{p}{\pi} E_m \sin\left(\frac{\pi}{p}\right) \cos\alpha - \frac{p\omega L_s}{2\pi} I_L$$

For the given system

$$\frac{p\omega L_s}{2\pi} I_L = 6 \times 100\pi \times 1.146 \times 10^{-3} \times \frac{28}{2\pi} = 9.63$$

and $\dfrac{p}{\pi} E_m \sin\left(\dfrac{\pi}{p}\right) = 6 \times 660 \times \sqrt{2}\ \sin(30°) = 891.3.$

(i) For $\alpha = 0°,$

$$E_{mean} = 891.3 - 9.63 = 881.7 \text{ V}$$

$$= E_{max} \text{ for cycloconverter.}$$

Hence, the RMS voltage of the cycloconverter is
$$E_{rms} = E_{max}/\sqrt{2} = 623.4 \text{ V}.$$

(ii) For $\quad \alpha = 30°, E_{mean} = 891.3 \cos(30°) - 9.63 = 762.3 \text{ V}$
$$= E_{max} \text{ for cycloconverter.}$$

Hence, the RMS voltage of the cycloconverter is
$$E_{rms} = E_{max}/\sqrt{2} = 539 \text{ V}$$

(iii) For $\quad \alpha = 60°, E_{mean} = 891.3 \cos(60°) - 9.63 = 436 \text{ V}.$
$$= E_{max} \text{ for cycloconverter.}$$

Hence, the RMS voltage of the cycloconverter is
$$E_{rms} = \frac{E_{max}}{\sqrt{2}} = 308.3 \text{ V}.$$

Example 10.3 A three phase to single phase cycloconverter employs three pulse positive and negative group converters. Each converter is supplied from delta/star transformer with per phase turns ratio of 3:1. The supply voltage is 410 V, 50 Hz. The R_L load has $R = 4\ \Omega$ and at low output frequency, $\omega oL = 3.0\ \Omega$. The commutation overlap and thyristor turn-off time limit the firing in the inversion mode to 160°. Determine:
   (i) the value of the fundamental RMS output voltage,
   (ii) RMS output current, and
   (iii) output power.

Solution:
   (i) Per phase input voltage to transformer = 410 V.

   $\therefore$ Per phase input voltage to converter, $E_{ph} = \dfrac{410}{2} = 205$ V.

   Now, voltage reduction factor, $r = \cos(180° - 160°) = \cos 20° = 0.9396$

   $\therefore$ From Eq. (10.5), the fundamental RMS output voltage from Eq. (10.5) becomes
   $$E_{or} = 0.9396\ [205\ (3/\pi) \sin \pi/3] = 159.30 \text{ V}.$$

   (ii) RMS output current, $I_{or} = \dfrac{159.30}{\sqrt{4^2 + 3^2}} \angle -\tan^{-1}\dfrac{3}{4} = 31.86\ \angle -36.87°$ A

   (iii) Output power = $I_{or}^2 R = (31.86)^2 \times 4 = 4060.24$ W.

Example 10.4 Repeat Example 10.3 in case of three phase to single phase cycloconverter employing six pulse bridge converter.

Solution:
   (i) Per phase input voltage to converter = 205 V.

   Line voltage input to bridge converter = $205\sqrt{3}$ V.
   Voltage reduction factor $\qquad r = \cos 20°$
   Now, for six pulse device, $\qquad m = 6.$

∴ From Eq. (10.5)., $E_{or} = \cos 20° \left[205\sqrt{3}(6/11)\sin \pi/6\right] = 318.61$ V.

(ii) RMS output current $= \dfrac{318.61}{\sqrt{4^2+3^2}} \angle -\tan^{-1}\dfrac{3}{4}$

$= 63.72 \angle -36.87°$ A.

(iii) RMS output power $= (63.72)^2 \times 4 = 16241.97$ W.

Example 10.5 A three phase six-pulse, 50 kVA, 415 V cycloconverter is operating at a firing angle of 45° and supplying load of 0.8 power factor. Determine input current to the converters.

Solution: For a three phase, six-pulse cycloconverter circuit, the input power is given by

$$p_i = 3 \cdot E \cdot I_{mphase} \text{ component of input current} = 3 E \cdot I \dfrac{\cos\theta}{\sqrt{2}} \cos \phi° \quad (10.9)$$

where θ is the angle to which firing angle α is restricted for voltage control, i.e. $\theta \le \alpha \le \pi - \theta$ and ϕ_0 is load power factor.

$$50 \times 10^3 = 3 \times 415 \times I \times \dfrac{1}{\sqrt{2}} \times \dfrac{1}{\sqrt{2}} \times 0.8 \therefore I = 100.40 \text{ A.}$$

REVIEW QUESTIONS

10.1 Explain the basic principle of operation of a cycloconverter with a neat equivalent circuit diagram. Also draw and discuss the waveforms illustrating the operation of the idealized cycloconverter circuit with loads of various displacement angles.

10.2 A single-phase to single-phase centre-tapped cycloconverter is delivering power to a resistive load. The supply transformer has turns ratio of 1 : 1 : 1. The frequency ratio is $f_o/f_s = 1/3$. The firing delay angle α for all the thyristors are the same. Sketch the time variations of the following waveforms for $\alpha = 0°$ and 45°.
   (a) Supply voltage   (b) Output current   (c) Supply current.
   Indicate the conduction of all the thyristors also.

10.3 Describe the basic principle of working of a single-phase to single-phase cycloconverter for both continuous and discontinuous conductions for a bridge-type cycloconverter. Mark the condition of various thyristors also.

10.4 Describe how single-phase, low frequency output voltage can be fabricated from the segments of three-phase input voltage waveform through the use of a three-phase half-wave circuit. Show a complete cycle of low frequency output voltage. In case load current lags the low frequency output voltage, discuss the operation of positive and negative group phase-controlled converters.

10.5 Discuss why a three-phase to single-phase cycloconverter requires positive and negative group phase-controlled converters. Under what conditions, the group work as inverters or rectifiers? How should the firing angles of the two converters be controlled?

10.6 Describe three-phase to three-phase cycloconverter with relevant circuit arrangements using 18 thyristors and 36 thyristors. What are the advantages of three-phase bridge cycloconverter circuit over three-phase to three-phase cycloconverter circuit consisting of 18 SCRs.

10.7 Show that the fundamental RMS value of per phase output voltage of low-frequency for an m-pulse cycloconverter is given by

$$E_{or} = E_{ph} \left(\frac{m}{\pi}\right) \sin\left(\frac{m}{\pi}\right)$$

Also, express E_{or} in terms of voltage reduction factor r.

10.8 Draw and explain the control circuit block diagram for a cycloconverter with non-circulating current mode.

10.9 Briefly discuss the suitable converter group selection scheme in non-circulating current mode of operation of a cycloconverter.

10.10 Describe any suitable reference voltage generator block diagram for a non-circulating cycloconverter (three phase to three-phase case).

10.11 Discuss the block diagram of logic and trigger circuit suitable for cycloconverters.

10.12 Describe the control scheme for a cycloconverter using voltage-sensing principle of converter group selection.

Also, draw and discuss the various voltage waveforms of a control scheme.

10.13 Give the comparison between a cycloconverter and a d.c. link converter.

10.14 What is a load commutated cycloconverter? How does it differ from line-commutated cycloconverters?

Problems

10.1 A three-pulse cycloconverter is supplying a single-phase load of 480 V, 72 A at a power factor of 0.85 lagging and a frequency of 25 Hz. Estimate the minimum supply voltage required, the ratings of the thyristors and the power factor of the a.c. supply. Neglect losses and device voltage drops.

[Ans $e = 998$ V (line), Peak current = 101.8 A, Thyristor RMS current = 58.8 A, Load-power = 29.4 kW, Input power factor = 0.41]

10.2 A three-phase to single-phase cycloconverter employs three-pulse positive and negative group converters. Each converter is supplied from delta/star transformer with per phase turns ratio of 2:1. The supply voltage is 415 V, 50 Hz. The RL load has $R = 3$ Ω and at low output frequency, $\omega oL = 2$ Ω. The commutation overlap and thyristor turn-off time limit the firing in the inversion mode to 170°. Determine:

(i) the value of fundamental RMS output voltage,

(ii) RMS output current, and
(iii) output power.

[*Ans* (i) 169 V, (ii) 46.87 ∠ − 33.69° A, (iii) 6590.39 W]

10.3 Repeat Problem 10.2 in case of three-phase to single-phase cycloconverter employing six-pulse bridge converter.

[*Ans* (i) 337.99 V, (ii) 93.74 ∠ − 33.69°, (iii) 26361.56 W]

10.4 A three-phase, six-pulse 50 kVA, 415 V cycloconverter is operating at a firing angle of 60° and supplying load having power factor of 0.8. Determine input current to cycloconverter.

[*Ans* 245.5 A.]

REFERENCES

1. B R Pelley, *Thyristor-phase controlled converters and cycloconverters*, New-York, Wiley-Interscience, 1971.
2. S B Dewan, *Power semiconductor circuits*, New York, John-Wiley, 1975.
3. G K Dubey et al., *Thyristorised power controllers*, Wiley Eastern, New Delhi.
4. S B Dewan and M D Kankam, *A method of harmonic analysis of cycloconverters*, IEEE Trans. 1970, p. 455.

Chapter 11
A.C. Regulators

LEARNING OBJECTIVES:
- To describe the need and function of a.c. regulators.
- To consider the operation of half-wave and full-wave a.c. voltage regulator with various loads.
- To develop the general equations describing regulator behaviour.
- To define power factor in relation to a.c. regulator circuit.
- To introduce the sequence control of a.c. regulators.
- To examine the operation of three-phase a.c. regulators.

11.1 INTRODUCTION

By connecting a reverse parallel pair of thyristors or Triac between a.c. supply and load, the voltage applied to the load can be controlled. This type of power controller is known as an a.c. voltage controller or a.c. regulators. Therefore, a.c. voltage regulators converts fixed mains voltage directly to variable alternating voltage without a change in the frequency. The important applications where a.c. voltage controllers are widely used are: speed control of polyphase induction motors, domestic and industrial heating, light controls, on-load transformer tap changing, static reactive power compensators, etc. Earlier, the devices used for these applications were auto transformers, tap-changing transformers, magnetic amplifiers, saturable reactors, etc. Now, thyristor and Triac a.c. regulators have replaced them in most of the applications because of high efficiency, compact size, flexibility in control, etc. The a.c. regulators are also suitable for closed loop control because of low control power and fast response. Since the a.c. regulators are phase-controlled controllers, thyristors and Triacs are line commutated and as such no complex commutation circuitry is required in these controllers. The main disadvantage of these regulators is the presence of harmonics in the supply current and load voltage waveforms, particularly at lower output voltage levels.

The a.c. voltage controllers can be classified as single-phase controllers and three-phase controllers.

Each type of controllers can be subdivided into
(a) unidirectional or half-wave control, and
(b) bidirectional or full wave control.

This chapter describes the single-phase and three-phase a.c. voltage controllers and their performance.

11.2 SINGLE-PHASE A.C. REGULATORS

11.2.1 Half-Wave A.C. Voltage Regulator

Figure 11.1(a) shows the power circuit diagram of a single-phase half-wave a.c. voltage regulator using one thyristor in antiparallel with one diode. The power flow to the load is controlled by delaying the firing angle of thyristor T_1. The output voltage and current waveforms obtained from this type of controller is shown in Fig. 11.1(b).

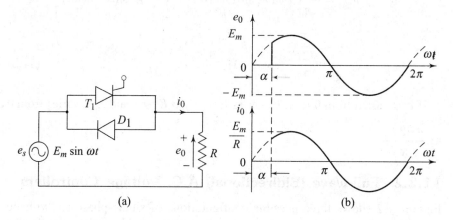

Fig. 11.1 Single-phase half-wave a.c. voltage controller. **(a)** power circuit diagram, and **(b)** voltage and current waveforms

Due to the presence of diode D_1, the control range is limited and the effective RMS output voltage can only be varied between 70.7 and 100%. It can be observed from Fig. 11.1 that positive half-cycle is not identical with negative half-cycle for both voltage and current waveforms. As a result, d.c. component is introduced in the supply and load circuits, which is undesirable. Since the power flow is controlled during the positive half-cycle of input voltage, this type of controller is also known as a unidirectional controller. This type of controller is only suitable for low power resistive loads, such as heating and lighting.

If $e_s = E_m \sin \omega t = \sqrt{2} E_s \sin \omega t$ is the input voltage and the delay angle of thyristor T_1 is $\omega t = \alpha$, the RMS output voltage can be obtained from

$$E_0 = \left\{ \frac{1}{2\pi} \left[\int_\alpha^\pi 2E_s^2 \sin^2 \omega t \, d(\omega t) + \int_\pi^{2\pi} 2E_s^2 \sin^2 \omega t \, d(\omega t) \right] \right\}^{1/2}$$

$$= \left\{ \frac{2E_s^2}{4\pi} \left[\int_\alpha^\pi (1 - \cos 2\omega t) \, d(\omega t) + \int_\pi^{2\pi} (1 - \cos 2\omega t) \, d(\omega t) \right] \right\}^{1/2}$$

$$= E_s \left[\frac{1}{2\pi} \left(2\pi - \alpha + \frac{\sin 2\alpha}{2} \right) \right]^{1/2} \quad (11.1)$$

The average value of output voltage is given by

$$E_{0\mathrm{av}} = \frac{1}{2\pi} \left[\int_\alpha^\pi \sqrt{2} E_s \sin \omega t \, d(\omega t) + \int_\pi^{2\pi} \sqrt{2} E_s \sin \omega t \, d(\omega t) \right]$$

$$= \frac{\sqrt{2} E_s}{2\pi} (\cos \alpha - 1) \quad (11.2)$$

If α is varied from 0 to π, E_0 varies from E_s to $E_s/\sqrt{2}$ and $E_{0\mathrm{av}}$ varies from 0 to $\dfrac{-\sqrt{2} E_s}{\pi}$.

11.2.2 Full-wave (Bidirectional) A.C. Voltage Controllers

Figure 11.2 shows three possible configurations of single-phase a.c. voltage controllers. Figure 11.2(a) uses two thyristors connected in antiparallel. In this circuit, isolation between control and power circuit is a must because the cathodes of two thyristors are not connected to a common point. Figure 11.2(b) employs four diodes and one thyristor. In this circuit, isolation between control and power circuit is not required. This scheme, therefore, offers a cheap a.c. voltage controller. Figure 11.2(c) shows the Triac based a.c. voltage regulator. This circuit configuration is suitable for low power applications where the load is resistive or has only a small inductance. The triggering circuit for the Triac need not be isolated from the power circuit.

1. Single-phase a.c. voltage controller with resistive (R) load Figure 11.3(a) shows the single-phase a.c. voltage controller with resistive load. As shown, two thyristors connected in antiparallel have been employed. Waveforms for source voltage e_s, gating pulses i_{g_1}, i_{g_2}, load-current i_o, load voltage e_0, voltage across T_1 as V_{T_1}, and voltage across T_2 as V_{T_2} are shown in Fig. 11.3(b).

A.C. Regulators

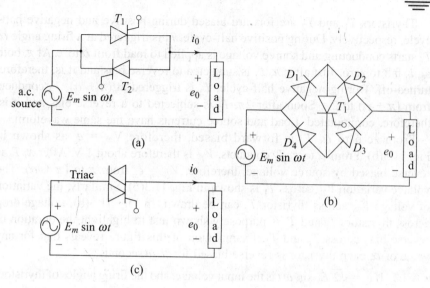

Fig. 11.2 Single-phase a.c. regulators

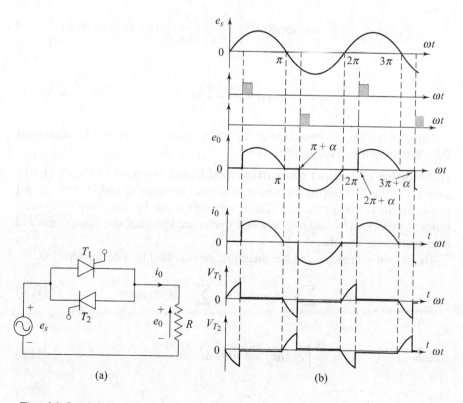

Fig. 11.3 *(a)* Single-phase a.c. voltage controller with R load *(b)* voltage and current waveforms

Thyristors T_1 and T_2 are forward biased during positive and negative half-cycle, respectively. During positive half-cycle, T_1 is triggered at a firing angle α. T_1 starts conducting and source voltage is applied to load from α to π. At π, both e_0, i_0 fall to zero. Just after π, T_1 is subjected to reverse bias and it is, therefore, turned-off. During negative half-cycle, T_2 is triggered at $(\pi + \alpha)$. T_2 conducts from $(\pi + \alpha)$ to 2π. Soon after 2π, T_2 is subjected to a reverse bias and it is, therefore, commutated. Load and source currents have the same waveform.

From zero to α, T_1 is forward biased, therefore $V_{T_1} = e_s$ as shown in Fig. 11.3(b). From α to π, T_1 conducts, V_{T_1} is therefore about 1 V. After π, T_1 is reverse biased by source voltage, therefore, $V_{T_1} = e_s$ from π to $(\pi + \alpha)$. The voltage variation V_{T_1} across T_1 is shown in Fig. 11.3(b). Similarly, the variation of voltage V_{T_2} across thyristor T_2 can be drawn. In Fig. 11.3(b), voltage drop across thyristors T_1 and T_2 is purposely shown just to highlight the duration of reverse bias across T_1 and T_2. Examination of this figure reveals that for any value of α, each thyristor is reverse biased for π/ω seconds.

If $e_s = \sqrt{2}\, E_s \sin \omega t$ is the input voltage, and the firing angles of thyristors T_1 and T_2 are equal ($\alpha_1 = \alpha_2 = \alpha$), the RMS output voltage can be obtained from

$$E_0 = \left[\frac{2}{2\pi}\int_\alpha^\pi 2E_s^2 \sin^2 \omega t\, d(\omega t)\right]^{1/2} = \left[\frac{4E_s^2}{4\pi}\int_\alpha^\pi (1 - \cos 2\omega t)d(\omega t)\right]^{1/2}$$

$$= E_s \left[\frac{1}{\pi}\left(\pi - \alpha + \frac{\sin 2\alpha}{2}\right)\right]^{1/2} \qquad (11.3)$$

Thus, by varying α from 0 to π, the RMS output voltage can be controlled from RMS input voltage E_s to zero.

Harmonics of output quantities and input current : Figure 11.3(b) shows that the waveforms for output quantities (voltage e_0 and current i_0) and input current is non-sinusoidal. These waveforms can be described by Fourier series. As the positive and negative half-cycles are identical, d.c. component and even harmonics are absent.

The output voltage e_0 can be, therefore, represented by Fourier series as

$$e_0 = \sum_{n=1,3,5}^{\infty} A_n \sin \omega t + \sum_{n=1,3,5}^{\infty} B_n \cos \omega t \qquad (11.4)$$

where
$$A_n = \frac{2}{\pi}\int_0^\pi e_0(\omega t) \sin n\omega t\, d(\omega t) \qquad (11.5)$$

A.C. Regulators

and
$$B_n = \frac{2}{\pi} \int_0^\pi e_0(\omega t) \cos n\omega t \, d(\omega t) \qquad (11.6)$$

The load voltage e_0 during the first half-cycle is
$$e_0 = E_m \sin \omega t, \quad \alpha < \omega t < \pi \qquad (11.7)$$

Substituting e_0 from Eq. (11.7) in Eqs (11.5) and (11.6) gives

$$A_n = \frac{2E_m}{\pi} \int_\alpha^\pi \sin \omega t \sin n\omega t \, d(\omega t)$$

$$= \frac{E_m}{\pi} \int_\alpha^\pi \left[\cos(n-1)\omega t - \cos(n+1)\omega t\right] d(\omega t)$$

$$= \frac{E_m}{\pi} \left[\frac{\sin(n+1)\alpha}{(n+1)} - \frac{\sin(n-1)\alpha}{(n-1)}\right] \qquad (11.8)$$

and
$$B_n = \frac{2E_m}{\pi} \int_\alpha^\pi \sin \omega t \cdot \cos n\omega t \cdot d(\omega t)$$

$$= \frac{E_m}{\pi} \int_\alpha^\pi \left[\sin(n+1)\omega t - \sin(n-1)\omega t\right] d(\omega t)$$

$$= \frac{E_m}{\pi} \left[\frac{\cos(n+1)\alpha - 1}{(n+1)} - \frac{\cos(n-1)\alpha - 1}{(n-1)}\right] \qquad (11.9)$$

where $E_m = \sqrt{2}\, E_s$ and E_s = RMS value of source voltage.
For obtaining Eq. (11.9), note that for $n = 1, 3, 5, \ldots$ $\cos(n+1)\pi = 1$ and $\cos(n-1)\pi = 1$.

The peak amplitude of the nth harmonic output voltage E_{nm}, and its phase ϕ_n are given by, $E_{nm} = \sqrt{A_n^2 + B_n^2}$ (11.10)

and
$$\phi_n = \tan^{-1} \frac{B_n}{A_n} \qquad (11.11)$$

Also, $\qquad I_{nm} = \dfrac{E_{nm}}{R} = n^{\text{th}}$ harmonic load current $\qquad (11.12)$

Equations (11.8) to (11.11) can be used to evaluate the magnitude of harmonics for which $n = 3, 5, \ldots$ It cannot be used to calculate the fundamental component because substitution of $n = 1$ in Eqs (11.8) and (11.9) leads to undefined expressions. The fundamental component which has the same frequency as the supply voltage, can, however, be obtained from Eqs (11.5) and (11.6) by substituting $n = 1$ and for e_0 from Eq. (11.7). This yields,

$$A_1 = \frac{2}{\pi} \int_\alpha^\pi E_m \sin^2 \omega t \, d(\omega t) = \frac{E_m}{\pi} \left[\frac{\sin 2\alpha}{2} + \pi - \alpha \right] \tag{11.13}$$

and $$B_1 = \frac{2}{\pi} \int_\alpha^\pi E_m \sin \omega t \cos \omega t \, d(\omega t) = \frac{E_m}{\pi} \left[\frac{\cos 2\alpha - 1}{2} \right] \tag{11.14}$$

Therefore, the value of E_{1m} and ϕ_1 for the fundamental frequency is given by

$$E_{1m} = \sqrt{A_1^2 + B_1^2} = \frac{E_m}{\pi} \left[\left\{ \frac{\sin 2\alpha}{2} + (\pi - \alpha) \right\}^2 + \left\{ \frac{\cos 2\alpha - 1}{2} \right\}^2 \right]^{1/2} \tag{11.15}$$

$I_{1m} = \dfrac{E_{1m}}{R}$ is the amplitude of fundamental component of load or source current,

and $$\phi_1 = \tan^{-1} \frac{B_1}{A_1} = \tan^{-1} \left[\frac{\cos 2\alpha - 1}{\sin 2\alpha + 2(\pi - \alpha)} \right] \tag{11.16}$$

When a.c. voltage controller is used for the speed control of a single phase induction motor, only fundamental component is useful in producing the torque. The harmonics in the motor current merely increase the losses and therefore heating of the induction motor.

For heating and lighting loads, however, both fundamental and harmonics are useful in producing the a.c. controlled power. In such applications, RMS value of the output voltage E_0 is of interest.

Figure 11.3(b) shows that source current waveform is identical with load current waveform. This shows that the expressions for both load and source currents for the appropriate harmonics are the same.

Power factor : Assuming that the source voltage remains sinusoidal even though non-sinusoidal current is drawn from it, the power factor is given by

$$PF = \frac{\text{real power}}{\text{apparent power}} = \frac{E_s I_1 \cos \phi_1}{E_s \cdot I_{\text{rms}}} = \frac{I_1 \cdot \cos \phi_1}{I_{\text{rms}}} \tag{11.17}$$

where $I_1 = \dfrac{I_{1m}}{\sqrt{2}}$ = RMS value of fundamental component of source current,

A.C. Regulators

I_{rms} = RMS value of source current, and ϕ_1 = phase angle between E_s and I_1, Eq. (11.16).

Equation (11.17) gives the definition of power factor when the source voltage is sinusoidal but the current is non-sinusoidal. For the present case, another expression for PF can be obtained as follows:

Real power delivered to load = $\dfrac{E_0^2}{R}$

Apparent power delivered to load = $E_s \cdot I_0 = E_s \left(\dfrac{E_0}{R} \right)$

$\therefore \qquad \text{PF} = \dfrac{E_0^2 / R}{E_s \cdot (E_0 / R)} = \dfrac{E_0}{E_s}$

From Eq. (11.3), $\text{PF} = \dfrac{E_0}{E_s} = \left[\dfrac{1}{\pi} \left\{ (\pi - \alpha) + \dfrac{1}{2}\sin 2\alpha \right\} \right]^{1/2}$ \qquad (11.18)

The maximum values of RMS output voltage and current occurs at $\alpha = 0$ and are given by E_s and E_s/R, respectively. Since harmonics are absent at $\alpha = 0$, these are also the maximum values of fundamental RMS voltage and current. If these quantities are used to normalise various voltages and currents, then from Eqs (11.18) and (11.3),

PF = normalised output voltage
   = normalised source (or load) current \qquad (11.19)

Figure 11.4 shows the plot of normalised fundamental component of load current, normalised values of harmonics in load current and power factor against normalised RMS load current. The power factor is poor for low values of the load current and appreciable amount of harmonics is present. For example, at a per unit value of 0.4 of the load current, third harmonic becomes comparable to the fundamental and at a p.u. value of 0.2, fifth and seventh harmonics also have comparable values. Since the source current and load current waveforms are identical, the curves of harmonics components and fundamental of Fig. 11.4 are also applicable to the source current.

2. Single-phase a.c. voltage controller with inductive (RL) load

Figure 11.5(a) shows a single phase a.c. voltage controller with RL load. The waveforms for source voltage E_s, gate currents i_{g1} and i_{g2}, load and source currents i_0 and i_s, load voltage e_0, and thyristor voltages are shown in Fig. 11.5(b).

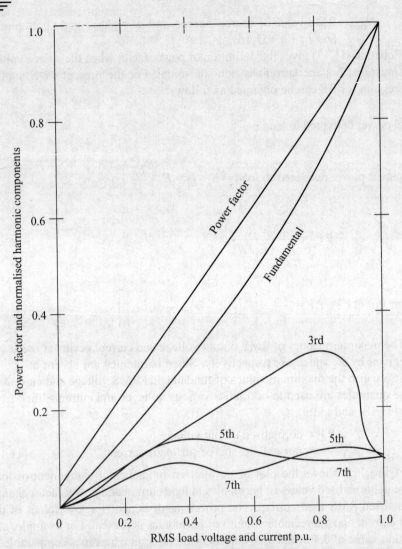

Fig. 11.4 *Power factor and normalised harmonic components versus RMS load voltage and current p.u.*

During the interval zero to π, thyristor T_1 is forward biased. At $\omega t = \alpha$, T_1 is triggered and $i_0 = i_{T_1}$ starts building up through the load. At π, load and source voltages are zero but the current is not zero because of the presence of inductance in the load circuit. Thyristor T_1 will continue to conduct until its current falls to zero at $\omega t = \beta$. Angle β is called as the extinction angle. The load is subjected to the source voltage from α to β. At β, when i_0 is zero, T_1 is turned-off as it is already reversed biased. After the commutation of T_1 at β, a voltage of magnitude $E_m \sin \beta$ at once appears as a reverse bias across T_1 and as a forward bias across

T_2, as shown in Fig. 11.5(b). From β to $\pi + \alpha$, no current exists in the power circuit. Thyristor T_2 is turned-on at $(\pi + \alpha) > \beta$. Current $i_0 = i_{T_2}$ starts building up in the reversed direction through the load. At 2π, e_s and e_0 are zero but $i_{T_2} = i_0$ is not zero. At $(\pi + \alpha + \gamma)$, $i_{T_2} = 0$ and T_2 is turned-off because it is already reverse biased. At $(\pi + \alpha + \gamma)$, $E_m \sin(\pi + \alpha + \gamma)$ appears as a forward bias across T_1 and as a reverse bias across T_2, as shown in Fig. 11.5(b). From $(\pi + \alpha + \gamma)$ to $(2\pi + \alpha)$ no current exists in the power circuit. At $(2\pi + \alpha)$, T_1 is turned-on and current starts building up as before.

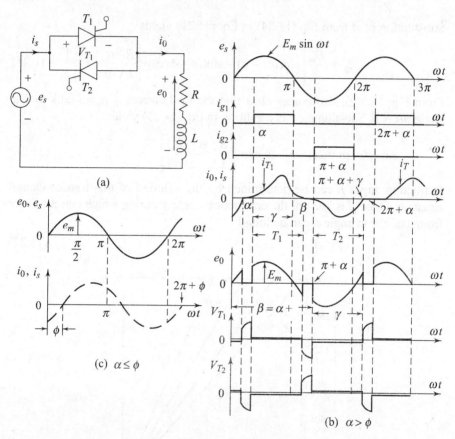

Fig. 11.5 *Single-phase a.c. regulator with RL-load*

The expression for load current i_s and β can be obtained as follows:
For $\alpha \leq \omega t \leq \beta$, the KVL for the circuit of Fig. 11.5(a) gives

$$e_s = E_m \sin \omega t = R \cdot i_0 + L \frac{di_0}{dt} \qquad (11.20)$$

The solution of this equation is of the form

$$i_0 = \frac{E_m}{Z} \sin(\omega t - \phi) + A \cdot e^{-(R/L)t} \qquad (11.21)$$

where
$$Z = (R^2 + \omega^2 L^2)^{1/2} \tag{11.22}$$

and
$$\phi = \tan^{-1} \frac{\omega L}{R} \tag{11.23}$$

Constant A can be obtained from the initial condition according to which $i_0 = 0$ at $\omega t = \alpha$, i.e. at $t = \alpha/\omega$. Therefore,

$$0 = \frac{E_m}{Z} \sin(\alpha - \phi) + A \cdot e^{-R\alpha/\omega L} \text{ or } A = \frac{-E_m}{Z} \sin(\alpha - \phi) e^{R\alpha/\omega L} \tag{11.24}$$

Substitution of A from Eq. (11.24) in Eq. (11.21) yields

$$i_0 = \frac{E_m}{Z} \left[\sin(\omega t - \phi) - \sin(\alpha - \phi) \exp\left\{ \frac{R}{L}\left(\frac{\alpha}{\omega} - t\right)\right\} \right] \tag{11.25}$$

From Fig. 11.5(b), it becomes clear that the load current i_0 again falls to 0 at angle $\omega t = \beta$. Substituting this condition in Eq. (11.25) yields

$$\sin(\beta - \phi) = \sin(\alpha - \phi) \exp\left[\frac{R}{L}\left(\frac{\alpha - \beta}{\omega}\right)\right] \tag{11.26}$$

Extinction angle β can be determined by the solution of this transcendental equation. Once β is known, the conduction angle γ during which current flows from angle α to angle β can be found:

$$\gamma = \beta - \alpha. \tag{11.27}$$

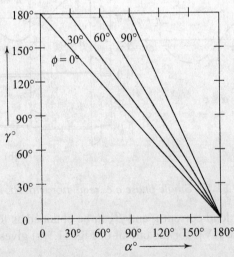

Fig. 11.6 γ versus α curves for a.c. regulator with RL load

From Eqs (11.26) and (11.27), one can obtain a relationship between γ and α for a given value of ϕ. γ versus α curves for various values of ϕ are shown in Fig. 11.6.

A.C. Regulators

Figure 11.6 shows that as α is decreased, the conduction angle γ increases. The waveforms of current i_0 in Fig. 11.5(b) reveals that for $\gamma < \pi$, current i_{T_1} through T_1 flows from α to $(\alpha + \gamma) = \beta$, and T_1 remains OFF from $(\alpha + \gamma)$ up to $(\pi + \alpha)$. At $(\pi + \alpha)$, current i_{T_2} through T_2 flows from $(\pi + \alpha)$ to $(\pi + \alpha + \gamma)$. T_2 remains OFF from $(\pi + \alpha + \gamma)$ to $(2\pi + \alpha)$. At $(2\pi + \alpha)$, T_1 is turned-on. With progressive decrease in α, γ may become equal to π. Under this condition, when γ is just equal to π, T_1 will be ON from α to $(\pi + \alpha)$ and i_{T_1} flows from α to $(\pi + \alpha)$. Further, T_2 will be ON from $(\pi + \alpha)$ to $(2\pi + \alpha)$ and current i_{T_2} flows from $\pi + \alpha$ to $(2\pi + \alpha)$. Thus, when $\gamma = \pi$, from 0 to $\alpha \to T_2$ conducts.

$$\text{from } \alpha \text{ to } (\pi + \alpha) \to T_1 \text{ conducts} \tag{11.28}$$

$$\text{from } (\pi + \alpha) \text{ to } (2\pi + \alpha) \to T_2 \text{ conducts, and so on}$$

This shows that load current will never become zero for any segment of time and therefore, for all the time load is connected to source. Hence, for $\gamma = \pi$, the load voltage is equal to sinusoidal source voltage provided the voltage drop in thyristors is neglected. Under these conditions, load behaves as if it is being fed directly by the a.c. source.

Now, we determine the value of α for which $\gamma = \pi$ and load is directly connected to a.c. source. For this, consider that the RL load, with load phase angle ϕ, is connected directly to a.c. source. Under steady state, the load current will be a sine wave and lag behind the voltage wave by an angle ϕ as shown in Fig. 11.5(c). The current is positive from ϕ to $(\pi + \phi)$ and negative from $(\pi + \phi)$ to $(2\pi + \phi)$. If it is required to obtain the current waveform of Fig. 11.5(c), through the operation of power circuit of Fig. 11.5(a), then

From 0 to $\phi \to T_2$ conducts

$$\phi \text{ to } (\pi + \phi) \to T_1 \text{ conducts} \tag{11.29}$$

$$(\pi + \phi) \text{ to } (2\pi + \phi) \to T_2 \text{ conducts, and so on}$$

Comparison of expressions (11.28) and (11.29) reveals that when $\alpha = \phi$, $\gamma = \pi$.

This can be verified by referring to Eqs (11.26) and (11.27). When $\alpha = \phi$, Eq. (11.26) gives

$$\sin(\beta - \alpha) = 0 = \sin \pi \quad \text{or} \quad (\beta - \alpha) = \pi$$

From Eq. (11.27), $\gamma = \beta - \alpha = \pi$.

This shows that for a single-phase a.c. voltage controller, waveforms of Fig. 11.5(b) are applicable only when $\alpha > \phi$ and that of Fig. 11.5(c) for $\alpha \leq \phi$.

The RMS output voltage can be found from

$$E_0 = \left[\frac{2}{2\pi} \int_\alpha^\beta 2E_s^2 \sin^2 \omega t \, d(\omega t) \right]^{1/2}$$

$$= \left[\frac{4E_s^2}{4\pi} \int_\alpha^\beta (1 - \cos 2\omega t) \, d(\omega t) \right]^{1/2} \tag{11.30}$$

$$= E_s \left[\frac{1}{\pi} \int_\alpha^\beta (\beta - \alpha + \frac{\sin 2\alpha}{2} - \frac{\sin 2\beta}{2}) \right]^{1/2} \qquad (11.31)$$

The RMS thyristor current can be found from Eq. (11.25) as

$$I_{\text{rms}} = \left[\frac{1}{2\pi} \int_\alpha^\beta i_0^2 \, d(\omega t) \right]^{1/2}$$

$$= \frac{E_s}{Z} \left[\frac{1}{\pi} \int_\alpha^\beta \left\{ \sin(\omega t - \phi) - \sin(\alpha - \phi) e^{R/L \left(\frac{\alpha}{\omega} - t \right)} \right\}^2 d(\omega t) \right]^{1/2} \qquad (11.32)$$

and the RMS output current can then be determined by combining the RMS current of each thyristor as

$$I_0 = \sqrt{I_{\text{rms}}^2 + I_{\text{rms}}^2} = \sqrt{2} I_{\text{rms}} \qquad (11.33)$$

The average value of thyristor current can also be found from Eq. (11.25) as

$$I_{\text{av}} = \frac{1}{2\pi} \int_\alpha^\beta i_0 \, d(\omega t)$$

$$= \frac{\sqrt{2} E_s}{2\pi Z} \int_\alpha^\beta \left[\sin(\omega t - \phi) - \sin(\alpha - \phi) e^{R/L \left(\frac{\alpha}{\omega} - t \right)} d(\omega t) \right] \qquad (11.34)$$

Operation with $\alpha \leq \phi$: Assume that a.c. voltage controller is working under steady-state with $\alpha = \phi$. From zero to ϕ, T_2 conducts and from ϕ to $(\pi + \phi)$, T_1 conducts; from $(\pi + \phi)$ to $(2\pi + \phi)$, T_2 conducts, and so on. Now, let α be decreased below ϕ. When t_1 is triggered at $\alpha < \phi$, T_1 will not get turned-on because it is reverse biased by voltage drop irn T_2 which is conducting current i_{T_2}. T_1 will get turned-on only at ϕ when $i_{T_2} = 0$ and reverse bias due to voltage drop in T_2 vanishes. Now T_1 will conduct from ϕ to $(\pi + \phi)$. T_2 will be triggered at an angle $(\pi + \alpha) < (\pi + \phi)$. As T_1 is conducting, a voltage drop in T_1 will apply a reverse bias across T_2, as a result T_2 will not get turned-on at $(\pi + \alpha)$ but only at $(\pi + \phi)$, when $i_{T_1} = 0$. Now T_2 will conduct from $(\pi + \phi)$ to $(2\pi + \phi)$ and so on. This shows that load voltage and current waveforms will not change from what they are at $\alpha = \phi$. Thus, the reduction of α below ϕ is not able to control the load voltage and load current. The a.c. output power can be controlled only for $\alpha > \phi$. Note that for $\alpha \leq \phi$, γ remains equal to π. Thus, the control range of firing angle is $\phi < \alpha < 180°$.

Gating Signal Requirements: For resistive load as in Fig. 11.3, thyristor T_1 stops conducting at π, and thyristor T_2 becomes forward biased after π. When thyristor T_2 is triggered at $\pi + \alpha$, it gets turned-on as it is already forward biased by source voltage. Thus, pulse gating is suitable for resistive load as shown in Fig. 11.3.

However, pulse gating is not suitable for *RL* loads. The reason for this can be explained by referring to Fig. 11.7. Thyristor T_1 is triggered at angle α and the current flows as shown in Fig. 11.7. At period $(\pi + \alpha)$, SCR T_2 is triggered. As SCR T_1 is still conducting, voltage drop across T_1 reverse biases T_2 at $\pi + \alpha$, T_2 is therefore not turned-on at $\pi + \alpha$. At $(\alpha + \gamma)$, current i_{T_1} decays to zero and T_1 stops conducting, as a result T_2 gets forward biased but the gate pulse I_{g_2} applied to SCR T_2 at $(\pi + \alpha)$ is already zero and therefore, T_2 does not get turned-on. At period $(2\pi + \alpha)$, when gate pulse is applied to T_1, it gets turned-on because it is already forward biased by source voltage. At $(3\pi + \alpha)$, when T_2 is pulse gated, it

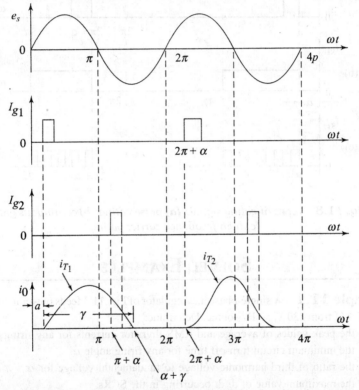

Fig. 11.7 *Single-phase regulator with R-L load with pulse gating*

will not turn-on as explained earlier. Thus, the a.c. voltage controller gives asymmetrical output voltage waveform due to the conduction of T_1 alone. This half-wave rectifier operation of the a.c. voltage controller is undesirable. This difficulty can be overcome by applying a continuous gate signal to the SCRs T_1 and T_2 so that when i_{T_1} becomes zero at $(\alpha + \gamma)$, T_2 gets turned-on due to the presence of continuous signal. A continuous gate signal is shown in Fig. 11.8(b). The duration of the continuous gate signal should last for a period of $(\pi - \alpha)/\omega$ seconds. Strictly speaking, sustained gate pulse may not last from α to π as shown. The minimum duration of gate pulse should be equal to ϕ plus the angle required for the current to reach latching current value.

In practice, continuous gating is undesirable as it leads to more heating of the SCR gate and at the same time, it increases the size at the pulse transformer. The technique that avoids the above disadvantages of continuous gate signal and ensures that thyristor is turned-on is to use a train of firing pulses from α to π as shown in Fig. 11.8(c). This type of signal is also termed as high frequency carrier gating.

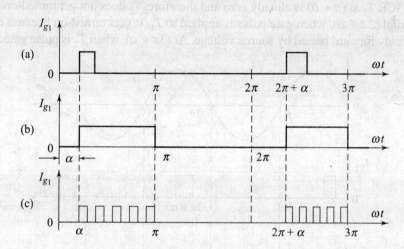

Fig. 11.8 *Types of gating signals* **(a)** *pulse gating* **(b)** *continuous gating* **(c)** *high frequency carrier gating*

SOLVED EXAMPLES

Example 11.1 A single-phase a.c. regulator of Fig. 11.3 feeds power to a resistive load of 4 Ω from 230 V, 50 Hz source. Determine:
  (i) the peak values of average and RMS thyristor currents for any firing angle α.
  (ii) the minimum circuit turn-off time for any firing angle α.
  (iii) the ratio of third harmonic voltage to fundamental voltage for $\alpha = \pi/3$.
  (iv) the maximum value of di/dt occurring in the SCRs.
  (v) the angle α at which the greatest forward or reverse voltage is applied to either of the thyristors and the magnitude of these voltages.

Solution:
(i) From Fig. 11.3(b), it can be observed that the current through the thyristor flows from α to π for the first cycle of 2π radians. Therefore, average thyristor current is

given by, $I_{TAV} = \dfrac{1}{2\pi}\displaystyle\int_{\alpha}^{\pi}\dfrac{E_m}{R}\sin\omega t\, \mathrm{d}(\omega t) = \dfrac{E_m}{2\pi R}(1+\cos\alpha)$ \hfill (11.35)

A.C. Regulators

This current has the maximum value when $\alpha = 0$.

$$\therefore \quad I_{TAVM} = \frac{E_m}{\pi R} = \frac{\sqrt{2} \times 230}{\pi \times 4} = 25.88 \text{ A}.$$

Now, RMS thyristor current is

$$I_{rmsm} = \left[\frac{1}{2\pi}\int_{\alpha}^{\pi}\left(\frac{E_m}{R}\sin\omega t\right)^2 d(\omega t)\right]^{1/2} = \frac{E_m}{2R\sqrt{\pi}}\left[(\pi - \alpha) + \frac{1}{2}\sin 2\alpha\right]^{1/2}$$

Its maximum value occurs at $\alpha = 0$.

$$\therefore \quad I_{rmsm} = \frac{E_m}{2R} = \frac{\sqrt{2} \times 230}{2 \times 4} = 40.66 \text{ A}.$$

(ii) Waveforms for V_{T_1} and V_{T_2} in Fig. 11.3(b) shows that for any value of firing angle α, the circuit turn-off time is always π radians.

$$\therefore \quad \text{Circuit turn-off time} = \frac{\pi}{\omega} = \frac{\pi}{2\pi f} = \frac{1}{2f} = \frac{1}{2 \times 50} = 10 \text{ ms}$$

(iii) For third harmonics, from Eq. (11.8),

$$A_3 = \frac{E_m}{\pi}\left[\frac{\sin 240°}{4} - \frac{\sin 120°}{2}\right] = 0.6495 \frac{E_m}{\pi}$$

From Eq. (11.9),

$$B_3 = \frac{E_m}{\pi}\left[\frac{\cos 240° - 1}{4} - \frac{\cos 120° - 1}{2}\right] = 0.375 \frac{E_m}{\pi}$$

The amplitude of third harmonic voltage, from Eq. (11.10) is

$$E_{3m} = \left[A_3^2 + B_3^2\right]^{1/2} = 0.75 \frac{E_m}{\pi}$$

The amplitude of fundamental frequency voltage from Eq. (11.15) is

$$E_{1m} = \frac{E_m}{\pi}\left[\left\{\frac{\sin 120°}{2} + (\pi - \pi/3)\right\}^2 + \left\{\frac{\cos 120° - 1}{2}\right\}^2\right]^{1/2} = 2.64 \frac{E_m}{\pi}$$

$$\therefore \quad \frac{E_{3m}}{E_{1m}} = \frac{0.75}{2.64} = 0.284.$$

(iv) As there is sudden rise of current from zero to $E_m/R \sin\alpha$ at firing angle α, di/dt is infinity.

(v) The waveforms of V_{T_1}, V_{T_2} in Fig. 11.3(b) reveals that the greatest forward or reverse voltage would appear across either of the thyristor when $\alpha = \pi/2$ or $\alpha > \pi/2$.

The magnitude of these voltages is, $E_m = \sqrt{2} E_s$.

Example 11.2 A single-phase half-wave a.c. regulator of Fig. Ex. 11.2 feeds power to a resistive load of 6 Ω from 230 V, 50 Hz source. The firing angle of SCR is $\pi/2$. Calculate:

(a) the RMS value of output voltage
(b) the input power factor p_f and
(c) the average input current.

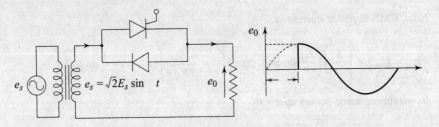

Fig. Ex. 11.2 Half-wave a.c. regulator

Solution:
(a) The RMS output voltage is given by the expression

$$E_{0(rms)} = \left\{ \frac{1}{2\pi} \left[\int_\alpha^\pi 2E_s^2 \cdot \sin^2 \omega t \, d(\omega t) + \int_\pi^{2\pi} 2E_s^2 \cdot \sin^2 \omega t \, d(\omega t) \right] \right\}^{1/2}$$

$$= \left\{ \frac{2E_s^2}{4\pi} \left[\int_\alpha^\pi (1 - \cos 2\omega t) d(\omega t) + \int_\pi^{2\pi} (1 - \cos 2\omega t) d(\omega t) \right] \right\}^{1/2}$$

$$= E_s \left[\frac{1}{2\pi} \left(2\pi - \alpha + \frac{\sin 2\alpha}{2} \right) \right]^{1/2} \tag{11.36}$$

$\therefore \quad E_{0(rms)} = 230\sqrt{3/4} = 199.19$ V.

(b) The RMS load current is

$$I_{0(rms)} = \frac{E_{0(rms)}}{R} = \frac{199.19}{6} = 33.198 \text{ A}.$$

The load power, $p_0 = I_{0(rms)}^2 \cdot R = (33.198)^2 \times 6 = 6612.64$ W.

Since the input current is the same as the load current, the input volt-ampere rating is

$$E_A = E_s I_s = E_s I_0 = 230 \times 33.190 = 7635.54 \text{ EA}$$

The input power factor

$$p_f = \frac{p_0}{E_A} = \frac{E_0}{E_s} = \left[\frac{1}{2\pi} \left(2\pi - \alpha + \frac{\sin 2\alpha}{2} \right) \right]^{1/2} = \sqrt{\frac{3}{4}} = \frac{6612.64}{7635.54} = 0.866 \text{ (logging)}$$

(c) The average value of output voltage can be obtained as

$$E_{dc} = \frac{1}{2\pi} \left[\int_\alpha^\pi \sqrt{2} E_s \sin \omega t \, d(\omega t) + \int_\pi^{2\pi} \sqrt{2} E_s \sin \omega t \, d(\omega t) \right] = \frac{\sqrt{2} E_s}{2\pi} (\cos \alpha - 1) \tag{11.37}$$

$\therefore \quad E_{dc} = \frac{\sqrt{2} \times 230}{2\pi} \left(\cos \frac{\pi}{2} - 1 \right) = -230 \times \frac{\sqrt{2}}{2\pi} = -51.77$ V

A.C. Regulators

$\therefore$ Average input current, $I_{dc} = \dfrac{E_{dc}}{R} = \dfrac{-51.77}{6} = -8.62$ A

The negative sign of I_{dc} signifies that the input current during the positive half-cycle is less than that during the negative half-cycle. If there is an input transformer, the transformer core may be saturated. This type of unidirectional control is not normally used in practice.

Example 11.3 A single phase a.c. voltage regulator with R–L load has the following details:

Supply voltage = 230 V, 50 Hz; details: $R = 4\,\Omega$ and $\omega L = 3\,\Omega$. Calculate:
(a) the control range of firing angle,
(b) the maximum value of RMS load current,
(c) the maximum power and power factor,
(d) the maximum values of average and RMS thyristor current,
(e) the maximum possible value of di/dt that may occur in the SCR, and
(f) the conduction angle for $\alpha = 0°$ and $\alpha = 120°$, assuming a gate pulse of duration π radian.

Solution:

(a) For controlling the load, the minimum firing angle α = load-phase angle ϕ

$$= \tan^{-1}\dfrac{\omega L}{R} = \tan^{-1}\dfrac{3}{4} = 36.87°.$$

The maximum possible value of α is 180°.
$\therefore$ The control range of firing angle is $36.87° \leq \alpha \leq 180°$.

(b) The maximum value of RMS load current occurs when $\alpha = 36.87°$. This value is given by

$$\therefore \quad I_{0(\text{rms } m)} = \dfrac{230}{\sqrt{R^2 + (\omega L)^2}} = \dfrac{230}{\sqrt{4^2 + 3^2}} = 46 \text{ A}.$$

(c) Maximum power,

$$P_0 = I_0^2 \cdot R = (46)^2 \times 4 = 8464 \text{ W}.$$

Power factor $= \dfrac{I_0^2 R}{E_s I_0} = \dfrac{46^2 \times 4}{230 \times 46} = 0.8$.

(d) From Fig. 11.5 average thyristor current is maximum when $\alpha = \phi$ and conduction angle $\gamma = \pi$.

$$\therefore \quad I_{T(AV)} = \dfrac{1}{2\pi}\int_{\alpha}^{\pi+\pi} \dfrac{E_m}{Z}\sin(\omega t - \phi)\,\mathrm{d}(\omega t) = \dfrac{E_m}{\pi Z} = \dfrac{\sqrt{2}\times 230}{\pi \cdot \sqrt{4^2 + 3^2}} = 20.707 \text{ A}.$$

Similarly, maximum value of RMS thyristor current is,

$$I_{(\text{rms } m)} = \left[\dfrac{1}{2\pi}\int_{\alpha}^{\alpha+\pi}\left\{\dfrac{E_m}{Z}\sin(\omega t - \alpha)\right\}^2 \mathrm{d}(\omega t)\right]^{1/2} = \dfrac{E_m}{2Z} = \dfrac{\sqrt{2}\times 230}{2\times\sqrt{4^2 + 3^2}} = 32.527 \text{ A}$$

(e) Maximum value of di/dt occurs when $\alpha = \phi$. From Eq. (11.25),

$$\dfrac{di}{dt} = \dfrac{\omega \cdot E_m}{Z}\cos(\omega t - \phi) - 0$$

Its value is maximum when $\cos(\omega t - \phi) = 1$.

$$\therefore \left(\frac{di}{dt}\right)_{max} = \frac{\sqrt{2} \times 230 \times 2\pi \times 50}{\sqrt{4^2 + 3^2}} \cdot 1 = 2.0437 \times 10^4 \text{ A/s}.$$

From Fig. 11.6, it can be observed that, for $\alpha = 0°$, the conduction angle is 180°. Similarly, for $\alpha = 120°$, the conduction angle is about 95°.

Example 11.4 The power in a resistor is to be controlled by a tapped transformer as shown in Fig. 11.9. Taking the load as 10 Ω, the full transformer secondary as 100 V, with the tapping at 70.7 V, plot a curve of load power against firing delay angle. Determine the required thyristor ratings. Neglect losses.

Solution: Assuming that the thyristors T_3 and T_4 are fired at voltage zero, and the firing delay angle relates to thyristors T_1 and T_2, the mean power is given by

$$\text{Power}, \quad P_0 = \frac{E_{rms}^2}{R} = \frac{1}{10\pi}\left[\int_0^\alpha (70.7\sqrt{2}\sin\omega t)^2 d(\omega t) + \int_\alpha^\pi (100\sqrt{2}\sin\omega t)^2 d(\omega t)\right]$$

$$= 1000 - (500\alpha - 250\sin 2\alpha)/\pi \cdot \omega.$$

The plot of mean load-power against firing delay angle is shown in Fig. Ex. 11.4.

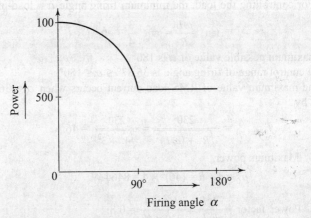

Fig. Ex. 11.4

With the circuit connected, but no firing pulses to the SCRs gates, the peak voltages (PV) experiences by the thyristors are:

Thyristors T_1, T_2: PRV = PFV = $100\sqrt{2}$ = 141 V.

Thyristors T_3, T_4: PRV = PFV = $70.7\sqrt{2}$ = 100 V.

At zero firing delay, thyristors T_1 and T_2 will carry a half-cycle of sinusoidal current of maximum value $\frac{100\sqrt{2}}{10}$, that is an RMS value of 70.7 A.

At 180° firing angle, thyristors T_3 and T_4 will carry a half-cycle of maximum value $70.7\sqrt{2}/10$, that is, an RMS value of 5 A.

Example 11.5 The single-phase transformer tap changer of Fig. 11.9 has the primary voltage of 260 V (RMS) 50 Hz. The secondary voltages are $e_1 = 130$ V and $e_2 = $

A.C. Regulators

130 V. If the load resistance is $R = 6 \, \Omega$, the RMS load voltage is 195 V, and the firing angle of thyristors T_1 and T_2 is 90°, determine:
  (a) the RMS current of thyristors T_1 and T_2,
  (b) the RMS current of thyristors T_3 and T_4, and
  (c) the input power factor.

Solution: Given:
$$e_1 = 130 \text{ V}, \ e_2 = 130 \text{ V}, \ E_p = 260 \text{ V}, \ E_0 = 195 \text{ V and } R = 6 \, \Omega$$

(a) The RMS current of thyristors T_1 and T_2 can be obtained as

$$I_1 = \left[\frac{1}{2\pi R^2} \int_\alpha^\pi 2(e_1+e_2)^2 \sin^2 \omega t \, d(\omega t)\right]^{1/2} = \frac{e_1+e_2}{\sqrt{2}R}\left[\frac{1}{\pi}\left(\pi - \alpha + \frac{\sin 2\alpha}{2}\right)\right]^{1/2} \quad (11.38)$$

$$= \frac{130+130}{\sqrt{2} \times 6}\left[\frac{1}{\pi}\left(\pi - 90° + \frac{\sin 180°}{2}\right)\right]^{1/2} \quad I_1 = 21.67 \text{ A.}$$

(b) The RMS current of thyristors T_3 and T_4 can be obtained as

$$I_2 = \left[\frac{1}{2\pi R^2} \int_0^\alpha 2e_1^2 \sin^2 \omega t \, d(\omega t)\right]^{1/2} = \frac{e_1}{\sqrt{2}R}\left[\frac{1}{\pi}\left(\alpha - \frac{\sin 2\alpha}{2}\right)\right]^{1/2} \quad (11.39)$$

$$= \frac{130}{\sqrt{2} \times 6}\left[\frac{1}{\pi}\left(90° - \frac{\sin 180°}{2}\right)\right]^{1/2} = 10.83 \text{ A}$$

(c) The RMS current of the second (top) secondary winding is $I_{\omega_2} = \sqrt{2} \, I_1 = 30.65$ A.

The RMS current of the first (lower) secondary winding, which is the total RMS current of thyristors T_1, T_2, T_3 and T_4, is

$$I_{\omega_1} = \left[\left(\sqrt{2}I_1\right)^2 + \left(\sqrt{2} \cdot I_2\right)^2\right]^{1/2} = \left[\left(\sqrt{2} \times 21.67\right)^2 + \left(\sqrt{2} \times 10.83\right)^2\right]^{1/2}$$

$$= [939.18 + 234.58]^{1/2} = 34.26 \text{ A.}$$

The volt-ampere rating of primary or secondary,

$$E_A = e_1 \cdot I_{\omega_1} + e_2 \cdot I_{\omega_2} = 130 \times 34.26 + 130 \times 30.65$$
$$= 4453.8 + 3984.5 = 8438.3$$

The load power, $P_0 = \dfrac{E_0^2}{R} = \dfrac{(195)^2}{6} = 6337.5$ W.

$\therefore$ The power factor, $\text{PF} = \dfrac{P_0}{E_A} = \dfrac{6337.5}{8438.3} = s0.751$ (lagging).

11.3 SEQUENCE CONTROL OF A.C. REGULATORS

For the improvement of system power factor, and reduction of harmonics in the source current and the load voltage, sequence control of a.c. regulators is employed. Sequence control of a.c. regulators means the use of two or more stages of voltage controllers in parallel for the regulation of output voltage. The term sequence control means that the stages of voltage controllers in parallel are triggered in a proper sequence one after the other so as to obtain a variable output with low harmonic content. Sequence controlled a.c. regulators can be used as voltage regulators in supply systems and for the speed control of induction motors. Such type of controllers are also known as synchronous tap changers or transformer tap changers.

11.3.1 Two Stage Sequence Control

Two stage sequence control of a.c. regulators employs two stages in parallel as shown in Fig. 11.9. The turns ratio from primary to each secondary is taken as unity for the sake of simplicity. This means that for source voltage $e_s = E_m \sin \omega t$, $e_1 = e_2 = E_m \sin \omega t$ and the sum of two secondary voltages is $2 E_m \sin \omega t$. The load may be R or RL. For both types of loads, for obtaining output voltage control from zero to RMS value E, use only thyristor pair T_3, T_4 in Fig. 11.9. For zero output voltage, α is 180° for T_3, T_4 and for E, α is zero. For output voltage control from E to $2E$, α for thyristor pair T_3, T_4 is always zero and for thyristor pair T_1, T_2; α is varied from zero to 180°.

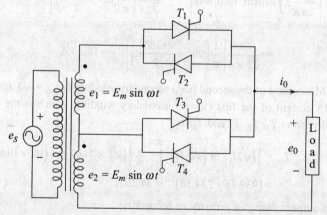

Fig. 11.9 *Two stage sequence control of a.c. regulator*

(A) Resistance load For resistance load, the load current waveform is identical with output voltage waveform. When thyristor pair T_3, T_4 is in operation, then the output voltage and current waveforms are as shown in Fig. 11.10. When both pairs T_1, T_2 and T_3, T_4 are in operation, then firing angle for T_3, T_4 is always zero, whereas firing angle for pair T_1, T_2 is varied from 180° to zero for obtaining output voltage from E to $2E$.

The output voltage, when SCR T_3 is triggered at $\omega t = 0$, follows $e_2 = E_m \sin \omega t$ curve. When SCR T_1 is triggered at $\omega t = \alpha$, voltage e_1 reverse biases T_3, it is, therefore, turned-off. After this, T_1 begins conduction and the output voltage jumps from e_2 to $(e_1 + e_2)$ and follows $2E_m \sin \omega t$ curve. At $\omega t = \pi$, output voltage and current are zero. At this instant, T_4 is triggered and output voltage

A.C. Regulators

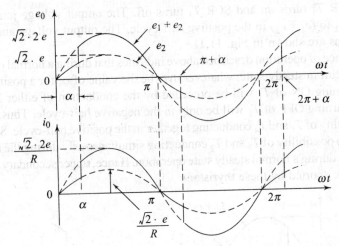

Fig. 11.10 *Output voltage and current waveforms with resistive load*

follows $E_m \sin \omega t$ curve. At $\omega t = \pi + \alpha$, when forward biased SCR T_2 is triggered, T_4 is reverse biased by $E_m \sin \alpha$, it is, therefore, turned-off. When T_2 begins conduction, output voltage follows $2E_m \sin \omega t$ curve as shown by the negative half-cycle in Fig. 11.10. In this figure, output current waveform i_o is shown identical with output voltage waveform e_0.

(B) Resistance–inductance (RL) load When thyristor pair T_3, T_4 alone is in operation, then the waveforms of output voltage and current are as shown in Fig. 11.5(b) for $\alpha > \phi$ and in Fig. 11.5 (c) for $\alpha \leq \phi$. For the control of output voltage from E to $2E$, thyristor pair T_3, T_4 is operated with a firing angle equal to zero, whereas firing angle for T_1, T_2 is varied from 180° to zero. The operation of the circuit is considered assuming that the duration of the firing pulse is chosen such that during positive half-cycle, firing pulses for T_1, T_3 last from $\omega t = 0$ to $\omega t = \pi$ and during negative half-cycle, firing pulse for thyristor pair T_3, T_4 extend from $\omega t = \pi$ to 2π.

Thyristor T_3 is conducting during positive half-cycle and a voltage $e_2 = \sqrt{2} E \sin \omega t$ is applied to the load. When SCR T_1 is triggered at $\omega t = \pi$, then SCR T_3 is turned-off and the reverse voltage e_1 and output voltage jump to $(e_1 + e_2)$, as shown. At $\omega t = \pi$, $(e_1 + e_2)$ reaches zero but output current i_o is not zero because of the presence of inductance L in the load. Thus, T_1 continues conducting until $\omega t = \beta$ where current i_o decays to zero and T_1, already reverse biased by $(e_1 + e_2)$, is turned-off.

Thyristor T_4, already gated at $\omega t = \pi$, starts conducting lowering the voltage to e_2 as shown. At $\omega t = (\pi + \alpha)$, SCR T_2 is triggered, voltage e_1 turns-off T_4 and output voltage in the negative half-cycle jumps to voltage $(e_1 + e_2)$ as shown. At $\omega t = 2\pi$, voltage $(e_1 + e_2)$ reaches zero but i_o is not zero because of inductance L. At instant $\omega t = (\pi + \beta)$, current i_o reaches zero. SCR T_2, already reverse biased by $(e_1 + e_2)$, is turned-off and SCR T_3, already gated at $\omega t = 2\pi$, is turned-on, lowering the voltage to e_2 at $\omega t = (\pi + \beta)$. At $\omega t = (2\pi + \alpha)$, already gated SCR T_1 turns-on and SCR T_3 turns-off. The output voltage jumps from voltage e_2 to $(e_1 + e_2)$ in the positive half-cycle. The output voltage and current waveforms are shown in Fig. 11.11.

Sequence of operation described above indicates that during a normal operation of the circuit in steady state, whatever may be the value of α, for a positive half-cycle, turning ON of T_1 will be preceded by the conduction of either T_2 or T_3, and the turning OFF of T_1 will be only in the negative half-cycle. Thus, there is no possibility of T_1 and T_4 conducting together in the positive half-cycle. Similarly, there is no possibility of T_3 and T_2 conducting simultaneously during the negative half-cycle during a normal steady state operation. Hence, upper secondary winding will not be shorted by these thyristors.

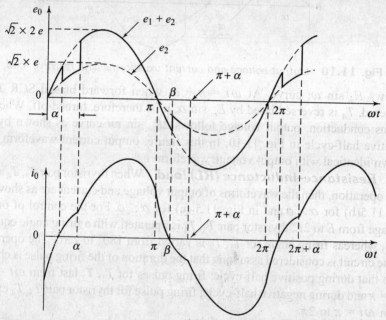

Fig. 11.11 *Output voltage and current waveforms with R–L load*

11.3.2 Multistage Sequence Control of A.C. Regulators

By using more than two stages of sequence control, it is possible to have further improvement in power factor and reduction in harmonics than that in a two-stage sequence control. Figure 11.12 shows the power circuit for n-stage sequence control of a.c. regulators. As shown, the transformer has n secondary windings. Each secondary is rated for e_s/n, where e_s is the source voltage. The voltage of node P with respect to K is e_s. Voltage of terminal q is $(n-1)e_s/n$ and so on. If voltage control from $e_{sK} = (n-3)e_s/n$ to $e_{rk} = (n-2)e_s/n$ is required, then thyristor pair 4 is triggered at $\alpha = 0°$ and the firing angle of thyristor pair 3 is controlled from $\alpha = 0°$ to $180°$, whereas all other thyristor pairs are kept OFF. Similarly, for controlling the voltage from $V_{qK} = (n-1)e_s/n$ to $e_{pk} = e_s$, thyristor pair 2 is triggered at $\alpha = 0°$, whereas for pair 1, firing angle α is varied from $0°$

to 180° keeping the remaining $(n-2)$ SCR pairs OFF. Thus, the load voltage can be controlled from e_s/n to e_s by an appropriate control of triggering the adjacent thyristor pairs.

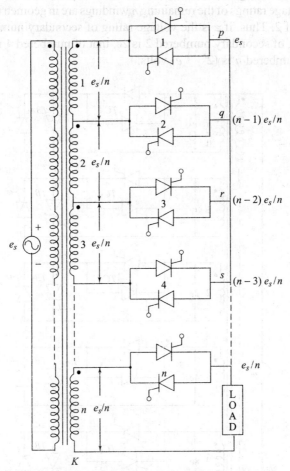

Fig. 11.12 *Multistage sequence control of a.c. regulators*

The presence of harmonics in the output voltage depends upon the magnitude of voltage variation. If this voltage variation is a small fraction of the total output voltage, the harmonic content in the output voltage is small. For example, for voltage control from $(n-2)\,e_s/n$ to $(n-1)\,e_s/n$, if voltage variation $e_s/n \ll (n-1)\,e_s/n$, then the harmonic content in the output voltage would be small.

11.3.3 Single-Phase Sinusoidal Voltage Regulator

If continuous voltage control over a wide range with low harmonic content and improved power factor is required, then a large number of stages will have to be used in a multistage sequence voltage controller. However, this is an expensive proposition. An alternative circuit requiring less number of stages is shown in Fig. 11.13.

This voltage controller, as shown in Fig. 11.13, has one primary winding and $(n + 1)$ secondary windings, i.e. it employs $(n + 1)$ stage; 0, 1, 2, ..., n. The top secondary winding, numbered p, is called as *vernier winding*. Its rating is e volts. The voltage ratings of the remaining n windings are in geometric progression with a ratio of 2. Thus, if e is the voltage rating of secondary numbered 1, then voltage rating of secondary numbered 2 is $2e$, that of numbered 4 is $8e$ ($2^{4-1} e$) and that of numbered n is ($2^{n-1} e$) volts.

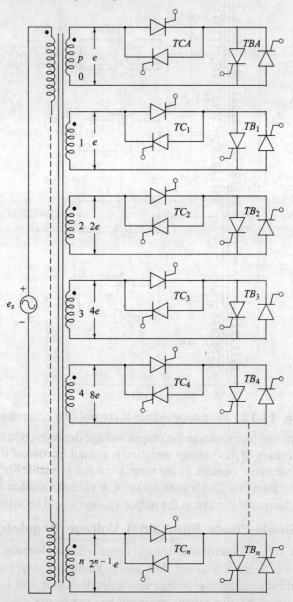

Fig. 11.13 *Single-phase sinusoidal voltage regulator*

As shown, the power circuit uses two sets of thyristors, named as *control thyristors* (TC_1, TC_2, ... TC_n) and by-pass thyristors. (TB_1, TB_2, ... TB_n). SCRs pertaining to n stages, i.e. TC_1, TC_2, ... TC_n and TB_1, TB_2, ... TB_n are either ON or OFF throughout a cycle. This means that control and by-pass thyristors are made to act as switches which remain either ON or OFF during a cycle. When control SCR pair for any stage is ON and its by-pass thyristor pair is OFF, then voltage of that stage would appear across the load and a load current would flow accordingly. On the other hand, if the control SCR pair is off and by-pass pair is ON for any stage, then this particular stage could be by-passed and will not contribute any voltage across the load. Thus, with an appropriate series combination of secondaries from 1 to n (without vernier winding), the load voltage can be varied from e to $(2^n - 1)e$ in discrete steps of e. This special feature of choosing any series combination of secondary windings is, however, not available in the multistage sequence control of thyristors in Fig. 11.12.

In the power circuit of Fig. 11.13, an additional stage A is employed as a vernier to permit continuous control of voltage from zero to e. It is a phase-controlled secondary winding. This winding contributes harmonics to line and load currents. The harmonic content would, however, be much lower because contribution of voltage by vernier winding to the load voltage is only a small fraction of the total load voltage.

The operation of the power circuit of Fig. 11.13 can now be explained for a resistance load. Let the load needs a stepless variation of voltage from $10e$ to $11e$. Stages 2 and 4 together can be used to get a voltage $10e$ and the variation from $10e$ to $11e$ is obtained by using stage 0. For stages 2 and 4, by-pass SCRs are kept OFF but their control SCRs are kept ON all the time. For the remaining stages, all by-pass SCRs are kept ON while their control SCRs are kept OFF all the time. With this, the circuit of Fig. 11.13 reduces to that shown in Fig. 11.14(a).

During the positive half-cycle, SCR T_3 is turned on at $\omega t = 0°$. A voltage equal to $10e$ is applied to the load. At $\omega t = \alpha$, T_1 is turned on and this turns off T_3, the voltage now jumps from $10e$ to $11e$, i.e. from $(10\,E_m \sin \alpha)$ to $(11\,E_m \sin \alpha)$, as shown in Fig. 11.14(b). After $\omega t = \alpha$, output voltage follows $(11\,E_m \sin \omega t)$ curve. At the end of positive half-cycle, T_4 is turned on at $\omega t = \pi$, the load voltage is now $10e$. At $\omega t = (\pi + \alpha)$, SCR T_2 is turned on and with this SCR T_4 is turned-off and load voltage jumps from $10e$ $(10\,E_m \sin \alpha)$ to $11e$ $(11\,E_m \sin \alpha)$ in the negative half-cycle. In this manner, load voltage can be continuously controlled from zero to $2^n e$ by a suitable choice of output voltage range.

11.4 THREE-PHASE A.C. REGULATORS

There are many three-phase versions of a.c. thyristor controllers and Fig. 11.15 shows a selection of them.

They all operate in slightly different ways. The following are the important points of comparison between the circuits of Fig. 11.15.
1. In circuits C and D, the individual phase controllers control their own loads independently of the other. They can, therefore, be studied as three single-phase controllers.

2. In other circuits, the individual phase controllers affect the other phase loads also and they have to be studied as complete three-phase circuits.
3. The peak voltages occur across thyristors at or near to the fully OFF state. In the case of circuits D, E and F the maximum thyristor voltage is

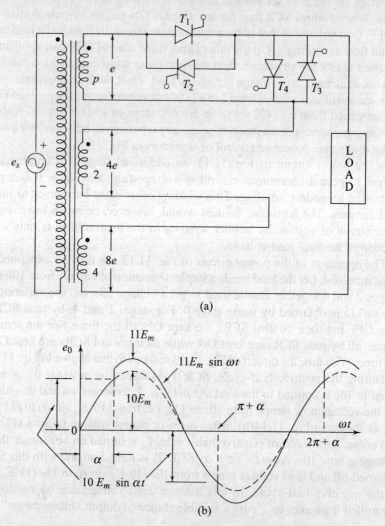

Fig. 11.14 *Operation of voltage regulator of Fig. 11.13*

the peak of the line voltage, whereas in circuit C it is the peak of the phase voltage; in circuits A and B the maximum thyristor voltage will be somewhere between the peak of the phase and line voltage depending on the leakage current of the thyristors, the method of firing, and the presence of voltage-sharing resistors across the thyristors.

4. All these circuits can be used under phase control.
5. The range of phase angle required to achieve full output range from zero to maximum varies between the circuits and these are given in Table 11.1.
6. The maximum current flow in the thyristors is decided from the fully on condition and the size of thyristors to be used should be chosen from

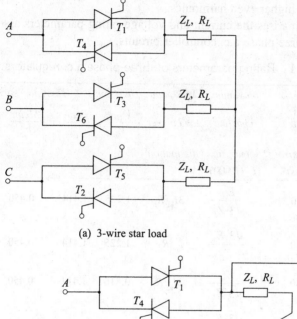

(a) 3-wire star load

(b) 3-wire delta load

Fig. 11.15 *(a) and (b)*

this condition. The peak, mean, and RMS thyristor currents given in Table 11.1 are related to the RMS a.c. input current which should be found by applying the full supply voltage to the load-circuit impedances.

7. Although only three thyristors are required in circuit F, they each have to carry a higher current than the other circuits. Under fully conduction conditions, the current flows in each thyristor for 240 degrees in every cycle.

8. The thyristor/diode versions of circuits A and B are shown in Fig. 11.15 (g and h, respectively). These circuits are most useful when integral cycle firing (Chapter 16) is adopted as they only then work in either the fully ON or fully OFF conditions.

When they are used with phase control as only three points per cycle are controllable, the load current and voltage will contain additional second and fourth harmonics and the higher even harmonics.

Table 11.1 summarises the current and voltage rating parameters associated with all of these three-phase a.c. controller circuits.

Table 11.1 Rating parameters of three-phase a.c. regulators

Circuits (Fig. 11.15)	Delay angle α for full control (degrees)	Maximum input line current (I_{ac} (RMS))	Maximum load power dissipation	$\dfrac{V_{RWm}}{V_{ac}}$	Thyristor Peak I_{ac}	Thyristor current Mean I_{ac}	RMS I_{ac}
A	150	$\dfrac{E_{ac}}{\sqrt{3} \cdot Z_L}$	$3I_{ac}^2 R_L$	1.225	1.414	0.450	0.707
B	150	$\dfrac{\sqrt{3} \cdot E_{ac}}{Z_L}$	$I_{ac}^2 R_L$	1.225	1.414	0.450	0.707
C	180	$\dfrac{E_{ac}}{\sqrt{3} \cdot Z_L}$	$3I_{ac}^2 R_L$	0.816	1.414	0.450	0.707
D	180	$\dfrac{\sqrt{3} \cdot E_{ac}}{Z_L}$	$I_{ac}^2 R_L$	1.414	0.816	0.260	0.408
E	150	$\dfrac{E_{ac}}{\sqrt{3} \cdot Z_L}$	$3I_{ac}^2 R_L$	1.414	0.816	0.260	0.408
F	210	$\dfrac{E_{ac}}{\sqrt{3} \cdot Z_L}$	$3I_{ac}^2 R_L$	1.414	1.414	0.675	0.766
G	210	$\dfrac{E_{ac}}{\sqrt{3} \cdot Z_L}$	$3I_{ac}^2 R_L$	1.225	0.414	0.450	0.707
H	210	$\dfrac{\sqrt{3} \cdot E_{ac}}{Z_L}$	$I_{ac}^2 R_L$	1.225	1.414	0.450	0.707

* E_{ac} = RMS line voltage; I_{ac} = RMS line current; Z_L = load impedance per phase; R_L = load resistance per phase.

11.4.1 Three-phase, Three-wire A.C. Regulator with Resistive Load

As an example of the method of analysing three-phase a.c. regulator circuits, circuit A of Fig. 11.15 will be considered in detail. The power diagram of a three-

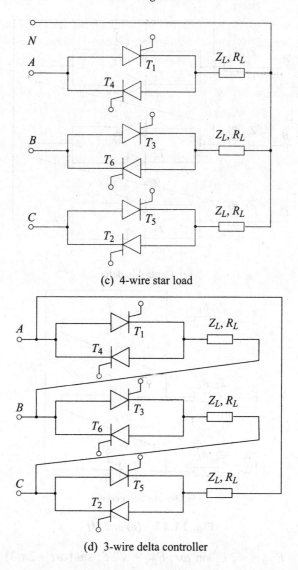

(c) 4-wire star load

(d) 3-wire delta controller

Fig. 11.15 *(c) and (d)*

phase full-wave (or bidirectional) a.c. regulator is shown in Fig. 11.16 with a star-connected resistive load. The current flow to the load is controlled by thyristors T_1, T_3, and T_5; and thyristors T_2, T_4, and T_6 provide the return current path. The firing sequence of thyristor is T_1, T_2, T_3, T_4, T_5, T_6. We may recall that a thyristor will conduct if its anode voltage is higher than that of its cathode and it is triggered. Once a thyristor starts conducting, it would be turned-off only when its current falls to zero.

If E_s is the RMS value of the input phase voltage and we define the instantaneous input phase voltages as

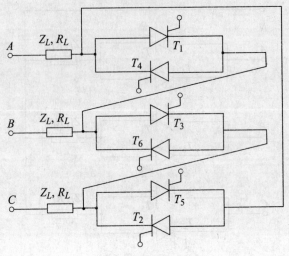

(e) Control in delta

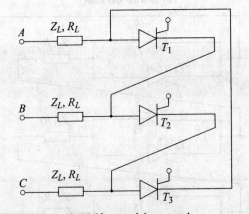

(f) Half-wave delta control

Fig. 11.15 *(e) and (f)*

$$E_{AN} = \sqrt{2}\,E_s \sin \omega t,\ E_{BN} = \sqrt{2}\,E_s \sin(\omega t - 2\pi/3)$$
$$E_{CN} = \sqrt{2}\,E_s \sin(\omega t - 4\pi/3)$$

then the instantaneous input line voltages are

$$E_{AB} = \sqrt{6}\,E_s \sin(\omega t + \pi/6),\ E_{BC} = \sqrt{6}\,E_s \sin(\omega t - \pi/2)$$
$$E_{CA} = \sqrt{6}\,E_s \sin(\omega t - 7\pi/6)$$

Figure 11.17 shows the waveforms for the input voltages, conduction angles of thyristors, and output phase voltages for $\alpha = 60°$ and $\alpha = 120°$. For $0 \leq \alpha \leq 60°$, immediately before the triggering of SCR T_1, two thyristors conduct. Once SCR T_1 is triggered, three thyristors conduct. A thyristor turns-off when its current attempts to reverse. The conditions alternate between two and three conducting thyristors.

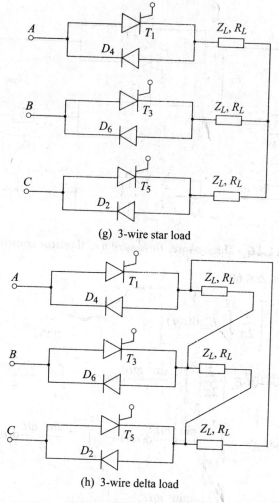

(g) 3-wire star load

(h) 3-wire delta load

Fig. 11.15 *(g) and (h)*

At anytime, only two thyristors conduct for $60° \leq \alpha \leq 90°$. For $90° \leq \alpha < 150°$, although two thyristors conduct at anytime, there are periods when no thyristors are ON. For $\alpha \geq 150°$, there is no period for two conducting thyristors and the output voltage becomes zero at $\alpha = 150°$. The range of delay angle is

$$0 \leq \alpha \leq 150° \qquad (11.40)$$

The expression for the RMS output phase voltage depends on the range of delay angle. The RMS output voltage for a star connected load can be obtained as follows.

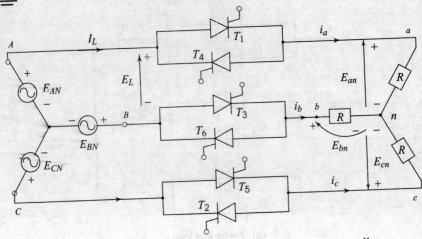

Fig. 11.16 *Three-phase, three-wire a.c. thyristor controller*

For $0 \leq \alpha \leq 60°$:

$$E_0 = \left[\frac{1}{2\pi}\int_0^{2\pi} E_{an}^2 \, d(\omega t)\right]^{1/2}$$

$$= \sqrt{6}E_s \left\{\frac{2}{2\pi}\left[\int_\alpha^{\pi/3} \frac{\sin^2 \omega t}{3} d(\omega t) + \int_{\pi/4}^{\pi/2+\alpha} \frac{\sin^2 \omega t}{4} d(\omega t)\right.\right.$$

$$+ \int_{\pi/3+\alpha}^{2\pi/3} \frac{\sin^2 \omega t}{3} d(\omega t) + \int_{\pi/2}^{\pi/2+\alpha} \frac{\sin^2 \omega t}{4} d(\omega t)$$

$$\left.\left.+ \int_{2\pi/3+\alpha}^{\pi} \frac{\sin^2 \omega t}{3} d(\omega t)\right]\right\}^{1/2}$$

$$= \sqrt{6}E_s \left[\frac{1}{\pi}\left(\frac{\pi}{6} - \frac{\alpha}{4} + \frac{\sin 2\alpha}{8}\right)\right]^{1/2} \tag{11.41}$$

For $60° \leq \alpha \leq 90°$:

$$E_0 = \sqrt{6}E_s \left[\frac{2}{2\pi}\left\{\int_{\pi/2-\pi/3+\alpha}^{5\pi/6-\pi/3+\alpha} \frac{\sin^2 \omega t}{4} d(\omega t)\right.\right.$$

$$\left.\left.+ \int_{\pi/2-\pi/3+\alpha}^{5\pi/6-\pi/3+\alpha} \frac{\sin^2 \omega t}{4} d(\omega t)\right\}\right]^{1/2}$$

$$= \sqrt{6}E_s \left[\frac{1}{\pi} \left(\frac{\pi}{12} + \frac{3\sin 2\alpha}{16} + \frac{\sqrt{3} \cdot \cos 2\alpha}{16} \right) \right]^{1/2} \quad (11.42)$$

For $90° \leq \alpha \leq 150°$:

$$E_0 = \sqrt{6}E_s \left\{ \frac{2}{\pi} \left[\int_{\pi/2 - \pi/3 + \alpha}^{\pi} \frac{\sin^2 \omega t}{4} d(\omega t) + \int_{\pi/2 - \pi/3 + \alpha}^{\pi} \frac{\sin^2 \omega t}{4} d(\omega t) \right] \right\}^{1/2}$$

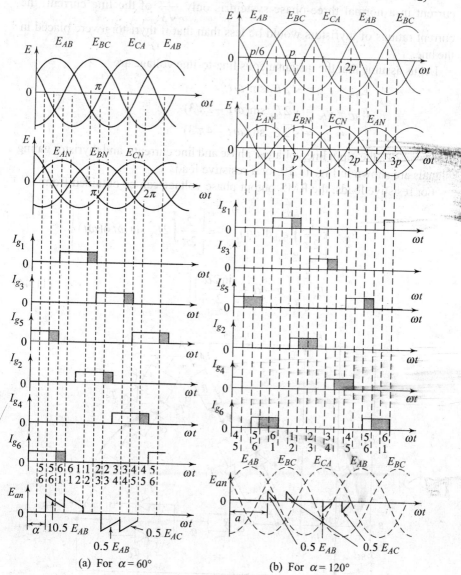

Fig. 11.17 *Waveforms for three-phase three-wire a.c. thyristor controller*

$$= \sqrt{6}E_s \left[\frac{1}{\pi} \left(\frac{5\pi}{24} - \frac{\alpha}{4} + \frac{\sin 2\alpha}{16} + \frac{\sqrt{3}\cos 2\alpha}{16} \right) \right]^{1/2} \qquad (11.43)$$

11.4.2 Three-phase Bidirectional Delta-Connected Regulators

Figure 11.18 shows the delta-connected three-phase regulator. Since the phase current in a normal three-phase system is only $\frac{1}{\sqrt{3}}$ of the line current, the current ratings of thyristors would be less than that if thyristors were placed in the line.

Let us assume that the instantaneous line-to-line voltage are

$$E_{AB} = e_{ab} = \sqrt{2}E_s \sin \omega t.$$
$$E_{BC} = e_{bc} = \sqrt{2}E_s \sin(\omega t - 2\pi/3).$$
$$E_{CA} = e_{ca} = \sqrt{2}E_s \sin(\omega t - 4\pi/3).$$

For $\alpha = 120°$, the input line voltages, phase and line currents, and thyristor gating signals are shown in Fig. 11.19 for a resistive load.

For resistive loads, the RMS output phase voltage can be obtained from

$$E_0 = \left[\frac{1}{2\pi} \int_{\alpha}^{2\pi} e_{ab}^2 \, d(\omega t) \right]^{1/2} = \left[\frac{2}{2\pi} \int_{\alpha}^{\pi} 2E_s^2 \sin \omega t \, d(\omega t) \right]^{1/2}$$

$$= E_s \left[\frac{1}{\pi} \left(\pi - \alpha + \frac{\sin 2\alpha}{2} \right) \right]^{1/2} \qquad (11.44)$$

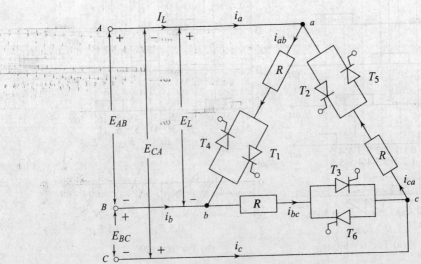

Fig. 11.18 *Delta connected three-phase a.c. regulator*

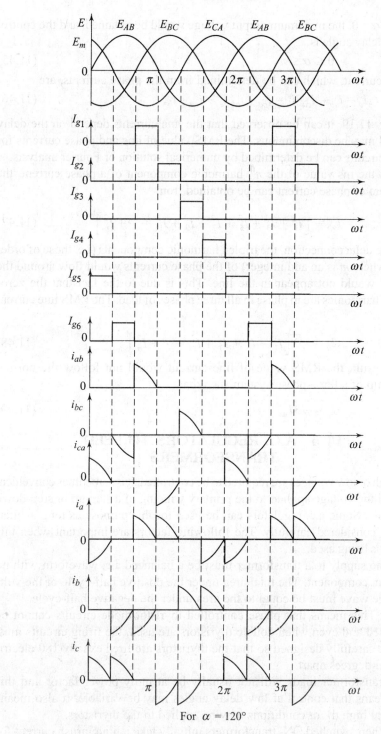

Fig. 11.19 *Waveforms for three-phase delta-connected regulator*

When $\alpha = 0$, the maximum output voltage would be obtained, and the control range of delay angle is

$$0 \le \alpha \le \pi \qquad (11.45)$$

The line current, which can be determined from the phase currents, are

$$i_a = i_{ab} - i_{ca} \quad i_b = i_{bc} - i_{ab} \quad i_c = i_{ca} - i_{bc} \qquad (11.46)$$

From Fig. 11.19, it can be observed that the line currents depend on the delay angle and may be discontinuous. The RMS value of line and phase currents for the load circuits can be determined by numerical solution or Fourier analysis.

If I_n is the ms value of the nth harmonic component of a phase current, the RMS value of phase current can be obtained from

$$I_{ab} = \left[I_1^2 + I_3^2 + I_5^2 + I_7^2 + I_9^2 + I_{11}^2 + \cdots + I_n^2 \right]^{1/2} \qquad (11.47)$$

Due to the delta connection, the triplen harmonic components (i.e. those of order $n = 3m$, where m is an add integer) of the phase currents would flow around the delta and would not appear in the line. This is due to the fact that the zero-sequence harmonics are in phase in all three phases of load. The RMS line current becomes

$$I_a = \sqrt{3} \left[I_1^2 + I_5^2 + I_7^2 + I_{11}^2 + \cdots + I_n^2 \right]^{1/2} \qquad (11.48)$$

As a result, the RMS value of line current would not follow the normal relationship of a three-phase system such that

$$I_a < \sqrt{3} I_{ab} \qquad (11.49)$$

11.5 A.C. REGULATORS TO FEED TRANSFORMERS

When high or low voltages are required to be controlled, it is sometimes convenient to control the voltage applied to the primary winding of a step-up or step-down transformer. Some a.c. regulators can be used for this purpose as long as their design is considered carefully. The following factors are important when this principle is being used:

1. The supply to a transformer must be a balanced a.c. waveform with no d.c. component. The total area under the positive half-cycle of the voltage wave must be equal to the area under the negative half-cycle.

    This means that phase controlled thyristor/diode circuits cannot be used and even when double thyristors are used, the firing circuits must be carefully designed so that the thyristors are fired exactly 180 electrical degrees apart.

2. Transformer loads will not usually be at unity power factor and this means that control at low delay angles may be variable. It also means that high dV/dt conditions may be applied to the thyristors.

3. When switched ON, transformers initially take a high inrush current for a few cycles while the correct core flux is being established. This means

that integral cycles control (Chapter 16) will not be acceptable in its normal form. Phase-controlled a.c. thyristor controllers, however, do provide a means of gradually increasing the applied voltage, so removing the high inrush currents.
4. If misfiring occurs in the controller, the transformer may saturate and take large currents sufficient to cause overheating (if prolonged), or operation of the protective fuses or switches.
5. The a.c. controllers will normally incorporate voltage-transient "snubber circuits" or their equivalent in parallel with the thyristors. These may provide a sufficient path for transformer magnetization and so voltage will be measured even when the thyristors are fully OFF.

Single and multiphase controllers are in use for transformer loads, and circuits A, B, and C of Fig. 11.15 can be employed. It is not possible to include thyristors within a three-phase primary delta winding as they will suppress the required flow of third harmonic currents.

SOLVED EXAMPLES

Example 11.6 A three-phase, three-wire bidirectional controller supplies a star-connected resistive load of $R = 5\ \Omega$ and the line-to-line input voltage is 210 V (RMS) 50 Hz. The firing angle is $\alpha = \pi/3$. Determine:
(i) the RMS output phase voltage, E_0. (ii) the input power factor, P_f and
(iii) expression for the instantaneous output voltage of phase a.

Solution: $E_L = 210$ V. $\therefore\ E_s = \dfrac{E_L}{\sqrt{3}} = \dfrac{210}{\sqrt{3}} = 121.24$ V.

(i) From Eq. (11.36), the RMS output voltage

$$E_0 = \sqrt{6} \times 121.24 \left[\frac{1}{\pi}\left(\frac{\pi}{6} - \frac{\pi}{12} + \frac{\sin 2 \cdot \pi/3}{8}\right)\right]^{1/2} = 101.924\ \text{V}$$

(ii) Thus RMS phase current of the load is

$$I_a = 101.924/5 = 20.3848\ \text{A}$$

and the output power is

$$P_0 = 3 I_a^2 R = 3 \times (20.3848)^2 \times 5 = 6233.101\ \text{W}.$$

Since the load is connected in star, the phase current is equal to the line current,

$\therefore\qquad I_L = I_a = 20.3848$ A.

The input volt-ampere, $E \cdot A = 3\ E_s \cdot I_L = 3 \times 121.24 \times 20.3848 = 7414.36$ VA

$\therefore$ The power factor $P_F = \dfrac{P_0}{E_A} = \dfrac{6233.101}{7414.36} = 0.84$ (lagging)

(iii) If the input phase voltage is taken as the reference and is

$$E_{AN} = 121.24\ \sqrt{2}\ \sin \omega t = 171.46 \sin \omega t,$$

the instantaneous input line voltages are

$$E_{AB} = 210\sqrt{2} \sin\left(\omega t + \frac{\pi}{6}\right) = 296.98 \sin\left(\omega t + \frac{\pi}{6}\right)$$

$$E_{BC} = 296.98 \sin\left(\omega t - \frac{\pi}{2}\right)$$

$$E_{CA} = 296.98 \sin\left(\omega t - \frac{7\pi}{6}\right)$$

The instantaneous output phase voltage, e_{an}, which depends on the number of conducting devices, can be determined from Fig. 11.17(a) as follows:

For $\quad 0 \le \omega t < \pi/3, e_{an} = 0$

For $\quad \pi/3 \le \omega t < 2\pi/3, e_{an} = E_{AB}/2 = 148.49 \sin(\omega t + \pi/6)$

For $\quad 2\pi/3 \le \omega t < \pi, e_{an} = E_{AC}/2 = -E_{CA}/2 = 148.49 \sin\left(\omega t - \frac{7\pi}{6} - \pi\right)$

For $\quad \pi \le \omega t < \frac{4\pi}{3}, e_{an} = 0$

For $\quad \frac{4\pi}{3} \le \omega t < \frac{5\pi}{3}, e_{an} = E_{AB}/2 = 148.49 \sin(\omega t + \pi/6)$

For $\quad \frac{5\pi}{3} \le \omega t < 2\pi, e_{an} = \frac{E_{AC}}{2} = 148.49 \sin\left(\omega t - \frac{7\pi}{6} - \pi\right)$.

Example 11.7 A three-phase resistive load is to be controlled by three Triacs from a 415 V supply. If the load is 15 kW, determine the required ratings of Triacs. If thyristors were used instead of Triacs, determine their rating.

Solution: The RMS line current = $\dfrac{15000}{\sqrt{3} \times 415}$ = 20.9 A.

Hence the required Triac RMS current rating = 20.9 A. In the OFF state, differences in leakage current should result in the line voltage appearing across the Triac, hence the required peak off-state voltage = $415 \times \sqrt{2}$ = 587 V.

If thyristors were used, they would conduct for only one half-cycle of the sine-wave, hence their required RMS current rating = $\dfrac{20.9}{\sqrt{2}}$ = 14.8 A.

The voltage requirement would be identical to that for the Triac.

REVIEW QUESTIONS

11.1 Describe the operation of single phase half-wave a.c. voltage regulator with the help of voltage and current waveforms. Also, derive the expression for average value of output voltage.

11.2 List the advantages and disadvantages of single-phase half-wave (unidirectional) a.c. regulator.

A.C. Regulators

11.3 For a single-phase a.c. voltage regulator feeding a resistive load, draw the waveforms of source voltage, gating signals, output voltage, source and output currents and voltage across SCRs. Describe its working with reference to the waveforms drawn.

11.4 Explain why the single-phase a.c. regulator using two SCRs must have its trigger sources isolated from each other?

11.5 Derive the expression for amplitude of fundamental component of load or source current for single-phase a.c. voltage regulator.

11.6 For a single-phase a.c. regulator feeding a resistive load, show that power factor is given by the expression

$$\left[\frac{1}{\pi}\{\pi - \alpha\} + \frac{\sin 2\alpha}{2}\right]^{1/2}$$

11.7 Define the term 'power-factors'. Derive its expression for single-phase a.c. voltage regulator feeding a resistive load circuit and show that
$$PF = [\text{per unit power}]^{1/2}$$

11.8 Analyse the output voltage waveform of single phase a.c. regulator into various harmonics with Fourier series and find expressions for the amplitude of nth harmonic, E_{nm} and its phase ϕ_n.

11.9 A single-phase a.c. voltage regulator, with two thyristors arranged in antiparallel, is connected to RL load. Discuss its working when firing angle is more than the load P_f angle. Illustrate your answer with waveforms of source voltage, gate signals, load and source currents, output voltage and voltage across the SCRs.

11.10 Derive an expression for the output current in terms of source voltage, load impedance and firing angle for a single-phase a.c. regulator with RL load.

11.11 For a single-phase a.c. voltage regulator, develop a relation between conduction angle γ and firing angle α, and plot their variation as a function of load phase angle ϕ. Under what conditions conduction angle γ becomes equal to π?

11.12 Discuss the operation of a single-phase a.c. regulator with RL load when α is less than, or equal to, load phase angle ϕ. Hence show that for α less than ϕ, output voltage of the a.c. regulator cannot be regulated.

11.13 For a single phase a.c. regulator, explain how pulse gating is suitable for R load and not for RL load. Hence, show that high frequency carrier gating is essential for RL loads.

11.14 Discuss the working of a two-stage sequence control of voltage regulators for both R and RL loads. What is the advantage of this regulator over a single-phase full-wave voltage regulator?

11.15 Describe the operation of multistage sequence control of a.c. voltage regulators with suitable power diagram.

11.16 Distinguish between two-stage and multi-stage sequence control of a.c. voltage regulators. What are the advantages of multistage over two-stage sequence control.

11.17 Describe a single-phase sinusoidal voltage regulator with vernier winding. What are the functions of controlled and by-pass thyristors? Describe how output voltage waveform from 8e to 9e can be obtained from this voltage regulator.

11.18 Give the various configurations of three-phase a.c. regulators. List the important points of comparison between these circuits.

11.19 Describe the operation of a 3-ϕ, three-wire a.c. thyristor controller with neat power-diagram and voltage and current waveforms.

11.20 Draw the waveforms for three-phase a.c. thyristor controller for $\alpha = 60°$, $90°$ and $120°$.

11.21 Discuss the working of delta connected three-phase a.c. regulator with suitable power diagram. Also, draw the various waveforms for $\alpha = 120°$.

11.22 Discuss the various important factors to be considered while feeding transformers through a.c. regulators.

PROBLEMS

11.1 A single-phase a.c. regulator of Fig. 11.3 with a resistive load has the following data:

Supply mains: 230 V, 50 Hz, $R = 4\ \Omega$. Determine:
  (i) the firing angle α at which, at greatest forward or reverse voltage is applied to either of the thyristors and the magnitude of these voltages;
  (ii) the greatest forward or reverse voltage that appears across either of the thyristors for firing angles of 120° and 60°.
  (iii) the RMS value of fifth harmonic current and its phase for $\alpha = \pi/4$.

[*Ans* (i) $\alpha \geq \pi/2$, 325.27 V. (ii) 325.27 V, 281.69 V. (iii) 27.28 V, $-63.435°$.]

11.2 A single-phase a.c. regulator of Fig. 11.4 controls the load power. The input to the controller is 230 V, 50 Hz, sinusoidal. The load circuit consists of $R = 3\ \Omega$ and $\omega_L = 4\ \Omega$. Determine:
  (a) the control range of firing angle.
  (b) the maximum value of RMS load current.
  (c) the maximum power and power factor.
  (d) the maximum values of average and RMS thyristor currents.
  (e) the maximum possible value of $\dfrac{di}{dt}$ that may occur in the SCR, and
  (f) the conduction angle for $\alpha = 0°$, and $\alpha = 135°$, assuming a gate pulse of duration π.

[*Ans* (a) $53.13° \leq \alpha \leq 180°$, (b) 46 A, (c) maximum power = 6.35 kW.
Power factor = 0.6, (d) 20.707 A, 32.527 A,
(e) 2.0437×10^4 A/s, (f) 180°, 100°.]

11.3 A single phase-load of resistance 12 Ω in series with an inductance of 24 mH is fed from a 240 V (RMS), 50 Hz supply by a pair of inverse parallel thyristors. Find the mean power in the load at firing angles of (a) 0°, (b) 90° and (c) 120°. Ignore source inductance and device voltage drops.

[*Ans* (a) 3441 V, (b) 1370 W, (c) 177.2 W]

A.C. Regulators

11.4 A single-phase, half-wave a.c. voltage regulator, using one SCR in antiparallel with a diode, feeds 1 kW, 230 V heater. Find the load power for a firing angle of (i) 0°, (ii) 180° (iii) 70°.

[Ans (i) 1 kW (ii) 500 W (iii) 856.71 W]

11.5 A three-phase resistive heating load is controlled by Triacs from a 415 V (line), 50 Hz supply. (a) If the maximum load is 24 kW, (i) determine the rating of the Triacs and their firing angles for loads of (ii) 16 kW and (iii) 8 kW. (b) If the Triacs are replaced by thyristors, how would the current rating change?

[Ans (a) Triac rating (RMS) = 33.39 A (i) 16 kW, α = 74.49°
(ii) 8 kW, α = 105.29°, (b) thyristor rating (RMS) = 23.61 A.

11.6 A single-phase load of resistance 12 Ω in series with an inductance of 24 mH is fed from a 240 V (RMS), 50 Hz supply by a pair of inverse parallel thyristors. Find the mean power in the load at firing angles of, (a) 0°, (b) 90° and (c) 120°. Ignore source inductance and device voltage drops.

[Ans (a) 3441 W, (b) 1370 W, (c) 177.2 W.]

11.7 The single-phase transformer tap-changer of Fig. 11.9 has the primary voltage of 220 V (RMS), 50 Hz. The secondary voltages are e_1 = 110 V and e_2 = 110 V. If the load resistance is 8 Ω, the RMS load voltage is 170 V and the firing angle of thyristors T_1, T_2 is 72°, determine:
(a) the RMS currents of SCRs T_1 and T_2,
(b) the RMS currents of SCRs T_3 and T_4, and
(c) the input power factor.

[Ans (a) 16.194 A, (b) 5.382 A, (c) 0.698 (lagging)]

REFERENCES

1. S B Dewan and A Straughen, *Power semiconductor circuit*, John Wiley, New York, 1975.
2. G K Dubey, S R Daradla, A Joshi and RMK Sinha, *Thyristorised power controllers*, Wiley, 1986.
3. B M Bird and KG King, *An introduction to power electronics*, John Wiley, New York, 1983.
4. W Shepered, *Thyristor control of a.c. circuits*, Bradford University Press and Crosby LockWood, 1976.
5. B J Chalmers and SA Hamed, "New method of analysis and performance prediction for thyristor voltage controlled RL loads," *IEEE proceedings*, Vol. 134, pp. 339–347.

Chapter 12
Resonant Converters

LEARNING OBJECTIVES:
- To introduce basic resonant circuit concepts.
- To consider the operation of series and parallel load resonant converters.
- To examine the operation of class-E resonant inverter and rectifier.
- To explain the working principles of resonant switch converters.
- To consider the operation of zero-voltage switch three-level PWM converter.
- To introduce the operation of resonant dc link inverter.

12.1 INTRODUCTION

Different converter topologies; ac–dc, ac–ac, dc–ac, and dc–dc were discussed in previous chapters. Basically, there are two circuit schemes for processing of power electronic technology. They are pulse width modulation (PWM) and resonance. During each switching in all the pulse-width modulated dc–dc and dc–ac converter topologies, the controllable switches are operated in a switch mode where they are required to turn-on and turn-off the entire load current. In the switch mode operation, the switches are subjected to high switching stresses and high switching power-loss that increases linearly with the switching frequency of the PWM. Also, one more problem with this operation is the electromagnetic interference (EMI) which is due to large di/dt and dv/dt.

Using high switching frequency in all the four types of converters will be desirable because it will provide tremendous savings in component sizes (reduce converter size) and weights and hence achieves high power density.

High-frequency resonant converters have become increasingly popular among power supply designers, especially at frequencies above 100 kHz because they offer small size, good reliability and reduced EMI/RFI.

Resonant converter is defined as a class of converters in which the topology constitutes of at least one resonant tank circuit as a subcircuit. A resonant tank is a subcircuit consisting of at least one inductor and one capacitor. The above-mentioned disadvantages of PWM control can be eliminated or minimized if the switching devices are turned 'on' and 'off' when the voltage across a device and/or its current become zero. The voltage and current are forced to pass through zero crossing by creating an L–C resonant circuit.

The resonant technique process power in sinusoidal form. The power switches are often turned-off under zero current and turned-on with large increase of device current. The resonant converter can be operated either below resonant frequency or above resonant frequency. If it is operated at frequencies above the resonant frequency, the switches are turned-on at zero voltage across them but turned-off abruptly. In these conveters, the switching losses and device stresses are lower compared to PWM converters but the conduction losses are higher because the peak and rms device currents are much higher in resonant converter. The control of PWM technique is simpler and is largely used in power conversion today, however, it is limited to low and medium power applications. The resonant converter can be used in low, medium and high power applications using high power switches but the control circuit is rather complex.

Two methods are employed in resonant converters. First, the switch generates a square wave of voltage that is applied to a series LC circuit. This configuration is known as series-resonant converters. Second, the switch generates a square-wave of current that is applied to a parallel LC circuit. It is known as parallel-resonant converter. The output of these converters is supplied to the load or processed further for the desired output, e.g. dc–dc, dc–ac or ac–ac conversion using conventional converters.

12.2 BASIC RESONANCE CIRCUIT CONCEPTS

In resonant converters, both series and parallel LC circuits are used. These circuits are discussed here briefly.

12.2.1 Series Resonant Circuits

Series resonant RLC circuit is shown in Fig. 12.1. Here the resistance R forms a series circuit with the resonating components L_r and C_r. The input voltage is a square-wave. The voltage equation is given by

$$R \cdot i + L_r \cdot \frac{di}{dt} + \frac{1}{C_r} \int i \, dt = 0 \tag{12.1}$$

The solution for this equation is given by

$$i(t) = \frac{E_{dc}}{w_o L_r} e^{\frac{-R_t}{2L_r}} \sin \omega_r \cdot t \tag{12.2}$$

and

$$Ec(t) = E_{dc} + (E_0 - E_{dc}) e^{\frac{-Rt}{2L_r}} \left(\frac{R}{2L_r \omega_r} \sin \omega_r t + \cos \omega_r t \right)$$

$$= E_{dc} + (E_0 - E_{dc})\, e^{\frac{-Rt}{2L_r}} \left\{ \left[1 + \left(\frac{R}{2L_r \omega_r} \right)^2 \right]^{1/2} \cdot \cos(\omega_r t + \theta) \right\} \quad (12.3)$$

where, angular undamped (resonance) frequency $= W_0 = \dfrac{1}{\sqrt{L_r C_r}}$ \hfill (12.4)

$$\text{Angular damped frequency} = W_r = \sqrt{\frac{1}{L_r \cdot C_r} - \frac{R \cdot L_r}{2 L_r}} \quad (12.5)$$

$$\theta = \tan^{-1}\left(\frac{R}{2 L_r W_r} \right) \quad (12.6)$$

If $R = 0$, then the circuit of Fig. 12.1 becomes an LC resonating circuit with undamped oscillation. Frequency domain analysis is important in order to determine the effectiveness of the LC resonating circuit. For Fig. 12.1, we can write

$$Z(s) = \frac{1}{S C_r} + s L_r + R$$

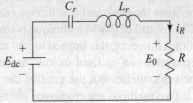

Fig. 12.1 *Series RLC circuit*

$$\therefore\ Y(s) = \left(\frac{1}{S C_r} + S L_r + R \right)^{-1} = \frac{(R/L_r) S}{R\left(S^2 + R/L_r\, S + \dfrac{1}{L_r C_r} \right)}$$

Let us define: $\omega_0 = \dfrac{1}{\sqrt{L_r \cdot C_r}}$,

Characteristic impedance, $Z_0 = \sqrt{\dfrac{L_r}{C_r}}$ \hfill (12.7)

$$\varepsilon \cdot \omega_0 = \alpha = R/2 L_r, \quad \therefore\ \varepsilon = \frac{R}{2}\sqrt{\frac{C_r}{L_r}} \quad (12.8)$$

$$\therefore\quad Y(s) = \frac{2\alpha \cdot S}{\left(S^2 + 2\alpha S + \omega_0^2 \right)} \cdot \frac{1}{R} \quad (12.9)$$

The roots of the above equation are

$$S_{1,2} = -\alpha \pm j\, \omega_0 \sqrt{1 - \varepsilon^2} = -\alpha \pm j\, \sqrt{\omega_0^2 - \alpha^2}$$

At resonance, $S = j\omega_0$, which gives, $Y(S) = \dfrac{1}{R}$ \hfill (12.10)

In the presence of the load resistance R, quality factor Q is defined as

$$Q = \frac{\omega_0 L_r}{R} = \frac{1}{\omega_0 C_r R} = \frac{Z_0}{R} \quad (12.11)$$

The voltage stress on the L_r and C_r depends upon Q.

The voltage gain of the RLC series circuit in frequency domain is given by

$$G(S) = \frac{E_0(j\omega_s)}{E_1(j\omega_s)} = \frac{R}{R + j\omega_s L_r - (j/\omega_s C_r)}$$

$$= \frac{1}{1 + jQ[(\omega_s/\omega_0) - (\omega_0/\omega_s)]} \quad (12.12)$$

and

$$|G(S)| = \left\{ \frac{1}{1 + Q^2 \left[\mu - (1/\mu)\right]^2} \right\}^{1/2}, \text{ where } \mu = \frac{\omega_s}{\omega_0} \quad (12.13)$$

or

$$|G(j\omega)| = \frac{1}{\left[1 + Q^2 (\mu - 1/\mu)^2\right]^{1/2}} \quad (12.14)$$

The magnitude plot for Eq. (12.14) is shown in Fig. 12.2. Since these converters are often switched at a variable frequency condition, the impedance of the series RLC circuit varies and depends upon the switching frequency. It becomes minimum at ω_0. If the switching frequency is greater than the resonant frequency, then the output voltage becomes continuous one. The current will rise to a high value, especially at a high load current, if the inverter operates near resonance and a short circuit occurs at the load. However, the output current can be controlled by increasing the switching frequency. At resonance, the maximum output occurs and the maximum gain for $\mu = 1$ is $|G(j\omega)|_{max} = 1$. Under no load conditions, $R = \infty$ and $Q = 0$. Thus, the curve would be a horizontal line. The characteristic has a poor selectivity for $Q = 1$ and the output voltage will change significantly from no load to full-load conditions, thereby yielding poor regulation.

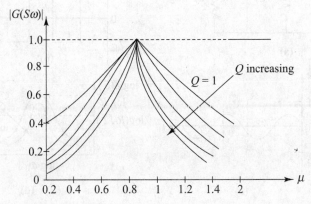

Fig. 12.2 *Frequency response for RLC series circuit*

12.2.2 Parallel Resonant Circuits

Figure 12.3 shows an undamped parallel resonant circuit supplied by a dc current I_{dc}. The circuit equations are

$$i_L + C_r \frac{dE_c}{dt} = I_{dc} \quad (12.15)$$

and

$$E_c L_c = L_r \frac{di_L}{dt} \quad (12.16)$$

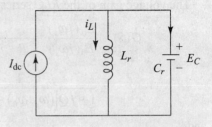

Fig. 12.3 Undamped parallel-resonant circuit

The solution of the above equations is given by

$$i_{L(t)} = I_{dc} + (I_{L0} - I_{dc}) \cos \omega_0 (t - t_0) + \frac{E_{C0}}{Z_0} \sin \omega_0 (t - t_0) \quad (12.17)$$

and

$$E_C(t) = Z_0(I_d - I_{L0}) \sin \omega_0 (t - t_0) + E_{C0} \cos (t - t_0) \quad (12.18)$$

where I_{L0} and V_{C0} are the initial conditions of the components at time $t = t_0$.

$$W_0 = \frac{1}{\sqrt{L_r C_r}} \quad (12.19)$$

and

$$Z_0 = \sqrt{\frac{L_r}{C_r}} \quad (12.20)$$

The parallel LC network offers infinite reactance or impedance at the resonance condition. The impedance becomes maximum at the switching frequency, $W_S = \omega_0$ and the load voltage wave-form becomes almost sinusoidal, i.e. ($E_R = E_C$) (Fig. 12.4). Also, E_R is in phase with i.

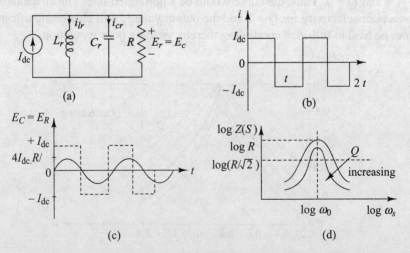

Fig. 12.4 Parallel Resonant circuit and responses

In the presence of load resistor R, the quality factor Q is defined as

$$Q = \omega_0 \cdot R \cdot C_r = \frac{R}{\omega_0 L_r} = \frac{R}{Z_0} \quad (12.21)$$

Impedance is given by

$$Z(S) = \frac{R \cdot S/(Q \cdot \omega_0)}{1 + \frac{s^2}{\omega_0^2} + \frac{s}{Q\omega_0}} \quad (12.22)$$

It can be observed from Fig. 12.4(d) that $Z(S)$ varies with the switching frequency ω_s, hence the load power. The inductive reactance is less than the capacitive reactance for $\omega_s < \omega_0$. Therefore, i_{L_r} dominates over i_{C_r} and the parallel LC circuit behaves as an equivalent inductor in parallel with the load resistor, R. Similarly for $\omega_s > \omega_0$, the circuit behaves as an equivalent capacitor in parallel with R. The gain is given by

$$G(j\omega) = \frac{R/j\omega C_r/(R + 1/j\omega C_r)}{j\omega L_r + \frac{R/j\omega C_r}{R + (1/j\omega C_r)}} \quad (12.23)$$

$$= \frac{1}{1 - \omega^2 L_r C_r + (j\omega L_r/R)} = \frac{1}{1 - (\omega/\omega_0)^2 + \frac{j(\omega/\omega_0)}{Q}}$$

$$|G(j\omega)| = \frac{1}{\sqrt{(1-\mu^2)^2 + (\mu/Q)^2}} \quad (12.24)$$

Also, $|G(j\omega)|_{max} = Q$ appears at $\mu = 1$ or $\omega_s = \omega_0$.

12.2.3 Parallel Loaded Resonant LC Converters

A parallel-loaded resonant converter is shown in Fig. 12.5(a). Here, the load is connected across the capacitor C directly. The voltage gain is given by

$$G(j\omega) = \frac{1}{1 - \omega_s^2 L_r C_r + j\omega_s \frac{L_r}{R}} \quad (12.25)$$

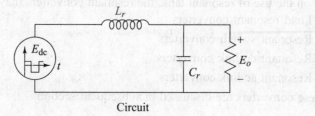

Circuit

Fig. 12.5(a)

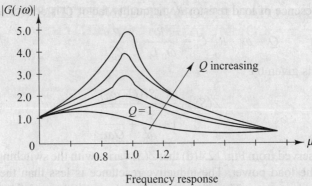

Fig. 12.5(b) *Parallel load resonant LC converter*

For the circuit of Fig. 12.5, the resonant frequency is given by

$$\omega_0 = \frac{1}{\sqrt{L_r \cdot C_r}} \qquad (12.26)$$

Also, the quality factor $(Q) = \dfrac{R}{\omega_0 L_r}$ \qquad (12.27)

Substituting ω_0 and Q in Eq. (12.25), we get

$$G(j\omega) = \frac{1}{\left[1-(\omega_s/\omega_0)^2\right]+j(\omega_s/\omega_0)Q}$$

Let us define, $\quad \mu = \dfrac{\omega_s}{\omega_0}, \quad \therefore \; G(j\omega) = \dfrac{1}{\left[(1-\mu^2)^2+(\mu/Q)^2\right]^{1/2}}$ \qquad (12.28)

The magnitude plot of the voltage-gain of Eq. (12.28) is shown in Fig. 12.5(b). The maximum gain occurs near resonance.

12.3 CLASSIFICATION OF RESONANT CONVERTERS

Based on the use of resonant tank, the resonant converters may be classified as
1. Load resonant converters
2. Resonant switch converters
3. Resonant-dc-link converters
4. Resonant ac link converters

All these converters are discussed in subsequent sections.

12.4 LOAD RESONANT CONVERTERS (SELF-COMMUTATING CONVERTERS)

In these converters, both series and parallel LC resonating circuits are used. Oscillating voltage and current due to LC resonance are applied to the load. The power flow to the load is controlled by the resonant tank impedance which in turn is controlled by the switching frequency (f_s) in comparison to the resonant frequency (f_0) of the tank. There are several types of self-or load resonating converters, used in both dc-to-dc and dc-to-ac converters. The series and parallel load resonating converters can be further classified as

  (i) Series/parallel resonant inverters with unidirectional switch.
  (ii) Series/parallel resonant inverters with bidirectional switches
  (iii) Class E resonant converter
  (iv) Class E resonant rectifier

12.4.1 Series Resonant Inverters with Unidirectional Switches

There are various configurations of series resonant inverters, depending on the connections of the switching devices and load. Series inverters discussed in Chapter 9 comes under this category of converters. Various circuit topologies such as basic series inverter, modified series inverter and half-bridge improved series inverter were discussed in Chapter 9. Resonant inverters were realized with SCR switches because the resonant load can be used to commutate the SCR. The operating frequency in an SCR inverter cannot be sufficiently high because of its high turn-off time. In modern converters, high frequency high power switching devices such as MOSFET, IGBTs or MCTs can be used.

Figure 12.6 shows the full-bridge series resonant inverter. For higher output power full-bridge circuit is normally used. When T_1 and T_2 are triggered simultaneously, the equation of current in the resonant circuit is given by

$$i(t) = \frac{E_{dc} + E_c}{\omega_r L_r} e^{-Rt/2L_r} \sin \omega_r t \qquad (12.29)$$

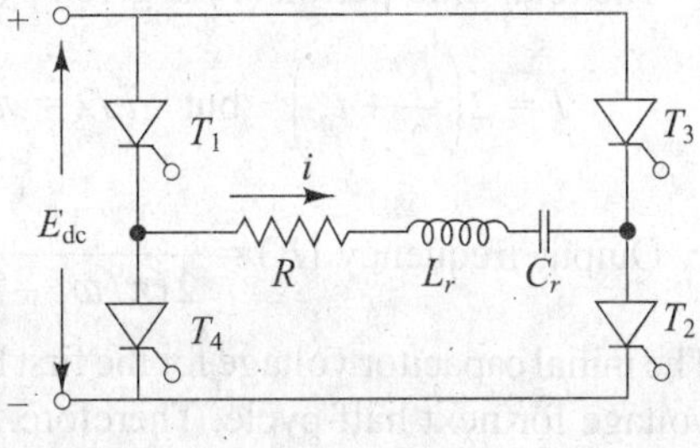

Fig. 12.6 *Full-bridge series resonant inverter*

where $-E_c$ is the initial voltage of the capacitor. This current flows for a half-cycle of the resonant frequency and becomes zero at $t = \dfrac{t_r}{2} = \pi/\omega_r$ and both T_1 and T_2 are turned-off.

The final capacitor voltage at $t = \dfrac{\pi}{\omega_r}$ is given by

$$E_{C1} = E_{dc} + (E_{dc} + E_c) \frac{-R\pi}{e^{2L_r\omega_r}} \qquad (12.30)$$

The final capacitor voltage in this half-cycle is the initial value for the next half-cycle.

At $t = t_2$, SCRs T_3 and T_4 are triggered. As shown in Fig. 12.7, the next half cycle begins after the expiry of the first half-cycle. This half cycle continues for T_1/ω_r when both T_3 and T_4 are turned-off.

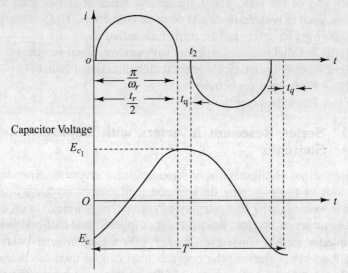

Fig. 12.7 *Current and voltage waveforms*

The total time-period (T) is given by

$$T = 2\left(\frac{t_r}{2} + t_q\right) \quad \text{but} \quad t_r/2 = \pi/\omega_r \therefore T = 2\left(\frac{\pi}{\omega_r} + t_q\right) \qquad (12.31)$$

$\therefore$ Output frequency $(F) = \dfrac{1}{2(\pi/\omega_r + t_q)}$ \hfill (12.32)

The initial capacitor voltage for the first half-cycle is the same as the final capacitor voltage for next half-cycle. Therefore, in Eq. (12.30), $E_{c_1} = E_c$ and hence

$$E_c = E_{dc} + (E_{dc} + E_c) \frac{-R\pi}{e^{2L_r\omega_r}}$$

Or, $\quad E_c = E_{dc} \dfrac{\left(1 + e^{\frac{-R\pi}{2L_r\omega_r}}\right)}{\left(1 - e^{\frac{R\pi}{2L_r\omega_r}}\right)} = E_{dc} \dfrac{\left(e^{\frac{R\pi}{2L_r\omega_r}} + 1\right)}{\left(e^{\frac{R\pi}{2L_r\omega_r}} - 1\right)}$ \hfill (12.33)

The average load current is given by

$$I_{L(\text{avg})} = \frac{2F(E_{\text{dc}}+E_c)}{\left[\omega_r^2 + (R/2L_r)^2\right]L_r}\left[1 + e^{\frac{-R\pi}{4\omega_r L_r}}\right] \quad (12.34)$$

Average thyristor current is given by

$$I_{\text{av(SCR)}} = \frac{1}{2} I_{L(\text{av})} \quad (12.35)$$

The rms load current is given by

$$I_{L(\text{rms})} = \frac{E_{\text{dc}}+E_c}{\omega_r L_r}\left[F\left(1 - e^{\frac{-R\pi}{2\omega_r L_r}}\right)\left\{\frac{L_r}{R} - \left(\frac{R}{L_r}\right)\frac{1}{(2\omega_r)^2 + \left(\frac{R}{L_r}\right)^2}\right\}\right]^{1/2} \quad (12.36)$$

The rms thyristor current is given by

$$I_{\text{rms(SCR)}} = \frac{I_{L(\text{rms})}}{\sqrt{2}} \quad (12.37)$$

Since the capacitor voltage (E_c) in full-bridge circuit is larger than that in half-bridge circuit, therefore, output power in this case is more than that of half-bridge circuit for the same input voltage and resonant frequency.

Solved Example

Example 12.1 The full-bridge series resonant inverter has $L = 40$ μH, $C = 6.8$ μF and $R = 3$ Ω. The frequency of output voltage is 6 kHz and the dc input is 200 V. SCR turn-off time is 15 μs. Calculate:
  (i) circuit turn-off time
  (ii) maximum possible output frequency
  (iii) capacitor voltage
  (iv) rms load current
  (v) output power
  (vi) average and rms SCR currents

Solution:
Given: $E_{\text{dc}} = 200$ V, $L_r = 40$ μH, $C_r = 6.8$ μF
$F = 6$ kHz, $t_q = 15$ μs, $R = 3$ Ω.
(i) For series resonant inverters (Chapter 9), we have

$$\omega_r = \sqrt{1/L_r C_r - R^2/4L_r^2} = \sqrt{\frac{1}{40 \times 6.8 \times 10^{-12}} - \frac{9}{4 \times 40 \times 40 \times 10^{-12}}} = 47{,}646 \text{ rad/s}.$$
$$(12.38)$$

But $\quad \omega_r = 2\pi f_r \therefore f_r = 7.58$ kHz.
and $\quad \omega_0 = 2\pi f_0 = 2 \cdot \pi \times 6000 = 37{,}699$ rad/s.

Also, circuit turn-off time (T_{off}) must be greater than the SCR turn-off time (t_q) and is given by (from Chapter 9),

$$T_{\text{off}} > t_q = \frac{\pi}{\omega_0} - \frac{\pi}{\omega_r} \qquad (12.39)$$

$$= \frac{\pi}{37{,}699} - \frac{\pi}{47{,}646} = 17.4 \ \mu\text{sec}.$$

(ii) From Eq. (12.32), maximum possible frequency is given by

$$F_{\max} = \frac{1}{2(t_q + \pi/\omega_r)} = \frac{1}{2\left(15 \times 10^{-6} + \dfrac{\pi}{47{,}846}\right)} = F_{\max} = 6.178 \text{ kHz}$$

(iii) Capacitor voltage from Eq. (12.33) is given by

$$E_c = 200 \left(\frac{e^{\frac{3\pi}{2 \times 40 \times 10^{-6} \times 47646}} + 1}{e^{\frac{3\pi}{2 \times 40 \times 10^{-6} \times 47646}} - 1} \right) = 235.77 \text{ V}$$

(iv) RMS load current from Eq. (12.36) is given by

$$I_{L\text{rms}} = \left(\frac{200 + 235.77}{47646 \times 40 \times 10^{-6}}\right) \left[(6.178 \times 10^3) \left(\frac{-3\pi}{1 - e^{2 \times 47646 \times 40 \times 10^{-6}}} \right) \right.$$

$$\left. \left\{ \left(\frac{40 \times 10^{-6}}{3}\right) - \left(\frac{3}{40 \times 10^{-6}}\right) \frac{1}{(2 \times 47646)^2 + \left(\dfrac{3}{40 \times 10^{-6}}\right)^2} \right\} \right]$$

$$= 49.38 \text{ A}$$

(v) Output power $P_0 = I_{L(\text{rms})}^2 \times R = (49.38)^2 \times 3 = 7315.15$ W

(vi) RMS thyristor current is given by (Eq. 12.37),

$$I_{rms(\text{SCR})} = \frac{49.38}{\sqrt{2}} = 34.91 \text{ A}$$

Average load current from Eq. (12.34) is given by

$$I_{L(\text{avg})} = \frac{2 \times 6.178 \times 10^3 (200 + 235.77)}{\left[(47{,}646)^2 + \left(\dfrac{3}{2 \times 40 \times 10^{-6}}\right)^2\right] \times 40 \times 10^{-6}} \left[1 + \frac{-3\pi}{e^{4 \times 47646 \times 40 \times 10^{-6}}}\right] = 47.65 \text{ A}$$

Average SCR current $I_{\text{avg(SCR)}} = \dfrac{1}{2} I_{L(\text{av})} = 1/2 \ (47.65) = 23.825$

12.4.2 Series Resonant Inverters with Bidirectional Switches

Series resonant inverters with unidirectional switches has number of drawbacks. As discussed, the power switches have to be turned-on in every half-cycle of output voltage. This limits the inverter frequency and the amount of energy transfer from the source to the load. Also, the major drawback is the high peak reverse voltage developed across the switches. This drawback can be overcome by connecting an antiparallel diode across a thyristor.

12.4.2.1 Half-Bridge Series Resonance Inverters with Bidirectional Switches

Figure 12.8 shows the half bridge circuit with antiparallel diodes. When thyristor T_1 is fired, current flows resonantly through the path $E_{dc+} - T_{1(A-k)} - L_r - R - C_{2+}$. With this capacitor C_2 charges and C_1 discharges. This current rises from zero to a peak value and falls to zero again at time t_1 (Fig. 12.9). The value of which depends on the resonant frequency of the circuit. Current flows in the opposite direction due to the oscillatory nature of the resonant circuit and this current flows through diode D_1 and reduces to zero at the end of the cycle at t_2. At t_1, SCR T_1 is self commutated.

The reverse voltage across the SCR is limited by the diode drop of about 1.0 V. Here, for successful operation, the conducting time of the diodes should be greater than the turn-off time of the SCRs: The output frequency can be made equal to half the resonant frequency. If the turn-off time of the SCR is t_q, the maximum resonant frequency is given by

$$F_{0max} = \frac{1}{2t_q} \qquad (12.40)$$

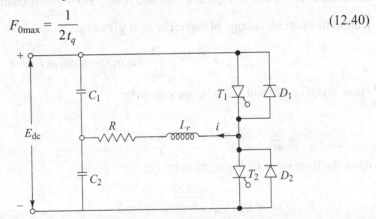

Fig. 12.8 Half-bridge inverter with antiparallel diodes

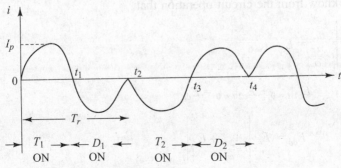

Fig. 12.9 Current waveforms

Similarly, SCR T_2 is triggered at t_2 and it becomes off at t_3. From period t_3 to t_4, diode D_2 conducts as shown in Fig. 12.9. Here, there are two current pulses per half-cycle instead of one through the load. Now, the load current waveform for T_2 conducting and T_1 off is given by

$$i = \frac{E_{dc} + E_{c1}(t=0)}{\omega_r L_r} \sin \omega_r \cdot t \cdot e^{-Rt/2L_r} \quad (12.41)$$

where $-E_{c1}|(t=0)$ is the initial voltage at capacitor c_1

For equilibrium condition, the initial voltage of C_1 is equal to the final voltage of C_2 and the initial voltage of c_2 is equal to the final voltage of C_1. At any instant, the sum of E_{c1} and E_{c2} will always be equal to supply voltage E_{dc}. Capacitor voltage E_{c1} is given by

$$E_{c1} = E_{dc} - \frac{(E_{dc} + E_c)}{\omega_r} e^{\frac{-Rt}{2L_r}} \left[\frac{R}{2L_r} \sin \omega_r t + \omega_r \cos \omega_r t \right] \quad (12.42)$$

This equation is valid for $0 < t \leq \dfrac{2\pi}{\omega_r}$ because the load current flows for one complete cycle before it becomes zero and finally at $t = \dfrac{\pi}{\omega_r}$ it changes its sign. During this period, voltage of capacitor c_2 is given by

$$E_{c2} = E_{c2}|_{t=0} \cdot e^{-Rt/2L_r} \left(\frac{R}{2L_r} \sin \omega_r t + \omega_r \cos \omega_r t \right) \Big/ \omega_r \quad (12.43)$$

From Eqs. (12.42) and (12.43), we can write

$$E_{c1}\big|_{t=\frac{2\pi}{\omega_r}} = E_{dc} - (E_{dc} + E_{c1}|_{t=0})\, e^{\frac{-R 2\pi}{2L_r \omega_r}} \quad (12.44)$$

Also, the final value of E_{c2} is given by

$$E_{c2}\big|_{t=\frac{2\pi}{\omega_r}} = E_{c2}|_{t=0} \cdot e^{\frac{-R 2\pi}{2L_r \omega_r}} \quad (12.45)$$

Now, we know from the circuit operation that

$$E_{c_2}\big|_{t=\frac{2\pi}{\omega_r}} = E_{c_1} \cdot t_{=0} \quad (12.46)$$

and
$$E_{c1}|_{t=0} + E_{c2}|_{t=0} = E_{dc} \quad (12.47)$$

$\therefore$
$$E_{dc} = E_{c_2}\big|_{t=\frac{2\pi}{\omega_r}} + E_{c_2}\big|_{t=0} \quad (12.48)$$

Or,
$$E_{c_2}|_{t=0}\left[1+e^{\frac{-R2\pi}{2L_r\omega_r}}\right]=E_{dc}$$

Or,
$$E_{c_2}|_{t=0}=\frac{E_{dc}}{1+e^{\frac{-R2\pi}{2L_r\omega_r}}} \quad (12.49)$$

Final value of E_{c_2}, i.e. $E_{c_2}\big|_{t=\frac{2\pi}{\omega_r}}$ becomes

$$E_{c_2}\bigg|_{t=\frac{2\pi}{\omega_r}}=E_{dc}-E_{c_2}|_{t=0}=E_{dc}\left[1-\frac{1}{e^{\frac{-R2\pi}{2L_r\omega_r}}}\right]$$

$$=E_{dc}\left[\frac{e^{\frac{-R2\pi}{2L_r\omega_r}}}{1+e^{\frac{-R2\pi}{2L_r\omega_r}}}\right] \quad (12.50)$$

Similarly,

$$E_{c_1}|_{(t=0)}=E_{dc}\left[\frac{e^{\frac{-R2\pi}{2L_r\omega_r}}}{1+e^{\frac{-R2\pi}{2L_r\omega_r}}}\right] \quad (12.51)$$

and
$$E_{c_1}\bigg|_{\left(t=\frac{2\pi}{\omega_r}\right)}=E_{dc}\left[\frac{1}{1-e^{\frac{-R2\pi}{2L_r\omega_r}}}\right] \quad (12.52)$$

In this case $E_{c_1}|_{(t=0)}$ is positive and not negative as assumed in Eq. (12.41)

If the conducting time of the diode is less than the turn-off time of the SCR, then a dead zone (t_d) should be provided as shown in Fig. 12.10 to allow the SCR

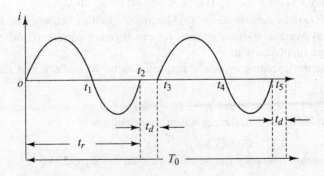

Fig. 12.10 *Current waveform with dead zone*

T_1 to reach the off-state. Normally, the reverse-conducting thyristor (RCT), where an asymmetric SCR and fast-recovery diode are fabricated together, is best for series resonant converters. Here, dead zone period is not necessary. Due to the turn-off time (12 μs to 25 μs) requirement of SCR, the maximum frequency of resonant inverters is restricted up to 10 kHz.

Figure 12.11 shows the half-bridge transistorized resonant inverter. This inverter can operate at the resonant frequency because the transistor switch needs microsecond or less to turn-off. Here switch S_2 can be turned-on almost instantaneously after switch S_1 is turned-off. With bidirectional switches, the current ratings of the devices are reduced. The average device current is half and the rms current is $1/\sqrt{2}$ of that for an inverter with unidirectional switches for the same output power.

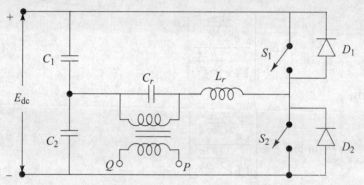

Fig. 12.11 *Half-bridge transistorised resonant inverter*

SOLVED EXAMPLES

Example 12.2 Half bridge series inverter with bidirectional switches (Fig. 12.8) has the following circuit parameters:
  (i) $C_1 = C_2 = 4\ \mu F$ (ii) $L_r = 40\ \mu H$ (iii) $R = 2\ \Omega$
  (iv) $E_{dc} = 120$ V.dc   (v) SCR turn-off time = 20 μs

Determine: (a) Resonant frequency, (b) Maximum operating frequency, (c) Peak thyristor current, (d) Average thyristor current, (e) rms thyristor current, (f) rms load current, (g) average supply current.

Assume the operating or output frequency to be 40% of resonant frequency.

Solution:
Given: $E_{dc} = 120$ V, $L_r = 40\ \mu H$, $R = 2\ \Omega$, $t_q = 20\ \mu s$
In the half-bridge circuit, the equivalent capacitance is

$$C_{eq} = C_1 + C_2 = 8\ \mu F$$

(a) Resonant frequency can be obtained from the Eq. (12.38)

$$\omega_r = \sqrt{\frac{1}{L_r C_{eq}} - \frac{R^2}{4L_r^2}} \quad \therefore\ f_r = \frac{1}{2\pi}\sqrt{\frac{1}{40\times 10^{-6} \times 8 \times 10^{-6}} - \frac{4}{4\times 40 \times 40 \times 10^{-12}}}$$

Resonant Converters 761

$= 7949.8 \text{ Hz} = 7.949 \text{ kHz}$

Period of resonant waveform, $t_r = 1/f_r = 0.126$ ms

(b) Maximum resonant frequency is obtained from Eq. (12.40), as

$$f_{r(max)} = \frac{1}{2 \times 20 \times 10^{-6}} = 25 \text{ kHz}$$

Now, output frequency $(f_0) = 0.4 f_r = 0.4 \times 7.949 \times 10^3 = 3.179$ kHz

$\therefore$ Output period $T_0 = \dfrac{1}{f_0} = \dfrac{1}{3.17 \times 10^3} = 0.315$ ms

Now, the period for the resonant waveform $(t_r) = \dfrac{1}{f_r} = \dfrac{1}{7.949 \times 10^3} = 0.125$ ms

Delay time $t_d = \dfrac{T_0}{2} - t_r$ (Fig. 12.10) $= \left(\dfrac{0.315}{2} - 0.125\right)$ ms $= 0.0325$ ms $= 32.5\ \mu s$

Since delay time is greater than zero, therefore this inverter circuit would operate in non-overlapping mode.

(c) Peak SCR current: This current is obtained by differentiating Eq. (12.41) and equating it to zero.

$$\therefore \quad \frac{di}{dt} = \omega_r \cos \omega_r \cdot t\ e^{\frac{-RT}{2L_r}} - \frac{R}{2L_r} \sin \omega_r \cdot t\ e^{\frac{-RT}{2L_r}} = 0$$

Or, $\tan \omega_r \cdot t = \dfrac{2\omega_r L_r}{R}$

$\therefore$ The instant at which the current reaches the peak $(t = t_p)$ is

$$t_p = \frac{1}{\omega_r} \tan^{-1}\left(\frac{2\omega_r L_r}{R}\right) = \frac{1}{2\pi \times 7949} \tan^{-1}\left(\frac{2 \times 2\pi \times 7949 \times 40 \times 10^{-6}}{2}\right)$$

$= (2.002 \times 10^{-5}) \tan^{-1}(1.9978) = 2.002 \times 10^{-5} \times \dfrac{63.409}{180} \times \pi = 22.15\ \mu s$

Now, initial capacitor voltage from Eq. (12.51) can be written as:

$$E_{c_1} = \frac{E_{dc}}{e^{\frac{R 2\pi}{2L_r \omega_r}+1}} = \frac{120}{e^{\frac{2 \times 2 \times \pi}{2 \times 40 \times 10^{-6} \times 2 \times \pi \times 7949}}+1} = \frac{120}{e^{3.145}+1} = 4.954 \text{ V}$$

Now, substitute t_p, and E_{c_1} in Eq. (12.41) to obtain peak current.

$$\therefore \quad I_p = \frac{E_{dc} + E_{c_1}}{\omega_r L_r} \sin(\omega_r \cdot t_p)\ e^{\frac{-Rt_p}{2L_r}}$$

$$= \frac{120 + 4.954}{2 \cdot \pi \times 7.949 \times 10^3 \times 40 \times 10^{-6}} \sin(2\pi \times 7.949 \times 10^3 \times 22.15 \times 10^{-6})(0.575)$$

$= 32.15$ A

(d) Average SCR current is obtained by integrating Eq. (12.41) we get

$$I_{av(SCR)} = \frac{1}{T_0} \int_0^{t_1} \frac{E_{dc} + E_{c_1}}{\omega_r L_r} \sin \omega_r t\ e^{\frac{-RT}{2L_r}}\ dt$$

$$= \frac{-(E_{dc}+E_{c_1})f_0}{\omega_r L_r}\left[e^{\frac{-RT}{2L_r}}\left(\frac{R}{2L_r}\sin\omega_r t\right)+\frac{\omega_r\cos\omega_r t}{\omega r^2+\left(\frac{R}{2L_r}\right)^2}\right]_0^{t_1}$$

Here $\quad t_1 = t_r/2$

$$\therefore \quad I_{av(SCR)} = \frac{(E_{dc}+E_{c_1})f_0}{\left[\omega_r^2+\left(\frac{R}{2L_r}\right)^2\right]L_r}\left[1+e^{\frac{-Rt_r}{4L_r}}\right]$$

$$= \frac{(120+4.954)\times 3.179\times 10^3}{\left[(49.945)^2\times 10^6+\left(\frac{2\times 10^6}{2\times 40}\right)^2\right]\times 40\times 10^{-6}}\left[1+e^{\frac{-2\times 0.126\times 10^{-3}}{40\times 4\times 10^{-6}}}\right]$$

$$= 3.842 \text{ A}$$

(e) rms thyristor current : rms current through the SCR is given by

$$I_{rms(SCR)} = \left[\frac{1}{T_o}\int_0^{t_1} i^2\,dt\right]^{1/2} = \frac{E_{dc}+E_{c_1}}{\omega_r L_r}\left[\frac{1}{T_o}\int_0^{\frac{t_r}{2}} \sin^2\omega_r t\, e^{\frac{-Rt}{L_r}}\,dt\right]^{1/2}$$

$$= \left(\frac{E_{dc}+E_{c_1}}{\omega_r L_r}\right)\left[\frac{1}{2T_0}\left(1-e^{\frac{-Rt_r}{2L_r}}\right)\times\left\{\frac{L_r}{R}-\frac{R}{L_r}\frac{1}{(2\omega_r)^2+\left(\frac{R}{L_r}\right)^2}\right\}\right]^{1/2}$$

$$= \left(\frac{120+4.954}{49.945\times 10^3\times 40\times 10^{-6}}\right)\left[\frac{1}{2\times 0.315\times 10^{-3}}\left(1-e^{\frac{-2\times 0.126\times 10^{-3}}{2\times 40\times 10^{-6}}}\right)\times\right.$$

$$\left.\left\{\frac{40\times 10^{-6}}{2}-\frac{2}{40\times 10^{-6}}\frac{1}{(2\times 49.945\times 10^3)^2+\left(\frac{2}{40\times 10^{-6}}\right)^2}\right\}\right]$$

$= 9.75 \text{ A}$

(f) RMS load current $= 2 \cdot I_{rms(SCR)}$
$$I_0 = 2\times 9.75 = 19.5 \text{ A}$$

(g) Average supply current,
Output power $P_o = I_0^2\times R = (19.5)^2\times 2 = 760.5$ W

$\therefore$ Average supply current, $I_s = \dfrac{P_o}{E_{dc}} = \dfrac{760.5}{120} = 6.34$ A

12.4.2.2 Full Bridge Series Resonant Inverters with Bidirectional Switches

Full-bridge series resonant inverters with bidirectional switches is shown in Fig. 12.12. When T_1 and T_2 are triggered, resonant current flows through the path E_{dc+}—T_1—R—L_r—C_r—T_2—E_{dc-}.
The instantaneous current is given by

$$L_r \frac{di}{dt} + R \cdot i + \frac{1}{C_r} \int i \, dt + E_{c(t=0)} = E_{dc} \tag{12.53}$$

with initial condition $i(t = 0) = 0$, $E_{c_1(t=0)} = -E_c$
The solution for the current is

$$i(t) = \frac{E_{dc} + E_c}{\omega_r L_r} \sin \omega_r \cdot t \, e^{-Rt/2L_r} \tag{12.54}$$

Capacitor voltage follows the equation

$$V_{c(t)} = E_{dc} - \frac{(E_{dc} + E_c)}{\omega_r} e^{-Rt/2L_r} \left(\frac{R}{2L_r} \sin \omega_r t + \omega_r \cos \omega_r t \right) \tag{12.55}$$

The SCRs T_1 and T_2 are turned-off at $t_1 = \dfrac{\pi}{\omega_r}$ when the circuit current becomes zero, and at t_1, the capacitor voltage E_c becomes

$$E_{c_1} = E_{c(t1)} = (E_{dc} + E_c) \, e^{\frac{-R \cdot \pi}{2L_r \, \omega_r}} + E_{dc} \tag{12.56}$$

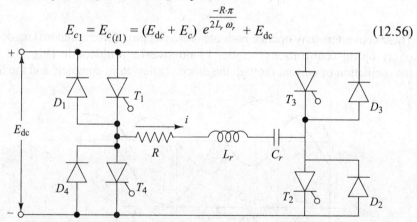

Fig. 12.12 *Full-bridge series inverter*

The second mode of operation begins when SCRs T_3 and T_4 are triggered. A negative load current (reverse resonant current) begins to flow from $t = t_3$. SCR current becomes zero at t_4 and diode current flows in the positive direction due to the oscillatory circuitry. In this half cycle, the current equation is

$$L_r \frac{di}{dt} + Ri + \frac{1}{C_r} \int i \, dt + E_{c(t=0)} = -E_{dc} \tag{12.57}$$

with the initial conditions at $i(t=0) = 0$ and $E_{c(t=0)} = E_{c_1}$ and the load current is

$$i(t) = -\left(\frac{E_{dc} + E_c}{\omega_r L_r}\right) \cdot \sin \omega_r t \cdot e^{\frac{-Rt}{2L_r}} \tag{12.58}$$

The capacitor voltage is

$$E_c = \frac{E_{dc} + E_{c_1}}{\omega_r} e^{\frac{-Rt}{2L_r}} \left(\frac{R}{2L_r} \sin \omega_r t + \omega_r \cos \omega_r t\right) - E_{dc} \tag{12.59}$$

SCRs T_3 and T_4 are turned-off at $t = \dfrac{\pi}{\omega_r}$ when current $i(t)$ becomes, zero.

∴ Substituting $t = \pi/\omega_r$ in Eq. (12.59), gives

$$E_c = -E_c(t = t_1) = (E_{dc} + E_{c_1}) e^{\frac{-R\pi}{2L_r \omega_r}} + E_{dc} \tag{12.60}$$

The initial voltage of capacitor is obtained from the condition that the capacitor voltage in the positive half-cycle $\left(\text{at } t = \dfrac{\pi}{\omega_r}\right)$ is the initial voltage of the capacitor for negative half-cycle. Thus, from Eqs (12.56) and (12.60), we can write

$$E_c = E_{c_1} = E_{dc} \frac{e^{\frac{R\pi}{2L_r \omega_r}} + 1}{e^{\frac{R\pi}{2L_r \omega_r}} - 1} \tag{12.61}$$

These converters may operate both in nonoverlapping (discontinuous) mode and overlapping (continuous) mode. In nonoverlapping mode (Fig. 12.13), the oscillation of current through the diode is allowed to complete and the SCR

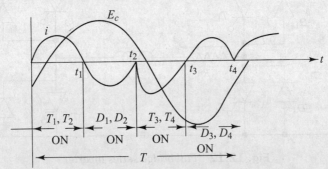

Fig. 12.13 Current and voltage waveforms (Non-overlapping mode)

(or switch) is triggered. In case of overlap mode, an SCR is triggered before the current through the diode ceases to zero, i.e. the current flowing through the diodes cannot complete the cycle and their current is transferred to the SCRs of the other branch (Fig. 12.14). In this case due to overlapping, the time period of one complete cycle is less than that at the resonance condition. Therefore, the operating frequency of series resonance inverter and its output power increases. Compared to half-bridge inverters, the output power is four-times and the device currents are twice in case of full-bridge inverters.

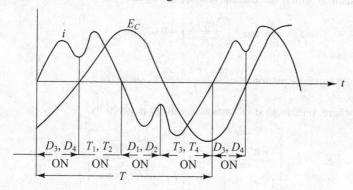

Fig. 12.14 *Capacitor voltage and current waveforms (Overlapping mode)*

SOLVED EXAMPLE

Example 12.3 Full-bridge series resonant inverter with bidirectional switches has the following circuit parameters:
  (i) $C_r = 8\ \mu F$ 　　　　　(ii) $L_r = 40\ \mu H$ 　　　　(iii) $R = 2\Omega$
  (iv) $E_{dc} = 120$ V dc 　　(v) SCR turn-off time = $20\ \mu s$

Determine: (a) Resonant frequency, (b) Maximum operating frequency, (c) Peak thyristor current, (d) Average thyristor current, (e) rms thyristor current, (f) rms load current, (g) average supply current.

Assume the operating or output frequency to be 40% of resonant frequency.

Solution:

Given: $C_r = 8\ \mu F$, $L_r = 40\ \mu H$, $R = 2\ \Omega$, $E_{dc} = 120$ V, $t_q = 20\ \mu s$.

(a) Resonant frequency f_r:

$$= \frac{1}{2\pi}\left(\frac{1}{L_r C_r} - \frac{R^2}{4L_r^2}\right)^{1/2}$$

$$= \frac{1}{2\pi}\left(\frac{10^{12}}{40 \times 8} - \frac{4 \times 10^{12}}{4 \times 40 \times 40}\right)^{1/2} = 7.96\ \text{kHz}$$

$\omega_r = 2\pi f_r = 50$ K rad/s and $f_0 = 0.4 \times f_r = 0.4 \times 7.96 \times 10^3 = 3.184$ kHz

∴ 　　$T_0 = 1/f_0 = 0.314$ msec and $t_r = 1/f_r = 0.126$ msec

∴ 　　$t_d = \dfrac{T_0}{2} - t_r = 0.031$ msec

Thus the mode of operation is nonoverlapping.

(b) Maximum operating frequency

$$f_{0(\max)} = 1/2\ (f_{r(\max)})\text{ where } f_{r\max} = \frac{1}{2t_q} = \frac{1}{2 \times 20 \times 10^{-6}} = 25\ \text{kHz}$$

$$\therefore \quad f_{0max} = \frac{1}{2} \times 25 \times 10^3 = 12.5 \text{ kHz}$$

(c) Instant at which the current reaches the peaks

$$t_p = \frac{1}{\omega_r} \tan^{-1}\left(\frac{2\omega_r L_r}{R}\right) = 0.022 \text{ ms}$$

the current equation, $i = \dfrac{E_{dc} + E_{c_1}}{\omega_r L_r} e^{\frac{-Rt}{2L_r}} \sin \omega_r t$

where, initial value of capacitor voltage is given by

$$E_{c_1} = E_{dc} \left[\frac{1 - e^{\frac{-R\pi}{L_r \omega_r}}}{1 + e^{\frac{-R\pi}{L_r \omega_r}}}\right] = 120 \left[\frac{1 - e^{\frac{-2\pi}{40 \times 10^{-6} \times 50 \times 10^3}}}{1 + e^{\frac{-2\pi}{40 \times 10^{-6} \times 50 \times 10^3}}}\right] = 110.11 \text{ V}$$

$\therefore$ peak current

$$I_p = \frac{E_{dc} + E_{c_1}}{\omega_r L_r} e^{\frac{-Rt_p}{2L_r}} \sin(\omega_r t_p)$$

$$= \left(\frac{120 + 110.11}{50 \times 10^3 \times 40 \times 10^{-6}}\right) e^{\frac{-2 \times 0.022 \times 10^{-3}}{2 \times 40 \times 10^{-6}}} \sin(50 \times 10^3 \times 0.022 \times 10^{-3}) = 59.084 \text{ A}$$

(d) Average SCR current:

$$I_{av(SCR)} = \frac{(E_{dc} + E_{c_1}) f_0}{\left[\omega_r^2 + \left(\frac{R}{2L_r}\right)^2\right] L_r} \left[1 + e^{\frac{-Rt_r}{4L_r}}\right]$$

$$= \frac{(120 + 59.084)(3.184 \times 10^3)}{\left[(50 \times 10^3)^2 + \left(\frac{2}{2 \times 40 \times 10^{-6}}\right)^2\right] \times 40 \times 10^{-6}} \left[1 + e^{\frac{-2 \times 0.126 \times 10^{-3}}{4 \times 40 \times 10^{-6}}}\right] = 5.47 \text{ A}$$

(e) Now, $I_{rms(SCR)} = \dfrac{E_{dc} + E_{c_1}}{\omega_r L_r} \left[\left(1 - e^{\frac{-Rt_r}{2L_r}}\right)\left(\dfrac{f_0}{2}\right)\left\{\left(\dfrac{L_r}{R}\right) - \left(\dfrac{R}{L_r}\right)\left(\dfrac{1}{(2\omega_r)^2 + \left(\dfrac{R}{L_r}\right)^2}\right)\right\}\right]^{\frac{1}{2}}$

$$= \left(\frac{120 + 59.084}{50 \times 10^3 \times 40 \times 10^{-6}}\right) \left[\left(1 - e^{\frac{-2 \times 0.126 \times 10^{-3}}{2 \times 40 \times 10^{-6}}}\right)\left(\frac{3.184}{2} \times 10^3\right)\right.$$

$$\left\{\left(\frac{40\times10^{-6}}{2}\right)-\left(\frac{2}{40\times10^{-6}}\right)\right\}\left\{\frac{1}{(2\times50\times10^3)^2+\left(\frac{2}{40\times10^{-6}}\right)^2}\right\}\right]^{\frac{1}{2}}=13.96\text{ A}$$

(f) RMS load current $I_0 = 2 \times I_{\text{rms(SCQ)}} = 27.92$ A

(g) Output power, $P_0 = I_0^2 R = 1559$ W ∴ Average supply current, $I_s = \dfrac{1559}{120} = 13$ A

12.5 PARALLEL RESONANT INVERTERS

A parallel resonant inverter circuit is shown in Fig. 12.15(a). Equivalent circuit Fig. 12.15(b) is obtained by referring the load resistance R_L into the primary side and also neglecting the leakage inductance of the transformer. The gating signals for the two switches are shown in Fig. 12.15(c). As shown, parallel resonance circuit is supplied by a dc current I_{dc}. Inductor L_s acts as a constant current source. Mutual inductance of the transformer (L_m) acts as a resonating inductor and capacitor C_r acts as a resonating element. A constant current is switched by switches S_1 and S_2 alternatively into the resonant circuit.

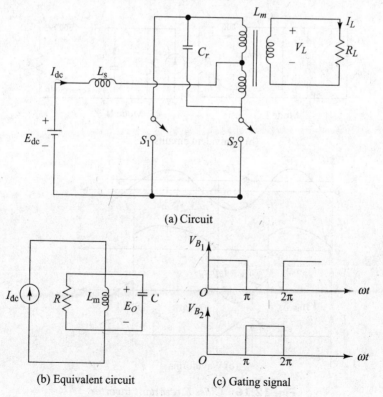

(a) Circuit

(b) Equivalent circuit

(c) Gating signal

Fig. 12.15 Parallel-resonant inverter

12.6 CLASS-E RESONANT INVERTERS

Class-E resonant inverter circuit is shown in Fig. 12.16(a). This converter uses only one switch. As shown, the input to the converter is through a sufficiently large inductor so that under steady state, the input to the converter is a dc current source (I_{dc}). The load is supplied through the sharply tuned series resonant circuit. This results in a sinusoidal current i_0. The topology without the diode across the switch is termed as the optimum and with the diode, suboptimum. The optimum switching scheme is described as follows:

Mode I : When the switch is turned-on, the current ($I_{dc} + i_0$) flows through the switch as shown in the equivalent circuit diagram (Fig. 12.16(b)). The converter is designed for a low-damping ratio ($\varepsilon \leq 0.07$) and a high-quality factor ($Q \geq 7$).

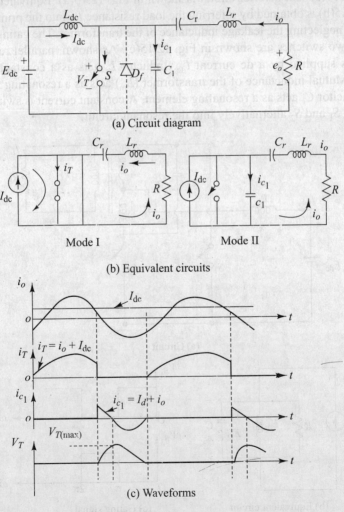

Fig. 12.16 Class-E resonant inverter

which results in almost sinusoidal load current (i_0). Now, the switch is turned-off at zero voltage. A slight variation (increase) in switching frequency ($\omega_s > \omega_0$), reduces the output voltage significantly.

Mode II : Switch S is turned-off during this mode of operation. Equivalent circuit is shown in Fig. 12.16(b). Current ($i_{c_1} = I_{dc} + i_0$) flows immediately through the capacitor c_1 when the switch is turned-off. The switch voltage rises from zero to a maximum value and falls to zero again. When $i_{c_1} = c_1 \, dv_T/dt$ becomes negative, the switch voltage falls to zero. Hence, switch voltage would tend to be negative. Antiparallel diode D_f will limit this negative voltage. It has been observed that the switch duty ratio $D = 0.5$ results in a maximum power capability, i.e. the maximum switch utilization ratio.

Switch utilization ratio is defined as the ratio of the output power P_0 to the product of the peak switch voltage and the peak switch current. Also, it has been found that the peak switch voltage is 3.5 E_{dc} and peak switch current is 3 I_{dc}.

Various current and voltage waveforms are shown in Fig. 12.16(c).

Mode III : During this mode, the switch voltage falls to zero with a finite negative slope. The equivalent circuit under this mode is similar to that for mode I, except the initial conditions. At the end of this mode, the load current falls to zero. When the switch voltage falls to zero with a zero slope ($i_{c_1} = 0$) then there will be no need for antiparallel diode and this mode would not exist, i.e., $V_T = 0$ and $dv_T/d_T = 0$.

In optimum mode, the design equations for the various circuit components with maximum efficiency is given by

$$L_{dc} = 0.4 \, \frac{R}{\omega_s} \tag{12.62}$$

$$c_1 = \frac{2.165}{R \cdot \omega_s} \tag{12.63}$$

and $\quad \omega_s \cdot L_r - \dfrac{1}{\omega_s \, c_r} = 0.353 \, R \tag{12.64}$

Class-E inverter is operated at a slightly higher frequency than the resonance frequency of the series resonant link. Output current is almost sinusoidal due to series resonance link. The main advantage of this inverter is the elimination of switching losses and the reduction in EMI. High peak voltage and current associated with the switch and large voltages and currents through the resonance LC elements are the significant disadvantages. Class-E inverter finds applications in high-frequency electronic lamp ballast.

SOLVED EXAMPLE

Example 12.4 Design a class-E resonance inverter for optimum values with following specifications:

(i) D.C. voltage E_{dc} = 24 V (ii) Load resistance R = 12 Ω
(iii) Switching frequency f_s = 20 kHz

Solution:
Given: $E_{dc} = 24$ V, $R = 12$ Ω, $\omega_s = 2\pi \times 20 \times 10^3 = 125663$ rad/s.

(i) *Selection of source inductance* (L_{dc}):

We have from Eq. (12.62), $L_{dc} = 0.4 \times R/\omega_s = \dfrac{0.4 \times 12}{125663} = 38.2\ \mu H$.

(ii) *Selection of resonance element:*

From Eq. (12.63), $C_1 = \dfrac{2.165}{R \cdot \omega_s} = \dfrac{2.165}{12 \times 125663} = 1.4\ \mu F$

Now, inductance L from series resonance operation is given by, $Q = \dfrac{\omega_s \cdot L_r}{R}$.

Let us assume $Q = 7$

$\therefore\ L_r = \dfrac{7 \times 12}{125663} = 668.5\ \mu H$

Capacitor C_r can be obtained from Eq. (12.64),

$$\omega_s \cdot L_r - \dfrac{1}{\omega_s C_r} = 0.353\ R$$

Or, $\omega_s \left(\dfrac{QR}{\omega_s}\right) - \dfrac{1}{\omega_s C_r} = 0.353\ R$

$7 \times 12 - \dfrac{1}{125663 \times C_r} = 0.353 \times 12$

$\therefore\ C_r = 0.10\ \mu F$

(iii) *Selection of resonance frequency:*

The damping factor is given by, $\varepsilon = R/2\ \sqrt{C_r/L_r}$

$= \left(\dfrac{12}{2}\right) \sqrt{\dfrac{0.10}{668.5}} = 0.0733$

The damping factor is very small. Hence, the output current will essentially be sinusoidal.

Now, resonant frequency, $f_0 = \dfrac{1}{2\pi\sqrt{L_r C_r}} = \dfrac{1}{2 \cdot \pi \sqrt{668.5 \times 10^{-6} \times 0.10 \times 10^{-6}}}$

$= 19.47$ kHz

12.7 CLASS-E RESONANT RECTIFIER

DC to DC converters basically consist of two conversion stages, namely, dc-to-ac resonant inverters and ac to dc rectifier.

Here, the high frequency diode rectifier suffers from the following drawbacks:
(i) High conduction and switching losses
(ii) High harmonic content of the input current and
(iii) high parasitic oscillations

All these limitations are overcome in a class-E resonant rectifier (Fig. 12.17a). In this rectifier, zero voltage switching principle of diode is used, i.e. in this circuit diode turns-off at zero voltage. In the resonant capacitance C_r, the diode junction capacitance is added and hence it does not adversely affect the circuit operation. As shown, input voltage $e_s = E_m \sin \omega t$ is applied to the circuit. For generating a constant dc output voltage E_0, capacitor C_1 of large value is selected. The circuit operation is explained in the following operating modes:

(i) **Mode I :** Diode D is OFF during this mode, components L_r and C_r are selected in such a way that at the operating frequency f,

$\omega \cdot L_r = 1/\omega \cdot C_r$. Equivalent circuit for this mode is shown in Fig. 12.17(b). Associated waveforms are shown in Fig. 12.17(c).

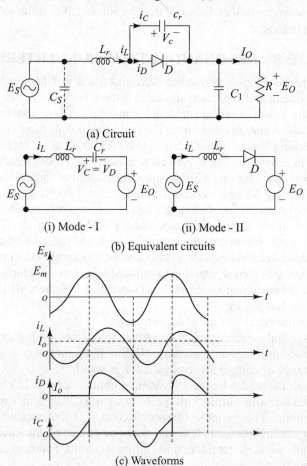

Fig. 12.17 *Resonant rectifier (class-E)*

(ii) **Mode II :** During this mode, diode D is ON. Now, when the diode current $(i_D = i_L)$ reaches zero, the diode turns-off. At turn-off, $i_D = i_L = 0$ and $V_D = V_C$

= 0. That is, $i_c = C_r \dfrac{dV_c}{dt} = 0$, which gives $\dfrac{dV_{C_r}}{dt} = 0$. Hence, at turn-off, diode voltage is zero which results in reduced switching losses.

Inductor current is given by, $i_L = I_m \sin(\omega_t - \phi) - I_0$ \hfill (12.65)

where: $\quad I_0 = \dfrac{V_0}{R}$ and $I_m = \dfrac{E_m}{R}$

Phase shift (ϕ) will be 90° when the diode is ON. Also, it will be 0° when the diode is OFF, provided that $\omega \times L_r = 1/\omega C_r$. Phase-shift value lies in the range 0 and 90° and its value depends on the load resistance (R). Peak to peak current is given by $2E_m/R$. Input current has a dc component I_0 and phase delay ϕ. An input capacitor (shown by dashed lines) C_s may be connected in order to improve the input power-factor.

12.8 RESONANT-SWITCH CONVERTERS

PWM square-wave dc-dc converters were discussed in Chapter 8. If resonant elements (inductor and capacitor) are added to these converters, then the resulting converters are known as resonant switch converters. These converters are also known as *quasi-resonant converters*. The main function of these converters is to reduce switching losses of the devices (MOSFETs, IGBTs, etc) by operating them either in zero voltage or in zero current switching mode. Hence, the current or voltage waveform becomes quasi-sinusoidal instead of square-wave in the dc-to-dc or dc-to-ac PWM converters. The other favourable features are:

(i) Since the switching losses are less, these circuits can be operated at comparatively high-frequencies than a PWM converter.
(ii) Also, since the operating frequency is high, the size and cost of the possible circuit elements are small leading to overall reduction of size.
(iii) Efficiency becomes high due to reduced losses in switching device.
(iv) Since the switching losses are less (virtually zero), the EMI/RFI problems are also less severe.

The main drawbacks of these converters are:

(i) Voltage and current stresses on switching devices are higher.
(ii) Conduction losses are also more (Higher peak currents).
(iii) The range of control for load variation is small.
(iv) In most cases, the frequency variation control scheme has to be adapted to regulate the output voltage both against load and supply voltage variations. The variable frequency control is undesirable because the ratings of the resonant elements should be guided by the lowest switching frequency and as a result optimal utilization of the resonant components is not possible.
(v) Radiated noise created by them is not easy to suppress.

In these converters, constant-frequency square wave or PWM control can be used with some additional constraints to provide zero-voltage and/or zero current switchings. Broadly, these converters can be classified in the following switching topologies:

(i) Zero-Current-Switching (ZCS) Topology : Here the switch turns-on and turns-off at zero current. Figure 12.18(a) shows switch types. As shown, in both types the circuit consists of switch, inductor and capacitor. In both types, inductor L_r limits the di/dt of the switch current, and L_r and C_r constitute a series-resonant circuit. When the switch current is zero, there will be a current $i = C_j \, dv_T/dt$ flowing through the internal capacitance c_j due to a finite slope of the switch voltage at turn-off. This current flow will cause power dissipation in the switch and limits the high switching frequency.

Since the devices do not turn-off at zero-current due to their recovery times, some energy will be trapped in the inductor of type-I circuit and voltage transients will appear across the switch. Hence type-I is preferred.

Figure 12.18(b) shows two circuit configurations. In half-wave configuration, unidirectional current flow is maintained by diode D whereas in full-wave configuration, switch current can flow bidirectionally.

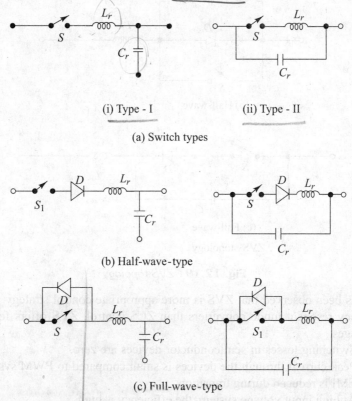

Fig. 12.18 *Zero-current switch converters*

(ii) Zero-Voltage Switching (ZVS) Topology : Here the switch turns-on and turns-off at zero voltage. ZVS circuit is shown in Fig. 12.19(a). To achieve ZVS, capacitor C_r is connected in parallel with the switch S. With the capacitor, C_r, the internal switch capacitance C_j is added and it affects the resonant frequency only. It is not contributing power dissipation in the switch. The ZVS can also be either of half-wave (unidirectional) or full-wave (bidirectional) type. These two types are shown in Fig. 12.19(b) and (c) respectively. In Fig. 12.19 (b), diode D is connected in antiparallel to switch. The voltage across C_r is clamped by D and the switch operates in half-wave configuration. In Fig. 12.19(c), diode D is connected in series with switch. The voltage across C_r can oscillate freely and the switch operates in full-wave configuration.

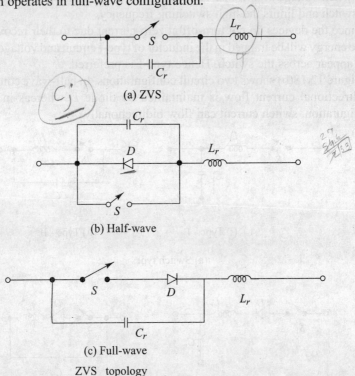

Fig. 12.19 *ZVS topology*

It has been observed that ZVS is more appropriate control strategy for high frequency resonant mode converters than ZCS control. ZVS offers following advantages:
  (i) Switching losses in semiconductor devices are zero.
  (ii) Peak current through the devices is small compared to PWM system.
  (iii) EMI is reduced during transition.
  (iv) At high input voltage system, the efficiency is high.
  (v) System can incorporate parasitic circuit and component inductance and capacitance.
  (vi) It can withstand short circuit conditions.
   Few disadvantages of ZVS are.
    (i) The off-state voltage across the switch is high.
    (ii) In general, they are regulated with variable frequency controller.

(iii) Zero-Voltage Multi-Resonant Switch : This configuration consists of more than one ZVS topology.

12.8.1 Zero-Current Switching Resonant Converters

In ZCS techniques, the turn-off losses of the switching devices are almost eliminated. Therefore, the converters can be operated at higher frequencies, in the range 1 MHz to 2 MHz. Figure 12.20(a) shows a ZCS resonant buck converter, where the switch of the simple buck converter is replaced by a type-I switch configuration of Fig. 12.18(a). Here the following assumptions are made for operating the load as a constant current sink (i.e. output voltage constant):
 (i) the filter inductance must be large ($L_f \gg L_r$).
 (ii) time constant of the output filter must be larger than the operating time period of the converter.

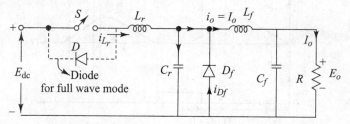

Fig. 12.20(a) *ZCS resonant buck converter*

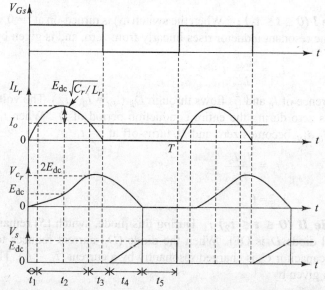

Fig. 12.20(b) *Waveforms for half-wave configuration*

The circuit operates in the half-wave mode if the switch is unilateral, i.e. there is no antiparallel diode across the switch S. It operates in full-wave mode if the switch is bilateral (i.e. diode is available across the switch).

A. Half-wave Mode : The associated waveforms for half-wave mode is shown in Fig. 12.20(b). The circuit operation is described in following operating modes:

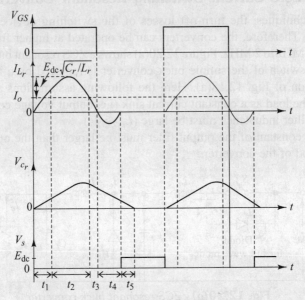

Fig. 12.20(c) *Waveforms for full-wave configuration*

(i) Mode I $(0 \leq t \leq t_1)$: When the switch (S) is turned-on at $t = 0$, the current through the resonant inductor rises linearly from zero, and is given by

$$i_{L_r} = \frac{E_{dc}}{L_r} t \qquad (12.66)$$

The difference of I_0 and i_{L_r} flows through D_F ($i_{Df} = I_0 - i_{Lr}$). The voltage across C remains zero during the entire conduction period of D_F. When i_{L_r} becomes equal to I_0, i_{DF} becomes zero and D_f turns-off at $t = t_1$.

$$\therefore \qquad t_1 = \frac{I_0 L_r}{E_{dc}} \qquad (12.67)$$

and $\qquad V_{c_r} = 0 \qquad (12.68)$

(ii) Mode II $(0 \leq t \leq t_2)$: During this mode, switch (S) remains ON but freewheel diode D_f is OFF. When the diode (D_f) current reduces to zero, the resonant capacitor C_r is charged resonantly by a current ($I_{L_r} - I_0$). The inductor current is given by

$$i_{Lr} = I_0 + \frac{E_{dc}}{Z_r} \sin \omega_r t \qquad (12.69)$$

where $\qquad Z_r = \sqrt{\frac{L_r}{C_r}} \qquad (12.70)$

and $\omega_r = \dfrac{1}{\sqrt{L_r \cdot C_r}}$ (12.71)

The capacitor voltage V_{Cr} is given by

$$V_{Cr} = E_{dc}(1 - \cos \omega_r t)$$ (12.72)

The peak switch current which occurs at $t = \dfrac{\pi}{2}\sqrt{L_r C_r}$ is given by

$$I_p = E_{dc}\sqrt{\dfrac{C_r}{L_r}} + I_0$$ (12.73)

The peak capacitor voltage is

$$V_{c(peak)} = 2 E_{dc}$$ (12.74)

The condition for zero current switching is

$$Z_r \leq \dfrac{E_{dc}}{I_0}$$ (12.75)

(iii) Mode III ($0 \leq t \leq t_3$) : During t_3 period, the voltage across capacitor and the current through the inductor are given by

$$i_{L_r} = I_0 - E_{dc}\sqrt{\dfrac{C_r}{L_r}} \sin \omega_r \cdot t$$ (12.76)

$$V_{c_r} = 2 E_{dc} \cdot \cos \omega_r \cdot t$$ (12.77)

At the end of this mode ($t = t_3$), $i_{L_r} = 0$ and $V_{c_r} = V_{c_3}$.

$\therefore$ From Eq. (12.76), we can write, $I_0 = E_{dc}\sqrt{\dfrac{C_r}{L_r}} \sin \omega_r \cdot t_3$

or $\quad t_3 = \sqrt{L_r c_r}\ \sin^{-1}\left(\dfrac{I_0}{E_{dc}}\sqrt{\dfrac{L_r}{c_r}}\right)$ (12.78)

(iv) Mode iv ($0 \leq t \leq t_4$) : During this period, the switch is OFF. The capacitor C_r begins to discharge through the output with constant output current (I_0) and V_{c_r} decreases linearly. Thus, the capacitor voltage is given by

$$V_{c_r} = V_{c_{r_3}} - \dfrac{I_0}{c_r} t$$ (12.79)

This mode ends at $t = t_4$ when $V_{c_r} = 0$

$\therefore$ From Eq. (12.79), we get

$\therefore \quad t_4 = \dfrac{V_{c_{r_3}} \cdot C_r}{I_0}$ (12.80)

(v) Mode V ($0 \leq t \leq t_5$) : At the begining of t_5, V_{c_r} tends to become negative due to the resonating $L_r \cdot C_r$ circuit. Therefore, D_f conducts and the load current I_0 flows through D_f. The peak resonating current must be higher than I_0 for a half-wave ZCS converter. This mode ends at time $t = t_5$ when the switch S is turned-ON again and the next cycle starts.

(B) Full-wave Mode Normally, an antiparallel diode (D) is also fabricated across the switching device (Fig. 12.20(a)). When this type of switch is used, then it is called a full-wave configuration, where the current reversal is possible. In the half-wave mode, we have seen that the switch does not allow negative current to flow and the inductor current i_{L_r} becomes zero when switch is turned-off, whereas, in the full-wave mode, current can flow in the negative direction and the current becomes oscillatory and the energy is fed-back to the source through the antiparallel diode (D). This current reduces to zero at $t = t_4$ when the negative half-cycle is complete. The current cannot flow further because the switch is open by that time. The associated waveforms are shown in Fig. 12.20 (c). Because of antiparallel diode across the switch, output voltage is insensitive to load variations.

In zero current switching, the turn-off losses of the power switches are almost eliminated but the turn-on losses are quite large. This is because when the device is turned-on, the energy stored by the output capacitance of the power-MOSFET is dissipated inside the device.

This loss is proportional to the switching frequency and therefore, at higher frequencies the loss becomes significantly large and overall efficiency is reduced. Due to large rms current flowing through the tank circuit and the semiconductor switch, the conduction losses are large. While designing the circuit, care must be taken for these losses.

ZCS technique can also be implemented in various converter circuit configurations, such as, forward, flyback and half-bridge. Normally, electrical isolation is required between the source and the load and this is done by an isolating transformer. The resonant circuit can be added either in the primary or in the secondary of the isolating transformer. All these configurations can operate either in half or in full-wave mode depending on whether the power switch is unidirectional or bidirectional. The half-bridge ZCS and full-wave half-bridge ZCS circuit topologies are shown in Fig. 12.21.

SOLVED EXAMPLE

Example 12.5 ZCS resonant converter operates with $E_{dc} = 18$ V, $E_0 = 12$ V, $f_r = 1$ MHz and $P_0 = 12$ W. Determine the following:
   (a) values of resonant components L_r and C_r (b) peak switch current
   (c) peak voltage rating of the resonant capacitor C_r

Solution:
Given: Unregulated d.c. (E_{dc}) = 18 V, Output voltage (E_0) = 12 V, $f_r = 1$ MHz and $P_0 = 12$ W

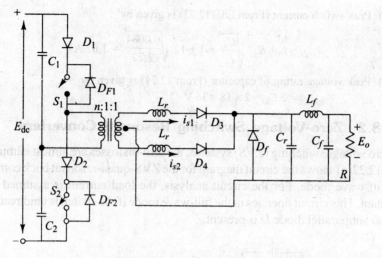

Fig. 12.21(a) *Half-bridge ZCS converter*

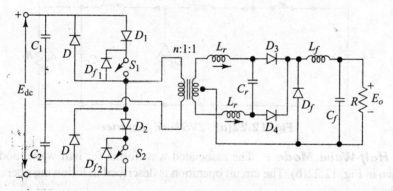

Fig. 12.21(b) *Full-wave half-bridge ZCS converter*

(a) We have $P_0 = E_0 \cdot I_0$; $12 = 12 \times I_0$ ∴ $I_0 = 1$ A
The condition of zero current switching (Eq. 12.75) is

$$Z_r \leq \frac{E_{dc}}{I_0} \leq \frac{18}{1} = 18$$

Also, from Eq. (12.70), $Z_r = \sqrt{\dfrac{L_r}{C_r}} = 18$

and from Eq. (12.71), $f_r = \dfrac{1}{2\pi\sqrt{L_r \cdot C_r}}$ ∴ $1 \times 10^6 = \dfrac{1}{2\pi\sqrt{L_r \cdot C_r}}$

From above, $L_r = 2.72\ \mu\text{H.}$, $C_r = 0.0084\ \mu\text{F}$

(b) Peak switch current (From Eq. (12.73)) is given by

$$i_{s(p)} = I_0 + E_{dc}\sqrt{\frac{C_r}{L_r}} = 1 + 12\sqrt{\frac{0.0084}{2.72}} = 1.67 \text{ A}$$

(c) Peak voltage rating of capacitor (From 12.74) is given by
$$V_{c(peak)} = 2E_{dc} = 2 \times 18 = 36 \text{ V}$$

12.8.2 Zero-Voltage Switching Resonant Converters

In zero voltage switching (ZVS) systems, the turn-on losses are almost eliminated. Fig. 12.22(a) shows the circuit diagram for the ZVS-quasi-resonant buck converter in half-wave mode. For the circuit analysis, the load current is assumed to be constant. This circuit operates in the full-wave mode if the switch is unidirectional, i.e. no antiparallel diode D is present.

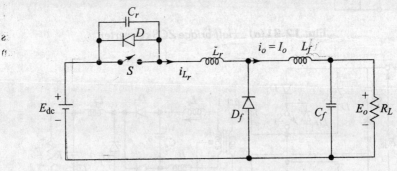

Fig. 12.22(a) *ZVS buck converter*

(A) Half-Wave Mode : The associated waveforms for half-wave mode is shown in Fig. 12.22(b). The circuit operation is described in following operating modes:

(i) Mode I $(0 \leq t \leq t_1)$: In this mode, when dc input E_{dc} is applied, capacitor C_r gets charged through the path $E_{dc+} - C_{r+} - C_{r-} - L_r - L_f - R_L - E_{dc-}$, at a constant load current I_0. The capacitor charging voltage is given by $V_{c_r} =$

$$\frac{1}{C_r}\int i\,dt \quad \therefore \quad V_{c_r} = \frac{I_0}{C_r}t \tag{12.81}$$

At the end of this mode, $t = t_1$, and $V_{c_r}' = E_{dc}$

$\therefore$ From equation (12.81), $t_1 = \dfrac{E_{dc}\,C_r}{I_0}$ \hfill (12.82)

(ii) Mode II $(0 \leq t \leq t_2)$: During this period, diode D_f turns-on and L_r and C_r starts resonating. The tank circuit current follows a cosine function and is given by

$$i_{L_r} = I_0 \cos \omega_r \cdot t = i_{C_r} \tag{12.83}$$

The capacitor voltage V_{c_r} is given by

$$V_{c_r} = E_m \sin \omega_r t + E_{dc} \tag{12.84}$$

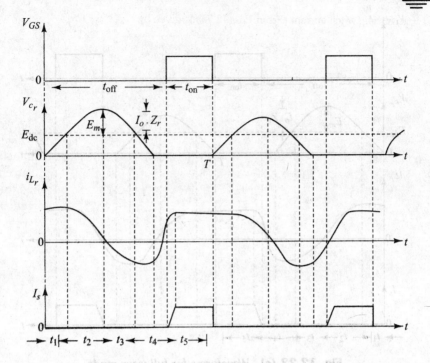

Fig. 12.22(b) *Waveforms for half-wave mode*

where $\qquad E_m = I_0 \sqrt{L_r / C_r} \qquad$ (12.85)

At $t = \dfrac{\pi}{2} \sqrt{L_r C_r}$ peak switch voltage occurs and is given by

$$V_{s(p)} = V_{c_r(p)} = E_{dc} + I_0 \cdot Z_r \qquad (12.86)$$

where Z_r = characteristic impedance of the tank circuit = $\sqrt{\dfrac{L_r}{C_r}}$

At the end of this mode, $t = t_2$ and $V_{c_r} = E_{dc}$ and $i_{Lr} = -I_0$.

Therefore, $t_2 = \pi \sqrt{L_r C_r}$.

(iii) Mode III ($0 \le t \le t_3$) : During this mode, the capacitor discharge from voltage E_{dc} to zero and is given by

$$V_{c_r} = E_{dc} - E_m \sin \omega_r \cdot t \qquad (12.87)$$

Also, the inductor current i_{L_r} is given by

$$i_{L_r} = -I_0 \cos \omega_r \cdot t \qquad (12.88)$$

At the end of this mode, $t = t_3$ and $V_{c_r} = 0$ and $i_{L_r} = I_{L_{r3}}$

∴ From Eq. 12.87, $0 = E_{dc} - E_m \sin \omega_r \cdot t_3$ ∴ $E_{dc} = E_m \sin \omega_r \cdot t_3$

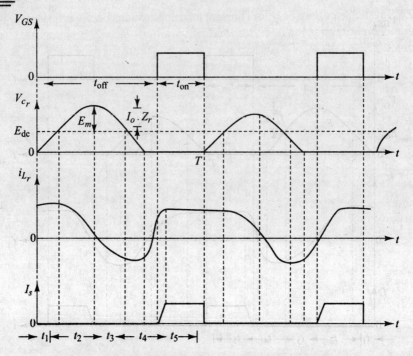

Fig. 12.22 (c) *Waveforms for full wave mode*

But $E_m = I_0 \sqrt{\dfrac{L_r}{C_r}}$ and $\omega_r = \sqrt{L_r \cdot C_r}$

$$\therefore E_{dc} = I_0 \sqrt{\dfrac{L_r}{C_r}} \sin \sqrt{L_r C_r}\; t_3 \text{ or } t_3 = \sqrt{L_r C_r}\; \sin^{-1}\left(\dfrac{E_{dc}}{I_0}\sqrt{\dfrac{C_r}{L_r}}\right) \quad (12.89)$$

(iv) Mode IV ($0 \le t \le t_4$): During this mode, switch S is turned-on and diode D_f remains ON. Inductor current i_{L_r} rises linearly from $I_{L_{r3}}$ to I_0 and is given by

$$i_{L_r} = i_{L_{r3}} + \dfrac{E_{dc}}{L_r} t \quad (12.90)$$

At the end of this mode, $t = t_4$ and $i_{L_r} = 0$.

Therefore, from Eq. (12.90), we get, $0 = (i_{L_{r3}} - I_0) + \dfrac{E_{dc}}{L_r} t_4$

$$\therefore \quad t_4 = (I_0 - i_{L_{r3}}) \dfrac{L_r}{E_{dc}} \quad (12.91)$$

(v) Mode V ($0 \le t \le t_5$): During this mode, switch S is ON and diode D_f is OFF. The load current I_0 flows through the switch S. At the end of this mode, $t = t_5$ and switch S is turned off and cycle is repeated.

ZVS systems are used only for constant load applications because the variations in the load current results in corresponding variations in switch voltages. Here, switch is turned-on at zero voltage only. Before making the switch-on, diode D must conduct.

(B) Full Wave-Mode : In this mode, the capacitor voltage (V_{cr}) goes negative for certain period as shown in Fig. 12.22(c). Here, the switch should be turned-on as long as the voltage across it is negative to ensure zero-voltage switching. In the half-wave mode of operation, the parasitic antiparallel diode of the MOSFET-comes into conduction in the negative half-cycle of the resonant oscillation, i.e., when the capacitor voltage reduces to zero then the antiparallel diode conducts. In the full-wave mode, the conduction of the switch in the negative half-cycle should be stopped and hence a diode in series with the MOSFET has to be incorporated to provide reverse blocking capability. The diode will prevent discharging of the switch output capacitance during the resonance. As a result, the energy stored in the junction capacitances of the semiconductor switch is trapped during the off-time and is dissipated internally after the switch turns-off. Hence, the full-wave mode of ZVS system suffers from the turn-on losses and dv/dt noise as in PWM converters.

ZVS technique can also be implemented in various converter circuit configurations, such as forward, flyback, half-bridge and full bridge.

12.8.2.1 ZVS Forward Converter

Figure 12.23 shows the ZVS forward converter. The switching frequency varies according to the requirement satisfying the zero-voltage switching condition. The values of the resonating components for ZVS are:

$$L_r \geq \frac{1}{2\pi f_r} \frac{N}{I_{0(\min)}} E_{dc(\max)} \qquad (12.92)$$

$$C_r \leq \frac{1}{2\pi \cdot f_r} \frac{I_{0(\min)}}{N} \frac{1}{E_{dc(\max)}} \qquad (12.93)$$

where N = transformer turns ratio.

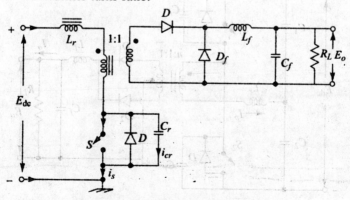

Fig. 12.23 *ZVS forward converter*

12.8.2.2 ZVS Flyback Converter

The basic circuit diagram of the flyback ZVS converter is shown in Fig. 12.24. This circuit is similar to that of flyback converter except for the mutual polarity of the transformer and the output rectifier connection. L_r includes transformer leakage inductance and capacitor C_r includes the capacitance of the device. The values of L_r and C_r can be obtained from the following equations:

$$Z_r = \text{Characteristic impedance of resonant circuit} = \sqrt{\frac{L_r}{C_r}} \quad (12.94)$$

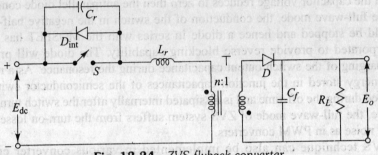

Fig. 12.24 *ZVS flyback converter*

and
$$f_r = \frac{1}{2\pi\sqrt{L_r C_r}} \quad (12.95)$$

The major limitation of these single-ended converter configurations is that the main switching device has to suffer from high-voltage stress.

12.8.2.3 ZVS Half-Bridge Converter

The circuit diagram of the half-bridge ZVS converter is shown in Fig. 12.25. The major drawback of previous circuit configurations is overcome by the use of half-bridge, full-bridge and push-pull circuit configurations.

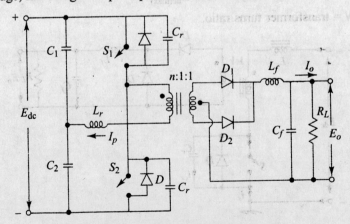

Fig. 12.25 *ZVS half-bridge converter*

The ZVS condition for this configuration is given by

$$\frac{E_{dc}}{2} \leq \frac{I_0 \cdot Z_r}{N} \qquad (12.96)$$

$$\omega_r = \frac{1}{\sqrt{2L_r \cdot C_r}} \qquad (12.97)$$

$$Z_r = \sqrt{\frac{L_r}{2C_r}} \qquad (12.98)$$

12.8.2.4 ZVS Full-Bridge PWM Converter

The circuit diagram of the full-bridge ZVS PWM converter is shown in Fig. 12.26. This topology has the advantages of both the ZVS converters and PWM techniques, therefore it is used in many applications. Here, the switching losses are minimum because of ZVS and relatively low current stress of the power switches. This circuit operates at a fixed frequency and output regulation is achieved by PWM technique.

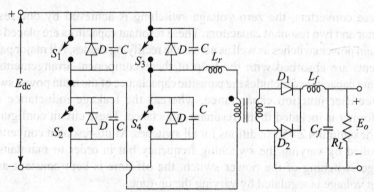

Fig. 12.26 Full-bridge ZVS PWM converter

SOLVED EXAMPLE

Example 12.6 ZVS buck converter operates in half-wave mode with $E_{dc} = 35$ V, $E_0 = 24$ V, $f_r = 500$ kHz and $P_o = 24$ W.

Determine the following:
(i) Values of resonant component L_r and C_r
(ii) Peak voltage rating of the resonant capacitor
(iii) Charging and discharging period of capacitor

Solution:
Given: $E_{dc} = 35$ V, $E_0 = 24$ V, $f_r = 500$ kHz and $P_o = 24$ W
(i) We have $P_o = E_0 \cdot I_0$ ∴ $24 = 24 I_0$, $I_0 = 1$A

Now, $Z_r = \sqrt{\dfrac{L_r}{C_r}} \leq \dfrac{E_{dc}}{I_0} = \dfrac{35}{1} = 35$

Also, $f_r = \dfrac{1}{2\pi\sqrt{L_r C_r}} = 500 \times 10^3$, $\therefore L_r = 11.03\,\mu\text{H}$, $C_r = 0.0090\,\mu\text{F}$

(ii) From Eq. (12.86), peak voltage rating of resonant capacitor is given by:

$$V_{S(p)} = E_{dc} + I_0 \cdot Z_r = 35 + 1 \times 35 = 70\text{ V}$$

(iii) From Eq. (12.82), the linear charging period of the capacitor is given by

$$t_1 = \dfrac{E_{dc(min)} C_r}{I_0} = \dfrac{35 \times 0.0090 \times 10^{-6}}{1} = 0.315\,\mu\text{sec}.$$

From Eq. (12.89), the constant discharging period of capacitor is given by

$$t_3 = \sqrt{L_r C_r}\,\sin^{-1}\left(\dfrac{E_{dc(min)}}{I_0}\sqrt{\dfrac{C_r}{L_r}}\right)$$

$$= \sqrt{11.03 \times 10^{-6} \times 0.0090 \times 10^{-6}}\,\sin^{-1}\left(\dfrac{35}{1}\sqrt{\dfrac{0.0090 \times 10^{-6}}{11.03 \times 10^{-6}}}\right) = 0.315\,\mu\text{sec}$$

12.8.3 Zero Voltage Switched Multi-resonant Converter

In these converters, the zero voltage switching is achieved by one resonant inductor and two resonant capacitors. These resonant capacitors are placed across the main power switches as well as across the rectifying diodes. All major parasitic elements are absorbed with the help of these component-arrangements. The resonant capacitance includes the parasitic capacitance of the main power switches and rectifier junction capacitance, whereas the leakage inductance of the transformer is included in the resonant inductance. These circuit configurations provide optimum ZVS conditions for all switches. Multi-resonant converters are controlled by varying the switching frequency but in order to maintain zero-voltage switching of the power switch, the off-time is kept constant and the output voltage is regulated by varying the on-time.

12.8.3.1 Buck-Converter

Figure 12.27 shows the basic circuit arrangement of ZVS multi-resonant converter. The voltage stress on the power switches is much less in this configuration compared to ZVS buck converter of Fig. 12.22. The output filter has a large time constant.

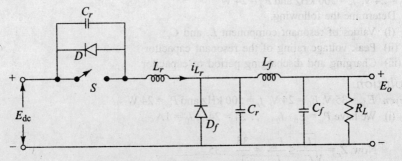

Fig. 12.27 *ZVS-multi-resonant converter*

12.8.3.2 Forward ZVS Multi-Resonant Converter

The basic circuit configuration for the forward ZVS multi-resonant converter is shown in Fig. 12.28. This circuit is similar to ZVS with the addition of second resonant capacitor C_{r2} across the secondary of the transformer. Here, power switch S and rectifiers D_2 and D_3 operate with zero voltage switching. This configuration is useful as the on-card regulator in a distributed power system where the main supply is provided by a separate source.

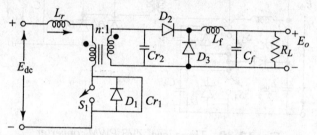

Fig. 12.28 Forward ZVS multi-resonant converter

12.8.3.3 Half-Bridge ZVS Multi-Resonant Converter

The basic circuit diagram of this topology is shown in Fig. 12.29. This circuit diagram is similar to that of Fig. 12.25, with the addition of two resonant capacitors across the rectifying diodes D_1 and D_2. These capacitors include the junction capacitors of the diodes. The topology is useful for high-power density converters and wide variations of supply voltage and load current.

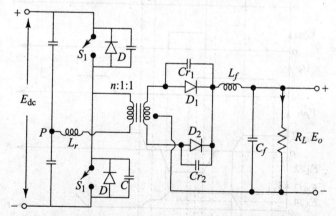

Fig. 12.29 Half-bridge ZVS multi-resonant converter

12.9 ZVS THREE-LEVEL PWM-CONVERTER

The circuit diagram of the three-level ZVS PWM converter is shown in Fig. 12.30. Phase-shifted PWM control is used to achieve zero voltage switching and output voltage regulation. In this configuration, the voltage stress on the main switch is just the half of the supply voltage, hence it is used in high-power

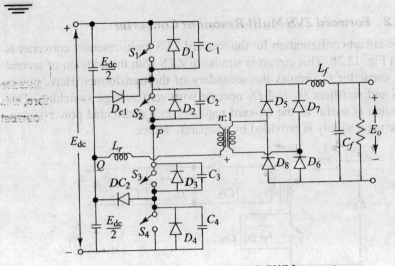

Fig. 12.30 *Three-level ZVS PWM-converter*

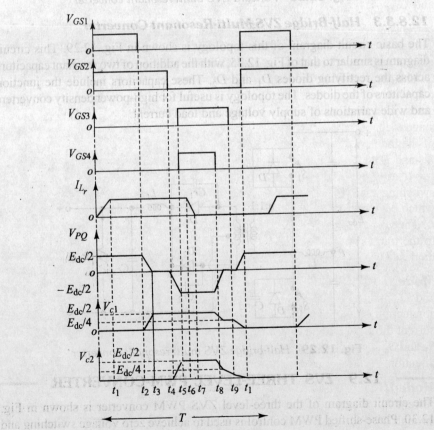

Fig. 12.31 *Voltage and current waveforms*

applications. Resonant commutation inductor L_r also includes the leakage inductance of the power transformer. Bridge rectifier consists of diodes D_5–D_8. Diodes D_1, D_2, D_3, D_4 are the body diodes of the switch (MOSFET). Since the switching frequency is high, no separate diodes or capacitors are required for zero voltage switching. The associated voltage and current waveforms are shown in Fig. 12.31 capacitors C_1, C_2, C_3 and C_4 are the intrinsic capacitors of the switches (MOSFETs). The operation of this circuit is explained in the following modes:

(i) Mode I : At t_1 switches S_1 and S_2 are turned-on simultaneously. Current will be drawn from $E_{dc}/2$ through S_1, S_2, transformer T and inductor L_r. This supply voltage ($E_{dc}/2$) is impressed on the transformer primary and load current flows through rectifier diode D_5 and D_6. This way, the power is transferred from supply to load. When S_1 and S_2 are ON, the voltage across them is zero and consequently the voltage across C_1 and C_2 is also zero. During this mode, S_3 and S_4 are OFF. Therefore, voltage across them will be E_{dc}. Also, voltage across C_3 and C_4 will be $\dfrac{E_{dc}}{2}$ each. Transformer primary current is equal to the load current referred to the primary and becomes I_0/N, where N is the turns ratio. This is because the output filter is sufficiently large to be approximated by a current source with a value equal to load current.

The above operation continues till the time $t = t_3$ at which switch S_1 is turned-off. With this, capacitor C_1 get charged to $E_{dc}/2$ from zero voltage and voltage across C_3 and C_4 ($V_{C3} + V_{C4}$) gradually drops from E_{dc} to $E_{dc}/2$. At $t = t_4$, this stage is complete and clamping diode D_{C1} starts conducting and the stored energy of the inductor L_r is discharged and load current flows through rectifier diodes D_5 and D_6. At this instant free-wheeling operation starts and continues up to $t = t_5$ at which switch S_2 is turned-off. Switch S_2 is also turned-off in soft switching manner (similar to S_1) because current is transferred from S_2 to C_2 which begins to charge up gradually increasing its voltage V_{c2} up to $E_{dc}/2$ at $t = t_6$. The voltage across C_3 and C_4 ($V_{C3} + V_{C4}$) decreases towards zero. This is due to the transition of the primary voltage of the transformer and all the rectifier diodes (D_5–D_8) conducts and the output is short circuited.

Mode II : During the period, $t_6 - t_7$, the inductor current I_{L_r} flows through D_4 and D_3 and decreases linearly. Switches S_3 and S_4 are turned-on during the conduction period of D_3 and D_4 at zero voltage and zero current. After $t = t_7$, current through L_r reduces to zero and diodes D_3 and D_4 turns-off and switches S_3 and S_4 start conducting and the current I_{L_r} increases linearly in the reverse-direction. At $t = t_8$, this condition is complete.

Mode III : After the instant t_7, the inductor current reaches the reflected load current I_0. Rectifier diodes D_5–D_6 turns-off and the load current is supplied through D_7 and D_8. Through S_3 and S_4, the load current is supplied by the lower half of the supply voltage $E_{dc}/2$ and power is transferred from the source to the load. This operation continues till t_9 when switch S_4 is turned-off and the process of the second half-period starts which is identical to the first one. It has been observed that the output current freewheels through switch S_2, D_{C_1}, L_r and the output rectifiers during the period $(t_4 - t_5)$. Thus, the output voltage depends on the conduction period of S_1 and S_4. Therefore, the output voltage can be regulated by altering the duration of gate commands of S_1 and S_4 keeping those of S_2 and S_3 unaltered, i.e. by PWM method.

Mathematical Analysis:

(i) Average Output Voltage Average output voltage (E_0) can be obtained from Fig. 12.25, is

$$E_0 = \frac{2 \frac{E_{dc}}{2}(t_3 - t_2)}{n T_s} = \frac{E_{dc}(t_3 - t_2)}{n T_s} \qquad (12.99)$$

where: n = Transformer turns ratio, T_s = Switching period

The inductor current during the time period $(t_2 - t_1)$ is given by

$$I_{L_r} = \frac{E_{dc}}{2 L_r} t - I_{L_0} \qquad (12.100)$$

where $\qquad I_{L_0} = I_0/n \qquad (12.101)$

Now, at $t = t_2$, $I_{L_r} = I_{L_0}$. Substituting in Eq. (12.101), we get,

$$t_2 = \frac{2 L_r \cdot I_{L_0}}{E_{dc}/2} = \frac{4 L_r \cdot I_{L_0}}{E_{dc}} \qquad (12.102)$$

Substituting Eq. (12.102) in Eq. (12.99), we get

$$E_0 = \frac{E_{dc}}{n T_s}\left[t_3 - \frac{4 L_r \cdot I_{L_0}}{E_{dc}}\right] = \frac{E_{dc}}{2\eta}\left[\frac{2 t_3}{T_s} - \frac{4 L_r I_{L_0}}{\frac{E_{dc}}{2} \cdot T_s}\right] \qquad (12.103)$$

(ii) Capacitor Voltage Equations : We have seen that during the interval $(t_4 - t_3)$, V_{C_1} rises and V_{C_3}, V_{C_4} decreases,

$$\therefore \qquad V_{C_1} = \frac{I_{L_0} \cdot t}{1.5 C} \qquad (12.104)$$

and $\qquad V_{C_3} + V_{C_4} = E_{dc} - \frac{I_{L_0}}{1.5 C} \cdot t \qquad (12.105)$

where $\qquad C_1 = C_2 = C_3 = C_4 = C$

Also, it has been discussed that during the interval $(t_6 - t_5)$, V_{C_2} increases from zero to $\frac{E_{dc}}{2}$ and $(V_{C_3} + V_{C_4})$ reduces from $E_{dc}/2$ to zero. The voltage across C_2 is given by, $V_{C_2} = \sqrt{\frac{L_r}{1.5 C}} \cdot I_{L_0} \sin \omega_r \cdot t \qquad (12.106)$

Resonant Converters

where:
$$\omega_r = \frac{1}{\sqrt{1.5\, C\, L_r}} \tag{12.107}$$

and
$$z_r = \sqrt{\frac{L_r}{1.5\, C}} \tag{12.108}$$

From Eq. (12.106), at $\omega_r \cdot t = \pi/2$, voltage V_{C_2} is maximum,

$$\therefore \quad V_{C_2} = I_{L_0} \cdot \sqrt{\frac{L_r}{1.5\, C}} \tag{12.109}$$

At minimum load current, $V_{C_2} = \dfrac{E_{dc}}{2}$

$$\therefore \quad \frac{E_{dc}}{2} = I_{L_0(min)} \cdot \sqrt{\frac{L_r}{1.5\, C}}$$

But $I_{L_0(min)} = I_{L_0(min)}/n \quad \therefore \quad I_{0(min)} = \dfrac{n \cdot E_{dc}}{2} \sqrt{\dfrac{1.5\, C}{L_r}} \tag{12.110}$

To ensure proper commutation, the minimum operating load current should be greater than $I_{0(min)}$.

(iii) Duty Ratio Let us define the duty ratio as

$$D = \frac{2(t_3 - t_1)}{T_s} = \frac{2 t_3}{T_s} \quad (\because\ t_1 = 0) \tag{12.111}$$

From Eq. (12.103), the reduction in duty-ratio is given by

$$\Delta D = \frac{4\, L_r \cdot I_{L_0}}{E_{dc}/2}\, f_s \tag{12.112}$$

Now, effective duty ratio may be defined as

$$D_{eff} = \frac{E_0}{E_{dc}/2n} \tag{12.113}$$

$$\therefore \quad D_{eff} = D - \Delta D \tag{12.114}$$

SOLVED EXAMPLE

Example 12.7 Design a ZVS three-level PWM converter to meet the following specifications:
(i) $E_{dc} = 300$ V d.c.   (ii) $E_0 = 60$ V   (iii) $I_0 = 10$ A
(iv) $f_s = 120$ kHz   (v) $D_{eff} = 0.5$

Solution: Given: $E_{dc} = 300$ V, $E_0 = 60$ V, $I_0 = 10$ A, $f_s = 120$ kHz and $D_{eff.} = 0.5$

(i) Calculation of duty ratio:

From Eq. (12.113), effective duty ratio is given by

$$D_{eff.} = \frac{2n \cdot E_0}{E_{dc}} \quad \therefore \quad 0.5 = \frac{2n \cdot 60}{300} \quad \therefore \quad n = 1.25$$

Let us assume the reduction of duty ratio at full load equal to 25% of duty ratio D_r

$\therefore$ From Eq. (12.114), $D_{eff.} = D - 0.25 D$ $\therefore$ $0.5 = 0.75 D$

$$\therefore \quad D = \frac{0.5}{0.75} = 0.67$$

(ii) Selection of resonant inductor L_r:

From Eq. (12.112), we get $L_r = \dfrac{\Delta D \cdot E_{dc}/2}{4\left(\dfrac{I_0}{n}\right) \cdot f_s}$, But $\Delta D = 0.25 D$

$$\therefore \quad L_r = \frac{0.25 \times 0.67 \times 150}{4 \times 8 \times 120 \times 10^3} = 6.54 \, \mu H$$

The effective inductance $L_{r(eff)} = L_r$ + transformer leakage inductance

$\qquad = 6.54 \, \mu H + 3.46 \, \mu H \text{ (say)} = 10 \, \mu H$

Now, minimum load current from Eq. (12.110), is given by

$$I_{0(min)} = \frac{1.25 \times 300}{2} \sqrt{\frac{1.5 \, C}{10 \times 10^{-6}}}$$

Let us select MOSFET switch with intrinsic capacitance $C = 500$ PF to ensure ZVS.

$$\therefore \quad I_{0(min)} = 187.5 \sqrt{\frac{1.5 \times 500 \times 10^{-12}}{10 \times 10^{-6}}} = 1.63 \text{ A}$$

Inductor current $I_{L_r} = I_0 = \dfrac{I_0}{n} = \dfrac{10}{1.25} = 8$ A

Voltage drop across L_r at 8 A rms current

$$V_{L_r} = I_{L_r} \times X_{L_r} = 8 \times 2 \times \pi \times 120 \times 10^3 \times 10 \times 10^{-6} = 60.32 \text{ V}$$

(iii) Transformer selection:

We have, $E_0 = D_{eff.} E_s$ $\therefore$ $E_s = \dfrac{E_0}{D_{eff.}} = \dfrac{60}{0.5} = 120$

Now, transformer primary voltage $E_p = nE_s = 1.25 \times 120 = 150$

Transformer Volt-ampere,

$$VA = 150 \times 8 = 1200 \text{ VA}$$

Transformer and inductor should be designed with ferrite core.

12.10 RESONANT DC LINK INVERTERS

The switching losses in a PWM inverter can be avoided by introducing a resonant circuit in between the dc input voltage and the PWM inverter. With this configuration, the input voltage to the inverter oscillates between zero and a value greater than twice the dc input voltage. The inverter switches are turned-on and turned-off at zero voltage.

The basic resonant dc-link inverter circuit is shown in Fig. 13.32(a). Here, the resonant circuit consists of L_r, C_r and a switch with an antiparallel diode. The circuit load current is represented by I_0 [i.e. current drawn by the inverter]. It is worth to assume I_0 to be constant in magnitude because of internal inductance of the load during a resonant frequency cycle. The associated voltage and current waveforms are shown in Fig. 12.32(b).

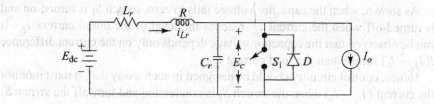

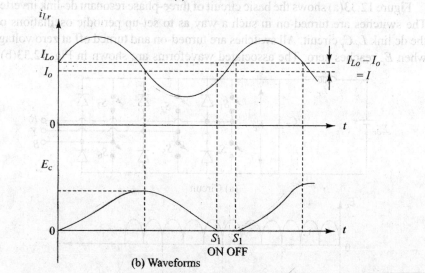

Fig. 12.32 dc-link resonance

Assuming a lossless circuit and $R = 0$, the link voltage is given by
$$E_c = E_{dc} (1 - \cos \omega_r \cdot t) \tag{12.115}$$
and the inductor current is given by
$$i_{L_r} = E_{dc} \sqrt{\frac{C_r}{L_r}} \sin \omega_r \cdot t + I_0 \tag{12.116}$$

These oscillations will continue under lossless conditions and hence switch S_1 is not turned-on.

Practically, there will be a power loss in R and i_{L_r} will be damped sinusoidal and therefore switch S_1 is turned-on to bring the current to the initial level. Circuit is underdamped since the R is very small. With this situation, i_{L_r} and E_c is given by

$$i_{L_r} = I_0 + e^{-\alpha t} \left[\frac{E_{dc}}{\omega \cdot L_r} \sin \omega t + (I_{L_0} - I_0) \cos \omega t \right] \quad (12.117)$$

and capacitor voltage E_c becomes

$$E_c = E_{dc} + e^{-\alpha t} [\omega \cdot L_r \cdot (I_{L_0} - I_0) \sin \omega t - E_{dc} \cos \omega t] \quad (12.118)$$

As shown, when the capacitor voltage falls to zero, switch S_1 is turned-on and is turned-off when the current i_{L_r} reaches the level of the initial current i_{L_0}. It can be observed that the capacitor voltage depends only on the current difference $I(I_{L_0} - I_0)$ rather than the load current I_0.

Hence, control circuitry should be designed in such a way that, it must monitor the current $(i_{L_r} - I_0)$ when the switch S_1 is conducting and turn-off the switch S_1 when the designed value of current I is reached.

Figure 12.33(a) shows the basic circuit of three-phase resonant dc-link inverter. The switches are turned-on in such a way as to set-up periodic oscillations on the dc link $L_r C_r$ circuit. All switches are turned-on and turned off at zero voltage when E_c reaches zero. The associated waveforms are shown in Fig. 12.33(b).

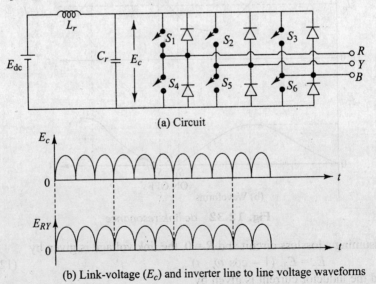

(a) Circuit

(b) Link-voltage (E_c) and inverter line to line voltage waveforms

Fig. 12.33 *Three-phase resonant dc link inverter*

REVIEW QUESTIONS

12.1 Draw the circuit diagram of series RLC circuit and derive the expression for the voltage gain. Also, draw and explain the magnitude plot.

12.2 Draw and explain the circuit of parallel resonant circuit. Also, derive the expression for voltage gain and plot its magnitude response.

12.3 Draw the circuit diagram and derive the expression for voltage gain of parallel-loaded resonant LC circuit. Also, plot its frequency response.

12.4 Describe the operation of full-bridge series resonant inverter with unidirectional switch. Also, draw and explain the current and voltage waveforms.

12.5 Why is a switch with an antiparallel diode called a bidirectional switch? How does it effect the operation of the inverter?

12.6 Draw and explain the non-overlapping and overlapping mode of operation of a half-bridge inverter with bidirectional switch.

12.7 Derive the expression for the voltage across the capacitor in case of half-bridge series resonant inverter with bidirectional switches.

12.8 With the help of neat circuit diagram and waveforms, explain the operation of full-bridge series resonant inverters with bidirectional switches.

12.9 Derive the expression for the voltage across the capacitor in case of full-bridge series resonant inverter.

12.10 Briefly explain the operation of parallel resonant inverter.

12.11 Briefly explain the difference between series and parallel resonant inverter.

12.12 With the help of neat circuit diagram and associated waveforms, explain the operation of class-E resonant inverters.

12.13 List the advantages and limitations of class-E resonant inverters.

12.14 List the advantages and disadvantages of resonant inverters with unidirectional switches.

12.15 List the advantages and disadvantages of resonant inverters with bidirectional switches.

12.16 What are the advantages of reverse-conducting thyristors in resonant inverters.

12.17 With the help of neat circuit diagram and waveforms, explain the operation of class-E resonant rectifier.

12.18 List the advantages and limitations of class E resonant rectifiers.

12.19 What is a quasi-resonant converter? Differentiate briefly between resonant and quasi-resonant converter.

12.20 What do you mean by resonant switch? List the favourable features over the PWM switches.

12.21 Draw and explain the various types of zero current switch topology.

12.22 Draw and explain the various versions of zero-voltage switch topology.

12.23 With the help of neat circuit diagram and associated waveforms, explain the operation for half-wave mode of ZCS resonant buck converter.

12.24 With the help of neat circuit diagram and associated waveforms, explain the operation for full-wave mode of ZCS resonant buck converter.

12.25 List the advantages and limitations of ZCS converters.

12.26 With the help of neat circuit diagram and associated waveforms, explain the operation of ZVS resonant converters in half-wave mode.

12.27 What is the principle of ZVS resonant converters.
12.28 With the help of neat circuit diagram and associated waveforms, explain the operation of ZVS resonant converters in full-wave mode.
12.29 List the advantages and limitations of ZVS converters.
12.30 Draw and explain the operation of ZVS forward converter.
12.31 Draw and explain the operation of ZVS flyback converter.
12.32 Draw and explain the operation of ZVS half-bridge converter.
12.33 Draw and explain the operation of full-bridge ZVS PWM converter.
12.34 What is the difference between single resonant and multiresonant converter? What are the advantages of multi-resonant converter over the single resonant converter topology?
12.35 With the help of neat circuit diagram and associated voltage and current waveforms, explain the operation of ZVS multi-resonant converter.
12.36 With the help of neat circuit diagram and associated waveforms, explain the operation of ZVS-three-level PWM converter.
12.37 List the advantages and limitations of ZVS-three-level PWM converter.
12.38 Draw and explain briefly the operation of resonant dc link inverter.

Problems

12.1 The full-bridge series resonant inverter has $L = 30\ \mu H$, $C = 5.2\ \mu F$ and $R = 4\ \Omega$. The frequency of output voltage is 5 kHz and the dc input is 150 V. SCR turn-off time is 20 μs. Determine:
  (i) circuit turn-off time
  (ii) maximum possible output frequency
  (iii) capacitor voltage
  (iv) rms load current
  (v) output power
  (vi) average and rms SCR currents
  Ans. (i) 28.94 μs
  (ii) 5.496 kHz
  (iii) 152.671
  (iv) 25.58 A
  (v) 2617.35 W
  (vi) 9.46 A, 18.08 A

12.2 Half-bridge series resonant inverter with bidirectional switches has the following circuit parameters:
  (i) $C_1 = C_2 = 6\ \mu F$
  (ii) $L_r = 25\ \mu H$
  (iii) $R = 1\ \Omega$
  (iv) $E_{dc} = 150$ V dc
  (v) SCR turn-off time = 25 μs
  Calculate: (a) resonant frequency
  (b) maximum operating frequency
  (c) peaks thyristor current
  (d) average thyristor current
  (e) rms thyristor current
  (f) rms load current
  (g) average supply current
  Assuming operating frequency to be 30% of resonant frequency
  Ans. (a) 6.62 kHz
  (b) 20 kHz
  (c) 73.78 A
  (d) 9.54 A
  (e) 19.71 A
  (f) 39.42
  (g) 5.179 A

12.3 Full bridge series resonant inverter with bidirectional switches has the following circuit parameters:
  (i) $C_r = 6\ \mu F$
  (ii) $L_r = 30\ \mu H$
  (iii) $R = 3\ \Omega$
  (iv) $E_{dc} = 180\ V_{dc}$
  (v) SCR turn-off time = 18 μs

Determine:
(a) resonant frequency
(b) maximum operating frequency
(c) peak thyristor current
(d) average SCR current
(e) rms thyristor current
(f) rms load current
(g) average supply current.

Assume the output frequency to be 30% of the resonant frequency.

Ans. (a) $f_r = 8.76$ kHz (b) $f = 13.89$ kHz
(c) $I_p = 75.18$ A (d) $I_{avg(SCR)} = 5.98$ A
(e) $I_{rms} = 25.34$ A (f) $I_0 = 50.68$ A
(g) 42.80 A

12.4 A series resonant inverter with unidirectional switch has the following parameters:
$L_r = 0.1\ \mu$H, $C_r = 10\ \mu$F, $R = 1\ \Omega$ and $t_q = 12\ \mu$sec.
Determine the maximum switching frequency for nonoverlap operation of the inverter
Ans. $f_{s(max)} = 4.44$ kHz

12.5 Determine the values of resonant components for class-E type of resonance inverter for the following specifications: $E_{dc} = 12$ V, $R = 8\ \Omega$, $F_s = 15$ kHz
Ans. $L_{dc} = 33.95\ \mu$H, $C_1 = 2.87\ \mu$F
$L_r = 594.18\ \mu$H, $C_r = 0.2\ \mu$F

12.6 ZCS step-down resonant converter operates with $E_{dc} = 24$ V, $f_r = 1.2$ MHz, $E_0 = 15$ V, $p_0 = 14$ W
Determine the following:
(i) Values of resonant component L_r and C_r (ii) Peak switch current
(iii) peak voltage rating of resonant capacitor
Ans. (i) $L_r = 3.44\ \mu$H, $C_r = 0.0052\ \mu$F (ii) 1.863 (iii) 48 V

12.7 Design a ZVS three-level PWM converter to meet the following specifications:
(i) $E_{dc} = 200\ V_{dc}$ (ii) $E_0 = 30$ V
(iii) $I_0 = 4$A (iv) $f_s = 100$ kHz
(v) $D_{eff.} = 0.5$
Ans. $n = 1.67$, $D = 0.67$, $L_r = 17.48\ \mu$H, $L_{reff.} = 22$ mH,
$I_{0min} = 0.98$ A, $V_{LR} = 33.18$ V, $V_{L_r} = 240$ V

REFERENCES

1. Mohan/Undeland/Robins, *Power-Electronics*, John Wiley and Sons, Inc. 1995.
2. M.H. Rashid, *Power-Electronics*, Pearson education, 2003.
3. Lee, F.C. "High frequency quasi-resonant converter technologies", *Proc. IEEI*, Vol. 76, pp. 377–390, 1988.
4. Toabisz, W.A., J.M.M., and Lee F.C., "High-frequency multi-resonant converter technology and its applications", *IEEE conf. rec./July 1990*.
5. Kang, Y.U. and Upadhyay A.K., "Analysis and design of half-bridge parallel resonant converter", *IEEE PESC rec.*, pp. 231–243, 1987.
6. Murai, Y. and Lipo, T.A., "High-frequency series resonant dc link power conversion", *IEEE Trans. and applns*, Vol. IA-28, No. 6, pp. 1277–1285, Nov/Dec. 1992.s

Chapter 13

Protection and Cooling of Power Switching Devices

LEARNING OBJECTIVES:

- To define the overvoltage conditions.
- To examine the various means of overvoltage protection.
- To define the overcurrent fault conditions.
- To consider the various means of overcurrent protection.
- To understand the effect of Radio Interference phenomenon on gate circuits.
- To consider the function and operation of a gate protection circuit.
- To consider losses and heat transfer properties.
- To introduce the major thyristor mounting techniques.
- To understand what is meant by SCR reliability.

13.1 INTRODUCTION

All power devices have limited operating capabilities. Reliable and satisfactory use of devices depends on ensuring that the circuit conditions imposed on them are always within their capabilities. To achieve this, device has to be surrounded by components chosen to protect it against the extreme conditions, enabling an economic and easily obtainable device to be used. If a device experiences overvoltage, overload, high current or voltage transients, high device (junction) temperature or other abnormal operating conditions, the device may be degraded in performance or destroyed permanently. The device should be protected against abnormal conditions for satisfactory and reliable operation so that it faithfully adheres to the specified characteristics of the manufacturer. To operate the device within its upper temperature limit, the heat produced by losses in a device must be dissipated sufficiently and effectively. Therefore, heat sink and cooling arrangement for devices are employed.

13.2 OVERVOLTAGE CONDITIONS

Devices can be damaged by excessive voltage applied even for very short periods of time. There are many such transient conditions in all power electronic circuits and it is necessary to understand them to ensure satisfactory protection. The conditions most important to power-electronic circuits are:
  (i) Lightning surges
 (ii) Transformer switching
(iii) Thyristor turn-off, and
(iv) Load switching, etc.

(i) Lightning Surges Lightning strikes on overhead power-lines can be passed through the supply network and appear on all power-electronic circuits which are directly connected to the network. They are usually attenuated by supply system transformers and lightning arrestors but can still be many times the normal voltage level lasting for periods up to tens of microseconds. Fortunately, their magnitude tends to reduce as the time of the transient increases. In some applications, the output connections from the thyristor circuit may be exposed to lightning, e.g. d.c. transmission and traction, and output voltage suppression circuits will then be needed.

(ii) Transformer Switching Transformer switching is a regular and significant source of transient overvoltages, particularly when a thyristor equipment is supplied by its own transformer. Transients will occur on the secondary windings when its primary current is opened or closed. These occur even when the equipment and transformer are unloaded due to the magnetizing conditions within the transformer. When the supply is closed on to the transformer, the inrush magnetizing current causes voltage of up to twice the normal to occur transiently. Capacitive coupling between the primary and secondary can temporarily boost the secondary voltage if the transformer has a large step-down ratio. The voltage-transients due to transformer switching are shown in Figs 13.1 to 13.3.

(iii) Power Device Turn-off When any device turn-off at a relatively high rate of change of current, a reverse current will flow to sweep away the stored charge. Once this has been achieved the current quickly reduces to zero, inducing high voltages in the circuit inductances. These voltages can be extremely high if no protection is included to limit them; they appear as reverse voltage across the device which are turning OFF, and they are reflected on to other devices in the circuit in both polarities. They occur everytime any thyristor turns-off and so becomes respective transients in nearly all practical applications. The stored charge and hence, the level of reverse 'charge recovery' current varies between thyristors

and due to temperature, such maximum values have to be used in assessing protection requirements. Cyclical commutation transient due to reverse recovery of SCRs is shown in Fig. 13.4.

(iv) Load Switching Load switching will result in overvoltages being induced in circuit inductances whether these be on the load or supply side of the thyristor circuit. Fuse blowing and the operation of protective circuit breakers are probably the most severe examples of this effect. The energy contained in circuit inductances will be given by

$$\text{Energy} = \frac{1}{2} LI^2$$

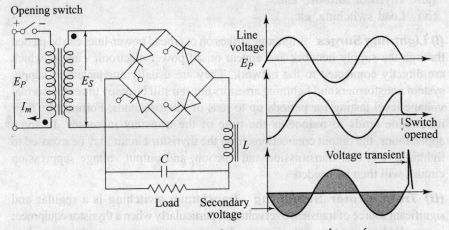

Fig. 13.1 *Voltage transient due to interruption of transformer magnetizing current*

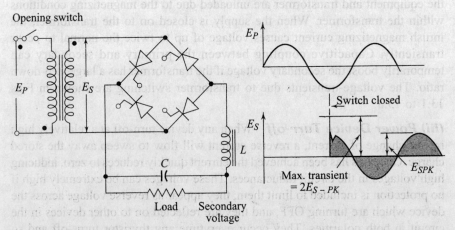

Fig. 13.2 *Voltage transient due to energizing transformer primary*

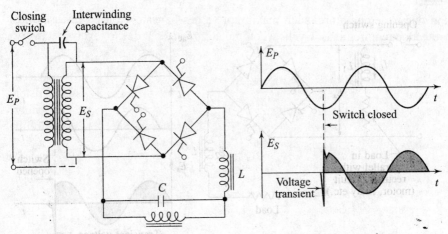

Fig. 13.3 *Voltage transient due to energizing step-down transformers*

where I is the current flowing and this will need to be dissipated in the protective components used, without exceeding the device voltage capabilities.

In general, devices which open the circuit slowly, dissipate this energy slowly within themselves by arcing or fast switches, i.e. fuses or high speed switches, will usually leave most of the energy to be absorbed by other circuit components. Load switching conditions significantly depend on the characteristics of the switch opening the circuit. The voltage transients due to load switching are described in Figs 13.5 and 13.6.

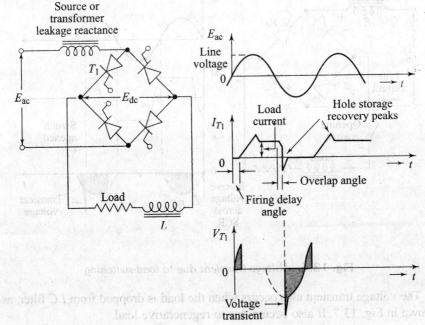

Fig. 13.4 *Transients due to reverse-recovery of SCRs*

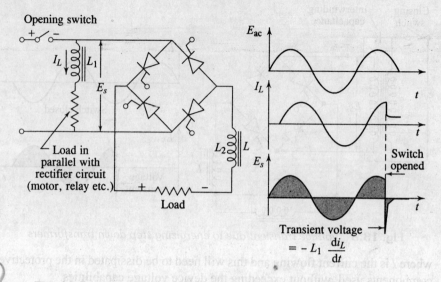

Fig. 13.5 *Voltage transient due to switching circuit with inductive load across input*

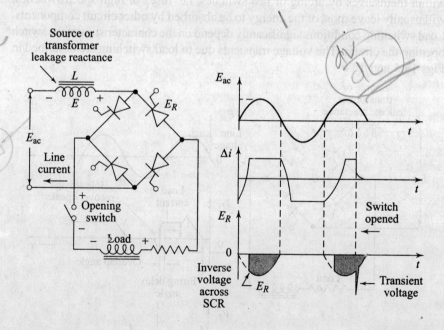

Fig. 13.6 *Voltage transient due to load-switching*

The voltage transient also occurs when the load is dropped from *LC* filter, as shown in Fig. 13.7. It also occurs due to regenerative load.

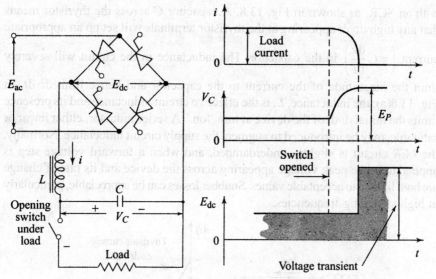

Fig. 13.7 *Voltage transient due to dropping load from LC-type filter with high L/C ratio*

13.3 OVERVOLTAGE PROTECTION

Various sources of voltage transients have been discussed in above section. Therefore, it is almost always necessary in a power semiconductor equipment to provide a means of limiting transient overvoltages which could otherwise overstress the semiconductor devices. To protect against all sources of voltage transients, it is necessary to protect each device individually. A simple means of protecting the devices against overvoltages is to introduce a safety factor V_f. In order to keep the protective components to a minimum, devices are selected with their peak voltage ratings of 1.5 to 2.5 times their normal peak working voltage. The effect of overvoltages is usually minimized by using the following protective elements and circuits.

 (i) Snubber circuits
 (ii) Nonlinear-surge suppressors
 (iii) Crowbar-circuits

13.3.1 Snubber Circuits (dv/dt Suppression)

In most power electronic circuits, protection is necessary against the effects of excessive rate of rise of forward voltage (dv/dt) across the devices, which can otherwise cause unintended breakover, leading to a malfunction of the circuit and possible to failure of the devices. The tendency to excessive dv/dt may arise from external causes such as the closing of main supply contactor, or from the operation of the circuit itself.

dv/dt suppression is achieved by means of snubber circuit. The snubber circuit basically consists of a series-connected resistor and capacitor placed in shunt with an SCR, as shown in Fig. 13.8. A capacitor C across the thyristor means that any high dv/dt appearing at the thyristor terminals will set up an appropriate current $\left(=C\dfrac{dv}{dt}\right)$ in the capacitor. The inductance in the circuit will severely limit the magnitude of the current to the capacitor and hence limit dv/dt. In Fig. 13.8(a), the inductance, L, is the effective circuit inductance and its presence limits the initial di/dt of the device at turn "on." A series inductor, either linear or saturable, may be introduced to augment the supply circuit inductance. Normally, the LCR circuit is slightly underdamped, and when a forward voltage step is applied to it, the peak voltage appearing across the device and its rate of change are both limited to acceptable values. Snubber losses can be appreciable, particularly at high-switching frequencies.

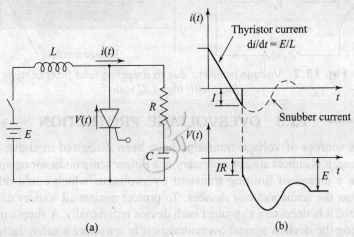

Fig. 13.8 *(a) Thyristor with snubber circuit (b) reverse recovery waveforms*

When the thyristor turn-off, there is a brief pulse of reverse recovery current that rises to a peak value, at which time the device blocks. In the absence of an RC snubber, the abrupt interruption of the reverse recovery current in the series inductance, L, will cause transient $L\,di/dt$ overvoltages that may destroy the device itself or other semiconductors in the converter circuit.

If an RC snubber is connected across the thyristor, the reverse recovery current can transfer to the RC path when the device blocks, as shown in Fig. 13.8(b). The voltage across the RC path appears as an oscillatory reverse voltage across the semiconductor. A correctly designed snubber will limit the amplitude of this reverse-recovery voltage and will also limit its rate of rise. By delaying the build-up of reverse voltage across the semiconductor, the recovery losses in the device are also reduced. If the snubber resistance is too small, excessive ringing will occur in the LCR circuit, and the reverse blocking voltage capability of the semiconductor may be exceeded.

When thyristor T_1 in Fig. 13.8(a) turns-on, the snubber capacitor, C, discharges into the thyristor, but resistor R limits the discharge current and prevents excessive di/dt at turn-on. However, at turn-off, a forward IR voltage drop suddenly appears across the thyristor due to reverse recovery of the diode, as indicated in Fig. 13.8(b). Thus, the presence of the resistor impairs the forward dv/dt limiting performance of the snubber.

If the primary function of the snubber is to limit forward dv/dt on the thyristor, the polarized snubber of Fig. 13.9 is often used. Resistor R_1 is a small damping resistor giving enhanced forward dv/dt protection, while R_2 is a much larger resistor for limiting the snubber discharge current when the thyristor is gated.

In a.c. utility-powered circuits, snubber networks are often connected between the a.c. supply lines to limit incoming utility-borne transients. RC snubbers are also connected across transformer secondary winding and other potential sources of inductive voltage transients. However, snubber networks are only partially effective in suppressing transient overvoltages and hence, large safety margins must be applied to the semiconductor blocking voltages. Voltage-damping devices are more effective in limiting transient overvoltages than RC snubbers and have lower losses for normal operating conditions.

(a) Design of Snubber Network for d.c. Circuit The snubber circuit used in d.c. circuit is shown in Fig. 13.10. In practice, a d.c. voltage is switched across an SCR, and the rate of change of voltage across the device must be below the dv/dt rating of the device, otherwise the device would start conducting even without the application of a gate trigger pulse.

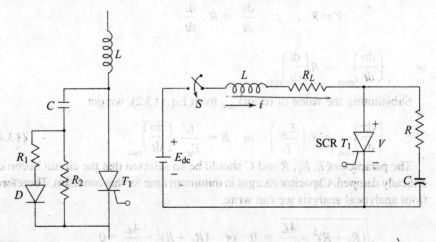

Fig. 13.9 *Polarized snubber*   **Fig. 13.10** *The snubber circuit*

When switch S is closed, the capacitor behaves like a short circuit and SCR in the forward blocking state offers a very high resistance. In the figure, R_L is the load resistance and L the source inductance.

For this circuit, the voltage equation is

$$E_{dc} = (R + R_L)\,i + L\,\frac{di}{dt}.$$

The solution of the above equation is

$$i = I(1 - e^{-t/\tau}) \tag{13.1}$$

where $I = \dfrac{E_{dc}}{R + R_L}$ and $\tau = \dfrac{L}{R + R_L}$

In Eq. (13.1), t is the time in seconds measured from the instant of closing the switch.

Differentiating Eq. (13.1) with respect to t, we get

$$\frac{di}{dt} = I e^{-t/\tau} \cdot \frac{1}{\tau} = \frac{E_{dc}}{R + R_L} \cdot \frac{R + R_L}{L} e^{-t/\tau} = \frac{E_{dc}}{L} e^{-t/\tau}$$

The value of $\dfrac{di}{dt}$ is maximum when $t = 0$.

$$\therefore \quad \left(\frac{di}{dt}\right)_{max} = \frac{E_{dc}}{L} \tag{13.2}$$

or

$$L = \frac{E_{dc}}{(di/dt)_{max}} \tag{13.3}$$

Now, voltage across the thyristor is given by

$$V = R \cdot i \quad \therefore \quad \frac{dv}{dt} = R \cdot \frac{di}{dt}$$

or

$$\left(\frac{dv}{dt}\right)_{max} = R \left(\frac{di}{dt}\right)_{max}$$

Substituting the value of $(di/dt)_{max}$ from Eq. (13.2), we get

$$\left(\frac{dv}{dt}\right)_{max} = R \left(\frac{E_{dc}}{L}\right) \quad \text{or} \quad R = \frac{L}{E_{dc}} \left(\frac{dv}{dt}\right)_{max} \tag{13.4}$$

The parameters L, R_L, R and C should be so selected that the circuit becomes critically damped. Capacitor charges in minimum time for this condition. Therefore, from analytical analysis we can write

$$\sqrt{(R_L + R)^2 - \frac{4L}{C}} = 0 \quad \text{or} \quad (R_L + R)^2 - \frac{4L}{C} = 0$$

or

$$(R_L + R)^2 = \frac{4L}{C} \quad \therefore \quad R_L + R = 2\sqrt{\frac{L}{C}} \tag{13.5}$$

Using the permissible value of dv/dt as the basis for avoiding any maloperation of the SCR, and from the given values of L and R_L, we can determine the snubber components R and C from Eqs (13.4) and (13.5), respectively.

(b) Snubber Design for a.c. Circuit (i) The following equation has proved useful in selecting the value of the capacitance required for keeping the voltage transients within the device rating:

$$C = 10 \cdot \frac{VA}{V_s^2} \frac{60}{f} \tag{13.6}$$

where
C = the minimum capacitance required (in microfarads)
(VA) = the transformer volt–ampere rating
V_s = the transformer secondary RMS voltage
f = operating frequency.

The required resistance to ensure adequate damping can be calculated from the following relationship:

$$R = 2\delta \sqrt{L/C} \tag{13.7}$$

where, δ = damping factor, normally taken to be about 0.65
R = the required resistance to damp the transient voltage to a desired level
L = effective circuit inductance and C = minimum required capacitance.

(b) If the maximum dv/dt for the thyristor is specified, the equation used in place of Eq. (13.6) to calculate the required value of the capacitance is

$$C = \frac{1}{2L} \left(\frac{0.564 \, E_m}{dv/dt} \right)^2 \tag{13.8}$$

where E_m is the peak-input-line-to-line voltage.

SOLVED EXAMPLES

Example 13.1 The SCR in Fig. 13.11 is used to control power in resistance R. The supply is 400 V, and the maximum allowable di/dt and dv/dt the SCR are 50 A/μsec and 200 V/μsec, respectively. Compute the values of the di/dt inductor and the snubber circuit components R_s and C_s.

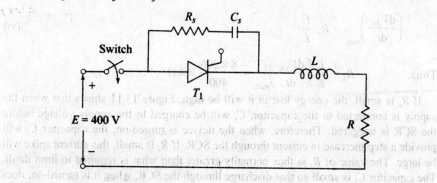

Fig. 13.11 *Circuit with dv/dt and di/dt protection*

Solution: The equivalent circuit immediately after closing the switch is shown in Fig. 13.12.

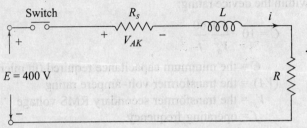

Fig. 13.12 *Equivalent circuit*

The voltage equation is

$$E = R_s i + L \frac{di}{dt} + R i \tag{i}$$

$$\therefore \quad \frac{di}{dt} = \frac{E - (R_s + R)i}{L} \tag{ii}$$

At $i = 0$, the $\frac{di}{dt}$ is maximum. Therefore, $\left(\frac{di}{dt}\right)_{max} = \frac{E}{L}$ (iii)

$$\therefore \quad L = \frac{E}{(di/dt)_{max}} = \frac{400}{50} \, \mu H = 8 \, \mu H.$$

Now, the voltage across the SCR is

$$V_{AK} = R_s i \tag{iv}$$

$$\therefore \quad \frac{dV_{AK}}{dt} = R_s \cdot \frac{di}{dt}$$

Therefore, $\left(\frac{dV_{AK}}{dt}\right)_{max} = R_s \cdot \left(\frac{di}{dt}\right)_{max}$ (v)

From Eqs (iii) and (v),

$$\left(\frac{dV_{AK}}{dt}\right)_{max} = R_s \frac{E}{L} \tag{vi}$$

Thus, $R_s = \frac{L}{E}\left(\frac{dV_{AK}}{dt}\right)_{max} = \frac{8 \times 200}{400} = 4 \, \Omega.$

If R_s is small, the energy lost in it will be high. Figure 13.11 shows that when the supply is connected to the capacitor, C_s will be charged to the supply voltage before the SCR is triggered. Therefore, when the device is turned-on, the capacitor C_s will provide a step increase in current through the SCR. If R_s is small, this current spike will be large. The value of R_s is thus normally greater than what is required to limit dv/dt. The capacitor C_s is small so that discharge through the SCR, when it is turned-on, does not harm the device. Typical values of R_s and C_s are 10 Ω and 0.1 μF, respectively. If this value of R_s is used, the value of L has to increase to limit dv/dt to the specified value.

From Eq. (vi),

$$L = \frac{(E \cdot R_s)}{(dv/dt)_{max}} = \frac{400 \times 10}{200} \mu H = 20 \mu H$$

The inductor value is not too high, and it is more than is required for di/dt protection.

Example 13.2 To provide reliable dv/dt protection to an SCR used in a single-phase fully controlled bridge, compute the required parameters for a snubber circuit. The SCR has maximum dv/dt capability of 50 V/μs. The input line-to-line voltage has a peak value 380 V and the source inductance is 0.1 mH.

Solution: Refer to Eq. (13.8).

$$C = \frac{10^4}{2} \times \left(\frac{0.564 \times 380 \times 10^{-6}}{50}\right)^2 \quad \therefore \quad C = 0.092 \mu F.$$

From Eq. (13.7), and assuming $\delta = 0.65$, we get

$$R = 2 \times 0.65 \times \sqrt{\frac{0.1 \times 10^{-3}}{0.092 \times 10^{-6}}} \quad \therefore \quad R = 42.86 \, \Omega.$$

Example 13.3 Design a snubber circuit to meet the following data:
 (i) Supply transformer is rated 5 kVA with secondary voltage of 120V RMS.
 (ii) Switching frequency = 400 Hz.
 (iii) Circuit inductance = 100 μH
 (iv) Peak transient voltage is to be limited at 200 volts, (v) $\delta = 0.75$

Solution:

(i) From Eq. (13.6), $C = 10 \cdot \dfrac{5000}{(120)^2} \times \dfrac{60}{400} = 0.52 \, \mu F.$

(ii) Now, calculate peak transient voltage V_s peak switching voltage ratio,

$$\frac{E_p}{E_{sp}} = \frac{200}{120\sqrt{2}} = 1.18.$$

(iii) From Eq. (13.7), $R = 2(0.75) \sqrt{\dfrac{100 \times 10^{-6}}{0.52 \times 10^{-6}}} = 21.2 \cong 20 \, \Omega.$

Example 13.4 A capacitor connected across the secondary of an input transformer, as shown in Fig. 13.13, with zero damping resistance, $R = 0$. The secondary voltage is $E_s = 140$ V, 50 Hz. If the magnetizing inductance referred to the secondary is $L_m = 3$ mH and the input supply to the transformer primary is disconnected at an angle of $\theta = 180°$ of the input a.c. voltage. Compute:
 (i) The initial capacitor voltage, $V_{c(0-)}$
 (ii) The magnetizing current, I_m, and
 (iii) The capacitor value to limit the transient capacitor voltage to $E_p = 330$ V.

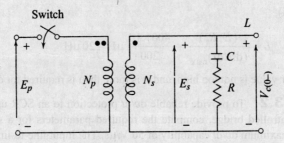

Fig. 13.13 *Switching of transients*

Solution:

$$E_m = \sqrt{2} \cdot E_s = \sqrt{2} \times 140 = 197.99 \text{ V}$$

(i) The capacitor voltage at the beginning of switch-off is given by

$$V_{c(0-)} = E_m \sin \theta = 197.99 \sin 180 = 0$$

(ii) The magnetizing current, I_m, is given by

$$I_m = \frac{-E_m}{W L_m} \cdot \cos \theta \, I_m = \frac{-197.99}{2 \times \pi \times 50 \times 3 \times 10^{-3}} \cdot \cos 180 = 210 \text{ A}$$

(iii) Given, $E_p = 330$ V.

The capacitor value to limit the maximum transient capacitor voltage to $E_p = 330$ V, is given by the equation

$$C = \frac{I_m \cdot E_m}{E_p^2 \, W} = \frac{210 \times 197.99}{330 \times 330 \times 2\pi \times 50} = 1215.3 \text{ µF.}$$

Example 13.5 A d.c. chopper (Fig. 13.14), uses a MOSFET as a switch. The input voltage $E_{dc} = 30$ V and the chopper operates at a switching frequency $f_s = 40$ kHz, supplying a load current of 30 A. The switching times MOSFET are: $t_r = 80$ ns and $t_f = 30$ ns. Determine:

(i) the snubber components value for critical damping
(ii) value of R if the maximum discharge current is limited to 10% of the load current.
(iii) Power loss due to RC snubber circuit.

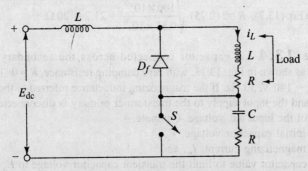

Fig. 13.14 *DC chopper circuit*

Solution:

From Fig. 13.14, when switch is turned-on, $\dfrac{di}{dt} = \dfrac{E_{dc}}{L}$ (a)

Also, $\dfrac{di}{dt} = \dfrac{i_L}{t_r}$ (b)

Now, when switch is turned-off, $i_L = C \cdot \dfrac{dV}{dt}$ (c)

and $\dfrac{dV}{dt} = \dfrac{E_{dc}}{t_f}$ (d)

(i) Equating (a) and (b), we get

$$L = \dfrac{E_{dc} \cdot t_r}{i_L} = \dfrac{30 \times 80 \times 10^{-9}}{30} = 80 \text{ nH}$$

Also, from Eqs. (c) and (d), we can write

$$C = \dfrac{i_L \cdot t_f}{E_{dc}} = \dfrac{30 \times 30 \times 10^{-9}}{30} = 30 \text{ nf}$$

From Eq. (13.7), for critical damping, we can write

$$R = \sqrt{\dfrac{4L}{C}} = \sqrt{\dfrac{4 \times 80 \times 10^{-9}}{30 \times 10^{-9}}} = 3.26 \text{ }\Omega$$

(ii) Now, $R = \dfrac{E_{dc}}{0.1 \times i_L} = \dfrac{30}{0.1 \times 30} = 10 \text{ }\Omega$

(iii) Power loss, $P_{loss} = \dfrac{1}{2} C E_{dc}^2 \cdot f = 0.54 \text{ W}$

13.3.2 Non-linear Surge Suppressors

Non-linear devices having the kind of characteristics illustrated in Fig. 13.15 are frequently used in place of, or in conjunction with, relative networks as a mean of limiting voltage surges, being so selected that they pass an acceptably small current at the normally working voltage but a sufficiently large current at higher voltages to limit surges at a tolerable level. Devices of this kind are:
  (i) Varistors
  (ii) Selenium surge-suppressors

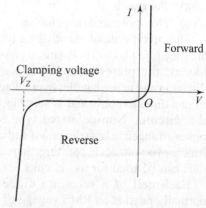

Fig. 13.15 Characteristics of non-linear surge suppressor

1. Voltage Dependent Resistors or Varistors *(Metal Oxide Varistor (MOV)]* These have been made for many years from a thyrite or metrosil material, which provides a current variation in proportion to the fourth or fifth power of the voltage. They are not normally satisfactory as the sole means of protection semiconductor circuits because surge voltage cannot be limited adequately without excessive dissipation under normal conditions.

New types of so-called metal oxide varistors are now available with a very much better characteristic, corresponding to thirteenth or higher power law, and can be used effectively within their limitations of size. The metal oxide varistor (MOV) consists of a pair of metal contact plates separated by a thin layer of small zinc oxide granules bonded in an amorphous mixture of other oxides. The boundaries of the granules have p-n junction characteristics so that the device has a symmetrical V–I characteristic similar to that of the avalanche diode suppressor. Varistors are now available with a.c. operating voltages from 6 to 2800 V.

2. Selenium Surge Suppressors: The selenium metal rectifier used extensively in the past for rectification may be used for protection. The metal-to-selenium junction has a low forward breakover voltage but, more importantly, a well-defined reverse-breakdown voltage. The characteristics of selenium diode are shown in Fig. 13.15. Normally, the operating point lies before the knee of the characteristic curve and draws very small current from the circuit. However, when an overvoltage appears, the knee point is crossed and the reverse current flowing through the selenium increases suddenly, thereby typically limiting the transient voltage to twice the normal-voltage.

The selenium diode (or suppressor) must be capable of absorbing the surge energy without undue temperature rise.
Selenium voltage-limiter can be of two types:
  (a) Polarized
  (b) Non-polarized (unpolarized).

The circuit symbol of both these types are shown in Fig. 13.16. Polarised version has all the plates arranged in same direction and hence handles overvoltages in one direction only and is suitable for d.c. circuits. Non-polarized type has plates arranged in both directions and can handle overvoltages in both directions and thus can be used for a.c. circuits.

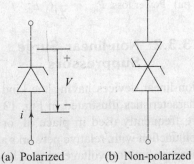

Fig. 13.16 *Circuit symbols*

Each cell of a selenium diode is normally rated at an RMS voltage of 25 V, with a clamping voltage of typically 72 V. Figure 13.17 shows the positioning of selenium suppression devices. If a d.c. circuit of 210 V is to be protected with 25-V selenium cells, then $210/25 \approx 9$ cells would be required and the clamping voltage would be $9 \times 72 = 648$ V. To protect a single-phase a.c. circuit of 230 V,

50 Hz, with 25 V selenium cells, $230/25 \approx 10$ cells would be required in each direction and total of $2 \times 10 = 20$ cells would be necessary for non-polarized suppression.

Due to low internal capacitance, the selenium diodes do not limit the dv/dt to the same extent as compared to the RC snubber circuits. However, they limit the transient voltages to well-defined magnitudes. In protecting a device, the reliability of an RC circuit is better than that of selenium diodes.

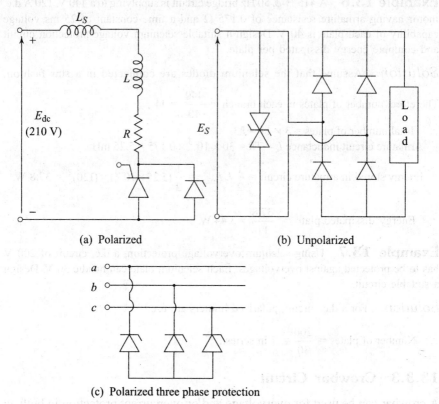

Fig. 13.17   *Voltage suppression diodes (thyrectors)*

3. Avalanche Diodes: Avalanche diodes are semiconductor diodes which can operate in the reverse-breakdown region with a high reverse current without damage. They can be used in a similar way to varistors by bypassing the current through them during the transient. They usually have the capability of absorbing significant energy in this avalanche breakdown condition. The limitations on the use of an avalanche diode are the reverse-breakdown voltage, the peak power loss, and the steady-state power loss.

Non-linear devices have the following advantages in comparison with capacitive networks. Firstly, they are free from certain inconvenient tendencies associated with stored energy in reactive suppression circuits and, secondly, that in the event of a square of unexpectedly high magnitude they will normally fail to a short-circuit rather than permit a damaging voltage to be applied to the more expensive equipment to which they are connected.

SOLVED EXAMPLES

Example 13.6 A 415, 3-ϕ, 50 Hz bridge circuit is supplying to a 440 V, 120 A d.c. motor having armature resistance of 0.175 Ω and a time constant of 25 ms voltage capability of each plate is 40 V. Design a suitable selenium voltage protection circuit and compute energy dissipated per plate.

Solution: Assume that the selenium diodes are connected in a star fashion.

Therefore, number of plates in each branch = $\dfrac{440}{40}$ = 11.

$\therefore$ Total number of plates = 3 × 11 = 33.
Armature circuit inductance $L = TR = 30 \times 10^{-3} \times 0.175 = 5.25$ mH

Energy stored in armature circuit = $\dfrac{1}{2} L I_{max}^2 = \dfrac{1}{2}(5.25 \times 10^{-3}) \times (120)^2 = 37.8$ W.

$\therefore$ Energy dissipated/plate = $\dfrac{37.8}{11} = 3.44$ W.

Example 13.7 Using selenium overvoltage protection, a d.c. circuit of 200 V has to be protected against overvoltages. Each selenium plate can handle 30 V. Design a suitable circuit.

Solution: For a d.c. circuit, polarized limiters are used.

$\therefore$ Number of plates = $\dfrac{200}{30} \cong 7$ in series.

13.3.3 Crowbar Circuit

A crowbar can be used for overvoltage and/or overcurrent protection in both ac and dc circuits. Figure 13.18 shows how an SCR can be used to provide fault protection for sensitive dc power electronic circuits and loads. Whenever a fault condition occurs the crowbar SCR is triggered, shorting the supply. The resultant high-supply current flowing blows the fuse, or initiates a fast-acting circuit breaker, thereby isolating the load from the supply. Diode D_f provides a current path for inductive load energy.

Load current is measured by the voltage across the sense resistor R. When this voltage reaches a preset limit, that is the load current has reached the fault-level, the SCR is triggered. The load or dc link voltage is measured from the resistor divider $R_2 - R_3$. When this voltage exceeds the predetermined limit, the SCR is triggered and the fuse is blown by the crowbar short-circuit current, isolating the sensitive load from the supply.

Crowbar SCR can conduct many times its average current rating. For the few milliseconds in which the fuse is isolating, the SCR I^2t surge current feature can be exploited. The SCR I^2t rating must be larger than the fuse total I^2t rating. If the SCR crowbar is fuse-link protected then the total I^2t of the dc-link fuse link must be less than the pre-arcing I^2t of the SCR crowbar fuse-link.

An ac crowbar can comprise two antiparallel-connected SCRs across the fuse-protected a.c. line, or alternatively one SCR in a four diode rectifying bridge.

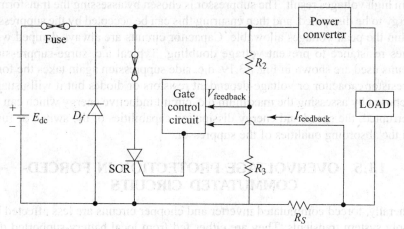

Fig. 13.18 *SCR crowbar for overvoltage and overcurrent protection*

13.4 PRACTICAL OVERVOLTAGE PROTECTION IN NATURALLY-COMMUTATED CIRCUITS

In simple application circuits, it is possible for one-voltage suppression circuit to cope with all the possible transient conditions which can exist. However, in the majority of complicated applications, a number of suppressors are included so as to optimize the overall system. For example:

(i) Surge voltage suppression circuit may be fitted to the input connections to limit incoming transients due to lightning strikes, and transformer and inductive switching in the input circuits.

(ii) A majority of thyristors used will have a snubber circuit connected to them to cope with thyristor switching effects, e.g. turn-off recovery.

(iii) Surges induced in the load circuit would be attenuated by suppressors connected to the output.

Although the principal aim of each of these suppression circuits may be clear, they all will, in fact, affect many of the transient conditions and all of them are to be considered when assessing their individual size for the magnitude of the voltage surges.

Even with all of these facilities, it is still necessary to use thyristors having nonrepetitive peak voltage ratings (V_{RSM} and V_{DSM}) of at least twice the normal crest working voltage.

The exact size of components used for surge-suppression circuits varies significantly among manufactures, the decisions being based on practical tests and experience. Nevertheless, they are all based on the same principles.

Supply-site suppression will usually be decided by supply-transformer switching, the worst case normally being opening the transformer primary circuit when no-load current is flowing. Under this condition, the transformer magnetizing energy has nowhere to go unless suppression circuits are included, and without them high voltages result. The suppressor is chosen by assessing the transformer energy to be dissipated, and then ensuring this can be accepted by the suppressor within the peak voltages allowable. Capacitor circuits are always damped with series resistance to prevent voltage doubling. Typical a.c. surge-suppression circuits used are shown in Fig. 13.19. d.c. side suppression again takes the form of resistor/capacitor or voltage-dependent resistors or diodes but it will usually be chosen by assessing the maximum d.c. circuit inductive energy which can be interrupted, the speed and energy dissipation capabilities of the switch or fuse, and the absorbing qualities of the suppressor.

13.5 OVERVOLTAGE PROTECTION IN FORCED-COMMUTATED CIRCUITS

Generally, forced commutated inverter and chopper circuits are less affected by supply system transients. They are either fed from local battery-supported d.c. systems or they have large values of d.c. capacitance which prevent any short-time transients from getting through to the thyristors. Consequently, as long as the internally generated thyristor turns-off voltages can be limited by correctly designed snubber circuits, it is possible to use the thyristors at normal peak voltages much nearer to their repetitive voltage ratings. However, the normal applied peak voltages can be well above the d.c. supply voltage level. Where load-switching transients can occur, capacitor/resistor circuits or nonlinear devices are the most effective when mounted near to the inductances in which the voltages are generated.

13.6 OVERCURRENT FAULT CONDITIONS

Most thyristor circuits are fast operating and can be controlled in such a way as to prevent the load current from rising too high to cause circuit damage or maloperation. However, there are a number of component failures and circuit maloperation conditions where the current can rise out of control to many times the normal rated value, and steps have to be taken to limit these conditions and to protect the other circuit components against the effects of the high current. An overcurrent can occur due to any of the following causes:

(i) Output short-circuit.
(ii) Internal faults in a thyristor circuit.

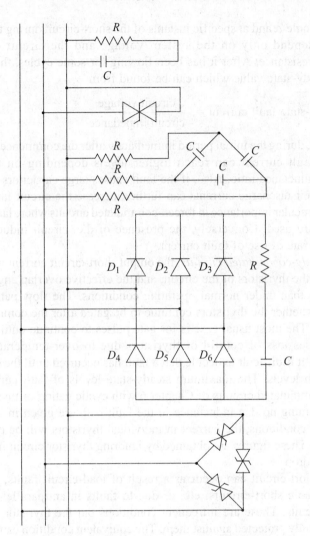

Fig. 13.19 *Typical a.c. surge voltage suppression circuits*

(iii) Inversion failure in naturally commutated circuits.
(iv) Commutation failure in a forced-commutation circuits, and
(v) Short circuit between one of the phases of mains and the bridge.

The following paragraphs cover most of these overcurrent fault conditions.

13.6.1 Output Short-circuits

On any system, output short circuit can occur and in many situations gate control of the thyristors cannot prevent high currents flow flowing; this will quickly result in loss of control of the thyristors due to over-temperature or commutation failure. The maximum level of the fault current during short-circuit will occur at zero delay angle α and at specific instants of the short-circuit during the cycle. It will then depend only on the system voltage and the circuit equivalent impedance/resistance. After it has been flowing for some cycles, the fault will reach a steady-state value which can be found from

$$\text{Steady-state fault current} = \frac{\text{circuit voltage}}{\text{circuit impedance}}$$

However, during the initial period immediately after the commencement of the fault, the fault current can reach higher levels depending on the circuit resistance/inductance ratio. Also, if the fault causes large capacitors to be short circuited, their discharge currents can further increase the circuit fault current. This is of particular importance in forced-commutated circuits where large filtering capacitors are used. Conversely, the presence of d.c. circuit inductance will slow up the rate of rise of fault current.

In naturally-commutated circuits, the output short-circuit current will be split up between the thyristors of the circuit, and the effective overlap angles will be much larger than under normal operating conditions. The flow path will also depend on whether the thyristors continue to be gated after the commencement of the fault. The most usual case is for gate pulses to continue until the circuit current causes loss of control of thyristors due to over-temperature of the junctions, as it is difficult to decide that a fault has occurred until the current has risen to high levels. The maximum steady-state levels of fault current in the naturally commutated circuits of Chapter 6 with cyclic gating of thyristors at α = zero, assuming no d.c. inductance in the fault path are given in Table 13.1. Under these conditions, the current in individual thyristors will be of half-sine wave shape. These figures are obtained by ignoring thyristor circuit impedances and voltage drop.

Output short-circuit can occur as a result of load-circuit faults, e.g. motor flashover, cable short-circuits, etc. or due to faults in antiparallel-connected thyristor circuits. These are infrequent conditions but the thyristor equipment needs to be fully protected against them. The equivalent condition can also occur if a d.c. motor load is inadvertently connected to the converter while the motor is at rest and the converter is controlled to give maximum output voltage. It can also be caused by control system maloperation suddenly resulting in operation at low-delay angles when the motor load is running at low speed.

13.6.2 Internal Faults in Power-Electronic Circuit

Internal faults within a thyristor circuit can also result in a high current flow. They can be caused by the failure of a thyristor or incorrect firing of it due to interference or a faulty firing circuit. Failure of voltage protection components, e.g., a snubber capacitor, can also cause over-currents due to thyristor voltage or dv/dt breakover.

If a fault results in a thyristor failing to block forward voltage, this thyristor will tend to carry the high-fault current on its own, even during periods when other thyristors would normally carry current. The worst possible case of this condition would mean the thyristor carrying the full short circuit current of the circuit, but in many cases, the level would be less than this until such a time as other thyristors fail to block voltage due to excessive junction temperature.

Table 13.1 Steady-state fault currents in naturally commutated rectifying circuits

Circuit	d.c. short-circuit mean current	a.c. RMS short circuit current	Thyristor peak current	Thyristor mean current	Half-cycle I^2t thyristor (Amp-sec^2) 50 Hz
1. Single-phase half wave	I_F	$\dfrac{\pi}{2\sqrt{2}} I_F$	$\dfrac{\pi}{2} I_F$	$\dfrac{1}{2} I_F$	$\dfrac{\pi^2}{8} 10^{-3} I_F^2$
Half-controlled bridge					
Fully-controlled bridge					
2. Three-phase	I_F	$\dfrac{\pi}{3\sqrt{2}} I_F$	$\dfrac{\pi}{3} I_F$	$\dfrac{1}{3} I_F$	$\dfrac{\pi^2}{18} 10^{-3} I_F^2$
Half-controlled bridge					
Fully-controlled bridge					
3. Six-phase half wave	I_F	$\dfrac{\pi}{6\sqrt{2}} I_F$	$\dfrac{\pi}{6} I_F$	$\dfrac{1}{6} I_F$	$\dfrac{\pi^2}{72} 10^{-3} I_F^2$

$$I_F = \frac{\text{rated d.c. current}}{\text{per unit reactance}} = \frac{I_{dw}}{dt}$$

If the fault is reverse-blocking failure of a thyristor, other thyristors within the circuit may feed into the fault, and as far as the good thyristors are concerned it can look like an output short-circuit. In naturally-commutated circuits, however, the fault will tend to start again every cycle and be continually asymmetric, i.e. a repetitive sequence of asymmetrical fault currents.

13.6.3 Inversion Failure in Naturally-Commutated Circuit

Inversion failure in naturally-commutated circuits can result in a high fault current during power flow from the d.c. to the a.c. side, i.e. inversion. The condition is normally only significant with motor loads where sustained inversion can occur. This power-flow condition can only exist as long as the converter is able to provide a voltage to balance the generated voltage of the d.c. machine. If the converter voltage disappears or reverses, a short-circuit current will flow (see Fig. 13.20).

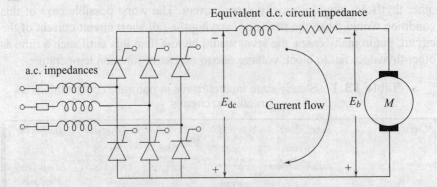

Fig. 13.20 *Inversion failure*

If the converter voltage, E_{dc}, disappears, it is equivalent to short-circuiting the motor through the circuit impedance (including the motor). This could be caused by loss of a.c. supply. A higher fault condition may result from a firing or control circuit fault which causes the converter suddenly to switch into the fully rectifying condition, in which case the total circuit voltage will be the sum of E_{dc} and E_b. The total fault circuit will now, however, include both the a.c. circuit impedances as well as the d.c. impedance, and its level and shape will depend on the relative values of the a.c. and d.c. impedances. An intermediate condition occurs if all the firing pulses suddenly disappear as those thyristors in the circuit which are carrying current at that instant continue to do so. The total circuit voltage under this conditions will be single-phase of the supply voltage, i.e. a sinusoidally changing voltage, plus the motor voltage; the fault current will flow through both the a.c. and d.c. circuits.

13.6.4 Commutation Failure in Forced Commutation Circuit

Correct operation of all forced-commutation inverter circuits depends on the ability of the commutating components to provide sufficient time for the thyristors to turn-off. If, due to excessive load-current, or insufficient charge on commutating capacitors, or incorrect firing of the thyristors, a thyristor does not have time to turn-off, it will stay in the on-state and this will very soon, if not instantly, result in the d.c. supply being short-circuited by the thyristors. The fault current then flowing will depend on the circuit inductance, resistance, and capacitance, as shown in Fig. 13.21. This figure shows that capacitance close to the inverter circuit will cause an initial high discharge current through the thyristors limited only by the local resistance and inductance near to the thyristors. An inductive input circuit, i.e. significant L and no C_2, will result in a comparatively slow build-up of fault current.

If the inverter has more phases and thyristors, other conditions of commutation failure may be possible but they will all result in either the full d.c. short-circuit current flowing through an individual thyristor or it will be shared with the other phases.

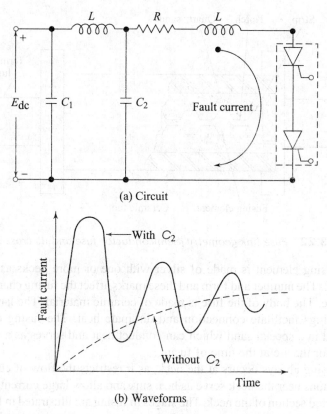

Fig. 13.21 *Fault current caused by commutation failure*

13.7 OVERCURRENT PROTECTION

As the thyristors have a restricted overcurrent capacity, special fast-acting fuses and circuit breakers are usually provided for overcurrent protection. As the fault current increases, the fuse opens and clears the fault current in few milliseconds. Before describing the fuse, it is necessary to consider carefully the requirement demands of it, which are:

(1) It must carry continuously the device rated current.
(2) Its thermal storage capacity must be less than that of the device being protected, that is I^2t let-through value of the fuse before the fault-current is cleared must be less than the rated I^2t of the device to be protected.
(3) The fuse voltage during arcing must be high enough to force the current down and dissipate the circuit energy.
(4) After breaking the current, the fuse must be able to withstand any re-striking voltage which appears across it.

Figure 13.22 shows the fuse-link geometry. It can be seen that the link is a strip with several narrow notches. In general, the fuse consists of (i) fusing element, (ii) ceramic-tube, and (iii) lugs.

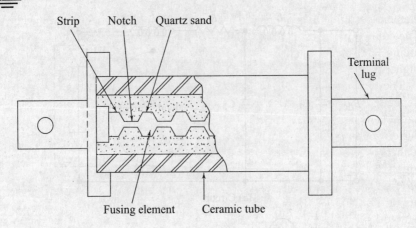

Fig. 13.22 *Fuse link-geometry [Semiconductor fuse and its cross section]*

The fusing element is made of silver with one or more necks as shown in Fig. 13.22. The number and form and these marks affect the fusing characteristics of the fuse. The body of the fuse is made of ceramic material. The large contact blade or lugs facilitate connection and dissipate heat. The fusing element is embedded in a special sand which can conduct heat and serves as a quenching medium for the arc at the time of fusing.

The fusing always occurs at the neck, as it restricts the flow of current. The large sections near the neck serve as heat sink and allow large current density in the restricted section of the neck. The stages in fusing are illustrated in Fig. 13.23.

When the fault current rises the fuse temperature also rises until $t = t_m$, at which time the fuse melts and arcs are developed across the fuse. Due to the arc, the impedance of the fuse is increased, thereby reducing the current. However, an arc voltage is formed across the fuse the generated heat vaporizes the fuse

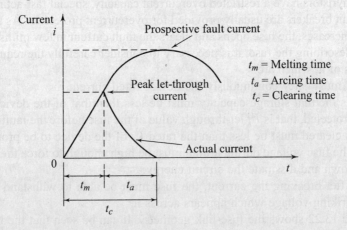

Fig. 13.23 *Fuse current*

element, resulting in an increased arc length and further reduction of the current. The cumulative effect is the extinction of the arc in a very short time. When the arcing is complete in time t_a, the fault is clear. The faster the fuse clears, the higher is the arc voltage.

The clearing time t_c is the sum of *melting time* t_m and *arc time* $t_a \cdot t_m$ is dependent on load current, whereas t_a is dependent on the power-factor or parameters of the fault-circuit. The fault is normally cleared before the fault current reaches its first peak, and the fault current, which might have blown if there was no fuse, is called the *prospective fault current*. This is shown in Fig. 13.23.

It may give adequate protection to place the fuses in the supply lines as shown in Fig. 13.24, particularly with a passive load. However, with a live load, that is, a motor, it is possible for the load to provide the fault current, in which case each individual thyristor can be fused as shown in Fig. 13.25. In general, larger the equipment rating, the more tendency is there to fuse the devices individually. It may be necessary to add inductance to limit the rate of rise of the fault current and hence avoid excessive stress on the fuse and device. The added inductance may be in the line or in series with the individual device. However, this inductance may affect the normal performance of the converter.

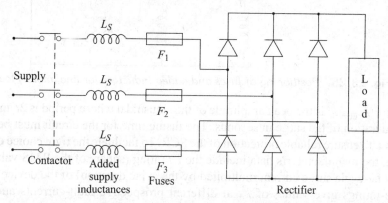

Fig. 13.24 *Positioning of fuses and added inductors for passive load*

13.7.1 Fuse–SCR Co-ordination in A.C. Circuits

A fast acting fuse can be used for protecting thyristors against large surge currents of very short duration, called *subcycle surge currents*. The coordination of the fusing time with the subcycle duration rating of the device is essential. Figure 13.26 shows the desired coordination between the ratings for SCRs and the fuses. The one-cycle surge rating $\hat{I}$ of an SCR is defined as the peak amplitude of the sinusoidal current which the SCR can carry for one half-cycle (10 ms on a 50 Hz basis). The subcycle-surge current rating may be obtained from

$$\hat{I}_{\text{subcycle}(t)} = \sqrt{\frac{\hat{I}^2 \times 1/100}{t}} \tag{13.9}$$

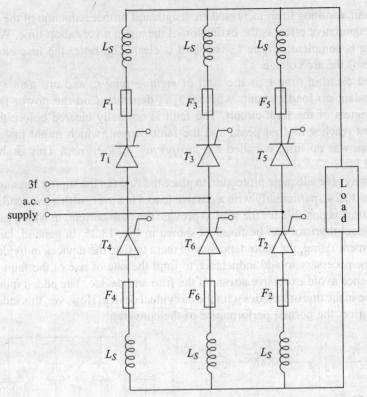

Fig. 13.25 *Positioning of fuses and added inductors for line (motor) load*

where $\hat{I}_{subcycle}$ is the peak amplitude of the sinusoidal whose period is $2t$ and t is the duration of the surge in seconds. The fusing time for the circuit must be less than t to ensure reliable protection of the SCR. To facilitate the right choice of the fuse, the manufacturer's data include the I^2t rating (square of the RMS value of the one-cycle surge current multiplied by the cycle duration) of the device. The fuse-ratings give values of I^2t at different prospective fault-currents and the corresponding peak let-through current.

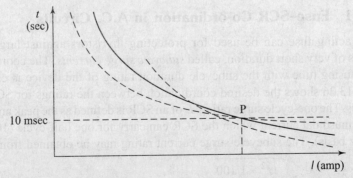

Fig. 13.26 *Fuse and circuit breaker coordination*

From these data, the fusing time for a given subcycle surge current can be computed. Assuming that the fault current waveform is triangular, the fusing time t_c will be

$$t_c = \frac{3.I_t^2}{I_p^2} \qquad (13.10)$$

where I_p is the peak magnitude of the let-through fault current.

The manufacturer's data on peak surge ratings of SCRs corresponds to peak value of rectified sinusoidal waveforms in a half-wave circuit operating at 50 Hz or 60 Hz. The one-cycle point (P of Fig. 13.26) will therefore give peak values of the noncurrent half-size wave of duration 10 ms. The fuse will provide protection when the current is higher than that corresponding to point P. For lower-currents, the device will usually be protected by the circuit-breaker.

13.7.2 Fuses in D.C. Circuits

Fuse protection in d.c. circuits presents greater problems than in a.c. circuits. Since natural periodic current zeros do not occur, there is no equivalent to the long-term current/time characteristics for fuses, and moderate or slowly rising fault-currents can produce a dangerous condition by allowing the fuse to arc continuously without breaking the current. The use of fuses in d.c. circuits is therefore subject to important restrictions regarding minimum prospective fault current, the maximum circuit time constant (L/R) under fault conditions, and the supply voltage. It is, therefore, necessary for the fuse manufacturers to supply separate d.c. peak let-through current values in relation to the fuse clearing time in order to coordinate with the SCR's sub-cycle capability.

13.7.3 Complete Protection of Converter

To illustrate the complete protection of a converter, Fig. 13.27 shows elements giving complete protection to surge currents and voltages, dv/dt and di/dt, and the contractors for the long-time low level overload. The value of R, C and reactors which may be satisfactory are indicated in Fig. 13.27.

SOLVED EXAMPLES

Example 13.8 A circuit having a prospective fault-current of 1.5 kA is protected by a fuse with I^2t rating of 150 A^2 sec on a 50 Hz basis. The faulted circuit is opened in 6 ms. Calculate the peak value of the fault current.

Solution: Assuming that the fault current waveform is triangular, the fusing time t_c will be given by

$$t_c = \frac{3I^2t}{I_p^2}$$

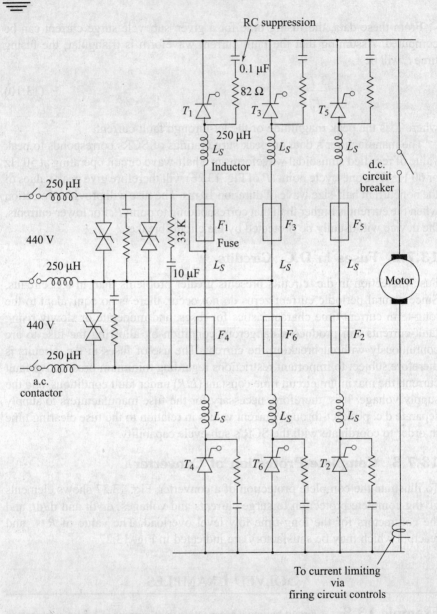

Fig. 13.27 *Complete protection of a converter*

$$\therefore \quad I_p^2 = \frac{3\,I_L^2}{t_c} = \frac{3 \times 150 \times 10^3}{6} \quad \therefore I_p = 273.86 \text{ A}.$$

Example 13.9 A particular SCR has a one-cycle non-repetitive surge current rating of 250 A on a 50 Hz basis. If the circuit is interrupted by a fuse before the SCR gets damaged, compute the required fusing time when the peak short-circuit current is 400 A. Assume the waveform to be sinusoidal.

Solution: From Eq. (13.9),

$$\hat{I}_{subcycle} = \sqrt{\frac{250^2 \times 1/100}{0.02}} = 176.78 \text{ A}^2\text{-s}.$$

Now, from Eq. (13.10), $t_c = \dfrac{3 \times 176.78}{(400)^2} = 3.31 \text{ m sec}.$

Example 13.10 An SCR has a di/dt_{max} rating of 15 A/μs. It is operated from a 150 V supply.
 (i) What is the minimum value of load inductance that will protect the device?
 (ii) If the recharge resistor of the snubber is 620 Ω and the load is 62 Ω, what will be the new value of L?

Solution:
(a) From Eq. (13.3), the inductance must satisfy the condition

$$L = \frac{E_{dc}}{(di/dt)_{max}} = \frac{150}{15} \times 10^6 = 10 \, \mu\text{H}$$

(b) In the presence of a recharge resistor R of 620 Ω, while $R_L = 62 \, \Omega$, the minimum inductance becomes,

$$L = \frac{E_{dc} \cdot \left(\dfrac{R+R_L}{R}\right)}{(di/dt)_{max}} = \frac{150\left(\dfrac{620+62}{620}\right)}{15} = 11 \, \mu\text{H}$$

13.8 GATE PROTECTION

13.8.1 Radio Interference Phenomenon

Generally, a number of devices are mounted close together in many power-control applications. When anyone of the devices is turned-on, the voltage across it collapses and current is established and due to this, associated electric field collapses and magnetic field is established. Now, when any-one of the devices is turned-off, the current across it collapses and voltage is established and due to this, associated magnetic field collapses and electric field is established. These changes in electric and magnetic fields give rise to induced voltages in the gate circuits of the nearby devices. These induced voltages may turn on the thyristors at wrong instances, causing maloperation of the entire control scheme. This phenomenon is known as *radio interference phenomenon*. Hence, the gate cathode circuit of the thyristor must be protected against this phenomenon.

Because of rapid changes in voltages and currents within a switching converter, power-electronic equipment is a source of electromagnetic interference (EMI) with other equipment as well as with its own proper operation. The EMI is transmitted in two forms: radiated and conducted. The switching converters supplied by the power lines generated conducted noise into the power lines that is usually several orders of magnitude higher than the radiated noise into free-space. Metal cabinets used for housing power converters reduce the radiated component of the EMI.

Figure 13.28 shows the conducted interference which consists of two categories commonly known as the differential mode and the common mode. The differential-mode noise is a current or a voltage measured between the lines of the source, that is, a line-to-line voltage or the line current I_{dm}. The common-mode noise is a voltage or current measured between the power lines and ground, such as I_{cm}. Both noises are generally present on input-lines as well as on the output lines. Any filter design has to take into account both of these modes of noise.

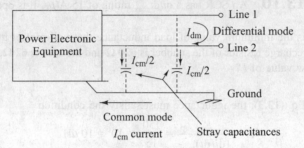

Fig. 13.28 *Conducted Interference*

Switching waveforms are inherent in all switching converters. These waveforms contain significant energy levels at harmonic frequencies in the radio-frequency region because of short-rise and fall times, several orders above the fundamental frequency. Transmission of the differential-mode noise is through the input-line to the utility system and through the dc side network to the load on the power converter. Moreover, conduction paths through stray capacitances between components and due to magnetic coupling between circuits must also be considered.

Transmission of common-mode noise is entirely through 'parasitic' or stray capacitors and stray electric and magnetic fields. These stray capacitances exist between various system components and between components and ground. For safety reasons, most power-electronic equipment has a grounded cabinet. The noise appearing on the ground line-contributes significantly to the EMI.

EMI Reduction: The most cost-effective-way of dealing with EMI is to prevent the EMI from being generated at its source, which can significantly reduce radiated and conducted interference. The effect of radio interference phenomenon is usually minimized by using the following elements and circuits:
  (i) Shielding
  (ii) R-F filters
  (iii) Zero-voltage switching.

(i) Shielding: The interference by external electric and magnetic fields induces spurious signals in the gate to cathode circuit. Hence, it is very important to protect the gate control by using shielded cables, as shown in Fig. 13.29. Generally, the radiated noise is effectively shielded by the metal cabinets used for housing the power-electronic equipment. Additional steps may be necessary if the power electronic equipment is operating near sensitive communication or medical equipment.

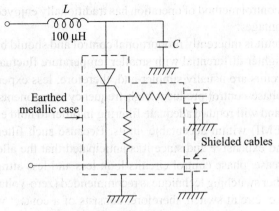

Fig. 13.29 *R-F filter and shielding*

(ii) R-F Filters: To minimize the radio interference of SCRs, an R-F filter, such as the one shown in Fig. 13.29, may be used. This filter consists of a small inductance which is connected in series of the SCR and shunt capacitor. Inductor slows down the rate of change of forward current and capacitor reduces the rate of decay of the forward-voltage. The values of L and C are small and can be determined for required attenuation in the ratio frequency band using the following relation:

$$X_L = R_L = X_C$$

where R_L is the load-resistance.

Now, for a radio-frequency band, the values of L and C will be

$$2\pi f_0 L = R_L = \frac{1}{2\pi f_0 C}$$

or
$$L = \frac{R_L}{2\pi f_0} \tag{13.11}$$

and
$$C = \frac{1}{2\pi R_L f_0} \tag{13.12}$$

where f_0 is the corner or breakpoint frequency, usually taken as 50 kHz, for the filter.

(iii) Zero Voltage Switching: As discussed above, the *RF* noise contribution of device is primarily due to sudden step in current as the device switches. In some applications, particularly in electric heating, satisfactory control may be contained by turning the devices ON at line voltage zeros, giving only complete half-cycles of current to the load. By eliminating the sudden steps of current, the *RF* noise contribution is brought to an absolute minimum. This eliminates the need for *RF* filter components, which, for longer heating load, can become quite large and costly.

The phase-control method of operation has traditionally enjoyed the following two main advantages:

(i) The circuit is inherently proportional control and should be able to maintain a tighter differential with smaller temperature fluctuations.

(ii) The circuits are usually simpler and, therefore, less expensive.

However, phase-control produces high-frequency components each time the switch closes and will require adequate filtering in order to hold electromagnetic interference (EMI) within acceptable limits. Because such filters can become rather large and expensive and since it is anticipated that the allowed limits on EMI will decrease, phase control circuits look less and less attractive for large loads and another switching technique is recommended (zero-voltage-switching).

The ideal a.c. circuit switch, therefore, consists of a contact which closes at the instant when voltage across it is zero, and opens at the instant when current through it is zero. This has become known as 'zero-voltage-switching'.

The half-wave, zero voltage switching circuit is shown in Fig. 13.30. This zero-voltage-switching circuit applies the voltage to the load in a.c. circuits when the instantaneous value of this voltage is going through the zero value. This avoids high rate of rise of current in purely resistive loads (e.g., furnace and lighting loads) and thereby reduces the generation of radio noise and hot-spot temperatures in the device carrying the load.

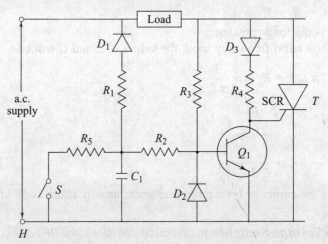

Fig. 13.30 *Ideal half-wave zero voltage switch*

Circuit Operation: The ON-OFF action of thyristor T_1 depends upon the state of transistor Q_1 (cut-off and conducting state) and the state of transistor Q_1 is controlled by the switch S. When transistor Q_1 is cut-off, positive-anode voltage on SCR T_1 causes the gate current to flow through D_3 and R_4, triggering SCR T_1 into conduction. When transistor Q_1 is biased into conduction, the current through

R_4 is shunted away from the gate of SCR T_1 through the collector of Q_1. Two following modes describe the circuit operation.

Mode 1 Consider the positive half-cycle and switch S closed. With this condition, the supply current has the first path, L-Load-R_3-R_2-R_5-N. This current causes a potential drop across R_3, R_2 and R_5. This potential drop is sufficient to drive transistor Q_1 in the saturation state, which offers negligible resistance. The supply current has the second path, L-Load-D_3-R_4-Q_1-N. Hence, SCR T_1 will not get the gate pulse and becomes off. Therefore, no voltage will be applied to the load.

Mode 2 Consider the negative half-cycle and switch S opened. With this condition, capacitor C_1 will charge to the peak of the supply voltage through R_1 and D_1. The charging path becomes N-G_+-C_1-R_1-D_1-L. As the a.c. supply voltage drops from its negative-peak, capacitor C_1 discharges through D_2 and R_2. This makes diode D_2 forward biased and this diode applies negative bias to transistor Q_1. The time constant for the discharge path is so chosen that D_2 will conduct for a part of the following positive half-cycle. Thus, the base–emitter junction of Q_1 is reverse biased in the beginning of the positive half-cycle and during this period, SCR T_1 will be triggered because all the current through R_4 will now flow through the gate. Therefore, voltage will be applied to the load.

Hence, it becomes clear that when switch S is closed, SCR T_1 will stop conducting at the end of the present or previous positive half-cycle and will not get triggered again. Similarly, when switch S is opened, SCR T_1 will be turned-on, only at the beginning of the following positive half-cycle of the applied-voltage. The associated load-voltage and capacitor-voltage waveforms are shown in Fig. 13.31.

Design Considerations:

(1) Resistance R_4 must be selected less than the ratio of supply voltage at which switching should occur (typically 3 to 5 V) to the maximum gate current required to trigger SCR T_1.

$$\therefore \quad R_4 < \frac{V_s}{I_{g\,max}}$$

Let us select a thyristor C106B having a maximum-gate current $I_{gmax} = 200\,\mu A$.

$$\therefore \quad R_4 = \frac{3}{200\,\mu A} = 15\,k\Omega$$

The lower limit of R_4 is determined by the collector current limit of Q_1.

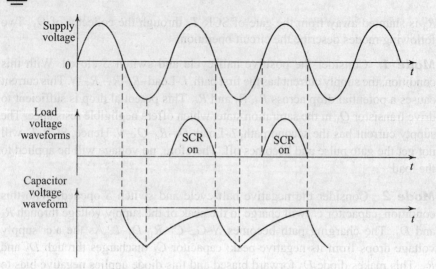

Fig. 13.31 *Load and capacitor voltage waveforms*

(2) The resistor R_3 is used to provide sufficient base drive to Q_1, to keep Q_1 in saturation throughout the cycle when C_1 is in the discharged state.
For transistor Q_1, we have the relation

$$I_C = \beta I_\beta$$

Also, from Fig. 13.30, $I_C = I_{gmax}$.
Let us select a transistor 2N5172 having a current gain of 15

$$\therefore \quad I_B = \frac{200\ \mu A}{15} = 13.33\ \mu A$$

Also, $\quad R_3 \approx \dfrac{V_s}{I_B} \approx \dfrac{3}{13.33\ \mu A} \approx 225\ k\Omega$

(3) Resistance R_2 should be substantially less than R_3.
$R_2 \approx 47\ k\Omega$.

(4) The time constant $R_1 C_1$ must be sufficient to extent bias current for Q_1 into the positive half-cycle, and hence should be approximately $\dfrac{1}{2}f$, where f is the supply-frequency.

For 50 Hz frequency, $R_2 C_1 = \dfrac{T}{2} = 0.01$

$$\therefore \quad C_1 = \frac{0.01}{47 K} = 0.22\ \mu F.$$

(5) The resistor R_5, which limits the capacitor discharge current through the contacts when S is closed, must be low compared to R_1. This will prevent C_1 from charging when the control contacts are closed.

$$\therefore \quad R_5 \approx 10\ K.$$

Protection and Cooling of Power Switching Devices

(iv) Use of Snubber: From the point of view of EMI reduction, a properly designed snubber is quite effectively, since it reduces both the dv/dt and di/dt of the circuit. The snubber must be connected directly on the component being snubbed—with as short leads as possible.

(v) Other Preventive Measures: The magnitudes of coupling fields should be reduced by proper mechanical layout, wiring and shielding. To reduce magnetic fields, it is important to minimize the net area enclosed by a current loop. All current loops with switching transients should be made to have as small an area as possible. All current carrying conductors should be run in close proximity to the return wire, such as by copper strips. A twisted pair of wires will reduce the generated external field to a minimum.

To reduce stray capacitances, the area of exposed metal at the switching potential should be minimized and kept as far from the ground as possible by proper mechanical design.

SOLVED EXAMPLE

Example 13.11 R-F filter is used to protect the device from R-F phenomenon, with load resistance of 120 Ω. Compute the values of RF filter components by assuming the operating frequency of 50 kHz.

Solution: From Eq. (13.11), $L = \dfrac{120}{2\pi \times 50 \times 10^3} = 381.97\,\mu H$.

and from Eq. (13.12), $C = \dfrac{1}{2\pi \times 120 \times 50 \times 10^3} = 0.026\,\mu F$.

13.9 HEAT SINKS

13.9.1 Temperature Control in Semiconductor Devices

Power dissipation in electrical components raises the internal temperature and affects performance and reliability. A high internal temperature may be detrimental to the physical structure of the component. In the designing of the small-size lightweight power electronic systems, removal of heat and control of internal temperature profile are very important considerations.

The theoretical upper limit on the internal temperature of a semiconductor device is the temperature at which the intrinsic carrier density of the most lightly doped region is equal to the majority carrier doping density in that region. The p-n junction would lose rectifying property at this limit temperature. The maximum allowed temperature is far below this theoretical limit manufacturers specify a maximum temperature, known as maximum junction temperature – T_{jmax}. This is also known as the worst-case temperature in the design procedure. Typical semiconductor packages are designed for T_{jmax} of 125°C.

13.9.2 Heat Transfer

Power-losses in semiconductor devices appear in the form of heat. The accumulation of heat energy increases the temperature of the internal structure of devices. The heat must be removed from the body of the device by transferring it to the ambient environment. Heat transfer takes place in three ways:

(i) Conduction (ii) Convection and/or (iii) Radiation.

Conduction is the heat transfer among stationary interfaces by the vibratory motion of atoms or molecules. Convection is the mechanical transport of heat by a moving fluid or gas. The fluid flow may be natural or forced. Fluid-flow in natural convection is caused by temperature gradients. The forced convection is created by pumps or fans.

Heat energy is converted into electromagnetic radiation in heat transfer by radiation which is absorbed by other components in the vicinity. In enclosed packages the radiation from hot devices may be absorbed by neighbouring devices at the lower temperatures. As a consequence, the neighbouring components are required to function at elevated temperatures something they were not planned to do. Shiny sheet metal partitions serve as radiation shields.

The heat transfer in power-electronic systems is dominated by conduction and convection processes.

13.9.3 Thermal Resistance and Thermal Model

Thermal energy flow takes place from a region of high temperature to a region of lower temperature. Similarly, in an electric circuit, current flows from a point of high potential to a point of lower potential. Heat transfer is therefore analogous to current flow. Thus, there is an analogy between thermal-power flow and current flow as given in Table 13.2. The thermal resistance is denoted by Θ and is a measure of the temperature difference per watt of heat flow (°C/W), just as electrical resistance, R, is a measure of the voltage difference per ampere of current flow in ohms.

Table 13.2

S.No.	Electrical quantities	Thermal quantities
1.	Potential difference, (V)	Temperature difference (°C)
2.	Current (A)	Thermal power, or rate of heat transfer (W)
3.	Electrical resistance (ohms)	Thermal-resistance (°C/W).

By analogy with Ohm's law, the steady-state temperature difference $(T_1 - T_2)$ when a constant thermal power of P_{av} watts flows through a thermal resistance Θ_{12} °C/W is given by

$$\Theta_{12} = \frac{T_1 - T_2}{P_{av}} \text{ °C/W}, \quad T_1 > T_2 \qquad (13.13)$$

Using the electrical analogy, heat flow from semiconductor junction to ambient fluid can be represented by the thermal equivalent circuit of Fig. 13.32 in which the thermal power of P_{av} W at the junction is analogous to a constant current source.

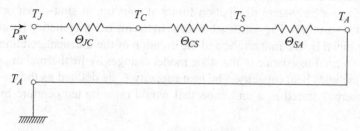

Fig. 13.32 *Thermal equivalent circuit*

As shown in Fig. 13.32, the heat generated at the junction flows from the junction to the case, from case to the heat-sink, and from the heat-sink to the surrounding atmosphere. The steady-state temperature difference between junction and ambient is given by

$$T_J - T_A = P_{av}(\Theta_{JC} + \Theta_{CS} + \Theta_{SA}) \qquad (13.14)$$

where

- Θ_{JC} = thermal resistance from junction to case, °C/W.
- Θ_{CS} = thermal resistance from case to sink, °C/W.
- Θ_{SA} = thermal resistance from sink to ambient, °C/W.
- T_J = ambient temperature, °C.

Also,

$$P = \frac{T_J - T_C}{\Theta_{JC}} = \frac{T_C - T_S}{\Theta_{CS}} = \frac{T_S - T_A}{\Theta_{SA}} = \frac{T_J - T_A}{\Theta_{JA}} \qquad (13.15)$$

where $\Theta_{JA} = \Theta_{JC} + \Theta_{CS} + \Theta_{SA}$, the total thermal resistance between junction and ambient.

The junction-to-case thermal resistance, Θ_{JC}, is specified in the deivce data sheet. The thermal resistance across the case-to-sink interface, Θ_{CS}, depends on the size of the device case because the size affects the contact area between the device and its heat-sink. The quality of the contact, also important, is dependent on the flatness of the surfaces, the clamping pressures and whether a thermally conducting grease is used between the surfaces. Practical values of Θ_{CS} vary between 0.05°C/W and 0.5°C/W, and the device data sheet may specify a typical value, assuming correct installation procedures and the use of interface thermal grease. If the case is electrically isolated from the heat sink with a thin insulating wafer, Θ_{CS} may be substantially increased.

The sink-to-ambient thermal resistance, Θ_{SA}, is measured from the device mounting area, where the heat-sink is hottest, to cooling fluid. Its value depends on parameters such as heat-sink material, surface area and finish, volume occupied, and air flow. For a natural convection heat-sink, the thermal resistance can be as low as 0.5°C/W, but for lower values, heat-sink size is excessive and forced convection cooling or liquid cooling is necessary.

13.9.4 Transient Thermal Impedance

The instantaneous dissipation in the semiconductor devices may greatly exceed the average steady-state dissipation limits at start-up, at shut-down or during steep load variation. The permissible surge in the maximum junction temperature is determined by the magnitude and the duration of the transient condition.

The thermal resistance of the static model changes to the thermal impedance, which includes heat capacity. The heat capacity C_s is defined as the amount of heat (energy) stored in a unit mass that would raise its temperature by a unit degree:

$$C_s = \frac{dH}{dt} \quad \text{where } dH = Q \cdot dt,$$

Rate of flow of heat is termed as the thermal power Q, $Q = \dfrac{dH}{dt}$.

$\therefore$ Thermal power $\quad Q = C_s \cdot \dfrac{dT_A}{dt} \quad$ (13.16)

Here, the temperature T_A is referenced to ambient temperature. This equation is analogous to i-v relation of a capacitor in electrical systems:

$$i = C \cdot \frac{dv}{dt} \quad (13.17)$$

Heat capacity C_s is analogous to the electrical capacitance C_s is reference to the ambient. Masses in a thermal system constitute the heat storage devices. Figure 13.33 shows a thermal mass and its equivalent thermal circuit with heat circuit included.

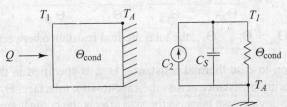

Fig. 13.33 *Thermal equivalent circuit in transient heat transfer*

From Fig. 13.33, thermal impedance using Laplace-transform is

$$T_1 = Z_\theta \cdot Q \quad \text{where } Z_\theta = \frac{\theta}{1 + S_\theta \cdot C_s} \quad (13.18)$$

For a step-change in the amount of heat imparted, that is to say, upon application of a step power input Q_0, the steady-state rise in the temperature is given by

$$T_{1,ss} = T_{1,0} + \theta \cdot Q_0 \quad (13.19)$$

The instantaneous temperature follows an exponential path,

$$T_1 = T_{1,0} + \theta \cdot Q_0 (1 - e^{-t/\tau})$$

where $\quad \tau = \theta \cdot C_s \quad$ (13.20)

Rearranging the variables in the above equation,

$$T_1 = T_{1,0} + Z_{\theta(t)} \cdot Q_0 \quad (13.21)$$

and
$$Z_\theta = \Theta \cdot (1 - e^{-t/\tau}) \quad (13.22)$$

Impedance Z_θ can be said to be an exponential function of time, as shown in Fig. 13.34.

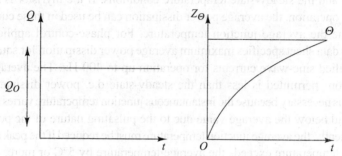

Fig. 13.34 Transient thermal impedance

13.9.5 Heat-Sink Specifications

The thyristor data sheet specifies the maximum allowable junction temperature, $T_{J(\max)}$ and Θ_{JC}. When Θ_{CS}, is specified or estimated, Eq. (13.14) can be used to determine the largest permissible heat-sink thermal resistance, Θ_{SA}, and hence the size of heat-sink required for a given steady-state power dissipation, P_{av}, and ambient temperature, T_A. The continuous d.c. current rating of the device can then be evaluated with the use of the on-state voltage current characteristic. Conversely, if the heat-sink has been specified, the junction temperatures can be determined for a given on-state current and ambient temperature.

As an example, consider a particular SCR, whose data sheet specifies $T_{J(\max)} = 125°C$, $\Theta_{JC} = 0.4°C/W$ and $\Theta_{CS} = 0.1 \ °C/W$. It is proposed to use this device in an application where the thyristor dissipates 100 W and the ambient temperature is 35°C. Substituting in Eq. (13.14) gives the required value of Θ_{SA} as 0.4°C/W. The circuit designer must now examine a heat-sink manufacturer's catalogue and select a heat-sink that has this value of thermal resistance or less, at power dissipation of 100 W. The heat-sink catalogue shows that natural convection thermal resistance decreases somewhat with increasing power dissipation and temperature rise above ambient.

The junction time, T_J, is not accessible to the user; consequently, the device manufacturer may define the allowable power dissipation and current capability as a function of the maximum allowable case temperature, $T_{C(\max)}$, which results in a junction temperature of $T_{J(\max)}$. For a given power dissipation, P_{av} and ambient temperature, T_A, the user now selects a heat-sink which maintains the case temperature at, or below, the specified value. Neglecting Θ_{CS}, the heat-sink thermal resistance is given by

$$\Theta_{SA} = \frac{T_C - T_A}{P_{av}} \quad (13.23)$$

As an example, assume a power semiconductor dissipates 100 W for operating conditions for which the data sheet specifies a maximum case temperature of 95°C. If the ambient temperature is 35°C, the maximum permissible heat-sink thermal resistance is given by Eq. (13.23) as 0.6°C/W.

The above equations and calculations have assumed d.c. current flow in the thyristor and the steady-state temperature conditions. If the thyristor is subjected to cyclic operation, the average power dissipation can be used in these equations to determine the average junction temperature. For phase-control applications, a thyristor data sheet specifies maximum average power dissipation for square-wave and rectified sine-wave currents for operation up to 400 Hz. The average power dissipation permitted is less than the steady-state d.c. power dissipation. This derating is necessary because the instantaneous junction temperature varies cyclically above and below the average value due to the pulsating nature of the power loss. Consequently, the average junction temperature must be reduced if the peak allowable junction temperature exceeds the average temperature by 5°C or more.

13.10 THYRISTOR MOUNTING TECHNIQUES

The internal power losses in a thyristor cause high thermal stresses which further give rise to mechanical forces. Therefore, a thyristor must be braced to withstand such mechanical forces. In addition, the SCR mounting must be so designed as to facilitate heat flow from junction to the case. In the following paragraphs, five major mounting techniques have been discussed, which depend upon the low or high power ratings of the thyristors:

1. Lead Mounting Figure 13.35 shows a lead-mounted thyristor. For load current rating of about one ampere, lead mounted SCRs are used. Such SCRs do not require any additional cooling or heat-sink. Their housings dissipate sufficient heat by radiation and convection.

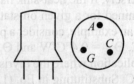

Fig. 13.35 *Lead mounting*

2. Stud Mounting Figure 13.36 shows a stud-mounted thyristor. The stud mounted SCR is a flexible component and has wide acceptance. This type of SCR uses a copper or aluminium stud with a machine thread for making mechanical electrical and thermal contact to a heat exchanger of the user's choice. The threaded stud can have different sizes. The SCR is attached to heat-sink by means of threaded stud and nut. Thus anode gets electrically connected to the heat-sink. This method of mounting is common with small and medium sized thyristors.

Fig. 13.36 *Stud mounting*

3. Bolt-down Mounting Figure 13.37 shows a bolt-down mounted thyristor. The device has flanges or tabs which usually contain one or more holes. Bolts are placed through these holes in order to attach the device to the heat-sink. This type of mounting is common in medium and small ratings. This is also called as flat pack mounting.

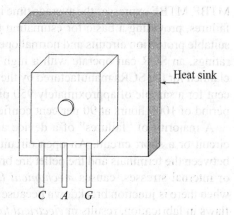

Fig. 13.37 *Bolt-down mounting*

4. Press-fit Mounting Figure 13.38 shows an SCR press-fit package. This package is designed primarily for forced insertion into a slightly undersized hole in the heat-exchanger. When properly mounted, this type of SCR has a lower-thermal drop to the heat exchanger than the stud type mounting. Also, in high volume applications the cost of this type of mounting is generally less than that for the stud-type of SCR. This type of mounting is employed for thyristors of a large rating.

5. Pressure Mounting This mounting technique is shown in Fig. 13.39. This device is clamped under a very large external force. Such devices are also called Hockey-puk SCRs because of their shape. The clamping force is applied smoothly, evenly and perpendicularly to Hockey-puk to insure that there is no deformation to either the pole faces or the heat-exchangers during mounting. It is possible to have double side cooling using air cooled or liquid cooled heat exchangers. This type of mounting is used for thyristors of very high-current ratings.

Fig. 13.38 *Press-fit mounting* Fig. 13.39 *Pressure mounting*

13.11 SCR RELIABILITY

Reliability may be defined as the probability of performing a specific function under given conditions for a specific period of time. In the case of large systems, the common unit of reliability measurement is Mean Time Between Failures, or

MTBF. MTBF expresses the average time in hours that a system operates between failures, providing a basic for estimating the cost of system maintenance. With suitable protection circuits and normal operation within the permissible limits of ratings, an SCR can operate with a high degree of reliability. The failure rate claimed for the SCRs manufactured by the General Electric Company is 0.41 per cent for a sample of approximately 950 pieces of C35 type SCRs, in a working period of 1000 hours at 90 per cent confidence limits.

A majority of "failures" of a device are due to the occurrence of an open-circuit or a short circuit. An open circuit, which occurs when ohmic contacts between the terminals and the pellet are broken due to mishandling of the device or internal stresses, causes *mechanical failure*. A short circuit, which occurs when there is junction breakdown because the device ratings are exceeded due to flaws in fabrication, results in *electrical failure*. Sometimes, failure is caused by inaccurate design of the heat-sink or inadequate cooling arrangements, this is called *thermal failure*. Since a device that has failed will have no switching ability, it cannot be used for power control. Therefore, care should be taken first in the selection of the device and then for its necessary protection.

SOLVED EXAMPLES

Example 13.12 A thyristor with a steady power loss of 30 W has a junction to heat-sink thermal resistance of 0.7°C/W. Determine the value of the thermal resistance, the heat sink can have if the ambient temperature is 40°C and the junction temperature is limited to 125°C. Give the base temperature at this condition.

Solution: The total thermal resistance = $(125 - 40)/30 = 2.83$°C/W.
The thermal resistance of the heat-sink = $2.83 - 0.7 = 2.13$°C/W.
Temperature at the base = $40 + (30 \times 2.13) = 104$°C.

Example 13.13 A 440 V, 960 A, thyristor converter is supplying to a load continuously. Determine junction-temperature with following details:
Ambient temperature = 80°C
On-state power loss = 150 watts.
Thermal resistance:
$$Q_{JC} = 0.15°C/W, Q_{CS} = 0.075°C/W, Q_{SA} = 0.45°C/W.$$

Solution: From Eq. (13.14),
$$T_J = 80 + 150(0.15 + 0.075 + 0.45) = 181.25°C.$$

Example 13.14 For a thyristor, maximum junction temperature is 150°C. The thermal resistances for the thyristor–sink combination are $\Theta_{JC} = 0.015$°C/W and $\Theta_{CS} = 0.08$°C/W. For a heat-sink temperature of 60°C, determine the total average power loss in the thyristor-sink combination.

In case the heat-sink temperature is brought down to 50 °C by forced cooling, find the percentage increase in the device rating.

Solution: From Eq. (13.14), $P_{av_1} = \dfrac{150° - 60}{0.015 + 0.08} = 947.37$ W

with improved cooling, $P_{av_2} = \dfrac{150 - 50}{0.095} = 1052.63$ W

Thyristor rating is proportional to the square root of average power loss.

$\therefore$ Percentage increase in thyristor rating = $\dfrac{\sqrt{1052.63} - \sqrt{947.37}}{\sqrt{947.37}} = 5.39\ \%$

Example 13.15 The power loss of a device is shown in Fig. 13.40. Plot the instantaneous junction temperature rise above the case. $p_2 = p_4 = p_6 = 0$, $p_1 = 1000$ W, $p_3 = 1400$ W, and $p_5 = 700$ W. For $t_1 = t_3 = t_5 = 1$ ms, the data sheet gives:
$\theta(t = t_1) = \theta_1 = \theta_3 = \theta_5 = 0.035°$C/W.
For $t_2 = t_4 = t_6 = 0.5$ ms, $\theta(t = t_2) = \theta_2 = \theta_4 = \theta_6 = 0.025$ °C/W.

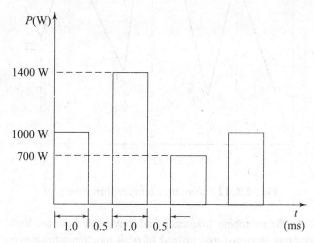

Fig. 13.40 *Thyristor power loss*

Solution: We calculate the junction temperature rise as:

$\therefore\ \Delta T_J(t = 1\text{ ms}) = T_J(t = \text{ms}) - T_{JO} = \theta_1 P_1 = 0.035 \times 1000 = 35°$C

$\Delta T_J(t = 1.5\text{ ms}) = 35 - \theta_2 P_1 = 35 - 0.025 \times 1000 = 10°$C.

$\Delta T_J(t = 2.5\text{ ms}) = 10 + \theta_3 P_3 = 10 + 0.035 \times 1400 = 59°$C.

$\Delta T_J(t = 3\text{ ms}) = 59 - \theta_4 P_3 = 50 - 0.025 \times 1400 = 15$ °C.

$\Delta T_J(t = 4\text{ ms}) = 15°\text{C} + \theta_5 P_5 = 15 + 0.035 \times 700 = 39.5°$C

$\Delta T_J(t = 4.5\text{ ms}) = 39.5 - \theta_6 P_5 = 39.5 - 0.025 \times 700 = 22°$C

The junction temperature rise above the case is plotted in Fig. 13.41.

Example 13.16 A power transistor is switching a highly inductive load. Determine the heat-sink required if the circuit has following data:

(i) $V_{CC} = 100$ V, $V_{CE(on)} = 1$ V, $I_{on} = 20$ A
(ii) $t_{on} = 1$ μs, $t_{off} = 2$ μsec, $f_s = 10$ kHz
(iii) Duty cycle $\delta = 0.9$
(iv) $T_A = 35°C$, $T_{J(max)} = 125°C$, $\theta_{jc} = 0.7°C/W$, and $\theta_{CS} = 0.1°C/W$.

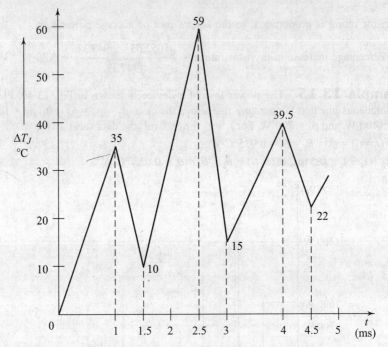

Fig. 13.41 *Junction temperature rise plot*

Solution: As the switching frequency and duty cycle both are high, the average junction temperature is considered instead of peak junction temperature.

$$\therefore \quad P_{cond.} = V_{CE(on)} \cdot I_C \cdot \delta = 1 \times 20 \times 0.9 = 18 \text{ W}$$

The switching loss for power-transistor switching inductive load is given by

$$P_{loss} = \frac{V_{CC} \cdot I_{on}}{2}(t_{on} + t_{off})f_s = \frac{100 \times 20}{2}(3 \times 10^{-6}) \times 10 \times 10^3 = 30 \text{ W}$$

Now, Total power loss, $P_{diss} = P_{cond} + P_{loss} = 48$ W

$$\therefore \quad P_{diss} = \frac{T_{J(max)} - T_A}{\theta_{jc} + \theta_{cs} + \theta_{sa}} = \frac{125 - 35}{0.7 + 0.1 + \theta_{sa}} \quad \therefore \theta_{sa} = 1.1°C/W$$

Example 13.17 Determine the thermal resistance of the heat sink needed to operate a MOSFET conducting a repetitive 20 A rectangular current waveform. On-time is 10 μs, duty-cycle is 0.1% and the ambient temperature is 40°C. Assume $R_{ds(on)}$ at 150°C and 20 A is 5 Ω, and $\theta_{js} = 1.5$ °C/W.

Solution: Since the on-state duty cycle and switching frequency are both very low, therefore switching losses can be neglected.

$$\therefore \quad P_{diss} = P_{cond.} = I_D^2 R_{ds(on)} = (20)^2 \times 5 = 2 \text{ kW}$$

Now, $\quad P_{peak} = \dfrac{T_{j(max)} - T_c}{Z_{\theta jc}} = \dfrac{T_{j(max)} - T_c}{T_{pr} \cdot \theta_{jc}} \quad \therefore \quad P_{peak} = \dfrac{150 - T_C}{0.03 \times 1.5} = 2 \times 10^3$

$\therefore \quad T_C = 60°C$

Now, $\quad P_{diss} = \dfrac{T_C - T_A}{\theta_{c-a}} = \delta \cdot \text{Peak} \quad \therefore \quad 0.001 \times 2 \times 10^3 = \dfrac{60 - 40}{\theta_{c-a}}$

$\therefore \quad \theta_{c-a} = 10°C/W$

Example 13.18 A power device has a thermal capacity of 0.2 $J°/C$ and a thermal resistance of 0.7°C/W. Determine the maximum power dissipation the power device can withstand for 0.1 second for a temperature not exceeding 40°C.

Solution: Given: Thermal storage capacity, $C_s = 0.2 J/°C$
Thermal resistance, $\theta = 0.7°C/W$
Junction temp., $T_J = 40°C$
During short period transients, the temperature rise of the junction is given by,

$$T_J = T_{J\,max}(1 - e^{-t/\tau}) \tag{13.24}$$

where thermal time constant, $\tau = \dfrac{\text{thermal storage capacity }(C_s)}{\text{power dissipated per °C rise}} \tag{13.25}$

Now, power dissipated per °C rise $= \dfrac{1}{\text{thermal resistance}} \tag{13.26}$

$= 1/0.7 = 1.428$ W/°C.

$\therefore \quad \tau = \dfrac{0.2}{1.428} = 0.14 \text{ sec}$

If a temperature increases to 40°C after 0.1 seconds, then from Eq. (13.24),

$$40 = T_{j(max)}(1 - e^{-0.1/0.14}) \quad \therefore \quad T_{J\,max} = 78.43°C$$

$\therefore \quad$ Maximum junction temperature, the device can withstand, is 78.43°C.
$\therefore \quad$ The power the device can withstand

$= T_{J(max)} \times P_{diss}$ per 1°C rise

$= 78.43 \times 1.428$

$= 112$ Watts

REVIEW QUESTIONS

13.1 What are the causes of voltage-transients on load side and supply side? How are these transients suppressed?

13.2 What happens if di/dt applied to a device exceeds its di/dt rating? How can this di/dt be limited below rated value?

13.3 What are the causes of overvoltages and overcurrent in thyristor circuits.

13.4 Explain in detail the various overvoltage conditions of an SCR.

13.5 What do you mean by snubber circuit. Draw and explain the function of each component.

13.6 Give the design details of snubber network for d.c. circuit.

13.7 Explain in detail the following overvoltage protecting devices:
(i) Metal oxide varistors
(ii) Selenium surge-suppressors
(iii) Avalanche diodes.

13.8 Explain in detail, the practical overvoltage protection in naturally commutated circuits.

13.9 Explain clearly the overvoltage protection in forced-commutated circuits.

13.10 Give the details of polarised and nonpolarised selenium protections. How they can be used for SCR protection?

13.11 What is meant by radio interference of SCRs? How is it minimized?

13.12 Explain in detail the various overcurrent conditions.

13.13 What is heat-sink? What is its necessity? What are the different type of heat-sink? How is a heat-sink mounted? Why is it coated black?

13.14 Explain how fault-current is caused by the commutation failure?

13.15 List the various demanding requirements of a fuse.

13.16 Explain the following terms in relation to fuse:
(1) Prospective fault current
(2) Melting time
(3) Arcing time, and
(4) Cleaning time.

13.17 Write a detailed note on fuse–SCR coordination in a.c. circuits.

13.18 Define one-cycle surge current rating and subcycle surge current rating of an SCR.

13.19 Explain the heat transfer process in thyristor?

13.20 Draw the thermal equivalent circuit for an SCR and discuss the various parameters involved in it.

13.21 Describe the terms electrical failure, mechanical failure and thermal failure as related to thyristors.

13.22 Explain the various thyristor mounting techniques in detail.

13.23 What are the problems of fusing d.c. circuits?

13.24 Explain heat-sink specifications in detail.

PROBLEMS

13.1 A circuit having a prospective fault current of 2kA is protected by a fuse with I^2t rating of 150 A^2– sec on a 50 Hz basis. The faulted circuit is opened in 4 msec. Compute the peak value of the fault current.

[*Ans* 335.41 A.]

13.2 (a) The maximum dissipation in a thyristor in a converter circuit is 30 W. From the data, the maximum case temperature for this load is specified to be 130°C. If the maximum ambient temperature is likely to be 50°C, what is the thermal resistance of the heat-sink?

(b) In part (a), if the maximum case temperature rise above the ambient is specified to be 100°C, can the circuit supply 330 kW load for 5 minutes, other conditions being same? Assume the thyristor loss proportional to the load. [Ans (a) 2.9°C/W, (b) Yes, temp. rise = 90°.]

13.3 A thyristor circuit is supplying constant load of 100 kW at a constant conduction angle with square waveforms. The case temperature rise after 5 minutes was 20°C and after 10 minutes was observed to be 30°C. Estimate the temperature rise of the heat-sink after the steady-state condition is reached.
[Ans = 40°C.]

13.4 An a.c. single phase supply of 230 V, 50 Hz has to be protected against overvoltages using selenium diodes. Each selenium plate can handle 30 V. Design a suitable circuit. [Ans Total number of plates = 16.]

13.5 A thyristor of thermal resistance 1.8°C/W, is mounted on a heat sink of thermal resistance 2.8°C/W. Compute the maximum power loss of the thyristor if the junction temperature is not to exceed 125°C in an ambient of 40°C.
[Ans 22.4 W.]

13.6 A thyristor has with its heat-sink a thermal resistance of 0.2°C/W steady-state and a 100 ms value of 0.05°C/W. What power loss can the thyristor tolerate for 100 ms if the junction temperature is not to exceed 125°C following a steady power loss of 300 W, the ambient temperature being 30°C? [Ans 1000 W.]

13.7 Compute the on-state power loss of a 440 V, 800 A thyristor converter operating continuously at rated load with following data:
$T_A = 45°C$, $T_J = 120°C$, $Q_{JS} = 0.1°C/W$, $Q_{SA} = 0.08°C/W$.
[Ans P = 416.67 W.]

REFERENCES

1. C W Lander, *Power Electronics*, McGraw-Hill (U.K. Ltd. 1981).
2. Ramshaw, *Power Electronics*, ELBs Chapman and Hall London. 1973.
3. R K Sugandhi and K K Sugandhi, *Thyristors Theory and Applications*, Wiley Eastern Ltd., New Delhi 1984.
4. Ramamoorthy, *An Introduction to Thyristors of Their Applications*, East-West Press, New Delhi 1978.
5. F Csaki, *Power Electronics* Akademia Kiado, Budapest 1980.
6. Mohan/undeland/Robbius/ *Power Electronics,* Wiley 1995.
7. M.H. Rashid, *Power-Electronics* Pearson Education, 2002.

Chapter **14**

Control of D.C. Drives

LEARNING OBJECTIVES:
- To introduce the basic principle of d.c. machines.
- To consider the various schemes of a d.c. motor speed control.
- To examine the operation of single-phase and three-phase half-wave, semiconverter and full converter controlled drives.
- To examine the operation of d.c. chopper drives.
- To examine the operation of converter-controlled and chopper-controlled d.c. drives.
- To introduce the phase-locked loop control of d.c. drives.
- To introduce the microcomputer control of d.c. drives.

14.1 INTRODUCTION

Today, industries are increasingly demanding process automation in all sectors. Automation results into better quality, increased production and reduced costs. The variable speed drives, which steplessly control speed of a.c./d.c. motors are indispensable controlling elements in automation systems. Depending on the application, some of them are fixed speed and some of them variable speed. The variable speed drives, till a couple of decades back, had various limitations, such as poor efficiencies, larger space, lower speeds, etc. However, the advent of power electronics transformed the scene completely and today we have variable drive systems which are not only smaller in size but also very efficient, highly reliable and meeting all the stringent demands of the various industries of modern era.

Direct current d.c. motors have been used in variable speed drives for a long time. The versatile control characteristics of d.c. motors have contributed to their extensive use in industry. d.c. motors can provide high starting torques which are required for traction drives. Control over a large speed range, both below and above the rated speed can be easily achieved. The methods of speed-control are simpler and less expensive than those of alternating current motors.

The invention of power semiconductors saw the advent of d.c. drive systems for most of the early variable speed requirements. Based on the advantages of simple construction and ease of control, this technology continued to be taken up for further improvements all over the world till very recently. No wonder that today this technology is well understood and proven which resulted in its popularity the world over. Almost perfected as far as the controller is concerned, this technology has most of its drawbacks due to d.c. motor. As is well known, d.c. motors have inherent disadvantages in that they need regular maintenance, they are tailor-made hence not readily available for replacements and are bulky in size. Added to this, due to the commutator sparking, they are simply not suitable in hazardous areas like chemical and petrochemical plants or in mines. Most of the variable-speed drives installed at present in works are d.c. drives.

For ratings above 500 kW, manufacture of d.c. motor itself poses problems. With these serious limitations, d.c. drive systems become unsuitable for energy saving applications of pumps or fans.

Both series and separately excited d.c. motors are normally used in variable-speed drives but series motors are traditionally employed for traction applications. At present, separately excited d.c. motors controlled by thyristor converters are the most widely used motor drive systems in industry. The thyristor converter provides variable armature voltage for the drive motor. The three basic methods for obtaining a variable d.c. output voltage from a fixed supply voltage ac or d.c. are phase control, integral cycle control, and chopper control. In all these methods, thyristors connect the supply to and disconnect it from the motor terminals. The frequency of switching is rapid. Therefore, the motor responds to the average output voltage level and not to the individual voltage pulses.

Thyristor d.c. drives frequently require sophisticated control systems. Both analog and digital feedback controls are used. Phase-locked loop control techniques are employed in some d.c. drives to provide precise speed control and essentially zero speed regulation. Microcomputer control of complex drive systems can provide great operating flexibility when required.

14.2 BASIC MACHINE EQUATIONS

The d.c. machine consists of a stationary field winding and a rotating armature winding as shown in Fig. 14.1. The field winding is supplied with an a.c. current to produce a static magnetic field pattern within the machine. This magnetic field then interacts with the current in the armature conductors to produce a torque. In order to sustain this torque the armature current distribution must be maintained relative to the field, irrespective of the actual rotor position. This is achieved by the action of the commutator, which reverses the current in the armature conductors as they pass from under one field pole to the next. Also, as the armature conductors are moving through the magnetic field produced by the field winding they have induced in them a back E_{mf} (E_b) which appears at the commutator.

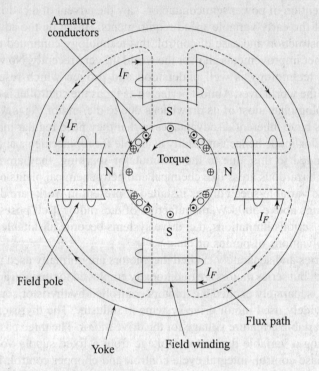

Fig. 14.1 4-pole d.c. machine

When motoring, the d.c. machine draws power from the d.c. source and the torque developed by the machine acts to rotate the armature against the applied mechanical load. In the generating mode, the torque developed opposes the applied mechanical torque driving the armature. The defining steady-state equations are, with reference to Fig. 14.2:

Generating $E_b = E + I_a R_a$ (14.1)

Motoring $E_b = E - I_a R_a$ (14.2)

$E_b = K \phi \omega$ (14.3)

where ϕ = flux per pole

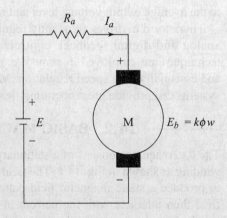

Fig. 14.2 Steady-state equivalent circuit of the armature circuit of a d.c. motor

and ω = rotational speed in radians per second.

Internal mechanical power = $T.W = E_b T_a$ (14.4)

Therefore, Torque $T = k \phi I_a$ (14.5)

The field winding can either be supplied from a separate d.c. source (shunt connection) or be connected as part of the armature circuit (series connection). Figure 14.3 shows these connections together with the motoring torque/speed characteristics that result.

Motoring Neglecting armature resistance, from Eqs (14.2) and (14.4),

$$E = E_b = K \phi \omega \quad (14.6)$$

If E is held constant it can be seen from Eq. (14.6) that the motor speed (ω) can be controlled by varying the field (ϕ) such that,

$$\omega \propto 1/\phi \quad (14.7)$$

The limiting condition for continuous operation is the maximum continuous current that can be carried by the armature winding. Using Eqs (14.4) and (14.6) with E (and hence, E_b) constant, the machine is seen to operate with a constant power limit. This condition is shown by Fig. 14.4. The maximum speed obtainable using field control occurs at conditions of maximum applied armature voltage and maximum field. Speed can be varied over a range of about 2 to 1 for large machines and 4 or 5 to 1 for small machines.

Referring again to Eq. (14.6), if the field (ϕ) is held constant then the motor speed can be controlled by varying the voltage applied to the machine armature when

$$E \propto \omega \quad (14.8)$$

Applying the same restriction on armature current as before, then by reference to Eq. (14.3) the motor is found to operate with a constant torque limit as in Fig. 14.5. The maximum operating speed occurs with maximum armature voltage and maximum field with an operating speed range of the order of 100 to 1.

14.3 SCHEMES FOR D.C. MOTOR SPEED CONTROL

The most common form of variable-speed d.c. drive is based on the control of armature voltage. The speed of a d.c. motor has to be controlled from an a.c. or d.c. source. The two basic schemes of d.c. motor speed control is shown in Fig. 14.6. The layout shown in Fig. 14.6(a) is more common, where a controlled rectifier is used to supply the motor armature. Any of the various controlled rectifier configurations described in Chapter 6 can be used. If the armature current is continuous, then the voltage waveform at the motor will be identical to those developed in Chapter 6. Typically, a motor rated above 2 kW will have sufficient inductance to maintain continuous current. The speed of the motor is determined by its mean armature voltage, any oscillating torque produced by the harmonic voltage (current) components being heavily damped by the motor

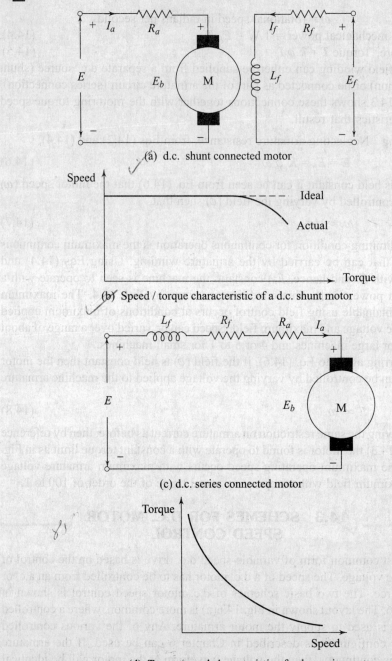

Fig. 14.3 *d.c. motor connections and characteristics*

inertia. Hence, the motor speed is dependent on the firing delay angle of the rectifier. If required, the field can be supplied via a controlled rectifier. Where

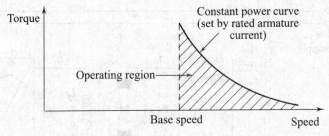

Fig. 14.4 *Operating region of a d.c. motor with field control*

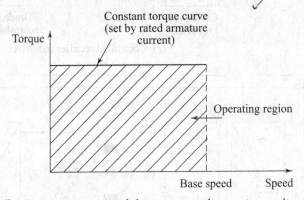

Fig. 14.5 *Operating region of d.c. motor with armature voltage control*

the machine is required to generate, (say) if rapid braking is required, then the converter must be capable of operation in the inverting mode.

An alternative is to use a diode rectifier to supply a d.c. chopper as in Fig. 14.6(b). A chopper would also be used where a d.c. supply is already available.

Thus, d.c. drives can be classified, in general, into following three types.
1. Single-phase drives
2. Three-phase drives
3. Chopper drives.

14.4 SINGLE-PHASE SEPARATELY EXCITED DRIVES

In phase-controlled d.c. drives, an a.c. to d.c. phase-controlled converter is used to control the d.c. drive motor. Controlled rectifier for d.c. drives are widely used in applications requiring a wide range of speed control and/or frequent starting, braking and reversing. Some prominent applications are in rolling mills, paper mills, printing presses, mine winders, machine tools, etc. The various configurations of single-phase and three-phase converters have been discussed in Chapter 6. The converter used for a particular application depends on such factors as supply available ($1-\phi$ or $3-\phi$), rating of the drive, amount of voltage ripple to be tolerated, reversible or nonreversible drive, need for regeneration, etc.

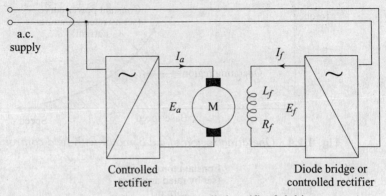

(a) Controlled rectifier fed-drive

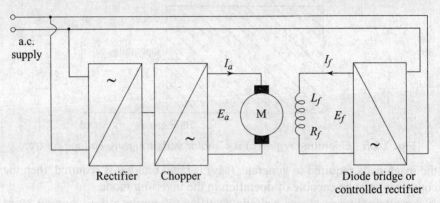

(b) Armature voltage control using d.c. chopper

Fig. 14.6 *Variable speed drives using armature voltage control*

The basic circuit arrangement for a single-phase separately excited d.c. motor drive is shown in Fig. 14.7. The armature voltage is controlled by a semiconverter or full-converter and the field circuit is fed from the a.c. supply through a diode bridge. The motor current cannot reverse due to the thyristors in the converters.

The armature circuit of the d.c. motor is represented by its back emf, e_b, armature resistance, R_a, and armature inductance, L_a, as shown in Fig. 14.7.

The back emf of the motor is given by

$$e_b = K_a \phi n \qquad (14.9)$$

where e_b is the instantaneous back emf and n is the instantaneous speed, under steady-state condition,

$$E_b = K_a \phi N \qquad (14.10)$$

where
    E_b = average back emf of the motor
    K_a = motor speed constant
    N = steady-state speed
    ϕ = flux in the machine

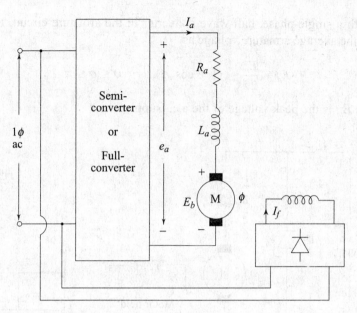

Fig. 14.7 *Basic single-phase circuit for the speed control of a separately excited d.c. motor*

The torque developed by the motor is given by

$$t = K_a \phi i_a \tag{14.11}$$

where i_a is the instantaneous current into the armature of the motor and t is the torque. The motor and t is the torque. The average developed torque is

$$T = K_a \phi i_a \tag{14.12}$$

where i_a is average d.c. current into the armature of the motor.
The voltage equation of the armature circuit is given by

$$e_a = R_a i_a + L_a \cdot \frac{di_a}{dt} + e_b \tag{14.13}$$

The average armature voltage is given by

$$E_a = R_a I_a + E_b \tag{14.14}$$

Note that the inductance L_a does not absorb any average voltage.
From Eqs (14.9) and (14.14), the average speed is

$$N = \frac{E_a - I_a R_a}{K_a \phi} \tag{14.15}$$

14.4.1 Single-Phase Half-Wave Converter Drives

Figure 14.8 shows a single-phase, half-wave converter for controlling a separately excited d.c. motor. It requires a single thyristor and a freewheeling diode. In this circuit the motor current is always discontinuous, resulting in poor motor performance. This type of converter is employed only for motors below 400 W.

With a single-phase, half-wave converter in the armature circuit, Eq. (9.1) gives the average armature voltage as

$$E_a = \frac{E_m}{2\pi} (1 + \cos \alpha), \quad 0 \leq \alpha \leq \pi \quad (14.16)$$

where E_m is the peak voltage of the a.c. supply.

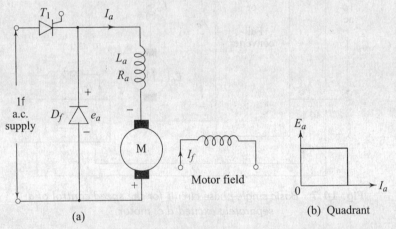

Fig. 14.8 *Single-phase, half-wave converter drive*

14.4.2 Single-phase Semiconverter Drives

Semiconverters are one-quadrant converters, that is, they have one polarity of voltage and current at the d.c. terminals. The circuit arrangement for a single-phase semiconverter drive is shown in Fig. 14.9. The armature voltage is controlled by a semiconverter and the field circuit is fed from the a.c. supply through a diode bridge. The motor current cannot reverse due to the thyristors in the converters. Due to the semiconverters, the average d.c. output voltage (E_a) is always positive. Therefore, power flow $(E_a I_a)$ is always positive, that is, from the a.c. supply to the d.c. load. Hence, in drive system using semiconverters, regeneration or reverse power flow from motor to a.c. supply is not possible.

In single-phase converters, the armature voltage, e_a, and current, i_a, change with time. The armature current may be continuous or discontinuous depending on the operating conditions and circuit parameters. The torque–speed characteristic would be different for two types of conduction.

1. Continuous Armature Current Typical voltage and current waveforms are shown in Fig. 14.10. Here, it has been assumed that the armature current is continuous over the whole range of operation. As shown in Fig. 14.10(a), SCR T_1 is triggered at an angle α and t_2 at an angle $(\pi + \alpha)$ with respect to the supply voltage e. For the period $\alpha < \omega t < \pi$, the motor is connected to the input supply through T_1 and D_2, and the motor terminal voltage e_a is the same as the supply input voltage e. Beyond period π, e_a tends to reverse as the input voltage changes polarity. This will forward bias the freewheeling diode D_f and it starts conducting. The motor current i_a, which was flowing from the supply through T_1, is transferred to D_F (i.e. T_1 commutates). Therefore, during period $\pi < \omega t < (\pi + \alpha)$, the motor terminals are shorted through D_F, making e_a zero.

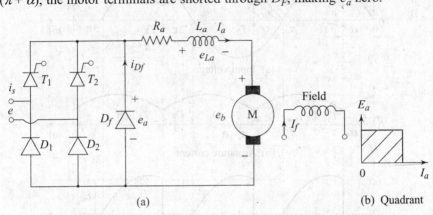

Fig. 14.9 *Semi-converter feeding a separately excited d.c. motor*

As discussed above, when the thyristor conducts from period α to π, energy from the supply is delivered to the armature circuit. This energy is partially stored in the inductance, partially stored in the kinetic energy of the moving system, and partially used to supply the mechanical load. During the freewheeling period, π to $(\pi + \alpha)$, energy is recovered from the inductance and is converted to mechanical form to supplement the kinetic energy in supplying the mechanical load. The freewheeling armature current continues to produce electromagnetic torque in the motor. During this period, no energy is fed back to the supply.

Torque–speed characteristics: The armature voltage equations are as follows:

$$e_a = e = R_a i_a + L_a \frac{di_a}{dt} + e_b, \quad \alpha < \omega t < \pi \tag{14.17}$$

$$e_a = 0 = R_a i_a + L_a \frac{di_a}{dt} + e_b, \quad \pi < \omega t < \pi + \alpha \tag{14.18}$$

With a single-phase semiconverter in the armature circuit, Eq. (6.16) gives the average armature voltage as

$$E_a = \frac{E_m}{\pi}(1 + \cos \alpha) \tag{14.19}$$

The steady-state speed equation is given by

$$N = \frac{E_a - I_a R_a}{K_a \phi} = \frac{E_m (1 + \cos \alpha)}{\pi K_a \phi} - \frac{R_a \cdot T}{(K_a \phi)^2} \tag{14.20}$$

where $I_a = \dfrac{T}{K_a \phi}$ from Eq. (14.12)

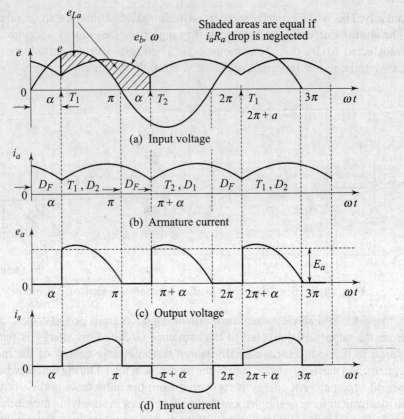

Fig. 14.10 *Voltage and current waveforms for continuous motor current*

The no-load speed of the motor is given by

$$N_{N-L} = \frac{E_m(1 + \cos \alpha)}{\pi K_a \phi} \quad \text{where } T = 0 \tag{14.21}$$

The torque–speed characteristics based on Eq. (14.20) is shown in Fig. 14.18. This characteristic shows that the good speed regulation is obtained with continuous armature current.

2. Discontinuous Armature Current The armature current becomes discontinuous for large values of the firing angle, high speed and low values of torque. If the armature current is discontinuous, the no-load speeds will be higher than those shown in Fig. 14.11, and the speed regulation will be significantly poor in the region of discontinuous armature current. The motor performance deteriorates with discontinuous armature current. The ratio of peak to average and RMS to average armature current increases. It is, therefore, desirable to operate the motor in the continuous current mode. To achieve this, an external armature circuit choke may be used which decreases the rate of current decay during the freewheeling operation.

Control of D.C. Drives

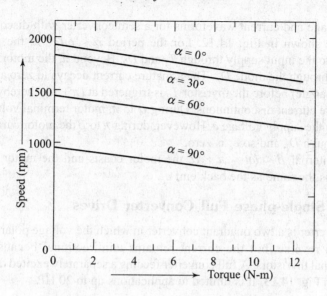

Fig. 14.11 *Torque–speed characteristics*

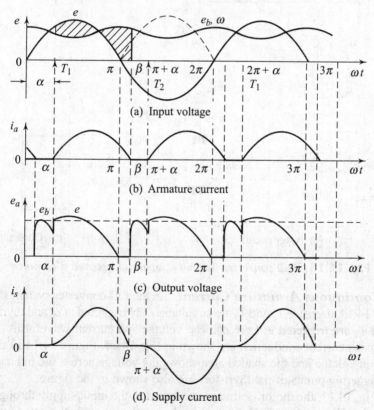

Fig. 14.12 *Voltage and current waveforms illustrating the effects of discontinuous motor current in semiconverter operation*

The voltage and current waveforms for a semiconverter with discontinuous current are shown in Fig. 14.12. For the period $\alpha < \omega t < \pi$, the motor is connected to the input supply through T_1 and D_2. Beyond π, the motor terminal is shorted through the diode D_F. The armature current decays to zero at angle β (extinction angle) before the thyristor T_2 is triggered at $(\pi + \alpha)$, thereby making the armature current discontinuous. During α to π, motor terminal voltage e_a is the same as the supply voltage e. However, during π to β the motor current freewheels through D_F and so e_a is zero.

In the internal $\beta < \omega t < \pi + \alpha$, the motor coasts and the motor terminal voltage e_a is the same as the back emf e_b.

14.4.3 Single-phase Full Converter Drives

A full converter is a two quadrant converter in which the voltage polarity of the output can reverse, but the current remains unidirectional because of the unidirectional thyristors. A full converter feeding a separately excited d.c. motor is shown in Fig. 14.13. It is limited to applications up to 20 HP.

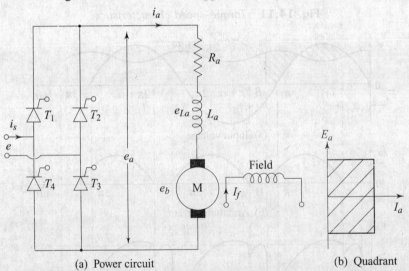

(a) Power circuit   (b) Quadrant

Fig. 14.13 *Full converter feeding a separately excited d.c. motor*

1. Continuous Armature Current In the full-converter system shown in Fig. 14.13. thyristor T_1 and T_3 are simultaneously triggered at α, and thyristors T_2 and T_4 are triggered at $(\pi + \alpha)$. The voltage and current waveforms under continuous current conduction are shown in Fig. 14.14. Figure 14.14(a) shows the input voltage and the shaded area shows the voltage across the inductance. The triggering points of the thyristors are also shown in the figure.

In Fig. 14.13, the motor is always connected to the input supply through the thyristors. Thyristors T_1 and T_3 conduct during the interval $\alpha < \omega t < (\pi + \alpha)$ and connect the motor to the supply. At $(\pi + \alpha)$, when thyristors T_2 and T_4 are

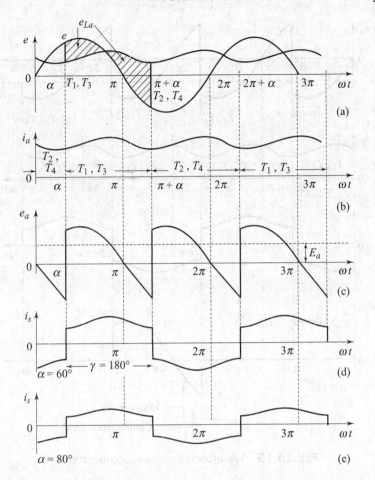

Fig. 14.14 *Voltage and current waveforms for continuous motor current (motoring operation)*

triggered, immediately the supply voltage appears across thyristors T_1 and T_3 as a reverse-bias voltage and turns them OFF. This is called natural or line commutation. The motor current, i_a, which was flowing from the supply through T_1 and T_3, is transferred to T_2 and T_4. During α to π, energy flows from the input supply to the motor (both e and i_s are positive, and e_a and i_a are positive, signifying positive power flow). However, during π to $\pi + \alpha$, some of the motor system energy is fed back to the input system (e and i_s have opposite polarities and likewise e_a and i_a, signifying reverse power flow).

The voltage and current waveforms for firing angle greater than 90° are shown in Fig. 14.15. The average motor terminal voltage E_a is now negative. If the motor back emf E_b is reversed, it will behave as a d.c. generator and will feed power back to the a.c. supply. This is known as the inversion operation of the converter, and this mode of operation is used in the regenerative braking of the motor.

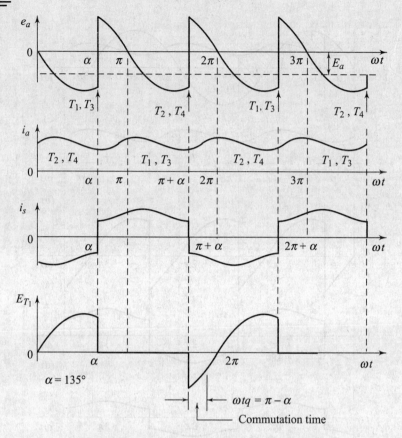

Fig. 14.15 *Waveforms for inversion operation*

Torque–speed characteristics: With a single-phase, full converter in the armature circuit, Eq. (6.16) gives the average armature voltage as

$$E_a = \frac{2E_m}{\pi} \cos \alpha \text{ for } 0 \leq \alpha < \pi \tag{14.22}$$

The steady-state speed equation is given by

$$N = \frac{E_a - I_a R_a}{K_a \phi} = \frac{E_m \cos \alpha}{\pi K_a \phi} - \frac{R_a \cdot T}{(K_a \phi)^2} \tag{14.23}$$

where $I_a = \dfrac{T}{K_a \phi}$, from Eq. (14.12)

The no-load speed of the motor is given by

$$N_{N-L} = \frac{2 E_m \cos \alpha}{\pi K_a \phi}, \quad \text{where } T = 0 \tag{14.24}$$

Control of D.C. Drives

The torque-speed characteristics based on Eq. (14.23) is shown in Fig. 14.16. This characteristic shows that good speed regulation is obtained with continuous armature current.

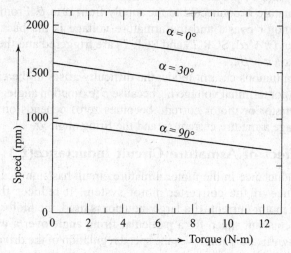

Fig. 14.16 *Torque-speed characteristic at different firing angles*

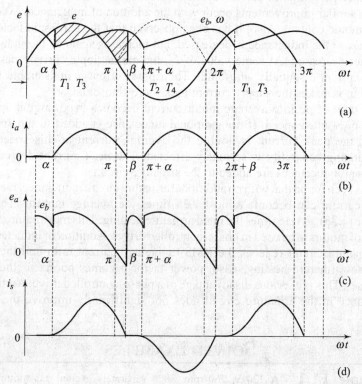

Fig. 14.17 *Voltage and current waveform illustrating the effects of discontinuous motor current in full-converter operation*

2. Discontinuous Armature Current The voltage and current waveforms for full converter with discontinuous armature current are shown in Fig. 14.17, SCR T_1 and T_3 are fired at α and armature current flows from α to β. The motor armature is connected to the supply from α to β. From period β to $(\pi + \alpha)$, the motor coasts and the armature voltage is the back emf of the motor. Again at $(\pi + \alpha)$, SCR T_2 and SCR T_4 are triggered and the conduction continues up to $(\pi + \beta)$.

In the discontinuous current mode, the difficulty arises in the calculation of the average motor terminal voltage E_a, because β (extinction angle, the instant at which the thyristor or motor current. becomes zero) depends on the average speed N, average armature current I_a, and the firing angle α.

14.4.4 Effect of Armature Circuit Inductance(L_a)

Addition of inductance in the motor armature circuit has significant effects on the performance of the converter motor system. It reduces the region of discontinuous motor current. If a large inductor is used, the motor will operate essentially at constant speed for a particular firing angle over a wide range of torque. This significantly improves the speed regulation of the drive. It has been found that additional armature circuit inductance improves the motor as well as the supply performance for the semiconverter system. For the full-converter system, similar improvements occur with the addition of inductance, except for the harmonic content. Also, it has been observed that the harmonic factor first decreases as the inductance is decreased, but increases again as the inductance is further decreased. At some value of L_a, the motor ripple current makes the supply current essentially sinusoidal. The ripple current is undesirable for the motor but it makes the supply current look more sinusoidal.

The ratio of peak-to-average motor current becomes maximum at approximately 63% rated speed. If no additional inductance is added to the armature circuit, the peak current is double the average current at this speed. The commutating capability is seriously effected. The peak current decreases as additional inductances are added to the armature circuit.

It has been found that with no additional armature circuit inductance, the RMS motor current can become almost 1.25 times the average current. This will produce $(1.25)^2 = 1.56$ times the rated armature heating. This may require special design of motors to be used in phase-controlled drives. Addition of inductances in the armature circuit reduces the RMS-to-average current ratio and therefore, decreases armature heating. Also, power factor becomes poorer as the speed decreases. This is a serious disadvantage of a phase-controlled drive. Addition of inductance in the armature circuit does not significantly improve the power factor.

SOLVED EXAMPLES

Example 14.1 A 220 V, 960 rpm, 80 A separately excited d.c. motor has an armature resistance 0.06 Ω. Under rated conditions, the motor is driving a load whose torque is constant and independent of speed. The speeds below the rated speed are

Control of D.C. Drives

obtained with armature voltage control (with full field), and the speeds above the rated speed are obtained by field control (with rated armature voltage). Determine:
(i) The motor terminal voltage when the speed is 620 rpm.
(ii) The value of flux as a percentage of rated flux if the motor speed is 1200 rpm.
Neglect the motor's rotational losses.

Solution: The back emf at 960 rpm,

$$E_{b_1} = E - I_a R_a = 220 - (80 \times 0.06) = 215.2 \text{ V}$$

The rated speed $= \omega_1 = \dfrac{960}{60} \times 2\pi = 100.53$ rad/s

Let us assume that the flux at the rated conditions be ϕ_1
Therefore,

$$K_a \phi_1 \times 100.53 = E_{b_1} = 215.2. \quad \therefore K_a \phi_1 = \dfrac{215.2}{100.53} = 2.14 \tag{i}$$

(i) Now, back emf at 620 rpm is,

$$\dfrac{E_{b_2}}{E_{b_1}} = \dfrac{620}{960} \quad \therefore E_{b_2} = E_{b_1} \times \dfrac{620}{960} = 215.2 \times \dfrac{620}{960} = 138.98 \text{ V}$$

Since the load torque is constant I_a 80 A. The motor terminal voltage

$$= E_{b_2} + I_a R_a = 138.98 + (80 \times 0.06) = 143.78 \text{ V} \tag{ii}$$

(ii) Let us take the new flux $\phi_2 = K \phi_1$.

Since $E_b = K_a \phi \omega \quad \therefore E_{b_3} = K_a \phi_2 \times \dfrac{1200 \times 2\pi}{60} = K_a \cdot k \cdot \phi_1 \times 125.67$

Substituting from Eq. (i) gives

$$E_{b_3} = K \times 2.14 \times 125.67 = 268.93 K \tag{iii}$$

Since $T = K_a \phi I_a$, we have,

$$K_a \phi_1 I_{a_1} = K_a \phi_2 I_{a_2}$$

or $\quad I_{a_2} = \dfrac{\phi_1}{\phi_2} I_{a1} \quad$ or $\quad I_{a_2} = \dfrac{I_{a_1}}{K} = \dfrac{80}{K} \tag{iv}$

$\therefore \quad E = E_{b_3} + I_{a_2} R_a \quad$ or $\quad 220 = E_{b_3} + 0.06 I_{a_2} \tag{v}$

Substituting from Eqs (iii), and (iv) into Eq. (v) gives

$$220 = 268.93 K + \dfrac{80}{K} \times 0.06 \quad \text{or} \quad 268.93 K^2 - 220 K + 4.8 = 0$$

or $\quad K = 0.8$ or 0.022. The feasible value of $K = 0.8$.
Hence, the flux must be reduced to 0.8 of its rated value.

Example 14.2 A separately excited motor of Example 14.1 is coupled to an overhauling load with a torque of 100 Nm. Compute the speed at which the motor can hold the load by regenerative braking. Source voltage is 220 V. Neglect the motors rotational losses.

Solution: From Example 14.1, $E_{b_1} = 215.2$ V. $\omega_1 = 100.53$ rad/s

$$\therefore \text{Rated motor torque} = T_1 = \frac{E_{b_1} \times I_{a_1}}{\omega_1} = \frac{215.2 \times 80}{100.53} = 171.25 \text{ Nm}.$$

Now, $T = K_a \phi I_a$.

$\therefore \qquad 171.25 = K_a \phi I_{a_1} = K_a \phi \times 80.$ \hfill (i)

and $\qquad 1000 = K_a \phi I_{a_2}$ \hfill (ii)

where I_{a_2} is the current under regenerative braking. From Eqs (i) and (ii),

$$I_{a_2} = \frac{1000 \times 80}{171.25} = 467.15 \text{ A}$$

Back emf, $\qquad E_{b_2} = E + I_{a_2} R_a = 220 + 467.15 \times 0.06 = 248.02$ V.

The new speed, $\omega_2 = \dfrac{E_{b_2}}{E_{b_1}} \times \omega_1 = \dfrac{248.02}{215.2} \times 100.53 = 115.86$ rad/s or 1106 rpm

Example 14.3 A separately excited d.c. motor is fed from a 230 V, 50 Hz supply via a single-phase, half-controlled bridge rectifier. Armature parameters are: inductance 0.06 H, resistance 0.3 Ω, the motor voltage constant is $K_a = 0.9$ V/A rad/s and the field resistance is $R_F = 104$ Ω. The field current is also controlled by a semiconverter and is set to the maximum possible value. The load torque is $T_L = 50$ N-m at 800 rpm. The inductances of the armatures and field circuits are sufficient enough to make the armature and field currents continuous and ripple free compute:
(i) the field current I_f,
(ii) the firing angle of the converter in the armature circuit, and
(iii) the input power factor of the armature circuit converter.
Neglect the system losses.

Solution:
(i) For single-phase semiconverter controlled d.c. drive, we can write for field circuit as, $E_f = \dfrac{E_m}{\pi}(\cos\alpha + 1)$

Now, the maximum field voltage and current are obtained when firing angle $\alpha = 0$.

$\therefore$ Field voltage, $E_f = \dfrac{2E_m}{\pi} = \dfrac{2 \times \sqrt{2} \times 230}{\pi} = 207.07$ V.

Field current, $\qquad I_f = \dfrac{E_f}{R_f} = \dfrac{207.07}{104} = 1.99$ A

(ii) Now, we have the relation

$$I_a = \frac{T}{K_a I_f} = \frac{50}{0.9 \times 1.99} = 27.92 \text{ A}$$

Also, $\qquad E_b = K_a \omega I_f = 0.9 \times \left(800 \times \dfrac{2\pi}{60}\right) \times 1.99 = 150.04$ V.

Now, armature voltage is, $E_a = E_b + I_a R_a = 150.04 + 27.92 \times 0.3 = 158.42$ V.
The armature voltage is given by the equation.

$$E_a = \frac{E_m}{\pi}(1 + \cos\alpha), \quad 158.42 = \frac{\sqrt{2} \times 230}{\pi}(1 + \cos\alpha) \therefore \alpha = 58°$$

(iii) For the constant and ripple free armature current, the output power is
$$p_0 = E_a I_a = 158.42 \times 27.92 = 4423 \text{ W}.$$

By neglecting the losses in the system, we can write
Power from the supply, $P_a = p_0 = 4423$ W.
The RMS input current of the armature converter as shown in Fig. 14.10 is

$$I_{rms} = \left[\frac{2}{2\pi}\int_\alpha^\pi I_a^2 d(\omega t)\right]^{1/2} = I_a\left(\frac{\pi-\alpha}{\pi}\right)^{1/2} = 27.92\left(\frac{\pi-58}{\pi}\right)^{1/2} = 22.99 \text{ A}.$$

Now, the input voltage-ampere rating, is
$$EI = E I_{rms} = 230 \times 22.99 = 5287.7$$

∴ Input power factor assuming negligible harmonics is given by

$$P_F = \frac{P_a}{EI} = \frac{4423}{5287.7} = 0.84 \text{ (lag)}$$

Example 14.4 The speed of a 10 HP, 210 V, 1000 rpm separately excited d.c. motor is controlled by a single-phase, full-converter as shown in Fig. 14.13. The rated motor armature current is 30 A, and the armature resistance is $R_a = 0.25$ Ω. The a.c. supply voltage is 230 V. The motor voltage constant is $K_a\phi = 0.172$ V/rpm. Assume that sufficing inductance is present in the armature circuit to make the motor current continuous and ripple free.

(a) *Rectifier operation (motoring action):* For a firing angle $\alpha = 45°$, and rated motor armature current, determine:
(1) The motor torque   (2) The speed of the motor
(3) The supply power factor.

(b) *Inverter operation (regenerative action):* The motor back emf polarity is reversed by reversing the field excitation. Determine:
(1) The firing angle to keep the motor current at its rated value.
(2) The power fed back to the supply.

Solution:

(a) (1) $K_a\phi = 0.172$ V/rpm

$$= \frac{0.172 \times 60}{2\pi} \text{ V-s/rad} = 1.64 \text{ V-s/rad}.$$

Now $\quad T = K_a\phi I_a = 1.64 \times 30 = 49.2$ N-m.

(2) The armature voltage, $E_a = \dfrac{2E_m}{\pi}\cos\alpha = \dfrac{2\sqrt{2}\times 230}{\pi}\cos 45° = 146.42$ V.

$$E_b = E_a - I_a R_a = 146.42 - (30 \times 0.25) = 138.92 \text{ V}.$$

Speed, $N = \dfrac{E_b}{K_a\phi} = \dfrac{138.92}{0.172} = 807.67$ rpm.

(3) If the motor current is constant and ripple free, the supply current is a square wave (Fig. 14.14) of amplitude 30 A. The rms supply current is $I = 30$ A

$\therefore$ Supply volt-ampere $= EI = 230 \times 30 = 6900$ VA.

If losses in the converter are neglected, the power from the supply is (real-power),

$$P_s = E_a I_a = 146.42 \times 30 = 4392.6 \text{ W}.$$

Therefore, the supply power factor is $P_F = \dfrac{4392.6}{6900} = 0.64$.

(b) (1) At the time of polarity reversal, the back emf is

$$E_b = 138.92 \text{ V}$$

Now, $E_a = E_b + I_a R_a = -138.92 + (30 \times 0.25) = -131.42$ V.

Also, $E_a = \dfrac{2\sqrt{2} \times 230}{\pi} \cos\alpha = -131.42 \quad \therefore \alpha = 129.39°$

(2) Power from the d.c. machine is

$$P_g = 138.92 \times 30 = 4167.6$$

Power lost in the armature resistance is

$$P_R = I_a^2 R_a = 30^2 \times 0.25 = 225 \text{ W}$$

Power fed back to the a.c. supply is

$$P_s = 4167.6 - 225 = 3942.6 \text{ W}.$$

or $\quad P_s = E_a I_a = 131.42 \times 30 = 3942.6$ W.

Example 14.5 A 210, 1200 rpm, 10 A separately excited motor is controlled by a 1-phase fully controlled converter with an a.c. source voltage of 230 V, 50 Hz. Assume that sufficient inductance is present in the armature circuit to make the motor current continuous and ripple free for any torque greater than 25 per cent of rated torque $R_a = 1.5\,\Omega$.
(a) What should be the value of the firing angle to get the rated torque at 800 rpm?
(b) Compute the firing angle for the rated braking torque at -1200 rpm.
(c) Calculate the motor-speed at the rated torque and $\alpha = 165°$ for the regenerative braking in the second quadrant?

Solution: $\quad E_m = 230 \times \sqrt{2} = 325.27$ V

Now, $\quad E_b = E_a - I_a R_a = 210 - (10 \times 1.5) = 195$ V

$$\omega = \dfrac{1200 \times 2\pi}{60} = 125.66 \text{ rad/s.}$$

$$K_a\phi = \dfrac{E_b}{\omega} = \dfrac{195}{125.66} = 1.55$$

Now, for continuous conduction, we can write.

$$E_a = E_b + I_a R_a$$

or $\dfrac{2E_m}{\pi} \cos \alpha = I_a R_a + E_b$ (i)

(a) At rated torque, $I_a = 10$ A.

Back emf at 800 rpm $= E_{b_1} = \dfrac{800}{1200} \times 195 = 130$ V

From Eq. (i)

$\dfrac{2 \times 325.27}{\pi} \cos \alpha = (10 \times 1.5) + 130 \quad \therefore \quad \alpha = 45.55°$.

(b) For the speed of –1200 rpm, $E_b = -195$ V. From Eq. (i)

$\dfrac{2 \times 325.27}{\pi} \cos \alpha = (10 \times 1.5) - 195 \quad \therefore \quad \alpha = 150.37°$.

(c) From Eq. (i), for $\alpha = 165°$, and $I_a = 10$ A

$\dfrac{2 \times 325.27}{\pi} \cos 165° = 10 \times 1.5 + E_b \quad \text{or} \quad E_b = -215.02$ V

Forward regeneration is obtained either by the field reversal or the armature reversal, for which,

$$K_a \phi = -1.55$$

Now, $\quad \omega = \dfrac{E_b}{K_a \phi} = \dfrac{-215.02}{-1.55} = 138.72$ rad/s $= 1324.7$ rpm.

Example 14.6 A small separately excited d.c. motor is supplied via a half controlled, single-phase bridge rectifier. The supply is 240 V, the thyristors are triggered at 110°, and the armature current continues for 50° beyond the voltage zero. Determine the motor speed at a torque of 1.8 N-m, given the motor torque characteristics is 1.0 N m/A and its armature resistance is 6 Ω. Neglect all converter losses.

Solution: The voltage is supplied to motor between 110° to 180°, zero voltage for 50°, and the motor back emf E_b from 50° to 110° (that is, a period of 60°).

$\therefore \quad E_a = \dfrac{1}{\pi} \left[\displaystyle\int_{110°}^{180°} 240\sqrt{2} \sin \omega t \, d(\omega t) + E_b \cdot \pi \cdot \dfrac{60°}{180°} \right] = 71.1 + 0.333 E_b$.

Average current, $I_a = \dfrac{1.8}{1.0} = 1.8$ A.

Now, $\quad E_b = E_a - I_a R_a \quad$ or $\quad E_b = 71.1 + 0.333 E_b - (1.8 \times 6) \quad \therefore \quad E_b = 90.43$ V.

From the relation of gross-mechanical power,

$$T \cdot N = E_b I_a \quad \text{or} \quad T/I_a = E_b/N,$$

Therefore voltage characteristics is 1.0 V/rad/s.

$$\therefore \quad \text{Speed}, N = \frac{90 \cdot 43}{1} = 90.43 \text{ rad/s} = 864 \text{ rpm}.$$

Example 14.7 An 80 kW, 440V, 800 rpm d.c. motor is operating at 600 rpm and developing 75% rated torque is controlled by 3-ϕ, six-pulse thyristor converter. If the back emf at rated speed is 410 V, determine the triggering angle of the converter. The input to the converter is 3-ϕ, 415 V, 50 Hz a.c. supply.

Solution: Given: $E_{b_1} = 410$ V, $N_1 = 800$ rpm, $N_2 = 600$ rpm.

Now, we have the relation: $\dfrac{E_{b_2}}{E_{b_1}} = \dfrac{N_2}{N_1}$

$$\therefore \quad E_{b_2} = 410 \times \frac{600}{800} = 307.5 \text{ V}$$

Now, $\quad E_{b_1} = E - I_a R_a$, $410 = 440 - I_a R_a$. $\therefore I_a R_a = 30$ V

But, $\quad I_a = \dfrac{80 \times 1000}{440} = 181.82$ A.

$$\therefore \quad R_a = \frac{30}{181.82} = 0.165 \, \Omega.$$

Now, terminal voltage of d.c. motor at 800 rpm and 75% rated torque,

$$= E_{b_2} + I_a R_a = 307.5 + (0.5 \times 181.82 \times 0.165) = 330 \text{ V}.$$

Neglecting voltage drop in the converter system,

now $\quad E_a = \dfrac{3\sqrt{3}}{\pi} E_m \cos \alpha$, $330 \text{ V} = \dfrac{3\sqrt{3}}{\pi} \times \dfrac{\sqrt{2}}{\sqrt{3}} E_{ac} \cos \alpha$

where $\quad E_{ac}$ = RMS value of the a.c. voltage.

$$\therefore \quad 330 = \frac{3\sqrt{2}}{\pi} E_{ac} \cos \alpha, \quad 330 = 1.35 \times 415 \times \cos \alpha \therefore \alpha = 53.91°$$

Example 14.8 The speed of a 150 HP, 650 V, 1750 rpm, separately excited d.c. motor is controlled by a 3-ϕ, full converter. The converter is operating from a 3-ϕ, 460 V, 50 Hz supply. The rated armature current of the motor is 170 A. The motor parameters are $R_a = 0.099 \, \Omega$, $L_a = 0.73$ mH, and $K_a \phi = 0.33$ V/rpm. Neglect the losses in converter system. Determine:
  (a) No-load speeds at firing angles $\alpha = 0°$ and $\alpha = 30°$. Assume that at no-load, the armature current is 10% of the rated current and is continuous.
  (b) The firing angle to obtain rated speed of 1750 rpm at rated motor current. Also, compute the supply power factor.
  (c) The speed regulation for the firing angle obtained in part (b).

Control of D.C. Drives

Solution: *(a) No-load condition:*

The supply phase voltage is,

$$E = \frac{460}{\sqrt{3}} = 265.58 \text{ V}$$

Now, the motor terminal voltage is

$$E_a = \frac{3\sqrt{6} \times 265.58}{\pi} \cos\alpha = 621.22 \cos\alpha$$

For $\alpha = 0°$, $\quad E_a = 621.22$ V.

∴ $\quad E_b = E_a - I_a R_a = 621.22 - (170 \times 0.099) = 604.39$ V.

No-load speed is

$$N_{NL} = \frac{E_b}{K_a\phi} = \frac{604.39}{0.33} = 1831.48 \text{ rpm}.$$

For $\alpha = 30°$, $\quad E_a = 621.22 \cos\alpha = 621.22 \cos 30° = 537.99$ V.

$E_b = 537.99 - (170 \times 0.099) = 521.16$ V

No-load speed is

$$N_{NL} = \frac{521.16}{0.33} = 1579.27 \text{ rpm}.$$

(b) Full-load condition: Motor back emf E_b at 1750 rpm is

$$E_b = 0.33 \times 1750 = 577.5 \text{ V}$$

Motor terminal voltage at rated current is

$$E_a = 577.5 + (170 \times 0.099) = 594.33 \text{ V}.$$

Therefore, $594.33 = 621.22 \cos\alpha$, ∴ $\alpha = 16.92°$

The ripple in the motor current is negligible at full-load condition. Therefore, from Fig. 14.30 the supply current i_A is a square-wave of amplitude 170 A.

∴ The RMS value of the supply current is

$$I_A = \left(\frac{1}{\pi} \times 170^2 \times \frac{2\pi}{3}\right)^{1/2} = \sqrt{\frac{2}{3}} \times 170 = 138.80 \text{ A}.$$

Supply volt-ampere $= 3EI_a = 3 \times 265.58 \times 138.80 = 110587.51$ VA

For the lossless converter system,

Power from the supply (p_s) = Power input to motor

∴ $\quad P_s = E_a I_a = 594.33 \times 170 = 101036.1$ W.

∴ Supply power factor $= \dfrac{101036.1}{110587.51} = 0.91$

(c) Speed regulation: At full-load condition, motor current is 170 A and speed is 1750 rpm. If the load is thrown-off keeping the firing angle same at $\alpha = 16.92°$, motor current decreases to 10%, i.e. 17 A.

Therefore, $E_b = 594.33 - (17 \times 0.099) = 592.65$ V.

Now, no-load speed is

$$N_{NL} = \frac{592.65}{0.33} = 1795.91 \text{ rpm.}$$

The speed regulation $= \frac{1795.91 - 1750}{1750} \times 100 = 2.62\%$

14.5 BRAKING OPERATION OF RECTIFIER CONTROLLED SEPARATELY EXCITED MOTOR

A single-phase fully controlled converter feeds the separately excited d.c. motor as shown in Fig. 14.18(a). The polarities of output voltage, back emf, and armature current shown are for the motoring operation in the forward direction. The rectifier output voltage is positive and the firing angle lies in the range $0 \leq \alpha \leq 90°$. The polarities of rectifier output voltage, back emf, and armature current show that the rectifier supplies power to the motor which is converted into mechanical power. With these polarities of the rectifier output voltage and the motor back emf, the direction of power flow can be reversed and thus, the motor can be made to work under regenerative braking if the armature current can reverse. This is not possible because the rectifier can carry current only in one direction. The only alternative available for the reversal of the flow of power is to reverse both the rectifier output voltage E_a and the motor back emf with respect to the rectifier terminals and make $|E_b| > |E_a|$, as shown in Fig. 14.18(b). The rectifier output voltage can be reversed by making $\alpha > 90°$, as shown in Fig. 14.18(b). Under this condition, the rectifier works as a line commutated inverter, transferring power from the d.c. side to the a.c. mains. The condition $|E_b| > |E_a|$ can be satisfied for any motor speed by choosing an appropriate value of α in the range $90° < \alpha < 180°$. The reversal of the motor emf with respect to the rectifier terminals can be done by any of the following changes:

1. An active load coupled to the motor shaft may drive it in the reverse direction. This gives reverse regeneration. In this case no changes are required in the armature connection with respect to the rectifier terminals.
2. The field current may be reversed with the motor running in the forward direction. This gives forward regeneration. In this case also, no changes are required in the armature connection.
3. The motor armature connections may be reversed with respect to the rectifier output terminals, with the motor still running in the forward direction. This will give forward regeneration.

If the drive shown in Fig. 14.18(a) runs only in the forward direction and if there is no arrangement for the reversal of either field or armature, regenerative braking cannot be obtained. The drive then works essentially as a single-quadrant drive.

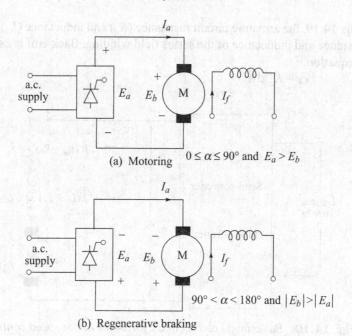

Fig. 14.18 *Two-quadrant operation of fully controlled rectifier fed separately excited motor*

Regenerative braking cannot be obtained with a half controlled rectifier because the output voltage cannot be reversed. The plugging operation can be obtained both with half-controlled and fully-controlled rectifiers by reversing the back emf by any of the three methods just stated and keeping the rectifier voltage still positive. An external resistance must then be included to limit the current. Because of the poor efficiency and the need for external resistance to limit the armature current, plugging is not employed with rectifier drives. While operating in regenerative braking, care should be taken to avoid accidental plugging.

14.6 SINGLE-PHASE SERIES D.C. MOTOR DRIVES

Figure 14.19 shows the basic single phase power circuit for series motor speed control. As shown, the field circuit is connected in series with the armature, and motor terminal voltage is controlled by a semi-converter or full-converter. Series motors are particularly suited for applications that require high starting torque, such as cranes, hoists, elevators, vehicles, etc. Inherently, series motors can provide essentially constant power output and are therefore particularly suitable for traction drives.

Speed control is very difficult with the series motor because any change in the load current will immediately be reflected in a speed change and hence, all speed-control systems will use separately excited motors.

In Fig. 14.19, the armature circuit resistance (R_a) and inductance (L_a) include the resistance and inductance of the series field winding. Back emf is expressed by the equation,

$$e_b = K_a \phi n \tag{14.25}$$

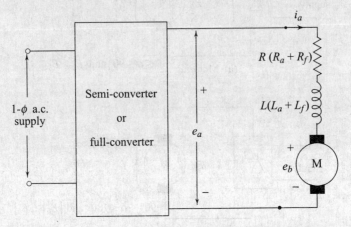

Fig. 14.19 Basic single-phase circuit for series motor speed control

The flux ϕ has two components. One component, say ϕ_a, is produced by the armature current flowing through the series field winding. The other component, say ϕ_{res}, is due to the residual magnetism. The latter is small and can be assumed constant.

$$\therefore \quad \phi = \phi_a + \phi_{res} \tag{14.26}$$

If magnetic linearity is assumed,

$$\phi_a = K_f i_a \tag{14.27}$$

Substituting Eqs (14.26) and (14.27) into Eq. (14.25), we get

$$e_b = K_a (K_f i_a + \phi_{res}) \cdot n = K_a K_f i_a n + K_a \phi_{res} \cdot n$$

$$= K_{af} i_a n + K_{res} \cdot n \tag{14.28}$$

The back emf due to the residual magnetism is very small and is proportional to speed. The back emf due to the flux produced by the armature current is the major voltage, and it is present when both i_a and n are present. Average back emf is given by the equation,

$$E_b = K_{af} \cdot N I_a + K_{res} \cdot N \tag{14.29}$$

The torque developed by the motor is given by

$$t = K_a \phi i_a \tag{14.30}$$

If the flux ϕ_{res} is neglected, then from Eqs (14.26), (14.27) and (14.30),

$$t \approx K_{af} i_a^2 \tag{14.31}$$

Torque is, therefore, developed in the same direction for either direction of current. Hence, speed reversal in the series motor can be achieved by reversing either the field winding or the armature terminals but not both. The average developed torque is

$$T = K_{af} \cdot i_a^2 \big|_{\text{average}} = K_{af} \cdot I_{ar}^2 \qquad (14.32)$$

The voltage equation of the armature circuit is given by

$$e_a = R \cdot i_a + L \cdot \frac{di_a}{dt} + e_b \qquad (14.33)$$

The average armature voltage is given by

$$E_a = R_a I_a + E_b \qquad (14.34)$$

Thus, although the time variations of voltages, currents, torque, and speed can assume various forms, the basic d.c. motor Eqs (14.29), (14.32), and (14.34) hold good for the phase-controlled series d.c. motor drives in terms of average quantities.

14.6.1 Single-phase Semiconverter Drives

Figure 14.20(a) shows the power circuit for the speed control of a d.c. series motor by a single-phase semiconverter. Current and voltage waveforms for discontinuous motor armature current and continuous motor armature current are shown in Figs 14.20(b) and (c), respectively. In the series motor drive, the current flow through the armature and the field when the SCR is triggered at an angle α. The current continuous up to β for discontinuous conduction and up to $(\pi + \alpha)$ for continuous conduction.

In separately excited motors, a large back emf, E_b, is always present, even when the motor armature current is absent. The back voltage, E_b, tends to oppose the motor current and so the motor current decays rapidly. This leads to discontinuous motor current over a wide range of operation. In series motors, the back emf is proportional to the motor current (speed is assumed constant and back emf due to residual magnetism is very small and neglected). Therefore, the back emf decreases (unlike the separately excited motor where E_b stays constant and accelerates the decay of i_a) as the motor current decreases and so the motor current tends to be continuous. In fact, motor current is continuous over a wide range of operation in d.c. series motor drives. Only at high speed and low current is the motor current likely to become discontinuous. In the discontinuous current mode, the motor terminal voltage e_a is the same as the back emf due to residual magnetism, which is very small.

Freewheeling action in a semiconverter system takes place during the internal

$\pi < \omega t < \beta$ for discontinuous current, and

$\pi < \omega t < (\pi + \alpha)$ for continuous current.

Freewheeling action will be absent if $\beta < \pi$.

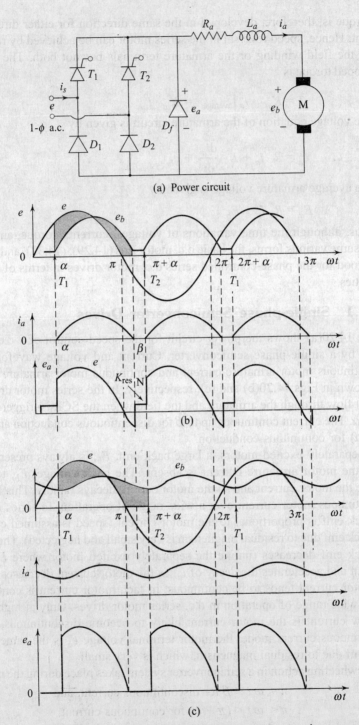

Fig. 14.20 *Speed control of a d.c.-series motor by a single-phase semiconverter*

$$e_a = 0 = R \cdot i_a + L \frac{di_a}{dt} + e_b$$

and $t = K_{af} i_a^2$

Substituting the value of e_b from Eq. (14.28),

$$e_a = 0 = R i_a + L \frac{di_a}{dt} + K_{af} \cdot i_a n + K_{res} \cdot n \quad (14.35)$$

In terms of average values,

$$E_q = R I_a + K_{af} \cdot I_a N + K_{res} \cdot N \quad (14.36)$$

and

$$T = K_{af} \cdot I_{ar}^2 \quad (14.37)$$

In phase-controlled series motor drives, the motor current is mostly continuous. For a continuous motor current, the motor terminal voltage can be written as

$$E_a = \frac{E_m}{\pi}(1 + \cos \alpha) = R I_a + K_{af} \cdot I_a N + K_{res} \cdot N \quad (14.38)$$

Hence, the average speed from the above equation becomes

$$N = \frac{(E_m/\pi)(1 + \cos \alpha) - R \cdot I_a}{K_{af} \cdot I_a + K_{res}} \quad (14.39)$$

This expression relates average speed to average motor current. However, we are more interested in the torque-speed characteristics of the drive. From Eq. (14.32), it becomes clear that in series motors, the torque developed is proportional to the square of the motor RMS current. The relationship between the average and RMS motor currents is not straight forward and therefore, an explicit expression relating torque and speed cannot be easily obtained.

However, if the ripple in the motor current can be neglected, then $I_a \approx I_{ar}$. This will provide a well-defined expression relating speed and torque.

$$\therefore \quad T = K_{af} \cdot I_{ar}^2 \approx K_{af} \cdot I_a^2 \quad (14.40)$$

From Eq. (14.38),

$$I_a = \frac{E_m/\pi(1 + \cos \alpha) - K_{res} \cdot N}{R + K_{af} \cdot N}$$

Substitute the above value of I_a in Eq. (14.40)

$$\therefore \quad T \approx K_{af} \cdot \left[\frac{E_m/\pi(1 + \cos \alpha) - K_{res} \cdot N}{R + K_{af} \cdot N}\right]^2 \quad (14.41)$$

The torque–speed characteristics under the assumption of continuous and ripple-free motor current can be obtained from Eq. (14.41) for different values of the firing angle α and are shown in Fig. 14.21.

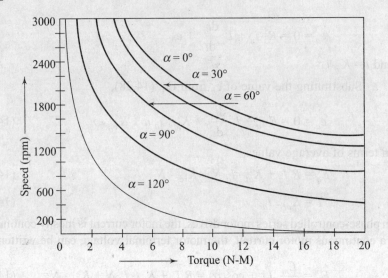

Fig. 14.21 *Torque–speed characteristics of d.c. series motor controlled by a single-phase semiconverter*

14.6.2 Single-phase Full Converter Drives

Figure 14.22(a) shows the power circuit for a speed control of a d.c. series motor by a single-phase full converter. Voltage and current waveforms are shown in Figs 14.22(b) and (c) for discontinuous and continuous motor armature current, respectively. As shown, thyristors T_1 and T_3 are simultaneously triggered at firing angle α, and thyristors T_2 and T_4 are triggered at $(\pi + \alpha)$. The current continues up to period β for discontinuous conduction and up to $(\pi + \alpha)$ for continuous conduction. In the discontinuous mode, the armature receives energy from the supply and accelerates during the conduction period α to β. From β to $(\pi + \alpha)$, the motor coasts and the load is supplied by the kinetic energy of the motor. For a continuous motor current, the motor terminal voltage can be written as

$$E_q = \frac{2E_m}{\pi} \cos \alpha = RI_a + K_{af} \cdot I_a N + K_{res} \cdot N \qquad (14.42)$$

Hence, the average speed from above equation becomes

$$N = \frac{(2E_m/\pi)\cos \alpha - RI_a}{K_{af} \cdot I_a + K_{res}} \qquad (14.43)$$

By making a similar assumption as in the case of semiconverter drive, the well-defined expression relating speed and torque can be written as

$$T \approx K_{af} \left[\frac{(2E_m/\pi)\cos \alpha - K_{res} \cdot N}{R + K_{af} \cdot N} \right]^2 \qquad (14.44)$$

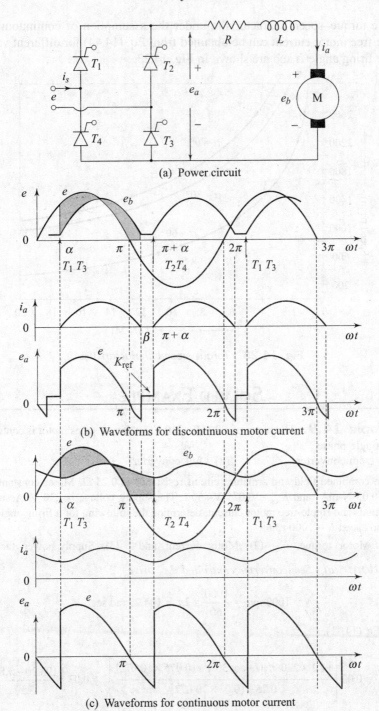

Fig. 14.22 *Single-phase full converter drive*

The torque–speed characteristic under the assumption of continuous and ripple-free motor current can be obtained from Eq. (14.44) for different values of the firing angle α and are shown in Fig. 14.23.

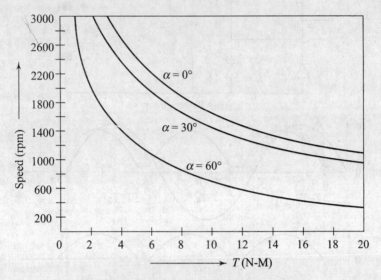

Fig. 14.23 *Torque–speed characteristics*

SOLVED EXAMPLE

Example 14.9 The speed of a 20 HP, 210 V, 1000 rpm series motor is controlled by a single-phase
   (a) semiconverter   (b) full converter.

The combined field and armature circuit resistance is 0.25 Ω. Motor constants are K_{af} = 0.03 N-mA^2 and K_{res} = 0.075 V-s/rad. The supply voltage is 230 V. Assuming continuous and ripple-free motor current, determine the following for a firing angle α = 30° and speed N = 1000 rpm;
   (1) Motor torque,   (2) Motor current, and   (3) Supply power, factor.

Solution: *(a)* *Semiconverter controlled d.c. drive*

(1) $\qquad N = 1000 \text{ rpm} = \dfrac{1000}{60} \times 2\pi = 104.72 \text{ rad/sec.}$

From Eq. (14.41),

$$T = 0.03 \left[\frac{\left(\sqrt{2} \times 230/\pi\right)(1+\cos 30°) - (0.075 \times 104.72)}{0.25 + (0.03 \times 104.72)} \right]^2 = 0.03 \left[\frac{193.20 - 7.85}{3.39} \right]^2$$

$\qquad$ = 89.68 N-m.

Control of D.C. Drives

(2) From Eq. (14.40),

$$I_a = \left(\frac{89.68}{0.03}\right)^{1/2} = 54.67 \text{ A.}$$

(3) Motor terminal voltage is given by

$$E_a = \frac{\sqrt{2} \times 230}{\pi}(1 + \cos 30°) = 193.20 \text{ V.}$$

If losses in the converter system are neglected, then input power is given by

$$P_s = E_a I_a = 193.20 \times 54.67 = 10562.24 \text{ W.}$$

Input volt-ampere $= 230 \times 54.67 \times \left(\frac{5}{6}\right)^{1/2} = 11478.53$ VA

Supply power-factor is $P_F = \dfrac{10562.24}{11478.53} = 0.92$

(b) **Full converter controlled d.c. series motor**

(1) From Eq. (14.41),

$$T = 0.03\left[\frac{\left(2\sqrt{2} \times 230/\pi\right)\cos 30° - 0.075 \times 104.72}{0.25 + (0.03 \times 104.72)}\right]^2 = 0.03\left(\frac{179.33 - 7.85}{3.39}\right)^2$$

$= 76.76$ N-m.

(2) From Eq. (14.40),

$$I_a = \left(\frac{76.76}{0.03}\right)^{1/2} = 50.58 \text{ A.}$$

(3) Now, $E_a = \dfrac{2\sqrt{2} \times 230}{\pi}\cos 30° = 179.33$ V

∴ $P_s = 179.33 \times 50.58 = 9070.51$ W.

Input volt-ampere $= 230 \times 50.58 = 11633.4$ VA

∴ $P_F = \dfrac{9070.51}{11633.4} = 0.78.$

Note that for the same firing angle, the power factor is better in the semiconverter system.

14.7 THREE-PHASE SEPARATELY EXCITED DRIVES

Three-phase drives are used for high power applications, up to megawatts power level. The three phase-controlled rectifiers provide power to these large horsepower d.c. drives. Three-phase drive is better compared to single-phase drive because the output ripple is small, the ripple frequency is large and filtering requirement is less. Moreover, in a three-phase drives, the armature current is mostly continuous and therefore, the motor performance is better compared to that of single-phase drives.

14.7.1 Three-phase Half-Wave Converter Drives

The basic power circuit of a three-phase half-wave converter drives is shown in Fig. 14.24. This type of converter is employed for motors in the typical range of 10–50 HP. This drive is impractical for most industrial applications because the supply currents would contain d.c. components.

With a three-phase half-wave converter in the armature circuit, Eq. (4.45) gives the armature voltage as

$$E_d = \frac{3\sqrt{3}}{2\pi} E_m \cos \alpha \qquad (14.45)$$

where E_m is the peak phase voltage of a star-connected three-phase a.c. supply.

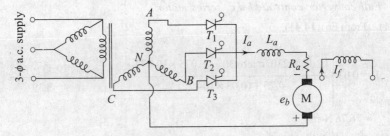

Fig. 14.24 Power circuit

14.7.2 Three-phase Semiconverter Drives

Figure 14.25 shows the power-circuit and the voltage and current waveforms for a three-phase semiconverter drive. It is a one-quadrant drive with field reversal and is limited to applications in the range of 15–150 HP. The field converter should also be a single-phase or three-phase semiconverter.

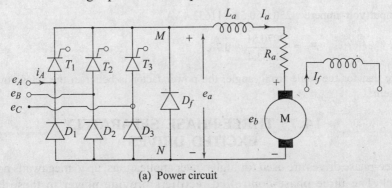

(a) Power circuit

Fig. 14.25(a)

The conduction periods of the diodes and the thyristors are shown in Fig. 14.25. As shown, the diodes D_1, D_2 and D_3 conduct during the intervals t_4 to t_6, t_6 to t_8, and t_2 to t_4, respectively. If thyristors T_1, T_2, and T_3 were diodes, they would conduct during the intervals t_1 to t_3, t_3 to t_5, and t_5 to t_7, respectively. Therefore, the references for the triggering angles for T_1, T_2 and T_3 are the instants t_1, t_3, and t_5, respectively; that is these are the crossing points of the phase voltages e_A, e_B and e_C.

The firing instants of the thyristors are marked for the triggering angle $\alpha = 90°$. The interval at which SCRs are triggered is 120°. Recall that the firing interval is 180° in single-phase converters. Thyristors are switched faster in three-phase converters and therefore, the time available for any current decay is less compared to the single-phase case. This tends to make the motor current continuous.

As shown, thyristor T_1 and diode D_3 conduct during the interval $(\pi/6 + \alpha) < \omega t < \omega t_4$. Therefore, motor terminal M is connected to phase voltage e_A, and terminal N is connected to phase voltage e_C. Thus, the motor terminal voltage during this period is $e_a = e_A - e_C = e_{AC}$. At ωt_4, e_a is zero, and from this time onwards e_a tends to be negative. The freewheeling diode D_F thus becomes forward biased at ωt_4, and motor current flows through it until the next thyristor T_2 is turned-on at $(\pi/6 + \alpha + 2\pi/3)$. In the absence of the freewheeling diode, free-wheeling action would have taken place through T_1 and D_1. At large firing angles the motor current can be continuous or discontinuous, depending on the current demand and speed.

The motor current may be discontinuous at large firing angles if the current demand is low and the speed is not low. Waveforms of voltages and currents are shown in Fig. 14.25(b) at $\alpha = 120°$ for both continuous and discontinuous motor currents.

In terms of average voltages,

$$E_a(\alpha) = I_a R_a + E_b = R_a I_a + K_a \phi N.$$

$\therefore$
$$N = \frac{E_a(\alpha) - R_a I_a}{K_a \phi} \tag{14.46}$$

If the motor current is continuous, then the motor armature voltage from Eq. (4.52) is given by

$$E_a(\alpha) = \frac{3\sqrt{3} E_m}{2\pi}(1 + \cos \alpha) \tag{14.47}$$

In Fig. 14.25(b), the fundamental component of the input current is also shown. The displacement angle ϕ_1 increases as the firing angle increases. Thus, the input power factor will decrease as the firing angle increases.

14.7.3 Three-phase Full Converter Drives

Figure 14.26 shows the full converter drive circuit, and the voltage and current waveforms. It is a two-quadrant drive without any field reversal, and is limited to applications in the range of 100–150 HP. The instants of firing the thyristors are worked for $\alpha = 60°$. The thyristors are fired at an interval of 60° and the ripple in the motor terminal voltage is six pulses per cycle. Since the SCRs are triggered at a faster rate, the motor current is mostly continuous. Therefore, the filtering requirement is less than that in the semiconverter system.

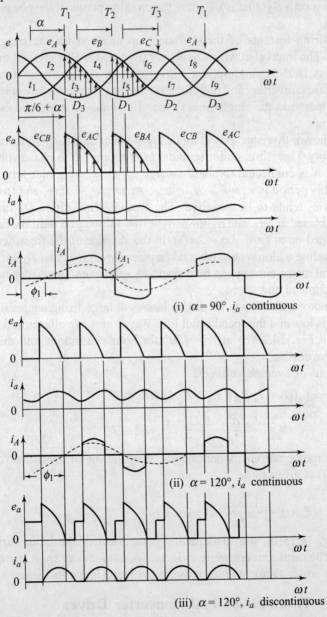

Fig. 14.25(b), (c) *Three-phase semiconverter controlled d.c. motor drive*

Thyristor T_1 turns-on at $\omega t = (\pi/6 + \alpha)$. Prior to this instant, SCR t_6 was turned-on. Therefore, during the interval $(\pi/6 + \alpha) < \omega t < (\pi/6 + \alpha + \pi/3)$, thyristors T_1 and T_6 conduct, and the motor terminals are connected to phase A and phase B, making $e_a = e_{AB}$. Thyristor T_2 is triggered at $\omega t = (\pi/6 + \alpha + \pi/3)$, and immediately SCR T_6 is reversed biased and turns-off. The current from T_6 is transferred to T_2 and therefore, the motor terminals are connected to phase A through T_1 and phase C through T_2, making $e_a = e_{AC}$. This process repeats after every 60° whenever an SCR is triggered. The thyristors are numbered in the sequence in which they are triggered.

For the triggering angle $\alpha = 120°$, the motor terminal voltage becomes negative as shown in Fig. 14.26(d). This is the inversion mode of operation of the converter. Power can be transferred from the motor to the a.c. supply if the motor voltage is reversed with a reversing contractor or by reversing the field current, which is commonly known as regeneration. The motor will slow down due to power feedback and thus, the motor voltage will decrease. Therefore, as the motor slows down, the firing angle of the converter has to be adjusted to keep the current up and to regenerate power.

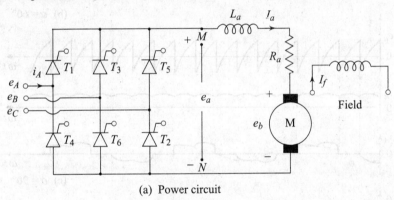

(a) Power circuit

Fig. 14.26(a)

The average motor terminal voltage from Eq. (4.51) is given by

$$E_{a(\alpha)} = \frac{3\sqrt{3}}{\pi} E_m \cos \alpha \tag{14.48}$$

The average speed is

$$N = \frac{E_{a(\alpha)} - R_a I_a}{K_a \phi} \tag{14.49}$$

In a separately excited motor, $T = K_a \phi I_a$. Therefore,

$$N = \frac{E_{a(\alpha)}}{K_a \phi} - \frac{R_a}{(K_a \phi)^2} T \tag{14.50}$$

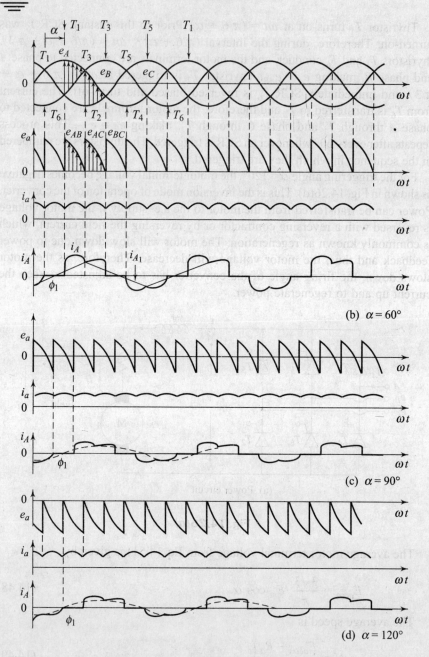

Fig. 14.26(b)–(d) *Three-phase full converter drive*

The first term in Eq. (14.50) represents the ideal no-load ($T \approx 0$) speed, which therefore, depends on $E_a(\alpha)$. If the motor current is assumed continuous, motor terminal voltage E_a depends only on the firing angle α and is given by Eqs (14.48) and (14.49) for the semiconverter and the full converter, respectively.

The variations of E_a with α for continuous motor current are shown in Fig. 14.27. These curves also represent the theoretical no-load speed as a function of firing angle. The second term in Eq. (14.50) represents the decrease in speed as the motor torque increases. Since the armatures circuit resistance R_a is small, the decrease in speed is small (i.e. good speed regulation). In large motors, the motor current at no-load is not small, and if a three-phase converter is used the motor current is likely to be continuous even at no-load condition. Therefore, three-phase drives provide better speed regulation and improved performance compared to single-phase drives.

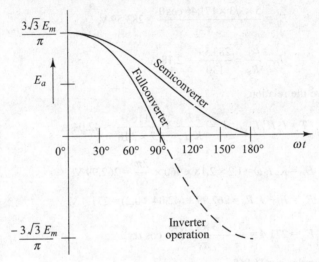

Fig. 14.27 *Average output voltage e_a as a function of firing angle in three-phase converters for continuous motor current*

SOLVED EXAMPLES

Example 14.10 The speed of a 25 HP, 320 V, 960 rpm separately excited d.c. motor is controlled by a 3-ϕ full converter. The field current is also controlled by a three-phase full converter and is set to the maximum possible value. The a.c. input is a 3-ϕ, star-connected 210 V, 50 Hz supply. The armature resistance is $R_a = 0.2\ \Omega$, the field resistance is $R_f = 130\ \Omega$, and the motor voltage constant is $K_a = 1.2$ V/A rad/s. The armature and field currents are continuous and ripple free. Determine:
  (a) The firing angle of the armature converter if the free converter is operated at the maximum field current and the developed torque is 110 N-m at 960 rpm.
  (b) The speed of the motor if the field circuit converter is set for the maximum field current, the developed torque is 110 N-m and the firing angle of the armature converter is 0°.
  (c) The firing angle of the field converter if the speed has to increase to 1750 rpm, for the same load requirement in part (b). Neglect the system losses.

Solution:

(a) Phase voltage, $E_p = \dfrac{210}{\sqrt{3}} = 121.24$ V $\therefore E_m = \sqrt{2} \times 121.24 = 171.46$ V

The field controlled converter Eq. is

$$E_f = \dfrac{3\sqrt{3} E_m}{\pi} \cos \alpha$$

For maximum field current, $\alpha = 0$,

$\therefore \qquad E_f = \dfrac{3 \times \sqrt{3} \times 171.46 \cos 0}{\pi} = 283.59$ V.

Also, $\qquad I_f = \dfrac{E_f}{R_f} = \dfrac{283.59}{130} = 2.18$ A.

Now, we have the relation,

$$T = I_a K_a I_f \quad \therefore I_a = \dfrac{T}{K_a I_f} = \dfrac{110}{1.2 \times 2.18} = 42.04 \text{ A.}$$

$$E_b = K_a I_f \omega = 1.2 \times 2.18 \times 960 \times \dfrac{2\pi}{60} = 262.99 \text{ V.}$$

$$E_a = E_b + I_a R_a = 262.99 + (42.04 \times 0.2) = 271.4 \text{ V.}$$

Now, $\qquad E_a = 271.4 = \dfrac{3 \times \sqrt{3} \times 171.46}{\pi} \cos \alpha.$

$\therefore$ Firing angle, $\alpha = 16.86°.$

(b) Given $\quad \alpha = 0, \quad \therefore E_a = \dfrac{3 \times \sqrt{3} \times 171.46}{\pi} = 283.59.$

$E_b = 283.59 - (42.04 \times 0.2) = 275.18$ V

$\therefore \quad$ Speed $= \dfrac{E_b}{K_a \cdot I_f} = \dfrac{275.18}{1.2 \times 2.18} = 105.19$ rad/s $= 1004.50$ rpm.

(c) $\omega = 1750 \times \dfrac{2\pi}{60} = 183.26$ rad/s

$E_b = 275.18 = 1.2 \times 183.26 \times I_f \quad \therefore I_f = 1.25$ A.

$\therefore \qquad E_f = 1.25 \times 130 = 162.5$ V

Now, $E_f = 162.5 = \dfrac{3 \times \sqrt{3} \times 171.46}{\pi} \cos \alpha_f, \quad \therefore \alpha_f = 55.04°.$

Example 14.11 The speed of a 100 kw, 1000 rpm, separately excited d.c. motor is controlled by 3-ϕ full converter. The specifications of the converter are 460 V, 300 A. The input to the converter is a 3-ϕ, 415 V, 50 Hz a.c. supply. Determine
  (a) Firing angle of the converter and power factor at rated speed.
  (b) Firing angle and power factor at 10% rated speed.

(c) Active and reactive power drawn from the system at rated speed.
(d) Active and reactive power drawn from the system at 10% rated speed.
(e) Ratio of reactive power drawn at 10% and rated speed.

Neglect the system losses and effect of commutation angle.

Solution:

(a) For 3-ϕ, fully controlled converter, we have the relation

$$E_a = \frac{3\sqrt{3}}{\pi} E_m \cos \alpha = \frac{3\sqrt{3}}{\pi} \times \frac{\sqrt{2}}{\sqrt{3}} E \cos \alpha$$

$$460 = 1.35 \times 415 \times \cos \alpha \therefore \alpha = 34.81°.$$

Since the effect of commutation angle is to be neglected, therefore,

Power factor $\cos \phi = \cos \alpha = 0.82$.

(b) At 10% rated speed,

$$E_{a\alpha} = 0.1 \times 460 = 46 \text{ V} \therefore 46 = 1.35 \times 415 \times \cos \alpha$$

or $\cos \alpha = 0.082$, or $\alpha = 85.2°$

Also, power factor, $\cos \phi = \cos \alpha = 0.082$

(c) At rated speed, active power drawn from the system is $\sqrt{3} \, EI \cos \phi$

Now, $I = \sqrt{\frac{2}{3}} \times I_a = 0.817 \, I_a$

But, $I_a = \dfrac{100 \times 1000}{460} = 217.39 \text{ A}$

$\therefore$ $I_a = 0.817 \times 217.39 = 177.61 \text{ A}$

$\therefore$ $\cos \phi = 0.821$

$\therefore$ Active power = $\sqrt{3} \times 415 \times 177.61 \times 0.821 = 104.7$ kW

Reactive power = $\sqrt{3} \, E I \sin \phi$

where $\sin \phi = \sqrt{1 - \cos^2 \phi} = \sqrt{1 - (0.821)^2} = 0.571$

Hence, reactive power = $\sqrt{3} \times 415 \times 177.61 \times 0.571 = 72.79$ kVAR.

(d) At 10% speed, active power = $\sqrt{3} \times E \times I \cos \phi$

$$= \sqrt{3} \times 415 \times 177.61 \times 0.082 = 10.47 \text{ kW}.$$

Reactive power at 10% speed = $\sqrt{3} \, EI \sin \phi$.

But, $\sin \phi = \sqrt{1 - \cos^2 \phi} = \sqrt{1 - (0.82)^2} = 0.9966$

Hence reactive power,

$$Q_i = \sqrt{3} \times 415 \times 177.61 \times 0.9966 \text{ VAR} = 127.08 \text{ kVAR}.$$

(e) $\dfrac{\text{Reactive power at 10\% speed}}{\text{Reactive power at rated speed}} = \dfrac{127.08}{72.79} = 1.75$

Hence, due to reduction in motor speed the reactive power has increased by 74.6%.

14.8 D.C. CHOPPER DRIVES

Control of a d.c. motor's speed by a chopper is required where the supply is d.c. (as from a battery) or an a.c. voltage that has already been rectified to a d.c. voltage. The most important applications of choppers are in the speed control of d.c. motors used in industrial or traction drives. Choppers are used for the control of d.c. motors because of a number of advantages, such as high efficiency, flexibility in control, light weight, small size, quick response, and regeneration down to very low speeds. Chopper controlled d.c. drives have also applications in servos in battery operated vehicles such as forklift trucks, trolleys, and so on.

Because of the flexible control characteristics, separately excited d.c. motors or permanent magnet field d.c. motors are used in servo applications. In the past, the series motor was mainly used in traction. Presently, the separately excited motor is also employed in traction. The high starting torque was the main reason for using a series motor. However, the series motor has a number of limitations. The field of the series motor cannot be easily controlled by static means. If field control is not employed, the series motor must be designed with its base speed equal to the highest desired speed of the drive. The higher base speeds are obtained using fewer turns in the field windings. However, this reduces the torque per ampere at zero and low speeds. Further, there are a number of problems with regenerative braking of a series motor. On the other hand, regenerative braking of a separately excited motor is fairly simple and can be carried out down to very low speeds. Because of the limitations of series motors, separately excited motors are now preferred even for traction applications.

The choppers offer a number of advantages over a controlled rectifiers for a d.c. motor control in open-loop and closed-loop configurations. Because of the higher frequency of the output voltage-ripple, the ripple in the motor armature current is less and the region of discontinuous conduction in the speed–torque plane is smaller. A reduction in the armature current ripple reduces the machine losses and its derating. A reduction or elimination of discontinuous conduction region improves speed regulation and transient response of a drive.

To realize a higher frequency of output voltage ripple, it is customary to use a rectifier with a higher pulse number. Use of a rectifier with a higher pulse number results in a low utility factor for thyristors and a relatively high cost. On the other hand, a chopper can be operated at comparatively high frequencies. For example, it is possible to operate a chopper at 300 Hz even with converter-grade thyristors. With inverter-grade thyristors the frequency can be increased to 600 Hz. If the output voltage range can be lowered, corresponding frequencies can be increased to 400 Hz and 800 Hz, respectively. When power transistors are employed, frequencies can be higher than 2.5 kHz. For low power applications, power MOSFET can be used and the frequency can be higher than 200 kHz. The rectifier output voltage and current have a much lower frequency, 100 Hz in the case of a single-phase rectifier and 300 Hz in the case of a three-phase fully controlled rectifier when the a.c. source frequency is 50 Hz.

Control of D.C. Drives

In Chapter 8, the principles and circuits used for chopper control were explained. In this section chopper-controlled d.c. motors are analyzed. The performance characteristics of series d.c. motors and separately excited d.c. motors are also discussed.

14.8.1 Chopper-fed D.C. Series Motor

Figure 14.28(a) shows the basic circuit and relevant waveforms for a chopper-driven series d.c. motor. Assume that the motor current i_a is continuous as shown in Fig. 14.28(b) and that the speed is constant.

During the ON period of the chopper, we can write

$$E_{dc} = R \cdot i_a + L \cdot \frac{di_a}{dt} + K_{af} \cdot i_a N + K_{res} \cdot N, \quad 0 < t < t_{on} \tag{14.51}$$

During the OFF period of the chopper, we can write

$$0 = R \, i_a + L \, \frac{di_a}{dt} + K_{af} \cdot i_a N + K_{res} N, \quad t_{on} < t < T \tag{14.52}$$

In the steady-state, the current will be minimum at the instant the chopper is turned-on and maximum at the instant the chopper is turned-off, that is,

$$i_a(0) = I_{min} \quad \text{and} \quad i_a(t_{on}) = I_{max}.$$

The solutions for i_a are

$$i_a = \frac{E_{dc} - K_{res} N}{R + K_{af} N}\left(1 - e^{-t/\tau}\right) + I_{min} e^{-t/\tau} \quad 0 < t < T_{on} \tag{14.53}$$

$$i_a = \frac{-K_{res} N}{R + K_{af} N}\left(1 - e^{-t'/\tau}\right) + I_{max} e^{-t'/\tau}, \quad t_{on} < t < T \tag{14.54}$$

where $\quad t' = t - t_{on}$

and $\quad \tau = \dfrac{L}{(R + K_{af} N)}$

From Eq. (14.53),

$$I_{max} = i_a(t_{on}) = \frac{E_{dc} - K_{res} N}{R + K_{af} N}\left(1 - e^{-t_{on}/\tau}\right) + \left(I_{min} e^{-t_{on}/\tau}\right) \tag{14.55}$$

Now $\quad i_a(0) = i_a(T) = i_a(t' = T - t_{on}).$

Therefore, from Eq. (14.54),

$$I_{min} = i_a(t' = T - t_{on})$$

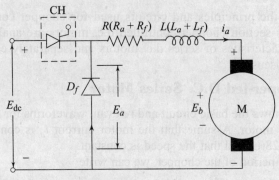

(a) Basic power circuit

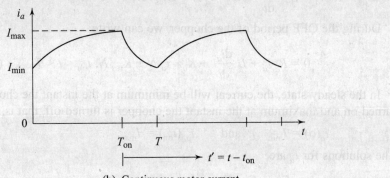

(b) Continuous motor current

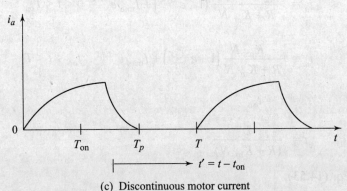

(c) Discontinuous motor current

Fig. 14.28 Chopper-fed d.c. series motor

$$= \frac{-K_{res} N}{R + K_{af} N}\left(1 - e^{-(T - t_{on}/\tau)}\right) + \left(I_{max} e^{-(T - t_{on}/\tau)}\right) \quad (14.56)$$

From Eqs (14.55) and (14.56),

$$I_{min} = \frac{E_{dc}}{R + K_{af} N} \frac{\left(e^{t_{on}/\tau} - 1\right)}{\left(e^{T/\tau} - 1\right)} - \frac{K_{res} N}{R + K_{af} N} \quad (14.57)$$

$$I_{\max} = \frac{E_{dc}}{R + K_{af} N} \frac{\left(1 - e^{-t_{on}/\tau}\right)}{\left(1 - e^{-T/\tau}\right)} - \frac{K_{res} N}{R + K_{af} N} \quad (14.58)$$

Discontinuous Armature Current In the case of discontinuous armature current, as shown in Fig. 14.28(c), $I_{\min} = 0$.
From Eq. (14.55),

$$I_{\max} = \frac{E_{dc} - K_{res} N}{R + K_{af} N} \left(1 - e^{-t_{on}/\tau}\right) \quad (14.59)$$

To find the instant $t = t_p$ or $t' = t_p - t_{on}$ at which the current becomes zero, the value of $I_{\max}$ obtained from Eq. (14.59) is substituted in Eq. (14.54) and i_a is made zero. Therefore,

$$0 = \frac{-K_{res} N}{R + K_{af} N}\left[1 - e^{-(t_p - t_{on}/\tau)}\right]$$

$$+ \frac{E_{dc} - K_{res} N}{R + K_{af} N}\left(1 - e^{-t_{on}/\tau}\right)\left(e^{-(t_p - t_{on}/\tau)}\right) \quad (14.60)$$

From Eq. (14.60),

$$t_p = \tau \ln \left\{ e^{t_{on}/\tau}\left[1 + \frac{E_{dc} - K_{res} N}{K_{res} N}\left(1 - e^{-t_{on}/\tau}\right)\right]\right\} \quad (14.61)$$

From Eq. 14.61, the continuity of the motor current can be determined. Once the nature of the current is known, the motor current is given by Eq. (14.53) when the chopper is ON and by Eq. (14.54) when the chopper is OFF. Equation (14.54) is valid for $t_{on} < t < T$ for continuous motor current and $t_{on} < t < t_p$ for discontinuous motor current.

To study the performance characteristics of the motor, expressions for average and RMS motor current and average motor voltage have to be determined. These are obtained as follows:

Average Motor Current I_a is given by the expression

$$I_a = \frac{1}{T}\left[\int_0^{T_{on}} i_{a(t)}\, dt + \int_{t_{on}}^{t_p} i_{a(t)}\, dt\right] \quad (14.62)$$

Let us put

$$I_1 = \frac{E_{dc} - K_{res} N}{R + K_{af} N} \quad (14.63)$$

and

$$I_2 = \frac{-K_{res} N}{R + K_{af} N} \quad (14.64)$$

in Eqs (14.53) and (14.54) respectively.

Therefore, from Eqs (14.53), (14.54), (14.62), (14.63), and (14.64) we can write

$$I_a = \frac{1}{T}\Big[I_1 t_{on} + \tau(I_{min} - I_1)(1 - e^{-t_{on}/\tau})$$

$$- I_2(t_p - t_{ON}) + \tau(I_{max} + I_2)(1 - e^{-(t_p - t_{ON})/\tau})\Big] \quad (14.65)$$

Equation (14.65) is an expression for average motor current in closed form.

RMS Motor Current I_{ar}, RMS motor current is given by

$$I_{ar} = \left\{\frac{1}{T}\left[\int_0^{t_{on}} i_a^2(t)dt + \int_{t_{on}}^{t_p} i_a^2(t)dt\right]\right\}^{1/2} \quad (14.66)$$

$$= \left\{\frac{1}{T}\Big[I_1^2 t_{on} + 2\tau I_1(I_{min} - I_1)(1 - e^{-t_{on}/\tau})\right.$$

$$+ \frac{\tau}{2}(I_{min} - I_1)^2(1 - e^{-2t_{on}/\tau}) + I_2^2(t_p - t_{on})$$

$$- 2\tau I_2(I_{max} + I_2)(1 - e^{-(t_p - t_{on})/\tau})$$

$$\left.+ \frac{\tau}{2}(I_{max} + I_2)^2\Big(1 - e^{-2(t_p - t_{on})/\tau}\Big)\Big]\right\}^{1/2} \quad (14.67)$$

Equation (14.67) is also in closed form.

Torque The average motor torque is given by

$$T = k_{af} I_{ar}^2 \quad (14.68)$$

Average Motor Voltage
(i) The average motor voltage for continuous current is,

$$E_a = E_{dc}\alpha \quad (14.69)$$

and $\quad t_p = T$

(ii) The average motor voltage for discontinuous current is,

$$E_a = E_{dc} \cdot \alpha + K_{res} \cdot N \cdot \frac{(T - t_p)}{T}$$

and $\quad t_p < T \quad (14.70)$

14.8.2 Chopper-Fed Separately Excited d.c. Motor

The above analysis for chopper fed-d.c. series motor is also valid for separately excited d.c. motors with the following exceptions.
(1) The term $k_{af} N$ is not present.

(2) $K_{res} N$ is replaced by a back emf $K_a \phi N$.

Therefore, $I_1 = \dfrac{E_{dc} - K_a \phi \cdot N}{R}$ \hfill (14.71)

and $I_2 = \dfrac{-K_a \phi \cdot N}{R}$ \hfill (14.72)

(3) Average torque, $T = k_a \phi I_a$ \hfill (14.73)

SOLVED EXAMPLE

Example 14.12 A d.c. shunt motor takes a current of 80 A on a 480 V supply and runs at 960 rpm. The armature resistance is 0.25 Ω and the field resistance is 120 Ω. A chopper is used to control the speed of the motor in the range of 400–750 rpm having constant torque. The on-period of the chopper is 3 ms. The field is supplied directly from 480 V supply. Determine, the range of frequencies of the chopper.

Solution: Field current is the motor $(I_f) = \dfrac{480}{120} = 4$ A.

∴ Armature current = 80 – 4 = 76 A.

Now, back emf at 960 rpm = $480 - 76 \times 0.25 = 461$ V.

Back emf at 400 rpm = $\dfrac{400}{960} \times 461 = 191.88$ V

Terminal voltage required = $191.88 + (76 \times 0.25) = 210.88$ V.

∴ Chopping period, $T = \dfrac{480}{210.88} \times 3 \times 10^{-3}$

∴ Chopping frequency = $\dfrac{210.88 \times 10^3}{480 \times 3} = 146.44$ Hz

Now, back emf at 750 rpm = $\dfrac{750}{960} \times 461 = 360.16$ V

∴ Terminal voltage required = $360.16 + (76 \times 0.25) = 379.16$ V.

∴ Chopper period = $\dfrac{480 \times 3 \times 10^{-3}}{379.16}$

Chopper frequency = $\dfrac{379.16 \times 10^3}{480 \times 3} = 263.31$ Hz

Therefore, the range of frequencies of the chopper is $146.44 \text{ Hz} < f \leq 263.31$ Hz

14.9 PHASE-LOCKED LOOP (PLL) CONTROL OF D.C. DRIVES

The degree of speed control required in industrial drives depends on the application. In some applications, open-loop regulation of the drive motor may be adequate. In others, closed-loop feedback control is required for precise speed control and fast response. Conventionally, this is achieved by an analog servo feedback system in which any change in speed is sensed by a tachometer and compared with a fixed reference voltage to generate an error signal. These analog devices for speed sensing and comparing signals are not ideal and the speed regulation is more than 0.2%. With a proportional-integral controller, when a torque is applied or changed, there is a transient speed dip. In addition, the steady-state speed is reached after a certain time interval. This type of operation may not be satisfactory in certain drives where high quality products are desired.

These shortcomings of analog feedback control system can be overcome by using a digital phase-looked loop control system. In a PLL control system, the motor speed is converted to a digital pulse train, which is synchronized with a reference digital pulse train. In this way, by locking onto a reference frequency precise control of speed is achieved.

Figure 14.29(a) shows the block diagram of a converter-fed d.c. motor drive with phase-locked loop control and the transfer function block diagram is shown in Fig. 14.29(b). In this PLL system, speed encoder is used to convert the motor speed to a digital pulse-train. The output of the encoder acts as the speed feedback signal of frequency f_0. The phase detector compares the reference pulse-train

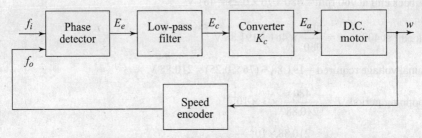

(a) Motor speed control using PLL-technique

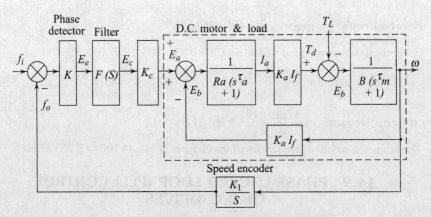

(b) Transfer function model

Fig. 14.29 *Phase-locked loop control system*

(or frequency) f_r with the feedback frequency f_0 and provides a pulse-width modulated output voltage, E_e, which is proportional to the difference in phases and frequencies of the reference and feedback pulse trains. The phase detector (or comparator) is available in integrated circuits. A low-pass loop filter converts the pulse train, E_e, to a continuous d.c. level E_c which varies the output of the power converter and, in turn, the motor speed.

When the motor runs at the same speed as the reference pulse train, the two frequencies would be synchronized (or locked) together with a phase difference. The output of the phase detector would be a constant voltage proportional to the phase-difference and the steady-state motor speed would be maintained at a fixed value irrespective of the load on the motor. Any disturbance contributing to the speed change would result in a phase difference and the output of the phase detector would respond immediately to vary the speed of the motor in such as direction and magnitude as to retain the locking of the reference and feedback frequencies. The response of the phase detector is very fast. As long as the two frequencies are locked, the speed regulation should ideally be zero. However, in practice, the speed regulation is limited to 0.002% and this represents a significant improvement over the analog speed control system.

PLL techniques have been used primarily in communication systems to synchronize signals. PLL are now available as inexpensive digital integrated circuits. With the development of both integrated circuits and solid-state power circuits, the use of PLL techniques in motor speed control has generated considerable interest.

14.10 MICROCOMPUTER CONTROL OF D.C. DRIVES

Microcomputers have profoundly influenced the technology of power electronics and drives systems. Their implementation has not only brought about simplification of hardware and improvement of reliability, but also permitted performance optimization and powerful diagnostic capability which was not possible before by dedicated hardware control. The performance of microcomputers in terms of functional integration, improvement of architecture and speed of computation has continually improved since their introduction at the beginning of the 1970s, and this trend will continue. Practically, all newly designed d.c. and a.c. drives use microprocessors for some part, or all of the control electronics function. The microprocessors will be supplemented by custom LSI, hard-wired digital, or analog circuits, that result in less cost, superior performance, and safer operation. The microprocessor makes possible monitoring and diagnostic functions that were too costly to provide with hard-wired circuits. These additional functions justify the cost of the microprocessor controller.

A conventional analog control scheme implemented in d.c. drives consists of an outer speed control loop and an inner current control loop. The analog control method has several disadvantages, such as nonlinearity in the analog speed transducer, difficulty in accurately transmitting the analog signal, errors due to temperature, component aging, drift and offset of the analog components, and so on. Once a control circuit is built to meet certain performance criteria, it may require major changes in the hardware logic circuits to meet other performance requirements. However, with the microcomputer control, the control scheme is implemented in the software and is flexible to change the control strategy to meet different performance characteristics or to add extra control features.

Figure 14.30 shows a schemating diagram for a microcomputer control of a converter-fed four-quadrant d.c. drive. The speed information is fed into the microcomputer using an A/D (analog-to-digital) converter. The motor current data is usually fed into the computer through a fast A/D converter or by sampling the armature current. A synchronizing circuit interface is required so that the microprocessor can synchronize the generation of the firing pulse data with the supply line frequency. Although the microcomputer can perform the functions of gate pulse generator and logic circuit, these are shown outside the microcomputer. The pulse amplifier provides the necessary isolation and produces gate pulse of required magnitude and duration.

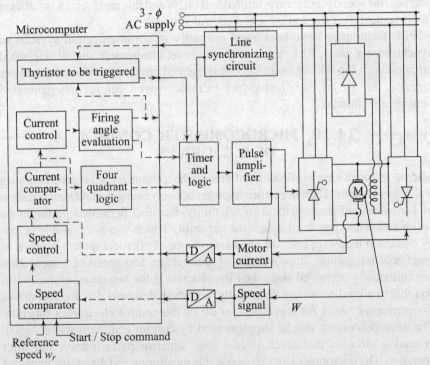

Fig. 14.30 *Schematic 4-quadrant microcomputer controlled d.c. drive system*

A set of instructions is stored in the memory, and those instructions are executed by the microprocessor for proper functioning of the drive. A typical program flow chart for this drive system is shown in Fig. 14.31. When the

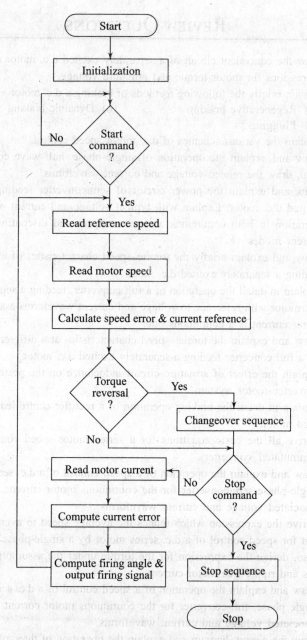

Fig. 14.31 *Program flow chart for a drive*

program starts, the first step is to initialise the internal registers and the interface circuits. The microprocessor is now ready to receive the starting command. After receiving the start signal, it begins executing the program in a continuous cyclic manner. The sequence of instructions allows the computer to process data for speed regulation, current regulation, and changeover reversal operation.

REVIEW QUESTIONS

14.1 Draw the equivalent circuit of a separately excited d.c. motor and derive the expressions for motor torque and armature voltage.

14.2 Explain briefly the following methods of braking a d.c. motor
   (a) Regenerative braking   (b) Dynamic braking
   (c) Plugging

14.3 Explain the various schemes of d.c. motor speed control.

14.4 Draw and explain the operation of single-phase, half-wave converter drive. Also, draw the related voltage and current waveforms.

14.5 Draw and explain the power circuit of semiconverter feeding a separately excited d.c. motor. Explain with typical voltage and current waveforms, the operation in both continuous armature current and discontinuous armature current modes.

14.6 Draw and explain briefly the torque–speed characteristics of a semiconverter feeding a separately excited d.c. motor.

14.7 Explain in detail the operation of a full converter, feeding a separately excited d.c. motor with reference to voltage and current waveforms assuming that the motor current is a continuous one.

14.8 Draw and explain the torque–speed characteristics at a different firing angles for a full converter feeding a separately excited d.c. motor.

14.9 Explain the effect of armature circuit inductance on the performance of the converter–motor system.

14.10 Explain in detail the braking operation of a rectifier controlled separately excited motor.

14.11 Derive all the basic equations for a series motor speed control using line commutated converters.

14.12 Draw and explain the operation of a speed control of a d.c. series motor by a single-phase semiconverter for the continuous motor current. Draw also the associated voltage and current waveforms.

14.13 Derive the expression which relates the average speed to average motor current for speed control of a d.c. series motor by a single-phase semiconverter. Also, derive the expression for the torque under the assumption of continuous and ripple-free motor-current.

14.14 Draw and explain the operation of a speed control of a d.c. series motor by a single phase, full-converter for the continuous motor current. Draw also the associated voltage and current waveforms.

14.15 Draw the circuit diagram and explain the operation of three-phase, half-wave converter drives.

14.16 Draw the circuit diagram and explain the operation of a three-phase semiconverter drive. Also, sketch and explain the following waveforms:
   (i) Output voltage and output current at $\alpha = 90°$
   (ii) Output voltage and output current at $\alpha = 120°$.

14.17 Draw the power circuit diagram and explain the operation of a three-phase full converter drive. Also, sketch and explain the output voltage and output current waveforms at firing angle of 60°, 90° and 120°.

14.18 List the advantages offered by d.c. chopper drives over line-commutated converter controlled d.c. drives.

14.19 Draw the circuit-diagram and explain the operation of chopper fed d.c. series motor. Also, derive the expressions for I_{max} and I_{min} assuming a continuous armature current.

14.20 Derive the expressions for average motor current, RMS motor current, torque, and average motor voltage, for chopper-fed d.c. series motor.

14.21 Derive the expressions for average motor current, currents I_1 and I_2, and average torque for chopper-fed d.c. separately excited motor.

14.22 Explain the principle of closed-loop control of d.c. drive using suitable block-diagram.

14.23 Draw the circuit diagram and explain the operation of closed-loop speed control with inner-current loop and field weakening.

14.24 What is the principle of phase-locked loop control of d.c. drives? Also, list the advantages of phase-locked loop control of d.c. drives.

14.25 Draw the block diagram and explain the operation of a phase-locked loop-control system.

14.26 Explain the principle of microcomputer control of d.c. drives. List the advantages of microcomputer control of d.c. drives.

14.27 Draw and explain the block schematic of a 4-quadrant, microcomputer controlled d.c. drive system. Also, draw and explain the suitable flow chart for the same scheme.

PROBLEMS

14.1 A 230 V, 800 rpm, 7 A separately excited d.c. motor has an armature resistance 0.2 Ω. Under rated conditions, the motor is driving a load whose torque is constant and independent of speed. The speeds below the rated speed are obtained with armature voltage control (with full field), and the speeds above the rated speed are obtained by field control (with rated armature voltage). Compute:
   (i) The motor terminal voltage when the speed is 500 rpm.
   (ii) The value of flux as a percentage of rated flux if the motor speed is 1100 rpm. Neglect the motor's rotational losses. [*Ans* (i) 149V
   (ii) 0.71 of its rated value.]

14.2 A separately excited motor of Example 14.1 is coupled to an overhauling load with a torque of 800 N-m. Determine the speed at which the motor can hold the load by regenerative braking the source voltage is 230 V. Neglect the motor's rotational losses. [*Ans* 1081 rpm.]

14.3 The speed of a separately excited d.c. motor is controlled by a single-phase semiconverter in Fig. 14.9. The field current is also controlled by a

semiconverter and field current is set to the maximum possible value. The a.c. supply voltage to the armature and field converter is single-phase, 210 V, 50 Hz. The armature and field parameters are $R_a = 0.25\ \Omega, R_f = 150\ \Omega, K_a = 1.055$ V/A rad/s. The load torque is TL = 80 N-m at a speed of 960 rpm. The armature and field currents are continuous and ripple-free. Determine:
(a) the field current I_f.
(b) the firing angle of the converter in the armature circuit, and
(c) the input power factor of the armature circuit converter.
Neglect the system losses. [Ans (a) 1.26A, (b) 55°, (c) 0.85.]

14.4 The speed of a 20 Hp, 460 V, 1000 rpm separately excited d.c. motor is controlled by a single-phase full wave bridge circuit as shown in Fig. 14.13. The rated motor armature current is 35 A, and the armature resistance is 0.15 Ω. The a.c. supply voltage is 480 V. The motor back emf constant is $K_a \phi = 0.45$ V/rpm. Assume that sufficient inductance is present in the armature circuit to make the motor current continuous and ripple free.
 (a) *Rectifier operation (motoring action)*: For a firing angle $\alpha = 60°$ and rated armature current, determine
   (1) the motor torque.
   (2) the speed of the motor.
   (3) the supply power factor.
 (b) *Inverter operation (regenerative action)*: The motor back emf polarity is reversed by reversing the field excitation. Determine:
   (1) the firing angle to keep the motor current at its rated value.
   (2) the power fed back to the supply. [Ans (a) (1) 150.4 N-m.
      (2) 468.6 rpm. (3) 0.45, (b) (1) 118.4° (2) 7196 W.]

14.5 A 220 V, 1500 rpm, 12 A separately excited motor is controlled by a 1-phase fully controlled converter with an a.c. source voltage of 230 V, 50 Hz. Assume that sufficient inductance is present in the armature circuit to make the motor current continuous and ripple free for any torque greater than 25 per cent of rated torque, $R_a = 2.5\ \Omega$.
 (a) What should be the value of the firing angle to get the rated torque at 1000 rpm?
 (b) Determine the firing angle for the rated braking torque at 1500 rpm.
 (c) Determine the motor speed at the rated torque and $a = 150°$ for the regenerative braking in the second-quadrant.
     [Ans (a) 40.44° (b) 140.56° (c) 1652.03 rpm.]

14.6 The speed of a 15 HP, 220V, 1200 rpm series motor is controlled by a single-phase
 (a) semiconverter
 (b) fullconverter
 The combined field and armature circuit resistance is 0.3 Ω. Motor constants are $K_{af} = 0.02$ N-m/A^2 and $K_{res} = 0.08$ V-s/rad. The supply voltage is 230 V. Assuming continuous and ripple-free motor current, compute the following for a firing angle $\alpha = 30°$ and speed $N = 1200$ rpm.
 (1) Motor torque
 (2) Motor current

(3) Supply power factor. [*Ans* (a) (1) 84.96 N-m (2) 65.18A (3) 0.92
(b) (1) 72.58 N-m (2) 60.24A (3) 0.78]

14.7 A 100 kW, 440V, 960 rpm d.c. motor is operating at 750 rpm and developing 75% rated torque is controlled by 3-ϕ, six-pulse thyristor converter. If the back emf at rated speed is 405 V, determine the triggering angle of the converters. The input to the converter is 3-ϕ, 415 V, 50 Hz a.c. supply.[*Ans* $\alpha = 53.26°$.]

14.8 The speed of a separately excited d.c. motor is controlled by a single-phase full wave converter. The field circuit is also controlled by a full-converter and the field current is set to the maximum possible value. The a.c. supply voltage to the armature and field converters is single-phase, 410 V, 50 Hz. The armature resistance is 0.3 Ω, R_f = 150 Ω, and the motor voltage constant is K_a = 1.2 V/A-rad/s. The armature current is 30 A. Assume that the sufficient inductance is present in the armature and field circuits to make the armature and field currents continuous and ripple-free. The firing angle of the armature controlled converter is 45°. Determine:
(a) the torque developed by the motor,
(b) the speed, ω, and
(c) the supply power factor.
Neglect the system losses.   [*Ans* (a) 88.56 N-m. (b) 815.22 rpm. (c) 0.71.]

14.9 The speed of a 25 HP, 380 V, 1800 rpm, separately excited d.c. motor is controlled by a 3-ϕ full converter. The field circuit is also controlled by 3-ϕ full converter and the field current is set to the maximum possible value. The a.c. input is a 3-ϕ, star-connected 210 V, 50 Hz supply. The armature and field parameters are R_a = 0.15 Ω, R_f = 250 Ω, and K_a = 1.2 V/A rad/s. Assume that the armature and field currents are continuous and ripple free. By neglecting the system losses, determine the following:
(a) the firing angle of the armature converter, α, if the motor supplies the rated power at the rated speed,
(b) the no-load speed, if the firing angles are the same in part (a) and the armature current at no-load is 10% of the rated value, and
(c) the speed regulation.   [*Ans* (a) 26.44. (b) 1848.42 rpm. (c) 2.69%.]

14.10 A d.c. chopper is used to control the speed of d.c. shunt motor. The supply voltage to the chopper is 220V. The on-time and off-time of the chopper is 10 ms and 12 msec, respectively. Assuming continuous conduction of the motor current, and neglecting the armature inductance, determine the average load current when the motor runs at a speed of 146.60 rad/s and has a voltage constant K_a of 0.495 V/A rad/s.   [*Ans* 9.2 A.]

14.11 A d.c. shunt motor has the parameters: R_a = 0.78 Ω, L_a = 16 mH, torque constant = 2.1 N.m/A and voltage constant 0.209 V/r/min. It is supplied by a fully controlled, 3-ϕ bridge converter from a 220 V/(line), 50 Hz a.c. source. Find expressions for the armature current and hence, determine the values for mean torque, and speed as appropriate for the following conditions:
(a) Motoring with a firing angle of 22.5°, a continuous current and 70 Nm torque.
(b) Generating with a firing advance angle of 22.5°, and continuous current and 70 Nm torque.

(c) Motoring with a firing angle of 67.5° at a speed of 720 r/min.
(d) Generating at a firing advance angle of 67.5° at a speed of 720 r/min.

[*Ans* (a) 1267 rpm (b) 1599 rpm (c) 107 Nm (d) 56.7 N-m.]

14.12 A 60 kW, 230 V, 1800 rpm, separately excited d.c. motor is controlled by a converter as shown in Fig. 14.35. The field current is kept constant at 1.3 A and the machine back emf constant is $K_a = 0.8$ V/A rad/s. The armature resistance is $R_a = 0.03$ Ω and the viscous friction constant is B = 0.3 N-m/rad/s. The amplification of the speed sensor is $K_s = 92$ mV/rad/s and the gain of the power control is $K_c = 140$. Determine the following:
(a) The rated torque of the motor.
(b) The reference voltage E_r to drive the motor at the rated speed.
(c) The speed at which the motor develops the rated torque if the reference voltage is kept unchanged.
(d) The motor speed, if the load torque is increased by 15% of the rated value.
(e) The motor speed, if the reference voltage is reduced by 15%.
(f) The motor speed, if the load torque is increased by 15% of the rated value and the reference voltage is reduced by 15%.
(g) The speed regulation for a reference voltage of 1.4V if there was no feedback as in an open-loop control.
(h) Speed regulation, with a closed-loop control. [*Ans* (a) 318.30 N-m. (b) 18.58 V (c) 1793.23 rpm. (d) 17.9280 rpm. (e) 1532.38 rpm. (f) 1525.14 rpm. (g) 4.87% (b) 0.38%.]

REFERENCES

1. G K Dubey, *Power semiconductor controlled drives*, Englewood Cliffs, NJ, Prentice Hall, 1989.
2. S B Dewan, G R Slemmon and A Stranghen, *Power Semiconductor drives*, Wiley Interscience, 1984
3. B R Pelly, *Thyristor phase controlled converter and cycloconverters*, Wiley-Interscience, 1971.
4. P C Sen, *Thyristor D.C. drives*, Wiley Interscience, 1981.
5. P C Sen and M L McDonald, "Thyristorized d.c. drives with regenerative braking and speed reversal," *IEEE Trans. on IECI*, Vol. IECI 25, No. 4, Nov. 1978.
6. K Ishida, K Nakamura, T Izumi and M Ohara, "Microprocessor control of converter fed d.c. motor drives", *Conference Record IEEE*, Industry applications society annual meeting, 1982.
7. W F Ray and A G Potamianos, "Microprocessor control of d.c. motor drive, Conference-Publication number 234 IEEE.
8. G H Khan, G K Dubey and S R Doradla, An economical four-quadrant GTO converter and its application to d.c. Drive", *IEEE Transactions on Power-Electronics*, Vol 8, No. 1, January 1993.
9. Clemente S and B Pelley, "A 48 V, 200 A chopper for motor-speed control, with regenerative braking capability using power-HEXFETS," *IEEE Industry applications society meeting*, 1980.

Chapter 15

Control of A.C. Drives

LEARNING OBJECTIVES:
- To introduce the basic principles of a.c. machines.
- To consider the various schemes of a.c. motor speed control.
- To introduce the basis principles of stator voltage control method.
- To consider the basic principles of variable frequency control method.
- To examine the operation of PWM voltage-source inverter drive using GTOs.
- To examine the operation of PWM current-source inverter drives using GTOs.
- To consider the operation of chopper-controlled wound rotor induction motor.
- To examine the operation of static Scherbius drives and Kramer drives.
- To introduce the operating principles of various synchronous machines.
- To introduce basic principles of self-controlled synchronous motor drives.
- To examine the operation of three-phase, half wave and full wave brushless d.c. motor drives.

15.1 INTRODUCTION

In the world which is facing severe energy crisis, the golden rule is "energy saved is energy generated." Use of variable speed drives for industrial applications is one way to generate energy and lots of it. With wide options which are open to engineers for selecting proper drive system, one can look forward to an era where every application in the industry will be driven by highly efficient and reliable drives.

Most of the drives used in the industries today are electrical. Depending on the application, some of them are fixed speed and some of them variable speed. The variable speed drives, till a couple of decades back, had various limitations such as poor efficiencies, larger space, lower speeds, etc. However, the advent of power electronics transformed the scene completely and today we have variable drive systems which are not only smaller in size but also very efficient, highly reliable and meeting all the stringent demands of the various industries of modern era.

Modern technology offers various alternatives in the selection of a drive system. Depending on the process requirements, environmental conditions and financial objectives of a company, choice of a drive can be made. Primarily, the drive systems can be divided into two groups, d.c. drive systems and a.c. drive systems. The invention of power semiconductors saw the advent of d.c. drive systems for most of the early variable speed requirements. Based on the advantages of simple construction and ease of control, this technology continued to be taken up for further improvements all over the world till very recently. No wonder that today this technology is well understood and proven which resulted in its popularity world the over. Almost perfected, as far as the controller is concerned, this technology has most of its drawbacks due to d.c. motor. As is well known, d.c. motors have inherent disadvantages in that they need regular maintenance, they are tailor made hence not readily available for replacements, and are bulky in size. Added to this, due to the commutator sparking, they are simply not suitable in hazardous areas like chemical and petrochemical plants or in mines. Also, for ratings above 500 kW, manufacture of d.c. motor itself poses problems. With these serious limitations, d.c. drive systems have become unsuitable for energy saving applications (for example pumps or fans).

A.C. drive systems use the a.c. motor as the driven element–either induction or synchronous type. Since most of the motors in industries are only of induction type, development in this field took place rapidly. Induction motors, particularly squirrel-cage type induction motors, have a number of advantages when compared with d.c. motors. Some of these are ruggedness; lower maintenance requirements, better reliability; lower cost, weight, volume, and inertia; higher efficiency; and the ability to operate in dirty and explosive environments.

When new innovations have becomes practically a day-to-day matter, the entire drive technology has undergone tremendous change in last few decades. However, the two most significant developments that are revolutionising the drive concepts are:
1. Development of new power devices like power transistors, GTOs, IGBTs, etc. and
2. Development of microprocessor based control systems.

These two factors will enable even the large drive systems to have—
- Very compact size
- Highly reliable operation with increased flexibility of control
- Communication with higher order automation systems
- Extensive fault diagnostics
- Ease of commissioning due to self-optimisation.

15.2 BASIC PRINCIPLE OF OPERATION

Since the introduction by Nikola Tesla in the 1880s of the concept of polyphase alternating current, the entire industrial world has witnessed a continued growth in the use of three-phase, squirrel-cage induction motors. These motors are now almost a commodity item. Approximately, 60% of the world's consumption of electrical energy passes through the windings of squirrel-cage induction motors in the range 1 to 125 horsepower. Therefore, the discussion here also assumes a three-phase, squirrel-cage induction motor.

A three-phase induction motor consists of a balanced three-phase winding on the stator. These windings are distributed in the stator slots. Also, these three windings are displaced by 120° in space with respect to each other. The squirrel-cage rotor consists of a stack of insulated laminations. It has electrically conducted bars inserted through it, close to the periphery in the axial direction, which are electrically shorted at each end of the rotor by end rings, thus producing a cage-like structure. This also illustrates the simple, low-cost, and rugged nature of the rotor.

15.2.1 Equivalent Circuit

The equivalent circuit of the polyphase induction motor is very similar to the usual transformer equivalent circuit because the induction motor is essentially a transformer with a rotating secondary winding. As in a static transformer, the primary or stator current establishes a mutual flux that links the secondary or rotor winding, and also a leakage flux that links only the primary winding. This leakage flux induces a primary emf which is proportional to the rate of change of primary current. Its effect may be represented, in the usual manner, by a series leakage reactance X_1, in each stator phase (Fig. 15.1). R_1 is the stator resistance per phase and $(R_1 + jx_1)$ is termed the stator leakage impedance. The mutual flux in the air gap induces slip-frequency emfs in the rotor and supply frequency emfs in the stator. The voltage drop across the stator leakage impedance causes the air gap emf per phase, E_1, and the mutual flux per pole, ϕ_1, to decrease slightly as load is applied to the motor. The resulted stator current, I_1, is composed of the magnetizing current, I_m, and the load component of stator current, I_2, which cancels the magneto-motive force (mmf) due to rotor current. Core losses and saturation effects are neglected.

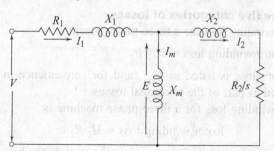

Fig. 15.1 *Single-phase equivalent circuit of the polyphase induction motor*

In deriving the rotor equivalent circuit, the actual cage or phase-wound rotor winding is considered to be replaced by an equivalent short circuited rotor winding having the same number of turns and the same winding arrangement as the stator. This is equivalent to the usual transformer procedure of referring secondary quantities to the primary.

At standstill, the induced emf per phase in the equivalent rotor is equal to the stator emf, E_1, and the rotor frequency equals the supply frequency, F_1. If the rotor slip with respect to the fundamental rotating field is denoted by S, the rotor emf, E_2, equals SE_1 and the rotor frequency, F_2 equals SF_1.

If R_2 is the equivalent rotor resistance per phase and X_2 is the equivalent rotor leakage reactance per phase at standstill, then the rotor current is given by

$$I_2 = \frac{E_2}{R_2 + jsx_2} = \frac{SE_1}{R_2 + jsx_2} \quad (15.1)$$

and hence,
$$I_2 = \frac{E_1}{(R_2/s) + jx_2} \quad (15.2)$$

In Eq. (15.1), all rotor quantities are at slip frequency but in Eq. (15.2), they are at supply frequency. These equations show that the rotor current, I_2, is unaltered in magnitude if the rotor is brought to standstill and the resistance increased from R_2 to R_2/s. The rotor equivalent circuit may, therefore, be joined directly to the stator circuit, as in Fig. 15.1, to give the complete equivalent circuit for one-phase of the motor.

15.2.2 Induction Motor Equations

There are several methods to evaluate machine performance used by numerous authors and machine designers. It has been reported that even though the Steinmetz model of Fig. 15.1 is an approximate circuit, the results are more realistic than many of the more refined equivalent circuits. Let us begin with the equation for power input to the motor.

1. **Power in**

$$p_{in} = \sqrt{3} V_1 I_1 \cos\theta_1 = \sqrt{3} \cdot \sqrt{3} V_{1ph} I_1 \cos\theta_1$$
$$= 3 V_{1ph} I_1 \cos\theta_1 \quad (15.3)$$

2. **There are five categories of losses**
   (a) Stator winding loss for a three-phase machine is
   $$\text{Stator winding loss} = 3 I_1^2 R_1 \quad (15.4)$$
   (b) The core loss is listed as P_{core} and for convenience in computation is included as part of the rotational losses.
   (c) Rotor winding loss for a three-phase machine is
   $$\text{Rotor winding loss} = 3 I_2^2 R_2 \quad (15.5)$$
   (d) Friction and windage loss = $p_{f\omega}$, is the sum of the bearing friction and the windage reaction of the rotor and rotor cooling fins.

Control of A.C. Drives

(e) Stray load loss = P_{stray} is caused by the magnetic fields in the end zone of the windings and harmonic magnetic fields. The stray loss is assumed to be 1.0 per cent of motor output when not available by other tests.

(f) The friction, windage, and stray losses included with the hysteresis and eddy current losses are called rotational losses.

3. Air gap power A convenient quantity in determining motor performance is the power transferred across the air gap and is the input power less the stator winding loss. Thus

$$P_g = \text{(power crossing the air gap)} = P_{in} - \text{stator winding loss}$$

$$P_g = \sqrt{3}\, V_1 I_1 \cos\theta_1 - 3I_1^2 R_1 = 3I_1^2 R_f \qquad (15.6)$$

4. Developed mechanical power The air gap power is partly consumed as rotor winding loss, and the remainder is the gross mechanical power (also called developed mechanical power (also called developed mechanical power). Thus,

$$P_{gross} = P_g - 3I_2^2 R_2 = P_g(1-s) \qquad (15.7)$$

5. Finally, the useful power available at the shaft is the gross mechanical power less the rotational losses.

$$P_{shaft} = P_{gross} - (P_{f\omega} + P_{core} + P_{stray}) \qquad (15.8)$$
$$= P_{gross} - P_{rot}.$$

6. The torque available to the machine shaft is

$$T_{shaft} = \frac{P_{shaft}}{\omega_r} = \frac{P_{gross} - P_{rot}}{\omega_s(1-s)} \text{ N-m} \qquad (15.9)$$

where ω_r is the rotor speed in rad/s, ω_s is the synchronous speed in rad/s, and s is the per unit slip.

There is a torque called the developed torque, also called internal torque or electromagnetic torque, that is developed by the electromagnetic energy conversion process and is defined by

$$T_d = \frac{P_{gross}}{\omega_r} \qquad (15.10)$$

or

$$T_d = \frac{P_{gross}}{\omega_s(1-s)} = \frac{P_g(1-s)}{\omega_s(1-s)} = \frac{P_g}{\omega_s} \qquad (15.11)$$

This means that three of the motor values may be determined directly from the air gap power.

Rotor winding loss = $s P_g$ \qquad (15.12)

$$P_{gross} = (1-s)P_g \qquad (15.13)$$

and

$$T_d = \frac{P_g}{\omega_s}$$

15.2.3 Induction Motor Performance Characteristics

When the stator is supplied by a balanced three-phase a.c. source of frequency ω rad/s (or f Hz), a rotating field moving at a synchronous speed, ω_s rad/s is produced.

(1)
$$\omega_s = \frac{2\pi/(P/2)}{1/f} = \frac{2}{P}(2\pi f) = \frac{2}{P}\omega \text{ rad/s} \tag{15.14}$$

where P = number of poles.
In terms of the revolutions per minute (rpm), the synchronous speed is

$$n_s = 60 \cdot \frac{\omega_s}{2\pi} = \frac{120}{p}f \tag{15.15}$$

(2) **If the rotor speed** is ω_r rad/s, then the relative speed between the stator rotating field and the rotor is given by

$$\omega_{ss} = S\omega_s = \omega_s - \omega_r \tag{15.16}$$

where ω_{ss} is called the slip-speed. The parameter S is known as slip and is given by

or
$$S = \frac{\omega_s - \omega_r}{\omega_s} \tag{15.17}$$

or
$$S = \frac{n_s - n_r}{n_s} \tag{15.18}$$

or
$$\omega_r = (1-s)\omega_s \tag{15.19}$$

(3) **Input impedance**

$$Z_{in} = (R_1 + jx_1) + Z_f \tag{15.20}$$

where $Z_f = (R_2/s + jx_2)$ in parallel with jx_m.

or
$$Z_f = \frac{(R_2/s + jx_2)jx_m}{R_2/s + j(x_2 + x_m)} \tag{15.21}$$

$Z_f = R_f + jx_f$

(4) **The stator current is then**

$$I_1 = \frac{V_{1ph}}{Z_{in}} \tag{15.22}$$

This means that the input power is:

$$P_{in} = 3V_{1pb} I_1 \cos\theta_1 = \sqrt{3} V_1 I_1 \cos\theta_1 \tag{15.23}$$

where θ_1 is the angle related to the input impedance, Z_{in}.

(5) Stator winding loss = $3I_1^2 R_1$.
(6) Air gap power, $P_g = 3I_2^2 R_2/s = 3I_1^2 R_F$

Control of A.C. Drives

(7) Rotor winding loss $= 3I_2^2 R_2 = S P_g$

(8) Gross developed mechanical power,

$$P_{gross} = P_g - 3I_2^2 R_2 = 3I_2^2 R_2 (1-s)/s = P_g (1-s)$$

(9) Developed torque $T_d = P_g/\omega_s$

(10) Net shaft power, $P_{shaft} = P_{gross} - P_{rot}.$

$$P_{shaft} = P_{gross} - (P_{f\omega} + P_{core} + P_{stray}) \text{ and}$$

$$HP = \frac{P_{shaft}}{746}$$

(11) Net shaft torque, $T_{shaft} = P_{shaft}/\omega_r$

(12) Motor efficiency, $E_{ff} = P_{shaft}/P_{in}.$

15.2.4 Maximum Torque (Breakdown) and Power

Two important quantities that explain the induction motor performance are:
(1) The maximum torque, and
(2) The slip at maximum torque.

The equations used in the previous sections do not directly lead to the desired form of torque equation. For this reason, consider the cantilever approximate circuit of Fig. 15.2, which is a circuit similar to the transformer cantilever circuit.

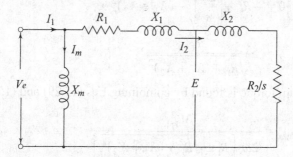

Fig. 15.2 *Cantilever equivalent circuit, Form I*

Begin by solving for the rotor current,

$$I_2 = \frac{V_1}{(R_1 + R_2/s) + j(X_1 + X_2)} \tag{15.24}$$

$$I_2 = \frac{V_1}{\sqrt{(R_1 + R_2/s)^2 + (X_1 + X_2)^2}} \tag{15.25}$$

Now, the gross developed mechanical power:

$$P_{gross} = 3I_2^2 R_2 (1-s)/s = \frac{3V_1^2 R_2 (1-s)/s}{(R_1 + R_2/s)^2 + (X_1 + X_2)^2} \tag{15.26}$$

The developed torque is then:

$$T_d = \frac{P_{gross}}{\omega_r} = \frac{P_{gross}}{\omega_s(1-s)} = \frac{3V_1^2 R_2/s}{\omega_s[(R_1 + R_2/s)^2 + (X_1 + X_2)^2]} \quad (15.27)$$

or

$$T_d = \frac{3}{\omega_s} I_2^2 \frac{R_2}{s} \text{ N-m} \quad (15.28)$$

The maximum torque is obtained by taking the derivative of the torque with respect to the slip and then setting the derivative to zero. Thus,

$$\frac{dT_d}{ds} = \left[\frac{3V_1^2 R_2}{\omega_s}\right]$$

$$\times \left[\frac{(R_1 + R_2)^2 (X_1 + X_2)^2 (-s^{-2}) - (2/s)(R_1 + R_2/s)(-R_2 s - 2)}{(R_1 + R_2/s)^2 (X_1 + X_2)^2}\right]$$

$$0 = \frac{-(R_1 + R_2/s)^2 + (X_1 + X_2)^2}{s^2} + (2R_2/s^3)\left(R_1 + \frac{R_2}{s}\right)$$

$$0 = -R_1^2 + \left[\frac{R_2}{s}\right]^2 - (X_1 + X_2)^2$$

$$\therefore \quad S_{max\,T} = \pm \frac{R_2}{\sqrt{R_1^2 + (X_1 + X_2)^2}} \quad (15.29)$$

The maximum torque is found by combining Eqs (15.29) and (15.28) for

$$T_{max} = \frac{3V_{1ph}^2}{2\omega_s\left[R_1 \pm \sqrt{R_1^2 + (X_1 + X_2)^2}\right]} \quad (15.30)$$

Note that the magnitude of maximum torque is independent of the rotor resistance, while the slip at maximum torque is directly related to the rotor resistance. The plus sign represents motoring and the minus sign the generating condition.

Maximum power output A second quality often referred to is that of maximum gross power output. This is readily found by using the maximum power transfer-theorem, where

$$\frac{R_2(1-s)}{s} = \sqrt{(R_1 + R_2)^2 + (X_1 + X_2)^2} = Z \quad (15.31)$$

Rearranging terms produces:

$$R_2 - SR_2 = SZ$$

or

$$\frac{R_2}{R_2 + Z} = S$$

$$\therefore \quad S_{\max P} = \frac{R_2}{R_2 + \sqrt{(R_1 + R_2)^2 + (X_1 + X_2)^2}} \quad (15.32)$$

The maximum power output corresponds to the slip defined by Eq. (15.31).

15.2.5 Typical Torque-speed Characteristic of a Polyphase Induction Motor

The general shape of the torque–speed or torque–slip curve with the motor connectted to a constant voltage, constant frequency source is shown in Fig. 15.3.

1. **Motoring region: $0 \le S \le 1$** For this range of slip, the mechanical power is outputted or torque developed is in the direction in which the rotor rotates. Also:
   (a) Torque is zero at $s = 0$.
   (b) The torque has a maximum value, called the breakdown torque, (T_{BD}), at slip, $S_{\max T}$. The motor would decelerate to a halt if it is loaded with more than the breakdown torque.
   (c) At $S = 1$, i.e. when the rotor is stationary, the torque corresponds to the starting torque, T_s. In a normally designed motor T_s is much less than T_{BD}.
   (d) The normal operating point is located well below T_{BD}. The full-load slip is usually 2–8%.
   (e) The torque–slip characteristic from no-load to somewhat beyond full-load is almost linear.

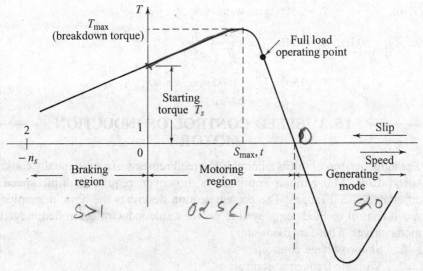

Fig. 15.3 *Torque–speed (slip) characteristics*

2. Generating region: $S < 0$ Negative slip implies rotor running at super-synchronous speed ($n > n_s$). The load resistance is negative, which means that mechanical power must be put in while electrical power is put out at the machine terminals.

3. Braking region: $S > 1$ The motor runs in opposite direction to the rotating field (i.e. n is negative), absorbing mechanical power (braking action) which is dissipated as heat in the rotor copper.

15.2.6 Starting Torque

Letting $S = 1$ in Eq. (15.27),

$$T_{start} = \frac{3V_1^2 R_2}{\omega_s \left[(R_1 + R_2)^2 + (X_1 + X_2)^2 \right]} \qquad (15.33)$$

The starting torque increases by adding resistance in the rotor circuit.

SOLVED EXAMPLE

Example 15.1 A three-phase, squirrel-cage induction motor is developing torque of 1500 sync. watts at 50 Hz and 1440 rpm (synchronous speed is 1500 rpm). If the motor frequency is now increased to 75 Hz using constant power mode, determine the new value of torque developed by motor at constant slip.

Solution: At a constant slip,

$$T \propto \frac{1}{f_1} \quad \text{where } f_1 \text{ is stator frequency}$$

Given, $T_1 = 1500$ sync. watts.

∴ $f_1 = 50$ Hz, $f_2 = 75$ Hz, ∴ $\dfrac{T_2}{T_1} = \dfrac{f_1}{f_2}$

∴ $T_2 = 1500 \times \dfrac{50}{75} = 1000$ sync. watts.

15.3 SPEED CONTROL OF INDUCTION MOTORS

The induction motor fulfils admirably the requirements of a substantially constant speed drive. Many motor applications, however, require multiple speed or adjustable speed ranges. The present section describes the basic principles of speed control methods employed in power semiconductor controlled induction motor drives. These methods are:
1. Stator voltage control.
2. Variable frequency control.
3. Rotor resistance control.
4. Slip-energy recovery scheme.

Methods 1 and 2 are applicable to both squirrel-cage and wound rotor motors, and methods 3 and 4 can be used only for wound rotor motors. A description of these methods is given by in the following to bring out their salient features with regard to ranges of speed control, areas of application, etc.

15.4 STATOR VOLTAGE CONTROL

From Eq. (15.27) it becomes clear that, the torque developed by an induction motor is proportional to the square of terminal voltage at a constant value of the supply frequency and the slip. When the applied voltage is varied, a set of speed torque curves can be obtained as shown in Fig. 15.4(a). When the voltage applied is n times the rated value, the torque ordinates will be n^2 times the ordinates corresponding to full voltage. If constant torque is required at different voltages, the slip of the motor increases when the voltage is reduced. To accommodate the required rotor current there is a consequent increase in the slip of the motor. The power factor of the motor deteriorates at low voltages. Figure 15.4(b) shows the torque–speed curve of the load. From the figure, it is clear that depending upon the type of load, speed control of the motor in a limited range is possible.

This method of speed control is not suitable for normal mains fed three-phase induction motor whose typical torque–speed curves are as shown in Fig. 15.4. The portion of the curve beyond the point of maximum torque is unstable. The normal cage motors have small resistance and therefore, the unstable portion is large. The speed control is possible only in a narrow band of speeds. The starting current of these motors is also very high. The equipment used to control the speed must be able to withstand this current. The power factor is very poor at large slips. Therefore, special rotor design with fairly high resistance is required to be able to take the advantage of speed control by voltage variation. The effect of variation of rotor resistance is to shift the slip for maximum torque towards unity. The portion of unstable region can be reduced or even eliminated by properly designing the rotor. This increases the range of speed control. This is helpful in reducing the starting current and improving the power factor. The method is highly suitable for speed control of solid induction motors which have inherent high rotor resistance.

However, in cage rotors with high inherent resistance for the purpose of the speed control by stator voltage variation, the rotor losses at large slips are dissipated in the rotor itself causing heating of the rotor. A slip-ring rotor, on the other hand, allows insertion of the required external resistance. The losses at large slips will heat only the resistance and rotor heating can be avoided.

An a.c. *voltage controller* (Chapter 11) can be used for the purpose of varying the applied voltage to achieve a speed control. By controlling the firing angle of the thyristors connected antiparallel in each phase, the RMS value of the voltage applied to the motor can be varied. This brings in speed control. Various types of a.c. voltage controllers discussed in Chapter 11 can be used for the speed control

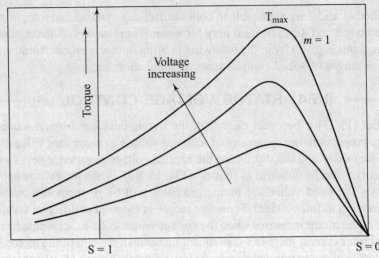

(a) Typical speed-torque curves for variation in stator voltage (low-resistance rotor)

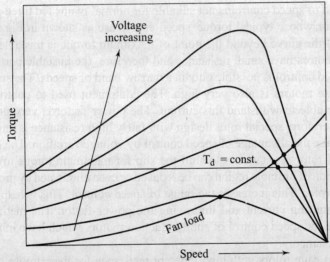

(b) Operating points and speed range for constant torque and fan type load (rotor resistance low)

Fig. 15.4 *Torque-speed characteristics*

purpose. Sometimes, half controlled converters are also used. But these introduce even harmonics, imbalance and result in noisy operation. In the case of fully controlled converters, the fifth harmonic is the lowest harmonic. The induction motors controlled by a.c. voltage controllers find wide applications in fan, pump and crane drives.

15.4.1 Four-Quadrant A.C. Voltage Controllers

The four-quadrant operation with plugging is obtained by the use of the circuit of Fig. 15.5. Thyristor pairs A, B and C provide operation in quadrants I and IV. The speed-torque curve at a fixed stator voltage and for operation in quadrants I and IV is shown by a solid-line in Fig. 15.6. Use of thyristor pairs A', B' and C' changes the phase sequence and thus, gives operation in quadrants II and III. The speed–torque curve for the same stator voltage and operation in quadrants II and III is shown by a dotted line in Fig. 15.6.

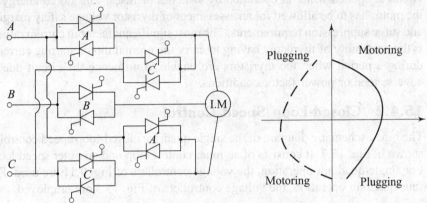

Fig. 15.5

Fig. 15.6 *Speed–torque curves for a fixed stator voltage and +ve and – ve phase sequence*

While changing from one set of thyristor pairs to another, that is from ABC to $A'BC'$ and *vice-versa*, care should be taken to ensure that the incoming pair is activated only after the outgoing pair is fully turned-off. Failure to satisfy this condition will cause short-circuiting of the supply by the conducting thyristors of the two pairs. The protection against such fault can be provided only by the fuse links and not by the current control.

This method of speed control has a few limitations, however:
(a) Reference back to Chapter 11 will show that the output voltage from an a.c. controller is dependent not only on the delay angle of the gate firing pulses but also on the periods of current flow which are dictated by the load power factor. An induction motor will draw a varying power factor and this will influence the voltage being applied to it. Whenever the load current is continuous, the a.c. controller will have no influence on the circuit conditions at all.
(b) Control is achieved by distortion of the voltage waveforms and by the reduction of current flow periods. Significant amounts of stator and rotor harmonic currents will flow and eddy currents will be induced in the iron core. These will cause additional motor heating and alter the motor's performance compared with sinusoidal operation. The practical results of these limitations are:

1. The motor's performance can only be predicted after a full understanding of the motor, thyristor converter and the load.
2. A closed-loop speed control based on a tachogenerator speed measurement is essential to ensure stable performance.
3. The system gains most practical application where the load is predictable and where the load torque required at reduced speed is relatively low.

As far as the thyristor voltage ratings are concerned, the normal crest working voltage is the peak of the supply line voltage, but high transients can occur if the circuit is opened while in operation by switches or fuses. The stored energy in the motor has to be allowed for an assessment of thyristor voltage safety margins and surge suppression requirements. The most significant factor in current ratings is the possibility of thyristors having to carry the normal motor starting currents during a period when the thyristors are unable to influence the circuit due to adverse load or power factor conditions.

15.4.2 Closed-Loop Speed Control

The block schematic diagram of the single quadrant closed-loop speed control is shown in Fig. 15.7. It consists of an inner control loop and an outer speed loop. For single quadrant operation, the voltage controllers of Fig. 9.11 are used. For four-quadrant operation, the voltage controllers of Fig. 15.5 are employed.

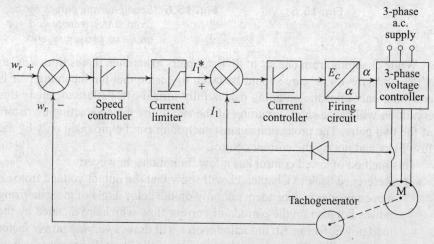

Fig. 15.7 *Single quadrant closed-loop speed control*

A four-quadrant, closed-loop drive can be realised using the voltage controller of Fig. 15.5 and the drive of Fig. 15.8. Now, let us consider the operation of the drive for speed reversal. When the speed command is set for the reverse direction, the speed error e_ω reverses and exceeds a prescribed limit. The master controller, on sensing this, withdraws the gate pulses to force the current to zero. The master controller provides a delay of 5 to 10 ms after the zero current is sensed for ensuring that the outgoing thyristors have turned-off. The gate pulses are now released to the other set of thyristors. The drive first decelerates and then accelerates in the reverse direction at a constant maximum allowable current and finally settles at the desired speed.

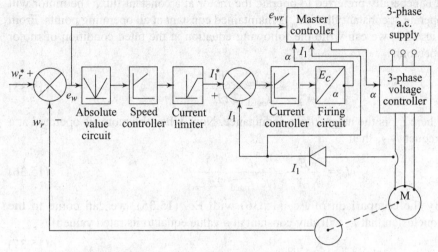

Fig. 15.8 *Four-quadrant closed-loop speed control*

15.5 VARIABLE FREQUENCY CONTROL

From Eq. (15.14) it becomes clear that the synchronous speed is directly proportional to the supply frequency. Hence, by changing the supply frequency, the synchronous speed and motor speed can be controlled below and above the normal full-load speed. The voltage induced in the stator, E, is proportional to the product of the slip frequency and air gap flux. The motor terminal voltage can be considered proportional to the product of the frequency and the flux, if the stator drop is neglected. Any reduction in the supply frequency without a change in the terminal voltage causes an increase in the air-gap flux. Induction motors are designed to operate at the knee point of the magnetization characteristics to make full use of the magnetic material. Therefore, the increase in flux will saturate the motor. This will increase the magnetizing current, distort the line current and voltage, increase the core loss and the stator copper loss, and produce a higher pitch acoustic noise.

While any increase in flux beyond the rated value is undesirable from the consideration of saturation effects, a decrease in flux is also avoided to retain the torque capability of motor. Therefore, the variable frequency control below the rated frequency is generally carried out by reducing the machine phase voltage, V, along with the frequency in such a manner that flux is maintained constant. Above the rated frequency, the motor is operated at a constant voltage because of the limitation imposed by stator insulation or by supply voltage limitations. Now, let us define a variable 'K' as,

$$K = f/f_{\text{rated}} \tag{15.34}$$

where f is the operating frequency and f_{rated} is the rated frequency of the motor, variable K is called as the per unit frequency.

(a) Operation Below the Rated Frequency ($K < 1$) As described above, it is generally preferred to operate the motor at a constant flux. The motor will operate at constant flux if I_m is maintained constant at all operating points. From Fig. 15.1, we can write the following equation at the rated condition of motor operation :

$$I_m = \frac{E_{rated}}{X_m} = \frac{E_{rated}}{f_{rated}} \cdot \frac{1}{2\pi L_m} \tag{15.35}$$

where L_m is the magnetizing inductance. Now, when the motor is operated at a frequency f, then

$$I_m = \frac{E}{K \cdot X_m} = \frac{E}{K \cdot f_{rated}} \cdot \frac{1}{2\pi L_m} \tag{15.36}$$

By the comparison of Eq. (15.36) with Eq. (15.35), we can come to the conclusion that I_m will stay constant at a value equal to its rated value if

$$E = K \cdot E_{rated} \tag{15.37}$$

Equation (15.37) suggests that the flux will remain constant if the back emf changes in the same ratio as the frequency, in other words, *when (E/f) ratio is maintained constant.*

Motor operation for a constant (E/f) ratio From Fig. 15.1, and at a frequency f, we can write

$$I_2 = \frac{K \cdot E_{rated}}{\sqrt{\left(\frac{R_2}{S}\right)^2 + (K X_2)^2}} = \frac{E_{rated}}{\sqrt{R_2^2/(KS)^2 + X_2^2}} \tag{15.38}$$

where

$$S = \frac{K \cdot \omega_s - \omega_r}{K \omega_s} \tag{15.39}$$

Note that ω_s is the synchronous speed at the rated frequency. Now, we can write the developed torque from Eq. (15.28) as

$$T = \frac{3}{K \omega_s} I_2^2 R_2/S \tag{15.40}$$

$$= \frac{3}{\omega_s} \left[\frac{E_{rated}^2 R_2/(KS)}{R_2^2/(KS)^2 + X_2^2} \right] \tag{15.41}$$

Now, E is maintained constant for a given frequency. The power transferred across the air-gap will be maximum at a slip S_m, for which

$$K \cdot X_2 = \pm R_2/S_m$$

or $$S_m = \pm \frac{R_2}{K X_2} \tag{15.42}$$

Substituting in Eq. (15.41) gives

$$T_{max} = \pm \frac{3}{2\omega_s} \frac{E_{rated}^2}{X_2} \tag{15.43}$$

From Eq. (15.43), it becomes clear that for a variable frequency control at a constant flux, the breakdown torque remains constant for all frequencies, both during motoring and regenerative braking. Also, the examination of Eqs (15.38) and (15.41) shows that for a constant (SK), the rotor current I_2 and torque T are constant. Now, if $\bar{E}$ is taken as a reference vector, then the phase lag of $\bar{I}_2$ is given by

$$Q_r = \tan^{-1}(K \cdot s \cdot X_2/R_2) \tag{15.44}$$

Since Q_r is also constant for a given (SK), the motor current will also be constant. Thus, the motor operates at constant value of torque, I_1, and I_2 when the flux and (KS) are maintained constant.

Let us now examine the physical significance of SK from Eq. (15.39)

$$SK = \frac{K\omega_s - \omega_r}{\omega_s} = \frac{\omega_{ss}}{\omega_s} \tag{15.45}$$

where $\qquad \omega_{ss} = K\omega_s - \omega_r \tag{15.46}$

Note that ω_{ss} is the slip speed, which is the difference in the field speed at frequency f (or synchronous speed $K\omega_s$) and the rotor speed ω_r. W_{ss} is the drop in motor speed from its no-load speed $(K\omega_s)$ when the machine is loaded. From Eq. (15.45), a constant value of (KS) implies the motor operation at a constant slip speed ω_{ss}.

From the above discussion, now, it becomes clear that for any value of T, the drop in the motor speed from its no-load speed $(K\omega_s)$ is the same for all frequencies. Hence, the machine speed–torque characteristics for $0 < s < S_m$ are parallel curves. The nature of speed–torque curves for the variable frequency operation at a constant flux are shown in Fig. 15.9 both for motoring and braking operations.

The operation of the machine at a constant slip speed also implies the operation at a constant rotor frequency as shown by Eq. (15.47).

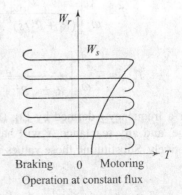

Fig. 15.9 *Speed–torque curves with variable frequency control*

$$SK = \frac{S \cdot f}{f_{rated}} = \frac{f_r}{f_{rated}} = \frac{\omega_r}{\omega_{rated}} \qquad (15.47)$$

where ω_r and f_r are the rotor frequency in rad/s and Hz, respectively. For $S < S_{max}$, $\left(\frac{R_2}{S}\right) \gg \cdot (K\,X_2)$, hence from Eqs (15.41) and Eq. (15.45).

$$T = \frac{3\,E_{rated}^2}{\omega_s\,R_2}\,(KS) = \text{constant} \times \omega_{ss} \qquad (15.48)$$

Equation (15.48) suggests that for $S < S_m$, the speed torque curves are nearly straight lines. Since they are also parallel, the speed–torque characteristics are approximately parallel straight lines for $S < S_m$, when flux is maintained constant.

The operation of the machine at constant flux requires a closed-loop control of flux. When the operating point changes, the closed-loop control adjusts the motor voltage to maintain a constant flux. The closed-loop control becomes complicated because the measurement of flux is always difficult. Hence, the flux is controlled indirectly by operating the machine at a constant (V/f) ratio for most of the frequency range, except at low frequencies, where the (V/f) ratio is increased to compensate for the stator resistance drop.

The (V/f) ratio is chosen equal to its value at the rated voltage and frequency. As the load on the machine is increased, the stator resistance drop increases and the back emf decreases and the flux reduces. Consequently, the machine does not operate exactly at a constant flux.

We will now examine *the motor operation when the (V/f) ratio is held* constant. For simplicity, the equivalent circuit of Fig. 15.2 is used here. From this equivalent circuit, at rated motor terminal voltage (V_{rated}) and frequency (ω_{rated}), we have from Eqs (15.37) and (15.30),

$$T = \frac{3}{\omega_s}\left[\frac{V_{rated}^2\,(R_2/s)}{(R_1 + R_2/s)^2 + (X_1 + X_2)^2}\right] \qquad (15.49)$$

and

$$T_{max} = \frac{3}{2\,\omega_s}\left[\frac{V_{rated}^2}{R_1 \pm \sqrt{R_1^2 + (X_1 + X_2)^2}}\right] \qquad (15.50)$$

For a frequency f defined by Eq. (15.34), the synchronous speed, terminal voltage, and any reactance X will have the values $K\,\omega_s$, $K\,V_{rated}$, and $K\,X$, respectively. Substituting these values in Eqs (15.49) and (15.50) yields

$$T = \frac{3}{\omega_s}\left[\frac{V_{rated}^2\,R_2/(KS)}{\left(\frac{R_1}{K} + \frac{R_2}{KS}\right)^2 + (X_1 + X_2)^2}\right],\ K < 1 \qquad (15.51)$$

$$T_{max} = \frac{3}{2\omega_s} \left[\frac{V_{rated}^2}{(R_1/K) \pm \sqrt{(R_1/K)^2 (X_1 + X_2)^2}} \right], \quad K < 1 \quad (15.52)$$

When f is large, $(R_1/K) \ll (X_1 + X_2)$ giving a constant value of T_{max}, both for motoring and regenerative braking. However, for low values of f, the maximum torque capability is altered. It decreases for motoring and increases for braking. What is true for the maximum torque is also true for the rated torque. This behaviour can also be explained from consideration of flux.

When the motor operates at a frequency f with a constant (V/f) control, the terminal voltage and all reactance are reduced by a factor K but the stator resistance remains fixed. The resistances drop, which is negligible for high values of f, becomes appreciable in comparison with the terminal voltage at low values of f. As a result, the (E/f) ratio decreases, decreasing flux and the motor torque capability. The lower the frequency, the greater the reduction in the torque capability.

When working in the regenerative braking mode, the rotor current direction is opposite to that in motoring. Hence, the stator resistance drop has an opposite effect that is the flux and braking torque have higher value at low frequencies.

Figure 15.10 shows the nature of the speed–torque characteristics for constant (V/f) control and for $f < f_{rated}$. The decrease in motoring torque and increase in braking torque at low frequencies have higher values for motors of low power rating.

To make full use of the motor's torque capability at the start and for low speeds, the (V/f) ratio is increased to compensate for the stator resistance drop at low frequencies as shown in Fig. 15.11. This allows a constant maximum torque to be obtained for motoring operation at all frequencies.

The motoring speed–torque characteristics become similar to those shown in Fig. 15.9. The braking torque which is already high at low frequencies is increased further. The large increase in braking torque may cause severe mechanical stress on the motor and the load.

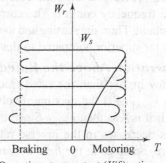

Operation at a constant (V/f) ratio

Fig. 15.10 Speed–torque curves with variable frequency control

For a given frequency, the exact compensation for the stator resistance drop can be done only for a particular operating point. This point is chosen either at the rated torque or the breakdown torque. Then, for lower torques, the motor saturates due to a large flux, particularly at low frequencies. It may be saturated to such an extent that the no-load current may be greater than rated current with the reduced cooling at low speeds, and this may lead to overheating.

To obtain a linear relation between V and f, V may also be varied as $V = V_0 + k \cdot f$, where V_0 is chosen to produce the nominal flux at zero speed and the constant K is chosen to get the rated terminal voltage at the rated frequency.

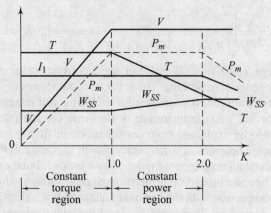

Fig. 15.11 V, T, P_m, I_1, and ω_{ss} versus per unit frequency, K_1, plots

To get a high torque to current ratio, and a high efficiency and power factor, the motor is operated for $S < S_m$ that is on the portion of speed–torque curves with a negative slope. Therefore, in Figs (15.9) and (15.10), only the portions with negative slope are shown. However, a complete characteristics is shown for the rated frequency to provide a comparison between the starting and low-speed torque available with variable frequency control and constant frequency operation. There is a large increase in the starting and low-speed torques with variable frequency control. The corresponding currents are also reduced by a large amount. Thus, the starting and low-speed performance of a variable frequency drive is far superior compared to that with the fixed frequency operation.

(b) Operation Above the Rated Frequency ($k > 1$) The operation at a frequency higher than the rated frequency takes place at a constant terminal voltage V_{rated} or at the maximum voltage available from the variable frequency source if it is less than V_{rated}. Since the terminal voltage is maintained constant, the flux decreases in the inverse ratio of per unit frequency 'K'. The motor, therefore, operates in the *field weaking mode*.

The torque expressions for the operation in this frequency range are obtained by the substitution of $k\,\omega_s$ for ω_s and $k\,(X_1 + X_2)$ for $(X_1 + X_2)$ in Eqs (15.49) and (15.50), giving

$$T = \frac{3}{\omega_s}\left[\frac{V_{\text{rated}}^2\, R_2/(KS)}{\left(R_1 + \dfrac{R_2}{s}\right)^2 + K^2\,(X_1 + X_2)^2}\right],\, K > 1 \qquad (15.53)$$

$$T_{\max} = \frac{3}{2\,\omega_s K}\left[\frac{V_{\text{rated}}^2}{R_1 \pm \sqrt{R_1^2 + K^2\,(X_1 + X_2)^2}}\right],\, K > 1 \qquad (15.54)$$

Since $k > 1$, the breakdown torque decreases with the increase in frequency and speed. The speed–torque curves for operation for field weakening mode of operation are shown in Fig. 15.12. Here, also the motor is made to operate for $S < S_m$ to get high torque per ampere, high efficiency, and a good power factor.

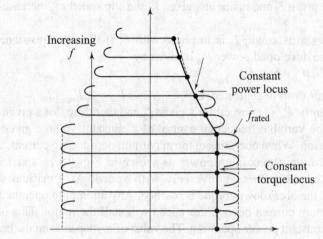

Fig. 15.12 *Speed–torque curves for variable frequency control of induction motor*

Torque and Power Capabilities : The torque and power variations for a given stator current and for frequencies below and above the rated frequency are shown by dots in Fig. 15.12. When the stator current has the maximum permissible value, these will represent the maximum torque and power capabilities of the machine. The variations of various variables, such as developed torque and power (P_m), slip speed, ω_{ss}, and terminal voltage, V, with per unit frequency 'K', for the motor operation at a given stator current are shown in Fig. 15.11. These curves will give the maximum torque and power capabilities of the machine when the stator current has the maximum permissible value. These variations can be explained as follows.

It is shown in the previous section that when the motor operates at a constant flux and a given stator current, the developed torque and slip speed have constant values at all frequencies. Thus, for $K < 1$, the variable frequency control with a constant flux gives constant torque operation.

When constant (V/f) control is used and the (V/f) ratio is increased at low frequencies to compensate for the stator resistance drop, at the maximum permissible current the drive operates essentially at a constant flux, providing constant torque operation.

For $K > 1$, $\quad I_2 = \dfrac{V_{rated}}{\sqrt{(R_1 + R_2/S)^2 + K^2(X_1 + X_2)^2}}$ \hfill (15.55)

Since the slip is small,

$$I_2 = \frac{S \cdot V_{rated}}{R_2} = \frac{V_{rated}}{R_2}\left(\frac{K\omega_s - \omega_r}{K\omega_s}\right)$$

or $(K\omega_s - \omega_r) = \omega_{ss} = \dfrac{R_2 \cdot \omega_s}{V_{rated}}(K I_2)$

Thus, at a given I_2 and hence at a given I_1, the slip speed ω_{ss} increases linearly with 'K'.

Since the slip is small, I_2 is in phase with E. If the rotor resistance loss is neglected, the developed power P_m is given by

$P_m = 3\,E\,I_2$

If the stator drop is neglected, $E = V_{rated}$ and $P_m = 3\,V_{rated}\,I_2$.

Consequently, P_m is constant for a given I_2 and therefore, for a given I_1. Thus, for $K > 1$, the variable frequency control at a constant voltage gives constant power operation. When operating at the maximum permissible current, the motor develops a constant maximum power as shown in Figs 15.11 and 15.12. The maximum torque decreases inversely with speed. At a critical speed ω_c (Fig. 15.12), the breakdown torque is reached. Any attempt to operate the motor at the maximum current beyond this speed will stall the motor. This is also the limit of the constant power operation. The value of ω_c depends on the breakdown torque of the machine. The range of constant power operation is higher for a motor with higher breakdown torque.

The speed and frequency at the transition from constant torque to constant power operation are called base speed and base frequency, respectively. They will usually be equal to ω_s and f_{rated}, respectively, but this will not always be so.

There are some applications, like traction, where speed control in a wide range is required and the torque demand in the high-speed range is low. For such applications, control beyond the constant power range is required. To prevent the torque from exceeding breakdown torque, the machine is operated at a constant slip speed and the machine current and power are allowed to decrease as shown in Fig. 15.11. Now, the motor current reduces inversely with speed; and the torque decreases inversely as the speed squared. This characteristic is often referred to as the series motor characteristics. The torque produced in this region is somewhat higher than that produced by a d.c. series motor.

Control and Advantages : The motor is always operated on the portion of the speed–torque curves with a negative slope, by limiting either the slip speed or the current, for getting the advantages of high torque to current ratio, high efficiency and a good power factor. Now, let us consider the operation of the drive for a change in speed command.

(1) When motoring, a decrease in the speed command decreases the supply frequency. This shifts the operation to regenerative braking (Fig. 15.12). The drive decelerates under the influence of braking torque and load torque. For speeds below ω_s, the voltage and frequency are reduced with speed to maintain the desired (V/f) ratio or constant flux, and to keep the operation on the portion of the speed–torque curves with a negative slope by limiting the slip speed. For speeds above ω_s, the frequency alone is reduced with speed to maintain the operation on the portion of speed torque curves with a negative slope. When

close to the desired speed, the operation shifts to the motoring operation and the drive settles at the desired speed. When regenerative braking is not possible due to the inability of the source to accept the energy, the motor torque can be made zero, at most. Then the deceleration occurs mainly due to the load torque.

(2) When motoring, an increase in the speed command increases the supply frequency. The motor torque exceeds the load torque and the drive accelerates. Again, the operation is maintained on the portion of the speed–torque curves with a negative slope by limiting the slip speed. The drive finally settles at the desired speed.

Usually, a Class B squirrel-cage motor is used. Some energy efficient applications may also employ the Class A design. It may be recalled that these designs have low full-load slips, which results in high running efficiency and good speed regulation. Even with these designs, variable frequency control gives large torques with reduced currents for the complete range of speeds. Thus, variable frequency control allows simultaneous realization of good running of transient performance from squirrel-cage motor which is cheap, rugged, reliable, longer lasting, and maintenance free. Since regenerative braking is also possible down to zero speed, the variable frequently control provides a highly efficient variable speed drive with excellent running and transient performance.

In special applications requiring maintenance free operation, such as underground and underwater installations, and also in applications involving explosive and contaminated environments, such as in mines and the chemical industry, variable frequency induction motor drives have already gained popularity. Because of the advantages of squirrel-cage motors and variable frequency control, the variable frequency a.c. drives, find applications in traction, mill runout tables, pumps, fans, blowers, compressors, spindle drives, conveyors, machine tools, and so on. Due to the availability of power transistors with improved ratings and characteristics, general purpose low-power variable frequency drives are now available with a cost comparable to that of d.c. drives. The recent progress in GTOs may provide variable frequency drives which will complete very well and probably replace d.c. drives in medium and lower range high power drives. The variable frequency supply to an induction motor for speed control can be made available using;

(i) Voltage source inverter
(ii) Current source inverter, and
(iii) Cycloconverter.

15.5.1 Control of Induction Motor by Voltage Source Inverters

An inverter belongs to the voltage source category if, viewed from the load side, the a.c. terminals of the inverter function as a voltage source. Because of a low internal impedance, the terminal voltage of a voltage source inverter remains substantially constant with variations in load. It is, therefore, equally suitable to single-motor and multi-motor drives. Any short circuit across its terminal causes current to rise very fast, due to the low time constant of its internal impedance. The fault current cannot be regulated by current control and must be cleared by fast acting fuse links.

The many inverter circuits referred to in Chapter 9 can be used to provide variable frequency, variable voltage supplies to alter the speed of induction motors. The speed of an induction motor has to be controlled from an a.c. or d.c. source. Figure 15.13 shows the variables schemes for the control of induction-motor by voltage source inverters.

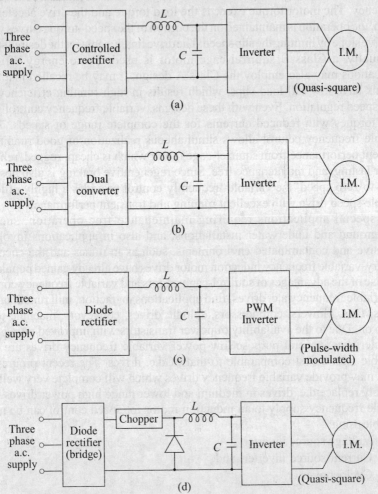

Fig. 15.13 *Schemes for induction motor speed control by voltage source inverters*

Two waveforms are in regular use for practical a.c. motor drives: the quasi-square wave and the synthesized sine wave, pulse-width modulated (PWM) waveform. Motors will operate quite successfully at high speeds with quasi-square applied waveforms but at low speeds the rotating fields within the machine will be stepped around rather than moving smoothly. The PWM waveform allows sinusoidal currents to flow in the motor even at low frequencies giving smooth rotation of the motor field flux and the smooth performance of the motor. In general, the applied voltage to the motor must increase with increase in frequency. When the motor is supplied with the same waveform at any frequency, e.g. the quasi-square waveform, then the level of the d.c. voltage will need to change to control the motor voltage.

In Fig. 15.13(a), the controlled rectifier varies the d.c. input to the inverter at the same time, as the inverter output frequency is varied. The rectifier is line-commutated and the inverter is forced commutated. The section between the rectifiers and the inverter is known as d.c. link. The d.c. link includes a series inductance and a capacitor, which smooths the d.c. input voltage to the inverter to an effectively constant value, E_{dc}. This system is not able to regenerate, because a reversal of i_0 would be required. If regeneration is necessary, it may be obtained by replacing the phase-controlled rectifier with a dual converter, see Fig. 15.13(b). A system in which the d.c. link voltage is constant is shown in Fig. 15.13(c). Here, the PWM techniques are applied to vary both the voltage and the frequency within the inverter. Due to diode rectifier, regeneration is not possible and the inverter would generate harmonics into the a.c. supply.

A fourth scheme is illustrated in Fig. 15.13(d), in which the variation of d.c. voltage is obtained by a chopper. Due to the chopper, the harmonic injection into the a.c. supply is reduced. This scheme is a combination that is used when a high frequency output is required; and PWM in the inverter is therefore not possible. In addition, a high input power factor is obtained from the diode rectifier. This arrangement may also be used with d.c. distribution for a transportation systems because the chopper will exclude from the distribution system the wide range of harmonics that would otherwise be produced in it by the inverter. These harmonics may interfere with signalling or communication systems in which the known harmonic frequencies produced by the chopper can be suppressed.

As discussed in Chapter 9, the inverter may be operated as a six-step inverter or a pulse-width modulated (PWM) inverter. The six-step inverter using thyristors was described in Chapter 9, whereas with power transistors it was discussed in Chapter 5.

PWM Voltage Source Inverter Drives Using GTO Figure 15.14 shows the power-circuit of a pulse width modulated inverter drive system using gate-turn-off thyristors. The six-step inverter circuit control the motor voltage magnitude by adjustment of the d.c. link voltage. The adaptation of the PWM strategy for the inverter, means that it is possible to control both voltage magnitude and frequency within the inverter, allowing the d.c. link voltage to be maintained constant. The emergence of the GTO & MOSFET has enabled switching within the inverter to be both faster and more efficient, thereby eliminating the need for the commutating circuits required by the convention thyristor. The variable d.c. link voltage system involved extra losses because thyristors were used in the rectifier rather than diodes, or where a chopper is used the losses within the chopper devices. The use of PWM results in an overall reduction of the losses because only a diode rectifier is required where GTOs are used there is an absence of commutation component losses.

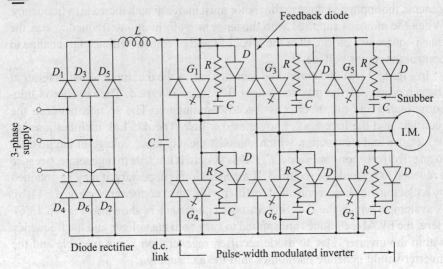

Fig. 15.14 *Pulse-width modulated inverter drive using GTO*

Typical waveforms associated with PWM waves are shown in Fig. (15.15). The frequency ratio is the ratio of the inverter switching frequency to the inverter output, that is, to the motor frequency. The current waveforms clearly show the reduction in harmonic current compared to the quasi-square-wave inverter-waveform, as well as the desirability to raise the frequency ratio.

The choice of the switching frequency with PWM inverter fed motor is a compromise between conflicting considerations. High switching frequency within the inverter will increase the inverter switching losses but reduce the harmonic content of the current waveforms and hence reduce the motor conductor losses giving a smoother torque. However, the magnetic circuit of the motor when having to respond to a high frequency voltage component will show increased magnetic losses and becomes a source of acoustic noise. Likewise, rapid switching of the inverter devices can and does generate high levels of acoustic noise which can be a nuisance in certain quiet locations.

It is clear that, technically the variable-speed cage induction motor drive is more complex than the controlled-rectifier fed d.c. motor drive. However, the constant improvement in the switching devices and the advance of the microprocessor control systems has made the variable speed PWM inverter fed cage induction motor an increasingly attractive proposition in areas which were traditionally those of the d.c. motor drive.

15.5.2 Control of Induction Motor by Current Source Inverters

As discussed in Chapter 9, an inverter which behaves as a current source at its a.c. terminals is called a current source. Inverter because of large internal impedance, the terminal voltage of a current source inverter changes substantially with a change in load. Therefore, if used in a multi-motor drive, a change in load on any motor affects other motors. Hence, current source inverters are not

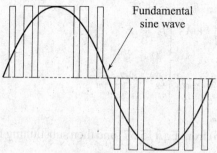

(a) Ideal line voltage with low frequency ratio

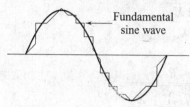

(b) Typical motor-line current with low frequency ratio

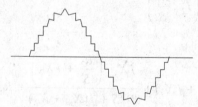

(c) Illustrating motor-line current with higher frequency ratio

Fig. 15.15 *Waveforms of three-phase PWM inverter*

suitable for multi-motor drives. Since the inverter current is independent of load impedance, it has inherent protection against short-circuits across its terminals.

The analysis and performance of an induction motor by a voltage source are presented in previous sections. This section now considers the case of a current source.

1. Operation at a Fixed Frequency The equivalent circuit of Fig. 15.1 is also applicable here. The only change is that the motor is now fed by a current source I_1 instead of a voltage source V. The motor input current will be independent of motor parameters and the terminal voltage V will change due to the change in the motor impedance. Now, the input current I_1 is shared between the magnetizing reactance X_m and the rotor impedance.

The rotor current is small and the magnetizing current I_m is nearly equal to I_1 for low values of S. The motor operates under saturation for low values of slip since I_1 is usually much higher than the normal magnetizing current. Here, motor should be analyzed taking the saturation into account.

From equivalent circuit of Fig. 15.1, we can write

$$\left[\left(\frac{R_2}{S}\right)^2 + X_2^2\right] I_2^2 = E^2 \tag{15.56}$$

$$\left[\left(\frac{R_2}{S}\right)^2 + (X_2 + X_m)^2\right] I_2^2 = I_1^2 X_m^2 \tag{15.57}$$

and
$$E = I_m X_m \tag{15.58}$$

Subtracting Eq. (15.56) from Eq. (15.57) and then substituting from Eq. (15.58) gives

$$I_2^2 = \frac{I_1^2 - I_m^2}{1 + \frac{2 X_2}{X_m}} \tag{15.59}$$

Also, from Eq. (15.56),

$$S = \frac{R_2}{\sqrt{(E/I_2)^2 - X_2^2}} \tag{15.60}$$

Now,
$$T = \frac{3}{w_s}\left(I_2^2 \cdot \frac{R_2}{S}\right) \tag{15.61}$$

motor input impedance

$$Z_{in} = (R_s + j X_s) + \frac{j X_s \cdot \left(\frac{R_2}{S} + j X_2\right)}{R_2/S + j(X_m + X_2)} = R_{in} + j X_{in} \tag{15.62}$$

where
$$R_{in} = R_s + \frac{\frac{R_2}{S} X_m^2}{(R_2/S)^2 + (X_m + X_2)^2} \tag{15.63}$$

and
$$X_{in} = X_s + \frac{X_m\left[(R_2/S)^2 + X_2(X_m + X_2)\right]}{(R_2/S)^2 + (X_m + X_2)^2} \tag{15.64}$$

$$V = I_S (R_{in}^2 + X_{in}^2)^{1/2} \tag{15.65}$$

Equations (15.56) to (15.62) are nonlinear algebraic equations due to the nonlinear relations of E and X_m with I_m. To avoid the need for a numerical solution, the calculations can be done in the following sequence.

A suitable value (less than I_1) is assumed for I_m for a given I_1, E and X_m are obtained from the magnetization characteristic, I_2 is obtained from Eq. (15.59); S is determined from Eq. (15.60) and then T and V are calculated from Eqs (15.61) and (15.65), respectively.

Figure 15.16 shows the speed–torque curves of a 2.8 kW, 400 V, 50 Hz star connected induction motor at different per unit values of I_1 and the rated frequency. The per unit value of I_1 is determined by dividing the actual value by the rated stator current. For the sake of comparison, the speed–torque and speed–current curves of the motor when fed by a voltage-source of rated voltage and frequency are also shown by dash-dots. The variation of the terminal–voltage for low value of slip and two values of I_1 is also shown. The motor operates nearly at the nominal value of flux and I_m for all loads, when fed by a voltage-source of rated voltage and frequency.

The starting torque is low due to the low values of flux (as I_m is low) and rotor current compared to their values at the rated voltage with a constant current source. Due to the increase in flux, torque increases with speed. With the rated voltage speed–torque curve, the flux and I_m have nominal values at the intersections. A further increase in speed increases the terminal voltage beyond the rated value. The flux and the magnetizing current are also increased beyond nominal values, and the motor saturates. Because of the saturation, the increase in terminal voltage and torque is much lower than what would be predicted if the saturation is neglected.

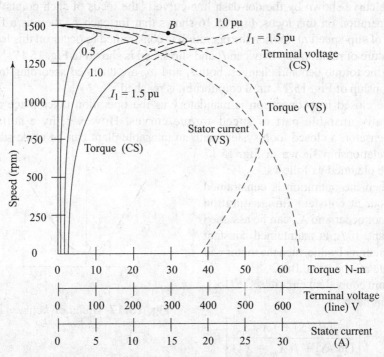

Fig. 15.16 *Induction motor operation with a current source*

For the constant current, $I_1 = 1.5$ per unit, now let us examine points A and B on the speed–torque curve. For the rated terminal voltage and frequency, point A also lies on the motor speed–torque curve. Hence, the motor operates at nominal flux and rated terminal voltage at point A. The motor terminal voltage will be higher than the rated value and the flux will be more than the nominal value at point B. Also, the machine will operate under saturation and core loss will be higher than at point A. With most loads, point B will also provide stable operation.

On the other hand, at point A, the rotor copper loss will be slightly higher. But, at both the points, the stator copper loss will be the same. On the whole, if the machine is operated at point A, then the losses will be less and saturation will be avoided. Even though point A lies on the statically unstable part of the speed–torque curve, the operation at this point is preferred because of these advantages.

For the constant current I_1 higher than 1.5 per unit, let us now consider the motor operation for the same developed torque. At this value of current, point B will shift up and A will shift down. Because of a further increase in flux, the operation at the new location of B will be worse than was before. At the new location of A, the machine will operate at a flux lower than the nominal value, thus not allowing full use of the motor torque capability.

The above discussion suggests that for a given I_1, operation is preferred at a point A which lies on the intersection of the speed–torque curve for a given I_1 and the speed curve for the rated terminal voltage and frequency. The speed–torque curve shown by the dot-dash line curve is the locus of such points "A". For operation on this locus, Fig. 15.16 shows that for each I_1, there is a fixed value of slip-speed ω_{ss}. When the operation is constrained to occur on this locus, the nature of relationship between I_1 and slip speed is shown in Fig. 15.17. Now, when the torque demand changes, both I_1 and ω_{ss} are changed according to the relationship of Fig. 15.17, until equilibrium is reached.

The closed loop operation is mandatory as the operation takes place on a statically unstable part of speed–torque curves. However, by a suitable compensator, a closed-loop system with an unusable plant can be made stable. The relationship shown in Fig. 15.17 can be obtained as follows.

When the operation is constrained to occur at constant flux, saturation does not occur and X_m can be assumed constant. If I_m is maintained constant at the nominal value, then the motor will operate at a nominal flux.

From equivalent circuit of Fig. 15.1, we can have

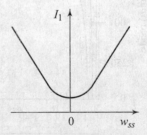

Fig. 15.17 Relation between I_1 and ω_{ss}

$$\bar{I}_m = \left[\frac{(R_2/S) + jX_2}{(R_2/S) + j(X_m + X_2)}\right]\bar{I}_1$$

or
$$I_m^2 = \left[\frac{(R_2/S)^2 + X_2^2}{(R_2/S) + (X_m + X_2)^2}\right]I_1^2 \qquad (15.66)$$

Now, for a given I_1, the value of 'S' can be determined which will provide operation of a flux of the nominal value from Eq. (15.66).

We have $\qquad T = \dfrac{3}{\omega_s} I_2^2 \dfrac{R_2}{S}$

From Eq. (15.57) put the value of I_2 in above equation, which gives

$$T = \frac{3}{\omega_s}\left[\frac{I_1^2 X_m^2 R_2/S}{(R_2/S)^2 + (X_2 + X_m)^2}\right] \qquad (15.67)$$

2. Variable Frequency Control

(a) Operation at and below rated frequency: As discussed previously, when the drive operates at a per unit frequency 'K' $\left(=\dfrac{f}{f_\text{rated}}\right)$, any reactance X will become KX. Hence, from equivalent circuit of Fig. 15.1.

$$I_m^2 = \left[\frac{(R_2/S)^2 + K^2/X_2^2}{(R_2/S)^2 + K^2(X_m + X_2)^2}\right] I_1^2$$

or

$$I_m^2 = \left[\frac{R_2^2/(SK)^2 + X_2^2}{R_2^2/(SK)^2 + (X_m + X_2)^2}\right] I_1^2 \qquad (15.68)$$

Now, let us consider the operation at a given I_1 – say at I_{11} – and variable frequency. The value of slip for $K = 1$ can be obtained from Eq. (15.66), which will give operation at a nominal value of I_m for $I_1 = I_{11}$. Let this slip be S_1. Then from Eq. (15.66),

$$I_m^2 = \left[\frac{(R_2/S_1)^2 + X_1^2}{(R_2/S_1)^2 + (X_m + X_2)^2}\right] I_{11}^2 \qquad (15.69)$$

Now, to have the operation at a nominal value of I_m for all value of per-unit frequency K, with I_1 remaining fixed at I_{11}, the following condition must be satisfied according to Eqs (15.68) and (15.69):

$$\frac{R_2^2/(SK)^2 + X_2^2}{R_2^2/(SK)^2 + (X_m + X_2)^2} = \frac{R_2^2/S_1^2 + X_2^2}{R_2^2/S_1^2 + (X_m + X_2)^2}$$

This equation yields,

$$SK = S_1 \qquad (15.70)$$

Now, multiplying both sides of Eq. (15.70) by ω_s, gives $SK\,\omega_s = S_1\omega_s$

or   Slip speed at per unit frequency = Slip speed at rated frequency   (15.71)

From Eq. (15.71), it becomes clear that, for a given I_1, the slip speed which provides motor operation at nominal flux at rated frequency also give operation at nominal flux at all frequencies. Hence, the relationship obtained between I_1 and ω_{ss} at rated frequency for the operation at nominal flux is valid for all frequencies.

From Fig. 15.1, the relation of Eq. (15.70) can also be derived. When the motor operates at a given I_1 and variable frequency, I_m will be held constant if the ratio between the rotor impedance and the magnetizing reactance is kept fixed. Since the reactance change in proportion to K, slip must change in inverse proportion to K. This gives Eq. (15.70).

When operating at per unit frequency K, ω_s and any reactance X in Eq. (15.67) should be replaced by $k\omega_s$ and KX, respectively giving

$$T = \frac{3}{\omega_s}\left[\frac{I_1^2 X_m^2 R_2/(SK)}{(R_2/SK)^2 + (X_2+X_m)^2}\right] \quad (15.72)$$

Hence, for a given I_1, if slip speed is maintained constant torque will also be constant for all frequencies. The motor can be made to operate at a constant flux, by controlling slip speed as a function of I_1. Hence, though it is fed by a current source, performance is identical to that on a variable frequency voltage source discussed in Section (15.5). It can be shown that Eqs (15.41) and (15.72) are also identical. The speed–torque curves for a constant flux operation are shown in Fig. 15.18. Also, this figure shows that how these curves are obtained by selecting constant flux operating points for different I_1.

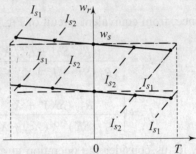

—— Locus of operating points at constant flux
------ Speed / torque curve for specific I_S

Fig. 15.18 Speed–torque curves at a constant flux for variable frequency current source controlled induction motor

(b) Operation above rated frequency: The motor terminal voltage changes when operating at a nominal flux or nominal I_m for $0 < K < 1$, as shown in Fig. 15.10. At the rated frequency, it becomes equal to the rated value. Thus, operation above the rated frequency is carried out with the terminal voltage held constant at the rated value. With the operation contained to occur at a fixed voltage, the machine behaviour is identical to that when fed by a constant voltage variable frequency source, discussed in Section 15.5 and shown in Figs 15.10 and 15.11.

The operation at the maximum permissible current gives operation at a constant maximum power. The maximum torque decreases inversely with speed. The machine impedance must be held constant as frequency is increased to get a constant terminal voltage. This is achieved by increasing the slip speed to compensate for an increase in reactances. The breakdown torque is reached at a certain speed. Now, the machine should be operated at a constant slip speed and the current should decrease with the increase in frequency to keep the terminal voltage constant.

Variable frequency current sources are obtained by using:
(i) current source inverters, and
(ii) cycloconverters.

The different schemes for the control of induction motor by current source inverters is shown in Fig. 15.19. In scheme (a), the inverter is fed by a d.c. source through a large inductance, L_d. The current I_d can be assumed to be ripple free d.c. because of the large value of L_d. With the six-step current source inverter, the waveforms of Fig. 9.70 (Chapter 9) is obtained. Therefore, the combination of L_d and the inverter is known as a current source inverter. This scheme, strictly speaking, does not act as a current source. Any change in the machine impedance, with a change in slip, changes the magnitudes of I_d and machine phase currents.

The magnitude of I_d could be maintained constant if both the waveform and the magnitudes of machine currents are to be made independent of changes in machine operation. This is obtained by a closed-loop control of I_d. Figures 15.19(b) and (c) show the current source inverter schemes incorporating closed-loop current control. Induction motor drives employs schemes of Figs. 15.19(b) or (c). Single-phase induction motor drives sometimes use the scheme of Fig. 15.19(a).

In Fig. 15.19(b), the inductor acts as a current source and the controlled rectifier controls the current source. As shown, the actual current I_d is compared with the reference value I_{dr}. The error is processed in a controller to adjust the rectifier-firing angle α to make I_d equal to I_{dr}. The operation of the scheme of Fig. 15.19(c) which is employed with a d.c. supply, can be similarly explained.

The major drawback of the rectifier current source is the poor power factor at low d.c. link voltages, whereas the power factor is higher with chopper controlled current source scheme.

The current source inverter controlled induction motor scheme of Fig. 15.19, generally uses the following inverter circuits.

(1) Six-step current source inverter
(2) Autosequentially commutated current source inverter
(3) PWM current source inverters.

The six-step current source inverter and autosequentially commutated current source inverter (ASCI) circuits were described in Chapter 9. Pulse-width modulated current source inverter circuit is now described in the next sub-section.

Pulse-Width Modulated GTO Current-source Inverter : Figure (15.20) shows the power circuit of a GTO current source inverter. This current source inverter produces input and output waveforms by PWM control through advantageously using the high-speed switching characteristics of the GTO as a switching device in the inverter. This current source inverter is composed of a rectifier section to convert constant-frequency a.c. power to d.c. power, a d.c. reactor to smooth the d.c. link current, and an inverter section to convert the d.c. power to variable voltage and variable frequency a.c. power. Both the rectifier and inverter sections are three phase bridge circuits—each composed of six GTOs. To absorb overvoltages produced when the GTO current is cut-off, three capacitors each are connected to the a.c. input and output terminals. A sinusoidally pulse width current flows in each GTO. The overvoltage absorption capacitors connected to the a.c. input and output terminals also functions as filters so that the input and output currents become sinusoidal.

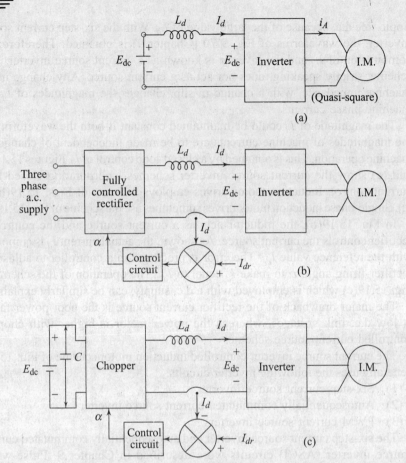

Fig. 15.19 *Current source inverter motor drive*

Figure (15.21) shows the modulation of line-current i_A, in its positive half-cycle. A carrier wave e_C is compared with a modulating reference wave e_R. When $e_R > e_c$, a pulse of current is produced. In this case, the waveform has quarter wave-symmetry. In a current source inverter, the fundamental component of the machine current can be varied by controlling I_d. Therefore, the pulse-width modulation is only required to improve the current waveforms. The two factors that change the PWM controlled current pattern are the modulation index A/A_m (amplitude ratio of e_R to e_C) and the number of pulses M per half-cycle of rectifier or inverter operation. Changing these values will change the harmonics in the input or output current. To make the input or output waveforms sinusoidal, both modulation index and number of pulses must be selected to minimize the harmonics. In determining these values, however, consideration must be given to GTO switching characteristic such as the turn-on/off time, the turn-on/off loss, etc. Figure 15.21(b) shows the modulating waveforms of line currents i_A, i_B and i_C. The sinusoidal machine currents will lag behind the respective line currents, due to a small leading current drawn by the capacitors.

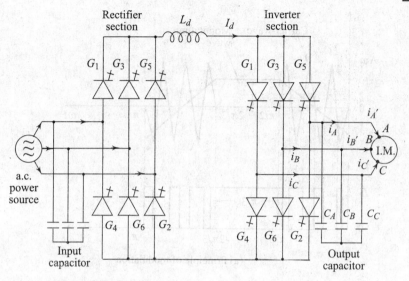

Fig. 15.20 *GTO current-source inverter power circuit*

Now, let us consider the operation of the GTO. Inverter at $\omega t = 0$ in Fig. 15.21(b). Since i_C is positive and i_B is negative, GTOs G_5 G_6 are ON, and the source current I_d is flowing through a path consisting of G_5, phase C phase B, and G_6 as shown in the equivalent circuit of Fig. 15.22(a). Let this path be designated loop 1. The machine current $i_{A'}$, which is sinusoidal and lags behind i_A, is still negative. Hence, another relatively small current is flowing through loop 2 formed by phase A, capacitor C_A, C_B and phase B. At angle α_1 (Fig. 15.29 (b)), I_d is transferred from line C to line A, by turning-off G_5 and turning-on G_1. Since the machine phase current $i_{A'}$ is still negative, the source current I_d, flows through a loop comprised of G_1, capacitor C_A, capacitor C_C, phase C, phase B, and G_6, as shown in the equivalent circuit of Fig. 15.22(b). Let us call this loop as loop 3. The current through loop 2 continues to flow as before. Now, capacitor C_A is charged both by the source current I_d and the loop 2 current, and its voltage shoots up. The loop 2 current charges capacitor C_B in the negative direction. The voltage C_{AB}, which is the sum of the voltages across capacitors C_A and C_B also shoots up. The modulated waveform of Fig. 15.21(b) shows only a few pulse, but in actual practice, there would be many more pulses. As the source current is transferred back and forth between loop 1 and loop 3, by alternate conductions of G_5 and G_1, the voltages across capacitors C_A and C_B build up. Consequently, voltage e_{AB} rises to a very high value. The build-up of voltage e_{AB} continues as long as the machine phase-current $i_{A'}$ remains negative. After $i_{A'}$ reverses, the loop 2 current flows to discharge capacitors C_A and C_B, while the loop 3 current continues to charge them. At a certain value of i_A, the build-up of the capacitor voltage is checked and e_{AB} decreases after attaining a maximum value. This is the maximum spike in the voltage e_{AB} and it occurs soon after the reversal of the machine current $i_{A'}$ since, the machine works symmetrically identical spikes will be produced at the reversal of currents $i_{B'}$ and $i_{C'}$.

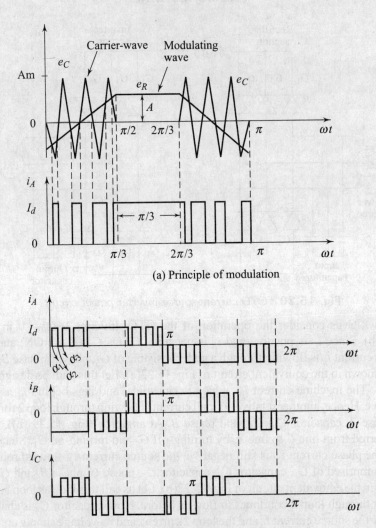

Fig. 15.21 *PWM control current waveform*

Even after the build-up of e_{AB} has been checked, the voltage spikes continues to be produced at each switching. For example, the transfer of the source current from G_5 to G_1 causes a current $(I_d - I_A')$ to flow through capacitor C_A. Consequently, the capacitor voltage shoots-up producing, a spike. Until $\omega t = \pi/2$, the pulse duration increases hence, the period of charging of capacitor C_A increases. Consequently, a voltage spike of appreciable magnitude is produced even though charging current $(I_d - i_{A'})$ is low.

The above discussion, shows that, the voltage spikes are produced primarily from capacitors being charged by the source current. In a particular case we have just examined, this happens due to the charging of C_A by the source current when it flows through loop 3. This suggests that the voltage spike can be reduced

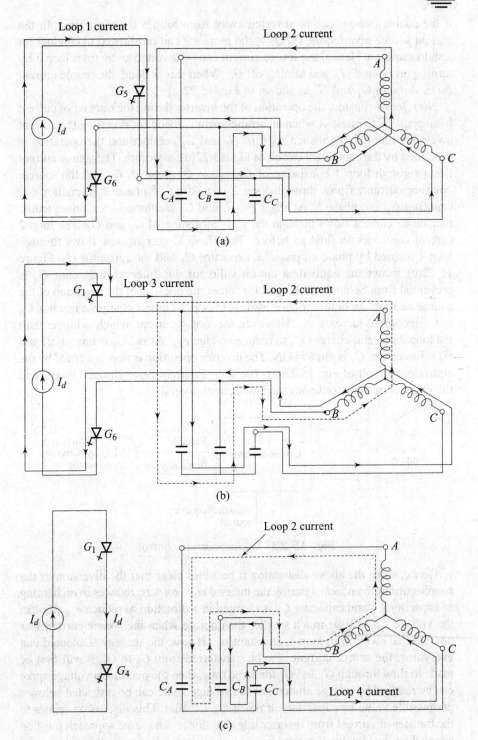

Fig. 15.22 *Equivalent circuits of the inverter of Fig. 15.20*

if the source current can be diverted away from loop 3 for some time. In the current source inverter, the GTOs in the same leg can be allowed to conduct for a short duration. Hence, the source current can be diverted away from loop 3 by turning on G_1 and G_4, and turning-off G_6. When this is done, the source current flows through G_1 and G_4 as shown in Fig. 15.22(c).

Now, let us consider the operation of the inverter during the transfer of current from phase C to phase A when an arrangement is made for diverting the current away from loop 3. At instant $\omega t = 0$, G_5 and G_6 conduct and the operation is described by the equivalent circuit of Fig. 15.22(a), as before. The source current flows through loop 1 consisting of G_5, phase C, phase B, G_6 and the source. Another current, flows through loop 2 consisting of phase A, capacitor C_A, capacitor C_B and phase B. At angle α_1, G_1 and G_4 are turned-on. Consequently, the source current flows through the path consisting of G_1 and G_4. The loop 2 current continues to flow as before. The phase C current now flows through loop 4 formed by phase C, phase B, capacitor C_B and the capacitor C_C. Figure 15.22(c) shows the equivalent circuit valid for this interval. Capacitor C_A is prevented from being over-charged for sometime because of the diversion of the source current. As before, notice also that, loop 2 current charges capacitor C_B in a direction to increase e_{AB}. However, the loop 4 current, which is higher than the loop 2 current, charges C_B to reduce voltage e_{AB}. At α_2, G_4 is turned-off and G_6 is turned-on. G_1 is already ON. The inverter operation is now governed by the equivalent circuit of Fig. 15.22(b). The source current flows through loop 3 and the phase A current continues to flow through loop 2.

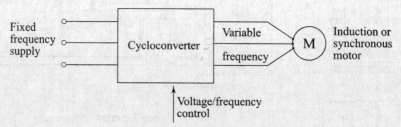

Fig. 15.23 *Cycloconverter control*

Hence, from the above discussion it becomes clear that the diversion of the source current from loop 3 during the interval $\alpha_1 < \omega t < \alpha_2$ reduces overcharging of capacitor C_A and capacitor C_B is charged in a direction to reduce e_{AB}. Hence, the voltage e_{AB} will be much smaller than before when the source current was not diverted away from loop 3, at instant α_3. If now, the strategy is adopted that everytime; the source current is to be transferred from G_5 to G_1, it will first be made to flow through G_1 and G_4 for some time, then the maximum voltage spike can be reduced by a large amount, or the voltage spike can be restricted below a permissible value by capacitors of much lower value. This discussion relates to the transfer of current from inverter line C to line A. The same approach must be adopted for the transfer of current from any one inverter line to another. Obviously,

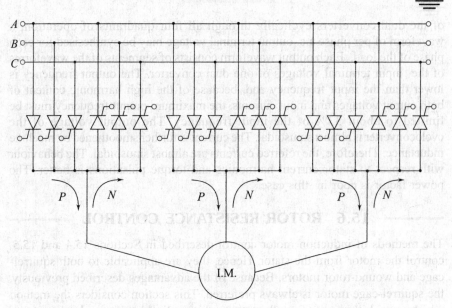

Fig. 15.24 *A three-phase, three-pulse cycloconverter induction motor drive (non-circulating current mode of operation)*

the control becomes complex. The motor operates with sinusoidal voltage and current, hence the problems of derating and torque-pulsations are eliminated. However, the diversion of the source current derates the inverter.

15.5.3 Control of Induction Motor by Cycloconverters

The principle and operation of the cycloconverter was explained in Chapter 10, where it was shown that a low-frequency supply can be directly synthesized from a higher frequency source by suitable switching of the cycloconverter elements. A major limitation of the cycloconverter is that its output frequency is limited to (say) one-third of the input frequency, possibly slightly better for a higher pulse configuration. If the input is 50 Hz or 60 Hz, the maximum output frequency is around 20 Hz, the next result being that the cycloconverter application is limited to low speed drives. If the power-source is (say) 400 Hz, then clearly higher speeds are possible.

Figure (15.23) shows that both voltage and frequency levels are directly controllable in the cycloconverter. However, although the cycloconverter is technically attractive for some applications, its use is severely limited on economic grounds compared to the inverter schemes. The cycloconverter is expensive in the number of thyristors and the extensive electronic control circuitry which is required. The system is inherently capable of braking by regeneration back into the a.c. source, thus four quadrant operation is possible. Technically, cycloconverter use is limited by the low-output frequency range and by the quite severe harmonic and power factor demands made on the a.c.

A useful power-circuit arrangement is shown in Fig. (15.24). It consists of three-dual-converters, each supplying one phase of the load. By switching each of the dual-converters cyclically through all four-quadrants of operation, a waveform of per phase a.c. output terminal voltage may be synthesised for each phase of the load. Each output waveform consists of segments of the waveforms of the input terminal voltages of one dual converter. The output frequency is lower than the input frequency and, because of the high harmonic content of both output voltages and input currents, the maximum output frequency must be limited to about 33% of the input frequency. The output voltage of the cycloconverter is almost sinusoidal. The current is further smoothened by machine inductance. Therefore, the referred currents are almost sinusoidal. The behaviour with respect to stator current harmonics and torque pulsations is better. The power factor is poor in this case.

15.6 ROTOR RESISTANCE CONTROL

The methods of induction motor control described in Sections 15.4 and 15.5, control the motor from the stator. Hence, they are applicable to both squirrel-cage and wound-rotor motors. Because of the advantages described previously, the squirrel-cage motor is always preferred. This section considers the method which controls the motor from the rotor, and is applicable to wound-rotor motors only. The wound-rotor motor has a number of disadvantages compared to squirrel-cage motor such as higher cost, weight, volume and inertia, and frequent maintenance due to the presence of brushes and slip-rings. However, the control of a wound-rotor motor from the rotor allows cheaper drives to be obtained for a few specific applications.

One of the important features of the wound-rotor induction motor is that unlike the squirrel-cage motor, it need not be designed to obtain a compromise between the normal running performance and the starting performance. The rotor winding is designed to have a low resistance so that running efficiency is high and the full-load slip is low.

In a wound-rotor motor, the improved starting performance is obtained by connecting an external resistance in series with the rotor winding, as shown in Fig. 15.25(a). The speed–torque and speed–rotor current characteristics of a wound rotor induction motor with different values of external resistances are shown in Fig. 15.25(b). The increase in the rotor resistance does not affect the value of maximum torque but increases the slip at maximum torque. When high starting torque is needed, the rotor resistance can be chosen to obtain the maximum torque at standstill. This also reduces the starting current substantially.

External resistances can be decreased, as the rotor speeds up, making the maximum torque available throughout the accelerating range. Since most of the rotor copper loss occurs in the external resistors, the rotor temperature rise during starting is substantially lower than it would be if the resistance were incorporated in the rotor winding, as in the case of squirrel-cage motors. This allows optimum use of the motor torque capabilities. Hence, the wound rotor motor is widely used in applications requiring frequent starting and braking with large motor torques.

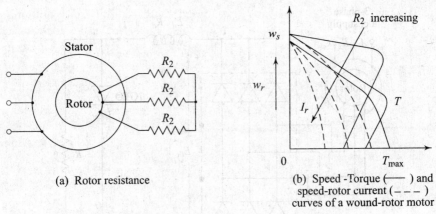

(a) Rotor resistance

(b) Speed-Torque (———) and speed-rotor current (- - -) curves of a wound-rotor motor

Fig. 15.25 *Speed control by rotor resistance*

From the foregoing discussion, it becomes clear that for a given load–torque, the motor speed is reduced as the rotor resistance is increased and the no-load speed, however, remains unaffected by the variation of the rotor resistance. Also, the motor efficiency decreases and the rotor copper loss increases with the decrease in speed. Thus, rotor resistance control method is an inefficient method of speed control like terminal voltage control. However, it has a number of advantages over terminal voltage control. It provides a constant torque operation with a high torque to current ratio. Though, the rotor copper loss increases with the decrease in speed, most of it is dissipated in the external resistors. The copper loss inside the motor in fact remains constant for a fixed torque. Because of this, a motor of smaller size can be employed.

15.6.1 Chopper Controlled Wound-Rotor Induction Motor

Instead of mechanically varying the resistance, the rotor circuit resistance can be varied statically by using the principle of a chopper. This gives stepless and smooth variation of resistance and consequently of motor speed. The external resistor may be replaced by a three-phase diode rectifier and a chopper as shown in Fig. (15.26), where the GTO operates as a chopper switch. As shown in Fig. (15.26), the slip frequency a.c. rotor voltages are converted into d.c. by a three-phase diode bridge and applied across an external resistance R. The GTO switch which is connected in parallel with R is operated periodically with a period T and remains ON for an interval t_{on} in each period. The effective value of resistance R changes from R to 0 as t_{on} changes from 0 to T. The filter inductor L_d is provided to minimize the ripple in current I_d. A high ripple in I_d produces high harmonic content in the rotor, increasing copper losses and causing derating of the motor. The filter inductor also helps in dominating discontinuous conduction at light loads. The diode bridge is the main contributor to the ripple and not the semiconductor switch because it operates at a sufficiently high frequency.

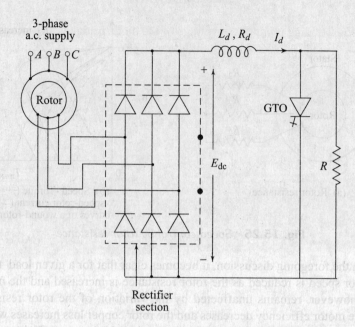

Fig. 15.26 *Induction motor speed control using chopper*

At standstill, the diode-bridge output voltage E_{dc} changes from its maximum value to nearly 5 per cent of the maximum value at near rated motor speed. Here, a thyristor is not suitable for this application because a reliable commutation can only be obtained either by using a bulky commutation capacitor or an auxiliary source for charging the commutating capacitor.

The voltage E_{dc} is small because induction motors are usually designed with stator-to-rotor turns ratio greater than 1. Hence, a transistor is suitable for low-power drives and GTO may be employed for ratings beyond the capability of transistors. The self-commutation capability of these devices ensures reliable commutation at all operating points and makes the semiconductor switch compact.

An alternative static rotor resistance control circuit is obtained by using either a six-pulse or three-pulse controlled rectifier instead of the diode bridge and GTO. By controlling rectifier firing angle, the power consumed by R is then controlled. As the firing angle is increased from 0 to the maximum, the effective rotor resistance increases from R to a maximum value, controlling the speed.

1. Equivalent Circuits of Wound-rotor Motor with Static Resistance Control For the analysis of the power-circuit of Fig. 15.26, the following assumptions have been made:

(i) The filter inductor current is assumed to be ripple-free d.c.

(ii) Due to the motor leakage inductances, commutation overlap in the diode bridge is ignored.

(iii) The rotor phase current will have a six-step waveform shown in Fig. 15.27, under assumptions (i) and (ii) above. In the same Fig. 15.27, the waveform of the corresponding input phase voltage of the diode bridge is also shown. The fundamental rotor current is in phase with the phase voltage. The rotor phase current waveform is similar to the phase current waveform of CSI (Fig. 9.58). As far as harmonics are concerned, the motor can be considered fed from the rotor by a current source. The harmonics in rotor current cause only a small harmonic current to flow in the stator. Therefore, the machine-induced emf and hence flux can be assumed sinusoidal. When the flux is sinusoidal, the torque is produced only by the fundamental. The harmonics produce only pulsating torques.

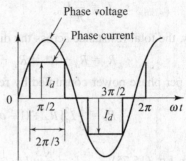

Fig. 15.27 *Rotor phase voltage and phase-current waveforms*

(iv) The losses in the diode bridge and the semiconductor switch are neglected. Let us define duty ratio α again as in the case of chopper (Chapter 8). Therefore,

$$\alpha = \frac{t_{on}}{T} \qquad (15.73)$$

The energy absorbed by resistance R during a period of operation of the GTO is given by

$$E_R = I_d^2 R (T - t_{on})$$

Now, the average power absorbed by resistance R during a period T is

$$P_r = \frac{1}{T} [I_d^2 R (T - t_{on})] = [I_d^2 R (1 - t_{on}/T)] = I_d^2 R (1 - \alpha)$$

Therefore, the effective value of resistance R is given by

$$R^* = (1 - \alpha) R \qquad (15.74)$$

Now, the RMS value of the rotor phase current from Fig. (15.26) is given by

$$I_{rms} = \left[\frac{1}{\pi} \int_0^{2\pi/3} I_d^2 \, d(\omega t) \right]^{1/2} = \frac{\sqrt{2}}{\sqrt{3}} I_d \qquad (15.75)$$

Because of the quarter-wave symmetry of the rotor phase current i_r, Fourier component $b_1 = 0$, and

$$a_1 = \frac{4}{\pi} \int_{\pi/6}^{\pi/2} I_d \sin \omega t \, d(\omega t) = \frac{2\sqrt{3}}{\pi} I_d$$

Therefore, the fundamental rotor current is

$$I_r = \frac{a_1}{\sqrt{2}} = \frac{2\sqrt{3}}{\pi} \frac{I_d}{\sqrt{2}} = \frac{\sqrt{6} I_d}{\pi} \qquad (15.76)$$

From Eqs (15.75) and (15.76)

$$I_r = \frac{3}{\pi} I_{rms} \tag{15.77}$$

Now, the total resistance across the diode bridge is

$$R_e = R_d + R^* = R_d + (1 - \alpha)R.$$

The per phase power consumed by resistance R_e,

$$P_e = \frac{1}{3} I_d^2 [R_d + (1 - \alpha)R] \tag{15.78}$$

From Eq. (15.75), $I_d = \frac{\sqrt{3}}{\sqrt{2}} I_{rms}$

Substituting in above equation, P_e

$$\therefore \quad P_e = \frac{1}{2}[R_d + (1-\alpha)R]I_{rms}^2 \tag{15.79}$$

This is equivalent to the power dissipation in a resistance of $\frac{1}{2}[R_d + (1 - \alpha)R]$ ohms, caused by the rms rotor current I_{rms}. Hence, the effective per-phase value of resistance R_e is given by

$$R_e^* = 0.5 [R_d + (1 - \alpha)R] \tag{15.80}$$

The power transferred across the air gap in a fundamental equivalent circuit is given by,

$$p_{ag} = 3EI_2 \cos \theta_r \tag{15.81}$$

where θ_r is the phase angle between phases E and I_2.

The total power consumed in the rotor circuit (P_{ag}') for the drive under consideration is given by

$$P_{ag}' = 3 I_{rms}^2 (R_r + R_e^*) + P_m \tag{15.82}$$

Substituting the value of I_r from Eq. (15.77), yields

$$P_{ag}' = \frac{\pi}{3} I_r^2 (R_r + R_e^*) + P_m \tag{15.83}$$

But, the fundamental equivalent circuit of the drive must satisfy the condition $P_{ag} = P_{ag}'$. Hence, from Eqs (15.82) and (15.83).

$$EI_2 \cos \theta_r = \frac{\pi^2}{9} I_r^2 (R_r + R_e^*) + \frac{P_m}{3} \tag{15.84}$$

Now, the slip power (the portion of the air gap power, which is not converted into mechanical power, is called slip power) due to the fundamental rotor current for the drive under consideration becomes

$$S \cdot p_{ag_1} = 3 I_r^2 (R_r + R_e *) \tag{15.85}$$

where p_{ag_1} is the fundamental air-gap power in the drive.

The mechanical power developed by the fundamental rotor current is given by

$$P_m = (1-s) p_{ag_1}$$

Substituting from Eq. (15.85) gives

$$P_m = 3 I_r^2 (R_r + R_e *) \frac{(1-S)}{s} \tag{15.86}$$

Now, substituting from Eq. (15.86) into Eq. (15.84) and rearranging the terms give

$$EI_2 \cos \theta_r = \left[\left(\frac{\pi^2}{9} - 1 \right)(R_r + R_e *) + \frac{(R_r + R_e *)}{s} \right] I_r^2$$

$$= \left(R_K + \frac{R_m}{s} \right) I_r^2 \tag{15.87}$$

where

$$R_K = \left(\frac{\pi^2}{9} - 1 \right)(R_r + R_e *) \tag{15.88}$$

$$R_m = (R_r + R_e *) \tag{15.89}$$

Referring all parameters on the right side of Eq. (15.87) also to be stator side gives

$$EI_2 \cos \theta_r = \left(R_K' + \frac{R_m'}{s} \right) I_r^2 \tag{15.90}$$

where R_K' and R_M' are respectively the values of R_K and R_M referred to the stator side.

Thus, $\qquad R_K' = a_{T_1}^2 \cdot R_K$

and $\qquad R_M' = a_{T_1}^2 \cdot R_m$ $\qquad (15.91)$

where a_{T_1} is the stator to rotor turns ratio.

The per phase fundamental equivalent circuit of the drive referred to the stator, as obtained from Eq. (15.90), is shown in Fig. 15.28(a). Resistance $\left(\frac{R_m'}{s} \right)$ accounts for the developed mechanical power and the fundamental rotor copper low, whereas, R_K' accounts for the rotor harmonic copper loss. The equivalent circuit of Fig. 15.28(a) can be simplified to that of Fig. 15.28(b).

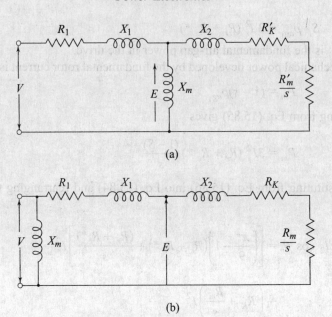

Fig. 15.28 Equivalent circuits

Now, from the equivalent circuit of Fig. 15.28,

$$\bar{I}_2 = \frac{V}{(R_1 + R'_K + R'_m/s) + j(X_1 + X_2)} \quad (15.92)$$

$$T = \frac{3}{\omega_s} I_2^2 \left(\frac{R'_m}{s}\right) \text{ N-m} \quad (15.93)$$

Substituting the value of I_2 in Eq. (15.93), yields

$$T = \frac{3}{\omega_s} \left[\frac{V^2 (R'_m/s)}{(R_1 + R'_K + R'_m/s)^2 + (X_1 + X_2)^2} \right] \quad (15.94)$$

For a given value of α and s, the rotor current and torque can be obtained from Eqs (15.92) and (15.94).

In the above analysis, the energy loss in the GTO and the diodes have been ignored. For low values of α, this loss is negligible compared to the total rotor loss. For α close to unity this loss forms a significant portion of the total rotor loss. Thus, for the values of a close to unity, appreciable error may be caused in the calculation of speed-torque curves.

Figure 15.29 shows the nature of speed-torque curves for different values of α. Now, R is fully bypassed by the GTO for $\alpha = 1$. However, the speed–torque curves for $\alpha = 1$ lies below the natural speed torque curve, due to the additional losses in the switch GTO, diodes, and resistances R_d and R'_K. As shown, for a given torque, speed reduces with α. The control region consists of the area enclosed in *PQRS*. By controlling α, any operating point in this region can be obtained. The operation is not possible in the area *PSO*. Now, when *R* is increased, the control region is increased and the area *PSO* is decreased.

Compared to conventional rotor resistance control methods, employing contractors, slip regulators and so on, the chopper controlled method has the disadvantage of requiring motor derating. But it has many more advantages, such as smooth and stepless control, fast response, less maintenance, longer life, compact size, assured balance between rotor phase currents, simple closed loop control, and so on.

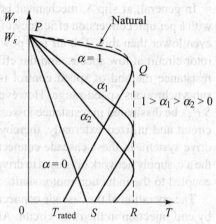

Fig. 15.29 Speed–torque curves for chopper controlled induction-motor

15.7 SLIP POWER RECOVERY SCHEME

In the method described in Section 15.6, the slip power is dissipated in the resistance and this effectively reduces the efficiency of the motor at low speeds. However, instead of dissipating it in the resistance, the slip power can be conveniently returned to the mains or used in a useful manner making the drive more efficient. This is achieved by means of slip energy recovery schemes widely known as Kramer and Scherbius drives.

As discussed previously, the fundamental power delivered to the rotor across the air gap, P_{ag}, is divided between the mechanical power output, P_m, and the rotor copper loss P_c. The rotor copper loss is given by

$$P_C = P_{ag} - P_m = T(\omega_s - \omega) = ST\omega s = SP_{ag} \quad (15.95)$$

and
$$P_m = (1-s)P_{ag} \quad (15.96)$$

Also
$$P_{ag} = T\omega_s \quad (15.97)$$

where *T* is the electromagnetic torque developed by the motor, and ω_s is the synchronous angular velocity. From these equations, it is obvious that the air gap power is constant when the induction motor drives a constant torque load and hence, the rotor copper loss is proportional to the slip. At half synchronous speed, the air gap power is divided equally between the mechanical power output and rotor I_R^2 loss, giving an overall efficiency when stator losses are taken into consideration, of less than 50 per cent.

In general, at slip S, mechanical power is obtained from the air gap power with a per unit conversion efficiency of $(1-S)$, and the overall motor efficiency is even lower than this. The air gap power is almost completely dissipated in the rotor circuit at low speeds, and the efficiency is very poor. Therefore, the rotor resistance method of speed control is uneconomical except for a very small subsynchronous speed range. However, it is not essential that the slip power, $S P_{ag}$, be dissipated in resistance losses because it can be removed from the rotor circuit and utilized externally, thereby improving the overall efficiency of the drive system. In these cascade connections, the slip power is either returned to the a.c. supply network or is used to drive an auxiliary motor which is mechanically coupled to the induction motor shaft.

The operation of the cascade connection may also be regarded as speed control by emf injection in the rotor circuit. Assume the motor is operating normally at slip S, and an external voltage is applied at the slip-rings in phase opposition to the rotor emf the resultant decrease in rotor current causes a reduction in motor torque, because torque is proportional to in-phase rotor current, assuming constant air gap flux. Therefore, the motor speed decreases due to the braking action of the load. However, as the slip increases, the rotor emf and current also increases, and stable operation is obtained at some reduced speed when the motor torque again equals the load torque.

When rotor resistance control is adopted, the slip-frequency voltage drop in the external resistor constitutes the injected voltage, but an external emf source of low impedance is equally effective and does not introduce excessive heat loss. The main problem in providing a suitable emf source is that the frequency of the injected emf must match the rotor slip frequency at all speeds.

Historically, mechanical frequency conversion methods have been employed. In the traditional Scherbius system shown in Fig. 15.30, a rotary converter rectifies the slip power, and the rectified output drives a d.c. motor which is mechanically coupled to a squirrel-cage induction generator. The induction generator is driven at supersynchronous speeds and returns the slip power to the a.c. supply. In the traditional Kramer system, the slip-power is also rectified by a rotary converter, and the output supplies a d.c. motor which is mechanically coupled to the main induction motor. Thus, the slip power is converted to mechanical power at the induction motor shaft.

Now, static frequency converters have replaced the auxiliary machines in the Scherbius system, and this change has resulted in a more compact adjustable speed drive with an improved operating efficiency and a better dynamic response. The Kramer drive can also be modified using a diode rectifier bridge in place of the rotory converter but a d.c. motor is still required to convert the rectified slip power to mechanical power.

Control of A.C. Drives

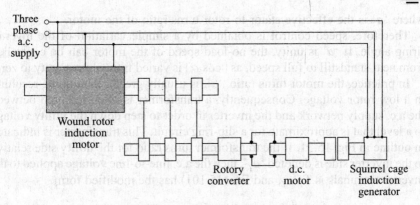

Fig. 15.30 *Traditional Scherbius system*

15.7.1 Static Scherbius Drives

Figure 15.31 shows the static Scherbius scheme for speed control of a wound-rotor induction motor. This drive scheme is also known as *subsynchronous converter* cascade because it is capable of providing speed control only in the subsynchronous speed range ($\omega_c < \omega_r \leq \omega_s$). The d.c. link converter consists of a three-phase diode rectifier bridge which operates at slip frequency and feeds rectified slip power through the smoothing inductor to the phase-controlled thyristor inverter. The inverter returns the rectified slip power to the a.c. supply network. The rectifier and inverter are both naturally commutated by the alternating emfs appearing at the slip-rings and supply busbars, respectively.

The problem of matching the frequencies of the injected emf and the rotor emf is eliminated by the rectification of the slip-ring voltages because an adjustable d.c. back emf can now be used as the injected voltage for speed control. In Fig. (15.30), the average back emf of the inverter is the injected d.c. emf opposing the rectified rotor voltage.

If commutation overlap is negligible, the direct output voltage of the uncontrolled three-phase bridge rectifier is obtained as

$$E_{dc} = 1.35 \, E_r \cdot S \tag{15.98}$$

where E_r is the line-to-line rotor voltage at standstill and S is the fractional-slip.

For a line-commutated three-phase bridge inverter with negligible overlap, the average back-emf is given by

$$E_1 = 1.35 \, E_L \cos \alpha \tag{15.99}$$

where α is the inverter firing angle ($\alpha > 90°$), and E_L is the a.c. line-to-line voltage.

At no-load the motor torque is negligible and the rectified rotor current is almost zero. Consequently, the two direct voltages of Eqs (15.98) and (15.99) must balance. Thus, $E_{dc} + E_1 = 0$

or $\quad 1.35 \, E_r \, S + 1.35 \, E_L \cos \alpha = 0 \tag{15.100}$

and hence, $\quad S = -\left(\dfrac{E_L}{E_r}\right) \cos \alpha = a|\cos \alpha| \tag{15.101}$

where 'a' is the effective stator-to-rotor turns ratio of the motor.

Therefore, speed control is obtained by a simple variation of the inverter firing angle. If 'a' is unity, the no-load speed of the motor can be controlled from near standstill to full speed, as $|\cos \alpha|$ is varied from almost unity to zero.

In practice, the motor turns ratio, 'a', is usually greater than unity, resulting in a low rotor voltage. Consequently, a transformer is often required between the a.c. supply network and the inverter in order to step down the utility voltage to a level that is approximate for a slip-ring circuit. This transformer is indicated in outline in Fig. 15.31. If the transformer turns ratio for the utility side relative to the inverter side is denoted 'a_T', then the a.c. line-to-line voltage applied to the inverter terminals is E_L/a_T, and Eq. (15.101) has the modified form

$$s = \frac{a}{a_T}|\cos \alpha| \qquad (15.102)$$

Clearly, if a cascade transformer is not used then a_T is unity, and Eq. (15.102) reduces to Eq. (15.101).

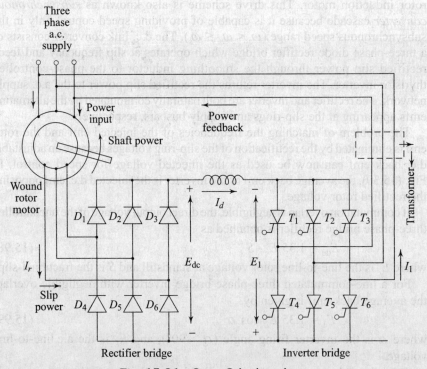

Fig. 15.31 Static Scherbius drive

In order to develop motor torque, a rotor current is required, and the rectified rotor voltage must force current flow against the inverter back emf. As the induction motor is loaded, the speed falls slightly so that the resulting increase in the rotor voltage can overcome the voltage drops in the rotor windings and in the d.c. link circuit.

Now, the fundamental rotor slip power, SP_{ag}, is approximately equal to the d.c. link power, if the rotor resistance is small. Thus,

$$S \cdot P_{ag} = E_1 I_d \qquad (15.103)$$

or

$$P_{ag} = E_1 I_d / S$$

But, as before,

$$P_{ag} = T \cdot \omega_s$$

and hence,

$$T = \frac{E_1 I_d}{s\omega_s} \qquad (15.104)$$

Substituting from Eq. (15.99), yields

$$T = \frac{1.35 E_L I_d}{a\omega_s} \qquad (15.105)$$

This equation is also valid when a transformer is present in the cascade circuit. Thus, the steady-state torque is proportional to the rectified rotor current, I_d, which in turn, is equal to the difference between the rectified rotor voltage and the average back emf of the inverter divided by the resistance of the d.c. link inductor. The inverter emf is constant for a fixed firing angle, and hence, the rotor slip increases linearly with load torque, giving a torque-speed characteristic similar to that of a separately excited d.c. motor with armature-voltage control. In practice, the complete open-loop torque-speed characteristics have the form shown in Fig. 15.32.

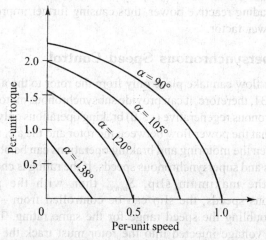

Fig. 15.32 Open loop torque–speed characteristics for the induction motor and static converter cascade

The principal disadvantage of the subsynchronous cascade drive is its low fundamental power factor, or displacement factor, particularly at reduced speeds. This is owing to the lagging reactive power drawn by the machine to maintain the air gap flux. The reactive power consumption of the line commutated inverter is largely responsible for the low power factor of the cascade drive. The reactive power drawn by the inverter increases as α advances from 180° to 90°. At $\alpha = 90°$, the inverter kVA is almost reactive. However, the total reactive kVA is the sum of the reactive powers absorbed by the motor and inverter and consequently, the system power factor is poor at low speeds when the active power consumption is small.

Several modifications have been reported in the literature to improve the power-factor of the slip energy recovery schemes:

1. Using a motor having a turns ratio 'a' less than 1. If 'a' is greater than 1 then a step-down transformer can be interposed.
2. The reactive power drawn by the inverter can be kept low and this is possible if $\alpha \approx 180°$ at the lowest value of the controllable speed range.
3. Use the forced-commutated inverters in place of link-commutated inverters. When forced-commutated, the inverter does not draw the reactive power from the line.
4. Capacitive compensation of the converter cascade. The capacitors may be connected to improve the power factor in the complete speed range. The compensation can be either on the stator side or on the rotor side. The latter has been found more advantageous in view of torque, current ratio, better regulation, and good power factor. However, the power capacitors required are of larger size and hence costly.
5. By operating the inverter with pulse-width modulation. For this, thyristors in the inverter are replaced by self-commutated switches. When pulse-width modulation is employed, the inverter can also be operated with a leading reactive power, thus causing further improvement in the drive power factor.

15.7.2 Supersynchronous Speed Control

Since the power flow can take place only from the rotor to the a.c. mains in the drive of Fig. 15.31, therefore, it can provide subsynchronous motoring ($1 \geq s \geq 0$) and supersynchronous regenerative ($s < 0$) braking operations only. If the changes are made such that the power flow between the rotor and the a.c. mains becomes bidirectional, then the motoring and braking operations can be obtained both for subsynchronous and supersynchronous speeds. If the rating is chosen to provide operation at the maximum slip, S_{max}, then with the provision for supersynchronous speeds, the slip can be controlled from $-S_{max}$ to $+S_{max}$, consequently doubling the speed range for the same rating. To realize these operations, the voltage injected into the rotor must track the frequency and phase-sequence of the rotor induced voltages. Thus, the phase sequence of the injected voltage must be reversed for operation at supersynchronous speeds. Here we consider two schemes for speed control at supersynchronous speeds.

1. D.C. Link Converter Cascade In order to reverse the power flow in the converter circuit of Fig. 15.31, the rectifier must operate as an inverter and *vice versa*. For speed control on either side of synchronism, the converter cascade therefore needs two phase controlled thyristor bridges, one operating at slip-frequency as a rectifier or inverter, while the other at network frequency as an inverter or rectifier. The circuit diagram is shown in Fig. 15.33. The replacement of six diodes by six thyristors increases the converter cost and also necessitates the introduction of slip frequency gating circuits for the new converter. Difficulty is experienced near synchronism when the slip frequency emfs are insufficient for natural commutation, and special circuit configurations employing forced commutation or devices with a self turn-off capability are necessary for the passage through synchronism. Thus, the provision of supersynchronous speed control complicates the static converter cascade system and nullifies the advantages of simplicity and economy which are inherent in a purely synchronous drive.

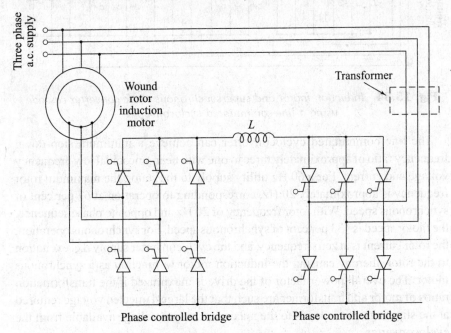

Fig. 15.33 *Induction motor and supersynchronous static converter cascade using a naturally commutated d.c. link converter*

2. Cycloconverter Cascade A second approach for achieving bidirectional power flow in the cascade circuit is through the use of a line-commutated cycloconverter. An 18-SCR circuit is shown in Fig. 15.34, but other circuit configurations can be employed and a voltage matching transformer may be introduced between the cycloconverter and the a.c. supply network. The cycloconverter must be controlled so that the frequency of the injected slip-ring voltages tracks the rotor slip frequency. As before, for subsynchronous motoring, the slip frequency rotor emfs deliver slip power to the cycloconverter, and this power is returned to the a.c. utility supply. At supersynchronous speeds, the slip-ring voltages have opposite phase sequence, and the cycloconverter delivers slip power to the rotor. Thus, for supersynchronous motoring the a.c. utility network must supply both the stator input power and the slip-power.

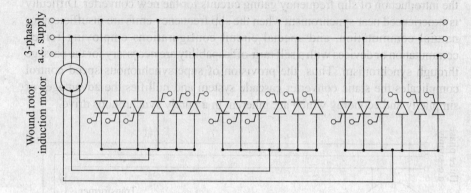

Fig. 15.34 *Induction motor and supersynchronous static converter cascade using a line-commutated cycloconverter*

The line commutated cycloconverter can achieve a minimum step-down frequency ratio of approximately three to one with near-sinusoidal low-frequency voltage and current. For a 60 Hz utility supply to the stator, the maximum rotor frequency is approximately 20 Hz, corresponding to operation at 67 per cent of synchronous speed. With rotor frequency of 20 Hz and opposite phase sequence, the motor speed is 133 per cent of synchronous speed. For synchronous operation, the rotor current is at zero frequency, and the cycloconverter supply d.c. excitation to the rotor, thereby causing the induction motor to function as a synchronous motor. The overall power factor of the drive is maximized if the transformation ratios of motor and transformer are such that the largest injected voltage required at the slip-rings corresponds to the maximum output voltage available from the cycloconverter.

The cycloconverter cascade is expensive, and it introduces additional control complexity, but the near sinusoidal rotor currents minimize harmonic heating effects and low frequency torque pulsations. Commutation problems near synchronous speed are also eliminated. Since the cycloconverter employs a large number of thyristors, the drive is suitable only for very large capacity drives.

15.7.3 Modified Kramer Drives

Kramer suggested that the slip power taken from the rotor for speed control can be usefully employed by converting it to mechanical power in an auxiliary motor mounted on the induction motor shaft. The mechanical power produced by the auxiliary motor supplements the main motor power, thus allowing the same power to be delivered to the load at different speeds.

In the traditional Kramer system, the slip power is also rectified by a rotary converter, and the output supplies a d.c. motor which is mechanically coupled to the main induction motor. Thus, the slip power is converted to mechanical power at the induction motor shaft.

In the earlier static Kramer drive, the slip frequency rotor voltages were converted into d.c. by a diode bridge. The rectified d.c. voltage was fed to the armature of a d.c. motor mounted on the induction motor shaft. If the armature resistance drop is neglected, then the d.c. machine voltage is equal to the d.c. output voltage of the diode bridge, which, in turn, depends on the induction motor slip. By controlling the field, the d.c. motor induced voltage, and consequently, the induction motor speed, can be controlled. This static Kramer drive is not employed any more because of problems associated with d.c. motors, particularly with high power levels. Instead, the drive of Fig. 15.35 is employed.

In this drive, the d.c. machine is replaced by a commutator less d.c. motor, which consists of a synchronous motor fed by a load commutated inverter. The speed is controlled by varying the commutation lead angle. The speed can also be controlled by the field current control. However, it is not preferred because of the following problems associated with it. To drive the system at synchronous speed, the field current must be reduced to zero to reduce the inverter d.c. terminal voltage to zero. Now, the induced voltage will not be sufficient to obtain load commutation. Two other problems associated with the field current control are the slow response of the field circuit and increased armature reaction at low field currents. However, the field current control can be used to control flux and the synchronous terminal voltage within the ratings of the inverter thyristors.

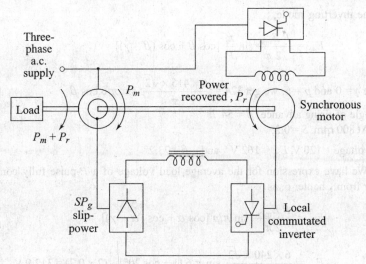

Fig. 15.35 *Commutatorless Kramer drive*

Compared to the Scherbius drive, this drive has a better power factor and lower harmonic content in the line current. In the static Scherbius drives, reactive power and harmonics are associated with the power feedback to the line. In the static Kramer drive since the power is not fed back to the line, problems associated with the feedback of power are also eliminated.

SOLVED EXAMPLE

Example 15.2 (a) A 6-pole, 50 Hz slip-ring induction motor is controlled by a static Scherbius drive. Determine the angle of firing advance in the inverter at (i) 600 rpm (ii) 800 rpm, if the open circuit standstill slip-ring voltage is 600 V, and the inverter is connected to a 415 V, 3-phase system. Neglect overlap and losses.

(b) Recalculate the firing advance in part (a(i)) if there is an overlap of 10° in the rectifier and 5° in the inverter. Allow for diode and thyristor voltage drops of 1.5 V and 0.7 V respectively.

(c) Use the data of part (a), and taking the minimum required speed to be 600 rpm, estimate the voltage ratio of a transformer to be interposed between the inverter and the supply. Also, specify the power-flow through the d.c. link as a ratio to the power input to the stator.

Solution: (a) Synchronous speed $= \left(\dfrac{50}{3}\right) \times 60 = 1000$ rpm.

(i) Slip $= \dfrac{1000 - 600}{1000} = 0.4$.

Now, rotor voltage at 600 rpm $= 600 \times 0.4 = 240$ V

Assuming 3-phase converter bridge with $\alpha = 0$, $\gamma = 0$

$\therefore$ The d.c. link voltage $= \dfrac{6 \times 240 \sqrt{2}}{\pi} \sin \pi/6$ $\therefore$ $E_{dc} = 324$ V

In the inverting mode,

$$E_{dc} = \dfrac{P E_{max}}{2\pi} \sin \dfrac{\pi}{P} [\cos \beta + \cos (\beta - \gamma)] \qquad (i)$$

With $\gamma = 0$ and $p = 6$, we get $324 = \dfrac{6 \times 415 \times \sqrt{2}}{\pi} \sin \pi/6 \cos \beta$

$\therefore$ Angle of firing advance, $\beta = 54.7°$

(ii) At 800 rpm, $S = 0.2$,

Rotor voltage $= 120$ V, $E_{dc} = 162$ V and $\beta = 73.2°$

(b) We have expression for the average load voltage of a P-pulse fully controlled rectifier from Chapter 6, as

$$E_{dc} = \dfrac{P E_{max}}{2\pi} \sin \pi/p [\cos \alpha + \cos (\alpha + \gamma)]$$

$\therefore \qquad E_{dc} = \dfrac{6 \times 240 \times \sqrt{2}}{2\pi} \sin \pi/6 \, [1 + \cos 20°] - (2 \times 0.7) = 312.9$ V.

Using Eq. (i),

$$312.9 - (2 \times 1.5) = \frac{6 \times 415 \times \sqrt{2}}{2\pi} \sin\frac{\pi}{6} \left[\cos\beta + \cos(\beta - 5°)\right] \therefore \beta = 58.9°$$

(c) As both rectifier and inverter are the same 3-phase configurations, then the same relationship exists between the a.c. side and the d.c. link voltage when the firing angle advance is zero.

Hence, the voltage required at the transformer is 240 V, the slip-ring voltage at 600 rpm. In practice, the firing advance will not be less than (say) 15° to allow for some overlap. The transformer ratio = $\frac{415}{240}$ = 1.73.

Taking the input stator power to be 100, the power out of the rotor is $100 \times 0.4 = 40$, and the shaft output power is $100 - 40 = 60$, neglecting losses within the motor.

15.8 SYNCHRONOUS MOTOR DRIVES

A synchronous machine is one in which alternating current flows in the armature winding and d.c. excitation is supplied to the field winding. The armature winding is almost invariably on the stator and is usually a three-phase winding. The field winding is on the rotor. The speed of synchronous machine under steady-state conditions is proportional to the frequency of the current in its armature. The magnetic field created by the armature currents rotates at the same speed as that created by the field current on the rotor (which is rotating at synchronous speed), and a steady, torque results. The armature is identical to the stator of induction motors, but there is no induction in the rotor.

A synchronous motor, therefore, is a constant speed machine which always rotates with zero slip at the synchronous speed, which depends on the frequency and the number of poles, as given by Eq. (15.14). A synchronous motor can be operated as a motor or generator. By varying the field current, the power-factor can be controlled. The synchronous motors can be of the following types:

1. Round or cylindrical rotor motors
2. Salient or projecting rotor pole-motors
3. Reluctance motors
4. Permanent magnet motors
5. Switched reluctance motors.
6. Brushless d.c. and a.c. motors.

In this section, the operating principles and phasor diagrams for the above mentioned machines are presented. The cylindrical rotor motors and salient pole motors are two distinct categories of wound-field machine. The wound-field machine has a distributed polyphase armature winding on the stator and a d.c.-excited field winding on the rotor. An inverted construction, with a d.c.-excited stator field winding is possible but the rotating three-phase armature winding would require four slip-rings (including a neutral connection), and the slip-ring currents and insulation voltage ratings would be quite large. The rotating-field system is preferred because the d.c. excitation current is relatively small and can be supplied by only two slip-rings. A brushless construction is obtained when the rotor field excitation is provided by permanent magnets.

The synchronous reluctance motor is a rugged brushless motor that uses an unexcited rotor having salient or projecting poles. The wound field and permanent magnet machines can operate at unity power factor at full load and thus minimize the stator current and inverter volt-ampere rating for a given shaft-power. These machines can also operate at a leading power factor and thereby provide load commutation for the inverter. The synchronous reluctance motor, on the other hand, always operates at a relatively low lagging power factor. The permanent magnet and reluctance machines are restricted to lower power ratings but the wound-field motor can be of very high power rating.

15.8.1 Round (Cylindrical) Rotor Motor

The round rotor or cylindrical rotor, synchronous machine has a uniform air gap between a slotted stator and rotor. This machine has its rotor in cylindrical form. The stator is composed of iron laminations stacked together. The rotor is a solid forging with rotor slots milled into its surface. A single field winding is placed in the rotor slots, and a conventional three-phase distributed armature winding is placed in the stator slot. For synchronous motor operation, the three-phase armature winding is supplied with balanced three-phase currents, and d.c. excitation is supplied to the rotor field winding. The balanced three-phase armature currents establish a component flux wave in the air gap, which has an approximately sinusoidal spatial distribution with a constant amplitude and which rotates at synchronous speed. If the rotor also rotates at synchronous speed, the magnetic fields of stator and rotor are stationary relative to one another, and a steady electromagnetic torque is developed because of the tendency of the two magnetic fields to align their axes.

The speed n_1 (in revolutions per minute) of the synchronous motor is related to the a.c. supply frequency f_1, and the number of pole pairs P, by $n_1 = 60 \cdot f_1/p$. On 50 or 60 Hz a.c. utility supplies, the synchronous motor has no starting torque; it must be brought up to synchronous speed by induction motor action or by an auxiliary motor. Induction motor torque is developed when the rotor is fitted with a squirrel-cage winding or when an unlaminated steel rotor is used.

The rotor field winding can be excited with direct current supplied through slip-rings and brushes from a static phase-controlled rectifier exciter or from a d.c. generator exciter on the shaft of the main synchronous machine. The field current of the synchronous machine is controlled by varying the field current of the d.c. generator exciter. A common approach, up to quite large ratings, is to use a shaft-coupled a.c. generator exciter with rectification of the a.c. output.

If the exciter armature winding is placed on the rotor and the rectifier diodes are mounted on the rotating shaft, the rectified output can be fed directly to the field winding of the main synchronous machine without any sliding contacts. This is the common brushless excitation system. An alternate approach is to fit a shaft-mounted exciter generator with three-phase windings on both stator and rotor. The three-phase stator-winding is energized by the a.c. power supply system and establishes a rotating air gap field. The rotor is driven against this rotating field so that alternating voltages are generated in the three-phase rotor winding. These a.c. voltages are rectified by a rotating diode rectifier bridge to provide the d.c. excitation for the main synchronous machine. The field current of the synchronous machine is varied by controlling the stator voltage of the exciter with a three-phase thyristor voltage controller.

Equivalent Circuit : The equivalent circuit per phase, neglecting the no-load loss, is shown in Fig. 15.35(a), where R_a is the armature resistance per phase and X_s is the synchronous reactance per-phase. The reactance are independent on the rotor position. E_f is known as excitation or field voltage and is dependent on the field current.

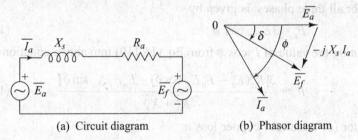

(a) Circuit diagram (b) Phasor diagram

Fig. 15.36 *Equivalent circuit of synchronous motors*

In Fig. 15.37 the E curves shows the typical variations of the armature current against the excitation current. The power factor depends on the field current. For the same armature current, the power factor could be leading or lagging, depending on the excitation current, I_f.

Now, if ϕ is the lagging power factor angle of the motor, then from Fig. 15.33(a), we have

$$\overline{E}_f = E_a \angle 0 - \overline{I}_a (R_a + jX_s) \tag{15.106}$$

$$= E_a \angle 0 - I_a (\cos\phi - j\sin\phi)(R_a + jX_s)$$

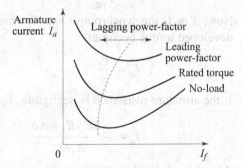

Fig. 15.37 *Typical E curves of synchronous motors*

$$= E_a - I_a X_s \sin\phi - I_a R_a \cos\phi - jI_a (X_s \cos\phi - R_a \sin\phi) \tag{15.107}$$
$$= E_f \angle \delta \tag{15.108}$$

where $\delta = \tan^{-1}\left[\dfrac{-(I_a X_s \cos\phi - I_a R_a \sin\phi)}{E_a - I_a X_s \sin\phi - I_a R_a \cos\phi}\right] \tag{15.109}$

and $E_f = [(E_a - I_a X_s \sin\phi - I_a R_a \cos\phi)^2$
$\quad\quad + (I_a X_s \cos\phi - I_a R_a \sin\phi)^2]^{1/2} \tag{15.110}$

From phasor diagram of Fig. 15.36(b), we can write

$$\overline{E}_f = E_f (\cos\delta + j\sin\delta) \tag{15.111}$$

and $\bar{I}_a = \dfrac{\bar{E}_a - \bar{E}_f}{R_a + jX_s} = \dfrac{\left[E_a - E_f(\cos\delta + j\sin\delta)\right](R_a - jX_s)}{R_a^2 + X_s^2}$ (15.112)

From Eq. (15.112) the real part is

$$I_a \cos\phi = \dfrac{R_a(E_a - E_f \cos\delta) - E_f jX_s \sin\delta}{R_a^2 + X_s^2}$$ (15.113)

Torque/Load-angle Characteristic : The shaft torque can be expressed in terms of the *load angle*, δ. Thus, the electrical power input to the synchronous motor for all three phases, is given by

$$P_i = 3E_a I_a \cos\phi$$ (15.114)

Substituting the value of $I_a \cos\phi$ from Eq. (15.110) into above equation yields

$$P_i = \dfrac{3[R_a(E_a^2 - E_a E_f \cos\delta) - E_a E_f X_s \sin\delta]}{R_a^2 + X_s^2}$$ (15.115)

The stator (or armature) copper loss is

$$P_c = 3I_a^2 R_a$$ (15.116)

The gap power, which is the same as the developed power, is

$$P_d = P_{ag} = P_i - P_c$$ (15.117)

Now, if ω_s is the synchronous speed, which in the same as the rotor speed, the developed torque becomes

$$T_d = \dfrac{P_d}{\omega_s}$$ (15.118)

If the armature resistance is negligible, T_d in Eq. (15.118) becomes

$$T_d = \dfrac{-3E_a \cdot E_f \sin\delta}{X_s \omega_s}$$ (15.119)

Also, Eq. 15.109 becomes

$$\delta = -\dfrac{\tan^{-1} I_a X_s \cos\phi}{E_a - I_a X_s \sin\phi}$$ (15.120)

A negative sign is introduced into Eq. (15.119) because δ is negative for *motoring operation* (E_f lagging E_a) and torque becomes positive. In the case of generating, δ is positive and the power (and torque) becomes negative. The angle δ is also called as the torque angle.

For a fixed voltage and frequency, the torque depends on the angle, δ, and is proportional to the excitation voltage E_f. For fixed values of E_f and δ, the torque depends on the voltage-to-frequency ratio and a constant volts/hertz control will provide speed control at a constant torque. If E_a, E_f, and δ remain fixed, the torque decreases with the speed and the motor operates in the field weakening mode.

If the $\delta = 90°$, then the torque becomes maximum and the maximum developed torque, which is called as the push–pull torque, becomes

$$T_p = T_m = \frac{-3E_a \cdot E_f}{X_s \cdot \omega_s} \tag{15.121}$$

Figure 15.38 shows the plot of developed torque against the angle, δ. For stability considerations, the motor is operated in the positive slope of $T_d - \delta$ characteristics and this limits the range of torque angle, $-90° \leq \delta \leq 90°$.

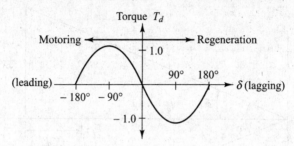

Fig. 15.38 *Torque versus torque angle with cylindrical rotor*

Round-rotor synchronous machines are used for steam and gas turbine driven generators in utility and industrial generating stations. Ratings can exceed 1500 MW for a single unit.

15.8.2 Salient-Pole Motors

In a salient-pole synchronous machine, the stator has a conventional three-phase distributed armature winding placed in slots, as in the round rotor motor. The field winding consists of a number of concentrated field coils placed around projecting poles on the rotor. To produce alternate north and south rotor poles, the field winding is excited with direct current. As before, synchronous motor torque is developed by the interaction between stator and rotor magnetic field when the stator winding is energized with balance three-phase currents. The salient-pole construction results in a nonuniform air gap. The air gap length is a minimum in the polar, or direct axis and is a maximum in the inter-polar, quadrature axis. Because of the variation in magnetic circuit reluctance, the stator winding mmf will establish a larger air gap flux when the mmf is centred on the direct axis than when centred on the quadrature axis. This variation in air gap flux implies a corresponding variation in stator phase reactance. The field current is supplied through slip rings and brushes or by a brushes exciter as previously described. A squirrel-cage damper winding is often embedded in the rotor pole faces to dampen rotor oscillations following sudden load changes and to provide induction motor torque for starting the motor.

In general, the armature current, I_a, can be resolved into two components, one in phase with, and the other in phase quadrature with, the generated emf, E_f, as shown in the phasor diagram of Fig. 15.39. The q-axis component of armature current, I_q, is in phase with E_f and establishes an armature reaction flux that is centred on the q-axis, or interpolar space between the field poles. The d-axis component of armature current, I_d, is in phase-quadrature with the generated emf and establishes an armature reaction flux along the polar axis or d-axis. Because of the different air gap reluctances in the d and q axes, a different synchronous reactance must be assigned to each axis. Thus, this machine has d-axis and q-axis synchronous reactances, X_d and X_q. The fundamental flux due to the given armature mmf is smaller in the q-axis than in the d-axis, and hence X_q is less than X_d. Typically, X_q is about 0.6 X_d.

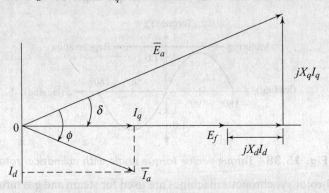

Fig. 15.39 *Phasor diagram for a salient-pole synchronous motor*

Now using Eq. (15.106), the excitation voltage becomes

$$\overline{E}_f = \overline{E}_a - jX_d \cdot \overline{I}_d - jX_q \cdot \overline{I}_q - R_a \overline{I}_a$$

For negligible armature resistance, the phasor diagram is shown in Fig. 15.39.
From the phasor diagram,

$$I_d = I_a \sin(\phi - \delta) \tag{15.122}$$

$$I_q = I_a \cos(\phi - \delta) \tag{15.123}$$

$$I_d X_d = E_a \cos \delta - E_f \tag{15.124}$$

$$I_q X_q = E_a \sin \delta \tag{15.125}$$

Substituting I_q from Eq. (15.123) in Eq. (15.125), we get

$$E_a \sin \delta = X_q I_a \cos(\phi - \delta)$$
$$= X_q I_a (\cos \delta \cos \phi + \sin \delta \sin \phi) \tag{15.126}$$

Dividing both sides by $\cos \delta$ and solving for δ gives

$$\delta = -\tan^{-1}\left[\frac{I_a X_q \cos \phi}{E_a - I_a X_q \sin \phi}\right] \tag{15.127}$$

where the negative sign signifies that E_f lags E_a.
Now, if the terminal voltage is resolved into a d-axis and a q-axis, then

$$E_{ad} = -E_a \sin \delta, \text{ and } E_{aq} = E_a \cos \delta$$

Control of A.C. Drives

∴ The input power becomes

$$P_i = -3(I_d E_{ad} + I_q E_{aq}) = 3I_d E_q \sin \delta - 3I_q E_a \cos \delta \quad (15.128)$$

Substituting the values of I_d from Eq. (15.124) and I_q from Eq. (15.125) in Eq. (15.128).

$$P_i = \frac{-3E_a E_f}{X_d} \sin \delta - \frac{3E_a^2}{2}\left(\frac{X_d - X_q}{X_d \cdot X_q}\right) \sin 2\delta \quad (15.129)$$

Neglecting stator losses, the air gap torque in newton-meters is again equal to the input power divided by the mechanical speed in radians/second. Thus,

$$T_d = \frac{-3E_a E_f}{X_d \omega_s} \sin \delta - \frac{3E_a^2}{2\omega_s}\left(\frac{X_d - X_q}{X_d \cdot X_q}\right) \sin 2\delta \quad (15.130)$$

As shown, the torque in Eq. (15.130) has two components. The first term is the same as that obtained for the round rotor machine. The second term is independent of field excitation and represents the reluctance torque due to the tendency for the salient poles to align themselves with the rotating air gap flux in the position of minimum reluctance. In a round rotor machine, $X_d = X_q = X_s$ and the reluctance torque term vanishes. The torque–load angle characteristics for a salient pole motor are shown in Fig. 15.40. The torque has a maximum value at $\delta = \pm \delta m$. For stability, the torque angle is limited in the range of $-\delta m \leq \delta \leq \delta m$ and in this stable range; the slope of $T_d - \delta$ characteristics is higher than that of cylindrical rotor motor.

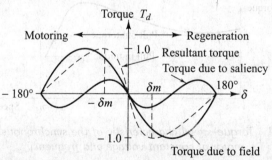

Fig. 15.40 *Torque versus torque angle with salient pole rotor*

Salient pole machines have been built in unit sizes of up to 500 MW for hydrogenerator utility applications.

15.8.3 Synchronous Reluctance Motors

The elementary synchronous reluctance motor has an unexcited ferromagnetic rotor with polar projections. Therefore, the reluctance motor differs from the conventional synchronous machine, which has either a d.c.-excited winding or permanent magnets on the rotor. In general, reluctance torque is developed by the tendency of a ferromagnetic material to align itself with a magnetic field.

If a synchronously rotating stator field is established by means of a conventional polyphase winding excited by a balanced polyphase a.c. supply, then the rotor runs in exact synchronism with this field as the salient poles seek to maintain the minimum reluctance position with respect to the stator flux. On a fixed frequency a.c. supply, the motor is not self-starting unless the rotor is fitted with a squirrel-cage winding to permit starting by induction motor action.

When the rotor speed approaches synchronous speed, the reluctance torque is superimposed on the induction motor torque, and as a result, the rotor speed oscillates above and below its average value. Provided the load torque and inertia are not excessive, the instantaneous rotor speed can increase to synchronous speed and the rotor locks into synchronism with the stator field. Figure 15.41 shows a typical torque–speed characteristics at a fixed supply frequency and voltage. At subsynchronous speeds, the torque is pulsating. Average values are shown in Fig. 15.41.

Once the rotor is synchronized, the cage winding rotates synchronously with the stator field. Thus, the rotor winding plays no part in the steady state synchronous operation of the motor. The machine continues to operate synchronously, provided the *pull-out torque* of the motor is not exceeded. This is the load torque required to pull the rotor out of synchronism.

The *pull-in torque* is defined as the maximum load torque which the rotor can pull into synchronism with a specified load inertia. The pull-in torque may be increased at the expense of a larger starting current, but it is always less than the pull-out torque.

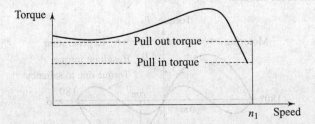

Fig. 15.41 *Torque–speed characteristic of the synchronous reluctance motor at constant voltage and frequency*

When the supply voltage and frequency are varied, a family of torque–speed curves is obtained as in the case of the induction motor. For constant-torque applications, the V/Hz ratio must again be held approximately constant but a voltage boost is necessary below 20 Hz. Reluctance motors have been widely used in adjustable-speed multimotor drives requiring exact speed coordination between individual motors. If all motors are accelerated simultaneously from standstill by increasing the supply frequency, the machines operate synchronously at all times, and they can be designed for optimum synchronous performance without regard to the pull-in torque requirements. Unfortunately, the reluctance motor also exhibits a tendency towards instability at lower supply frequencies, but it forms a low cost, robust, and reliable synchronous machine.

Equation (15.130) can be used to determine the reluctance torque with $E_f = 0$.

$$\therefore \quad T_d = \frac{-3E_a^2}{2\omega_s}\left(\frac{X_d - X_q}{X_d \cdot X_q}\right)\sin 2\delta \qquad (15.131)$$

where
$$\delta = -\tan^{-1}\frac{I_a X_q \cos\phi}{E_a - I_a X_q \sin\phi} \qquad (15.132)$$

The pull-out torque for $\delta = -45°$ is

$$T_p = \frac{3E_a^2}{2\omega_s}\left(\frac{X_d - X_q}{X_d \cdot X_q}\right) \qquad (15.133)$$

15.8.4 Permanent Magnet Motors

In a permanent magnet motor operating on a fixed frequency a.c. supply, the constant rotor flux produced by the permanent magnets generates a constant value of excitation emf, E_f. The actual value of excitation emf depends on the magnet material, its physical dimensions, the rotor design, and the air gap length. The machine designer must ensure that the magnets are not demagnetized by the application of normal or overload currents.

The excitation voltage cannot be varied. The equations for the salient pole motors may be applied to the permanent magnet motors if the excitation voltage, E_f, is assumed constant. This emf has a value such that the motor operates near unity power factor at rated load. For the same frame size, permanent magnet motors have higher pull-out torque.

The elimination of field coil, d.c. supply, and slip rings reduce the motor loss and the complexity. These motors are also known as brushless motors and finding increasing applications in robots and machine tools. A permanent magnet motor can be fed by either rectangular current or sinusoidal current. The rectangular current-fed motors, which have concentrated windings on the stator inducing a square or trapezoidal voltage, are normally used in low power drives. The sinusoidal current fed motors, which have distributed windings on the stator, provide smoother torque and are normally used in high power drives.

15.8.5 Switched Reluctance Motors

The switched reluctance motor (SRM) is a member of the class of variable reluctance machines (VRM), but somewhat different from the commonly known stepping motors. This type of motor has a doubly salient construction with projecting poles on both stator and rotor. The machine rotor construction is the same as that of a synchronous reluctance motor, i.e. it does not have permanent magnets or any winding, but the stator poles have a concentrated winding (instead of the sinusoidal winding in a synchronous reluctance motor).

In the doubly salient variable reluctance motor, various combinations of stator and rotor pole numbers are possible. A four-phase machine with eight stator poles and six rotor poles is most common, but a three-phase 6/4 pole machine is also used. Figure 15.42 shows a cross-sectional view of an SRM with 8 stator poles and 6 rotor poles. Windings on diametrically opposite stator poles are connected in series (or parallel) so that one stator pole acts as a north pole and the other acts as a south pole. Thus, there are four phases, A, B, C and D, giving a four-phase 8/6 pole motor.

When phase A of the stator is energized with direct current, a single excited magnetic system is formed. In such a system, torque is developed by the tendency for the magnetic circuit to adopt the configuration of minimum reluctance. This reluctance torque, which is independent of current direction, causes the rotor to turn until a pair of its poles (1 and 4) are exactly aligned with the stator poles of phase A, as shown in Fig. 15.42. The rotor is stable in this position and cannot move until phase A is de-energized. If phase B is then energized, the rotor turns clockwise through the step-angle of 15 degrees until rotor poles 3 and 6 are aligned with phase B. Energization of phases C and D in sequence causes further clockwise rotation through two more 15 degree steps. After the sequence $ABCDA$, the rotor has turned through four steps, or 60 degrees, and hence six cycles of this sequence are required for one revolution of the rotor.

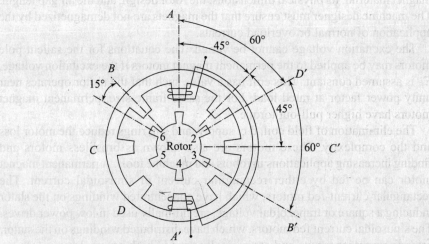

Fig. 15.42 *A single-stack, four-phase, 8/6 pole SRM. Only phase A winding is shown for clarity*

The SRM has a number of inherent advantages that have sparked interest in its use as an adjustable speed drive. These features are:
(1) The machine has a simple construction, making it a potentially cheaper alternative.
(2) There is independent and uncoupled operation of motor phases so that the machine will continue to operate (with higher pulsating torque) if a phase fails.
(3) There is no possibility of converter shoot-through fault because the winding is always in series with the devices. Also, no-short circuit fault current is possible in the stator due to rotor induced emf.

(4) Its robust rotor construction makes it more reliable and suitable for high-speed operation.
(5) Efficient motor cooling because all windings are on the stator.
(6) Low rotor inertia and high torque inertia ratio.

The principle demerits of the drive are that the machine pulsating torque is high, giving high acoustic noise, and the machine is somewhat bulky (comparable to the synchronous reluctance machine).

1. Power Converter Circuits As explained above, each stator phase of the switched-reluctance motor must be energized with a unidirectional current pulse while the rotor is appropriately positioned relative to the stator. A rapid response is necessary for positive and negative changes in the demanded phase current level. This function requires a two-quadrant power converter that is capable of applying equal positive and negative phase voltages to produce approximately equal rates of current increase and decrease. The required converter characteristics is provided by each of the circuit configurations of Fig. 15.43. Because the inverter phases are completely independent of each other, Fig. 15.43 shows the circuitry required for just one phase of the SRM. Thus, a four-phase machine will require four of these circuits.

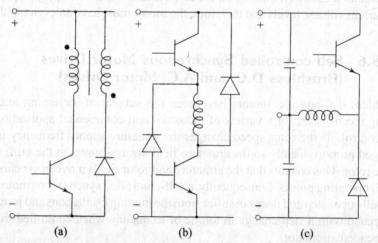

Fig. 15.43 *Power converter configurations for one-phase of an SRM*

The d.c. supply is usually obtained from a diode rectifier bridge with an LC filter, power transistors are shown in Fig. 15.43, but GTOs and MOSFETs can also be used. Each circuit topology shown in Fig. 15.43 has the advantage that there is always a motor winding in series with a main power device. Consequently, there is no shoot-through fault current path across the d.c. bus, giving added ruggedness and reliability to the power-electronic converter.

In Fig. 15.43(a), the phase current is controlled by a single power transistor in conjunction with a closed coupled bifilar motor winding and feedback diode. When the transistor is switched-off, freewheeling current continues to flow through the secondary winding and the feedback diode, and the inductive energy is returned to the d.c. supply as the current collapses. At the instant of turn-off, the transistor C–E voltage rises to twice the d.c. link voltage, assuming a turns ratio of unity and perfect coupling between the bifilar windings. In practice, there is always some uncoupled leakage inductance giving rise to higher voltage transients and requiring large snubber circuits to protect the devices. The bifilar winding also results in a poor utilization of copper in the motor and doubles the number of motor connections as compared with alternative circuits.

The circuit of Fig. 15.43(b) dispenses with the bifilar winding but requires two power transistors and two feedback diodes in each phase. The upper and lower power transistors are turned-on and off simultaneously. At turn-off, the two diodes conduct, thereby reversing the voltage across the motor phase and allowing the return of inductive energy to the d.c. supply.

In Fig. 15.43(c), only one power transistor and one diode are used per phase but centre-tapped d.c. supply is required. Note that the d.c. link voltage must be doubled in this connection, as compared with the other circuits, in order to apply the same voltage across the motor phase winding. The d.c. centre-tap can be created by a split capacitor circuit, as shown in Fig. 15.43(c). It is important to maintain a balanced loading on the two valves of the d.c. supply to preserve the current voltage levels and therefore, the motor must have an even number of phases.

15.8.6 Self-controlled Synchronous Motor Drives (Brushless D.C. and A.C. Motor Drives)

Brushless d.c. and a.c. motors have been the subject of increasing attention during recent years for a variety of industrial and commercial applications. In self-control, as the rotor speed changes the armature supply frequency is also changed proportionately so the armature field always moves at the same speed as the rotor. This ensures that the armature and rotor fields move in synchronism for all operating points. Consequently, a self-controlled synchronous motor does not pull out of step and does not suffer from the hunting oscillations and instability associated with a step change in torque or frequency when controlled from an independent oscillator.

The term "brushless d.c. motor" is used to identify the combination of a.c. machine, solid-state inverter, and rotor position sensor that results in a drive system having a linear torque–speed characteristic, as in a conventional d.c. machine. The a.c. motor has a polyphase winding on the stator and permanent magnets on the rotor. Motor operation is made self-synchronous by the addition of a rotor position sensor that controls the firing signals for the solid-state inverter. In response to these firing signals, the inverter directs current through the stator phase windings in a controlled sequence.

If the stator is fitted with a conventional three-phase winding, the motor has the construction of a standard permanent magnet synchronous machine and operates as a self-controlled synchronous motor, or inverted d.c. motor with an electronic commutator. In self-synchronous machine, the air gap flux distribution and counter emf, or back emf, waveform are approximately trapezoidal, as in a conventional d.c. machine. In the brushless d.c. motor, the torque function is trapezoidal, whereas in the permanent magnet synchronous motor, the torque function is sinusoidal.

In a self-controlled synchronous motor drive, the motor is only a synchronous machine as regards constructions because its mode of operation is like that of a d.c. motor with an electronic commutator. In an electronic commutator, the number of 'segments' is reduced causing the output torque to pulsate above and below its mean value. When a mechanical commutator is replaced by its electronic counterpart, the number of switching operations is reduced in order to limit the number of semiconductors and improve their utilization. The commutator in a conventional d.c. motor is placed on the rotor to avoid the need for rotating brushes but the construction is inverted when an electronic commutator is employed.

1. Brushless d.c. Motor : The block-diagram of a self-controlled synchronous motor fed from a three-phase inverter is shown in Fig. 15.44. As shown, when an inverter is employed, the drive is controlled from a d.c. source. The inverter may be a current source inverter, current controlled PWM inverter or a voltage source inverter. Depending on the type of inverter, the d.c. source may be a controllable current source or a constant or a controllable voltage source. Operation of the drive of Fig. 15.44 is similar to that of a d.c. motor. The inverter's frequency is changed in proportion to the speed so that the armature and rotor mmf waves revolve at the same speed, thus producing a steady torque for all speeds, as in a d.c. motor. The rotor position sensor and inverter now perform the same function as the brushes and commutator in a d.c. motor.

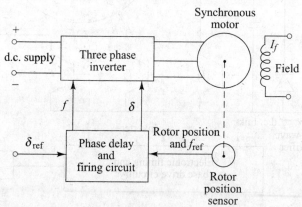

Fig. 15.44 *Self-controlled synchronous motor (brushless d.c. motor)*

Due to the similarity in operation with a d.c. motor, the inverter fed self-controlled synchronous motor drive of Fig. 15.44 is known as a *commutatorless d.c. motor*. If the synchronous motor is a permanent magnet motor or a reluctance motor or a wound field motor with *a* brushless excitation, it is known as a *brushless* and *commutatorless d.c. motor* or simply a *brushless d.c. motor*. By connecting the field in series with the d.c. supply gives the characteristics of a d.c. series motor. The brushless d.c. motors offer the characteristics of d.c. motors and do not have limitations such as frequent maintenance and inability to operate in explosive environments. They are finding increasing applications in servo drives.

(1) The Three-phase Half-wave Brushless d.c. Motor: Figure 15.45 shows a basic three-phase half-wave brushless d.c. motor with its electronic controller. The three stator phases are star-connected with the neutral point joined to the positive terminal of the d.c. supply. Transistors T_1, T_2 and T_3 deliver unidirectional phase currents in response to base-drive signals that are under the control of rotor position sensor. The simple half-wave circuit is unusual in that there are no freewheeling or feedback diodes to provide an alternatives path for the inductive winding current when a transistor is turned-off. Nevertheless, the circuit is satisfactory for small, low-cost systems in which the inductive winding energy is so small that it cannot cause destructive breakdown of the transistor at turn-off.

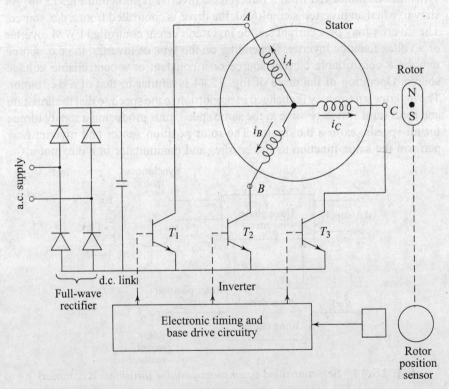

Fig. 15.45 *A three-phase, half-wave brushless d.c. motor drive*

Figures 15.46(a), (b) and (c) show the idealized static torque/angle characteristics for the individual phases. This trapezoidal torque function has a 120-degree flat topped region that can be utilized to produce useful shaft torque.

Thus, if the motor phases are energized sequentially with a constant current during each 120-degree interval of constant positive torque, the motor develops a steady positive torque that is independent of shaft position. The corresponding transistor, or phase currents are shown in Figs. 15.46(d), (e) and (f). Each motor phase has a conduction angle of 120 electrical degrees, and the current amplitude is I, giving a resultant motor torque of

$$T = K_T \cdot I \tag{15.134}$$

where K_T, the torque constant, is the torque per ampere over the flat portion of the trapezoid.

From the static-torque angle characteristics, it is clear that a torque reversal is achieved when the transistor conduction interval is delayed by 180°. The current direction is unchanged but each phase now conducts during the 120° interval of negative torque, as shown in Figs. 15.46(g), (h) and (i). Thus, torque-reversal is obtained in a brushless d.c. motor by shifting the firing signals by 180°, whereas in a conventional brush type d.c. motor, a reversal of the supply voltage polarity and armature current direction is necessary.

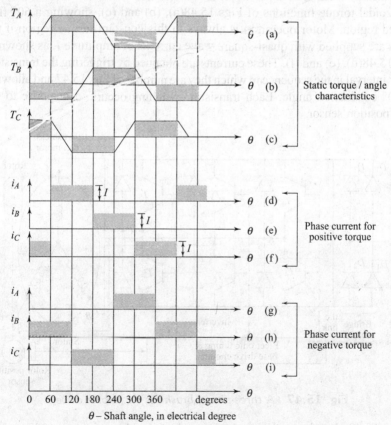

Fig. 15.46 *Idealized waveforms for the three-phase, half-wave brushless d.c. motor*

This circuit configuration is suitable for small brushless d.c. motors in a wide variety of applications at power levels from a few watts to 100 W. Typical applications for these small brushless d.c. motors include turntable drives in record players, spindle drives in hard-disk drives for computers, and various applications in low-coast instruments and computer peripheral equipment.

When the drive power level is increased, the stored energy in the motor winding inductance is more significant, and it must be returned to the d.c. supply to prevent destructive breakdown of the transistor at turn-off. This energy feedback can be accomplished in a half-wave system using a centre-tapped d.c. supply and feedback diodes. However, a full-wave system permits the use of a single d.c. supply and gives improved winding utilization.

(2) The Three-phase Full-wave Brushless d.c. Motor : A three-phase full-wave brushless motor system with feedback diodes for energy recovery is shown in Figs. 15.47. The star-connected stator winding has no neutral connection, and two of the three-phase are active at all times. For line-to-line d.c. currents from A to B, B to C, and C to A, the motor has the idealized trapezoidal torque functions of Figs 15.48(a), (b) and (c), showing a 60° flat-topped region. Motor operation is always in this constant-torque region if the phase are supplied with quasi-square wave currents of amplitude I, as shown in Figs 15.48(d), (e) and (f). These currents are obtained by triggering the transistors at 60° interval in the sequence in which they are numbered of Fig. 15.47 and allowing a 120° conduction angle. Each transistor switching occurs in response to the rotor position sensor.

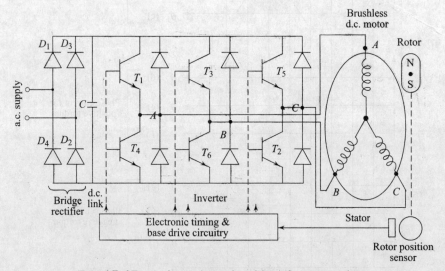

Fig. 15.47 *A three-phase, brushless d.c. motor drive*

Figure 15.48 also shows that each motor phase conducts for two 120° periods in each cycle, giving twice the winding utilization of the previous three-phase half-wave system. A steady nonpulsating torque is developed of magnitude $T =$

$K_T I$, and as before, a torque reversal is achieved by phase-shifting the transistor base-drive signals by 180°. The idealized quasi-square wave currents of Fig. 15.48 imply instantaneous switching of current from one phase combination to the next. In a practical voltage fed system, the inductive load will delay the build-up of current and will also prolong conduction after the theoretical turn-off instant. In an actual motor, the torque function will also depart somewhat from the ideal trapezoidal waveshape. Despite these practical imperfections, the commercial brushless d.c. motor can achieve a very low torque ripple and is eminently suitable for use in a high-performance servo drive.

2. Brushless a.c. Motor : The block diagram of a self-controlled synchronous motor fed from a three-phase cycloconverter is shown in Fig. 15.49. As shown, when a cycloconverter is employed, the drive is fed directly from the a.c. main. The cycloconverter can be controlled to produce either a voltage source or a current source. The drive of Fig. 15.49 is also known as *brushless* and *commutatorless a.c. motor* or simply a *brushless a.c. motor*. These a.c. motors are used for high-power applications (up to megawatts range), such as compressors, blowers, fans, conveyors, steel rolling mills and cement plants.

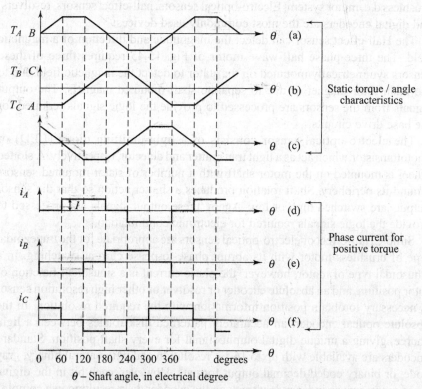

Fig. 15.48 *Idealized waveforms for three-phase full-wave brushless d.c. motor*

The self-control is also used for starting large synchronous motor in gas turbine and pump storage power plants.

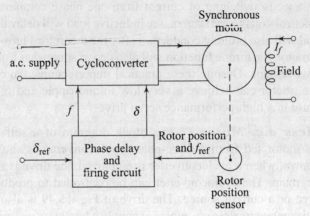

Fig. 15.49 *Brushless a.c. motor*

3. Position Sensors : The rotor position sensor is an integral part of the brushless d.c. motor system. Electro-optical sensors, hall-effect sensors, resolvers, and digital encoders are the most commonly used devices.

The Hall-effect sensor can detect the magnitude and direction of a magnetic field. The three-phase half-wave motor of Fig. 15.45 requires three of these sensors symmetrically mounted on the stator to detect the magnetic field due to the main rotor magnet or due to separate shaft mounted magnets. The output signals from the sensors are processed to provide the logic signals required for the base drive circuits.

The electro-optical sensor consists of a light-emitting diode (LED) of phototransistor,which act as a light transmitter and detector, respectively. A slotted wheel is mounted on the motor shaft with a number of stator mounted sensors around its periphery. Shaft rotation produces a shutter action so that the sensor outputs are switched high or low. Again, these output signals are processed to provide the logic signals required for electronic commutation.

Simple Hall-effect or electro-optical sensors are appropriate for the trapezoidal type of brushless motor with its abrupt phase-to-phase current switching. In a sinusoidal type of motor, however, the phase current is a sinusoidal function of rotor position, and an absolute encoder or resolver or other high resolution sensor is necessary to obtain position information with the required resolution. In the absolute optical encoder, an accurately patterned disk rotates between a light source, giving a unique digital output signal for every shaft position. Standard encoders are available with up to 16-bit resolution and with natural binary, gray code, or binary coded decimal output formats. However, each bit in the digital word represents an independent track on the encoder disk, resulting in a complex and costly sensor.

Brushless resolver operation is based on inductive coupling between stator and rotor windings. The resolver with its resolver-to-digital (R/D) converter also gives precise absolute digital position information, but again, the cost is often prohibitive.

In a sinusoidal type of brushless motor, absolute rotor position information is required to at least a 9-or-10 bit resolution. Sinusoidal reference current waveforms are generated with this precise position information, and the actual phase currents track the reference currents in a current controlled PWM inverter.

The provision of a high-resolution sensor adds significantly to the cost of the sinusoidal system. On the other hand, the trapezoidal motor has a rugged and inexpensive sensor and simple control logic. It also has excellent performance characteristics and is the preferred brushless motor in a wide variety of applications.

SOLVED EXAMPLE

Example 15.3 A three-phase, 400 V, 50 Hz, 6-pole, star connected round rotor synchronous motor has a synchronous reactance of $X_s = 2.0\ \Omega$ per phase and the armature resistance is negligible. The load torque which is proportional to the speed squared, is $T_L = 300$ N-m at 100 rpm. The power factor is maintained at unity by field control and the voltage-to-frequency ratio is kept constant at the rated value. If the inverter frequency is 40 Hz and the motor speed is 560 rpm, determine
(a) the input voltage E_a,   (b) the armature current I_a,
(c) the excitation voltage E_f, (d) the torque angle δ, and
(e) the pull-out torque T_p.

Solution: $P_F = \cos\phi = 1.0,\ \phi = 0$.

$\therefore\qquad E_{a(\text{rated})} = E_b = E_s = 400/\sqrt{3} = 230.94$ V

Poles, $P = 6$, $\omega = 2\pi \times 50 = 314.16$ rad/s.

Base speed, $\omega_b,\ \omega_s = \dfrac{2 \times 314.16}{6} = 104.72$ rad/s $= 993.13$ rpm.

Now, ratio $K = E_b/\omega_b = \dfrac{230.94}{104.72} = 2.21$

At 560 rpm, $T_L = 300 \times \left(\dfrac{560}{1000}\right)^2 = 94.08$ N-m.

$\omega_s = \omega_m = 560 \times \pi/30 = 58.64$ rad/s.

$P_0 = 94.08 \times 58.64 = 5516.85$ W.

(a) $E_a = K\omega_s = 2.21 \times 58.64 = 129.59$ V

(b) $P_0 = 3\,E_a\,I_a\,P_F = 5516.85$ $\therefore\ I_a = \dfrac{5516.85}{(3 \times 129.59)} = 14.19$ A

(c) From Eq. (15.108),

$\overline{E}_f = 129.59 - 14.19 \times (1 + j0)(j2.0) = 132.66\ \angle -12.35°$

(d) The torque angle, $\delta = -12.35°$
(e) From Eq. (15.121),

$$T_p = \frac{3 \times 129.59 \times 132.66}{2 \times 58.64} = 439.75 \text{ N-m}$$

15.9 THE DRIVE SELECTION

Modern technology offers various alternatives in the selection of a drive system. Depending on the process requirements, environmental conditions and financial objectives of a company, choice of a drive can be made. The drive systems can be divided into two groups, d.c. drive systems and a.c. drive systems.

1. D.C. Drive Systems The invention of power semiconductors saw the advent of d.c. drive systems for most of the early variable speed requirements. Based on the advantages of simple construction and ease of control, this technology continued to be taken up for further improvements all over the world till very recently. The d.c. drives offer the following:

Advantages
- No start-up problems with high torque.
- Equipment required for the converter in the power section and for the electronic control circuits reduced to a minimum.
- Large speed setting range without additional equipment.
- Multimotor operation with load sharing or precision coordinated speed and tension control.
- Fast control response and rotational accuracy even at very low speeds.
- Speed stability high.
- Insensitivity to system voltage drops.
- Precise replication of load torque for process control.
- Converter losses reduced to a minimum.

Disadvantages
- Generally not suitable for harsh explosive, or corrosive environments.
- Controls are relatively complex, particularly for high performance drives.
- Motor requires maintenance.
- May create line harmonics.

Applications
- Log carriages.
- Veneer lathes.
- Machine tools.
- Mine hoists.
- Paper making and finishing machinery.
- Metals industry primary reduction mills, cold mills, and processing lines.
- Shovels and draglines.
- Core-driven reels with constant tension over the range of roll diameters, both wind-ups and up-winds.
- Plastic processing lines.
- Textile range drives.

Control of A.C. Drives

2. A.C. Drive System The motor selection for a particular application is important as it would lead to overheating of the motor if it is underrated and it would put extra stress on the converter if the motor selected is overrated. a.c. drive systems use the a.c. motor as the drive element of either induction or synchronous type.

The four important considerations that enter into motor and inverter selection are:
   (i)   actual motor capabilities of ampere versus torque
   (ii) thermal (low speed) limits
   (iii) high speed limitations, and
   (iv) starting and peak running torque.

(a) A.C. Drive with Current Source d.c. Link Converter The main advantageous features of a.c. drives with a current source d.c.-link converter are as follows:
- For large ratings (> 400 kW), it is easier to handle the current in the d.c. link.
- Robust, favourably priced squirrel-cage motors can be used.
- Wear and maintenance reduces to a minimum.
- High degrees of motor protection can be easily obtained.
- Mainly suitable for individual drive duty.
- Start-up at full-rated torque at only rated current.
- Increased starting torque where the converter is rated accordingly.
- Speed setting range up to 1:10.
- Stable load angle control without tachogenerator.
- Speed stability ± 100%.
- About the same overall efficiency as with d.c. drives.
- System power factor, depending on firing angle setting, is approximately 0.85 at rated speed.

Disadvantages
- Converter costs increased compared to single-quadrant d.c. drives.
- Control structure more complex compared to d.c. drives.

(b) A.C. Drives with PWM Inverters a.c. drives with PWM inverters differ from drives with current-source d.c. link converters in the following main features:
- Multi-motor operation from the same converter.
- Speed setting range down to zero speed.
- Full motor torque even at standstill.
- Converter and motor need not be matched to each other.
- System failures can be bridged by standby supply to d.c. link from capacitors or storage batteries.
- Efficiency is slightly improved by special design.
- System power factor approximately 1 over the entire speed setting range.
- Most suitable for pump/fan ratings up to about 400 kW.

Disadvantages
- Power and control section structure more complex.
- Operation above 150 Hz is difficult.

Applications
- Fans, pumps, compressors
- Machine tools
- Conveyors
- Centrifuges
- Retrofits for simple d.c. drives
- Retrofits for mechanical speed changes
- Retrofits for eddy current coupling
- Retrofits for wound rotor controllers that do not require high starting torque or controlled slow speed operation
- Paper machine etc.

(c) A.C. Drives with Cycloconverters

The main features of a.c. drives with cycloconverters are:
- There is only one conversion stage.
- Capable of transferring power in either direction without addition of any other circuitry.
- Output voltage and frequency can be controlled independent of each other.
- Popular for drive ratings in MW range which require low speeds.

Disadvantages
- Requires large number of power devices and hence, power and control circuits are complicated.
- Harmonics are high at frequencies near supply frequency and hence, waveform is distorted.
- Input power factor is poor at low output voltage.
- Maximum output frequency can be only 50% of the mains frequency.

Applications
- Rolling-mill drives
- Conveyor drives
- Tube mills
- Diesel electric locomotives
- Gearless cement mills etc.

(d) Microprocessor Controlled a.c. Drives

Advantages
- Better solutions to process engineering problems
- More adaptable dynamic characteristics
- High operational reliability
- High product consistency
- Low maintenance and low wear and tear
- Smaller dimensions
- Fast communication with higher order automation systems
- Quick-fault diagnosis
- Accurate monitoring of protective functions such as phase loss, overcurrent, and high and low voltages without the use of discrete relays.

(e) Subsynchronous Converter Cascade (SEC) Drive
The main features are:
- Competitive alternative to the d.c. drive in the power range above 50 kW.
- For small speed setting ranges, it is in most cases available at a lower price.
- The induction motor can be operated as a conventional drive in the event of converter failure.
- Starting and running up to the lowest speed of the control range is normally performed by means of a starter, hence, high starting torque.
- Motor with brushes but without a commutator.
- Speed stability ±1 %.

Disadvantages
- System power factor around 0.7 depending on the speed setting range and the type of converter used
- Torque, active power and reactive power pulsations.

Hence, depending upon the application requirement, the choice of a drive can be made.

Review Questions

15.1 Why is it essential to operate an induction motor between the synchronous speed and the breakdown speed when it is used to holding an active load by the regenerative braking?

15.2 Explain the basic principle of operation of an induction motor with reference to its equivalent circuit.

15.3 Explain in detail the induction motor performance characteristics.

15.4 Draw the torque speed characteristics of the polyphase induction motor. Also, explain the following operating regions:
   (i) Motoring region.   (ii) Generating region.
   (iii) Braking region

15.5 State and discuss the various methods of speed control of induction motor.

15.6 Discuss briefly the stator voltage control scheme of induction motor. Also, draw and explain the speed torque curves.

15.7 Draw and explain the operation of four-quadrant a.c. voltage controller.

15.8 Discuss the variable-frequency control method of an induction motor. Also, explain operation for two different modes,
   (a) operation below the rated frequency
   (b) operation above the rated frequency.

15.9 Explain the induction motor operation when the V/f ratio is held constant. Also, derive the expression for the maximum torque.

15.10 Draw and explain the speed–torque curves with variable frequency control for two different modes:
   (i) operation at constant flux
   (ii) operation at constant (V/f) ratio.

15.11 Explain the advantages of variable frequency induction motor drives.

15.12 Sate and explain the various schemes for induction motor speed control by voltage source inverters.

15.13 Explain the operation of induction motor by current source inverters.

15.14 Derive the expression for the torque and, when the induction motor is controlled by current-source inverters at a fixed-frequency.

15.15 Derive an expression for the optimum slip which maximizes the efficiency of an induction motor fed by a current source of variable magnitude and fixed frequency.

15.16 Explain the operation of induction motor for two different cases when fed by current source inverters,
(a) operation at and below rated frequency
(b) operation above rated frequency.

15.17 Explain the method of controlling induction motors using cycloconverters.

15.18 Draw the circuit diagram and explain the operation of rotor-resistance control using chopper.

15.19 Derive an expression for the torque T for the rotor resistance control using the chopper scheme.

15.20 Discuss the speed–torque curves for a chopper-controlled induction motor.

15.21 Discuss briefly the traditional Scherbius system.

15.22 Draw the circuit diagram and explain the working of slip power recovery system using solid-state Scherbius system.

15.23 Suggest the certain modifications for improving the power factor of the slip energy recovery scheme. Draw a suitable diagram and explain the working of slip-power recovery system using commutatorless Kramer drive.

15.24 List the various types of synchronous motors.

15.25 Explain briefly the operation of round-rotor synchronous motor. Also, derive the expression for field excitation.

15.26 What is the torque angle of synchronous motors. Also, derive the expression for δ.

15.27 What are the differences between salient pole motors and reluctance motors.

15.28 Derive an expression for torque angle and developed torque in case of salient-pole synchronous motors.

15.29 Explain the following terms with respect to synchronous motors:
(i) Pull-out torque.   (ii) Starting torque
(iii) E curves.

15.30 Draw and explain the torque versus torque-angle characteristic curves with salient pole rotor.

15.31 Draw and explain briefly the torque speed characteristic of the synchronous reluctance motor at constant voltage and frequency.

15.32 What are the differences between salient pole motor and permanent magnet motors?

15.33 Explain the operation of SRMs. Also, list the advantages of SRM that have sparked interest in its use as an adjustable-speed drive.

15.34 Draw and explain the various power converter configurations for one-phase of an SRM.

15.35 What is a self-control mode of synchronous motors?

15.36 Draw and explain the block diagram of a self-controlled synchronous motor fed from a three-phase inverter.

15.37 Explain with a neat diagram the operation of a three-phase, half-wave brushless d.c. motor drive. Also, draw and explain the associated waveforms.

15.38 Draw and explain the operation of a three-phase brushless d.c. motor drive. Also, explain the related waveforms.

15.39 What is a brushless a.c. motor?

15.40 Discuss briefly about the position sensors.

15.41 Draw and explain the servo-control and a permanent magnet synchronous motor with a PWM inverter.

15.42 Explain with a neat block diagram the operation of a high-performance brushless d.c. motor servo drive.

15.43 Explain in detail how choice of a drive can be made.

PROBLEMS

15.1 For a three-phase induction motor, 350 bp, 1100 rev/min, 2300 V, design B, with the following performance parameters:

$$R_1 = 0.15\,\Omega, R_2 = 0.95\,\Omega,$$
$$X_1 = X_2 = 1.63\,\Omega, X_m = 81.4\,\Omega$$
$$P_{fw} + P_{core} = 8300\,W$$

Assume $P_{stray} = 1$ per cent of p_{gross}.
Determine the following for a rotor slip of 4 per cent :
 (a) the stator current and input power,
 (b) the shaft power output and torque,
 (c) the efficiency.   [Ans (a) 57.28 A, 209.2 kW (b) 189.1 kW, 1568 N-m. (c) 90.4 %]

15.2 For the motor in Problem 15.1, determine the following for the starting condition:
 (a) stator current
 (b) input power, and
 (c) starting developed torque (T_d).
      [Ans (a) 389.6 A (b) 484 kW (c) 3310 N-m]

15.3 For Problem 15.1, determine the slip when the machine is developing maximum torque and also the magnitude of the maximum torque
      [Ans $S_{max}T = 0.291$ per unit slip, $T_{max} = 6165$ N-m.]

15.4 A three-phase, 1500 rpm, induction motor is developing torque of 2000 syn. watts at a input frequency of 50 Hz. If the motor torque is now reduced to 1500 syn. watts, determine new value of stator frequency. The motor is operating in constant HP region. Assume constant rotor frequency and neglect effect of rotor resistance.   [Ans $f_2 = 57.74$ Hz.]

15.5 A three-phase, star-connected, 60 Hz, 4-pole induction motor has the following constants in ohms per phase referred to the stator:

$$R_1 = R_2 = 0.024$$
and $$X_1 = X_2 = 0.18$$

The motor is controlled by the variable frequency control with a constant (V/f). Determine the following for an operating frequency of 0.012 kHz:
 (a) The breakdown torque as a ratio of its value at rated frequency for motoring and braking.
 (b) The starting torque and rotor current in terms of their values at the rated frequency.   [Ans (a) 0.68, 1.46. (b) 2.6, 0.72.]

15.6 For the motor, Example (15.5), if the rated slip is 4%, then determine the motor speed for rated torque and $f = 30$ Hz. The motor is controlled with a constant (V/f) ratio. [Ans 820 rpm.]

15.7 A three-phase, 400 V, 50 Hz, 6-pole, 960 rpm, star-connected induction motor has the following parameters per phase referred to the stator:

$$R_1 = 0.4\,\Omega, R_2 = 0.20, x_1 = x_2 = 1.5\,\Omega, x_m = 30\,\Omega.$$

The motor is controlled by variable frequency control at a constant flux of rated value.

Determine the following:

(a) The motor speed and the stator current at half the rated torque and 25 Hz.

(b) By assuming the speed–torque curves to be straight lines, solve for part (a), for $s < s_m$.

(c) The frequency, stator current, and voltage at the rated braking torque and 800 rpm. [Ans (a) 481.3 rpm, 19.85 A (b) 480 rpm, 20.6 A.
(c) 38.Hz, 38.52 ∠– 154°, 156 ∠– 173° V.]

15.8 The motor in Problem 15.7 is fed by a variable frequency current source. The motor is made to operate at the rated flux, at all operating points. Compute the following:

(a) Slip speed for $I_1 = 60$ A

(b) The frequency and stator current for operation at 500 rpm for the following torque values:

   (i) 139 N-m.
   (ii) –188 N-m.

(c) Also obtain the solution of b(i) assuming speed–torque curves to be straight lines in the region of interest.

[Ans (a) 67 rpm (b) (i) 26.42 Hz, 29.7 A
(ii) 23 Hz, 38.5 A (c) 26.48 Hz, 29.7 A.]

15.9 A three-phase, 460 V, 60 Hz, six-pole, 1180 rpm, star-connected, squirrel-cage induction motor has the following parameters per phase referred to the stator:

$$R_1 = 0.19\,\Omega, R_2 = 0.07\,\Omega,$$
$$X_1 = 0.75\,\Omega, X_2 = 0.67\,\Omega \text{ and } X_m = 20\,\Omega.$$

The motor is controlled by a six-step inverter. The d.c. input to the inverter is fed by a six-pulse, fully controlled rectifier. (a) What should be the rectifier firing angle for getting the rated fundamental voltage across the motor if the rectifier is fed by an a.c. source of 460 V, 50 Hz.

(b) If the machine is operated at a constant flux then determine:

   (i) the inverter frequency at 600 rpm and rated torque.
   (ii) the inverter frequency at 500 rpm and half the rated torque. Also, determine the motor current. [Ans (a) 18.25° (b) (i) 31 Hz,
(ii) 25.5 Hz, 332.7/2 N-m, 34.32 A]

15.10 A three-phase, 460 V, 60 Hz, six-pole, 1176 rpm, star-connected, squirrel-cage induction motor has the following parameters per phase referred to the stator.

$$R_1 = 0.29\,\Omega, R_2 = 0.145\,\Omega, X_1 = 0.21\Omega, X_2 = 0.5\,\Omega \text{ and } X_m = 15.3\,\Omega.$$

The current source inverter controls the motor. At the rated value, flux is maintained constant. Determine the following.

(a) The stator current and d.c. link current when the machine operates at rated torque and 60 Hz.
(b) The inverter frequency and d.c. link current for a speed of 600 rpm and rated torque.
(c) The motor speed, stator current, and d.c. link current for half for the rated torque and inverter frequency of 30 Hz.

[*Ans* (a) 42.4A, 51.94 A., 206.6 N-m.
(b) 31.2 Hz, 51.94 A. (c) 588 rpm, 27.33 A, 33.47 A.]

15.11 A three-phase, 460V, 60 Hz, 1164 rpm, six-pole, star-connected, wound-round induction motor has the following parameters per phase referred to the stator:

$R_1 = 0.4\ \Omega, R_2 = 0.6\ \Omega, X_1 = X_2 = 1.8\ \Omega, X_m = 40\ \Omega$, stator-to rotor turns ratio is 2.5.

The motor speed is controlled by static rotor resistance control. The filter resistance is 0.02 Ω. The value of external resistance is chosen such that $\alpha = 0$, and the breakdown torque is obtained at standstill.

Determine the following:
(a) the value of the external resistance
(b) α for a speed of 960 rpm at 1.5 times the rated torques.
(c) the speed for $\alpha = 0.6$ and 1.5 times the rated torque. Neglect the friction and windage loss.   [*Ans*: (a) 0.96 Ω. (b) 0.42 (c) 1020 rpm.]

15.12 A three-phase, 460 V, 60 Hz, 6-pole, star-connected cylindrical rotor synchronous motor has synchronous reactance of $X_s = 2.5\ \Omega$ per phase and the armature resistance is negligible. The load torque, which is proportional to the speed squared, is $I_L = 398$ N-m at 1200 rpm. The power factor is maintained at unity by field control and the voltage-to-frequency ratio is kept constant at the rated value. If the inverter frequency is 36 Hz, and the motor speed is 720 rpm, determine
(a) the input voltage, E_a (b) the armature current, I_a (c) the excitation voltage, E_f (d) the torque-angle, δ, and (e) the pull-out torque, T_p.

[*Ans* (a) 159.34 V (b) 22.6 A (c) 169.1 $\angle$–19.52°
(d) –19.52° (e) 428.82 N–m.]

REFERENCES

1. S B Dewan, G B Slemon, and A Straughen, *Power-semiconductor drives*, John Wiley, New York, 1984.
2. A E Fitzgerald, C Kingsley, and A Kushko, *Electric machinery*, McGraw-Hill, 1971.
3. W Sheperd, *Thyristor control of a.c. circuits*, Crosby Lockwood staples, London, 1975
4. B K Bose, *Power electronics and a.c. Drives*. Prentice Hall, 1986.
5. B K Bose *Adjustable a.c. drives*, IEEE Press, New York; 1980.
6. G K Dubey, *Power semiconductor controlled drives*, Prentice Hall, Englewood Cliffs, N. J: 1989.
7. K P Phillips, "Current source converter for a.c. motor drives," *IEEE Trans. on Ind. Appl.* Vol. IA-8, Nov-Dec, 1972, pp. 679–683.
8. M Hombu, S Ueda, A Veda, and Y Matsuda, "A new current source inverter with sinusoidal output voltage and current," *IEEE Trans. on Ind. Appl.* Vol. IA-21, Sept–Oct 1985, pp. 1192–1198.

9. S Boloznani and G S Buja; "Control system design of a current inverter induction motor drive," *IEEE Trans. on Ind. Appl.*, Vol. IA - 21, Sept-Oct, 1955. pp. 1145–1153.
10. J M D Murphy and V B Hosinger, "Efficiency optimisation of inverter-fed induction motor drives," *Conf. Rec. IEEE and Appl. Soc. Annual meeting*, 1982, pp. 544–552.
11. P C Sen and K H Ma, "Rotor chopper control of induction motor drive; TRC strategy," *IEEE Trans. on Ind. Appl.*, Vol. IA-11, 1975, pp. 43–49.
12. W Drury, B L Jones and J E Brown, "Application of controlled flywheeling to the recovery bridge of a static Kramer drive", *IEE Proc.*, Vol. 130 Pt 13, No. 2, March 1983, pp. 73–85.
15. T Wakabayashi, T Hori, K Shimzu and Yoshioka, "Commutatorless Kramer control system for large capacity induction motors for driving water service pumps" *Conference RCC, IEEE IAS Annual meeting*, 1976, pp. 822–828.
14. B K Bose, "Sliding mode control of induction motor," *IEEE Industry Applications Society Conference Record*, 1985, pp. 155–161.
15. S Nishikata, S Muto and T Kataoka, "Dynamic performance analysis of self-controlled synchronous motor speed control systems," *IEEE IAS Annual meeting*, 1981, pp. 671–677.
16. A Smith, "Static Scherbius system of induction motor speed control," *proceedings IEE*, Vol. 124, 1977, pp. 557–565.
17. M F Brosnan and B Brown, "Closed loop speed control using an a.c. synchronous motor," *IEE Conf. pub. No. 234, Power electronics and variable-speed drives* 1984 pp. 373–376.
18. R Gabriel and W Lecnhard, "Microprocessor Control of Induction motor," *IEEE/IAS 7th International semiconductor power converter conference*, 1982.

Chapter 16

Power Electronic Applications

LEARNING OBJECTIVES:

- To describe the need and function of uninterruptible and switch mode power supplies.
- To examine the application of power-semiconductors to high voltage d.c. transmission.
- To consider the operation of static VAR compensators.
- To describe the operation of induction and dielectric heating.
- To introduce switch-mode-welding.
- To consider the operation of battery charge systems, circuit breakers, timing circuits and flasher circuits.

16.1 INTRODUCTION

Power electronics and motor drives constitute a vast, complex and interdisciplinary subject that has gone through rapid technological evolution during the last four decades. As the technology is advancing and apparatus cost is decreasing along with the improvement of reliability, their applications are expanding in industrial, commercial, residential, military, aerospace and utility environments. Many innovations in power semiconductor devices, converter topologies, analytical and simulation techniques, electrical machine drives, and control and estimation techniques are contributing to this advancement. The frontier of the technology has been further advanced by the artificial intelligence (AI) techniques such as fuzzy logic and artificial neural networks, thus bringing more challenge to power-electronic engineers. In the global industrial automation, energy generation, conservation of the 21st century, the widespread impact of power-electronics is inevitable.

As the technology is maturing and cost is decreasing, power-electronics is expanding in applications, such as switching mode power supplies (SMPS), UPS systems, electrochemical processes, heating and lighting, static VAR compensation, active filtering, high voltage dc system, photo-voltaic system and variable frequency motor drives. The importance of power electronics is being increasingly visible now-a-days in the energy saving of electrical apparatus by more efficient use of electricity. The energy consumption in the world is increasing by leaps and bounds to improve the human living standard, particularly in industrialized countries. The major amount of this energy comes by burning fossil fuels, such as coal, natural gas and oil which create global warming effect besides urban pollution problem. The energy efficiency improvement of electrical apparatus with the help of power electronics not only reduces electricity consumption but the corresponding reduced power generation indirectly helps reduction of environmental pollution problem.

16.2 UNINTERRUPTIBLE POWER SUPPLY

In applications, such as medical intensive care systems, chemical plant process control, safety monitors or a major computer installation, where even a temporary loss of supply could have severe consequence, there is need to provide an uninterruptible power supply system which can maintain the supply under all conditions. Therefore, the function of a UPS is to provide an interrupted free supply of power to the a.c. load, which cannot be directly fed from dc source and dc is required to be converted into a.c. A UPS is a power conditioner which:
  (i) Provides good quality power to the load at all conditions of supply power.
  (ii) Regulates the load voltage when the mains voltage fluctuates.
  (iii) Provides complete isolation between the load and the mains.
  (iv) Suppresses the line transient (voltage spikes) and minimizes EMI (RFI) problems.
  (v) Provides a constant voltage and constant frequency supply to the critical load.

Figure 16.1 shows the block diagram of a typical UPS system. A rectifier converts a single-phase or three-phase a.c. voltage into dc, which supplies power to the inverter as well as the battery bank (to charge it). The inverter gets a d.c. input voltage from the rectifier when the ac mains is ON, and from the battery bank when the a.c. mains is OFF. Inverter converts this d.c. voltage into a.c. voltage and through a suitable filter applies it to the load. If the PWM inverter is used, then the filter can be eliminated. A static switch will connect or disconnect the battery from the input of the inverter depending on the status of a.c. mains.

16.2.1 UPS Configurations

Depending on the arrangement of the basic blocks, UPS systems are classified as:
  (i) On-line or inverter preferred UPS system.

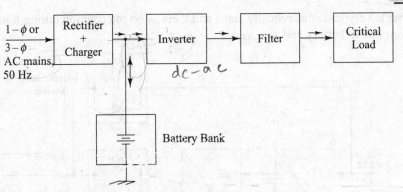

Fig. 16.1 *Block diagram of typical UPS system*

(ii) Off-line or line-preferred UPS system.
(iii) Line-interactive UPS system.

These configurations have been discussed in the following sections.

16.2.1.1 On-Line UPS (Inverter-Preferred) Figure 16.2 shows the block diagram of the on-line UPS systems. On this mode of operation, the load is always connected to the inverter through the UPS static switch. The UPS static switch (Fig. 16.2) is a normally 'ON' switch. It turns-off only when the UPS system fails. In that case the 'Mains Static Off' switch and used only when UPS is to be bypassed. The various operating modes are:

(i) Mode I: When the AC mains is ON, the rectifier circuit will supply the power to the inverter as well as to the battery. Therefore, it acts as a rectifier cum charger. Hence, its ratings are usually higher. The inverter output is connected to the load via UPS Static Switch. Battery will be charged in this mode.

(ii) Mode II: If the supply power fails suddenly, the rectifier output will be zero and hence the battery-bank now supplies power to the inverter without any interruption and delay. There will not be any change in inverter as well as the load.

After restoration of the line supply, the charger supplies the inverter and recharges the battery automatically first in constant current mode and then in constant potential mode. Various rates of battery recharge may be set depending upon the application. The inverter has to be designed carefully because it supplies the load continuously. There should not be frequent failure of the inverter-system.

(iii) Mode III: In case if the UPS fails (inverter fails), then the normally "OFF" mains static switch is turned-on which automatically transfers the a.c. line to the load in less than 1/4 cycle period with no phase discontinuity. This not only maintains power to the load, but also actuates the failure alarm signal to draw attention of the attendant.

This type of system is more popular because it can provide full isolation of the critical load from the a.c. line and also provides power conditioning. Also, the mode of operation does not change during the failure of power. Its changeover time is very less (theoretically zero) and there is no interruption during transfer from line to battery and vice versa.

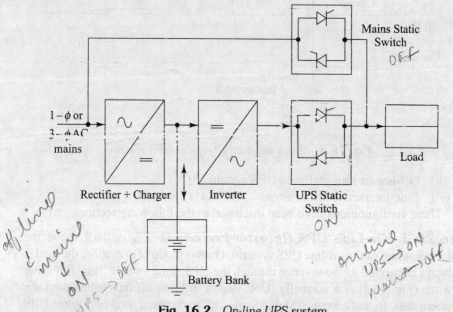

Fig. 16.2 *On-line UPS system*

This system protects the critical load against surges, spikes, line noise, frequency and voltage variation, brownout and outages. All these protections are not available in the off line systems. Table 16.1 list the important specifications of an on-line UPS system.

Table 16.1 Specifications of on-line-UPS

S.N.	Parameter	Specifications
1.	Power rating	500 VA, 1 kVA, 5 kVA, 50 kVA etc.
2.	Output voltage	230 V ± 0.1%
3.	Output frequency	50 Hz ± 0.1 Hz
4.	Input voltage	230 V ± 15%
5.	Output voltage waveform	Sine wave
6.	Power-factor	> 0.8 Lagging
7.	Back-up time	30 min. to 4 hours
8.	Total Harmonic Distortion (THD)	< 3%
9.	Efficiency	> 85%
10.	Protections provided	(i) Overvoltage/undervoltage cutout (ii) Overcurrent trip with reset

16.2.1.2 Off-Line UPS (Line Preferred)

The block diagram for the off-line UPS is shown in Fig. 16.3. As can be observed, the only difference between the off-line and on-line UPS is that the mains-static switch here is a 'normally ON' switch. It connects the a.c. mains directly to the load when the mains is ON. The battery charger is a stabilized one which maintains the battery on float at fully charged condition and at the same time provides stabilized power to the inverter. The other static switch, i.e. 'UPS Static Switch' is normally OFF switch. It is closed only when the mains fails. Thus, in the off-line UPS, the inverter comes into the circuit only when the mains fails.

The rectifier/charger has to do only one function, to charge the battery bank. Therefore, its size and power rating is lower than that of an on-line UPS system charger. Under the condition of mains failure, the static switch operate to disconnect the mains from the load and connect the load to the UPS output. The battery will then supply the power to the load via the inverter. The total time taken to sense the power failure and make a changeover from mains to UPS is about 5 ms (1/4 cycle).

Here no isolation is provided between the load and mains and hence this UPS is not recommended for highly critical loads. The specifications of the UPS are listed in Table 16.2.

Table 16.2 Off-line UPS specifications

S.N.	Parameter	Specifications
1.	Power rating	500 VA, 1kVA, 2 kVA etc.
2.	Input voltage	(i) Mains: 170-270 V, 50 Hz
		(ii) Battery: 24 V/6.5 AH, 48 V/6.5 AH etc.
3.	Output voltage	(i) Mains on: 230 V ± 5% Volts.
		(ii) UPS on: 230 V ± 0.5% Volts
4.	Output voltage waveform	Sinusoidal
5.	Output frequency	(i) Main on: 50 Hz ± 3%
		(ii) UPS on: 50 Hz ± 2Hz
6.	Battery recharge time	4 hours.
7.	Back-up time	30 minutes or more.
8.	Transfer time	5 ms (typical)
9.	Efficiency	> 85%
10.	Protections	(i) Output overload
		(ii) Mains overvoltage
		(iii) Low battery

6.2.1.3 Line Interactive UPS The block-diagram of a line interactive UPS is shown in Fig. 16.4(a). The heart of this system is the battery-charger-cum-inverter which acts as an inverter. In this system normally the critical load is supplied from the commercial supply through the static switch and an inductance L. The operation of the system can be divided into the following two modes.

(i) Mode I (Mains ON): When mains is ON, the static switch is closed and the load gets connected directly to the ac mains through the inductance L. The inverter/charger block operates as a charger and charges the battery bank. The equivalent circuit for this mode is shown in Fig. 16.4(b).

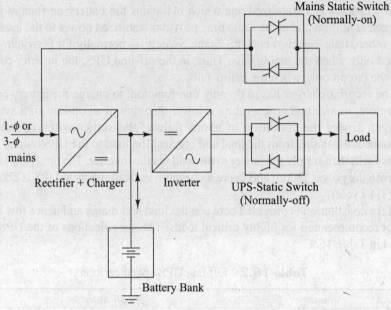

Fig. 16.3 *Off-line UPS system.*

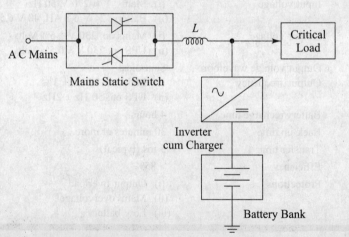

Fig. 16.4(a) *Off-line interactive UPS*

(ii) Mode II (Mains Off) As soon as the mains fails, the static switch is turned-off and the inverter/charger block operates as an inverter and the battery supplies power to the load through the inverter. The equivalent circuit for this mode is shown in Fig. 16.4(c).

The total time taken for sensing and changeover after the failure of mains is less than 5 msec. This configuration does not provide any isolation between the load and the a.c. mains. However, it is possible to give a pure, clean voltage to the load with good voltage regulation by using power conditioners like ac stabilizer or a constant voltage transformer (CVT).

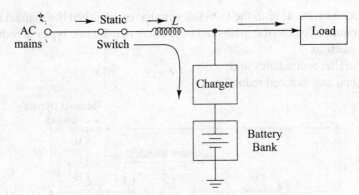

Fig. 16.4(b) *Equivalent circuit when mains ON.*

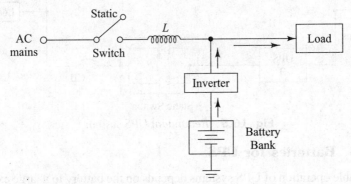

Fig. 16.4(c) *Equivalent circuit when mains fails*

16.2.2 Reliability of UPS System

Reliability in power conditioning systems is very important. Reliability is measured in term of mean time between failures or MTBF. It is expressed in thousands of hours. As the UPS system is an alternative for the ac mains supply and the critical loads like computers or medical equipments are being operated on the UPS, the UPS system needs to be extremely reliable, failsafe and easy to maintain.

Reliability can be greatly increased by the use of the redundancy in UPS. Redundancy means to provide facilities in the power conditioning system to transfer the critical load from a failed conditioner (UPS) to an alternate or by-pass power source, i.e. the use of more number of UPS than are actually required for the critical load is called redundancy. If one UPS fails, the other one is still there to supply the critical load, thus increasing the reliability. A typical redundant UPS system is shown in Fig. 16.5. Two UPS are connected in parallel using the two static switches. The circuit breakers are used to isolate the load from the two UPS systems, and the mains.

Two UPS are connected in parallel using load sharing circuitry not shown here. The critical load is supplied by both UPS in parallel. If one UPS fails, then it is isolated for servicing using the circuit breaker. The load continues to work

on the second UPS. If both the UPS fail simultaneously, then the manual bypass switch connects load to the mains directly. There may be two types of redundant systems such as
(i) Parallel redundancy and
(ii) Dual and isolated redundancy.

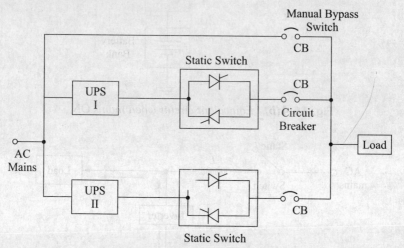

Fig. 16.5 *Redundant UPS system*

16.2.3 Batteries for UPS

The reliable operation of UPS systems depends on the battery to a large extent. If the battery is not properly selected the overall life is adversely affected. There are different types of batteries available in the market namely the lead-acid batteries, nickel–cadmium batteries and the sealed maintenance free (SMF) batteries. However, the conventional lead acid batteries are most commonly used in UPS system.

(A) Capacity of the Battery The capacity of a battery is expressed in ampere hour (AH). This rating tells us about, how much amount of current that battery can supply for 1 hour. For example, 10 AH battery can supply 10 A of current for 1 hour. Capacity of a battery depends on the following factors:
  (i) *Rate of Discharge:* AH rating decreases with increase in rate of discharge. Due to rapid rate of discharge cell potential falls significantly, due to internal losses. Weakening of acid at higher discharge rate in porous plate is also greater at higher discharge rates. This also effects the capacity adversely.
  (ii) *Temperature:* Capacity of a battery increases with increase in temperature.
  (iii) *Density of Electrolyte:* As the density of electrolyte effects the internal resistance and the vigour of the chemical reaction, it has an important effect on the capacity. Capacity increases with density.

Battery capacity can be calculated from the following equations:

(i) $\text{Battery (kW)} = \dfrac{\text{Load kVA} \times \text{Power-factor}}{\text{Inverter Efficiency}}$ \hfill (16.1)

(ii) $\text{Number of cells} = \dfrac{\text{Minimum allowable battery voltage}}{\text{Final voltage per cell}}$ \hfill (16.2)

(iii) $\text{Cells capacity (kW/cell)} = \dfrac{\text{Battery kW}}{\text{Number of cells}}$ \hfill (16.3)

Unless the UPS has protection for excess discharge, <u>1.75 V per cell</u> should be selected to prevent battery damage. Battery discharge rates in kW at 25°C are given in Table 16.3.

Table 16.3 Battery discharge rates in kW at 25°C.

Type	5 min	10 min	15 min	20 min	30 min	60 min
SB-45	0.201	0.155	0.128	0.108	0.086	0.53
SB-60	0.268	0.206	0.170	0.144	0.114	0.070
SB-75	0.335	0.258	0.213	0.180	0.143	0.088
SB-90	0.402	0.309	0.255	0.216	0.171	0.105

(B) Efficiency of the Battery There are two different values of the battery efficiency:
(i) Ampere-hour efficiency (ii) Watt-hour efficiency.

(i) AH Efficiency: The ampere-hour efficiency is defined as the ratio of ampere hours taken from battery to the ampere hours supplied to it while charging.

$$\therefore \quad \text{AH efficiency} = \dfrac{\text{A-H during discharge}}{\text{A-H input while charging}} \qquad (16.4)$$

The typical value of A-H efficiency is <u>90 to 95%</u>. 5 to 10% reduction is due to the losses taking place in the battery. The ampere hour efficiency takes into account only the current and time but it does not consider the battery terminal voltage at all.

(ii) Watt-hour (WH) (Energy) Efficiency: The WH efficiency is defined as follows:

$$\text{WH-efficiency} = \text{AH-efficiency} \times \dfrac{\text{Average cell voltage while discharging}}{\text{Average cell voltage while charging}}$$

\hfill (16.5)

= <u>70 – 80%</u> usually.

If the charge volts increases or discharge volts decreases then WH-efficiency will also decrease. High charging and discharging rates will usually do this and hence are not recommended.

SOLVED EXAMPLES

Example 16.1 A UPS is driving a 600 W load which has a lagging power factor of 0.8. The efficiency of the inverter is 80%. The battery voltage is 24 volts dc. Assume that there is a separate charger for the battery. Determine the following:
 (i) kVA rating of the inverter.   (ii) Wattage of the rectifier.
 (iii) A-H rating of the battery for a back-up time of 30 minutes.

Solution: *Given:* Load = 600 W, PF = 0.8, Efficiency = 80%
E_{dc} = 24 V, Back-up time = 30 min.
 (i) *kVA rating of inverter:*

$$PF = \frac{\text{Active power in Watts}}{\text{Total RMS power (kVA)}}, \quad 0.8 = \frac{600}{\text{Total RMS power}}$$

∴ Total RMS power = 750 VA ∴ The VA rating of the inverter = 0.75 kVA
 (ii) *Wattage of the rectifier:*

$$\text{Rectifier wattage} = \frac{\text{kVA rating of UPS} \times \text{p.f.}}{\text{Inverter efficiency}} = \frac{0.75 \text{ kVA} \times 0.8}{0.8} = 0.75 \text{ kW}$$

 (iii) *AH Rating of Battery:*
  When mains fails, the battery supplies power to the inverter. The power is 0.75 kW. Now, battery kW rating = Battery voltage × current

$$0.75 \text{ kW} = 48 \times I_{dc}, \quad \therefore I_{dc} = 15.63 \text{ Amp} \approx 16 \text{ A}.$$

Assuming battery to be 100% efficient, the A-H rating of the battery is $16 \times \frac{1}{2} = 8$ AH

Example 16.2 For a UPS system, select a suitable battery for following specifications:
 (i) UPS rating = 20 kVA.   (ii) Back-up time = 15 min.
 (iii) Inverter efficiency at full load = 85%. (iv) Type of load = inductive with a lagging PF = 0.8
 (v) Type of battery = Lead acid.   (vi) Battery voltage range = 147–190 V.
 (vii) Final volts on cell = 1.75 V/cell. (viii) Cell groups = Six per jar.

Solution:

 (i) From Eq. (16.1), Battery (kW) = $\dfrac{\text{Load kVA} \times \text{PF}}{\text{Efficiency}} = \dfrac{20 \times 0.8}{0.85} = 18.82$ kW

  ∴ Battery needs to supply 18.82 kW power.

 (ii) From Eq. (16.2), No. of cells = $\dfrac{\text{Minimum allowable battery voltage}}{\text{Final voltage per cell}}$

  $= \dfrac{147}{1.75} = 84$

 (iii) No. of jars = $\dfrac{\text{No. of cells}}{\text{No. of cells per jar}} = \dfrac{84}{6} = 14$

 (iv) From, Eq. (16.3) cell size (kW) = $\dfrac{\text{Battery (kW)}}{\text{No. of cells}} = \dfrac{18.82 \text{ kW}}{84} = 0.224$ kW.

For battery selection, refer Table 16.3 (Battery discharge rate table). From this table, the 90-AH battery is selected which has 0.255 kW discharge rate at 15 min. back-up time which is greater than required 0.224 kW.

Example: 16.3 A 2.0 volts lead acid cell is discharged at a uniform rate of 35 A for 5 hours at an average emf of 1.95 V. If it is then charged at a rate of 45 A for 4 hours to restore its original voltage. Determine:
(i) A-H efficiency   (ii) W-H efficiency.

Solution: From Eq. (16.4), A-H efficiency = $\dfrac{35 \times 5}{45 \times 4}$ = 97.22%

From Eq. (16.5), W-H efficiency = $\dfrac{97.22 \times 1.95}{2}$ = 94.79%.

Example 16.4 18 lead-acid cells, each having a discharge capacity of 90 A-H at 10 hours rate are being charged at constant current continuously for 8 hours. The A-H efficiency is 85% and terminal voltage per cell at the beginning of charging is 2.4 volts. Determine the dc supply voltage required to charge the battery and charging current if the internal resistance of the battery is 0.1 Ω.

Solution:
(i) From the definition of A-H efficiency,

$$\text{A-H efficiency} = \frac{\text{A-H output/cell}}{\text{A-H input/cell}}, \quad 0.85 = \frac{90}{\text{A-H input/cell}}$$

∴ A-H input/cell = 105.88 A-H.
But A-H input/cell = charging current × charging time

∴ Charging current = $\dfrac{105.88}{8}$ = 13.23 A.

(ii) Total terminal voltage of 18 cells = 18 × 2.4 = 43.20 V
Voltage drop across internal resistance = 0.1 × 13.23 = 1.323 V.
∴ Charger supply voltage = terminal voltage + Drop = 43.20 + 1.323 = 44.523 V.

Example 16.5 A 6 kVA, 230 V UPS operates from a 144V DC link and feeds a load having 0.8 PF. If the inverter efficiency is 0.85, calculate the back-up time available if the battery bank has capacity of 500 AH. Also, determine charger peak output power if batteries must be restored within four hours when power returns. Assume the following: (a) Battery voltage varies from 10.6 V to 13.4 V, Normal voltage = 12 V. (b) Capacity derating of the battery up to 8 hour discharge is 0.5.

Solution:
(i) *Back-up time:*

From Eq. (16.1), Battery kW = $\dfrac{6 \times 0.8}{0.85}$ = 5.647 kW.

No. of batteries to be connected in series = $\dfrac{144}{12}$ = 12.

Now let us consider the worst case battery voltage = 10.6 V for calculation of battery discharge current.

∴ Battery kW = Battery voltage × Battery discharge current.
Now, total battery voltage = $10.6 \times 12 = 127.2$ V.

∴ Battery discharge current, $I_{dc} = \dfrac{\text{Battery kW}}{127.2} = \dfrac{5.647 \times 10^3}{127.2} = 44.39$ Amp.

Capacity derating of battery up to 8 hr discharge is 0.5.

∴ Back-up time = $\dfrac{\text{A-H capacity}}{\text{Discharge current}} = \dfrac{250 \text{ A-H}}{44.39} = 4.864$ hours.

(ii) *Charger Peak Rating*
Assume that the battery is completely discharged during the back-up time of 4.864 hrs.

∴ AH output from the battery = 250 A-H
Now, in order to charge the battery completely,
A-H input = A-H output
i.e. charging current × No. of charging hours = 250.

∴ charging current = $\dfrac{250}{4} = 62.50$ Amp.

∴ Charger peak power = $144 \times 62.5 = 9$ kW.

16.3 SWITCHED MODE POWER SUPPLIES (SMPS)

In the low to medium power range, say, upwards from 50 W, a d.c. source of supply is often required which contains negligible a.c. ripple which can be controlled in magnitude. For this application, switched mode power supplies circuits are used. Though the SMPS was developed originally by NASA in the 1960s to provide a compact power supply system for space vehicles, it had no significant impact on the general power supply market until the 1970s. From that point development was rapid and the SMPS now accounts for some 75% of the power supplies being produced.

An SMPS is based on a d.c. chopper with a rectified and possibly transformed output. The output voltage amplitude is controlled by varying the mark–space ratio of the chopper. This may be achieved by means of pulse-width control or frequency variation with constant pulse width, the former being the more common. The circuit techniques used for SMPS can be separated into following four broad categories.

(i) Flyback SMPS
(ii) Feed-forward SMPS
(iii) Push-pull SMPS
(iv) Bridge SMPS

Figure 16.6 Shows the building blocks of a typical high-frequency off-the line switching power-supply. As the name implies, the input rectification in a switching power supply is done directly off-the line, without the use of the low-frequency isolation power transformer, as in the case of linear power supply. The a.c line is directly rectified and filtered to produce a raw high-voltage d.c., which in turn is fed into a switching element. The switch is operating at the high-frequencies of 20 kHz to 1 MHz, chopping the dc voltage into a high-frequency square-wave. This square-wave is fed into the power isolation transformer, stopped down to a predetermined value and then rectified and filtered to produce the required dc output.

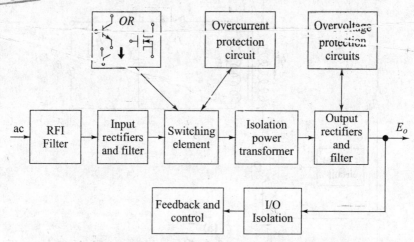

Fig. 16.6 *Block diagram of off-line high frequency SMPS*

A portion of this output is monitored and compared against the fixed reference voltage, and the error signal is used to control the on–off times of the switch, thus regulating the output. Since the switch is either ON or OFF, it is dissipating very little energy, resulting in a very high overall power supply efficiency of about 70 to 80%. Another advantage is the power-transformer size which can be quite small due to high operating frequency. Hence the combination of high efficiency (i.e. no large heat sinks) and relatively small magnetics, results in compact, light weight power supplies, with power densities up to 30 W/in^3 versus 0.3 W/in^3 for linears.

16.3.1 Isolated Flyback Converters

Although there are numerous converter circuits described by number of authors and researchers, basically all of them are related to three classical circuits known as the "flyback or buck-boost", the "forward or buck", and the "push-pull or buck derived" converter. The buck-boost regulators (Chapter 8) comes under the category of nonisolated SMPS. In these nonisolated regulators, the isolation between the load and source is provided only by the power-semiconductor devices, therefore if these devices fail then the load will be directly connected across the source and may damage it completely.

In isolated converters, the ferrite core high-frequency transformer is used for electrical isolation between the load and the source. Another advantage of using a transformer is the possibility to get multiple outputs.

16.3.1.1 Discontinuous Mode (Flyback Converter)

The circuit diagram for the discontinuous flyback converter is shown in Fig. 16.7 and associated steady-state waveforms are shown in Fig. 16.8. The circuit operates as follows:

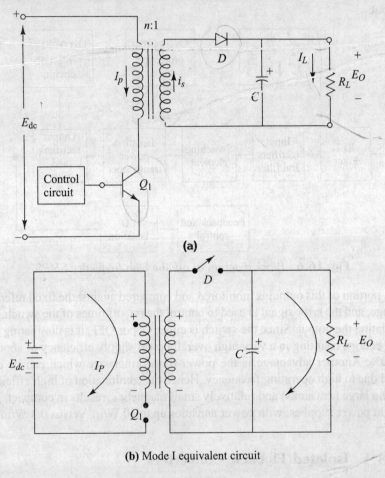

(a)

(b) Mode I equivalent circuit

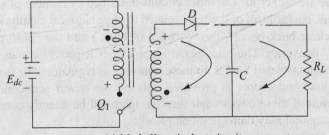

(c) Mode II equivalent circuit

Fig 16.7 *Flyback converters*

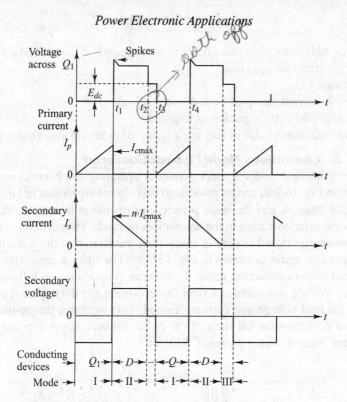

Fig. 16.8 *Flyback converter associated waveforms.*

Mode I ($t_0 - t_1$): When transistor Q_1 is ON at $t = 0$, primary current starts to build-up in the primary winding, storing energy. Due to the opposite primary arrangement between the input and output windings of the transformer choke, there is no energy transferred to the load since diode D is reverse-biased. Transistor Q_1 is turned-off abruptly at instant $t = t_1$ The equivalent circuit for this mode is shown in Fig. 16.7 (b).

Mode II ($t_1 - t_2$): When the transistor is turned-off at $t = t_1$, the polarity of the windings reverses due to the collapsing magnetic field. Now, diode D is conducting, charging the output capacitor C and delivering current I_L to load.

The voltage across Q_1 is the sum of input supply voltage (E_{dc}) and the self induced voltage across the primary winding (Ldi/dt), therefore it is higher than the supply voltage (E_{dc}). The equivalent circuit for this mode is shown in Fig. 16.7(c). The secondary current goes to zero at $t = t_2$. Thus, the stored energy in the transformer core is delivered to the load during this mode of operation.

(iii) **Mode III ($t_2 - t_3$):** In this mode, transistor and diode both are in the off-state. Therefore, primary and secondary currents are zero. As there is no voltage drop across the primary winding of the transformer, the voltage across the transistor Q_1 is equal to the dc supply voltage (E_{dc}). The secondary voltage is zero. The one-cycle operation completes in this mode and repeats itself.

Advantages of Discontinuous Mode
  (i) Slower diodes can be used on the secondary side for rectification.
  (ii) Size of transformer is smaller than that in the continuous mode.

(iii) An additional filter inductance is not required on the secondary side.
(iv) Fast transient response.

Disadvantages
(i) Peak current rating of both devices is high.
(ii) Large size filter capacitor is required.
(iii) The maximum value of duty cycle is limited to 50% to avoid core saturation.

16.3.1.2 Continuous Mode Flyback Converter

The circuit diagram for the flyback converter operating in a continuous mode is as shown in Fig. 16.9(a) and its associated waveforms are shown in Fig. 16.9(b). The circuit diagram and the basic principle of operation is same as that of the flyback converter operating in the discontinuous mode. However, in continuous mode, the energy stored in core is completely transferred to the load, transistor Q1 is turned-on again as shown in Fig. 16.9(b). The flyback converter operates in the continuous conduction mode if converter is operating on full-load and if the supply voltage is minimum. Under this condition, the duty cycle is increased to meet the load voltage and current demand. This increases the on-time of the pulse and decreases the off-time of the pulse. Flyback converters are used in computers, printers, video games, CATV, etc.

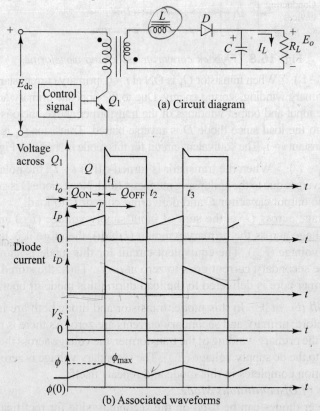

Fig. 16.9 *Flyback converter continuous mode.*

(i) Advantages of Continuous Mode:
  (i) The size of the filter capacitor required for this mode in the secondary side is nearly half the size of the filter capacitor used in discontinuous mode.
  (ii) The peak current rating of the diode and the transistor reduces to 50% of that used in the discontinuous mode.

(ii) Disadvantages of Continuous Mode:
  (i) Rectifier diodes should be nearly four times faster than discontinuous current mode (typical t_{rr} = 25–100 ns).
  (ii) Size of the transformer is larger than that of the discontinuous mode.

(iii) Expression for Voltage Transfer Ratio:
Let, E_{dc} be the dc input voltage to the converter,
N_1/N_2 be the transformer turns ratio,
E_0 be the output voltage and
α be the duty ratio of the transistor.

As shown in Fig. 16.9, the inductor core flux increases linearly from its initial value $\phi(0)$, which is finite and positive and is given by

$$\phi_{(t)} = \phi_{(0)} + \frac{E_{dc}}{N_1} t, \text{ for } 0 < t < t_{on} \qquad (16.6)$$

The value of peak flux at the end of this interval is given by

$$\phi_{max} = \phi(t_{on}) = \phi_{(0)} + \frac{E_{dc}}{N_1} t_{on} \qquad (16.7)$$

Transistor is turned-off after t_{on} and the energy stored in the transformer core causes the current to flow in the secondary winding. The voltage across the secondary $e_2 = -E_0$, and therefore the flux decreases linearly during the off-time of switch, i.e. t_{off}.

$$\therefore \quad \phi_{(t)} = \phi_{max} - \frac{E_o}{N_2}(t - t_{on}) \text{ for } t_{on} < t < T \qquad (16.8)$$

and

$$\phi_{(T)} = \phi_{max} - \frac{E_o}{N_2}(T - t_{on}) \qquad (16.9)$$

Substituting ϕ_{max} from (16.7) in (16.9), we get

$$\phi_{(T)} = \phi_{(0)} + \frac{E_{dc}}{N_1} t_{on} - \frac{E_o}{N_2}(T - t_{on}) \qquad (16.10)$$

In steady state, since the net change of flux through the core over one time period must be zero,
$\phi_{(T)} = \phi_{(0)}$, substitute in Eq. (16.10),

$$\therefore \quad \phi_{(0)} = \phi_{(0)} + \frac{E_{dc}}{N_1} t_{on} - \frac{E_o}{N_2}(T - t_{on})$$

$$\frac{E_{dc}}{N_1} t_{on} = \frac{E_o}{N_2}(T - t_{on})$$

$$\therefore \quad \frac{E_o}{E_{dc}} = \frac{t_{on}}{(T-t_{on})} \frac{N_2}{N_1} = \frac{t_{on}/T}{(1-t_{on}/T)} \frac{N_2}{N_1}$$

but, $\quad \alpha = t_{on}/T,$

$$\therefore \quad \frac{E_o}{E_{dc}} = \frac{\alpha}{1-\alpha} \frac{N_2}{N_1} \tag{16.11}$$

Equation (16.11) shows that the voltage transfer ratio in flyback converter depends on α in an identical manner as in the buck-boost converter.

(iv) Selection Criteria for Switching Transistor:

(a) The switching transistor used in the flyback converter must be chosen to handle peak collector voltage at turn-off and peak collector currents at turn-on. The peak collector voltage which the transistor must sustain at turn-off is

$$V_{CE,max} = \frac{E_{dc}}{1-\alpha_{max}} \tag{16.12}$$

From Eq. (16.12), it becomes clear that in order to limit the collector voltage to a safe value, the duty cycle must be kept relatively low, normally below 50 per cent, i.e. $\alpha_{max} < 0.5$.

In practice, α_{max} is taken at about 0.4, which limits the peak collector voltage $V_{CE,max} < 2.2\, E_{dc}$ and, therefore, transistors with working voltages above 800 V are usually used in the off-line flyback converter designs.

(b) The Selected transistor must have,

$$I_c = \frac{I_L}{n} = I_p \tag{16.13}$$

where I_p is the primary transformer-choke peak current,
n is the primary to secondary turns ratio,
and I_L is the output load current.

(v) Selection of Transformer-Choke:

Since the transformer-choke of the flyback converter is driven in one direction only of the B-H characteristics curve, it has to be designed so that it will not saturate. Therefore, a core with a relatively large volume and air-gap must be used. The effective transformer-choke volume is given by

$$\text{Volume }(V) = \frac{\mu_o \cdot \mu_e \cdot I_{L,max}^2 \cdot L_{out}}{\beta_{max}^2} \tag{16.14}$$

where, $I_{L,max}$ = determined by load current.
μ_e = relative permeability of the chosen core material.
β_{max} = maximum flux density of the core.

The relative permeability μ_e must be chosen to be large enough to avoid excessive temperature rise in core due to restricting core and wire size and therefore copper and core-losses.

(vi) Expression for Peak Working Collector Current:
The energy transferred in the choke is given by

$$P_{out} = \left(\frac{L I_p^2}{2T}\right) n \tag{16.15}$$

where $\quad n =$ efficiency of the converter.

The voltage across the transformer may be expressed as

$$E_{dc} = L \frac{di}{dt} \tag{16.16}$$

Let us take $\quad di = I_p$ and $\dfrac{1}{dt} = \dfrac{f}{\alpha_{max}}$

Substitute in Eq. (16.16), we get

$$E_{dc} = \frac{L I_p \cdot f}{\alpha_{max}} \tag{16.17}$$

or

$$L = \frac{E_{dc} \cdot \alpha_{max}}{I_p \cdot f} \tag{16.18}$$

Substituting Eq. (16.18) into Eq. (16.15), we get

$$P_o = \left(\frac{E_{dc} \cdot f \cdot \alpha_{max} \cdot I_p^2}{2 \cdot f \cdot I_L}\right) \eta = \frac{1}{2} \eta\, E_{dc}\, \alpha_{max}\, I_p$$

or,

$$I_p = \frac{2 P_o}{\eta\, E_{dc} \cdot \alpha_{max}} \tag{16.19}$$

Substituting Eq. (16.19), in Eq. (16.13), we get

$$I_c = \frac{2 P_o}{\eta\, E_{dc} \cdot \alpha_{max}} \tag{16.20}$$

For converter with efficiency 80% and duty-cycle $\alpha_{max} = 0.4$, we get

$$I_c = \frac{6.2 \cdot P_o}{E_{dc}} \tag{16.21}$$

16.3.1.3 Variations of Basic Flyback Converter

We have seen that with the basic flyback circuit, the collector voltage of the switching transistor must sustain at least twice the input voltage at turn-off. In cases where the voltage value is too high to use a commercial transistor type, the two-transistor flyback converter may be used, as shown in Fig. 16.10. This circuit uses two transistors, which are switched ON or OFF simultaneously. Diodes D_1 and D_2 act as clamping diodes restricting the maximum collector voltage of the transistor to E_{dc}. Thus, lower voltage transistors may be used to realize this design, but at the expense of three-extra components, i.e. Q_2, D_1 and D_2.

An advantage of the flyback circuit is the simplicity by which a multiple output switching power supply may be realized. This is because the isolation element acts as a common choke to all outputs, this only a diode and a capacitor are needed for an extra output voltage. Figure 16.11 illustrates a practical circuit.

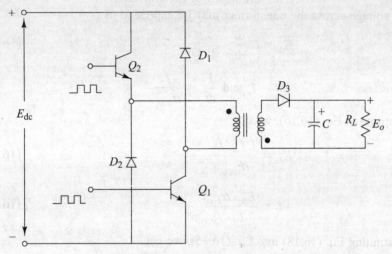

Fig. 16.10 *Two transistor flyback circuit*

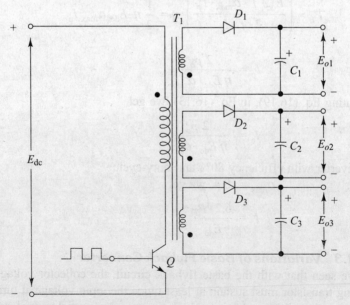

Fig. 16.11 *Flyback converter with multiple outputs*

Solved Examples

Example 16.6 An isolated flyback converter operating in continuous conduction mode at a frequency of 60 kHz is fed from the supply mains via a full-wave bridge rectifier-cum-capacitor filter and uses a 1200 V power-MOSFET as switching element. Assuming a ripple-free output of capacitor filter and neglecting rectifier drops, determine:

(i) Maximum duty-cycle, allowing a margin of 180 V for voltage spikes on the drain of power MOSFET for a nominal mains voltage of 230 V, 50 Hz.

(ii) The flyback converter turns ratio required to obtain a maximum output voltage of 12 V for a minimum mains voltage of 180 V.

Solution: Given: Switching frequency, $f_s = 60$ kHz., E_{sp} = Spike voltage = 180 V.
Mains = 230 V, 50 Hz, $E_0 = 12$ V

(i) *Calculation of* α_{max}:
From Eq. 16.12, voltage across switch is given by

$$E_{DS} = \frac{E_{dc}}{1-\alpha_{max}} + \text{spike voltage}$$

Voltage across the MOSFET should not exceed 1200 V.
∴ $E_{DS} < 1200$ Volts.

∴ $$E_{DS} = \frac{E_{dc}}{1-\alpha_{max}} + E_{sp} < 1200 \text{ V}$$

E_{dc} at 230V mains voltage with capacitor filter = $230 \times \sqrt{2}$

∴ $$\frac{230 \times \sqrt{2}}{1-\alpha_{max}} + 180 < 1200 \text{ V}, \quad \frac{325.26}{1-\alpha_{max}} < 1020$$

∴ $\alpha_{max} = 0.68$

(ii) *Calculation of turns-ratio* (N_1/N_2):

From Eq. 16.11, $E_0 = \frac{\alpha}{1-\alpha}\left(\frac{N_2}{N_1}\right)E_{dc}$

∴ $$\frac{N_1}{N_2} = \frac{\alpha}{1-\alpha}\frac{E_{dc}}{E_o} = \left(\frac{0.68}{1-0.68}\right)\left(\frac{230 \times \sqrt{2}}{12}\right) = 57.6.$$

Example 16.7 A flyback converter is operated in a continuous mode from a supply of 14V to 30V with two outputs 12 V at 0.6A and –12 V at 0.6 A. The switching power supply is used to power some drivers that have intermittent load demands. The load can vary from 0.1 to 0.5 Amp. Assuming the efficiency of the converter to be 80% and switching frequency to be 50 kHz, determine:

(i) The average input power and current. (ii) Ratings of the transistor.

(iii) The primary widing inductance and the number of turns of the primary and the secondary if the core exhibits 80 MH per 1100 turns.

(iv) Ratings of the rectifying diode.

Solution:

(i) *Average input power and current:*
The full-load input power is given by
$P_o = (12 \times 0.6) + (12 \times 0.6) = 14.4$ W.

Input power $P_i = \dfrac{P_o}{\eta} = \dfrac{14.4}{0.8} = 18$ W

Now, maximum value of average input current,

$$I_{(avg)max} = \dfrac{P_i}{E_{dc_{min}}} = \dfrac{18}{14} = 1.285 \text{ Amp.}$$

Similarly, minimum value of average input current,

$$I_{i(av)min} = \dfrac{P_i}{E_{dc_{max}}} = \dfrac{18}{30} = 0.6 \text{ A}$$

(ii) *Ratings of transistor:*

From Eq. 16.20, $I_{c(max)} = \dfrac{2P_o}{\eta E_{dc_{min}} \cdot \alpha_{max}}$ Assume, $\alpha_{max} = 0.5$

$$I_{c_{max}} = \dfrac{2 \times 14.4}{0.8 \times 14 \times 0.5} = 5.14 \text{ Amp.}$$

Now, $\quad V_{CE_{max}} = \dfrac{E_{dc_{max}}}{1 - \alpha_{max}} = \dfrac{30}{1 - 0.5} = 60$ V

$P_{d_{max}} = V_{CE_{max}} \times I_{c_{max}} = 60 \times 5.14 = 308$ W.

(iii) *Primary winding inductance and turns:*
The value of primary inductance is given by

$$L_{prim} = \dfrac{E_{dc_{(min)}}}{I_{c_{max}} \cdot f_s} \alpha_{max} \qquad (16.22)$$

$$= \dfrac{14}{5.14 \times 50 \times 10^3} \times 0.5 = 27 \, \mu H.$$

Since the core exhibits 80 MH per 1100 turns,

$$\therefore N_{prim(N1)} = 1100 \sqrt{\dfrac{L_{prim.}}{L_{1100}}} = 1100 \sqrt{\dfrac{27 \times 10^{-6}}{80 \times 10^{-3}}}$$

$= 21$ turns.

From Eq. (16.11),
$\quad N_2 = 18 \qquad E_o = 12$

(iv) *Diode ratings:*

Peak diode current $I_D \geq \left(\dfrac{N_1}{N_2}\right) \cdot I_p \geq \left(\dfrac{N_1}{N_2}\right) \cdot I_{c_{max}}$

$$\geq \left(\frac{21}{18}\right) \cdot (5.14) \geq 5.6 \text{ Amp}$$

$$\text{PIV} \geq E_0 + \frac{E_{dc(max)}}{\eta} \geq 12 + \frac{30}{(21/18)} \geq 37.71 \text{ V.} = 40 \text{ V}$$

16.3.2 Isolated Forward Converter

Figure 16.12(a) shows the basic circuit diagram of a single ended isolated forward converter. At first glance, this power circuit resembles that of the flyback converter. However there are some distinct differences between the two circuits:

(i) The dot polarities are on the same side of the transformer. This means that the two windings are wound in the same sense and hence carry the current simultaneously. Thus, the transformer of the forward converter is a pure transformer.

(ii) Since the transformer is merely a transformer, this suggests a need of a filter inductance on the secondary side as shown in Fig. 16.12(a). The circuit operates as follows:

(i) Mode I (Q_1 ON) As soon as Q_1 is turned-on, the supply voltage E_{dc} is applied across the primary winding of the transformer. Due to this constant voltage, the primary current increases at a constant rate. Due to the winding polarity as shown in Fig. 16.12(a), the induced voltage in the secondary winding will forward bias diode D_1 and the secondary current starts flowing.

(ii) Mode II (Q_1 OFF) When Q_1 is turned-off, the primary voltage will change its polarity as shown in Fig. 16.12(b). The secondary voltage also will change its polarity. Diode D_1 is reversed-biased and D_2 is forward biased due to the induced voltage in the filter inductance and the current flows through the load as shown in Fig. 16.12(c).

Disadvantage: With basic configuration of Fig. 16.12(a), some residual energy remains in the transformer core. Due to this residual energy, the transformer core will saturate after a few cycles of operation. This will lead to overcurrent through Q_1 and may damage Q_1. The core saturation can be avoided only by removing the residual energy from the core. This is achieved by adding a "tertiary" or "demagnetizing" winding into the basic configuration of forward-converter.

16.3.2.1 Forward Converter with Tertiary Winding Figure 16.13(a) shows the circuit diagram for the forward converter with tertiary winding added to the basic configuration. Operation of the circuit is same as the basic forward converter, except for the operation of the demagnetizing winding. Such converters are used in computers, wordprocessor, televisions, etc.

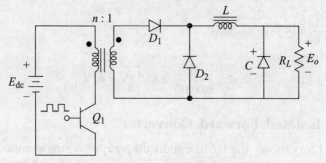

Fig. 16.12(a) *Basic single-ended forward converter*

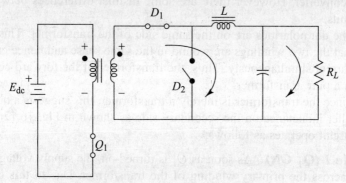

Fig. 16.12(b) Mode I (Q1—ON) operation

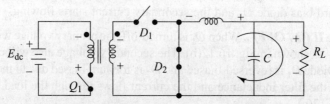

Fig. 16.12(c) Mode-II (Q_1—OFF) operation

Operation of Tertiary Winding: When the transistor Q_1 is turned-on, due to the winding polarities, diode D_m is reverse-biased and does not conduct. However, when Q_1 is turned-OFF, D_m is forward-biased and the current flows through the tertiary winding as shown in Fig. 16.13(b). The residual energy in the transformer core is returned back to the dc source via diode D_m and the tertiary winding. This is how the tertiary winding helps to demagnetize the core and avoids the core saturation. Due to the dot convention as shown in Fig. 16.13(a), the primary winding and tertiary winding will never carry current simultaneously. Associated waveforms are shown in Fig. 16.14. The dark-areas on the waveforms of Fig. 16.14 shows the magnetizing-demagnetizing current, given as,

$$I_{mag} = \frac{t_{on} \cdot E_{dc}}{L} \qquad (16.23)$$

where L is the output inductance in microhenries.

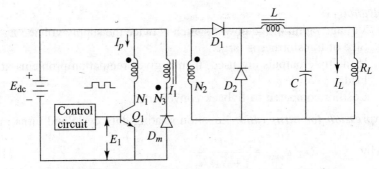

Fig. 16.13(a) *Forward converter with tertiary winding*

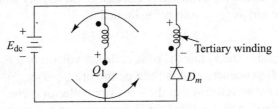

Fig. 16.13(b) *Demagnetization of transformer core by tertiary winding*

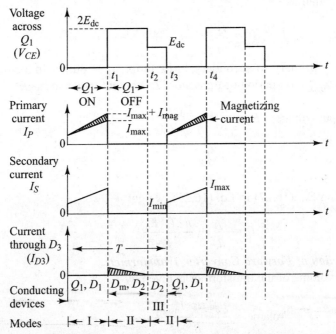

Fig. 16.14 *Waveforms with tertiary winding*

Advantages:
  (i) It needs only one power transistor as switching element.
 (ii) Simple driver circuits are required as compared to forward converters.

Disadvantages:
(i) Voltage rating of the power switch is twice the supply voltage E_{dc}.
(ii) Size of transformer is large.
(iii) If multiple outputs are used, then it gives regulation problems at light loads.
(iv) Costlier compared to flyback converters.

(i) Expression for turns ratio: Equation relating duty cycle and turns ratio is given by

$$\alpha_{max} = \frac{1}{1 + N_3/N_1} \qquad (16.24)$$

(ii) Selection of Switching Transistor: Because of the tertiary winding and diode D_3, the voltage across Q_1 at turn-off is limited to

$$V_{CE,max} = 2\,E_{dc} \qquad (16.25)$$

The waveforms also shows that the peak collector voltage of $2\,E_{dc}$ is maintained for as long as D_m conducts, that is for a period of t_{on} (=$T\cdot\alpha_{max}$). Also, it can be observed from the waveforms that the transistor collector current at turn-on will have a value equal to that derived for flyback converter plus the net amount of the magnetization current. Therefore, peak collector current in the transistor may be written as,

$$I_{C_{max}} = \frac{I_L}{n} + \frac{T\cdot\alpha_{max}\cdot E_{dc}}{L} \qquad (16.26)$$

where, n = primary to secondary turns ratio., I_L = output inductor current, Amp., $T\cdot\alpha_{max}$ = on-period of transistor., L = output inductance, μH.

Also, output voltage, $\quad E_0 = \dfrac{\alpha_{max}\cdot E_{dc}}{n}$

or, $\quad E_{dc} = \dfrac{n\cdot E_o}{\alpha_{max}} \qquad (16.27)$

Substituting Eq. (16.27) in Eq. (16.25), we get

$$I_c = \frac{I_L}{n} + \frac{n\cdot T\cdot E_o}{L} \qquad (16.28)$$

(iii) Selection of Forward Converter Transformer:
The core volume of the transformer is given by

$$\text{Volume} = \frac{\mu_o\cdot\mu_e\cdot I_{mag.}^2\cdot L}{B_{max}^2} \qquad (16.29)$$

where, $\quad I_{mag} = \dfrac{n\cdot T\cdot E_o}{L} \qquad (16.30)$

It should be noted that the α_{max} must be kept below 50%, so that when the transformer voltage is clamped through the tertiary winding, the integral of the volt-seconds between the input voltage, when Q_1 is ON, and the clamping level, when Q_1 is OFF, amounts to zero. Duty cycles above 50 per cent, will upset the volt-seconds balance, driving the transformer into saturation, which in turn produces high collector current spikes that may destroy the switching transistor.

Also, care must be taken during construction to couple the tertiary winding tightly to the primary (bifilar wound) to eliminate fatal voltage spikes caused by leakage inductance.

16.3.2.2 Two-Transistor Forward Converters

Figure 16.15(a) shows the circuit diagram of the two-transistor forward converter, and the associated waveforms are shown in Fig. 16.15(b). This configuration reduces the voltage ratings of the transistor to $E_{dc(max)}$ instead of $2 E_{dc(max)}$ (Single-ended configuration). The circuits operate as follows:

(i) Mode I (Q_1, Q_2 and D_3 ON): At $t = 0$, transistors Q_1 and Q_2 are turned on simultaneously. The supply voltage E_{dc} is connected across the primary winding. The primary current starts increasing linearly from I_{min} to $I_{max} + I_{mg}$, where I_{mg} is the magnetizing component shown by shaded area in the primary current waveform. Due to the specific winding directions, the induced voltage in the tertiary winding will reverse bias diode D_m and the induced voltage in the secondary winding will forward bias the rectifying diode D_3. The secondary will deliver power to the inductance L, capacitor C and the load as shown in Fig. 16.15(b).

(ii) Mode II (Q_1, Q_2, OFF and D_4. D_m): At $t = t_1$, both the transistors Q_1 and Q_2 are turned-off simultaneously. Due to the sudden interruption of primary current, the induced voltage across the primary winding will change its polarities as shown in Fig. 16.15(b). This voltage will forward bias the diodes D_1, D_2 and they will clamp the primary voltage to E_{dc} volts.

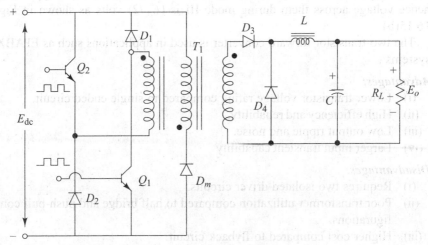

Fig. 16.15(a) *Two-transistor forward converter*

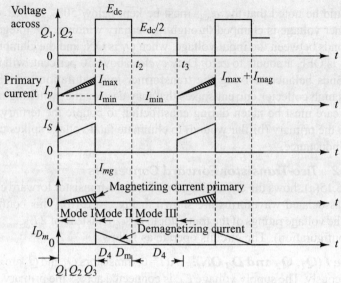

Fig. 16.15(b) *Voltage and current waveforms*

The induced voltages across the secondary and tertiary windings will also change their polarities. This will forward bias diode D_m and reverse bias D_3. The demagnetizing current starts flowing through the tertiary winding and D_m as shown in Fig. 16.15(b). Due to the induced voltage across the filter inductance L, diode D_4 is forward biased and the load current is maintained by the inductance L and filter capacitor C as shown in Fig. 16.15(b).

(iii) Mode III (D_4 ON): At $t = t_2$, the demagnetizing current reduces to zero so diode D_m is turned-off. All the other devices except D_4 are in the off-state. D_4, however continues to conduct and the load current is maintained. The supply voltage E_{dc} gets divided equally across the nonconducting transistors Q_1 and Q_2 hence voltage across them during mode III is ($E_{dc}/2$) volts as shown in Fig. 16.15(b).

The two-transistor forward converter is used in applications such as EPABX systems.

Advantages:
  (i) Lower transistor voltage rating compared to single ended circuit.
  (ii) High efficiency and reliability.
  (iii) Low output ripple and noise.
  (iv) Larger input transient capability.

Disadvantages:
  (i) Requires two isolated driver circuits.
  (ii) Poor transformer utilization compared to half bridge and push-pull configurations.
  (iii) Higher cost compared to flyback circuit.
  (iv) Regulation problems at light loads for multiple outputs.

SOLVED EXAMPLE

Example 16.8 A switch mode power supply is to be designed with following specifications: $E_0 = 12$ V, $I_0 = 12$ A, $f_s = 60$ kHz
AC rectified mains with LC filter: 230 V, 50Hz

A forward converter operating in continuous-conduction mode with a demagnetizing winding is chosen. Assume all components to be ideal except for the presence of transformer magnetization inductance. Determine:
 (a) turns ratio of demagnetizing winding with primary winding at maximum duty cycle of 0.6.
 (b) switch voltage rating allowing for 50% voltage of input voltage as spike.
 (c) d.c supply current at full load for input reduced by 20%.

Solution:
Given : $\alpha_{max} = 0.6$, spike voltage = 50% E_{dc}.
 (i) *Calculation of turns ratio:*

From Eq. (16.25), $\alpha_{max} = \dfrac{1}{1 + N_3/N_1}$, $0.6 = \dfrac{1}{1 + N_3/N_1}$

$\therefore \qquad \dfrac{N_3}{N_1} = 0.67$ Or $\dfrac{N_1}{N_3} = 1.5$

 (ii) *Voltage rating of switch:*
Voltage across the switch in its off-state is given by

$$V_{sw} \geq E_{dc} + \dfrac{N_1}{N_3} \cdot E_{dc} + \text{Spike voltage.}$$

Now, $\qquad E_{dc} = 230 \times \sqrt{2} = 325.3$ V.

Also, spike voltage = $\dfrac{1}{2}$ (325.3) = 162.64 V.

$\therefore \qquad V_{sw} \geq 325.3 + (1.5 \times 325.3) + 162.64 = 975.88$ V. $= 1000$ V

 (iii) Assuming the transformer efficiency to be 100%

$\therefore \qquad N_1 i_1 = N_2 i_2, \quad \therefore i_1 = \dfrac{N_2}{N_1} \cdot i_2$

Also, $\qquad \dfrac{N_1}{N_2} = \dfrac{V_1}{V_2} = \dfrac{0.8 \times 230 \times \sqrt{2}}{12} = 21.68$ V.

$i_1 = \dfrac{12}{12.68} = 0.55$ Amp. $\therefore$ DC supply current = 0.55 Amp.

16.3.3 Half Bridge Converter

Half bridge converter circuit is shown in Fig. 16.16. This converter is derived from the step-down converter configuration. If you observe the circuit diagram, you will notice that it uses a transistorised half bridge inverter discussed in Chapter 9. The inverter is dc-ac converter, the ac output of which is rectified by rectifier block which consists of diodes D_3 and D_4 and filter components L and C. The isolation is provided by the transformer.

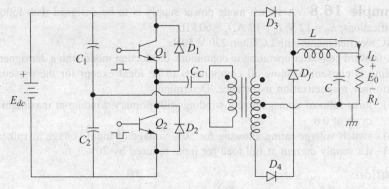

Fig. 16.16 *Half-bridge converter*

The large value capacitors (C_1 and C_2) on the input-side generates the centre-tap input dc supply required for the operation of half-bridge inverter. Diodes (D_1 and D_2) across Q_1 and Q_2 clamps the forward voltage across them to E_{dc} and reverse voltage to $(-V_D)$ where V_D is a diode drop. The capacitors (C_1 and C_2) divides the supply E_{dc} into two voltages each of $E_{dc}/2$ across each capacitor.

In order to avoid core saturation, the volt-second integral of each switching transistor is balanced automatically by inserting a coupling capacitor C_c in series with the transformer primary. Also, base drives to two transistors Q_1 and Q_2 should be 180° phase-shifted with a delay between them to avoid shoot through fault or cross conduction. The circuit operates as follows:

(i) Mode I (Q_1 ON): As soon as transistor Q_1 is turned-on, the entire capacitor voltage ($E_{dc}/2$) is placed across the primary of the transformer. This will make the dotted-terminal of the primary and secondary windings positive. Diode D_3 is forward biased and D_4 is reverse biased. Thus, the energy is delivered to the load and the LC filter through the upper half of the centre-tap secondary and diode D_3. During this period, the voltage across transistor Q_2 is E_{dc}.

(ii) Mode II (Q_2 ON): During the on-period of Q_2 entire lower capacitor voltage ($E_{dc}/2$) is placed across the primary winding of the transformer with dot terminal being at negative potential. The induced voltage on the secondary side with dot terminal negative will reverse bias D_3 and forward-bias diode D_4. Thus, the energy is delivered to the load and LC filter through the lower half of the centre-tap secondary and diode D_4. During this period, voltage across Q_1 is E_{dc}. The associated waveforms are shown in Fig. 16.17.

When both Q_1 and Q_2 are off, the load current is maintained by diode D_f.

A half-bridge converter is widely used in personal computer, EPABX systems and video games where the converter has to satisfy the following requirements:
  (i) High output power (500 W).
  (ii) Optimizing transformer utilization by operating in first and third quadrant.
  (iii) Efficient design.

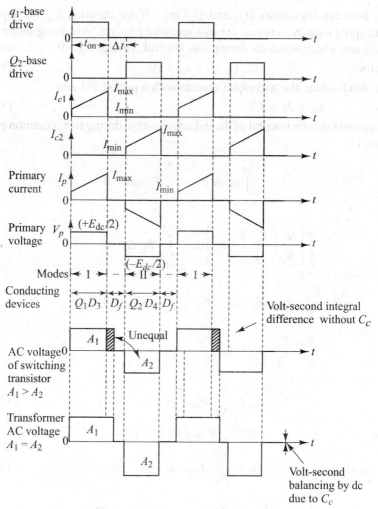

Fig. 16.17 *Associated voltage-current waveforms*

(i) Expression for output voltage (E_0):

Refer to Fig. 16.16 and 16.17.

Let, E_0 be the output voltage,

E_L be the voltage across the filter inductance L.

E_2 be the rectifier output voltage.

When each transistor remains ON for period t_{on}, the rectifier output voltage is given by

$$E_2 = (N_2/N_1)\, E_{dc}/2 \qquad (16.31)$$

where N_2/N_1 = Secondary to primary transformer turns ratio.

Now, when Q_1 is conducting, E_L is given by

$$E_L = \frac{N_2}{N_1}\left(\frac{E_{dc}}{2}\right) - E_0,\ 0 < t < t_{on} \qquad (16.32)$$

When both the transistors (Q_1 and Q_2) are off for duration t_{on}, the inductor current splits equally between the two secondary halves. Assuming diode D_f to be ideal one which conducts during this interval, voltage $E_2 = 0$.

Therefore, $\quad\quad E_L = -E_0, \; t_{on} < t < t_{on} + \Delta t \quad\quad$ (16.33)

In the steady-state, the waveform repeats with a period $T/2$ and

$$t_{on} + \Delta t = T/2 \quad\quad (16.34)$$

Now, equate the time integral of the inductor voltage during one repetition period to zero, i.e.

$$\int_0^{t_{on}} E_L \, dt + \int_{t_{on}}^{t_{on}+\Delta t} E_L \cdot dt = 0.$$

$$\therefore \quad \int_0^{t_{on}} \left[\frac{N_2}{N_1} \left(\frac{E_{dc}}{2} \right) - E_o \right] dt - \int_{t_{on}}^{T/2} E_o \, dt = 0$$

$$\therefore \quad \frac{N_2}{N_1} \frac{E_{dc}}{2} t_{on} - E_o \cdot t_{on} - E_0 \frac{T}{2} + E_0 \cdot t_{on} = 0$$

$$\therefore \quad \frac{N_2}{N_1} \frac{E_{dc}}{2} t_{on} = E_0 \frac{T}{2}$$

$$\therefore \quad E_o = \frac{N_2}{N_1} E_{dc} \frac{t_{on}}{T} \quad\quad (16.35)$$

but $\quad\quad \alpha = t_{on}/T$ and $0 < \alpha < 0.5$

$$\therefore \quad E_0 = \frac{N_2}{N_1} E_{dc} \cdot \alpha \quad\quad (16.36)$$

or, $\quad\quad \dfrac{E_o}{E_{dc}} = \dfrac{N_2}{N_1} \alpha \quad\quad$ (16.37)

(ii) Series coupling capacitor:

Referring to Fig 16.16, assume that the two switching power transistors have unmatched switching characteristics, that is, transistor Q_1 has a slow turn-off while transistor Q_2 has a fast turn-off. Figure 16.17 shows the effect of slow-turn- off of Q_1 upon the a.c. voltage waveforms at the junction of Q_1 and Q_2. A volt-seconds-unbalance, depicted by the cross-hatched area, is added to one-side of the a.c. voltage waveform. If this unbalanced waveform is allowed to drive the power transformer, flux walking will occur, resulting in core saturation, producing transistor collector current spiking, which will lower converter efficiency and may also drive the transistor into thermal runaway to destruction.

By inserting a coupling capacitor in series with the primary transformer winding, a d.c. bias proportional to the volt-seconds unbalance is picked up by this capacitor, shifting the d.c. level as shown in Fig. 16.17, thus balancing the volt-seconds integrals of the two switching periods.

It can be observed from the Fig. 16.16, that the coupling capacitor and the output filter inductor forms a series resonant circuit. The resonant frequency is given by

$$f_r = \frac{1}{2\pi\sqrt{L_R C_c}} \qquad (16.38)$$

The reflected filter inductance to the transformer primary is

$$L_R = \left(\frac{N_1}{N_2}\right)^2 \cdot L \qquad (16.39)$$

where, f_r = resonant frequency, kHz, C_c = coupling capacitor, μf L_R = reflected filter inductance, μH L = output inductance.

Substituting Eq. (16.39) in Eq. (16.38), we get

$$C_c = \frac{1}{4\pi^2 f_r^2 (N_1/N_2)^2 \cdot L_R} \qquad (16.40)$$

In order for the changing of the coupling capacitor, to be linear, the resonant frequency must be chosen to be well below the converter switching frequency.

(iii) Advantages:
  (i) Due to the coupling capacitor, no-flux symmetry problem.
  (ii) Due to the core utilization in either direction, the transformer utilization factor is much better than the flyback converter.
  (iii) Leakage inductance energy and magnetizing energy are pumped into the input and output filter capacitors and hence improves the efficiency of the converter.

(iv) Disadvantages:
  (i) Use of two capacitors (C_1 and C_2) adds to the size and cost of power supply.
  (ii) Difference in the turn-off time of two transistors.
  (iii) A delay must be provided between the base drives of the two transistors to avoid shoot-through fault.
  (iv) At turn-on, the transistor collector current doubles due to the decrease in its voltage rating.

SOLVED EXAMPLE

Example 16.9 For a half-bridge converter of 200 W, working at 20 kHz with a turns-ratio of 10, determine the value of coupling capacitor C_c. Also, verify if the calculated value is acceptable or not. If not then determine the new-value. Assume filter inductance of 20 μH and 80% efficiency.

Solution:
Given: $P_0 = 200$ W, $f_s = 20$ kHz, $= n\ 10$, $L = 20\ \mu$H, $e = 80\%$.
From Eq. (16.39),
$$L_R = (10)^2 \times (20 \times 10^{-6}) = 2\text{ mH}.$$
For practical purposes, we will assume that the resonant frequency must be about one-fourth the switching frequency, expressed as
$$f_R = 0.25 f_s$$
$$= 0.25 \times 20 \times 10^3 = 5\text{ kHz}.$$
From Eq. (16.40),
$$C_c = \frac{1}{4(3.14)^2 (25 \times 10^6)(2 \times 10^{-3})} \quad C_c = 0.50\ \mu\text{F}.$$

The transistor collector working current in terms of output power, efficiency and duty cycle is given by
$$I_c = \frac{2P_o}{n \cdot \alpha_{max} \cdot E_{dc}} \tag{16.41}$$

Let $\alpha_{max} = 0.8$ and input is mains rectified and filtered.
Here,
$$E_{dc} = \sqrt{2} \times 230 \approx 325\text{ V}$$
$$\therefore\quad I_c = \frac{2 \times 200}{(0.8)(0.8)(325)} = 1.923\text{ A}$$

Now, assuming that the converter has an input voltage tolerance of ±20 per cent, then the heaviest current of the transistor will occur at low line.
∴ Worst case collector current will be
$$\therefore\quad I_c = 1.923 + 0.2(1.923) = 2.31\text{ Amp}.$$
Coupling capacitor charging voltage is given by
$$V_c = \frac{I}{C}\ dt \tag{16.42}$$

where $dt = \frac{T}{2}\ \alpha_{max}$ and $T = 1/f_s$

$$\therefore\quad dt = \frac{\alpha_{max}}{2f_s} \quad \therefore\quad dt = \frac{0.8}{2 \times 20 \times 10^3} = 20\ \mu\text{S} \tag{16.43}$$

$$\therefore\quad V_c = \frac{2.31(20 \times 10^{-6})}{0.5 \times 10^{-6}} = 92.4\text{ V}.$$

Note V_c must have a reasonable value anywhere between 10 and 20% of $E_{dc}/2$ (6 to 32 V). A charging voltage of 92.4 V is too high and will interfere with converter regulation at low line; therefore a new value of coupling capacitance must be found from the equation, $C = I \cdot \dfrac{dt}{dV_c}$ (16.44)
where, I = average primary current, A., dt = charging interval, μs. dV_c = arbitrary number from 16 to 32 V.

Let us choose $V_c = 30$ V.

$$\therefore \quad C = \frac{2.31(20 \times 10^{-6})}{30} = 1.5 \ \mu F$$

16.3.4 Full-Bridge Converter

Figure 16.18 shows the circuit diagram of the full-bridge converter. This converter uses a full-bridge inverter to first convert the input d.c. voltage into a.c., then through a transformer it is applied to a full-wave rectifier and filter combination. At the output we get a filtered d.c voltage. Associated waveforms are shown in Fig. 16.19.

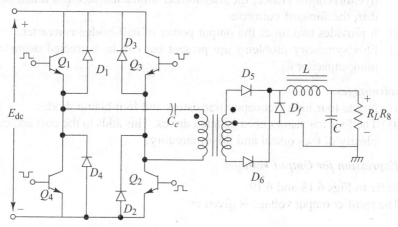

Fig. 16.18 *Full-bridge converter circuit*

In half-bridge circuit the voltage stress of the switching transistors at turn-off reduces to $\dfrac{E_{dc}}{2}$ but the collector current is doubled at turn-on. Therefore, half-bridge converter circuit is not suitable for high-current applications. This problem is solved in full-bridge converter circuit. The circuit operates as follows:

(i) Mode I (Q_1, Q_2 ON): When the transistors Q_1 and Q_2 are turned-on simultaneously, the entire d.c supply voltage appears across the primary winding with dotted terminal positive. This will induce voltage into the secondary winding in such a way that the diode D_5 is forward biased and D_6 is reverse biased. The energy is delivered to the LC filter and load from the upper half of the secondary winding. The voltage across nonconducting transistors Q_3 and Q_4 is E_{dc}, during this mode of operation.

(ii) Mode II (Q_3, Q_4 ON): As soon as transistors Q_3 and Q_4 are ON, the entire supply voltage appears across the primary with dot terminal negative. The negative primary current starts flowing. The induced voltage in the secondary winding now forward biases diode D_6 and reverse-biases diode D_5. The energy is delivered to the load and LC filter from the lower half of the secondary winding. The voltage across the nonconducting transistors Q_1 and Q_2 is E_{dc} volts.

During the time when all the transistors are nonconducting, diode D_f is forward biased and the LC filter will maintain the load current through this diode.

Full bridge converter is widely used in main frame computers, EPABX system and in applications where power rating greater than 500 W and up to 2 kW are required.

Advantages:
  (i) Low ripple and noise at the output.
  (ii) This converter drives the transformer flux in both the positive and negative directions. Hence, the transformer utilization factor is much better than the forward converter.
  (iii) It provides two times the output power of half-bridge converter.
  (iv) Flux symmetry problems are present but can be corrected using coupling capacitor C_c.

Disadvantages:
  (i) Needs four high-frequency transistors and four clamp diodes.
  (ii) The four transistors need isolated drives. This adds to the cost and complexity of the control and driver circuitry.

(i) Expression for Output Voltage:

Refer to Figs 6.18 and 6.19.

The rectifier output voltage is given by

$$E_2 = \frac{N_2}{N_1} \cdot E_{dc} \tag{16.45}$$

Voltage across the filter inductance L is given by

$$E_L = \left(\frac{N_2}{N_1} \cdot E_{dc}\right) - E_o, \quad 0 < t < t_{on} \tag{16.46}$$

When all transistors are nonconducting, D_f is ON and $E_2 = 0$

$\therefore \qquad E_L = -E_0, \quad t_{on} < t < (t_{on} + \Delta t)$

Equating the time integral of the inductor voltage over one-time period to zero in the steady-state and substituting $(t_{on} + \Delta t) = T/2$, we get

$$E_0 = 2 \frac{N_2}{N_1} \cdot E_{dc} \cdot \alpha_{max} \tag{16.47}$$

From Eq. (16.47) it becomes clear that the output voltage if the full-bridge converter is twice that of the half-bridge.

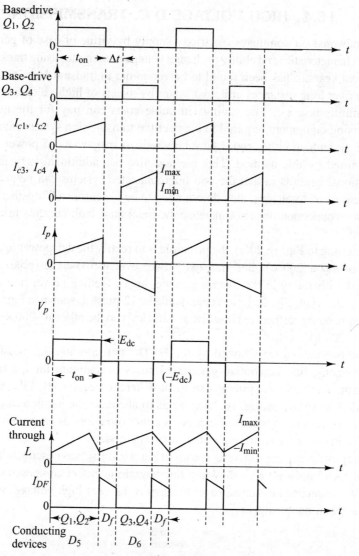

Fig. 16.19 *Voltage and current waveforms of full-bridge converter*

(ii) Transistor Working Current:
Assuming a converter efficiency of 80% and 0.8 duty-cycle, then the transistor working current is

$$I_c = \frac{1.6 P_o}{E_{dc}} \qquad (16.48)$$

All other properties of this converter remains the same as for the half-bridge converter and all the formulas developed for the calculation of other elements may be used here as well.

16.4 HIGH VOLTAGE D.C. TRANSMISSION

A.C transmission dominates electricity supply by virtue of ease of generation, motor characteristics and ability to change voltage magnitudes using transformers. In recent years, it has been found to be economic to transmit large amounts of power over long distances or across water by means of high voltage d.c. rather than multiphase a.c. The savings in cable costs can pay for the necessary conversion equipment required to connect the transmission system into the a.c. power systems at either end. Many hundreds of megawatts of power are now transmitted by this method. This method also has additional very important operational benefits in that the two interconnected systems can be at different frequencies and voltages, they do not need to be synchronized together, and the d.c. interconnection does not increase the maximum fault currents in either a.c system.

Referring to Fig. 16.20(a), the principle is to rectify the a.c. power to a voltage level of say ± 200 kV, and transmit power over a two-cable (pole) line to a converter operating in the inverting mode, hence feeding power into the other a.c. system. Typically, each converter would be 12-pulse as shown in Fig. 16.20(b), the centre being earthed and one line at +200 kV and the other at −200 kV to earth for a ± 200 kV system.

The two a.c. systems linked by the HVDC transmission line must each be synchronous, that is, contain generators so as to maintain the a.c frequency constant. The two a.c. systems can be at different frequencies. Filters must be included, as shown in Fig. 16.20(a), so as to attenuate the harmonics generated by the converters. Power flow can be in either direction, as each converter can operate in either the rectifying or the inverting mode.

The principle of the converter shown in Fig. 16.20(b), was described in Chapter 6, but each thyristor valve shown in the diagram is in fact composed of several thyristor connected in series, so as to support the very high voltages across the valves when nonconducting.

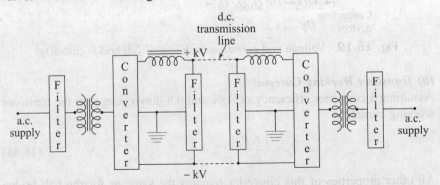

Fig. 16.20(a) *Overall layout*

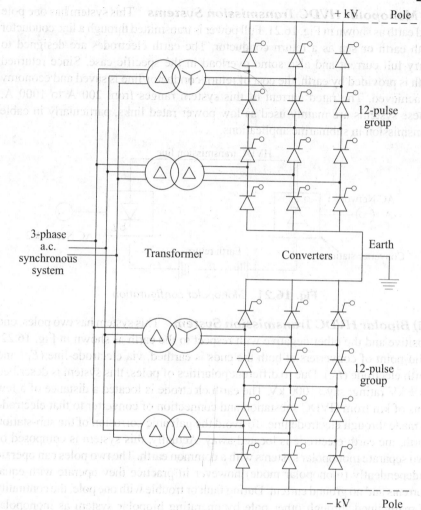

Fig. 16.20(b) *HVDC transmission*

HVDC system with converters is economical only above 800 kilometres. Additional cost of thyristors in rectifying and inverting equipment can only be justified for high power transmission over large distances. Pole refers to line or conductor carrying d.c. having fixed polarity with respect to earth.

16.4.1 Types of HVDC Transmission Systems

Various configurations of HVDC systems are:
  (i) Monoplor HVDC transmission systems.
  (ii) Bipolar HVDC transmission systems.
  (iii) Homopolar HVDC transmission systems.
  (iv) Back-to-back tie station.

(i) Monopolar HVDC Transmission Systems This system has one pole and earth as shown in Fig. 16.21. Full power is transmitted through a line conductor with earth or sea as a return conductor. The earth electrodes are designed to carry full current and also some overload in the specific case. Since returned path is provided by earth, the cost of return conductor/line is saved and economy is achieved. The rated current of this system ranges from 200 A to 1000 A. These systems are mainly used as low power rated links, particularly in cable transmission in submarine applications.

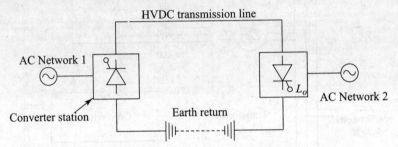

Fig. 16.21 *Monopolar configuration*

(ii) Bipolar HVDC Transmission System This system has two poles, one positive and the other negative with respect to the earth as shown in Fig. 16.22. Mid-point of converters at both the ends is earthed, via electrode-line (E_L) and earth electrode (E_E). Due to different polarities of poles, this system is described as ± kV rating, say ±1000 kV. The earth electrode is located a distance of a few tens of km from HVDC substation and connection of converter to that electrode is made through electrode line. To avoid the galvanic corrosion of the sub-station earth, the earth electrode is located away. Actually, this system is composed of two separate monopolar systems with a common earth. The two poles can operate independently (monopolar mode) however in practice they operate with equal currents and no ground current. During fault or trouble with one pole, the continuity is maintained through other pole by operating biopolar system as monopolar system, however power handling is reduced to half.

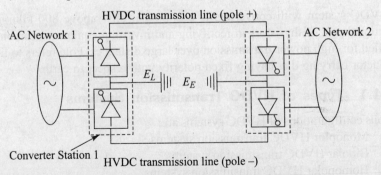

Fig. 16.22 *Bipolar configuration*

(iii) Homopolar HVDC Transmission System This system has two poles of same polarity and earth as a return path as shown in Fig. 16.23. Transmission between converter substations is carried out using two separate conductors/lines. However, at the ends they are connected together to common pole feeding converter substation (similar to monopolar system). The two poles/lines can be separate transmission lines having independent towers or can be supported by some towers.

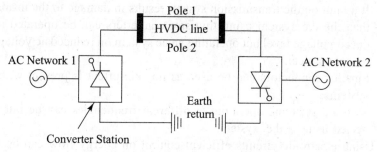

Fig. 16.23 *Homopolar configuration*

(iv) Back-to-Back HVDC Tie-Station This system has no d.c. line between converter stations (rectifier and inversion mode) are at the same locations as shown in Fig. 16.24. Rectification and inversion is done by back-to-back converter. Two adjacent a.c. networks which are geographically close to each other can be tied up using this system.

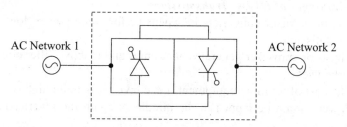

Fig. 16.24 *Back-to-back configuration*

16.4.2 Advantages and Disadvantages

(i) Advantages:
Transmission of power through high-voltage d.c. transmission line has always been advantageous compared to a.c transmission because of the following reasons:
   (i) Power on each d.c. pole is independently controllable.
   (ii) No reactive power transmission problem and hence ideally suited for cable transmission.
   (iii) Intermediate switching stations are not required.
   (iv) System is stable and its stability depends only on thermal capacity of the line equipments.

(v) Lightening never hits both poles of a bipolar transmission line but it can hit simultaneously both circuits of an a.c. double circuit line.

(vi) In d.c., a lightening fault can be quickly cleared as compared to comparatively large time taken a.c. systems.

(vii) A d.c. system offers more flexibility in power-flow.

(viii) A d.c. system offers high degree of reliability at reduced cost.

(ix) Fewer conductors and small towers are required.

(x) If a fault on the transmission system results in damage to the insulation then the d.c system using thyristor converters can be operated at reduced voltage level but operating an ac system on reduced ac voltages is un-economical.

(xi) Smaller conductors can be used as no skin-effect is present with d.c. systems.

(xii) Two a.c systems, operating at different frequencies can be interconnected using a d.c. system.

(xiii) Using electronic circuits efficient control on energy flow can be exercised.

(xiv) No undesirable oscillation in d.c. system can take place as a result of an event on ac system.

(xv) Less corona loss.

(xvi) Short circuit level of the system does not increase.

(xvii) Less radio and TV interference.

(ii) Disadvantages of HVDC Transmission:

(i) Use of thyristor converters is expensive for small power, low distance transmission.

(ii) Thyristor converter generates harmonics and reactive filter circuit have to be used.

(iii) The cost of terminal equipment, i.e. converter is quite high.

(iv) Power cannot be tapped in the middle so again the inverter has to be used.

SOLVED EXAMPLE

Example 16.10 An HVDC transmission system is rated at 500 mW ± 250 kV, and the converters are arranged as shown in Fig. 16.20. Determine the RMS current and peak reverse voltage of each thyristor value.

Solution: The direct current $= \dfrac{(500 \times 10^6)}{250 \times 2 \times 10^3} = 1000$ A

Each thyristor valve conducts 1000 A for one-third cycle, therefore,

RMS current is $= \dfrac{1000}{\sqrt{3}} = 577.35$ A.

Each transformer feeds a 6-pulse group of mean voltage,

$$= \frac{250}{2} = 125 \text{ kV}.$$

∴ Hence, peak reverse voltage $= E_{\text{line(max)}} = 125 \times \pi/3 = 131 \text{ kV}.$

16.5 STATIC VAR COMPENSATORS

The equivalent circuit of an electric utility for the fundamental harmonic components is given in Fig. 16.25. L_S is the source inductance due to long wires in the distribution system. I_L is the current drawn by a user and, in general, it has real and imaginary components. The voltage available to a user is given by

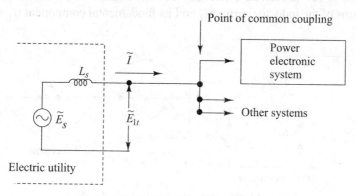

Fig. 16.25 *Electric-utility system*

$$\tilde{E}_{1t} = \tilde{E}_s - j\omega L_s \, \tilde{I}_1 \quad (16.49)$$

where
$$\tilde{I}_1 = \tilde{I}_{1P} + j\tilde{I}_{1Q} \quad (16.50)$$

$$\tilde{E}_{1t} = \tilde{E}_s + \omega L_s \cdot \tilde{I}_{1Q} - j\omega L_s \tilde{I}_{1P} \quad (16.51)$$

The subscript 's' refers to the line-source. The subscript '1' refers to the fundamental component. The imaginary (reactive) component, I_{1Q} is the main contributor to the line voltage regulation. Minimizing I_{1Q} is referred to as the static VAR control. The real component of the fundamental harmonic, I_{1P}, contributes little to the line voltage regulation. In electric utility network, it is desired to regulate the voltage within a narrow range, +5% to –10%, and have balanced loads on all phases. The static VAR controllers are used to prevent voltage flickers caused by the industrial loads that cause very rapid changes in the reactive power. Primarily, there are three types of static VAR controllers:

(i) Switched-Inductor Compensator.
(ii) Switched-Capacitor Compensator.
(iii) Switch-Mode Compensator.

16.5.1 Switched-inductor Compensator (SIC) [Thyristor-controlled Inductors TCIs]

This compensator is used for compensating negative VAR drawn by the load. Connecting a variable inductor across the utility terminals controls the reactive current drawn from the electric utility. As shown in Fig. 16.26, the variable inductor can be realized by controlled phase-delay switching of the inductor. Switch S_1 is turned-on at an angle α with respect to the line voltage. Nominally, the current in an ideal inductor lags the inductor voltage by $\pi/2$, hence, the phase-delay angle should be greater than $\pi/2$. After turning-on S_1, the inductor current increases sinusoidally and commutates naturally at an angle $(3\pi/2 - \alpha)$.

At $(\pi + \alpha)$, switch S_2 is turned-on and commutates at angle $(5\pi/2 - \alpha)$. Thyristor is the preferred switching device. Figure 16.26 also shows the waveforms of the inductor current I_L and its fundamental component i_{L_1} at delay angle α.

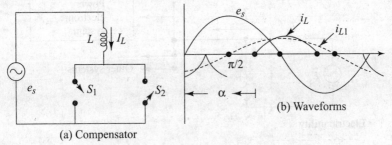

(a) Compensator

(b) Waveforms

Fig. 16.26 *Switched-inductor capacitor*

The fundamental harmonic of the inductor current waveform is given by

$$I_Q = I_{L1} = \frac{E_s}{\pi \omega L} (2\pi - 2\alpha + \sin 2\alpha) \qquad (16.52)$$

where $\quad \dfrac{\pi}{2} \leq \alpha \leq \pi$

The variable equivalent inductance is given by

$$L_{eq} = \frac{E_s}{\omega |I_{L_1}|} = L \frac{\pi}{(2\pi - 2\alpha + \sin 2\alpha)} \qquad (16.53)$$

16.5.2 Switched-Capacitor Compensator (SCC)

This compensator is used for compensating positive VARs in the load. Figure 16.27 shows a switched-capacitor compensator. Each switch is bi-directional such as two thyristors back to back. These compensators use integral half-cycle control, that is, the capacitors are fully in or fully out of the circuit in each half cycle of the line-frequency. The inductors in series with each capacitor branch are for limiting the current. Unlike the continuous control of the static VAR in the SIC, the VAR in the Switched-capacitor method is compensated in discrete steps.

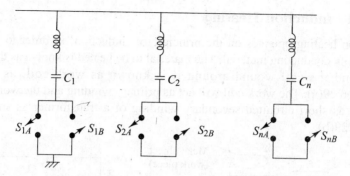

Fig. 16.27 Switched–capacitor compensator

16.5.3 Switch Mode Compensator (SMC)

The main drawback of the above methods is that the compensation is done at line frequency. Hence, the value and the size of the inductor and capacitor banks are large and expensive. Switch-mode control of reactive current is easier and less expensive. The switch mode of converter of Fig. 16.28 can be used as the pre-phase compensator at high switching frequencies. Current mode control is the preferred mode of control. Both the magnitude and the phase of line a.c. current can be controlled by this compensator.

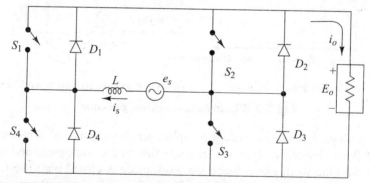

Fig. 16.28 Switch-mode converter for the a.c.-line interface

16.6 RF HEATING

Radio frequency heating is high frequency heating. The frequencies of operation ranges from hundreds of kilohertz to tens of megahertz, and is basically dependent upon the heating effect of alternating current. Based on the means of production of high frequency alternating current, RF heating is classified as:

(i) Induction heating, and (ii) dielectric heating. Induction heating is done by making use of eddy current loss, and is hence used for heating metallic objects. On the other hand, dielectric heating makes use of dielectric loss in imperfect dielectric materials, and is therefore, suitable for heating non-metals.

16.6.1 Induction Heating

Induction heating operates on the principle of 'induction' in order to heat an electrically conducting material. The material to be heated is known as the work piece and the coil wound around it is known as work coil, as shown in Fig. 16.29(a). The work coil will act as primary winding and the work piece will act as short circuited secondary winding of a transformer as shown in Fig. 16.29(b).

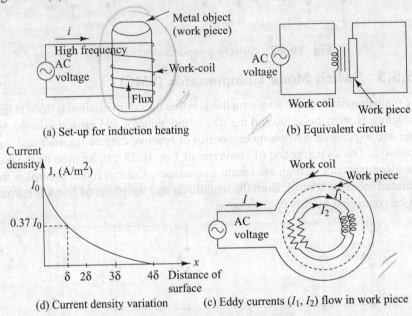

Fig. 16.29 *Induction heating principle*

When a high-frequency ac voltage is applied across the work coil, a magnetizing current flows through it. This will generate flux in the work piece and induce voltage into the work piece. Since the work piece is closed onto itself, eddy current flows into it as shown in Fig. 16.29(c). The work piece will be heated up due to the finite resistance offered by the work piece to the flow of the eddy current. The heat loss in the work piece is normally confined to surface of the work piece due to 'skin-effect'. It is the tendency of current to get concentrated at the surface. Skin-effect is observed at higher frequencies of operation. Therefore, as we go deeper into the work piece, the eddy current reduce as shown in Fig. 16.29(d).

From Fig 16.29(d) it can be observed that the magnitude of the inducted currents (current diversity) in work piece decreases exponentially with the distance x from the surface, and is given by

$$I(x) = I_0 \, e^{-x/\delta} \qquad (16.54)$$

where I_0 is the current at the surface (Amp), x is the distance from the surface in metres, δ is the penetration depth (skin depth) at which current is reduced to I_0 times a factor $1/e$ (0.368).

Eq.(16.54) can be written in terms of current density as

$$J = J_0 \cdot e^{-x/\delta} \tag{16.55}$$

The penetration depth is inversely proportional to the square root of the work piece resistivity ρ, given by

$$\delta = k \cdot \sqrt{\frac{\rho}{f}} \tag{16.56}$$

where, k is constant, and f is the frequency of the input current, (Hz)
The line frequency is used for heating large work pieces to greater depths. High-frequency up to 1 MHz are used for hardening and annealing of metal gears.

Skin depth can also be given by

$$\delta = \frac{1}{\sqrt{\pi \cdot f \cdot \mu \cdot \sigma}} \tag{16.57}$$

where, μ is permeability of the conducting workpiece, Wb/m^2 and σ is conductivity of work piece, mho/metre.

Also
$$\mu = \mu_r \cdot \mu_0 \tag{16.58}$$

where, μ_r is relative permeability of the material.
μ_0 is permeability of free-space. $= 4\pi \times 10^{-7}$ henry/m

Also, $\qquad \sigma = 1/\rho \tag{16.59}$
where ρ = resistivity of the material, ohm-metre.
Substituting the values of μ_0 and σ into Equation (16.57), urgent

$$\delta = \left[\pi.f.\mu_r (4\pi \times 10^{-7}) \times \frac{1}{\rho} \right]^{1/2}$$

i.e.
$$\delta = 503 \sqrt{\frac{\rho}{\mu_r \cdot f}}, \text{ metres} \tag{16.60}$$

Equation (16.60) shows that the depth of penetration increases by:
(i) decreasing the frequency, f
(ii) decrease in relative permeability of the work piece
(iii) increase in the resistivity of the material.
Substitute δ from Eq. (16.60) into Eq. (16.55), we get

$$J = J_0 \exp \left[\frac{-x}{503\sqrt{\dfrac{\rho}{\mu_r \cdot f}}} \right] \tag{16.61}$$

The volume power density can now be written as

$$p = \frac{J^2}{6}, \text{ Watt/m}^3 \quad (16.62)$$

i.e $\quad p = J^2 \cdot \rho, \text{ Watt/m}^3 \quad (16.63)$

where, J is expressed in A/m^2 and ρ in ohm-meter,

i.e $\quad p = (J_0 \, e^{-x/\delta})^2 \cdot \rho$

or $\quad p = J_o^2 \cdot \rho \cdot e^{-2x/\delta} \quad (16.64)$

Also, the total power entering the metallic work piece per sq. m. of surface is gives by

$$p_a = 2\pi \cdot \left(\frac{N \cdot I}{L}\right)^2 \cdot \sqrt{\frac{\delta \cdot \mu_r \cdot f}{10^7}}, \text{ W/m}^2 \quad (16.65)$$

This power is available in the form of heat per unit area of the surface, due to the circulating eddy current I passing through the work-coil of the length L(mt) and N number of turns.

Figure 16.30 shows two prominent induction heating circuits. In the voltage-source series resonant inverters, the output current is nearly sinusoidal at the switching frequency slightly below the resonance. The power is controlled by a variable switching frequency control. In the current source parallel resonant inverter, the output current is nearly sinusoidal at the switching frequency slightly above the resonance. The power is controlled by variable switching frequency control. Alternately, the current can be used as the control parameter while keeping the switching frequency constant.

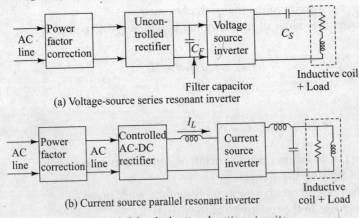

Fig. 16.30 Induction-heating circuits

16.6.1.1 Advantages of Induction Heating

(1) The depth of heat penetration depends on the supply frequency. Hence, depth of heating can be controlled by changing the frequency. It can be precisely controlled up to 0.001 mm, thus avoiding wastage of heating power.

(ii) In inducting heating, the current and therefore, the heat is concentrated at limited parts of the surface. Hence useful for surface treatment (Surface hardening applications).
(iii) High heating reduces the possibility of oxidation.
(iv) Heating can be done in vacuum or in any gas.
(v) Automatic control of temperature is easily possible.
(vi) Smaller portion of the surface can be heated without heating the entire work piece.
(vii) Very high heating rates are possible.
(viii) Since the heating is faster, the heating unit need not be kept energized continuously, the unit may be energized at the moment of heating. Thus, electricity can be saved.
(ix) Due to eddy currents, the stirring effect is automatic, hence during melting there is thorough mixing of metals to form alloys.
(x) Heating process is pollution free, hence quality of product is maintained.
(xi) Due to automatic control, unskilled labour can handle the operation.
(xii) Because of the absence of ash and flue gas, better working conditions are preserved.

16.6.1.2 Disadvantages of Induction Heating

(i) A high frequency power source is required, which is costly and complex. Thus, initial cost required is more.
(ii) The running cost or cost of operation is high.
(iii) Due to conversion of a.c. supply (50 Hz) into high frequency supply and low efficiency of induction coil, this heating process is not efficient. The overall efficiency is poor, less than 50% in many cases.

16.6.1.3 Application of Induction Heating

Induction heating is used in the following applications:

(i) Induction Cooking: In a standard electric or gas cooking range, a significant amount of heat escapes to the surroundings, thus resulting in poor thermal efficiency. This can be avoided by means of induction cooking, as shown in Fig. 16.31. As shown, the 50 Hz a.c. input is converted to a high-frequency a.c. in a range of 25–40 kHz, which is supplied to an induction coil. This induces circulating currents in the metal pan on the top of the induction coil, thus directly heating the pan.

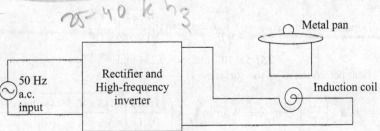

Fig. 16.31 *Induction cooking*

(ii) Surface Hardening of Metals Induction heating is found advantageous to harden the surfaces of steel, bearings of crank shafts, gear teeth, surgical instruments, etc., by using cooled copper coil operated at about 400 kHz. The heat penetration in the metal is then less than 1 mm at this frequency in a very short time duration of about 5 sec. This technique avoids the scaly surfaces encountered in other methods of surface hardening. The depth of the penetration can be very precisely controlled up to 0.1 mm by variation in frequency.

(iii) Annealing of Metals Induction heating is preferred for annealing of metals like brass, bronze, etc., which have a tendency of breaking when annealed in cold state, such metals are therefore first heated by induction heating up to about 750°C, within some seconds and then annealed.

(iv) Brazing The process of brazing requires heating of nonsimilar metals. The metals will have different values of resistivity and hence takes different amount of time to heat up to the desired temperature. Therefore in order to heat the slowly heating metal quickly, the magnetic flux of the work coil must be concentrated on the metal that heats up slowly. This will ensure proper joints at correct temperature before the brazing alloy melts.

(v) Soldering High quality soldering can be expeditiously done at high temperatures, by induction heating, when silver, copper and their alloys are used as solders.

(vi) Other Applications Induction heating finds its use in variety of industrial jobs such as welding, forging of bolt heads, drying of paints on metals, sintering-powdered metals, sterilizing surgical instruments, melting of metals, etc.

SOLVED EXAMPLES

Example 16.11 A metal cylinder 10 cm long and 1 cm diameter is placed inside a coil having 15 turns. If the coil carries current of 80 A at 200 kHz, determine the depth of heating and heat generated per unit surface area. Assume $\mu_r = 1$ and the resistivity of the metal to be 5×10^{-8} Ω-m.

Solution:
Given: $L = 1$ cm $= 0.01$ m, $N = 15$ turns., $i = 80$ A, $f = 200 \times 10^3$ Hz, $\rho = 5 \times 10^{-8}$ Ω-m

(i) Depth of heating $(\delta) = 503 \sqrt{\dfrac{\rho}{\mu_r \cdot f}} = 503 \sqrt{\dfrac{5 \times 10^{-8}}{1 \times 200 \times 10^3}}$

$\qquad = 251.5 \times 10^{-6}$ metres $= 0.251$ mm

(ii) Heat per unit area of the surface $= H$

$= 2 \cdot \pi \cdot \left(\dfrac{NI}{L}\right)^2 \cdot \dfrac{\rho \cdot \mu_r \cdot f}{10^7}$ W/m^2 $= 2 \cdot \pi \left(\dfrac{15 \times 80}{0.01}\right)^2 \cdot \sqrt{\dfrac{5 \times 10^{-8} \times 1 \times 200 \times 10^3}{10^7}}$

$\qquad = 2861.16$ kW/m^2

Example 16.12 Determine the supply frequency for the case hardening of shaft having specific resistivity of 5×10^{-5} Ω-cm and the relative permeability equal to 1 for depth of heating 2.5 mm.

Solution:
Given: $\delta = 2.5$ mm $= 2.5 \times 10^{-3}$ m, $\rho = 5 \times 10^{-5}$ Ω-cm $= 5 \times 10^{-7}$ ohm-m $\mu_r = 1$

$$\therefore f = \frac{\rho}{u_r}\left(\frac{503}{\delta}\right)^2 = \frac{5 \times 10^{-7}}{1}\left(\frac{503}{2.5 \times 10^{-3}}\right)^2 = 20.24 \text{ kHz}$$

16.6.2 Dielectric Heating

Dielectric heating is another type of *RF* heating. It makes use of the dielectric loss in imparted dielectric materials and is therefore suitable for heating non-metals.

When a.c. current of magnitude I at frequency f(Hz) is passed through a capacitor, then the voltage developed across a capacitor is given by:

$$V = \frac{1}{j \, 2\pi \cdot f \cdot c} I = -j \cdot \frac{1}{2\pi f.c} \cdot I \qquad (16.66)$$

Term $-j$ indicates that the ac voltage across the capacitor lags behind the current by 90°, if the capacitor is ideal one. At high frequencies, a practical capacitor can be represented by the equivalent circuit of Fig. 16.32(a), where, C is the rated value of the capacitance.

R_p is the equivalent parallel resistance.
R_s is the equivalent series resistance, and
L_s is the equivalent series lead inductance.

Figure 16.32(b) shows the corresponding phasor diagram. I_A a practical capacitor at high frequencies, the voltage across the capacitor lags behind the current phasor by an angle θ (power factor angle) which is less than 90°. The current can be resolved into two components, one in phase with the voltage and the other at 90° with respect to voltage as shown in Fig 16.32(b).

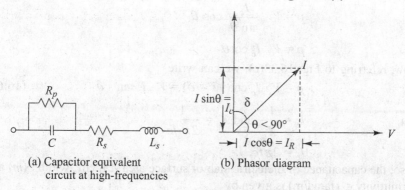

(a) Capacitor equivalent circuit at high-frequencies

(b) Phasor diagram

Fig. 16.32

The angle ($\delta = 90° - \theta$) is called as the loss angle and is a measure of the loss in the capacitor. The dissipation factor (D) of the dielectric material of the capacitor is given by

$$\tan \delta = D = 2 \cdot \pi \cdot f \cdot c \cdot R_s \tag{16.67}$$

At low frequencies, $\tan \delta$ is small giving low dissipation. However, at radio frequencies, $\tan \delta$ is quite large giving considerable dissipation across the capacitor. This, dissipation loss can be effectively utilized to heat the dielectric materials like wood, plastic, etc., by placing the same between two conducting electrodes, giving rise to the capacitive effect. This is the principle on which dielectric heating works.

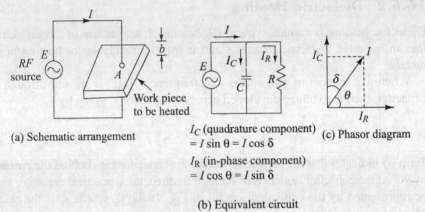

(a) Schematic arrangement

I_C (quadrature component)
$= I \sin \theta = I \cos \delta$

I_R (in-phase component)
$= I \cos \theta = I \sin \delta$

(c) Phasor diagram

(b) Equivalent circuit

Fig. 16.33

As shown in Fig. 16.33, the material to be heated in dielectric heating is placed between two metallic electrodes which in turn are connected across the tank circuit of an oscillator and forms part of the tank circuit capacitance. Then power dissipated in the dielectric slab is given by

$$p = V \cdot I_R = V \cdot I \cdot \cos \theta \tag{16.68}$$

(But $I_c = I_0 \sin \theta$)

$$\therefore \quad p = V \cdot \frac{I_c}{\sin \theta} \cos \theta$$

$$p = V \cdot I_c \cot \theta$$

Now, referring to Fig. 16.33(c), we can write

$$p = V \cdot I_c \cot(90° - \delta) = V \cdot I_c \tan \cdot \delta \tag{16.69}$$

But, $\quad I_c = \dfrac{V}{X_c}$ and $X_c = \dfrac{1}{2\pi fc}$

$\therefore \quad I_c = 2\pi fc \cdot V$

Also, the capacitance of dielectric slab of surface area AA, cm^2 width $b(m)$ and permittivity ϵ (farad/m) is given by

$$C = \frac{\epsilon \cdot A}{b} \text{ farad} \tag{16.70}$$

Substitute the values of C and I_c in Eq. (16.69), we get
$$p = 2\pi \cdot f \cdot c\, V^2 \tan \delta \qquad (16.71)$$
i.e
$$p = \frac{2\pi \cdot f \cdot \epsilon\, A}{b} V^2 \tan\delta \qquad (16.72)$$
but volume of dielectric slab = surface area × width
$$= A \cdot b \cdot (m^3)$$
Hence, power loss per unit volume of the dielectric is given by
$$p_o = 2\pi \cdot f \cdot \epsilon \left(\frac{V}{b}\right)^2 \tan\delta,\; Watts/m^3 \qquad (16.73)$$
Now, $\epsilon = \epsilon_o \cdot \epsilon_r$, where ϵ_o = permittivity of free-space
$$= \frac{1}{36 \cdot \pi} \times 10^{-9}\; farad/m \qquad (16.74)$$
and ϵ_r = relative permittivity of the dielectric – slab.
If b is expressed in cm., then
$$p_o = 2\pi f \left(\frac{1}{36\pi} \times 10^{-9}\, \epsilon_r\right) \cdot \left(\frac{V}{10^2 \cdot b}\right)^2 \tan\delta \cdot Watts/cm^3$$
$$= 0.555 \times 10^{-12} f\, \epsilon_r \cdot \left(\frac{V}{b}\right)^2 \tan\delta \cdot Watts/cm^3 \quad (16.75)$$
Note that power factor of the capacitor is given by
p.f. = $\cos\theta = \cos(90° - \delta)$, where δ is the loss angle
= $\sin\delta \approx \delta$, if the loss angle is small
= $\tan\delta$, the loss tangent
Thus, for small loss angles,
p.f. ≈ loss tangent (16.76)
Using Eq. (16.75) and (16.76), we got power loss per unit volume in the dielectric slab as
$$P_o = 0.555 \times 10^{-12} f \cdot \epsilon_r \left(\frac{V}{b}\right)^2 \times (p.f.) \qquad (16.77)$$
The factor $(\epsilon_r \tan\delta)$ is called the *loss-factor* of the dielectric, so that another expression for p_o is
$$P_o = 0.555 \times 10^{-12} f \cdot \left(\frac{V}{b}\right)^2 (loss\; factor) \qquad (16.78)$$
From Eq.(16.78), it is cleared that the power loss per unit volume (p_o) in a dielectric slab depends on the following factors:
  (i) It increases with the square of the voltage gradient, $(V/b)^2$,
  (ii) It increases with increase in frequency of operation, and
  (iii) It depends upon the constants of the dielectric material through the loss factor $(\epsilon_r \cdot \tan\delta)$.

16.6.2.1 Applications of Dielectric Heating

Dielectric heating is employed in industrial process for heating nonconducting objects. Some of the applications are:

(i) **Food Processing:** In food processing industries, dielectric heating is employed for the following purposes:
   (a) Defrosting of frozen foods which then proceeds to cook the food in few minutes.
   (b) Pasteurizing milk and bear inside the bottles or containers.
   (c) For disinfecting grains and cereals by killing eggs of insects and pests.
   (d) For cooking foods without removing the outer-shell.
   (e) For dehydrating fruits, vegetables, etc.
   (f) For sterilizing food while sealed in their final container.

(ii) **Plastic Processing:** Many articles are made by moulding or shaping plastic materials with large press machines. Only at higher temperature, the plastic can flow into the shape in press machines. To increase the production rate, the biscuit-size 'preform' of plastic is preheated by dielectric heater before the press machine receives it and moulds it into the desired shape.

(iii) **Wood Processing:** In the plywood manufacturing, the glue between the layers of wood is dried.

(iv) **Electronic Sewing:** For Sewing plastic fabrics, they are pressed and passed together between roller type electrodes of dielectric heater. Using such sewing many watertight articles such as raincoats, umbrellas, food packets, shower curtains, etc., can be produced.

(v) Drying and heat treatment of textiles such as rayon, nylon, terylene etc.

(vi) Processing of chemicals during their manufacture.

(vii) Producing artificial fever in human body for medical treatment.

16.6.2.2 Comparison between Induction and Dielectric Heating

Induction Heating	Dielectric Heating
(i) It can be employed for heating the obfjects of only conducting materials.	(i) It can be employed for heating the objects of only non-conducting materials.
(ii) Rate of heating is proportional to the square of current (I^2) and square root of frequency ($\sqrt{f}$).	(ii) Rate of heating is proportional to the square of applied voltage (V^2) and the frequency of supply (f).
(iii) In induction heating, the change in frequency f affect the depth of penetration of heat and also the rate of heating.	(iii) In dielectric heating, the change in frequency (f) affect only the rate of heating.
(iv) Heating is not uniform throughout the work piece.	(iv) Heating is uniform throughout the work piece.

SOLVED EXAMPLES

Example 16.13 A work piece of dielectric material with size 25 cm × 15 cm × 1cm (thick) is to be heated using 1 kW heating power. If relative permittivity of material is 4 and power factor is 0.6, determine voltage and current necessary at frequency of 40 MHz.

Solution: Area $(A) = (25 \times 15)$ cm^2 = 375 cm^2 = 0.0375 m^2.
b = distance between electrodes = 1 cm = 0.01 m.

$$\epsilon_r = 4, \cos\theta = 0.6, f = 4 \times 10^6 \text{ Hz}.$$

From Eq.(16.69), $C = \dfrac{\epsilon_0 \cdot \epsilon_r \cdot A}{b} = \dfrac{8.854 \times 10^{-12} \times 4 \times 0.0375}{0.01} = 132.81 \times 10^{-12}$ F

Power dissipated = Heating power = 1 kW.
From Eq. (16.70), $p = 2\pi f.c. V^2 \cos\theta$ [∵ $\tan\delta = \cos\theta$]

$$\therefore V^2 = \dfrac{1000}{(2\pi \times 40 \times 10^6 \times 132.81 \times 10^{-12} \times 0.6)}, V = 223.45 \text{ V}.$$

Now, current $I = \dfrac{V}{X_C} = 2\pi f.c.V = 2\pi \times 40 \times 10^6 \times 132.81 \times 10^{-12} \times 223.45$ ∴ $I = 7.46$ A.

Example 16.14 In a dielectric heating process, the capacitor formed between the two electrodes is 5.3 μF. If supply voltage is 600 V ac and current flowing through capacitor is 100 A, determine the supply frequency.

Solution:

We have, $X_c = \dfrac{V}{I}$ ∴ $f = \dfrac{I}{2\pi \cdot c \cdot v} = \dfrac{100}{2\pi \times 5.3 \times 10^{-6} \times 600} = 5$ kHz.

16.7 SWITCH-MODE WELDING

In arc welding, an arc is established between two electrodes, one of the electrodes being the metal work piece. In all welding applications, the output needs to be electrically isolated from the utility input. The voltage-current characteristic of the welder depends on the type of welding process employed. Also, another requirement is to maintain a low-ripple current after the arc is established. Figure 16.34 shows the block-diagram of the switch-mode welder. Here, the isolation is provided by a high frequency transformer. A small inductance is needed at the output to limit the output current ripple at high frequencies. The efficiency of such welder is in the 85–90% range. Also, these welders have smaller weight and size compared with the welders employing a 50Hz power transformer.

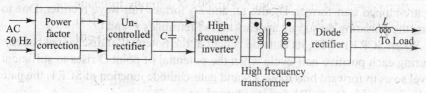

Fig. 16.34 *Switch-mode welder*

16.8 ELECTRONIC LAMP BALLAST

Lighting consumes nearly 15–20% of energy in residential buildings and nearly 30–35% in commercial buildings. Fluorescent lamps are three to four times more energy efficient than incandescent lamps. The energy efficiency of fluorescent lamps can be further boosted by 20% to 30% by operating at frequencies higher than 25 kHz. Conventional fluorescent lamps use an inductive ballast (also called as the choke) in series for stable operation with 50 Hz a.c. source shown in Fig. 16.35(a). Figure 16.35(b) shows the block-diagram of a modern high-frequency fluorescent system. The uncontrolled ac–dc converter with filter is followed by a high-frequency dc–ac inverter. Class-E resonant inverter can be used here. The switching frequency of the inverter lies in the range of 25 to less than 40 kHz and 60 kHz or higher. The current from the source has a very poor power factor, hence, the power factor correction is very necessary. The EMI filter prevents the conducted electromagnetic interference signal from being fed back to the ac line.

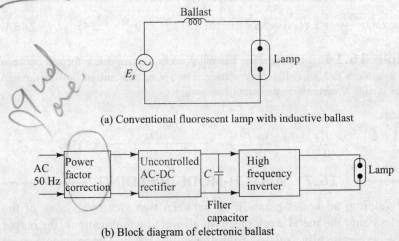

(a) Conventional fluorescent lamp with inductive ballast

(b) Block diagram of electronic ballast

Fig. 16.35

16.9 BATTERY CHARGER

The increase in the usage of portable appliances and communication equipment is responsible for a spurt in the demand of the low-power battery charger systems. Automatic battery charging circuit using SCR is shown in Fig. 16.36. A 12 V discharged battery is connected in the circuit as shown. When switch S_1 is closed, the single-phase 230V supply is stepped down to (15–0–15) V by a centre-tapped transformer. Diodes D_1 and D_2 formes full-wave rectifier. Due to this, the pulsating dc supply appears across terminals P and Q.

When SCR1 is OFF, its cathode is held at the potential of discharged battery. During each positive half-cycle when the potential of point O rises to sufficient level so as to forward bias diode D_3 and gate-cathode junction of SCR1, the gate pulse is provided to SCR1 and it is turned-on.

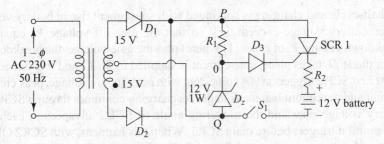

Fig. 16.36 *Battery charger using SCR*

When SCR1 is turned-on, the charging current flows through battery. Thus, during each positive half-cycle of pulsating dc supply voltage across P–Q, SCR1 is triggered and charging current is passed till the end of that half-cycle. Due to zener diode D_z, the maximum voltage of point O is held at 12 V. Due to the charging process, the battery voltage rises and finally attains full-value of 12 V. Thus, when the battery is fully charged, the cathode of SCR1 is held at 12 V. Therefore, diode D_3 and gate-cathode junction of SCR1 cannot be forward biased, since the potential of point O can reach up to 12 V. Hence, no gate-current is supplied and SCR1 is not triggered. In this way, after full charging further charging is automatically stopped.

Automatic Battery Charger with Trickle Charging Arrangement Figure 16.37 shows the circuit arrangement for automatic battery charger with trickle charging. Using centre-tap transformer, the input 230 V AC supply is stepped-down to 15–0–15 V AC supply using centre-tapped transformer. Diodes D_3 and D_4 act as rectifier and generate pulsating dc supply. This dc voltage is applied to battery charger circuit between points P and Q. The discharged battery is connected between terminals O and Q. When battery voltage is less than 11V, the peak-off voltage V_{R_3} at the wiper of pot R_3 is less than the breakdown voltage V_z (here 11V) of zener diode D_z. Hence-gate-current is not injected into SCR 2 and it is not turned-on. Thus, with SCR 2 OFF, during each positive half-cycle of pulsating dc supply voltage, SCR1 receives gate-current through R_1 and D_1 and turned-on.

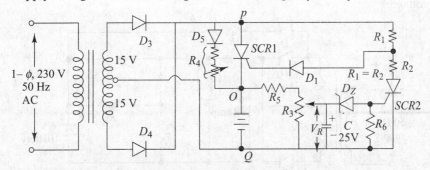

Fig. 16.37 *Battery charger with trickle charging arrangement*

After turning-on SCR1, charging current flows through battery. In this way, during each positive half-cycle, SCR1 is triggered and it conducts till the end of the half-cycle and charging is continued with subsequent rise in battery voltage. When battery voltage-exceeds 11V so that the pick-off voltage V_R equals the breakdown voltage V_z of zener (11V here) plus the gate-voltage, the breakdown of zener diode D_z takes place, gate-current is supplied to SCR2 and it is turned-on.

At first SCR2 triggers at 90° coincident with peak supply voltage, peak charging current and maximum battery voltage. As charging continues through SCR1, the battery voltage rises and the triggering angle of SCR2 advances in each half-cycle until it triggers before main SCR1. When this happens, with SCR2 ON, the voltage divider arrangement of R_1 and R_2 keeps voltage at anode of D_1 less than voltage at cathode of SCR1. Therefore, D_1 gets reverse biased and gate current is not supplied to SCR1. Thus, SCR1 is prohibited from turning-on and heavy charging current through SCR1 is stopped.

For trickle charging, a branch containing diode D_s and register R_4 can be added. So, when SCR1 is OFF, the low charging current is maintained through D_5 and R_4, keeping slow charging of battery. The heavy charging can start only after the battery voltage drops so that V_R drops below V_Z and triggering of SCR2 is stopped in each cycle.

16.10 EMERGENCY LIGHTING SYSTEM

Figure 16.38 shows a very simple single-source emergency lighting system which can be most suitable in household application. The input 230 V ac supply is stepped down to 6–0–6 V ac supply by centre tapped transformer. The diodes D_1 and D_2 form full-wave rectifier and converts 6–0–6 V ac supply to 6V DC supply for 6V lamp. When ac supply is available, 6 V dc supply appears across lamp and it glows. Pulsating current also flows through D_3, R_1 to trickle charge the battery. Thus battery charging is carried-out. The capacitor C gets charged with upper plate positive to some voltage less than 6 V. Due to capacitor voltage, gate-cathode junction of SCR1 gets reverse-biased. The anode is at battery voltage and cathode is at rectifier output voltage, which is slightly higher, hence SCR1 is reverse-biased and cannot conduct. The lamp glows due to rectifier output dc voltage.

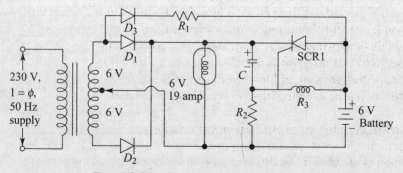

Fig. 16.38 *Emergency lighting system*

16.11 STATIC CIRCUIT BREAKER

16.11.1 Static DC Circuit Breaker

Figure 16.39 shows circuit configuration of static circuit breaker using SCR. The circuit is basically a parallel capacitor commutated power flip-flop. SCR1 receives gate-current through R_3 when the start button is momentarily depressed and device starts conducting. Turning-on of SCR1 causes major part of supply voltage to appear across load and hence power is delivered to load. The capacitor C gets charged to load voltage with right-plate positive, as shown in Fig. 16.39 through R_4 and SCR1.

Stop-button is momentarily depressed to switch-off load. SCR2 receives gate-current through R_5 and it is turned-on. Turning-on of SCR2 causes charged capacitor C to place across conducting SCR1. Capacitor C applies reverse-bias for SCR1 and starts discharging. Due to this, SCR1 turns-off and load current is continued through C and SCR2. Capacitor C gets fully discharged and then charges with reverse polarity to supply voltage. Load current at this instant falls to zero. The current flowing through R_4 and SCR2 is below holding valve of SCR2, hence SCR2 is turned-off. Thus, manual firing of SCR2 by pressing stop-button interrupts load current through SCR1 and finally opens the circuit.

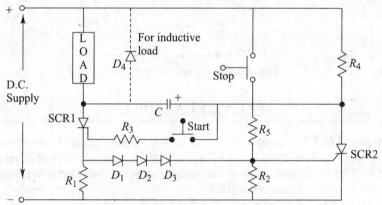

Fig. 16.39 *DC circuit breaker*

Under overload condition, the load current can be automatically interrupted. Consider that SCR1 is ON and carrying load current. If overload occurs, the voltage drop across R_1 exceeds the forward voltage drop of string of series connected diodes D_1, D_2, D_3 and gate-cathode junction. Gate-current is therefore supplied to SCR2 and it is turned-on. The turning-on of SCR2 results in turning-off of SCR1 immediately due to capacitor voltage and load is switched-off from the circuit. The circuit can be made to trip by adjustment of the value of R_1 and by selecting the proper number of series diodes $D_1, D_2, D_3 \ldots$ etc., and interrupt overload or fault current at any predetermined level.

16.11.2 Static AC Circuit Breaker

Figure 16.40(a) shows the circuit configuration of static ac circuit breaker. SCRs1 and 2 are triggered in positive and negative half-cycles respectively when switch

S is closed. During positive half-cycle of the input, SCR1 receives gate current through $(D_2 \parallel R_2)$, switch S and R_3 and it conducts. At the end of positive half-cycle, SCR1 is turned-off due to natural current zero.

SCR2 receives gate-current through $(D_1 \parallel R_1)$, R_3 and switch S during negative half-cycle and conducts. It is turned-off at the end of this negative half-cycle due to natural current zero value. When the load current is required to be interrupted, the switch S is opened. Opening of switch S results in blocking of gate currents of both SCRs and hence both SCRs are maintained-off. When switch S is opened at any instant in a particular half-cycle, the load current continues to flow through conducting SCR till the end of this half-cycle, however in the next half-cycle the other SCR is not triggered due to nonavailability of gate current. Thus, the maximum time delay for breaking the circuit is one half-cycle.

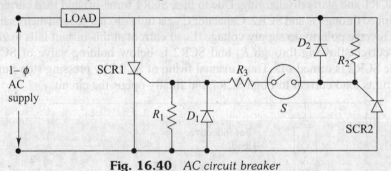

Fig. 16.40 *AC circuit breaker*

SOLVED EXAMPLE

Example 16.15 A static d.c. circuit breaker employs Class C turn-off method using auxiliary thyristor. If the supply voltage is 50 V d.c., load current is 10 A, calculate the value of load resistance, value of commutating capacitor and value of the resistance connected in series of auxiliary thyristor.

Given: $t_{off} = 20$ μs for main thyristor, $I_H = 5$ mA for auxiliary SCR

Solution: Circuit diagram is given in Fig. Ex. 16.15

(i) $R_2 = E_{dc}/I_H = 50/5 \times 10^3 = 10$ kΩ

(ii) Now load resistance

$$R_1 = \frac{E_{dc}}{I_L} = \frac{50}{10} = 5 \, \Omega$$

(iii) $C \geq \dfrac{1.5 \, t_q \, I}{E}$

$\geq \dfrac{1.5 \times 20 \times 10^{-6} \times 10}{50}$

$\geq 6 \, \mu F$

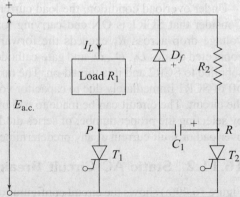

Fig. Ex. 16.15 *D.C. circuit breaker*

16.12 TIME-DELAY CIRCUIT

Figure 16.41 shows a simple but accurate and versatile time-delay circuit using SCR and UJT. When switch S_1 is closed, UJT relaxation oscillator gets energized and capacitor C starts charging through R_1 and R_2. During charging, since charging current is small, the voltage drop across R_5 is small, and negligible voltage appears across external load.

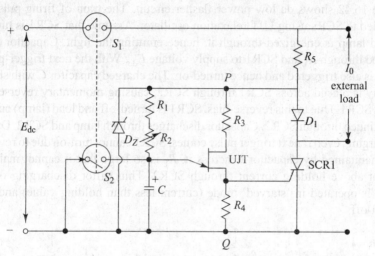

Fig. 16.41 *Time delay circuit using SCR and UJT*

When the capacitor C charges to peak point voltage of UJT, the UJT is turned-on and pulse voltage appears across R_4. SCR1 receives this pulse and remains-on. Once SCR1 turns-on, the supply voltage appears across R_5 assuming negligible voltage drop across D_1 and SCR1. Thus, turning-on of SCR1 results in application of voltage and supply of power to external load. After firing of SCR1, the voltage across relaxation oscillator circuit (PQ), falls to very low value (less than 2 V). Due to this, a low voltage is maintained across C and reasonable accuracy is achieved if the circuit is rapidly recycled. From the above discussion, it is seen that the voltage is applied across load after a time delay from the instant of closing switch S_1. Time delay depends on the time constant $(R_1 + R_2)$. C and can be set to any desired value by proper selection of R_1, R_2 and C. The maximum value $(R_1 + R_2)$ is determined by the requirement that the emitter current into UJT be large enough to turn it ON. The circuit can be reset by opening switch S_1. Switchs S_1 and S_2 are mechanically engaged, hence when S_1 is closed, S_2 gets opened and short-circuit across C is removed.

16.13 FLASHER CIRCUITS

Flasher circuits are basically power-flip-flop circuits. The output can be turned on or off by switching the power device. SCR and triac are ideally suited for this type of application since they can function over a wide range of voltage and current with much higher degree of reliability than the commonly used electromechanical systems. Also, SCR and triac can handle the inrush currents of some loads such as incandescent lamps, hence they do not require large current derating. Flasher circuits are used in traffic lights, aircraft beacons, navigational beacons and illuminated signs.

16.13.1 DC Low Power Flasher

Figure 16.42 shows dc low power flasher circuit. The train of firing pulses is provided to SCRs using UJT relaxation oscillator. Assume that SCR1 is turned-on and lamp is energized through it, hence emitting the light. Capacitor C_1 is charged through R_1 and SCR1 to supply voltage E_{dc}. With the next trigger-pulse, SCR2 is also triggered and hence turned-on. The charged capacitor C with shown polarity is placed across SCR1 through SCR2, causing momentary reverse bias across SCR1. Due to this reverse-bias, SCR1 is turned-off and load (lamp) current is continued through SCR2. Capacitor discharges through lamp and SCR2. During discharging, even if next trigger pulse comes, SCR1 cannot turn-on due to reverse-bias maintained by capacitor C across it. R_1 is so large that it cannot maintain current above holding current through SCR2. Thus, after discharging of C_1, SCR2 is operated in 'starved' mode (current less than holding value) and it is turned-off.

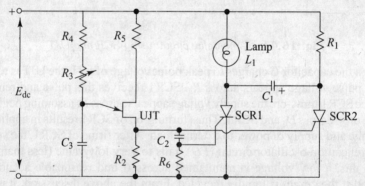

Fig. 16.42 *DC low power flasher*

When next trigger pulse comes, SCR1 is turned-on, lamp glows and capacitor C_1 gets charged through R_1 so as to develop commutating voltage for SCR1. Switching of SCR1 gives us flashes from lamp. The flash rate thus depends on triggering of SCR1, which depends on the firing pulse frequency of UJT relaxation oscillator. Therefore, by adjusting pot R_3, the required flash rate can be obtained.

16.13.2 Sequential Flasher

A sequential flasher using SCR and PUT is shown in Fig. 16.43. Lamp L_1 glows and timer 1 gets energized when switch S_1 is closed. Capacitor C_1 charges through R_2 and lamp L_2. Since the charging current through lamp L_2 is not sufficient, lamp L_2 cannot glow. When C_1 charges to a voltage one diode drop higher than the gate voltage of PUT1, it is triggered. The turning-on of PUT1 causes firing pulses for SCR1 and it is turned-on. When SCR1 turns-on, supply voltage appears across lamps L_2 and it glows. Thus lamp L_2 glows after lamp L_1 with some time delay. The turning-on of SCR1 also results in energizing of timer 2. Similar sequence of operations as in timer 1 takes place in timer 2 and after some time delay lamp L_3 glows. Thus, lamp L_1, L_2 and L_3 glow in a particular sequence and time delay depends on the delay of firing pulse provided by PUT relaxation oscillator to SCR1 and SCR2. After glowing of lamp L_3, thermal flasher switch gets opened, both SCR1 and SCR2 are commutated and circuit becomes ready for next-cycle. Such a sequential flasher is used in automotive turn-signals.

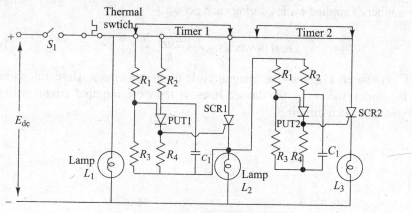

Fig. 16.43 *Sequential flasher*

16.14 INTEGRAL CYCLE TRIGGERING (OR BURST FIRING)

The integral cycle control, which is sometimes known as burst-firing, means that where full cycles of the supply voltage are applied to the load, control is affected by changing the number of full cycles applied compared to the number of cycles when the supply is not connected to the load at all. Any particular control condition result in a number of full "on" cycles followed by a number of full "off" cycles. Figure 16.44 shows this generalised condition where the time-period of T cycles is split up into N "on" cycles and $(T - N)$ "off" cycles.

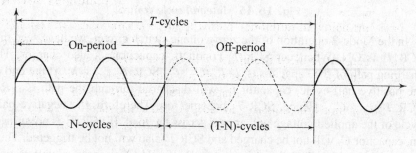

Fig. 16.44 *Generalized integral-cycle control*

The value of N and T may vary over the control range, both N and $(T-N)$ may vary from zero to T cycles. In this way the effective voltage, current, and power applied to the load can be controlled.

The principle advantage of this on-off method of control is that sinusoidal currents will flow during the on-periods, so avoiding the high-frequency harmonic currents which will exist with phase control in every cycle.

In practical systems of integral cycle control, the thyristors are usually switched ON and OFF, at, or very near to, the voltage zero points. The load power passed to a resistance load with integral cycle control can be calculated simply from the number of applied cycles during each period:

$$\text{Load power } P_{\text{Load}} = \frac{E_{\text{load}}^2}{R} \frac{N}{T} \qquad (16.78)$$

Figure 16.45 shows the integral cycle triggering circuit. Here, the portion of the circuit shown by the dashed lines, is the extra-required circuit with zero-voltage switch circuit.

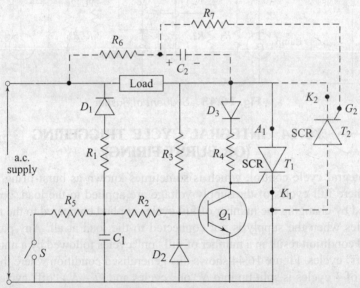

Fig. 16.45 *Integral-cycle control*

In the Mode-2 operation of the zero-voltage switch circuit, we have seen that SCR T_1 is ON. Whenever SCR T_1 conducts, capacitor C_2 also charges. The charging path of the capacitor is, $L-R_6-C_{2+}-C_2-\text{SCR } T_{1(A_1-K_1)}-N$. At the end of the positive half-cycle, capacitor C_2 will discharge through the path $C_{2+}-R_7-$ SCR $T_{2(G_2K_2)}-C_{2-}$. Hence, SCR T_2 becomes on. Therefore, the negative half-cycle of the applied voltage will appear across the load. If SCR T_1 is not turned-on, capacitor C_2 will not be charged and SCR T_2 also will not be triggered. Thus, the negative half-cycle follows the positive half-cycle of the applied voltage only if SCR T_1 is turned on, or if the switch S is opened. Therefore, the load will carry an integral number of full-cycles of applied voltage. If switch S is closed, the load-current will stop at the end of the present or the following negative half cycle. hence, the maximum interrupting time will be one cycle.

Heating loads are suitable for integral cycle control but lighting and motor loads cannot, in general, operate satisfactorily in this pulsating-manner.

SOLVED EXAMPLE

Example 16.16 A single-phase resistive heating load is to be controlled from a single phase, 50 Hz a.c. supply be means of an inverse parallel pair of SCRs. What will be the firing-angle of the thyristors when the load is 60% of its maximum value? If this controller is replaced by an integral triggering system operating with a constant repetition period of 0.24 s, what will be range of output powers available?

Solution:

(i) We can write, $P_{max} = E_m^2/2R$

Hence, $\dfrac{P_{load}}{P_{max}} = (\pi - \alpha + 1/2 \sin^2 \alpha)/\pi$

When $P_{load} = 0.6\, P_{max}$, $0.6 = (\pi - \alpha + 1/2 \sin 2\alpha)/\pi$ ∴ $\alpha = 80.55°$

(ii) For integral cycle triggering, the repetition frequency of 0.24 s represents 12 cycles at 50 Hz or 24 half-cycles.

Therefore, using Eq. (16.78), $T = 24$

and $\dfrac{P_{load}}{P_{max}} = \dfrac{N}{T} = \dfrac{N}{24}$.

Available power range from 4.17 % of P_{max} ($N = 1$) to 100% of P_{max} ($N = 24$), varying in steps of 4.17% of P_{max}.

REVIEW QUESTIONS

16.1 With the help of block diagram, explain the operation of solid-state UPS system.
16.2 Describe the operation of on-line UPS system with the help of neat block diagram. Also, list the important specifications of on-line UPS.
16.3 Why an on-line UPS is costlier?
16.4 Draw a schematic diagram of off-line UPS and its operation. Also list the specifications of off-line UPS.
16.5 List the merits and demerits of on-line UPS and off-line UPS.
16.6 Why an off-line UPS is comparatively cheaper?
16.7 With the help of block diagram explain the operation of off-line interactive UPS system.
16.8 What do you mean by redundancy in UPS system? How many types of redundant systems are there?
16.9 Explain the following terms:-
　(i) Capacity of the battery in UPS system.
　(ii) Efficiency of the battery

16.10 With the help of block-diagram explain the basic principle of operation of SMPS.

16.11 With the help of circuit diagram and waveforms explain the operation of flyback converter in discontinuous mode. Also, list the advantages and disadvantages.

16.12 With the help of heat circuit diagram and relevant waveforms, explain the operation of continuous mode flyback converters. Also, derive the expression for voltage transfer ratio in terms of transformer turns ratio and duty ratio.

16.13 Discuss the selection criterion for various components of flyback converter.

16.14 with the help of neat circuit diagram explain the operation of isolated forward converter. Also, explain the disadvantages of the basic circuitry.

16.15 List the distinct differences between flyback converter and forward converter.

16.16 With the help of neat circuit diagram and waveforms explain the operation of forward converter with tertiary winding. Also list the advantages and disadvantages of the same.

16.17 Explain the following design issues related to forward converter:
  (i) Selection of switching transistor
  (ii) Selection of converter transformer.

16.18 What is the need of demagnetizing winding in transformer for forward converter?

16.19 With the help of circuit diagram and waveforms explain the working of two transistor forward converter.

16.20 With the help of neat circuit diagram and waveforms explain the working of half-bridge converter. List the advantages and the disadvantages of the same.

16.21 Derive the expression for output voltage of half bridge converter.

16.22 Explain the effect of series coupling capacitor on the performance of half-bridge converter.

16.23 Draw and explain the working of full-bridge converter.

16.24 Explain the basic principle of HVDC transmission.

16.25 List the advantages of HVDC transmission.

16.26 List the drawbacks of the HVDC transmission.

16.27 List the various types of HVDC transmission systems and explain any two.

16.28 Explain the basic principle of static VAR compensators.

16.29 With the help of neat diagram, explain the operation of switched inductor compensator.

16.30 Briefly explain switched capacitor compensator.

16.31 Describe briefly the switch mode compensator.

16.32 Explain the principle of induction heating. Enlist the merits of induction heating over conventional methods.

16.33 Explain briefly at least four applications of induction heating.

16.34 Explain the basic principle of high-frequency dielectric heating. Derive expression for heating power per unit volume. Give two applications.

16.35 Compare dielectric heating with induction heating.

16.36 Explain briefly the operation of switch mode welding.

16.37 With the help of block diagram, explain the operation of electronic lamp ballast.

16.38 Draw and discuss the operation of automatic battery charger with trickle charging arrangement.

16.39 Briefly explain the operation of single source emergency lighting system.

16.40 Draw and explain the operation of d.c. circuit breaker. List its applications.

16.41 Briefly explain the operation of a.c. circuit breaker.
16.42 Draw and explain the operation of time delay circuit using SCR and UJT.
16.43 Discuss the operation of d.c. low-power flasher circuit.
16.44 Draw and explain the operation of sequential flasher.

PROBLEMS

16.1 A UPS is driving an 800W load which has a lagging power factor of 0.8. The efficiency of the inverter is 85%. The battery voltage is 48 V d.c. Assume that there is a separate charger for the battery. Determine the following:
  (i) kVA rating of inverter
  (ii) Wattage of rectifier
  (iii) AH rating for 30min back-up time
  [*Ans:* (i) 1 kVA, (ii) 1 kW, (iii) 10.5 AH].

16.2 Select a suitable battery for the following specifications:
  (i) UPS rating = 50 kVA
  (ii) Type of load: Inductive with a lagging P.F. = 0.8.
  (iii) Back-up time = 15min. (iv) Battery type = lead-acid
  (v) Final Volts = 1.75 V/Cell (vi) Battery voltage range = 210–280 V.
  (vii) Efficiency = 90% (viii) Cell group = Six per jar.
  [*Ans:* Battery kW = 44.44 kW, No. of cells = 120, Two 75 AH cell, Capacity = 0.426 kW.]

16.3 A single transistor flyback SMPS operating at 16 kHz is supplying a mean load power of 120 W at a mean voltage of 80 V from a d.c source of 110 V. Estimate the mark/space ratio of the output voltage and value of inductance required in the circuit.
  [*Ans:* 0.559 mH].

16.4 A brass tube 25 cm long is placed within a coil. It has 20 turns and carries 200A current at a frequency of 450 kHz. If resistivity of brass is 6.8×10^{-8} Ω m. Determine: (i) depth of penetration, and (ii) heat per unit area of the surface.
  [*Ans:* (i) δ = 0.19mm., (ii) H = 88.92 kW/m^2]

16.5 A work piece of dielectric material with size 20 cm × 15 cm × 1 cm (thick) is heated using frequency of 20 MHz and applied voltage of 750 V. If relative permittivity of material is 3 and power-factor is 0.4, calculate power developed per unit volume.
  [*Ans:* $p = 7.51 \times 10^6$ W/m^3]

REFERENCES

1. Mohan/Undeland/Robbins, *Power Electronics*, John Wiley and Sons, INC.
2. C.W Lander, *Power Electronics*, McGraw-Hill 1981.
3. B.K. Bose, *Modern Power Electronics*, B.K. Bose, 2000.

Appendix A

Simulation Tools for Power Electronic Circuits

Simulation tools are used both for research and teaching, to allow a good understanding of the structures that are studied before practical tests. The computer simulation of power electronics systems has presented many challenges and opportunities over the years. Fortunately, the general nature of power electronics systems remained relatively the same for a long period of time. This allowed system engineers to improve modeling techniques progressively and to apply computer hardware and software technology to design study tools that met the analysis requirements. The models were based on fundamental frequency responses. However, with the widespread use of processor-based controls and the associated advances in power electronic devices over the past 10 years, the nature of modern power electronic systems has significantly changed.

Computer simulation of power electronic circuits provides a sound solution for circuit design and analysis, as computers are well capable of performing the tedious, lengthy and repetitive computations. From a circuit designer and practitioner viewpoint, this allows an investigation of the circuit behavior prior to its fabrication and hence resulting in a considerable saving in material and labor. From an educational viewpoint, the student would gain some additional insight from this strong waveform oriented approach, which otherwise might be overlooked.

Currently, there are a number of powerful and flexible circuit simulation packages in the market; some at no cost to the academic community and some at reasonable cost for given features. Adoption of these circuit simulators for classroom applications is considered to be an effective educational tool for teaching a wide range of core and elective subjects: from elementary circuit analysis to the advance topic such as Fourier analysis, active filter design, and switched-mode and resonant-mode power electronic circuits.

The widespread acceptance of simulation packages for classroom applications is mainly related to the availability of digital computer facilities in tertiary institutions worldwide. Hardware requirement for running these packages is normally kept minimum and there are always options for selecting a preferred host computer. Most currently available software packages (commercial and public domain) are ported to be run on almost all popular personal computers as well as workstations and mainframes. Moreover, computer circuit simulation for classroom purposes will free the student of lengthy and repetitive computation. This allows more concentration on (a) circuit topology and its modification if needed, (b) design of the required control circuitry in a well defined situation, (c) making simulated fault measurements in a safe environment, and (d) interpreting the resulting waveforms rather than dealing with equations which often fail to provide a satisfactory explanation for the operation of the circuit it hand.

A 1.1 WHY SIMULATIONS!

Simulation is a very important means in the field of power electronics. There are a number of reasons for this fact, as there are:

- Saving of development time
- Saving of costs ('burnt power circuits tend to be expensive')
- Better understanding of the function
- Testing and finding of critical states and regions of operation
- Fast optimisation of system and control

Today it is difficult to imagine the task of power electronics development without help of a simulation!

A 1.2 SIMULATION TOOLS

The best available simulation tool is the tool you are used to! (Provided, that it can solve the task.)

There are a number of powerful simulation tools available. All of them have advantages and disadvantages. They can be grouped into three parts: according to the way the system to be simulated is 'entered into the computer':

(i) Mathematical input

In this case an exact mathematical description of the circuit to be simulated is entered into the machine. This can be done with various programming languages as there are basic, Fortran, ..., or the very practical MATLAB. (MATLAB with the toolbox 'Simulink' has an additional feature: the possibility to enter the mathematical description in graphic form.).

(ii) Netlist input

This form of an input was used for example by former versions of Spice. The physical elements of a circuit, i.e. resistors, capacitors, active elements were entered on the keyboard as a description list. This form of entering a circuit can still be used in Spice. But, it is no longer necessary. Nowadays, the circuit can be entered into the computer graphically.

(iii) Graphical input

This group contains the tools with the most comfortable form of entering the system into the machine. Most of today's simulation packages for power electronic provide graphic input.

A 1.3 FEATURES OF A GOOD SIMULATION TOOL

To address the challenges of the modern power electronic systems, relevant simulation technology is a must. From a general perspective, the two most important aspects of any power electronic system simulation toolset are the accuracy of the simulation and the ease-of-use of the tools. Some properties of a good tool are:

- Comfortable, intuitive input of the circuit.
- Correct models for the elements (as simple as possible but, as good as necessary).
- Correct error messages.
- Robust execution of the simulation.
- Sophisticated integration algorithm.
  (various algorithms to be chosen for the type of model)
- Good output of the simulation results (formats which can be exported to other programs).
- Support from the manufacturer.
- Portability of models from one program version to following ones.

More specific critical requirements of simulation tools are:

- **Accuracy:** accurate system solutions, particularly where power electronic devices are involved.
- **Intuitive user interface:** a user interface and navigation system geared towards efficient power system analysis providing immediate visual feedback.
- **Size and duration:** the ability to model large systems and for a longer period of time than in the past.
- **Component library:** a comprehensive library of power system components, including all modern power system devices and controls.

It is the responsibility of simulation developers to provide tools that can add a valuable extension to end-suer capability. The tools must incorporate the ability to simulate the elements of a power electronic system, such as, diodes, thyristors, GTOs, IGBTs, as well as complex control and protection systems.

A 1.4 CRITICAL TASKS AND OTHER CRITICAL POINTS

Difficult for digital simulation tools are:

- fast events like switching actions.
- differentiation

Some tools do have electrical elements, which are not good enough modeled (some had even been wrong...). To find 'bad' elements in a simulation can be very time consuming.

A 1.5 SELECTION OF SIMULATION TOOLS

There are a number of simulation tools available. All of them have advantages and disadvantages (one of the latter being often the costs).

Computer circuit simulation generally provides an overall waveform-oriented solution for circuit design and analysis. This allows incorporation and testing of any additional circuitry for harmonic suppression as these changes can easily be submitted to the computer for another look at the resulting waveforms and post-processing manipulation. Adoption of these circuit simulators for classroom design and analysis applications will enhance the understanding of power electronic loads and their impact on the utility. This can be regarded as an effective demonstration tool for illustrating the advantages and pitfalls of power electronic circuit technology.

Apart from the complexities of semiconductor switching device modeling most power electronic circuits are characterized by extreme nonlinearities presented by their switches and also by their widely varied time constants. Noting that because of the presence of a wide range of time constants, the numerical integration algorithm used for solving the circuit equations must be carefully chosen. In addition, most integration methods require small step sizes to ensure stability. This implies that many steps are needed for a typical simulation resulting in very long simulation run time. Better simulation packages overcome this problem by means of utilizing a variable step size approach.

Most widely used integration methods for power electronic circuit simulation were compared in; using a circuit with very large and very small time constant. The results indicated that there is no single integration method algorithm applicable to a wide variety of power electronic circuits while maintaining the numerical stability, simulation accuracy and optimum computational effort simultaneously.

In addition to parameters outlined above, the use of a circuit simulator for educational purposes adds a new scope to the selection of a general purpose power electronic circuit simulator. In this respect, among others, the following points are considered important: (a) cost of software and hardware requirements, (b) ease of use (input file format and syntax, schematic facility, output graphical interface, error reporting and on-line help facility), (c) data modularization feature of the software, and (d) provision for simple functional macro models for controllers.

The following selection of tools provides graphical input.

(a) PSpice

PSpice exists for a long time on the market. It started as a simulation tool for low power electronic circuits. SPICE, which is mainly intended for low power analog electronic and digital circuits simulation appears to be the most popular circuit analysis program in both industry and educational institutions. A large library with PSpice models for various electronic components exists. Further models can be added. Today it can simulate analog and digital circuits with a lot of features. The representation of numerical blocks and controllers is difficult.

PSpice was the first SPICE-based circuit simulator for IBM-PC and since 1986 its free classroom (evaluation) version has been widely used in schools throughout the world. PSpice is very powerful and flexible. The free evaluation version of PSpice is sufficient to perform almost all classroom problems in circuit theory, power electronics, and so forth.

Student licenses/demo version available (reduced model sizes)
Information:
www.logmatic.ch; www.orcad.com; www.orcadpcp.com

(b) MATLAB/Simulink/SimPowerSystems/PLECS

MATLAB is a mathematical tool that is established for a long time. Toolboxes for various applications exist. One of them is Simulink, a graphic tool for the entering of functions.

Simulink itself can be expanded with another toolbox: SimPowerSystem. This toolbox is designed for the simulation of electrical power systems including power electronics. The elements of the various toolboxes can be combined.

Analysis of new converter topologies, large converters, multi-converter systems, line commutated converters and electrical drive systems requires an appropriate global simulation tool. Because the dynamic model of AC machines is very complex due to the nonlinearities of the system, it is necessary to simulate the whole system containing an AC machine. A user friendly simulator must be able to handle complex systems as such electrical drives by dividing them in small interconnected sub-systems (converter-machine-regulator) assembled in a modular structure. The interconnection of these subsystems must be done in a simple manner, without additional programming, using a graphic interface. The built-in functions and tool boxes of SIMULINK facilitate the design and the analysis of high performance AC drive systems.

Demo version/Student licenses available for a small amount (reduced model sizes)
Information:
www.mathworks.ch; www.mathworks.com
An additional toolbox for the simulation of power electronics is PLECS.
This is a fast and reliable power toolbox for matlab.
www.plexim.com

(c) PSIM

PSIM is one of the tools that had been developed specifically for power electronics. Therefore, it is optimized for the tasks arising in this field. This results in fast simulation runs. PSIM offers some add-ons, one of them being an interface to Matlab/Simulink. With that interface the full mathematical power of Matlab is accessible.
Demo version/Student licenses available for a small amount (reduced model sizes) information:
www.powersimtech.com; www.powersys.fr

(d) Simplorer

Basically Simplorer consists of four modeling languages:

- VHDL-AMS for analog-mixed-signal-design
- Circuit Simulator for the simulation of power electronic circuits
- Block diagram simulator for the simulation of controllers and similar tasks
- State machine simulator for even driven systems

These features enable the engineer to choose the language most appropriate to the task. Simplorer can be interfaced to a number of other Simulation tools.
Information:
www.ansoft.com

(e) CASPOC

This tool is designed for the simulation of power electronics and electrical drives. It provides a large library of blocks for both topics. Further, code in Pascal and C can be included.
A freeware version of CASPOC is available.
www.integratedsoft.com

(f) Saber

Saber is a tool that has been developed for a wide range of applications, including power electronics. Saber can handle analogue, digital, mixed and event driven devices. It can be linked to digital simulations to handle models written in Verilog or VHDL.
Information:
www.avanticorp.com

Appendix B

High Voltage Gate Driver ICs (HVICs), Power Modules and Intelligent Power Modules

MOSFET and IGBT gate driver ICs are the simplest, smallest and lowest cost solution to drive MOSFETs or IGBTs up to 1200V in applications up to 12 kW, and can save over 30% in part count in a 50% smaller PCB area compared to a discrete optocoupler or transformer based solution. With the addition of few external components, gate driver ICs provide full driver capability with extremely fast switching speeds, designed-in ruggedness and low-power dissipation.

Gate driver IC's generate the current and voltage necessary to turn MOSFETs or IGBTs on and off from the logic output of a DSP, micro-controller or other logic device. The input is typically a 3.3-volt logic-level signal. Gate driver ICs are either CMOS compatible, or are TTL compatible.

B 1.1 ADVANTAGES OF HVICs

- Dead-time as low as 500 ns allows frequency up to 100 kHz
- Enable rugged gate drive design
- Low power dissipation
- Compared with opto-coupler based solutions: 30% fewer parts and 50% smaller PCB.
- Doesn't need auxiliary power supply

Appendix B 1061

- 10X faster delay matching (±50 ns)
- No degradation of performance over time
- Shorter time to signal over-current 1.5 μs
- Reduced EMI and voltage spikes

B 1.2 FEATURES OF HVICs

- 600 V and 1200 V gate driver in a single IC for MOSFET and IGBTs
- Multiple Configurations
- Single high side
- Half-bridge
- Three phase inverter driver
- Output source/sink current enables fast switching
- Integrated protection and feedback functions
- Optional deadtime control
- Tolerant to negative voltage transient
- Up to 50 V/ns dV/dt immunity
- Optional soft turn-on
- Uses low cost bootstrap power supply
- CMOS and LSTTL input compatible

B 1.3 APPLICATIONS

- Motor Drive
- Lighting Ballast
- Switched Mode Power Supplies
- Automotive
- Plasma Display Panels

B 1.4 GATE DRIVER ICS SIMPLIFY DESIGN

Driving a MOSFET or IGBT in the high side position of a half-bridge topology or three phase inverter leg offers the additional challenge that the gate voltage is referenced to the source rather than to ground. The source voltage is a floating point at up to the maximum bus voltage, or voltage rating of the MOSFET or IGBT, 600 V and up for motor drive, lighting or SMPS applications. Gate driver uses a level shifter technology for high voltage application and offers the only 1200 V rating in the industry.

These ICs simplify circuit designs by integrating extensive functionality. They use a low cost bootstrap supply, while opto-coupler-based circuits typically require an auxiliary power supply. Gate Driver ICs offer optional single input or dual input programmable deadtime control for low side and high side drivers as well as for 3 phase drivers to provide design flexibility and allows to minimize cross-conduction. Unique three-phase drivers allow driving a three-phase inverter using a single IC.

B 1.5 GATE DRIVER ICS ENABLE RUGGED DRIVER DESIGNS

Gate Driver ICs are specifically designed with motor drive applications in mind. The newest soft-turn-on limits voltage and current spike and reduce EMI. In addition, they have up to 50V/ns dV/dt immunity and are tolerant to negative voltage transient. The under-voltage lockout available for most drivers prevents shoot-through currents and device failures during power-up and power-down without any additional circuitry. The output drivers feature a high pulse current buffer stage designed for minimum driver cross-conduction.

Noise immunity is important for the high-side position, which has a floating voltage and is susceptible to high noise levels, particularly in motor drive applications. Noise immunity ensures that the MOSFET or IGBT doesn't turn on accidentally. Noise immunity is obtained by using Schmitt-triggered input with pull-down. Additional noise immunity is obtained with separate logic and ground pins in some ICs, such as the 600V ICs in 14-pin packages.

B 1.6 GATE DRIVER ICS ENABLE FAST SWITCHING SPEEDS

Gate Drive ICs have ten times better delay matching performance than optocoupler-based solutions. Delay matching between the low-side and high-side driver is typically within ±50 ns (and as low as ±10 ns for some specialty products), allowing complete dead-time control for better speed range and torque control in motor drive applications. Fast switching also reduces switching power losses and allows leveraging the full benefits of the fastest IGBTs available on the market today for better torque control over a wider speed range.

B 2.1 INTELLIGENT POWER MODULES

Integrating industry benchmark three-phase high voltage ICs and rugged trench IGBTs in a sleek and innovative single in-line package (SIP), intelligent power modules (IPMs) deliver a complete power stage solution for today's energy-efficient appliance and light industrial equipment driven by variable speed motors ranging from 400 W to 2500 W. Together with a few external components and particular controllers, they form a complete motor drive system, greatly accelerating the design path compared to a multidiscrete solution.

Built-in over-temperature/over-current protection, along with short-circuit rated IGBTs, an integrated under-voltage lockout function, and built-in temperature monitor provide a high level of protection and fail-safe operation. Other integrated features, such as bootstrap diodes for the highside drive function and the single polarity power supply, simplify overall system design.

These modules represent a sophisticated, integrated solution for three phase motor drives used in a variety of appliances, such as washing machines, energy efficient refrigerators and air conditioning compressor drives. They utilize NPT (non-punch through) IGBTs matched with Ultra-soft recovery diodes to minimize EMI generation. In addition to the IGBT power switches, the modules contain a six output monolithic driver chip, matched to the IGBTs to generate the most efficient power switch consistent with minimum noise generation and maximum ruggednes.

- Packaging options include staggered pinout for maximum creepage distances and straight or 90° bend options for heat-sinks parallel or perpendicular to the circuit board.
- Capacitive switch noise coupling to the mounting surface is prevented by a ground plane isolated to 2000 Vrms connected to the Vss pin.
- Insulated Metal Substrate technology ensures low thermal resistance and less stringent flatness requirements for the heat-sink. It also offers significant flexibility in the module layout and internal electrical system.

B 2.2 IPM SOLUTION

Apart from the better known advantages of modules (smaller, more reliable, single component) compared with discrete solutions; the IPM modules relieve the designer from several pitfalls often associated with IGBT/MOSFET inverter designs:

- Lower circuit inductance than discrete solutions results in voltage spike reduction and the ability to operate at higher switching frequency with lower switch losses.
- Simple power connection
- The integrated driver requires only six logic level inputs. (includes 3.3V logic) and three bootstrap capacitors selected for the switching frequency.
- Propagation delays for all low-side and high-side IGBTs are matched to prevent DC core flux from being applied to the motor.
- Built in dead time control prevents conduction overlap between high-side and low-side IGBTs.
- Fail-safe operation is ensured by built in shut down features for over current and over temperature.
- Analog temperature monitor and phase leg current pins are provided.

B 2.3 IPM SYSTEM DESCRIPTION

The primary advantage in using the IPM modules is the ease in which an optimized, reliable motor drive system can be implemented. The designer is relieved of the following headaches:

- How to provide sufficient dead time to prevent shoot through failures.
- How to design an overcurrent protection circuit to protect the IGBT switches.
- How to design an over-temperature detection circuit that actually monitors IGBT temperature.

- How to match propagation delay times in the drive circuits to prevent DC current flow in the motor windings.
- How to select the optimum switch times to minimize EMI generation and maximize efficiency.
- How to minimize inductive loop size for minimum turn-off voltage overshoots in the IGBTs.

The IPM provides answers to all the above questions in a compact, electrically isolated package.

B 2.4 INTERNAL CIRCUITRY

The 600 V IRAMS module contains six IGBT die each with its own discrete gate resistor, six commutation diode die, one three phase monolithic, level shifting driver chip, three bootstrap diodes with current limiting resistor and an NTC thermistor/resistor pair for over temperature trip. The over current trip circuit responds to an input voltage generated from an external sense element such as a current transformer or sense resistor. The input pin for the trip circuit performs a dual function:

- Input pin for over current trip voltage.
- Output pin for module analog temperature sensing thermistor.

Because of the dual function requirements, an external circuit similar to the diagram below is recommended.

B 2.5 INPUT SIGNALS

The complete, open loop motor drive system comprises a signal source, a drive stage an a power stage. The three-phase motor can be a simple induction type, or a permanent magnet synchronous type. The IRAMS module integrates the driver and power stages into an isolated module but the 'brains' of the system must generate timing, speed and direction PWM or PFM information to complete the motor drive function. 5 volt logic systems are generally preferred from a noise immunity standpoint but the module can also accept 3.3 V logic or any pulse input up to the Vcc level (+15 V).

Further detailed information is available on company's, websites, given in Appendix C

Index

A.C chopper 504
  drives 903
  gate trigerring 35
  line commutation 53
  link chopper 435
  regulators 12
  voltage controllers 12
  to DC Converters 11
  regulators 704
Active power input 306
Air gap power 907
All difused 25
Alloy diffused 25
Ampere-hour 994
Amplitude modulation index 558
Annealing 1033, 1036
Armature circuit inductance 862
Artificial Intelligence 2
ASCR 142
Assymetric firing 383
Assymetric IGBTS 191
Attenuation factor 600
Auxillary commutated inverter 621
Auxillary commutation 49
Avalanche Breakdown 19
Avalanche diode 813

Back end 852
Basic series inverter 594
Batteries 994
Battery charger 1042
Bidirectional AC voltage controller 706
Bidirectional Diode Thyristor
  (DIAC) 145
Bipolar voltage switching 571
Block converter 437

Block Stage 18
Boost chopper 441
Boost converter 515
Braking operation 870
Brazing 1036
Brushless AC motor 975
Brushless DC motor 971, 972
Buck-boost converter 519
Buffer layer-233
Burst firing 1049

Capacitor commutated current source
  inverter 644
Carrier frequency ratio 566
Carrier wave 938
Characteristic impedence 748, 781
Chopper controlled induction
  motor 943
Chopper drives 881
Chopper 11, 434
Chopping period 485
Circuit Breaker 1045
Circuit turn off time 484
Circulating current 421
Class A
  chopper 448
  commutation 40
Class B
  chopper 466
  commutation 45
Class C
  chopper 468
  commutation 46
Class D
  chopper 471
  commutation 49

Index

Class E-*52*
Class E Chopper-*478*
Class E Resonant rectifier-*770*
Class F-*53*
CMOS drive circuits *180*
Commutating
   capacitor *52, 484*
   inductance *485*
   inductor *52*
Commutation *38*
   failure *38, 820*
   of triac *151*
   overlap *362*
Commutatorless kramer drives *957*
Complementary
   SCR *101*
   commutated inverter *630*
   commutation *46*
Computer Simulation *14*
Conduction
   angle *714*
   loss *199*
   mode *575, 583, 586*
Conduction *834*
Conductivity modulation *192*
Constant frequency system *444*
Continuous conduction *297, 309, 333*
Control techniques *259*
Convection *834*
Coupled inductors *601*
Cross conduction *543*
Crossover distortion *696*
Crowbar circuit *814*
Cuk converter *524*
Current commutated chopper *486*
Current
   density *1033*
   detection *431*
   force motor drive *936*
   limit control *445*
   rating *57*
   ripple factor *303*
   sharing *126*
   source inverter *643*
CVT *992*
Cycloconverter
   cascade *955*
   control *940*
Cycloconverters *12, 670*

Damped frequency *748*
Damping Ratio *42*
DC chopper *435*
DC
   drive *846*
   link converter *697, 954*
   voltage ratio *302, 305*
Dead Zone *760*
Delay
   control *397*
   time *35, 230*
Depletion enhancement *160*
Depth of penetration *1033*
   derating *131*
   di/dt *59, 64*
Diac *145*
Dielectric
   heating *1037*
   Slab *1038*
Discontinuous conduction *298, 1332*
Displacement
   angle *302, 387*
   factor *302, 387*
Dissipation factor *1038*
Distortion factor *302, 552*
Distribution of voltage *116*
Drive-losses *199*
Dual converter *412*
Dual mode dual converter *426*
Duty Cycle *456*
   dv/dt triggering *34*
   dv/dt *57, 64*
Dynamic Equalising *119*
Dynamic latch up *195*

Edge detector *401*
Electronic lamp Ballast *1042*
Electronic swing *1040*
Emergancy lightings System *1044*
EMI *827, 828*
Emitter follower driving circuit *178*
Energy loss *199*
Energy recovery circuit *636*
Equilising
   circuit *117*
   reactor *128*
ETO turn off *243*
ETO *236*
External
   control of A.C voltage *555*

Index

pulse commutation 52
Extinction
   control 261, 385
   angle 714

Fall time 197, 230
Fault current 817
Field weakening mode 922
Filter design impedence 661
Filter input impedence 662
Firing of thyristor 71
Flasher circuit 1047, 1048
Flux 923
Flyback converter 510, 1006
Food processing 1040
Forced commutated inverter 538
Forced commutation 39
Forward blocking
   losses 61
   region 20
Forward conducting region 20
Forward conduction loss 60
Forward voltage triggering 33
Four quadrant AC voltage
   controller 915
Four quadrant chopper 478
Freewheeling diode 267
Frequency modulation controller 444
Frequency modulation ratio 560
Full bridge inverter 454, 570, 618
Full bridge converter 1024
Full converter drives 858, 881
Fully controlled converter 363
Fusing 822
Fuzzy logic 2

Gate characteristic 28
Gate circuit parameter 29
Gate
   current 29
   amplitude 71
Gate drive design consideration 176, 204
Gate
   power losses 61
   protection 827
   pulse duration 72
   trigger voltage 56
   triggering 34
   turn off mechanism 216

Gate voltage 29
Gradient 1039
GTO current source inverter 937
GTO 212

Half bridge
   converter 1015
   inverter 538
Half wave
   AC regulator 705
   converter drives 853, 880
Half-waving effect 354
Hardening 1033
Harmonic
   distortion 552
   factor 302, 305, 388, 393
   filters 657
   injection 927
   reduction 653
   spectrum 573
   voltage 454
Heat Sink 833,
HEXFET 184
High
   bridge inverter 615
   frequency series inverter 604
   voltage DC transmission 1024
Hockey puck 27
Holding current 21, 59, 63
Hote diffusion coefficient 238
HSPICE 14
HVIC's 246

I2t rating 59
IGBT
   driver circuit 187, 207
IGCT 231
Impulse commutated chopper 481
Induction
   cooking 1035
   heating 1032
Inductive switching 165
Input power factor 302
Integral cycle triggering 1049
Integrated buffer 179
Internal
   control of inverter 556
   faults 818
Interphase reactor 326
Intregal cycle triggering 1049

Index

Intrinsic stand off ratio 89
Inversion
    failure *819*
    layer *192*
Inverter
    gain *567*
    grade thyristor *141*
Inverter *11, 35, 767*
Inverting mode *284, 337*
IR21*10-210*
IRED *79*
Isolated flyback converter *999*
    forward converter *1009*

Jones chopper *496*
Junction temp *62*

Kramer drives *949*

Lagging power factor *961*
LASCR *158*
Latch-up *194*
Latching current *21, 63, 159*
LC filter *506, 658*
Lead mounting *838*
Lightening surges *799*
Line commutated inverter *538*
Line interactive UPS-*991*
Line synchronization *396*
Load commutated
    chopper *492*
    cycloconverter *698*
Load commutation *481*
Load current polarity *419*
Load resonant converter *753*
Load switching *800*
Loss
    angle *1038*
    factor *1039*

Magnetic Amplifier *74*
MATLAB *15*
McMurray Bedford inverter *641*
McMurray inverter *621*
MCT *224*
Mechanical power *907*
Mesa-type *26*
Microcomputercontrol DC drives *895*
Microprocessor based firing *395*
Microprocessor interfacing *108*
Mid point converter *213*

Modified kramer drives *956*
Modified McMurray Bedford
    inverter *641*
Modified McMurray inverter *626, 635*
Modulating wave *938*
Modulation index *558*
morgan chopper *502*
MOSFET losses *181*
MOSFET *159*
Motoring *848, 911*
Mounting *838*
MTBF *840*
MTO *235*
Multiphase chopper *508*
Multiple
    modulation index *560*
    pulse width modulation *561*

Natural commutation *39*
N-MCT *224*
Non-circulating current mode *428*
Non-punch through IGBT'S *188*

Off state loss *199*
Offline UPS *990*
On state current *57*
OPAMP emitter follower *180*
Optical
    isolator *78*
    triggering *124*
Oscillating frequency *52*
OTT filter *660*
Over voltage condition *799*
Overcurrent
    fault condition *816*
    protection *208, 221, 821*
Overlap angle *362*

Parallel
    convertion *173*
    inverter *609*
    operation *124*
Parasitic thyristor *193*
Peak point current *90*
Pedestal triggering *106*
Penetration depth *1033*
Permanent magnet motors *967*
Permeability *1033*
Phase
    angle control *153, 260, 380*
    controlled rectifier *11, 258*

Index

locked loop 893
shifted operation 509
Photo
  darlington 80
  transistor 79
PICs 244
Planner diffused 25
PMCT 226
PNPN 17
Polarised snubber 805
Position sensors 976
Power electronics
  application 13
  convertors 11
  system 2
Power factor
  improvement 380, 710
Power modulator 3
Power rating 60
Power semiconductor devices 4
Press
  fit mounting 839
  pack 27
Pressure mounting 839
Protection 798
PSPICE 14
Pull
  in torque 966
  out torque 966
Pulse
  gate triggering 35
  gating 29
  transformer 76
  width 29
Punch-through IGBTs 188
PUT relaxation oscillator 101
PUT 100
PWM
  control 261, 390, 557
  drives 927
  inverters 565

Quality factor 748, 751

Radian frequency 661
Radiation triggering 33
Radiation 834
Radio interference phenomenon 827
Ramp triggering 98
RC full wave trigger circuit 86

RCT 144
Reactive power input 306
Rectifying mode 284
Reflected load current 614
Regenerative breaking 859
Reliability 993
Reluctance motors 967
Residual magnetism 872
Resistance capacitance (RC) firing circuit 84
Resistance firing circuit 83
Resistivity 1033
Resonant arm filter 659
  multi resonant 787
Resonant circuits 747
  series 747,
  parallel 750, 751
Resonant converters-746,
  multiresonant 787
Resonant
  DC link inverter 793, 794
  inverters-parallel 767
  class E 768
Resonant pulse commutated chopper 486
Resonant switch converter 772
Reverse
  blocking region 19
  conduction 234
  recovery time 116
  recovery 801, 804
RF filters 828, 829
RF Heating 1031, 1037
Ringing frequency 484, 597
Ripple factor 455
Rise time 36, 71, 230
RLC load 616
Rotar
  speed 908
  current 945
  resistance control 942
Round rotor motor 960

Safe operating area 172, 203, 231
Salient pole motor 963
Scherbius drives 949
SCR reliability 839
SCR 18
SCS 156
Second quadrant chopper 467

Selenium *812*
Self commutated inverter *606*
Self commutation *39, 45*
Self controlled synchronous motor drives *970*
Semiconductor operation *381*
Semiconverter drives *854, 880*
Sequence control of A.C regulators *723*,
 two stage *724*
 multistage *726*
Sequence control of A.C regulators *723*
 two stage *724*
 multistage *726*
Sequencial flasher *1048*
Sequential triggering *122*
Series
 DC motor drives *871*
 inverter control *556*
 resonant inverter *594, 753, 756, 757, 760, 763*
Shielding *828*
Shoot through fault *543*
Silicon carbide devices *250*
Simultaneous triggering *121*
Single phase
 drives *851*
 sinusoidal voltage regulator *727*
 width modulation *558*
Single quadrant closed loop *916*
Sinusoidal pulse width modulation *563*
SITHs *223*
SITs *222*
Six-pulse converter *323*
Skin depth *1033*
 effect *1032*
Slip power recovery scheme *949*
Slip *909*
Smart power *244*
Snubbers *204, 220, 803, 806, 807*
Soldering *1036*
Source filter *505*
 impedance *361*
Space-charge region *238*
Speed control *849*
 Torque curves *919*
Spread time *36*
Spurious triggering *73*
Static
 circuit breaker *1045*
 equalising *118*

latchup *194*
resistance control *944*
scherbius drives *951*
VAR compensators *1029*
Stator
 current *908*
 voltage control *913*
Steady stage ripple *453*
Step-down
 chopper *437*
 converter *510*
Step up
 chopper *441*
 converter *515*
Step up/down *443*
Storage time *184*
String efficiency *130*
SUB *156*
Subcycle surge *58, 823*
Subsynchronous converter *951*
Supersynchronous speed control *954*
Supply power factor *387*
Surface
 hardening *1036*
 current *58*
Surge suppressors *811*
SUS *156*
Switch mode
 compensator (SMC) *1031*
 welding *1041*
Switch reluctance motors *967*
Switched capacitor compensator (SCC) *1030*
Switched inductor compensator (SIC) *1030*
Switched mode power supply (SMPS) *998*
Switching losses *199*
Switching Regulators *510*
Symmetric IGBTs *190*
Symmetricangle control *338*
Synchronous reluctance motor *965*

Tertiary winding *1009*
Thermal
 ratings *62*
 triggering *33*
 resistance *834*
 model *834*
Three phase AC regulators *729*